CRC Handbook of Tables for Order Statistics from Inverse Gaussian Distributions with Applications

N. Balakrishnan
Department of Mathematics and Statistics
McMaster University
Hamilton, Ontario, Canada

William W.S. Chen
Internal Revenue Service
Washington, DC

CRC Press is an imprint of the
Taylor & Francis Group, an **informa** business

CRC Press
Taylor & Francis Group
6000 Broken Sound Parkway NW, Suite 300
Boca Raton, FL 33487-2742

First issued in paperback 2019

ISBN 13: 978-0-367-44815-8 (pbk)
ISBN 13: 978-0-8493-3118-3 (hbk)

Visit the Taylor & Francis Web site at
http://www.taylorandfrancis.com

and the CRC Press Web site at
http://www.crcpress.com

Library of Congress Cataloging-in-Publication Data

Catolog record is available from the Library of Congress

To Sudha, Shubha, Ravi and Aditya

To Heng Chen

Preface

Inverse Gaussian distributions have recently become one of the most commonly studied models in the theoretical literature while being commonly used in the applied literature. Though these distributions have been around for over fifty years, much of the work concerning inferential methods for the parameters of inverse Gaussian distributions has been carried out only in the recent past. Most of these methods, particularly those based on censored samples, do involve the use of numerical methods to solve some equations.

Order statistics and their moments have been studied quite extensively for numerous distributions. However, moments of order statistics from inverse Gaussian distributions have not been discussed in the statistical literature until now primarily due to the extreme computational complexity in their numerical determination. It is this challenge that motivated us to enter into this tremendous task of computing the means, variances and covariances of order statistics for all sample sizes up to twenty five and for many choices of the shape parameter. Another major reason for taking up this study is, of course, to make use of the tabulated values of means, variances and covariances of order statistics in order to derive the best linear unbiased estimators of the location and scale parameters based on complete as well as Type-II censored samples. Both these goals have been achieved successfully and the extensive tables that emerged out of our efforts have been presented in this volume. We have also presented a few examples to illustrate some practical applications of these tables.

Our sincere thanks go to Mr. Robert Stern and Mr. Tim Pletscher (both of CRC Press, Boca Raton) for taking interest in this project and also providing great support during the course of preparation of this volume. Thanks are due to Ms. Nora Konopka and Ms. Suzanne Lassandro (both of CRC Press, Boca Raton) for helping in the production of the volume. We also thank Ms. Debbie Iscoe (Hamilton, Ontario, Canada) for the fine typesetting of the entire text. Finally, the first author acknowledges the research support received from the Natural Sciences and Engineering Research Council of Canada.

N. BALAKRISHNAN
Hamilton, Ontario, Canada

W. W. S. CHEN
Washington, DC

May 1997

Contents

Preface v

List of Figures ix

List of Tables xi

1. Introduction 1

2. Inverse Gaussian Distributions and Properties 5

3. Order Statistics and Moments 9

4. Best Linear Unbiased Estimation of Location and Scale Parameters 13

5. Illustrative Examples 17

6. Best Linear Unbiased Prediction 29

7. Quantile-Quantile Plots and Goodness-of-Fit Test 37

Bibliography 39

Figures 41

Tables 71

List of Figures

Figure 1.	Plot of IG(0.1) and N(0,1) density functions	**43**
Figure 2.	Plot of IG(0.2) and N(0,1) density functions	**44**
Figure 3.	Plot of IG(0.3) and N(0,1) density functions	**45**
Figure 4.	Plot of IG(0.4) and N(0,1) density functions	**46**
Figure 5.	Plot of IG(0.5) and N(0,1) density functions	**47**
Figure 6.	Plot of IG(0.6) and N(0,1) density functions	**48**
Figure 7.	Plot of IG(0.7) and N(0,1) density functions	**49**
Figure 8.	Plot of IG(0.8) and N(0,1) density functions	**50**
Figure 9.	Plot of IG(0.9) and N(0,1) density functions	**51**
Figure 10.	Plot of IG(1.0) and N(0,1) density functions	**52**
Figure 11.	Plot of IG(1.1) and N(0,1) density functions	**53**
Figure 12.	Plot of IG(1.2) and N(0,1) density functions	**54**
Figure 13.	Plot of IG(1.3) and N(0,1) density functions	**55**
Figure 14.	Plot of IG(1.4) and N(0,1) density functions	**56**
Figure 15.	Plot of IG(1.5) and N(0,1) density functions	**57**
Figure 16.	Plot of IG(1.6) and N(0,1) density functions	**58**
Figure 17.	Plot of IG(1.7) and N(0,1) density functions	**59**
Figure 18.	Plot of IG(1.8) and N(0,1) density functions	**60**

Figure 19. Plot of IG(1.9) and N(0,1) density functions **61**

Figure 20. Plot of IG(2.0) and N(0,1) density functions **62**

Figure 21. Plot of IG(2.1) and N(0,1) density functions **63**

Figure 22. Plot of IG(2.2) and N(0,1) density functions **64**

Figure 23. Plot of IG(2.3) and N(0,1) density functions **65**

Figure 24. Plot of IG(2.4) and N(0,1) density functions **66**

Figure 25. Plot of IG(2.5) and N(0,1) density functions **67**

Figure 26. Quantile-Quantile plots for Example 1 **68**

Figure 27. Quantile-Quantile plots for Example 3 **69**

List of Tables

Table 1. Means of order statistics for $n = 1(1)25$, $i = 1(1)n$ and $k = 0.0(0.1)2.5$ **73**

Table 2. Variances of order statistics for $n = (1)25$, $i = 1(1)n$ and $k = 0.0(0.1)2.5$ **95**

Table 3. Covariances of order statistics for $n = 2(1)25$, $1 \leq i < j \leq n$ and $k = 0.0(0.1)2.5$ **117**

Table 4. Coefficients a_i for the BLUE of the location parameter μ for $n = 2(1)25$ and the censoring number $s = 0(1)\left[\frac{n+1}{2}\right]$ **287**

Table 5. Coefficients b_i for the BLUE of the scale parameter σ for $n = 2(1)25$ and the censoring number $s = 0(1)\left[\frac{n+1}{2}\right]$ **457**

Table 6. Variances and covariance factors, V_1, V_2 and V_3, of the BLUEs of μ and σ for $n = 2(1)25$ and the censoring number $s = 0(1)\left[\frac{n+1}{2}\right]$ **627**

Table 7. Simulated values of $100p$-percentage points of the correlation goodness-of-fit statistic R for $n = 20$ (based on 5,000 Monte Carlo simulations) **673**

Table 8. Simulated values of $100p$-percentage points of the correlation goodness-of-fit statistic R for $n = 23$ (based on 5,000 Monte Carlo simulations) **677**

Table 9. Simulated values of $100p$-percentage points of the correlation goodness-of-fit statistic R for $n = 25$ (based on 5,000 Monte Carlo simulations) **681**

Table 10. Values of the correlation statistic R and the corresponding p-values for Example 1 **685**

Table 11. Values of the correlation statistic R and the corresponding p-values for Example 3 **687**

1

Introduction

Inverse Gaussian distributions have recently become one of the commonly used statistical models in various problems. Much of this is due to pioneering contributions made by numerous researchers; in particular, the early work by Halphen[1], Tweedie, Wald and Wasan, and more recent work by Chhikara, Folks, Seshadri, Whitmore, Cohen and Jørgensen. Due to the extensive developments made on the Inverse Gaussian distributions during the past thirty years or so, this wide family of distributions started getting due attention from theoreticians as well as applied researchers. As a result, numerous theoretically sophisticated methodologies have been developed based on the Inverse Gaussian distributions. Simultaneously, the Inverse Gaussian distributions were also being used as models for analyzing data in a variety of applied problems.

Tweedie (1947) first noted the inverse relationship between the cumulant generating functions of a wide class of distributions and those of Gaussian (normal) distributions and consequently termed that class of distributions as **Inverse Gaussian distributions**. Independently, Wald (1947) derived the same class of distributions as an asymptotic form of distributions of average sample number (ASN) in sequential analysis. As a result, the name **Wald distributions** is also used for some members of the class of Inverse Gaussian distributions. Wasan (1968) observed the Inverse Gaussian distributions while studying the first passage time of Brownian motion and hence called them **first passage time distributions of Brownian motion with positive shift**, and denoted them by *T.B.M.P. distributions*.

Due to the prominence and wide applicability that the Inverse Gaussian distributions have achieved over the years and especially during the recent times, two books have been devoted completely to the theory and applications of Inverse Gaussian distributions; see Chhikara and Folks (1989) and Seshadri (1993). For a concise review of developments on this class of distributions, one may also refer to Chapter 15 of Johnson, Kotz and Balakrishnan (1994). Koziol (1988) has prepared a handbook of tables giving percentile points of Inverse Gaussian distributions.

As mentioned earlier, the class of Inverse Gaussian distributions is currently being used quite commonly as a life-time model in reliability studies. Because of this particular application, several authors have paid special attention to inferential issues concerning these distributions. Both the two-parameter (scale and shape parameters) and the three-parameter (location, scale and shape parameters) models have been discussed in this context. The books by Chhikara and Folks (1989) and Seshadri (1993) present extensive discussions on classical inference for the parameters of Inverse Gaussian distributions. Some simple and efficient methods of estimation and their relative merits have been

[1]Thanks are due to Professor V. Seshadri for bringing to our attention some historical facts about the original work of Etienne Halphen on the Inverse Gaussian and generalized Inverse Gaussian distributions.

discussed in the books by Cohen and Whitten (1988) and Balakrishnan and Cohen (1991).

In this book, we are primarily concerned with the three-parameter form of Inverse Gaussian distributions although the tables of moments of order statistics included here can be easily used to discuss the estimation for the two-parameter form as well. As can be noted from the books by Cohen and Whitten (1988) and Balakrishnan and Cohen (1991), the estimation problem associated with the three-parameter Inverse Gaussian distribution is a difficult and challenging one. In this connection, we have undertaken in this book a two-fold study—firstly to study order statistics from standard Inverse Gaussian distributions and their moments, and secondly to apply the computed moments of order statistics in order to derive the best linear unbiased estimators of the location and scale parameters (with the shape parameter being fixed) based on complete as well as Type-II censored samples. In order to achieve these goals, we had to resolve several numerical complications and also employ many checks in order to verify the accuracy of the computations performed. These will be explained in Chapter 3 along with other pertinent details concerning the computations carried out for this book.

In Chapter 2, we start with a brief presentation of different forms of Inverse Gaussian distributions and some basic properties. We then describe a reparametrization suggested by Chan, Cohen and Whitten (1983) using which a three-parameter form of Inverse Gaussian distributions is presented. Plots of the standard form of this density function are also presented for various values of the shape parameter along with an indication of its departure from the standard normal density function.

In Chapter 3, we start with an introduction to order statistics and then present the necessary formulas for the marginal and joint densities of order statistics and the computational formulas for the means, variances and covariances of order statistics. Next, we describe the computational procedure used for the determination of the means, variances and covariances of order statistics and explain some of the numerical complications that had to be resolved while doing these computations. We also present the relations and identities that we used for checking the accuracy of the computations, and demonstrate them with a few examples. Values of means of order statistics for sample sizes up to twenty five and for values of the shape parameter being 0.0(0.1)2.5 computed by a numerical (single) integration procedure are presented in Table 1. Similarly, values of variances of order statistics for sample sizes up to twenty five and for values of the shape parameter being 0.0(0.1)2.5 computed by a numerical (single) integration procedure are presented in Table 2. Finally, values of covariances of order statistics for sample sizes up to twenty five and for values of the shape parameter being 0.0(0.1)2.5 computed by a numerical (double) integration procedure are presented in Table 3.

One of the primary applications of means, variances and covariances of order statistics is in deriving best linear unbiased estimators of the location and scale parameters; see, for example, David (1981), Balakrishnan and Cohen (1991), and Arnold, Balakrishnan and Nagaraja (1992). In Chapter 4, we first present the necessary formulas for the best linear unbiased estimators and then describe the computational procedure employed for the derivation of these best linear unbiased estimators. We discuss the best linear unbiased estimation of the location and scale parameters (assuming the shape parameter to be fixed) based on complete as well as Type-II right-censored samples which are most likely to be encountered in life-testing experiments. The coefficients in the best linear unbiased estimator of the location parameter for sample sizes up to twenty five, for values of the shape parameter being 0.0(0.1)2.5, and for various levels of censoring are presented in Table 4. Similarly, the coefficients in the best linear unbiased estimator of the scale parameter for sample sizes up to twenty five, for values of the shape parameter being 0.0(0.1)2.5, and for various levels of censoring are presented in Table 5. Finally, the variances and covariance of these best linear unbiased estimators for sample sizes up to twenty five, for values of the shape parameter being 0.0(0.1)2.5, and for various levels of censoring are presented in Table 6.

In Chapter 5, we present some numerical examples in order to illustrate a direct application of these tables. The best linear unbiased estimators so derived can also be used successfully to develop best linear unbiased prediction of the failure times coorresponding to the censored units. This is demonstrated in Chapter 6 and is also illustrated with some numerical examples. Another very useful application of means of order statistics is in constructing Q-Q plots and in developing formal correlation-type goodness-of-fit tests in order to verify the validity [see D'Agostino and Stephens (1986)] of Inverse Gaussian distributions for the observed data. In Chapter 7, we illustrate this use of means of order statistics presented in Table 1 through the numerical examples considered earlier. Through Monte Carlo simulations, we also demonstrate the correlation-type goodness-of-fit test with the help of the previously considered numerical examples.

2

Inverse Gaussian Distributions and Properties

The standard or canonical form of the two-parameter inverse Gaussian distribution has the probability density function

$$f(y|\delta,\lambda) = \left(\frac{\lambda}{2\pi y^3}\right)^{1/2} exp\left\{-\frac{\lambda}{2\delta^2 y}(y-\delta)^2\right\}, \quad y > 0. \tag{2.1}$$

It is known that, in this form of the distribution, δ is the mean and δ^3/λ is the variance. Furthermore, δ/λ is the square of the coefficient of variation.

As Tweedie (1957a,b) has shown, the above distribution can be reparametrized and written in the following three equivalent forms as well:

$$f(y|\delta,\phi) = \left(\frac{\delta\phi}{2\pi y^3}\right)^{1/2} e^{\phi}\, exp\left\{-\frac{\phi}{2}\left(\frac{y}{\delta}+\frac{\delta}{y}\right)\right\}, \ y > 0, \tag{2.2}$$

$$f(y|\phi,\lambda) = \left(\frac{\lambda}{2\pi y^3}\right)^{1/2} e^{\phi}\, exp\left\{-\frac{1}{2}\left(\frac{\phi^2 y}{\lambda}+\frac{\lambda}{y}\right)\right\}, \ y > 0, \tag{2.3}$$

and

$$f(y|\alpha,\lambda) = \left(\frac{\lambda}{2\pi y^3}\right)^{1/2} exp\left[-\frac{\lambda}{2}\left\{\alpha y-(2\alpha)^{1/2}+\frac{1}{2y}\right\}\right], \ y > 0, \tag{2.4}$$

In the above forms, $\phi = \frac{\lambda}{\delta}$, $\alpha = \surd(2\delta)$, and the parameters λ, δ, ϕ, and α are all positive. The Wald distribution, derived by Wald (1947) as an asymptotic form of the average sample number in sequential analysis, is obtained from Eq. (2.1) by taking $\delta = 1$ and $\lambda = \phi$. It is also easy to observe that if the random variable Y is distributed as inverse Gaussian in (2.1), then a scalar multiple of it (say, cY) is distributed once again as inverse Gaussian of the form (2.1) with δ and λ replaced by $c\delta$ and $c\lambda$, respectively. Also, the cumulative distribution function corresponding to the density in (2.1) can be shown to be

$$F(y|\delta,\lambda) = \Phi\left\{\left(\frac{\lambda}{y}\right)^{1/2}\left(\frac{y}{\delta}-1\right)\right\} + e^{2\lambda/\delta}\,\Phi\left\{-\left(\frac{\lambda}{y}\right)^{1/2}\left(\frac{y}{\delta}+1\right)\right\}, \ y > 0, \tag{2.5}$$

where $\Phi(\cdot)$ is the cumulative distribution function of a standard normal variable.

For the inverse Gaussian density function in (2.1), we can show that the cumulant generating function is

$$K_Y(t) = \frac{\lambda}{\delta}\left\{1-\left(1-\frac{2\delta^2 t}{\lambda}\right)^{1/2}\right\} \tag{2.6}$$

from which, upon differentiating successively with respect to t and then setting $t = 0$, we obtain the first four cumulants as

$$\kappa_1 = \delta \quad \text{(is the mean)},$$

$$\kappa_2 = \frac{\delta^3}{\lambda} \quad \text{(is the variance)},$$
$$\kappa_3 = \frac{3\delta^5}{\lambda^2},$$
$$\kappa_4 = \frac{15\delta^7}{\lambda^3}. \tag{2.7}$$

In general, the r-th cumulant of the distribution in (2.1) can be shown to be

$$\kappa_r = 1 \times 3 \times 5 \times \cdots \times (2r-3) \times \frac{\delta^{2r-1}}{\lambda^{r-1}}. \tag{2.8}$$

Furthermore, from Eq. (2.7), we obtain the coefficients of skewness and kurtosis of the distribution as

$$\sqrt{\beta_1} = 3\left(\frac{\delta}{\lambda}\right)^{1/2} = 3/\sqrt{\phi},$$
$$\beta_2 = 3 + \frac{15\delta}{\lambda} = 3 + 15/\phi. \tag{2.9}$$

From (2.9), we note the following relationship between the coefficients of skewness and kurtosis:

$$\beta_2 = 3 + \frac{5}{3}\beta_1. \tag{2.10}$$

As Johnson, Kotz and Balakrishnan (1994) have mentioned, the relationship in (2.10) shows that in the Pearson (β_1,β_2) plane the inverse Gaussian points lie between the gamma (Pearson's Type III) and the lognormal lines. It is also of interest to note from (2.9) that, as the parameter $\phi \to \infty$, the coefficients of skewness and kurtosis approach the values 0 and 3, respectively, which are the values corresponding to a normal distribution.

From the standard two-parameter form of the inverse Gaussian distribution in (2.1), a three-parameter inverse Gaussian distribution can be proposed easily by introducing a threshold (location) parameter, γ; see Cheng and Amin (1981). In this case, the probability density function is easily seen to become

$$f(y|\gamma,\delta,\lambda) = \left(\frac{\lambda}{2\pi(y-\gamma)^3}\right)^{1/2} exp\left\{-\frac{\lambda}{2\delta^2(y-\gamma)}(y-\gamma-\delta)^2\right\}, \; y > \gamma. \tag{2.11}$$

Upon reparametrizing the above model by setting $\sigma^2 = \delta^3/\lambda$, we obtain the reparametrized probability density function as

$$f(y|\gamma,\delta,\sigma) = \frac{\delta}{\sigma}\left\{\frac{\delta}{2\pi(y-\gamma)^3}\right\}^{1/2} exp\left\{-\frac{\delta}{2\sigma^2(y-\gamma)}(y-\gamma-\delta)^2\right\}, \; y > \gamma. \tag{2.12}$$

Then, using the formulas in (2.7) and (2.9), we readily obtain the mean, variance and the coefficient of skewness for the reparametrized distribution in (2.12) as

$$\text{Mean, } E(Y) = \gamma + \delta = \mu \; (say),$$
$$\text{Variance, } Var(Y) = \sigma^2,$$
$$\text{Coe. of skewness, } \sqrt{\beta_1}(Y) = 3\sigma/\delta = k \; (say). \tag{2.13}$$

That is, γ is the threshold parameter for the density function in (2.12), $\gamma + \delta = \mu$ is the mean, σ is the standard deviation, and $k = 3\sigma/\delta$ is the shape parameter (denoting the coefficient of skewness); see Chan, Cohen and Whitten (1983). Now, if we consider the standardized random variable $Z = (Y - \gamma - \delta)/\sigma$, we obtain the probability density function of Z from (2.12) easily as

$$f(z|k) = \frac{1}{\sqrt{(2\pi)}} \left(\frac{3}{3+kz} \right)^{3/2} exp\left\{ - \frac{3z^2}{2(3+kz)} \right\}, \ -\frac{3}{k} < z < \infty. \quad (2.14)$$

It is clear from (2.14) that $f(0|k) = 1/\sqrt{(2\pi)}$ for any value of the shape parameter k. It is also clear from (2.14) that the limiting form of this standard inverse Gaussian distribution, as k approaches 0, is

$$\frac{1}{\sqrt{(2\pi)}} e^{-z^2/2}, \ -\infty < z < \infty,$$

which is just the standard normal density function. Thus, in (2.14), as the shape parameter k moves away from 0, we get a family of skewed distributions which provide a moderate to large departures from normality. Further, from (2.14) we observe that the mode of the standardized inverse Gaussian distribution is given by the solution of the equation

$$k + \frac{6z + kz^2}{3 + kz} = 0$$

which is given by

$$Mode(Z) = -\frac{3}{k} - \frac{k}{2} + \frac{(36 + k^4)^{1/2}}{2k}. \quad (2.15)$$

The expression for the mode of Z in (2.15) readily reveals that $Mode(Z) > -\frac{3}{k}$; hence, the standardized inverse Gaussian distribution in (2.14), unlike the gamma and the Weibull distributions, does not become reverse J-shaped and retains the bell shape with a discernible mode for all values of the shape parameter k. In this respect, it resembles the lognormal distribution more than either the Weibull or the gamma distributions. Furthermore, from (2.14), it can also be shown that the cumulative distribution function of the standardized inverse Gaussian distribution is given by [also see Eq. (2.5)]

$$\begin{aligned} F(z|k) &= \Phi\left\{ \frac{z}{\left(1 + \frac{kz}{3}\right)^{1/2}} \right\} + e^{18/k^2}\, \Phi\left\{ - \frac{z + \frac{6}{k}}{\left(1 + \frac{kz}{3}\right)^{1/2}} \right\}, \ -\frac{3}{k} < z < \infty \\ &= 0, \ \text{otherwise}. \end{aligned} \quad (2.16)$$

In Figs. 1–25, the standardized probability density function $f(z|k)$ in (2.14) has been plotted along with the standard normal density function for the value of the shape parameter $k = 0.1(0.1)2.5$. From these figures, it is once again clear that when the shape parameter k is small (close to 0) the standardized inverse Gaussian density function is very close (almost indistinguishable) to the standard normal density function. However, as the shape parameter k gets large (away from 0), the standardized inverse Gaussian density function becomes highly skewed and is significantly different from the standard normal density function.

3

Order Statistics and Moments

Let $Z_1, Z_2, \ldots, Z_n$ be a random sample of size n from the standard inverse Gaussian distribution in (2.14). Let $Z_{1:n} \le Z_{2:n} \le \ldots \le Z_{n:n}$ be the order statistics obtained by arranging this sample in increasing order of magnitude.

Then, it is well-known that the marginal density function of the i-th order statistic, $Z_{i:n}$, is given by [see Arnold, Balakrishnan and Nagaraja (1992, p. 10), Balakrishnan and Cohen (1991, p. 12) and David (1981, p. 9)]

$$f_{i:n}(z|k) = \frac{n!}{(i-1)!(n-i)!}\{F(z|k)\}^{i-1}\{1-F(z|k)\}^{n-i}f(z|k), \quad -\frac{3}{k} < z < \infty, \tag{3.1}$$

where $f(z|k)$ and $F(z|k)$ are the standardized inverse Gaussian density and cumulative distribution functions as given in (2.14) and (2.16), respectively. In the above equation, we have deliberately kept the parameter k in all the notations in order to emphasize that the distribution of $Z_{i:n}$ involves the shape parameter k.

Similarly, it is known that the joint density function of the i-th and j-th order statistics, $Z_{i:n}$ and $Z_{j:n}$, is given by [see Arnold, Balakrishnan and Nagaraja (1992, p. 16), Balakrishnan and Cohen (1991, p. 9) and David (1981, p. 10)]

$$\begin{aligned} f_{i,j:n}(z_1, z_2|k) = {} & \frac{n!}{(i-1)!(j-i-1)!(n-j)!}\{F(z_1|k)\}^{i-1}\{F(z_2|k)-F(z_1|k)\}^{j-i-1} \\ & \times \{1-F(z_2|k)\}^{n-j}f(z_1|k)f(z_2|k), \quad -\frac{3}{k} < z_1 < z_2 < \infty. \end{aligned} \tag{3.2}$$

From the expressions of the marginal and joint density functions of order statistics presented in Eqs. (3.1) and (3.2), the single and the product moments of order statistics can be computed by employing one- and two-dimensional numerical integration routines. First, let us denote the single moments of order statistics by $\mu_{i:n}^{(\ell)}$, which are given by the single integral

$$\mu_{i:n}^{(\ell)} = \int_{-3/k}^{\infty} z^{\ell} f_{i:n}(z|k)\, dz, \ i = 1, 2, \ldots, n, \ \ell \ge 0. \tag{3.3}$$

When $\ell = 1$, we simply obtain the mean of the i-th order statistic, and in this case we will denote it by $\mu_{i:n}$ for simplicity in notation. From the first two single moments, we can readily find the variance of the i-th order statistic, denoted by $\sigma_{i,i:n}$, as

$$\sigma_{i,i:n} = \mu_{i:n}^{(2)} - \mu_{i:n}^2, \qquad i = 1, 2, \ldots, n. \tag{3.4}$$

Next, let us denote the product moments of order statistics by $\mu_{i,j:n}$, which are given by the double integral

$$\mu_{i,j:n} = \int_{-3/k}^{\infty} \int_{z_1}^{\infty} z_1 z_2 f_{i,j:n}(z_1, z_2|k)\, dz_2\, dz_1, \quad 1 \le i < j \le n. \tag{3.5}$$

From the product moments and the means of order statistics, we can readily find the covariance of the i-th and j-th order statistics, denoted by $\sigma_{i,j:n}$, as

$$\sigma_{i,j:n} = \mu_{i,j:n} - \mu_{i:n}\mu_{j:n}. \tag{3.6}$$

Realize that the single as well as the product moments of order statistics computed from (3.3) and (3.5), respectively, will depend on the particular value of k; yet, for convenience in notation, we are not directly incorporating the parameter k in our notations of these moments and also in variances and covariances. The main reason for doing so will become clear in the following chapter when we discuss the best linear unbiased estimators of the location and scale parameters.

Since the integration required in (3.3) and (3.5) cannot be performed algebraically, numerical integration techniques have been adopted. Specifically, the single integrals needed for the computation of the means and variances of order statistics were computed by using the Gaussian quadrature with 512 lattice points over differing lengths of intervals. Since the single integrals involved only finite lower limits (being $-3/k$), except in the case of the standard normal distribution (case $k = 0$), determination of an appropriate upper termination point for the integrals posed quite a challenge as it very much depended on the values of n, i and k. For this purpose, we tabulated the value of the cumulative distribution function, $F(z|k)$, in (2.16) and chose the upper termination point for the integrals to be that value of z which yielded the cumulative distribution function value to be as close to 1 as we desired. Naturally, as the shape parameter k increased, the upper termination point also became large as the distribution becomes highly right skewed. For example, when $k = 2.5$, the upper termination point turned out to be as large as 55 in many cases (keep in mind that the lower limit for the integral in this case is -1.2). Thus, after determining the suitable upper termination points in each case by examining the cumulative distribution function, the Gaussian quadrature with 512 lattice points was employed for the computation of the single integrals (now with finite lower and upper limits).

The values of the means of order statistics so computed for the choices $n = 1(1)25$, $i = 1(1)n$, and the shape parameter $k = 0.0(0.1)2.5$ are presented in Table 1. Similarly, the values of the variances of order statistics so computed for the choices $n = 1(1)25$, $i = 1(1)n$, and the shape parameter $k = 0.0(0.1)2.5$ are presented in Table 2. It is important to mention here that all the means and variances of order statistics computed for the standard normal case ($k = 0$) agreed with the previously published tables; see, for example, Parrish (1992a,b) and Harter and Balakrishnan (1996). In the case of the computation of the integrals in (3.5), determining an appropriate range of integration became immensely complicated due to the fact that the lower limit of integration for the integral for z_2 is z_1 while the lower limit of integration for the integral for z_1 is $-3/k$. Hence, extra care had to be taken in fixing the limits of integration appropriately and once again the upper termination points were determined appropriately for each case by looking at the cumulative distribution function. In this case, we employed the Gaussian quadrature with 96 lattice points. Then, the values of the product moments of order statistics were computed from (3.5) by carefully avoiding the multiplication of large integral constants (arising from the combinatorial terms) at the last stage in order to preserve the high-degree of accuracy in the computed values.

Using the values of the product moments of order statistics so computed, we then computed the covariances of order statistics from (3.6) for the choices $n = 2(1)25$, $1 \leq i < j \leq n$, and the shape parameter $k = 0.0(0.1)2.5$. These values are presented in Table 3. Once again, it was verified that all the covariances of order statistics computed for the standard normal case ($k = 0$) agreed with the previously published tables; see, for example, Tietjen, Kahaner and Beckman (1977), Parrish (1992b) and Harter and Balakrishnan (1996).

For this entire laborious computational process, we wrote one short calling program with eight subfunctions or subroutines. In the main program, we only read in the preassigned values of n, i, j and pass on these values to the subfunctions in order to compute each possible combination of the power of the cumulative distribution function and the density function. All these computations were carried out in double precision format in order to ensure that at least the first ten digits of the computed results were accurate. In addition to carrying out all the computations with double precision, we also employed at least fifteen different recurrence relations and identities to check the accuracy of the computed values of the means, variances and covariances of order statistics presented in Tables 1–3 and verified that they were all satisfied. Some of these relations and identities involved only the single moments of order statistics, while some involved only the product moments and the covariances of order statistics. For an exhaustive review of all the different recurrence relations and identities, interested readers may refer to Arnold and Balakrishnan (1989).

First of all, it is well-known that

$$\sum_{i=1}^{n} \mu_{i:n} = n\mu_{1:1} = 0, \tag{3.7}$$

since the mean of the standardized inverse Gaussian distribution is 0. This identity was satisfied to the desired accuracy in all the cases listed in Table 1. Next, we used the *triangle rule* given by

$$i\mu_{i+1:n} + (n-i)\mu_{i:n} = n\mu_{i:n-1}; \tag{3.8}$$

this was also satisfied to the desired accuracy in all the cases listed in Table 1. In addition to the above two, we also applied Joshi's (1973) identities

$$\sum_{i=1}^{n} \frac{1}{i}\mu_{i:n} = \sum_{i=1}^{n} \frac{1}{i}\mu_{1:i} \tag{3.9}$$

and

$$\sum_{i=1}^{n} \frac{1}{n-i+1}\mu_{i:n} = \sum_{i=1}^{n} \frac{1}{i}\mu_{i:i} \tag{3.10}$$

and verified them to be satisfied for all the cases to the desired accuracy. Two more relations, due to Sillitto (1964) and Downton (1966), given by

$$\sum_{r=0}^{n-\ell} \binom{i+r-1}{r}\binom{n-i-r}{\ell-i}\mu_{i+r:n} = \binom{n}{\ell}\mu_{i:\ell},\ 1 \le i \le \ell < n \tag{3.11}$$

and

$$\sum_{i=k+1}^{n-\ell} (i-1)^{(k)}(n-\ell)^{(\ell)}\mu_{i:n} = k!\ell!\binom{n}{k+\ell+1}\mu_{k+1:k+\ell+1},\ k+1 \le n-\ell, \tag{3.12}$$

respectively, were also successfully satisfied to the desired level of accuracy by the values of the means of order statistics presented in Table 1.

Similarly, the values of the variances and covariances of order statistics presented in Tables 2 and 3 were checked for their accuracy through a relation given by Balakrishnan

(1989) as

$$(i-1)\sigma_{i,j:n} + (j-i)\sigma_{i-1,j:n} + (n-j+1)\sigma_{i-1,j-1:n}$$
$$= n\{\sigma_{i-1,j-1:n-1} + (\mu_{i-1:n-1} - \mu_{i-1:n})(\mu_{j-1:n-1} - \mu_{j:n})\},$$
$$2 \le i < j \le n. \tag{3.13}$$

All the values presented in Tables 2 and 3 satisfied the above relation to the desired level of accuracy.

For example, we have presented in the following table some particular cases and the results obtained from the above recurrence relations and identities in order to give a glimpse of the nature of this computational accuracy verification process:

Relation	k	n	Index	Both sides equal
(3.8)	1.2	15	i = 8	-1.21918
(3.8)	1.5	15	i = 8	-1.75030
(3.8)	2.0	20	i = 10	-4.92265
(3.8)	2.5	25	i = 20	15.33447
(3.9)	0.1	10	***	-1.93385
(3.9)	0.7	12	***	-2.02600
(3.9)	1.0	8	***	-1.44699
(3.10)	0.3	8	***	1.67709
(3.10)	0.6	10	***	2.05767
(3.11)	1.4	18	i=9, ℓ=11	18982
(3.11)	1.8	20	i=10, ℓ=15	1281
(3.11)	2.3	25	i=17, ℓ=20	37854
(3.12)	0.5	15	k=5, ℓ=9	-17164891
(3.12)	1.5	20	k=3, ℓ=7	-2784894178
(3.12)	0.9	18	k=4, ℓ=8	3925019802
(3.13)	0.2	19	i=5, j=6	1.46562
(3.13)	0.6	20	i=7, j=8	1.17180
(3.13)	1.2	10	i=5, j=7	0.87064
(3.13)	1.5	15	i=10, j=13	1.45910
(3.13)	2.1	20	i=8, j=14	0.52172

4

Best Linear Unbiased Estimation of Location and Scale Parameters

Let us now assume that we have a random sample of size n, say $X_1, X_2, \ldots, X_n$, from a three-parameter inverse Gaussian distribution [obtained by introducing location and scale parameters in (2.14)] with probability density function

$$f(x|\mu,\sigma,k) = \frac{1}{\sqrt{(2\pi)}\sigma}\left(\frac{3\sigma}{3\sigma+k(x-\mu)}\right)^{3/2} exp\left\{-\frac{3(x-\mu)^2}{2\sigma\left\{3\sigma+k(x-\mu)\right\}}\right\},$$

$$\mu - \frac{3\sigma}{k} < x < \infty. \tag{4.1}$$

Upon comparing this with the density function in Eq. (2.12), it is then clear that the location parameter μ is simply the mean (viz., $\delta + \gamma$) and the scale parameter σ is simply the standard deviation. Note that, since the shape parameter k (being the coefficient of skewness) equals $3\sigma/\delta$, the lower endpoint of the support of the above density function is simply γ which agrees with that of (2.12).

Let $X_{1:n} \leq X_{2:n} \leq \ldots \leq X_{n:n}$ be the order statistics obtained by arranging the above sample in increasing order of magnitude. Then, due to the obvious relationship $Z_{i:n} = (X_{i:n} - \mu)/\sigma$ between the order statistics from the standard inverse Gaussian distribution in (2.14) and those from the above three-parameter inverse Gaussian distribution in (4.1), we have the relationships

$$\begin{aligned} E(X_{i:n}) &= \mu + \sigma\ \mu_{i:n}, \\ Var(X_{i:n}) &= \sigma^2\ \sigma_{i,i:n}, \\ Cov(X_{i:n}, X_{j:n}) &= \sigma^2\ \sigma_{i,j:n}, \end{aligned} \tag{4.2}$$

where the expressions for $\mu_{i:n}$ and $\sigma_{i,j:n}$ are presented in Eqs. (3.3)-(3.6). Remember that these are precisely the quantities that have been tabulated in Tables 1-3.

Let us now denote the column vector of order statistics by $\boldsymbol{X}$, the column vector of $\mu_{i:n}$ by $\boldsymbol{\mu}$, the column vector of 1's by $\mathbf{1}$, the matrix of $\sigma_{i,j:n}$ by $\boldsymbol{\Sigma}$, and its inverse by $\boldsymbol{\Sigma}^{-1}$.

The Best Linear Unbiased Estimators (BLUEs) of μ and σ are estimators which are linear functions of the components of $\boldsymbol{X}$ which are unbiased and which minimize the generalized variance. This method of estimation based on the least-squares theory was originally developed by Aitken (1935) and was used in the present framework of order statistics by Lloyd (1952); see also Balakrishnan and Rao (1997a,b) for some insights into this method of estimation. The generalized variance that has to be minimized is given by

$$Q(\mu,\sigma) = (\boldsymbol{X} - \mu\mathbf{1} - \sigma\boldsymbol{\mu})'\ \boldsymbol{\Sigma}^{-1}\ (\boldsymbol{X} - \mu\mathbf{1} - \sigma\boldsymbol{\mu}).$$

The estimators so derived by minimizing the above generalized variance are given by [see David (1981, pp. 129-130), Balakrishnan and Cohen (1991, pp. 80-81) and Arnold, Balakrishnan and Nagaraja (1992, pp. 171-173)]

$$\mu^* = -\boldsymbol{\mu}'\boldsymbol{\Gamma}\boldsymbol{X} = \sum_{i=1}^{n-s} a_i X_{i:n} \tag{4.3}$$

and

$$\sigma^* = \mathbf{1}'\boldsymbol{\Gamma}\boldsymbol{X} = \sum_{i=1}^{n-s} b_i X_{i:n}, \tag{4.4}$$

where $\boldsymbol{\Gamma}$ is a skew-symmetric matrix of order $n-s$ given by

$$\boldsymbol{\Gamma} = \frac{1}{\Delta}\boldsymbol{\Sigma}^{-1}(\mathbf{1}\boldsymbol{\mu}' - \boldsymbol{\mu}\mathbf{1}')\boldsymbol{\Sigma}^{-1} \tag{4.5}$$

and

$$\Delta = (\boldsymbol{\mu}'\boldsymbol{\Sigma}^{-1}\boldsymbol{\mu})(\mathbf{1}'\boldsymbol{\Sigma}^{-1}\mathbf{1}) - (\boldsymbol{\mu}'\boldsymbol{\Sigma}^{-1}\mathbf{1})^2. \tag{4.6}$$

In the above formulas, it has been assumed that the observed sample is a Type-II right-censored sample of size $n-s$; that is, of the n items placed on a life-test, the smallest $n-s$ failures were observed and the experimentation was terminated as soon as the $(n-s)$-th failure occurred with the largest s order statistics having been censored. As a result, $\boldsymbol{\mu}$, $\mathbf{1}$ and $\boldsymbol{\Sigma}$ all are of appropriate dimension. Naturally, the complete-sample results will be obtained from the above formulas if we set $s = 0$.

In addition, the variances and covariance of these BLUEs of μ and σ are given by [see David (1981, p. 130), Balakrishnan and Cohen (1991, p. 81) and Arnold, Balakrishnan and Nagaraja (1992, pp. 172-173)]

$$\begin{aligned} Var(\mu^*) &= \sigma^2(\boldsymbol{\mu}'\boldsymbol{\Sigma}^{-1}\boldsymbol{\mu})/\Delta = \sigma^2 V_1, \\ Var(\sigma^*) &= \sigma^2(\mathbf{1}'\boldsymbol{\Sigma}^{-1}\mathbf{1})/\Delta = \sigma^2 V_2, \\ Cov(\mu^*, \sigma^*) &= -\sigma^2(\boldsymbol{\mu}'\boldsymbol{\Sigma}^{-1}\mathbf{1})/\Delta = \sigma^2 V_3. \end{aligned} \tag{4.7}$$

In the above equations, $V_1 = \frac{1}{\Delta}(\boldsymbol{\mu}'\boldsymbol{\Sigma}^{-1}\boldsymbol{\mu})$, $V_2 = \frac{1}{\Delta}(\mathbf{1}'\boldsymbol{\Sigma}^{-1}\mathbf{1})$, and $V_3 = -\frac{1}{\Delta}(\boldsymbol{\mu}'\boldsymbol{\Sigma}^{-1}\mathbf{1})$.

By making use of the tabulated values of means, variances and covariances of order statistics from the standard inverse Gaussian distribution (see Tables 1-3), we computed the coefficients a_i and b_i for the BLUEs of μ and σ, respectively. Once again, these computations were carried out for all sample sizes up to 25 and for various choices of censoring (corresponding to different values of s). The coefficients a_i required for the computation of the BLUE of the location parameter μ have been presented in Table 4. Similarly, the coefficients b_i required for the computation of the BLUE of the scale parameter σ have been presented in Table 5. Once again, these values have been presented for the choices of the shape parameter $k = 0.0(0.1)2.5$. Because of increasing number of choices for the censoring number as the sample size increases, we deliberately restricted the choice of the censoring number s to be at most $[(n+1)/2]$. Thus, the entries in Tables 4 and 5 can be successfully used to compute the BLUEs of the parameters μ and σ based on samples of size at most 25 with at most 50% censoring.

Finally, the variances and covariance factors (V_1, V_2 and V_3) of the BLUEs were also computed for all sample sizes up to 25, for all choices of s up to $[(n+1)/2]$, and for the choices of the shape parameter $k = 0.0(0.1)2.5$. These values are presented in Table 6. Using these values, one can easily compute the standard errors of the best linear unbiased estimates of μ and σ computed for any given observed Type-II censored data.

The values of a_i and b_i reported in Tables 4 and 5 as well as those of V_1, V_2 and V_3 reported in Table 6 for the case $k = 0$ agreed with the previously published results for the case of the normal distribution. In addition, a few more checks were employed in order to test for the validity and accuracy of the values reported in Tables 4 and 5. For example, it is well-known from the formulae of the BLUEs that

$$\sum_{i=1}^{n-s} a_i = 1 \text{ and } \sum_{i=1}^{n-s} b_i = 0. \tag{4.8}$$

These were used for the verification of the computation process.

5

Illustrative Examples

In this chapter, we shall present three examples in order to illustrate the use of Tables 4–6 for the computation of the best linear unbiased estimates of the mean and standard deviation of the assumed inverse Gaussian distribution for the observed data.

Example 1: The following data give the maximum flood levels of Susquehanna river at Harrisburg, Pennsylvania, over four-year periods (during the period 1890–1969) in millions of cubic feet per second. The observed sample of size 20 is presented here in ordered form:

0.265 0.269 0.297 0.315 0.3225 0.338 0.379 0.379 0.392 0.402
0.412 0.416 0.418 0.423 0.449 0.484 0.494 0.613 0.654 0.740

Let us now assume the three-parameter inverse Gaussian distribution in (4.1) for this data. We then present below details of the computation of the best linear unbiased estimates of the mean and standard deviation.

Assuming that the shape parameter $\mathbf{k} = \mathbf{0.5}$, we have from Tables 4 and 5

$$\begin{aligned}\mu^* &= (0.04429 \times 0.265) + (0.04922 \times 0.269) + (0.05013 \times 0.297) + (0.05060 \times 0.315)\\ &+ (0.05087 \times 0.3225) + (0.05103 \times 0.338) + (0.05111 \times 0.379) + (0.05113 \times 0.379)\\ &+ (0.05112 \times 0.392) + (0.05107 \times 0.402) + (0.05099 \times 0.412) + (0.05089 \times 0.416)\\ &+ (0.05075 \times 0.418) + (0.05059 \times 0.423) + (0.05039 \times 0.449) + (0.05015 \times 0.484)\\ &+ (0.04986 \times 0.494) + (0.04947 \times 0.613) + (0.04891 \times 0.654) + (0.04742 \times 0.740)\\ &= 0.42250\end{aligned} \tag{5.1}$$

and

$$\begin{aligned}\sigma^* &= -(0.18567 \times 0.265) - (0.09212 \times 0.269) - (0.06350 \times 0.297) - (0.04475 \times 0.315)\\ &- (0.03072 \times 0.3225) - (0.01946 \times 0.338) - (0.01000 \times 0.379) - (0.00179 \times 0.379)\\ &+ (0.00551 \times 0.392) + (0.01213 \times 0.402) + (0.01822 \times 0.412) + (0.02393 \times 0.416)\\ &+ (0.02935 \times 0.418) + (0.03457 \times 0.423) + (0.03968 \times 0.449) + (0.04479 \times 0.484)\\ &+ (0.05002 \times 0.494) + (0.05559 \times 0.613) + (0.06201 \times 0.654) + (0.07221 \times 0.740)\\ &= 0.11577.\end{aligned} \tag{5.2}$$

From Table 6, we then compute the standard errors of the above BLUEs as

$$S.E.(\mu^*) = \sigma^*(0.04999)^{1/2} = 0.11577(0.04999)^{1/2} = 0.02588,$$
$$S.E.(\sigma^*) = \sigma^*(0.02773)^{1/2} = 0.11577(0.02773)^{1/2} = 0.01928. \tag{5.3}$$

Next, assuming that the shape parameter $\mathbf{k = 1.0}$, we have from Tables 4 and 5

$$\begin{aligned}\mu^* &= (0.01758 \times 0.265) + (0.04866 \times 0.269) + (0.05278 \times 0.297) + (0.05452 \times 0.315)\\ &+ (0.05526 \times 0.3225) + (0.05549 \times 0.338) + (0.05542 \times 0.379) + (0.05515 \times 0.379)\\ &+ (0.05476 \times 0.392) + (0.05426 \times 0.402) + (0.05368 \times 0.412) + (0.05304 \times 0.416)\\ &+ (0.05234 \times 0.418) + (0.05157 \times 0.423) + (0.05072 \times 0.449) + (0.04979 \times 0.484)\\ &+ (0.04874 \times 0.494) + (0.04750 \times 0.613) + (0.04592 \times 0.654) + (0.04282 \times 0.740)\\ &= 0.42202\end{aligned} \tag{5.4}$$

and

$$\begin{aligned}\sigma^* &= -(0.30407 \times 0.265) - (0.10494 \times 0.269) - (0.05921 \times 0.297) - (0.03273 \times 0.315)\\ &- (0.01491 \times 0.3225) - (0.00190 \times 0.338) + (0.00808 \times 0.379) + (0.01600 \times 0.379)\\ &+ (0.02245 \times 0.392) + (0.02781 \times 0.402) + (0.03232 \times 0.412) + (0.03615 \times 0.416)\\ &+ (0.03944 \times 0.418) + (0.04227 \times 0.423) + (0.04472 \times 0.449) + (0.04684 \times 0.484)\\ &+ (0.04865 \times 0.494) + (0.05016 \times 0.613) + (0.05135 \times 0.654) + (0.05153 \times 0.740)\\ &= 0.11892.\end{aligned} \tag{5.5}$$

From Table 6, we then compute the standard errors of the above BLUEs as

$$S.E.(\mu^*) = \sigma^*(0.04984)^{1/2} = 0.11892(0.04984)^{1/2} = 0.02655,$$
$$S.E.(\sigma^*) = \sigma^*(0.03066)^{1/2} = 0.11892(0.03066)^{1/2} = 0.02082. \tag{5.6}$$

Similarly, by assuming that the shape parameter $\mathbf{k = 1.5}$, we have from Tables 4 and 5

$$\begin{aligned}\mu^* &= -(0.04685 \times 0.265) + (0.05235 \times 0.269) + (0.06203 \times 0.297) + (0.06507 \times 0.315)\\ &+ (0.06565 \times 0.3225) + (0.06510 \times 0.338) + (0.06399 \times 0.379) + (0.06257 \times 0.379)\\ &+ (0.06099 \times 0.392) + (0.05931 \times 0.402) + (0.05758 \times 0.412) + (0.05582 \times 0.416)\\ &+ (0.05403 \times 0.418) + (0.05223 \times 0.423) + (0.05039 \times 0.449) + (0.04851 \times 0.484)\\ &+ (0.04654 \times 0.494) + (0.04443 \times 0.613) + (0.04200 \times 0.654) + (0.03826 \times 0.740)\\ &= 0.42321\end{aligned} \tag{5.7}$$

and

$$\begin{aligned}
\sigma^* &= -(0.48194 \times 0.265) - (0.11057 \times 0.269) - (0.04483 \times 0.297) - (0.01138 \times 0.315) \\
&+ (0.00859 \times 0.3225) + (0.02154 \times 0.338) + (0.03034 \times 0.379) + (0.03648 \times 0.379) \\
&+ (0.04082 \times 0.392) + (0.04387 \times 0.402) + (0.04597 \times 0.412) + (0.04736 \times 0.416) \\
&+ (0.04818 \times 0.418) + (0.04855 \times 0.423) + (0.04852 \times 0.449) + (0.04816 \times 0.484) \\
&+ (0.04746 \times 0.494) + (0.04640 \times 0.613) + (0.04482 \times 0.654) + (0.04166 \times 0.740) \\
&= 0.13231.
\end{aligned} \tag{5.8}$$

From Table 6, we then compute the standard errors of the above BLUEs as

$$\begin{aligned}
S.E.(\mu^*) &= \sigma^*(0.04932)^{1/2} = 0.13231(0.04932)^{1/2} = 0.02938, \\
S.E.(\sigma^*) &= \sigma^*(0.03603)^{1/2} = 0.13231(0.03603)^{1/2} = 0.02511.
\end{aligned} \tag{5.9}$$

Similarly, by assuming that the shape parameter $\mathbf{k = 2.0}$, we have from Tables 4 and 5

$$\begin{aligned}
\mu^* &= -(0.16520 \times 0.265) + (0.06518 \times 0.269) + (0.08192 \times 0.297) + (0.08509 \times 0.315) \\
&+ (0.08384 \times 0.3225) + (0.08089 \times 0.338) + (0.07728 \times 0.379) + (0.07345 \times 0.379) \\
&+ (0.06962 \times 0.392) + (0.06589 \times 0.402) + (0.06230 \times 0.412) + (0.05886 \times 0.416) \\
&+ (0.05558 \times 0.418) + (0.05244 \times 0.423) + (0.04944 \times 0.449) + (0.04654 \times 0.484) \\
&+ (0.04372 \times 0.494) + (0.04091 \times 0.613) + (0.03799 \times 0.654) + (0.03425 \times 0.740) \\
&= 0.42733
\end{aligned} \tag{5.10}$$

and

$$\begin{aligned}
\sigma^* &= -(0.73942 \times 0.265) - (0.10364 \times 0.269) - (0.01596 \times 0.297) + (0.02248 \times 0.315) \\
&+ (0.04201 \times 0.3225) + (0.05245 \times 0.338) + (0.05795 \times 0.379) + (0.06055 \times 0.379) \\
&+ (0.06137 \times 0.392) + (0.06105 \times 0.402) + (0.05998 \times 0.412) + (0.05842 \times 0.416) \\
&+ (0.05651 \times 0.418) + (0.05438 \times 0.423) + (0.05208 \times 0.449) + (0.04965 \times 0.484) \\
&+ (0.04709 \times 0.494) + (0.04438 \times 0.613) + (0.04138 \times 0.654) + (0.03727 \times 0.740) \\
&= 0.15147.
\end{aligned} \tag{5.11}$$

From Table 6, we then compute the standard errors of the above BLUEs as

$$\begin{aligned}
S.E.(\mu^*) &= \sigma^*(0.04824)^{1/2} = 0.15147(0.04824)^{1/2} = 0.03327, \\
S.E.(\sigma^*) &= \sigma^*(0.04378)^{1/2} = 0.15147(0.04378)^{1/2} = 0.03169.
\end{aligned} \tag{5.12}$$

Finally, by assuming that the shape parameter $\mathbf{k} = \mathbf{2.5}$, we have from Tables 4 and 5

$$\begin{aligned}\mu^* &= -(0.35144 \times 0.265) + (0.09197 \times 0.269) + (0.11595 \times 0.297) + (0.11680 \times 0.315)\\ &+ (0.11110 \times 0.3225) + (0.10346 \times 0.338) + (0.09546 \times 0.379) + (0.08771 \times 0.379)\\ &+ (0.08046 \times 0.392) + (0.07377 \times 0.402) + (0.06765 \times 0.412) + (0.06207 \times 0.416)\\ &+ (0.05698 \times 0.418) + (0.05235 \times 0.423) + (0.04814 \times 0.449) + (0.04428 \times 0.484)\\ &+ (0.04073 \times 0.494) + (0.03744 \times 0.613) + (0.03429 \times 0.654) + (0.03083 \times 0.740)\\ &= 0.43527\end{aligned} \tag{5.13}$$

and

$$\begin{aligned}\sigma^* &= -(1.10707 \times 0.265) - (0.07662 \times 0.269) + (0.03402 \times 0.297) + (0.07395 \times 0.315)\\ &+ (0.08914 \times 0.3225) + (0.09363 \times 0.338) + (0.09301 \times 0.379) + (0.08983 \times 0.379)\\ &+ (0.08540 \times 0.392) + (0.08040 \times 0.402) + (0.07522 \times 0.412) + (0.07008 \times 0.416)\\ &+ (0.06508 \times 0.418) + (0.06031 \times 0.423) + (0.05577 \times 0.449) + (0.05149 \times 0.484)\\ &+ (0.04744 \times 0.494) + (0.04358 \times 0.613) + (0.03980 \times 0.654) + (0.03552 \times 0.740)\\ &= 0.18019.\end{aligned} \tag{5.14}$$

From Table 6, we then compute the standard errors of the above BLUEs as

$$\begin{aligned}S.E.(\mu^*) &= \sigma^*(0.04660)^{1/2} = 0.18019(0.04660^{1/2} = 0.03890,\\ S.E.(\sigma^*) &= \sigma^*(0.05344)^{1/2} = 0.18019(0.05344)^{1/2} = 0.04165.\end{aligned} \tag{5.15}$$

Example 2: Twenty three ball bearings were placed on a life-test and the following data give the number of million revolutions before failure for each of these ball bearings. The life-test experiment was terminated as soon as the twentieth ball bearing failed with three bearings still surviving at the time of termination of the experiment. The observed Type-II right-censored sample of size 20 (obtained from a sample of size 23) is presented here in ordered form, with the largest three order statistics censored:

17.88	28.92	33.00	41.52	42.12	45.60	48.48	51.84	51.96	54.12
55.56	67.80	68.64	68.64	68.88	84.12	93.12	96.64	105.12	105.84

Let us now assume the three-parameter inverse Gaussian distribution in (4.1) for this data. We then present below details of the computation of the best linear unbiased estimates of the mean and standard deviation.

Assuming that the shape parameter $\mathbf{k} = \mathbf{0.3}$, we have from Tables 4 and 5

$$\begin{aligned}\mu^* &= (0.02661 \times 17.88) + (0.03471 \times 28.92) + (0.03733 \times 33.00) + (0.03908 \times 41.52)\\ &+ (0.04042 \times 42.12) + (0.04151 \times 45.60) + (0.04244 \times 48.48) + (0.04325 \times 51.84)\\ &+ (0.04398 \times 51.96) + (0.04465 \times 54.12) + (0.04526 \times 55.56) + (0.04583 \times 67.80)\end{aligned}$$

$$
\begin{aligned}
&+ (0.04637 \times 68.64) + (0.04689 \times 68.64) + (0.04738 \times 68.68) + (0.04785 \times 84.12) \\
&+ (0.04831 \times 93.12) + (0.04874 \times 96.64) + (0.04914 \times 105.12) + (0.18024 \times 105.84) \\
&= 69.694
\end{aligned}
\tag{5.16}
$$

and

$$
\begin{aligned}
\sigma^* &= -(0.16035 \times 17.88) - (0.09274 \times 28.92) - (0.06957 \times 33.00) - (0.05371 \times 41.52) \\
&- (0.04140 \times 42.12) - (0.03122 \times 45.60) - (0.02242 \times 48.48) - (0.01460 \times 51.84) \\
&- (0.00750 \times 51.96) - (0.00093 \times 54.12) + (0.00524 \times 55.56) + (0.01110 \times 67.80) \\
&+ (0.01674 \times 68.64) + (0.02223 \times 68.64) + (0.02763 \times 68.88) + (0.03300 \times 84.12) \\
&+ (0.03842 \times 93.12) + (0.04395 \times 96.64) + (0.04970 \times 105.12) + (0.24643 \times 105.84) \\
&= 32.003.
\end{aligned}
\tag{5.17}
$$

From Table 6, we compute the standard errors of the above BLUEs as

$$
\begin{aligned}
S.E.(\mu^*) &= \sigma^*(0.04536)^{1/2} = 32.003(0.04536)^{1/2} = 6.816, \\
S.E.(\sigma^*) &= \sigma^*(0.02806)^{1/2} = 32.003(0.02806)^{1/2} = 5.361.
\end{aligned}
\tag{5.18}
$$

Next, assuming that the shape parameter $\mathbf{k = 0.5}$, we have from Tables 4 and 5

$$
\begin{aligned}
\mu^* &= (0.01905 \times 17.88) + (0.03324 \times 28.92) + (0.03704 \times 33.00) + (0.03941 \times 41.52) \\
&+ (0.04111 \times 42.12) + (0.04242 \times 45.60) + (0.04348 \times 48.48) + (0.04436 \times 51.84) \\
&+ (0.04510 \times 51.96) + (0.04575 \times 54.12) + (0.04631 \times 55.56) + (0.04681 \times 67.80) \\
&+ (0.04725 \times 68.64) + (0.04764 \times 68.64) + (0.04799 \times 68.68) + (0.04829 \times 84.12) \\
&+ (0.04855 \times 93.12) + (0.04876 \times 96.64) + (0.04889 \times 105.12) + (0.17853 \times 105.84) \\
&= 69.949
\end{aligned}
\tag{5.19}
$$

and

$$
\begin{aligned}
\sigma^* &= -(0.19500 \times 17.88) - (0.09948 \times 28.92) - (0.07050 \times 33.00) - (0.05157 \times 41.52) \\
&- (0.03746 \times 42.12) - (0.02617 \times 45.60) - (0.01672 \times 48.48) - (0.00856 \times 51.84) \\
&- (0.00135 \times 51.96) + (0.00514 \times 54.12) + (0.01108 \times 55.56) + (0.01657 \times 67.80) \\
&+ (0.02172 \times 68.64) + (0.02659 \times 68.64) + (0.03126 \times 68.88) + (0.03576 \times 84.12) \\
&+ (0.04017 \times 93.12) + (0.04452 \times 96.64) + (0.04886 \times 105.12) + (0.22513 \times 105.84) \\
&= 32.575.
\end{aligned}
\tag{5.20}
$$

From Table 6, we compute the standard errors of the above BLUEs as

$$S.E.(\mu^*) = \sigma^*(0.04563)^{1/2} = 32.575(0.04563)^{1/2} = 6.958,$$
$$S.E.(\sigma^*) = \sigma^*(0.02814)^{1/2} = 32.575(0.02814)^{1/2} = 5.464. \tag{5.21}$$

Next, assuming that the shape parameter $\mathbf{k} = \mathbf{1.0}$, we have from Tables 4 and 5

$$\begin{aligned}\mu^* &= -(0.01911 \times 17.88) + (0.02998 \times 28.92) + (0.03869 \times 33.00) + (0.04321 \times 41.52)\\ &+ (0.04591 \times 42.12) + (0.04763 \times 45.60) + (0.04876 \times 48.48) + (0.04950 \times 51.84)\\ &+ (0.04996 \times 51.96) + (0.05022 \times 54.12) + (0.05032 \times 55.56) + (0.05030 \times 67.80)\\ &+ (0.05018 \times 68.64) + (0.04996 \times 68.64) + (0.04966 \times 68.68) + (0.04928 \times 84.12)\\ &+ (0.04881 \times 93.12) + (0.04825 \times 96.64) + (0.04756 \times 105.12) + (0.17092 \times 105.84)\\ &= 70.889\end{aligned} \tag{5.22}$$

and

$$\begin{aligned}\sigma^* &= -(0.31748 \times 17.88) - (0.11527 \times 28.92) - (0.06866 \times 33.00) - (0.04157 \times 41.52)\\ &- (0.02326 \times 42.12) - (0.00984 \times 45.60) + (0.00049 \times 48.48) + (0.00873 \times 51.84)\\ &+ (0.01548 \times 51.96) + (0.02110 \times 54.12) + (0.02586 \times 55.56) + (0.02993 \times 67.80)\\ &+ (0.03345 \times 68.64) + (0.03651 \times 68.64) + (0.03918 \times 68.88) + (0.04152 \times 84.12)\\ &+ (0.04356 \times 93.12) + (0.04533 \times 96.64) + (0.04683 \times 105.12) + (0.18811 \times 105.84)\\ &= 35.572\end{aligned} \tag{5.23}$$

From Table 6, we compute the standard errors of the above BLUEs as

$$S.E.(\mu^*) = \sigma^*(0.04607)^{1/2} = 35.572(0.04607)^{1/2} = 7.635,$$
$$S.E.(\sigma^*) = \sigma^*(0.03000)^{1/2} = 35.572(0.03000)^{1/2} = 6.161. \tag{5.24}$$

Next, assuming that the shape parameter $\mathbf{k} = \mathbf{1.5}$, we have from Tables 4 and 5

$$\begin{aligned}\mu^* &= -(0.09919 \times 17.88) + (0.03034 \times 28.92) + (0.04688 \times 33.00) + (0.05385 \times 41.52)\\ &+ (0.05706 \times 42.12) + (0.05843 \times 45.60) + (0.05880 \times 48.48) + (0.05857 \times 51.84)\\ &+ (0.05797 \times 51.96) + (0.05713 \times 54.12) + (0.05612 \times 55.56) + (0.05499 \times 67.80)\\ &+ (0.05379 \times 68.64) + (0.05252 \times 68.64) + (0.05121 \times 68.68) + (0.04984 \times 84.12)\\ &+ (0.04843 \times 93.12) + (0.04696 \times 96.64) + (0.04541 \times 105.12) + (0.16089 \times 105.84)\\ &= 72.659\end{aligned} \tag{5.25}$$

and

$$\begin{aligned}\sigma^* &= -(0.50490 \times 17.88) - (0.12647 \times 28.92) - (0.05812 \times 33.00) - (0.02282 \times 41.52)\\ &- (0.00139 \times 42.12) + (0.01275 \times 45.60) + (0.02257 \times 48.48) + (0.02961 \times 51.84)\\ &+ (0.03474 \times 51.96) + (0.03852 \times 54.12) + (0.04130 \times 55.56) + (0.04333 \times 67.80)\\ &+ (0.04476 \times 68.64) + (0.04572 \times 68.64) + (0.04629 \times 68.88) + (0.04654 \times 84.12)\\ &+ (0.04651 \times 93.12) + (0.04622 \times 96.64) + (0.04568 \times 105.12) + (0.16913 \times 105.84)\\ &= 41.538.\end{aligned} \tag{5.26}$$

From Table 6, we compute the standard errors of the above BLUEs as

$$\begin{aligned}S.E.(\mu^*) &= \sigma^*(0.04596)^{1/2} = 41.538(0.04596)^{1/2} = 8.905,\\ S.E.(\sigma^*) &= \sigma^*(0.03448)^{1/2} = 41.538(0.03448)^{1/2} = 7.713.\end{aligned} \tag{5.27}$$

Next, assuming that the shape parameter $\mathbf{k} = \mathbf{2.0}$, we have from Tables 4 and 5

$$\begin{aligned}\mu^* &= -(0.23871 \times 17.88) + (0.03877 \times 28.92) + (0.06558 \times 33.00) + (0.07429 \times 41.52)\\ &+ (0.07660 \times 42.12) + (0.07614 \times 45.60) + (0.07435 \times 48.48) + (0.07192 \times 51.84)\\ &+ (0.06918 \times 51.96) + (0.06633 \times 54.12) + (0.06345 \times 55.56) + (0.06060 \times 67.80)\\ &+ (0.05783 \times 68.64) + (0.05513 \times 68.64) + (0.05252 \times 68.68) + (0.04999 \times 84.12)\\ &+ (0.04754 \times 93.12) + (0.04516 \times 96.64) + (0.04283 \times 105.12) + (0.15050 \times 105.84)\\ &= 75.750\end{aligned} \tag{5.28}$$

and

$$\begin{aligned}\sigma^* &= -(0.78100 \times 17.88) - (0.12812 \times 28.92) - (0.03417 \times 33.00) + (0.00843 \times 41.52)\\ &+ (0.03101 \times 42.12) + (0.04382 \times 45.60) + (0.05122 \times 48.48) + (0.05540 \times 51.84)\\ &+ (0.05755 \times 51.96) + (0.05837 \times 54.12) + (0.05828 \times 55.56) + (0.05759 \times 67.80)\\ &+ (0.05647 \times 68.64) + (0.05504 \times 68.64) + (0.05340 \times 68.88) + (0.05161 \times 84.12)\\ &+ (0.04970 \times 93.12) + (0.04770 \times 96.64) + (0.04563 \times 105.12) + (0.16207 \times 105.84)\\ &= 50.366.\end{aligned} \tag{5.29}$$

From Table 6, we compute the standard errors of the above BLUEs as

$$\begin{aligned}S.E.(\mu^*) &= \sigma^*(0.04517)^{1/2} = 50.366(0.04517)^{1/2} = 10.704,\\ S.E.(\sigma^*) &= \sigma^*(0.04141)^{1/2} = 50.366(0.04141)^{1/2} = 10.249.\end{aligned} \tag{5.30}$$

Next, assuming that the shape parameter $\mathbf{k} = \mathbf{2.5}$, we have from Tables 4 and 5

$$\begin{aligned}\mu^* &= -(0.45339 \times 17.88) + (0.05983 \times 28.92) + (0.09845 \times 33.00) + (0.10706 \times 41.52)\\ &+ (0.10616 \times 42.12) + (0.10169 \times 45.60) + (0.09589 \times 48.48) + (0.08972 \times 51.84)\\ &+ (0.08359 \times 51.96) + (0.07772 \times 54.12) + (0.07218 \times 55.56) + (0.06702 \times 67.80)\\ &+ (0.06222 \times 68.64) + (0.05778 \times 68.64) + (0.05368 \times 68.68) + (0.04989 \times 84.12)\\ &+ (0.04639 \times 93.12) + (0.04317 \times 96.64) + (0.04019 \times 105.12) + (0.14076 \times 105.84)\\ &= 80.575\end{aligned} \tag{5.31}$$

and

$$\begin{aligned}\sigma^* &= -(1.18096 \times 17.88) - (0.11309 \times 28.92) + (0.01018 \times 33.00) + (0.05793 \times 41.52)\\ &+ (0.07849 \times 42.12) + (0.08690 \times 45.60) + (0.08921 \times 48.48) + (0.08829 \times 51.84)\\ &+ (0.08562 \times 51.96) + (0.08200 \times 54.12) + (0.07791 \times 55.56) + (0.07363 \times 67.80)\\ &+ (0.06932 \times 68.64) + (0.06508 \times 68.64) + (0.06098 \times 68.88) + (0.05705 \times 84.12)\\ &+ (0.05331 \times 93.12) + (0.04976 \times 96.64) + (0.04640 \times 105.12) + (0.16201 \times 105.84)\\ &= 62.755.\end{aligned} \tag{5.32}$$

From Table 6, we compute the standard errors of the above BLUEs as

$$\begin{aligned}S.E.(\mu^*) &= \sigma^*(0.04375)^{1/2} = 62.755(0.04375)^{1/2} = 13.126,\\ S.E.(\sigma^*) &= \sigma^*(0.05027)^{1/2} = 62.755(0.05027)^{1/2} = 14.070.\end{aligned} \tag{5.33}$$

Example 3: Let us consider the life-test experiment taken in the last example. Suppose the complete data on all twenty three ball bearings were available; that is, the number of million revolutions before the failure of each of the 23 ball bearings were available. The so observed complete sample of size 23 is presented here in ordered form:

17.88 28.92 33.00 41.52 42.12 45.60 48.48 51.84 51.96 54.12
55.56 67.80 68.64 68.64 68.88 84.12 93.12 96.64 105.12 105.84
127.92 128.04 173.40

Let us now assume the three-parameter inverse Gaussian distribution in (4.1) for this complete data. We then present below details of the computation of the best linear unbiased estimates of the mean and standard deviation.

Assuming that the shape parameter $\mathbf{k} = \mathbf{0.3}$, we have from Tables 4 and 5

$$\begin{aligned}\mu^* &= (0.04182 \times 17.88) + (0.04311 \times 28.92) + (0.04339 \times 33.00) + (0.04355 \times 41.52)\\ &+ (0.04365 \times 42.12) + (0.04372 \times 45.60) + (0.04376 \times 48.48) + (0.04379 \times 51.84)\\ &+ (0.04381 \times 51.96) + (0.04382 \times 54.12) + (0.04381 \times 55.56) + (0.04380 \times 67.80)\\ &+ (0.04379 \times 68.64) + (0.04376 \times 68.64) + (0.04373 \times 68.68) + (0.04369 \times 84.12)\end{aligned}$$

$$
\begin{aligned}
&+ (0.04364 \times 93.12) + (0.04358 \times 96.64) + (0.04351 \times 105.12) + (0.04341 \times 105.84) \\
&+ (0.04328 \times 127.92) + (0.04308 \times 128.04) + (0.04248 \times 173.40) \\
&= 72.069
\end{aligned}
\tag{5.34}
$$

and

$$
\begin{aligned}
\sigma^* &= -(0.13630 \times 17.88) - (0.07947 \times 28.92) - (0.05998 \times 33.00) - (0.04664 \times 41.52) \\
&- (0.03629 \times 42.12) - (0.02772 \times 45.60) - (0.02032 \times 48.48) - (0.01374 \times 51.84) \\
&- (0.00775 \times 51.96) - (0.00222 \times 54.12) + (0.00298 \times 55.56) + (0.00792 \times 67.80) \\
&+ (0.01268 \times 68.64) + (0.01732 \times 68.64) + (0.02189 \times 68.88) + (0.02645 \times 84.12) \\
&+ (0.03106 \times 93.12) + (0.03581 \times 96.64) + (0.04081 \times 105.12) + (0.04622 \times 105.84) \\
&+ (0.05238 \times 127.92) + (0.06006 \times 128.04) + (0.07485 \times 173.40) \\
&= 35.735.
\end{aligned}
\tag{5.35}
$$

From Table 6, we have the standard errors of the above BLUEs to be

$$
\begin{aligned}
S.E.(\mu^*) &= \sigma^*(0.04348)^{1/2} = 35.735(0.04348)^{1/2} = 7.451, \\
S.E.(\sigma^*) &= \sigma^*(0.02339)^{1/2} = 35.735(0.02339)^{1/2} = 5.465
\end{aligned}
\tag{5.36}
$$

Next, assuming that the shape parameter $\mathbf{k = 0.5}$, we have from Tables 4 and 5

$$
\begin{aligned}
\mu^* &= (0.03806 \times 17.88) + (0.04253 \times 28.92) + (0.04339 \times 33.00) + (0.04384 \times 41.52) \\
&+ (0.04412 \times 42.12) + (0.04428 \times 45.60) + (0.04439 \times 48.48) + (0.04444 \times 51.84) \\
&+ (0.04446 \times 51.96) + (0.04446 \times 54.12) + (0.04443 \times 55.56) + (0.04438 \times 67.80) \\
&+ (0.04431 \times 68.64) + (0.04422 \times 68.64) + (0.04412 \times 68.68) + (0.04399 \times 84.12) \\
&+ (0.04385 \times 93.12) + (0.04367 \times 96.64) + (0.04346 \times 105.12) + (0.04321 \times 105.84) \\
&+ (0.04287 \times 127.92) + (0.04239 \times 128.04) + (0.04112 \times 173.40) \\
&= 71.993
\end{aligned}
\tag{5.37}
$$

and

$$
\begin{aligned}
\sigma^* &= -(0.16822 \times 17.88) - (0.08639 \times 28.92) - (0.06155 \times 33.00) - (0.04532 \times 41.52) \\
&- (0.03322 \times 42.12) - (0.02354 \times 45.60) - (0.01543 \times 48.48) - (0.00843 \times 51.84) \\
&- (0.00224 \times 51.96) + (0.00334 \times 54.12) + (0.00843 \times 55.56) + (0.01316 \times 67.80) \\
&+ (0.01759 \times 68.64) + (0.02179 \times 68.64) + (0.02582 \times 68.88) + (0.02972 \times 84.12) \\
&+ (0.03355 \times 93.12) + (0.03736 \times 96.64) + (0.04123 \times 105.12) + (0.04524 \times 105.84) \\
&+ (0.04959 \times 127.92) + (0.05465 \times 128.04) + (0.06286 \times 173.40)
\end{aligned}
$$

$$= 35.658. \tag{5.38}$$

From Table 6, we have the standard errors of the above BLUEs to be

$$\begin{aligned} S.E.(\mu^*) &= \sigma^*(0.04347)^{1/2} = 35.658(0.04347)^{1/2} = 7.435, \\ S.E.(\sigma^*) &= \sigma^*(0.02387)^{1/2} = 35.658(0.02387)^{1/2} = 5.509. \end{aligned} \tag{5.39}$$

Next, assuming that the shape parameter $\mathbf{k = 1.0}$, we have from Tables 4 and 5

$$\begin{aligned} \mu^* &= (0.01215 \times 17.88) + (0.04099 \times 28.92) + (0.04506 \times 33.00) + (0.04688 \times 41.52) \\ &+ (0.04777 \times 42.12) + (0.04816 \times 45.60) + (0.04827 \times 48.48) + (0.04820 \times 51.84) \\ &+ (0.04800 \times 51.96) + (0.04771 \times 54.12) + (0.04735 \times 55.56) + (0.04694 \times 67.80) \\ &+ (0.04647 \times 68.64) + (0.04597 \times 68.64) + (0.04543 \times 68.68) + (0.04484 \times 84.12) \\ &+ (0.04420 \times 93.12) + (0.04351 \times 96.64) + (0.04275 \times 105.12) + (0.04189 \times 105.84) \\ &+ (0.04088 \times 127.92) + (0.03958 \times 128.04) + (0.03700 \times 173.40) \\ &= 71.980 \end{aligned} \tag{5.40}$$

and

$$\begin{aligned} \sigma^* &= -(0.28131 \times 17.88) - (0.10252 \times 28.92) - (0.06129 \times 33.00) - (0.03731 \times 41.52) \\ &- (0.02110 \times 42.12) - (0.00923 \times 45.60) - (0.00007 \times 48.48) + (0.00723 \times 51.84) \\ &+ (0.01321 \times 51.96) + (0.01820 \times 54.12) + (0.02242 \times 55.56) + (0.02604 \times 67.80) \\ &+ (0.02917 \times 68.64) + (0.03190 \times 68.64) + (0.03428 \times 68.88) + (0.03638 \times 84.12) \\ &+ (0.03823 \times 93.12) + (0.03985 \times 96.64) + (0.04127 \times 105.12) + (0.04250 \times 105.84) \\ &+ (0.04353 \times 127.92) + (0.04433 \times 128.04) + (0.04430 \times 173.40) \\ &= 36.965. \end{aligned} \tag{5.41}$$

From Table 6, we have the standard errors of the above BLUEs to be

$$\begin{aligned} S.E.(\mu^*) &= \sigma^*(0.04334)^{1/2} = 36.965(0.04334)^{1/2} = 7.695, \\ S.E.(\sigma^*) &= \sigma^*(0.02635)^{1/2} = 36.965(0.02635)^{1/2} = 6.000. \end{aligned} \tag{5.42}$$

Next, assuming that the shape parameter $\mathbf{k = 1.5}$, we have from Tables 4 and 5

$$\begin{aligned} \mu^* &= -(0.05157 \times 17.88) + (0.04214 \times 28.92) + (0.05223 \times 33.00) + (0.05587 \times 41.52) \\ &+ (0.05706 \times 42.12) + (0.05711 \times 45.60) + (0.05655 \times 48.48) + (0.05567 \times 51.84) \\ &+ (0.05459 \times 51.96) + (0.05339 \times 54.12) + (0.05212 \times 55.56) + (0.05082 \times 67.80) \end{aligned}$$

$$+ (0.04949 \times 68.64) + (0.04814 \times 68.64) + (0.04678 \times 68.68) + (0.04540 \times 84.12)$$
$$+ (0.04401 \times 93.12) + (0.04259 \times 96.64) + (0.04113 \times 105.12) + (0.03959 \times 105.84)$$
$$+ (0.03793 \times 127.92) + (0.03600 \times 128.04) + (0.03297 \times 173.40)$$
$$= 72.691 \tag{5.43}$$

and

$$\sigma^* = -(0.45397 \times 17.88) - (0.11385 \times 28.92) - (0.05240 \times 33.00) - (0.02066 \times 41.52)$$
$$- (0.00139 \times 42.12) + (0.01133 \times 45.60) + (0.02017 \times 48.48) + (0.02650 \times 51.84)$$
$$+ (0.03112 \times 51.96) + (0.03453 \times 54.12) + (0.03703 \times 55.56) + (0.03886 \times 67.80)$$
$$+ (0.04016 \times 68.64) + (0.04103 \times 68.64) + (0.04156 \times 68.88) + (0.04180 \times 84.12)$$
$$+ (0.04179 \times 93.12) + (0.04155 \times 96.64) + (0.04111 \times 105.12) + (0.04044 \times 105.84)$$
$$+ (0.03951 \times 127.92) + (0.03819 \times 128.04) + (0.03559 \times 173.40)$$
$$= 41.567. \tag{5.44}$$

From Table 6, we have the standard errors of the above BLUEs to be

$$S.E.(\mu^*) = \sigma^*(0.04286)^{1/2} = 41.567(0.04286)^{1/2} = 8.605,$$
$$S.E.(\sigma^*) = \sigma^*(0.03094)^{1/2} = 41.567(0.03094)^{1/2} = 7.312. \tag{5.45}$$

Next, assuming that the shape parameter $\mathbf{k} = \mathbf{2.0}$, we have from Tables 4 and 5

$$\mu^* = -(0.17018 \times 17.88) + (0.05026 \times 28.92) + (0.06883 \times 33.00) + (0.07379 \times 41.52)$$
$$+ (0.07410 \times 42.12) + (0.07250 \times 45.60) + (0.07005 \times 48.48) + (0.06724 \times 51.84)$$
$$+ (0.06430 \times 51.96) + (0.06136 \times 54.12) + (0.05848 \times 55.56) + (0.05569 \times 67.80)$$
$$+ (0.05301 \times 68.64) + (0.05043 \times 68.64) + (0.04795 \times 68.68) + (0.04558 \times 84.12)$$
$$+ (0.04329 \times 93.12) + (0.04108 \times 96.64) + (0.03892 \times 105.12) + (0.03680 \times 105.84)$$
$$+ (0.03466 \times 127.92) + (0.03241 \times 128.04) + (0.02946 \times 173.40)$$
$$= 74.658 \tag{5.46}$$

and

$$\sigma^* = -(0.70707 \times 17.88) - (0.11572 \times 28.92) - (0.03067 \times 33.00) + (0.00788 \times 41.52)$$
$$+ (0.02831 \times 42.12) + (0.03989 \times 45.60) + (0.04658 \times 48.48) + (0.05035 \times 51.84)$$
$$+ (0.05229 \times 51.96) + (0.05301 \times 54.12) + (0.05293 \times 55.56) + (0.05229 \times 67.80)$$
$$+ (0.05127 \times 68.64) + (0.04997 \times 68.64) + (0.04848 \times 68.88) + (0.04685 \times 84.12)$$
$$+ (0.04512 \times 93.12) + (0.04330 \times 96.64) + (0.04141 \times 105.12) + (0.03943 \times 105.84)$$

$$
\begin{aligned}
&+ (0.03734 \times 127.92) + (0.03501 \times 128.04) + (0.03177 \times 173.40) \\
&= 49.188.
\end{aligned}
\tag{5.47}
$$

From Table 6, we have the standard errors of the above BLUEs to be

$$
\begin{aligned}
S.E.(\mu^*) &= \sigma^*(0.04188)^{1/2} = 49.188(0.04188)^{1/2} = 10.066, \\
S.E.(\sigma^*) &= \sigma^*(0.03758)^{1/2} = 49.188(0.03758)^{1/2} = 9.535.
\end{aligned}
\tag{5.48}
$$

Next, assuming that the shape parameter $\mathbf{k} = \mathbf{2.5}$, we have from Tables 4 and 5

$$
\begin{aligned}
\mu^* &= -(0.35856 \times 17.88) + (0.06972 \times 28.92) + (0.09837 \times 33.00) + (0.10307 \times 41.52) \\
&+ (0.10044 \times 42.12) + (0.09524 \times 45.60) + (0.08920 \times 48.48) + (0.08305 \times 51.84) \\
&+ (0.07710 \times 51.96) + (0.07148 \times 54.12) + (0.06624 \times 55.56) + (0.06139 \times 67.80) \\
&+ (0.05691 \times 68.64) + (0.05278 \times 68.64) + (0.04898 \times 68.68) + (0.04548 \times 84.12) \\
&+ (0.04226 \times 93.12) + (0.03928 \times 96.64) + (0.03651 \times 105.12) + (0.03394 \times 105.84) \\
&+ (0.03151 \times 127.92) + (0.02915 \times 128.04) + (0.02647 \times 173.40) \\
&= 78.294
\end{aligned}
\tag{5.49}
$$

and

$$
\begin{aligned}
\sigma^* &= -(1.07222 \times 17.88) - (0.10176 \times 28.92) + (0.01008 \times 33.00) + (0.05335 \times 41.52) \\
&+ (0.07194 \times 42.12) + (0.07950 \times 45.60) + (0.08154 \times 48.48) + (0.08065 \times 51.84) \\
&+ (0.07817 \times 51.96) + (0.07485 \times 54.12) + (0.07109 \times 55.56) + (0.06717 \times 67.80) \\
&+ (0.06322 \times 68.64) + (0.05935 \times 68.64) + (0.05559 \times 68.88) + (0.05199 \times 84.12) \\
&+ (0.04856 \times 93.12) + (0.04530 \times 96.64) + (0.04219 \times 105.12) + (0.03923 \times 105.84) \\
&+ (0.03638 \times 127.92) + (0.03355 \times 128.04) + (0.03026 \times 173.40) \\
&= 60.138.
\end{aligned}
\tag{5.50}
$$

From Table 6, we have the standard errors of the above BLUEs to be

$$
\begin{aligned}
S.E.(\mu^*) &= \sigma^*(0.04040)^{1/2} = 60.138(0.04040)^{1/2} = 12.088, \\
S.E.(\sigma^*) &= \sigma^*(0.04586)^{1/2} = 60.138(0.04586)^{1/2} = 12.879.
\end{aligned}
\tag{5.51}
$$

6

Best Linear Unbiased Prediction

Prediction problems arise naturally in life-testing experiments. For example, let us consider the life-test experiment described in Example 2 of the last Chapter. In this case, twenty three ball bearings were placed on a life-test and the data on the number of million revolutions before failure of each of these ball bearings were observed. The experiment itself was terminated as soon as the twentieth ball bearing failed with three ball bearings still surviving at the time of termination of the experiment. It is, therefore, natural for the experimenter to be interested in predicting the number of million revolutions before failure of the remaining three surviving ball bearings. In particular, the experimenter may be interested in predicting the very next failure or the very last failure.

Clearly, the best unbiased predictor of $X_{\ell:n}$, based on the observed Type-II right-censored sample of size $n-s$ given by $X_{1:n}, X_{2:n}, \ldots, X_{n-s:n}$, is

$$E[X_{\ell:n}|X_{1:n}, X_{2:n}, \ldots, X_{n-s:n}] = E[X_{\ell:n}|X_{n-s:n}]$$

by the Markovian property of order statistics from absolutely continuous distributions. However, this conditional expectation involves the unknown parameters of the underlying distribution that need to be estimated. From the results on the general linear model [see Goldberger (1962)], it is well-known that the Best Linear Unbiased Predictor (BLUP) of $X_{\ell:n}$ is

$$X^*_{\ell:n} = \mu^* + \sigma^* \mu_{\ell:n} + \boldsymbol{\omega}'\boldsymbol{\Sigma}^{-1}\left(\boldsymbol{X} - \mu^*\mathbf{1} - \sigma^*\boldsymbol{\mu}\right), \tag{6.1}$$

where μ^* and σ^* are the BLUEs of μ and σ, respectively, based on the observed Type-II right-censored sample of size $n-s$, and $\boldsymbol{\omega}' = (\sigma_{1,\ell:n}, \ldots, \sigma_{n-s,\ell:n})$. Using the above formula, and the values of means and variances and covariances of order statistics tabulated in Tables 1—3, one can compute the BLUP of $X_{\ell:n}$ based on the observed values of the smallest $n-s$ order statistics.

Recently, Doganaksoy and Balakrishnan (1997) presented another simple way of getting the BLUP simply from the tables of BLUEs of either the location parameter μ or the scale parameter σ. These authors showed that the BLUEs of μ and σ remain unchanged when the BLUP value of the order statistic $X_{n-s+1:n}$ is used in turn as an observed value of that order statistic. Based on this observation, the BLUP of $X_{n-s+1:n}$ can then be simply calculated by solving for $X_{n-s+1:n}$ from the equation $\mu^*_{n-s+1} = \mu^*_{n-s}$ or $\sigma^*_{n-s+1} = \sigma^*_{n-s}$, where μ^*_{n-s} and μ^*_{n-s+1} are the BLUEs of μ based on the smallest $n-s$ and $n-s+1$ order statistics, respectively, and similarly σ^*_{n-s} and σ^*_{n-s+1} are the BLUEs of σ based on the smallest $n-s$ and $n-s+1$ order statistics, respectively. As Doganaksoy and Balakrishnan (1997) have shown, this procedure may be applied repeatedly in order to find the BLUP of any order statistic $X_{\ell:n}$ for $\ell \geq n-s+1$.

For the purpose of illustration, let us consider the data presented in Example 2 of the last Chapter. In this case, we have $n = 23$ and $s = 3$. The Type-II right-censored sample of size 20 so observed has been presented in Example 2 of the last Chapter. Now, suppose we are interested in finding the best linear unbiased predictor of $X_{21:23}$.

Assuming that the shape parameter $\mathbf{k} = \mathbf{0.3}$, we computed earlier the BLUE of μ based on the smallest 20 order statistics to be $\mu_{20}^* = 69.694$. Now, assuming that the smallest 21 order statistics are available, we have from Table 4 the BLUE of μ to be

$$\begin{aligned}
\mu_{21}^* &= (0.03260 \times 17.88) + (0.03799 \times 28.92) + (0.03968 \times 33.00) + (0.04080 \times 41.52) \\
&\quad + (0.04164 \times 42.12) + (0.04233 \times 45.60) + (0.04290 \times 48.48) + (0.04341 \times 51.84) \\
&\quad + (0.04385 \times 51.96) + (0.04426 \times 54.12) + (0.04462 \times 55.56) + (0.04497 \times 67.80) \\
&\quad + (0.04529 \times 68.64) + (0.04559 \times 68.64) + (0.04588 \times 68.88) + (0.04615 \times 84.12) \\
&\quad + (0.04641 \times 93.12) + (0.04666 \times 96.64) + (0.04688 \times 105.12) + (0.04706 \times 105.84) \\
&\quad + (0.13104 \times X_{21:23}) \\
&= 54.921 + (0.13104 \times X_{21:23}).
\end{aligned} \tag{6.2}$$

When the above expression is equated to $\mu_{20}^* = 69.694$ and solved for $X_{21:23}$, we obtain the BLUP of $X_{21:23}$ to be $X_{21:23}^* = 112.74$.

Next, assuming that the shape parameter $\mathbf{k} = \mathbf{0.5}$, we computed earlier the BLUE of μ based on the smallest 20 order statistics to be $\mu_{20}^* = 69.949$. Now, assuming that the smallest 21 order statistics are available, we have from Table 4 the BLUE of μ to be

$$\begin{aligned}
\mu_{21}^* &= (0.02648 \times 17.88) + (0.03684 \times 28.92) + (0.03948 \times 33.00) + (0.04110 \times 41.52) \\
&\quad + (0.04224 \times 42.12) + (0.04310 \times 45.60) + (0.04378 \times 48.48) + (0.04433 \times 51.84) \\
&\quad + (0.04479 \times 51.96) + (0.04518 \times 54.12) + (0.04551 \times 55.56) + (0.04580 \times 67.80) \\
&\quad + (0.04604 \times 68.64) + (0.04624 \times 68.64) + (0.04642 \times 68.88) + (0.04655 \times 84.12) \\
&\quad + (0.04666 \times 93.12) + (0.04672 \times 96.64) + (0.04674 \times 105.12) + (0.04666 \times 105.84) \\
&\quad + (0.12935 \times X_{21:23}) \\
&= 55.279 + (0.12935 \times X_{21:23}).
\end{aligned} \tag{6.3}$$

When the above expression is equated to $\mu_{20}^* = 69.949$ and solved for $X_{21:23}$, we obtain the BLUP of $X_{21:23}$ to be $X_{21:23}^* = 113.41$.

Next, assuming that the shape parameter $\mathbf{k} = \mathbf{1.0}$, we computed earlier the BLUE of μ based on the smallest 20 order statistics to be $\mu_{20}^* = 70.889$. Now, assuming that the smallest 21 order statistics are available, we have from Table 4 the BLUE of μ to be

$$\begin{aligned}
\mu_{21}^* &= -(0.00709 \times 17.88) + (0.03418 \times 28.92) + (0.04110 \times 33.00) + (0.04457 \times 41.52) \\
&\quad + (0.04657 \times 42.12) + (0.04779 \times 45.60) + (0.04852 \times 48.48) + (0.04895 \times 51.84) \\
&\quad + (0.04916 \times 51.96) + (0.04921 \times 54.12) + (0.04913 \times 55.56) + (0.04896 \times 67.80) \\
&\quad + (0.04871 \times 68.64) + (0.04838 \times 68.64) + (0.04799 \times 68.88) + (0.04754 \times 84.12) \\
&\quad + (0.04701 \times 93.12) + (0.04640 \times 96.64) + (0.04570 \times 105.12) + (0.04485 \times 105.84) \\
&\quad + (0.12236 \times X_{21:23})
\end{aligned}$$

$$= 56.747 + (0.12236 \times X_{21:23}). \tag{6.4}$$

When the above expression is equated to $\mu^*_{20} = 70.889$ and solved for $X_{21:23}$, we obtain the BLUP of $X_{21:23}$ to be $X^*_{21:23} = 115.58$.

Next, assuming that the shape parameter $\mathbf{k} = \mathbf{1.5}$, we computed earlier the BLUE of μ based on the smallest 20 order statistics to be $\mu^*_{20} = 72.659$. Now, assuming that the smallest 21 order statistics are available, we have from Table 4 the BLUE of μ to be

$$\begin{aligned}
\mu^*_{21} &= -(0.08121 \times 17.88) + (0.03477 \times 28.92) + (0.04887 \times 33.00) + (0.05458 \times 41.52)\\
&\quad + (0.05703 \times 42.12) + (0.05790 \times 45.60) + (0.05792 \times 48.48) + (0.05745 \times 51.84)\\
&\quad + (0.05667 \times 51.96) + (0.05569 \times 54.12) + (0.05458 \times 55.56) + (0.05339 \times 67.80)\\
&\quad + (0.05214 \times 68.64) + (0.05085 \times 68.64) + (0.04952 \times 68.88) + (0.04815 \times 84.12)\\
&\quad + (0.04675 \times 93.12) + (0.04530 \times 96.64) + (0.04379 \times 105.12) + (0.04218 \times 105.84)\\
&\quad + (0.11367 \times X_{21:23})\\
&= 59.201 + (0.11367 \times X_{21:23}).
\end{aligned} \tag{6.5}$$

When the above expression is equated to $\mu^*_{20} = 72.659$ and solved for $X_{21:23}$, we obtain the BLUP of $X_{21:23}$ to be $X^*_{21:23} = 118.40$.

Next, assuming that the shape parameter $\mathbf{k} = \mathbf{2.0}$, we computed earlier the BLUE of μ based on the smallest 20 order statistics to be $\mu^*_{20} = 75.750$. Now, assuming that the smallest 21 order statistics are available, we have from Table 4 the BLUE of μ to be

$$\begin{aligned}
\mu^*_{21} &= -(0.21339 \times 17.88) + (0.04301 \times 28.92) + (0.06678 \times 33.00) + (0.07411 \times 41.52)\\
&\quad + (0.07568 \times 42.12) + (0.07479 \times 45.60) + (0.07276 \times 48.48) + (0.07019 \times 51.84)\\
&\quad + (0.06738 \times 51.96) + (0.06449 \times 54.12) + (0.06161 \times 55.56) + (0.05879 \times 67.80)\\
&\quad + (0.05605 \times 68.64) + (0.05339 \times 68.64) + (0.05083 \times 68.88) + (0.04836 \times 84.12)\\
&\quad + (0.04597 \times 93.12) + (0.04366 \times 96.64) + (0.04140 \times 105.12) + (0.03917 \times 105.84)\\
&\quad + (0.10496 \times X_{21:23})\\
&= 62.953 + (0.10496 \times X_{21:23}).
\end{aligned} \tag{6.6}$$

When the above expression is equated to $\mu^*_{20} = 75.750$ and solved for $X_{21:23}$, we obtain the BLUP of $X_{21:23}$ to be $X^*_{21:23} = 121.92$.

Finally, assuming that the shape parameter $\mathbf{k} = \mathbf{2.5}$, we computed earlier the BLUE of μ based on the smallest 20 order statistics to be $\mu^*_{20} = 80.575$. Now, assuming that the smallest 21 order statistics are available, we have from Table 4 the BLUE of μ to be

$$\begin{aligned}
\mu^*_{21} &= -(0.41925 \times 17.88) + (0.06345 \times 28.92) + (0.09847 \times 33.00) + (0.10568 \times 41.52)\\
&\quad + (0.10414 \times 42.12) + (0.09940 \times 45.60) + (0.09352 \times 48.48) + (0.08735 \times 51.84)\\
&\quad + (0.08128 \times 51.96) + (0.07550 \times 54.12) + (0.07007 \times 55.56) + (0.06501 \times 67.80)\\
&\quad + (0.06033 \times 68.64) + (0.05600 \times 68.64) + (0.05200 \times 68.88) + (0.04832 \times 84.12)
\end{aligned}$$

$$
\begin{aligned}
&+ (0.04492 \times 93.12) + (0.04178 \times 96.64) + (0.03887 \times 105.12) + (0.03618 \times 105.84) \\
&+ (0.09698 \times X_{21:23}) \\
&= 68.335 + (0.09698 \times X_{21:23}).
\end{aligned}
\tag{6.7}
$$

When the above expression is equated to $\mu^*_{20} = 80.575$ and solved for $X_{21:23}$, we obtain the BLUP of $X_{21:23}$ to be $X^*_{21:23} = 126.21$.

Proceeding similarly, and using the BLUEs of μ presented in Table 4 for the case $n = 23$ and $s = 1$, we can determine the Best Linear Unbiased Predictor of $X_{22:23}$, the twenty second failure. For example, let us assume that the shape parameter $\mathbf{k} = \mathbf{0.3}$. Then, treating the BLU predicted value of 112.74 for $X_{21:23}$ as its observed value, we readily have the BLUE of μ from (6.2) as $\mu^*_{21} = 54.921 + (0.13104 \times 112.74) = 69.694$. Now, assuming that the smallest 22 order statistics are available (with the BLUP value as the observed value of $X_{21:23}$), we have from Table 4 the BLUE of μ to be

$$
\begin{aligned}
\mu^*_{22} &= (0.03766 \times 17.88) + (0.04079 \times 28.92) + (0.04170 \times 33.00) + (0.04229 \times 41.52) \\
&+ (0.04272 \times 42.12) + (0.04306 \times 45.60) + (0.04335 \times 48.48) + (0.04359 \times 51.84) \\
&+ (0.04380 \times 51.96) + (0.04398 \times 54.12) + (0.04415 \times 55.56) + (0.04429 \times 67.80) \\
&+ (0.04443 \times 68.64) + (0.04455 \times 68.64) + (0.04467 \times 68.88) + (0.04477 \times 84.12) \\
&+ (0.04486 \times 93.12) + (0.04494 \times 96.64) + (0.04501 \times 105.12) + (0.04504 \times 105.84) \\
&+ (0.04502 \times 112.74) + (0.08534 \times X_{22:23}) \\
&= 59.2660 + (0.08534 \times X_{22:23}).
\end{aligned}
\tag{6.8}
$$

When the above expression is equated to $\mu^*_{21} = 69.694$ and solved for $X_{22:23}$, we obtain the BLUP of $X_{22:23}$ to be $X^*_{22:23} = 122.19$.

Next, let us assume that the shape parameter $\mathbf{k} = \mathbf{0.5}$. Then, treating the BLU predicted value of 113.41 for $X_{21:23}$ as its observed value, we readily have the BLUE of μ from (6.3) as $\mu^*_{21} = 55.279 + (0.12935 \times 113.41) = 69.949$. Now, assuming that the smallest 22 order statistics are available (with the BLUP value as the observed value of $X_{21:23}$), we have from Table 4 the BLUE of μ to be

$$
\begin{aligned}
\mu^*_{22} &= (0.03282 \times 17.88) + (0.03994 \times 28.92) + (0.04159 \times 33.00) + (0.04257 \times 41.52) \\
&+ (0.04324 \times 42.12) + (0.04372 \times 45.60) + (0.04408 \times 48.48) + (0.04436 \times 51.84) \\
&+ (0.04458 \times 51.96) + (0.04475 \times 54.12) + (0.04489 \times 55.56) + (0.04499 \times 67.80) \\
&+ (0.04506 \times 68.64) + (0.04511 \times 68.64) + (0.04513 \times 68.88) + (0.04512 \times 84.12) \\
&+ (0.04509 \times 93.12) + (0.04503 \times 96.64) + (0.04493 \times 105.12) + (0.04476 \times 105.84) \\
&+ (0.04449 \times 113.41) + (0.08375 \times X_{22:23}) \\
&= 59.596 + (0.08375 \times X_{22:23}).
\end{aligned}
\tag{6.9}
$$

When the above expression is equated to $\mu^*_{21} = 69.949$ and solved for $X_{22:23}$, we obtain the BLUP of $X_{22:23}$ to be $X^*_{22:23} = 123.62$.

Next, let us assume that the shape parameter $\mathbf{k} = \mathbf{1.0}$. Then, treating the BLU predicted value of 115.58 for $X_{21:23}$ as its observed value, we readily have the BLUE of μ from (6.4) as $\mu_{21}^* = 56.747 + (0.12236 \times 115.58) = 70.889$. Now, assuming that the smallest 22 order statistics are available (with the BLUP value as the observed value of $X_{21:23}$), we have from Table 4 the BLUE of μ to be

$$\begin{aligned}\mu_{22}^* &= (0.00333 \times 17.88) + (0.03785 \times 28.92) + (0.04322 \times 33.00) + (0.04580 \times 41.52)\\ &+ (0.04720 \times 42.12) + (0.04797 \times 45.60) + (0.04836 \times 48.48) + (0.04852 \times 51.84)\\ &+ (0.04851 \times 51.96) + (0.04837 \times 54.12) + (0.04814 \times 55.56) + (0.04784 \times 67.80)\\ &+ (0.04748 \times 68.64) + (0.04706 \times 68.64) + (0.04658 \times 68.88) + (0.04606 \times 84.12)\\ &+ (0.04547 \times 93.12) + (0.04483 \times 96.64) + (0.04410 \times 105.12) + (0.04325 \times 105.84)\\ &+ (0.04222 \times 115.58) + (0.07786 \times X_{22:23})\\ &= 60.686 + (0.07786 \times X_{22:23}).\end{aligned} \tag{6.10}$$

When the above expression is equated to $\mu_{21}^* = 70.889$ and solved for $X_{22:23}$, we obtain the BLUP of $X_{22:23}$ to be $X_{22:23}^* = 131.04$.

Next, let us assume that the shape parameter $\mathbf{k} = \mathbf{1.5}$. Then, treating the BLU predicted value of 118.40 for $X_{21:23}$ as its observed value, we readily have the BLUE of μ from (6.5) as $\mu_{21}^* = 59.201 + (0.11367 \times 118.40) = 72.659$. Now, assuming that the smallest 22 order statistics are available (with the BLUP value as the observed value of $X_{21:23}$), we have from Table 4 the BLUE of μ to be

$$\begin{aligned}\mu_{22}^* &= -(0.06537 \times 17.88) + (0.03869 \times 28.92) + (0.05065 \times 33.00) + (0.05525 \times 41.52)\\ &+ (0.05703 \times 42.12) + (0.05746 \times 45.60) + (0.05718 \times 48.48) + (0.05648 \times 51.84)\\ &+ (0.05554 \times 51.96) + (0.05445 \times 54.12) + (0.05326 \times 55.56) + (0.05201 \times 67.80)\\ &+ (0.05071 \times 68.64) + (0.04939 \times 68.64) + (0.04805 \times 68.88) + (0.04668 \times 84.12)\\ &+ (0.04528 \times 93.12) + (0.04385 \times 96.64) + (0.04237 \times 105.12) + (0.04080 \times 105.84)\\ &+ (0.03909 \times 118.40) + (0.07113 \times X_{22:23})\\ &= 62.697 + (0.07113 \times X_{22:23}).\end{aligned} \tag{6.11}$$

When the above expression is equated to $\mu_{21}^* = 72.659$ and solved for $X_{22:23}$, we obtain the BLUP of $X_{22:23}$ to be $X_{22:23}^* = 140.05$.

Next, let us assume that the shape parameter $\mathbf{k} = \mathbf{2.0}$. Then, treating the BLU predicted value of 121.92 for $X_{21:23}$ as its observed value, we readily have the BLUE of μ from (6.6) as $\mu_{21}^* = 62.953 + (0.10496 \times 121.92) = 75.750$. Now, assuming that the smallest 22 order statistics are available (with the BLUP value as the observed value of $X_{21:23}$), we have from Table 4 the BLUE of μ to be

$$\begin{aligned}\mu_{22}^* &= -(0.19068 \times 17.88) + (0.04683 \times 28.92) + (0.06786 \times 33.00) + (0.07394 \times 41.52)\\ &+ (0.07485 \times 42.12) + (0.07359 \times 45.60) + (0.07134 \times 48.48) + (0.06864 \times 51.84)\\ &+ (0.06577 \times 51.96) + (0.06285 \times 54.12) + (0.05997 \times 55.56) + (0.05717 \times 67.80)\\ &+ (0.05445 \times 68.64) + (0.05184 \times 68.64) + (0.04932 \times 68.88) + (0.04690 \times 84.12)\end{aligned}$$

$$+ (0.04457 \times 93.12) + (0.04230 \times 96.64) + (0.04010 \times 105.12) + (0.03793 \times 105.84)$$
$$+ (0.03575 \times 121.92) + (0.06472 \times X_{22:23})$$
$$= 65.943 + (0.06472 \times X_{22:23}). \tag{6.12}$$

When the above expression is equated to $\mu_{21}^* = 75.750$ and solved for $X_{22:23}$, we obtain the BLUP of $X_{22:23}$ to be $X_{22:23}^* = 151.53$.

Next, let us assume that the shape parameter $\mathbf{k} = \mathbf{2.5}$. Then, treating the BLU predicted value of 126.21 for $X_{21:23}$ as its observed value, we readily have the BLUE of μ from (6.8) as $\mu_{21}^* = 68.335 + (0.09698 \times 126.21) = 80.575$. Now, assuming that the smallest 22 order statistics are available (with the BLUP value as the observed value of $X_{21:23}$), we have from Table 4 the BLUE of μ to be

$$\mu_{22}^* = -(0.38795 \times 17.88) + (0.06672 \times 28.92) + (0.09845 \times 33.00) + (0.10436 \times 41.52)$$
$$+ (0.10226 \times 42.12) + (0.09728 \times 45.60) + (0.09131 \times 48.48) + (0.08515 \times 51.84)$$
$$+ (0.07914 \times 51.96) + (0.07344 \times 54.12) + (0.06811 \times 55.56) + (0.06316 \times 67.80)$$
$$+ (0.05858 \times 68.64) + (0.05435 \times 68.64) + (0.05045 \times 68.88) + (0.04686 \times 84.12)$$
$$+ (0.04355 \times 93.12) + (0.04049 \times 96.64) + (0.03766 \times 105.12) + (0.03502 \times 105.84)$$
$$+ (0.03256 \times 126.21) + (0.05902 \times X_{22:23})$$
$$= 70.785 + (0.05902 \times X_{22:23}). \tag{6.13}$$

When the above expression is equated to $\mu_{21}^* = 80.575$ and solved for $X_{22:23}$, we obtain the BLUP of $X_{22:23}$ to be $X_{22:23}^* = 165.88$.

Finally, upon using Table 4 for the case $n = 23$ and $s = 0$, we can similarly determine the Best Linear Unbiased Predictor of $X_{23:23}$, the final failure. For example, let us assume that the shape parameter $\mathbf{k} = \mathbf{0.3}$. Then, treating the BLU predicted value of 122.19 for $X_{22:23}$ as its observed value, we readily have the BLUE of μ from (6.8) as $\mu_{22}^* = 59.2660 + (0.08534 \times 122.19) = 69.694$. Now, assuming that all 23 order statistics are available (with the BLUP values as the observed values of $X_{21:23}$ and $X_{22:23}$), we have from Table 4 the BLUE of μ to be

$$\mu_{23}^* = (0.04182 \times 17.88) + (0.04311 \times 28.92) + (0.04339 \times 33.00) + (0.04355 \times 41.52)$$
$$+ (0.04365 \times 42.12) + (0.04372 \times 45.60) + (0.04376 \times 48.48) + (0.04379 \times 51.84)$$
$$+ (0.04381 \times 51.96) + (0.04382 \times 54.12) + (0.04381 \times 55.56) + (0.04380 \times 67.80)$$
$$+ (0.04379 \times 68.64) + (0.04376 \times 68.64) + (0.04373 \times 68.88) + (0.04369 \times 84.12)$$
$$+ (0.04364 \times 93.12) + (0.04358 \times 96.64) + (0.04351 \times 105.12) + (0.04341 \times 105.84)$$
$$+ (0.04328 \times 112.74) + (0.04308 \times 122.19) + (0.04248 \times X_{23:23})$$
$$= 63.7940 + (0.04248 \times X_{23:23}). \tag{6.14}$$

When the above expression is equated to $\mu_{22}^* = 69.694$ and solved for $X_{23:23}$, we obtain the BLUP of $X_{23:23}$ to be $X_{23:23}^* = 138.89$.

Next, let us assume that the shape parameter $\mathbf{k} = \mathbf{0.5}$. Then, treating the BLU predicted value of 123.62 for $X_{22:23}$ as its observed value, we readily have the BLUE of μ from (6.9) as $\mu_{22}^* = 59.596 + (0.08375 \times 123.62) = 69.949$. Now, assuming that all 23 order statistics are available (with the BLUP values as the observed values of $X_{21:23}$ and $X_{22:23}$), we have from Table 4 the BLUE of μ to be

$$\begin{aligned}
\mu_{23}^* &= (0.03806 \times 17.88) + (0.04253 \times 28.92) + (0.04339 \times 33.00) + (0.04384 \times 41.52) \\
&+ (0.04412 \times 42.12) + (0.04428 \times 45.60) + (0.04439 \times 48.48) + (0.04444 \times 51.84) \\
&+ (0.04446 \times 51.96) + (0.04446 \times 54.12) + (0.04443 \times 55.56) + (0.04438 \times 67.80) \\
&+ (0.04431 \times 68.64) + (0.04422 \times 68.64) + (0.04412 \times 68.88) + (0.04399 \times 84.12) \\
&+ (0.04385 \times 93.12) + (0.04367 \times 96.64) + (0.04346 \times 105.12) + (0.04321 \times 105.84) \\
&+ (0.04287 \times 112.74) + (0.04239 \times 123.62) + (0.04112 \times X_{23:23}) \\
&= 64.025 + (0.04112 \times X_{23:23}).
\end{aligned} \tag{6.15}$$

When the above expression is equated to $\mu_{22}^* = 69.949$ and solved for $X_{23:23}$, we obtain the BLUP of $X_{23:23}$ to be $X_{23:23}^* = 144.07$.

Next, let us assume that the shape parameter $\mathbf{k} = \mathbf{1.0}$. Then, treating the BLU predicted value of 131.04 for $X_{22:23}$ as its observed value, we readily have the BLUE of μ from (6.10) as $\mu_{22}^* = 60.686 + (0.07786 \times 131.04) = 70.889$. Now, assuming that all 23 order statistics are available (with the BLUP values as the observed values of $X_{21:23}$ and $X_{22:23}$), we have from Table 4 the BLUE of μ to be

$$\begin{aligned}
\mu_{23}^* &= (0.01215 \times 17.88) + (0.04099 \times 28.92) + (0.04506 \times 33.00) + (0.04688 \times 41.52) \\
&+ (0.04777 \times 42.12) + (0.04816 \times 45.60) + (0.04827 \times 48.48) + (0.04820 \times 51.84) \\
&+ (0.04800 \times 51.96) + (0.04771 \times 54.12) + (0.04735 \times 55.56) + (0.04694 \times 67.80) \\
&+ (0.04647 \times 68.64) + (0.04597 \times 68.64) + (0.04543 \times 68.88) + (0.04484 \times 84.12) \\
&+ (0.04420 \times 93.12) + (0.04351 \times 96.64) + (0.04275 \times 105.12) + (0.04189 \times 105.84) \\
&+ (0.04088 \times 112.74) + (0.03958 \times 131.04) + (0.03700 \times X_{23:23}) \\
&= 65.062 + (0.03700 \times X_{23:23}).
\end{aligned} \tag{6.16}$$

When the above expression is equated to $\mu_{22}^* = 70.889$ and solved for $X_{23:23}$, we obtain the BLUP of $X_{23:23}$ to be $X_{23:23}^* = 157.49$.

Next, let us assume that the shape parameter $\mathbf{k} = \mathbf{1.5}$. Then, treating the BLU predicted value of 140.05 for $X_{22:23}$ as its observed value, we readily have the BLUE of μ from (6.11) as $\mu_{22}^* = 62.697 + (0.07113 \times 140.05) = 72.659$. Now, assuming that all 23 order statistics are available (with the BLUP values as the observed values of $X_{21:23}$ and $X_{22:23}$), we have from Table 4 the BLUE of μ to be

$$\begin{aligned}
\mu_{23}^* &= -(0.05157 \times 17.88) + (0.04214 \times 28.92) + (0.05223 \times 33.00) + (0.05587 \times 41.52) \\
&+ (0.05706 \times 42.12) + (0.05711 \times 45.60) + (0.05655 \times 48.48) + (0.05567 \times 51.84) \\
&+ (0.05459 \times 51.96) + (0.05339 \times 54.12) + (0.05212 \times 55.56) + (0.05082 \times 67.80) \\
&+ (0.04949 \times 68.64) + (0.04814 \times 68.64) + (0.04678 \times 68.88) + (0.04540 \times 84.12)
\end{aligned}$$

$$+ (0.04401 \times 93.12) + (0.04259 \times 96.64) + (0.04113 \times 105.12) + (0.03959 \times 105.84)$$
$$+ (0.03793 \times 112.74) + (0.03600 \times 140.05) + (0.03297 \times X_{23:23})$$
$$= 66.831 + (0.03297 \times X_{23:23}). \qquad (6.17)$$

When the above expression is equated to $\mu_{22}^* = 72.659$ and solved for $X_{23:23}$, we obtain the BLUP of $X_{23:23}$ to be $X_{23:23}^* = 176.77$.

Next, let us assume that the shape parameter $\mathbf{k} = \mathbf{2.0}$. Then, treating the BLU predicted value of 151.53 for $X_{22:23}$ as its observed value, we readily have the BLUE of μ from (6.12) as $\mu_{22}^* = 65.943 + (0.06472 \times 151.53) = 75.750$. Now, assuming that all 23 order statistics are available (with the BLUP values as the observed values of $X_{21:23}$ and $X_{22:23}$), we have from Table 4 the BLUE of μ to be

$$\mu_{23}^* = -(0.17018 \times 17.88) + (0.05026 \times 28.92) + (0.06883 \times 33.00) + (0.07379 \times 41.52)$$
$$+ (0.07410 \times 42.12) + (0.07250 \times 45.60) + (0.07005 \times 48.48) + (0.06724 \times 51.84)$$
$$+ (0.06430 \times 51.96) + (0.06136 \times 54.12) + (0.05848 \times 55.56) + (0.05569 \times 67.80)$$
$$+ (0.05301 \times 68.64) + (0.05043 \times 68.64) + (0.04795 \times 68.88) + (0.04558 \times 84.12)$$
$$+ (0.04329 \times 93.12) + (0.04108 \times 96.64) + (0.03892 \times 105.12) + (0.03680 \times 105.84)$$
$$+ (0.03466 \times 112.74) + (0.03241 \times 151.53) + (0.02946 \times X_{23:23})$$
$$= 69.785 + (0.02946 \times X_{23:23}). \qquad (6.18)$$

When the above expression is equated to $\mu_{22}^* = 75.750$ and solved for $X_{23:23}$, we obtain the BLUP of $X_{23:23}$ to be $X_{23:23}^* = 202.48$.

Finally, let us assume that the shape parameter $\mathbf{k} = \mathbf{2.5}$. Then, treating the BLU predicted value of 165.88 for $X_{22:23}$ as its observed value, we readily have the BLUE of μ from (6.13) as $\mu_{22}^* = 70.785 + (0.05902 \times 165.88) = 80.575$. Now, assuming that all 23 order statistics are available (with the BLUP values as the observed values of $X_{21:23}$ and $X_{22:23}$), we have from Table 4 the BLUE of μ to be

$$\mu_{23}^* = -(0.35856 \times 17.88) + (0.06972 \times 28.92) + (0.09837 \times 33.00) + (0.10307 \times 41.52)$$
$$+ (0.10044 \times 42.12) + (0.09524 \times 45.60) + (0.08920 \times 48.48) + (0.08305 \times 51.84)$$
$$+ (0.07710 \times 51.96) + (0.07148 \times 54.12) + (0.06624 \times 55.56) + (0.06139 \times 67.80)$$
$$+ (0.05691 \times 68.64) + (0.05278 \times 68.64) + (0.04898 \times 68.88) + (0.04548 \times 84.12)$$
$$+ (0.04226 \times 93.12) + (0.03928 \times 96.64) + (0.03651 \times 105.12) + (0.03394 \times 105.84)$$
$$+ (0.03151 \times 112.74) + (0.02915 \times 165.88) + (0.02647 \times X_{23:23})$$
$$= 74.329 + (0.02647 \times X_{23:23}). \qquad (6.19)$$

When the above expression is equated to $\mu_{22}^* = 80.575$ and solved for $X_{23:23}$, we obtain the BLUP of $X_{23:23}$ to be $X_{23:23}^* = 235.97$.

7

Quantile–Quantile Plots and Goodness-of-Fit Test

In any statistical analysis based on the assumption of a particular distribution for the data at hand, one will naturally be interested in assessing the validity of that assumption; more specifically, one will be interested in testing for the hypothesis that the data has come from that specific distribution wherein only the functional form of the distribution is assumed to be known while it may involve some unknown parameters. For example, we may be interested in testing whether the data at hand has possibly arisen from the three-parameter inverse Gaussian distribution in (4.1), wherein we may assume that all three parameters μ, σ and k are unknown.

One of the simple and most commonly used methods for this task is called *Quantile–Quantile Plot*; see, for example, the book by D'Agostino and Stephens (1986). A Quantile–Quantile plot simply plots the order statistics $X_{i:n}$ (sample quantiles) obtained from the sample at hand against the values $\mu_{i:n}$, which are the expected values of order statistics from the standard distribution (population quantiles). Note that if the sample had in fact come from the assumed distribution (with some location parameter and scale parameter), then the plot of $X_{i:n}$ against $\mu_{i:n}$ will be nearly linear. Naturally, based on this fact, one could propose the correlation coefficient R between these two sets of values as a possible goodness-of-fit test statistic with values of R close to 1 indicating the suitability of the assumed distribution while smaller values of R indicating the inappropriateness of the assumed distribution. Realize that, however, though this procedure could be used when the location parameter μ and the scale parameter σ are both unknown, one has to assume the shape parameter k to be known since the values of $\mu_{i:n}$ depend on the particular choice of k. For this reason, one could repeat the Quantile–Quantile plot for various choices of the shape parameter k and determine a reasonable range of values of k for which the assumed distributional family fit the data well.

The exact values of the means of order statistics presented in Table 1 can be used in the construction of the Quantile-Quantile plot as well as in the determination of the correlation goodness-of-fit statistic R. In order to carry out the goodness-of-fit test at a pre-specified level of significance, we need the percentage points of the null distribution of the statistic R, which is nothing but the sampling distribution of R when the sample actually arises from the three-parameter inverse Gaussian distribution in (4.1) with the shape parameter k being known, but the location parameter μ and the scale parameter σ both being unknown. For the determination of these percentage points, we used Monte Carlo simulations. We simulated the required sample observations from the inverse Gaussian distribution in (4.1) with a specific value of the shape parameter k (with $\mu = 0$ and $\sigma = 1$, without loss of any generality), ordered this sample data, and then determined the value of the correlation coefficient R between this ordered data and the values of $\mu_{i:n}$ taken from Table 1 for the corresponding value of k. This Monte Carlo process was repeated 5,000 times. The values of R so determined were then ordered, and from this ordered string of 5,000 values of R, we determined out the 50, 125, 250, 375, 500, 750, 1000, 1250, 1500, 2000, 2500, 3000, 3500, 3750, 4000, 4250, 4500, 4625, 4750, 4875, 4950-th values as simulated estimates of the lower and upper 1%, 2.5%, 5%, 7.5%, 10%, 15%, 20%, 25%, 30%, 40% and 50% values of the null distribution of the test statistic R. These values are presented in Tables 7–9 for sample sizes $n = 20, 23, 25$, respectively. These tables and some additional tables, including power comparisons, have been provided by Asha (1996).

For the purpose of illustration, let us consider the data giving the maximum flood levels of Susquehanna river, which was considered earlier in Chapter 5 as the first example there. In this case, we have the sample size $n = 20$. Then, using the values of means of order statistics presented in Table 1 corresponding to $n = 20$ and for various values of the shape parameter k, Quantile–Quantile plots were constructed taking $k = 0.1, 0.5(0.5)2.5$. These plots are presented in Fig. 26. Just a casual look through these figures reveals that the inverse Gaussian model with $k = 0.1$ does not fit the data well at all, while the models with values of $k = 0.5(0.5)2.5$ all fit the data well. Also, from the simulated percentage points of the correlation statistic R presented in Table 7, we found the p-values through interpolation. The values of the correlation statistic R as well as the corresponding p-values are presented in Table 10 for all values of the shape parameter k. This table also supports the comment made above with regard to the suitability of the inverse Gaussian model for this data.

Next, let us consider the data giving the number of million revolutions before failure of the 23 ball bearings, which was considered earlier in Chapter 5 as the third example there. In this case, we have the sample size $n = 23$. Then, upon using the values of means of order statistics presented in Table 3 corresponding to $n = 23$ and for various values of the shape parameter k, Quantile–Quantile plots were constructed taking $k = 0.1, 0.5(0.5)2.5$. These plots are presented in Fig. 27. Once again, a casual look through these figures reveals that the inverse Gaussian model fits the data with any value of $k \geq 0.3$. Also, from the simulated percentage points of the correlation statistic R presented in Table 8, we found the p-values through interpolation. The values of the correlation statistic R as well as the corresponding p-values are presented in Table 11 for all values of the shape parameter k. This table clearly supports the comment made above with regard to the suitability of the inverse Gaussian model for this data.

Bibliography

1. Aitken, A. C. (1935). On least squares and linear combinations of observations. *Proceedings of the Royal Society of Edinburgh* **55**, 42-48.
2. Arnold, B. C. and Balakrishnan, N. (1989). *Relations, Bounds and Approximations for Order Statistics*, Lecture Notes in Statistics **53**. Springer-Verlag, New York.
3. Arnold, B. C., Balakrishnan, N., and Nagaraja, H. N. (1992). *A First Course in Order Statistics*. John Wiley & Sons, New York.
4. Asha, M. M. (1996). Correlation type goodness-of-fit test for the inverse Gaussian distribution. *M. Sc. Project*, McMaster University, Hamilton, Ontario, Canada.
5. Balakrishnan, N. (1989). A relation for the covariances of order statistics from n independent and non-identically distributed random variables. *Statistische Hefte* **30**, 141-146.
6. Balakrishnan, N. and Cohen, A. C. (1991). *Order Statistics and Inference: Estimation Methods*. Academic Press, San Diego.
7. Balakrishnan, N. and Rao, C. R. (1997a). A note on the best linear unbiased estimation based on order statistics. *The American Statistician* (to appear).
8. Balakrishnan, N. and Rao, C. R. (1997b). On the efficiency properties of BLUEs. *Journal of Statistical Planning and Inference* (to appear).
9. Chan, M. Y., Cohen, A. C., and Whitten, B. J. (1983). The standardized inverse Gaussian distribution Tables of the cumulative probability function. *Communications in Statistics - Simulation and Computation* **12**, 423-442.
10. Cheng, R. C. H. and Amin, N. A. K. (1981). Maximum likelihood estimation of parameters in the inverse Gaussian distribution with unknown origin. *Technometrics* **23**, 257-263.
11. Chhikara, R. S. and Folks, J. L. (1989). *The Inverse Gaussian Distribution: Theory, Methodology and Applications*. Marcel Dekker, New York.
12. Cohen, A. C. and Whitten, B. J. (1988). *Parameter Estimation in Reliability and Life Span Models*. Marcel Dekker, New York.
13. D'Agostino, R. B. and Stephens, M. A. (Eds.) (1986). *Goodness-of-Fit Techniques*. Marcel Dekker, New York.
14. David, H. A. (1981). *Order Statistics*, Second edition. John Wiley & Sons, New York.
15. Doganaksoy, N. and Balakrishnan, N. (1997). A useful property of best linear unbiased predictors with applications to life-testing. *The American Statistician* (to appear).
16. Downton, F. (1966). Linear estimates with polynomial coefficients. *Biometrika* **53**,

129-141.

17. Goldberger, A. S. (1962). Best linear unbiased prediction in the generalized linear regression model. *Journal of the American Statistical Association* **57**, 369-375.

18. Harter, H. L. and Balakrishnan, N. (1996). *CRC Handbook of Tables for the Use of Order Statistics in Estimation.* CRC Press, Boca Raton.

19. Johnson, N. L., Kotz, S., and Balakrishnan, N. (1994). *Continuous Univariate Distributions Vol. 1*, Second edition. John Wiley & Sons, New York.

20. Joshi, P. C. (1973). Two identities involving order statistics, *Biometrika* **60**, 428-429.

21. Koziol, J. A. (Ed.) (1988). *CRC Handbook of Percentile Points of the Inverse Gaussian Distribution.* CRC Press, Boca Raton.

22. Lloyd, E. H. (1952). Least-squares estimation of location and scale parameters using order statistics. *Biometrika* **39**, 88-95.

23. Parrish, R. S. (1992a). Computing expected values of normal order statistics. *Communications in Statistics - Simulation and Computation* **21**, 57-70.

24. Parrish, R. S. (1992b). Computing variances and covariances of normal order statistics. *Communications in Statistics - Simulation and Computation* **21**, 71-101.

25. Seshadri, V. (1993). *The Inverse Gaussian Distribution: A Case Study in Exponential Families.* Clarendon Press, Oxford, England.

26. Sillitto, G. P. (1964). Some relations between expectations of order statistics in samples of different sizes. *Biometrika* **51**, 259-262.

27. Tietjen, G. L., Kahaner, D. K., and Beckman, R. J. (1977). Variances and covariances of the normal order statistics for sample sizes 2 to 50. *Selected Tables in Mathematical Statistics* **5**, 1-73.

28. Tweedie, M. C. K. (1947). Functions of a statistical variate with given margins, with special reference to Laplacian distributions. *Proceedings of the Cambridge Philosophical Society* **43**, 41-49.

29. Tweedie, M. C. K. (1957a). Statistical properties of inverse Gaussian distributions, I. *Annals of Mathematical Statistics* **28**, 362-377.

30. Tweedie, M. C. K. (1957b). Statistical properties of inverse Gaussian distributions, II. *Annals of Mathematical Statistics* **28**, 696-705.

31. Wald, A. (1947). *Sequential Analysis.* John Wiley & Sons, New York.

32. Wasan, M. T. (1968). First passage time distribution of Brownian motion. *Monograph*, Department of Mathematics, Queen's University, Kingston, Ontario, Canada.

Figures

Figure 1. Plot of IG(0.1) and N(0,1) density functions

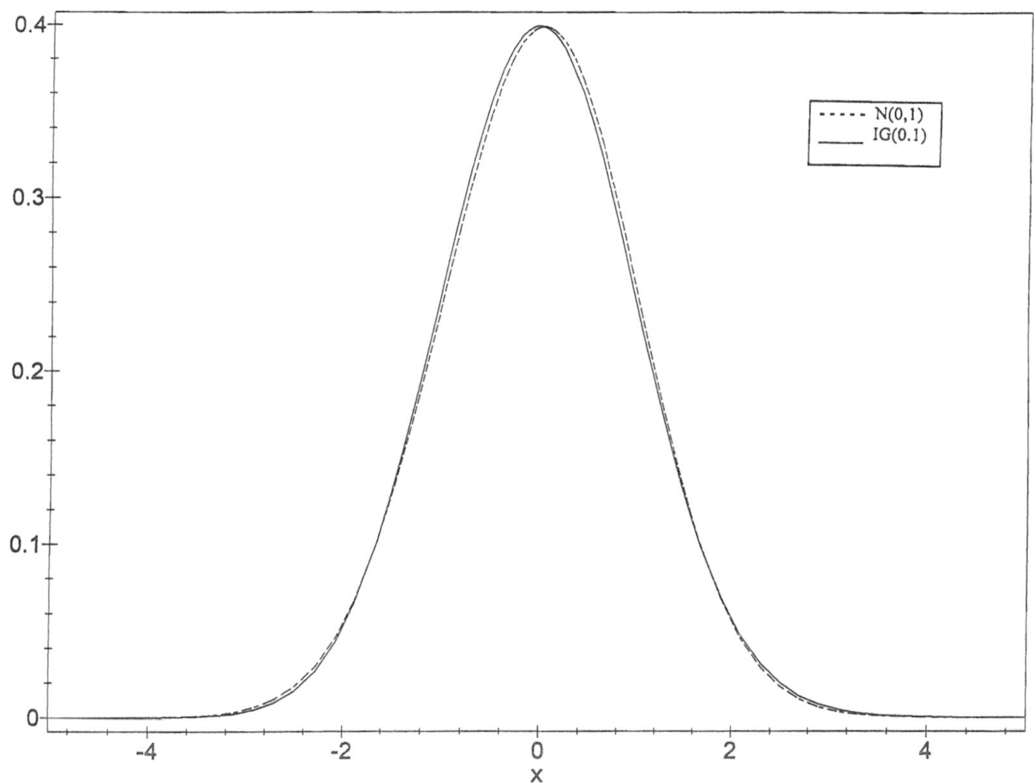

Figure 2. Plot of IG(0.2) and N(0,1) density functions

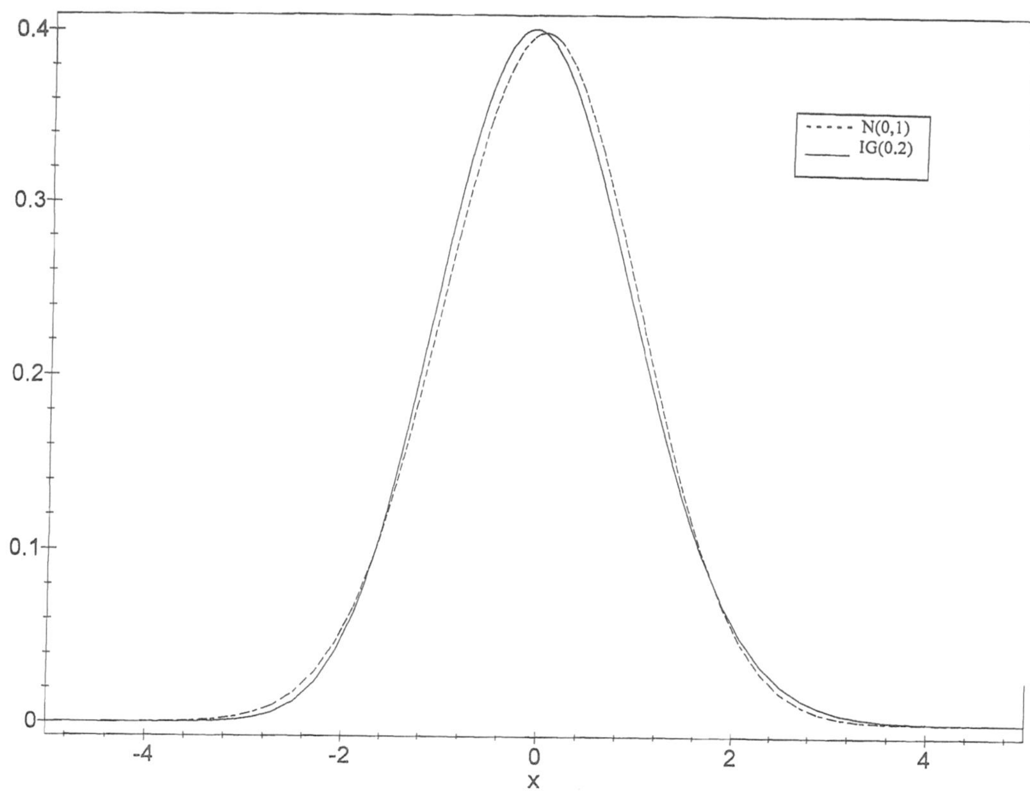

Figure 3. Plot of IG(0.3) and N(0,1) density functions

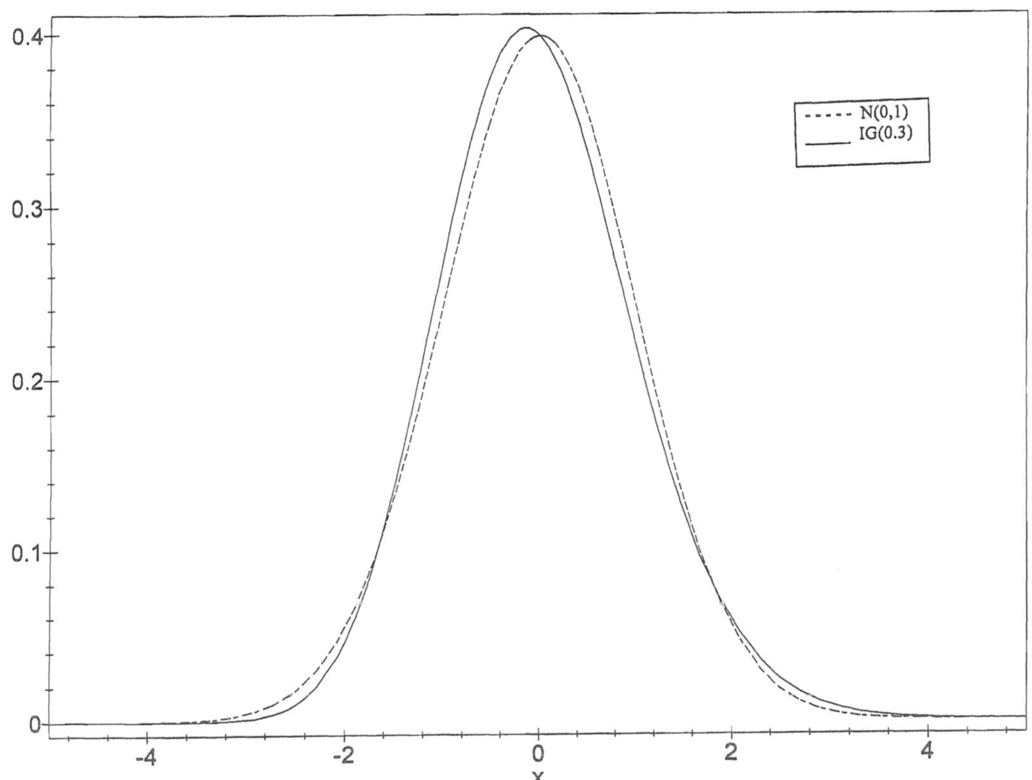

Figure 4. Plot of IG(0.4) and N(0,1) density functions

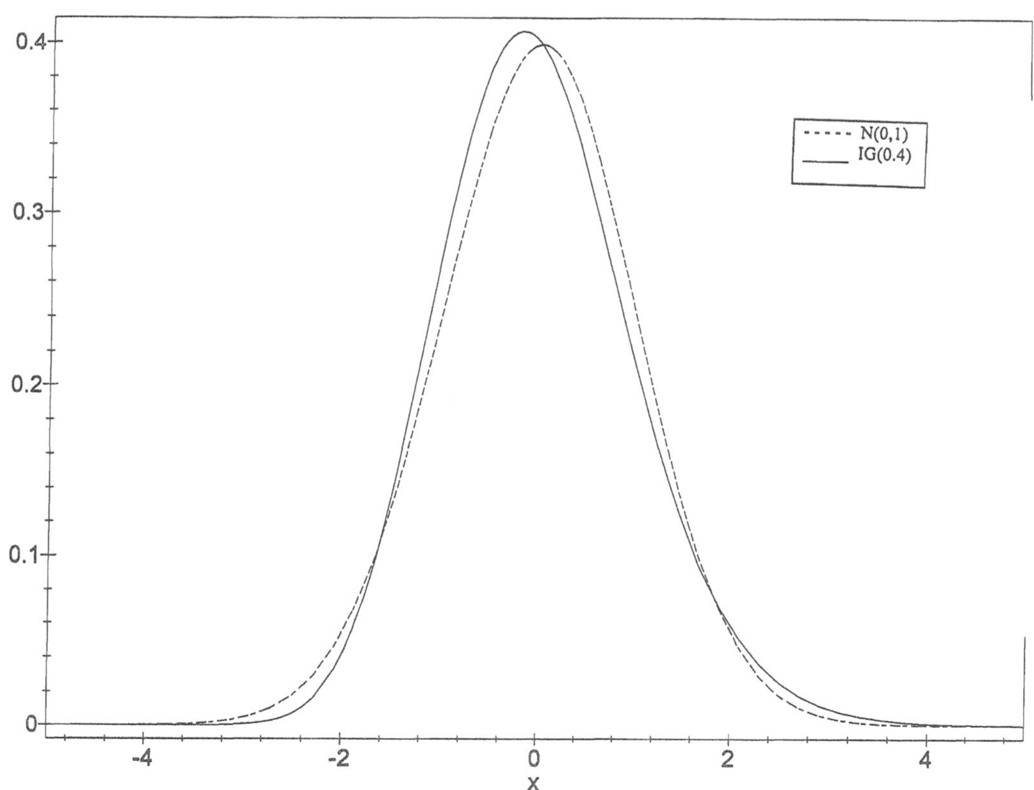

Figure 5. Plot of IG(0.5) and N(0,1) density functions

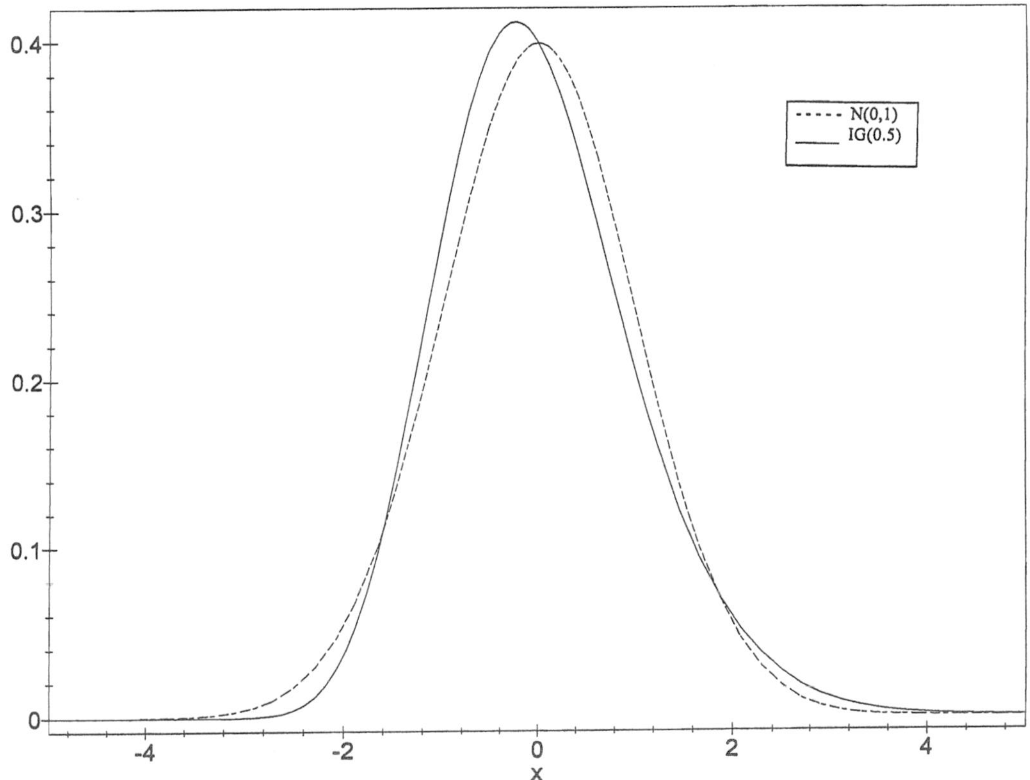

Figure 6. Plot of IG(0.6) and N(0,1) density functions

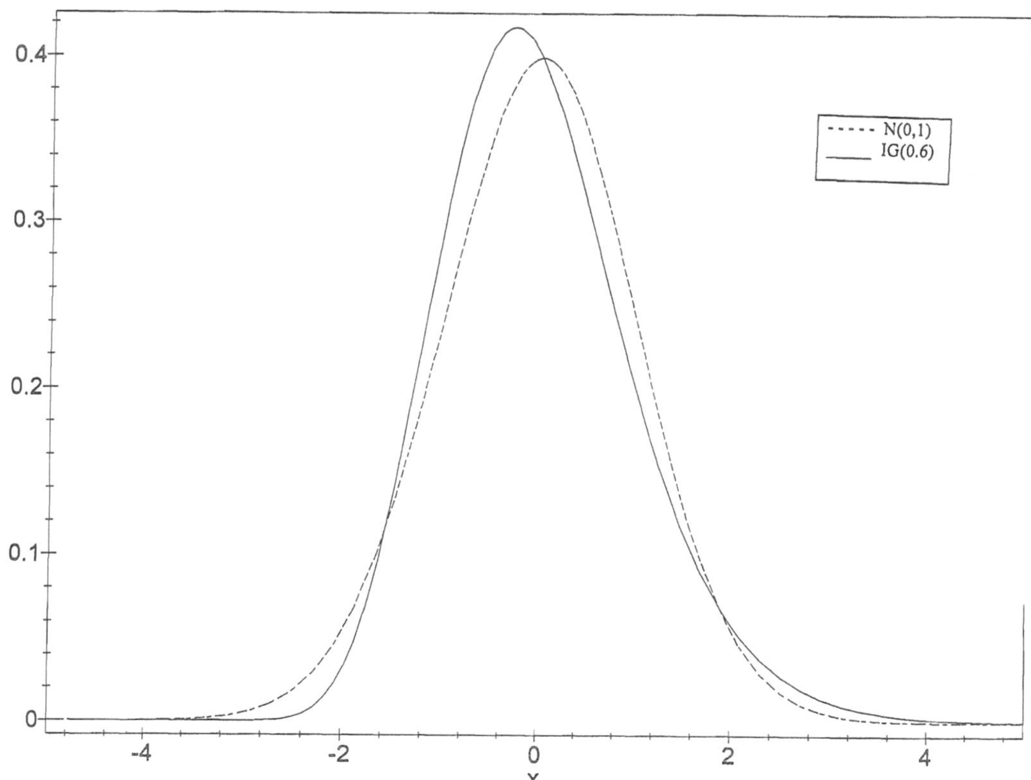

Figure 7. Plot of IG(0.7) and N(0,1) density functions

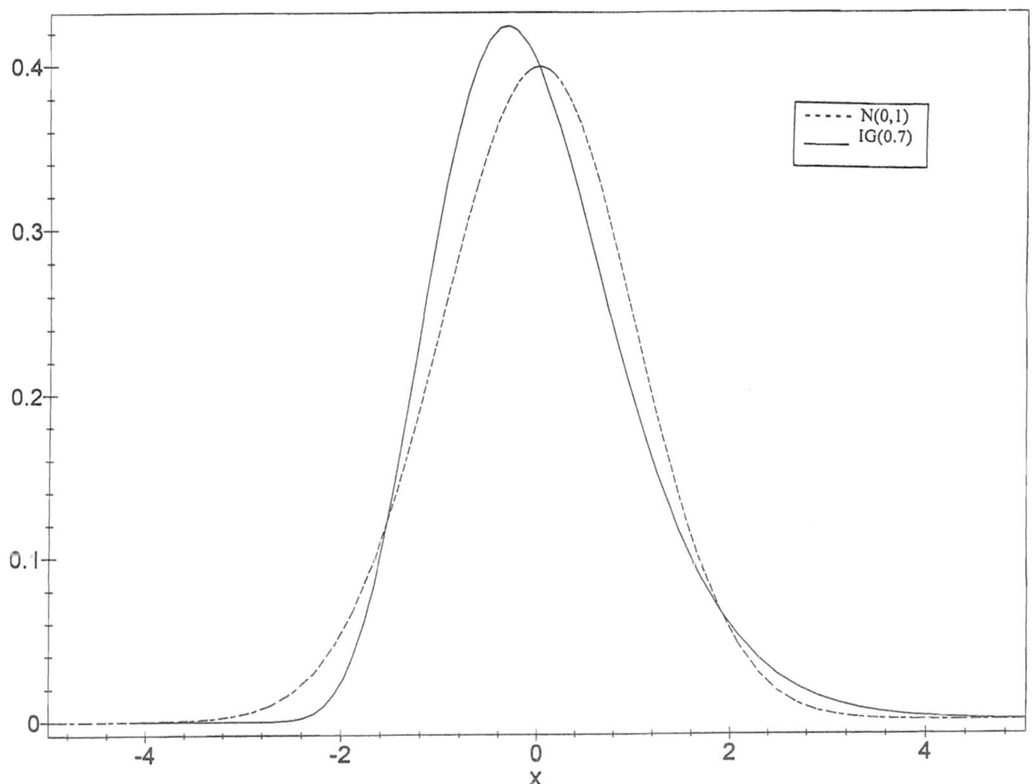

Figure 8. Plot of IG(0.8) and N(0,1) density functions

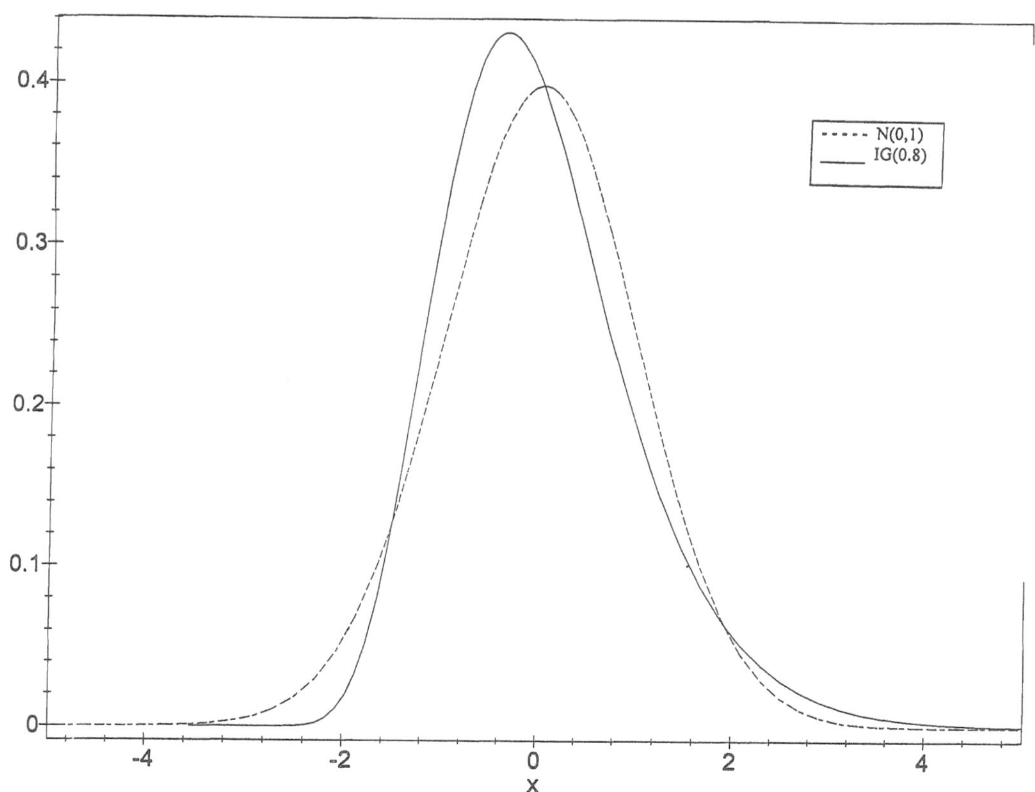

Figure 9. Plot of IG(0.9) and N(0,1) density functions

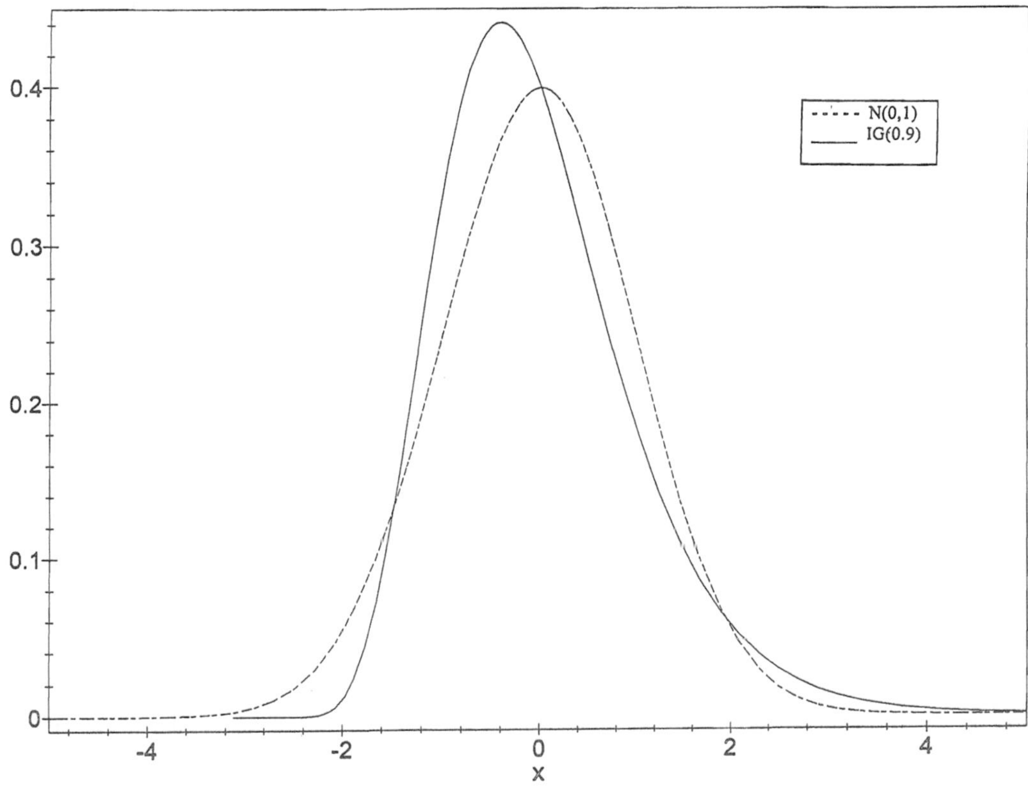

Figure 10. Plot of IG(1.0) and N(0,1) density functions

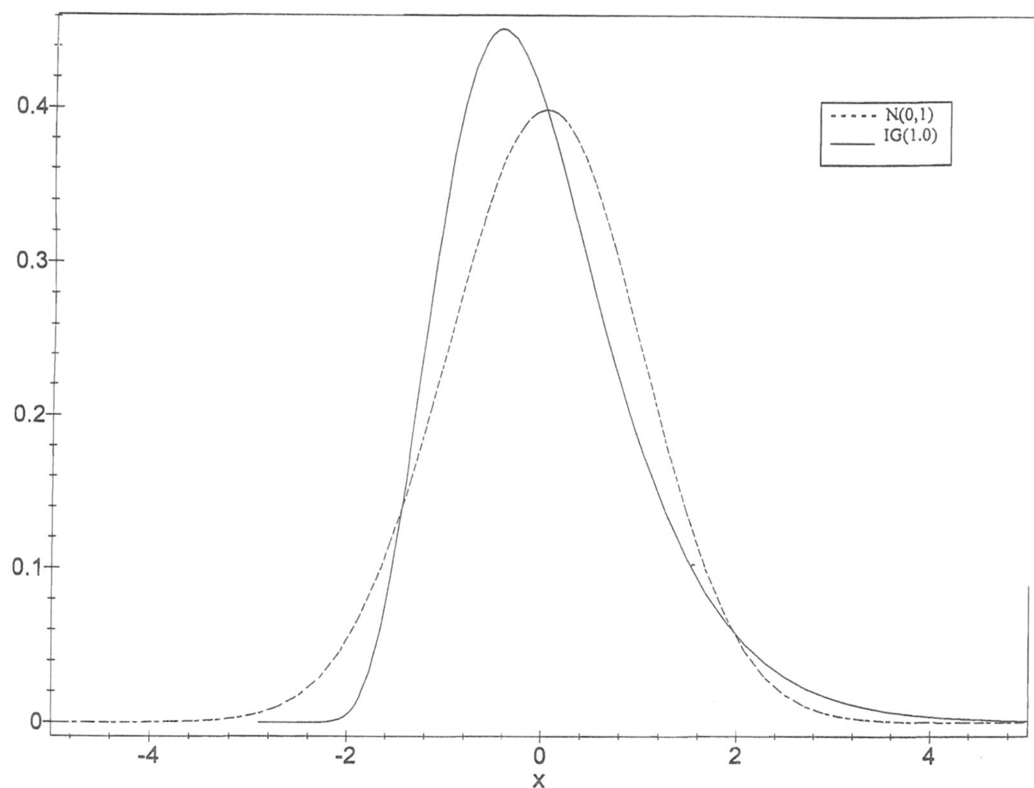

Figure 11. Plot of IG(1.1) and N(0,1) density functions

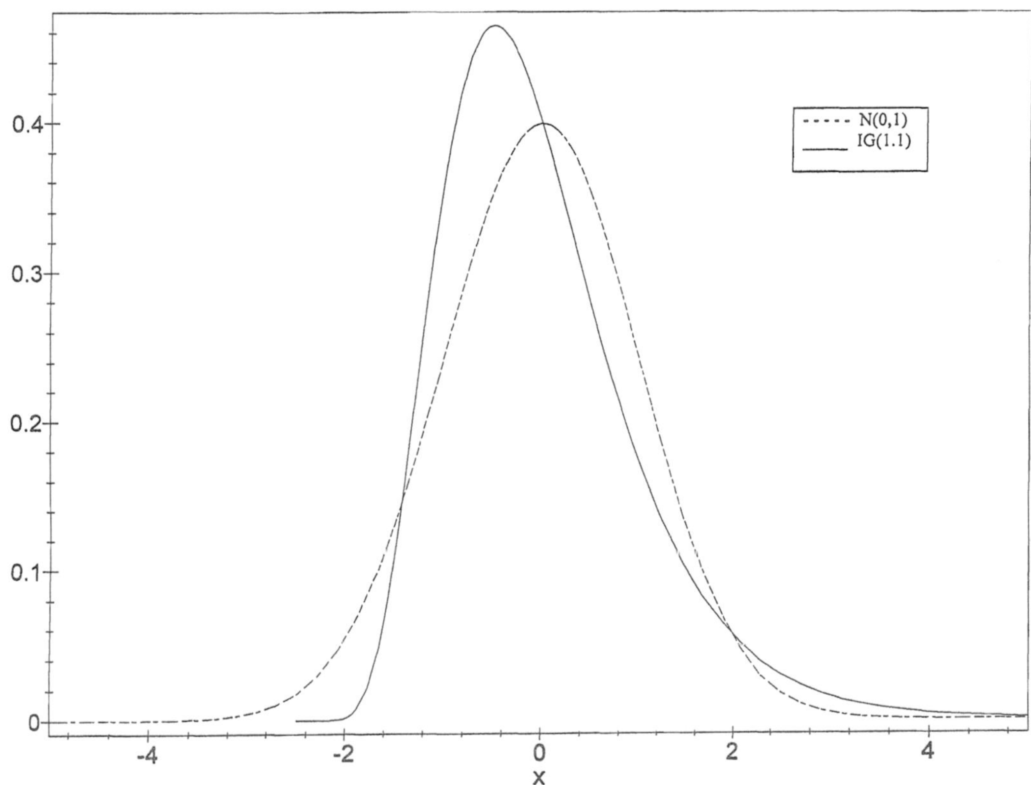

Figure 12. Plot of IG(1.2) and N(0,1) density functions

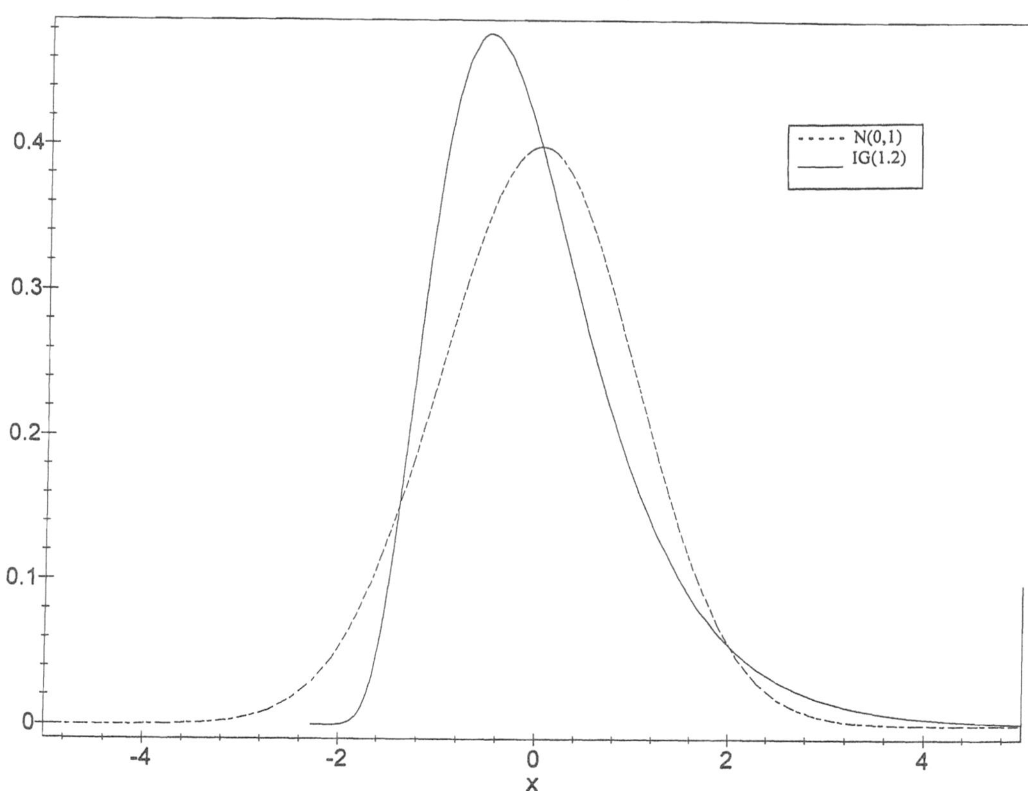

Figure 13. Plot of IG(1.3) and N(0,1) density functions

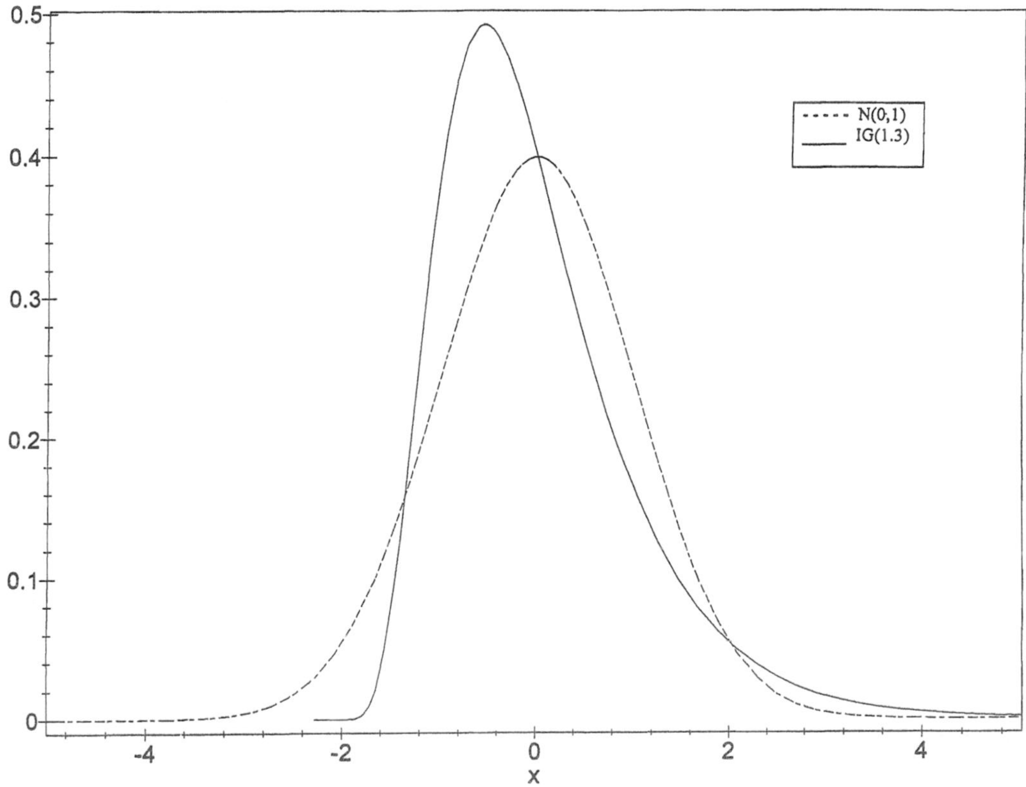

Figure 14. Plot of IG(1.4) and N(0,1) density functions

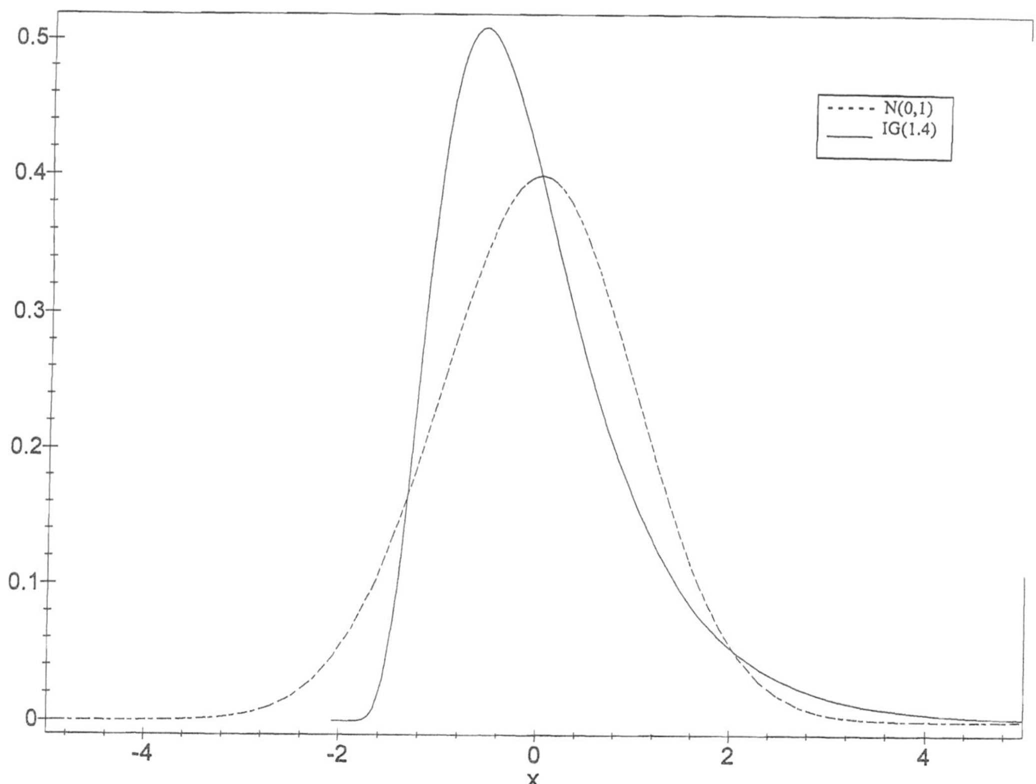

Figure 15. Plot of IG(1.5) and N(0,1) density functions

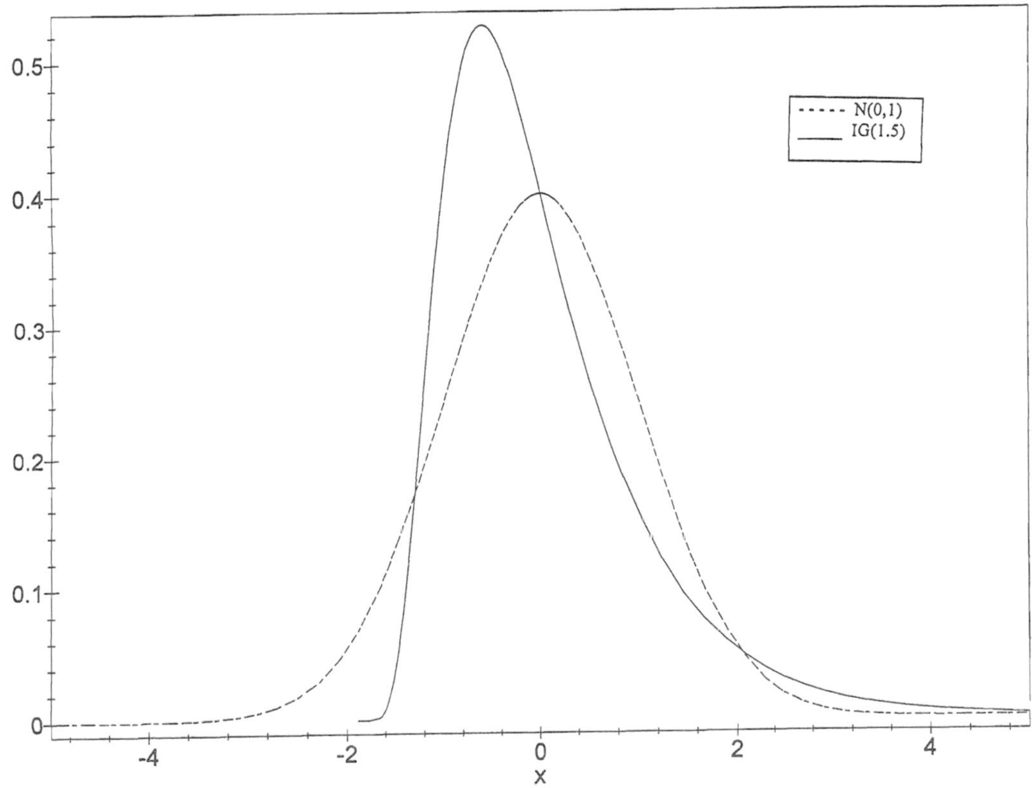

Figure 16. Plot of IG(1.6) and N(0,1) density functions

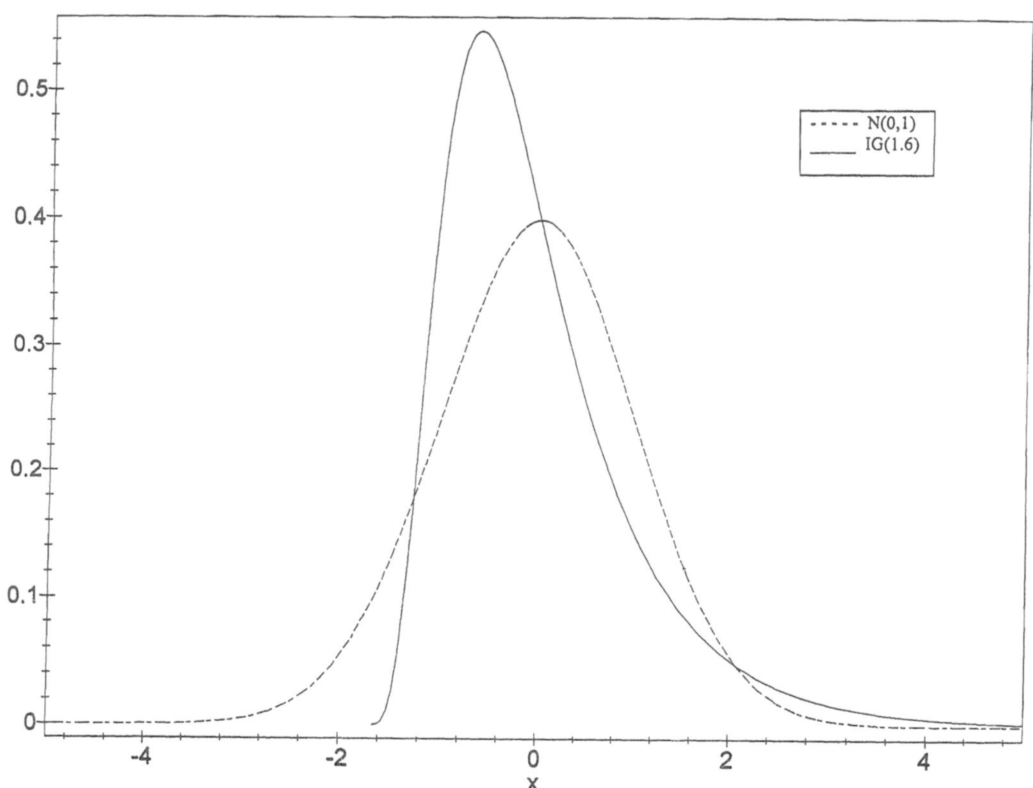

Figure 17. Plot of IG(1.7) and N(0,1) density functions

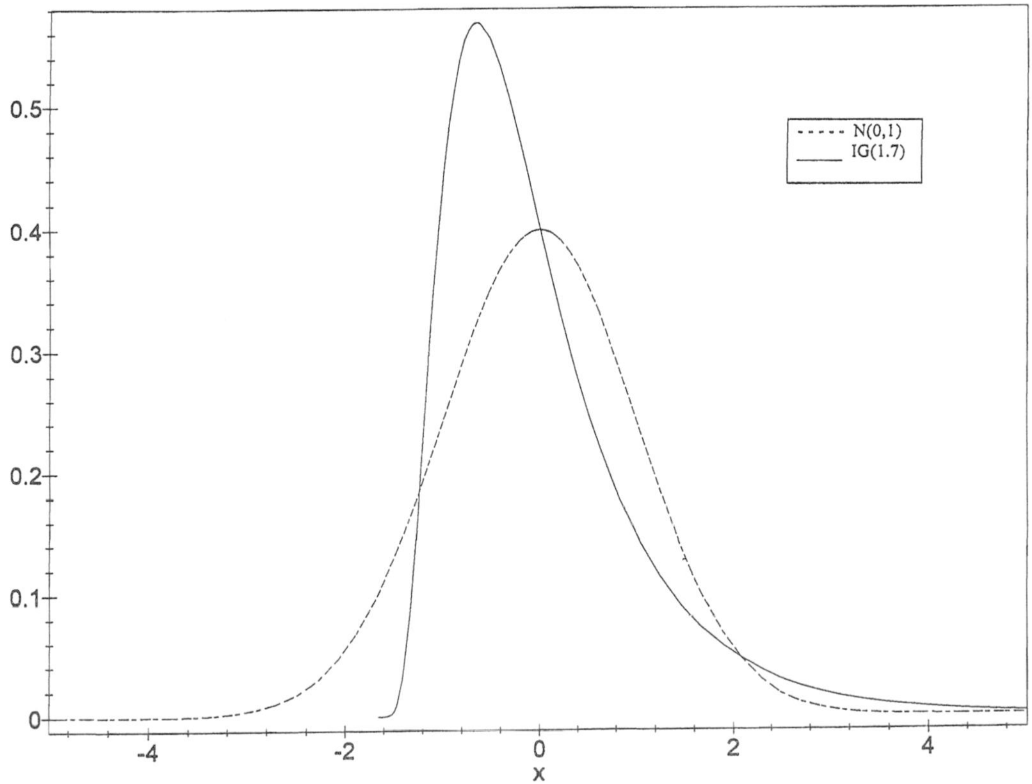

Figure 18. Plot of IG(1.8) and N(0,1) density functions

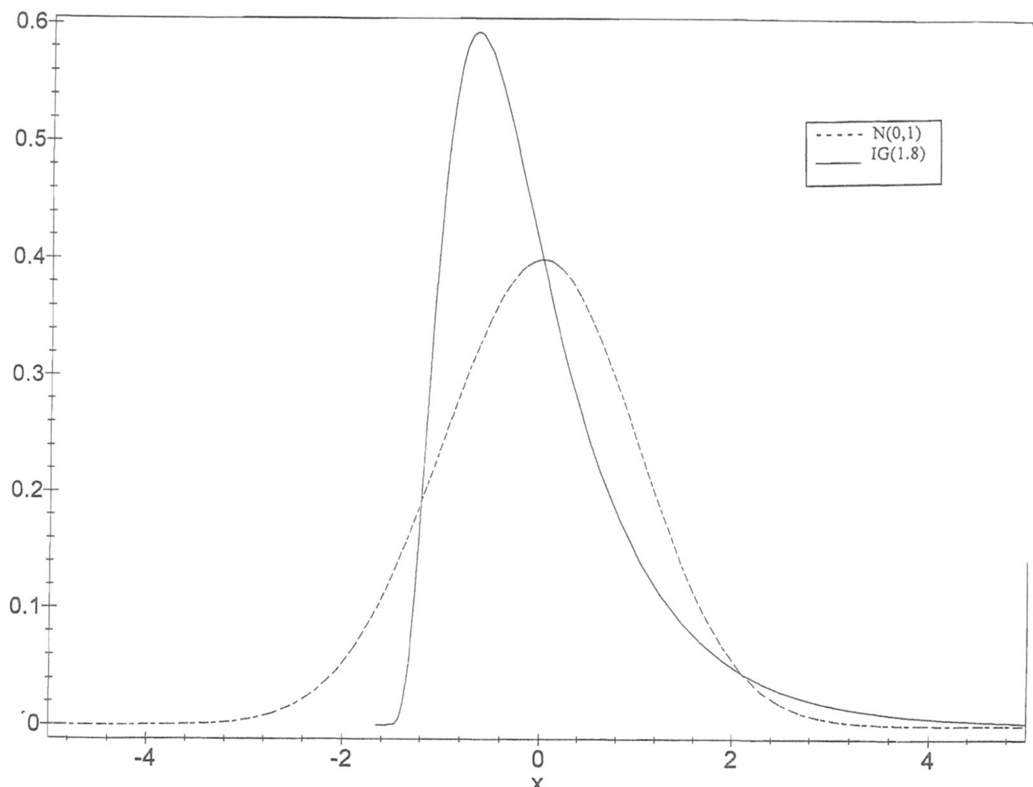

Figure 19. Plot of IG(1.9) and N(0,1) density functions

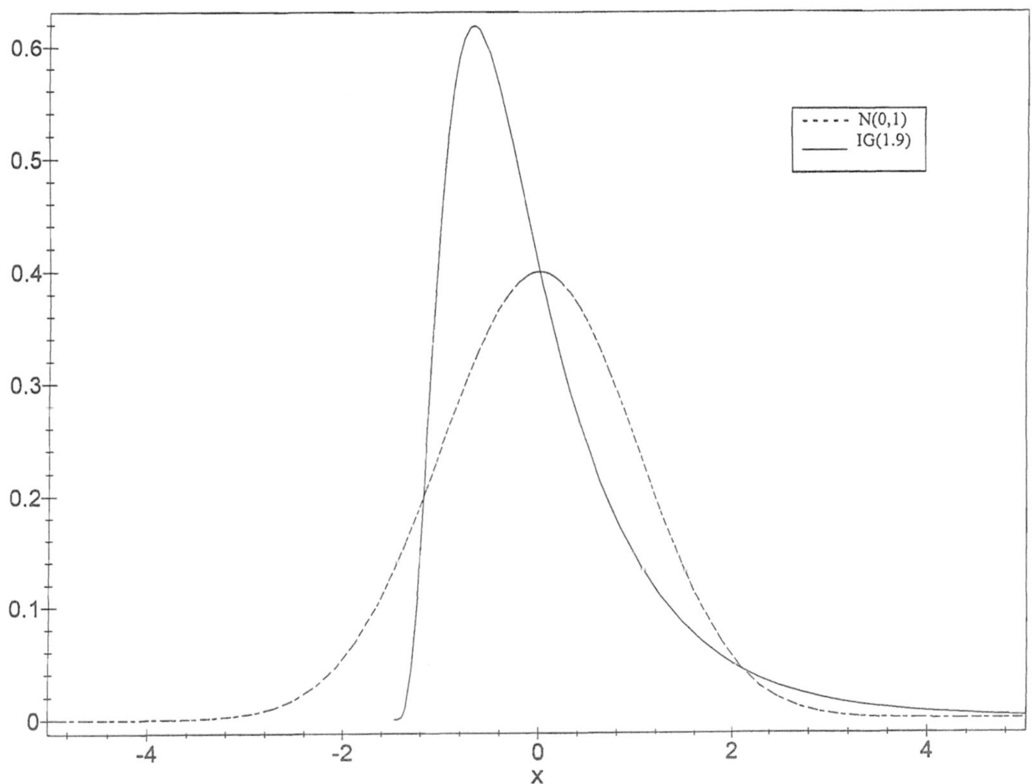

Figure 20. Plot of IG(2.0) and N(0,1) density functions

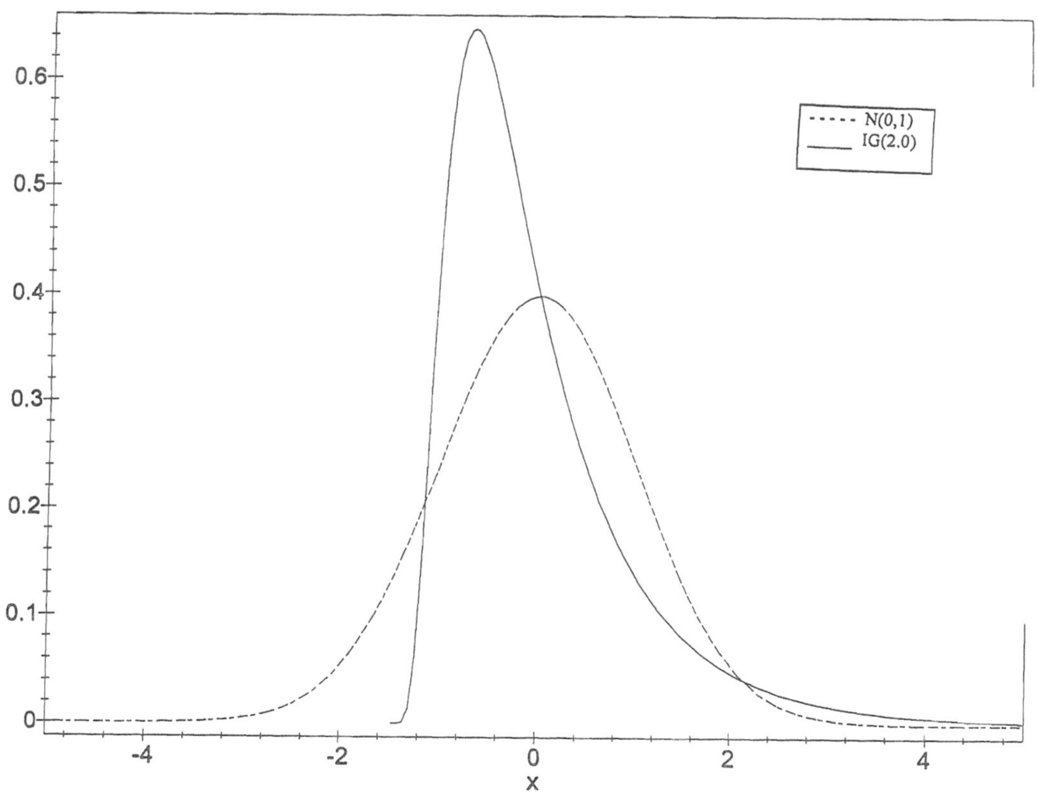

Figure 21. Plot of IG(2.1) and N(0,1) density functions

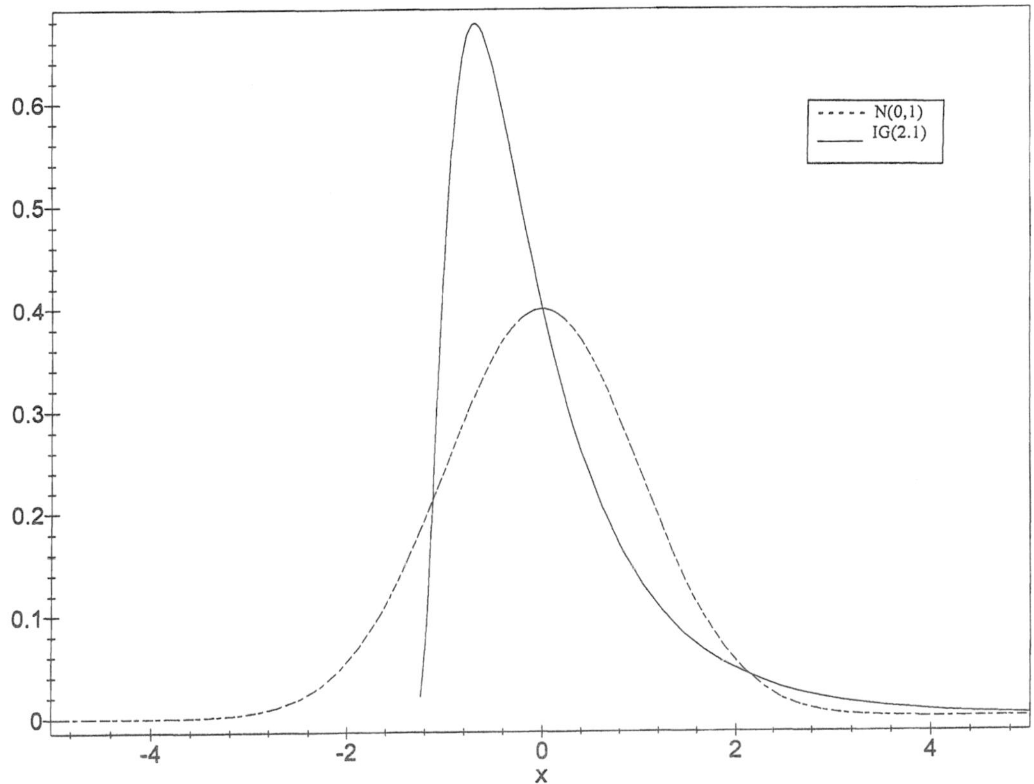

Figure 22. Plot of IG(2.2) and N(0,1) density functions

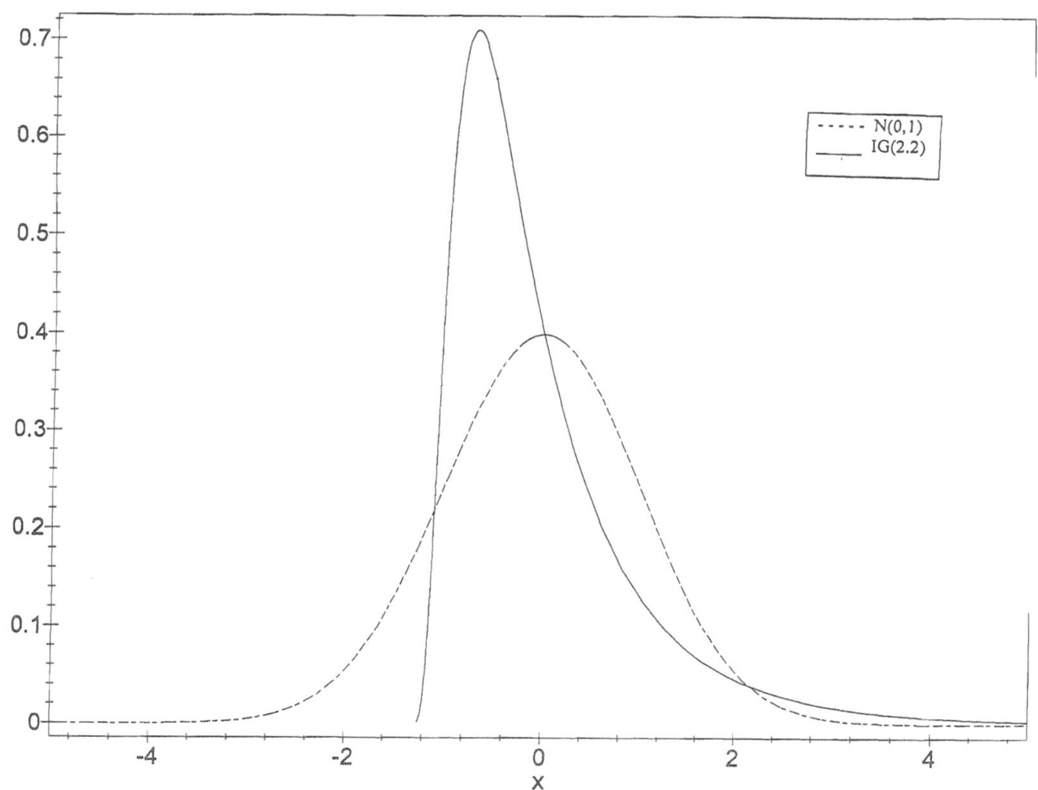

Figure 23. Plot of IG(2.3) and N(0,1) density functions

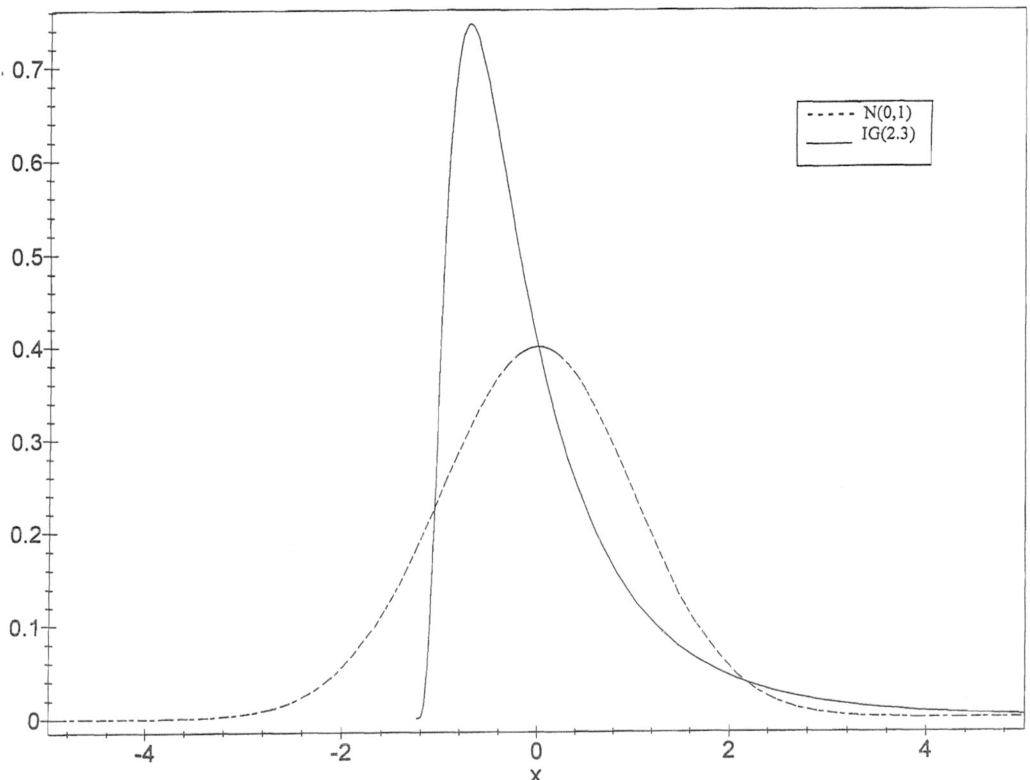

Figure 24. Plot of IG(2.4) and N(0,1) density functions

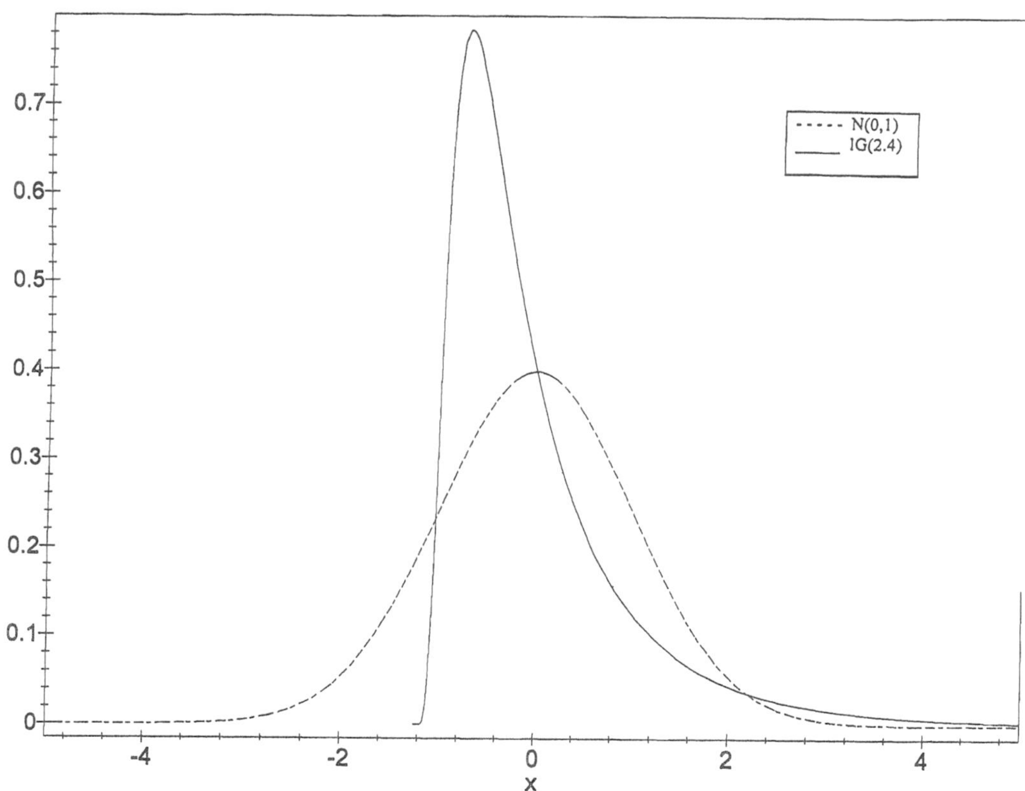

Figure 25. Plot of IG(2.5) and N(0,1) density functions

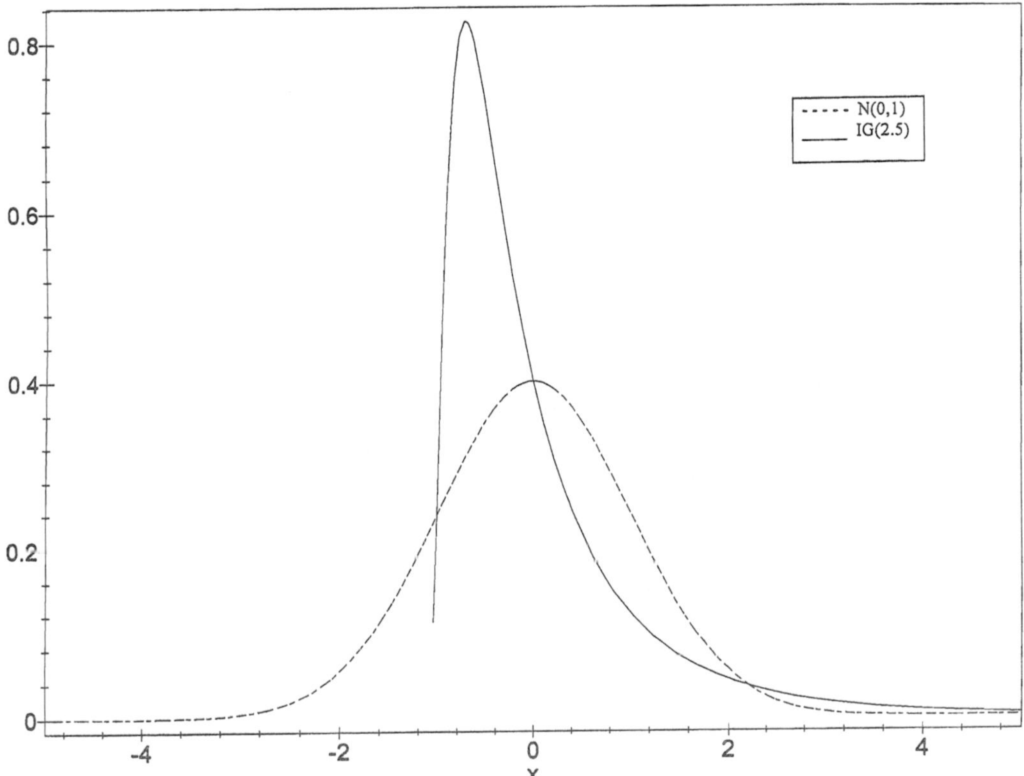

Figure 26. Quantile-Quantile plots for Example 1

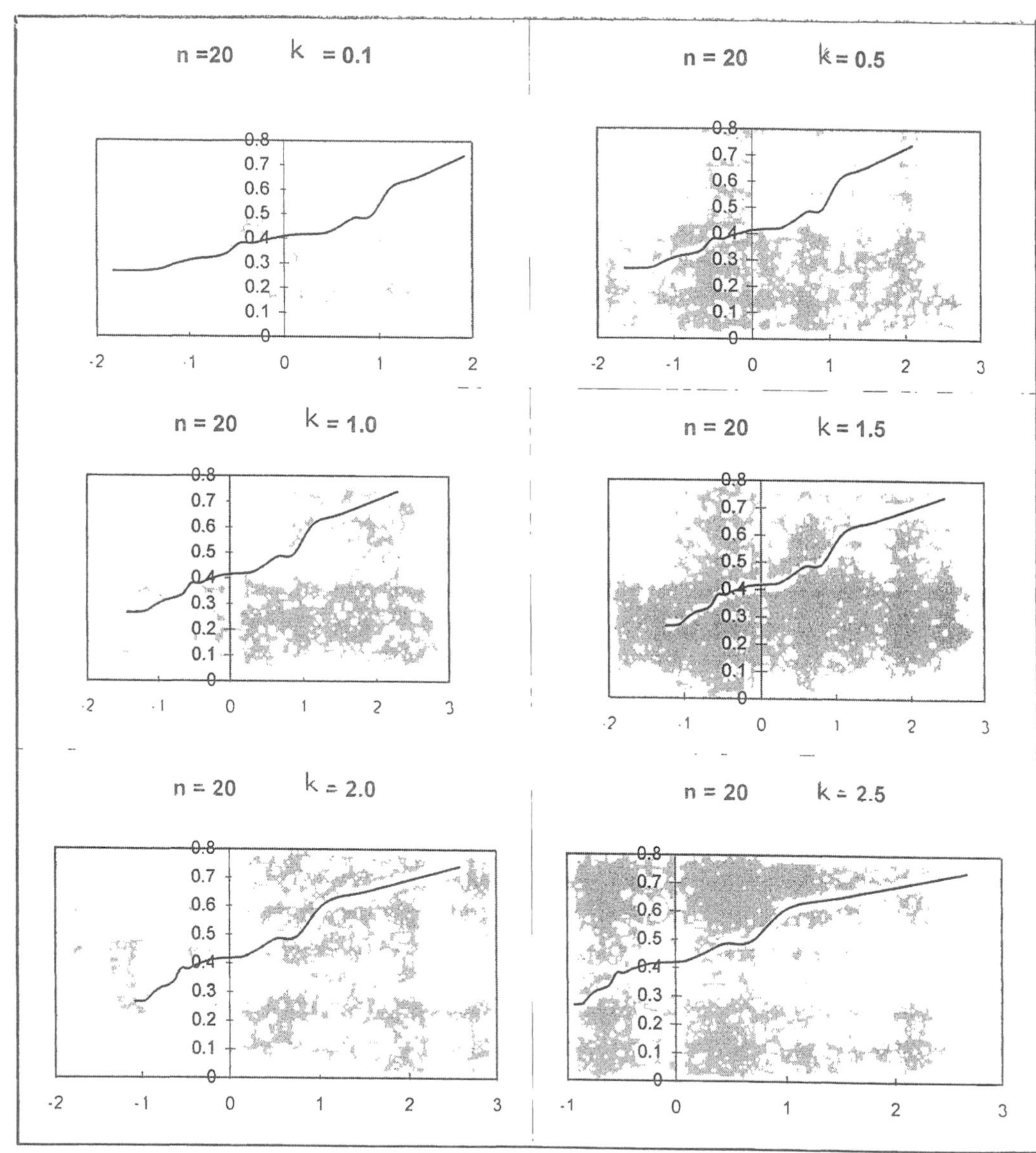

Figure 27. Quantile-Quantile plots for Example 3

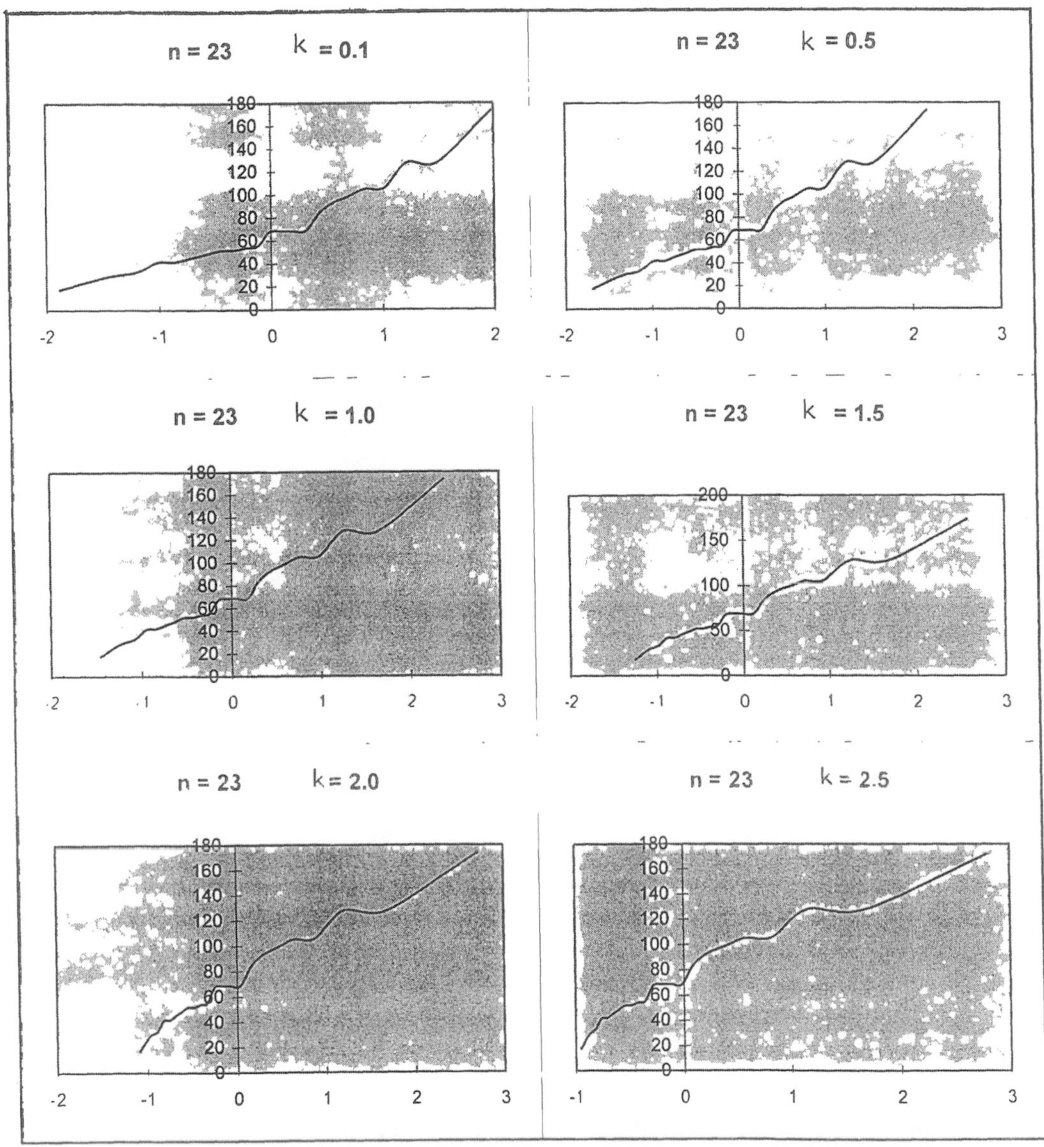

Tables

Table 1. Means of order statistics for $n = 1(1)25$, $i = 1(1)n$ and $k = 0.0(0.1)2.5$

Expected Values of Inverse Gaussian Order Statistics
Shape Parameter k

n	i	0.0	0.1	0.2	0.3	0.4	0.5	0.6	0.7
1	1	.00000	.00000	.00000	.00000	.00000	.00000	.00000	.00000
2	1	-.56419	-.56399	-.56341	-.56244	-.56110	-.55940	-.55735	-.55498
2	2	.56419	.56399	.56341	.56244	.56110	.55940	.55735	.55498
3	1	-.84628	-.84140	-.83595	-.82996	-.82346	-.81649	-.80908	-.80129
3	2	.00000	-.00918	-.01833	-.02741	-.03638	-.04522	-.05389	-.06238
3	3	.84628	.85058	.85428	.85736	.85984	.86171	.86298	.86367
4	1	-1.02938	-1.01985	-1.00969	-.99894	-.98764	-.97585	-.96363	-.95105
4	2	-.29701	-.30604	-.31472	-.32302	-.33093	-.33840	-.34544	-.35202
4	3	.29701	.28767	.27806	.26821	.25816	.24797	.23765	.22727
4	4	1.02938	1.03822	1.04635	1.05375	1.06040	1.06629	1.07142	1.07580
5	1	-1.16296	-1.14928	-1.13493	-1.11998	-1.10450	-1.08856	-1.07223	-1.05558
5	2	-.49502	-.50214	-.50873	-.51474	-.52017	-.52501	-.52926	-.53291
5	3	.00000	-.01188	-.02371	-.03545	-.04706	-.05849	-.06971	-.08068
5	4	.49502	.48737	.47924	.47065	.46165	.45227	.44256	.43256
5	5	1.16296	1.17593	1.18813	1.19952	1.21008	1.21979	1.22863	1.23661
6	1	-1.26721	-1.24982	-1.23178	-1.21315	-1.19403	-1.17448	-1.15461	-1.13449
6	2	-.64176	-.64656	-.65069	-.65413	-.65688	-.65894	-.66032	-.66103
6	3	-.20155	-.21331	-.22479	-.23595	-.24675	-.25715	-.26713	-.27666
6	4	.20155	.18955	.17737	.16504	.15263	.14017	.12772	.11531
6	5	.64176	.63628	.63017	.62345	.61616	.60832	.59998	.59119
6	6	1.26721	1.28386	1.29972	1.31474	1.32887	1.34208	1.35436	1.36569
7	1	-1.35218	-1.33149	-1.31016	-1.28828	-1.26594	-1.24324	-1.22028	-1.19714
7	2	-.75737	-.75982	-.76149	-.76240	-.76254	-.76193	-.76060	-.75857
7	3	-.35271	-.36343	-.37369	-.38347	-.39273	-.40145	-.40960	-.41718
7	4	.00000	-.01315	-.02625	-.03925	-.05210	-.06475	-.07717	-.08931
7	5	.35271	.34158	.33008	.31826	.30618	.29387	.28138	.26878
7	6	.75737	.75417	.75021	.74553	.74015	.73410	.72742	.72015
7	7	1.35218	1.37214	1.39131	1.40961	1.42699	1.44342	1.45885	1.47328
8	1	-1.42360	-1.39994	-1.37566	-1.35086	-1.32566	-1.30016	-1.27447	-1.24868
8	2	-.85222	-.85238	-.85170	-.85020	-.84789	-.84480	-.84097	-.83642
8	3	-.47282	-.48214	-.49087	-.49900	-.50649	-.51333	-.51952	-.52504
8	4	-.15251	-.16557	-.17839	-.19093	-.20314	-.21498	-.22641	-.23741
8	5	.15251	.13927	.12589	.11243	.09894	.08547	.07207	.05879
8	6	.47282	.46296	.45259	.44176	.43052	.41891	.40697	.39478
8	7	.85222	.85124	.84942	.84678	.84336	.83916	.83423	.82861
8	8	1.42360	1.44656	1.46872	1.49001	1.51037	1.52974	1.54808	1.56538
9	1	-1.48501	-1.45864	-1.43169	-1.40427	-1.37649	-1.34848	-1.32034	-1.29218
9	2	-.93230	-.93027	-.92737	-.92361	-.91902	-.91364	-.90750	-.90065
9	3	-.57197	-.57975	-.58685	-.59324	-.59892	-.60388	-.60811	-.61161
9	4	-.27453	-.28692	-.29893	-.31051	-.32163	-.33225	-.34234	-.35188
9	5	.00000	-.01389	-.02772	-.04145	-.05502	-.06839	-.08150	-.09433
9	6	.27453	.26180	.24879	.23554	.22211	.20856	.19493	.18127
9	7	.57197	.56354	.55450	.54488	.53472	.52408	.51300	.50153
9	8	.93230	.93343	.93368	.93304	.93154	.92919	.92602	.92206
9	9	1.48501	1.51070	1.53560	1.55963	1.58272	1.60481	1.62584	1.64579
10	1	-1.53875	-1.50991	-1.48051	-1.45070	-1.42058	-1.39029	-1.35994	-1.32965
10	2	-1.00136	-.99728	-.99229	-.98642	-.97971	-.97219	-.96393	-.95498
10	3	-.65606	-.66226	-.66771	-.67239	-.67629	-.67941	-.68176	-.68335
10	4	-.37576	-.38721	-.39816	-.40856	-.41840	-.42764	-.43626	-.44424
10	5	-.12267	-.13648	-.15009	-.16343	-.17647	-.18916	-.20147	-.21334
10	6	.12267	.10870	.09464	.08052	.06642	.05238	.03846	.02469
10	7	.37576	.36386	.35155	.33888	.32591	.31268	.29924	.28567
10	8	.65606	.64912	.64148	.63316	.62421	.61468	.60460	.59404
10	9	1.00136	1.00451	1.00673	1.00802	1.00837	1.00781	1.00637	1.00406
10	10	1.53875	1.56694	1.59436	1.62092	1.64654	1.67114	1.69467	1.71710

Expected Values of Inverse Gaussian Order Statistics
Shape Parameter k

n	i	0.0	0.1	0.2	0.3	0.4	0.5	0.6	0.7
11	1	-1.58644	-1.55531	-1.52367	-1.49166	-1.45940	-1.42702	-1.39466	-1.36243
11	2	-1.06192	-1.05590	-1.04895	-1.04111	-1.03242	-1.02293	-1.01271	-1.00182
11	3	-.72884	-.73349	-.73732	-.74032	-.74250	-.74387	-.74442	-.74420
11	4	-.46198	-.47234	-.48210	-.49124	-.49972	-.50753	-.51464	-.52106
11	5	-.22489	-.23824	-.25125	-.26388	-.27609	-.28783	-.29908	-.30980
11	6	.00000	-.01437	-.02869	-.04289	-.05693	-.07076	-.08433	-.09760
11	7	.22489	.21126	.19740	.18337	.16922	.15501	.14078	.12660
11	8	.46198	.45106	.43964	.42775	.41544	.40277	.38980	.37656
11	9	.72884	.72339	.71716	.71019	.70250	.69414	.68516	.67559
11	10	1.06192	1.06698	1.07108	1.07420	1.07634	1.07752	1.07775	1.07706
11	11	1.58644	1.61693	1.64669	1.67559	1.70356	1.73050	1.75637	1.78110
12	1	-1.62923	-1.59598	-1.56227	-1.52823	-1.49399	-1.45970	-1.42549	-1.39148
12	2	-1.11573	-1.10788	-1.09909	-1.08940	-1.07887	-1.06755	-1.05551	-1.04282
12	3	-.79284	-.79597	-.79823	-.79962	-.80016	-.79985	-.79872	-.79679
12	4	-.53684	-.54605	-.55459	-.56242	-.56953	-.57591	-.58155	-.58644
12	5	-.31225	-.32491	-.33713	-.34887	-.36009	-.37075	-.38083	-.39031
12	6	-.10259	-.11690	-.13102	-.14490	-.15849	-.17174	-.18462	-.19708
12	7	.10259	.08815	.07364	.05911	.04462	.03021	.01595	.00188
12	8	.31225	.29920	.28581	.27213	.25823	.24414	.22994	.21568
12	9	.53684	.52700	.51655	.50555	.49405	.48209	.46972	.45701
12	10	.79284	.78886	.78404	.77840	.77199	.76483	.75697	.74845
12	11	1.11573	1.12261	1.12849	1.13336	1.13721	1.14006	1.14191	1.14278
12	12	1.62923	1.66187	1.69380	1.72489	1.75504	1.78418	1.81223	1.83913
13	1	-1.66799	-1.63277	-1.59713	-1.56120	-1.52513	-1.48907	-1.45316	-1.41751
13	2	-1.16408	-1.15450	-1.14397	-1.13255	-1.12029	-1.10726	-1.09352	-1.07917
13	3	-.84983	-.85149	-.85224	-.85210	-.85106	-.84917	-.84644	-.84291
13	4	-.60285	-.61088	-.61818	-.62472	-.63048	-.63546	-.63965	-.64307
13	5	-.38833	-.40018	-.41150	-.42226	-.43241	-.44194	-.45081	-.45902
13	6	-.19052	-.20448	-.21814	-.23146	-.24438	-.25686	-.26887	-.28037
13	7	.00000	-.01471	-.02936	-.04390	-.05828	-.07243	-.08632	-.09990
13	8	.19052	.17632	.16193	.14741	.13281	.11820	.10362	.08912
13	9	.38833	.37599	.36323	.35008	.33661	.32286	.30890	.29477
13	10	.60285	.59411	.58470	.57465	.56402	.55285	.54120	.52911
13	11	.84983	.84728	.84384	.83953	.83438	.82842	.82170	.81425
13	12	1.16408	1.17267	1.18024	1.18678	1.19227	1.19672	1.20012	1.20251
13	13	1.66799	1.70264	1.73659	1.76973	1.80194	1.83313	1.86323	1.89218
14	1	-1.70338	-1.66632	-1.62887	-1.59118	-1.55341	-1.51570	-1.47820	-1.44103
14	2	-1.20790	-1.19669	-1.18452	-1.17147	-1.15758	-1.14294	-1.12763	-1.11172
14	3	-.90113	-.90136	-.90067	-.89904	-.89652	-.89312	-.88888	-.88384
14	4	-.66176	-.66862	-.67469	-.67995	-.68439	-.68801	-.69082	-.69283
14	5	-.45557	-.46653	-.47690	-.48663	-.49569	-.50406	-.51173	-.51868
14	6	-.26730	-.28073	-.29378	-.30639	-.31851	-.33012	-.34117	-.35164
14	7	-.08816	-.10281	-.11730	-.13155	-.14553	-.15919	-.17248	-.18535
14	8	.08816	.07339	.05857	.04374	.02898	.01432	-.00017	-.01445
14	9	.26730	.25352	.23945	.22516	.21069	.19610	.18146	.16680
14	10	.45557	.44404	.43199	.41948	.40656	.39328	.37970	.36587
14	11	.66176	.65414	.64578	.63672	.62700	.61668	.60580	.59441
14	12	.90113	.89995	.89785	.89484	.89094	.88617	.88058	.87421
14	13	1.20790	1.21812	1.22731	1.23544	1.24249	1.24847	1.25338	1.25723
14	14	1.70338	1.73991	1.77577	1.81083	1.84497	1.87811	1.91015	1.94102
15	1	-1.73591	-1.69711	-1.65797	-1.61863	-1.57926	-1.54001	-1.50103	-1.46245
15	2	-1.24794	-1.23517	-1.22145	-1.20685	-1.19144	-1.17529	-1.15850	-1.14114
15	3	-.94769	-.94656	-.94447	-.94144	-.93750	-.93267	-.92700	-.92054
15	4	-.71488	-.72058	-.72544	-.72945	-.73262	-.73493	-.73640	-.73706
15	5	-.51570	-.52575	-.53513	-.54382	-.55178	-.55900	-.56547	-.57119

Expected Values of Inverse Gaussian Order Statistics
Shape Parameter k

n	i	0.0	0.1	0.2	0.3	0.4	0.5	0.6	0.7
15	6	-.33530	-.34810	-.36044	-.37225	-.38351	-.39418	-.40423	-.41365
15	7	-.16530	-.17968	-.19380	-.20759	-.22101	-.23402	-.24657	-.25863
15	8	.00000	-.01496	-.02987	-.04466	-.05927	-.07367	-.08780	-.10161
15	9	.16530	.15070	.13595	.12109	.10620	.09132	.07651	.06182
15	10	.33530	.32206	.30846	.29454	.28035	.26596	.25142	.23679
15	11	.51570	.50502	.49376	.48196	.46967	.45694	.44384	.43041
15	12	.71488	.70836	.70106	.69299	.68422	.67477	.66469	.65404
15	13	.94769	.94785	.94705	.94530	.94262	.93902	.93456	.92925
15	14	1.24794	1.25970	1.27042	1.28007	1.28863	1.29608	1.30243	1.30769
15	15	1.73591	1.77421	1.81187	1.84874	1.88471	1.91968	1.95355	1.98626
16	1	-1.76599	-1.72556	-1.68481	-1.64392	-1.60305	-1.56236	-1.52199	-1.48209
16	2	-1.28474	-1.27050	-1.25532	-1.23926	-1.22240	-1.20483	-1.18664	-1.16791
16	3	-.99027	-.98783	-.98441	-.98003	-.97473	-.96854	-.96152	-.95371
16	4	-.76317	-.76773	-.77142	-.77422	-.77615	-.77720	-.77739	-.77676
16	5	-.57001	-.57912	-.58751	-.59515	-.60203	-.60812	-.61343	-.61795
16	6	-.39622	-.40833	-.41990	-.43087	-.44123	-.45094	-.45998	-.46833
16	7	-.23375	-.24772	-.26134	-.27455	-.28732	-.29958	-.31133	-.32251
16	8	-.07729	-.09220	-.10695	-.12149	-.13576	-.14972	-.16331	-.17650
16	9	.07729	.06227	.04722	.03218	.01721	.00237	-.01228	-.02672
16	10	.23375	.21948	.20496	.19025	.17541	.16050	.14557	.13068
16	11	.39622	.38362	.37056	.35711	.34331	.32924	.31493	.30046
16	12	.57001	.56021	.54976	.53871	.52710	.51499	.50243	.48947
16	13	.76317	.75775	.75149	.74442	.73659	.72803	.71878	.70890
16	14	.99027	.99172	.99219	.99166	.99016	.98772	.98435	.98010
16	15	1.28474	1.29798	1.31017	1.32128	1.33127	1.34013	1.34787	1.35449
16	16	1.76599	1.80596	1.84531	1.88390	1.92161	1.95832	1.99393	2.02838
17	1	-1.79394	-1.75196	-1.70970	-1.66735	-1.62506	-1.58301	-1.54134	-1.50019
17	2	-1.31878	-1.30314	-1.28655	-1.26911	-1.25088	-1.23197	-1.21245	-1.19244
17	3	-1.02946	-1.02575	-1.02106	-1.01539	-1.00880	-1.00132	-.99301	-.98393
17	4	-.80738	-.81084	-.81338	-.81501	-.81574	-.81559	-.81456	-.81270
17	5	-.61946	-.62763	-.63503	-.64164	-.64745	-.65244	-.65661	-.65997
17	6	-.45133	-.46270	-.47346	-.48358	-.49302	-.50176	-.50979	-.51708
17	7	-.29519	-.30865	-.32169	-.33425	-.34629	-.35777	-.36866	-.37894
17	8	-.14599	-.16068	-.17513	-.18928	-.20307	-.21647	-.22942	-.24190
17	9	.00000	-.01516	-.03025	-.04523	-.06004	-.07463	-.08894	-.10293
17	10	.14599	.13110	.11609	.10099	.08589	.07082	.05585	.04103
17	11	.29519	.28134	.26717	.25273	.23808	.22327	.20837	.19343
17	12	.45133	.43940	.42695	.41404	.40072	.38704	.37306	.35884
17	13	.61946	.61055	.60093	.59065	.57976	.56830	.55633	.54390
17	14	.80738	.80304	.79781	.79174	.78484	.77717	.76876	.75967
17	15	1.02946	1.03216	1.03384	1.03450	1.03416	1.03283	1.03055	1.02734
17	16	1.31878	1.33343	1.34702	1.35951	1.37088	1.38111	1.39018	1.39811
17	17	1.79394	1.83549	1.87645	1.91668	1.95603	1.99439	2.03167	2.06777
18	1	-1.82003	-1.77658	-1.73289	-1.68915	-1.64552	-1.60218	-1.55928	-1.51697
18	2	-1.35041	-1.33343	-1.31551	-1.29675	-1.27722	-1.25703	-1.23627	-1.21504
18	3	-1.06573	-1.06081	-1.05488	-1.04798	-1.04016	-1.03145	-1.02192	-1.01163
18	4	-.84813	-.85049	-.85193	-.85243	-.85201	-.85069	-.84849	-.84545
18	5	-.66479	-.67203	-.67846	-.68406	-.68881	-.69273	-.69580	-.69805
18	6	-.50158	-.51218	-.52212	-.53137	-.53989	-.54767	-.55470	-.56097
18	7	-.35084	-.36374	-.37614	-.38800	-.39927	-.40994	-.41996	-.42932
18	8	-.20774	-.22210	-.23613	-.24979	-.26302	-.27578	-.28804	-.29976
18	9	-.06880	-.08391	-.09888	-.11363	-.12813	-.14232	-.15615	-.16958
18	10	.06880	.05360	.03837	.02316	.00804	-.00694	-.02172	-.03627
18	11	.20774	.19310	.17826	.16326	.14816	.13303	.11791	.10286
18	12	.35084	.33749	.32375	.30966	.29530	.28070	.26594	.25107

Expected Values of Inverse Gaussian Order Statistics
Shape Parameter k

n	i	0.0	0.1	0.2	0.3	0.4	0.5	0.6	0.7
18	13	.50158	.49036	.47856	.46623	.45343	.44020	.42662	.41272
18	14	.66479	.65677	.64800	.63851	.62835	.61757	.60622	.59436
18	15	.84813	.84483	.84062	.83552	.82956	.82277	.81520	.80690
18	16	1.06573	1.06962	1.07248	1.07430	1.07508	1.07485	1.07362	1.07143
18	17	1.35041	1.36640	1.38133	1.39516	1.40786	1.41939	1.42975	1.43894
18	18	1.82003	1.86309	1.90558	1.94735	1.98827	2.02822	2.06707	2.10476
19	1	-1.84448	-1.79963	-1.75458	-1.70952	-1.66463	-1.62007	-1.57600	-1.53258
19	2	-1.37994	-1.36167	-1.34248	-1.32246	-1.30170	-1.28029	-1.25835	-1.23597
19	3	-1.09945	-1.09336	-1.08626	-1.07818	-1.06917	-1.05929	-1.04859	-1.03715
19	4	-.88586	-.88718	-.88754	-.88695	-.88542	-.88298	-.87966	-.87550
19	5	-.70661	-.71293	-.71839	-.72300	-.72674	-.72961	-.73162	-.73279
19	6	-.54771	-.55753	-.56664	-.57502	-.58263	-.58947	-.59551	-.60077
19	7	-.40164	-.41393	-.42566	-.43679	-.44729	-.45713	-.46628	-.47472
19	8	-.26374	-.27769	-.29124	-.30434	-.31696	-.32904	-.34056	-.35149
19	9	-.13072	-.14566	-.16036	-.17478	-.18886	-.20255	-.21582	-.22863
19	10	.00000	-.01531	-.03056	-.04569	-.06065	-.07539	-.08984	-.10397
19	11	.13072	.11562	.10041	.08514	.06987	.05467	.03958	.02466
19	12	.26374	.24945	.23488	.22007	.20510	.19002	.17488	.15974
19	13	.40164	.38884	.37559	.36192	.34791	.33360	.31906	.30434
19	14	.54771	.53721	.52608	.51437	.50213	.48941	.47626	.46274
19	15	.70661	.69947	.69154	.68284	.67343	.66335	.65264	.64137
19	16	.88586	.88359	.88037	.87623	.87119	.86528	.85855	.85104
19	17	1.09945	1.10450	1.10850	1.11143	1.11331	1.11414	1.11394	1.11275
19	18	1.37994	1.39721	1.41343	1.42854	1.44251	1.45530	1.46690	1.47732
19	19	1.84448	1.88897	1.93292	1.97618	2.01859	2.06004	2.10042	2.13962
20	1	-1.86748	-1.82128	-1.77494	-1.72863	-1.68252	-1.63681	-1.59164	-1.54716
20	2	-1.40760	-1.38811	-1.36770	-1.34648	-1.32454	-1.30198	-1.27891	-1.25544
20	3	-1.13095	-1.12373	-1.11549	-1.10628	-1.09614	-1.08513	-1.07333	-1.06079
20	4	-.92098	-.92128	-.92059	-.91894	-.91635	-.91283	-.90843	-.90320
20	5	-.74538	-.75079	-.75532	-.75895	-.76170	-.76356	-.76455	-.76468
20	6	-.59030	-.59934	-.60763	-.61514	-.62185	-.62775	-.63284	-.63712
20	7	-.44833	-.45998	-.47102	-.48141	-.49112	-.50013	-.50841	-.51595
20	8	-.31493	-.32841	-.34142	-.35393	-.36589	-.37727	-.38803	-.39816
20	9	-.18696	-.20161	-.21597	-.22996	-.24356	-.25670	-.26936	-.28149
20	10	-.06200	-.07727	-.09239	-.10732	-.12200	-.13637	-.15039	-.16402
20	11	.06200	.04665	.03127	.01593	.00069	-.01440	-.02929	-.04392
20	12	.18696	.17206	.15697	.14175	.12647	.11117	.09592	.08077
20	13	.31493	.30105	.28682	.27229	.25752	.24258	.22751	.21238
20	14	.44833	.43612	.42338	.41019	.39658	.38262	.36836	.35386
20	15	.59030	.58054	.57010	.55902	.54736	.53517	.52250	.50940
20	16	.74538	.73912	.73202	.72411	.71545	.70607	.69602	.68535
20	17	.92098	.91971	.91746	.91426	.91012	.90509	.89919	.89246
20	18	1.13095	1.13712	1.14221	1.14623	1.14917	1.15103	1.15184	1.15162
20	19	1.40760	1.42611	1.44357	1.45991	1.47510	1.48911	1.50191	1.51351
20	20	1.86748	1.91333	1.95868	2.00335	2.04720	2.09009	2.13192	2.17257
21	1	-1.88917	-1.84170	-1.79412	-1.74661	-1.69935	-1.65253	-1.60631	-1.56084
21	2	-1.43362	-1.41294	-1.39137	-1.36900	-1.34593	-1.32227	-1.29812	-1.27361
21	3	-1.16047	-1.15217	-1.14284	-1.13254	-1.12131	-1.10923	-1.09636	-1.08278
21	4	-.95380	-.95310	-.95141	-.94874	-.94511	-.94056	-.93513	-.92887
21	5	-.78150	-.78601	-.78962	-.79231	-.79409	-.79497	-.79497	-.79411
21	6	-.62982	-.63808	-.64555	-.65221	-.65804	-.66303	-.66719	-.67052
21	7	-.49148	-.50248	-.51281	-.52246	-.53138	-.53956	-.54698	-.55364
21	8	-.36203	-.37499	-.38743	-.39932	-.41061	-.42126	-.43126	-.44058
21	9	-.23841	-.25272	-.26666	-.28018	-.29324	-.30579	-.31779	-.32923
21	10	-.11836	-.13347	-.14838	-.16301	-.17732	-.19126	-.20478	-.21784

Expected Values of Inverse Gaussian Order Statistics
Shape Parameter k

n	i	0.0	0.1	0.2	0.3	0.4	0.5	0.6	0.7
21	11	.00000	-.01544	-.03081	-.04607	-.06115	-.07600	-.09058	-.10482
21	12	.11836	.10308	.08771	.07230	.05692	.04161	.02643	.01144
21	13	.23841	.22379	.20891	.19384	.17864	.16335	.14804	.13277
21	14	.36203	.34860	.33476	.32056	.30607	.29133	.27641	.26136
21	15	.49148	.47988	.46770	.45500	.44183	.42826	.41434	.40011
21	16	.62982	.62080	.61106	.60063	.58958	.57794	.56576	.55312
21	17	.78150	.77609	.76982	.76270	.75479	.74611	.73673	.72668
21	18	.95380	.95350	.95220	.94992	.94667	.94249	.93741	.93147
21	19	1.16047	1.16772	1.17388	1.17895	1.18292	1.18579	1.18758	1.18831
21	20	1.43362	1.45331	1.47196	1.48949	1.50586	1.52104	1.53500	1.54774
21	21	1.88917	1.93633	1.98301	2.02904	2.07427	2.11855	2.16176	2.20382
22	1	-1.90969	-1.86100	-1.81224	-1.76358	-1.71523	-1.66735	-1.62013	-1.57371
22	2	-1.45816	-1.43635	-1.41366	-1.39019	-1.36603	-1.34132	-1.31615	-1.29064
22	3	-1.18824	-1.17889	-1.16851	-1.15716	-1.14489	-1.13178	-1.11789	-1.10331
22	4	-.98459	-.98293	-.98025	-.97659	-.97197	-.96643	-.96000	-.95275
22	5	-.81527	-.81891	-.82162	-.82339	-.82424	-.82417	-.82321	-.82139
22	6	-.66667	-.67415	-.68081	-.68663	-.69159	-.69570	-.69895	-.70135
22	7	-.53157	-.54190	-.55152	-.56042	-.56856	-.57592	-.58250	-.58829
22	8	-.40559	-.41801	-.42986	-.44111	-.45171	-.46164	-.47087	-.47939
22	9	-.28579	-.29969	-.31318	-.32618	-.33867	-.35060	-.36194	-.37266
22	10	-.16997	-.18486	-.19946	-.21373	-.22761	-.24105	-.25403	-.26649
22	11	-.05642	-.07181	-.08708	-.10215	-.11697	-.13150	-.14567	-.15946
22	12	.05642	.04094	.02545	.01001	-.00533	-.02051	-.03548	-.05019
22	13	.16997	.15487	.13959	.12421	.10879	.09337	.07803	.06280
22	14	.28579	.27150	.25691	.24205	.22699	.21180	.19652	.18121
22	15	.40559	.39265	.37925	.36543	.35125	.33678	.32206	.30717
22	16	.53157	.52058	.50897	.49680	.48411	.47095	.45740	.44349
22	17	.66667	.65838	.64934	.63957	.62913	.61805	.60640	.59423
22	18	.81527	.81071	.80525	.79892	.79175	.78378	.77506	.76563
22	19	.98459	.98523	.98486	.98347	.98110	.97776	.97349	.96833
22	20	1.18824	1.19653	1.20373	1.20981	1.21478	1.21863	1.22138	1.22305
22	21	1.45816	1.47899	1.49878	1.51745	1.53497	1.55128	1.56636	1.58021
22	22	1.90969	1.95811	2.00607	2.05340	2.09995	2.14556	2.19012	2.23351
23	1	-1.92916	-1.87930	-1.82940	-1.77965	-1.73024	-1.68136	-1.63318	-1.58585
23	2	-1.48137	-1.45847	-1.43470	-1.41017	-1.38499	-1.35926	-1.33311	-1.30665
23	3	-1.21445	-1.20408	-1.19269	-1.18032	-1.16705	-1.15295	-1.13808	-1.12255
23	4	-1.01356	-1.01096	-1.00734	-1.00273	-.99714	-.99064	-.98326	-.97506
23	5	-.84697	-.84975	-.85159	-.85247	-.85241	-.85142	-.84953	-.84679
23	6	-.70115	-.70787	-.71373	-.71873	-.72284	-.72608	-.72845	-.72996
23	7	-.56896	-.57862	-.58754	-.59569	-.60306	-.60962	-.61536	-.62030
23	8	-.44609	-.45795	-.46920	-.47979	-.48970	-.49890	-.50737	-.51510
23	9	-.32965	-.34312	-.35611	-.36858	-.38048	-.39177	-.40243	-.41243
23	10	-.21755	-.23214	-.24638	-.26023	-.27364	-.28656	-.29896	-.31079
23	11	-.10813	-.12340	-.13846	-.15327	-.16776	-.18190	-.19562	-.20889
23	12	.00000	-.01554	-.03102	-.04638	-.06156	-.07652	-.09119	-.10553
23	13	.10813	.09272	.07722	.06170	.04622	.03083	.01559	.00054
23	14	.21755	.20267	.18757	.17229	.15691	.14148	.12606	.11069
23	15	.32965	.31575	.30148	.28689	.27204	.25700	.24181	.22655
23	16	.44609	.43367	.42072	.40731	.39350	.37933	.36486	.35016
23	17	.56896	.55860	.54758	.53595	.52375	.51104	.49788	.48432
23	18	.70115	.69360	.68525	.67615	.66632	.65582	.64470	.63302
23	19	.84697	.84324	.83858	.83302	.82659	.81932	.81127	.80247
23	20	1.01356	1.01512	1.01565	1.01515	1.01363	1.01112	1.00764	1.00324
23	21	1.21445	1.22374	1.23194	1.23901	1.24495	1.24976	1.25344	1.25602
23	22	1.48137	1.50330	1.52419	1.54397	1.56259	1.57999	1.59616	1.61108

Expected Values of Inverse Gaussian Order Statistics
Shape Parameter k

n	i	0.0	0.1	0.2	0.3	0.4	0.5	0.6	0.7
23	23	1.92916	1.97878	2.02797	2.07656	2.12437	2.17127	2.21711	2.26180
24	1	-1.94767	-1.89669	-1.84569	-1.79489	-1.74447	-1.69463	-1.64553	-1.59733
24	2	-1.50338	-1.47943	-1.45463	-1.42908	-1.40290	-1.37620	-1.34911	-1.32174
24	3	-1.23924	-1.22789	-1.21552	-1.20218	-1.18794	-1.17288	-1.15708	-1.14063
24	4	-1.04091	-1.03740	-1.03286	-1.02732	-1.02081	-1.01338	-1.00508	-.99597
24	5	-.87682	-.87877	-.87975	-.87976	-.87881	-.87694	-.87416	-.87052
24	6	-.73354	-.73950	-.74458	-.74877	-.75205	-.75445	-.75596	-.75661
24	7	-.60399	-.61298	-.62119	-.62861	-.63521	-.64098	-.64591	-.65002
24	8	-.48391	-.49519	-.50582	-.51576	-.52498	-.53346	-.54117	-.54812
24	9	-.37047	-.38347	-.39595	-.40785	-.41914	-.42979	-.43977	-.44905
24	10	-.26163	-.27588	-.28973	-.30313	-.31603	-.32840	-.34020	-.35139
24	11	-.15583	-.17090	-.18570	-.20018	-.21430	-.22799	-.24122	-.25395
24	12	-.05176	-.06726	-.08264	-.09783	-.11277	-.12742	-.14173	-.15565
24	13	.05176	.03618	.02060	.00507	-.01035	-.02561	-.04064	-.05541
24	14	.15583	.14055	.12513	.10962	.09409	.07858	.06317	.04789
24	15	.26163	.24704	.23216	.21706	.20179	.18641	.17098	.15555
24	16	.37047	.35698	.34307	.32879	.31420	.29935	.28432	.26914
24	17	.48391	.47201	.45955	.44658	.43314	.41931	.40513	.39067
24	18	.60399	.59426	.58383	.57275	.56106	.54881	.53607	.52288
24	19	.73354	.72672	.71906	.71061	.70141	.69149	.68092	.66973
24	20	.87682	.87390	.87003	.86523	.85953	.85296	.84557	.83740
24	21	1.04091	1.04337	1.04477	1.04513	1.04445	1.04275	1.04006	1.03641
24	22	1.23924	1.24951	1.25868	1.26671	1.27360	1.27933	1.28392	1.28739
24	23	1.50338	1.52637	1.54833	1.56918	1.58886	1.60733	1.62455	1.64051
24	24	1.94767	1.99845	2.04883	2.09862	2.14766	2.19579	2.24288	2.28882
25	1	-1.96531	-1.91324	-1.86120	-1.80938	-1.75800	-1.70723	-1.65725	-1.60822
25	2	-1.52430	-1.49934	-1.47354	-1.44701	-1.41987	-1.39224	-1.36425	-1.33601
25	3	-1.26275	-1.25046	-1.23714	-1.22286	-1.20769	-1.19171	-1.17501	-1.15768
25	4	-1.06679	-1.06240	-1.05697	-1.05053	-1.04312	-1.03480	-1.02561	-1.01562
25	5	-.90501	-.90614	-.90629	-.90545	-.90365	-.90091	-.89727	-.89277
25	6	-.76405	-.76927	-.77358	-.77698	-.77946	-.78103	-.78171	-.78152
25	7	-.63690	-.64522	-.65274	-.65943	-.66528	-.67027	-.67442	-.67772
25	8	-.51935	-.53005	-.54006	-.54935	-.55788	-.56564	-.57261	-.57880
25	9	-.40860	-.42112	-.43306	-.44440	-.45508	-.46508	-.47437	-.48294
25	10	-.30268	-.31654	-.32996	-.34289	-.35527	-.36707	-.37826	-.38881
25	11	-.20006	-.21488	-.22937	-.24349	-.25718	-.27040	-.28311	-.29527
25	12	-.09953	-.11492	-.13012	-.14507	-.15972	-.17402	-.18791	-.20136
25	13	.00000	-.01563	-.03119	-.04664	-.06191	-.07695	-.09170	-.10612
25	14	.09953	.08401	.06841	.05280	.03724	.02178	.00649	-.00860
25	15	.20006	.18499	.16970	.15427	.13876	.12321	.10770	.09228
25	16	.30268	.28841	.27380	.25892	.24381	.22854	.21316	.19774
25	17	.40860	.39555	.38203	.36809	.35379	.33919	.32434	.30931
25	18	.51935	.50799	.49603	.48351	.47049	.45702	.44316	.42896
25	19	.63690	.62781	.61798	.60745	.59628	.58451	.57220	.55941
25	20	.76405	.75795	.75098	.74319	.73461	.72528	.71525	.70457
25	21	.90501	.90289	.89980	.89574	.89076	.88488	.87815	.87060
25	22	1.06679	1.07013	1.07239	1.07359	1.07372	1.07282	1.07090	1.06800
25	23	1.26275	1.27397	1.28408	1.29305	1.30085	1.30749	1.31297	1.31731
25	24	1.52430	1.54832	1.57131	1.59319	1.61390	1.63340	1.65164	1.66861
25	25	1.96531	2.01721	2.06873	2.11968	2.16990	2.21922	2.26751	2.31466

Expected Values of Inverse Gaussian Order Statistics
Shape Parameter k

n	i	0.8	0.9	1.0	1.1	1.2	1.3	1.4	1.5
1	1	.00000	.00000	.00000	.00000	.00000	.00000	.00000	.00000
2	1	-.55231	-.54936	-.54614	-.54269	-.53903	-.53517	-.53115	-.52697
2	2	.55231	.54936	.54614	.54269	.53903	.53517	.53115	.52697
3	1	-.79315	-.78469	-.77598	-.76704	-.75792	-.74865	-.73926	-.72980
3	2	-.07065	-.07868	-.08647	-.09400	-.10125	-.10822	-.11491	-.12131
3	3	.86379	.86338	.86245	.86104	.85917	.85687	.85417	.85111
4	1	-.93815	-.92500	-.91165	-.89816	-.88457	-.87093	-.85729	-.84367
4	2	-.35813	-.36379	-.36897	-.37370	-.37797	-.38180	-.38520	-.38818
4	3	.21684	.20642	.19603	.18570	.17547	.16536	.15539	.14557
4	4	1.07944	1.08236	1.08459	1.08615	1.08706	1.08737	1.08710	1.08629
5	1	-1.03869	-1.02163	-1.00446	-.98724	-.97003	-.95289	-.93585	-.91896
5	2	-.53598	-.53847	-.54041	-.54182	-.54270	-.54310	-.54303	-.54253
5	3	-.09137	-.10175	-.11181	-.12152	-.13087	-.13985	-.14845	-.15667
5	4	.42232	.41187	.40126	.39052	.37970	.36883	.35794	.34706
5	5	1.24372	1.24999	1.25542	1.26005	1.26391	1.26701	1.26939	1.27109
6	1	-1.11421	-1.09384	-1.07346	-1.05314	-1.03294	-1.01290	-.99309	-.97353
6	2	-.66110	-.66056	-.65942	-.65773	-.65552	-.65282	-.64967	-.64610
6	3	-.28573	-.29431	-.30239	-.30998	-.31707	-.32366	-.32977	-.33538
6	4	.10299	.09080	.07877	.06694	.05533	.04396	.03286	.02205
6	5	.58198	.57240	.56250	.55231	.54189	.53127	.52048	.50957
6	6	1.37607	1.38551	1.39401	1.40160	1.40831	1.41415	1.41917	1.42339
7	1	-1.17393	-1.15072	-1.12760	-1.10463	-1.08189	-1.05942	-1.03727	-1.01549
7	2	-.75588	-.75256	-.74864	-.74419	-.73922	-.73381	-.72797	-.72177
7	3	-.42416	-.43056	-.43637	-.44159	-.44625	-.45035	-.45390	-.45694
7	4	-.10115	-.11264	-.12376	-.13450	-.14484	-.15476	-.16425	-.17331
7	5	.25610	.24338	.23068	.21802	.20545	.19300	.18070	.16857
7	6	.71233	.70401	.69523	.68603	.67647	.66657	.65639	.64597
7	7	1.48669	1.49909	1.51047	1.52086	1.53028	1.53875	1.54630	1.55297
8	1	-1.22289	-1.19720	-1.17169	-1.14643	-1.12149	-1.09693	-1.07278	-1.04911
8	2	-.83121	-.82538	-.81898	-.81205	-.80466	-.79686	-.78868	-.78019
8	3	-.52989	-.53409	-.53765	-.54058	-.54291	-.54466	-.54585	-.54651
8	4	-.24795	-.25801	-.26756	-.27661	-.28515	-.29316	-.30066	-.30765
8	5	.04566	.03273	.02004	.00761	-.00453	-.01635	-.02784	-.03897
8	6	.38236	.36977	.35706	.34427	.33144	.31861	.30582	.29310
8	7	.82232	.81542	.80795	.79995	.79147	.78256	.77325	.76359
8	8	1.58160	1.59676	1.61083	1.62385	1.63583	1.64678	1.65674	1.66573
9	1	-1.26411	-1.23622	-1.20860	-1.18133	-1.15446	-1.12807	-1.10219	-1.07686
9	2	-.89315	-.88505	-.87641	-.86729	-.85773	-.84781	-.83758	-.82708
9	3	-.61441	-.61652	-.61795	-.61874	-.61891	-.61850	-.61754	-.61606
9	4	-.36085	-.36925	-.37705	-.38427	-.39091	-.39697	-.40246	-.40741
9	5	-.10682	-.11895	-.13070	-.14203	-.15294	-.16340	-.17341	-.18296
9	6	.16764	.15408	.14063	.12732	.11420	.10129	.08862	.07622
9	7	.48972	.47762	.46528	.45274	.44006	.42728	.41442	.40154
9	8	.91735	.91194	.90586	.89916	.89188	.88407	.87577	.86703
9	9	1.66464	1.68236	1.69896	1.71444	1.72882	1.74212	1.75436	1.76557
10	1	-1.29952	-1.26966	-1.24016	-1.21109	-1.18252	-1.15450	-1.12708	-1.10031
10	2	-.94539	-.93523	-.92457	-.91346	-.90197	-.89016	-.87809	-.86582
10	3	-.68420	-.68434	-.68379	-.68260	-.68080	-.67842	-.67551	-.67211
10	4	-.45158	-.45827	-.46432	-.46973	-.47451	-.47869	-.48227	-.48529
10	5	-.22477	-.23571	-.24616	-.25609	-.26550	-.27439	-.28275	-.29058
10	6	.01112	-.00220	-.01524	-.02798	-.04038	-.05241	-.06407	-.07534
10	7	.27199	.25826	.24454	.23085	.21725	.20376	.19042	.17727
10	8	.58303	.57162	.55988	.54784	.53555	.52307	.51042	.49766
10	9	1.00093	.99702	.99235	.98699	.98096	.97432	.96711	.95938
10	10	1.73838	1.75851	1.77747	1.79527	1.81192	1.82743	1.84183	1.85515

Expected Values of Inverse Gaussian Order Statistics
Shape Parameter k

n	i	0.8	0.9	1.0	1.1	1.2	1.3	1.4	1.5
11	1	-1.33044	-1.29880	-1.26760	-1.23691	-1.20681	-1.17734	-1.14855	-1.12048
11	2	-.99031	-.97828	-.96577	-.95286	-.93962	-.92612	-.91241	-.89855
11	3	-.74323	-.74153	-.73916	-.73614	-.73252	-.72835	-.72368	-.71854
11	4	-.52678	-.53181	-.53616	-.53983	-.54286	-.54527	-.54707	-.54830
11	5	-.31997	-.32957	-.33860	-.34704	-.35490	-.36218	-.36888	-.37502
11	6	-.11052	-.12308	-.13523	-.14695	-.15823	-.16904	-.17938	-.18925
11	7	.11250	.09853	.08474	.07116	.05783	.04477	.03202	.01958
11	8	.36313	.34954	.33585	.32210	.30834	.29461	.28094	.26737
11	9	.66549	.65491	.64389	.63249	.62076	.60874	.59648	.58402
11	10	1.07548	1.07304	1.06979	1.06576	1.06100	1.05556	1.04947	1.04279
11	11	1.80467	1.82705	1.84824	1.86822	1.88701	1.90462	1.92107	1.93638
12	1	-1.35780	-1.32453	-1.29178	-1.25963	-1.22813	-1.19735	-1.16733	-1.13810
12	2	-1.02956	-1.01579	-1.00160	-.98705	-.97222	-.95718	-.94198	-.92669
12	3	-.79411	-.79070	-.78663	-.78192	-.77664	-.77083	-.76455	-.75783
12	4	-.59059	-.59402	-.59675	-.59878	-.60016	-.60091	-.60107	-.60066
12	5	-.39916	-.40739	-.41498	-.42193	-.42826	-.43397	-.43907	-.44359
12	6	-.20909	-.22063	-.23167	-.24220	-.25221	-.26168	-.27061	-.27901
12	7	-.01196	-.02552	-.03878	-.05169	-.06424	-.07640	-.08816	-.09949
12	8	.20139	.18714	.17297	.15892	.14503	.13133	.11786	.10463
12	9	.44399	.43074	.41729	.40369	.39000	.37624	.36248	.34874
12	10	.73932	.72963	.71943	.70876	.69768	.68624	.67448	.66244
12	11	1.14271	1.14172	1.13986	1.13716	1.13367	1.12942	1.12447	1.11886
12	12	1.86485	1.88936	1.91263	1.93468	1.95549	1.97509	1.99348	2.01070
13	1	-1.38226	-1.34750	-1.31333	-1.27984	-1.24707	-1.21510	-1.18396	-1.15367
13	2	-1.06427	-1.04891	-1.03316	-1.01711	-1.00083	-.98438	-.96783	-.95125
13	3	-.83862	-.83363	-.82797	-.82171	-.81489	-.80757	-.79981	-.79166
13	4	-.64572	-.64763	-.64882	-.64932	-.64915	-.64837	-.64699	-.64507
13	5	-.46655	-.47340	-.47958	-.48509	-.48994	-.49415	-.49774	-.50074
13	6	-.29135	-.30176	-.31161	-.32089	-.32957	-.33768	-.34521	-.35216
13	7	-.11313	-.12598	-.13841	-.15041	-.16194	-.17301	-.18359	-.19368
13	8	.07476	.06058	.04662	.03291	.01950	.00640	-.00636	-.01876
13	9	.28054	.26625	.25194	.23768	.22349	.20941	.19549	.18175
13	10	.51664	.50385	.49077	.47747	.46400	.45039	.43670	.42295
13	11	.80612	.79736	.78802	.77815	.76779	.75699	.74581	.73429
13	12	1.20391	1.20433	1.20383	1.20244	1.20019	1.19714	1.19332	1.18878
13	13	1.91993	1.94644	1.97170	1.99570	2.01843	2.03992	2.06016	2.07920
14	1	-1.40433	-1.36819	-1.33272	-1.29799	-1.26406	-1.23099	-1.19882	-1.16758
14	2	-1.09531	-1.07847	-1.06129	-1.04384	-1.02622	-1.00848	-.99069	-.97292
14	3	-.87805	-.87157	-.86444	-.85673	-.84849	-.83978	-.83066	-.82119
14	4	-.69405	-.69452	-.69426	-.69330	-.69170	-.68948	-.68669	-.68337
14	5	-.52491	-.53042	-.53523	-.53935	-.54279	-.54558	-.54775	-.54932
14	6	-.36151	-.37077	-.37941	-.38742	-.39480	-.40157	-.40772	-.41328
14	7	-.19779	-.20975	-.22122	-.23218	-.24261	-.25250	-.26185	-.27066
14	8	-.02847	-.04220	-.05560	-.06863	-.08128	-.09351	-.10532	-.11669
14	9	.15219	.13767	.12328	.10908	.09508	.08133	.06786	.05469
14	10	.35184	.33768	.32342	.30912	.29482	.28057	.26640	.25234
14	11	.58256	.57031	.55771	.54482	.53167	.51832	.50482	.49120
14	12	.86710	.85929	.85083	.84178	.83218	.82209	.81154	.80059
14	13	1.26004	1.26184	1.26266	1.26255	1.26153	1.25965	1.25695	1.25348
14	14	1.97069	1.99910	2.02624	2.05209	2.07666	2.09994	2.12195	2.14271
15	1	-1.42440	-1.38699	-1.35030	-1.31443	-1.27943	-1.24535	-1.21223	-1.18010
15	2	-1.12330	-1.10508	-1.08657	-1.06784	-1.04897	-1.03004	-1.01111	-.99225
15	3	-.91333	-.90545	-.89695	-.88789	-.87833	-.86833	-.85796	-.84728
15	4	-.73692	-.73602	-.73439	-.73208	-.72913	-.72557	-.72146	-.71685
15	5	-.57616	-.58038	-.58388	-.58667	-.58878	-.59023	-.59106	-.59130

Expected Values of Inverse Gaussian Order Statistics
Shape Parameter k

n	i	0.8	0.9	1.0	1.1	1.2	1.3	1.4	1.5
15	6	-.42240	-.43050	-.43793	-.44470	-.45082	-.45629	-.46113	-.46537
15	7	-.27017	-.28118	-.29162	-.30149	-.31078	-.31948	-.32761	-.33516
15	8	-.11506	-.12813	-.14077	-.15297	-.16470	-.17595	-.18670	-.19696
15	9	.04730	.03299	.01893	.00516	-.00828	-.02138	-.03411	-.04645
15	10	.22212	.20746	.19285	.17835	.16399	.14981	.13584	.12212
15	11	.41670	.40279	.38870	.37451	.36024	.34595	.33167	.31746
15	12	.64287	.63123	.61917	.60675	.59401	.58100	.56778	.55438
15	13	.92315	.91630	.90875	.90054	.89173	.88236	.87248	.86214
15	14	1.31187	1.31500	1.31711	1.31824	1.31842	1.31769	1.31610	1.31369
15	15	2.01775	2.04797	2.07689	2.10451	2.13082	2.15581	2.17951	2.20193
16	1	-1.44278	-1.40417	-1.36636	-1.32942	-1.29342	-1.25841	-1.22441	-1.19146
16	2	-1.14875	-1.12924	-1.10948	-1.08954	-1.06952	-1.04949	-1.02950	-1.00964
16	3	-.94518	-.93599	-.92619	-.91587	-.90508	-.89389	-.88236	-.87055
16	4	-.77533	-.77313	-.77022	-.76663	-.76240	-.75760	-.75227	-.74645
16	5	-.62169	-.62468	-.62692	-.62845	-.62929	-.62948	-.62905	-.62805
16	6	-.47598	-.48293	-.48919	-.49475	-.49964	-.50387	-.50747	-.51045
16	7	-.33311	-.34312	-.35251	-.36128	-.36944	-.37698	-.38390	-.39023
16	8	-.18925	-.20154	-.21333	-.22460	-.23535	-.24556	-.25523	-.26435
16	9	-.04087	-.05472	-.06822	-.08133	-.09405	-.10633	-.11817	-.12956
16	10	.11587	.10120	.08671	.07244	.05842	.04469	.03127	.01819
16	11	.28587	.27121	.25654	.24190	.22733	.21289	.19859	.18448
16	12	.47618	.46259	.44878	.43478	.42065	.40643	.39217	.37790
16	13	.69844	.68744	.67597	.66407	.65179	.63919	.62631	.61320
16	14	.97501	.96911	.96247	.95511	.94709	.93847	.92929	.91959
16	15	1.35999	1.36441	1.36778	1.37012	1.37146	1.37187	1.37136	1.36999
16	16	2.06160	2.09354	2.12417	2.15347	2.18144	2.20807	2.23339	2.25739
17	1	-1.45970	-1.41998	-1.38111	-1.34318	-1.30625	-1.27036	-1.23555	-1.20184
17	2	-1.17203	-1.15131	-1.13038	-1.10932	-1.08823	-1.06716	-1.04619	-1.02539
17	3	-.97414	-.96371	-.95271	-.94120	-.92925	-.91694	-.90433	-.89147
17	4	-.81003	-.80661	-.80247	-.79767	-.79226	-.78629	-.77981	-.77288
17	5	-.66254	-.66434	-.66538	-.66571	-.66536	-.66436	-.66275	-.66058
17	6	-.52365	-.52949	-.53460	-.53901	-.54273	-.54577	-.54818	-.54997
17	7	-.38858	-.39758	-.40593	-.41362	-.42066	-.42706	-.43283	-.43798
17	8	-.25387	-.26531	-.27620	-.28652	-.29627	-.30543	-.31401	-.32201
17	9	-.11655	-.12979	-.14259	-.15495	-.16683	-.17822	-.18910	-.19948
17	10	.02640	.01201	-.00210	-.01590	-.02935	-.04244	-.05512	-.06740
17	11	.17850	.16364	.14888	.13428	.11986	.10567	.09174	.07810
17	12	.34443	.32989	.31526	.30060	.28596	.27137	.25687	.24250
17	13	.53107	.51789	.50441	.49069	.47677	.46271	.44854	.43432
17	14	.74993	.73961	.72876	.71742	.70565	.69350	.68101	.66824
17	15	1.02324	1.01829	1.01255	1.00604	.99883	.99097	.98249	.97345
17	16	1.40489	1.41056	1.41514	1.41866	1.42115	1.42265	1.42321	1.42286
17	17	2.10264	2.13622	2.16848	2.19940	2.22896	2.25716	2.28402	2.30955
18	1	-1.47536	-1.43459	-1.39473	-1.35587	-1.31806	-1.28136	-1.24579	-1.21138
18	2	-1.19345	-1.17160	-1.14957	-1.12746	-1.10535	-1.08332	-1.06143	-1.03976
18	3	-1.00065	-.98905	-.97690	-.96428	-.95125	-.93789	-.92427	-.91044
18	4	-.84162	-.83703	-.83174	-.82580	-.81926	-.81219	-.80464	-.79666
18	5	-.69949	-.70015	-.70006	-.69925	-.69776	-.69563	-.69291	-.68964
18	6	-.56647	-.57123	-.57524	-.57853	-.58113	-.58305	-.58433	-.58501
18	7	-.43801	-.44601	-.45332	-.45996	-.46592	-.47122	-.47588	-.47991
18	8	-.31091	-.32148	-.33144	-.34080	-.34954	-.35766	-.36518	-.37209
18	9	-.18258	-.19511	-.20715	-.21868	-.22968	-.24014	-.25005	-.25941
18	10	-.05053	-.06446	-.07803	-.09121	-.10397	-.11629	-.12816	-.13955
18	11	.08794	.07318	.05864	.04435	.03034	.01665	.00330	-.00968
18	12	.23614	.22120	.20631	.19150	.17683	.16232	.14802	.13396

Expected Values of Inverse Gaussian Order Statistics
Shape Parameter k

n	i	0.8	0.9	1.0	1.1	1.2	1.3	1.4	1.5
18	13	.39857	.38423	.36974	.35515	.34052	.32589	.31129	.29677
18	14	.58203	.56930	.55621	.54282	.52917	.51533	.50133	.48722
18	15	.79791	.78828	.77806	.76731	.75607	.74440	.73235	.71996
18	16	1.06830	1.06430	1.05944	1.05379	1.04739	1.04028	1.03252	1.02415
18	17	1.44697	1.45385	1.45960	1.46427	1.46787	1.47045	1.47205	1.47270
18	18	2.14121	2.17636	2.21018	2.24264	2.27373	2.30344	2.33179	2.35877
19	1	-1.48992	-1.44816	-1.40736	-1.36763	-1.32900	-1.29154	-1.25526	-1.22018
19	2	-1.21327	-1.19033	-1.16727	-1.14416	-1.12111	-1.09817	-1.07543	-1.05294
19	3	-1.02504	-1.01234	-.99911	-.98543	-.97139	-.95705	-.94247	-.92773
19	4	-.87054	-.86484	-.85846	-.85144	-.84385	-.83574	-.82718	-.81821
19	5	-.73315	-.73272	-.73154	-.72964	-.72708	-.72389	-.72012	-.71582
19	6	-.60525	-.60895	-.61191	-.61413	-.61566	-.61651	-.61673	-.61634
19	7	-.48246	-.48948	-.49579	-.50140	-.50631	-.51056	-.51415	-.51712
19	8	-.36180	-.37148	-.38053	-.38892	-.39667	-.40378	-.41026	-.41613
19	9	-.24094	-.25272	-.26395	-.27463	-.28473	-.29425	-.30318	-.31154
19	10	-.11774	-.13111	-.14404	-.15652	-.16852	-.18002	-.19101	-.20149
19	11	.00996	-.00448	-.01863	-.03244	-.04589	-.05894	-.07159	-.08381
19	12	.14465	.12967	.11483	.10019	.08578	.07163	.05777	.04423
19	13	.28950	.27459	.25967	.24477	.22994	.21523	.20067	.18630
19	14	.44891	.43483	.42054	.40610	.39156	.37696	.36234	.34776
19	15	.62957	.61732	.60466	.59164	.57832	.56475	.55097	.53703
19	16	.84280	.83386	.82430	.81415	.80347	.79231	.78072	.76874
19	17	1.11059	1.10750	1.10353	1.09872	1.09312	1.08677	1.07973	1.07204
19	18	1.48654	1.49459	1.50150	1.50727	1.51196	1.51559	1.51820	1.51984
19	19	2.17758	2.21424	2.24955	2.28349	2.31605	2.34721	2.37698	2.40538
20	1	-1.50352	-1.46081	-1.41914	-1.37857	-1.33918	-1.30099	-1.26404	-1.22834
20	2	-1.23167	-1.20772	-1.18368	-1.15963	-1.13568	-1.11190	-1.08835	-1.06509
20	3	-1.04761	-1.03385	-1.01960	-1.00493	-.98992	-.97465	-.95919	-.94359
20	4	-.89718	-.89042	-.88300	-.87495	-.86636	-.85728	-.84776	-.83788
20	5	-.76400	-.76252	-.76030	-.75738	-.75379	-.74960	-.74484	-.73957
20	6	-.64060	-.64330	-.64524	-.64644	-.64694	-.64677	-.64598	-.64459
20	7	-.52275	-.52882	-.53415	-.53876	-.54267	-.54590	-.54848	-.55043
20	8	-.40763	-.41643	-.42455	-.43200	-.43879	-.44491	-.45040	-.45525
20	9	-.29307	-.30407	-.31449	-.32429	-.33349	-.34208	-.35006	-.35744
20	10	-.17722	-.18995	-.20219	-.21392	-.22512	-.23578	-.24589	-.25545
20	11	-.05826	-.07226	-.08589	-.09912	-.11191	-.12426	-.13614	-.14754
20	12	.06577	.05097	.03640	.02211	.00813	-.00550	-.01878	-.03166
20	13	.19724	.18213	.16712	.15224	.13754	.12305	.10880	.09482
20	14	.33919	.32438	.30950	.29459	.27970	.26487	.25014	.23555
20	15	.49594	.48216	.46813	.45389	.43950	.42500	.41043	.39584
20	16	.67412	.66237	.65017	.63756	.62460	.61134	.59782	.58409
20	17	.88496	.87674	.86783	.85830	.84819	.83755	.82644	.81490
20	18	1.15040	1.14823	1.14513	1.14115	1.13634	1.13075	1.12443	1.11741
20	19	1.52389	1.53308	1.54109	1.54795	1.55369	1.55835	1.56195	1.56455
20	20	2.21199	2.25009	2.28684	2.32221	2.35617	2.38873	2.41988	2.44964
21	1	-1.51625	-1.47265	-1.43015	-1.38880	-1.34868	-1.30981	-1.27223	-1.23594
21	2	-1.24884	-1.22392	-1.19895	-1.17402	-1.14922	-1.12463	-1.10032	-1.07634
21	3	-1.06857	-1.05381	-1.03859	-1.02298	-1.00706	-.99092	-.97461	-.95820
21	4	-.92183	-.91406	-.90564	-.89663	-.88709	-.87708	-.86666	-.85591
21	5	-.79242	-.78995	-.78674	-.78283	-.77827	-.77312	-.76742	-.76124
21	6	-.67303	-.67476	-.67571	-.67594	-.67546	-.67432	-.67256	-.67023
21	7	-.55953	-.56466	-.56904	-.57269	-.57564	-.57790	-.57951	-.58049
21	8	-.44921	-.45713	-.46436	-.47089	-.47674	-.48191	-.48643	-.49031
21	9	-.34006	-.35028	-.35986	-.36881	-.37712	-.38480	-.39184	-.39827
21	10	-.23041	-.24247	-.25398	-.26494	-.27533	-.28513	-.29436	-.30301

Expected Values of Inverse Gaussian Order Statistics
Shape Parameter k

n	i	0.8	0.9	1.0	1.1	1.2	1.3	1.4	1.5
21	11	-.11870	-.13218	-.14522	-.15780	-.16989	-.18148	-.19257	-.20313
21	12	-.00331	-.01779	-.03196	-.04577	-.05921	-.07224	-.08484	-.09700
21	13	.11759	.10254	.08767	.07303	.05864	.04455	.03077	.01734
21	14	.24625	.23112	.21602	.20099	.18610	.17136	.15682	.14251
21	15	.38565	.37101	.35624	.34138	.32650	.31162	.29681	.28208
21	16	.54005	.52663	.51289	.49890	.48470	.47034	.45588	.44135
21	17	.71601	.70479	.69307	.68089	.66832	.65539	.64217	.62870
21	18	.92472	.91719	.90895	.90004	.89051	.88041	.86980	.85871
21	19	1.18802	1.18673	1.18449	1.18134	1.17732	1.17248	1.16687	1.16053
21	20	1.55925	1.56954	1.57863	1.58654	1.59331	1.59896	1.60354	1.60708
21	21	2.24462	2.28412	2.32225	2.35899	2.39432	2.42822	2.46070	2.49176
22	1	-1.52822	-1.48378	-1.44048	-1.39839	-1.35757	-1.31807	-1.27989	-1.24304
22	2	-1.26491	-1.23907	-1.21322	-1.18744	-1.16185	-1.13649	-1.11146	-1.08681
22	3	-1.08812	-1.07241	-1.05627	-1.03976	-1.02298	-1.00600	-.98889	-.97173
22	4	-.94473	-.93601	-.92664	-.91670	-.90626	-.89537	-.88411	-.87253
22	5	-.81874	-.81531	-.81115	-.80630	-.80082	-.79476	-.78817	-.78112
22	6	-.70294	-.70372	-.70374	-.70302	-.70161	-.69955	-.69687	-.69364
22	7	-.59329	-.59751	-.60098	-.60371	-.60572	-.60705	-.60773	-.60780
22	8	-.48719	-.49426	-.50061	-.50624	-.51118	-.51543	-.51902	-.52198
22	9	-.38274	-.39217	-.40093	-.40903	-.41647	-.42325	-.42939	-.43490
22	10	-.27841	-.28977	-.30054	-.31072	-.32029	-.32925	-.33761	-.34536
22	11	-.17282	-.18571	-.19812	-.21001	-.22137	-.23220	-.24247	-.25219
22	12	-.06459	-.07864	-.09232	-.10558	-.11841	-.13077	-.14266	-.15407
22	13	.04775	.03292	.01835	.00407	-.00987	-.02346	-.03666	-.04945
22	14	.16594	.15074	.13567	.12076	.10607	.09162	.07745	.06359
22	15	.29214	.27705	.26193	.24684	.23182	.21692	.20217	.18760
22	16	.42929	.41486	.40025	.38550	.37068	.35582	.34097	.32617
22	17	.58159	.56854	.55513	.54142	.52746	.51329	.49897	.48454
22	18	.75555	.74487	.73364	.72191	.70975	.69719	.68429	.67110
22	19	.96231	.95549	.94791	.93962	.93068	.92113	.91102	.90041
22	20	1.22366	1.22324	1.22184	1.21950	1.21626	1.21216	1.20726	1.20160
22	21	1.59280	1.60417	1.61431	1.62325	1.63102	1.63764	1.64317	1.64763
22	22	2.27566	2.31649	2.35596	2.39403	2.43066	2.46586	2.49963	2.53196
23	1	-1.53950	-1.49425	-1.45020	-1.40741	-1.36593	-1.32581	-1.28707	-1.24970
23	2	-1.28001	-1.25329	-1.22659	-1.20002	-1.17366	-1.14758	-1.12187	-1.09657
23	3	-1.10643	-1.08981	-1.07278	-1.05542	-1.03782	-1.02005	-1.00219	-.98430
23	4	-.96610	-.95646	-.94619	-.93537	-.92406	-.91234	-.90027	-.88791
23	5	-.84322	-.83887	-.83379	-.82805	-.82168	-.81475	-.80733	-.79945
23	6	-.73064	-.73052	-.72963	-.72802	-.72571	-.72277	-.71922	-.71513
23	7	-.62444	-.62779	-.63037	-.63220	-.63332	-.63376	-.63356	-.63274
23	8	-.52208	-.52831	-.53381	-.53857	-.54263	-.54600	-.54870	-.55077
23	9	-.42176	-.43040	-.43836	-.44562	-.45221	-.45812	-.46338	-.46800
23	10	-.32204	-.33269	-.34271	-.35211	-.36087	-.36901	-.37651	-.38340
23	11	-.22169	-.23397	-.24571	-.25690	-.26752	-.27757	-.28703	-.29592
23	12	-.11950	-.13307	-.14619	-.15886	-.17103	-.18270	-.19385	-.20448
23	13	-.01425	-.02876	-.04293	-.05675	-.07017	-.08317	-.09574	-.10785
23	14	.09545	.08036	.06548	.05085	.03651	.02248	.00879	-.00453
23	15	.21125	.19598	.18078	.16571	.15079	.13608	.12159	.10737
23	16	.33528	.32028	.30521	.29011	.27504	.26004	.24514	.23039
23	17	.47042	.45624	.44183	.42724	.41252	.39772	.38289	.36807
23	18	.62083	.60817	.59512	.58172	.56802	.55408	.53994	.52565
23	19	.79298	.78284	.77212	.76086	.74912	.73694	.72439	.71150
23	20	.99796	.99183	.98492	.97726	.96890	.95990	.95031	.94018
23	21	1.25751	1.25795	1.25738	1.25584	1.25336	1.25000	1.24581	1.24082
23	22	1.62474	1.63714	1.64830	1.65824	1.66698	1.67456	1.68101	1.68637

Expected Values of Inverse Gaussian Order Statistics
Shape Parameter k

n	i	0.8	0.9	1.0	1.1	1.2	1.3	1.4	1.5
23	23	2.30525	2.34737	2.38813	2.42747	2.46538	2.50183	2.53684	2.57040
24	1	-1.55017	-1.50415	-1.45937	-1.41591	-1.37381	-1.33311	-1.29383	-1.25596
24	2	-1.29422	-1.26666	-1.23917	-1.21183	-1.18474	-1.15798	-1.13162	-1.10571
24	3	-1.12361	-1.10613	-1.08825	-1.07008	-1.05170	-1.03318	-1.01460	-.99602
24	4	-.98611	-.97559	-.96445	-.95279	-.94066	-.92815	-.91531	-.90221
24	5	-.86606	-.86083	-.85488	-.84827	-.84106	-.83331	-.82508	-.81642
24	6	-.75642	-.75543	-.75367	-.75118	-.74802	-.74423	-.73986	-.73495
24	7	-.65332	-.65582	-.65754	-.65851	-.65878	-.65836	-.65731	-.65566
24	8	-.55430	-.55972	-.56438	-.56830	-.57151	-.57402	-.57587	-.57709
24	9	-.45763	-.46550	-.47266	-.47911	-.48487	-.48994	-.49435	-.49812
24	10	-.36197	-.37190	-.38118	-.38980	-.39777	-.40509	-.41176	-.41780
24	11	-.26615	-.27780	-.28886	-.29934	-.30922	-.31849	-.32716	-.33523
24	12	-.16914	-.18217	-.19471	-.20674	-.21824	-.22920	-.23961	-.24946
24	13	-.06987	-.08397	-.09768	-.11097	-.12382	-.13620	-.14810	-.15950
24	14	.03281	.01796	.00339	-.01087	-.02477	-.03830	-.05144	-.06415
24	15	.14019	.12493	.10984	.09494	.08028	.06589	.05181	.03805
24	16	.25389	.23861	.22335	.20817	.19310	.17819	.16346	.14897
24	17	.37598	.36112	.34613	.33108	.31601	.30096	.28598	.27111
24	18	.50931	.49541	.48123	.46683	.45226	.43756	.42280	.40799
24	19	.65800	.64576	.63308	.62002	.60661	.59292	.57899	.56487
24	20	.82850	.81891	.80870	.79792	.78662	.77484	.76265	.75009
24	21	1.03185	1.02642	1.02016	1.01312	1.00536	.99691	.98784	.97820
24	22	1.28975	1.29103	1.29127	1.29051	1.28879	1.28616	1.28266	1.27833
24	23	1.65519	1.66860	1.68076	1.69167	1.70136	1.70987	1.71723	1.72347
24	24	2.33351	2.37689	2.41888	2.45946	2.49859	2.53627	2.57247	2.60722
25	1	-1.56027	-1.51352	-1.46806	-1.42395	-1.38125	-1.34000	-1.30020	-1.26186
25	2	-1.30765	-1.27929	-1.25102	-1.22296	-1.19518	-1.16777	-1.14079	-1.11429
25	3	-1.13980	-1.12148	-1.10280	-1.08385	-1.06472	-1.04549	-1.02622	-1.00700
25	4	-1.00491	-.99353	-.98157	-.96910	-.95619	-.94292	-.92935	-.91555
25	5	-.88745	-.88137	-.87458	-.86715	-.85914	-.85060	-.84160	-.83220
25	6	-.78049	-.77865	-.77606	-.77274	-.76876	-.76416	-.75900	-.75331
25	7	-.68020	-.68187	-.68277	-.68292	-.68235	-.68112	-.67926	-.67681
25	8	-.58419	-.58881	-.59266	-.59577	-.59815	-.59985	-.60088	-.60128
25	9	-.49079	-.49790	-.50428	-.50993	-.51488	-.51914	-.52273	-.52568
25	10	-.39870	-.40792	-.41646	-.42433	-.43152	-.43804	-.44390	-.44913
25	11	-.30687	-.31786	-.32825	-.33802	-.34715	-.35566	-.36355	-.37081
25	12	-.21433	-.22680	-.23874	-.25012	-.26094	-.27118	-.28085	-.28993
25	13	-.12018	-.13382	-.14702	-.15975	-.17199	-.18372	-.19494	-.20562
25	14	-.02343	-.03795	-.05213	-.06594	-.07935	-.09233	-.10486	-.11692
25	15	.07699	.06188	.04701	.03241	.01811	.00414	-.00946	-.02268
25	16	.18232	.16697	.15172	.13663	.12173	.10706	.09265	.07854
25	17	.29414	.27891	.26364	.24841	.23324	.21819	.20330	.18858
25	18	.41449	.39981	.38495	.36999	.35496	.33991	.32489	.30994
25	19	.54618	.53258	.51867	.50449	.49010	.47554	.46087	.44613
25	20	.69331	.68150	.66922	.65650	.64341	.62999	.61629	.60237
25	21	.86229	.85327	.84358	.83328	.82242	.81106	.79924	.78701
25	22	1.06415	1.05940	1.05379	1.04738	1.04020	1.03232	1.02377	1.01461
25	23	1.32051	1.32262	1.32366	1.32367	1.32269	1.32078	1.31796	1.31430
25	24	1.68429	1.69869	1.71181	1.72367	1.73429	1.74371	1.75195	1.75905
25	25	2.36056	2.40514	2.44834	2.49012	2.53044	2.56929	2.60666	2.64256

Expected Values of Inverse Gaussian Order Statistics
Shape Parameter k

n	i	1.6	1.7	1.8	1.9	2.0	2.1	2.2	2.3
1	1	.00000	.00000	.00000	.00000	.00000	.00000	.00000	.00000
2	1	-.52266	-.51824	-.51373	-.50914	-.50448	-.49978	-.49503	-.49026
2	2	.52266	.51824	.51373	.50914	.50448	.49978	.49503	.49026
3	1	-.72029	-.71075	-.70121	-.69169	-.68222	-.67280	-.66345	-.65420
3	2	-.12742	-.13324	-.13878	-.14404	-.14902	-.15374	-.15820	-.16240
3	3	.84770	.84399	.83999	.83573	.83124	.82654	.82165	.81659
4	1	-.83013	-.81667	-.80334	-.79015	-.77713	-.76429	-.75164	-.73920
4	2	-.39077	-.39297	-.39481	-.39630	-.39747	-.39832	-.39889	-.39918
4	3	.13594	.12649	.11725	.10823	.09942	.09084	.08250	.07439
4	4	1.08496	1.08315	1.08090	1.07823	1.07518	1.07177	1.06803	1.06400
5	1	-.90225	-.88576	-.86951	-.85352	-.83781	-.82239	-.80728	-.79248
5	2	-.54162	-.54032	-.53868	-.53671	-.53444	-.53189	-.52910	-.52608
5	3	-.16450	-.17194	-.17901	-.18569	-.19201	-.19797	-.20357	-.20883
5	4	.33622	.32545	.31476	.30417	.29371	.28339	.27321	.26320
5	5	1.27214	1.27258	1.27243	1.27175	1.27054	1.26886	1.26674	1.26419
6	1	-.95427	-.93534	-.91675	-.89854	-.88070	-.86327	-.84624	-.82961
6	2	-.64216	-.63788	-.63329	-.62842	-.62331	-.61799	-.61248	-.60681
6	3	-.34053	-.34522	-.34947	-.35328	-.35669	-.35971	-.36235	-.36463
6	4	.01154	.00134	-.00855	-.01811	-.02733	-.03623	-.04480	-.05304
6	5	.49857	.48750	.47641	.46531	.45423	.44320	.43222	.42132
6	6	1.42686	1.42959	1.43164	1.43303	1.43381	1.43400	1.43364	1.43277
7	1	-.99411	-.97316	-.95265	-.93262	-.91306	-.89398	-.87540	-.85731
7	2	-.71523	-.70841	-.70134	-.69406	-.68660	-.67899	-.67126	-.66345
7	3	-.45948	-.46153	-.46314	-.46432	-.46509	-.46548	-.46552	-.46522
7	4	-.18194	-.19013	-.19790	-.20524	-.21216	-.21867	-.22479	-.23051
7	5	.15665	.14494	.13347	.12224	.11129	.10060	.09019	.08007
7	6	.63534	.62453	.61359	.60254	.59141	.58023	.56903	.55782
7	7	1.55878	1.56377	1.56798	1.57145	1.57421	1.57629	1.57774	1.57859
8	1	-1.02592	-1.00326	-.98114	-.95957	-.93857	-.91813	-.89826	-.87895
8	2	-.77142	-.76243	-.75325	-.74392	-.73448	-.72495	-.71538	-.70579
8	3	-.54668	-.54638	-.54564	-.54449	-.54296	-.54109	-.53890	-.53642
8	4	-.31414	-.32013	-.32565	-.33070	-.33529	-.33946	-.34321	-.34656
8	5	-.04974	-.06013	-.07015	-.07978	-.08903	-.09789	-.10637	-.11447
8	6	.28048	.26798	.25564	.24346	.23147	.21969	.20813	.19680
8	7	.75362	.74338	.73291	.72223	.71139	.70041	.68933	.67816
8	8	1.67380	1.68097	1.68728	1.69276	1.69747	1.70142	1.70466	1.70722
9	1	-1.05212	-1.02798	-1.00447	-.98159	-.95935	-.93776	-.91679	-.89646
9	2	-.81637	-.80549	-.79449	-.78340	-.77227	-.76112	-.74998	-.73889
9	3	-.61411	-.61171	-.60891	-.60573	-.60221	-.59838	-.59428	-.58993
9	4	-.41181	-.41570	-.41910	-.42201	-.42448	-.42651	-.42814	-.42938
9	5	-.19205	-.20067	-.20884	-.21655	-.22381	-.23064	-.23704	-.24303
9	6	.06411	.05229	.04080	.02963	.01880	.00831	-.00183	-.01163
9	7	.38866	.37583	.36305	.35038	.33781	.32539	.31311	.30101
9	8	.85790	.84840	.83858	.82848	.81813	.80756	.79682	.78592
9	9	1.77579	1.78504	1.79336	1.80080	1.80738	1.81315	1.81814	1.82239
10	1	-1.07420	-1.04877	-1.02405	-1.00003	-.97672	-.95412	-.93221	-.91100
10	2	-.85339	-.84086	-.82827	-.81565	-.80305	-.79050	-.77801	-.76563
10	3	-.66826	-.66400	-.65936	-.65440	-.64914	-.64361	-.63786	-.63191
10	4	-.48776	-.48971	-.49117	-.49216	-.49271	-.49285	-.49260	-.49200
10	5	-.29789	-.30469	-.31099	-.31680	-.32213	-.32701	-.33145	-.33546
10	6	-.08620	-.09665	-.10669	-.11630	-.12549	-.13427	-.14263	-.15059
10	7	.16432	.15159	.13912	.12692	.11500	.10336	.09203	.08101
10	8	.48481	.47193	.45903	.44614	.43330	.42054	.40786	.39530
10	9	.95117	.94251	.93347	.92406	.91433	.90432	.89406	.88357
10	10	1.86741	1.87865	1.88891	1.89822	1.90661	1.91413	1.92082	1.92670

Expected Values of Inverse Gaussian Order Statistics
Shape Parameter k

n	i	1.6	1.7	1.8	1.9	2.0	2.1	2.2	2.3
11	1	-1.09316	-1.06659	-1.04079	-1.01577	-.99152	-.96803	-.94530	-.92332
11	2	-.88460	-.87060	-.85659	-.84263	-.82874	-.81495	-.80130	-.78780
11	3	-.71298	-.70705	-.70080	-.69425	-.68745	-.68043	-.67323	-.66589
11	4	-.54900	-.54918	-.54888	-.54813	-.54697	-.54543	-.54352	-.54130
11	5	-.38060	-.38565	-.39017	-.39420	-.39775	-.40083	-.40348	-.40572
11	6	-.19864	-.20754	-.21596	-.22391	-.23140	-.23843	-.24501	-.25116
11	7	.00749	-.00425	-.01562	-.02662	-.03724	-.04747	-.05731	-.06677
11	8	.25393	.24065	.22755	.21466	.20199	.18955	.17738	.16546
11	9	.57140	.55865	.54583	.53295	.52005	.50716	.49429	.48149
11	10	1.03556	1.02782	1.01961	1.01097	1.00195	.99258	.98289	.97293
11	11	1.95059	1.96374	1.97584	1.98694	1.99708	2.00629	2.01461	2.02208
12	1	-1.10968	-1.08209	-1.05534	-1.02942	-1.00433	-.98006	-.95660	-.93393
12	2	-.91137	-.89605	-.88080	-.86563	-.85060	-.83572	-.82103	-.80655
12	3	-.75074	-.74331	-.73559	-.72763	-.71946	-.71113	-.70265	-.69408
12	4	-.59972	-.59829	-.59641	-.59410	-.59140	-.58835	-.58498	-.58132
12	5	-.44754	-.45095	-.45383	-.45621	-.45812	-.45958	-.46062	-.46126
12	6	-.28688	-.29423	-.30106	-.30739	-.31323	-.31859	-.32350	-.32797
12	7	-.11039	-.12085	-.13087	-.14044	-.14957	-.15826	-.16652	-.17435
12	8	.09169	.07903	.06669	.05468	.04300	.03167	.02069	.01006
12	9	.33505	.32146	.30798	.29465	.28148	.26850	.25572	.24316
12	10	.65018	.63772	.62511	.61238	.59957	.58671	.57382	.56093
12	11	1.11264	1.10584	1.09851	1.09069	1.08243	1.07375	1.06471	1.05533
12	12	2.02677	2.04173	2.05560	2.06842	2.08023	2.09106	2.10096	2.10996
13	1	-1.12427	-1.09575	-1.06814	-1.04141	-1.01557	-.99060	-.96648	-.94321
13	2	-.93468	-.91818	-.90179	-.88554	-.86948	-.85363	-.83802	-.82266
13	3	-.78317	-.77438	-.76535	-.75611	-.74672	-.73720	-.72759	-.71793
13	4	-.64264	-.63974	-.63641	-.63268	-.62860	-.62421	-.61952	-.61458
13	5	-.50316	-.50504	-.50640	-.50728	-.50769	-.50768	-.50726	-.50647
13	6	-.35855	-.36440	-.36971	-.37450	-.37880	-.38262	-.38598	-.38891
13	7	-.20327	-.21237	-.22097	-.22909	-.23673	-.24390	-.25060	-.25687
13	8	-.03078	-.04241	-.05363	-.06445	-.07486	-.08486	-.09444	-.10362
13	9	.16822	.15493	.14190	.12914	.11667	.10450	.09265	.08112
13	10	.40920	.39547	.38180	.36821	.35473	.34139	.32820	.31518
13	11	.72247	.71040	.69811	.68564	.67303	.66031	.64751	.63466
13	12	1.18357	1.17773	1.17131	1.16434	1.15686	1.14893	1.14056	1.13181
13	13	2.09704	2.11373	2.12929	2.14376	2.15717	2.16957	2.18100	2.19148
14	1	-1.13727	-1.10792	-1.07951	-1.05206	-1.02553	-.99993	-.97523	-.95140
14	2	-.95522	-.93764	-.92022	-.90300	-.88602	-.86929	-.85284	-.83670
14	3	-.81142	-.80140	-.79117	-.78078	-.77028	-.75970	-.74907	-.73842
14	4	-.67957	-.67532	-.67067	-.66566	-.66033	-.65471	-.64884	-.64276
14	5	-.55032	-.55079	-.55075	-.55025	-.54930	-.54794	-.54621	-.54414
14	6	-.41826	-.42269	-.42657	-.42993	-.43280	-.43519	-.43715	-.43867
14	7	-.27894	-.28668	-.29389	-.30060	-.30680	-.31252	-.31777	-.32257
14	8	-.12760	-.13806	-.14805	-.15758	-.16665	-.17527	-.18344	-.19117
14	9	.04184	.02933	.01718	.00539	-.00602	-.01705	-.02770	-.03796
14	10	.23844	.22471	.21118	.19788	.18482	.17203	.15950	.14727
14	11	.47750	.46377	.45004	.43634	.42269	.40913	.39568	.38235
14	12	.78928	.77766	.76576	.75363	.74130	.72881	.71619	.70347
14	13	1.24929	1.24441	1.23890	1.23279	1.22612	1.21894	1.21129	1.20320
14	14	2.16225	2.18060	2.19778	2.21383	2.22879	2.24270	2.25559	2.26750
15	1	-1.14897	-1.11884	-1.08972	-1.06160	-1.03445	-1.00827	-.98303	-.95871
15	2	-.97351	-.95494	-.93659	-.91848	-.90065	-.88312	-.86592	-.84907
15	3	-.83633	-.82517	-.81385	-.80242	-.79091	-.77936	-.76781	-.75628
15	4	-.71178	-.70629	-.70043	-.69425	-.68778	-.68105	-.67412	-.66700
15	5	-.59098	-.59014	-.58881	-.58703	-.58484	-.58226	-.57934	-.57610

Expected Values of Inverse Gaussian Order Statistics
Shape Parameter k

n	i	1.6	1.7	1.8	1.9	2.0	2.1	2.2	2.3
15	6	-.46901	-.47210	-.47464	-.47667	-.47822	-.47931	-.47997	-.48022
15	7	-.34214	-.34857	-.35445	-.35981	-.36466	-.36902	-.37291	-.37636
15	8	-.20670	-.21594	-.22468	-.23292	-.24068	-.24795	-.25475	-.26109
15	9	-.05839	-.06990	-.08099	-.09166	-.10189	-.11168	-.12105	-.12998
15	10	.10866	.09549	.08263	.07009	.05789	.04603	.03453	.02339
15	11	.30333	.28932	.27546	.26178	.24829	.23502	.22199	.20920
15	12	.54084	.52721	.51352	.49981	.48611	.47244	.45883	.44531
15	13	.85139	.84027	.82882	.81708	.80510	.79290	.78053	.76800
15	14	1.31051	1.30659	1.30199	1.29674	1.29090	1.28449	1.27756	1.27016
15	15	2.22309	2.24302	2.26176	2.27934	2.29579	2.31114	2.32545	2.33874
16	1	-1.15957	-1.12874	-1.09896	-1.07022	-1.04250	-1.01579	-.99006	-.96528
16	2	-.98994	-.97046	-.95124	-.93232	-.91372	-.89546	-.87758	-.86008
16	3	-.85851	-.84631	-.83399	-.82159	-.80916	-.79673	-.78433	-.77201
16	4	-.74021	-.73358	-.72661	-.71934	-.71182	-.70409	-.69618	-.68813
16	5	-.62650	-.62444	-.62192	-.61897	-.61563	-.61194	-.60792	-.60362
16	6	-.51284	-.51467	-.51597	-.51676	-.51709	-.51697	-.51644	-.51553
16	7	-.39597	-.40114	-.40577	-.40986	-.41344	-.41654	-.41917	-.42136
16	8	-.27293	-.28097	-.28848	-.29547	-.30195	-.30793	-.31344	-.31849
16	9	-.14048	-.15092	-.16089	-.17038	-.17940	-.18796	-.19605	-.20370
16	10	.00546	-.00689	-.01886	-.03043	-.04159	-.05235	-.06271	-.07265
16	11	.17058	.15692	.14352	.13040	.11758	.10507	.09288	.08102
16	12	.36367	.34950	.33544	.32150	.30771	.29409	.28067	.26747
16	13	.59990	.58645	.57289	.55925	.54558	.53189	.51822	.50460
16	14	.90943	.89884	.88788	.87658	.86499	.85313	.84106	.82879
16	15	1.36780	1.36484	1.36115	1.35677	1.35174	1.34611	1.33992	1.33321
16	16	2.28011	2.30157	2.32180	2.34084	2.35872	2.37548	2.39115	2.40577
17	1	-1.16924	-1.13775	-1.10736	-1.07806	-1.04982	-1.02262	-.99644	-.97124
17	2	-1.00481	-.98449	-.96447	-.94480	-.92549	-.90656	-.88805	-.86996
17	3	-.87844	-.86527	-.85202	-.83873	-.82545	-.81222	-.79905	-.78599
17	4	-.76554	-.75785	-.74985	-.74159	-.73311	-.72445	-.71565	-.70674
17	5	-.65788	-.65469	-.65106	-.64703	-.64264	-.63792	-.63291	-.62764
17	6	-.55118	-.55184	-.55198	-.55162	-.55082	-.54959	-.54797	-.54599
17	7	-.44254	-.44652	-.44995	-.45285	-.45525	-.45717	-.45864	-.45969
17	8	-.32945	-.33632	-.34264	-.34843	-.35371	-.35848	-.36278	-.36662
17	9	-.20935	-.21870	-.22754	-.23588	-.24372	-.25107	-.25794	-.26435
17	10	-.07925	-.09067	-.10164	-.11216	-.12224	-.13186	-.14104	-.14979
17	11	.06476	.05176	.03909	.02679	.01486	.00330	-.00787	-.01865
17	12	.22829	.21428	.20048	.18691	.17361	.16058	.14783	.13539
17	13	.42008	.40585	.39167	.37757	.36358	.34973	.33603	.32250
17	14	.65523	.64202	.62864	.61515	.60157	.58794	.57428	.56063
17	15	.96390	.95388	.94343	.93260	.92143	.90996	.89822	.88625
17	16	1.42166	1.41963	1.41684	1.41332	1.40911	1.40427	1.39882	1.39280
17	17	2.33376	2.35669	2.37836	2.39881	2.41807	2.43618	2.45317	2.46908
18	1	-1.17812	-1.14602	-1.11506	-1.08523	-1.05650	-1.02886	-1.00226	-.97667
18	2	-1.01835	-.99725	-.97650	-.95612	-.93616	-.91662	-.89752	-.87888
18	3	-.89646	-.88240	-.86829	-.85418	-.84012	-.82614	-.81226	-.79853
18	4	-.78830	-.77962	-.77067	-.76149	-.75213	-.74261	-.73299	-.72330
18	5	-.68587	-.68163	-.67697	-.67194	-.66657	-.66090	-.65497	-.64881
18	6	-.58510	-.58466	-.58371	-.58229	-.58043	-.57818	-.57555	-.57259
18	7	-.48334	-.48620	-.48851	-.49030	-.49159	-.49242	-.49281	-.49279
18	8	-.37842	-.38417	-.38936	-.39402	-.39815	-.40179	-.40495	-.40767
18	9	-.26823	-.27650	-.28424	-.29145	-.29815	-.30435	-.31006	-.31530
18	10	-.15047	-.16090	-.17085	-.18031	-.18928	-.19779	-.20582	-.21339
18	11	-.02228	-.03448	-.04627	-.05765	-.06860	-.07913	-.08923	-.09890
18	12	.12015	.10663	.09342	.08053	.06797	.05576	.04390	.03241

Expected Values of Inverse Gaussian Order Statistics
Shape Parameter k

n	i	1.6	1.7	1.8	1.9	2.0	2.1	2.2	2.3
18	13	.28236	.26810	.25400	.24011	.22643	.21298	.19980	.18688
18	14	.47304	.45883	.44462	.43044	.41634	.40232	.38842	.37466
18	15	.70728	.69435	.68122	.66793	.65450	.64097	.62738	.61376
18	16	1.01522	1.00578	.99587	.98554	.97482	.96376	.95239	.94075
18	17	1.47246	1.47137	1.46946	1.46679	1.46340	1.45933	1.45462	1.44931
18	18	2.38443	2.40877	2.43183	2.45364	2.47423	2.49364	2.51191	2.52907
19	1	-1.18631	-1.15363	-1.12215	-1.09183	-1.06265	-1.03458	-1.00760	-.98166
19	2	-1.03076	-1.00893	-.98749	-.96647	-.94589	-.92578	-.90615	-.88701
19	3	-.91288	-.89797	-.88307	-.86820	-.85341	-.83874	-.82421	-.80985
19	4	-.80891	-.79931	-.78947	-.77943	-.76924	-.75894	-.74856	-.73814
19	5	-.71104	-.70581	-.70019	-.69423	-.68795	-.68140	-.67462	-.66764
19	6	-.61539	-.61391	-.61194	-.60953	-.60669	-.60348	-.59993	-.59606
19	7	-.51948	-.52128	-.52253	-.52327	-.52353	-.52334	-.52274	-.52174
19	8	-.42139	-.42607	-.43019	-.43377	-.43683	-.43940	-.44151	-.44318
19	9	-.31933	-.32655	-.33323	-.33936	-.34497	-.35007	-.35469	-.35884
19	10	-.21145	-.22089	-.22982	-.23823	-.24613	-.25354	-.26047	-.26693
19	11	-.09558	-.10691	-.11777	-.12818	-.13812	-.14760	-.15663	-.16521
19	12	.03103	.01819	.00572	-.00636	-.01804	-.02932	-.04020	-.05068
19	13	.17214	.15822	.14457	.13121	.11814	.10539	.09297	.08088
19	14	.33323	.31881	.30451	.29037	.27640	.26264	.24910	.23580
19	15	.52297	.50884	.49466	.48047	.46631	.45221	.43818	.42426
19	16	.75643	.74383	.73097	.71791	.70468	.69131	.67784	.66429
19	17	1.06374	1.05490	1.04554	1.03572	1.02547	1.01484	1.00387	.99259
19	18	1.52055	1.52036	1.51934	1.51751	1.51492	1.51162	1.50765	1.50304
19	19	2.43242	2.45813	2.48252	2.50565	2.52753	2.54820	2.56770	2.58607
20	1	-1.19389	-1.16068	-1.12870	-1.09792	-1.06832	-1.03987	-1.01252	-.98625
20	2	-1.04219	-1.01967	-.99759	-.97596	-.95482	-.93418	-.91405	-.89443
20	3	-.92791	-.91223	-.89657	-.88099	-.86553	-.85021	-.83507	-.82014
20	4	-.82767	-.81721	-.80654	-.79571	-.78475	-.77371	-.76263	-.75154
20	5	-.73383	-.72768	-.72117	-.71433	-.70720	-.69984	-.69227	-.68454
20	6	-.64265	-.64020	-.63728	-.63393	-.63018	-.62608	-.62166	-.61696
20	7	-.55179	-.55258	-.55284	-.55259	-.55189	-.55075	-.54921	-.54730
20	8	-.45949	-.46315	-.46625	-.46882	-.47087	-.47245	-.47357	-.47426
20	9	-.36423	-.37044	-.37609	-.38119	-.38577	-.38984	-.39343	-.39655
20	10	-.26445	-.27292	-.28084	-.28823	-.29510	-.30147	-.30734	-.31274
20	11	-.15845	-.16887	-.17879	-.18822	-.19716	-.20562	-.21360	-.22111
20	12	-.04415	-.05621	-.06785	-.07905	-.08981	-.10014	-.11002	-.11946
20	13	.08115	.06779	.05477	.04211	.02981	.01788	.00634	-.00482
20	14	.22114	.20692	.19293	.17918	.16570	.15251	.13961	.12703
20	15	.38128	.36676	.35233	.33802	.32385	.30984	.29602	.28241
20	16	.57020	.55619	.54210	.52796	.51380	.49966	.48556	.47154
20	17	.80299	.79074	.77819	.76540	.75240	.73922	.72591	.71248
20	18	1.10976	1.10151	1.09272	1.08342	1.07366	1.06348	1.05292	1.04202
20	19	1.56619	1.56690	1.56674	1.56574	1.56395	1.56142	1.55817	1.55427
20	20	2.47801	2.50503	2.53072	2.55511	2.57824	2.60013	2.62083	2.64037
21	1	-1.20095	-1.16724	-1.13479	-1.10358	-1.07359	-1.04477	-1.01708	-.99050
21	2	-1.05276	-1.02960	-1.00692	-.98473	-.96305	-.94191	-.92131	-.90127
21	3	-.94176	-.92533	-.90897	-.89273	-.87663	-.86072	-.84501	-.82955
21	4	-.84487	-.83360	-.82214	-.81056	-.79889	-.78717	-.77543	-.76372
21	5	-.75461	-.74759	-.74023	-.73257	-.72466	-.71653	-.70823	-.69979
21	6	-.66735	-.66399	-.66017	-.65594	-.65135	-.64642	-.64120	-.63572
21	7	-.58089	-.58073	-.58005	-.57888	-.57727	-.57524	-.57284	-.57008
21	8	-.49359	-.49628	-.49841	-.50002	-.50112	-.50176	-.50196	-.50174
21	9	-.40409	-.40933	-.41400	-.41812	-.42172	-.42481	-.42743	-.42960
21	10	-.31108	-.31859	-.32554	-.33195	-.33783	-.34320	-.34808	-.35248

Expected Values of Inverse Gaussian Order Statistics
Shape Parameter k

n	i	1.6	1.7	1.8	1.9	2.0	2.1	2.2	2.3
21	11	-.21316	-.22267	-.23166	-.24013	-.24810	-.25556	-.26253	-.26903
21	12	-.10871	-.11996	-.13073	-.14103	-.15086	-.16022	-.16912	-.17756
21	13	.00428	-.00840	-.02068	-.03256	-.04402	-.05507	-.06569	-.07589
21	14	.12845	.11468	.10121	.08806	.07524	.06278	.05066	.03892
21	15	.26748	.25304	.23879	.22474	.21093	.19738	.18409	.17108
21	16	.42679	.41225	.39775	.38333	.36901	.35483	.34080	.32694
21	17	.61502	.60118	.58721	.57315	.55905	.54492	.53080	.51672
21	18	.84721	.83534	.82313	.81064	.79790	.78494	.77181	.75854
21	19	1.15352	1.14587	1.13765	1.12888	1.11962	1.10990	1.09977	1.08927
21	20	1.60963	1.61122	1.61191	1.61173	1.61072	1.60894	1.60643	1.60321
21	21	2.52143	2.54972	2.57666	2.60228	2.62662	2.64969	2.67155	2.69223
22	1	-1.20754	-1.17335	-1.14047	-1.10886	-1.07849	-1.04932	-1.02132	-.99444
22	2	-1.06258	-1.03882	-1.01557	-.99284	-.97067	-.94906	-.92803	-.90757
22	3	-.95455	-.93744	-.92042	-.90355	-.88686	-.87039	-.85415	-.83819
22	4	-.86069	-.84866	-.83648	-.82419	-.81185	-.79948	-.78714	-.77484
22	5	-.77365	-.76581	-.75765	-.74923	-.74058	-.73174	-.72275	-.71366
22	6	-.68988	-.68565	-.68099	-.67594	-.67054	-.66484	-.65887	-.65266
22	7	-.60728	-.60622	-.60465	-.60262	-.60016	-.59730	-.59408	-.59054
22	8	-.52433	-.52610	-.52731	-.52801	-.52823	-.52798	-.52731	-.52625
22	9	-.43979	-.44410	-.44783	-.45102	-.45370	-.45587	-.45758	-.45885
22	10	-.35252	-.35911	-.36512	-.37059	-.37553	-.37995	-.38389	-.38735
22	11	-.26135	-.26997	-.27804	-.28558	-.29260	-.29910	-.30511	-.31064
22	12	-.16497	-.17538	-.18528	-.19469	-.20360	-.21201	-.21995	-.22741
22	13	-.06183	-.07377	-.08527	-.09632	-.10692	-.11706	-.12676	-.13601
22	14	.05005	.03685	.02403	.01158	-.00048	-.01215	-.02341	-.03427
22	15	.17326	.15915	.14531	.13176	.11852	.10559	.09300	.08074
22	16	.31145	.29685	.28241	.26813	.25406	.24021	.22660	.21324
22	17	.47005	.45552	.44100	.42653	.41212	.39781	.38362	.36958
22	18	.65766	.64402	.63021	.61628	.60226	.58819	.57409	.56000
22	19	.88934	.87785	.86600	.85383	.84137	.82867	.81575	.80266
22	20	1.19523	1.18819	1.18054	1.17231	1.16355	1.15431	1.14462	1.13452
22	21	1.65107	1.65352	1.65504	1.65567	1.65544	1.65441	1.65261	1.65008
22	22	2.56288	2.59240	2.62055	2.64736	2.67286	2.69709	2.72007	2.74186
23	1	-1.21371	-1.17908	-1.14578	-1.11379	-1.08307	-1.05358	-1.02528	-.99813
23	2	-1.07173	-1.04741	-1.02362	-1.00039	-.97775	-.95570	-.93426	-.91342
23	3	-.96644	-.94867	-.93103	-.91357	-.89632	-.87932	-.86259	-.84616
23	4	-.87533	-.86257	-.84970	-.83675	-.82378	-.81081	-.79790	-.78505
23	5	-.79118	-.78256	-.77366	-.76451	-.75517	-.74566	-.73604	-.72632
23	6	-.71054	-.70549	-.70003	-.69420	-.68805	-.68162	-.67494	-.66806
23	7	-.63136	-.62945	-.62705	-.62419	-.62093	-.61729	-.61331	-.60902
23	8	-.55223	-.55312	-.55347	-.55331	-.55268	-.55161	-.55013	-.54827
23	9	-.47201	-.47542	-.47827	-.48058	-.48237	-.48368	-.48453	-.48496
23	10	-.38968	-.39537	-.40049	-.40505	-.40909	-.41262	-.41566	-.41824
23	11	-.30423	-.31197	-.31915	-.32579	-.33190	-.33749	-.34258	-.34720
23	12	-.21458	-.22415	-.23320	-.24172	-.24972	-.25723	-.26424	-.27077
23	13	-.11950	-.13067	-.14137	-.15158	-.16131	-.17057	-.17935	-.18767
23	14	-.01747	-.03000	-.04212	-.05381	-.06508	-.07591	-.08631	-.09627
23	15	.09345	.07983	.06655	.05361	.04104	.02884	.01702	.00558
23	16	.21582	.20145	.18732	.17344	.15984	.14652	.13352	.12083
23	17	.35329	.33859	.32401	.30956	.29529	.28120	.26732	.25368
23	18	.51126	.49679	.48230	.46781	.45335	.43896	.42467	.41049
23	19	.69833	.68491	.67130	.65752	.64363	.62964	.61560	.60153
23	20	.92955	.91847	.90699	.89516	.88300	.87057	.85789	.84501
23	21	1.23508	1.22865	1.22157	1.21389	1.20564	1.19687	1.18763	1.17795
23	22	1.69068	1.69399	1.69633	1.69774	1.69828	1.69798	1.69689	1.69505

Expected Values of Inverse Gaussian Order Statistics
Shape Parameter k

n	i	1.6	1.7	1.8	1.9	2.0	2.1	2.2	2.3
23	23	2.60252	2.63324	2.66256	2.69053	2.71716	2.74250	2.76658	2.78944
24	1	-1.21951	-1.18445	-1.15076	-1.11841	-1.08736	-1.05756	-1.02898	-1.00157
24	2	-1.08030	-1.05543	-1.03114	-1.00744	-.98435	-.96189	-.94006	-.91887
24	3	-.97751	-.95912	-.94090	-.92288	-.90511	-.88761	-.87042	-.85354
24	4	-.88892	-.87548	-.86195	-.84838	-.83481	-.82128	-.80783	-.79447
24	5	-.80739	-.79804	-.78843	-.77860	-.76860	-.75847	-.74824	-.73795
24	6	-.72956	-.72374	-.71752	-.71097	-.70411	-.69699	-.68965	-.68213
24	7	-.65345	-.65073	-.64754	-.64391	-.63988	-.63551	-.63081	-.62584
24	8	-.57771	-.57777	-.57729	-.57632	-.57489	-.57304	-.57079	-.56819
24	9	-.50128	-.50384	-.50584	-.50730	-.50827	-.50876	-.50880	-.50844
24	10	-.42323	-.42807	-.43232	-.43603	-.43921	-.44189	-.44409	-.44583
24	11	-.34270	-.34959	-.35591	-.36168	-.36692	-.37164	-.37586	-.37961
24	12	-.25876	-.26751	-.27571	-.28337	-.29050	-.29713	-.30325	-.30889
24	13	-.17040	-.18080	-.19069	-.20007	-.20895	-.21733	-.22523	-.23265
24	14	-.07643	-.08826	-.09963	-.11055	-.12101	-.13100	-.14053	-.14961
24	15	.02465	.01162	-.00103	-.01328	-.02513	-.03656	-.04757	-.05817
24	16	.13472	.12076	.10710	.09375	.08074	.06808	.05577	.04383
24	17	.25637	.24180	.22743	.21329	.19938	.18575	.17239	.15932
24	18	.39320	.37845	.36377	.34921	.33477	.32050	.30641	.29253
24	19	.55061	.53624	.52181	.50734	.49288	.47845	.46409	.44981
24	20	.73720	.72404	.71064	.69704	.68330	.66943	.65547	.64146
24	21	.96802	.95736	.94620	.93470	.92294	.91079	.89837	.88572
24	22	1.27323	1.26741	1.26090	1.25376	1.24602	1.23774	1.22895	1.21969
24	23	1.72863	1.73277	1.73591	1.73810	1.73939	1.73982	1.73943	1.73826
24	24	2.64052	2.67239	2.70285	2.73193	2.75967	2.78610	2.81124	2.83514
25	1	-1.22498	-1.18951	-1.15545	-1.12276	-1.09139	-1.06131	-1.03246	-1.00481
25	2	-1.08834	-1.06296	-1.03818	-1.01404	-.99053	-.96768	-.94549	-.92395
25	3	-.98787	-.96889	-.95011	-.93157	-.91330	-.89533	-.87770	-.86040
25	4	-.90158	-.88749	-.87335	-.85919	-.84506	-.83099	-.81703	-.80320
25	5	-.82245	-.81240	-.80212	-.79165	-.78103	-.77030	-.75951	-.74868
25	6	-.74717	-.74060	-.73367	-.72642	-.71890	-.71114	-.70318	-.69506
25	7	-.67382	-.67032	-.66637	-.66201	-.65727	-.65220	-.64684	-.64121
25	8	-.60110	-.60035	-.59909	-.59735	-.59517	-.59258	-.58962	-.58631
25	9	-.52802	-.52976	-.53095	-.53162	-.53180	-.53151	-.53080	-.52968
25	10	-.45374	-.45775	-.46118	-.46407	-.46643	-.46830	-.46970	-.47066
25	11	-.37747	-.38354	-.38904	-.39398	-.39838	-.40227	-.40566	-.40859
25	12	-.29844	-.30638	-.31375	-.32058	-.32688	-.33266	-.33793	-.34273
25	13	-.21578	-.22540	-.23449	-.24305	-.25110	-.25863	-.26568	-.27223
25	14	-.12852	-.13963	-.15025	-.16039	-.17004	-.17920	-.18789	-.19611
25	15	-.03550	-.04789	-.05986	-.07139	-.08248	-.09313	-.10333	-.11308
25	16	.06474	.05129	.03819	.02546	.01311	.00116	-.01040	-.02156
25	17	.17409	.15984	.14586	.13217	.11878	.10572	.09299	.08061
25	18	.29509	.28037	.26582	.25146	.23731	.22340	.20975	.19637
25	19	.43135	.41659	.40187	.38722	.37268	.35826	.34401	.32992
25	20	.58827	.57403	.55968	.54527	.53084	.51641	.50201	.48767
25	21	.77443	.76154	.74838	.73499	.72141	.70768	.69384	.67991
25	22	1.00489	.99466	.98396	.97283	.96133	.94948	.93733	.92492
25	23	1.30983	1.30460	1.29867	1.29207	1.28484	1.27705	1.26871	1.25989
25	24	1.76505	1.77000	1.77393	1.77689	1.77892	1.78006	1.78036	1.77986
25	25	2.67699	2.70999	2.74155	2.77173	2.80054	2.82801	2.85420	2.87911

Expected Values of Inverse Gaussian Order Statistics
Shape Parameter k

n	i	2.4	2.5
1	1	.00000	.00000
2	1	-.48548	-.48068
2	2	.48548	.48068
3	1	-.64504	-.63598
3	2	-.16635	-.17007
3	3	.81139	.80606
4	1	-.72697	-.71497
4	2	-.39922	-.39903
4	3	.06652	.05888
4	4	1.05968	1.05512
5	1	-.77800	-.76385
5	2	-.52286	-.51946
5	3	-.21376	-.21837
5	4	.25337	.24371
5	5	1.26126	1.25797
6	1	-.81340	-.79760
6	2	-.60101	-.59509
6	3	-.36658	-.36821
6	4	-.06094	-.06853
6	5	.41052	.39983
6	6	1.43141	1.42960
7	1	-.83971	-.82259
7	2	-.65556	-.64764
7	3	-.46462	-.46373
7	4	-.23586	-.24085
7	5	.07024	.06071
7	6	.54663	.53548
7	7	1.57887	1.57862
8	1	-.86021	-.84202
8	2	-.69619	-.68662
8	3	-.53367	-.53068
8	4	-.34953	-.35214
8	5	-.12220	-.12956
8	6	.18571	.17487
8	7	.66694	.65569
8	8	1.70915	1.71046
9	1	-.87675	-.85765
9	2	-.72786	-.71691
9	3	-.58537	-.58061
9	4	-.43027	-.43081
9	5	-.24861	-.25380
9	6	-.02107	-.03017
9	7	.28910	.27738
9	8	.77490	.76377
9	9	1.82593	1.82880
10	1	-.89046	-.87059
10	2	-.75338	-.74126
10	3	-.62579	-.61953
10	4	-.49106	-.48982
10	5	-.33907	-.34229
10	6	-.15814	-.16530
10	7	.07031	.05992
10	8	.38287	.37058
10	9	.87290	.86207
10	10	1.93182	1.93621
11	1	-.90206	-.88151
11	2	-.77448	-.76135
11	3	-.65842	-.65086
11	4	-.53877	-.53598
11	5	-.40756	-.40903
11	6	-.25689	-.26221
11	7	-.07585	-.08455
11	8	.15382	.14247
11	9	.46876	.45613
11	10	.96271	.95228
11	11	2.02873	2.03461
12	1	-.91204	-.89089
12	2	-.79229	-.77827
12	3	-.68543	-.67673
12	4	-.57740	-.57324
12	5	-.46153	-.46145
12	6	-.33201	-.33565
12	7	-.18176	-.18876
12	8	-.00020	-.01011
12	9	.23084	.21876
12	10	.54807	.53525
12	11	1.04564	1.03568
12	12	2.11810	2.12542
13	1	-.92074	-.89907
13	2	-.80757	-.79277
13	3	-.70823	-.69853
13	4	-.60942	-.60407
13	5	-.50534	-.50388
13	6	-.39143	-.39355
13	7	-.26270	-.26810
13	8	-.11239	-.12075
13	9	.06991	.05904
13	10	.30236	.28974
13	11	.62178	.60891
13	12	1.12271	1.11328
13	13	2.20105	2.20976
14	1	-.92843	-.90628
14	2	-.82087	-.80536
14	3	-.72779	-.71720
14	4	-.63649	-.63007
14	5	-.54174	-.53906
14	6	-.43981	-.44056
14	7	-.32693	-.33087
14	8	-.19846	-.20534
14	9	-.04783	-.05732
14	10	.13532	.12368
14	11	.36917	.35616
14	12	.69067	.67784
14	13	1.19471	1.18586
14	14	2.27846	2.28852
15	1	-.93527	-.91270
15	2	-.83257	-.81644
15	3	-.74481	-.73341
15	4	-.65974	-.65236
15	5	-.57257	-.56878
15	6	-.48010	-.47962
15	7	-.37937	-.38198

n	i	2.4	2.5
15	8	-.26699	-.27247
15	9	-.13850	-.14660
15	10	.01262	.00220
15	11	.19668	.18442
15	12	.43190	.41861
15	13	.75537	.74264
15	14	1.26230	1.25404
15	15	2.35105	2.36242
16	1	-.94143	-.91846
16	2	-.84297	-.82626
16	3	-.75977	-.74765
16	4	-.67997	-.67172
16	5	-.59907	-.59428
16	6	-.51427	-.51268
16	7	-.42314	-.42453
16	8	-.32309	-.32727
16	9	-.21090	-.21767
16	10	-.08219	-.09132
16	11	.06950	.05832
16	12	.25449	.24174
16	13	.49104	.47757
16	14	.81637	.80381
16	15	1.32601	1.31836
16	16	2.41938	2.43202
17	1	-.94700	-.92367
17	2	-.85229	-.83506
17	3	-.77306	-.76027
17	4	-.69776	-.68872
17	5	-.62214	-.61645
17	6	-.54369	-.54108
17	7	-.46033	-.46061
17	8	-.37001	-.37299
17	9	-.27030	-.27583
17	10	-.15809	-.16597
17	11	-.02905	-.03906
17	12	.12325	.11144
17	13	.30917	.29604
17	14	.54700	.53343
17	15	.87409	.86175
17	16	1.38627	1.37924
17	17	2.48395	2.49782
18	1	-.95207	-.92842
18	2	-.86071	-.84300
18	3	-.78496	-.77157
18	4	-.71356	-.70380
18	5	-.64245	-.63594
18	6	-.56933	-.56579
18	7	-.49240	-.49165
18	8	-.40995	-.41183
18	9	-.32009	-.32445
18	10	-.22051	-.22720
18	11	-.10815	-.11699
18	12	.02129	.01053
18	13	.17424	.16189
18	14	.36106	.34764
18	15	.60013	.58651
18	16	.92888	.91680
18	17	1.44344	1.43705
18	18	2.54516	2.56022

n	i	2.4	2.5
19	1	-.95672	-.93276
19	2	-.86836	-.85020
19	3	-.79569	-.78174
19	4	-.72771	-.71729
19	5	-.66050	-.65322
19	6	-.59192	-.58753
19	7	-.52038	-.51869
19	8	-.44443	-.44529
19	9	-.36254	-.36582
19	10	-.27293	-.27849
19	11	-.17334	-.18104
19	12	-.06074	-.07041
19	13	.06914	.05774
19	14	.22274	.20996
19	15	.41046	.39681
19	16	.65070	.63710
19	17	.98104	.96924
19	18	1.49784	1.49208
19	19	2.60334	2.61956
20	1	-.96101	-.93676
20	2	-.87535	-.85679
20	3	-.80544	-.79097
20	4	-.74047	-.72944
20	5	-.67667	-.66869
20	6	-.61201	-.60683
20	7	-.54506	-.54250
20	8	-.47455	-.47447
20	9	-.39924	-.40151
20	10	-.31769	-.32219
20	11	-.22817	-.23479
20	12	-.12848	-.13706
20	13	-.01559	-.02597
20	14	.11476	.10282
20	15	.26902	.25587
20	16	.45760	.44379
20	17	.69898	.68543
20	18	1.03081	1.01933
20	19	1.54973	1.54461
20	20	2.65879	2.67613
21	1	-.96497	-.94046
21	2	-.88177	-.86283
21	3	-.81434	-.79940
21	4	-.75205	-.74045
21	5	-.69125	-.68262
21	6	-.63001	-.62411
21	7	-.56701	-.56365
21	8	-.50115	-.50019
21	9	-.43134	-.43268
21	10	-.35643	-.35995
21	11	-.27506	-.28065
21	12	-.18554	-.19309
21	13	-.08568	-.09504
21	14	.02754	.01653
21	15	.15837	.14597
21	16	.31328	.29983
21	17	.50270	.48877
21	18	.74516	.73170
21	19	1.07842	1.06727
21	20	1.59934	1.59486

n	i	2.4	2.5
21	21	2.71177	2.73020
22	1	-.96865	-.94389
22	2	-.88770	-.86840
22	3	-.82250	-.80712
22	4	-.76262	-.75050
22	5	-.70448	-.69525
22	6	-.64625	-.63967
22	7	-.58670	-.58259
22	8	-.52483	-.52306
22	9	-.45971	-.46017
22	10	-.39037	-.39296
22	11	-.31571	-.32034
22	12	-.23442	-.24097
22	13	-.14482	-.15319
22	14	-.04473	-.05478
22	15	.06883	.05728
22	16	.20016	.18735
22	17	.35571	.34201
22	18	.54594	.53193
22	19	.78943	.77609
22	20	1.12405	1.11324
22	21	1.64687	1.64302
22	22	2.76248	2.78197
23	1	-.97208	-.94709
23	2	-.89319	-.87356
23	3	-.83003	-.81423
23	4	-.77231	-.75970
23	5	-.71656	-.70677
23	6	-.66100	-.65379
23	7	-.60447	-.59967
23	8	-.54607	-.54356
23	9	-.48499	-.48464
23	10	-.42038	-.42211
23	11	-.35135	-.35507
23	12	-.27683	-.28245
23	13	-.19553	-.20295
23	14	-.10580	-.11491
23	15	-.00547	-.01613
23	16	.10846	.09643
23	17	.24027	.22713
23	18	.39645	.38256
23	19	.58747	.57343
23	20	.83195	.81876
23	21	1.16786	1.15741
23	22	1.69249	1.68927
23	23	2.81111	2.83163
24	1	-.97529	-.95008
24	2	-.89830	-.87835
24	3	-.83700	-.82081
24	4	-.78125	-.76818
24	5	-.72764	-.71732
24	6	-.67446	-.66666
24	7	-.62061	-.61517
24	8	-.56527	-.56204
24	9	-.50769	-.50658
24	10	-.44715	-.44807
24	11	-.38290	-.38577
24	12	-.31406	-.31879
24	13	-.23960	-.24611

n	i	2.4	2.5
24	14	-.15824	-.16643
24	15	-.06835	-.07811
24	16	.03226	.02106
24	17	.14656	.13412
24	18	.27886	.26543
24	19	.43564	.42161
24	20	.62742	.61338
24	21	.87286	.85983
24	22	1.21001	1.19993
24	23	1.73636	1.73375
24	24	2.85784	2.87937
25	1	-.97830	-.95288
25	2	-.90306	-.88282
25	3	-.84347	-.82692
25	4	-.78952	-.77601
25	5	-.73785	-.72704
25	6	-.68681	-.67846
25	7	-.63535	-.62930
25	8	-.58271	-.57882
25	9	-.52820	-.52639
25	10	-.47121	-.47137
25	11	-.41107	-.41313
25	12	-.34705	-.35094
25	13	-.27832	-.28396
25	14	-.20386	-.21117
25	15	-.12240	-.13129
25	16	-.03231	-.04266
25	17	.06858	.05691
25	18	.18326	.17046
25	19	.31604	.30236
25	20	.47341	.45926
25	21	.66592	.65191
25	22	.91228	.89944
25	23	1.25060	1.24090
25	24	1.77860	1.77661
25	25	2.90281	2.92532

Table 2. Variances of order statistics for $n = 1(1)25$, $i = 1(1)n$ and $k = 0.0(0.1)2.5$

Variances of Inverse Gaussian Order Statistics
Shape Parameter k

n	i	j	0.0	0.1	0.2	0.3	0.4	0.5	0.6	0.7
1	1	1	1.00000	1.00000	1.00000	1.00000	1.00000	1.00000	1.00000	1.00000
2	1	1	.68169	.65372	.62628	.59947	.57335	.54801	.52348	.49981
2	2	2	.68169	.71010	.73886	.76785	.79698	.82614	.85524	.88417
3	1	1	.55947	.52555	.49293	.46169	.43187	.40353	.37668	.35132
3	2	2	.44867	.44833	.44732	.44564	.44332	.44038	.43686	.43279
3	3	3	.55947	.59459	.63082	.66802	.70608	.74485	.78421	.82402
4	1	1	.49172	.45581	.42172	.38949	.35915	.33069	.30411	.27936
4	2	2	.36046	.35261	.34431	.33562	.32660	.31729	.30776	.29807
4	3	3	.36046	.36781	.37463	.38088	.38653	.39156	.39596	.39972
4	4	4	.49172	.52936	.56864	.60946	.65170	.69520	.73984	.78545
5	1	1	.44753	.41089	.37641	.34412	.31401	.28607	.26023	.23642
5	2	2	.31152	.30050	.28928	.27794	.26654	.25513	.24379	.23257
5	3	3	.28683	.28655	.28571	.28431	.28238	.27994	.27701	.27362
5	4	4	.31152	.32228	.33273	.34281	.35245	.36163	.37030	.37842
5	5	5	.44753	.48630	.52711	.56988	.61447	.66078	.70864	.75791
6	1	1	.41593	.37903	.34456	.31251	.28287	.25556	.23052	.20764
6	2	2	.27958	.26692	.25428	.24172	.22930	.21709	.20515	.19351
6	3	3	.24621	.24251	.23837	.23381	.22887	.22359	.21802	.21218
6	4	4	.24621	.24944	.25218	.25442	.25614	.25735	.25804	.25823
6	5	5	.27958	.29218	.30466	.31695	.32899	.34072	.35208	.36302
6	6	6	.41593	.45523	.49689	.54082	.58691	.63505	.68510	.73691
7	1	1	.39192	.35499	.32069	.28900	.25986	.23320	.20891	.18687
7	2	2	.25673	.24314	.22972	.21654	.20368	.19117	.17907	.16743
7	3	3	.21972	.21415	.20826	.20209	.19568	.18908	.18233	.17548
7	4	4	.21045	.21022	.20953	.20839	.20682	.20483	.20245	.19969
7	5	5	.21972	.22493	.22976	.23416	.23812	.24162	.24464	.24719
7	6	6	.25673	.27044	.28419	.29791	.31153	.32498	.33821	.35115
7	7	7	.39192	.43148	.47362	.51830	.56542	.61485	.66649	.72017
8	1	1	.37290	.33605	.30199	.27068	.24205	.21599	.19238	.17108
8	2	2	.23940	.22523	.21136	.19788	.18482	.17224	.16019	.14868
8	3	3	.20077	.19405	.18713	.18004	.17283	.16554	.15824	.15094
8	4	4	.18719	.18501	.18245	.17952	.17625	.17267	.16880	.16468
8	5	5	.18719	.18896	.19032	.19125	.19177	.19187	.19155	.19083
8	6	6	.20077	.20723	.21340	.21923	.22470	.22978	.23444	.23866
8	7	7	.23940	.25382	.26842	.28313	.29786	.31257	.32716	.34159
8	8	8	.37290	.41255	.45500	.50018	.54802	.59843	.65127	.70640
9	1	1	.35735	.32064	.28685	.25593	.22777	.20226	.17926	.15862
9	2	2	.22570	.21115	.19703	.18339	.17028	.15775	.14581	.13451
9	3	3	.18638	.17892	.17134	.16368	.15601	.14836	.14078	.13331
9	4	4	.17056	.16716	.16345	.15945	.15520	.15072	.14606	.14124
9	5	5	.16610	.16591	.16534	.16439	.16308	.16142	.15943	.15713
9	6	6	.17056	.17362	.17632	.17865	.18059	.18213	.18329	.18404
9	7	7	.18638	.19369	.20078	.20762	.21418	.22041	.22629	.23179
9	8	8	.22570	.24060	.25580	.27122	.28678	.30243	.31808	.33366
9	9	9	.35735	.39702	.43964	.48518	.53356	.58471	.63850	.69480
10	1	1	.34434	.30779	.27428	.24372	.21601	.19101	.16856	.14850
10	2	2	.21452	.19974	.18547	.17177	.15868	.14624	.13447	.12338
10	3	3	.17500	.16702	.15900	.15100	.14305	.13520	.12750	.11998
10	4	4	.15794	.15372	.14925	.14456	.13970	.13469	.12958	.12439
10	5	5	.15105	.14961	.14783	.14573	.14333	.14065	.13771	.13455
10	6	6	.15105	.15215	.15290	.15329	.15333	.15301	.15236	.15138
10	7	7	.15794	.16189	.16553	.16885	.17183	.17445	.17670	.17857
10	8	8	.17500	.18290	.19067	.19826	.20564	.21275	.21957	.22607
10	9	9	.21452	.22976	.24540	.26135	.27755	.29393	.31042	.32694
10	10	10	.34434	.38397	.42669	.47248	.52128	.57302	.62757	.68483

Variances of Inverse Gaussian Order Statistics
Shape Parameter k

n	i	j	0.0	0.1	0.2	0.3	0.4	0.5	0.6	0.7
11	1	1	.33325	.29687	.26363	.23342	.20612	.18158	.15963	.14008
11	2	2	.20520	.19026	.17591	.16220	.14917	.13685	.12526	.11440
11	3	3	.16572	.15737	.14906	.14082	.13271	.12476	.11702	.10952
11	4	4	.14795	.14315	.13816	.13301	.12775	.12241	.11703	.11164
11	5	5	.13964	.13734	.13474	.13189	.12879	.12547	.12196	.11829
11	6	6	.13716	.13700	.13651	.13570	.13458	.13316	.13146	.12949
11	7	7	.13964	.14164	.14333	.14468	.14571	.14639	.14675	.14677
11	8	8	.14795	.15254	.15688	.16095	.16471	.16816	.17126	.17403
11	9	9	.16572	.17406	.18233	.19050	.19851	.20632	.21390	.22120
11	10	10	.20520	.22068	.23664	.25300	.26971	.28668	.30385	.32113
11	11	11	.33325	.37281	.41558	.46155	.51068	.56289	.61808	.67614
12	1	1	.32364	.28744	.25446	.22458	.19766	.17354	.15204	.13296
12	2	2	.19726	.18222	.16783	.15415	.14121	.12902	.11761	.10697
12	3	3	.15798	.14936	.14083	.13245	.12424	.11626	.10853	.10109
12	4	4	.13981	.13458	.12921	.12374	.11822	.11267	.10713	.10164
12	5	5	.13061	.12769	.12453	.12114	.11757	.11384	.10997	.10599
12	6	6	.12664	.12560	.12427	.12265	.12078	.11865	.11629	.11373
12	7	7	.12664	.12738	.12781	.12793	.12775	.12727	.12651	.12546
12	8	8	.13061	.13327	.13565	.13774	.13952	.14098	.14212	.14295
12	9	9	.13981	.14488	.14975	.15439	.15877	.16287	.16667	.17015
12	10	10	.15798	.16664	.17531	.18392	.19244	.20082	.20901	.21698
12	11	11	.19726	.21292	.22913	.24582	.26293	.28039	.29812	.31605
12	12	12	.32364	.36311	.40590	.45200	.50139	.55400	.60972	.66846
13	1	1	.31521	.27919	.24646	.21688	.19032	.16659	.14550	.12685
13	2	2	.19041	.17530	.16091	.14728	.13443	.12239	.11115	.10072
13	3	3	.15139	.14257	.13390	.12541	.11716	.10918	.10149	.09413
13	4	4	.13301	.12745	.12181	.11612	.11041	.10473	.09911	.09357
13	5	5	.12325	.11987	.11628	.11253	.10863	.10461	.10050	.09634
13	6	6	.11832	.11664	.11471	.11253	.11013	.10753	.10474	.10179
13	7	7	.11680	.11666	.11623	.11552	.11454	.11331	.11182	.11011
13	8	8	.11832	.11972	.12084	.12168	.12222	.12247	.12244	.12213
13	9	9	.12325	.12641	.12933	.13198	.13435	.13644	.13822	.13970
13	10	10	.13301	.13845	.14373	.14883	.15371	.15835	.16271	.16679
13	11	11	.15139	.16031	.16928	.17826	.18719	.19603	.20474	.21327
13	12	12	.19041	.20620	.22260	.23955	.25700	.27486	.29306	.31154
13	13	13	.31521	.35458	.39736	.44357	.49317	.54610	.60229	.66162
14	1	1	.30773	.27189	.23940	.21011	.18388	.16050	.13979	.12152
14	2	2	.18442	.16927	.15489	.14132	.12858	.11667	.10561	.09537
14	3	3	.14570	.13673	.12795	.11941	.11114	.10318	.09555	.08827
14	4	4	.12723	.12142	.11557	.10971	.10389	.09812	.09246	.08691
14	5	5	.11710	.11336	.10946	.10543	.10130	.09708	.09283	.08855
14	6	6	.11153	.10938	.10700	.10441	.10164	.09870	.09562	.09243
14	7	7	.10903	.10824	.10719	.10589	.10436	.10260	.10064	.09849
14	8	8	.10903	.10955	.10980	.10979	.10950	.10896	.10816	.10713
14	9	9	.11153	.11344	.11510	.11649	.11761	.11845	.11901	.11930
14	10	10	.11710	.12065	.12399	.12710	.12995	.13254	.13485	.13687
14	11	11	.12723	.13296	.13857	.14404	.14933	.15441	.15925	.16383
14	12	12	.14570	.15482	.16404	.17332	.18259	.19182	.20097	.20998
14	13	13	.18442	.20030	.21685	.23402	.25174	.26995	.28856	.30751
14	14	14	.30773	.34700	.38976	.43604	.48581	.53903	.59561	.65546
15	1	1	.30104	.26538	.23311	.20410	.17817	.15512	.13475	.11684
15	2	2	.17912	.16395	.14960	.13610	.12346	.11169	.10079	.09074
15	3	3	.14073	.13165	.12279	.11421	.10595	.09802	.09046	.08328
15	4	4	.12223	.11623	.11022	.10425	.09834	.09253	.08685	.08132
15	5	5	.11187	.10786	.10371	.09948	.09517	.09082	.08646	.08212

Variances of Inverse Gaussian Order Statistics
Shape Parameter k

n	i	j	0.0	0.1	0.2	0.3	0.4	0.5	0.6	0.7
15	6	6	.10587	.10334	.10062	.09773	.09468	.09150	.08822	.08486
15	7	7	.10269	.10141	.09990	.09817	.09623	.09410	.09180	.08936
15	8	8	.10169	.10157	.10119	.10056	.09970	.09860	.09729	.09577
15	9	9	.10269	.10372	.10451	.10503	.10530	.10532	.10508	.10461
15	10	10	.10587	.10818	.11026	.11210	.11368	.11500	.11606	.11685
15	11	11	.11187	.11573	.11941	.12289	.12614	.12915	.13191	.13439
15	12	12	.12223	.12820	.13408	.13986	.14548	.15094	.15618	.16120
15	13	13	.14073	.15001	.15943	.16895	.17851	.18808	.19760	.20703
15	14	14	.17912	.19507	.21174	.22909	.24705	.26554	.28451	.30387
15	15	15	.30104	.34020	.38293	.42926	.47918	.53264	.58957	.64987
16	1	1	.29501	.25951	.22746	.19870	.17306	.15032	.13027	.11268
16	2	2	.17439	.15922	.14491	.13148	.11894	.10731	.09656	.08669
16	3	3	.13634	.12716	.11826	.10966	.10141	.09353	.08604	.07896
16	4	4	.11787	.11171	.10558	.09952	.09355	.08772	.08204	.07655
16	5	5	.10735	.10312	.09879	.09439	.08996	.08552	.08109	.07672
16	6	6	.10105	.09823	.09524	.09211	.08886	.08551	.08209	.07862
16	7	7	.09740	.09574	.09387	.09181	.08957	.08717	.08464	.08198
16	8	8	.09572	.09510	.09425	.09317	.09187	.09038	.08870	.08684
16	9	9	.09572	.09611	.09625	.09615	.09582	.09526	.09447	.09348
16	10	10	.09740	.09884	.10004	.10101	.10173	.10220	.10243	.10242
16	11	11	.10105	.10368	.10610	.10831	.11028	.11200	.11348	.11469
16	12	12	.10735	.11146	.11543	.11921	.12280	.12617	.12930	.13218
16	13	13	.11787	.12402	.13013	.13616	.14208	.14785	.15344	.15884
16	14	14	.13634	.14574	.15533	.16505	.17487	.18472	.19456	.20435
16	15	15	.17439	.19039	.20716	.22466	.24281	.26156	.28084	.30057
16	16	16	.29501	.33406	.37675	.42312	.47315	.52682	.58406	.64478
17	1	1	.28953	.25420	.22235	.19383	.16846	.14600	.12625	.10895
17	2	2	.17014	.15498	.14070	.12735	.11492	.10341	.09282	.08311
17	3	3	.13242	.12318	.11424	.10564	.09741	.08958	.08217	.07519
17	4	4	.11401	.10772	.10150	.09537	.08938	.08353	.07788	.07243
17	5	5	.10340	.09899	.09451	.08999	.08546	.08096	.07650	.07211
17	6	6	.09688	.09383	.09063	.08732	.08391	.08043	.07691	.07337
17	7	7	.09290	.09094	.08879	.08647	.08400	.08140	.07869	.07589
17	8	8	.09074	.08973	.08850	.08708	.08546	.08367	.08172	.07963
17	9	9	.09005	.08993	.08959	.08903	.08826	.08727	.08609	.08473
17	10	10	.09074	.09153	.09209	.09242	.09252	.09240	.09206	.09150
17	11	11	.09290	.09467	.09622	.09754	.09864	.09950	.10012	.10051
17	12	12	.09688	.09978	.10249	.10500	.10729	.10936	.11119	.11277
17	13	13	.10340	.10772	.11191	.11596	.11983	.12351	.12697	.13019
17	14	14	.11401	.12031	.12661	.13286	.13902	.14507	.15097	.15670
17	15	15	.13242	.14193	.15165	.16155	.17158	.18168	.19181	.20192
17	16	16	.17014	.18617	.20302	.22064	.23897	.25795	.27750	.29755
17	17	17	.28953	.32848	.37112	.41751	.46765	.52150	.57902	.64010
18	1	1	.28453	.24935	.21770	.18941	.16428	.14210	.12262	.10560
18	2	2	.16629	.15114	.13691	.12363	.11131	.09992	.08947	.07992
18	3	3	.12890	.11960	.11064	.10205	.09386	.08608	.07875	.07186
18	4	4	.11057	.10418	.09789	.09171	.08569	.07986	.07423	.06882
18	5	5	.09991	.09535	.09075	.08614	.08154	.07699	.07251	.06813
18	6	6	.09324	.08999	.08663	.08317	.07964	.07607	.07248	.06889
18	7	7	.08902	.08681	.08443	.08191	.07926	.07651	.07367	.07077
18	8	8	.08650	.08517	.08366	.08196	.08010	.07809	.07594	.07368
18	9	9	.08531	.08481	.08409	.08317	.08206	.08076	.07929	.07766
18	10	10	.08531	.08560	.08568	.08554	.08518	.08462	.08386	.08291
18	11	11	.08650	.08761	.08852	.08921	.08967	.08992	.08995	.08976
18	12	12	.08902	.09105	.09288	.09451	.09592	.09711	.09807	.09880

Variances of Inverse Gaussian Order Statistics
Shape Parameter k

n	i	j	0.0	0.1	0.2	0.3	0.4	0.5	0.6	0.7
18	13	13	.09324	.09635	.09930	.10207	.10464	.10700	.10914	.11105
18	14	14	.09991	.10440	.10879	.11306	.11717	.12112	.12486	.12839
18	15	15	.11057	.11700	.12346	.12989	.13627	.14256	.14873	.15475
18	16	16	.12890	.13849	.14833	.15838	.16859	.17891	.18930	.19970
18	17	17	.16629	.18235	.19926	.21699	.23547	.25464	.27443	.29478
18	18	18	.28453	.32337	.36597	.41237	.46259	.51661	.57437	.63578
19	1	1	.27994	.24491	.21344	.18536	.16047	.13854	.11931	.10255
19	2	2	.16279	.14765	.13347	.12027	.10804	.09678	.08646	.07705
19	3	3	.12571	.11638	.10741	.09883	.09067	.08295	.07569	.06890
19	4	4	.10747	.10101	.09466	.08845	.08242	.07660	.07100	.06565
19	5	5	.09679	.09211	.08742	.08273	.07809	.07351	.06902	.06465
19	6	6	.09002	.08661	.08311	.07953	.07591	.07227	.06864	.06502
19	7	7	.08562	.08320	.08065	.07796	.07518	.07231	.06938	.06640
19	8	8	.08283	.08126	.07951	.07760	.07554	.07335	.07106	.06867
19	9	9	.08129	.08047	.07945	.07825	.07687	.07533	.07365	.07183
19	10	10	.08079	.08069	.08038	.07987	.07917	.07828	.07721	.07597
19	11	11	.08129	.08191	.08232	.08253	.08253	.08232	.08192	.08132
19	12	12	.08283	.08422	.08541	.08640	.08717	.08773	.08808	.08822
19	13	13	.08562	.08787	.08995	.09183	.09351	.09498	.09624	.09727
19	14	14	.09002	.09331	.09646	.09946	.10227	.10489	.10730	.10948
19	15	15	.09679	.10143	.10599	.11045	.11477	.11895	.12295	.12675
19	16	16	.10747	.11402	.12061	.12720	.13377	.14027	.14668	.15296
19	17	17	.12571	.13537	.14531	.15549	.16586	.17638	.18699	.19765
19	18	18	.16279	.17885	.19582	.21363	.23225	.25159	.27160	.29221
19	19	19	.27994	.31867	.36121	.40762	.45792	.51208	.57007	.63178
20	1	1	.27570	.24081	.20952	.18165	.15698	.13528	.11630	.09978
20	2	2	.15957	.14446	.13034	.11721	.10508	.09393	.08374	.07447
20	3	3	.12281	.11345	.10447	.09591	.08779	.08014	.07295	.06624
20	4	4	.10468	.09814	.09175	.08552	.07949	.07369	.06813	.06283
20	5	5	.09400	.08921	.08444	.07970	.07502	.07042	.06594	.06159
20	6	6	.08715	.08361	.07999	.07632	.07263	.06894	.06527	.06164
20	7	7	.08261	.08003	.07732	.07451	.07161	.06865	.06565	.06263
20	8	8	.07963	.07785	.07590	.07382	.07161	.06929	.06688	.06440
20	9	9	.07781	.07673	.07547	.07405	.07246	.07073	.06888	.06692
20	10	10	.07695	.07653	.07591	.07511	.07414	.07299	.07169	.07024
20	11	11	.07695	.07717	.07720	.07703	.07667	.07612	.07539	.07450
20	12	12	.07781	.07870	.07940	.07990	.08020	.08029	.08020	.07991
20	13	13	.07963	.08124	.08268	.08391	.08495	.08579	.08642	.08684
20	14	14	.08261	.08505	.08733	.08944	.09135	.09307	.09458	.09588
20	15	15	.08715	.09060	.09392	.09710	.10013	.10297	.10562	.10806
20	16	16	.09400	.09875	.10346	.10808	.11259	.11697	.12120	.12524
20	17	17	.10468	.11131	.11802	.12475	.13148	.13818	.14479	.15131
20	18	18	.12281	.13253	.14255	.15285	.16336	.17405	.18486	.19575
20	19	19	.15957	.17565	.19266	.21055	.22928	.24878	.26899	.28983
20	20	20	.27570	.31433	.35682	.40323	.45359	.50788	.56607	.62806
21	1	1	.27177	.23702	.20590	.17822	.15376	.13228	.11352	.09723
21	2	2	.15662	.14153	.12746	.11440	.10237	.09133	.08125	.07212
21	3	3	.12016	.11077	.10180	.09326	.08518	.07758	.07047	.06385
21	4	4	.10213	.09554	.08911	.08287	.07685	.07107	.06554	.06030
21	5	5	.09146	.08660	.08176	.07697	.07226	.06766	.06319	.05887
21	6	6	.08457	.08091	.07720	.07346	.06971	.06598	.06229	.05867
21	7	7	.07993	.07721	.07437	.07145	.06847	.06544	.06239	.05933
21	8	8	.07680	.07484	.07274	.07051	.06817	.06575	.06325	.06070
21	9	9	.07477	.07348	.07202	.07041	.06866	.06678	.06480	.06272
21	10	10	.07363	.07295	.07209	.07105	.06986	.06851	.06702	.06542

Variances of Inverse Gaussian Order Statistics
Shape Parameter k

n	i	j	0.0	0.1	0.2	0.3	0.4	0.5	0.6	0.7
21	11	11	.07326	.07317	.07288	.07242	.07178	.07096	.06998	.06885
21	12	12	.07363	.07413	.07443	.07455	.07448	.07423	.07379	.07319
21	13	13	.07477	.07589	.07683	.07757	.07813	.07849	.07866	.07864
21	14	14	.07680	.07860	.08024	.08170	.08296	.08404	.08491	.08558
21	15	15	.07993	.08253	.08499	.08728	.08941	.09134	.09308	.09461
21	16	16	.08457	.08815	.09162	.09497	.09818	.10122	.10408	.10675
21	17	17	.09146	.09633	.10115	.10592	.11060	.11516	.11959	.12385
21	18	18	.10213	.10884	.11565	.12251	.12938	.13624	.14305	.14978
21	19	19	.12016	.12992	.14002	.15041	.16105	.17189	.18289	.19399
21	20	20	.15662	.17270	.18974	.20770	.22653	.24617	.26655	.28762
21	21	21	.27177	.31030	.35274	.39914	.44956	.50397	.56234	.62459
22	1	1	.26811	.23350	.20254	.17505	.15079	.12951	.11097	.09489
22	2	2	.15388	.13883	.12481	.11182	.09988	.08894	.07898	.06997
22	3	3	.11772	.10832	.09935	.09083	.08280	.07526	.06822	.06168
22	4	4	.09980	.09317	.08671	.08046	.07445	.06869	.06321	.05802
22	5	5	.08916	.08422	.07933	.07451	.06978	.06518	.06072	.05643
22	6	6	.08224	.07848	.07469	.07089	.06710	.06334	.05964	.05602
22	7	7	.07752	.07468	.07174	.06873	.06567	.06259	.05950	.05642
22	8	8	.07427	.07216	.06993	.06758	.06515	.06264	.06007	.05747
22	9	9	.07209	.07061	.06899	.06722	.06534	.06334	.06126	.05910
22	10	10	.07073	.06983	.06876	.06754	.06616	.06466	.06303	.06130
22	11	11	.07008	.06972	.06919	.06848	.06761	.06658	.06541	.06411
22	12	12	.07008	.07025	.07025	.07007	.06971	.06918	.06848	.06763
22	13	13	.07073	.07145	.07200	.07236	.07255	.07254	.07237	.07201
22	14	14	.07209	.07340	.07454	.07550	.07628	.07687	.07727	.07749
22	15	15	.07427	.07624	.07806	.07970	.08117	.08245	.08354	.08444
22	16	16	.07752	.08026	.08287	.08533	.08763	.08976	.09170	.09345
22	17	17	.08224	.08593	.08953	.09303	.09639	.09961	.10266	.10554
22	18	18	.08916	.09411	.09905	.10394	.10877	.11350	.11810	.12256
22	19	19	.09980	.10658	.11347	.12044	.12745	.13446	.14144	.14836
22	20	20	.11772	.12752	.13768	.14816	.15891	.16989	.18105	.19235
22	21	21	.15388	.16996	.18703	.20505	.22397	.24373	.26428	.28555
22	22	22	.26811	.30655	.34893	.39533	.44579	.50031	.55884	.62133
23	1	1	.26470	.23022	.19941	.17209	.14802	.12695	.10860	.09272
23	2	2	.15135	.13632	.12235	.10944	.09758	.08674	.07690	.06800
23	3	3	.11546	.10605	.09709	.08860	.08061	.07313	.06616	.05970
23	4	4	.09766	.09099	.08451	.07826	.07226	.06654	.06110	.05596
23	5	5	.08705	.08205	.07712	.07227	.06753	.06294	.05850	.05423
23	6	6	.08011	.07628	.07242	.06857	.06475	.06097	.05727	.05365
23	7	7	.07534	.07239	.06936	.06628	.06317	.06005	.05693	.05384
23	8	8	.07200	.06977	.06742	.06497	.06245	.05988	.05726	.05463
23	9	9	.06969	.06806	.06629	.06441	.06241	.06032	.05816	.05593
23	10	10	.06816	.06708	.06584	.06446	.06294	.06131	.05957	.05774
23	11	11	.06729	.06672	.06597	.06507	.06402	.06282	.06150	.06006
23	12	12	.06701	.06693	.06667	.06624	.06565	.06489	.06399	.06296
23	13	13	.06729	.06770	.06793	.06799	.06788	.06759	.06714	.06653
23	14	14	.06816	.06908	.06983	.07041	.07081	.07103	.07108	.07095
23	15	15	.06969	.07117	.07249	.07364	.07461	.07540	.07602	.07644
23	16	16	.07200	.07411	.07608	.07789	.07954	.08100	.08229	.08339
23	17	17	.07534	.07820	.08094	.08355	.08601	.08831	.09043	.09237
23	18	18	.08011	.08390	.08762	.09124	.09476	.09813	.10136	.10441
23	19	19	.08705	.09208	.09711	.10212	.10708	.11195	.11672	.12136
23	20	20	.09766	.10449	.11146	.11853	.12565	.13280	.13994	.14703
23	21	21	.11546	.12529	.13551	.14606	.15692	.16803	.17934	.19081
23	22	22	.15135	.16742	.18451	.20258	.22158	.24146	.26216	.28361

Variances of Inverse Gaussian Order Statistics
Shape Parameter k

n	i	j	0.0	0.1	0.2	0.3	0.4	0.5	0.6	0.7
23	23	23	.26470	.30305	.34537	.39176	.44226	.49687	.55556	.61827
24	1	1	.26151	.22715	.19649	.16934	.14545	.12456	.10640	.09071
24	2	2	.14899	.13399	.12007	.10723	.09546	.08471	.07497	.06619
24	3	3	.11337	.10395	.09501	.08655	.07860	.07118	.06427	.05789
24	4	4	.09568	.08898	.08249	.07624	.07026	.06456	.05916	.05408
24	5	5	.08511	.08006	.07509	.07022	.06548	.06089	.05648	.05224
24	6	6	.07817	.07426	.07035	.06646	.06261	.05882	.05512	.05152
24	7	7	.07336	.07032	.06722	.06407	.06092	.05776	.05462	.05153
24	8	8	.06995	.06760	.06516	.06263	.06004	.05742	.05476	.05210
24	9	9	.06753	.06577	.06389	.06190	.05981	.05764	.05541	.05314
24	10	10	.06588	.06464	.06325	.06174	.06010	.05836	.05653	.05462
24	11	11	.06484	.06407	.06315	.06209	.06088	.05955	.05810	.05655
24	12	12	.06433	.06403	.06355	.06292	.06214	.06121	.06015	.05896
24	13	13	.06433	.06448	.06445	.06426	.06390	.06339	.06273	.06192
24	14	14	.06484	.06544	.06587	.06614	.06624	.06617	.06594	.06554
24	15	15	.06588	.06696	.06789	.06865	.06924	.06966	.06991	.06998
24	16	16	.06753	.06916	.07063	.07194	.07309	.07407	.07486	.07548
24	17	17	.06995	.07218	.07428	.07624	.07804	.07968	.08114	.08242
24	18	18	.07336	.07632	.07918	.08192	.08452	.08697	.08926	.09138
24	19	19	.07817	.08204	.08586	.08960	.09324	.09676	.10014	.10337
24	20	20	.08511	.09021	.09533	.10044	.10551	.11052	.11543	.12023
24	21	21	.09568	.10257	.10960	.11675	.12398	.13125	.13853	.14579
24	22	22	.11337	.12323	.13349	.14411	.15506	.16628	.17774	.18937
24	23	23	.14899	.16505	.18215	.20027	.21934	.23933	.26017	.28179
24	24	24	.26151	.29976	.34203	.38841	.43894	.49364	.55248	.61539
25	1	1	.25851	.22426	.19375	.16676	.14304	.12233	.10435	.08884
25	2	2	.14678	.13182	.11795	.10518	.09348	.08283	.07319	.06451
25	3	3	.11143	.10201	.09307	.08464	.07674	.06937	.06253	.05622
25	4	4	.09385	.08712	.08062	.07438	.06841	.06275	.05739	.05236
25	5	5	.08332	.07823	.07323	.06834	.06360	.05903	.05463	.05043
25	6	6	.07638	.07241	.06846	.06453	.06066	.05687	.05317	.04959
25	7	7	.07154	.06842	.06526	.06207	.05887	.05569	.05254	.04945
25	8	8	.06808	.06564	.06311	.06051	.05787	.05520	.05252	.04984
25	9	9	.06558	.06371	.06172	.05964	.05748	.05525	.05297	.05066
25	10	10	.06382	.06244	.06094	.05931	.05758	.05575	.05385	.05188
25	11	11	.06264	.06172	.06065	.05945	.05812	.05667	.05513	.05349
25	12	12	.06197	.06147	.06082	.06002	.05908	.05801	.05682	.05551
25	13	13	.06175	.06167	.06143	.06103	.06048	.05978	.05895	.05799
25	14	14	.06197	.06230	.06248	.06249	.06235	.06205	.06159	.06099
25	15	15	.06264	.06341	.06402	.06447	.06476	.06488	.06484	.06464
25	16	16	.06382	.06505	.06614	.06706	.06782	.06841	.06884	.06909
25	17	17	.06558	.06733	.06894	.07040	.07170	.07284	.07380	.07459
25	18	18	.06808	.07042	.07264	.07473	.07667	.07845	.08007	.08151
25	19	19	.07154	.07459	.07756	.08041	.08315	.08574	.08817	.09045
25	20	20	.07638	.08033	.08424	.08808	.09184	.09549	.09901	.10239
25	21	21	.08332	.08848	.09367	.09887	.10405	.10918	.11423	.11918
25	22	22	.09385	.10077	.10787	.11509	.12242	.12980	.13722	.14462
25	23	23	.11143	.12130	.13161	.14229	.15332	.16465	.17623	.18801
25	24	24	.14678	.16283	.17995	.19810	.21724	.23733	.25829	.28007
25	25	25	.25851	.29667	.33888	.38525	.43581	.49059	.54956	.61267

Variances of Inverse Gaussian Order Statistics
Shape Parameter k

n	i	j	0.8	0.9	1.0	1.1	1.2	1.3	1.4	1.5
1	1	1	1.00000	1.00000	1.00000	1.00000	1.00000	1.00000	1.00000	1.00000
2	1	1	.47703	.45516	.43422	.41419	.39507	.37684	.35949	.34299
2	2	2	.91287	.94125	.96924	.99678	1.02383	1.05034	1.07628	1.10161
3	1	1	.32744	.30501	.28400	.26435	.24602	.22893	.21303	.19825
3	2	2	.42821	.42316	.41770	.41185	.40568	.39923	.39253	.38563
3	3	3	.86415	.90447	.94486	.98522	1.02544	1.06543	1.10511	1.14441
4	1	1	.25639	.23514	.21552	.19745	.18084	.16560	.15164	.13886
4	2	2	.28828	.27843	.26859	.25879	.24908	.23949	.23007	.22084
4	3	3	.40284	.40532	.40719	.40846	.40914	.40927	.40887	.40797
4	4	4	.83190	.87903	.92670	.97477	1.02312	1.07163	1.12017	1.16866
5	1	1	.21457	.19456	.17630	.15967	.14456	.13085	.11843	.10720
5	2	2	.22152	.21068	.20011	.18982	.17987	.17025	.16100	.15212
5	3	3	.26981	.26562	.26108	.25624	.25113	.24579	.24026	.23458
5	4	4	.38597	.39294	.39930	.40506	.41020	.41475	.41869	.42206
5	5	5	.80844	.86006	.91263	.96598	1.01999	1.07450	1.12940	1.18454
6	1	1	.18682	.16792	.15081	.13538	.12147	.10898	.09776	.08770
6	2	2	.18224	.17135	.16088	.15085	.14128	.13218	.12354	.11537
6	3	3	.20614	.19992	.19358	.18715	.18067	.17417	.16770	.16127
6	4	4	.25794	.25717	.25595	.25430	.25225	.24984	.24708	.24402
6	5	5	.37351	.38351	.39298	.40191	.41026	.41805	.42524	.43185
6	6	6	.79032	.84518	.90132	.95858	1.01682	1.07588	1.13562	1.19590
7	1	1	.16694	.14899	.13286	.11840	.10548	.09395	.08367	.07453
7	2	2	.15626	.14560	.13547	.12587	.11680	.10827	.10027	.09278
7	3	3	.16858	.16166	.15476	.14792	.14117	.13455	.12807	.12175
7	4	4	.19659	.19318	.18949	.18556	.18141	.17708	.17259	.16799
7	5	5	.24925	.25083	.25194	.25260	.25280	.25258	.25195	.25094
7	6	6	.36375	.37596	.38774	.39905	.40986	.42015	.42991	.43910
7	7	7	.77576	.83308	.89197	.95228	1.01384	1.07650	1.14010	1.20452
8	1	1	.15194	.13479	.11948	.10584	.09372	.08298	.07346	.06505
8	2	2	.13775	.12740	.11765	.10849	.09993	.09193	.08450	.07760
8	3	3	.14371	.13657	.12956	.12271	.11605	.10959	.10336	.09736
8	4	4	.16035	.15583	.15117	.14639	.14152	.13661	.13167	.12674
8	5	5	.18974	.18827	.18647	.18434	.18192	.17923	.17630	.17316
8	6	6	.24245	.24577	.24864	.25105	.25300	.25452	.25560	.25626
8	7	7	.35579	.36970	.38328	.39647	.40924	.42156	.43339	.44471
8	8	8	.76369	.82296	.88406	.94683	1.01109	1.07670	1.14350	1.21133
9	1	1	.14016	.12371	.10909	.09615	.08471	.07461	.06572	.05790
9	2	2	.12384	.11382	.10445	.09572	.08760	.08008	.07314	.06675
9	3	3	.12598	.11883	.11188	.10517	.09871	.09251	.08659	.08095
9	4	4	.13631	.13130	.12623	.12114	.11607	.11103	.10606	.10116
9	5	5	.15454	.15170	.14862	.14534	.14188	.13827	.13454	.13072
9	6	6	.18441	.18440	.18402	.18329	.18223	.18086	.17919	.17725
9	7	7	.23689	.24156	.24581	.24962	.25300	.25593	.25842	.26048
9	8	8	.34912	.36439	.37941	.39414	.40852	.42251	.43608	.44919
9	9	9	.75346	.81433	.87725	.94206	1.00859	1.07667	1.14615	1.21689
10	1	1	.13063	.11479	.10078	.08843	.07756	.06802	.05965	.05232
10	2	2	.11299	.10329	.09427	.08592	.07820	.07111	.06459	.05862
10	3	3	.11267	.10560	.09879	.09227	.08605	.08013	.07452	.06922
10	4	4	.11916	.11392	.10871	.10355	.09847	.09349	.08863	.08390
10	5	5	.13118	.12764	.12395	.12015	.11626	.11231	.10831	.10431
10	6	6	.15008	.14849	.14663	.14451	.14216	.13960	.13686	.13396
10	7	7	.18007	.18120	.18196	.18235	.18240	.18211	.18150	.18058
10	8	8	.23221	.23798	.24335	.24831	.25286	.25697	.26067	.26393
10	9	9	.34342	.35980	.37602	.39202	.40775	.42317	.43822	.45287
10	10	10	.74464	.80685	.87131	.93784	1.00630	1.07650	1.14830	1.22154

Variances of Inverse Gaussian Order Statistics
Shape Parameter k

n	i	j	0.8	0.9	1.0	1.1	1.2	1.3	1.4	1.5
11	1	1	.12275	.10744	.09396	.08212	.07174	.06267	.05475	.04785
11	2	2	.10427	.09487	.08617	.07816	.07081	.06407	.05793	.05233
11	3	3	.10229	.09534	.08871	.08239	.07640	.07075	.06543	.06044
11	4	4	.10627	.10096	.09572	.09060	.08560	.08075	.07606	.07154
11	5	5	.11449	.11058	.10660	.10256	.09851	.09445	.09042	.08643
11	6	6	.12728	.12485	.12222	.11942	.11646	.11338	.11020	.10693
11	7	7	.14648	.14587	.14497	.14380	.14236	.14068	.13877	.13666
11	8	8	.17643	.17848	.18016	.18149	.18246	.18309	.18338	.18336
11	9	9	.22819	.23486	.24117	.24710	.25263	.25777	.26249	.26680
11	10	10	.33847	.35578	.37301	.39010	.40699	.42362	.43996	.45595
11	11	11	.73692	.80027	.86605	.93408	1.00420	1.07625	1.15007	1.22550
12	1	1	.11611	.10127	.08825	.07686	.06692	.05825	.05072	.04417
12	2	2	.09709	.08797	.07957	.07187	.06483	.05842	.05259	.04731
12	3	3	.09396	.08715	.08069	.07458	.06882	.06342	.05836	.05365
12	4	4	.09622	.09090	.08571	.08067	.07579	.07110	.06659	.06229
12	5	5	.10194	.09784	.09372	.08960	.08551	.08147	.07749	.07360
12	6	6	.11098	.10807	.10503	.10187	.09863	.09532	.09197	.08859
12	7	7	.12415	.12260	.12081	.11882	.11664	.11429	.11179	.10916
12	8	8	.14346	.14365	.14355	.14315	.14248	.14154	.14036	.13895
12	9	9	.17330	.17611	.17857	.18068	.18244	.18386	.18495	.18570
12	10	10	.22469	.23211	.23921	.24597	.25236	.25838	.26401	.26923
12	11	11	.33410	.35221	.37031	.38834	.40623	.42394	.44140	.45858
12	12	12	.73008	.79442	.86135	.93069	1.00228	1.07596	1.15157	1.22894
13	1	1	.11042	.09600	.08339	.07240	.06284	.05453	.04733	.04110
13	2	2	.09108	.08220	.07407	.06665	.05990	.05377	.04823	.04322
13	3	3	.08711	.08045	.07416	.06825	.06271	.05753	.05272	.04825
13	4	4	.08815	.08287	.07776	.07282	.06808	.06355	.05923	.05513
13	5	5	.09215	.08795	.08378	.07966	.07559	.07162	.06774	.06398
13	6	6	.09872	.09553	.09226	.08892	.08555	.08216	.07877	.07540
13	7	7	.10818	.10606	.10377	.10133	.09876	.09607	.09330	.09046
13	8	8	.12154	.12070	.11962	.11830	.11677	.11504	.11314	.11108
13	9	9	.14087	.14173	.14230	.14256	.14255	.14226	.14170	.14090
13	10	10	.17056	.17401	.17714	.17993	.18238	.18449	.18626	.18771
13	11	11	.22159	.22965	.23744	.24492	.25206	.25885	.26528	.27132
13	12	12	.33021	.34901	.36787	.38672	.40550	.42415	.44261	.46085
13	13	13	.72396	.78918	.85712	.92761	1.00051	1.07564	1.15285	1.23196
14	1	1	.10548	.09144	.07921	.06857	.05934	.05135	.04445	.03849
14	2	2	.08595	.07731	.06943	.06226	.05575	.04988	.04459	.03983
14	3	3	.08138	.07486	.06874	.06301	.05767	.05270	.04810	.04386
14	4	4	.08152	.07631	.07128	.06646	.06186	.05749	.05334	.04943
14	5	5	.08429	.08006	.07588	.07179	.06779	.06391	.06015	.05653
14	6	6	.08914	.08579	.08239	.07898	.07556	.07216	.06880	.06550
14	7	7	.09617	.09370	.09111	.08841	.08563	.08278	.07989	.07698
14	8	8	.10587	.10439	.10272	.10088	.09887	.09672	.09445	.09208
14	9	9	.11932	.11907	.11857	.11783	.11686	.11568	.11430	.11274
14	10	10	.13860	.14003	.14117	.14201	.14257	.14285	.14285	.14260
14	11	11	.16814	.17214	.17584	.17922	.18227	.18500	.18739	.18945
14	12	12	.21882	.22744	.23583	.24394	.25174	.25922	.26636	.27314
14	13	13	.32672	.34613	.36565	.38523	.40479	.42428	.44365	.46283
14	14	14	.71845	.78443	.85327	.92481	.99887	1.07531	1.15396	1.23465
15	1	1	.10114	.08745	.07555	.06523	.05631	.04860	.04196	.03624
15	2	2	.08152	.07310	.06544	.05850	.05223	.04658	.04151	.03697
15	3	3	.07650	.07013	.06417	.05861	.05345	.04867	.04427	.04022
15	4	4	.07598	.07083	.06590	.06120	.05674	.05252	.04854	.04480
15	5	5	.07783	.07359	.06945	.06541	.06150	.05772	.05408	.05060

Variances of Inverse Gaussian Order Statistics
Shape Parameter k

n	i	j	0.8	0.9	1.0	1.1	1.2	1.3	1.4	1.5
15	6	6	.08145	.07800	.07455	.07111	.06769	.06433	.06104	.05782
15	7	7	.08678	.08409	.08132	.07848	.07559	.07268	.06976	.06685
15	8	8	.09406	.09219	.09016	.08799	.08572	.08334	.08089	.07837
15	9	9	.10389	.10296	.10182	.10048	.09896	.09728	.09546	.09351
15	10	10	.11737	.11763	.11764	.11740	.11692	.11622	.11530	.11419
15	11	11	.13660	.13852	.14015	.14150	.14256	.14334	.14385	.14409
15	12	12	.16596	.17045	.17466	.17855	.18214	.18541	.18836	.19098
15	13	13	.21632	.22544	.23435	.24302	.25142	.25951	.26729	.27473
15	14	14	.32356	.34350	.36361	.38384	.40411	.42436	.44454	.46458
15	15	15	.71343	.78011	.84977	.92223	.99736	1.07498	1.15493	1.23705
16	1	1	.09731	.08393	.07233	.06230	.05365	.04620	.03979	.03429
16	2	2	.07766	.06944	.06198	.05525	.04918	.04374	.03887	.03452
16	3	3	.07230	.06607	.06025	.05485	.04986	.04526	.04103	.03716
16	4	4	.07126	.06620	.06137	.05679	.05245	.04837	.04455	.04097
16	5	5	.07242	.06821	.06411	.06014	.05631	.05264	.04913	.04578
16	6	6	.07513	.07163	.06816	.06472	.06134	.05804	.05482	.05170
16	7	7	.07923	.07641	.07353	.07061	.06768	.06476	.06185	.05898
16	8	8	.08484	.08270	.08044	.07809	.07565	.07316	.07062	.06805
16	9	9	.09228	.09090	.08935	.08764	.08580	.08383	.08176	.07961
16	10	10	.10218	.10170	.10102	.10012	.09903	.09777	.09634	.09477
16	11	11	.11565	.11635	.11680	.11699	.11695	.11668	.11618	.11547
16	12	12	.13480	.13715	.13922	.14101	.14252	.14375	.14471	.14540
16	13	13	.16400	.16892	.17356	.17793	.18199	.18575	.18920	.19233
16	14	14	.21405	.22361	.23299	.24216	.25109	.25974	.26811	.27615
16	15	15	.32068	.34109	.36174	.38255	.40346	.42440	.44531	.46614
16	16	16	.70885	.77616	.84654	.91986	.99595	1.07465	1.15579	1.23923
17	1	1	.09388	.08079	.06947	.05970	.05130	.04408	.03788	.03258
17	2	2	.07426	.06622	.05895	.05241	.04653	.04128	.03658	.03241
17	3	3	.06864	.06254	.05686	.05161	.04677	.04233	.03826	.03455
17	4	4	.06720	.06222	.05749	.05302	.04881	.04487	.04118	.03775
17	5	5	.06782	.06365	.05960	.05571	.05197	.04840	.04501	.04179
17	6	6	.06983	.06632	.06285	.05944	.05611	.05288	.04974	.04672
17	7	7	.07303	.07011	.06717	.06423	.06129	.05838	.05552	.05271
17	8	8	.07742	.07510	.07270	.07023	.06771	.06516	.06259	.06003
17	9	9	.08320	.08152	.07970	.07776	.07571	.07358	.07138	.06913
17	10	10	.09073	.08978	.08864	.08733	.08587	.08426	.08254	.08071
17	11	11	.10066	.10059	.10029	.09979	.09908	.09819	.09712	.09589
17	12	12	.11411	.11519	.11603	.11662	.11696	.11707	.11695	.11662
17	13	13	.13318	.13590	.13836	.14055	.14246	.14410	.14547	.14657
17	14	14	.16222	.16751	.17255	.17733	.18183	.18603	.18994	.19353
17	15	15	.21198	.22192	.23173	.24135	.25076	.25992	.26882	.27741
17	16	16	.31804	.33888	.36001	.38135	.40284	.42441	.44599	.46754
17	17	17	.70464	.77251	.84357	.91766	.99463	1.07432	1.15657	1.24121
18	1	1	.09080	.07797	.06691	.05738	.04920	.04219	.03619	.03106
18	2	2	.07123	.06336	.05627	.04990	.04420	.03911	.03458	.03056
18	3	3	.06542	.05944	.05390	.04878	.04409	.03979	.03587	.03230
18	4	4	.06367	.05877	.05413	.04977	.04568	.04186	.03830	.03501
18	5	5	.06386	.05973	.05575	.05193	.04829	.04482	.04154	.03844
18	6	6	.06533	.06182	.05837	.05500	.05173	.04857	.04552	.04260
18	7	7	.06783	.06486	.06190	.05895	.05603	.05315	.05034	.04760
18	8	8	.07132	.06888	.06639	.06385	.06129	.05872	.05616	.05363
18	9	9	.07590	.07401	.07201	.06992	.06775	.06553	.06327	.06098
18	10	10	.08179	.08049	.07905	.07747	.07577	.07397	.07207	.07010
18	11	11	.08937	.08878	.08800	.08704	.08592	.08465	.08323	.08170
18	12	12	.09931	.09958	.09964	.09948	.09912	.09856	.09781	.09690

Variances of Inverse Gaussian Order Statistics
Shape Parameter k

n	i	j	0.8	0.9	1.0	1.1	1.2	1.3	1.4	1.5
18	13	13	.11272	.11414	.11532	.11626	.11695	.11741	.11764	.11764
18	14	14	.13170	.13476	.13756	.14011	.14239	.14440	.14614	.14762
18	15	15	.16058	.16621	.17161	.17677	.18166	.18627	.19059	.19461
18	16	16	.21007	.22037	.23055	.24059	.25043	.26006	.26944	.27855
18	17	17	.31560	.33683	.35840	.38022	.40224	.42439	.44660	.46881
18	18	18	.70075	.76914	.84081	.91561	.99340	1.07400	1.15727	1.24303
19	1	1	.08800	.07543	.06460	.05529	.04732	.04050	.03468	.02971
19	2	2	.06852	.06081	.05388	.04767	.04213	.03720	.03282	.02894
19	3	3	.06257	.05670	.05128	.04630	.04173	.03757	.03378	.03034
19	4	4	.06056	.05574	.05120	.04694	.04296	.03926	.03582	.03264
19	5	5	.06042	.05634	.05242	.04868	.04512	.04175	.03857	.03559
19	6	6	.06146	.05796	.05454	.05122	.04801	.04492	.04196	.03914
19	7	7	.06341	.06042	.05744	.05450	.05161	.04879	.04604	.04337
19	8	8	.06621	.06369	.06114	.05857	.05600	.05344	.05090	.04841
19	9	9	.06989	.06785	.06573	.06355	.06131	.05905	.05676	.05446
19	10	10	.07459	.07306	.07141	.06965	.06780	.06587	.06388	.06184
19	11	11	.08054	.07959	.07848	.07722	.07583	.07431	.07269	.07098
19	12	12	.08815	.08789	.08742	.08678	.08596	.08498	.08386	.08259
19	13	13	.09808	.09867	.09903	.09919	.09913	.09888	.09843	.09780
19	14	14	.11145	.11317	.11466	.11591	.11693	.11770	.11825	.11857
19	15	15	.13034	.13370	.13682	.13969	.14230	.14465	.14674	.14856
19	16	16	.15908	.16501	.17074	.17623	.18148	.18646	.19117	.19559
19	17	17	.20831	.21892	.22946	.23987	.25012	.26017	.27000	.27958
19	18	18	.31335	.33493	.35689	.37917	.40167	.42435	.44714	.46997
19	19	19	.69714	.76600	.83824	.91371	.99224	1.07370	1.15790	1.24471
20	1	1	.08546	.07312	.06250	.05340	.04562	.03898	.03331	.02849
20	2	2	.06608	.05852	.05174	.04568	.04028	.03549	.03124	.02749
20	3	3	.06001	.05425	.04895	.04409	.03965	.03561	.03194	.02862
20	4	4	.05780	.05306	.04861	.04445	.04057	.03698	.03365	.03059
20	5	5	.05739	.05336	.04950	.04584	.04237	.03909	.03601	.03313
20	6	6	.05808	.05460	.05122	.04796	.04481	.04180	.03892	.03619
20	7	7	.05961	.05660	.05363	.05072	.04786	.04509	.04240	.03981
20	8	8	.06186	.05930	.05671	.05413	.05156	.04902	.04653	.04409
20	9	9	.06485	.06272	.06052	.05827	.05600	.05372	.05143	.04916
20	10	10	.06866	.06697	.06517	.06330	.06135	.05935	.05730	.05523
20	11	11	.07344	.07223	.07088	.06942	.06785	.06618	.06443	.06262
20	12	12	.07943	.07878	.07796	.07699	.07587	.07462	.07325	.07178
20	13	13	.08706	.08707	.08690	.08653	.08599	.08528	.08441	.08340
20	14	14	.09696	.09783	.09848	.09891	.09914	.09916	.09899	.09863
20	15	15	.11028	.11228	.11405	.11559	.11689	.11796	.11880	.11941
20	16	16	.12909	.13272	.13612	.13929	.14221	.14487	.14728	.14942
20	17	17	.15768	.16389	.16991	.17572	.18130	.18663	.19169	.19648
20	18	18	.20667	.21758	.22843	.23918	.24980	.26025	.27050	.28052
20	19	19	.31125	.33316	.35549	.37817	.40113	.42430	.44762	.47102
20	20	20	.69377	.76308	.83584	.91192	.99116	1.07340	1.15848	1.24626
21	1	1	.08314	.07100	.06059	.05168	.04407	.03759	.03208	.02739
21	2	2	.06386	.05644	.04980	.04388	.03862	.03396	.02984	.02621
21	3	3	.05772	.05206	.04686	.04211	.03779	.03386	.03031	.02710
21	4	4	.05534	.05068	.04631	.04224	.03846	.03497	.03175	.02879
21	5	5	.05471	.05073	.04694	.04334	.03995	.03677	.03378	.03099
21	6	6	.05512	.05167	.04833	.04511	.04203	.03909	.03630	.03365
21	7	7	.05629	.05329	.05034	.04745	.04464	.04192	.03930	.03678
21	8	8	.05812	.05552	.05292	.05034	.04779	.04528	.04283	.04045
21	9	9	.06057	.05836	.05611	.05383	.05155	.04926	.04699	.04475
21	10	10	.06370	.06188	.05999	.05803	.05602	.05398	.05191	.04984

Variances of Inverse Gaussian Order Statistics
Shape Parameter k

n	i	j	0.8	0.9	1.0	1.1	1.2	1.3	1.4	1.5
21	11	11	.06759	.06619	.06468	.06308	.06138	.05962	.05780	.05593
21	12	12	.07241	.07148	.07041	.06921	.06789	.06646	.06494	.06334
21	13	13	.07843	.07805	.07749	.07677	.07590	.07490	.07376	.07250
21	14	14	.08606	.08633	.08641	.08630	.08601	.08555	.08492	.08414
21	15	15	.09594	.09705	.09796	.09865	.09913	.09941	.09949	.09938
21	16	16	.10921	.11145	.11348	.11527	.11684	.11818	.11929	.12017
21	17	17	.12793	.13181	.13547	.13891	.14211	.14506	.14776	.15020
21	18	18	.15639	.16285	.16914	.17524	.18112	.18677	.19216	.19729
21	19	19	.20515	.21632	.22747	.23854	.24950	.26032	.27095	.28138
21	20	20	.30929	.33150	.35417	.37723	.40061	.42424	.44805	.47199
21	21	21	.69063	.76034	.83359	.91024	.99013	1.07311	1.15902	1.24770
22	1	1	.08100	.06907	.05884	.05011	.04266	.03633	.03096	.02639
22	2	2	.06184	.05456	.04805	.04225	.03712	.03258	.02857	.02505
22	3	3	.05564	.05008	.04498	.04034	.03612	.03230	.02885	.02574
22	4	4	.05313	.04854	.04426	.04027	.03659	.03319	.03006	.02720
22	5	5	.05231	.04838	.04466	.04114	.03782	.03472	.03182	.02912
22	6	6	.05249	.04907	.04578	.04262	.03960	.03673	.03401	.03145
22	7	7	.05338	.05039	.04745	.04460	.04184	.03918	.03662	.03418
22	8	8	.05486	.05224	.04964	.04707	.04455	.04208	.03968	.03735
22	9	9	.05688	.05462	.05234	.05005	.04776	.04549	.04325	.04105
22	10	10	.05947	.05757	.05560	.05359	.05155	.04949	.04743	.04537
22	11	11	.06268	.06115	.05953	.05782	.05605	.05422	.05236	.05047
22	12	12	.06663	.06550	.06425	.06288	.06142	.05987	.05826	.05658
22	13	13	.07149	.07081	.06998	.06902	.06792	.06671	.06540	.06399
22	14	14	.07752	.07738	.07706	.07657	.07593	.07514	.07422	.07317
22	15	15	.08514	.08564	.08596	.08608	.08602	.08578	.08538	.08482
22	16	16	.09500	.09634	.09747	.09839	.09911	.09963	.09994	.10006
22	17	17	.10821	.11068	.11294	.11498	.11679	.11838	.11974	.12087
22	18	18	.12685	.13095	.13486	.13854	.14200	.14522	.14820	.15092
22	19	19	.15518	.16187	.16841	.17478	.18094	.18688	.19258	.19803
22	20	20	.20373	.21514	.22656	.23792	.24920	.26036	.27136	.28217
22	21	21	.30746	.32994	.35293	.37634	.40011	.42417	.44845	.47289
22	22	22	.68768	.75777	.83148	.90866	.98916	1.07283	1.15951	1.24905
23	1	1	.07903	.06728	.05723	.04866	.04137	.03518	.02993	.02548
23	2	2	.06000	.05283	.04644	.04077	.03575	.03132	.02742	.02400
23	3	3	.05374	.04828	.04328	.03873	.03461	.03089	.02754	.02453
23	4	4	.05113	.04661	.04240	.03850	.03490	.03159	.02855	.02578
23	5	5	.05016	.04628	.04262	.03916	.03592	.03290	.03008	.02747
23	6	6	.05015	.04676	.04351	.04040	.03745	.03464	.03200	.02951
23	7	7	.05080	.04782	.04492	.04210	.03939	.03678	.03428	.03191
23	8	8	.05199	.04937	.04677	.04422	.04173	.03930	.03695	.03468
23	9	9	.05367	.05138	.04908	.04678	.04450	.04225	.04004	.03789
23	10	10	.05583	.05386	.05185	.04981	.04775	.04569	.04364	.04160
23	11	11	.05851	.05688	.05516	.05339	.05157	.04971	.04783	.04594
23	12	12	.06179	.06050	.05912	.05764	.05608	.05445	.05278	.05106
23	13	13	.06578	.06488	.06385	.06271	.06145	.06011	.05867	.05717
23	14	14	.07066	.07020	.06959	.06884	.06795	.06694	.06582	.06459
23	15	15	.07669	.07676	.07665	.07638	.07595	.07537	.07464	.07378
23	16	16	.08429	.08501	.08553	.08587	.08602	.08599	.08580	.08544
23	17	17	.09412	.09567	.09701	.09815	.09909	.09982	.10035	.10070
23	18	18	.10728	.10996	.11243	.11469	.11673	.11855	.12014	.12152
23	19	19	.12584	.13015	.13427	.13819	.14189	.14536	.14859	.15158
23	20	20	.15405	.16095	.16773	.17434	.18076	.18698	.19297	.19872
23	21	21	.20239	.21404	.22570	.23734	.24892	.26039	.27173	.28290
23	22	22	.30574	.32848	.35176	.37550	.39963	.42409	.44881	.47372

Variances of Inverse Gaussian Order Statistics
Shape Parameter k

n	i	j	0.8	0.9	1.0	1.1	1.2	1.3	1.4	1.5
23	23	23	.68491	.75535	.82949	.90717	.98824	1.07256	1.15997	1.25032
24	1	1	.07720	.06563	.05574	.04733	.04018	.03412	.02899	.02465
24	2	2	.05830	.05125	.04497	.03941	.03450	.03018	.02638	.02305
24	3	3	.05201	.04663	.04173	.03728	.03325	.02962	.02635	.02343
24	4	4	.04931	.04486	.04073	.03691	.03339	.03016	.02720	.02451
24	5	5	.04821	.04439	.04078	.03739	.03422	.03127	.02853	.02599
24	6	6	.04804	.04469	.04148	.03843	.03553	.03279	.03022	.02780
24	7	7	.04850	.04553	.04266	.03988	.03722	.03466	.03223	.02992
24	8	8	.04945	.04683	.04425	.04172	.03926	.03687	.03457	.03236
24	9	9	.05084	.04853	.04622	.04393	.04167	.03945	.03728	.03516
24	10	10	.05266	.05064	.04860	.04654	.04448	.04243	.04039	.03839
24	11	11	.05491	.05320	.05143	.04961	.04776	.04588	.04400	.04212
24	12	12	.05766	.05626	.05478	.05321	.05159	.04992	.04821	.04647
24	13	13	.06098	.05992	.05875	.05747	.05611	.05466	.05316	.05160
24	14	14	.06500	.06431	.06349	.06254	.06148	.06032	.05906	.05772
24	15	15	.06989	.06964	.06923	.06867	.06798	.06715	.06620	.06515
24	16	16	.07592	.07619	.07628	.07620	.07596	.07557	.07502	.07435
24	17	17	.08351	.08442	.08513	.08566	.08601	.08618	.08618	.08601
24	18	18	.09330	.09504	.09658	.09792	.09906	.09999	.10073	.10128
24	19	19	.10642	.10928	.11195	.11441	.11666	.11870	.12051	.12211
24	20	20	.12490	.12940	.13372	.13786	.14178	.14548	.14895	.15219
24	21	21	.15298	.16009	.16707	.17391	.18058	.18706	.19332	.19935
24	22	22	.20114	.21299	.22489	.23678	.24864	.26041	.27207	.28357
24	23	23	.30412	.32710	.35065	.37470	.39918	.42401	.44914	.47449
24	24	24	.68229	.75307	.82761	.90575	.98737	1.07231	1.16040	1.25150
25	1	1	.07550	.06409	.05436	.04609	.03908	.03314	.02812	.02388
25	2	2	.05673	.04979	.04362	.03817	.03336	.02913	.02543	.02218
25	3	3	.05043	.04513	.04031	.03594	.03200	.02846	.02528	.02244
25	4	4	.04765	.04326	.03920	.03545	.03201	.02886	.02598	.02336
25	5	5	.04644	.04267	.03912	.03580	.03269	.02981	.02714	.02468
25	6	6	.04613	.04282	.03966	.03666	.03382	.03114	.02863	.02629
25	7	7	.04643	.04349	.04064	.03791	.03529	.03279	.03041	.02817
25	8	8	.04718	.04457	.04200	.03950	.03707	.03473	.03248	.03033
25	9	9	.04834	.04601	.04370	.04142	.03918	.03699	.03486	.03280
25	10	10	.04987	.04782	.04576	.04369	.04163	.03960	.03759	.03562
25	11	11	.05178	.05001	.04819	.04635	.04448	.04260	.04072	.03886
25	12	12	.05411	.05262	.05106	.04944	.04777	.04607	.04434	.04260
25	13	13	.05691	.05572	.05443	.05306	.05161	.05011	.04856	.04696
25	14	14	.06026	.05939	.05841	.05732	.05613	.05486	.05351	.05210
25	15	15	.06429	.06379	.06316	.06239	.06151	.06051	.05942	.05823
25	16	16	.06919	.06912	.06889	.06851	.06799	.06734	.06656	.06567
25	17	17	.07521	.07565	.07593	.07603	.07597	.07575	.07538	.07487
25	18	18	.08278	.08386	.08476	.08547	.08600	.08635	.08653	.08654
25	19	19	.09254	.09445	.09617	.09769	.09902	.10015	.10108	.10182
25	20	20	.10560	.10864	.11149	.11415	.11659	.11883	.12085	.12266
25	21	21	.12401	.12869	.13320	.13753	.14166	.14559	.14929	.15276
25	22	22	.15198	.15927	.16645	.17351	.18040	.18712	.19364	.19994
25	23	23	.19995	.21200	.22412	.23625	.24836	.26042	.27238	.28420
25	24	24	.30259	.32579	.34960	.37394	.39874	.42393	.44944	.47522
25	25	25	.67982	.75091	.82582	.90442	.98655	1.07206	1.16081	1.25263

Variances of Inverse Gaussian Order Statistics
Shape Parameter k

n	i	j	1.6	1.7	1.8	1.9	2.0	2.1	2.2	2.3
1	1	1	1.00000	1.00000	1.00000	1.00000	1.00000	1.00000	1.00000	1.00000
2	1	1	.32732	.31244	.29832	.28493	.27223	.26020	.24880	.23799
2	2	2	1.12633	1.15041	1.17384	1.19663	1.21876	1.24024	1.26109	1.28129
3	1	1	.18453	.17180	.15999	.14904	.13890	.12950	.12079	.11273
3	2	2	.37856	.37137	.36409	.35675	.34937	.34198	.33461	.32727
3	3	3	1.18326	1.22160	1.25939	1.29658	1.33315	1.36906	1.40429	1.43883
4	1	1	.12717	.11650	.10676	.09786	.08975	.08235	.07560	.06944
4	2	2	.21182	.20304	.19450	.18623	.17823	.17050	.16305	.15589
4	3	3	.40660	.40479	.40258	.39999	.39706	.39382	.39030	.38652
4	4	4	1.21699	1.26507	1.31284	1.36022	1.40716	1.45359	1.49947	1.54477
5	1	1	.09705	.08788	.07961	.07214	.06541	.05934	.05387	.04893
5	2	2	.14363	.13552	.12780	.12045	.11348	.10688	.10063	.09473
5	3	3	.22878	.22289	.21695	.21097	.20499	.19903	.19310	.18723
5	4	4	.42485	.42709	.42881	.43002	.43074	.43100	.43083	.43025
5	5	5	1.23983	1.29516	1.35042	1.40553	1.46042	1.51500	1.56921	1.62300
6	1	1	.07869	.07063	.06342	.05697	.05121	.04606	.04146	.03735
6	2	2	.10766	.10040	.09358	.08718	.08119	.07559	.07036	.06548
6	3	3	.15492	.14867	.14253	.13653	.13068	.12499	.11947	.11413
6	4	4	.24067	.23707	.23325	.22924	.22506	.22075	.21632	.21179
6	5	5	.43788	.44332	.44819	.45251	.45627	.45951	.46224	.46447
6	6	6	1.25660	1.31760	1.37879	1.44006	1.50132	1.56248	1.62347	1.68420
7	1	1	.06640	.05918	.05277	.04708	.04203	.03755	.03357	.03004
7	2	2	.08579	.07927	.07321	.06759	.06238	.05755	.05310	.04898
7	3	3	.11562	.10969	.10396	.09845	.09317	.08811	.08327	.07865
7	4	4	.16330	.15855	.15375	.14894	.14414	.13936	.13463	.12995
7	5	5	.24956	.24785	.24582	.24350	.24092	.23810	.23506	.23184
7	6	6	.44773	.45579	.46328	.47020	.47655	.48235	.48760	.49231
7	7	7	1.26959	1.33521	1.40125	1.46759	1.53413	1.60077	1.66742	1.73400
8	1	1	.05761	.05105	.04526	.04015	.03564	.03167	.02816	.02506
8	2	2	.07121	.06532	.05988	.05487	.05027	.04605	.04218	.03863
8	3	3	.09162	.08612	.08089	.07591	.07119	.06671	.06249	.05850
8	4	4	.12183	.11697	.11217	.10746	.10285	.09835	.09397	.08971
8	5	5	.16982	.16633	.16269	.15895	.15511	.15120	.14724	.14325
8	6	6	.25651	.25639	.25589	.25505	.25388	.25242	.25066	.24865
8	7	7	.45551	.46576	.47547	.48461	.49320	.50122	.50869	.51561
8	8	8	1.28005	1.34954	1.41965	1.49027	1.56129	1.63260	1.70409	1.77569
9	1	1	.05103	.04499	.03970	.03505	.03097	.02739	.02424	.02148
9	2	2	.06087	.05548	.05055	.04604	.04192	.03817	.03475	.03165
9	3	3	.07560	.07053	.06575	.06124	.05700	.05303	.04931	.04583
9	4	4	.09637	.09170	.08715	.08275	.07849	.07439	.07045	.06667
9	5	5	.12682	.12287	.11889	.11491	.11093	.10698	.10308	.09922
9	6	6	.17506	.17265	.17004	.16724	.16429	.16120	.15799	.15468
9	7	7	.26213	.26336	.26420	.26466	.26476	.26451	.26394	.26305
9	8	8	.46183	.47396	.48558	.49666	.50720	.51719	.52662	.53551
9	9	9	1.28872	1.36151	1.43512	1.50943	1.58432	1.65967	1.73538	1.81134
10	1	1	.04592	.04031	.03542	.03114	.02741	.02414	.02129	.01879
10	2	2	.05318	.04821	.04369	.03959	.03586	.03248	.02943	.02666
10	3	3	.06423	.05954	.05515	.05105	.04722	.04365	.04034	.03727
10	4	4	.07933	.07492	.07067	.06660	.06271	.05899	.05546	.05210
10	5	5	.10030	.09632	.09239	.08851	.08471	.08098	.07735	.07381
10	6	6	.13092	.12777	.12452	.12120	.11782	.11441	.11098	.10754
10	7	7	.17939	.17792	.17621	.17427	.17213	.16980	.16730	.16465
10	8	8	.26677	.26920	.27121	.27283	.27406	.27492	.27543	.27559
10	9	9	.46710	.48086	.49415	.50693	.51920	.53094	.54215	.55281
10	10	10	1.29606	1.37172	1.44839	1.52592	1.60420	1.68311	1.76254	1.84237

Variances of Inverse Gaussian Order Statistics
Shape Parameter k

n	i	j	1.6	1.7	1.8	1.9	2.0	2.1	2.2	2.3
11	1	1	.04183	.03659	.03203	.02806	.02461	.02161	.01899	.01671
11	2	2	.04724	.04263	.03846	.03469	.03128	.02821	.02545	.02296
11	3	3	.05578	.05142	.04737	.04361	.04012	.03690	.03392	.03117
11	4	4	.06721	.06306	.05911	.05535	.05179	.04842	.04524	.04224
11	5	5	.08250	.07864	.07487	.07120	.06764	.06420	.06087	.05767
11	6	6	.10361	.10024	.09686	.09347	.09009	.08674	.08342	.08015
11	7	7	.13438	.13192	.12933	.12662	.12380	.12090	.11793	.11491
11	8	8	.18302	.18240	.18150	.18034	.17894	.17731	.17548	.17347
11	9	9	.27069	.27417	.27723	.27989	.28215	.28402	.28552	.28665
11	10	10	.47157	.48678	.50154	.51584	.52966	.54297	.55577	.56804
11	11	11	1.30239	1.38058	1.45994	1.54034	1.62163	1.70371	1.78645	1.86974
12	1	1	.03849	.03356	.02928	.02557	.02236	.01957	.01715	.01504
12	2	2	.04254	.03823	.03435	.03086	.02772	.02490	.02238	.02011
12	3	3	.04927	.04520	.04144	.03797	.03477	.03183	.02913	.02665
12	4	4	.05819	.05430	.05062	.04714	.04387	.04079	.03790	.03520
12	5	5	.06980	.06612	.06255	.05911	.05580	.05263	.04959	.04670
12	6	6	.08521	.08184	.07851	.07521	.07197	.06879	.06569	.06267
12	7	7	.10643	.10361	.10073	.09779	.09482	.09182	.08883	.08584
12	8	8	.13732	.13550	.13350	.13135	.12905	.12664	.12411	.12150
12	9	9	.18613	.18625	.18608	.18563	.18492	.18395	.18276	.18134
12	10	10	.27405	.27846	.28247	.28606	.28926	.29206	.29447	.29651
12	11	11	.47543	.49192	.50801	.52367	.53888	.55362	.56787	.58161
12	12	12	1.30793	1.38838	1.47015	1.55310	1.63710	1.72202	1.80774	1.89415
13	1	1	.03570	.03104	.02701	.02352	.02051	.01790	.01565	.01369
13	2	2	.03872	.03467	.03104	.02779	.02487	.02227	.01995	.01787
13	3	3	.04412	.04031	.03680	.03357	.03062	.02791	.02544	.02318
13	4	4	.05125	.04759	.04414	.04091	.03789	.03507	.03243	.02998
13	5	5	.06035	.05684	.05348	.05026	.04719	.04426	.04149	.03886
13	6	6	.07207	.06878	.06557	.06242	.05936	.05639	.05352	.05074
13	7	7	.08756	.08464	.08169	.07875	.07581	.07290	.07003	.06720
13	8	8	.10887	.10655	.10412	.10160	.09901	.09637	.09369	.09098
13	9	9	.13987	.13862	.13716	.13553	.13372	.13176	.12967	.12745
13	10	10	.18882	.18962	.19012	.19032	.19024	.18989	.18928	.18844
13	11	11	.27697	.28223	.28708	.29153	.29559	.29924	.30250	.30538
13	12	12	.47880	.49644	.51372	.53062	.54710	.56314	.57871	.59381
13	13	13	1.31284	1.39532	1.47926	1.56452	1.65096	1.73846	1.82689	1.91614
14	1	1	.03335	.02891	.02509	.02180	.01896	.01651	.01440	.01257
14	2	2	.03556	.03174	.02832	.02527	.02256	.02013	.01798	.01606
14	3	3	.03995	.03636	.03307	.03006	.02731	.02481	.02253	.02046
14	4	4	.04575	.04230	.03907	.03605	.03325	.03064	.02822	.02598
14	5	5	.05306	.04973	.04656	.04355	.04069	.03798	.03543	.03303
14	6	6	.06226	.05909	.05602	.05304	.05017	.04740	.04474	.04220
14	7	7	.07405	.07113	.06824	.06537	.06255	.05978	.05707	.05443
14	8	8	.08962	.08709	.08451	.08189	.07925	.07660	.07396	.07132
14	9	9	.11101	.10913	.10712	.10500	.10278	.10047	.09809	.09566
14	10	10	.14210	.14136	.14041	.13925	.13790	.13638	.13469	.13286
14	11	11	.19118	.19260	.19370	.19450	.19501	.19523	.19518	.19487
14	12	12	.27954	.28556	.29119	.29643	.30127	.30571	.30976	.31342
14	13	13	.48178	.50046	.51883	.53685	.55449	.57172	.58852	.60487
14	14	14	1.31723	1.40155	1.48746	1.57482	1.66349	1.75334	1.84425	1.93610
15	1	1	.03133	.02710	.02346	.02033	.01765	.01533	.01334	.01163
15	2	2	.03291	.02928	.02606	.02318	.02063	.01836	.01635	.01457
15	3	3	.03651	.03311	.03001	.02719	.02462	.02229	.02018	.01827
15	4	4	.04130	.03804	.03500	.03217	.02956	.02714	.02490	.02284
15	5	5	.04728	.04412	.04113	.03830	.03563	.03312	.03077	.02856

Variances of Inverse Gaussian Order Statistics
Shape Parameter k

n	i	j	1.6	1.7	1.8	1.9	2.0	2.1	2.2	2.3
15	6	6	.05469	.05165	.04873	.04591	.04321	.04063	.03817	.03583
15	7	7	.06396	.06110	.05829	.05554	.05286	.05024	.04771	.04527
15	8	8	.07581	.07322	.07062	.06802	.06543	.06286	.06032	.05783
15	9	9	.09144	.08928	.08703	.08472	.08236	.07997	.07754	.07511
15	10	10	.11289	.11143	.10981	.10805	.10618	.10419	.10211	.09995
15	11	11	.14407	.14381	.14331	.14260	.14168	.14056	.13927	.13781
15	12	12	.19327	.19525	.19691	.19826	.19931	.20007	.20055	.20076
15	13	13	.28182	.28854	.29488	.30083	.30640	.31158	.31636	.32076
15	14	14	.48444	.50407	.52344	.54249	.56119	.57952	.59745	.61495
15	15	15	1.32119	1.40720	1.49491	1.58419	1.67491	1.76692	1.86011	1.95434
16	1	1	.02957	.02552	.02205	.01907	.01652	.01433	.01244	.01082
16	2	2	.03064	.02720	.02413	.02142	.01901	.01688	.01500	.01333
16	3	3	.03362	.03040	.02747	.02481	.02240	.02022	.01825	.01647
16	4	4	.03764	.03454	.03166	.02901	.02656	.02430	.02222	.02032
16	5	5	.04261	.03961	.03677	.03411	.03161	.02927	.02709	.02505
16	6	6	.04868	.04578	.04300	.04034	.03781	.03540	.03311	.03095
16	7	7	.05615	.05338	.05068	.04806	.04551	.04306	.04070	.03843
16	8	8	.06547	.06290	.06033	.05780	.05530	.05285	.05045	.04811
16	9	9	.07738	.07509	.07277	.07041	.06804	.06567	.06331	.06096
16	10	10	.09306	.09123	.08930	.08728	.08519	.08304	.08084	.07860
16	11	11	.11457	.11349	.11223	.11082	.10927	.10759	.10579	.10390
16	12	12	.14583	.14600	.14593	.14563	.14512	.14439	.14347	.14237
16	13	13	.19514	.19763	.19981	.20168	.20324	.20450	.20547	.20616
16	14	14	.28386	.29122	.29821	.30484	.31108	.31694	.32241	.32750
16	15	15	.48683	.50734	.52761	.54761	.56731	.58666	.60564	.62421
16	16	16	1.32480	1.41234	1.50172	1.59277	1.68537	1.77938	1.87467	1.97112
17	1	1	.02804	.02415	.02082	.01798	.01554	.01345	.01166	.01013
17	2	2	.02870	.02540	.02249	.01991	.01763	.01562	.01384	.01228
17	3	3	.03117	.02810	.02532	.02281	.02053	.01849	.01664	.01498
17	4	4	.03456	.03162	.02889	.02639	.02408	.02197	.02003	.01826
17	5	5	.03875	.03589	.03321	.03069	.02835	.02616	.02412	.02224
17	6	6	.04382	.04105	.03840	.03588	.03350	.03124	.02911	.02711
17	7	7	.04996	.04729	.04470	.04220	.03980	.03749	.03528	.03317
17	8	8	.05747	.05495	.05245	.05001	.04762	.04529	.04303	.04084
17	9	9	.06684	.06452	.06219	.05986	.05754	.05524	.05297	.05074
17	10	10	.07878	.07678	.07471	.07259	.07044	.06826	.06606	.06386
17	11	11	.09451	.09299	.09136	.08961	.08778	.08586	.08388	.08184
17	12	12	.11608	.11534	.11443	.11334	.11209	.11071	.10919	.10755
17	13	13	.14740	.14798	.14831	.14840	.14826	.14790	.14734	.14658
17	14	14	.19682	.19979	.20244	.20479	.20683	.20856	.21000	.21116
17	15	15	.28570	.29365	.30125	.30849	.31537	.32186	.32798	.33372
17	16	16	.48900	.51031	.53143	.55231	.57292	.59322	.61318	.63276
17	17	17	1.32810	1.41707	1.50798	1.60067	1.69502	1.79088	1.88812	1.98663
18	1	1	.02668	.02294	.01974	.01701	.01468	.01269	.01098	.00952
18	2	2	.02700	.02385	.02106	.01861	.01644	.01454	.01286	.01138
18	3	3	.02906	.02613	.02349	.02110	.01895	.01702	.01528	.01373
18	4	4	.03196	.02914	.02656	.02418	.02201	.02002	.01820	.01655
18	5	5	.03553	.03280	.03024	.02786	.02565	.02359	.02169	.01993
18	6	6	.03981	.03716	.03463	.03225	.03000	.02788	.02589	.02403
18	7	7	.04494	.04237	.03989	.03751	.03524	.03306	.03100	.02903
18	8	8	.05112	.04866	.04625	.04390	.04163	.03942	.03729	.03525
18	9	9	.05867	.05636	.05407	.05180	.04955	.04735	.04520	.04309
18	10	10	.06807	.06599	.06388	.06174	.05960	.05745	.05532	.05320
18	11	11	.08005	.07831	.07648	.07459	.07264	.07065	.06862	.06657
18	12	12	.09582	.09459	.09323	.09175	.09015	.08847	.08669	.08485

Variances of Inverse Gaussian Order Statistics
Shape Parameter k

n	i	j	1.6	1.7	1.8	1.9	2.0	2.1	2.2	2.3
18	13	13	.11744	.11703	.11643	.11564	.11469	.11359	.11233	.11095
18	14	14	.14883	.14978	.15048	.15093	.15115	.15114	.15092	.15049
18	15	15	.19833	.20175	.20485	.20764	.21013	.21231	.21419	.21579
18	16	16	.28737	.29586	.30403	.31185	.31931	.32641	.33313	.33948
18	17	17	.49097	.51303	.53493	.55663	.57810	.59928	.62015	.64068
18	18	18	1.33114	1.42143	1.51376	1.60798	1.70395	1.80154	1.90060	2.00103
19	1	1	.02547	.02186	.01878	.01616	.01392	.01202	.01038	.00899
19	2	2	.02551	.02248	.01982	.01747	.01541	.01359	.01200	.01060
19	3	3	.02723	.02443	.02190	.01963	.01759	.01576	.01412	.01265
19	4	4	.02972	.02703	.02456	.02231	.02025	.01837	.01666	.01511
19	5	5	.03279	.03018	.02774	.02548	.02339	.02146	.01967	.01802
19	6	6	.03645	.03391	.03150	.02924	.02711	.02511	.02325	.02151
19	7	7	.04080	.03832	.03595	.03368	.03153	.02948	.02754	.02571
19	8	8	.04596	.04358	.04126	.03901	.03684	.03476	.03275	.03084
19	9	9	.05218	.04991	.04767	.04547	.04332	.04122	.03917	.03719
19	10	10	.05976	.05766	.05555	.05344	.05134	.04926	.04721	.04520
19	11	11	.06919	.06733	.06542	.06347	.06150	.05950	.05750	.05550
19	12	12	.08120	.07970	.07811	.07643	.07468	.07286	.07100	.06910
19	13	13	.09701	.09605	.09495	.09371	.09235	.09088	.08932	.08766
19	14	14	.11867	.11857	.11826	.11777	.11710	.11625	.11526	.11412
19	15	15	.15012	.15142	.15247	.15327	.15382	.15414	.15424	.15413
19	16	16	.19972	.20354	.20706	.21027	.21318	.21579	.21809	.22010
19	17	17	.28889	.29790	.30659	.31495	.32296	.33062	.33792	.34485
19	18	18	.49278	.51553	.53816	.56063	.58290	.60491	.62664	.64806
19	19	19	1.33395	1.42547	1.51913	1.61478	1.71227	1.81146	1.91224	2.01446
20	1	1	.02439	.02090	.01793	.01540	.01325	.01142	.00985	.00852
20	2	2	.02419	.02128	.01872	.01647	.01450	.01277	.01125	.00993
20	3	3	.02563	.02294	.02052	.01835	.01641	.01467	.01312	.01173
20	4	4	.02778	.02520	.02284	.02069	.01874	.01696	.01534	.01388
20	5	5	.03044	.02794	.02561	.02346	.02148	.01965	.01796	.01642
20	6	6	.03360	.03116	.02886	.02671	.02469	.02280	.02105	.01941
20	7	7	.03732	.03494	.03267	.03051	.02847	.02653	.02471	.02299
20	8	8	.04171	.03940	.03717	.03501	.03295	.03097	.02909	.02729
20	9	9	.04691	.04469	.04252	.04040	.03833	.03633	.03440	.03254
20	10	10	.05315	.05106	.04898	.04692	.04489	.04289	.04093	.03902
20	11	11	.06075	.05885	.05691	.05496	.05300	.05105	.04910	.04718
20	12	12	.07021	.06856	.06685	.06507	.06326	.06141	.05954	.05766
20	13	13	.08225	.08098	.07960	.07813	.07656	.07493	.07322	.07147
20	14	14	.09809	.09739	.09653	.09553	.09439	.09313	.09176	.09030
20	15	15	.11980	.11998	.11995	.11973	.11932	.11874	.11799	.11708
20	16	16	.15131	.15293	.15430	.15543	.15630	.15694	.15735	.15753
20	17	17	.20098	.20519	.20910	.21271	.21601	.21902	.22172	.22413
20	18	18	.29028	.29977	.30895	.31782	.32636	.33455	.34239	.34986
20	19	19	.49445	.51785	.54116	.56435	.58736	.61016	.63270	.65495
20	20	20	1.33656	1.42924	1.52414	1.62112	1.72003	1.82074	1.92312	2.02703
21	1	1	.02341	.02003	.01716	.01472	.01265	.01088	.00938	.00810
21	2	2	.02301	.02020	.01774	.01558	.01370	.01204	.01060	.00933
21	3	3	.02422	.02163	.01931	.01723	.01537	.01372	.01224	.01093
21	4	4	.02608	.02360	.02134	.01929	.01743	.01574	.01421	.01283
21	5	5	.02840	.02600	.02377	.02172	.01983	.01810	.01651	.01505
21	6	6	.03116	.02881	.02661	.02455	.02263	.02085	.01919	.01766
21	7	7	.03438	.03208	.02991	.02785	.02590	.02407	.02235	.02073
21	8	8	.03814	.03590	.03376	.03170	.02973	.02786	.02608	.02439
21	9	9	.04255	.04039	.03829	.03625	.03428	.03237	.03054	.02879
21	10	10	.04777	.04572	.04368	.04168	.03972	.03781	.03595	.03414

Variances of Inverse Gaussian Order Statistics
Shape Parameter k

n	i	j	1.6	1.7	1.8	1.9	2.0	2.1	2.2	2.3
21	11	11	.05404	.05212	.05019	.04827	.04635	.04445	.04258	.04074
21	12	12	.06167	.05994	.05817	.05637	.05455	.05271	.05088	.04904
21	13	13	.07114	.06969	.06816	.06656	.06490	.06320	.06146	.05969
21	14	14	.08322	.08216	.08099	.07970	.07832	.07685	.07531	.07370
21	15	15	.09909	.09862	.09799	.09721	.09629	.09523	.09406	.09277
21	16	16	.12084	.12128	.12152	.12155	.12140	.12106	.12054	.11987
21	17	17	.15239	.15433	.15600	.15743	.15861	.15955	.16025	.16073
21	18	18	.20214	.20671	.21099	.21497	.21865	.22204	.22512	.22791
21	19	19	.29157	.30150	.31115	.32049	.32952	.33821	.34657	.35456
21	20	20	.49599	.52000	.54395	.56781	.59153	.61506	.63837	.66142
21	21	21	1.33900	1.43276	1.52883	1.62706	1.72731	1.82945	1.93333	2.03884
22	1	1	.02253	.01925	.01646	.01410	.01210	.01040	.00895	.00772
22	2	2	.02196	.01924	.01687	.01479	.01298	.01140	.01001	.00880
22	3	3	.02296	.02046	.01823	.01624	.01446	.01288	.01148	.01023
22	4	4	.02458	.02220	.02003	.01806	.01628	.01468	.01322	.01191
22	5	5	.02662	.02431	.02218	.02021	.01841	.01676	.01526	.01388
22	6	6	.02904	.02678	.02467	.02270	.02087	.01918	.01761	.01616
22	7	7	.03185	.02964	.02755	.02558	.02372	.02198	.02035	.01883
22	8	8	.03511	.03295	.03088	.02891	.02704	.02525	.02357	.02198
22	9	9	.03889	.03680	.03477	.03281	.03092	.02911	.02737	.02572
22	10	10	.04333	.04131	.03934	.03740	.03552	.03368	.03191	.03020
22	11	11	.04857	.04667	.04477	.04288	.04102	.03919	.03740	.03565
22	12	12	.05485	.05310	.05131	.04952	.04772	.04592	.04414	.04237
22	13	13	.06250	.06095	.05934	.05768	.05599	.05428	.05254	.05080
22	14	14	.07201	.07074	.06938	.06794	.06644	.06487	.06326	.06160
22	15	15	.08410	.08325	.08227	.08117	.07996	.07866	.07727	.07580
22	16	16	.10000	.09976	.09935	.09878	.09806	.09720	.09621	.09510
22	17	17	.12179	.12249	.12297	.12325	.12334	.12323	.12294	.12249
22	18	18	.15340	.15562	.15758	.15930	.16077	.16199	.16298	.16374
22	19	19	.20322	.20812	.21275	.21708	.22112	.22487	.22831	.23146
22	20	20	.29276	.30311	.31319	.32299	.33248	.34165	.35049	.35898
22	21	21	.49742	.52200	.54656	.57106	.59544	.61967	.64370	.66750
22	22	22	1.34129	1.43606	1.53323	1.63265	1.73416	1.83764	1.94295	2.04996
23	1	1	.02172	.01853	.01583	.01355	.01161	.00997	.00857	.00738
23	2	2	.02100	.01838	.01609	.01409	.01234	.01082	.00949	.00833
23	3	3	.02183	.01942	.01727	.01536	.01365	.01214	.01080	.00961
23	4	4	.02325	.02095	.01887	.01698	.01528	.01374	.01236	.01112
23	5	5	.02505	.02282	.02077	.01889	.01717	.01560	.01417	.01286
23	6	6	.02718	.02501	.02298	.02110	.01935	.01774	.01625	.01488
23	7	7	.02966	.02752	.02552	.02363	.02186	.02020	.01866	.01722
23	8	8	.03250	.03042	.02843	.02654	.02475	.02305	.02146	.01995
23	9	9	.03579	.03376	.03180	.02991	.02810	.02637	.02473	.02316
23	10	10	.03960	.03763	.03571	.03384	.03203	.03028	.02859	.02697
23	11	11	.04405	.04217	.04031	.03848	.03668	.03492	.03321	.03154
23	12	12	.04931	.04755	.04577	.04400	.04224	.04050	.03878	.03709
23	13	13	.05561	.05400	.05236	.05069	.04900	.04730	.04560	.04392
23	14	14	.06328	.06189	.06043	.05891	.05734	.05574	.05412	.05247
23	15	15	.07280	.07171	.07052	.06923	.06787	.06644	.06495	.06342
23	16	16	.08492	.08426	.08346	.08254	.08150	.08036	.07912	.07779
23	17	17	.10085	.10082	.10061	.10025	.09972	.09905	.09824	.09730
23	18	18	.12267	.12360	.12433	.12484	.12515	.12527	.12521	.12496
23	19	19	.15432	.15682	.15906	.16105	.16279	.16429	.16555	.16657
23	20	20	.20421	.20944	.21439	.21906	.22344	.22753	.23132	.23482
23	21	21	.29387	.30461	.31511	.32533	.33526	.34488	.35418	.36315
23	22	22	.49876	.52387	.54901	.57410	.59912	.62400	.64872	.67323

Variances of Inverse Gaussian Order Statistics
Shape Parameter k

n	i	j	1.6	1.7	1.8	1.9	2.0	2.1	2.2	2.3
23	23	23	1.34343	1.43917	1.53738	1.63791	1.74062	1.84537	1.95203	2.06046
24	1	1	.02098	.01788	.01526	.01304	.01116	.00957	.00822	.00707
24	2	2	.02014	.01760	.01538	.01345	.01176	.01030	.00902	.00791
24	3	3	.02082	.01849	.01641	.01457	.01293	.01148	.01019	.00906
24	4	4	.02206	.01984	.01783	.01602	.01439	.01292	.01160	.01041
24	5	5	.02366	.02151	.01953	.01773	.01608	.01458	.01322	.01197
24	6	6	.02555	.02345	.02150	.01969	.01802	.01649	.01507	.01377
24	7	7	.02774	.02568	.02375	.02194	.02024	.01867	.01720	.01584
24	8	8	.03025	.02823	.02631	.02450	.02278	.02117	.01965	.01823
24	9	9	.03312	.03115	.02926	.02744	.02571	.02406	.02250	.02102
24	10	10	.03643	.03451	.03265	.03085	.02911	.02743	.02583	.02429
24	11	11	.04025	.03841	.03659	.03481	.03308	.03139	.02975	.02817
24	12	12	.04472	.04297	.04122	.03948	.03777	.03608	.03443	.03282
24	13	13	.05000	.04837	.04671	.04505	.04339	.04173	.04008	.03845
24	14	14	.05631	.05485	.05333	.05178	.05020	.04860	.04699	.04538
24	15	15	.06400	.06276	.06144	.06006	.05862	.05713	.05560	.05405
24	16	16	.07354	.07261	.07158	.07045	.06922	.06793	.06656	.06513
24	17	17	.08568	.08520	.08458	.08382	.08295	.08196	.08086	.07967
24	18	18	.10163	.10180	.10180	.10162	.10128	.10079	.10016	.09939
24	19	19	.12349	.12465	.12559	.12633	.12686	.12720	.12734	.12730
24	20	20	.15519	.15793	.16043	.16269	.16469	.16645	.16797	.16925
24	21	21	.20511	.21067	.21593	.22091	.22562	.23003	.23415	.23798
24	22	22	.29490	.30602	.31690	.32753	.33787	.34793	.35767	.36709
24	23	23	.50001	.52563	.55130	.57697	.60258	.62809	.65346	.67865
24	24	24	1.34546	1.44210	1.54129	1.64288	1.74672	1.85268	1.96062	2.07041
25	1	1	.02030	.01728	.01473	.01257	.01075	.00921	.00790	.00679
25	2	2	.01935	.01689	.01474	.01287	.01124	.00983	.00860	.00753
25	3	3	.01990	.01764	.01564	.01386	.01228	.01089	.00965	.00856
25	4	4	.02099	.01884	.01691	.01516	.01359	.01218	.01092	.00979
25	5	5	.02241	.02033	.01843	.01670	.01512	.01368	.01238	.01119
25	6	6	.02410	.02207	.02019	.01846	.01686	.01539	.01404	.01280
25	7	7	.02605	.02406	.02220	.02046	.01884	.01733	.01593	.01464
25	8	8	.02827	.02632	.02447	.02273	.02109	.01955	.01810	.01675
25	9	9	.03081	.02890	.02707	.02532	.02366	.02209	.02060	.01919
25	10	10	.03370	.03184	.03003	.02829	.02662	.02502	.02349	.02203
25	11	11	.03702	.03522	.03345	.03173	.03005	.02843	.02687	.02537
25	12	12	.04086	.03913	.03742	.03572	.03406	.03244	.03085	.02931
25	13	13	.04535	.04371	.04207	.04043	.03880	.03719	.03559	.03403
25	14	14	.05064	.04913	.04760	.04604	.04446	.04289	.04131	.03975
25	15	15	.05697	.05564	.05425	.05281	.05134	.04983	.04831	.04677
25	16	16	.06467	.06357	.06239	.06114	.05982	.05844	.05701	.05555
25	17	17	.07422	.07345	.07257	.07158	.07050	.06933	.06808	.06677
25	18	18	.08639	.08608	.08562	.08503	.08431	.08347	.08251	.08146
25	19	19	.10236	.10272	.10290	.10291	.10275	.10244	.10197	.10136
25	20	20	.12425	.12562	.12678	.12773	.12847	.12901	.12936	.12952
25	21	21	.15599	.15898	.16173	.16423	.16648	.16850	.17027	.17180
25	22	22	.20600	.21182	.21737	.22266	.22767	.23240	.23684	.24099
25	23	23	.29586	.30734	.31859	.32960	.34034	.35081	.36097	.37082
25	24	24	.50118	.52729	.55347	.57968	.60586	.63197	.65796	.68379
25	25	25	1.34737	1.44488	1.54500	1.64759	1.75251	1.85962	1.96877	2.07985

Variances of Inverse Gaussian Order Statistics
Shape Parameter k

n	i	j	2.4	2.5
1	1	1	1.00000	1.00000
2	1	1	.22775	.21805
2	2	2	1.30087	1.31984
3	1	1	.10526	.09834
3	2	2	.31998	.31277
3	3	3	1.47265	1.50576
4	1	1	.06383	.05870
4	2	2	.14900	.14239
4	3	3	.38251	.37831
4	4	4	1.58944	1.63346
5	1	1	.04448	.04046
5	2	2	.08916	.08390
5	3	3	.18144	.17572
5	4	4	.42928	.42796
5	5	5	1.67631	1.72909
6	1	1	.03367	.03037
6	2	2	.06093	.05670
6	3	3	.10897	.10399
6	4	4	.20720	.20256
6	5	5	.46623	.46754
6	6	6	1.74462	1.80467
7	1	1	.02690	.02412
7	2	2	.04518	.04168
7	3	3	.07426	.07009
7	4	4	.12534	.12081
7	5	5	.22844	.22489
7	6	6	.49650	.50019
7	7	7	1.80043	1.86663
8	1	1	.02233	.01991
8	2	2	.03539	.03242
8	3	3	.05475	.05122
8	4	4	.[illegible]8559	.08161
8	5	5	.13925	.13525
8	6	6	.24640	.24393
8	7	7	.52198	.52782
8	8	8	1.84729	1.91882
9	1	1	.01906	.01692
9	2	2	.02882	.02625
9	3	3	.04258	.03955
9	4	4	.06305	.05960
9	5	5	.09543	.09171
9	6	6	.15129	.14784
9	7	7	.26188	.26044
9	8	8	.54385	.55164
9	9	9	1.88748	1.96369
10	1	1	.01661	.01470
10	2	2	.02416	.02190
10	3	3	.03442	.03178
10	4	4	.04892	.04591
10	5	5	.07039	.06707
10	6	6	.10411	.10070
10	7	7	.16188	.15899
10	8	8	.27543	.27496
10	9	9	.56292	.57250
10	10	10	1.92252	2.00289
11	1	1	.01472	.01298
11	2	2	.02072	.01870
11	3	3	.02864	.02631
11	4	4	.03942	.03677
11	5	5	.05460	.05166
11	6	6	.07694	.07380
11	7	7	.11185	.10876
11	8	8	.17128	.16895
11	9	9	.28744	.28788
11	10	10	.57978	.59099
11	11	11	1.95348	2.03758
12	1	1	.01322	.01163
12	2	2	.01808	.01627
12	3	3	.02437	.02229
12	4	4	.03268	.03032
12	5	5	.04394	.04132
12	6	6	.05974	.05689
12	7	7	.08287	.07992
12	8	8	.11881	.11606
12	9	9	.17973	.17793
12	10	10	.29818	.29949
12	11	11	.59484	.60755
12	12	12	1.98115	2.06863
13	1	1	.01200	.01053
13	2	2	.01601	.01436
13	3	3	.02112	.01925
13	4	4	.02771	.02559
13	5	5	.03637	.03402
13	6	6	.04807	.04551
13	7	7	.06442	.06170
13	8	8	.08825	.08552
13	9	9	.12513	.12272
13	10	10	.18737	.18609
13	11	11	.30788	.31001
13	12	12	.60841	.62250
13	13	13	2.00610	2.09666
14	1	1	.01099	.00962
14	2	2	.01435	.01283
14	3	3	.01857	.01687
14	4	4	.02391	.02200
14	5	5	.03077	.02865
14	6	6	.03977	.03745
14	7	7	.05187	.04939
14	8	8	.06871	.06613
14	9	9	.09318	.09067
14	10	10	.13090	.12883
14	11	11	.19433	.19355
14	12	12	.31670	.31960
14	13	13	.62074	.63612
14	14	14	2.02877	2.12216
15	1	1	.01015	.00887
15	2	2	.01299	.01158
15	3	3	.01653	.01497
15	4	4	.02095	.01920
15	5	5	.02650	.02458
15	6	6	.03361	.03151
15	7	7	.04291	.04064

n	i	j	2.4	2.5
15	8	8	.05538	.05299
15	9	9	.07267	.07024
15	10	10	.09772	.09543
15	11	11	.13620	.13446
15	12	12	.20071	.20041
15	13	13	.32477	.32840
15	14	14	.63201	.64859
15	15	15	2.04951	2.14551
16	1	1	.00943	.00822
16	2	2	.01185	.01055
16	3	3	.01487	.01343
16	4	4	.01857	.01697
16	5	5	.02316	.02140
16	6	6	.02891	.02699
16	7	7	.03626	.03418
16	8	8	.04583	.04362
16	9	9	.05864	.05635
16	10	10	.07634	.07407
16	11	11	.10191	.09985
16	12	12	.14110	.13967
16	13	13	.20659	.20675
16	14	14	.33220	.33652
16	15	15	.64237	.66008
16	16	16	2.06861	2.16703
17	1	1	.00881	.00767
17	2	2	.01090	.00968
17	3	3	.01349	.01215
17	4	4	.01664	.01516
17	5	5	.02049	.01887
17	6	6	.02523	.02347
17	7	7	.03117	.02926
17	8	8	.03873	.03669
17	9	9	.04856	.04642
17	10	10	.06167	.05949
17	11	11	.07976	.07764
17	12	12	.10581	.10397
17	13	13	.14563	.14452
17	14	14	.21203	.21264
17	15	15	.33907	.34404
17	16	16	.65195	.67071
17	17	17	2.08627	2.18694
18	1	1	.00827	.00719
18	2	2	.01008	.00893
18	3	3	.01233	.01108
18	4	4	.01504	.01367
18	5	5	.01831	.01681
18	6	6	.02228	.02066
18	7	7	.02718	.02542
18	8	8	.03328	.03140
18	9	9	.04104	.03906
18	10	10	.05111	.04905
18	11	11	.06451	.06244
18	12	12	.08295	.08099
18	13	13	.10944	.10783
18	14	14	.14986	.14905
18	15	15	.21709	.21813
18	16	16	.34545	.35104
18	17	17	.66084	.68060
18	18	18	2.10269	2.20547

n	i	j	2.4	2.5
19	1	1	.00780	.00677
19	2	2	.00937	.00829
19	3	3	.01134	.01017
19	4	4	.01370	.01242
19	5	5	.01651	.01512
19	6	6	.01988	.01838
19	7	7	.02398	.02235
19	8	8	.02901	.02727
19	9	9	.03528	.03343
19	10	10	.04322	.04130
19	11	11	.05351	.05153
19	12	12	.06717	.06522
19	13	13	.08593	.08414
19	14	14	.11284	.11145
19	15	15	.15381	.15329
19	16	16	.22182	.22326
19	17	17	.35140	.35758
19	18	18	.66913	.68983
19	19	19	2.11801	2.22277
20	1	1	.00738	.00640
20	2	2	.00876	.00774
20	3	3	.01050	.00939
20	4	4	.01256	.01136
20	5	5	.01500	.01370
20	6	6	.01790	.01650
20	7	7	.02138	.01900
20	8	8	.02559	.02397
20	9	9	.03075	.02903
20	10	10	.03716	.03536
20	11	11	.04528	.04341
20	12	12	.05577	.05388
20	13	13	.06968	.06785
20	14	14	.08874	.08711
20	15	15	.11603	.11485
20	16	16	.15751	.15728
20	17	17	.22625	.22808
20	18	18	.35697	.36371
20	19	19	.67689	.69847
20	20	20	2.13235	2.23898
21	1	1	.00701	.00607
21	2	2	.00822	.00725
21	3	3	.00976	.00872
21	4	4	.01158	.01046
21	5	5	.01372	.01250
21	6	6	.01624	.01492
21	7	7	.01922	.01781
21	8	8	.02279	.02129
21	9	9	.02711	.02551
21	10	10	.03239	.03070
21	11	11	.03894	.03719
21	12	12	.04722	.04542
21	13	13	.05790	.05611
21	14	14	.07204	.07033
21	15	15	.09138	.08991
21	16	16	.11904	.11806
21	17	17	.16099	.16104
21	18	18	.23040	.23261
21	19	19	.36220	.36947
21	20	20	.68417	.70659

n	i	j	2.4	2.5
21	21	21	2.14584	2.25421
22	1	1	.00667	.00578
22	2	2	.00775	.00682
22	3	3	.00912	.00813
22	4	4	.01074	.00967
22	5	5	.01262	.01148
22	6	6	.01483	.01360
22	7	7	.01742	.01610
22	8	8	.02048	.01907
22	9	9	.02414	.02264
22	10	10	.02855	.02697
22	11	11	.03395	.03230
22	12	12	.04063	.03893
22	13	13	.04906	.04733
22	14	14	.05992	.05822
22	15	15	.07427	.07269
22	16	16	.09388	.09256
22	17	17	.12187	.12110
22	18	18	.16427	.16458
22	19	19	.23432	.23689
22	20	20	.36712	.37490
22	21	21	.69102	.71424
22	22	22	2.15854	2.26858
23	1	1	.00637	.00551
23	2	2	.00732	.00644
23	3	3	.00855	.00762
23	4	4	.01000	.00899
23	5	5	.01167	.01060
23	6	6	.01362	.01246
23	7	7	.01589	.01465
23	8	8	.01854	.01722
23	9	9	.02168	.02028
23	10	10	.02542	.02394
23	11	11	.02993	.02838
23	12	12	.03543	.03382
23	13	13	.04224	.04059
23	14	14	.05081	.04915
23	15	15	.06184	.06023
23	16	16	.07639	.07492
23	17	17	.09625	.09508
23	18	18	.12455	.12398
23	19	19	.16737	.16794
23	20	20	.23802	.24093
23	21	21	.37177	.38004
23	22	22	.69749	.72148
23	23	23	2.17055	2.28217
24	1	1	.00610	.00527
24	2	2	.00694	.00610
24	3	3	.00805	.00716
24	4	4	.00935	.00840
24	5	5	.01085	.00983
24	6	6	.01258	.01148
24	7	7	.01458	.01341
24	8	8	.01690	.01566
24	9	9	.01962	.01830
24	10	10	.02283	.02143
24	11	11	.02665	.02519
24	12	12	.03125	.02972
24	13	13	.03685	.03527

n	i	j	2.4	2.5
24	14	14	.04377	.04217
24	15	15	.05248	.05089
24	16	16	.06366	.06214
24	17	17	.07840	.07705
24	18	18	.09849	.09748
24	19	19	.12709	.12671
24	20	20	.17031	.17113
24	21	21	.24152	.24477
24	22	22	.37617	.38491
24	23	23	.70362	.72833
24	24	24	2.18192	2.29504
25	1	1	.00585	.00505
25	2	2	.00660	.00579
25	3	3	.00760	.00675
25	4	4	.00877	.00787
25	5	5	.01012	.00915
25	6	6	.01167	.01063
25	7	7	.01344	.01234
25	8	8	.01549	.01432
25	9	9	.01787	.01663
25	10	10	.02065	.01934
25	11	11	.02393	.02255
25	12	12	.02782	.02638
25	13	13	.03250	.03101
25	14	14	.03820	.03667
25	15	15	.04523	.04369
25	16	16	.05406	.05255
25	17	17	.06539	.06397
25	18	18	.08031	.07908
25	19	19	.10062	.09976
25	20	20	.12950	.12932
25	21	21	.17309	.17416
25	22	22	.24485	.24841
25	23	23	.38034	.38953
25	24	24	.70942	.73483
25	25	25	2.19272	2.30727

Table 3. Covariances of order statistics for $n = 2(1)25$, $1 \leq i < j \leq n$ and $k = 0.0(0.1)2.5$

Covariances of Inverse Gaussian Order Statistics
Shape Parameter k

n	i	j	0.0	0.1	0.2	0.3	0.4	0.5	0.6	0.7
2	1	2	.31831	.31809	.31743	.31634	.31483	.31293	.31064	.30801
3	1	2	.27566	.26776	.25963	.25132	.24289	.23439	.22586	.21734
3	1	3	.16487	.16471	.16423	.16343	.16234	.16096	.15930	.15740
3	2	3	.27566	.28330	.29061	.29757	.30413	.31027	.31597	.32119
4	1	2	.24559	.23463	.22370	.21287	.20219	.19171	.18147	.17152
4	1	3	.15801	.15417	.15010	.14584	.14143	.13689	.13225	.12756
4	1	4	.10468	.10456	.10420	.10361	.10279	.10176	.10053	.09911
4	2	3	.23594	.23572	.23503	.23390	.23234	.23037	.22800	.22527
4	2	4	.15801	.16160	.16492	.16793	.17064	.17301	.17505	.17675
4	3	4	.24559	.25653	.26739	.27811	.28863	.29890	.30887	.31850
5	1	2	.22433	.21186	.19963	.18770	.17612	.16494	.15418	.14389
5	1	3	.14815	.14247	.13670	.13087	.12503	.11921	.11344	.10776
5	1	4	.10577	.10344	.10092	.09823	.09541	.09247	.08944	.08634
5	1	5	.07422	.07412	.07384	.07337	.07272	.07191	.07094	.06984
5	2	3	.20844	.20474	.20070	.19634	.19168	.18677	.18164	.17634
5	2	4	.14994	.14978	.14929	.14848	.14736	.14594	.14424	.14229
5	2	5	.10577	.10790	.10981	.11148	.11290	.11407	.11499	.11565
5	3	4	.20844	.21175	.21466	.21716	.21923	.22087	.22207	.22285
5	3	5	.14815	.15369	.15907	.16425	.16919	.17387	.17826	.18235
5	4	5	.22433	.23699	.24977	.26260	.27543	.28818	.30080	.31323
6	1	2	.20850	.19519	.18228	.16983	.15788	.14647	.13561	.12534
6	1	3	.13944	.13270	.12598	.11933	.11277	.10635	.10009	.09402
6	1	4	.10243	.09886	.09518	.09142	.08761	.08377	.07994	.07613
6	1	5	.07736	.07576	.07401	.07212	.07010	.06799	.06580	.06354
6	1	6	.05634	.05626	.05603	.05565	.05512	.05446	.05367	.05276
6	2	3	.18899	.18343	.17765	.17167	.16555	.15931	.15300	.14666
6	2	4	.13966	.13748	.13502	.13231	.12938	.12623	.12291	.11944
6	2	5	.10591	.10578	.10540	.10478	.10393	.10284	.10155	.10005
6	2	6	.07736	.07879	.08004	.08110	.08195	.08261	.08306	.08331
6	3	4	.18327	.18307	.18248	.18150	.18014	.17842	.17635	.17397
6	3	5	.13966	.14156	.14315	.14444	.14540	.14605	.14638	.14640
6	3	6	.10243	.10586	.10913	.11222	.11510	.11775	.12017	.12234
6	4	5	.18899	.19427	.19925	.20390	.20820	.21211	.21563	.21874
6	4	6	.13944	.14615	.15280	.15934	.16575	.17197	.17798	.18374
6	5	6	.20850	.22216	.23611	.25028	.26459	.27900	.29342	.30779
7	1	2	.19620	.18239	.16911	.15641	.14432	.13287	.12209	.11196
7	1	3	.13212	.12472	.11744	.11030	.10336	.09664	.09016	.08396
7	1	4	.09849	.09414	.08976	.08537	.08100	.07668	.07244	.06829
7	1	5	.07656	.07406	.07147	.06878	.06604	.06326	.06046	.05767
7	1	6	.05992	.05874	.05742	.05599	.05446	.05284	.05114	.04940
7	1	7	.04480	.04474	.04454	.04422	.04377	.04321	.04255	.04179
7	2	3	.17448	.16781	.16101	.15414	.14723	.14032	.13346	.12668
7	2	4	.13073	.12732	.12370	.11990	.11596	.11190	.10776	.10355
7	2	5	.10196	.10049	.09881	.09692	.09485	.09261	.09022	.08771
7	2	6	.07998	.07988	.07958	.07908	.07839	.07753	.07649	.07530
7	2	7	.05992	.06096	.06184	.06257	.06313	.06352	.06375	.06381
7	3	4	.16556	.16340	.16090	.15809	.15499	.15163	.14803	.14422
7	3	5	.12960	.12945	.12901	.12826	.12724	.12594	.12439	.12260
7	3	6	.10196	.10319	.10419	.10494	.10545	.10571	.10573	.10551
7	3	7	.07656	.07893	.08115	.08322	.08510	.08680	.08830	.08960
7	4	5	.16556	.16737	.16882	.16990	.17061	.17094	.17091	.17053
7	4	6	.13073	.13392	.13685	.13952	.14191	.14399	.14578	.14726
7	4	7	.09849	.10276	.10694	.11099	.11488	.11860	.12211	.12541
7	5	6	.17448	.18099	.18730	.19336	.19914	.20462	.20976	.21454
7	5	7	.13212	.13958	.14708	.15456	.16199	.16931	.17650	.18351

Covariances of Inverse Gaussian Order Statistics
Shape Parameter k

n	i	j	0.0	0.1	0.2	0.3	0.4	0.5	0.6	0.7
7	6	7	.19620	.21049	.22520	.24027	.25563	.27120	.28693	.30273
8	1	2	.18631	.17219	.15870	.14589	.13379	.12240	.11175	.10182
8	1	3	.12597	.11813	.11049	.10307	.09591	.08905	.08249	.07626
8	1	4	.09472	.08986	.08502	.08023	.07553	.07094	.06647	.06216
8	1	5	.07477	.07168	.06853	.06536	.06217	.05900	.05587	.05279
8	1	6	.06021	.05834	.05638	.05434	.05224	.05010	.04794	.04577
8	1	7	.04830	.04738	.04635	.04521	.04399	.04269	.04133	.03991
8	1	8	.03684	.03678	.03661	.03633	.03595	.03546	.03489	.03424
8	2	3	.16320	.15580	.14837	.14095	.13360	.12634	.11921	.11226
8	2	4	.12326	.11904	.11468	.11021	.10567	.10109	.09649	.09190
8	2	5	.09756	.09522	.09270	.09003	.08723	.08432	.08132	.07826
8	2	6	.07872	.07766	.07641	.07500	.07344	.07173	.06991	.06797
8	2	7	.06325	.06316	.06291	.06250	.06193	.06121	.06035	.05936
8	2	8	.04830	.04909	.04976	.05028	.05067	.05091	.05102	.05099
8	3	4	.15236	.14897	.14532	.14144	.13735	.13308	.12866	.12413
8	3	5	.12096	.11953	.11783	.11589	.11372	.11133	.10876	.10601
8	3	6	.09782	.09770	.09734	.09676	.09594	.09492	.09369	.09227
8	3	7	.07872	.07960	.08028	.08076	.08104	.08113	.08101	.08071
8	3	8	.06021	.06196	.06359	.06507	.06640	.06757	.06858	.06941
8	4	5	.14918	.14900	.14849	.14764	.14647	.14498	.14320	.14114
8	4	6	.12096	.12213	.12301	.12360	.12391	.12393	.12368	.12316
8	4	7	.09756	.09972	.10167	.10339	.10489	.10615	.10716	.10794
8	4	8	.07477	.07777	.08068	.08346	.08610	.08858	.09088	.09299
8	5	6	.15236	.15545	.15824	.16070	.16283	.16460	.16602	.16709
8	5	7	.12326	.12732	.13118	.13483	.13824	.14138	.14426	.14684
8	5	8	.09472	.09958	.10439	.10914	.11378	.11829	.12264	.12681
8	6	7	.16320	.17052	.17773	.18478	.19162	.19824	.20458	.21062
8	6	8	.12597	.13395	.14204	.15020	.15837	.16652	.17460	.18257
8	7	8	.18631	.20101	.21625	.23196	.24808	.26453	.28125	.29816
9	1	2	.17814	.16383	.15024	.13740	.12533	.11405	.10355	.09383
9	1	3	.12075	.11261	.10473	.09714	.08987	.08294	.07637	.07018
9	1	4	.09131	.08608	.08093	.07588	.07097	.06621	.06163	.05725
9	1	5	.07274	.06924	.06572	.06221	.05875	.05534	.05201	.04877
9	1	6	.05948	.05715	.05475	.05231	.04985	.04739	.04495	.04254
9	1	7	.04908	.04761	.04607	.04444	.04277	.04105	.03931	.03755
9	1	8	.04009	.03935	.03851	.03758	.03657	.03550	.03436	.03319
9	1	9	.03106	.03101	.03086	.03061	.03027	.02985	.02935	.02878
9	2	3	.15412	.14622	.13838	.13063	.12302	.11557	.10834	.10134
9	2	4	.11701	.11222	.10736	.10246	.09754	.09263	.08778	.08300
9	2	5	.09345	.09049	.08741	.08421	.08094	.07761	.07425	.07087
9	2	6	.07655	.07482	.07294	.07093	.06881	.06658	.06428	.06192
9	2	7	.06324	.06242	.06145	.06034	.05910	.05774	.05628	.05473
9	2	8	.05171	.05164	.05143	.05108	.05059	.04998	.04925	.04841
9	2	9	.04009	.04073	.04124	.04164	.04192	.04207	.04211	.04202
9	3	4	.14208	.13787	.13347	.12890	.12420	.11940	.11452	.10961
9	3	5	.11377	.11147	.10895	.10623	.10334	.10029	.09712	.09384
9	3	6	.09336	.09233	.09109	.08964	.08800	.08620	.08423	.08213
9	3	7	.07724	.07714	.07685	.07636	.07570	.07485	.07384	.07268
9	3	8	.06324	.06389	.06439	.06471	.06487	.06486	.06469	.06436
9	3	9	.04908	.05044	.05169	.05281	.05380	.05465	.05536	.05592
9	4	5	.13699	.13556	.13382	.13181	.12952	.12698	.12423	.12127
9	4	6	.11267	.11253	.11213	.11146	.11054	.10937	.10797	.10636
9	4	7	.09336	.09417	.09476	.09511	.09523	.09513	.09481	.09427
9	4	8	.07655	.07811	.07950	.08071	.08172	.08254	.08316	.08358
9	4	9	.05948	.06174	.06390	.06595	.06786	.06963	.07124	.07270

Covariances of Inverse Gaussian Order Statistics
Shape Parameter k

n	i	j	0.0	0.1	0.2	0.3	0.4	0.5	0.6	0.7
9	5	6	.13699	.13811	.13892	.13940	.13956	.13940	.13893	.13817
9	5	7	.11377	.11583	.11765	.11921	.12049	.12151	.12224	.12271
9	5	8	.09345	.09625	.09889	.10134	.10358	.10561	.10741	.10897
9	5	9	.07274	.07621	.07962	.08294	.08615	.08923	.09216	.09492
9	6	7	.14208	.14605	.14978	.15322	.15637	.15921	.16172	.16390
9	6	8	.11701	.12168	.12622	.13058	.13476	.13871	.14243	.14589
9	6	9	.09131	.09658	.10187	.10713	.11234	.11747	.12248	.12735
9	7	8	.15412	.16201	.16987	.17764	.18529	.19276	.20003	.20706
9	7	9	.12075	.12910	.13763	.14629	.15504	.16382	.17261	.18134
9	8	9	.17814	.19313	.20875	.22494	.24163	.25876	.27625	.29404
10	1	2	.17126	.15683	.14318	.13036	.11836	.10721	.09688	.08737
10	1	3	.11626	.10791	.09987	.09217	.08484	.07790	.07136	.06523
10	1	4	.08825	.08276	.07738	.07216	.06712	.06227	.05763	.05323
10	1	5	.07074	.06693	.06315	.05941	.05576	.05219	.04874	.04542
10	1	6	.05840	.05572	.05301	.05030	.04760	.04493	.04231	.03976
10	1	7	.04892	.04707	.04517	.04322	.04124	.03926	.03728	.03532
10	1	8	.04108	.03990	.03864	.03731	.03592	.03450	.03305	.03159
10	1	9	.03404	.03343	.03272	.03194	.03109	.03017	.02921	.02821
10	1	10	.02670	.02665	.02652	.02630	.02600	.02563	.02519	.02468
10	2	3	.14662	.13838	.13026	.12230	.11453	.10700	.09973	.09275
10	2	4	.11170	.10651	.10130	.09610	.09093	.08584	.08084	.07596
10	2	5	.08974	.08634	.08285	.07930	.07571	.07211	.06852	.06496
10	2	6	.07420	.07199	.06967	.06724	.06474	.06218	.05957	.05695
10	2	7	.06223	.06089	.05942	.05784	.05616	.05439	.05254	.05065
10	2	8	.05231	.05166	.05088	.04997	.04895	.04784	.04663	.04534
10	2	9	.04337	.04331	.04312	.04282	.04240	.04187	.04123	.04051
10	2	10	.03404	.03456	.03497	.03528	.03549	.03558	.03558	.03547
10	3	4	.13380	.12902	.12410	.11908	.11399	.10885	.10371	.09860
10	3	5	.10774	.10483	.10174	.09850	.09513	.09166	.08811	.08451
10	3	6	.08923	.08754	.08568	.08365	.08147	.07915	.07673	.07421
10	3	7	.07492	.07413	.07317	.07204	.07075	.06932	.06775	.06607
10	3	8	.06303	.06295	.06270	.06230	.06173	.06102	.06017	.05919
10	3	9	.05231	.05282	.05319	.05342	.05351	.05345	.05325	.05293
10	3	10	.04108	.04218	.04317	.04405	.04482	.04546	.04598	.04637
10	4	5	.12751	.12521	.12267	.11988	.11689	.11370	.11036	.10688
10	4	6	.10579	.10476	.10349	.10200	.10028	.09837	.09627	.09401
10	4	7	.08895	.08884	.08851	.08796	.08721	.08626	.08512	.08380
10	4	8	.07492	.07552	.07593	.07615	.07618	.07602	.07568	.07516
10	4	9	.06223	.06342	.06447	.06536	.06608	.06664	.06703	.06726
10	4	10	.04892	.05069	.05237	.05394	.05540	.05673	.05792	.05897
10	5	6	.12560	.12545	.12500	.12426	.12324	.12194	.12038	.11859
10	5	7	.10579	.10656	.10708	.10734	.10735	.10710	.10661	.10588
10	5	8	.08923	.09071	.09199	.09305	.09389	.09451	.09491	.09508
10	5	9	.07420	.07627	.07820	.07996	.08155	.08295	.08416	.08518
10	5	10	.05840	.06103	.06359	.06607	.06844	.07068	.07279	.07475
10	6	7	.12751	.12953	.13127	.13271	.13385	.13468	.13521	.13544
10	6	8	.10774	.11047	.11297	.11525	.11728	.11906	.12058	.12183
10	6	9	.08974	.09303	.09619	.09919	.10201	.10465	.10707	.10929
10	6	10	.07074	.07455	.07834	.08208	.08574	.08930	.09274	.09603
10	7	8	.13380	.13841	.14282	.14700	.15093	.15458	.15794	.16100
10	7	9	.11170	.11683	.12187	.12679	.13156	.13615	.14055	.14471
10	7	10	.08825	.09383	.09948	.10515	.11080	.11642	.12196	.12740
10	8	9	.14662	.15494	.16328	.17160	.17986	.18801	.19602	.20383
10	8	10	.11626	.12488	.13374	.14280	.15199	.16129	.17064	.18000
10	9	10	.17126	.18645	.20235	.21890	.23604	.25371	.27183	.29034

Covariances of Inverse Gaussian Order Statistics
Shape Parameter k

n	i	j	0.0	0.1	0.2	0.3	0.4	0.5	0.6	0.7
11	1	2	.16536	.15086	.13720	.12441	.11251	.10148	.09132	.08201
11	1	3	.11236	.10385	.09571	.08795	.08060	.07367	.06718	.06113
11	1	4	.08552	.07982	.07429	.06895	.06382	.05892	.05427	.04987
11	1	5	.06885	.06481	.06083	.05693	.05315	.04948	.04596	.04260
11	1	6	.05720	.05426	.05133	.04842	.04555	.04274	.04001	.03737
11	1	7	.04838	.04624	.04408	.04189	.03971	.03754	.03541	.03332
11	1	8	.04124	.03974	.03817	.03657	.03493	.03328	.03163	.03000
11	1	9	.03511	.03412	.03307	.03195	.03078	.02958	.02835	.02710
11	1	10	.02942	.02890	.02830	.02763	.02689	.02610	.02527	.02440
11	1	11	.02332	.02328	.02316	.02296	.02269	.02235	.02195	.02150
11	2	3	.14031	.13182	.12350	.11540	.10755	.09999	.09274	.08582
11	2	4	.10715	.10166	.09620	.09078	.08546	.08024	.07517	.07025
11	2	5	.08644	.08271	.07893	.07512	.07131	.06753	.06379	.06013
11	2	6	.07192	.06935	.06670	.06397	.06121	.05841	.05561	.05282
11	2	7	.06089	.05916	.05733	.05541	.05341	.05136	.04927	.04715
11	2	8	.05195	.05088	.04969	.04840	.04702	.04557	.04405	.04247
11	2	9	.04425	.04372	.04307	.04232	.04146	.04052	.03949	.03840
11	2	10	.03710	.03705	.03688	.03662	.03624	.03578	.03522	.03458
11	2	11	.02942	.02985	.03020	.03044	.03060	.03065	.03062	.03049
11	3	4	.12697	.12177	.11648	.11114	.10579	.10045	.09516	.08994
11	3	5	.10264	.09927	.09577	.09216	.08846	.08471	.08093	.07714
11	3	6	.08552	.08336	.08104	.07859	.07603	.07337	.07064	.06786
11	3	7	.07247	.07118	.06974	.06814	.06642	.06458	.06265	.06063
11	3	8	.06189	.06127	.06049	.05958	.05852	.05734	.05605	.05466
11	3	9	.05276	.05268	.05247	.05212	.05163	.05102	.05029	.04945
11	3	10	.04425	.04467	.04496	.04512	.04516	.04508	.04488	.04456
11	3	11	.03511	.03601	.03683	.03754	.03815	.03864	.03903	.03931
11	4	5	.11988	.11696	.11384	.11053	.10706	.10346	.09975	.09596
11	4	6	.10003	.09837	.09649	.09442	.09217	.08976	.08721	.08454
11	4	7	.08488	.08410	.08313	.08196	.08061	.07909	.07742	.07562
11	4	8	.07255	.07245	.07218	.07172	.07109	.07029	.06934	.06823
11	4	9	.06189	.06235	.06265	.06279	.06277	.06259	.06225	.06177
11	4	10	.05195	.05290	.05372	.05440	.05494	.05533	.05558	.05569
11	4	11	.04124	.04268	.04403	.04528	.04643	.04746	.04837	.04916
11	5	6	.11674	.11572	.11442	.11286	.11107	.10905	.10682	.10440
11	5	7	.09919	.09907	.09871	.09811	.09728	.09622	.09496	.09351
11	5	8	.08488	.08544	.08580	.08595	.08588	.08560	.08513	.08446
11	5	9	.07247	.07360	.07455	.07531	.07590	.07629	.07650	.07652
11	5	10	.06089	.06249	.06397	.06530	.06648	.06750	.06835	.06905
11	5	11	.04838	.05046	.05247	.05440	.05622	.05794	.05953	.06098
11	6	7	.11674	.11750	.11797	.11816	.11807	.11770	.11706	.11617
11	6	8	.10003	.10148	.10269	.10365	.10437	.10485	.10508	.10506
11	6	9	.08552	.08751	.08932	.09094	.09235	.09354	.09453	.09529
11	6	10	.07192	.07438	.07672	.07892	.08097	.08284	.08454	.08606
11	6	11	.05720	.06012	.06300	.06581	.06855	.07118	.07369	.07608
11	7	8	.11988	.12256	.12499	.12715	.12904	.13064	.13195	.13297
11	7	9	.10264	.10585	.10889	.11172	.11434	.11673	.11887	.12077
11	7	10	.08644	.09010	.09366	.09709	.10038	.10351	.10645	.10920
11	7	11	.06885	.07292	.07701	.08107	.08510	.08905	.09291	.09665
11	8	9	.12697	.13205	.13698	.14172	.14626	.15056	.15460	.15837
11	8	10	.10715	.11263	.11806	.12341	.12866	.13376	.13870	.14346
11	8	11	.08552	.09134	.09726	.10325	.10927	.11529	.12127	.12719
11	9	10	.14031	.14893	.15765	.16640	.17515	.18384	.19245	.20092
11	9	11	.11236	.12119	.13030	.13966	.14922	.15893	.16875	.17863
11	10	11	.16536	.18069	.19680	.21364	.23114	.24924	.26789	.28699

Covariances of Inverse Gaussian Order Statistics
Shape Parameter k

n	i	j	0.0	0.1	0.2	0.3	0.4	0.5	0.6	0.7
12	1	2	.16024	.14569	.13204	.11931	.10750	.09661	.08662	.07749
12	1	3	.10893	.10031	.09210	.08430	.07695	.07006	.06363	.05766
12	1	4	.08307	.07723	.07158	.06615	.06097	.05605	.05140	.04703
12	1	5	.06709	.06287	.05875	.05473	.05086	.04713	.04357	.04019
12	1	6	.05599	.05286	.04975	.04669	.04370	.04080	.03800	.03531
12	1	7	.04766	.04531	.04294	.04059	.03826	.03596	.03373	.03156
12	1	8	.04102	.03927	.03748	.03568	.03386	.03205	.03027	.02852
12	1	9	.03544	.03419	.03287	.03152	.03013	.02873	.02733	.02593
12	1	10	.03050	.02966	.02876	.02780	.02680	.02576	.02469	.02362
12	1	11	.02579	.02535	.02483	.02424	.02360	.02290	.02217	.02140
12	1	12	.02062	.02059	.02048	.02030	.02005	.01975	.01939	.01898
12	2	3	.13490	.12622	.11777	.10959	.10170	.09414	.08692	.08008
12	2	4	.10320	.09748	.09183	.08627	.08083	.07555	.07044	.06552
12	2	5	.08350	.07952	.07551	.07151	.06755	.06365	.05983	.05610
12	2	6	.06979	.06694	.06403	.06109	.05813	.05517	.05224	.04935
12	2	7	.05946	.05743	.05533	.05315	.05093	.04868	.04641	.04415
12	2	8	.05121	.04982	.04833	.04676	.04512	.04342	.04169	.03993
12	2	9	.04427	.04339	.04241	.04133	.04017	.03894	.03766	.03632
12	2	10	.03812	.03767	.03712	.03648	.03574	.03493	.03404	.03310
12	2	11	.03225	.03220	.03206	.03182	.03148	.03107	.03057	.03000
12	2	12	.02579	.02616	.02645	.02665	.02677	.02680	.02675	.02661
12	3	4	.12121	.11569	.11014	.10458	.09905	.09358	.08821	.08295
12	3	5	.09826	.09455	.09074	.08686	.08293	.07899	.07506	.07116
12	3	6	.08222	.07970	.07704	.07429	.07146	.06856	.06562	.06267
12	3	7	.07012	.06844	.06663	.06470	.06267	.06055	.05836	.05612
12	3	8	.06044	.05941	.05824	.05695	.05555	.05404	.05245	.05078
12	3	9	.05228	.05178	.05114	.05037	.04949	.04850	.04741	.04623
12	3	10	.04504	.04497	.04479	.04448	.04405	.04352	.04288	.04214
12	3	11	.03812	.03846	.03869	.03881	.03882	.03872	.03852	.03822
12	3	12	.03050	.03126	.03194	.03253	.03302	.03342	.03372	.03391
12	4	5	.11357	.11020	.10666	.10298	.09918	.09529	.09134	.08736
12	4	6	.09516	.09302	.09069	.08820	.08557	.08282	.07997	.07704
12	4	7	.08124	.07997	.07852	.07690	.07513	.07322	.07119	.06906
12	4	8	.07008	.06947	.06869	.06774	.06665	.06540	.06403	.06254
12	4	9	.06066	.06058	.06035	.05995	.05941	.05873	.05791	.05697
12	4	10	.05228	.05265	.05288	.05297	.05292	.05273	.05241	.05196
12	4	11	.04427	.04505	.04571	.04624	.04666	.04694	.04710	.04714
12	4	12	.03544	.03664	.03775	.03878	.03970	.04053	.04124	.04185
12	5	6	.10962	.10795	.10605	.10393	.10160	.09910	.09643	.09362
12	5	7	.09370	.09292	.09192	.09071	.08930	.08771	.08594	.08401
12	5	8	.08090	.08080	.08049	.07999	.07930	.07842	.07737	.07615
12	5	9	.07008	.07051	.07077	.07085	.07074	.07047	.07002	.06941
12	5	10	.06044	.06132	.06205	.06263	.06305	.06330	.06340	.06334
12	5	11	.05121	.05250	.05367	.05471	.05562	.05639	.05702	.05750
12	5	12	.04102	.04272	.04435	.04590	.04736	.04871	.04995	.05107
12	6	7	.10839	.10826	.10787	.10721	.10630	.10516	.10378	.10219
12	6	8	.09370	.09424	.09456	.09464	.09449	.09412	.09353	.09273
12	6	9	.08124	.08233	.08321	.08390	.08438	.08465	.08472	.08460
12	6	10	.07012	.07165	.07302	.07423	.07526	.07611	.07677	.07726
12	6	11	.05946	.06139	.06320	.06489	.06644	.06784	.06909	.07017
12	6	12	.04766	.04999	.05226	.05448	.05661	.05864	.06056	.06237
12	7	8	.10962	.11104	.11220	.11309	.11372	.11407	.11416	.11399
12	7	9	.09516	.09711	.09885	.10037	.10166	.10271	.10353	.10411
12	7	10	.08222	.08461	.08684	.08889	.09076	.09243	.09390	.09516
12	7	11	.06979	.07255	.07522	.07777	.08019	.08246	.08457	.08651

Covariances of Inverse Gaussian Order Statistics
Shape Parameter k

n	i	j	0.0	0.1	0.2	0.3	0.4	0.5	0.6	0.7
12	7	12	.05599	.05914	.06226	.06536	.06839	.07134	.07420	.07694
12	8	9	.11357	.11675	.11971	.12244	.12491	.12713	.12907	.13074
12	8	10	.09826	.10185	.10530	.10858	.11167	.11455	.11722	.11965
12	8	11	.08350	.08745	.09133	.09512	.09878	.10232	.10569	.10889
12	8	12	.06709	.07137	.07569	.08003	.08435	.08862	.09283	.09696
12	9	10	.12121	.12665	.13198	.13718	.14220	.14703	.15163	.15599
12	9	11	.10320	.10894	.11468	.12038	.12601	.13155	.13695	.14219
12	9	12	.08307	.08908	.09522	.10147	.10779	.11414	.12049	.12681
12	10	11	.13490	.14376	.15277	.16186	.17101	.18015	.18926	.19828
12	10	12	.10893	.11792	.12723	.13684	.14670	.15675	.16696	.17728
12	11	12	.16024	.17567	.19193	.20900	.22679	.24526	.26434	.28395
13	1	2	.15573	.14116	.12753	.11487	.10317	.09241	.08257	.07361
13	1	3	.10589	.09719	.08893	.08112	.07379	.06694	.06058	.05469
13	1	4	.08086	.07491	.06917	.06369	.05848	.05355	.04892	.04458
13	1	5	.06546	.06111	.05687	.05277	.04883	.04507	.04149	.03811
13	1	6	.05482	.05153	.04829	.04512	.04205	.03908	.03623	.03351
13	1	7	.04688	.04436	.04184	.03935	.03691	.03453	.03222	.03000
13	1	8	.04061	.03867	.03671	.03475	.03281	.03089	.02901	.02718
13	1	9	.03542	.03395	.03245	.03092	.02937	.02783	.02631	.02481
13	1	10	.03093	.02986	.02874	.02757	.02638	.02517	.02395	.02274
13	1	11	.02685	.02613	.02535	.02451	.02363	.02272	.02179	.02084
13	1	12	.02289	.02249	.02204	.02152	.02095	.02033	.01968	.01899
13	1	13	.01843	.01840	.01830	.01814	.01792	.01764	.01731	.01693
13	2	3	.13021	.12139	.11284	.10460	.09670	.08917	.08201	.07526
13	2	4	.09973	.09384	.08804	.08238	.07687	.07155	.06643	.06153
13	2	5	.08088	.07669	.07251	.06838	.06431	.06032	.05644	.05269
13	2	6	.06781	.06475	.06165	.05854	.05544	.05237	.04934	.04639
13	2	7	.05805	.05579	.05347	.05110	.04871	.04632	.04393	.04156
13	2	8	.05032	.04867	.04695	.04516	.04333	.04146	.03957	.03768
13	2	9	.04391	.04275	.04151	.04019	.03881	.03738	.03591	.03442
13	2	10	.03836	.03762	.03678	.03586	.03487	.03382	.03271	.03156
13	2	11	.03331	.03293	.03246	.03190	.03126	.03054	.02977	.02894
13	2	12	.02840	.02836	.02823	.02801	.02771	.02733	.02689	.02638
13	2	13	.02289	.02321	.02345	.02362	.02371	.02372	.02365	.02352
13	3	4	.11627	.11051	.10476	.09904	.09340	.08785	.08243	.07717
13	3	5	.09446	.09048	.08643	.08235	.07827	.07419	.07016	.06620
13	3	6	.07929	.07648	.07357	.07059	.06755	.06449	.06141	.05834
13	3	7	.06793	.06595	.06386	.06167	.05941	.05708	.05471	.05232
13	3	8	.05892	.05758	.05611	.05454	.05287	.05113	.04932	.04746
13	3	9	.05145	.05060	.04964	.04857	.04739	.04613	.04478	.04337
13	3	10	.04496	.04454	.04400	.04335	.04259	.04174	.04080	.03979
13	3	11	.03906	.03901	.03884	.03857	.03819	.03772	.03715	.03650
13	3	12	.03331	.03360	.03379	.03388	.03387	.03377	.03357	.03328
13	3	13	.02685	.02751	.02808	.02858	.02899	.02931	.02954	.02968
13	4	5	.10825	.10453	.10068	.09673	.09269	.08861	.08451	.08041
13	4	6	.09099	.08847	.08581	.08301	.08011	.07711	.07406	.07096
13	4	7	.07802	.07636	.07455	.07259	.07051	.06832	.06604	.06369
13	4	8	.06772	.06671	.06555	.06424	.06280	.06124	.05957	.05782
13	4	9	.05916	.05867	.05802	.05724	.05632	.05528	.05412	.05286
13	4	10	.05173	.05166	.05146	.05112	.05064	.05005	.04933	.04851
13	4	11	.04496	.04527	.04544	.04550	.04543	.04524	.04493	.04452
13	4	12	.03836	.03901	.03955	.03998	.04030	.04051	.04061	.04060
13	4	13	.03093	.03194	.03288	.03374	.03450	.03518	.03575	.03623
13	5	6	.10374	.10159	.09924	.09671	.09400	.09116	.08819	.08513
13	5	7	.08904	.08778	.08631	.08466	.08283	.08085	.07873	.07649

Covariances of Inverse Gaussian Order Statistics
Shape Parameter k

n	i	j	0.0	0.1	0.2	0.3	0.4	0.5	0.6	0.7
13	5	8	.07736	.07675	.07595	.07498	.07383	.07252	.07107	.06949
13	5	9	.06762	.06754	.06728	.06685	.06626	.06551	.06461	.06357
13	5	10	.05916	.05951	.05970	.05973	.05961	.05934	.05892	.05837
13	5	11	.05145	.05216	.05274	.05319	.05350	.05366	.05369	.05359
13	5	12	.04391	.04497	.04592	.04676	.04748	.04808	.04855	.04889
13	5	13	.03542	.03684	.03819	.03947	.04066	.04176	.04275	.04364
13	6	7	.10168	.10090	.09988	.09863	.09716	.09548	.09361	.09157
13	6	8	.08842	.08831	.08798	.08743	.08668	.08573	.08458	.08326
13	6	9	.07736	.07777	.07799	.07802	.07785	.07749	.07695	.07623
13	6	10	.06772	.06857	.06925	.06975	.07008	.07024	.07022	.07003
13	6	11	.05892	.06014	.06122	.06215	.06293	.06355	.06402	.06433
13	6	12	.05032	.05187	.05333	.05466	.05588	.05696	.05791	.05872
13	6	13	.04061	.04252	.04438	.04617	.04788	.04951	.05103	.05243
13	7	8	.10168	.10222	.10250	.10253	.10231	.10185	.10115	.10023
13	7	9	.08904	.09010	.09095	.09157	.09197	.09215	.09210	.09185
13	7	10	.07802	.07951	.08081	.08193	.08286	.08359	.08411	.08444
13	7	11	.06793	.06978	.07149	.07305	.07445	.07568	.07673	.07761
13	7	12	.05805	.06023	.06232	.06430	.06615	.06788	.06946	.07089
13	7	13	.04688	.04940	.05189	.05434	.05672	.05903	.06123	.06333
13	8	9	.10374	.10566	.10735	.10880	.11000	.11093	.11161	.11204
13	8	10	.09099	.09333	.09549	.09745	.09920	.10073	.10203	.10310
13	8	11	.07929	.08199	.08455	.08697	.08921	.09128	.09316	.09484
13	8	12	.06781	.07083	.07376	.07660	.07933	.08193	.08439	.08668
13	8	13	.05482	.05815	.06148	.06480	.06808	.07130	.07445	.07750
13	9	10	.10825	.11181	.11519	.11837	.12133	.12405	.12651	.12872
13	9	11	.09446	.09835	.10213	.10576	.10924	.11254	.11564	.11852
13	9	12	.08088	.08506	.08919	.09327	.09725	.10112	.10486	.10845
13	9	13	.06546	.06991	.07443	.07898	.08355	.08810	.09261	.09706
13	10	11	.11627	.12199	.12765	.13321	.13863	.14390	.14897	.15382
13	10	12	.09973	.10568	.11167	.11766	.12361	.12950	.13529	.14095
13	10	13	.08086	.08702	.09335	.09981	.10637	.11301	.11967	.12633
13	11	12	.13021	.13925	.14849	.15786	.16734	.17686	.18638	.19587
13	11	13	.10589	.11500	.12448	.13429	.14439	.15474	.16528	.17598
13	12	13	.15573	.17123	.18762	.20486	.22290	.24168	.26113	.28118
14	1	2	.15172	.13714	.12356	.11097	.09936	.08873	.07904	.07025
14	1	3	.10317	.09441	.08612	.07831	.07101	.06421	.05791	.05211
14	1	4	.07887	.07282	.06703	.06151	.05628	.05136	.04675	.04245
14	1	5	.06397	.05950	.05518	.05101	.04703	.04324	.03966	.03629
14	1	6	.05371	.05029	.04694	.04369	.04055	.03754	.03466	.03193
14	1	7	.04609	.04342	.04079	.03820	.03567	.03323	.03087	.02861
14	1	8	.04011	.03802	.03593	.03385	.03180	.02980	.02785	.02597
14	1	9	.03521	.03358	.03192	.03025	.02859	.02695	.02534	.02377
14	1	10	.03104	.02978	.02849	.02717	.02584	.02450	.02318	.02187
14	1	11	.02734	.02641	.02543	.02442	.02337	.02231	.02124	.02018
14	1	12	.02391	.02327	.02259	.02185	.02107	.02027	.01944	.01859
14	1	13	.02051	.02016	.01975	.01929	.01878	.01823	.01764	.01702
14	1	14	.01663	.01660	.01651	.01636	.01615	.01589	.01559	.01524
14	2	3	.12608	.11716	.10855	.10028	.09238	.08489	.07780	.07113
14	2	4	.09665	.09062	.08473	.07899	.07344	.06810	.06298	.05812
14	2	5	.07852	.07417	.06987	.06563	.06148	.05744	.05353	.04977
14	2	6	.06600	.06276	.05951	.05627	.05307	.04991	.04683	.04383
14	2	7	.05669	.05424	.05175	.04924	.04672	.04422	.04175	.03932
14	2	8	.04937	.04752	.04561	.04366	.04168	.03968	.03769	.03570
14	2	9	.04336	.04199	.04055	.03904	.03749	.03591	.03431	.03270
14	2	10	.03823	.03726	.03620	.03508	.03389	.03266	.03139	.03010

Covariances of Inverse Gaussian Order Statistics
Shape Parameter k

n	i	j	0.0	0.1	0.2	0.3	0.4	0.5	0.6	0.7
14	2	11	.03369	.03305	.03233	.03153	.03067	.02975	.02878	.02777
14	2	12	.02947	.02914	.02872	.02823	.02766	.02703	.02634	.02560
14	2	13	.02529	.02525	.02513	.02493	.02466	.02432	.02391	.02345
14	2	14	.02051	.02079	.02100	.02114	.02121	.02121	.02114	.02100
14	3	4	.11198	.10604	.10013	.09430	.08857	.08298	.07755	.07230
14	3	5	.09112	.08693	.08270	.07847	.07427	.07011	.06601	.06201
14	3	6	.07668	.07363	.07052	.06737	.06418	.06099	.05782	.05468
14	3	7	.06591	.06369	.06138	.05899	.05655	.05408	.05158	.04909
14	3	8	.05743	.05583	.05413	.05234	.05048	.04856	.04659	.04460
14	3	9	.05047	.04936	.04814	.04683	.04543	.04396	.04244	.04086
14	3	10	.04452	.04381	.04300	.04208	.04108	.04000	.03884	.03763
14	3	11	.03924	.03888	.03841	.03785	.03719	.03645	.03562	.03473
14	3	12	.03433	.03428	.03413	.03389	.03355	.03312	.03261	.03203
14	3	13	.02947	.02971	.02987	.02994	.02991	.02981	.02961	.02934
14	3	14	.02391	.02447	.02497	.02539	.02574	.02600	.02618	.02628
14	4	5	.10369	.09970	.09561	.09145	.08725	.08303	.07883	.07467
14	4	6	.08736	.08455	.08163	.07860	.07549	.07232	.06912	.06591
14	4	7	.07515	.07319	.07110	.06888	.06657	.06418	.06172	.05921
14	4	8	.06553	.06421	.06274	.06116	.05946	.05766	.05578	.05384
14	4	9	.05761	.05679	.05583	.05474	.05354	.05223	.05083	.04935
14	4	10	.05084	.05043	.04989	.04922	.04843	.04754	.04655	.04546
14	4	11	.04482	.04476	.04458	.04428	.04386	.04333	.04270	.04198
14	4	12	.03924	.03949	.03963	.03966	.03958	.03939	.03911	.03872
14	4	13	.03369	.03423	.03469	.03505	.03530	.03546	.03552	.03548
14	4	14	.02734	.02821	.02901	.02974	.03038	.03094	.03141	.03179
14	5	6	.09877	.09626	.09357	.09073	.08775	.08466	.08149	.07826
14	5	7	.08505	.08340	.08158	.07959	.07746	.07520	.07283	.07037
14	5	8	.07422	.07322	.07204	.07071	.06923	.06761	.06587	.06403
14	5	9	.06529	.06479	.06414	.06333	.06237	.06128	.06006	.05872
14	5	10	.05764	.05756	.05734	.05697	.05645	.05580	.05502	.05412
14	5	11	.05084	.05112	.05126	.05127	.05114	.05088	.05050	.04999
14	5	12	.04452	.04511	.04558	.04594	.04617	.04627	.04626	.04613
14	5	13	.03823	.03912	.03992	.04061	.04119	.04166	.04202	.04227
14	5	14	.03104	.03224	.03339	.03446	.03546	.03637	.03718	.03789
14	6	7	.09614	.09487	.09338	.09169	.08981	.08775	.08554	.08320
14	6	8	.08396	.08335	.08254	.08153	.08033	.07896	.07743	.07576
14	6	9	.07391	.07381	.07353	.07307	.07243	.07162	.07064	.06952
14	6	10	.06529	.06561	.06577	.06576	.06559	.06525	.06475	.06411
14	6	11	.05761	.05829	.05883	.05921	.05944	.05952	.05946	.05924
14	6	12	.05047	.05146	.05233	.05307	.05368	.05415	.05448	.05468
14	6	13	.04336	.04465	.04584	.04693	.04791	.04877	.04952	.05013
14	6	14	.03521	.03681	.03836	.03985	.04126	.04259	.04382	.04495
14	7	8	.09531	.09519	.09484	.09425	.09344	.09241	.09118	.08976
14	7	9	.08396	.08436	.08456	.08454	.08431	.08388	.08325	.08243
14	7	10	.07422	.07504	.07568	.07613	.07639	.07647	.07635	.07606
14	7	11	.06553	.06671	.06773	.06859	.06928	.06980	.07014	.07032
14	7	12	.05743	.05892	.06028	.06151	.06259	.06353	.06431	.06494
14	7	13	.04937	.05115	.05283	.05442	.05589	.05724	.05847	.05956
14	7	14	.04011	.04219	.04423	.04623	.04816	.05001	.05177	.05343
14	8	9	.09614	.09718	.09799	.09857	.09890	.09899	.09884	.09847
14	8	10	.08505	.08652	.08778	.08884	.08968	.09031	.09073	.09093
14	8	11	.07515	.07696	.07861	.08009	.08138	.08249	.08340	.08412
14	8	12	.06591	.06802	.07001	.07187	.07358	.07513	.07651	.07773
14	8	13	.05669	.05908	.06140	.06362	.06574	.06773	.06960	.07133
14	8	14	.04609	.04877	.05144	.05408	.05667	.05921	.06166	.06401

Covariances of Inverse Gaussian Order Statistics
Shape Parameter k

n	i	j	0.0	0.1	0.2	0.3	0.4	0.5	0.6	0.7
14	9	10	.09877	.10110	.10321	.10510	.10676	.10818	.10935	.11026
14	9	11	.08736	.09002	.09252	.09484	.09697	.09889	.10060	.10210
14	9	12	.07668	.07963	.08246	.08517	.08773	.09013	.09236	.09440
14	9	13	.06600	.06921	.07237	.07545	.07844	.08132	.08408	.08669
14	9	14	.05371	.05718	.06068	.06419	.06768	.07114	.07453	.07785
14	10	11	.10369	.10756	.11128	.11482	.11817	.12131	.12421	.12687
14	10	12	.09112	.09525	.09930	.10323	.10704	.11068	.11415	.11743
14	10	13	.07852	.08288	.08723	.09155	.09580	.09996	.10401	.10793
14	10	14	.06397	.06855	.07322	.07796	.08274	.08753	.09230	.09702
14	11	12	.11198	.11793	.12385	.12971	.13546	.14109	.14656	.15185
14	11	13	.09665	.10278	.10897	.11520	.12142	.12761	.13373	.13975
14	11	14	.07887	.08515	.09162	.09826	.10504	.11191	.11884	.12580
14	12	13	.12608	.13527	.14469	.15430	.16405	.17389	.18378	.19366
14	12	14	.10317	.11237	.12198	.13197	.14227	.15287	.16370	.17473
14	13	14	.15172	.16727	.18376	.20115	.21940	.23844	.25820	.27863
15	1	2	.14813	.13356	.12002	.10750	.09600	.08549	.07594	.06730
15	1	3	.10072	.09192	.08361	.07581	.06854	.06179	.05556	.04985
15	1	4	.07706	.07094	.06510	.05955	.05432	.04942	.04484	.04059
15	1	5	.06258	.05803	.05364	.04943	.04542	.04162	.03804	.03469
15	1	6	.05265	.04913	.04571	.04239	.03920	.03616	.03326	.03053
15	1	7	.04531	.04253	.03980	.03713	.03454	.03205	.02966	.02738
15	1	8	.03957	.03736	.03515	.03298	.03086	.02879	.02680	.02488
15	1	9	.03490	.03313	.03135	.02958	.02783	.02611	.02444	.02282
15	1	10	.03096	.02956	.02813	.02669	.02525	.02383	.02242	.02105
15	1	11	.02752	.02643	.02530	.02415	.02298	.02181	.02065	.01950
15	1	12	.02441	.02360	.02274	.02184	.02092	.01998	.01903	.01808
15	1	13	.02148	.02092	.02031	.01965	.01896	.01824	.01749	.01674
15	1	14	.01853	.01822	.01786	.01744	.01698	.01648	.01594	.01539
15	1	15	.01511	.01509	.01500	.01486	.01467	.01443	.01415	.01383
15	2	3	.12242	.11342	.10476	.09648	.08861	.08115	.07414	.06757
15	2	4	.09391	.08777	.08179	.07600	.07042	.06508	.05999	.05516
15	2	5	.07639	.07192	.06751	.06319	.05898	.05490	.05098	.04723
15	2	6	.06434	.06096	.05759	.05425	.05096	.04775	.04463	.04161
15	2	7	.05541	.05280	.05018	.04755	.04494	.04236	.03982	.03735
15	2	8	.04842	.04641	.04435	.04227	.04017	.03808	.03600	.03396
15	2	9	.04273	.04118	.03958	.03793	.03625	.03455	.03285	.03115
15	2	10	.03792	.03675	.03552	.03423	.03290	.03154	.03015	.02875
15	2	11	.03372	.03288	.03196	.03099	.02996	.02888	.02777	.02664
15	2	12	.02992	.02936	.02873	.02803	.02727	.02646	.02560	.02471
15	2	13	.02633	.02604	.02567	.02523	.02472	.02416	.02354	.02288
15	2	14	.02272	.02268	.02257	.02239	.02214	.02183	.02146	.02104
15	2	15	.01853	.01878	.01897	.01909	.01914	.01913	.01905	.01892
15	3	4	.10821	.10212	.09610	.09019	.08441	.07879	.07337	.06815
15	3	5	.08816	.08380	.07944	.07509	.07080	.06658	.06245	.05844
15	3	6	.07433	.07110	.06783	.06454	.06124	.05796	.05472	.05154
15	3	7	.06406	.06164	.05915	.05661	.05404	.05145	.04886	.04630
15	3	8	.05601	.05421	.05231	.05035	.04834	.04628	.04420	.04212
15	3	9	.04945	.04812	.04670	.04520	.04363	.04201	.04035	.03865
15	3	10	.04390	.04296	.04193	.04081	.03962	.03836	.03705	.03569
15	3	11	.03904	.03844	.03774	.03695	.03608	.03514	.03413	.03307
15	3	12	.03465	.03434	.03393	.03344	.03286	.03220	.03147	.03069
15	3	13	.03051	.03046	.03033	.03010	.02980	.02941	.02895	.02842
15	3	14	.02633	.02654	.02668	.02672	.02669	.02658	.02640	.02614
15	3	15	.02148	.02198	.02242	.02278	.02307	.02329	.02344	.02351
15	4	5	.09973	.09552	.09125	.08693	.08261	.07830	.07403	.06983

Covariances of Inverse Gaussian Order Statistics
Shape Parameter k

n	i	j	0.0	0.1	0.2	0.3	0.4	0.5	0.6	0.7
15	4	6	.08417	.08113	.07800	.07479	.07153	.06824	.06494	.06165
15	4	7	.07259	.07039	.06807	.06566	.06317	.06062	.05803	.05542
15	4	8	.06352	.06194	.06024	.05843	.05654	.05456	.05253	.05045
15	4	9	.05610	.05501	.05380	.05248	.05106	.04955	.04797	.04632
15	4	10	.04982	.04913	.04832	.04740	.04638	.04526	.04406	.04279
15	4	11	.04432	.04398	.04351	.04293	.04225	.04147	.04061	.03966
15	4	12	.03935	.03930	.03913	.03886	.03849	.03802	.03745	.03681
15	4	13	.03465	.03486	.03498	.03499	.03491	.03473	.03446	.03410
15	4	14	.02992	.03039	.03077	.03107	.03128	.03140	.03142	.03136
15	4	15	.02441	.02517	.02587	.02649	.02705	.02752	.02791	.02821
15	5	6	.09452	.09171	.08876	.08568	.08250	.07924	.07593	.07258
15	5	7	.08159	.07963	.07753	.07528	.07292	.07045	.06791	.06530
15	5	8	.07143	.07012	.06865	.06704	.06530	.06345	.06151	.05948
15	5	9	.06312	.06231	.06134	.06024	.05900	.05765	.05619	.05464
15	5	10	.05608	.05567	.05512	.05443	.05362	.05268	.05164	.05049
15	5	11	.04991	.04985	.04965	.04932	.04887	.04829	.04761	.04681
15	5	12	.04432	.04456	.04467	.04466	.04453	.04428	.04392	.04346
15	5	13	.03904	.03954	.03994	.04022	.04040	.04046	.04042	.04027
15	5	14	.03372	.03447	.03515	.03572	.03621	.03659	.03687	.03705
15	5	15	.02752	.02856	.02955	.03047	.03131	.03207	.03275	.03333
15	6	7	.09147	.08982	.08797	.08595	.08376	.08143	.07898	.07642
15	6	8	.08014	.07914	.07795	.07659	.07507	.07339	.07158	.06965
15	6	9	.07086	.07037	.06970	.06886	.06787	.06672	.06543	.06402
15	6	10	.06298	.06290	.06266	.06226	.06170	.06100	.06016	.05918
15	6	11	.05608	.05634	.05646	.05643	.05625	.05594	.05548	.05490
15	6	12	.04982	.05038	.05081	.05111	.05127	.05131	.05121	.05098
15	6	13	.04390	.04473	.04545	.04605	.04653	.04689	.04713	.04725
15	6	14	.03792	.03901	.04001	.04091	.04172	.04242	.04301	.04349
15	6	15	.03096	.03233	.03365	.03490	.03609	.03720	.03821	.03914
15	7	8	.09005	.08943	.08860	.08756	.08631	.08488	.08327	.08151
15	7	9	.07967	.07957	.07927	.07877	.07809	.07722	.07617	.07497
15	7	10	.07086	.07117	.07131	.07126	.07103	.07063	.07006	.06934
15	7	11	.06312	.06378	.06428	.06462	.06479	.06480	.06465	.06435
15	7	12	.05610	.05706	.05788	.05855	.05908	.05946	.05969	.05978
15	7	13	.04945	.05067	.05178	.05277	.05364	.05437	.05496	.05542
15	7	14	.04273	.04421	.04560	.04690	.04810	.04919	.05017	.05102
15	7	15	.03490	.03665	.03837	.04003	.04163	.04315	.04459	.04593
15	8	9	.09005	.09044	.09061	.09055	.09027	.08977	.08906	.08815
15	8	10	.08014	.08095	.08156	.08196	.08217	.08217	.08197	.08158
15	8	11	.07143	.07259	.07357	.07437	.07499	.07542	.07568	.07575
15	8	12	.06352	.06496	.06627	.06742	.06841	.06924	.06991	.07040
15	8	13	.05601	.05772	.05932	.06079	.06214	.06334	.06440	.06530
15	8	14	.04842	.05038	.05226	.05405	.05575	.05733	.05880	.06014
15	8	15	.03957	.04179	.04399	.04615	.04827	.05031	.05228	.05416
15	9	10	.09147	.09292	.09415	.09515	.09593	.09648	.09680	.09689
15	9	11	.08159	.08338	.08498	.08640	.08761	.08862	.08942	.09002
15	9	12	.07259	.07467	.07660	.07837	.07998	.08141	.08265	.08371
15	9	13	.06406	.06638	.06861	.07071	.07268	.07451	.07618	.07769
15	9	14	.05541	.05797	.06048	.06291	.06525	.06748	.06959	.07158
15	9	15	.04531	.04812	.05093	.05374	.05652	.05925	.06191	.06449
15	10	11	.09452	.09716	.09962	.10187	.10391	.10573	.10731	.10865
15	10	12	.08417	.08709	.08987	.09249	.09494	.09720	.09926	.10112
15	10	13	.07433	.07748	.08055	.08351	.08634	.08902	.09155	.09390
15	10	14	.06434	.06771	.07105	.07434	.07756	.08068	.08369	.08657
15	10	15	.05265	.05625	.05989	.06356	.06723	.07089	.07450	.07805

Covariances of Inverse Gaussian Order Statistics
Shape Parameter k

n	i	j	0.0	0.1	0.2	0.3	0.4	0.5	0.6	0.7
15	11	12	.09973	.10385	.10784	.11169	.11537	.11886	.12214	.12519
15	11	13	.08816	.09249	.09676	.10094	.10502	.10896	.11275	.11637
15	11	14	.07639	.08091	.08544	.08995	.09443	.09884	.10316	.10737
15	11	15	.06258	.06728	.07209	.07698	.08194	.08693	.09192	.09689
15	12	13	.10821	.11435	.12048	.12659	.13262	.13857	.14438	.15003
15	12	14	.09391	.10017	.10654	.11296	.11941	.12586	.13227	.13861
15	12	15	.07706	.08343	.09003	.09682	.10377	.11085	.11802	.12524
15	13	14	.12242	.13173	.14130	.15110	.16109	.17120	.18140	.19164
15	13	15	.10072	.11000	.11972	.12984	.14033	.15114	.16222	.17353
15	14	15	.14813	.16371	.18028	.19780	.21622	.23548	.25553	.27629
16	1	2	.14489	.13033	.11684	.10439	.09299	.08260	.07318	.06469
16	1	3	.09850	.08966	.08135	.07357	.06633	.05964	.05348	.04785
16	1	4	.07540	.06923	.06335	.05779	.05257	.04768	.04313	.03893
16	1	5	.06131	.05668	.05224	.04800	.04397	.04017	.03660	.03327
16	1	6	.05166	.04806	.04457	.04120	.03798	.03491	.03201	.02928
16	1	7	.04455	.04168	.03887	.03614	.03350	.03097	.02856	.02628
16	1	8	.03902	.03670	.03441	.03217	.02998	.02787	.02584	.02390
16	1	9	.03454	.03266	.03078	.02892	.02710	.02532	.02360	.02194
16	1	10	.03078	.02926	.02772	.02618	.02466	.02317	.02170	.02028
16	1	11	.02754	.02631	.02507	.02381	.02254	.02129	.02005	.01884
16	1	12	.02465	.02369	.02270	.02168	.02064	.01960	.01856	.01754
16	1	13	.02200	.02127	.02051	.01971	.01888	.01804	.01719	.01634
16	1	14	.01946	.01896	.01841	.01782	.01719	.01654	.01587	.01518
16	1	15	.01687	.01659	.01626	.01588	.01546	.01500	.01452	.01401
16	1	16	.01383	.01380	.01372	.01359	.01342	.01319	.01293	.01264
16	2	3	.11914	.11008	.10140	.09312	.08527	.07787	.07092	.06444
16	2	4	.09144	.08521	.07917	.07334	.06775	.06242	.05735	.05257
16	2	5	.07446	.06988	.06539	.06101	.05676	.05267	.04874	.04500
16	2	6	.06281	.05931	.05584	.05243	.04908	.04583	.04268	.03965
16	2	7	.05420	.05147	.04874	.04602	.04333	.04069	.03811	.03561
16	2	8	.04750	.04535	.04317	.04098	.03880	.03663	.03449	.03240
16	2	9	.04206	.04037	.03864	.03687	.03508	.03330	.03152	.02977
16	2	10	.03750	.03618	.03481	.03339	.03194	.03047	.02900	.02752
16	2	11	.03356	.03255	.03149	.03037	.02920	.02801	.02679	.02557
16	2	12	.03005	.02931	.02852	.02766	.02675	.02580	.02481	.02381
16	2	13	.02682	.02633	.02577	.02515	.02448	.02375	.02298	.02218
16	2	14	.02373	.02347	.02314	.02275	.02229	.02178	.02122	.02062
16	2	15	.02058	.02054	.02044	.02028	.02005	.01976	.01942	.01903
16	2	16	.01687	.01709	.01726	.01736	.01740	.01738	.01730	.01717
16	3	4	.10487	.09866	.09255	.08658	.08076	.07514	.06974	.06456
16	3	5	.08552	.08102	.07654	.07211	.06776	.06349	.05935	.05534
16	3	6	.07221	.06883	.06543	.06203	.05865	.05530	.05202	.04881
16	3	7	.06236	.05978	.05715	.05449	.05181	.04914	.04648	.04387
16	3	8	.05467	.05270	.05065	.04855	.04642	.04426	.04210	.03994
16	3	9	.04844	.04693	.04535	.04370	.04199	.04025	.03848	.03670
16	3	10	.04320	.04208	.04087	.03959	.03824	.03685	.03541	.03395
16	3	11	.03867	.03787	.03698	.03601	.03497	.03388	.03273	.03154
16	3	12	.03463	.03411	.03350	.03281	.03204	.03121	.03032	.02938
16	3	13	.03091	.03064	.03028	.02984	.02933	.02874	.02809	.02738
16	3	14	.02736	.02732	.02720	.02699	.02671	.02636	.02594	.02546
16	3	15	.02373	.02392	.02403	.02407	.02403	.02392	.02374	.02349
16	3	16	.01946	.01990	.02029	.02061	.02086	.02104	.02115	.02120
16	4	5	.09625	.09187	.08744	.08301	.07860	.07422	.06992	.06571
16	4	6	.08135	.07812	.07483	.07148	.06810	.06471	.06134	.05801
16	4	7	.07030	.06790	.06540	.06283	.06020	.05754	.05485	.05217

Covariances of Inverse Gaussian Order Statistics
Shape Parameter k

n	i	j	0.0	0.1	0.2	0.3	0.4	0.5	0.6	0.7
16	4	8	.06167	.05989	.05800	.05602	.05396	.05185	.04970	.04753
16	4	9	.05466	.05336	.05194	.05043	.04884	.04718	.04546	.04369
16	4	10	.04876	.04785	.04683	.04571	.04450	.04320	.04185	.04043
16	4	11	.04366	.04308	.04238	.04159	.04070	.03973	.03869	.03757
16	4	12	.03911	.03881	.03840	.03790	.03730	.03661	.03585	.03501
16	4	13	.03493	.03488	.03473	.03448	.03415	.03372	.03321	.03263
16	4	14	.03091	.03110	.03119	.03119	.03111	.03093	.03068	.03034
16	4	15	.02682	.02723	.02757	.02782	.02799	.02807	.02808	.02800
16	4	16	.02200	.02267	.02328	.02382	.02430	.02470	.02503	.02528
16	5	6	.09082	.08778	.08462	.08136	.07802	.07463	.07122	.06780
16	5	7	.07855	.07635	.07402	.07157	.06903	.06641	.06373	.06102
16	5	8	.06895	.06738	.06567	.06385	.06191	.05988	.05778	.05562
16	5	9	.06114	.06006	.05885	.05751	.05606	.05451	.05287	.05116
16	5	10	.05456	.05389	.05308	.05214	.05109	.04994	.04869	.04735
16	5	11	.04887	.04852	.04805	.04746	.04675	.04594	.04503	.04402
16	5	12	.04379	.04373	.04355	.04326	.04285	.04234	.04173	.04103
16	5	13	.03911	.03931	.03939	.03937	.03924	.03901	.03868	.03825
16	5	14	.03463	.03506	.03539	.03562	.03576	.03579	.03573	.03557
16	5	15	.03005	.03070	.03128	.03177	.03218	.03249	.03271	.03284
16	5	16	.02465	.02556	.02642	.02721	.02794	.02859	.02916	.02965
16	6	7	.08746	.08551	.08338	.08110	.07868	.07615	.07352	.07082
16	6	8	.07682	.07551	.07403	.07239	.07062	.06871	.06670	.06459
16	6	9	.06815	.06734	.06637	.06524	.06397	.06257	.06106	.05943
16	6	10	.06085	.06045	.05988	.05918	.05833	.05735	.05625	.05504
16	6	11	.05452	.05445	.05424	.05388	.05339	.05278	.05204	.05118
16	6	12	.04887	.04909	.04917	.04913	.04896	.04866	.04824	.04771
16	6	13	.04366	.04413	.04449	.04472	.04484	.04484	.04472	.04449
16	6	14	.03867	.03937	.03998	.04048	.04087	.04115	.04132	.04139
16	6	15	.03356	.03449	.03534	.03611	.03678	.03736	.03784	.03822
16	6	16	.02754	.02872	.02986	.03094	.03195	.03289	.03374	.03451
16	7	8	.08562	.08461	.08341	.08202	.08045	.07871	.07684	.07483
16	7	9	.07600	.07551	.07482	.07396	.07292	.07173	.07038	.06889
16	7	10	.06789	.06781	.06755	.06712	.06652	.06577	.06487	.06383
16	7	11	.06085	.06111	.06120	.06114	.06092	.06055	.06003	.05938
16	7	12	.05456	.05511	.05551	.05576	.05587	.05584	.05568	.05537
16	7	13	.04876	.04956	.05023	.05078	.05119	.05148	.05163	.05165
16	7	14	.04320	.04423	.04515	.04597	.04667	.04725	.04772	.04806
16	7	15	.03750	.03875	.03993	.04102	.04202	.04291	.04370	.04439
16	7	16	.03078	.03228	.03374	.03515	.03651	.03778	.03898	.04009
16	8	9	.08503	.08492	.08460	.08407	.08334	.08241	.08130	.08001
16	8	10	.07600	.07631	.07642	.07634	.07607	.07561	.07497	.07417
16	8	11	.06815	.06880	.06927	.06957	.06969	.06964	.06942	.06903
16	8	12	.06114	.06207	.06285	.06348	.06395	.06426	.06440	.06440
16	8	13	.05466	.05585	.05690	.05783	.05861	.05925	.05974	.06009
16	8	14	.04844	.04985	.05116	.05237	.05345	.05441	.05524	.05593
16	8	15	.04206	.04370	.04526	.04675	.04814	.04943	.05060	.05167
16	8	16	.03454	.03641	.03826	.04008	.04184	.04353	.04515	.04668
16	9	10	.08562	.08642	.08700	.08737	.08752	.08745	.08718	.08670
16	9	11	.07682	.07796	.07891	.07967	.08023	.08060	.08077	.08074
16	9	12	.06895	.07037	.07163	.07273	.07365	.07440	.07497	.07536
16	9	13	.06167	.06334	.06488	.06628	.06754	.06863	.06957	.07034
16	9	14	.05467	.05657	.05836	.06005	.06162	.06305	.06435	.06550
16	9	15	.04750	.04961	.05165	.05363	.05551	.05730	.05897	.06053
16	9	16	.03902	.04135	.04368	.04599	.04827	.05049	.05264	.05471
16	10	11	.08746	.08923	.09081	.09217	.09333	.09426	.09496	.09545

Covariances of Inverse Gaussian Order Statistics
Shape Parameter k

n	i	j	0.0	0.1	0.2	0.3	0.4	0.5	0.6	0.7
16	10	12	.07855	.08060	.08249	.08420	.08573	.08706	.08820	.08913
16	10	13	.07030	.07259	.07476	.07678	.07865	.08036	.08189	.08324
16	10	14	.06236	.06486	.06728	.06960	.07179	.07386	.07578	.07755
16	10	15	.05420	.05691	.05958	.06218	.06471	.06715	.06948	.07170
16	10	16	.04455	.04747	.05041	.05336	.05630	.05920	.06205	.06483
16	11	12	.09082	.09372	.09646	.09902	.10138	.10354	.10547	.10718
16	11	13	.08135	.08448	.08749	.09036	.09308	.09563	.09800	.10017
16	11	14	.07221	.07553	.07879	.08196	.08502	.08795	.09074	.09337
16	11	15	.06281	.06632	.06982	.07328	.07669	.08002	.08325	.08638
16	11	16	.05166	.05536	.05912	.06293	.06676	.07059	.07439	.07815
16	12	13	.09625	.10057	.10479	.10889	.11285	.11664	.12025	.12365
16	12	14	.08552	.09001	.09447	.09886	.10317	.10737	.11144	.11535
16	12	15	.07446	.07910	.08378	.08847	.09314	.09777	.10233	.10679
16	12	16	.06131	.06609	.07102	.07605	.08116	.08632	.09151	.09670
16	13	14	.10487	.11115	.11747	.12378	.13006	.13627	.14238	.14836
16	13	15	.09144	.09782	.10432	.11091	.11756	.12424	.13090	.13751
16	13	16	.07540	.08186	.08856	.09548	.10258	.10984	.11722	.12468
16	14	15	.11914	.12854	.13825	.14821	.15839	.16875	.17922	.18977
16	14	16	.09850	.10784	.11765	.12789	.13853	.14953	.16083	.17240
16	15	16	.14489	.16049	.17712	.19474	.21331	.23277	.25306	.27412
17	1	2	.14194	.12741	.11396	.10159	.09028	.08001	.07071	.06235
17	1	3	.09647	.08761	.07930	.07154	.06434	.05770	.05161	.04606
17	1	4	.07388	.06766	.06176	.05620	.05098	.04612	.04161	.03745
17	1	5	.06013	.05544	.05096	.04669	.04265	.03885	.03530	.03200
17	1	6	.05073	.04706	.04351	.04011	.03686	.03379	.03088	.02816
17	1	7	.04382	.04087	.03800	.03522	.03255	.03000	.02757	.02528
17	1	8	.03847	.03606	.03370	.03140	.02916	.02702	.02496	.02301
17	1	9	.03414	.03217	.03021	.02829	.02641	.02459	.02283	.02115
17	1	10	.03054	.02891	.02729	.02568	.02409	.02254	.02103	.01958
17	1	11	.02745	.02612	.02477	.02342	.02208	.02076	.01947	.01821
17	1	12	.02472	.02365	.02255	.02143	.02031	.01919	.01808	.01700
17	1	13	.02226	.02141	.02053	.01962	.01869	.01776	.01682	.01590
17	1	14	.01997	.01932	.01864	.01792	.01717	.01641	.01564	.01487
17	1	15	.01775	.01730	.01680	.01627	.01570	.01510	.01449	.01386
17	1	16	.01546	.01520	.01490	.01455	.01417	.01375	.01330	.01283
17	1	17	.01273	.01270	.01263	.01251	.01234	.01213	.01189	.01161
17	2	3	.11619	.10708	.09838	.09011	.08230	.07495	.06808	.06169
17	2	4	.08920	.08290	.07682	.07097	.06537	.06005	.05502	.05028
17	2	5	.07270	.06803	.06347	.05905	.05477	.05067	.04675	.04303
17	2	6	.06140	.05780	.05426	.05078	.04739	.04411	.04094	.03791
17	2	7	.05308	.05024	.04742	.04462	.04187	.03919	.03658	.03406
17	2	8	.04661	.04435	.04207	.03980	.03754	.03532	.03314	.03102
17	2	9	.04139	.03958	.03773	.03587	.03401	.03215	.03032	.02852
17	2	10	.03703	.03559	.03410	.03257	.03103	.02948	.02794	.02641
17	2	11	.03329	.03215	.03096	.02972	.02846	.02717	.02587	.02458
17	2	12	.03000	.02912	.02818	.02720	.02617	.02511	.02404	.02295
17	2	13	.02702	.02637	.02566	.02490	.02409	.02324	.02236	.02146
17	2	14	.02424	.02381	.02331	.02275	.02214	.02149	.02080	.02007
17	2	15	.02155	.02131	.02102	.02066	.02024	.01978	.01927	.01872
17	2	16	.01877	.01873	.01864	.01849	.01827	.01801	.01769	.01733
17	2	17	.01546	.01566	.01580	.01589	.01592	.01589	.01581	.01568
17	3	4	.10188	.09557	.08939	.08338	.07755	.07193	.06655	.06142
17	3	5	.08314	.07853	.07397	.06947	.06507	.06078	.05663	.05263
17	3	6	.07029	.06679	.06328	.05980	.05635	.05296	.04964	.04642
17	3	7	.06080	.05809	.05534	.05258	.04982	.04708	.04438	.04173

Covariances of Inverse Gaussian Order Statistics
Shape Parameter k

n	i	j	0.0	0.1	0.2	0.3	0.4	0.5	0.6	0.7
17	3	8	.05342	.05130	.04913	.04692	.04469	.04245	.04022	.03802
17	3	9	.04746	.04580	.04408	.04231	.04050	.03866	.03681	.03497
17	3	10	.04247	.04119	.03984	.03843	.03696	.03546	.03393	.03239
17	3	11	.03819	.03723	.03619	.03508	.03391	.03269	.03143	.03015
17	3	12	.03442	.03373	.03295	.03210	.03119	.03022	.02921	.02816
17	3	13	.03100	.03055	.03001	.02940	.02872	.02798	.02718	.02634
17	3	14	.02782	.02758	.02726	.02687	.02640	.02587	.02528	.02464
17	3	15	.02473	.02470	.02458	.02440	.02414	.02381	.02343	.02299
17	3	16	.02155	.02171	.02181	.02183	.02179	.02168	.02151	.02128
17	3	17	.01775	.01815	.01849	.01877	.01899	.01914	.01923	.01926
17	4	5	.09316	.08864	.08410	.07957	.07509	.07067	.06635	.06214
17	4	6	.07883	.07545	.07202	.06855	.06509	.06163	.05822	.05486
17	4	7	.06823	.06566	.06302	.06032	.05759	.05483	.05208	.04935
17	4	8	.05998	.05802	.05598	.05386	.05168	.04947	.04723	.04498
17	4	9	.05331	.05182	.05024	.04858	.04685	.04507	.04324	.04139
17	4	10	.04772	.04662	.04543	.04414	.04278	.04135	.03987	.03835
17	4	11	.04293	.04215	.04127	.04030	.03925	.03813	.03694	.03571
17	4	12	.03869	.03819	.03759	.03689	.03612	.03526	.03434	.03336
17	4	13	.03486	.03460	.03424	.03379	.03326	.03265	.03196	.03121
17	4	14	.03129	.03124	.03111	.03089	.03058	.03019	.02973	.02920
17	4	15	.02782	.02798	.02806	.02805	.02796	.02780	.02756	.02724
17	4	16	.02424	.02460	.02489	.02511	.02525	.02531	.02530	.02522
17	4	17	.01997	.02057	.02111	.02159	.02200	.02235	.02263	.02283
17	5	6	.08757	.08434	.08101	.07760	.07415	.07067	.06718	.06372
17	5	7	.07585	.07345	.07094	.06833	.06565	.06292	.06015	.05736
17	5	8	.06672	.06494	.06304	.06104	.05895	.05679	.05457	.05231
17	5	9	.05932	.05803	.05661	.05509	.05347	.05176	.04999	.04815
17	5	10	.05313	.05222	.05120	.05007	.04884	.04751	.04610	.04463
17	5	11	.04780	.04722	.04653	.04573	.04482	.04382	.04273	.04157
17	5	12	.04310	.04280	.04239	.04187	.04125	.04053	.03973	.03884
17	5	13	.03884	.03878	.03862	.03836	.03800	.03754	.03698	.03635
17	5	14	.03486	.03503	.03510	.03507	.03494	.03472	.03441	.03401
17	5	15	.03100	.03138	.03166	.03185	.03196	.03197	.03190	.03174
17	5	16	.02702	.02759	.02809	.02852	.02886	.02912	.02929	.02939
17	5	17	.02226	.02307	.02383	.02452	.02515	.02571	.02620	.02661
17	6	7	.08398	.08178	.07943	.07694	.07435	.07166	.06891	.06610
17	6	8	.07391	.07235	.07063	.06878	.06680	.06472	.06255	.06031
17	6	9	.06574	.06467	.06345	.06209	.06061	.05902	.05732	.05555
17	6	10	.05890	.05823	.05741	.05646	.05538	.05419	.05289	.05150
17	6	11	.05301	.05267	.05219	.05158	.05085	.05000	.04904	.04798
17	6	12	.04781	.04775	.04756	.04724	.04681	.04626	.04560	.04484
17	6	13	.04310	.04328	.04334	.04329	.04313	.04285	.04246	.04198
17	6	14	.03869	.03910	.03939	.03958	.03967	.03964	.03951	.03929
17	6	15	.03442	.03503	.03554	.03596	.03629	.03651	.03664	.03666
17	6	16	.03000	.03081	.03155	.03220	.03278	.03326	.03365	.03395
17	6	17	.02472	.02576	.02676	.02770	.02857	.02938	.03010	.03075
17	7	8	.08182	.08050	.07901	.07734	.07552	.07356	.07148	.06929
17	7	9	.07282	.07200	.07102	.06987	.06856	.06712	.06554	.06385
17	7	10	.06527	.06486	.06428	.06355	.06267	.06165	.06050	.05923
17	7	11	.05876	.05869	.05846	.05808	.05756	.05690	.05611	.05520
17	7	12	.05301	.05322	.05329	.05322	.05301	.05266	.05219	.05160
17	7	13	.04780	.04825	.04858	.04878	.04885	.04879	.04862	.04832
17	7	14	.04293	.04360	.04416	.04461	.04494	.04515	.04525	.04523
17	7	15	.03819	.03907	.03985	.04054	.04112	.04160	.04196	.04222
17	7	16	.03329	.03437	.03538	.03631	.03715	.03790	.03855	.03910

Covariances of Inverse Gaussian Order Statistics
Shape Parameter k

n	i	j	0.0	0.1	0.2	0.3	0.4	0.5	0.6	0.7
17	7	17	.02745	.02875	.03002	.03123	.03239	.03348	.03449	.03542
17	8	9	.08080	.08030	.07960	.07871	.07763	.07639	.07498	.07342
17	8	10	.07246	.07237	.07209	.07163	.07100	.07020	.06924	.06814
17	8	11	.06527	.06551	.06559	.06549	.06524	.06482	.06425	.06353
17	8	12	.05890	.05943	.05981	.06003	.06010	.06001	.05978	.05941
17	8	13	.05313	.05390	.05454	.05504	.05540	.05562	.05570	.05564
17	8	14	.04772	.04872	.04959	.05035	.05098	.05148	.05186	.05210
17	8	15	.04247	.04367	.04477	.04577	.04666	.04744	.04810	.04864
17	8	16	.03703	.03843	.03975	.04100	.04216	.04323	.04420	.04506
17	8	17	.03054	.03215	.03374	.03528	.03677	.03820	.03956	.04083
17	9	10	.08080	.08110	.08119	.08108	.08076	.08025	.07955	.07868
17	9	11	.07282	.07345	.07390	.07416	.07424	.07413	.07385	.07339
17	9	12	.06574	.06666	.06742	.06800	.06842	.06867	.06875	.06866
17	9	13	.05932	.06048	.06150	.06237	.06310	.06366	.06407	.06433
17	9	14	.05331	.05468	.05594	.05708	.05808	.05895	.05967	.06025
17	9	15	.04746	.04903	.05052	.05190	.05318	.05433	.05536	.05627
17	9	16	.04139	.04316	.04487	.04651	.04807	.04953	.05089	.05214
17	9	17	.03414	.03612	.03809	.04003	.04193	.04378	.04556	.04726
17	10	11	.08182	.08294	.08387	.08459	.08510	.08541	.08550	.08539
17	10	12	.07391	.07532	.07655	.07760	.07847	.07915	.07963	.07993
17	10	13	.06672	.06837	.06987	.07121	.07240	.07341	.07425	.07492
17	10	14	.05998	.06184	.06358	.06519	.06667	.06800	.06918	.07020
17	10	15	.05342	.05547	.05744	.05930	.06106	.06270	.06421	.06558
17	10	16	.04661	.04885	.05104	.05316	.05521	.05718	.05904	.06079
17	10	17	.03847	.04090	.04334	.04578	.04819	.05056	.05287	.05512
17	11	12	.08398	.08602	.08788	.08954	.09101	.09226	.09331	.09413
17	11	13	.07585	.07812	.08025	.08222	.08401	.08562	.08705	.08828
17	11	14	.06823	.07070	.07307	.07531	.07741	.07935	.08114	.08275
17	11	15	.06080	.06345	.06604	.06854	.07093	.07320	.07534	.07734
17	11	16	.05308	.05591	.05871	.06147	.06416	.06678	.06931	.07173
17	11	17	.04382	.04683	.04988	.05296	.05603	.05908	.06210	.06506
17	12	13	.08757	.09069	.09366	.09648	.09912	.10156	.10380	.10583
17	12	14	.07883	.08213	.08534	.08843	.09138	.09418	.09682	.09927
17	12	15	.07029	.07376	.07718	.08053	.08379	.08694	.08996	.09283
17	12	16	.06140	.06502	.06865	.07227	.07584	.07936	.08280	.08614
17	12	17	.05073	.05451	.05838	.06230	.06627	.07025	.07423	.07817
17	13	14	.09316	.09765	.10206	.10638	.11058	.11464	.11852	.12222
17	13	15	.08314	.08777	.09238	.09696	.10148	.10590	.11021	.11438
17	13	16	.07270	.07745	.08225	.08709	.09193	.09675	.10152	.10621
17	13	17	.06013	.06499	.07001	.07515	.08040	.08572	.09109	.09647
17	14	15	.10188	.10828	.11475	.12124	.12773	.13418	.14055	.14681
17	14	16	.08920	.09567	.10230	.10904	.11586	.12273	.12961	.13648
17	14	17	.07388	.08040	.08719	.09422	.10146	.10888	.11644	.12410
17	15	16	.11619	.12566	.13548	.14558	.15594	.16649	.17721	.18804
17	15	17	.09647	.10586	.11574	.12609	.13686	.14802	.15952	.17132
17	16	17	.14194	.15756	.17424	.19194	.21064	.23027	.25078	.27211
18	1	2	.13925	.12474	.11134	.09905	.08783	.07766	.06849	.06026
18	1	3	.09462	.08574	.07743	.06970	.06254	.05596	.04993	.04445
18	1	4	.07249	.06623	.06031	.05474	.04954	.04470	.04023	.03612
18	1	5	.05903	.05430	.04978	.04549	.04145	.03766	.03413	.03085
18	1	6	.04986	.04613	.04254	.03911	.03584	.03276	.02986	.02716
18	1	7	.04313	.04011	.03719	.03437	.03167	.02910	.02667	.02438
18	1	8	.03793	.03545	.03303	.03067	.02841	.02623	.02416	.02220
18	1	9	.03374	.03168	.02966	.02768	.02576	.02390	.02212	.02042
18	1	10	.03026	.02855	.02685	.02518	.02354	.02194	.02040	.01892

Covariances of Inverse Gaussian Order Statistics
Shape Parameter k

n	i	j	0.0	0.1	0.2	0.3	0.4	0.5	0.6	0.7
18	1	11	.02729	.02587	.02445	.02303	.02163	.02025	.01892	.01763
18	1	12	.02470	.02353	.02233	.02114	.01995	.01877	.01761	.01649
18	1	13	.02238	.02142	.02044	.01944	.01843	.01743	.01644	.01546
18	1	14	.02025	.01949	.01870	.01788	.01704	.01619	.01535	.01451
18	1	15	.01825	.01767	.01705	.01639	.01572	.01502	.01432	.01362
18	1	16	.01629	.01588	.01543	.01494	.01442	.01387	.01331	.01274
18	1	17	.01424	.01401	.01373	.01341	.01305	.01267	.01225	.01182
18	1	18	.01177	.01175	.01168	.01156	.01141	.01121	.01098	.01073
18	2	3	.11351	.10437	.09566	.08740	.07963	.07234	.06554	.05923
18	2	4	.08716	.08081	.07469	.06882	.06323	.05793	.05293	.04824
18	2	5	.07108	.06635	.06174	.05727	.05298	.04888	.04497	.04127
18	2	6	.06010	.05642	.05281	.04928	.04586	.04256	.03939	.03636
18	2	7	.05202	.04910	.04620	.04334	.04055	.03783	.03520	.03268
18	2	8	.04577	.04341	.04105	.03870	.03639	.03412	.03191	.02977
18	2	9	.04073	.03882	.03688	.03494	.03301	.03110	.02922	.02739
18	2	10	.03655	.03499	.03340	.03179	.03017	.02856	.02696	.02539
18	2	11	.03297	.03171	.03041	.02908	.02773	.02637	.02501	.02366
18	2	12	.02984	.02884	.02779	.02670	.02558	.02444	.02329	.02214
18	2	13	.02705	.02627	.02544	.02456	.02365	.02270	.02174	.02076
18	2	14	.02448	.02391	.02327	.02259	.02186	.02110	.02030	.01949
18	2	15	.02206	.02167	.02122	.02072	.02017	.01958	.01895	.01829
18	2	16	.01969	.01948	.01921	.01888	.01850	.01808	.01761	.01711
18	2	17	.01722	.01719	.01710	.01695	.01676	.01651	.01621	.01588
18	2	18	.01424	.01442	.01455	.01462	.01465	.01462	.01454	.01441
18	3	4	.09918	.09280	.08657	.08052	.07468	.06908	.06373	.05865
18	3	5	.08099	.07629	.07165	.06710	.06266	.05836	.05422	.05024
18	3	6	.06853	.06493	.06134	.05779	.05429	.05086	.04753	.04431
18	3	7	.05936	.05654	.05370	.05085	.04803	.04524	.04251	.03984
18	3	8	.05225	.05001	.04773	.04543	.04313	.04083	.03855	.03631
18	3	9	.04652	.04474	.04290	.04103	.03913	.03722	.03531	.03342
18	3	10	.04175	.04034	.03886	.03734	.03578	.03419	.03259	.03099
18	3	11	.03767	.03657	.03540	.03417	.03290	.03158	.03024	.02889
18	3	12	.03411	.03327	.03236	.03138	.03035	.02928	.02817	.02703
18	3	13	.03092	.03031	.02962	.02887	.02806	.02720	.02629	.02535
18	3	14	.02799	.02758	.02710	.02656	.02595	.02528	.02457	.02381
18	3	15	.02522	.02501	.02472	.02436	.02394	.02346	.02292	.02234
18	3	16	.02252	.02248	.02238	.02220	.02197	.02167	.02131	.02090
18	3	17	.01969	.01984	.01992	.01994	.01989	.01979	.01962	.01940
18	3	18	.01629	.01664	.01695	.01720	.01739	.01752	.01760	.01761
18	4	5	.09040	.08575	.08112	.07652	.07199	.06755	.06322	.05902
18	4	6	.07656	.07305	.06951	.06596	.06242	.05892	.05547	.05210
18	4	7	.06635	.06365	.06089	.05808	.05526	.05244	.04964	.04687
18	4	8	.05843	.05633	.05415	.05192	.04964	.04735	.04504	.04274
18	4	9	.05204	.05040	.04869	.04690	.04506	.04318	.04128	.03936
18	4	10	.04672	.04546	.04412	.04270	.04122	.03968	.03810	.03650
18	4	11	.04217	.04123	.04020	.03909	.03790	.03666	.03537	.03403
18	4	12	.03819	.03751	.03675	.03590	.03498	.03399	.03295	.03185
18	4	13	.03462	.03418	.03365	.03304	.03235	.03158	.03076	.02988
18	4	14	.03135	.03111	.03079	.03039	.02991	.02936	.02874	.02807
18	4	15	.02825	.02821	.02809	.02789	.02760	.02725	.02683	.02634
18	4	16	.02522	.02536	.02543	.02542	.02533	.02517	.02494	.02465
18	4	17	.02206	.02238	.02264	.02283	.02294	.02299	.02297	.02288
18	4	18	.01825	.01878	.01927	.01970	.02006	.02036	.02060	.02077
18	5	6	.08469	.08130	.07783	.07431	.07077	.06721	.06368	.06019
18	5	7	.07345	.07088	.06822	.06548	.06269	.05987	.05703	.05419

Covariances of Inverse Gaussian Order Statistics
Shape Parameter k

n	i	j	0.0	0.1	0.2	0.3	0.4	0.5	0.6	0.7
18	5	8	.06471	.06276	.06070	.05856	.05634	.05407	.05177	.04944
18	5	9	.05765	.05618	.05460	.05292	.05117	.04934	.04746	.04554
18	5	10	.05178	.05069	.04949	.04820	.04681	.04535	.04383	.04225
18	5	11	.04675	.04598	.04510	.04413	.04306	.04191	.04069	.03940
18	5	12	.04234	.04185	.04125	.04054	.03975	.03887	.03791	.03689
18	5	13	.03839	.03813	.03777	.03731	.03676	.03612	.03540	.03461
18	5	14	.03477	.03472	.03457	.03433	.03400	.03359	.03309	.03251
18	5	15	.03135	.03149	.03154	.03151	.03138	.03118	.03088	.03052
18	5	16	.02799	.02831	.02856	.02872	.02880	.02880	.02872	.02856
18	5	17	.02448	.02499	.02543	.02580	.02609	.02631	.02645	.02651
18	5	18	.02025	.02097	.02165	.02226	.02282	.02331	.02372	.02407
18	6	7	.08092	.07852	.07598	.07334	.07060	.06780	.06495	.06206
18	6	8	.07133	.06956	.06765	.06562	.06349	.06127	.05899	.05665
18	6	9	.06358	.06230	.06087	.05933	.05768	.05593	.05410	.05221
18	6	10	.05712	.05622	.05520	.05405	.05279	.05143	.04998	.04845
18	6	11	.05159	.05102	.05032	.04950	.04857	.04754	.04641	.04520
18	6	12	.04674	.04644	.04603	.04549	.04485	.04410	.04326	.04232
18	6	13	.04239	.04233	.04216	.04188	.04149	.04099	.04040	.03972
18	6	14	.03839	.03855	.03860	.03854	.03838	.03812	.03776	.03732
18	6	15	.03462	.03497	.03522	.03537	.03543	.03539	.03526	.03503
18	6	16	.03092	.03145	.03189	.03225	.03252	.03270	.03279	.03279
18	6	17	.02705	.02776	.02840	.02897	.02946	.02987	.03020	.03044
18	6	18	.02238	.02330	.02418	.02501	.02577	.02647	.02710	.02765
18	7	8	.07852	.07695	.07522	.07333	.07132	.06919	.06695	.06464
18	7	9	.07002	.06895	.06772	.06634	.06482	.06318	.06144	.05959
18	7	10	.06293	.06225	.06143	.06046	.05935	.05812	.05677	.05533
18	7	11	.05685	.05650	.05602	.05539	.05463	.05374	.05274	.05163
18	7	12	.05152	.05145	.05125	.05091	.05045	.04987	.04917	.04836
18	7	13	.04674	.04691	.04696	.04688	.04668	.04636	.04593	.04539
18	7	14	.04234	.04273	.04300	.04315	.04320	.04312	.04294	.04265
18	7	15	.03819	.03876	.03924	.03962	.03988	.04004	.04010	.04005
18	7	16	.03411	.03487	.03554	.03612	.03661	.03700	.03730	.03749
18	7	17	.02984	.03079	.03166	.03246	.03318	.03381	.03436	.03482
18	7	18	.02470	.02585	.02696	.02802	.02902	.02996	.03083	.03162
18	8	9	.07718	.07636	.07536	.07418	.07284	.07135	.06972	.06796
18	8	10	.06939	.06898	.06839	.06764	.06672	.06566	.06446	.06312
18	8	11	.06271	.06263	.06239	.06199	.06144	.06074	.05990	.05893
18	8	12	.05685	.05705	.05710	.05700	.05676	.05638	.05586	.05521
18	8	13	.05159	.05203	.05233	.05250	.05253	.05243	.05220	.05184
18	8	14	.04675	.04740	.04793	.04834	.04862	.04878	.04881	.04872
18	8	15	.04217	.04301	.04375	.04438	.04490	.04530	.04559	.04576
18	8	16	.03767	.03870	.03964	.04048	.04123	.04187	.04241	.04284
18	8	17	.03297	.03417	.03531	.03638	.03737	.03827	.03908	.03979
18	8	18	.02729	.02870	.03008	.03141	.03270	.03392	.03507	.03615
18	9	10	.07674	.07665	.07635	.07587	.07520	.07436	.07334	.07217
18	9	11	.06939	.06963	.06968	.06956	.06927	.06881	.06818	.06740
18	9	12	.06293	.06344	.06380	.06399	.06402	.06389	.06360	.06317
18	9	13	.05712	.05788	.05849	.05896	.05927	.05944	.05946	.05933
18	9	14	.05178	.05274	.05359	.05430	.05487	.05531	.05561	.05578
18	9	15	.04672	.04788	.04893	.04987	.05069	.05138	.05195	.05240
18	9	16	.04175	.04308	.04434	.04550	.04655	.04750	.04834	.04906
18	9	17	.03655	.03806	.03951	.04090	.04221	.04343	.04455	.04558
18	9	18	.03026	.03197	.03366	.03532	.03694	.03850	.04000	.04142
18	10	11	.07718	.07780	.07823	.07847	.07850	.07834	.07800	.07747
18	10	12	.07002	.07093	.07166	.07221	.07258	.07278	.07279	.07263

Covariances of Inverse Gaussian Order Statistics
Shape Parameter k

n	i	j	0.0	0.1	0.2	0.3	0.4	0.5	0.6	0.7
18	10	13	.06358	.06473	.06572	.06655	.06722	.06773	.06807	.06824
18	10	14	.05765	.05901	.06023	.06132	.06226	.06305	.06369	.06418
18	10	15	.05204	.05358	.05501	.05633	.05753	.05859	.05952	.06030
18	10	16	.04652	.04823	.04986	.05141	.05285	.05418	.05539	.05648
18	10	17	.04073	.04261	.04445	.04622	.04793	.04954	.05107	.05249
18	10	18	.03374	.03581	.03788	.03993	.04196	.04393	.04586	.04771
18	11	12	.07852	.07991	.08112	.08214	.08296	.08357	.08399	.08420
18	11	13	.07133	.07296	.07444	.07574	.07687	.07781	.07857	.07915
18	11	14	.06471	.06654	.06825	.06981	.07122	.07246	.07355	.07446
18	11	15	.05843	.06045	.06236	.06416	.06583	.06736	.06875	.06999
18	11	16	.05225	.05443	.05654	.05857	.06050	.06231	.06401	.06558
18	11	17	.04577	.04811	.05042	.05268	.05488	.05700	.05903	.06096
18	11	18	.03793	.04044	.04298	.04553	.04806	.05056	.05302	.05542
18	12	13	.08092	.08318	.08528	.08720	.08893	.09047	.09180	.09292
18	12	14	.07345	.07590	.07823	.08041	.08244	.08429	.08597	.08746
18	12	15	.06635	.06898	.07152	.07394	.07624	.07839	.08040	.08225
18	12	16	.05936	.06214	.06487	.06753	.07009	.07255	.07489	.07709
18	12	17	.05202	.05495	.05788	.06077	.06361	.06639	.06909	.07169
18	12	18	.04313	.04622	.04936	.05254	.05573	.05892	.06208	.06520
18	13	14	.08469	.08798	.09116	.09419	.09707	.09976	.10227	.10458
18	13	15	.07656	.08001	.08339	.08666	.08982	.09284	.09571	.09840
18	13	16	.06853	.07213	.07569	.07919	.08263	.08597	.08919	.09229
18	13	17	.06010	.06382	.06756	.07131	.07503	.07871	.08233	.08586
18	13	18	.04986	.05371	.05766	.06170	.06578	.06990	.07403	.07814
18	14	15	.09040	.09502	.09960	.10411	.10852	.11280	.11693	.12090
18	14	16	.08099	.08573	.09048	.09522	.09991	.10453	.10905	.11346
18	14	17	.07108	.07592	.08084	.08581	.09079	.09578	.10074	.10564
18	14	18	.05903	.06396	.06906	.07430	.07967	.08513	.09065	.09621
18	15	16	.09918	.10569	.11229	.11894	.12560	.13226	.13886	.14538
18	15	17	.08716	.09371	.10044	.10731	.11428	.12132	.12840	.13549
18	15	18	.07249	.07906	.08592	.09305	.10041	.10797	.11569	.12354
18	16	17	.11351	.12304	.13295	.14317	.15368	.16442	.17535	.18642
18	16	18	.09462	.10404	.11399	.12442	.13531	.14662	.15829	.17029
18	17	18	.13925	.15487	.17159	.18937	.20818	.22796	.24867	.27023
19	1	2	.13678	.12229	.10895	.09673	.08560	.07553	.06647	.05836
19	1	3	.09291	.08402	.07572	.06801	.06090	.05437	.04841	.04299
19	1	4	.07119	.06491	.05898	.05341	.04823	.04342	.03898	.03491
19	1	5	.05801	.05324	.04869	.04440	.04035	.03657	.03306	.02981
19	1	6	.04904	.04526	.04164	.03818	.03490	.03182	.02893	.02624
19	1	7	.04247	.03940	.03643	.03358	.03086	.02828	.02585	.02357
19	1	8	.03740	.03486	.03239	.03000	.02770	.02551	.02343	.02146
19	1	9	.03333	.03121	.02913	.02710	.02514	.02326	.02146	.01975
19	1	10	.02996	.02818	.02642	.02469	.02301	.02138	.01981	.01832
19	1	11	.02710	.02560	.02411	.02263	.02118	.01977	.01840	.01709
19	1	12	.02461	.02336	.02209	.02083	.01958	.01835	.01716	.01600
19	1	13	.02240	.02136	.02029	.01922	.01815	.01709	.01605	.01504
19	1	14	.02040	.01954	.01866	.01775	.01685	.01594	.01504	.01415
19	1	15	.01854	.01785	.01713	.01639	.01563	.01486	.01409	.01333
19	1	16	.01677	.01625	.01568	.01508	.01447	.01383	.01319	.01254
19	1	17	.01502	.01465	.01424	.01379	.01331	.01281	.01229	.01176
19	1	18	.01318	.01297	.01271	.01242	.01209	.01173	.01134	.01094
19	1	19	.01094	.01092	.01085	.01074	.01060	.01041	.01020	.00995
19	2	3	.11106	.10189	.09318	.08495	.07721	.06998	.06325	.05703
19	2	4	.08529	.07890	.07275	.06687	.06129	.05601	.05105	.04642
19	2	5	.06960	.06480	.06015	.05566	.05136	.04725	.04336	.03969

Covariances of Inverse Gaussian Order Statistics
Shape Parameter k

n	i	j	0.0	0.1	0.2	0.3	0.4	0.5	0.6	0.7
19	2	6	.05889	.05514	.05148	.04791	.04447	.04115	.03798	.03497
19	2	7	.05104	.04803	.04507	.04217	.03934	.03660	.03396	.03143
19	2	8	.04497	.04252	.04009	.03769	.03533	.03303	.03080	.02864
19	2	9	.04009	.03809	.03607	.03407	.03208	.03013	.02822	.02637
19	2	10	.03605	.03440	.03273	.03105	.02937	.02770	.02606	.02446
19	2	11	.03261	.03126	.02987	.02846	.02704	.02562	.02421	.02282
19	2	12	.02963	.02852	.02738	.02620	.02500	.02379	.02258	.02138
19	2	13	.02697	.02609	.02515	.02418	.02318	.02216	.02113	.02009
19	2	14	.02456	.02387	.02313	.02234	.02152	.02067	.01980	.01891
19	2	15	.02233	.02182	.02125	.02063	.01997	.01927	.01855	.01781
19	2	16	.02020	.01985	.01944	.01899	.01848	.01794	.01736	.01676
19	2	17	.01810	.01790	.01766	.01736	.01701	.01661	.01618	.01572
19	2	18	.01588	.01585	.01577	.01563	.01545	.01521	.01494	.01462
19	2	19	.01318	.01334	.01346	.01353	.01354	.01351	.01343	.01331
19	3	4	.09674	.09029	.08402	.07795	.07212	.06653	.06122	.05619
19	3	5	.07903	.07425	.06955	.06496	.06051	.05620	.05206	.04811
19	3	6	.06693	.06324	.05958	.05597	.05243	.04898	.04564	.04242
19	3	7	.05803	.05512	.05219	.04929	.04641	.04359	.04083	.03815
19	3	8	.05115	.04882	.04645	.04407	.04170	.03936	.03704	.03478
19	3	9	.04562	.04374	.04181	.03985	.03788	.03591	.03396	.03203
19	3	10	.04104	.03951	.03794	.03633	.03469	.03303	.03137	.02972
19	3	11	.03713	.03592	.03464	.03331	.03194	.03055	.02915	.02773
19	3	12	.03374	.03278	.03175	.03067	.02954	.02838	.02719	.02599
19	3	13	.03072	.02998	.02918	.02831	.02740	.02644	.02544	.02443
19	3	14	.02798	.02744	.02683	.02616	.02543	.02466	.02384	.02300
19	3	15	.02544	.02508	.02465	.02416	.02361	.02300	.02235	.02166
19	3	16	.02302	.02283	.02256	.02224	.02185	.02141	.02092	.02039
19	3	17	.02062	.02059	.02049	.02033	.02011	.01983	.01950	.01912
19	3	18	.01810	.01823	.01830	.01831	.01827	.01816	.01800	.01779
19	3	19	.01502	.01535	.01563	.01585	.01602	.01613	.01619	.01619
19	4	5	.08791	.08316	.07845	.07380	.06924	.06478	.06045	.05627
19	4	6	.07450	.07089	.06726	.06364	.06005	.05651	.05304	.04966
19	4	7	.06464	.06182	.05896	.05607	.05319	.05031	.04747	.04468
19	4	8	.05700	.05478	.05249	.05017	.04781	.04545	.04309	.04076
19	4	9	.05086	.04909	.04726	.04537	.04345	.04149	.03952	.03755
19	4	10	.04576	.04437	.04290	.04137	.03979	.03817	.03652	.03485
19	4	11	.04142	.04034	.03918	.03795	.03665	.03531	.03394	.03253
19	4	12	.03764	.03682	.03592	.03495	.03391	.03281	.03167	.03049
19	4	13	.03428	.03369	.03301	.03226	.03145	.03057	.02963	.02866
19	4	14	.03123	.03084	.03037	.02982	.02920	.02852	.02778	.02699
19	4	15	.02839	.02819	.02790	.02754	.02710	.02660	.02604	.02542
19	4	16	.02569	.02566	.02554	.02535	.02509	.02477	.02438	.02393
19	4	17	.02302	.02314	.02320	.02318	.02309	.02294	.02272	.02245
19	4	18	.02020	.02049	.02072	.02088	.02098	.02101	.02098	.02089
19	4	19	.01677	.01726	.01769	.01808	.01840	.01866	.01887	.01901
19	5	6	.08211	.07858	.07500	.07139	.06778	.06418	.06061	.05711
19	5	7	.07128	.06857	.06579	.06295	.06007	.05718	.05428	.05141
19	5	8	.06289	.06079	.05860	.05634	.05403	.05167	.04930	.04692
19	5	9	.05613	.05450	.05278	.05098	.04911	.04719	.04523	.04324
19	5	10	.05051	.04927	.04793	.04650	.04499	.04342	.04180	.04014
19	5	11	.04573	.04480	.04378	.04266	.04146	.04019	.03886	.03748
19	5	12	.04157	.04091	.04015	.03929	.03836	.03734	.03627	.03513
19	5	13	.03786	.03743	.03690	.03628	.03558	.03480	.03394	.03303
19	5	14	.03450	.03427	.03395	.03354	.03304	.03247	.03182	.03111
19	5	15	.03138	.03133	.03120	.03098	.03068	.03029	.02984	.02931

Covariances of Inverse Gaussian Order Statistics
Shape Parameter k

n	i	j	0.0	0.1	0.2	0.3	0.4	0.5	0.6	0.7
19	5	16	.02839	.02852	.02856	.02852	.02840	.02821	.02793	.02759
19	5	17	.02544	.02573	.02595	.02608	.02614	.02613	.02604	.02588
19	5	18	.02233	.02279	.02318	.02350	.02375	.02394	.02405	.02409
19	5	19	.01854	.01919	.01979	.02034	.02083	.02126	.02163	.02193
19	6	7	.07820	.07563	.07295	.07017	.06733	.06443	.06151	.05858
19	6	8	.06903	.06708	.06501	.06284	.06058	.05826	.05589	.05348
19	6	9	.06163	.06017	.05858	.05688	.05509	.05322	.05129	.04931
19	6	10	.05549	.05441	.05321	.05190	.05049	.04899	.04742	.04579
19	6	11	.05025	.04949	.04861	.04762	.04653	.04535	.04409	.04276
19	6	12	.04568	.04519	.04459	.04388	.04306	.04216	.04116	.04010
19	6	13	.04162	.04136	.04100	.04053	.03995	.03929	.03854	.03770
19	6	14	.03793	.03788	.03772	.03747	.03711	.03667	.03613	.03551
19	6	15	.03450	.03463	.03467	.03461	.03446	.03421	.03388	.03347
19	6	16	.03123	.03153	.03175	.03187	.03191	.03186	.03173	.03151
19	6	17	.02798	.02845	.02884	.02915	.02938	.02952	.02958	.02956
19	6	18	.02456	.02520	.02577	.02627	.02669	.02704	.02732	.02751
19	6	19	.02040	.02123	.02201	.02274	.02342	.02403	.02457	.02505
19	7	8	.07562	.07384	.07191	.06985	.06768	.06542	.06307	.06066
19	7	9	.06754	.06626	.06482	.06326	.06158	.05979	.05790	.05595
19	7	10	.06083	.05994	.05890	.05774	.05645	.05505	.05356	.05198
19	7	11	.05510	.05453	.05383	.05300	.05204	.05098	.04981	.04855
19	7	12	.05011	.04981	.04939	.04884	.04817	.04739	.04651	.04553
19	7	13	.04566	.04560	.04542	.04512	.04470	.04418	.04355	.04283
19	7	14	.04162	.04177	.04180	.04172	.04153	.04124	.04084	.04035
19	7	15	.03786	.03819	.03842	.03855	.03857	.03849	.03830	.03803
19	7	16	.03428	.03478	.03519	.03550	.03572	.03585	.03587	.03580
19	7	17	.03072	.03139	.03197	.03247	.03289	.03322	.03345	.03360
19	7	18	.02697	.02780	.02857	.02927	.02989	.03043	.03090	.03127
19	7	19	.02240	.02342	.02441	.02534	.02622	.02704	.02780	.02847
19	8	9	.07403	.07295	.07171	.07031	.06876	.06707	.06527	.06337
19	8	10	.06670	.06602	.06518	.06419	.06306	.06179	.06039	.05889
19	8	11	.06044	.06009	.05959	.05894	.05815	.05723	.05619	.05503
19	8	12	.05498	.05490	.05469	.05433	.05384	.05322	.05248	.05162
19	8	13	.05011	.05027	.05031	.05021	.04998	.04963	.04915	.04856
19	8	14	.04568	.04606	.04631	.04644	.04644	.04633	.04610	.04576
19	8	15	.04157	.04213	.04257	.04291	.04314	.04325	.04325	.04314
19	8	16	.03764	.03836	.03900	.03953	.03996	.04029	.04051	.04062
19	8	17	.03374	.03463	.03544	.03616	.03680	.03734	.03778	.03812
19	8	18	.02963	.03068	.03167	.03260	.03345	.03421	.03490	.03549
19	8	19	.02461	.02585	.02706	.02823	.02935	.03041	.03140	.03232
19	9	10	.07327	.07285	.07225	.07148	.07053	.06943	.06817	.06678
19	9	11	.06642	.06633	.06608	.06566	.06507	.06433	.06345	.06242
19	9	12	.06044	.06063	.06066	.06054	.06027	.05985	.05928	.05858
19	9	13	.05510	.05553	.05582	.05596	.05596	.05582	.05553	.05512
19	9	14	.05025	.05089	.05139	.05177	.05201	.05212	.05210	.05195
19	9	15	.04573	.04655	.04726	.04785	.04832	.04866	.04889	.04898
19	9	16	.04142	.04241	.04330	.04409	.04477	.04534	.04580	.04614
19	9	17	.03713	.03828	.03936	.04034	.04124	.04203	.04272	.04331
19	9	18	.03261	.03392	.03518	.03637	.03749	.03852	.03947	.04032
19	9	19	.02710	.02859	.03006	.03150	.03290	.03424	.03552	.03673
19	10	11	.07327	.07350	.07354	.07340	.07307	.07256	.07189	.07104
19	10	12	.06670	.06720	.06754	.06771	.06770	.06753	.06719	.06669
19	10	13	.06083	.06158	.06217	.06260	.06288	.06300	.06296	.06277
19	10	14	.05549	.05644	.05726	.05793	.05846	.05885	.05909	.05918
19	10	15	.05051	.05165	.05267	.05356	.05433	.05496	.05545	.05581

Covariances of Inverse Gaussian Order Statistics
Shape Parameter k

n	i	j	0.0	0.1	0.2	0.3	0.4	0.5	0.6	0.7
19	10	16	.04576	.04706	.04826	.04936	.05035	.05122	.05196	.05258
19	10	17	.04104	.04249	.04388	.04518	.04638	.04749	.04848	.04936
19	10	18	.03605	.03766	.03923	.04074	.04217	.04353	.04480	.04597
19	10	19	.02996	.03175	.03353	.03530	.03702	.03870	.04033	.04188
19	11	12	.07403	.07492	.07564	.07616	.07650	.07664	.07660	.07637
19	11	13	.06754	.06868	.06965	.07045	.07107	.07153	.07181	.07191
19	11	14	.06163	.06297	.06417	.06521	.06610	.06684	.06741	.06781
19	11	15	.05613	.05764	.05904	.06031	.06145	.06244	.06328	.06397
19	11	16	.05086	.05254	.05412	.05560	.05696	.05820	.05931	.06028
19	11	17	.04562	.04745	.04922	.05090	.05249	.05398	.05535	.05661
19	11	18	.04009	.04207	.04401	.04591	.04774	.04949	.05116	.05274
19	11	19	.03333	.03548	.03764	.03979	.04192	.04402	.04607	.04805
19	12	13	.07562	.07724	.07868	.07995	.08103	.08192	.08261	.08311
19	12	14	.06903	.07085	.07252	.07404	.07540	.07658	.07758	.07840
19	12	15	.06289	.06488	.06676	.06850	.07011	.07156	.07286	.07399
19	12	16	.05700	.05915	.06122	.06317	.06502	.06673	.06831	.06975
19	12	17	.05115	.05345	.05569	.05785	.05993	.06191	.06377	.06552
19	12	18	.04497	.04740	.04982	.05220	.05452	.05678	.05896	.06105
19	12	19	.03740	.03999	.04262	.04526	.04790	.05052	.05311	.05564
19	13	14	.07820	.08065	.08295	.08509	.08705	.08884	.09042	.09181
19	13	15	.07128	.07389	.07639	.07876	.08099	.08306	.08496	.08668
19	13	16	.06464	.06740	.07008	.07267	.07514	.07748	.07969	.08174
19	13	17	.05803	.06093	.06378	.06657	.06929	.07191	.07442	.07681
19	13	18	.05104	.05406	.05708	.06009	.06306	.06598	.06883	.07160
19	13	19	.04247	.04563	.04885	.05212	.05542	.05873	.06203	.06529
19	14	15	.08211	.08555	.08890	.09212	.09520	.09812	.10087	.10343
19	14	16	.07450	.07808	.08160	.08504	.08837	.09159	.09466	.09758
19	14	17	.06693	.07062	.07431	.07795	.08154	.08505	.08845	.09174
19	14	18	.05889	.06270	.06654	.07040	.07425	.07808	.08185	.08556
19	14	19	.04904	.05295	.05698	.06111	.06530	.06954	.07380	.07806
19	15	16	.08791	.09265	.09737	.10204	.10663	.11111	.11547	.11967
19	15	17	.07903	.08387	.08874	.09361	.09845	.10325	.10797	.11259
19	15	18	.06960	.07451	.07952	.08460	.08973	.09486	.09998	.10507
19	15	19	.05801	.06299	.06816	.07350	.07897	.08455	.09021	.09593
19	16	17	.09674	.10333	.11004	.11682	.12365	.13048	.13729	.14404
19	16	18	.08529	.09191	.09873	.10571	.11281	.12001	.12727	.13455
19	16	19	.07119	.07781	.08474	.09195	.09941	.10710	.11496	.12298
19	17	18	.11106	.12065	.13063	.14096	.15160	.16250	.17363	.18492
19	17	19	.09291	.10236	.11236	.12287	.13387	.14530	.15713	.16932
19	18	19	.13678	.15240	.16914	.18699	.20590	.22582	.24670	.26848
20	1	2	.13449	.12004	.10675	.09460	.08356	.07359	.06463	.05663
20	1	3	.09132	.08243	.07415	.06647	.05939	.05292	.04702	.04167
20	1	4	.06999	.06368	.05775	.05219	.04702	.04224	.03784	.03382
20	1	5	.05706	.05225	.04769	.04338	.03934	.03558	.03209	.02887
20	1	6	.04827	.04445	.04080	.03732	.03404	.03096	.02808	.02540
20	1	7	.04184	.03873	.03572	.03284	.03011	.02752	.02509	.02282
20	1	8	.03689	.03430	.03179	.02937	.02705	.02484	.02275	.02079
20	1	9	.03293	.03075	.02862	.02656	.02457	.02267	.02085	.01914
20	1	10	.02966	.02781	.02600	.02423	.02251	.02085	.01927	.01776
20	1	11	.02688	.02532	.02377	.02224	.02075	.01930	.01791	.01658
20	1	12	.02448	.02316	.02183	.02051	.01922	.01795	.01673	.01555
20	1	13	.02236	.02124	.02011	.01898	.01785	.01675	.01567	.01463
20	1	14	.02046	.01952	.01856	.01759	.01662	.01566	.01472	.01379
20	1	15	.01871	.01793	.01713	.01631	.01548	.01466	.01383	.01303
20	1	16	.01707	.01645	.01579	.01511	.01441	.01371	.01300	.01230

Covariances of Inverse Gaussian Order Statistics
Shape Parameter k

n	i	j	0.0	0.1	0.2	0.3	0.4	0.5	0.6	0.7
20	1	17	.01550	.01501	.01450	.01395	.01338	.01279	.01220	.01160
20	1	18	.01392	.01358	.01320	.01279	.01234	.01188	.01140	.01091
20	1	19	.01225	.01206	.01182	.01155	.01124	.01090	.01054	.01017
20	1	20	.01020	.01018	.01012	.01002	.00988	.00971	.00950	.00927
20	2	3	.10881	.09963	.09093	.08272	.07502	.06784	.06118	.05504
20	2	4	.08358	.07714	.07097	.06510	.05953	.05428	.04935	.04477
20	2	5	.06822	.06338	.05869	.05419	.04988	.04578	.04191	.03827
20	2	6	.05777	.05397	.05026	.04666	.04319	.03987	.03671	.03371
20	2	7	.05011	.04705	.04403	.04109	.03823	.03547	.03282	.03030
20	2	8	.04420	.04169	.03920	.03675	.03436	.03203	.02978	.02762
20	2	9	.03947	.03739	.03531	.03325	.03123	.02924	.02730	.02543
20	2	10	.03556	.03383	.03209	.03034	.02861	.02691	.02524	.02361
20	2	11	.03224	.03080	.02934	.02786	.02638	.02491	.02346	.02204
20	2	12	.02937	.02818	.02695	.02570	.02444	.02317	.02192	.02067
20	2	13	.02683	.02585	.02483	.02378	.02271	.02163	.02054	.01946
20	2	14	.02455	.02376	.02292	.02205	.02115	.02022	.01929	.01835
20	2	15	.02245	.02183	.02116	.02045	.01970	.01893	.01814	.01733
20	2	16	.02049	.02002	.01951	.01894	.01834	.01771	.01705	.01637
20	2	17	.01860	.01828	.01791	.01749	.01703	.01653	.01600	.01544
20	2	18	.01671	.01654	.01631	.01603	.01571	.01535	.01495	.01452
20	2	19	.01471	.01468	.01461	.01448	.01431	.01409	.01383	.01353
20	2	20	.01225	.01240	.01251	.01257	.01258	.01255	.01247	.01235
20	3	4	.09450	.08800	.08170	.07562	.06980	.06424	.05896	.05398
20	3	5	.07724	.07239	.06765	.06303	.05855	.05425	.05013	.04620
20	3	6	.06545	.06169	.05797	.05431	.05075	.04729	.04394	.04073
20	3	7	.05681	.05381	.05082	.04786	.04494	.04209	.03931	.03663
20	3	8	.05013	.04771	.04527	.04283	.04041	.03802	.03568	.03340
20	3	9	.04478	.04280	.04079	.03876	.03674	.03472	.03273	.03077
20	3	10	.04035	.03873	.03707	.03538	.03367	.03196	.03026	.02857
20	3	11	.03659	.03527	.03390	.03249	.03106	.02960	.02814	.02668
20	3	12	.03334	.03227	.03115	.02998	.02877	.02754	.02629	.02503
20	3	13	.03046	.02962	.02871	.02775	.02674	.02570	.02464	.02356
20	3	14	.02788	.02722	.02650	.02572	.02490	.02404	.02314	.02223
20	3	15	.02550	.02502	.02447	.02386	.02320	.02250	.02176	.02099
20	3	16	.02327	.02295	.02256	.02211	.02161	.02105	.02046	.01983
20	3	17	.02113	.02095	.02071	.02041	.02006	.01965	.01920	.01871
20	3	18	.01899	.01896	.01887	.01872	.01851	.01825	.01794	.01759
20	3	19	.01671	.01683	.01690	.01690	.01686	.01675	.01660	.01640
20	3	20	.01392	.01422	.01447	.01467	.01482	.01492	.01497	.01496
20	4	5	.08564	.08082	.07605	.07136	.06676	.06230	.05798	.05383
20	4	6	.07263	.06892	.06522	.06154	.05791	.05435	.05087	.04750
20	4	7	.06307	.06015	.05721	.05425	.05131	.04840	.04554	.04273
20	4	8	.05569	.05336	.05098	.04857	.04616	.04374	.04135	.03899
20	4	9	.04975	.04788	.04595	.04398	.04198	.03996	.03794	.03593
20	4	10	.04485	.04334	.04177	.04015	.03849	.03679	.03508	.03337
20	4	11	.04068	.03948	.03821	.03688	.03550	.03408	.03264	.03117
20	4	12	.03707	.03613	.03512	.03404	.03290	.03171	.03049	.02925
20	4	13	.03388	.03316	.03237	.03150	.03058	.02960	.02858	.02753
20	4	14	.03100	.03048	.02988	.02921	.02848	.02769	.02685	.02597
20	4	15	.02837	.02802	.02760	.02710	.02654	.02593	.02525	.02454
20	4	16	.02589	.02570	.02544	.02511	.02472	.02426	.02374	.02318
20	4	17	.02351	.02347	.02337	.02319	.02295	.02265	.02229	.02187
20	4	18	.02113	.02124	.02128	.02126	.02118	.02103	.02083	.02056
20	4	19	.01860	.01886	.01907	.01921	.01929	.01931	.01927	.01918
20	4	20	.01550	.01594	.01633	.01667	.01696	.01720	.01737	.01749

Covariances of Inverse Gaussian Order Statistics
Shape Parameter k

n	i	j	0.0	0.1	0.2	0.3	0.4	0.5	0.6	0.7
20	5	6	.07978	.07614	.07247	.06879	.06512	.06148	.05790	.05439
20	5	7	.06932	.06649	.06361	.06068	.05773	.05479	.05186	.04896
20	5	8	.06123	.05900	.05671	.05435	.05195	.04953	.04711	.04469
20	5	9	.05472	.05297	.05113	.04923	.04726	.04526	.04324	.04120
20	5	10	.04934	.04796	.04650	.04495	.04335	.04169	.03999	.03827
20	5	11	.04477	.04370	.04254	.04130	.04000	.03863	.03721	.03576
20	5	12	.04080	.04000	.03910	.03812	.03707	.03595	.03478	.03356
20	5	13	.03729	.03672	.03605	.03529	.03446	.03357	.03261	.03159
20	5	14	.03414	.03375	.03329	.03273	.03210	.03140	.03063	.02981
20	5	15	.03123	.03103	.03074	.03037	.02992	.02940	.02881	.02816
20	5	16	.02851	.02847	.02835	.02815	.02787	.02752	.02709	.02661
20	5	17	.02589	.02600	.02604	.02599	.02588	.02569	.02543	.02511
20	5	18	.02327	.02353	.02372	.02384	.02388	.02386	.02377	.02361
20	5	19	.02049	.02090	.02125	.02153	.02175	.02191	.02200	.02202
20	5	20	.01707	.01766	.01820	.01870	.01913	.01951	.01983	.02009
20	6	7	.07577	.07306	.07025	.06737	.06444	.06147	.05850	.05553
20	6	8	.06696	.06486	.06266	.06037	.05801	.05561	.05316	.05071
20	6	9	.05987	.05825	.05652	.05470	.05280	.05083	.04881	.04677
20	6	10	.05399	.05275	.05141	.04996	.04843	.04683	.04517	.04346
20	6	11	.04900	.04808	.04705	.04592	.04470	.04340	.04203	.04061
20	6	12	.04467	.04402	.04325	.04239	.04144	.04040	.03929	.03812
20	6	13	.04084	.04041	.03988	.03925	.03853	.03773	.03685	.03589
20	6	14	.03738	.03716	.03683	.03641	.03590	.03530	.03462	.03387
20	6	15	.03421	.03416	.03402	.03379	.03347	.03306	.03257	.03201
20	6	16	.03123	.03135	.03138	.03132	.03117	.03094	.03063	.03025
20	6	17	.02837	.02863	.02882	.02893	.02895	.02889	.02875	.02854
20	6	18	.02550	.02591	.02626	.02653	.02672	.02684	.02688	.02684
20	6	19	.02245	.02302	.02353	.02397	.02434	.02464	.02487	.02503
20	6	20	.01871	.01945	.02016	.02081	.02141	.02195	.02243	.02284
20	7	8	.07304	.07108	.06900	.06680	.06451	.06213	.05969	.05722
20	7	9	.06533	.06386	.06226	.06055	.05873	.05682	.05483	.05279
20	7	10	.05894	.05786	.05665	.05533	.05389	.05236	.05075	.04907
20	7	11	.05351	.05275	.05187	.05086	.04975	.04854	.04724	.04587
20	7	12	.04879	.04830	.04769	.04697	.04613	.04520	.04417	.04306
20	7	13	.04461	.04436	.04398	.04350	.04291	.04221	.04143	.04056
20	7	14	.04085	.04079	.04063	.04035	.03998	.03950	.03894	.03828
20	7	15	.03738	.03751	.03753	.03745	.03727	.03700	.03663	.03618
20	7	16	.03414	.03442	.03462	.03472	.03472	.03463	.03446	.03419
20	7	17	.03100	.03145	.03180	.03207	.03225	.03234	.03235	.03227
20	7	18	.02788	.02846	.02898	.02942	.02977	.03005	.03024	.03035
20	7	19	.02455	.02529	.02597	.02658	.02712	.02760	.02799	.02831
20	7	20	.02046	.02137	.02225	.02308	.02386	.02459	.02524	.02583
20	8	9	.07126	.06997	.06853	.06694	.06522	.06339	.06145	.05944
20	8	10	.06431	.06342	.06237	.06119	.05987	.05844	.05690	.05527
20	8	11	.05840	.05783	.05712	.05627	.05529	.05419	.05298	.05168
20	8	12	.05326	.05297	.05253	.05197	.05128	.05047	.04955	.04853
20	8	13	.04872	.04865	.04846	.04814	.04770	.04715	.04648	.04572
20	8	14	.04461	.04475	.04477	.04467	.04446	.04413	.04370	.04316
20	8	15	.04084	.04116	.04137	.04147	.04146	.04134	.04112	.04079
20	8	16	.03729	.03778	.03816	.03845	.03863	.03870	.03868	.03856
20	8	17	.03388	.03451	.03506	.03552	.03588	.03615	.03632	.03640
20	8	18	.03046	.03125	.03195	.03258	.03313	.03358	.03395	.03423
20	8	19	.02683	.02776	.02864	.02945	.03018	.03085	.03143	.03194
20	8	20	.02236	.02347	.02454	.02557	.02656	.02749	.02835	.02915
20	9	10	.07025	.06957	.06873	.06771	.06655	.06524	.06380	.06224

Covariances of Inverse Gaussian Order Statistics
Shape Parameter k

n	i	j	0.0	0.1	0.2	0.3	0.4	0.5	0.6	0.7
20	9	11	.06382	.06347	.06296	.06229	.06147	.06052	.05943	.05822
20	9	12	.05822	.05815	.05792	.05755	.05703	.05638	.05559	.05469
20	9	13	.05326	.05342	.05344	.05332	.05307	.05268	.05216	.05153
20	9	14	.04879	.04915	.04939	.04949	.04947	.04932	.04905	.04866
20	9	15	.04467	.04521	.04564	.04595	.04614	.04621	.04616	.04600
20	9	16	.04080	.04151	.04211	.04261	.04299	.04327	.04343	.04348
20	9	17	.03707	.03793	.03870	.03937	.03995	.04042	.04079	.04105
20	9	18	.03334	.03434	.03527	.03612	.03688	.03756	.03814	.03862
20	9	19	.02937	.03052	.03161	.03265	.03361	.03450	.03531	.03603
20	9	20	.02448	.02580	.02710	.02836	.02958	.03075	.03185	.03289
20	10	11	.06993	.06984	.06957	.06912	.06851	.06773	.06680	.06572
20	10	12	.06382	.06400	.06403	.06388	.06358	.06312	.06251	.06175
20	10	13	.05840	.05882	.05909	.05921	.05918	.05900	.05867	.05820
20	10	14	.05351	.05413	.05462	.05497	.05518	.05525	.05517	.05497
20	10	15	.04900	.04981	.05049	.05105	.05147	.05177	.05194	.05197
20	10	16	.04477	.04573	.04659	.04734	.04798	.04849	.04888	.04915
20	10	17	.04068	.04180	.04283	.04376	.04458	.04530	.04591	.04641
20	10	18	.03659	.03785	.03904	.04015	.04117	.04210	.04293	.04366
20	10	19	.03224	.03364	.03500	.03630	.03753	.03869	.03976	.04074
20	10	20	.02688	.02845	.03001	.03154	.03303	.03448	.03587	.03720
20	11	12	.07025	.07075	.07108	.07122	.07118	.07096	.07058	.07002
20	11	13	.06431	.06505	.06562	.06603	.06627	.06635	.06626	.06601
20	11	14	.05894	.05988	.06067	.06132	.06181	.06215	.06233	.06237
20	11	15	.05399	.05511	.05610	.05696	.05768	.05826	.05869	.05898
20	11	16	.04934	.05062	.05179	.05284	.05377	.05457	.05525	.05578
20	11	17	.04485	.04627	.04761	.04885	.04998	.05100	.05190	.05268
20	11	18	.04035	.04191	.04341	.04483	.04617	.04741	.04855	.04958
20	11	19	.03556	.03726	.03892	.04054	.04209	.04357	.04497	.04627
20	11	20	.02966	.03152	.03338	.03523	.03706	.03884	.04058	.04225
20	12	13	.07126	.07238	.07334	.07411	.07470	.07510	.07532	.07536
20	12	14	.06533	.06666	.06783	.06884	.06969	.07037	.07088	.07122
20	12	15	.05987	.06137	.06274	.06397	.06505	.06598	.06676	.06738
20	12	16	.05472	.05638	.05793	.05936	.06066	.06183	.06286	.06374
20	12	17	.04975	.05155	.05327	.05489	.05640	.05780	.05907	.06021
20	12	18	.04478	.04671	.04859	.05039	.05211	.05374	.05526	.05667
20	12	19	.03947	.04154	.04358	.04557	.04752	.04940	.05120	.05291
20	12	20	.03293	.03514	.03738	.03962	.04185	.04405	.04621	.04832
20	13	14	.07304	.07485	.07650	.07799	.07929	.08042	.08135	.08210
20	13	15	.06696	.06894	.07079	.07249	.07404	.07543	.07665	.07769
20	13	16	.06123	.06336	.06538	.06729	.06907	.07071	.07219	.07352
20	13	17	.05569	.05795	.06015	.06224	.06424	.06611	.06786	.06947
20	13	18	.05013	.05253	.05487	.05716	.05937	.06149	.06351	.06541
20	13	19	.04420	.04672	.04923	.05171	.05416	.05654	.05885	.06108
20	13	20	.03689	.03955	.04225	.04497	.04771	.05044	.05314	.05581
20	14	15	.07577	.07837	.08085	.08318	.08535	.08734	.08916	.09078
20	14	16	.06932	.07206	.07471	.07725	.07965	.08191	.08401	.08594
20	14	17	.06307	.06595	.06875	.07148	.07411	.07662	.07900	.08124
20	14	18	.05681	.05979	.06276	.06567	.06852	.07129	.07396	.07651
20	14	19	.05011	.05321	.05633	.05944	.06253	.06557	.06856	.07148
20	14	20	.04184	.04506	.04836	.05171	.05511	.05852	.06193	.06533
20	15	16	.07978	.08335	.08684	.09023	.09349	.09661	.09957	.10235
20	15	17	.07263	.07632	.07996	.08354	.08704	.09042	.09368	.09680
20	15	18	.06545	.06924	.07303	.07679	.08051	.08417	.08774	.09121
20	15	19	.05777	.06165	.06558	.06954	.07351	.07746	.08138	.08525
20	15	20	.04827	.05224	.05633	.06054	.06482	.06917	.07355	.07795

Covariances of Inverse Gaussian Order Statistics
Shape Parameter k

n	i	j	0.0	0.1	0.2	0.3	0.4	0.5	0.6	0.7
20	16	17	.08564	.09049	.09533	.10014	.10489	.10956	.11411	.11853
20	16	18	.07724	.08216	.08713	.09212	.09710	.10205	.10695	.11176
20	16	19	.06822	.07321	.07830	.08348	.08872	.09399	.09926	.10451
20	16	20	.05706	.06209	.06732	.07273	.07830	.08399	.08978	.09564
20	17	18	.09450	.10117	.10797	.11488	.12184	.12884	.13584	.14279
20	17	19	.08358	.09025	.09714	.10422	.11144	.11878	.12620	.13366
20	17	20	.06999	.07664	.08363	.09091	.09847	.10627	.11427	.12244
20	18	19	.10881	.11844	.12849	.13891	.14967	.16072	.17202	.18351
20	18	20	.09132	.10080	.11084	.12143	.13251	.14406	.15604	.16839
20	19	20	.13449	.15011	.16688	.18478	.20377	.22382	.24486	.26684
21	1	2	.13238	.11796	.10472	.09263	.08168	.07180	.06294	.05505
21	1	3	.08985	.08096	.07269	.06504	.05801	.05158	.04574	.04046
21	1	4	.06887	.06255	.05661	.05106	.04591	.04116	.03679	.03281
21	1	5	.05617	.05133	.04676	.04245	.03841	.03466	.03119	.02800
21	1	6	.04754	.04369	.04001	.03653	.03324	.03016	.02730	.02464
21	1	7	.04125	.03809	.03506	.03216	.02941	.02682	.02439	.02213
21	1	8	.03641	.03377	.03122	.02877	.02643	.02422	.02213	.02017
21	1	9	.03254	.03031	.02814	.02604	.02403	.02211	.02029	.01857
21	1	10	.02935	.02745	.02559	.02378	.02203	.02036	.01876	.01724
21	1	11	.02665	.02503	.02343	.02186	.02034	.01886	.01745	.01611
21	1	12	.02433	.02294	.02156	.02019	.01886	.01756	.01631	.01512
21	1	13	.02228	.02110	.01990	.01872	.01755	.01641	.01531	.01424
21	1	14	.02046	.01944	.01842	.01740	.01638	.01538	.01440	.01345
21	1	15	.01879	.01794	.01707	.01619	.01531	.01443	.01357	.01272
21	1	16	.01725	.01654	.01581	.01506	.01431	.01355	.01279	.01205
21	1	17	.01579	.01522	.01462	.01399	.01335	.01271	.01205	.01141
21	1	18	.01438	.01394	.01346	.01296	.01243	.01189	.01134	.01078
21	1	19	.01296	.01264	.01229	.01191	.01149	.01106	.01062	.01016
21	1	20	.01144	.01126	.01103	.01078	.01049	.01018	.00984	.00949
21	1	21	.00955	.00954	.00948	.00938	.00925	.00908	.00889	.00867
21	2	3	.10675	.09755	.08885	.08067	.07302	.06589	.05930	.05323
21	2	4	.08199	.07552	.06934	.06347	.05791	.05269	.04781	.04327
21	2	5	.06695	.06207	.05735	.05283	.04852	.04444	.04058	.03697
21	2	6	.05673	.05287	.04913	.04551	.04203	.03870	.03555	.03256
21	2	7	.04924	.04612	.04307	.04009	.03721	.03444	.03179	.02927
21	2	8	.04348	.04091	.03837	.03588	.03346	.03111	.02885	.02668
21	2	9	.03887	.03673	.03460	.03249	.03043	.02841	.02646	.02458
21	2	10	.03507	.03328	.03147	.02968	.02791	.02617	.02447	.02283
21	2	11	.03186	.03035	.02882	.02729	.02576	.02425	.02277	.02133
21	2	12	.02909	.02782	.02652	.02521	.02390	.02259	.02129	.02002
21	2	13	.02665	.02559	.02450	.02338	.02225	.02111	.01998	.01886
21	2	14	.02447	.02359	.02268	.02173	.02076	.01978	.01880	.01782
21	2	15	.02248	.02177	.02101	.02022	.01940	.01857	.01771	.01686
21	2	16	.02064	.02008	.01947	.01882	.01814	.01743	.01670	.01597
21	2	17	.01890	.01847	.01800	.01749	.01693	.01635	.01574	.01512
21	2	18	.01721	.01692	.01657	.01619	.01576	.01530	.01481	.01429
21	2	19	.01551	.01535	.01514	.01488	.01458	.01424	.01387	.01347
21	2	20	.01369	.01366	.01359	.01347	.01331	.01310	.01286	.01258
21	2	21	.01144	.01158	.01167	.01172	.01173	.01169	.01162	.01150
21	3	4	.09246	.08592	.07959	.07351	.06769	.06216	.05692	.05199
21	3	5	.07559	.07068	.06590	.06126	.05678	.05248	.04838	.04448
21	3	6	.06409	.06026	.05649	.05280	.04922	.04574	.04240	.03921
21	3	7	.05567	.05260	.04955	.04655	.04360	.04073	.03794	.03526
21	3	8	.04917	.04667	.04417	.04168	.03922	.03680	.03445	.03216
21	3	9	.04398	.04192	.03984	.03776	.03568	.03363	.03161	.02964

Covariances of Inverse Gaussian Order Statistics
Shape Parameter k

n	i	j	0.0	0.1	0.2	0.3	0.4	0.5	0.6	0.7
21	3	10	.03969	.03798	.03625	.03450	.03273	.03098	.02924	.02753
21	3	11	.03606	.03465	.03320	.03172	.03022	.02872	.02721	.02572
21	3	12	.03293	.03177	.03056	.02932	.02804	.02675	.02545	.02415
21	3	13	.03017	.02923	.02823	.02719	.02611	.02501	.02389	.02276
21	3	14	.02770	.02695	.02613	.02527	.02437	.02343	.02247	.02150
21	3	15	.02546	.02487	.02422	.02352	.02278	.02199	.02118	.02035
21	3	16	.02337	.02294	.02244	.02189	.02129	.02065	.01998	.01927
21	3	17	.02140	.02111	.02075	.02034	.01988	.01938	.01883	.01825
21	3	18	.01949	.01933	.01911	.01883	.01851	.01813	.01771	.01726
21	3	19	.01757	.01754	.01745	.01731	.01712	.01688	.01659	.01626
21	3	20	.01551	.01562	.01567	.01568	.01563	.01553	.01538	.01519
21	3	21	.01296	.01323	.01346	.01364	.01378	.01386	.01390	.01388
21	4	5	.08358	.07869	.07387	.06914	.06453	.06007	.05577	.05164
21	4	6	.07092	.06713	.06337	.05964	.05598	.05240	.04892	.04555
21	4	7	.06163	.05863	.05561	.05260	.04962	.04668	.04379	.04099
21	4	8	.05447	.05205	.04959	.04712	.04465	.04220	.03977	.03740
21	4	9	.04872	.04676	.04474	.04270	.04063	.03857	.03651	.03448
21	4	10	.04398	.04238	.04072	.03902	.03729	.03554	.03378	.03203
21	4	11	.03997	.03867	.03731	.03589	.03444	.03295	.03145	.02994
21	4	12	.03650	.03546	.03435	.03317	.03196	.03070	.02942	.02812
21	4	13	.03345	.03263	.03173	.03077	.02976	.02870	.02761	.02650
21	4	14	.03072	.03009	.02938	.02861	.02778	.02690	.02598	.02503
21	4	15	.02823	.02777	.02723	.02663	.02597	.02525	.02449	.02360
21	4	16	.02593	.02561	.02523	.02478	.02428	.02371	.02310	.02244
21	4	17	.02374	.02357	.02334	.02303	.02267	.02225	.02178	.02126
21	4	18	.02162	.02159	.02149	.02133	.02110	.02082	.02049	.02010
21	4	19	.01949	.01959	.01963	.01961	.01952	.01938	.01918	.01894
21	4	20	.01721	.01744	.01763	.01775	.01782	.01783	.01779	.01769
21	4	21	.01438	.01478	.01514	.01545	.01571	.01592	.01608	.01618
21	5	6	.07766	.07394	.07019	.06645	.06273	.05907	.05547	.05197
21	5	7	.06753	.06460	.06163	.05864	.05563	.05265	.04969	.04678
21	5	8	.05970	.05737	.05498	.05255	.05009	.04761	.04515	.04270
21	5	9	.05343	.05156	.04963	.04763	.04560	.04353	.04146	.03938
21	5	10	.04824	.04675	.04518	.04354	.04185	.04012	.03837	.03660
21	5	11	.04385	.04266	.04140	.04006	.03866	.03721	.03573	.03422
21	5	12	.04005	.03913	.03812	.03703	.03588	.03467	.03342	.03214
21	5	13	.03671	.03601	.03522	.03435	.03342	.03242	.03138	.03029
21	5	14	.03372	.03321	.03261	.03194	.03120	.03039	.02953	.02862
21	5	15	.03099	.03065	.03023	.02974	.02917	.02853	.02784	.02709
21	5	16	.02846	.02828	.02802	.02768	.02727	.02680	.02626	.02566
21	5	17	.02607	.02603	.02592	.02573	.02547	.02514	.02476	.02431
21	5	18	.02374	.02384	.02387	.02383	.02371	.02353	.02329	.02299
21	5	19	.02140	.02163	.02180	.02190	.02194	.02191	.02181	.02166
21	5	20	.01890	.01927	.01958	.01984	.02003	.02016	.02023	.02024
21	5	21	.01579	.01633	.01682	.01727	.01766	.01800	.01828	.01851
21	6	7	.07358	.07074	.06783	.06486	.06186	.05884	.05583	.05284
21	6	8	.06508	.06285	.06054	.05816	.05572	.05324	.05075	.04826
21	6	9	.05825	.05650	.05466	.05273	.05074	.04870	.04662	.04452
21	6	10	.05261	.05124	.04977	.04822	.04659	.04490	.04316	.04138
21	6	11	.04784	.04678	.04562	.04437	.04304	.04165	.04020	.03870
21	6	12	.04371	.04291	.04201	.04103	.03996	.03882	.03761	.03635
21	6	13	.04007	.03949	.03883	.03807	.03722	.03630	.03531	.03427
21	6	14	.03680	.03643	.03596	.03540	.03476	.03403	.03324	.03239
21	6	15	.03383	.03363	.03334	.03296	.03249	.03195	.03134	.03066
21	6	16	.03107	.03103	.03090	.03068	.03039	.03001	.02956	.02905

Covariances of Inverse Gaussian Order Statistics
Shape Parameter k

n	i	j	0.0	0.1	0.2	0.3	0.4	0.5	0.6	0.7
21	6	17	.02846	.02856	.02858	.02852	.02838	.02817	.02788	.02752
21	6	18	.02593	.02616	.02633	.02642	.02643	.02636	.02623	.02602
21	6	19	.02337	.02374	.02405	.02429	.02445	.02454	.02457	.02452
21	6	20	.02064	.02115	.02160	.02200	.02233	.02259	.02279	.02292
21	6	21	.01725	.01793	.01856	.01915	.01969	.02017	.02059	.02095
21	7	8	.07073	.06863	.06641	.06409	.06170	.05924	.05674	.05421
21	7	9	.06334	.06172	.05998	.05814	.05621	.05420	.05213	.05002
21	7	10	.05722	.05599	.05464	.05318	.05162	.04998	.04828	.04652
21	7	11	.05204	.05112	.05009	.04895	.04771	.04638	.04498	.04351
21	7	12	.04756	.04690	.04614	.04527	.04430	.04323	.04209	.04088
21	7	13	.04360	.04318	.04265	.04201	.04127	.04044	.03953	.03854
21	7	14	.04006	.03983	.03950	.03907	.03854	.03792	.03721	.03643
21	7	15	.03683	.03678	.03663	.03638	.03604	.03561	.03509	.03449
21	7	16	.03383	.03394	.03395	.03388	.03371	.03345	.03311	.03268
21	7	17	.03099	.03125	.03141	.03149	.03149	.03139	.03122	.03096
21	7	18	.02823	.02862	.02894	.02917	.02932	.02939	.02938	.02929
21	7	19	.02546	.02598	.02644	.02682	.02713	.02736	.02752	.02760
21	7	20	.02248	.02314	.02375	.02429	.02477	.02519	.02553	.02579
21	7	21	.01879	.01962	.02041	.02115	.02185	.02249	.02307	.02359
21	8	9	.06880	.06733	.06572	.06398	.06213	.06018	.05814	.05604
21	8	10	.06218	.06110	.05989	.05854	.05708	.05551	.05386	.05213
21	8	11	.05656	.05581	.05492	.05390	.05277	.05152	.05019	.04877
21	8	12	.05170	.05121	.05060	.04986	.04900	.04804	.04698	.04583
21	8	13	.04741	.04716	.04678	.04628	.04567	.04495	.04413	.04322
21	8	14	.04357	.04351	.04334	.04305	.04265	.04215	.04155	.04086
21	8	15	.04006	.04018	.04019	.04009	.03989	.03959	.03919	.03869
21	8	16	.03680	.03708	.03726	.03734	.03731	.03719	.03697	.03667
21	8	17	.03372	.03414	.03448	.03471	.03486	.03491	.03487	.03474
21	8	18	.03072	.03128	.03176	.03216	.03247	.03269	.03282	.03286
21	8	19	.02770	.02840	.02902	.02957	.03004	.03044	.03074	.03097
21	8	20	.02447	.02530	.02607	.02679	.02744	.02802	.02852	.02895
21	8	21	.02046	.02145	.02241	.02333	.02420	.02502	.02578	.02648
21	9	10	.06760	.06670	.06565	.06444	.06310	.06163	.06005	.05836
21	9	11	.06151	.06094	.06022	.05935	.05835	.05722	.05597	.05462
21	9	12	.05624	.05594	.05550	.05492	.05420	.05336	.05241	.05134
21	9	13	.05159	.05152	.05132	.05098	.05052	.04994	.04924	.04843
21	9	14	.04741	.04755	.04756	.04744	.04720	.04684	.04637	.04579
21	9	15	.04360	.04391	.04411	.04419	.04415	.04400	.04374	.04337
21	9	16	.04007	.04054	.04090	.04116	.04130	.04134	.04128	.04111
21	9	17	.03671	.03733	.03785	.03827	.03859	.03881	.03893	.03895
21	9	18	.03345	.03421	.03487	.03546	.03595	.03634	.03665	.03685
21	9	19	.03017	.03105	.03187	.03261	.03327	.03385	.03433	.03473
21	9	20	.02665	.02767	.02864	.02954	.03039	.03116	.03185	.03247
21	9	21	.02228	.02346	.02461	.02573	.02681	.02783	.02880	.02970
21	10	11	.06702	.06667	.06615	.06546	.06462	.06363	.06250	.06124
21	10	12	.06130	.06122	.06098	.06059	.06004	.05936	.05853	.05758
21	10	13	.05624	.05639	.05640	.05626	.05598	.05556	.05501	.05433
21	10	14	.05170	.05206	.05228	.05236	.05231	.05213	.05181	.05138
21	10	15	.04756	.04809	.04850	.04878	.04894	.04897	.04888	.04867
21	10	16	.04371	.04440	.04498	.04544	.04579	.04602	.04614	.04614
21	10	17	.04005	.04089	.04163	.04226	.04279	.04321	.04352	.04373
21	10	18	.03650	.03748	.03836	.03916	.03987	.04047	.04097	.04137
21	10	19	.03293	.03403	.03506	.03602	.03690	.03769	.03839	.03900
21	10	20	.02909	.03032	.03151	.03264	.03371	.03471	.03563	.03646
21	10	21	.02433	.02571	.02709	.02843	.02974	.03101	.03221	.03336

Covariances of Inverse Gaussian Order Statistics
Shape Parameter k

n	i	j	0.0	0.1	0.2	0.3	0.4	0.5	0.6	0.7
21	11	12	.06702	.06720	.06721	.06705	.06672	.06622	.06557	.06476
21	11	13	.06151	.06193	.06219	.06228	.06222	.06200	.06163	.06112
21	11	14	.05656	.05718	.05765	.05798	.05816	.05819	.05807	.05782
21	11	15	.05204	.05283	.05350	.05403	.05442	.05468	.05479	.05478
21	11	16	.04784	.04879	.04962	.05034	.05093	.05140	.05173	.05194
21	11	17	.04385	.04494	.04594	.04683	.04761	.04827	.04881	.04923
21	11	18	.03997	.04120	.04234	.04340	.04436	.04521	.04596	.04659
21	11	19	.03606	.03741	.03871	.03993	.04106	.04211	.04307	.04393
21	11	20	.03186	.03335	.03479	.03618	.03752	.03878	.03997	.04108
21	11	21	.02665	.02828	.02991	.03153	.03311	.03465	.03615	.03758
21	12	13	.06760	.06833	.06889	.06928	.06949	.06952	.06939	.06909
21	12	14	.06218	.06311	.06389	.06451	.06497	.06526	.06540	.06537
21	12	15	.05722	.05833	.05930	.06013	.06081	.06134	.06172	.06195
21	12	16	.05261	.05388	.05502	.05604	.05693	.05768	.05829	.05875
21	12	17	.04824	.04964	.05095	.05214	.05322	.05418	.05501	.05570
21	12	18	.04398	.04552	.04697	.04833	.04960	.05075	.05180	.05273
21	12	19	.03969	.04134	.04294	.04447	.04592	.04729	.04856	.04972
21	12	20	.03507	.03685	.03861	.04031	.04197	.04356	.04507	.04650
21	12	21	.02935	.03127	.03320	.03513	.03705	.03893	.04077	.04255
21	13	14	.06880	.07012	.07128	.07226	.07307	.07371	.07416	.07444
21	13	15	.06334	.06483	.06618	.06738	.06842	.06930	.07002	.07056
21	13	16	.05825	.05990	.06142	.06281	.06407	.06518	.06613	.06694
21	13	17	.05343	.05521	.05689	.05846	.05991	.06124	.06243	.06348
21	13	18	.04872	.05063	.05246	.05420	.05585	.05738	.05880	.06010
21	13	19	.04398	.04600	.04797	.04989	.05173	.05348	.05513	.05668
21	13	20	.03887	.04102	.04314	.04523	.04728	.04927	.05119	.05302
21	13	21	.03254	.03481	.03711	.03943	.04174	.04404	.04631	.04853
21	14	15	.07073	.07270	.07453	.07621	.07771	.07904	.08019	.08116
21	14	16	.06508	.06720	.06920	.07107	.07279	.07436	.07577	.07702
21	14	17	.05970	.06195	.06411	.06616	.06809	.06989	.07155	.07306
21	14	18	.05447	.05684	.05914	.06136	.06349	.06551	.06742	.06919
21	14	19	.04917	.05166	.05410	.05650	.05882	.06107	.06322	.06527
21	14	20	.04348	.04607	.04866	.05124	.05378	.05628	.05872	.06108
21	14	21	.03641	.03911	.04188	.04468	.04750	.05033	.05314	.05592
21	15	16	.07358	.07632	.07894	.08144	.08378	.08597	.08798	.08982
21	15	17	.06753	.07039	.07317	.07585	.07841	.08083	.08311	.08523
21	15	18	.06163	.06460	.06752	.07037	.07314	.07580	.07834	.08075
21	15	19	.05567	.05874	.06179	.06481	.06778	.07068	.07349	.07621
21	15	20	.04924	.05241	.05561	.05881	.06200	.06516	.06828	.07133
21	15	21	.04125	.04452	.04787	.05130	.05478	.05829	.06182	.06533
21	16	17	.07766	.08135	.08497	.08850	.09193	.09522	.09836	.10135
21	16	18	.07092	.07470	.07845	.08216	.08579	.08933	.09276	.09606
21	16	19	.06409	.06795	.07183	.07571	.07955	.08334	.08706	.09069
21	16	20	.05673	.06067	.06467	.06873	.07280	.07687	.08092	.08493
21	16	21	.04754	.05156	.05571	.05999	.06436	.06880	.07330	.07782
21	17	18	.08358	.08852	.09346	.09840	.10329	.10811	.11284	.11746
21	17	19	.07559	.08058	.08564	.09073	.09584	.10093	.10598	.11096
21	17	20	.06695	.07199	.07716	.08243	.08777	.09316	.09857	.10397
21	17	21	.05617	.06124	.06652	.07200	.07765	.08345	.08935	.09534
21	18	19	.09246	.09919	.10607	.11308	.12017	.12732	.13448	.14162
21	18	20	.08199	.08871	.09567	.10283	.11016	.11762	.12519	.13282
21	18	21	.06887	.07556	.08259	.08994	.09758	.10548	.11360	.12191
21	19	20	.10675	.11641	.12651	.13701	.14788	.15906	.17051	.18219
21	19	21	.08985	.09935	.10943	.12007	.13124	.14290	.15500	.16751
21	20	21	.13238	.14799	.16477	.18272	.20179	.22195	.24313	.26529

Covariances of Inverse Gaussian Order Statistics
Shape Parameter k

n	i	j	0.0	0.1	0.2	0.3	0.4	0.5	0.6	0.7
22	1	2	.13041	.11602	.10283	.09082	.07994	.07015	.06139	.05360
22	1	3	.08848	.07960	.07134	.06372	.05673	.05035	.04457	.03935
22	1	4	.06782	.06149	.05555	.05001	.04488	.04016	.03583	.03189
22	1	5	.05533	.05047	.04589	.04157	.03755	.03382	.03037	.02721
22	1	6	.04686	.04298	.03928	.03578	.03250	.02943	.02657	.02394
22	1	7	.04068	.03749	.03443	.03152	.02876	.02617	.02375	.02150
22	1	8	.03594	.03326	.03068	.02821	.02586	.02364	.02155	.01959
22	1	9	.03215	.02988	.02768	.02556	.02352	.02159	.01977	.01805
22	1	10	.02904	.02709	.02520	.02336	.02159	.01989	.01828	.01676
22	1	11	.02641	.02474	.02310	.02149	.01994	.01845	.01702	.01567
22	1	12	.02415	.02271	.02128	.01988	.01852	.01719	.01592	.01471
22	1	13	.02217	.02093	.01969	.01846	.01726	.01609	.01496	.01387
22	1	14	.02041	.01934	.01826	.01719	.01613	.01510	.01409	.01311
22	1	15	.01882	.01790	.01697	.01604	.01511	.01419	.01330	.01243
22	1	16	.01735	.01657	.01578	.01497	.01416	.01336	.01257	.01179
22	1	17	.01598	.01533	.01466	.01397	.01328	.01258	.01188	.01119
22	1	18	.01467	.01415	.01359	.01302	.01243	.01183	.01122	.01062
22	1	19	.01340	.01299	.01255	.01208	.01159	.01109	.01058	.01006
22	1	20	.01211	.01181	.01149	.01113	.01074	.01034	.00992	.00950
22	1	21	.01071	.01054	.01034	.01010	.00983	.00953	.00922	.00889
22	1	22	.00898	.00896	.00890	.00881	.00868	.00853	.00834	.00814
22	2	3	.10483	.09563	.08694	.07879	.07118	.06411	.05758	.05159
22	2	4	.08052	.07403	.06784	.06197	.05643	.05124	.04639	.04190
22	2	5	.06577	.06085	.05611	.05158	.04728	.04320	.03937	.03579
22	2	6	.05575	.05185	.04808	.04444	.04095	.03763	.03448	.03152
22	2	7	.04843	.04526	.04217	.03916	.03627	.03349	.03084	.02833
22	2	8	.04280	.04018	.03760	.03507	.03262	.03026	.02799	.02583
22	2	9	.03830	.03610	.03393	.03178	.02969	.02765	.02569	.02380
22	2	10	.03460	.03275	.03089	.02906	.02725	.02548	.02377	.02211
22	2	11	.03148	.02991	.02832	.02674	.02518	.02364	.02213	.02067
22	2	12	.02880	.02746	.02610	.02474	.02338	.02204	.02071	.01941
22	2	13	.02644	.02531	.02415	.02298	.02180	.02062	.01945	.01831
22	2	14	.02434	.02339	.02241	.02140	.02038	.01935	.01833	.01731
22	2	15	.02244	.02165	.02082	.01997	.01909	.01820	.01730	.01640
22	2	16	.02070	.02005	.01936	.01864	.01790	.01713	.01635	.01557
22	2	17	.01906	.01855	.01800	.01740	.01678	.01613	.01546	.01478
22	2	18	.01751	.01712	.01669	.01621	.01570	.01517	.01460	.01403
22	2	19	.01599	.01572	.01540	.01505	.01465	.01422	.01377	.01329
22	2	20	.01444	.01430	.01410	.01386	.01358	.01326	.01292	.01254
22	2	21	.01278	.01276	.01269	.01258	.01242	.01223	.01200	.01174
22	2	22	.01071	.01084	.01093	.01097	.01098	.01094	.01086	.01075
22	3	4	.09057	.08400	.07765	.07157	.06577	.06026	.05507	.05019
22	3	5	.07406	.06911	.06430	.05964	.05516	.05087	.04678	.04292
22	3	6	.06282	.05894	.05513	.05142	.04781	.04434	.04100	.03782
22	3	7	.05460	.05148	.04838	.04534	.04237	.03948	.03669	.03401
22	3	8	.04828	.04571	.04315	.04062	.03813	.03569	.03332	.03102
22	3	9	.04322	.04109	.03895	.03682	.03471	.03262	.03058	.02860
22	3	10	.03905	.03728	.03548	.03367	.03186	.03007	.02830	.02657
22	3	11	.03554	.03405	.03254	.03100	.02945	.02790	.02636	.02484
22	3	12	.03251	.03127	.02999	.02868	.02735	.02601	.02467	.02334
22	3	13	.02986	.02883	.02775	.02664	.02550	.02435	.02318	.02201
22	3	14	.02749	.02665	.02575	.02482	.02385	.02285	.02184	.02082
22	3	15	.02535	.02467	.02394	.02316	.02234	.02149	.02062	.01973
22	3	16	.02338	.02285	.02226	.02162	.02095	.02023	.01949	.01872
22	3	17	.02153	.02114	.02069	.02018	.01964	.01905	.01843	.01778

Covariances of Inverse Gaussian Order Statistics
Shape Parameter k

n	i	j	0.0	0.1	0.2	0.3	0.4	0.5	0.6	0.7
22	3	18	.01978	.01951	.01918	.01881	.01838	.01791	.01741	.01687
22	3	19	.01806	.01791	.01771	.01746	.01715	.01680	.01641	.01599
22	3	20	.01632	.01629	.01621	.01608	.01590	.01567	.01540	.01509
22	3	21	.01444	.01454	.01459	.01459	.01455	.01445	.01431	.01412
22	3	22	.01211	.01236	.01257	.01273	.01285	.01293	.01295	.01294
22	4	5	.08169	.07674	.07188	.06712	.06251	.05805	.05376	.04967
22	4	6	.06934	.06549	.06167	.05791	.05422	.05063	.04715	.04380
22	4	7	.06030	.05722	.05415	.05109	.04807	.04511	.04222	.03941
22	4	8	.05333	.05084	.04832	.04579	.04328	.04079	.03835	.03596
22	4	9	.04776	.04571	.04363	.04152	.03941	.03730	.03521	.03316
22	4	10	.04317	.04148	.03975	.03798	.03619	.03439	.03260	.03082
22	4	11	.03929	.03790	.03646	.03497	.03345	.03191	.03037	.02882
22	4	12	.03595	.03481	.03361	.03236	.03108	.02976	.02843	.02708
22	4	13	.03302	.03209	.03111	.03007	.02898	.02786	.02671	.02554
22	4	14	.03040	.02967	.02887	.02801	.02710	.02615	.02517	.02416
22	4	15	.02804	.02747	.02683	.02614	.02539	.02460	.02376	.02290
22	4	16	.02586	.02544	.02496	.02441	.02381	.02316	.02246	.02173
22	4	17	.02383	.02354	.02320	.02279	.02232	.02180	.02124	.02064
22	4	18	.02188	.02173	.02151	.02123	.02090	.02051	.02007	.01959
22	4	19	.01998	.01995	.01986	.01971	.01950	.01924	.01892	.01856
22	4	20	.01806	.01815	.01818	.01816	.01808	.01794	.01775	.01752
22	4	21	.01599	.01620	.01637	.01648	.01654	.01654	.01650	.01640
22	4	22	.01340	.01377	.01410	.01438	.01462	.01480	.01494	.01502
22	5	6	.07574	.07193	.06812	.06433	.06058	.05690	.05330	.04980
22	5	7	.06589	.06288	.05984	.05678	.05373	.05071	.04774	.04483
22	5	8	.05830	.05588	.05341	.05091	.04840	.04588	.04338	.04092
22	5	9	.05223	.05027	.04824	.04618	.04408	.04196	.03985	.03774
22	5	10	.04722	.04562	.04396	.04225	.04049	.03870	.03690	.03509
22	5	11	.04298	.04169	.04033	.03891	.03743	.03592	.03438	.03282
22	5	12	.03933	.03830	.03719	.03602	.03478	.03350	.03219	.03085
22	5	13	.03613	.03532	.03443	.03346	.03244	.03137	.03025	.02910
22	5	14	.03328	.03265	.03195	.03118	.03034	.02945	.02851	.02753
22	5	15	.03069	.03024	.02970	.02910	.02843	.02770	.02692	.02609
22	5	16	.02831	.02801	.02763	.02718	.02666	.02608	.02545	.02477
22	5	17	.02608	.02592	.02568	.02537	.02500	.02456	.02407	.02352
22	5	18	.02396	.02392	.02382	.02365	.02341	.02310	.02274	.02233
22	5	19	.02188	.02197	.02199	.02195	.02184	.02167	.02144	.02116
22	5	20	.01978	.01999	.02014	.02022	.02025	.02022	.02012	.01997
22	5	21	.01751	.01784	.01813	.01836	.01853	.01864	.01870	.01870
22	5	22	.01467	.01517	.01562	.01602	.01638	.01668	.01693	.01713
22	6	7	.07159	.06865	.06565	.06261	.05955	.05649	.05345	.05045
22	6	8	.06337	.06103	.05863	.05616	.05366	.05113	.04860	.04608
22	6	9	.05678	.05492	.05297	.05096	.04889	.04678	.04465	.04251
22	6	10	.05134	.04986	.04828	.04663	.04492	.04315	.04135	.03953
22	6	11	.04675	.04557	.04431	.04296	.04154	.04006	.03854	.03698
22	6	12	.04279	.04187	.04086	.03977	.03860	.03737	.03609	.03477
22	6	13	.03931	.03862	.03783	.03696	.03601	.03500	.03392	.03280
22	6	14	.03621	.03571	.03512	.03444	.03368	.03286	.03197	.03103
22	6	15	.03340	.03307	.03265	.03215	.03156	.03091	.03020	.02942
22	6	16	.03082	.03064	.03037	.03003	.02960	.02911	.02855	.02793
22	6	17	.02839	.02835	.02823	.02804	.02776	.02741	.02700	.02652
22	6	18	.02608	.02617	.02619	.02613	.02599	.02579	.02552	.02518
22	6	19	.02383	.02404	.02418	.02426	.02426	.02419	.02406	.02386
22	6	20	.02153	.02187	.02214	.02235	.02249	.02257	.02258	.02253
22	6	21	.01906	.01953	.01994	.02029	.02058	.02081	.02098	.02109

Covariances of Inverse Gaussian Order Statistics
Shape Parameter k

n	i	j	0.0	0.1	0.2	0.3	0.4	0.5	0.6	0.7
22	6	22	.01598	.01660	.01717	.01771	.01819	.01863	.01900	.01932
22	7	8	.06865	.06642	.06409	.06168	.05920	.05667	.05412	.05155
22	7	9	.06154	.05978	.05793	.05598	.05396	.05187	.04974	.04758
22	7	10	.05566	.05429	.05282	.05124	.04959	.04786	.04608	.04426
22	7	11	.05069	.04964	.04848	.04722	.04587	.04444	.04296	.04141
22	7	12	.04641	.04562	.04472	.04372	.04264	.04147	.04023	.03894
22	7	13	.04265	.04208	.04141	.04064	.03978	.03884	.03782	.03674
22	7	14	.03929	.03891	.03844	.03787	.03721	.03647	.03565	.03477
22	7	15	.03624	.03604	.03575	.03536	.03488	.03431	.03367	.03296
22	7	16	.03344	.03339	.03326	.03303	.03271	.03232	.03184	.03130
22	7	17	.03082	.03091	.03092	.03084	.03068	.03044	.03012	.02972
22	7	18	.02831	.02854	.02868	.02875	.02873	.02864	.02846	.02822
22	7	19	.02586	.02621	.02649	.02669	.02682	.02687	.02684	.02675
22	7	20	.02338	.02385	.02426	.02460	.02487	.02506	.02519	.02525
22	7	21	.02070	.02130	.02184	.02233	.02275	.02312	.02341	.02364
22	7	22	.01735	.01810	.01882	.01949	.02012	.02069	.02120	.02166
22	8	9	.06660	.06498	.06323	.06136	.05940	.05735	.05523	.05307
22	8	10	.06026	.05903	.05767	.05619	.05461	.05293	.05119	.04937
22	8	11	.05490	.05398	.05294	.05178	.05052	.04917	.04773	.04621
22	8	12	.05027	.04962	.04885	.04796	.04697	.04588	.04471	.04346
22	8	13	.04620	.04578	.04524	.04459	.04383	.04298	.04204	.04102
22	8	14	.04257	.04234	.04201	.04156	.04101	.04037	.03963	.03882
22	8	15	.03928	.03922	.03906	.03880	.03844	.03799	.03744	.03681
22	8	16	.03624	.03634	.03635	.03625	.03606	.03578	.03541	.03495
22	8	17	.03340	.03365	.03380	.03386	.03383	.03370	.03349	.03320
22	8	18	.03069	.03107	.03136	.03156	.03168	.03171	.03166	.03152
22	8	19	.02804	.02854	.02896	.02930	.02957	.02975	.02986	.02988
22	8	20	.02535	.02597	.02652	.02701	.02742	.02776	.02802	.02821
22	8	21	.02244	.02319	.02388	.02452	.02510	.02561	.02605	.02641
22	8	22	.01882	.01971	.02058	.02141	.02219	.02292	.02359	.02420
22	9	10	.06525	.06417	.06294	.06158	.06009	.05849	.05679	.05501
22	9	11	.05946	.05870	.05780	.05677	.05561	.05434	.05296	.05150
22	9	12	.05445	.05397	.05334	.05259	.05171	.05072	.04963	.04844
22	9	13	.05006	.04980	.04941	.04890	.04827	.04752	.04667	.04573
22	9	14	.04613	.04607	.04589	.04559	.04517	.04464	.04401	.04328
22	9	15	.04257	.04268	.04268	.04257	.04235	.04201	.04158	.04105
22	9	16	.03929	.03956	.03972	.03978	.03973	.03958	.03933	.03898
22	9	17	.03621	.03662	.03694	.03715	.03727	.03729	.03721	.03703
22	9	18	.03328	.03382	.03427	.03464	.03491	.03509	.03517	.03517
22	9	19	.03040	.03107	.03166	.03216	.03259	.03292	.03317	.03333
22	9	20	.02749	.02827	.02900	.02965	.03022	.03072	.03114	.03147
22	9	21	.02434	.02525	.02611	.02692	.02766	.02834	.02894	.02947
22	9	22	.02041	.02147	.02250	.02350	.02446	.02537	.02622	.02701
22	10	11	.06447	.06390	.06316	.06228	.06125	.06009	.05880	.05741
22	10	12	.05906	.05876	.05831	.05771	.05698	.05611	.05511	.05401
22	10	13	.05431	.05424	.05403	.05368	.05319	.05258	.05184	.05099
22	10	14	.05006	.05019	.05018	.05005	.04979	.04940	.04890	.04828
22	10	15	.04620	.04650	.04669	.04675	.04668	.04650	.04620	.04579
22	10	16	.04265	.04311	.04345	.04369	.04381	.04381	.04371	.04349
22	10	17	.03931	.03992	.04041	.04081	.04110	.04128	.04136	.04133
22	10	18	.03613	.03686	.03750	.03805	.03850	.03885	.03910	.03925
22	10	19	.03302	.03387	.03465	.03534	.03594	.03646	.03688	.03721
22	10	20	.02986	.03083	.03174	.03258	.03334	.03402	.03462	.03514
22	10	21	.02644	.02754	.02859	.02958	.03052	.03139	.03219	.03291
22	10	22	.02217	.02341	.02464	.02583	.02699	.02810	.02916	.03016

Covariances of Inverse Gaussian Order Statistics
Shape Parameter k

n	i	j	0.0	0.1	0.2	0.3	0.4	0.5	0.6	0.7
22	11	12	.06422	.06414	.06389	.06348	.06291	.06219	.06133	.06033
22	11	13	.05906	.05921	.05921	.05905	.05875	.05829	.05770	.05698
22	11	14	.05445	.05480	.05501	.05508	.05500	.05478	.05443	.05395
22	11	15	.05027	.05079	.05119	.05145	.05158	.05158	.05145	.05119
22	11	16	.04641	.04709	.04765	.04809	.04841	.04860	.04868	.04863
22	11	17	.04279	.04361	.04433	.04493	.04543	.04580	.04606	.04621
22	11	18	.03933	.04028	.04114	.04190	.04256	.04311	.04356	.04389
22	11	19	.03595	.03702	.03801	.03892	.03974	.04046	.04109	.04161
22	11	20	.03251	.03370	.03482	.03588	.03687	.03777	.03858	.03930
22	11	21	.02880	.03010	.03137	.03259	.03375	.03485	.03587	.03681
22	11	22	.02415	.02560	.02704	.02846	.02985	.03120	.03250	.03374
22	12	13	.06447	.06488	.06513	.06520	.06511	.06486	.06445	.06389
22	12	14	.05946	.06007	.06052	.06083	.06098	.06097	.06082	.06051
22	12	15	.05490	.05568	.05633	.05684	.05720	.05742	.05749	.05742
22	12	16	.05069	.05163	.05245	.05314	.05370	.05412	.05441	.05456
22	12	17	.04675	.04783	.04880	.04966	.05039	.05101	.05149	.05186
22	12	18	.04298	.04419	.04530	.04632	.04722	.04802	.04870	.04926
22	12	19	.03929	.04061	.04186	.04303	.04410	.04508	.04595	.04671
22	12	20	.03554	.03698	.03836	.03968	.04092	.04208	.04315	.04413
22	12	21	.03148	.03304	.03456	.03604	.03747	.03883	.04012	.04134
22	12	22	.02641	.02810	.02980	.03148	.03315	.03478	.03636	.03789
22	13	14	.06525	.06617	.06693	.06753	.06796	.06822	.06830	.06823
22	13	15	.06026	.06136	.06231	.06312	.06376	.06426	.06459	.06476
22	13	16	.05566	.05691	.05804	.05903	.05987	.06058	.06113	.06154
22	13	17	.05134	.05273	.05401	.05517	.05620	.05711	.05788	.05851
22	13	18	.04722	.04873	.05015	.05147	.05268	.05377	.05474	.05559
22	13	19	.04317	.04479	.04635	.04782	.04921	.05049	.05166	.05272
22	13	20	.03905	.04079	.04248	.04411	.04566	.04714	.04852	.04981
22	13	21	.03460	.03645	.03828	.04007	.04182	.04351	.04513	.04667
22	13	22	.02904	.03101	.03301	.03501	.03700	.03897	.04091	.04279
22	14	15	.06660	.06808	.06942	.07059	.07160	.07243	.07309	.07358
22	14	16	.06154	.06317	.06467	.06603	.06725	.06831	.06920	.06994
22	14	17	.05678	.05854	.06020	.06174	.06314	.06441	.06553	.06651
22	14	18	.05223	.05411	.05591	.05761	.05920	.06066	.06200	.06321
22	14	19	.04776	.04976	.05169	.05354	.[illegible]31	.05697	.05852	.05996
22	14	20	.04322	.04532	.04738	.04939	.[illegible]134	.05320	.05498	.05666
22	14	21	.03830	.04051	.04271	.04489	.04703	.04912	.05114	.05310
22	14	22	.03215	.03448	.03684	.03923	.04162	.04401	.04637	.04869
22	15	16	.06865	.07076	.07275	.07458	.07627	.07778	.07912	.08029
22	15	17	.06337	.06561	.06774	.06975	.07163	.07337	.07495	.07637
22	15	18	.05830	.06066	.06293	.06511	.06718	.06912	.07093	.07260
22	15	19	.05333	.05579	.05820	.06053	.06278	.06493	.06697	.06889
22	15	20	.04828	.05083	.05337	.05586	.05829	.06065	.06293	.06511
22	15	21	.04280	.04545	.04812	.05078	.05341	.05601	.05856	.06104
22	15	22	.03594	.03870	.04152	.04439	.04729	.05020	.05311	.05599
22	16	17	.07159	.07444	.07720	.07984	.08235	.08470	.08690	.08892
22	16	18	.06589	.06886	.07175	.07455	.07725	.07983	.08227	.08456
22	16	19	.06030	.06336	.06638	.06934	.07222	.07502	.07770	.08027
22	16	20	.05460	.05775	.06089	.06401	.06708	.07010	.07304	.07589
22	16	21	.04843	.05166	.05492	.05821	.06149	.06476	.06799	.07116
22	16	22	.04068	.04400	.04741	.05090	.05446	.05806	.06168	.06530
22	17	18	.07574	.07952	.08325	.08691	.09048	.09393	.09724	.10041
22	17	19	.06934	.07321	.07706	.08087	.08463	.08831	.09189	.09535
22	17	20	.06282	.06676	.07072	.07469	.07864	.08255	.08641	.09019
22	17	21	.05575	.05975	.06382	.06795	.07212	.07630	.08047	.08460

Covariances of Inverse Gaussian Order Statistics
Shape Parameter k

n	i	j	0.0	0.1	0.2	0.3	0.4	0.5	0.6	0.7
22	17	22	.04686	.05092	.05512	.05946	.06391	.06844	.07304	.07767
22	18	19	.08169	.08670	.09174	.09678	.10180	.10677	.11166	.11645
22	18	20	.07406	.07912	.08425	.08944	.09466	.09988	.10507	.11021
22	18	21	.06577	.07086	.07609	.08143	.08687	.09237	.09790	.10345
22	18	22	.05533	.06043	.06577	.07131	.07704	.08292	.08893	.09504
22	19	20	.09057	.09736	.10431	.11141	.11862	.12589	.13321	.14052
22	19	21	.08052	.08728	.09430	.10153	.10896	.11653	.12423	.13201
22	19	22	.06782	.07454	.08161	.08902	.09674	.10473	.11296	.12140
22	20	21	.10483	.11452	.12467	.13524	.14620	.15750	.16910	.18095
22	20	22	.08848	.09799	.10811	.11880	.13004	.14180	.15402	.16667
22	21	22	.13041	.14601	.16281	.18079	.19994	.22019	.24151	.26383
23	1	2	.12857	.11422	.10108	.08913	.07833	.06862	.05996	.05226
23	1	3	.08720	.07832	.07008	.06249	.05554	.04921	.04348	.03833
23	1	4	.06684	.06050	.05456	.04903	.04393	.03923	.03494	.03104
23	1	5	.05454	.04967	.04507	.04076	.03675	.03303	.02961	.02647
23	1	6	.04621	.04231	.03859	.03509	.03181	.02874	.02590	.02329
23	1	7	.04015	.03692	.03384	.03092	.02816	.02557	.02315	.02091
23	1	8	.03549	.03278	.03018	.02769	.02533	.02310	.02101	.01906
23	1	9	.03178	.02947	.02724	.02510	.02305	.02111	.01928	.01756
23	1	10	.02873	.02675	.02482	.02295	.02116	.01946	.01784	.01632
23	1	11	.02617	.02445	.02277	.02114	.01957	.01805	.01662	.01525
23	1	12	.02397	.02248	.02101	.01958	.01818	.01684	.01556	.01433
23	1	13	.02205	.02075	.01947	.01820	.01697	.01577	.01462	.01352
23	1	14	.02034	.01921	.01809	.01698	.01589	.01482	.01379	.01280
23	1	15	.01880	.01783	.01685	.01587	.01491	.01396	.01303	.01214
23	1	16	.01740	.01656	.01571	.01486	.01400	.01316	.01234	.01154
23	1	17	.01609	.01538	.01465	.01391	.01316	.01242	.01169	.01097
23	1	18	.01487	.01427	.01365	.01302	.01237	.01172	.01108	.01044
23	1	19	.01369	.01320	.01269	.01215	.01161	.01105	.01049	.00993
23	1	20	.01253	.01215	.01174	.01130	.01085	.01038	.00990	.00942
23	1	21	.01135	.01108	.01077	.01044	.01008	.00970	.00931	.00891
23	1	22	.01007	.00991	.00971	.00949	.00924	.00896	.00866	.00835
23	1	23	.00846	.00844	.00839	.00830	.00818	.00803	.00786	.00766
23	2	3	.10305	.09385	.08518	.07705	.06948	.06246	.05600	.05008
23	2	4	.07915	.07264	.06644	.06058	.05506	.04990	.04510	.04065
23	2	5	.06466	.05971	.05496	.05043	.04612	.04207	.03826	.03471
23	2	6	.05484	.05090	.04710	.04345	.03996	.03664	.03350	.03056
23	2	7	.04766	.04445	.04133	.03830	.03539	.03261	.02996	.02746
23	2	8	.04215	.03948	.03687	.03432	.03185	.02947	.02720	.02504
23	2	9	.03776	.03551	.03329	.03111	.02900	.02694	.02497	.02308
23	2	10	.03415	.03224	.03034	.02847	.02663	.02484	.02311	.02145
23	2	11	.03111	.02948	.02785	.02622	.02463	.02306	.02153	.02006
23	2	12	.02850	.02710	.02570	.02429	.02289	.02151	.02016	.01885
23	2	13	.02622	.02502	.02381	.02259	.02137	.02015	.01896	.01779
23	2	14	.02419	.02317	.02213	.02107	.02000	.01894	.01788	.01683
23	2	15	.02236	.02150	.02061	.01970	.01877	.01784	.01690	.01597
23	2	16	.02069	.01998	.01922	.01844	.01764	.01682	.01600	.01518
23	2	17	.01915	.01856	.01793	.01727	.01658	.01588	.01516	.01444
23	2	18	.01769	.01722	.01671	.01616	.01558	.01498	.01437	.01374
23	2	19	.01629	.01593	.01553	.01509	.01462	.01412	.01360	.01306
23	2	20	.01491	.01466	.01437	.01404	.01367	.01327	.01285	.01240
23	2	21	.01350	.01337	.01318	.01296	.01270	.01240	.01207	.01172
23	2	22	.01198	.01196	.01189	.01179	.01164	.01145	.01124	.01099
23	2	23	.01007	.01019	.01027	.01031	.01031	.01027	.01019	.01008
23	3	4	.08883	.08222	.07587	.06979	.06400	.05852	.05337	.04854

Covariances of Inverse Gaussian Order Statistics
Shape Parameter k

n	i	j	0.0	0.1	0.2	0.3	0.4	0.5	0.6	0.7
23	3	5	.07265	.06766	.06282	.05815	.05367	.04939	.04533	.04149
23	3	6	.06165	.05772	.05388	.05014	.04652	.04305	.03972	.03656
23	3	7	.05361	.05043	.04730	.04422	.04123	.03833	.03554	.03287
23	3	8	.04743	.04481	.04221	.03964	.03712	.03466	.03228	.02999
23	3	9	.04250	.04031	.03813	.03595	.03380	.03170	.02964	.02765
23	3	10	.03845	.03661	.03475	.03290	.03105	.02923	.02744	.02570
23	3	11	.03503	.03348	.03190	.03031	.02872	.02714	.02557	.02403
23	3	12	.03210	.03079	.02945	.02808	.02670	.02532	.02395	.02259
23	3	13	.02953	.02843	.02729	.02612	.02493	.02373	.02252	.02132
23	3	14	.02725	.02633	.02537	.02437	.02334	.02230	.02124	.02018
23	3	15	.02520	.02444	.02363	.02278	.02191	.02100	.02008	.01915
23	3	16	.02332	.02270	.02204	.02133	.02059	.01981	.01901	.01820
23	3	17	.02158	.02109	.02056	.01998	.01936	.01870	.01802	.01731
23	3	18	.01993	.01957	.01916	.01870	.01819	.01765	.01707	.01648
23	3	19	.01836	.01811	.01781	.01746	.01707	.01663	.01617	.01567
23	3	20	.01680	.01667	.01648	.01624	.01596	.01563	.01527	.01487
23	3	21	.01522	.01520	.01512	.01500	.01482	.01461	.01435	.01406
23	3	22	.01350	.01360	.01364	.01364	.01359	.01349	.01336	.01318
23	3	23	.01135	.01158	.01178	.01193	.01204	.01210	.01212	.01210
23	4	5	.07995	.07495	.07005	.06528	.06066	.05621	.05194	.04788
23	4	6	.06789	.06398	.06012	.05632	.05262	.04902	.04555	.04221
23	4	7	.05907	.05593	.05280	.04970	.04666	.04368	.04078	.03798
23	4	8	.05228	.04971	.04714	.04457	.04202	.03951	.03705	.03465
23	4	9	.04686	.04474	.04259	.04043	.03828	.03614	.03403	.03196
23	4	10	.04240	.04063	.03883	.03701	.03517	.03334	.03151	.02971
23	4	11	.03864	.03717	.03566	.03411	.03254	.03095	.02937	.02780
23	4	12	.03541	.03419	.03291	.03160	.03026	.02889	.02751	.02613
23	4	13	.03258	.03157	.03051	.02940	.02825	.02707	.02587	.02467
23	4	14	.03007	.02924	.02836	.02743	.02645	.02544	.02441	.02335
23	4	15	.02780	.02714	.02642	.02565	.02483	.02397	.02307	.02216
23	4	16	.02573	.02522	.02465	.02401	.02333	.02261	.02185	.02106
23	4	17	.02381	.02343	.02299	.02249	.02194	.02134	.02071	.02004
23	4	18	.02200	.02174	.02143	.02105	.02062	.02015	.01963	.01907
23	4	19	.02026	.02012	.01992	.01966	.01935	.01899	.01858	.01813
23	4	20	.01855	.01852	.01843	.01829	.01809	.01785	.01755	.01721
23	4	21	.01680	.01689	.01691	.01689	.01681	.01668	.01650	.01627
23	4	22	.01491	.01511	.01526	.01536	.01541	.01541	.01536	.01526
23	4	23	.01253	.01287	.01317	.01343	.01365	.01381	.01393	.01400
23	5	6	.07397	.07009	.06623	.06240	.05862	.05493	.05133	.04785
23	5	7	.06439	.06130	.05819	.05509	.05201	.04897	.04598	.04307
23	5	8	.05701	.05451	.05197	.04942	.04685	.04431	.04179	.03931
23	5	9	.05111	.04907	.04697	.04484	.04269	.04054	.03839	.03627
23	5	10	.04626	.04458	.04284	.04106	.03924	.03740	.03556	.03373
23	5	11	.04216	.04079	.03934	.03785	.03631	.03474	.03315	.03156
23	5	12	.03864	.03752	.03632	.03507	.03377	.03243	.03106	.02967
23	5	13	.03556	.03465	.03367	.03263	.03153	.03039	.02921	.02801
23	5	14	.03283	.03210	.03131	.03045	.02953	.02857	.02756	.02652
23	5	15	.03036	.02980	.02917	.02848	.02772	.02691	.02606	.02517
23	5	16	.02810	.02769	.02721	.02666	.02605	.02539	.02468	.02393
23	5	17	.02600	.02573	.02539	.02497	.02450	.02397	.02339	.02276
23	5	18	.02403	.02388	.02366	.02338	.02303	.02263	.02217	.02166
23	5	19	.02213	.02210	.02200	.02184	.02161	.02133	.02099	.02061
23	5	20	.02026	.02034	.02036	.02031	.02021	.02005	.01983	.01956
23	5	21	.01836	.01855	.01868	.01876	.01877	.01874	.01864	.01849
23	5	22	.01629	.01660	.01685	.01706	.01721	.01731	.01735	.01734

Covariances of Inverse Gaussian Order Statistics
Shape Parameter k

n	i	j	0.0	0.1	0.2	0.3	0.4	0.5	0.6	0.7
23	5	23	.01369	.01414	.01455	.01492	.01525	.01552	.01574	.01591
23	6	7	.06977	.06674	.06367	.06057	.05747	.05438	.05132	.04831
23	6	8	.06180	.05937	.05689	.05435	.05179	.04922	.04666	.04412
23	6	9	.05542	.05346	.05143	.04934	.04721	.04505	.04288	.04071
23	6	10	.05017	.04858	.04692	.04519	.04340	.04158	.03973	.03787
23	6	11	.04574	.04446	.04309	.04166	.04017	.03862	.03704	.03544
23	6	12	.04193	.04090	.03979	.03861	.03736	.03606	.03471	.03333
23	6	13	.03859	.03779	.03690	.03593	.03489	.03380	.03265	.03147
23	6	14	.03563	.03501	.03431	.03353	.03268	.03177	.03081	.02980
23	6	15	.03295	.03250	.03197	.03136	.03068	.02994	.02914	.02828
23	6	16	.03050	.03020	.02982	.02937	.02884	.02825	.02759	.02689
23	6	17	.02823	.02807	.02783	.02751	.02713	.02667	.02616	.02558
23	6	18	.02609	.02605	.02594	.02575	.02550	.02518	.02479	.02435
23	6	19	.02403	.02411	.02412	.02406	.02393	.02374	.02348	.02316
23	6	20	.02200	.02219	.02232	.02238	.02238	.02231	.02218	.02199
23	6	21	.01993	.02024	.02048	.02067	.02079	.02085	.02085	.02079
23	6	22	.01769	.01811	.01848	.01880	.01906	.01927	.01941	.01950
23	6	23	.01487	.01543	.01596	.01645	.01689	.01728	.01761	.01789
23	7	8	.06676	.06442	.06200	.05950	.05696	.05438	.05178	.04919
23	7	9	.05989	.05803	.05607	.05403	.05193	.04978	.04760	.04541
23	7	10	.05423	.05275	.05116	.04950	.04776	.04596	.04412	.04224
23	7	11	.04945	.04828	.04701	.04564	.04421	.04270	.04114	.03955
23	7	12	.04534	.04443	.04341	.04231	.04113	.03987	.03856	.03720
23	7	13	.04174	.04105	.04026	.03938	.03842	.03738	.03628	.03512
23	7	14	.03854	.03804	.03744	.03676	.03599	.03515	.03424	.03327
23	7	15	.03565	.03532	.03489	.03438	.03379	.03312	.03238	.03158
23	7	16	.03300	.03282	.03255	.03220	.03177	.03125	.03067	.03002
23	7	17	.03055	.03050	.03038	.03017	.02988	.02951	.02907	.02857
23	7	18	.02823	.02831	.02832	.02824	.02809	.02786	.02756	.02719
23	7	19	.02600	.02621	.02633	.02638	.02636	.02627	.02610	.02587
23	7	20	.02381	.02413	.02437	.02455	.02466	.02469	.02466	.02456
23	7	21	.02158	.02200	.02237	.02267	.02291	.02308	.02318	.02322
23	7	22	.01915	.01969	.02018	.02062	.02100	.02132	.02158	.02178
23	7	23	.01609	.01678	.01743	.01804	.01861	.01912	.01959	.01999
23	8	9	.06462	.06286	.06099	.05902	.05697	.05484	.05266	.05044
23	8	10	.05852	.05716	.05567	.05408	.05240	.05064	.04882	.04694
23	8	11	.05338	.05233	.05116	.04988	.04851	.04706	.04554	.04395
23	8	12	.04895	.04816	.04726	.04625	.04515	.04395	.04269	.04135
23	8	13	.04507	.04451	.04383	.04305	.04218	.04121	.04017	.03905
23	8	14	.04162	.04125	.04077	.04019	.03952	.03875	.03791	.03699
23	8	15	.03850	.03830	.03800	.03760	.03711	.03652	.03586	.03512
23	8	16	.03565	.03560	.03546	.03522	.03489	.03447	.03397	.03339
23	8	17	.03300	.03309	.03309	.03300	.03282	.03255	.03220	.03178
23	8	18	.03050	.03072	.03085	.03089	.03086	.03073	.03053	.03025
23	8	19	.02810	.02843	.02869	.02887	.02896	.02898	.02892	.02878
23	8	20	.02573	.02618	.02656	.02686	.02709	.02724	.02732	.02733
23	8	21	.02332	.02388	.02437	.02481	.02517	.02547	.02569	.02584
23	8	22	.02069	.02137	.02200	.02257	.02308	.02353	.02392	.02424
23	8	23	.01740	.01821	.01900	.01975	.02045	.02111	.02171	.02225
23	9	10	.06314	.06190	.06053	.05903	.05742	.05572	.05393	.05207
23	9	11	.05760	.05669	.05564	.05447	.05318	.05179	.05032	.04877
23	9	12	.05284	.05218	.05141	.05051	.04950	.04838	.04718	.04589
23	9	13	.04866	.04823	.04769	.04702	.04625	.04537	.04440	.04335
23	9	14	.04494	.04471	.04437	.04391	.04334	.04268	.04191	.04106
23	9	15	.04158	.04152	.04136	.04108	.04070	.04022	.03965	.03899

Covariances of Inverse Gaussian Order Statistics
Shape Parameter k

n	i	j	0.0	0.1	0.2	0.3	0.4	0.5	0.6	0.7
23	9	16	.03850	.03860	.03859	.03849	.03828	.03797	.03757	.03708
23	9	17	.03565	.03588	.03602	.03606	.03601	.03586	.03562	.03529
23	9	18	.03295	.03331	.03359	.03377	.03386	.03386	.03377	.03360
23	9	19	.03036	.03084	.03124	.03155	.03178	.03193	.03199	.03196
23	9	20	.02780	.02840	.02892	.02936	.02973	.03002	.03023	.03035
23	9	21	.02520	.02590	.02654	.02712	.02763	.02806	.02842	.02871
23	9	22	.02236	.02318	.02396	.02468	.02534	.02593	.02646	.02692
23	9	23	.01880	.01976	.02069	.02159	.02245	.02326	.02402	.02472
23	10	11	.06221	.06145	.06054	.05949	.05831	.05701	.05559	.05409
23	10	12	.05707	.05658	.05595	.05518	.05428	.05327	.05214	.05091
23	10	13	.05257	.05231	.05191	.05138	.05073	.04996	.04908	.04810
23	10	14	.04856	.04849	.04830	.04799	.04755	.04700	.04634	.04557
23	10	15	.04494	.04505	.04504	.04491	.04466	.04431	.04384	.04328
23	10	16	.04162	.04188	.04203	.04207	.04200	.04183	.04154	.04116
23	10	17	.03854	.03894	.03924	.03943	.03952	.03951	.03939	.03918
23	10	18	.03563	.03615	.03659	.03693	.03717	.03731	.03735	.03731
23	10	19	.03283	.03347	.03403	.03451	.03489	.03518	.03539	.03549
23	10	20	.03007	.03082	.03151	.03212	.03264	.03308	.03344	.03371
23	10	21	.02725	.02812	.02892	.02967	.03034	.03093	.03145	.03188
23	10	22	.02419	.02517	.02611	.02699	.02782	.02859	.02928	.02991
23	10	23	.02034	.02146	.02255	.02362	.02466	.02564	.02658	.02746
23	11	12	.06176	.06145	.06099	.06038	.05962	.05872	.05769	.05654
23	11	13	.05690	.05682	.05660	.05624	.05573	.05509	.05432	.05343
23	11	14	.05257	.05269	.05268	.05253	.05225	.05183	.05129	.05064
23	11	15	.04866	.04895	.04912	.04917	.04908	.04887	.04854	.04809
23	11	16	.04507	.04552	.04586	.04607	.04617	.04614	.04600	.04575
23	11	17	.04174	.04233	.04281	.04318	.04344	.04359	.04363	.04355
23	11	18	.03859	.03931	.03993	.04044	.04086	.04117	.04137	.04147
23	11	19	.03556	.03640	.03714	.03780	.03836	.03883	.03920	.03947
23	11	20	.03258	.03352	.03439	.03519	.03589	.03652	.03705	.03748
23	11	21	.02953	.03058	.03157	.03250	.03336	.03414	.03484	.03546
23	11	22	.02622	.02738	.02850	.02958	.03060	.03156	.03245	.03326
23	11	23	.02205	.02334	.02463	.02589	.02712	.02831	.02946	.03054
23	12	13	.06176	.06190	.06189	.06171	.06138	.06090	.06027	.05951
23	12	14	.05707	.05742	.05761	.05766	.05756	.05731	.05693	.05641
23	12	15	.05284	.05335	.05373	.05398	.05408	.05405	.05388	.05358
23	12	16	.04895	.04962	.05017	.05059	.05088	.05104	.05107	.05098
23	12	17	.04534	.04615	.04685	.04743	.04789	.04823	.04844	.04854
23	12	18	.04193	.04286	.04370	.04443	.04505	.04555	.04595	.04623
23	12	19	.03864	.03969	.04066	.04153	.04230	.04297	.04354	.04400
23	12	20	.03541	.03656	.03765	.03866	.03958	.04042	.04115	.04179
23	12	21	.03210	.03336	.03457	.03572	.03679	.03779	.03871	.03954
23	12	22	.02850	.02987	.03121	.03251	.03375	.03494	.03605	.03710
23	12	23	.02397	.02547	.02697	.02846	.02992	.03135	.03273	.03406
23	13	14	.06221	.06281	.06326	.06354	.06367	.06363	.06343	.06308
23	13	15	.05760	.05838	.05902	.05950	.05984	.06002	.06005	.05994
23	13	16	.05338	.05431	.05511	.05578	.05630	.05669	.05693	.05704
23	13	17	.04945	.05052	.05147	.05230	.05300	.05357	.05401	.05432
23	13	18	.04574	.04693	.04802	.04900	.04987	.05061	.05124	.05174
23	13	19	.04216	.04347	.04469	.04581	.04683	.04775	.04856	.04925
23	13	20	.03864	.04005	.04139	.04265	.04383	.04492	.04590	.04679
23	13	21	.03503	.03654	.03801	.03941	.04075	.04201	.04319	.04427
23	13	22	.03111	.03273	.03432	.03588	.03739	.03884	.04023	.04154
23	13	23	.02617	.02791	.02966	.03141	.03315	.03486	.03653	.03815
23	14	15	.06314	.06423	.06517	.06595	.06657	.06702	.06731	.06743

Covariances of Inverse Gaussian Order Statistics
Shape Parameter k

n	i	j	0.0	0.1	0.2	0.3	0.4	0.5	0.6	0.7
23	14	16	.05852	.05977	.06087	.06184	.06265	.06332	.06383	.06418
23	14	17	.05423	.05561	.05687	.05800	.05899	.05985	.06057	.06114
23	14	18	.05017	.05167	.05306	.05435	.05552	.05656	.05747	.05825
23	14	19	.04626	.04786	.04939	.05082	.05215	.05337	.05447	.05545
23	14	20	.04240	.04411	.04575	.04733	.04882	.05021	.05150	.05269
23	14	21	.03845	.04026	.04203	.04374	.04539	.04697	.04846	.04986
23	14	22	.03415	.03606	.03796	.03983	.04166	.04343	.04515	.04680
23	14	23	.02873	.03076	.03281	.03488	.03694	.03899	.04101	.04298
23	15	16	.06462	.06624	.06773	.06907	.07024	.07126	.07210	.07278
23	15	17	.05989	.06165	.06329	.06479	.06616	.06738	.06844	.06935
23	15	18	.05542	.05730	.05907	.06073	.06227	.06368	.06495	.06608
23	15	19	.05111	.05309	.05499	.05680	.05851	.06011	.06158	.06292
23	15	20	.04686	.04893	.05096	.05291	.05478	.05656	.05824	.05980
23	15	21	.04250	.04467	.04681	.04891	.05095	.05292	.05481	.05660
23	15	22	.03776	.04003	.04229	.04454	.04677	.04895	.05107	.05313
23	15	23	.03178	.03415	.03657	.03902	.04148	.04395	.04640	.04882
23	16	17	.06676	.06900	.07111	.07310	.07493	.07662	.07813	.07948
23	16	18	.06180	.06415	.06640	.06854	.07056	.07244	.07417	.07576
23	16	19	.05701	.05946	.06183	.06412	.06631	.06839	.07034	.07216
23	16	20	.05228	.05482	.05731	.05974	.06210	.06437	.06654	.06859
23	16	21	.04743	.05006	.05267	.05524	.05778	.06024	.06264	.06494
23	16	22	.04215	.04486	.04759	.05032	.05304	.05574	.05838	.06097
23	16	23	.03549	.03829	.04116	.04409	.04706	.05006	.05305	.05603
23	17	18	.06977	.07273	.07560	.07837	.08102	.08353	.08589	.08808
23	17	19	.06439	.06744	.07043	.07335	.07617	.07889	.08147	.08392
23	17	20	.05907	.06220	.06531	.06837	.07136	.07428	.07709	.07980
23	17	21	.05361	.05682	.06003	.06324	.06641	.06954	.07260	.07557
23	17	22	.04766	.05094	.05427	.05763	.06100	.06436	.06769	.07098
23	17	23	.04015	.04350	.04696	.05051	.05414	.05782	.06153	.06525
23	18	19	.07397	.07784	.08167	.08544	.08913	.09272	.09619	.09953
23	18	20	.06789	.07182	.07576	.07967	.08354	.08734	.09106	.09467
23	18	21	.06165	.06564	.06968	.07373	.07778	.08181	.08578	.08970
23	18	22	.05484	.05888	.06301	.06722	.07147	.07574	.08002	.08428
23	18	23	.04621	.05030	.05456	.05895	.06347	.06808	.07277	.07751
23	19	20	.07995	.08502	.09015	.09529	.10042	.10552	.11056	.11551
23	19	21	.07265	.07776	.08296	.08824	.09356	.09889	.10421	.10949
23	19	22	.06466	.06979	.07508	.08050	.08602	.09162	.09727	.10294
23	19	23	.05454	.05968	.06506	.07065	.07645	.08241	.08852	.09473
23	20	21	.08883	.09566	.10268	.10986	.11716	.12456	.13201	.13949
23	20	22	.07915	.08595	.09302	.10032	.10783	.11551	.12332	.13124
23	20	23	.06684	.07358	.08069	.08815	.09593	.10401	.11234	.12090
23	21	22	.10305	.11276	.12296	.13359	.14464	.15604	.16777	.17978
23	21	23	.08720	.09673	.10687	.11761	.12892	.14075	.15308	.16586
23	22	23	.12857	.14416	.16097	.17899	.19819	.21854	.23998	.26245
24	1	2	.12686	.11253	.09944	.08755	.07683	.06721	.05863	.05102
24	1	3	.08600	.07713	.06891	.06135	.05444	.04816	.04248	.03738
24	1	4	.06592	.05957	.05363	.04812	.04304	.03837	.03412	.03026
24	1	5	.05380	.04891	.04431	.04000	.03600	.03230	.02890	.02579
24	1	6	.04560	.04167	.03795	.03444	.03116	.02811	.02528	.02268
24	1	7	.03964	.03639	.03329	.03035	.02759	.02500	.02260	.02037
24	1	8	.03506	.03232	.02970	.02719	.02482	.02259	.02051	.01857
24	1	9	.03142	.02908	.02682	.02466	.02260	.02065	.01882	.01711
24	1	10	.02844	.02642	.02446	.02257	.02076	.01905	.01742	.01590
24	1	11	.02593	.02417	.02246	.02080	.01921	.01768	.01624	.01487
24	1	12	.02378	.02225	.02074	.01928	.01787	.01651	.01521	.01398

Covariances of Inverse Gaussian Order Statistics
Shape Parameter k

n	i	j	0.0	0.1	0.2	0.3	0.4	0.5	0.6	0.7
24	1	13	.02190	.02057	.01924	.01795	.01669	.01547	.01431	.01320
24	1	14	.02025	.01907	.01791	.01676	.01564	.01455	.01350	.01250
24	1	15	.01876	.01773	.01671	.01570	.01470	.01372	.01278	.01187
24	1	16	.01740	.01651	.01562	.01472	.01383	.01296	.01211	.01129
24	1	17	.01615	.01539	.01461	.01382	.01303	.01226	.01150	.01075
24	1	18	.01499	.01433	.01366	.01297	.01228	.01160	.01092	.01025
24	1	19	.01388	.01333	.01276	.01217	.01157	.01097	.01037	.00977
24	1	20	.01281	.01236	.01189	.01139	.01088	.01035	.00983	.00931
24	1	21	.01175	.01140	.01102	.01061	.01019	.00975	.00930	.00885
24	1	22	.01067	.01042	.01013	.00982	.00948	.00912	.00876	.00838
24	1	23	.00949	.00934	.00916	.00894	.00870	.00844	.00816	.00787
24	1	24	.00799	.00797	.00792	.00784	.00772	.00758	.00742	.00723
24	2	3	.10140	.09219	.08353	.07544	.06791	.06095	.05454	.04869
24	2	4	.07787	.07134	.06515	.05929	.05380	.04866	.04390	.03949
24	2	5	.06363	.05865	.05389	.04935	.04506	.04101	.03723	.03371
24	2	6	.05398	.05001	.04619	.04253	.03903	.03572	.03260	.02967
24	2	7	.04694	.04369	.04054	.03750	.03458	.03180	.02915	.02666
24	2	8	.04154	.03883	.03618	.03361	.03113	.02874	.02647	.02431
24	2	9	.03724	.03495	.03269	.03049	.02835	.02628	.02430	.02241
24	2	10	.03371	.03175	.02982	.02791	.02605	.02424	.02250	.02083
24	2	11	.03074	.02906	.02739	.02573	.02410	.02252	.02097	.01949
24	2	12	.02820	.02675	.02530	.02385	.02243	.02102	.01965	.01832
24	2	13	.02598	.02473	.02347	.02221	.02095	.01971	.01849	.01730
24	2	14	.02402	.02294	.02185	.02075	.01964	.01854	.01745	.01639
24	2	15	.02225	.02134	.02039	.01943	.01846	.01748	.01651	.01556
24	2	16	.02065	.01987	.01906	.01822	.01737	.01651	.01566	.01480
24	2	17	.01917	.01852	.01783	.01711	.01637	.01562	.01486	.01410
24	2	18	.01779	.01725	.01667	.01606	.01543	.01478	.01411	.01344
24	2	19	.01648	.01604	.01557	.01507	.01453	.01398	.01340	.01282
24	2	20	.01521	.01488	.01451	.01410	.01366	.01320	.01271	.01221
24	2	21	.01395	.01372	.01345	.01314	.01280	.01243	.01203	.01161
24	2	22	.01267	.01254	.01237	.01216	.01191	.01163	.01132	.01099
24	2	23	.01126	.01124	.01118	.01108	.01094	.01076	.01056	.01032
24	2	24	.00949	.00960	.00967	.00971	.00971	.00967	.00959	.00948
24	3	4	.08721	.08058	.07422	.06814	.06237	.05693	.05182	.04704
24	3	5	.07133	.06631	.06145	.05678	.05230	.04803	.04399	.04019
24	3	6	.06055	.05658	.05271	.04896	.04533	.04186	.03855	.03540
24	3	7	.05268	.04946	.04629	.04319	.04018	.03728	.03449	.03183
24	3	8	.04664	.04397	.04133	.03873	.03618	.03371	.03133	.02903
24	3	9	.04182	.03958	.03735	.03514	.03296	.03084	.02877	.02677
24	3	10	.03787	.03597	.03407	.03218	.03030	.02845	.02664	.02489
24	3	11	.03454	.03293	.03131	.02967	.02804	.02643	.02484	.02329
24	3	12	.03169	.03032	.02892	.02751	.02609	.02468	.02328	.02190
24	3	13	.02920	.02804	.02684	.02562	.02438	.02314	.02190	.02068
24	3	14	.02700	.02601	.02499	.02393	.02286	.02177	.02068	.01959
24	3	15	.02502	.02419	.02332	.02241	.02148	.02053	.01957	.01860
24	3	16	.02321	.02253	.02180	.02103	.02022	.01940	.01855	.01770
24	3	17	.02155	.02099	.02039	.01974	.01906	.01835	.01761	.01686
24	3	18	.02000	.01956	.01907	.01854	.01796	.01736	.01673	.01608
24	3	19	.01853	.01820	.01781	.01739	.01692	.01642	.01589	.01533
24	3	20	.01710	.01687	.01660	.01627	.01591	.01550	.01507	.01460
24	3	21	.01569	.01557	.01539	.01517	.01490	.01460	.01426	.01388
24	3	22	.01424	.01422	.01415	.01403	.01387	.01366	.01342	.01315
24	3	23	.01267	.01275	.01279	.01279	.01274	.01265	.01251	.01234
24	3	24	.01067	.01089	.01107	.01121	.01130	.01136	.01137	.01135

Covariances of Inverse Gaussian Order Statistics
Shape Parameter k

n	i	j	0.0	0.1	0.2	0.3	0.4	0.5	0.6	0.7
24	4	5	.07834	.07329	.06837	.06358	.05896	.05452	.05028	.04625
24	4	6	.06654	.06258	.05868	.05486	.05115	.04755	.04408	.04077
24	4	7	.05792	.05473	.05155	.04843	.04536	.04237	.03947	.03667
24	4	8	.05130	.04867	.04605	.04344	.04086	.03833	.03586	.03346
24	4	9	.04601	.04383	.04163	.03942	.03723	.03507	.03294	.03086
24	4	10	.04167	.03984	.03798	.03611	.03423	.03236	.03051	.02870
24	4	11	.03801	.03648	.03490	.03330	.03169	.03007	.02845	.02686
24	4	12	.03488	.03359	.03225	.03088	.02949	.02808	.02667	.02526
24	4	13	.03214	.03106	.02993	.02876	.02756	.02633	.02510	.02386
24	4	14	.02972	.02882	.02787	.02687	.02584	.02478	.02369	.02260
24	4	15	.02754	.02681	.02601	.02517	.02428	.02337	.02243	.02147
24	4	16	.02556	.02497	.02432	.02361	.02286	.02208	.02126	.02043
24	4	17	.02374	.02327	.02275	.02217	.02155	.02088	.02019	.01946
24	4	18	.02203	.02168	.02128	.02082	.02031	.01976	.01917	.01856
24	4	19	.02040	.02017	.01988	.01953	.01913	.01869	.01821	.01769
24	4	20	.01884	.01871	.01852	.01828	.01799	.01765	.01727	.01686
24	4	21	.01728	.01726	.01718	.01704	.01686	.01662	.01634	.01603
24	4	22	.01569	.01577	.01579	.01576	.01569	.01556	.01539	.01518
24	4	23	.01395	.01414	.01427	.01436	.01441	.01440	.01435	.01425
24	4	24	.01175	.01207	.01235	.01259	.01279	.01294	.01304	.01310
24	5	6	.07234	.06840	.06449	.06063	.05684	.05314	.04955	.04608
24	5	7	.06300	.05985	.05669	.05354	.05043	.04737	.04438	.04147
24	5	8	.05581	.05324	.05065	.04804	.04544	.04287	.04034	.03786
24	5	9	.05008	.04796	.04580	.04362	.04142	.03924	.03706	.03493
24	5	10	.04536	.04360	.04180	.03996	.03809	.03622	.03434	.03248
24	5	11	.04139	.03993	.03842	.03686	.03527	.03365	.03203	.03041
24	5	12	.03798	.03678	.03551	.03419	.03283	.03143	.03002	.02860
24	5	13	.03501	.03402	.03296	.03184	.03068	.02948	.02826	.02702
24	5	14	.03237	.03156	.03069	.02975	.02877	.02774	.02668	.02560
24	5	15	.03001	.02936	.02865	.02787	.02704	.02617	.02526	.02431
24	5	16	.02785	.02735	.02678	.02615	.02546	.02473	.02395	.02314
24	5	17	.02586	.02549	.02505	.02456	.02400	.02339	.02274	.02205
24	5	18	.02400	.02375	.02344	.02306	.02262	.02214	.02160	.02102
24	5	19	.02224	.02210	.02190	.02163	.02131	.02094	.02051	.02005
24	5	20	.02053	.02050	.02040	.02025	.02004	.01978	.01946	.01910
24	5	21	.01884	.01891	.01892	.01888	.01878	.01862	.01842	.01816
24	5	22	.01710	.01728	.01740	.01746	.01748	.01743	.01734	.01720
24	5	23	.01521	.01549	.01573	.01592	.01605	.01613	.01617	.01615
24	5	24	.01281	.01323	.01361	.01395	.01425	.01450	.01469	.01485
24	6	7	.06810	.06500	.06187	.05872	.05558	.05246	.04939	.04639
24	6	8	.06036	.05785	.05529	.05270	.05010	.04750	.04491	.04236
24	6	9	.05417	.05212	.05002	.04787	.04568	.04348	.04128	.03909
24	6	10	.04908	.04740	.04566	.04386	.04202	.04014	.03826	.03637
24	6	11	.04479	.04342	.04197	.04047	.03891	.03731	.03569	.03405
24	6	12	.04111	.03999	.03880	.03754	.03622	.03486	.03346	.03203
24	6	13	.03790	.03700	.03602	.03497	.03386	.03270	.03149	.03026
24	6	14	.03505	.03434	.03354	.03268	.03175	.03077	.02974	.02867
24	6	15	.03249	.03194	.03131	.03061	.02985	.02903	.02815	.02724
24	6	16	.03016	.02976	.02928	.02873	.02811	.02743	.02670	.02592
24	6	17	.02801	.02774	.02739	.02698	.02650	.02595	.02535	.02470
24	6	18	.02599	.02585	.02563	.02534	.02498	.02456	.02409	.02356
24	6	19	.02408	.02405	.02394	.02377	.02353	.02323	.02288	.02246
24	6	20	.02224	.02231	.02231	.02225	.02213	.02195	.02170	.02140
24	6	21	.02040	.02058	.02069	.02075	.02074	.02067	.02054	.02035
24	6	22	.01853	.01880	.01903	.01919	.01930	.01935	.01934	.01927

Covariances of Inverse Gaussian Order Statistics
Shape Parameter k

n	i	j	0.0	0.1	0.2	0.3	0.4	0.5	0.6	0.7
24	6	23	.01648	.01686	.01720	.01749	.01773	.01791	.01803	.01810
24	6	24	.01388	.01440	.01489	.01533	.01574	.01609	.01639	.01664
24	7	8	.06504	.06261	.06010	.05754	.05493	.05231	.04969	.04708
24	7	9	.05839	.05643	.05438	.05227	.05011	.04790	.04568	.04346
24	7	10	.05291	.05133	.04965	.04791	.04610	.04424	.04235	.04044
24	7	11	.04830	.04703	.04566	.04421	.04269	.04112	.03951	.03787
24	7	12	.04434	.04332	.04221	.04102	.03975	.03843	.03705	.03563
24	7	13	.04088	.04008	.03919	.03822	.03717	.03605	.03488	.03366
24	7	14	.03781	.03720	.03650	.03572	.03486	.03393	.03294	.03190
24	7	15	.03506	.03461	.03408	.03347	.03277	.03201	.03119	.03031
24	7	16	.03254	.03225	.03187	.03140	.03087	.03026	.02958	.02885
24	7	17	.03022	.03006	.02982	.02950	.02910	.02863	.02809	.02749
24	7	18	.02805	.02801	.02790	.02770	.02743	.02709	.02669	.02622
24	7	19	.02599	.02607	.02607	.02599	.02585	.02563	.02535	.02501
24	7	20	.02400	.02418	.02429	.02434	.02431	.02421	.02405	.02383
24	7	21	.02203	.02231	.02253	.02269	.02278	.02280	.02276	.02266
24	7	22	.02000	.02039	.02072	.02099	.02120	.02135	.02143	.02146
24	7	23	.01779	.01828	.01873	.01913	.01947	.01976	.01999	.02016
24	7	24	.01499	.01562	.01621	.01677	.01729	.01775	.01817	.01853
24	8	9	.06282	.06095	.05898	.05692	.05478	.05260	.05037	.04811
24	8	10	.05694	.05546	.05387	.05218	.05042	.04858	.04670	.04478
24	8	11	.05199	.05082	.04954	.04816	.04670	.04517	.04358	.04194
24	8	12	.04774	.04683	.04581	.04469	.04349	.04222	.04087	.03947
24	8	13	.04402	.04333	.04254	.04165	.04067	.03961	.03849	.03730
24	8	14	.04072	.04022	.03963	.03893	.03815	.03729	.03635	.03535
24	8	15	.03775	.03743	.03700	.03648	.03587	.03519	.03442	.03359
24	8	16	.03505	.03487	.03460	.03424	.03379	.03326	.03265	.03198
24	8	17	.03256	.03251	.03238	.03216	.03185	.03147	.03101	.03048
24	8	18	.03022	.03030	.03030	.03021	.03004	.02979	.02946	.02907
24	8	19	.02801	.02820	.02831	.02835	.02830	.02818	.02799	.02772
24	8	20	.02586	.02616	.02639	.02654	.02662	.02662	.02656	.02642
24	8	21	.02374	.02414	.02448	.02474	.02494	.02507	.02513	.02512
24	8	22	.02155	.02206	.02251	.02290	.02322	.02348	.02367	.02379
24	8	23	.01917	.01978	.02035	.02087	.02133	.02173	.02207	.02235
24	8	24	.01615	.01690	.01762	.01830	.01894	.01953	.02007	.02055
24	9	10	.06124	.05986	.05837	.05676	.05505	.05326	.05140	.04948
24	9	11	.05593	.05487	.05369	.05241	.05101	.04953	.04797	.04635
24	9	12	.05136	.05057	.04966	.04864	.04752	.04630	.04500	.04363
24	9	13	.04737	.04680	.04612	.04533	.04444	.04345	.04238	.04124
24	9	14	.04383	.04345	.04297	.04238	.04169	.04091	.04004	.03909
24	9	15	.04064	.04044	.04013	.03972	.03921	.03861	.03792	.03715
24	9	16	.03773	.03768	.03753	.03728	.03693	.03650	.03597	.03536
24	9	17	.03505	.03514	.03513	.03502	.03482	.03454	.03416	.03371
24	9	18	.03254	.03275	.03287	.03290	.03284	.03269	.03246	.03215
24	9	19	.03016	.03048	.03072	.03088	.03095	.03093	.03084	.03067
24	9	20	.02785	.02828	.02863	.02891	.02911	.02923	.02926	.02923
24	9	21	.02556	.02610	.02656	.02696	.02728	.02753	.02770	.02780
24	9	22	.02321	.02385	.02443	.02494	.02539	.02578	.02609	.02633
24	9	23	.02065	.02139	.02209	.02274	.02333	.02386	.02433	.02473
24	9	24	.01740	.01828	.01912	.01994	.02071	.02144	.02212	.02274
24	10	11	.06018	.05926	.05820	.05701	.05571	.05429	.05278	.05118
24	10	12	.05528	.05463	.05384	.05293	.05190	.05076	.04952	.04819
24	10	13	.05099	.05057	.05001	.04934	.04855	.04765	.04665	.04556
24	10	14	.04719	.04696	.04661	.04614	.04555	.04486	.04407	.04319
24	10	15	.04376	.04370	.04353	.04324	.04285	.04235	.04174	.04105

Covariances of Inverse Gaussian Order Statistics
Shape Parameter k

n	i	j	0.0	0.1	0.2	0.3	0.4	0.5	0.6	0.7
24	10	16	.04064	.04073	.04072	.04059	.04037	.04004	.03961	.03909
24	10	17	.03775	.03798	.03811	.03814	.03806	.03789	.03762	.03726
24	10	18	.03506	.03541	.03567	.03583	.03590	.03587	.03575	.03554
24	10	19	.03249	.03296	.03334	.03363	.03383	.03395	.03397	.03390
24	10	20	.03001	.03058	.03108	.03149	.03182	.03207	.03224	.03231
24	10	21	.02754	.02822	.02883	.02937	.02983	.03021	.03051	.03073
24	10	22	.02502	.02579	.02652	.02718	.02777	.02829	.02874	.02911
24	10	23	.02225	.02314	.02398	.02478	.02551	.02619	.02681	.02735
24	10	24	.01876	.01977	.02076	.02173	.02266	.02354	.02437	.02515
24	11	12	.05957	.05908	.05844	.05765	.05673	.05569	.05453	.05326
24	11	13	.05496	.05470	.05429	.05375	.05308	.05229	.05137	.05035
24	11	14	.05087	.05080	.05060	.05027	.04982	.04924	.04855	.04775
24	11	15	.04719	.04729	.04727	.04713	.04687	.04648	.04599	.04539
24	11	16	.04383	.04408	.04422	.04425	.04416	.04395	.04364	.04322
24	11	17	.04072	.04111	.04140	.04158	.04164	.04160	.04146	.04121
24	11	18	.03781	.03833	.03875	.03907	.03928	.03939	.03940	.03931
24	11	19	.03505	.03568	.03622	.03667	.03702	.03728	.03744	.03751
24	11	20	.03237	.03311	.03377	.03434	.03483	.03523	.03553	.03575
24	11	21	.02972	.03056	.03133	.03203	.03265	.03319	.03364	.03401
24	11	22	.02700	.02793	.02882	.02964	.03040	.03108	.03169	.03222
24	11	23	.02402	.02506	.02607	.02703	.02793	.02878	.02956	.03027
24	11	24	.02025	.02141	.02257	.02370	.02480	.02586	.02688	.02783
24	12	13	.05937	.05929	.05906	.05868	.05815	.05749	.05668	.05576
24	12	14	.05496	.05508	.05506	.05490	.05459	.05415	.05358	.05289
24	12	15	.05099	.05128	.05144	.05147	.05136	.05113	.05076	.05028
24	12	16	.04737	.04781	.04814	.04833	.04840	.04835	.04818	.04789
24	12	17	.04402	.04460	.04507	.04542	.04566	.04577	.04577	.04566
24	12	18	.04088	.04159	.04219	.04268	.04307	.04335	.04351	.04357
24	12	19	.03790	.03872	.03944	.04007	.04060	.04103	.04135	.04157
24	12	20	.03501	.03593	.03677	.03753	.03820	.03877	.03925	.03963
24	12	21	.03214	.03317	.03412	.03501	.03581	.03653	.03716	.03770
24	12	22	.02920	.03032	.03139	.03240	.03334	.03421	.03501	.03572
24	12	23	.02598	.02721	.02840	.02954	.03064	.03168	.03265	.03356
24	12	24	.02190	.02325	.02459	.02592	.02721	.02848	.02970	.03086
24	13	14	.05957	.05991	.06010	.06013	.06000	.05973	.05931	.05875
24	13	15	.05528	.05579	.05616	.05639	.05647	.05641	.05620	.05586
24	13	16	.05136	.05202	.05256	.05296	.05322	.05335	.05335	.05321
24	13	17	.04774	.04854	.04922	.04978	.05021	.05052	.05070	.05075
24	13	18	.04434	.04526	.04608	.04678	.04737	.04785	.04820	.04843
24	13	19	.04111	.04215	.04309	.04393	.04466	.04529	.04581	.04621
24	13	20	.03798	.03912	.04018	.04115	.04203	.04281	.04349	.04406
24	13	21	.03488	.03612	.03729	.03838	.03940	.04033	.04117	.04192
24	13	22	.03169	.03302	.03431	.03553	.03670	.03778	.03879	.03972
24	13	23	.02820	.02963	.03104	.03240	.03373	.03499	.03619	.03732
24	13	24	.02378	.02533	.02688	.02843	.02996	.03146	.03292	.03433
24	14	15	.06018	.06095	.06157	.06204	.06235	.06250	.06250	.06234
24	14	16	.05593	.05685	.05764	.05828	.05878	.05913	.05934	.05940
24	14	17	.05199	.05305	.05399	.05479	.05546	.05600	.05640	.05666
24	14	18	.04830	.04948	.05055	.05151	.05234	.05305	.05363	.05407
24	14	19	.04479	.04608	.04728	.04837	.04935	.05022	.05098	.05161
24	14	20	.04139	.04278	.04409	.04532	.04645	.04748	.04840	.04921
24	14	21	.03801	.03950	.04092	.04228	.04355	.04474	.04583	.04682
24	14	22	.03454	.03612	.03766	.03914	.04057	.04192	.04319	.04437
24	14	23	.03074	.03242	.03407	.03570	.03729	.03882	.04030	.04170
24	14	24	.02593	.02771	.02952	.03133	.03313	.03491	.03666	.03836

Covariances of Inverse Gaussian Order Statistics
Shape Parameter k

n	i	j	0.0	0.1	0.2	0.3	0.4	0.5	0.6	0.7
24	15	16	.06124	.06247	.06357	.06451	.06530	.06592	.06639	.06669
24	15	17	.05694	.05831	.05955	.06066	.06163	.06244	.06311	.06363
24	15	18	.05291	.05440	.05578	.05704	.05817	.05916	.06002	.06074
24	15	19	.04908	.05067	.05217	.05357	.05486	.05603	.05707	.05798
24	15	20	.04536	.04705	.04867	.05020	.05164	.05297	.05419	.05529
24	15	21	.04167	.04345	.04518	.04684	.04843	.04993	.05133	.05262
24	15	22	.03787	.03974	.04158	.04338	.04511	.04678	.04837	.04988
24	15	23	.03371	.03567	.03763	.03957	.04148	.04334	.04514	.04688
24	15	24	.02844	.03050	.03261	.03473	.03686	.03898	.04107	.04314
24	16	17	.06282	.06457	.06619	.06767	.06900	.07018	.07119	.07203
24	16	18	.05839	.06025	.06201	.06364	.06514	.06651	.06772	.06878
24	16	19	.05417	.05614	.05802	.05979	.06145	.06299	.06440	.06567
24	16	20	.05008	.05214	.05413	.05604	.05786	.05957	.06117	.06264
24	16	21	.04601	.04816	.05026	.05230	.05427	.05616	.05794	.05963
24	16	22	.04182	.04406	.04627	.04844	.05057	.05263	.05462	.05652
24	16	23	.03724	.03956	.04188	.04420	.04650	.04877	.05098	.05314
24	16	24	.03142	.03383	.03630	.03880	.04133	.04387	.04640	.04890
24	17	18	.06504	.06738	.06962	.07173	.07370	.07553	.07721	.07872
24	17	19	.06036	.06280	.06515	.06741	.06955	.07156	.07344	.07517
24	17	20	.05581	.05834	.06081	.06320	.06550	.06769	.06977	.07172
24	17	21	.05130	.05390	.05648	.05900	.06145	.06383	.06611	.06829
24	17	22	.04664	.04933	.05200	.05466	.05728	.05984	.06234	.06475
24	17	23	.04154	.04430	.04709	.04989	.05268	.05546	.05820	.06089
24	17	24	.03506	.03790	.04082	.04380	.04684	.04990	.05298	.05605
24	18	19	.06810	.07115	.07413	.07701	.07979	.08243	.08494	.08729
24	18	20	.06300	.06613	.06921	.07223	.07516	.07800	.08072	.08331
24	18	21	.05792	.06112	.06430	.06745	.07055	.07357	.07651	.07935
24	18	22	.05268	.05595	.05923	.06251	.06577	.06900	.07216	.07526
24	18	23	.04694	.05026	.05365	.05707	.06052	.06396	.06739	.07079
24	18	24	.03964	.04302	.04653	.05014	.05383	.05758	.06137	.06519
24	19	20	.07234	.07628	.08020	.08407	.08788	.09160	.09521	.09869
24	19	21	.06654	.07054	.07455	.07855	.08252	.08644	.09028	.09403
24	19	22	.06055	.06460	.06870	.07283	.07697	.08109	.08518	.08922
24	19	23	.05398	.05806	.06225	.06652	.07085	.07521	.07959	.08396
24	19	24	.04560	.04972	.05402	.05847	.06305	.06773	.07251	.07734
24	20	21	.07834	.08347	.08866	.09389	.09913	.10435	.10952	.11462
24	20	22	.07133	.07649	.08176	.08711	.09251	.09795	.10339	.10881
24	20	23	.06363	.06880	.07413	.07961	.08521	.09090	.09665	.10244
24	20	24	.05380	.05896	.06438	.07003	.07588	.08192	.08811	.09443
24	21	22	.08721	.09408	.10115	.10841	.11580	.12331	.13089	.13851
24	21	23	.07787	.08470	.09181	.09918	.10676	.11453	.12246	.13051
24	21	24	.06592	.07268	.07982	.08733	.09517	.10332	.11175	.12042
24	22	23	.10140	.11113	.12135	.13204	.14316	.15467	.16652	.17867
24	22	24	.08600	.09553	.10570	.11648	.12785	.13977	.15220	.16510
24	23	24	.12686	.14242	.15924	.17729	.19655	.21698	.23853	.26115
25	1	2	.12524	.11095	.09791	.08608	.07543	.06588	.05739	.04986
25	1	3	.08488	.07601	.06781	.06028	.05341	.04717	.04154	.03650
25	1	4	.06505	.05870	.05277	.04727	.04221	.03757	.03335	.02953
25	1	5	.05310	.04820	.04359	.03929	.03530	.03162	.02824	.02516
25	1	6	.04502	.04108	.03734	.03383	.03056	.02751	.02470	.02213
25	1	7	.03915	.03588	.03277	.02982	.02706	.02448	.02208	.01987
25	1	8	.03465	.03189	.02924	.02673	.02435	.02212	.02004	.01811
25	1	9	.03107	.02870	.02643	.02425	.02218	.02023	.01840	.01669
25	1	10	.02814	.02609	.02411	.02220	.02038	.01866	.01703	.01551
25	1	11	.02569	.02390	.02216	.02048	.01887	.01733	.01588	.01451

Covariances of Inverse Gaussian Order Statistics
Shape Parameter k

n	i	j	0.0	0.1	0.2	0.3	0.4	0.5	0.6	0.7
25	1	12	.02358	.02202	.02048	.01900	.01756	.01619	.01488	.01365
25	1	13	.02175	.02038	.01902	.01770	.01642	.01519	.01401	.01289
25	1	14	.02014	.01893	.01773	.01655	.01540	.01429	.01323	.01221
25	1	15	.01869	.01763	.01656	.01552	.01449	.01349	.01253	.01160
25	1	16	.01738	.01645	.01551	.01458	.01366	.01276	.01189	.01105
25	1	17	.01618	.01536	.01454	.01371	.01289	.01209	.01130	.01054
25	1	18	.01506	.01435	.01363	.01291	.01218	.01146	.01075	.01006
25	1	19	.01401	.01340	.01278	.01214	.01150	.01086	.01023	.00961
25	1	20	.01301	.01250	.01196	.01141	.01086	.01029	.00973	.00918
25	1	21	.01203	.01161	.01117	.01070	.01022	.00974	.00924	.00875
25	1	22	.01106	.01073	.01037	.00999	.00959	.00918	.00876	.00833
25	1	23	.01006	.00982	.00955	.00926	.00894	.00861	.00826	.00790
25	1	24	.00896	.00882	.00865	.00845	.00823	.00798	.00771	.00743
25	1	25	.00757	.00755	.00750	.00742	.00731	.00718	.00702	.00684
25	2	3	.09985	.09064	.08200	.07394	.06645	.05954	.05319	.04740
25	2	4	.07667	.07014	.06394	.05810	.05262	.04752	.04279	.03843
25	2	5	.06266	.05766	.05289	.04835	.04406	.04004	.03628	.03279
25	2	6	.05317	.04918	.04534	.04167	.03817	.03487	.03176	.02885
25	2	7	.04626	.04298	.03980	.03675	.03382	.03104	.02840	.02593
25	2	8	.04096	.03821	.03554	.03295	.03045	.02806	.02579	.02364
25	2	9	.03674	.03441	.03213	.02990	.02774	.02567	.02368	.02179
25	2	10	.03329	.03129	.02932	.02739	.02550	.02368	.02193	.02026
25	2	11	.03039	.02866	.02695	.02526	.02361	.02201	.02045	.01896
25	2	12	.02790	.02641	.02492	.02344	.02198	.02056	.01917	.01783
25	2	13	.02574	.02445	.02314	.02184	.02055	.01929	.01805	.01684
25	2	14	.02383	.02271	.02157	.02043	.01929	.01816	.01705	.01596
25	2	15	.02212	.02115	.02016	.01915	.01815	.01714	.01615	.01517
25	2	16	.02057	.01974	.01888	.01800	.01711	.01621	.01532	.01445
25	2	17	.01915	.01844	.01770	.01693	.01615	.01536	.01456	.01378
25	2	18	.01783	.01723	.01660	.01593	.01525	.01456	.01386	.01315
25	2	19	.01659	.01609	.01556	.01500	.01441	.01380	.01319	.01256
25	2	20	.01540	.01500	.01457	.01410	.01360	.01308	.01255	.01200
25	2	21	.01425	.01394	.01360	.01322	.01281	.01237	.01192	.01145
25	2	22	.01310	.01289	.01263	.01234	.01202	.01167	.01129	.01090
25	2	23	.01191	.01179	.01163	.01144	.01120	.01094	.01065	.01034
25	2	24	.01062	.01060	.01054	.01044	.01031	.01014	.00994	.00972
25	2	25	.00896	.00907	.00914	.00917	.00916	.00912	.00905	.00895
25	3	4	.08570	.07905	.07269	.06662	.06087	.05546	.05038	.04565
25	3	5	.07011	.06506	.06018	.05550	.05103	.04678	.04276	.03899
25	3	6	.05953	.05552	.05163	.04786	.04423	.04076	.03746	.03434
25	3	7	.05181	.04855	.04535	.04223	.03921	.03630	.03352	.03087
25	3	8	.04590	.04318	.04050	.03788	.03532	.03284	.03045	.02816
25	3	9	.04118	.03890	.03662	.03438	.03218	.03004	.02796	.02596
25	3	10	.03732	.03537	.03343	.03150	.02959	.02773	.02591	.02414
25	3	11	.03407	.03241	.03074	.02906	.02740	.02577	.02416	.02260
25	3	12	.03129	.02987	.02842	.02697	.02552	.02407	.02265	.02126
25	3	13	.02887	.02765	.02640	.02514	.02386	.02259	.02133	.02008
25	3	14	.02673	.02569	.02461	.02351	.02239	.02127	.02015	.01903
25	3	15	.02482	.02393	.02300	.02205	.02107	.02008	.01908	.01809
25	3	16	.02308	.02233	.02154	.02072	.01986	.01899	.01811	.01723
25	3	17	.02149	.02086	.02019	.01949	.01875	.01799	.01722	.01643
25	3	18	.02001	.01950	.01894	.01835	.01772	.01706	.01638	.01569
25	3	19	.01861	.01821	.01776	.01727	.01674	.01618	.01559	.01499
25	3	20	.01728	.01698	.01663	.01623	.01580	.01533	.01483	.01431
25	3	21	.01599	.01578	.01552	.01522	.01488	.01450	.01409	.01366

Covariances of Inverse Gaussian Order Statistics
Shape Parameter k

n	i	j	0.0	0.1	0.2	0.3	0.4	0.5	0.6	0.7
25	3	22	.01470	.01458	.01442	.01421	.01396	.01368	.01336	.01300
25	3	23	.01337	.01335	.01328	.01317	.01302	.01282	.01259	.01233
25	3	24	.01191	.01199	.01203	.01202	.01197	.01189	.01176	.01160
25	3	25	.01006	.01026	.01043	.01056	.01065	.01070	.01070	.01067
25	4	5	.07684	.07176	.06681	.06202	.05740	.05298	.04876	.04476
25	4	6	.06529	.06128	.05735	.05352	.04979	.04619	.04274	.03944
25	4	7	.05685	.05361	.05040	.04724	.04416	.04116	.03826	.03547
25	4	8	.05037	.04770	.04503	.04239	.03979	.03724	.03476	.03237
25	4	9	.04521	.04298	.04073	.03849	.03627	.03408	.03194	.02986
25	4	10	.04098	.03909	.03719	.03527	.03336	.03146	.02960	.02777
25	4	11	.03742	.03582	.03420	.03255	.03089	.02924	.02761	.02599
25	4	12	.03437	.03302	.03163	.03021	.02877	.02733	.02588	.02446
25	4	13	.03172	.03057	.02938	.02816	.02691	.02564	.02437	.02311
25	4	14	.02937	.02841	.02739	.02634	.02525	.02415	.02303	.02190
25	4	15	.02727	.02646	.02560	.02470	.02376	.02280	.02181	.02082
25	4	16	.02537	.02470	.02398	.02321	.02241	.02157	.02071	.01983
25	4	17	.02362	.02308	.02248	.02184	.02115	.02043	.01969	.01891
25	4	18	.02199	.02157	.02109	.02056	.01999	.01937	.01873	.01806
25	4	19	.02046	.02014	.01977	.01935	.01888	.01837	.01783	.01725
25	4	20	.01900	.01878	.01851	.01819	.01782	.01741	.01696	.01648
25	4	21	.01758	.01746	.01728	.01706	.01679	.01647	.01612	.01572
25	4	22	.01616	.01614	.01606	.01593	.01576	.01554	.01527	.01497
25	4	23	.01470	.01477	.01479	.01476	.01469	.01457	.01440	.01420
25	4	24	.01310	.01327	.01340	.01348	.01351	.01350	.01345	.01335
25	4	25	.01106	.01136	.01162	.01184	.01202	.01215	.01224	.01229
25	5	6	.07084	.06685	.06290	.05901	.05521	.05150	.04792	.04447
25	5	7	.06171	.05850	.05530	.05212	.04898	.04591	.04292	.04001
25	5	8	.05470	.05207	.04942	.04678	.04415	.04155	.03901	.03653
25	5	9	.04911	.04693	.04471	.04249	.04026	.03804	.03585	.03370
25	5	10	.04452	.04270	.04083	.03894	.03703	.03512	.03323	.03135
25	5	11	.04066	.03913	.03756	.03594	.03430	.03265	.03100	.02935
25	5	12	.03735	.03607	.03474	.03336	.03195	.03052	.02907	.02762
25	5	13	.03447	.03341	.03228	.03110	.02989	.02864	.02738	.02610
25	5	14	.03193	.03104	.03010	.02910	.02805	.02697	.02587	.02474
25	5	15	.02965	.02892	.02813	.02729	.02640	.02547	.02451	.02352
25	5	16	.02758	.02699	.02635	.02565	.02489	.02410	.02326	.02240
25	5	17	.02568	.02522	.02471	.02413	.02350	.02283	.02212	.02137
25	5	18	.02391	.02357	.02318	.02272	.02221	.02165	.02105	.02041
25	5	19	.02225	.02202	.02173	.02138	.02098	.02053	.02003	.01950
25	5	20	.02066	.02053	.02035	.02010	.01981	.01946	.01906	.01862
25	5	21	.01911	.01909	.01900	.01886	.01866	.01841	.01811	.01777
25	5	22	.01758	.01764	.01765	.01761	.01751	.01736	.01717	.01692
25	5	23	.01599	.01615	.01626	.01632	.01632	.01628	.01619	.01605
25	5	24	.01425	.01451	.01473	.01490	.01502	.01509	.01512	.01509
25	5	25	.01203	.01242	.01277	.01309	.01336	.01358	.01376	.01389
25	6	7	.06657	.06340	.06021	.05702	.05385	.05072	.04764	.04464
25	6	8	.05903	.05645	.05383	.05119	.04855	.04592	.04332	.04076
25	6	9	.05301	.05089	.04872	.04651	.04428	.04205	.03982	.03762
25	6	10	.04806	.04631	.04450	.04264	.04075	.03883	.03691	.03500
25	6	11	.04391	.04245	.04094	.03936	.03775	.03611	.03444	.03278
25	6	12	.04034	.03914	.03787	.03654	.03517	.03375	.03231	.03085
25	6	13	.03724	.03625	.03519	.03407	.03290	.03168	.03043	.02915
25	6	14	.03449	.03369	.03282	.03188	.03088	.02984	.02876	.02764
25	6	15	.03203	.03139	.03068	.02990	.02907	.02818	.02725	.02628
25	6	16	.02980	.02930	.02874	.02810	.02741	.02666	.02587	.02503

Covariances of Inverse Gaussian Order Statistics
Shape Parameter k

n	i	j	0.0	0.1	0.2	0.3	0.4	0.5	0.6	0.7
25	6	17	.02775	.02738	.02695	.02645	.02588	.02526	.02459	.02388
25	6	18	.02584	.02559	.02528	.02490	.02446	.02396	.02340	.02280
25	6	19	.02404	.02391	.02371	.02344	.02311	.02272	.02228	.02179
25	6	20	.02233	.02230	.02220	.02204	.02181	.02153	.02120	.02081
25	6	21	.02066	.02072	.02073	.02067	.02055	.02038	.02014	.01986
25	6	22	.01900	.01916	.01926	.01930	.01929	.01922	.01909	.01891
25	6	23	.01728	.01754	.01774	.01789	.01798	.01802	.01801	.01794
25	6	24	.01540	.01576	.01607	.01633	.01655	.01671	.01681	.01687
25	6	25	.01301	.01349	.01394	.01435	.01471	.01504	.01531	.01553
25	7	8	.06346	.06094	.05837	.05575	.05310	.05044	.04779	.04517
25	7	9	.05700	.05496	.05284	.05066	.04844	.04620	.04395	.04170
25	7	10	.05170	.05002	.04827	.04645	.04458	.04268	.04075	.03881
25	7	11	.04724	.04587	.04442	.04290	.04132	.03969	.03803	.03635
25	7	12	.04341	.04230	.04110	.03983	.03849	.03710	.03568	.03422
25	7	13	.04007	.03918	.03820	.03714	.03602	.03484	.03361	.03234
25	7	14	.03712	.03641	.03562	.03475	.03381	.03281	.03176	.03067
25	7	15	.03448	.03393	.03331	.03260	.03183	.03099	.03010	.02916
25	7	16	.03208	.03168	.03120	.03064	.03002	.02933	.02858	.02778
25	7	17	.02987	.02960	.02926	.02884	.02835	.02779	.02717	.02650
25	7	18	.02782	.02767	.02745	.02715	.02679	.02635	.02586	.02531
25	7	19	.02589	.02585	.02574	.02556	.02531	.02500	.02462	.02418
25	7	20	.02404	.02411	.02411	.02403	.02390	.02369	.02343	.02310
25	7	21	.02225	.02241	.02251	.02255	.02251	.02242	.02226	.02205
25	7	22	.02046	.02072	.02092	.02106	.02113	.02115	.02110	.02100
25	7	23	.01861	.01897	.01927	.01952	.01970	.01983	.01990	.01991
25	7	24	.01659	.01705	.01746	.01782	.01813	.01838	.01858	.01873
25	7	25	.01401	.01459	.01514	.01565	.01612	.01655	.01692	.01724
25	8	9	.06117	.05921	.05715	.05501	.05282	.05057	.04830	.04602
25	8	10	.05549	.05391	.05222	.05046	.04862	.04673	.04480	.04284
25	8	11	.05071	.04944	.04806	.04660	.04507	.04347	.04182	.04013
25	8	12	.04661	.04560	.04448	.04328	.04199	.04064	.03924	.03778
25	8	13	.04304	.04224	.04135	.04036	.03930	.03816	.03697	.03572
25	8	14	.03987	.03927	.03856	.03777	.03690	.03595	.03494	.03387
25	8	15	.03704	.03659	.03606	.03544	.03474	.03396	.03311	.03221
25	8	16	.03446	.03417	.03378	.03331	.03276	.03214	.03144	.03069
25	8	17	.03209	.03193	.03168	.03135	.03094	.03046	.02990	.02928
25	8	18	.02989	.02985	.02973	.02952	.02924	.02889	.02846	.02797
25	8	19	.02782	.02789	.02788	.02780	.02763	.02740	.02709	.02672
25	8	20	.02584	.02601	.02611	.02614	.02609	.02597	.02578	.02553
25	8	21	.02391	.02418	.02438	.02452	.02458	.02458	.02450	.02437
25	8	22	.02199	.02236	.02266	.02290	.02308	.02318	.02323	.02321
25	8	23	.02001	.02047	.02088	.02123	.02152	.02174	.02191	.02201
25	8	24	.01783	.01839	.01891	.01938	.01980	.02016	.02046	.02070
25	8	25	.01506	.01575	.01640	.01703	.01761	.01814	.01863	.01906
25	9	10	.05951	.05802	.05642	.05472	.05293	.05106	.04914	.04717
25	9	11	.05440	.05322	.05194	.05055	.04907	.04751	.04588	.04420
25	9	12	.05001	.04910	.04808	.04695	.04573	.04443	.04306	.04162
25	9	13	.04618	.04549	.04469	.04380	.04280	.04172	.04057	.03935
25	9	14	.04279	.04229	.04169	.04099	.04019	.03931	.03835	.03732
25	9	15	.03975	.03942	.03899	.03846	.03784	.03714	.03635	.03549
25	9	16	.03699	.03681	.03653	.03616	.03570	.03515	.03452	.03382
25	9	17	.03445	.03441	.03427	.03404	.03372	.03331	.03283	.03227
25	9	18	.03209	.03217	.03215	.03205	.03187	.03160	.03125	.03083
25	9	19	.02987	.03006	.03016	.03018	.03012	.02997	.02975	.02946
25	9	20	.02775	.02803	.02825	.02838	.02844	.02841	.02832	.02815

Covariances of Inverse Gaussian Order Statistics
Shape Parameter k

n	i	j	0.0	0.1	0.2	0.3	0.4	0.5	0.6	0.7
25	9	21	.02568	.02606	.02638	.02662	.02679	.02689	.02691	.02686
25	9	22	.02362	.02410	.02452	.02487	.02515	.02537	.02551	.02559
25	9	23	.02149	.02207	.02259	.02305	.02345	.02379	.02406	.02427
25	9	24	.01915	.01983	.02046	.02105	.02158	.02206	.02248	.02283
25	9	25	.01618	.01698	.01775	.01849	.01920	.01986	.02047	.02102
25	10	11	.05835	.05729	.05611	.05480	.05339	.05188	.05029	.04862
25	10	12	.05365	.05286	.05195	.05091	.04977	.04853	.04720	.04580
25	10	13	.04956	.04899	.04830	.04750	.04659	.04558	.04448	.04330
25	10	14	.04592	.04555	.04506	.04447	.04376	.04295	.04206	.04108
25	10	15	.04267	.04247	.04215	.04173	.04121	.04058	.03987	.03907
25	10	16	.03971	.03966	.03950	.03924	.03887	.03842	.03787	.03723
25	10	17	.03699	.03707	.03705	.03694	.03672	.03641	.03602	.03553
25	10	18	.03446	.03466	.03477	.03479	.03471	.03454	.03428	.03394
25	10	19	.03208	.03239	.03262	.03275	.03280	.03277	.03264	.03244
25	10	20	.02980	.03021	.03055	.03081	.03098	.03107	.03107	.03100
25	10	21	.02758	.02809	.02853	.02890	.02919	.02940	.02953	.02959
25	10	22	.02537	.02597	.02652	.02700	.02741	.02774	.02800	.02818
25	10	23	.02308	.02379	.02444	.02503	.02556	.02602	.02641	.02673
25	10	24	.02057	.02138	.02214	.02286	.02352	.02412	.02467	.02515
25	10	25	.01738	.01830	.01921	.02008	.02092	.02172	.02247	.02316
25	11	12	.05761	.05696	.05617	.05524	.05419	.05302	.05175	.05039
25	11	13	.05322	.05280	.05224	.05155	.05074	.04981	.04878	.04766
25	11	14	.04933	.04910	.04874	.04826	.04766	.04695	.04613	.04522
25	11	15	.04584	.04578	.04560	.04530	.04489	.04436	.04374	.04301
25	11	16	.04267	.04276	.04273	.04260	.04235	.04200	.04155	.04099
25	11	17	.03975	.03998	.04009	.04011	.04001	.03982	.03952	.03913
25	11	18	.03704	.03738	.03763	.03778	.03782	.03777	.03762	.03738
25	11	19	.03448	.03493	.03530	.03557	.03575	.03584	.03583	.03573
25	11	20	.03203	.03259	.03307	.03346	.03376	.03398	.03410	.03414
25	11	21	.02965	.03030	.03089	.03139	.03182	.03216	.03242	.03259
25	11	22	.02727	.02802	.02871	.02933	.02987	.03034	.03074	.03105
25	11	23	.02482	.02566	.02646	.02719	.02786	.02846	.02899	.02945
25	11	24	.02212	.02307	.02397	.02483	.02564	.02639	.02708	.02771
25	11	25	.01869	.01975	.02080	.02182	.02281	.02376	.02467	.02552
25	12	13	.05726	.05699	.05658	.05602	.05533	.05451	.05357	.05251
25	12	14	.05308	.05301	.05280	.05246	.05199	.05139	.05066	.04983
25	12	15	.04933	.04943	.04941	.04925	.04897	.04856	.04804	.04741
25	12	16	.04592	.04618	.04631	.04632	.04621	.04598	.04564	.04519
25	12	17	.04279	.04318	.04345	.04361	.04366	.04360	.04342	.04314
25	12	18	.03987	.04038	.04079	.04109	.04128	.04136	.04134	.04122
25	12	19	.03712	.03774	.03827	.03869	.03902	.03925	.03937	.03940
25	12	20	.03449	.03521	.03585	.03640	.03685	.03722	.03748	.03765
25	12	21	.03193	.03275	.03349	.03416	.03474	.03523	.03563	.03595
25	12	22	.02937	.03028	.03113	.03191	.03262	.03324	.03379	.03424
25	12	23	.02673	.02774	.02869	.02959	.03042	.03118	.03187	.03249
25	12	24	.02383	.02493	.02600	.02702	.02800	.02892	.02978	.03057
25	12	25	.02014	.02135	.02256	.02375	.02491	.02604	.02712	.02815
25	13	14	.05726	.05737	.05734	.05716	.05684	.05637	.05577	.05504
25	13	15	.05322	.05351	.05366	.05367	.05355	.05328	.05289	.05237
25	13	16	.04956	.05000	.05031	.05049	.05054	.05046	.05025	.04993
25	13	17	.04618	.04676	.04721	.04755	.04776	.04785	.04782	.04767
25	13	18	.04304	.04373	.04432	.04480	.04516	.04540	.04553	.04555
25	13	19	.04007	.04088	.04159	.04219	.04269	.04309	.04337	.04355
25	13	20	.03724	.03814	.03897	.03969	.04033	.04086	.04129	.04162
25	13	21	.03447	.03548	.03641	.03725	.03801	.03868	.03926	.03974

Covariances of Inverse Gaussian Order Statistics
Shape Parameter k

n	i	j	0.0	0.1	0.2	0.3	0.4	0.5	0.6	0.7
25	13	22	.03172	.03281	.03385	.03481	.03570	.03651	.03723	.03786
25	13	23	.02887	.03006	.03119	.03228	.03330	.03425	.03513	.03592
25	13	24	.02574	.02702	.02827	.02948	.03065	.03176	.03282	.03380
25	13	25	.02175	.02314	.02453	.02592	.02728	.02861	.02990	.03114
25	14	15	.05761	.05812	.05848	.05869	.05875	.05866	.05842	.05805
25	14	16	.05365	.05431	.05484	.05522	.05546	.05556	.05552	.05535
25	14	17	.05001	.05080	.05147	.05201	.05242	.05270	.05284	.05285
25	14	18	.04661	.04753	.04833	.04901	.04957	.05001	.05032	.05051
25	14	19	.04341	.04443	.04535	.04617	.04688	.04747	.04794	.04830
25	14	20	.04034	.04147	.04250	.04344	.04428	.04502	.04565	.04617
25	14	21	.03735	.03857	.03971	.04077	.04175	.04263	.04341	.04408
25	14	22	.03437	.03568	.03692	.03810	.03921	.04023	.04116	.04201
25	14	23	.03129	.03268	.03404	.03534	.03658	.03775	.03884	.03986
25	14	24	.02790	.02939	.03085	.03228	.03367	.03501	.03630	.03751
25	14	25	.02358	.02517	.02678	.02838	.02997	.03154	.03307	.03456
25	15	16	.05835	.05927	.06005	.06067	.06115	.06147	.06163	.06165
25	15	17	.05440	.05545	.05637	.05716	.05780	.05831	.05866	.05888
25	15	18	.05071	.05188	.05294	.05387	.05467	.05535	.05588	.05628
25	15	19	.04724	.04851	.04969	.05076	.05171	.05254	.05324	.05382
25	15	20	.04391	.04528	.04657	.04776	.04886	.04984	.05070	.05145
25	15	21	.04066	.04213	.04352	.04484	.04606	.04719	.04822	.04914
25	15	22	.03742	.03897	.04047	.04191	.04327	.04455	.04574	.04683
25	15	23	.03407	.03571	.03731	.03887	.04037	.04180	.04316	.04444
25	15	24	.03039	.03211	.03383	.03552	.03717	.03878	.04034	.04183
25	15	25	.02569	.02751	.02936	.03123	.03309	.03493	.03676	.03854
25	16	17	.05951	.06088	.06210	.06319	.06413	.06491	.06554	.06600
25	16	18	.05549	.05697	.05833	.05957	.06067	.06163	.06244	.06310
25	16	19	.05170	.05328	.05477	.05614	.05739	.05851	.05950	.06035
25	16	20	.04806	.04974	.05134	.05284	.05423	.05551	.05667	.05771
25	16	21	.04452	.04629	.04799	.04961	.05114	.05258	.05391	.05512
25	16	22	.04098	.04283	.04463	.04637	.04805	.04964	.05114	.05254
25	16	23	.03732	.03925	.04115	.04302	.04483	.04659	.04827	.04987
25	16	24	.03329	.03530	.03731	.03931	.04129	.04323	.04511	.04694
25	16	25	.02814	.03025	.03240	.03457	.03676	.03895	.04112	.04326
25	17	18	.06117	.06303	.06477	.06638	.06785	.06917	.07033	.07134
25	17	19	.05700	.05897	.06083	.06257	.06420	.06569	.06704	.06824
25	17	20	.05301	.05506	.05703	.05891	.06068	.06234	.06387	.06527
25	17	21	.04911	.05124	.05332	.05532	.05724	.05905	.06076	.06235
25	17	22	.04521	.04743	.04960	.05173	.05378	.05576	.05765	.05944
25	17	23	.04118	.04347	.04575	.04799	.05020	.05235	.05443	.05643
25	17	24	.03674	.03911	.04149	.04387	.04624	.04858	.05088	.05313
25	17	25	.03107	.03352	.03603	.03859	.04117	.04378	.04638	.04897
25	18	19	.06346	.06589	.06823	.07046	.07256	.07453	.07634	.07800
25	18	20	.05903	.06155	.06399	.06635	.06861	.07074	.07275	.07462
25	18	21	.05470	.05730	.05985	.06233	.06473	.06703	.06923	.07130
25	18	22	.05037	.05304	.05569	.05829	.06084	.06331	.06570	.06799
25	18	23	.04590	.04863	.05137	.05410	.05679	.05945	.06204	.06456
25	18	24	.04096	.04376	.04660	.04946	.05233	.05518	.05801	.06079
25	18	25	.03465	.03752	.04048	.04352	.04661	.04974	.05289	.05605
25	19	20	.06657	.06970	.07276	.07575	.07864	.08141	.08405	.08655
25	19	21	.06171	.06491	.06807	.07118	.07422	.07717	.08001	.08273
25	19	22	.05685	.06011	.06336	.06659	.06978	.07290	.07595	.07891
25	19	23	.05181	.05513	.05847	.06182	.06516	.06847	.07174	.07494
25	19	24	.04626	.04962	.05306	.05654	.06006	.06358	.06710	.07059
25	19	25	.03915	.04257	.04611	.04977	.05352	.05733	.06120	.06511

Covariances of Inverse Gaussian Order Statistics
Shape Parameter k

n	i	j	0.0	0.1	0.2	0.3	0.4	0.5	0.6	0.7
25	20	21	.07084	.07484	.07884	.08280	.08671	.09055	.09429	.09791
25	20	22	.06529	.06934	.07341	.07750	.08156	.08558	.08954	.09342
25	20	23	.05953	.06362	.06778	.07198	.07620	.08041	.08461	.08876
25	20	24	.05317	.05729	.06153	.06586	.07026	.07470	.07917	.08364
25	20	25	.04502	.04917	.05350	.05800	.06263	.06739	.07224	.07717
25	21	22	.07684	.08202	.08728	.09259	.09792	.10325	.10854	.11377
25	21	23	.07011	.07530	.08062	.08604	.09153	.09707	.10262	.10816
25	21	24	.06266	.06786	.07324	.07878	.08445	.09022	.09607	.10197
25	21	25	.05310	.05829	.06373	.06943	.07534	.08145	.08772	.09414
25	22	23	.08570	.09260	.09973	.10705	.11453	.12213	.12983	.13758
25	22	24	.07667	.08353	.09068	.09810	.10575	.11361	.12164	.12981
25	22	25	.06505	.07183	.07900	.08655	.09445	.10267	.11118	.11995
25	23	24	.09985	.10959	.11985	.13059	.14178	.15338	.16534	.17762
25	23	25	.08488	.09441	.10460	.11542	.12684	.13883	.15135	.16436
25	24	25	.12524	.14079	.15762	.17569	.19500	.21550	.23716	.25991

Covariances of Inverse Gaussian Order Statistics
Shape Parameter k

n	i	j	0.8	0.9	1.0	1.1	1.2	1.3	1.4	1.5
2	1	2	.30505	.30179	.29827	.29452	.29055	.28641	.28212	.27770
3	1	2	.20889	.20053	.19231	.18426	.17638	.16872	.16128	.15408
3	1	3	.15527	.15293	.15042	.14774	.14493	.14201	.13900	.13591
3	2	3	.32594	.33021	.33399	.33729	.34011	.34247	.34438	.34586
4	1	2	.16189	.15261	.14369	.13515	.12701	.11926	.11191	.10495
4	1	3	.12283	.11811	.11341	.10875	.10417	.09968	.09528	.09101
4	1	4	.09753	.09580	.09395	.09198	.08993	.08780	.08561	.08338
4	2	3	.22220	.21882	.21517	.21127	.20716	.20287	.19843	.19387
4	2	4	.17810	.17912	.17982	.18019	.18025	.18003	.17953	.17877
4	3	4	.32774	.33658	.34498	.35292	.36039	.36737	.37386	.37986
5	1	2	.13409	.12478	.11597	.10767	.09986	.09255	.08571	.07933
5	1	3	.10219	.09676	.09148	.08639	.08148	.07677	.07226	.06796
5	1	4	.08320	.08004	.07687	.07373	.07062	.06756	.06456	.06163
5	1	5	.06860	.06726	.06581	.06429	.06270	.06106	.05938	.05768
5	2	3	.17089	.16533	.15971	.15405	.14839	.14275	.13716	.13165
5	2	4	.14009	.13768	.13508	.13231	.12939	.12635	.12322	.12000
5	2	5	.11606	.11622	.11615	.11585	.11534	.11463	.11373	.11267
5	3	4	.22321	.22315	.22270	.22188	.22070	.21919	.21737	.21527
5	3	5	.18611	.18954	.19263	.19538	.19778	.19983	.20156	.20295
5	4	5	.32541	.33731	.34887	.36007	.37087	.38123	.39115	.40061
6	1	2	.11566	.10657	.09807	.09014	.08278	.07595	.06963	.06381
6	1	3	.08816	.08254	.07716	.07204	.06717	.06257	.05823	.05415
6	1	4	.07237	.06868	.06507	.06157	.05817	.05490	.05176	.04875
6	1	5	.06124	.05892	.05659	.05427	.05196	.04969	.04746	.04528
6	1	6	.05176	.05067	.04950	.04827	.04699	.04567	.04433	.04296
6	2	3	.14033	.13404	.12783	.12172	.11574	.10990	.10424	.09876
6	2	4	.11584	.11214	.10838	.10457	.10074	.09690	.09309	.08932
6	2	5	.09838	.09655	.09457	.09247	.09026	.08797	.08561	.08319
6	2	6	.08337	.08324	.08293	.08244	.08181	.08102	.08010	.07907
6	3	4	.17129	.16835	.16516	.16176	.15818	.15445	.15058	.14662
6	3	5	.14613	.14556	.14472	.14362	.14228	.14072	.13895	.13700
6	3	6	.12425	.12590	.12729	.12842	.12930	.12993	.13032	.13049
6	4	5	.22144	.22373	.22560	.22706	.22812	.22879	.22909	.22903
6	4	6	.18924	.19444	.19934	.20391	.20815	.21204	.21560	.21881
6	5	6	.32206	.33616	.35004	.36366	.37697	.38994	.40252	.41470
7	1	2	.10251	.09370	.08554	.07800	.07105	.06467	.05882	.05347
7	1	3	.07803	.07241	.06709	.06208	.05737	.05296	.04885	.04502
7	1	4	.06426	.06036	.05661	.05302	.04958	.04632	.04322	.04030
7	1	5	.05490	.05217	.04949	.04687	.04433	.04188	.03952	.03725
7	1	6	.04761	.04580	.04398	.04216	.04036	.03858	.03683	.03513
7	1	7	.04095	.04003	.03906	.03803	.03697	.03588	.03476	.03364
7	2	3	.12002	.11350	.10716	.10102	.09509	.08939	.08394	.07873
7	2	4	.09931	.09507	.09085	.08667	.08255	.07852	.07459	.07076
7	2	5	.08510	.08240	.07964	.07684	.07401	.07118	.06836	.06556
7	2	6	.07396	.07250	.07092	.06926	.06751	.06569	.06382	.06192
7	2	7	.06371	.06347	.06308	.06256	.06192	.06116	.06031	.05936
7	3	4	.14024	.13611	.13187	.12754	.12315	.11874	.11431	.10991
7	3	5	.12058	.11837	.11598	.11343	.11075	.10796	.10507	.10211
7	3	6	.10506	.10439	.10352	.10246	.10122	.09983	.09828	.09661
7	3	7	.09069	.09157	.09224	.09272	.09299	.09308	.09299	.09273
7	4	5	.16980	.16874	.16737	.16571	.16378	.16160	.15920	.15660
7	4	6	.14843	.14929	.14985	.15012	.15011	.14983	.14929	.14850
7	4	7	.12847	.13129	.13385	.13616	.13820	.13999	.14151	.14278
7	5	6	.21895	.22298	.22660	.22983	.23265	.23508	.23711	.23875
7	5	7	.19031	.19687	.20317	.20919	.21490	.22030	.22537	.23010

Covariances of Inverse Gaussian Order Statistics
Shape Parameter k

n	i	j	0.8	0.9	1.0	1.1	1.2	1.3	1.4	1.5
7	6	7	.31856	.33433	.35000	.36550	.38080	.39583	.41056	.42496
8	1	2	.09261	.08410	.07627	.06908	.06251	.05652	.05107	.04612
8	1	3	.07037	.06482	.05962	.05476	.05024	.04604	.04215	.03857
8	1	4	.05802	.05406	.05028	.04670	.04332	.04014	.03715	.03436
8	1	5	.04978	.04686	.04404	.04132	.03872	.03624	.03387	.03163
8	1	6	.04361	.04147	.03937	.03731	.03532	.03338	.03151	.02972
8	1	7	.03847	.03700	.03552	.03404	.03257	.03113	.02970	.02831
8	1	8	.03352	.03274	.03190	.03103	.03012	.02919	.02825	.02730
8	2	3	.10550	.09897	.09269	.08666	.08091	.07544	.07026	.06536
8	2	4	.08736	.08288	.07849	.07421	.07005	.06603	.06216	.05844
8	2	5	.07516	.07204	.06892	.06582	.06276	.05975	.05680	.05393
8	2	6	.06595	.06386	.06172	.05954	.05734	.05513	.05292	.05074
8	2	7	.05826	.05705	.05576	.05439	.05295	.05146	.04993	.04838
8	2	8	.05082	.05054	.05013	.04962	.04901	.04831	.04753	.04668
8	3	4	.11953	.11487	.11019	.10552	.10089	.09631	.09180	.08739
8	3	5	.10313	.10012	.09702	.09385	.09062	.08736	.08409	.08083
8	3	6	.09068	.08893	.08705	.08504	.08293	.08074	.07847	.07616
8	3	7	.08022	.07957	.07875	.07779	.07668	.07546	.07413	.07270
8	3	8	.07007	.07056	.07089	.07105	.07105	.07091	.07062	.07021
8	4	5	.13883	.13629	.13354	.13060	.12751	.12429	.12096	.11754
8	4	6	.12237	.12135	.12009	.11862	.11695	.11510	.11310	.11096
8	4	7	.10847	.10877	.10884	.10868	.10831	.10775	.10699	.10606
8	4	8	.09490	.09661	.09811	.09940	.10049	.10136	.10204	.10251
8	5	6	.16781	.16818	.16821	.16793	.16733	.16644	.16528	.16386
8	5	7	.14914	.15113	.15282	.15422	.15531	.15612	.15665	.15690
8	5	8	.13078	.13453	.13805	.14133	.14435	.14711	.14962	.15186
8	6	7	.21634	.22171	.22673	.23136	.23561	.23947	.24294	.24602
8	6	8	.19040	.19804	.20547	.21266	.21958	.22622	.23256	.23859
8	7	8	.31520	.33229	.34938	.36641	.38331	.40004	.41654	.43279
9	1	2	.08487	.07664	.06911	.06225	.05601	.05036	.04524	.04063
9	1	3	.06436	.05892	.05386	.04917	.04483	.04083	.03715	.03378
9	1	4	.05308	.04912	.04538	.04186	.03857	.03549	.03263	.02997
9	1	5	.04564	.04264	.03977	.03703	.03443	.03198	.02967	.02750
9	1	6	.04018	.03787	.03564	.03349	.03142	.02944	.02755	.02576
9	1	7	.03580	.03406	.03235	.03068	.02905	.02746	.02594	.02447
9	1	8	.03198	.03075	.02952	.02828	.02705	.02584	.02464	.02348
9	1	9	.02815	.02747	.02674	.02598	.02520	.02439	.02358	.02276
9	2	3	.09460	.08814	.08198	.07613	.07059	.06537	.06046	.05586
9	2	4	.07832	.07375	.06933	.06506	.06096	.05703	.05329	.04973
9	2	5	.06751	.06418	.06090	.05768	.05455	.05150	.04856	.04573
9	2	6	.05952	.05710	.05467	.05224	.04984	.04748	.04516	.04289
9	2	7	.05310	.05141	.04968	.04791	.04613	.04434	.04255	.04077
9	2	8	.04748	.04646	.04536	.04420	.04299	.04174	.04046	.03916
9	2	9	.04183	.04153	.04113	.04064	.04008	.03943	.03873	.03797
9	3	4	.10470	.09981	.09497	.09021	.08555	.08100	.07658	.07231
9	3	5	.09048	.08707	.08362	.08017	.07672	.07331	.06994	.06663
9	3	6	.07991	.07759	.07519	.07272	.07021	.06767	.06513	.06258
9	3	7	.07138	.06995	.06841	.06677	.06505	.06327	.06143	.05955
9	3	8	.06389	.06327	.06252	.06166	.06068	.05961	.05845	.05722
9	3	9	.05633	.05661	.05674	.05673	.05660	.05635	.05599	.05552
9	4	5	.11813	.11485	.11145	.10794	.10437	.10074	.09709	.09343
9	4	6	.10454	.10255	.10039	.09810	.09568	.09316	.09056	.08789
9	4	7	.09352	.09259	.09148	.09020	.08877	.08720	.08552	.08373
9	4	8	.08381	.08385	.08370	.08337	.08288	.08224	.08145	.08053
9	4	9	.07398	.07509	.07603	.07679	.07738	.07781	.07807	.07818

Covariances of Inverse Gaussian Order Statistics
Shape Parameter k

n	i	j	0.8	0.9	1.0	1.1	1.2	1.3	1.4	1.5
9	5	6	.13711	.13579	.13421	.13239	.13036	.12814	.12575	.12320
9	5	7	.12290	.12283	.12251	.12194	.12115	.12014	.11892	.11753
9	5	8	.11030	.11140	.11225	.11287	.11327	.11344	.11340	.11315
9	5	9	.09751	.09990	.10209	.10408	.10586	.10743	.10879	.10994
9	6	7	.16574	.16724	.16840	.16923	.16973	.16992	.16981	.16941
9	6	8	.14908	.15199	.15462	.15695	.15898	.16072	.16217	.16333
9	6	9	.13205	.13657	.14088	.14496	.14881	.15242	.15578	.15887
9	7	8	.21381	.22027	.22640	.23219	.23763	.24269	.24738	.25168
9	7	9	.18999	.19851	.20686	.21503	.22297	.23066	.23809	.24522
9	8	9	.31206	.33023	.34848	.36676	.38501	.40316	.42117	.43898
10	1	2	.07864	.07067	.06342	.05684	.05089	.04553	.04071	.03639
10	1	3	.05951	.05420	.04928	.04474	.04058	.03676	.03327	.03010
10	1	4	.04906	.04514	.04146	.03803	.03483	.03187	.02913	.02660
10	1	5	.04224	.03921	.03633	.03362	.03106	.02867	.02643	.02434
10	1	6	.03728	.03490	.03260	.03041	.02833	.02636	.02450	.02274
10	1	7	.03339	.03151	.02968	.02791	.02621	.02458	.02303	.02155
10	1	8	.03013	.02868	.02725	.02584	.02447	.02315	.02186	.02063
10	1	9	.02718	.02613	.02507	.02401	.02296	.02192	.02090	.01991
10	1	10	.02412	.02352	.02288	.02221	.02151	.02081	.02009	.01937
10	2	3	.08609	.07974	.07374	.06808	.06275	.05777	.05312	.04879
10	2	4	.07123	.06666	.06226	.05806	.05406	.05026	.04666	.04327
10	2	5	.06146	.05803	.05468	.05144	.04831	.04530	.04241	.03967
10	2	6	.05433	.05172	.04914	.04660	.04412	.04170	.03937	.03711
10	2	7	.04871	.04675	.04478	.04281	.04086	.03893	.03704	.03519
10	2	8	.04399	.04258	.04114	.03967	.03818	.03668	.03519	.03371
10	2	9	.03970	.03882	.03787	.03688	.03583	.03476	.03366	.03254
10	2	10	.03526	.03496	.03458	.03412	.03359	.03301	.03237	.03168
10	3	4	.09355	.08857	.08370	.07896	.07436	.06993	.06566	.06157
10	3	5	.08089	.07727	.07366	.07010	.06659	.06315	.05980	.05655
10	3	6	.07161	.06897	.06629	.06359	.06090	.05822	.05557	.05297
10	3	7	.06428	.06241	.06048	.05849	.05646	.05440	.05234	.05028
10	3	8	.05810	.05690	.05560	.05423	.05279	.05129	.04976	.04819
10	3	9	.05247	.05190	.05122	.05045	.04958	.04863	.04762	.04655
10	3	10	.04663	.04677	.04679	.04670	.04650	.04620	.04581	.04533
10	4	5	.10329	.09962	.09589	.09214	.08837	.08461	.08089	.07722
10	4	6	.09160	.08907	.08644	.08372	.08095	.07813	.07528	.07243
10	4	7	.08233	.08070	.07895	.07709	.07512	.07308	.07097	.06882
10	4	8	.07448	.07364	.07266	.07154	.07030	.06896	.06752	.06600
10	4	9	.06732	.06723	.06698	.06660	.06607	.06542	.06466	.06379
10	4	10	.05987	.06063	.06123	.06169	.06201	.06219	.06223	.06215
10	5	6	.11657	.11435	.11195	.10939	.10669	.10388	.10097	.09799
10	5	7	.10493	.10376	.10240	.10086	.09915	.09730	.09531	.09322
10	5	8	.09504	.09479	.09434	.09370	.09289	.09190	.09076	.08948
10	5	9	.08600	.08663	.08706	.08731	.08737	.08726	.08698	.08654
10	5	10	.07655	.07819	.07965	.08094	.08205	.08299	.08376	.08436
10	6	7	.13537	.13501	.13438	.13348	.13234	.13098	.12940	.12762
10	6	8	.12281	.12353	.12399	.12419	.12415	.12387	.12336	.12265
10	6	9	.11128	.11303	.11456	.11585	.11690	.11773	.11833	.11871
10	6	10	.09917	.10214	.10492	.10751	.10990	.11208	.11405	.11581
10	7	8	.16373	.16615	.16823	.16999	.17142	.17253	.17332	.17381
10	7	9	.14864	.15231	.15571	.15883	.16167	.16421	.16647	.16844
10	7	10	.13271	.13787	.14284	.14762	.15218	.15652	.16062	.16447
10	8	9	.21143	.21878	.22584	.23260	.23904	.24513	.25087	.25624
10	8	10	.18932	.19857	.20771	.21669	.22550	.23410	.24246	.25057
10	9	10	.30916	.32822	.34746	.36680	.38618	.40555	.42485	.44402

Covariances of Inverse Gaussian Order Statistics
Shape Parameter k

n	i	j	0.8	0.9	1.0	1.1	1.2	1.3	1.4	1.5
11	1	2	.07350	.06577	.05876	.05244	.04676	.04165	.03708	.03300
11	1	3	.05551	.05032	.04554	.04115	.03715	.03350	.03018	.02717
11	1	4	.04574	.04187	.03827	.03492	.03182	.02897	.02635	.02394
11	1	5	.03940	.03637	.03352	.03085	.02834	.02601	.02385	.02185
11	1	6	.03483	.03240	.03009	.02790	.02584	.02389	.02207	.02037
11	1	7	.03129	.02933	.02744	.02563	.02391	.02227	.02072	.01926
11	1	8	.02838	.02680	.02527	.02378	.02234	.02097	.01965	.01840
11	1	9	.02586	.02462	.02339	.02219	.02102	.01988	.01878	.01772
11	1	10	.02350	.02259	.02167	.02075	.01983	.01893	.01804	.01717
11	1	11	.02100	.02046	.01989	.01929	.01867	.01804	.01741	.01677
11	2	3	.07925	.07304	.06719	.06172	.05660	.05184	.04742	.04334
11	2	4	.06552	.06098	.05665	.05254	.04864	.04497	.04153	.03830
11	2	5	.05655	.05308	.04972	.04650	.04341	.04046	.03767	.03502
11	2	6	.05006	.04735	.04470	.04212	.03962	.03721	.03490	.03269
11	2	7	.04502	.04290	.04080	.03872	.03669	.03471	.03279	.03093
11	2	8	.04087	.03923	.03759	.03595	.03432	.03271	.03112	.02957
11	2	9	.03725	.03605	.03482	.03357	.03230	.03102	.02975	.02849
11	2	10	.03387	.03310	.03227	.03140	.03049	.02955	.02859	.02762
11	2	11	.03028	.02999	.02963	.02920	.02872	.02818	.02759	.02698
11	3	4	.08483	.07985	.07502	.07035	.06586	.06156	.05746	.05357
11	3	5	.07337	.06964	.06597	.06238	.05888	.05549	.05221	.04906
11	3	6	.06504	.06221	.05938	.05658	.05381	.05109	.04843	.04584
11	3	7	.05855	.05642	.05425	.05207	.04988	.04770	.04555	.04342
11	3	8	.05319	.05164	.05003	.04837	.04668	.04497	.04325	.04154
11	3	9	.04851	.04748	.04637	.04519	.04396	.04268	.04137	.04003
11	3	10	.04413	.04361	.04299	.04229	.04151	.04067	.03977	.03882
11	3	11	.03947	.03953	.03949	.03935	.03911	.03879	.03840	.03793
11	4	5	.09211	.08823	.08436	.08050	.07668	.07292	.06924	.06564
11	4	6	.08177	.07894	.07604	.07311	.07017	.06723	.06431	.06142
11	4	7	.07369	.07166	.06954	.06735	.06511	.06283	.06053	.05822
11	4	8	.06700	.06564	.06418	.06262	.06098	.05928	.05752	.05573
11	4	9	.06115	.06039	.05952	.05854	.05745	.05629	.05504	.05373
11	4	10	.05566	.05550	.05521	.05480	.05428	.05366	.05294	.05213
11	4	11	.04981	.05034	.05074	.05102	.05117	.05120	.05113	.05095
11	5	6	.10182	.09909	.09625	.09331	.09029	.08721	.08410	.08098
11	5	7	.09187	.09007	.08813	.08606	.08387	.08160	.07925	.07685
11	5	8	.08361	.08259	.08141	.08008	.07862	.07705	.07537	.07361
11	5	9	.07637	.07605	.07556	.07491	.07413	.07321	.07217	.07102
11	5	10	.06957	.06993	.07014	.07018	.07007	.06983	.06944	.06893
11	5	11	.06230	.06347	.06449	.06536	.06609	.06666	.06710	.06739
11	6	7	.11503	.11366	.11208	.11030	.10834	.10623	.10398	.10161
11	6	8	.10482	.10434	.10365	.10276	.10167	.10041	.09898	.09741
11	6	9	.09584	.09617	.09629	.09621	.09594	.09548	.09484	.09405
11	6	10	.08738	.08852	.08946	.09020	.09076	.09113	.09132	.09134
11	6	11	.07832	.08040	.08232	.08407	.08565	.08705	.08828	.08933
11	7	8	.13369	.13413	.13428	.13416	.13378	.13314	.13227	.13118
11	7	9	.12241	.12379	.12491	.12577	.12638	.12674	.12687	.12676
11	7	10	.11174	.11406	.11616	.11803	.11967	.12108	.12225	.12321
11	7	11	.10026	.10371	.10700	.11011	.11303	.11575	.11826	.12057
11	8	9	.16185	.16502	.16788	.17042	.17264	.17455	.17613	.17741
11	8	10	.14800	.15230	.15636	.16017	.16370	.16695	.16992	.17261
11	8	11	.13300	.13870	.14424	.14962	.15480	.15978	.16453	.16905
11	9	10	.20922	.21731	.22517	.23275	.24005	.24703	.25368	.25999
11	9	11	.18852	.19839	.20819	.21789	.22744	.23682	.24601	.25497
11	10	11	.30649	.32631	.34638	.36663	.38700	.40742	.42784	.44820

Covariances of Inverse Gaussian Order Statistics
Shape Parameter k

n	i	j	0.8	0.9	1.0	1.1	1.2	1.3	1.4	1.5
12	1	2	.06918	.06167	.05489	.04880	.04334	.03846	.03412	.03025
12	1	3	.05215	.04707	.04243	.03818	.03432	.03082	.02766	.02480
12	1	4	.04294	.03913	.03560	.03235	.02935	.02660	.02408	.02179
12	1	5	.03699	.03399	.03117	.02855	.02611	.02385	.02176	.01984
12	1	6	.03274	.03030	.02799	.02582	.02378	.02188	.02010	.01846
12	1	7	.02946	.02746	.02554	.02372	.02200	.02038	.01885	.01742
12	1	8	.02681	.02515	.02356	.02203	.02057	.01918	.01786	.01662
12	1	9	.02455	.02320	.02188	.02060	.01937	.01819	.01705	.01598
12	1	10	.02254	.02146	.02039	.01935	.01833	.01734	.01638	.01545
12	1	11	.02061	.01981	.01899	.01818	.01737	.01658	.01579	.01503
12	1	12	.01852	.01803	.01752	.01698	.01642	.01586	.01529	.01472
12	2	3	.07362	.06755	.06186	.05656	.05164	.04708	.04288	.03901
12	2	4	.06081	.05633	.05208	.04806	.04429	.04075	.03745	.03437
12	2	5	.05249	.04902	.04568	.04249	.03947	.03660	.03389	.03135
12	2	6	.04651	.04375	.04107	.03848	.03599	.03362	.03135	.02920
12	2	7	.04190	.03968	.03751	.03539	.03333	.03134	.02942	.02759
12	2	8	.03815	.03638	.03462	.03288	.03118	.02951	.02790	.02633
12	2	9	.03496	.03357	.03217	.03077	.02938	.02800	.02665	.02532
12	2	10	.03210	.03106	.03000	.02891	.02781	.02670	.02560	.02450
12	2	11	.02937	.02868	.02795	.02717	.02637	.02554	.02469	.02384
12	2	12	.02641	.02613	.02578	.02538	.02493	.02444	.02390	.02334
12	3	4	.07783	.07289	.06812	.06355	.05919	.05504	.05111	.04739
12	3	5	.06731	.06354	.05986	.05629	.05283	.04951	.04633	.04330
12	3	6	.05971	.05678	.05388	.05103	.04824	.04553	.04291	.04037
12	3	7	.05384	.05155	.04925	.04696	.04470	.04248	.04030	.03817
12	3	8	.04906	.04729	.04549	.04367	.04185	.04003	.03823	.03645
12	3	9	.04498	.04366	.04229	.04088	.03945	.03799	.03653	.03507
12	3	10	.04132	.04042	.03945	.03843	.03735	.03624	.03510	.03394
12	3	11	.03782	.03733	.03677	.03613	.03543	.03468	.03387	.03303
12	3	12	.03401	.03402	.03393	.03376	.03351	.03319	.03280	.03235
12	4	5	.08337	.07939	.07546	.07158	.06778	.06408	.06048	.05700
12	4	6	.07406	.07104	.06800	.06497	.06196	.05899	.05607	.05321
12	4	7	.06684	.06456	.06222	.05986	.05747	.05508	.05271	.05035
12	4	8	.06095	.05927	.05751	.05570	.05383	.05194	.05003	.04811
12	4	9	.05591	.05475	.05350	.05217	.05078	.04932	.04783	.04631
12	4	10	.05139	.05071	.04993	.04905	.04810	.04707	.04598	.04484
12	4	11	.04706	.04686	.04655	.04614	.04564	.04505	.04438	.04364
12	4	12	.04234	.04271	.04298	.04313	.04318	.04313	.04299	.04275
12	5	6	.09069	.08767	.08458	.08145	.07828	.07510	.07194	.06879
12	5	7	.08195	.07976	.07748	.07511	.07268	.07019	.06768	.06515
12	5	8	.07479	.07329	.07167	.06994	.06813	.06624	.06429	.06229
12	5	9	.06865	.06775	.06671	.06556	.06430	.06294	.06150	.05999
12	5	10	.06313	.06278	.06229	.06167	.06094	.06009	.05914	.05811
12	5	11	.05784	.05804	.05811	.05804	.05785	.05753	.05711	.05658
12	5	12	.05207	.05293	.05367	.05427	.05475	.05510	.05533	.05544
12	6	7	.10041	.09845	.09632	.09406	.09168	.08919	.08662	.08399
12	6	8	.09173	.09055	.08919	.08768	.08602	.08424	.08235	.08037
12	6	9	.08428	.08377	.08309	.08224	.08124	.08010	.07883	.07744
12	6	10	.07756	.07768	.07763	.07741	.07704	.07651	.07585	.07505
12	6	11	.07110	.07186	.07246	.07289	.07317	.07329	.07326	.07310
12	6	12	.06404	.06557	.06695	.06819	.06928	.07022	.07101	.07165
12	7	8	.11357	.11290	.11200	.11088	.10956	.10806	.10638	.10456
12	7	9	.10445	.10455	.10444	.10411	.10357	.10283	.10191	.10083
12	7	10	.09621	.09704	.09766	.09807	.09828	.09830	.09812	.09777
12	7	11	.08826	.08983	.09122	.09240	.09340	.09421	.09483	.09527

Covariances of Inverse Gaussian Order Statistics
Shape Parameter k

n	i	j	0.8	0.9	1.0	1.1	1.2	1.3	1.4	1.5
12	7	12	.07956	.08203	.08435	.08650	.08849	.09031	.09196	.09343
12	8	9	.13212	.13322	.13403	.13457	.13484	.13485	.13460	.13412
12	8	10	.12184	.12378	.12547	.12690	.12809	.12902	.12971	.13016
12	8	11	.11190	.11471	.11730	.11968	.12183	.12375	.12545	.12692
12	8	12	.10096	.10484	.10857	.11213	.11552	.11872	.12172	.12452
12	9	10	.16008	.16390	.16742	.17064	.17355	.17614	.17843	.18040
12	9	11	.14725	.15210	.15674	.16113	.16527	.16914	.17275	.17607
12	9	12	.13306	.13922	.14526	.15115	.15688	.16242	.16776	.17288
12	10	11	.20717	.21590	.22443	.23273	.24077	.24854	.25599	.26313
12	10	12	.18766	.19806	.20844	.21875	.22896	.23903	.24894	.25866
12	11	12	.30403	.32450	.34529	.36634	.38757	.40892	.43033	.45174
13	1	2	.06550	.05818	.05161	.04572	.04047	.03579	.03164	.02796
13	1	3	.04928	.04432	.03979	.03568	.03196	.02859	.02556	.02284
13	1	4	.04055	.03680	.03335	.03018	.02727	.02462	.02220	.02001
13	1	5	.03493	.03195	.02918	.02661	.02423	.02204	.02003	.01818
13	1	6	.03093	.02849	.02620	.02405	.02205	.02020	.01848	.01689
13	1	7	.02787	.02584	.02392	.02210	.02039	.01880	.01730	.01592
13	1	8	.02541	.02371	.02208	.02053	.01907	.01768	.01638	.01516
13	1	9	.02334	.02192	.02055	.01923	.01797	.01677	.01563	.01455
13	1	10	.02154	.02036	.01921	.01810	.01702	.01599	.01500	.01406
13	1	11	.01989	.01894	.01800	.01708	.01618	.01531	.01446	.01365
13	1	12	.01829	.01757	.01685	.01612	.01540	.01469	.01399	.01331
13	1	13	.01652	.01607	.01560	.01511	.01461	.01409	.01358	.01306
13	2	3	.06891	.06297	.05744	.05230	.04756	.04319	.03917	.03550
13	2	4	.05687	.05245	.04828	.04437	.04071	.03729	.03412	.03119
13	2	5	.04908	.04562	.04232	.03919	.03623	.03344	.03083	.02838
13	2	6	.04351	.04072	.03804	.03546	.03301	.03068	.02847	.02639
13	2	7	.03924	.03696	.03475	.03261	.03055	.02858	.02669	.02489
13	2	8	.03580	.03394	.03211	.03032	.02858	.02690	.02528	.02372
13	2	9	.03291	.03140	.02989	.02841	.02694	.02552	.02413	.02278
13	2	10	.03038	.02917	.02796	.02675	.02554	.02434	.02316	.02201
13	2	11	.02806	.02715	.02621	.02525	.02428	.02331	.02234	.02138
13	2	12	.02581	.02519	.02453	.02384	.02311	.02237	.02162	.02085
13	2	13	.02331	.02305	.02272	.02235	.02193	.02147	.02098	.02047
13	3	4	.07208	.06719	.06251	.05804	.05381	.04981	.04604	.04250
13	3	5	.06231	.05854	.05488	.05135	.04797	.04473	.04165	.03873
13	3	6	.05530	.05231	.04938	.04652	.04376	.04108	.03851	.03605
13	3	7	.04992	.04753	.04515	.04282	.04053	.03829	.03613	.03403
13	3	8	.04557	.04366	.04174	.03983	.03794	.03607	.03424	.03245
13	3	9	.04191	.04041	.03888	.03733	.03578	.03423	.03269	.03117
13	3	10	.03870	.03757	.03638	.03516	.03392	.03266	.03139	.03013
13	3	11	.03577	.03497	.03412	.03321	.03226	.03129	.03028	.02926
13	3	12	.03291	.03246	.03194	.03136	.03072	.03004	.02931	.02855
13	3	13	.02973	.02970	.02959	.02941	.02915	.02883	.02846	.02803
13	4	5	.07634	.07233	.06838	.06453	.06079	.05717	.05367	.05032
13	4	6	.06783	.06471	.06160	.05853	.05551	.05256	.04968	.04689
13	4	7	.06128	.05884	.05638	.05391	.05146	.04903	.04664	.04429
13	4	8	.05598	.05409	.05215	.05019	.04820	.04621	.04422	.04226
13	4	9	.05151	.05009	.04860	.04706	.04548	.04387	.04224	.04061
13	4	10	.04759	.04658	.04550	.04434	.04313	.04188	.04058	.03926
13	4	11	.04400	.04338	.04268	.04189	.04104	.04012	.03916	.03815
13	4	12	.04049	.04027	.03996	.03957	.03909	.03853	.03791	.03723
13	4	13	.03660	.03687	.03704	.03712	.03710	.03700	.03681	.03655
13	5	6	.08199	.07880	.07557	.07234	.06912	.06593	.06278	.05968
13	5	7	.07415	.07172	.06923	.06669	.06413	.06155	.05898	.05642

Covariances of Inverse Gaussian Order Statistics
Shape Parameter k

n	i	j	0.8	0.9	1.0	1.1	1.2	1.3	1.4	1.5
13	5	8	.06778	.06598	.06409	.06213	.06011	.05805	.05596	.05386
13	5	9	.06241	.06113	.05975	.05829	.05675	.05514	.05348	.05179
13	5	10	.05769	.05688	.05597	.05495	.05384	.05266	.05140	.05009
13	5	11	.05335	.05299	.05252	.05193	.05125	.05047	.04961	.04868
13	5	12	.04912	.04921	.04920	.04906	.04882	.04848	.04804	.04752
13	5	13	.04441	.04507	.04561	.04604	.04635	.04656	.04666	.04666
13	6	7	.08937	.08705	.08460	.08207	.07945	.07678	.07408	.07135
13	6	8	.08178	.08015	.07838	.07651	.07453	.07247	.07034	.06816
13	6	9	.07535	.07431	.07313	.07183	.07040	.06888	.06726	.06557
13	6	10	.06969	.06918	.06853	.06775	.06683	.06580	.06467	.06344
13	6	11	.06448	.06448	.06434	.06405	.06364	.06310	.06244	.06168
13	6	12	.05939	.05991	.06029	.06054	.06065	.06063	.06048	.06022
13	6	13	.05372	.05489	.05592	.05683	.05760	.05824	.05876	.05915
13	7	8	.09910	.09776	.09624	.09456	.09272	.09075	.08867	.08648
13	7	9	.09138	.09072	.08987	.08884	.08765	.08632	.08485	.08326
13	7	10	.08457	.08451	.08427	.08385	.08326	.08251	.08162	.08059
13	7	11	.07830	.07882	.07916	.07932	.07932	.07916	.07884	.07838
13	7	12	.07216	.07327	.07422	.07500	.07562	.07609	.07640	.07656
13	7	13	.06531	.06716	.06887	.07044	.07185	.07312	.07424	.07521
13	8	9	.11220	.11212	.11180	.11126	.11049	.10952	.10836	.10702
13	8	10	.10394	.10455	.10493	.10509	.10503	.10477	.10431	.10366
13	8	11	.09631	.09758	.09863	.09948	.10013	.10057	.10081	.10087
13	8	12	.08881	.09077	.09254	.09412	.09551	.09672	.09773	.09856
13	8	13	.08044	.08325	.08592	.08844	.09080	.09299	.09502	.09687
13	9	10	.13065	.13231	.13370	.13481	.13565	.13622	.13653	.13660
13	9	11	.12119	.12361	.12579	.12773	.12942	.13086	.13206	.13301
13	9	12	.11186	.11509	.11812	.12094	.12355	.12594	.12810	.13004
13	9	13	.10141	.10565	.10977	.11373	.11754	.12117	.12461	.12786
13	10	11	.15844	.16281	.16690	.17071	.17422	.17743	.18034	.18294
13	10	12	.14645	.15178	.15691	.16182	.16650	.17093	.17510	.17901
13	10	13	.13296	.13952	.14599	.15235	.15856	.16460	.17046	.17612
13	11	12	.20527	.21455	.22367	.23260	.24130	.24975	.25793	.26581
13	11	13	.18678	.19764	.20852	.21937	.23016	.24085	.25141	.26181
13	12	13	.30176	.32280	.34422	.36597	.38796	.41013	.43243	.45478
14	1	2	.06231	.05518	.04879	.04309	.03802	.03352	.02954	.02603
14	1	3	.04680	.04195	.03754	.03354	.02994	.02670	.02379	.02119
14	1	4	.03847	.03480	.03142	.02833	.02551	.02295	.02062	.01852
14	1	5	.03314	.03020	.02747	.02495	.02264	.02051	.01857	.01679
14	1	6	.02935	.02692	.02466	.02255	.02059	.01878	.01711	.01557
14	1	7	.02647	.02443	.02251	.02071	.01903	.01746	.01600	.01466
14	1	8	.02416	.02244	.02080	.01925	.01779	.01642	.01514	.01394
14	1	9	.02225	.02078	.01938	.01804	.01676	.01556	.01443	.01337
14	1	10	.02059	.01935	.01815	.01700	.01589	.01484	.01384	.01290
14	1	11	.01912	.01808	.01707	.01608	.01513	.01421	.01334	.01250
14	1	12	.01775	.01690	.01606	.01524	.01444	.01366	.01290	.01218
14	1	13	.01639	.01574	.01509	.01443	.01378	.01314	.01252	.01190
14	1	14	.01486	.01446	.01402	.01357	.01311	.01265	.01217	.01170
14	2	3	.06490	.05909	.05370	.04872	.04414	.03994	.03610	.03259
14	2	4	.05351	.04916	.04508	.04126	.03771	.03442	.03137	.02856
14	2	5	.04616	.04273	.03948	.03641	.03352	.03081	.02829	.02594
14	2	6	.04093	.03814	.03547	.03293	.03052	.02824	.02609	.02408
14	2	7	.03694	.03464	.03241	.03028	.02823	.02628	.02443	.02268
14	2	8	.03375	.03183	.02997	.02815	.02640	.02473	.02312	.02159
14	2	9	.03109	.02950	.02793	.02639	.02490	.02345	.02205	.02071
14	2	10	.02879	.02748	.02617	.02488	.02361	.02237	.02116	.01998

Covariances of Inverse Gaussian Order Statistics
Shape Parameter k

n	i	j	0.8	0.9	1.0	1.1	1.2	1.3	1.4	1.5
14	2	11	.02674	.02568	.02462	.02355	.02248	.02143	.02039	.01938
14	2	12	.02482	.02401	.02318	.02232	.02146	.02059	.01973	.01887
14	2	13	.02293	.02237	.02177	.02114	.02049	.01982	.01914	.01845
14	2	14	.02080	.02055	.02024	.01989	.01950	.01907	.01862	.01815
14	3	4	.06726	.06244	.05785	.05350	.04939	.04552	.04190	.03852
14	3	5	.05813	.05436	.05074	.04727	.04396	.04082	.03785	.03504
14	3	6	.05159	.04858	.04564	.04280	.04007	.03745	.03494	.03256
14	3	7	.04660	.04415	.04174	.03938	.03709	.03487	.03274	.03069
14	3	8	.04260	.04060	.03861	.03664	.03471	.03283	.03100	.02922
14	3	9	.03926	.03763	.03600	.03436	.03274	.03114	.02957	.02804
14	3	10	.03637	.03507	.03375	.03241	.03106	.02972	.02838	.02707
14	3	11	.03379	.03279	.03175	.03068	.02958	.02848	.02737	.02626
14	3	12	.03138	.03066	.02990	.02909	.02825	.02737	.02648	.02558
14	3	13	.02899	.02857	.02809	.02756	.02698	.02636	.02570	.02501
14	3	14	.02630	.02625	.02612	.02593	.02567	.02536	.02500	.02460
14	4	5	.07056	.06654	.06262	.05882	.05515	.05162	.04825	.04503
14	4	6	.06270	.05952	.05639	.05332	.05032	.04740	.04459	.04188
14	4	7	.05668	.05414	.05161	.04909	.04661	.04418	.04181	.03950
14	4	8	.05184	.04981	.04776	.04570	.04365	.04161	.03960	.03763
14	4	9	.04780	.04620	.04455	.04288	.04119	.03949	.03780	.03612
14	4	10	.04430	.04307	.04179	.04045	.03909	.03770	.03629	.03488
14	4	11	.04117	.04028	.03932	.03830	.03724	.03613	.03500	.03384
14	4	12	.03824	.03768	.03704	.03633	.03556	.03474	.03387	.03297
14	4	13	.03534	.03512	.03481	.03443	.03397	.03345	.03288	.03225
14	4	14	.03208	.03227	.03238	.03240	.03234	.03220	.03199	.03172
14	5	6	.07499	.07170	.06841	.06514	.06192	.05875	.05565	.05263
14	5	7	.06785	.06527	.06266	.06003	.05741	.05480	.05222	.04968
14	5	8	.06210	.06009	.05803	.05592	.05379	.05164	.04949	.04736
14	5	9	.05728	.05576	.05416	.05249	.05078	.04903	.04726	.04548
14	5	10	.05311	.05201	.05081	.04954	.04821	.04682	.04539	.04393
14	5	11	.04937	.04865	.04783	.04693	.04594	.04489	.04379	.04263
14	5	12	.04588	.04553	.04507	.04452	.04389	.04317	.04239	.04154
14	5	13	.04241	.04244	.04237	.04220	.04194	.04158	.04115	.04064
14	5	14	.03850	.03901	.03942	.03972	.03992	.04003	.04005	.03998
14	6	7	.08073	.07818	.07554	.07285	.07012	.06737	.06462	.06187
14	6	8	.07395	.07203	.07001	.06791	.06574	.06353	.06128	.05902
14	6	9	.06826	.06687	.06538	.06378	.06210	.06035	.05855	.05670
14	6	10	.06332	.06241	.06137	.06023	.05898	.05766	.05625	.05479
14	6	11	.05889	.05840	.05779	.05707	.05623	.05530	.05429	.05319
14	6	12	.05474	.05467	.05448	.05416	.05373	.05320	.05257	.05185
14	6	13	.05062	.05099	.05123	.05135	.05136	.05125	.05104	.05073
14	6	14	.04597	.04688	.04767	.04835	.04891	.04935	.04968	.04991
14	7	8	.08816	.08641	.08451	.08248	.08035	.07813	.07583	.07348
14	7	9	.08144	.08028	.07897	.07752	.07595	.07427	.07249	.07064
14	7	10	.07559	.07496	.07417	.07324	.07217	.07098	.06968	.06829
14	7	11	.07034	.07019	.06988	.06943	.06883	.06811	.06727	.06631
14	7	12	.06541	.06573	.06590	.06592	.06580	.06554	.06516	.06466
14	7	13	.06051	.06132	.06199	.06252	.06291	.06316	.06328	.06328
14	7	14	.05497	.05640	.05770	.05888	.05992	.06084	.06162	.06227
14	8	9	.09788	.09707	.09607	.09489	.09353	.09201	.09036	.08858
14	8	10	.09092	.09071	.09030	.08970	.08893	.08799	.08691	.08568
14	8	11	.08465	.08498	.08512	.08508	.08486	.08448	.08393	.08324
14	8	12	.07876	.07962	.08031	.08082	.08115	.08132	.08133	.08119
14	8	13	.07290	.07432	.07558	.07668	.07762	.07840	.07902	.07948
14	8	14	.06626	.06839	.07039	.07225	.07397	.07554	.07696	.07823

Covariances of Inverse Gaussian Order Statistics
Shape Parameter k

n	i	j	0.8	0.9	1.0	1.1	1.2	1.3	1.4	1.5
14	9	10	.11093	.11136	.11154	.11148	.11120	.11071	.11001	.10912
14	9	11	.10336	.10441	.10522	.10582	.10619	.10635	.10631	.10607
14	9	12	.09625	.09789	.09934	.10058	.10161	.10244	.10307	.10350
14	9	13	.08914	.09143	.09355	.09548	.09723	.09880	.10017	.10136
14	9	14	.08107	.08418	.08716	.09001	.09270	.09523	.09760	.09980
14	10	11	.12928	.13143	.13331	.13492	.13626	.13734	.13816	.13872
14	10	12	.12049	.12334	.12596	.12834	.13048	.13238	.13403	.13544
14	10	13	.11170	.11529	.11871	.12193	.12495	.12776	.13034	.13271
14	10	14	.10168	.10624	.11070	.11502	.11920	.12322	.12706	.13072
14	11	12	.15692	.16176	.16636	.17068	.17473	.17849	.18196	.18512
14	11	13	.14565	.15139	.15696	.16233	.16749	.17241	.17710	.18152
14	11	14	.13276	.13968	.14653	.15329	.15993	.16643	.17276	.17891
14	12	13	.20351	.21327	.22291	.23239	.24169	.25075	.25958	.26813
14	12	14	.18590	.19717	.20849	.21982	.23113	.24238	.25352	.26453
14	13	14	.29965	.32119	.34318	.36555	.38822	.41113	.43422	.45742
15	1	2	.05953	.05256	.04634	.04081	.03591	.03157	.02774	.02438
15	1	3	.04463	.03988	.03558	.03170	.02821	.02508	.02228	.01978
15	1	4	.03666	.03305	.02974	.02673	.02399	.02151	.01927	.01725
15	1	5	.03157	.02867	.02599	.02352	.02127	.01920	.01732	.01561
15	1	6	.02796	.02555	.02332	.02124	.01932	.01756	.01594	.01446
15	1	7	.02523	.02319	.02129	.01951	.01785	.01632	.01490	.01359
15	1	8	.02305	.02132	.01967	.01813	.01668	.01533	.01408	.01291
15	1	9	.02126	.01976	.01834	.01699	.01572	.01453	.01341	.01237
15	1	10	.01972	.01844	.01720	.01603	.01491	.01385	.01286	.01192
15	1	11	.01837	.01727	.01621	.01518	.01420	.01327	.01238	.01154
15	1	12	.01714	.01621	.01531	.01443	.01357	.01276	.01197	.01123
15	1	13	.01598	.01521	.01446	.01372	.01300	.01229	.01161	.01096
15	1	14	.01481	.01422	.01363	.01303	.01244	.01186	.01129	.01074
15	1	15	.01348	.01310	.01271	.01229	.01187	.01144	.01100	.01057
15	2	3	.06144	.05575	.05050	.04567	.04124	.03719	.03350	.03015
15	2	4	.05061	.04633	.04233	.03862	.03517	.03198	.02905	.02636
15	2	5	.04365	.04025	.03705	.03404	.03122	.02860	.02616	.02390
15	2	6	.03870	.03592	.03328	.03077	.02840	.02618	.02409	.02215
15	2	7	.03495	.03263	.03041	.02828	.02626	.02434	.02254	.02084
15	2	8	.03195	.03000	.02812	.02630	.02455	.02289	.02131	.01981
15	2	9	.02948	.02783	.02622	.02466	.02315	.02170	.02031	.01898
15	2	10	.02736	.02597	.02461	.02327	.02196	.02069	.01947	.01830
15	2	11	.02549	.02434	.02319	.02205	.02093	.01983	.01876	.01772
15	2	12	.02379	.02285	.02190	.02095	.02000	.01907	.01814	.01724
15	2	13	.02218	.02145	.02070	.01993	.01915	.01838	.01760	.01683
15	2	14	.02057	.02005	.01951	.01894	.01834	.01773	.01711	.01649
15	2	15	.01872	.01848	.01819	.01786	.01750	.01710	.01668	.01624
15	3	4	.06316	.05842	.05392	.04968	.04569	.04196	.03848	.03524
15	3	5	.05456	.05083	.04726	.04385	.04062	.03757	.03469	.03199
15	3	6	.04843	.04541	.04249	.03968	.03699	.03442	.03199	.02968
15	3	7	.04376	.04127	.03885	.03649	.03422	.03203	.02994	.02794
15	3	8	.04004	.03797	.03594	.03395	.03201	.03013	.02832	.02657
15	3	9	.03695	.03524	.03353	.03185	.03019	.02857	.02700	.02547
15	3	10	.03430	.03289	.03148	.03006	.02865	.02726	.02590	.02456
15	3	11	.03197	.03083	.02967	.02849	.02731	.02613	.02496	.02380
15	3	12	.02984	.02896	.02803	.02708	.02611	.02513	.02414	.02315
15	3	13	.02783	.02719	.02650	.02577	.02501	.02422	.02342	.02261
15	3	14	.02581	.02542	.02498	.02449	.02395	.02338	.02278	.02215
15	3	15	.02350	.02343	.02330	.02310	.02285	.02255	.02220	.02182
15	4	5	.06572	.06172	.05784	.05411	.05052	.04709	.04382	.04073

Covariances of Inverse Gaussian Order Statistics
Shape Parameter k

n	i	j	0.8	0.9	1.0	1.1	1.2	1.3	1.4	1.5
15	4	6	.05840	.05519	.05206	.04900	.04604	.04319	.04044	.03782
15	4	7	.05281	.05021	.04764	.04510	.04263	.04022	.03788	.03562
15	4	8	.04834	.04622	.04410	.04199	.03990	.03785	.03585	.03389
15	4	9	.04463	.04291	.04116	.03940	.03765	.03591	.03419	.03250
15	4	10	.04145	.04007	.03865	.03720	.03574	.03427	.03280	.03135
15	4	11	.03864	.03756	.03644	.03527	.03407	.03285	.03162	.03038
15	4	12	.03608	.03529	.03443	.03353	.03258	.03160	.03059	.02957
15	4	13	.03366	.03314	.03256	.03191	.03121	.03047	.02969	.02887
15	4	14	.03122	.03099	.03070	.03033	.02990	.02941	.02887	.02829
15	4	15	.02843	.02857	.02863	.02862	.02853	.02837	.02815	.02787
15	5	6	.06922	.06588	.06257	.05931	.05611	.05299	.04996	.04703
15	5	7	.06265	.05998	.05730	.05463	.05198	.04938	.04682	.04433
15	5	8	.05739	.05524	.05307	.05088	.04868	.04650	.04434	.04220
15	5	9	.05301	.05131	.04956	.04777	.04596	.04413	.04230	.04048
15	5	10	.04925	.04793	.04655	.04512	.04364	.04213	.04060	.03906
15	5	11	.04593	.04495	.04390	.04279	.04162	.04040	.03915	.03787
15	5	12	.04290	.04224	.04150	.04069	.03981	.03887	.03788	.03686
15	5	13	.04002	.03968	.03925	.03873	.03814	.03748	.03677	.03600
15	5	14	.03713	.03712	.03701	.03682	.03654	.03619	.03577	.03528
15	5	15	.03383	.03423	.03453	.03475	.03487	.03491	.03487	.03476
15	6	7	.07378	.07107	.06833	.06555	.06277	.06000	.05725	.05453
15	6	8	.06762	.06551	.06333	.06109	.05882	.05653	.05424	.05195
15	6	9	.06250	.06088	.05917	.05739	.05556	.05368	.05177	.04985
15	6	10	.05809	.05689	.05560	.05422	.05277	.05126	.04971	.04812
15	6	11	.05419	.05338	.05246	.05144	.05034	.04917	.04794	.04666
15	6	12	.05063	.05017	.04960	.04893	.04817	.04732	.04641	.04542
15	6	13	.04725	.04714	.04692	.04659	.04616	.04565	.04505	.04437
15	6	14	.04385	.04411	.04425	.04430	.04423	.04408	.04383	.04350
15	6	15	.03996	.04068	.04130	.04181	.04222	.04253	.04274	.04286
15	7	8	.07960	.07757	.07543	.07320	.07089	.06853	.06614	.06372
15	7	9	.07361	.07212	.07051	.06880	.06699	.06511	.06317	.06118
15	7	10	.06846	.06744	.06629	.06503	.06367	.06221	.06067	.05907
15	7	11	.06389	.06330	.06257	.06172	.06076	.05969	.05854	.05730
15	7	12	.05972	.05952	.05918	.05873	.05815	.05746	.05667	.05580
15	7	13	.05575	.05594	.05600	.05593	.05574	.05544	.05503	.05452
15	7	14	.05175	.05236	.05284	.05319	.05343	.05355	.05355	.05345
15	7	15	.04717	.04830	.04932	.05022	.05101	.05168	.05223	.05267
15	8	9	.08705	.08578	.08435	.08276	.08105	.07923	.07730	.07529
15	8	10	.08101	.08026	.07934	.07828	.07707	.07573	.07428	.07273
15	8	11	.07564	.07536	.07492	.07433	.07358	.07270	.07170	.07058
15	8	12	.07073	.07089	.07090	.07075	.07045	.07001	.06944	.06875
15	8	13	.06606	.06666	.06711	.06740	.06756	.06757	.06744	.06719
15	8	14	.06134	.06241	.06334	.06412	.06477	.06528	.06565	.06589
15	8	15	.05593	.05760	.05914	.06056	.06185	.06301	.06404	.06494
15	9	10	.09675	.09640	.09585	.09510	.09416	.09305	.09179	.09038
15	9	11	.09040	.09058	.09056	.09035	.08995	.08937	.08864	.08774
15	9	12	.08458	.08525	.08574	.08604	.08616	.08610	.08588	.08550
15	9	13	.07903	.08020	.08119	.08201	.08265	.08313	.08344	.08359
15	9	14	.07342	.07512	.07666	.07805	.07927	.08034	.08124	.08199
15	9	15	.06698	.06936	.07161	.07374	.07573	.07758	.07928	.08083
15	10	11	.10976	.11061	.11123	.11161	.11177	.11169	.11141	.11092
15	10	12	.10276	.10418	.10538	.10636	.10712	.10767	.10800	.10814
15	10	13	.09608	.09806	.09985	.10143	.10282	.10400	.10498	.10576
15	10	14	.08931	.09190	.09433	.09658	.09866	.10055	.10225	.10377
15	10	15	.08153	.08490	.08816	.09129	.09429	.09713	.09982	.10234

Covariances of Inverse Gaussian Order Statistics
Shape Parameter k

n	i	j	0.8	0.9	1.0	1.1	1.2	1.3	1.4	1.5
15	11	12	.12801	.13058	.13290	.13495	.13674	.13827	.13954	.14056
15	11	13	.11979	.12301	.12602	.12879	.13134	.13365	.13571	.13754
15	11	14	.11145	.11538	.11914	.12272	.12610	.12929	.13226	.13502
15	11	15	.10182	.10667	.11143	.11608	.12059	.12496	.12917	.13320
15	12	13	.15550	.16077	.16580	.17059	.17512	.17937	.18334	.18702
15	12	14	.14484	.15096	.15691	.16270	.16829	.17366	.17881	.18371
15	12	15	.13249	.13973	.14692	.15405	.16108	.16798	.17474	.18134
15	13	14	.20187	.21206	.22216	.23214	.24196	.25158	.26099	.27015
15	13	15	.18503	.19666	.20838	.22015	.23192	.24366	.25533	.26690
15	14	15	.29770	.31969	.34218	.36510	.38838	.41196	.43577	.45974
16	1	2	.05706	.05025	.04419	.03881	.03406	.02987	.02618	.02295
16	1	3	.04271	.03806	.03386	.03008	.02669	.02367	.02097	.01857
16	1	4	.03506	.03151	.02828	.02533	.02267	.02027	.01810	.01616
16	1	5	.03018	.02732	.02469	.02228	.02007	.01807	.01625	.01460
16	1	6	.02673	.02435	.02214	.02010	.01823	.01651	.01493	.01350
16	1	7	.02412	.02210	.02021	.01846	.01683	.01533	.01394	.01268
16	1	8	.02206	.02032	.01868	.01715	.01572	.01439	.01317	.01203
16	1	9	.02036	.01885	.01742	.01608	.01481	.01363	.01253	.01151
16	1	10	.01892	.01761	.01636	.01517	.01405	.01300	.01201	.01108
16	1	11	.01766	.01652	.01543	.01438	.01339	.01245	.01156	.01073
16	1	12	.01653	.01555	.01460	.01368	.01281	.01197	.01117	.01042
16	1	13	.01549	.01466	.01384	.01305	.01228	.01154	.01084	.01016
16	1	14	.01449	.01380	.01312	.01245	.01179	.01115	.01053	.00994
16	1	15	.01348	.01294	.01240	.01185	.01131	.01078	.01026	.00975
16	1	16	.01231	.01196	.01159	.01121	.01082	.01042	.01002	.00962
16	2	3	.05842	.05286	.04773	.04303	.03874	.03483	.03128	.02807
16	2	4	.04808	.04387	.03996	.03633	.03298	.02990	.02707	.02449
16	2	5	.04145	.03809	.03494	.03200	.02925	.02670	.02434	.02216
16	2	6	.03675	.03399	.03137	.02890	.02659	.02442	.02239	.02051
16	2	7	.03319	.03088	.02866	.02656	.02456	.02269	.02092	.01927
16	2	8	.03037	.02840	.02651	.02469	.02296	.02132	.01976	.01830
16	2	9	.02804	.02636	.02473	.02316	.02165	.02020	.01882	.01752
16	2	10	.02607	.02463	.02322	.02186	.02053	.01926	.01804	.01687
16	2	11	.02434	.02312	.02191	.02073	.01957	.01845	.01737	.01633
16	2	12	.02279	.02176	.02074	.01973	.01873	.01775	.01679	.01587
16	2	13	.02136	.02052	.01966	.01881	.01796	.01712	.01629	.01548
16	2	14	.01999	.01932	.01864	.01795	.01724	.01654	.01584	.01514
16	2	15	.01859	.01812	.01762	.01710	.01655	.01599	.01543	.01486
16	2	16	.01698	.01675	.01648	.01617	.01582	.01545	.01506	.01465
16	3	4	.05963	.05496	.05056	.04642	.04255	.03894	.03559	.03249
16	3	5	.05148	.04779	.04427	.04093	.03778	.03482	.03204	.02944
16	3	6	.04569	.04268	.03978	.03701	.03437	.03187	.02950	.02727
16	3	7	.04130	.03880	.03637	.03403	.03178	.02963	.02758	.02564
16	3	8	.03781	.03571	.03365	.03166	.02972	.02786	.02607	.02436
16	3	9	.03492	.03316	.03141	.02970	.02803	.02641	.02484	.02332
16	3	10	.03247	.03099	.02951	.02804	.02660	.02518	.02381	.02247
16	3	11	.03033	.02909	.02785	.02660	.02536	.02414	.02293	.02175
16	3	12	.02840	.02739	.02636	.02532	.02427	.02322	.02217	.02115
16	3	13	.02663	.02583	.02500	.02415	.02328	.02239	.02151	.02062
16	3	14	.02492	.02433	.02371	.02304	.02235	.02164	.02091	.02018
16	3	15	.02318	.02282	.02241	.02195	.02146	.02093	.02038	.01980
16	3	16	.02118	.02110	.02096	.02076	.02052	.02023	.01990	.01953
16	4	5	.06161	.05764	.05381	.05014	.04664	.04331	.04015	.03718
16	4	6	.05473	.05152	.04840	.04538	.04247	.03968	.03701	.03447
16	4	7	.04950	.04687	.04428	.04175	.03929	.03692	.03462	.03242

Covariances of Inverse Gaussian Order Statistics
Shape Parameter k

n	i	j	0.8	0.9	1.0	1.1	1.2	1.3	1.4	1.5
16	4	8	.04534	.04316	.04099	.03886	.03676	.03472	.03274	.03082
16	4	9	.04190	.04009	.03828	.03647	.03468	.03292	.03120	.02952
16	4	10	.03897	.03748	.03597	.03445	.03292	.03141	.02992	.02845
16	4	11	.03641	.03520	.03395	.03268	.03140	.03011	.02882	.02755
16	4	12	.03411	.03315	.03215	.03111	.03005	.02897	.02788	.02678
16	4	13	.03198	.03126	.03049	.02968	.02883	.02795	.02704	.02612
16	4	14	.02993	.02946	.02892	.02833	.02769	.02701	.02630	.02556
16	4	15	.02785	.02763	.02734	.02699	.02659	.02613	.02563	.02509
16	4	16	.02545	.02555	.02557	.02553	.02542	.02525	.02503	.02475
16	5	6	.06440	.06103	.05772	.05448	.05133	.04827	.04532	.04249
16	5	7	.05829	.05556	.05284	.05016	.04752	.04494	.04243	.03999
16	5	8	.05342	.05119	.04895	.04671	.04448	.04229	.04014	.03803
16	5	9	.04939	.04757	.04572	.04386	.04198	.04012	.03827	.03644
16	5	10	.04595	.04449	.04298	.04143	.03987	.03829	.03670	.03513
16	5	11	.04294	.04179	.04058	.03932	.03803	.03671	.03537	.03402
16	5	12	.04024	.03937	.03844	.03744	.03640	.03532	.03421	.03308
16	5	13	.03773	.03713	.03646	.03573	.03493	.03408	.03320	.03227
16	5	14	.03533	.03500	.03459	.03410	.03356	.03295	.03229	.03158
16	5	15	.03288	.03283	.03271	.03250	.03222	.03188	.03147	.03100
16	5	16	.03005	.03036	.03060	.03075	.03082	.03081	.03073	.03059
16	6	7	.06805	.06526	.06244	.05963	.05684	.05407	.05136	.04870
16	6	8	.06240	.06016	.05787	.05556	.05324	.05091	.04861	.04633
16	6	9	.05772	.05594	.05409	.05219	.05027	.04832	.04636	.04441
16	6	10	.05373	.05233	.05086	.04933	.04775	.04613	.04448	.04282
16	6	11	.05023	.04917	.04804	.04683	.04556	.04424	.04288	.04149
16	6	12	.04707	.04634	.04551	.04460	.04362	.04258	.04149	.04035
16	6	13	.04416	.04372	.04319	.04257	.04187	.04110	.04026	.03937
16	6	14	.04135	.04121	.04097	.04064	.04023	.03973	.03917	.03854
16	6	15	.03849	.03867	.03875	.03874	.03863	.03844	.03818	.03783
16	6	16	.03518	.03577	.03626	.03665	.03695	.03716	.03729	.03733
16	7	8	.07270	.07048	.06819	.06583	.06344	.06101	.05858	.05615
16	7	9	.06729	.06557	.06376	.06188	.05993	.05793	.05590	.05384
16	7	10	.06266	.06137	.05998	.05850	.05695	.05533	.05365	.05194
16	7	11	.05860	.05769	.05668	.05556	.05436	.05308	.05173	.05033
16	7	12	.05494	.05438	.05371	.05293	.05206	.05110	.05006	.04896
16	7	13	.05155	.05132	.05098	.05053	.04997	.04933	.04859	.04778
16	7	14	.04828	.04839	.04838	.04825	.04803	.04770	.04728	.04678
16	7	15	.04496	.04542	.04576	.04600	.04613	.04616	.04609	.04593
16	7	16	.04110	.04202	.04283	.04353	.04413	.04463	.04503	.04532
16	8	9	.07857	.07698	.07526	.07343	.07151	.06950	.06742	.06530
16	8	10	.07320	.07209	.07084	.06947	.06798	.06640	.06474	.06301
16	8	11	.06849	.06780	.06696	.06600	.06492	.06373	.06245	.06108
16	8	12	.06424	.06393	.06348	.06290	.06219	.06138	.06045	.05943
16	8	13	.06029	.06035	.06027	.06006	.05972	.05926	.05869	.05802
16	8	14	.05649	.05692	.05721	.05737	.05740	.05732	.05712	.05681
16	8	15	.05261	.05343	.05413	.05470	.05515	.05548	.05569	.05579
16	8	16	.04812	.04945	.05067	.05178	.05277	.05365	.05441	.05505
16	9	10	.08603	.08517	.08414	.08295	.08161	.08014	.07855	.07685
16	9	11	.08053	.08014	.07958	.07885	.07797	.07695	.07580	.07453
16	9	12	.07557	.07560	.07547	.07518	.07473	.07413	.07340	.07254
16	9	13	.07095	.07140	.07168	.07181	.07178	.07160	.07129	.07084
16	9	14	.06651	.06736	.06806	.06861	.06902	.06927	.06939	.06937
16	9	15	.06196	.06326	.06442	.06544	.06633	.06707	.06767	.06814
16	9	16	.05668	.05856	.06032	.06196	.06348	.06487	.06613	.06726
16	10	11	.09571	.09576	.09559	.09523	.09467	.09393	.09301	.09194

Covariances of Inverse Gaussian Order Statistics
Shape Parameter k

n	i	j	0.8	0.9	1.0	1.1	1.2	1.3	1.4	1.5
16	10	12	.08986	.09038	.09071	.09083	.09077	.09053	.09011	.08953
16	10	13	.08441	.08540	.08619	.08680	.08722	.08747	.08754	.08745
16	10	14	.07916	.08060	.08187	.08297	.08390	.08466	.08524	.08567
16	10	15	.07378	.07573	.07752	.07917	.08066	.08199	.08316	.08417
16	10	16	.06753	.07013	.07262	.07498	.07722	.07932	.08128	.08309
16	11	12	.10866	.10990	.11090	.11167	.11221	.11252	.11262	.11250
16	11	13	.10214	.10390	.10544	.10677	.10788	.10878	.10946	.10993
16	11	14	.09584	.09812	.10021	.10211	.10382	.10532	.10662	.10773
16	11	15	.08938	.09223	.09494	.09748	.09985	.10204	.10405	.10588
16	11	16	.08185	.08546	.08897	.09237	.09563	.09876	.10174	.10455
16	12	13	.12682	.12977	.13247	.13492	.13712	.13906	.14074	.14216
16	12	14	.11909	.12264	.12599	.12913	.13204	.13472	.13717	.13938
16	12	15	.11115	.11537	.11944	.12334	.12706	.13059	.13392	.13704
16	12	16	.10187	.10698	.11201	.11695	.12177	.12646	.13100	.13538
16	13	14	.15418	.15982	.16525	.17045	.17541	.18011	.18454	.18869
16	13	15	.14406	.15050	.15680	.16296	.16894	.17472	.18029	.18563
16	13	16	.13218	.13970	.14720	.15466	.16204	.16931	.17647	.18348
16	14	15	.20035	.21092	.22143	.23186	.24215	.25228	.26222	.27193
16	14	16	.18418	.19614	.20821	.22037	.23257	.24476	.25692	.26899
16	15	16	.29588	.31827	.34121	.36463	.38847	.41266	.43712	.46180
17	1	2	.05487	.04820	.04229	.03705	.03243	.02838	.02481	.02170
17	1	3	.04101	.03645	.03234	.02866	.02536	.02243	.01982	.01751
17	1	4	.03363	.03015	.02698	.02410	.02151	.01918	.01708	.01521
17	1	5	.02894	.02612	.02354	.02117	.01903	.01708	.01531	.01372
17	1	6	.02563	.02327	.02110	.01910	.01726	.01559	.01406	.01267
17	1	7	.02313	.02113	.01926	.01753	.01593	.01446	.01311	.01188
17	1	8	.02116	.01943	.01780	.01628	.01488	.01357	.01237	.01127
17	1	9	.01955	.01803	.01661	.01527	.01402	.01285	.01177	.01077
17	1	10	.01818	.01686	.01560	.01441	.01329	.01225	.01127	.01036
17	1	11	.01700	.01584	.01472	.01367	.01267	.01173	.01085	.01002
17	1	12	.01595	.01493	.01395	.01301	.01212	.01128	.01048	.00973
17	1	13	.01499	.01411	.01325	.01243	.01163	.01088	.01016	.00948
17	1	14	.01410	.01334	.01260	.01188	.01119	.01052	.00987	.00926
17	1	15	.01323	.01260	.01198	.01137	.01076	.01018	.00962	.00907
17	1	16	.01235	.01185	.01135	.01085	.01035	.00986	.00938	.00892
17	1	17	.01131	.01098	.01064	.01028	.00992	.00955	.00917	.00881
17	2	3	.05577	.05031	.04531	.04073	.03656	.03278	.02936	.02627
17	2	4	.04585	.04172	.03788	.03434	.03108	.02810	.02536	.02288
17	2	5	.03951	.03620	.03310	.03022	.02753	.02506	.02277	.02067
17	2	6	.03503	.03229	.02971	.02728	.02501	.02289	.02093	.01911
17	2	7	.03164	.02933	.02713	.02505	.02309	.02125	.01953	.01793
17	2	8	.02896	.02699	.02509	.02329	.02158	.01996	.01844	.01701
17	2	9	.02676	.02506	.02342	.02184	.02034	.01890	.01755	.01627
17	2	10	.02490	.02343	.02200	.02062	.01929	.01802	.01681	.01566
17	2	11	.02329	.02202	.02078	.01957	.01839	.01726	.01618	.01514
17	2	12	.02185	.02077	.01969	.01864	.01760	.01660	.01563	.01470
17	2	13	.02055	.01963	.01871	.01780	.01690	.01602	.01516	.01433
17	2	14	.01933	.01856	.01779	.01702	.01625	.01549	.01473	.01400
17	2	15	.01814	.01754	.01692	.01628	.01564	.01500	.01435	.01372
17	2	16	.01693	.01649	.01603	.01554	.01504	.01453	.01401	.01348
17	2	17	.01551	.01528	.01502	.01473	.01441	.01406	.01370	.01332
17	3	4	.05656	.05197	.04765	.04361	.03985	.03636	.03312	.03014
17	3	5	.04880	.04515	.04169	.03842	.03535	.03246	.02977	.02727
17	3	6	.04331	.04031	.03745	.03472	.03213	.02969	.02739	.02523
17	3	7	.03915	.03664	.03423	.03191	.02969	.02758	.02558	.02369

Covariances of Inverse Gaussian Order Statistics
Shape Parameter k

n	i	j	0.8	0.9	1.0	1.1	1.2	1.3	1.4	1.5
17	3	8	.03585	.03373	.03167	.02967	.02775	.02591	.02416	.02249
17	3	9	.03314	.03133	.02956	.02784	.02617	.02455	.02300	.02151
17	3	10	.03085	.02931	.02779	.02629	.02483	.02341	.02203	.02071
17	3	11	.02885	.02755	.02624	.02495	.02368	.02243	.02121	.02003
17	3	12	.02708	.02598	.02488	.02377	.02267	.02158	.02050	.01945
17	3	13	.02547	.02456	.02364	.02270	.02176	.02082	.01988	.01896
17	3	14	.02396	.02324	.02249	.02171	.02093	.02013	.01933	.01853
17	3	15	.02249	.02196	.02138	.02078	.02014	.01949	.01883	.01816
17	3	16	.02099	.02065	.02026	.01984	.01938	.01889	.01838	.01785
17	3	17	.01923	.01914	.01900	.01880	.01856	.01829	.01797	.01763
17	4	5	.05806	.05413	.05037	.04677	.04335	.04011	.03706	.03420
17	4	6	.05157	.04837	.04528	.04230	.03944	.03671	.03412	.03166
17	4	7	.04665	.04400	.04141	.03890	.03647	.03413	.03189	.02975
17	4	8	.04274	.04052	.03833	.03619	.03410	.03208	.03013	.02825
17	4	9	.03952	.03766	.03580	.03397	.03217	.03040	.02869	.02703
17	4	10	.03680	.03523	.03366	.03209	.03053	.02900	.02749	.02603
17	4	11	.03443	.03312	.03180	.03046	.02912	.02779	.02648	.02518
17	4	12	.03232	.03125	.03015	.02902	.02788	.02674	.02559	.02446
17	4	13	.03040	.02955	.02865	.02772	.02677	.02580	.02482	.02384
17	4	14	.02861	.02796	.02726	.02652	.02575	.02495	.02413	.02330
17	4	15	.02686	.02642	.02592	.02538	.02479	.02417	.02352	.02284
17	4	16	.02507	.02485	.02457	.02423	.02385	.02342	.02295	.02245
17	4	17	.02297	.02303	.02303	.02297	.02205	.02267	.02245	.02217
17	5	6	.06029	.05692	.05363	.05043	.04733	.04434	.04147	.03873
17	5	7	.05458	.05181	.04908	.04640	.04379	.04124	.03878	.03641
17	5	8	.05003	.04774	.04546	.04320	.04097	.03879	.03666	.03459
17	5	9	.04628	.04438	.04247	.04056	.03866	.03677	.03492	.03311
17	5	10	.04311	.04154	.03994	.03833	.03670	.03508	.03347	.03189
17	5	11	.04034	.03907	.03774	.03639	.03502	.03363	.03224	.03086
17	5	12	.03788	.03686	.03579	.03468	.03353	.03236	.03117	.02998
17	5	13	.03564	.03486	.03402	.03313	.03220	.03123	.03024	.02922
17	5	14	.03354	.03299	.03238	.03170	.03098	.03021	.02940	.02857
17	5	15	.03150	.03118	.03079	.03034	.02983	.02926	.02865	.02801
17	5	16	.02940	.02933	.02919	.02898	.02870	.02836	.02797	.02753
17	5	17	.02694	.02719	.02737	.02747	.02750	.02746	.02736	.02719
17	6	7	.06326	.06040	.05756	.05474	.05195	.04922	.04655	.04395
17	6	8	.05802	.05569	.05334	.05098	.04863	.04631	.04402	.04177
17	6	9	.05370	.05180	.04986	.04789	.04591	.04392	.04195	.04000
17	6	10	.05003	.04849	.04690	.04527	.04360	.04192	.04023	.03854
17	6	11	.04684	.04562	.04433	.04299	.04161	.04019	.03875	.03730
17	6	12	.04399	.04306	.04205	.04098	.03986	.03869	.03748	.03625
17	6	13	.04140	.04073	.03998	.03916	.03828	.03734	.03636	.03534
17	6	14	.03896	.03855	.03805	.03748	.03683	.03612	.03536	.03455
17	6	15	.03660	.03644	.03620	.03587	.03547	.03500	.03446	.03387
17	6	16	.03416	.03428	.03432	.03426	.03413	.03392	.03365	.03330
17	6	17	.03131	.03179	.03218	.03249	.03271	.03285	.03291	.03290
17	7	8	.06702	.06468	.06229	.05986	.05741	.05497	.05253	.05012
17	7	9	.06206	.06019	.05825	.05625	.05422	.05216	.05009	.04801
17	7	10	.05785	.05637	.05482	.05319	.05151	.04979	.04804	.04627
17	7	11	.05417	.05305	.05183	.05054	.04918	.04776	.04629	.04480
17	7	12	.05090	.05009	.04918	.04818	.04711	.04598	.04478	.04354
17	7	13	.04791	.04739	.04677	.04606	.04526	.04439	.04345	.04246
17	7	14	.04510	.04486	.04452	.04408	.04356	.04295	.04227	.04152
17	7	15	.04237	.04241	.04235	.04220	.04195	.04162	.04120	.04071
17	7	16	.03956	.03991	.04016	.04032	.04038	.04034	.04023	.04003

Covariances of Inverse Gaussian Order Statistics
Shape Parameter k

n	i	j	0.8	0.9	1.0	1.1	1.2	1.3	1.4	1.5
17	7	17	.03626	.03701	.03767	.03823	.03870	.03907	.03935	.03954
17	8	9	.07173	.06993	.06802	.06603	.06397	.06186	.05971	.05754
17	8	10	.06689	.06552	.06404	.06247	.06081	.05908	.05729	.05547
17	8	11	.06267	.06168	.06058	.05937	.05807	.05668	.05523	.05372
17	8	12	.05890	.05826	.05750	.05662	.05565	.05459	.05344	.05222
17	8	13	.05545	.05513	.05469	.05414	.05347	.05271	.05186	.05093
17	8	14	.05221	.05220	.05207	.05183	.05147	.05101	.05046	.04982
17	8	15	.04906	.04937	.04955	.04962	.04958	.04944	.04919	.04885
17	8	16	.04582	.04646	.04700	.04742	.04773	.04793	.04804	.04804
17	8	17	.04201	.04310	.04409	.04497	.04575	.04643	.04699	.04746
17	9	10	.07763	.07642	.07508	.07360	.07200	.07031	.06853	.06667
17	9	11	.07276	.07198	.07104	.06998	.06879	.06748	.06608	.06459
17	9	12	.06841	.06800	.06745	.06676	.06594	.06501	.06396	.06281
17	9	13	.06443	.06438	.06418	.06385	.06338	.06279	.06209	.06128
17	9	14	.06068	.06098	.06113	.06114	.06102	.06078	.06042	.05994
17	9	15	.05704	.05768	.05818	.05856	.05880	.05892	.05891	.05879
17	9	16	.05328	.05430	.05519	.05596	.05661	.05713	.05754	.05782
17	9	17	.04887	.05038	.05179	.05309	.05428	.05535	.05630	.05714
17	10	11	.08509	.08460	.08392	.08307	.08207	.08092	.07963	.07822
17	10	12	.08004	.07996	.07971	.07929	.07871	.07797	.07710	.07610
17	10	13	.07541	.07573	.07587	.07586	.07568	.07534	.07487	.07425
17	10	14	.07106	.07175	.07229	.07266	.07288	.07295	.07287	.07266
17	10	15	.06681	.06789	.06883	.06961	.07024	.07073	.07107	.07128
17	10	16	.06243	.06393	.06531	.06655	.06765	.06861	.06943	.07011
17	10	17	.05728	.05934	.06130	.06315	.06487	.06647	.06795	.06929
17	11	12	.09474	.09513	.09532	.09529	.09507	.09466	.09408	.09332
17	11	13	.08931	.09014	.09077	.09121	.09145	.09151	.09139	.09109
17	11	14	.08419	.08544	.08652	.08740	.08811	.08863	.08898	.08916
17	11	15	.07919	.08088	.08240	.08376	.08495	.08596	.08681	.08749
17	11	16	.07402	.07619	.07822	.08010	.08183	.08340	.08482	.08608
17	11	17	.06794	.07074	.07344	.07603	.07850	.08083	.08303	.08508
17	12	13	.10764	.10921	.11056	.11168	.11256	.11323	.11367	.11389
17	12	14	.10153	.10358	.10544	.10708	.10850	.10972	.11072	.11152
17	12	15	.09555	.09810	.10047	.10266	.10465	.10645	.10805	.10946
17	12	16	.08936	.09246	.09541	.09822	.10085	.10332	.10562	.10773
17	12	17	.08207	.08589	.08963	.09327	.09678	.10017	.10341	.10651
17	13	14	.12572	.12899	.13204	.13485	.13741	.13972	.14178	.14359
17	13	15	.11840	.12225	.12591	.12937	.13261	.13564	.13843	.14100
17	13	16	.11081	.11530	.11964	.12384	.12787	.13171	.13537	.13882
17	13	17	.10185	.10719	.11247	.11767	.12277	.12775	.13260	.13730
17	14	15	.15295	.15892	.16470	.17028	.17563	.18073	.18558	.19017
17	14	16	.14329	.15002	.15665	.16314	.16947	.17563	.18159	.18733
17	14	17	.13184	.13962	.14740	.15515	.16285	.17047	.17798	.18537
17	15	16	.19893	.20984	.22073	.23155	.24228	.25287	.26329	.27352
17	15	17	.18336	.19561	.20801	.22052	.23310	.24571	.25831	.27086
17	16	17	.29418	.31693	.34028	.36416	.38851	.41325	.43832	.46365
18	1	2	.05291	.04637	.04059	.03548	.03099	.02705	.02361	.02060
18	1	3	.03948	.03501	.03099	.02739	.02418	.02134	.01881	.01658
18	1	4	.03235	.02893	.02582	.02301	.02048	.01821	.01618	.01437
18	1	5	.02783	.02505	.02251	.02020	.01810	.01620	.01449	.01295
18	1	6	.02464	.02231	.02017	.01820	.01641	.01477	.01329	.01194
18	1	7	.02224	.02025	.01841	.01670	.01513	.01370	.01239	.01119
18	1	8	.02036	.01863	.01701	.01551	.01413	.01285	.01168	.01060
18	1	9	.01881	.01730	.01587	.01454	.01331	.01216	.01110	.01013
18	1	10	.01751	.01618	.01491	.01373	.01262	.01158	.01062	.00973

Covariances of Inverse Gaussian Order Statistics
Shape Parameter k

n	i	j	0.8	0.9	1.0	1.1	1.2	1.3	1.4	1.5
18	1	11	.01639	.01521	.01409	.01303	.01203	.01109	.01022	.00940
18	1	12	.01540	.01436	.01336	.01241	.01151	.01066	.00987	.00912
18	1	13	.01451	.01359	.01271	.01186	.01105	.01029	.00956	.00888
18	1	14	.01369	.01289	.01211	.01136	.01064	.00995	.00929	.00867
18	1	15	.01292	.01222	.01155	.01089	.01025	.00964	.00905	.00849
18	1	16	.01216	.01158	.01100	.01044	.00989	.00935	.00883	.00833
18	1	17	.01137	.01091	.01045	.00999	.00953	.00907	.00863	.00820
18	1	18	.01044	.01014	.00981	.00948	.00914	.00879	.00845	.00810
18	2	3	.05341	.04806	.04316	.03870	.03465	.03099	.02769	.02471
18	2	4	.04387	.03981	.03605	.03259	.02942	.02652	.02388	.02148
18	2	5	.03779	.03452	.03148	.02865	.02603	.02362	.02141	.01938
18	2	6	.03349	.03078	.02823	.02585	.02363	.02156	.01965	.01789
18	2	7	.03026	.02796	.02578	.02373	.02180	.02000	.01833	.01677
18	2	8	.02770	.02573	.02384	.02205	.02037	.01878	.01729	.01590
18	2	9	.02561	.02390	.02225	.02068	.01919	.01777	.01644	.01519
18	2	10	.02385	.02236	.02092	.01953	.01820	.01694	.01574	.01461
18	2	11	.02233	.02103	.01976	.01854	.01735	.01622	.01514	.01411
18	2	12	.02099	.01986	.01875	.01766	.01661	.01560	.01463	.01370
18	2	13	.01978	.01880	.01783	.01688	.01595	.01505	.01418	.01334
18	2	14	.01866	.01783	.01700	.01617	.01535	.01456	.01378	.01302
18	2	15	.01761	.01691	.01621	.01550	.01480	.01410	.01342	.01275
18	2	16	.01657	.01602	.01545	.01486	.01427	.01368	.01310	.01251
18	2	17	.01550	.01510	.01467	.01422	.01376	.01328	.01280	.01231
18	2	18	.01424	.01403	.01378	.01350	.01320	.01287	.01253	.01217
18	3	4	.05385	.04934	.04511	.04117	.03751	.03412	.03100	.02813
18	3	5	.04644	.04284	.03944	.03623	.03323	.03043	.02782	.02541
18	3	6	.04120	.03823	.03540	.03272	.03018	.02780	.02556	.02348
18	3	7	.03725	.03475	.03235	.03005	.02787	.02581	.02386	.02202
18	3	8	.03412	.03199	.02993	.02794	.02604	.02423	.02251	.02088
18	3	9	.03156	.02973	.02794	.02622	.02455	.02295	.02142	.01996
18	3	10	.02939	.02782	.02627	.02476	.02329	.02187	.02051	.01920
18	3	11	.02753	.02617	.02483	.02351	.02221	.02095	.01973	.01856
18	3	12	.02588	.02472	.02356	.02240	.02127	.02015	.01907	.01801
18	3	13	.02439	.02341	.02241	.02142	.02043	.01945	.01848	.01754
18	3	14	.02302	.02220	.02136	.02052	.01966	.01881	.01796	.01713
18	3	15	.02172	.02106	.02038	.01967	.01895	.01823	.01750	.01677
18	3	16	.02045	.01995	.01942	.01886	.01828	.01769	.01708	.01646
18	3	17	.01913	.01881	.01845	.01805	.01762	.01717	.01669	.01620
18	3	18	.01757	.01747	.01733	.01714	.01691	.01664	.01634	.01602
18	4	5	.05498	.05109	.04738	.04386	.04052	.03738	.03443	.03166
18	4	6	.04881	.04564	.04257	.03964	.03684	.03417	.03166	.02928
18	4	7	.04415	.04150	.03892	.03643	.03404	.03174	.02956	.02748
18	4	8	.04046	.03822	.03603	.03389	.03182	.02982	.02790	.02607
18	4	9	.03744	.03553	.03365	.03180	.03000	.02825	.02655	.02492
18	4	10	.03488	.03326	.03165	.03005	.02847	.02693	.02543	.02398
18	4	11	.03267	.03130	.02991	.02853	.02716	.02581	.02448	.02318
18	4	12	.03072	.02956	.02839	.02720	.02601	.02482	.02365	.02250
18	4	13	.02896	.02800	.02701	.02600	.02498	.02396	.02293	.02191
18	4	14	.02734	.02656	.02575	.02491	.02405	.02318	.02229	.02140
18	4	15	.02580	.02520	.02457	.02389	.02319	.02246	.02172	.02096
18	4	16	.02429	.02388	.02342	.02291	.02237	.02180	.02120	.02057
18	4	17	.02272	.02251	.02224	.02192	.02156	.02116	.02072	.02025
18	4	18	.02088	.02092	.02090	.02082	.02069	.02051	.02029	.02002
18	5	6	.05676	.05340	.05013	.04697	.04393	.04101	.03823	.03557
18	5	7	.05137	.04859	.04586	.04320	.04062	.03812	.03571	.03341

Covariances of Inverse Gaussian Order Statistics
Shape Parameter k

n	i	j	0.8	0.9	1.0	1.1	1.2	1.3	1.4	1.5
18	5	8	.04710	.04478	.04247	.04021	.03799	.03582	.03373	.03170
18	5	9	.04359	.04164	.03968	.03774	.03583	.03395	.03211	.03032
18	5	10	.04063	.03899	.03733	.03567	.03401	.03237	.03076	.02918
18	5	11	.03807	.03670	.03530	.03388	.03245	.03103	.02961	.02821
18	5	12	.03580	.03467	.03350	.03230	.03108	.02985	.02862	.02739
18	5	13	.03376	.03284	.03189	.03089	.02986	.02881	.02775	.02668
18	5	14	.03187	.03116	.03040	.02960	.02875	.02788	.02698	.02606
18	5	15	.03008	.02957	.02901	.02839	.02772	.02702	.02629	.02552
18	5	16	.02832	.02802	.02765	.02723	.02675	.02622	.02566	.02506
18	5	17	.02650	.02642	.02627	.02605	.02578	.02546	.02508	.02466
18	5	18	.02435	.02455	.02468	.02475	.02475	.02468	.02456	.02439
18	6	7	.05917	.05629	.05343	.05062	.04786	.04517	.04256	.04003
18	6	8	.05428	.05190	.04951	.04713	.04478	.04247	.04021	.03801
18	6	9	.05026	.04828	.04628	.04426	.04225	.04026	.03829	.03636
18	6	10	.04686	.04522	.04354	.04184	.04013	.03841	.03670	.03500
18	6	11	.04392	.04257	.04118	.03975	.03829	.03682	.03533	.03385
18	6	12	.04131	.04023	.03910	.03791	.03668	.03543	.03416	.03287
18	6	13	.03896	.03812	.03722	.03626	.03525	.03420	.03312	.03202
18	6	14	.03678	.03617	.03549	.03475	.03395	.03310	.03221	.03129
18	6	15	.03472	.03433	.03387	.03333	.03274	.03208	.03138	.03064
18	6	16	.03270	.03254	.03229	.03197	.03159	.03114	.03064	.03008
18	6	17	.03060	.03068	.03068	.03060	.03045	.03023	.02995	.02961
18	6	18	.02812	.02852	.02883	.02907	.02923	.02932	.02933	.02928
18	7	8	.06226	.05983	.05738	.05491	.05245	.05000	.04759	.04521
18	7	9	.05767	.05569	.05366	.05159	.04951	.04742	.04534	.04328
18	7	10	.05379	.05218	.05051	.04879	.04703	.04525	.04346	.04167
18	7	11	.05043	.04914	.04778	.04636	.04490	.04339	.04186	.04031
18	7	12	.04745	.04645	.04537	.04423	.04302	.04177	.04047	.03915
18	7	13	.04475	.04402	.04320	.04231	.04135	.04032	.03926	.03815
18	7	14	.04226	.04178	.04121	.04055	.03982	.03903	.03817	.03727
18	7	15	.03990	.03966	.03933	.03891	.03841	.03784	.03720	.03651
18	7	16	.03759	.03759	.03750	.03732	.03706	.03673	.03632	.03585
18	7	17	.03518	.03545	.03563	.03573	.03573	.03566	.03551	.03529
18	7	18	.03233	.03295	.03349	.03394	.03431	.03458	.03478	.03490
18	8	9	.06610	.06414	.06211	.06002	.05788	.05572	.05353	.05135
18	8	10	.06168	.06013	.05849	.05678	.05501	.05319	.05133	.04946
18	8	11	.05784	.05664	.05535	.05397	.05252	.05101	.04945	.04786
18	8	12	.05444	.05356	.05257	.05150	.05034	.04912	.04783	.04649
18	8	13	.05136	.05077	.05007	.04928	.04839	.04743	.04640	.04531
18	8	14	.04851	.04819	.04777	.04724	.04662	.04591	.04513	.04428
18	8	15	.04581	.04576	.04560	.04533	.04497	.04452	.04399	.04338
18	8	16	.04316	.04338	.04349	.04349	.04340	.04322	.04295	.04260
18	8	17	.04040	.04091	.04133	.04164	.04185	.04197	.04200	.04194
18	8	18	.03714	.03804	.03885	.03956	.04018	.04071	.04114	.04147
18	9	10	.07085	.06941	.06784	.06617	.06441	.06258	.06069	.05875
18	9	11	.06647	.06541	.06422	.06292	.06153	.06005	.05849	.05687
18	9	12	.06258	.06186	.06102	.06006	.05899	.05783	.05658	.05526
18	9	13	.05906	.05866	.05813	.05748	.05672	.05586	.05490	.05386
18	9	14	.05580	.05570	.05547	.05512	.05465	.05408	.05341	.05265
18	9	15	.05271	.05290	.05296	.05290	.05273	.05245	.05207	.05159
18	9	16	.04967	.05015	.05052	.05077	.05090	.05093	.05085	.05066
18	9	17	.04650	.04731	.04802	.04861	.04909	.04946	.04972	.04989
18	9	18	.04275	.04400	.04515	.04619	.04714	.04798	.04871	.04934
18	10	11	.07676	.07589	.07487	.07370	.07241	.07100	.06949	.06788
18	10	12	.07230	.07181	.07116	.07037	.06945	.06840	.06724	.06598

Covariances of Inverse Gaussian Order Statistics
Shape Parameter k

n	i	j	0.8	0.9	1.0	1.1	1.2	1.3	1.4	1.5
18	10	13	.06826	.06811	.06782	.06737	.06679	.06609	.06526	.06433
18	10	14	.06451	.06469	.06473	.06462	.06438	.06400	.06350	.06289
18	10	15	.06095	.06145	.06182	.06204	.06213	.06208	.06192	.06163
18	10	16	.05745	.05828	.05898	.05955	.05998	.06029	.06047	.06054
18	10	17	.05380	.05499	.05607	.05703	.05786	.05857	.05915	.05962
18	10	18	.04947	.05115	.05273	.05421	.05557	.05682	.05795	.05897
18	11	12	.08422	.08404	.08368	.08314	.08244	.08158	.08057	.07943
18	11	13	.07954	.07975	.07977	.07963	.07932	.07885	.07823	.07747
18	11	14	.07520	.07577	.07617	.07640	.07647	.07638	.07614	.07576
18	11	15	.07107	.07200	.07276	.07337	.07382	.07411	.07426	.07426
18	11	16	.06701	.06830	.06944	.07044	.07129	.07199	.07254	.07295
18	11	17	.06277	.06447	.06604	.06747	.06878	.06994	.07097	.07186
18	11	18	.05774	.05998	.06212	.06415	.06607	.06787	.06954	.07109
18	12	13	.09384	.09454	.09503	.09532	.09541	.09530	.09501	.09454
18	12	14	.08876	.08987	.09078	.09149	.09201	.09235	.09250	.09247
18	12	15	.08392	.08543	.08675	.08789	.08885	.08964	.09024	.09067
18	12	16	.07916	.08107	.08282	.08441	.08584	.08709	.08818	.08910
18	12	17	.07418	.07655	.07878	.08088	.08284	.08464	.08628	.08777
18	12	18	.06826	.07125	.07414	.07693	.07960	.08215	.08457	.08685
18	13	14	.10668	.10856	.11022	.11165	.11285	.11383	.11459	.11512
18	13	15	.10092	.10325	.10538	.10730	.10902	.11053	.11183	.11292
18	13	16	.09524	.09803	.10065	.10310	.10536	.10743	.10931	.11099
18	13	17	.08929	.09260	.09579	.09883	.10172	.10444	.10699	.10937
18	13	18	.08221	.08623	.09018	.09403	.09778	.10140	.10490	.10825
18	14	15	.12468	.12825	.13161	.13475	.13764	.14029	.14270	.14485
18	14	16	.11773	.12185	.12579	.12954	.13309	.13643	.13955	.14244
18	14	17	.11046	.11518	.11978	.12424	.12855	.13269	.13664	.14041
18	14	18	.10178	.10733	.11284	.11828	.12364	.12889	.13402	.13901
18	15	16	.15179	.15806	.16416	.17008	.17579	.18127	.18650	.19149
18	15	17	.14255	.14955	.15646	.16326	.16992	.17642	.18273	.18885
18	15	18	.13148	.13949	.14752	.15555	.16354	.17148	.17932	.18705
18	16	17	.19759	.20881	.22004	.23123	.24236	.25337	.26424	.27494
18	16	18	.18257	.19508	.20778	.22062	.23355	.24654	.25954	.27252
18	17	18	.29259	.31566	.33939	.36369	.38851	.41376	.43938	.46530
19	1	2	.05113	.04472	.03906	.03407	.02970	.02587	.02253	.01962
19	1	3	.03811	.03371	.02977	.02626	.02313	.02036	.01791	.01575
19	1	4	.03120	.02783	.02478	.02203	.01956	.01736	.01539	.01363
19	1	5	.02682	.02408	.02159	.01932	.01727	.01542	.01376	.01227
19	1	6	.02375	.02145	.01934	.01741	.01565	.01405	.01261	.01130
19	1	7	.02144	.01946	.01764	.01596	.01442	.01302	.01174	.01058
19	1	8	.01962	.01790	.01630	.01482	.01346	.01221	.01106	.01001
19	1	9	.01814	.01663	.01521	.01389	.01267	.01155	.01051	.00956
19	1	10	.01690	.01556	.01430	.01312	.01202	.01100	.01005	.00918
19	1	11	.01583	.01464	.01351	.01245	.01145	.01053	.00966	.00886
19	1	12	.01489	.01383	.01282	.01186	.01096	.01012	.00933	.00859
19	1	13	.01405	.01311	.01220	.01134	.01053	.00976	.00904	.00836
19	1	14	.01329	.01245	.01165	.01087	.01014	.00944	.00878	.00816
19	1	15	.01258	.01184	.01113	.01044	.00978	.00915	.00855	.00798
19	1	16	.01190	.01126	.01064	.01003	.00945	.00888	.00834	.00783
19	1	17	.01122	.01069	.01016	.00964	.00913	.00863	.00815	.00769
19	1	18	.01052	.01010	.00967	.00923	.00881	.00839	.00798	.00758
19	1	19	.00968	.00940	.00910	.00878	.00846	.00814	.00782	.00750
19	2	3	.05130	.04605	.04126	.03690	.03296	.02941	.02621	.02334
19	2	4	.04210	.03811	.03442	.03104	.02795	.02513	.02257	.02025
19	2	5	.03624	.03303	.03003	.02726	.02471	.02236	.02021	.01824

Covariances of Inverse Gaussian Order Statistics
Shape Parameter k

n	i	j	0.8	0.9	1.0	1.1	1.2	1.3	1.4	1.5
19	2	6	.03212	.02944	.02692	.02458	.02240	.02039	.01853	.01682
19	2	7	.02902	.02673	.02458	.02256	.02067	.01890	.01727	.01576
19	2	8	.02657	.02460	.02273	.02096	.01929	.01773	.01628	.01492
19	2	9	.02458	.02286	.02121	.01965	.01817	.01678	.01547	.01425
19	2	10	.02290	.02139	.01994	.01856	.01724	.01598	.01480	.01369
19	2	11	.02146	.02013	.01885	.01762	.01643	.01530	.01423	.01322
19	2	12	.02019	.01903	.01789	.01679	.01573	.01472	.01374	.01282
19	2	13	.01906	.01804	.01704	.01606	.01511	.01420	.01332	.01247
19	2	14	.01802	.01714	.01626	.01540	.01455	.01373	.01294	.01217
19	2	15	.01706	.01630	.01554	.01478	.01404	.01331	.01260	.01191
19	2	16	.01614	.01550	.01486	.01421	.01356	.01292	.01229	.01168
19	2	17	.01523	.01472	.01419	.01365	.01310	.01256	.01201	.01148
19	2	18	.01428	.01390	.01350	.01308	.01265	.01221	.01176	.01131
19	2	19	.01314	.01294	.01271	.01244	.01216	.01185	.01153	.01119
19	3	4	.05145	.04701	.04287	.03901	.03545	.03216	.02914	.02637
19	3	5	.04435	.04079	.03745	.03431	.03138	.02865	.02613	.02379
19	3	6	.03934	.03639	.03360	.03096	.02848	.02615	.02398	.02196
19	3	7	.03556	.03307	.03069	.02843	.02628	.02426	.02236	.02058
19	3	8	.03258	.03044	.02839	.02642	.02455	.02277	.02108	.01949
19	3	9	.03014	.02830	.02651	.02479	.02313	.02155	.02004	.01862
19	3	10	.02809	.02649	.02493	.02341	.02194	.02053	.01918	.01789
19	3	11	.02633	.02494	.02357	.02223	.02092	.01966	.01845	.01728
19	3	12	.02478	.02357	.02237	.02119	.02004	.01891	.01782	.01676
19	3	13	.02339	.02235	.02131	.02027	.01925	.01825	.01727	.01631
19	3	14	.02212	.02124	.02034	.01944	.01854	.01765	.01678	.01592
19	3	15	.02094	.02020	.01944	.01867	.01789	.01711	.01634	.01558
19	3	16	.01982	.01921	.01858	.01794	.01728	.01661	.01594	.01528
19	3	17	.01870	.01824	.01775	.01723	.01670	.01615	.01558	.01501
19	3	18	.01753	.01723	.01689	.01652	.01612	.01569	.01525	.01479
19	3	19	.01614	.01604	.01590	.01572	.01549	.01524	.01495	.01464
19	4	5	.05226	.04843	.04478	.04132	.03807	.03501	.03215	.02948
19	4	6	.04639	.04324	.04021	.03732	.03458	.03198	.02953	.02723
19	4	7	.04196	.03931	.03675	.03429	.03193	.02968	.02755	.02553
19	4	8	.03846	.03621	.03401	.03189	.02984	.02787	.02599	.02420
19	4	9	.03559	.03366	.03177	.02992	.02812	.02639	.02472	.02312
19	4	10	.03318	.03152	.02988	.02826	.02668	.02515	.02366	.02222
19	4	11	.03111	.02968	.02825	.02684	.02545	.02409	.02276	.02147
19	4	12	.02928	.02806	.02683	.02560	.02438	.02317	.02198	.02083
19	4	13	.02765	.02661	.02555	.02449	.02342	.02236	.02131	.02027
19	4	14	.02615	.02529	.02439	.02348	.02256	.02163	.02071	.01979
19	4	15	.02476	.02406	.02332	.02255	.02177	.02097	.02017	.01936
19	4	16	.02343	.02288	.02230	.02168	.02103	.02036	.01968	.01899
19	4	17	.02211	.02173	.02130	.02083	.02032	.01979	.01924	.01866
19	4	18	.02073	.02053	.02027	.01997	.01962	.01924	.01883	.01839
19	4	19	.01909	.01911	.01908	.01900	.01886	.01868	.01846	.01820
19	5	6	.05368	.05034	.04711	.04400	.04101	.03817	.03546	.03289
19	5	7	.04858	.04580	.04308	.04044	.03790	.03544	.03310	.03086
19	5	8	.04454	.04220	.03989	.03763	.03542	.03329	.03123	.02925
19	5	9	.04124	.03925	.03727	.03532	.03340	.03153	.02971	.02796
19	5	10	.03846	.03676	.03506	.03337	.03170	.03006	.02845	.02688
19	5	11	.03606	.03462	.03316	.03170	.03024	.02880	.02737	.02598
19	5	12	.03395	.03274	.03150	.03024	.02897	.02770	.02645	.02520
19	5	13	.03206	.03105	.03000	.02893	.02784	.02674	.02563	.02453
19	5	14	.03033	.02951	.02865	.02775	.02682	.02587	.02491	.02395
19	5	15	.02872	.02808	.02739	.02665	.02588	.02509	.02427	.02344

Covariances of Inverse Gaussian Order Statistics
Shape Parameter k

n	i	j	0.8	0.9	1.0	1.1	1.2	1.3	1.4	1.5
19	5	16	.02718	.02671	.02619	.02562	.02501	.02436	.02369	.02299
19	5	17	.02566	.02537	.02502	.02462	.02417	.02368	.02315	.02260
19	5	18	.02406	.02397	.02381	.02360	.02333	.02302	.02266	.02226
19	5	19	.02216	.02232	.02242	.02246	.02243	.02235	.02222	.02203
19	6	7	.05566	.05276	.04991	.04711	.04439	.04174	.03919	.03674
19	6	8	.05106	.04864	.04623	.04385	.04151	.03923	.03700	.03485
19	6	9	.04729	.04525	.04321	.04117	.03915	.03716	.03521	.03331
19	6	10	.04411	.04240	.04066	.03892	.03717	.03544	.03372	.03204
19	6	11	.04138	.03994	.03847	.03698	.03547	.03396	.03246	.03097
19	6	12	.03896	.03778	.03654	.03528	.03398	.03268	.03136	.03005
19	6	13	.03680	.03584	.03482	.03376	.03266	.03154	.03040	.02925
19	6	14	.03482	.03406	.03325	.03238	.03147	.03052	.02955	.02856
19	6	15	.03298	.03241	.03179	.03111	.03037	.02960	.02879	.02795
19	6	16	.03121	.03084	.03040	.02991	.02935	.02875	.02810	.02742
19	6	17	.02946	.02929	.02905	.02874	.02837	.02794	.02747	.02695
19	6	18	.02763	.02768	.02765	.02755	.02739	.02717	.02689	.02656
19	6	19	.02545	.02578	.02603	.02622	.02633	.02638	.02636	.02629
19	7	8	.05820	.05573	.05324	.05075	.04829	.04586	.04348	.04115
19	7	9	.05393	.05187	.04978	.04768	.04557	.04347	.04140	.03936
19	7	10	.05033	.04862	.04686	.04508	.04328	.04146	.03966	.03786
19	7	11	.04721	.04581	.04435	.04284	.04131	.03975	.03818	.03660
19	7	12	.04447	.04334	.04214	.04088	.03958	.03825	.03689	.03552
19	7	13	.04201	.04112	.04015	.03913	.03805	.03693	.03577	.03459
19	7	14	.03976	.03909	.03835	.03754	.03667	.03574	.03478	.03378
19	7	15	.03766	.03720	.03667	.03607	.03539	.03466	.03388	.03306
19	7	16	.03565	.03540	.03508	.03468	.03420	.03367	.03308	.03243
19	7	17	.03366	.03363	.03352	.03333	.03306	.03273	.03234	.03188
19	7	18	.03157	.03178	.03191	.03196	.03193	.03183	.03166	.03142
19	7	19	.02908	.02960	.03005	.03041	.03070	.03091	.03104	.03110
19	8	9	.06138	.05931	.05720	.05504	.05286	.05067	.04848	.04631
19	8	10	.05729	.05561	.05387	.05206	.05022	.04835	.04646	.04457
19	8	11	.05377	.05242	.05099	.04949	.04795	.04636	.04474	.04310
19	8	12	.05066	.04960	.04846	.04724	.04596	.04462	.04324	.04184
19	8	13	.04787	.04707	.04619	.04522	.04419	.04309	.04194	.04074
19	8	14	.04531	.04476	.04412	.04339	.04259	.04171	.04078	.03979
19	8	15	.04292	.04261	.04220	.04170	.04111	.04046	.03973	.03895
19	8	16	.04064	.04055	.04037	.04010	.03974	.03930	.03879	.03821
19	8	17	.03837	.03852	.03858	.03854	.03842	.03821	.03793	.03757
19	8	18	.03600	.03641	.03673	.03696	.03710	.03716	.03713	.03703
19	8	19	.03316	.03392	.03459	.03518	.03567	.03609	.03641	.03665
19	9	10	.06527	.06365	.06193	.06013	.05827	.05636	.05441	.05243
19	9	11	.06127	.06001	.05864	.05719	.05565	.05405	.05240	.05072
19	9	12	.05775	.05680	.05575	.05459	.05336	.05204	.05067	.04924
19	9	13	.05458	.05392	.05315	.05228	.05131	.05027	.04915	.04796
19	9	14	.05168	.05128	.05078	.05017	.04946	.04867	.04779	.04685
19	9	15	.04896	.04882	.04858	.04822	.04776	.04721	.04658	.04587
19	9	16	.04636	.04648	.04648	.04638	.04617	.04587	.04548	.04501
19	9	17	.04379	.04416	.04442	.04458	.04464	.04460	.04447	.04425
19	9	18	.04108	.04174	.04230	.04276	.04312	.04338	.04354	.04362
19	9	19	.03785	.03889	.03984	.04070	.04147	.04213	.04270	.04318
19	10	11	.07005	.06891	.06765	.06626	.06478	.06320	.06155	.05983
19	10	12	.06604	.06525	.06433	.06328	.06212	.06087	.05953	.05811
19	10	13	.06244	.06196	.06135	.06061	.05976	.05880	.05775	.05661
19	10	14	.05913	.05894	.05863	.05818	.05762	.05695	.05617	.05530
19	10	15	.05604	.05613	.05609	.05593	.05565	.05526	.05475	.05416

Covariances of Inverse Gaussian Order Statistics
Shape Parameter k

n	i	j	0.8	0.9	1.0	1.1	1.2	1.3	1.4	1.5
19	10	16	.05307	.05344	.05368	.05380	.05380	.05369	.05347	.05315
19	10	17	.05013	.05078	.05132	.05173	.05203	.05222	.05229	.05226
19	10	18	.04705	.04801	.04887	.04962	.05026	.05079	.05121	.05152
19	10	19	.04336	.04475	.04604	.04724	.04834	.04933	.05022	.05101
19	11	12	.07596	.07539	.07466	.07377	.07275	.07159	.07033	.06896
19	11	13	.07184	.07161	.07122	.07068	.07000	.06918	.06825	.06720
19	11	14	.06806	.06815	.06808	.06787	.06751	.06702	.06640	.06566
19	11	15	.06452	.06491	.06516	.06526	.06522	.06504	.06474	.06431
19	11	16	.06112	.06182	.06237	.06279	.06307	.06321	.06323	.06312
19	11	17	.05775	.05876	.05964	.06038	.06100	.06149	.06185	.06208
19	11	18	.05420	.05556	.05681	.05793	.05893	.05981	.06057	.06121
19	11	19	.04997	.05179	.05353	.05516	.05669	.05811	.05942	.06060
19	12	13	.08341	.08351	.08344	.08318	.08275	.08215	.08140	.08051
19	12	14	.07904	.07950	.07978	.07989	.07983	.07960	.07922	.07869
19	12	15	.07496	.07575	.07638	.07684	.07714	.07728	.07726	.07709
19	12	16	.07103	.07216	.07314	.07395	.07461	.07512	.07547	.07568
19	12	17	.06713	.06861	.06995	.07114	.07219	.07308	.07384	.07445
19	12	18	.06303	.06490	.06665	.06827	.06976	.07111	.07233	.07341
19	12	19	.05812	.06051	.06282	.06502	.06712	.06910	.07096	.07269
19	13	14	.09299	.09397	.09474	.09531	.09568	.09585	.09583	.09562
19	13	15	.08822	.08958	.09074	.09171	.09248	.09307	.09348	.09370
19	13	16	.08363	.08536	.08691	.08829	.08949	.09050	.09135	.09201
19	13	17	.07907	.08118	.08315	.08495	.08660	.08808	.08939	.09053
19	13	18	.07427	.07682	.07925	.08155	.08370	.08572	.08758	.08928
19	13	19	.06850	.07165	.07472	.07769	.08056	.08331	.08594	.08843
19	14	15	.10579	.10794	.10987	.11158	.11308	.11435	.11540	.11623
19	14	16	.10033	.10291	.10529	.10747	.10946	.11123	.11280	.11417
19	14	17	.09490	.09792	.10077	.10345	.10596	.10828	.11042	.11236
19	14	18	.08918	.09269	.09608	.09934	.10246	.10542	.10821	.11084
19	14	19	.08230	.08650	.09063	.09469	.09864	.10249	.10621	.10980
19	15	16	.12371	.12755	.13119	.13462	.13782	.14078	.14351	.14599
19	15	17	.11709	.12144	.12564	.12966	.13349	.13711	.14053	.14373
19	15	18	.11009	.11503	.11986	.12457	.12913	.13354	.13778	.14183
19	15	19	.10167	.10741	.11313	.11879	.12439	.12989	.13528	.14055
19	16	17	.15070	.15725	.16364	.16987	.17590	.18173	.18732	.19267
19	16	18	.14183	.14908	.15625	.16333	.17029	.17710	.18375	.19021
19	16	19	.13112	.13934	.14760	.15588	.16414	.17236	.18051	.18857
19	17	18	.19634	.20784	.21938	.23091	.24240	.25380	.26509	.27622
19	17	19	.18181	.19456	.20753	.22066	.23392	.24726	.26064	.27402
19	18	19	.29109	.31447	.33854	.36323	.38847	.41419	.44033	.46681
20	1	2	.04952	.04322	.03768	.03280	.02854	.02481	.02156	.01874
20	1	3	.03685	.03253	.02867	.02523	.02219	.01949	.01711	.01502
20	1	4	.03015	.02683	.02384	.02115	.01874	.01659	.01468	.01297
20	1	5	.02591	.02321	.02076	.01854	.01653	.01473	.01311	.01166
20	1	6	.02293	.02066	.01858	.01669	.01497	.01341	.01200	.01073
20	1	7	.02070	.01875	.01695	.01530	.01379	.01241	.01117	.01004
20	1	8	.01895	.01724	.01566	.01420	.01286	.01163	.01051	.00949
20	1	9	.01753	.01602	.01461	.01331	.01211	.01100	.00998	.00905
20	1	10	.01633	.01499	.01373	.01256	.01148	.01047	.00954	.00869
20	1	11	.01531	.01411	.01298	.01192	.01094	.01002	.00917	.00838
20	1	12	.01442	.01334	.01233	.01137	.01047	.00963	.00885	.00812
20	1	13	.01362	.01266	.01174	.01087	.01006	.00929	.00857	.00790
20	1	14	.01290	.01204	.01122	.01043	.00969	.00898	.00832	.00770
20	1	15	.01224	.01147	.01073	.01002	.00935	.00871	.00810	.00753
20	1	16	.01161	.01094	.01028	.00965	.00904	.00846	.00790	.00738

Covariances of Inverse Gaussian Order Statistics
Shape Parameter k

n	i	j	0.8	0.9	1.0	1.1	1.2	1.3	1.4	1.5
20	1	17	.01101	.01042	.00985	.00929	.00874	.00822	.00772	.00725
20	1	18	.01041	.00991	.00942	.00894	.00846	.00800	.00756	.00713
20	1	19	.00978	.00938	.00898	.00858	.00818	.00779	.00740	.00703
20	1	20	.00902	.00875	.00847	.00817	.00787	.00757	.00726	.00696
20	2	3	.04940	.04424	.03955	.03530	.03146	.02800	.02490	.02213
20	2	4	.04051	.03658	.03296	.02965	.02663	.02389	.02141	.01916
20	2	5	.03486	.03168	.02874	.02602	.02352	.02124	.01915	.01724
20	2	6	.03088	.02823	.02575	.02345	.02132	.01935	.01754	.01589
20	2	7	.02790	.02563	.02350	.02151	.01965	.01793	.01633	.01486
20	2	8	.02555	.02359	.02173	.01998	.01834	.01681	.01538	.01406
20	2	9	.02364	.02192	.02028	.01873	.01727	.01590	.01461	.01342
20	2	10	.02203	.02052	.01907	.01769	.01638	.01514	.01397	.01288
20	2	11	.02066	.01932	.01803	.01679	.01561	.01449	.01343	.01244
20	2	12	.01946	.01827	.01712	.01601	.01495	.01393	.01297	.01205
20	2	13	.01839	.01734	.01631	.01532	.01436	.01344	.01256	.01172
20	2	14	.01742	.01649	.01558	.01469	.01383	.01300	.01220	.01143
20	2	15	.01652	.01571	.01491	.01412	.01335	.01260	.01188	.01118
20	2	16	.01568	.01498	.01428	.01359	.01291	.01224	.01159	.01095
20	2	17	.01487	.01428	.01368	.01309	.01249	.01190	.01132	.01075
20	2	18	.01406	.01358	.01309	.01259	.01209	.01158	.01108	.01058
20	2	19	.01321	.01286	.01248	.01209	.01169	.01127	.01085	.01044
20	2	20	.01219	.01199	.01177	.01152	.01125	.01096	.01065	.01034
20	3	4	.04931	.04494	.04087	.03711	.03363	.03044	.02751	.02484
20	3	5	.04248	.03897	.03568	.03260	.02974	.02708	.02463	.02238
20	3	6	.03767	.03475	.03200	.02940	.02697	.02470	.02259	.02063
20	3	7	.03405	.03157	.02922	.02699	.02488	.02290	.02104	.01931
20	3	8	.03119	.02907	.02702	.02508	.02323	.02148	.01983	.01828
20	3	9	.02887	.02702	.02523	.02352	.02188	.02032	.01884	.01745
20	3	10	.02692	.02530	.02373	.02221	.02075	.01935	.01802	.01676
20	3	11	.02524	.02383	.02244	.02109	.01979	.01853	.01732	.01617
20	3	12	.02378	.02254	.02131	.02011	.01895	.01782	.01673	.01568
20	3	13	.02247	.02139	.02031	.01925	.01820	.01719	.01620	.01525
20	3	14	.02129	.02035	.01940	.01846	.01754	.01663	.01574	.01487
20	3	15	.02020	.01939	.01857	.01775	.01693	.01612	.01532	.01454
20	3	16	.01917	.01849	.01779	.01708	.01637	.01566	.01495	.01425
20	3	17	.01818	.01763	.01705	.01645	.01584	.01523	.01461	.01400
20	3	18	.01720	.01677	.01631	.01583	.01533	.01482	.01430	.01377
20	3	19	.01616	.01587	.01555	.01520	.01482	.01442	.01401	.01358
20	3	20	.01491	.01481	.01466	.01448	.01427	.01402	.01375	.01345
20	4	5	.04985	.04607	.04248	.03910	.03592	.03294	.03017	.02759
20	4	6	.04424	.04111	.03813	.03529	.03260	.03006	.02768	.02545
20	4	7	.04001	.03737	.03483	.03240	.03009	.02788	.02580	.02384
20	4	8	.03667	.03442	.03223	.03012	.02810	.02616	.02432	.02258
20	4	9	.03395	.03200	.03010	.02826	.02648	.02476	.02312	.02155
20	4	10	.03166	.02998	.02832	.02670	.02512	.02359	.02212	.02071
20	4	11	.02970	.02824	.02678	.02535	.02395	.02259	.02127	.01999
20	4	12	.02798	.02671	.02544	.02418	.02294	.02172	.02053	.01938
20	4	13	.02645	.02535	.02425	.02314	.02205	.02096	.01989	.01885
20	4	14	.02506	.02412	.02317	.02221	.02124	.02028	.01933	.01839
20	4	15	.02378	.02299	.02218	.02135	.02051	.01966	.01882	.01798
20	4	16	.02257	.02193	.02125	.02055	.01983	.01910	.01836	.01763
20	4	17	.02141	.02090	.02036	.01979	.01919	.01858	.01795	.01731
20	4	18	.02025	.01989	.01949	.01905	.01858	.01808	.01757	.01703
20	4	19	.01903	.01882	.01858	.01829	.01796	.01760	.01721	.01680
20	4	20	.01756	.01756	.01752	.01743	.01729	.01711	.01689	.01664

Covariances of Inverse Gaussian Order Statistics
Shape Parameter k

n	i	j	0.8	0.9	1.0	1.1	1.2	1.3	1.4	1.5
20	5	6	.05097	.04766	.04446	.04140	.03848	.03570	.03307	.03059
20	5	7	.04612	.04334	.04065	.03804	.03553	.03313	.03084	.02866
20	5	8	.04229	.03993	.03763	.03538	.03320	.03110	.02908	.02715
20	5	9	.03917	.03715	.03515	.03320	.03129	.02944	.02765	.02593
20	5	10	.03654	.03480	.03308	.03137	.02969	.02806	.02646	.02492
20	5	11	.03428	.03279	.03129	.02980	.02833	.02687	.02545	.02406
20	5	12	.03230	.03102	.02973	.02843	.02713	.02585	.02457	.02333
20	5	13	.03054	.02945	.02834	.02721	.02608	.02494	.02381	.02269
20	5	14	.02894	.02803	.02708	.02611	.02513	.02413	.02314	.02214
20	5	15	.02746	.02671	.02593	.02511	.02426	.02340	.02253	.02165
20	5	16	.02607	.02548	.02484	.02417	.02346	.02274	.02199	.02123
20	5	17	.02473	.02429	.02381	.02328	.02271	.02211	.02149	.02085
20	5	18	.02339	.02311	.02278	.02241	.02198	.02153	.02103	.02051
20	5	19	.02198	.02188	.02172	.02151	.02126	.02095	.02061	.02023
20	5	20	.02028	.02042	.02049	.02050	.02046	.02037	.02023	.02004
20	6	7	.05259	.04969	.04685	.04409	.04140	.03881	.03632	.03394
20	6	8	.04825	.04580	.04339	.04102	.03870	.03645	.03426	.03216
20	6	9	.04470	.04262	.04055	.03851	.03649	.03451	.03259	.03072
20	6	10	.04171	.03994	.03817	.03639	.03463	.03290	.03119	.02952
20	6	11	.03914	.03764	.03612	.03458	.03304	.03152	.03000	.02852
20	6	12	.03689	.03562	.03432	.03300	.03166	.03032	.02898	.02765
20	6	13	.03488	.03382	.03272	.03159	.03043	.02926	.02808	.02691
20	6	14	.03306	.03219	.03127	.03032	.02933	.02832	.02729	.02625
20	6	15	.03138	.03069	.02994	.02915	.02832	.02746	.02658	.02568
20	6	16	.02979	.02927	.02869	.02807	.02739	.02668	.02594	.02517
20	6	17	.02826	.02791	.02750	.02703	.02652	.02595	.02536	.02472
20	6	18	.02673	.02656	.02632	.02602	.02567	.02527	.02482	.02433
20	6	19	.02512	.02514	.02510	.02499	.02482	.02459	.02432	.02400
20	6	20	.02319	.02346	.02367	.02382	.02389	.02391	.02387	.02377
20	7	8	.05471	.05220	.04969	.04721	.04476	.04236	.04002	.03775
20	7	9	.05070	.04859	.04646	.04433	.04222	.04013	.03808	.03607
20	7	10	.04733	.04555	.04374	.04191	.04008	.03826	.03645	.03467
20	7	11	.04443	.04293	.04140	.03984	.03825	.03666	.03507	.03350
20	7	12	.04188	.04064	.03935	.03802	.03666	.03527	.03388	.03249
20	7	13	.03961	.03859	.03752	.03640	.03524	.03405	.03284	.03162
20	7	14	.03755	.03674	.03587	.03494	.03397	.03296	.03192	.03085
20	7	15	.03564	.03502	.03434	.03360	.03281	.03197	.03109	.03018
20	7	16	.03384	.03341	.03292	.03235	.03173	.03106	.03034	.02959
20	7	17	.03211	.03186	.03155	.03117	.03072	.03022	.02966	.02906
20	7	18	.03038	.03033	.03020	.03000	.02974	.02942	.02903	.02860
20	7	19	.02855	.02871	.02880	.02881	.02876	.02864	.02845	.02821
20	7	20	.02635	.02680	.02717	.02746	.02769	.02784	.02793	.02795
20	8	9	.05735	.05522	.05305	.05085	.04865	.04646	.04429	.04214
20	8	10	.05356	.05178	.04996	.04809	.04621	.04431	.04241	.04053
20	8	11	.05029	.04882	.04730	.04572	.04411	.04247	.04082	.03916
20	8	12	.04742	.04623	.04496	.04364	.04227	.04087	.03944	.03799
20	8	13	.04485	.04391	.04288	.04179	.04065	.03946	.03823	.03697
20	8	14	.04252	.04180	.04100	.04013	.03919	.03820	.03716	.03608
20	8	15	.04037	.03986	.03926	.03859	.03785	.03705	.03620	.03530
20	8	16	.03834	.03803	.03764	.03716	.03662	.03601	.03533	.03461
20	8	17	.03638	.03627	.03608	.03580	.03545	.03503	.03454	.03400
20	8	18	.03442	.03452	.03454	.03447	.03433	.03411	.03382	.03346
20	8	19	.03235	.03269	.03294	.03311	.03320	.03321	.03314	.03301
20	8	20	.02987	.03051	.03108	.03156	.03196	.03229	.03253	.03270
20	9	10	.06058	.05884	.05701	.05513	.05321	.05125	.04927	.04729

Covariances of Inverse Gaussian Order Statistics
Shape Parameter k

n	i	j	0.8	0.9	1.0	1.1	1.2	1.3	1.4	1.5
20	9	11	.05690	.05549	.05400	.05243	.05080	.04913	.04743	.04570
20	9	12	.05367	.05255	.05135	.05006	.04870	.04729	.04584	.04435
20	9	13	.05078	.04993	.04898	.04795	.04684	.04567	.04444	.04317
20	9	14	.04815	.04754	.04684	.04604	.04516	.04422	.04320	.04214
20	9	15	.04572	.04534	.04486	.04429	.04363	.04290	.04209	.04123
20	9	16	.04343	.04327	.04301	.04266	.04221	.04169	.04109	.04043
20	9	17	.04121	.04127	.04123	.04110	.04088	.04057	.04018	.03972
20	9	18	.03900	.03929	.03948	.03958	.03958	.03950	.03934	.03909
20	9	19	.03666	.03721	.03766	.03801	.03828	.03846	.03856	.03857
20	9	20	.03385	.03473	.03553	.03624	.03686	.03740	.03785	.03821
20	10	11	.06451	.06318	.06175	.06021	.05860	.05692	.05518	.05340
20	10	12	.06087	.05986	.05873	.05751	.05619	.05480	.05334	.05182
20	10	13	.05760	.05688	.05604	.05510	.05406	.05293	.05173	.05046
20	10	14	.05463	.05418	.05360	.05292	.05213	.05125	.05029	.04926
20	10	15	.05189	.05168	.05135	.05091	.05037	.04974	.04901	.04821
20	10	16	.04929	.04932	.04924	.04904	.04874	.04834	.04785	.04727
20	10	17	.04679	.04706	.04721	.04726	.04720	.04704	.04679	.04645
20	10	18	.04428	.04480	.04521	.04551	.04571	.04581	.04582	.04573
20	10	19	.04164	.04243	.04313	.04372	.04422	.04461	.04491	.04511
20	10	20	.03845	.03961	.04070	.04169	.04258	.04339	.04409	.04470
20	11	12	.06931	.06845	.06745	.06632	.06508	.06374	.06230	.06079
20	11	13	.06561	.06506	.06438	.06356	.06262	.06158	.06043	.05920
20	11	14	.06225	.06199	.06159	.06106	.06041	.05964	.05877	.05781
20	11	15	.05913	.05914	.05902	.05876	.05838	.05789	.05728	.05658
20	11	16	.05619	.05646	.05660	.05661	.05650	.05628	.05594	.05549
20	11	17	.05334	.05387	.05428	.05456	.05473	.05477	.05471	.05453
20	11	18	.05050	.05130	.05199	.05256	.05301	.05335	.05357	.05369
20	11	19	.04748	.04859	.04960	.05049	.05128	.05195	.05252	.05298
20	11	20	.04386	.04538	.04681	.04815	.04939	.05053	.05157	.05249
20	12	13	.07522	.07491	.07444	.07381	.07303	.07211	.07107	.06992
20	12	14	.07139	.07139	.07123	.07092	.07046	.06987	.06914	.06829
20	12	15	.06783	.06813	.06828	.06827	.06812	.06783	.06740	.06685
20	12	16	.06447	.06506	.06550	.06579	.06594	.06595	.06583	.06558
20	12	17	.06122	.06209	.06283	.06342	.06388	.06420	.06439	.06445
20	12	18	.05797	.05914	.06018	.06110	.06188	.06254	.06306	.06346
20	12	19	.05452	.05603	.05743	.05871	.05987	.06091	.06183	.06263
20	12	20	.05037	.05233	.05421	.05600	.05768	.05925	.06072	.06206
20	13	14	.08265	.08301	.08319	.08318	.08300	.08265	.08214	.08148
20	13	15	.07856	.07925	.07976	.08009	.08026	.08026	.08010	.07978
20	13	16	.07469	.07569	.07653	.07720	.07771	.07806	.07824	.07828
20	13	17	.07094	.07226	.07343	.07444	.07530	.07600	.07655	.07695
20	13	18	.06719	.06884	.07036	.07173	.07296	.07405	.07499	.07578
20	13	19	.06321	.06524	.06715	.06895	.07061	.07214	.07354	.07479
20	13	20	.05841	.06095	.06341	.06577	.06803	.07019	.07222	.07413
20	14	15	.09220	.09343	.09445	.09528	.09590	.09632	.09656	.09660
20	14	16	.08770	.08928	.09067	.09187	.09288	.09371	.09435	.09481
20	14	17	.08333	.08526	.08702	.08861	.09003	.09127	.09233	.09322
20	14	18	.07895	.08125	.08340	.08541	.08726	.08894	.09046	.09182
20	14	19	.07430	.07702	.07963	.08211	.08446	.08667	.08873	.09064
20	14	20	.06868	.07198	.07521	.07836	.08140	.08434	.08716	.08985
20	15	16	.10494	.10734	.10953	.11150	.11326	.11480	.11612	.11723
20	15	17	.09976	.10256	.10517	.10759	.10982	.11185	.11367	.11529
20	15	18	.09456	.09777	.10084	.10374	.10648	.10903	.11141	.11359
20	15	19	.08903	.09273	.09631	.09978	.10310	.10628	.10930	.11216
20	15	20	.08234	.08670	.09101	.09525	.09940	.10345	.10739	.11120

Covariances of Inverse Gaussian Order Statistics
Shape Parameter k

n	i	j	0.8	0.9	1.0	1.1	1.2	1.3	1.4	1.5
20	16	17	.12279	.12688	.13078	.13448	.13796	.14121	.14423	.14701
20	16	18	.11646	.12104	.12547	.12973	.13382	.13772	.14141	.14490
20	16	19	.10972	.11486	.11990	.12483	.12964	.13429	.13879	.14311
20	16	20	.10154	.10745	.11336	.11923	.12504	.13077	.13641	.14193
20	17	18	.14968	.15647	.16314	.16965	.17599	.18213	.18805	.19375
20	17	19	.14114	.14861	.15602	.16336	.17059	.17770	.18466	.19145
20	17	20	.13075	.13916	.14763	.15614	.16466	.17315	.18159	.18994
20	18	19	.19516	.20692	.21874	.23059	.24241	.25417	.26584	.27739
20	18	20	.18108	.19405	.20727	.22067	.23423	.24790	.26163	.27539
20	19	20	.28968	.31334	.33772	.36277	.38841	.41457	.44118	.46818
21	1	2	.04805	.04186	.03642	.03165	.02748	.02385	.02069	.01795
21	1	3	.03571	.03146	.02767	.02431	.02133	.01870	.01639	.01435
21	1	4	.02920	.02593	.02299	.02035	.01800	.01590	.01404	.01238
21	1	5	.02508	.02242	.02000	.01782	.01586	.01410	.01252	.01112
21	1	6	.02219	.01995	.01790	.01604	.01435	.01283	.01145	.01022
21	1	7	.02003	.01810	.01632	.01470	.01321	.01187	.01065	.00955
21	1	8	.01834	.01664	.01508	.01364	.01232	.01112	.01002	.00902
21	1	9	.01696	.01546	.01407	.01278	.01159	.01051	.00951	.00860
21	1	10	.01581	.01447	.01322	.01206	.01099	.01000	.00909	.00825
21	1	11	.01483	.01363	.01250	.01145	.01047	.00956	.00873	.00796
21	1	12	.01397	.01289	.01187	.01092	.01002	.00919	.00842	.00771
21	1	13	.01321	.01224	.01132	.01045	.00963	.00886	.00815	.00749
21	1	14	.01253	.01165	.01082	.01002	.00928	.00857	.00791	.00730
21	1	15	.01190	.01112	.01036	.00964	.00896	.00831	.00770	.00713
21	1	16	.01132	.01062	.00994	.00928	.00866	.00807	.00751	.00699
21	1	17	.01077	.01014	.00954	.00895	.00839	.00785	.00734	.00685
21	1	18	.01023	.00969	.00915	.00863	.00813	.00765	.00718	.00674
21	1	19	.00970	.00923	.00877	.00832	.00788	.00745	.00704	.00664
21	1	20	.00912	.00875	.00838	.00800	.00763	.00726	.00690	.00655
21	1	21	.00843	.00818	.00791	.00763	.00735	.00706	.00678	.00649
21	2	3	.04768	.04261	.03801	.03385	.03010	.02674	.02373	.02104
21	2	4	.03907	.03519	.03165	.02840	.02546	.02279	.02037	.01820
21	2	5	.03360	.03047	.02757	.02491	.02246	.02023	.01820	.01636
21	2	6	.02976	.02714	.02470	.02243	.02034	.01842	.01666	.01505
21	2	7	.02688	.02464	.02253	.02057	.01874	.01706	.01550	.01407
21	2	8	.02462	.02267	.02083	.01910	.01749	.01598	.01459	.01330
21	2	9	.02278	.02106	.01944	.01790	.01646	.01511	.01385	.01268
21	2	10	.02124	.01972	.01828	.01690	.01560	.01438	.01324	.01217
21	2	11	.01993	.01858	.01728	.01605	.01487	.01376	.01272	.01174
21	2	12	.01878	.01758	.01642	.01530	.01424	.01323	.01227	.01137
21	2	13	.01776	.01669	.01565	.01464	.01368	.01276	.01188	.01105
21	2	14	.01685	.01589	.01496	.01405	.01318	.01234	.01154	.01077
21	2	15	.01601	.01516	.01433	.01352	.01273	.01197	.01123	.01053
21	2	16	.01522	.01448	.01375	.01302	.01231	.01162	.01096	.01031
21	2	17	.01448	.01384	.01320	.01256	.01193	.01131	.01070	.01012
21	2	18	.01376	.01322	.01267	.01211	.01156	.01101	.01047	.00995
21	2	19	.01304	.01260	.01214	.01167	.01120	.01073	.01026	.00980
21	2	20	.01227	.01194	.01159	.01122	.01084	.01046	.01006	.00967
21	2	21	.01135	.01116	.01095	.01071	.01045	.01017	.00989	.00959
21	3	4	.04738	.04308	.03909	.03541	.03201	.02891	.02606	.02348
21	3	5	.04080	.03734	.03410	.03108	.02828	.02570	.02331	.02113
21	3	6	.03616	.03328	.03056	.02801	.02563	.02341	.02136	.01946
21	3	7	.03268	.03023	.02790	.02570	.02363	.02169	.01988	.01820
21	3	8	.02995	.02783	.02580	.02387	.02205	.02033	.01872	.01721
21	3	9	.02772	.02587	.02409	.02238	.02076	.01923	.01778	.01642

Covariances of Inverse Gaussian Order Statistics
Shape Parameter k

n	i	j	0.8	0.9	1.0	1.1	1.2	1.3	1.4	1.5
21	3	10	.02585	.02423	.02265	.02114	.01969	.01831	.01700	.01576
21	3	11	.02426	.02282	.02143	.02007	.01877	.01752	.01633	.01520
21	3	12	.02287	.02160	.02036	.01915	.01797	.01684	.01576	.01473
21	3	13	.02163	.02051	.01941	.01833	.01727	.01625	.01526	.01431
21	3	14	.02052	.01953	.01855	.01759	.01664	.01572	.01482	.01395
21	3	15	.01949	.01864	.01777	.01692	.01607	.01524	.01443	.01364
21	3	16	.01854	.01780	.01705	.01630	.01555	.01481	.01407	.01336
21	3	17	.01764	.01701	.01637	.01572	.01506	.01440	.01375	.01311
21	3	18	.01677	.01625	.01571	.01516	.01460	.01403	.01346	.01289
21	3	19	.01589	.01549	.01506	.01461	.01415	.01367	.01319	.01270
21	3	20	.01495	.01468	.01438	.01405	.01370	.01332	.01293	.01253
21	3	21	.01383	.01372	.01358	.01341	.01320	.01296	.01270	.01242
21	4	5	.04770	.04397	.04044	.03712	.03402	.03112	.02843	.02593
21	4	6	.04232	.03922	.03628	.03349	.03085	.02838	.02606	.02390
21	4	7	.03827	.03565	.03313	.03073	.02846	.02630	.02427	.02237
21	4	8	.03508	.03282	.03065	.02856	.02657	.02467	.02286	.02116
21	4	9	.03248	.03052	.02862	.02679	.02502	.02333	.02172	.02019
21	4	10	.03030	.02859	.02693	.02530	.02373	.02222	.02077	.01938
21	4	11	.02843	.02694	.02547	.02403	.02263	.02127	.01996	.01870
21	4	12	.02681	.02550	.02420	.02293	.02167	.02045	.01927	.01812
21	4	13	.02536	.02422	.02308	.02195	.02083	.01973	.01866	.01762
21	4	14	.02406	.02307	.02207	.02107	.02007	.01909	.01812	.01717
21	4	15	.02286	.02201	.02114	.02027	.01939	.01851	.01764	.01679
21	4	16	.02175	.02103	.02029	.01953	.01876	.01798	.01721	.01644
21	4	17	.02069	.02010	.01948	.01883	.01817	.01750	.01682	.01614
21	4	18	.01967	.01920	.01870	.01817	.01761	.01704	.01646	.01587
21	4	19	.01864	.01830	.01792	.01751	.01707	.01661	.01613	.01563
21	4	20	.01755	.01735	.01711	.01684	.01653	.01618	.01582	.01543
21	4	21	.01622	.01622	.01617	.01607	.01593	.01575	.01554	.01529
21	5	6	.04857	.04529	.04213	.03912	.03626	.03355	.03099	.02858
21	5	7	.04394	.04118	.03850	.03593	.03346	.03111	.02888	.02676
21	5	8	.04029	.03793	.03563	.03340	.03125	.02919	.02721	.02533
21	5	9	.03732	.03528	.03329	.03134	.02945	.02762	.02586	.02417
21	5	10	.03483	.03306	.03132	.02961	.02794	.02631	.02473	.02321
21	5	11	.03269	.03116	.02964	.02813	.02664	.02519	.02377	.02240
21	5	12	.03082	.02950	.02817	.02684	.02552	.02422	.02295	.02171
21	5	13	.02917	.02802	.02686	.02569	.02453	.02337	.02223	.02111
21	5	14	.02767	.02669	.02569	.02467	.02364	.02261	.02159	.02058
21	5	15	.02630	.02547	.02461	.02373	.02284	.02193	.02102	.02012
21	5	16	.02502	.02434	.02362	.02287	.02210	.02131	.02051	.01971
21	5	17	.02381	.02326	.02268	.02205	.02140	.02073	.02004	.01934
21	5	18	.02263	.02222	.02177	.02128	.02075	.02019	.01962	.01902
21	5	19	.02145	.02118	.02087	.02051	.02011	.01968	.01922	.01874
21	5	20	.02019	.02009	.01993	.01972	.01947	.01918	.01885	.01850
21	5	21	.01867	.01878	.01883	.01882	.01877	.01867	.01852	.01833
21	6	7	.04990	.04701	.04419	.04145	.03881	.03627	.03384	.03153
21	6	8	.04577	.04332	.04091	.03855	.03626	.03404	.03190	.02985
21	6	9	.04241	.04031	.03823	.03618	.03418	.03222	.03033	.02849
21	6	10	.03959	.03778	.03598	.03419	.03243	.03070	.02901	.02737
21	6	11	.03717	.03562	.03405	.03249	.03094	.02940	.02789	.02642
21	6	12	.03505	.03372	.03237	.03100	.02964	.02828	.02693	.02560
21	6	13	.03317	.03204	.03087	.02969	.02849	.02729	.02609	.02490
21	6	14	.03148	.03052	.02953	.02850	.02746	.02640	.02534	.02428
21	6	15	.02992	.02913	.02830	.02743	.02653	.02561	.02468	.02373
21	6	16	.02847	.02784	.02715	.02643	.02567	.02488	.02408	.02325

Covariances of Inverse Gaussian Order Statistics
Shape Parameter k

n	i	j	0.8	0.9	1.0	1.1	1.2	1.3	1.4	1.5
21	6	17	.02709	.02661	.02608	.02549	.02487	.02422	.02353	.02282
21	6	18	.02575	.02542	.02504	.02460	.02411	.02359	.02303	.02244
21	6	19	.02441	.02424	.02400	.02371	.02338	.02299	.02257	.02211
21	6	20	.02298	.02298	.02292	.02280	.02263	.02241	.02214	.02183
21	6	21	.02125	.02148	.02166	.02176	.02181	.02181	.02175	.02164
21	7	8	.05167	.04913	.04662	.04415	.04172	.03936	.03706	.03484
21	7	9	.04789	.04573	.04358	.04145	.03934	.03727	.03524	.03327
21	7	10	.04471	.04288	.04103	.03918	.03734	.03552	.03372	.03196
21	7	11	.04199	.04043	.03884	.03724	.03563	.03402	.03243	.03086
21	7	12	.03961	.03829	.03693	.03554	.03414	.03273	.03131	.02991
21	7	13	.03749	.03638	.03523	.03404	.03282	.03158	.03034	.02909
21	7	14	.03558	.03466	.03370	.03269	.03164	.03057	.02948	.02837
21	7	15	.03382	.03309	.03230	.03145	.03057	.02965	.02871	.02774
21	7	16	.03219	.03162	.03099	.03031	.02958	.02881	.02801	.02718
21	7	17	.03063	.03023	.02977	.02924	.02866	.02804	.02738	.02668
21	7	18	.02912	.02889	.02858	.02822	.02779	.02732	.02680	.02624
21	7	19	.02760	.02754	.02740	.02720	.02694	.02663	.02626	.02585
21	7	20	.02599	.02612	.02617	.02616	.02609	.02595	.02576	.02552
21	7	21	.02404	.02442	.02473	.02497	.02515	.02526	.02531	.02530
21	8	9	.05389	.05170	.04949	.04728	.04508	.04289	.04074	.03863
21	8	10	.05033	.04849	.04661	.04471	.04280	.04089	.03900	.03713
21	8	11	.04728	.04573	.04413	.04250	.04084	.03918	.03751	.03585
21	8	12	.04461	.04331	.04196	.04057	.03914	.03769	.03623	.03476
21	8	13	.04223	.04117	.04004	.03886	.03764	.03638	.03510	.03381
21	8	14	.04008	.03923	.03830	.03732	.03629	.03522	.03411	.03298
21	8	15	.03811	.03745	.03672	.03592	.03507	.03416	.03322	.03225
21	8	16	.03627	.03579	.03524	.03462	.03394	.03320	.03242	.03160
21	8	17	.03452	.03422	.03385	.03340	.03289	.03232	.03169	.03102
21	8	18	.03282	.03270	.03250	.03223	.03189	.03148	.03102	.03051
21	8	19	.03112	.03118	.03117	.03108	.03092	.03069	.03040	.03006
21	8	20	.02930	.02957	.02977	.02989	.02994	.02992	.02983	.02967
21	8	21	.02710	.02765	.02813	.02853	.02886	.02912	.02930	.02942
21	9	10	.05659	.05476	.05286	.05092	.04896	.04698	.04499	.04302
21	9	11	.05318	.05165	.05006	.04842	.04673	.04502	.04329	.04155
21	9	12	.05018	.04894	.04762	.04623	.04480	.04332	.04182	.04029
21	9	13	.04752	.04652	.04544	.04429	.04309	.04183	.04053	.03920
21	9	14	.04511	.04434	.04348	.04255	.04155	.04049	.03939	.03824
21	9	15	.04290	.04234	.04169	.04095	.04015	.03929	.03836	.03740
21	9	16	.04084	.04047	.04002	.03948	.03887	.03819	.03744	.03665
21	9	17	.03887	.03870	.03844	.03809	.03767	.03717	.03660	.03598
21	9	18	.03696	.03698	.03692	.03676	.03653	.03622	.03583	.03539
21	9	19	.03505	.03527	.03540	.03545	.03542	.03531	.03512	.03487
21	9	20	.03300	.03345	.03382	.03410	.03430	.03442	.03446	.03443
21	9	21	.03053	.03128	.03196	.03255	.03307	.03350	.03386	.03413
21	10	11	.05987	.05839	.05683	.05520	.05350	.05175	.04996	.04815
21	10	12	.05651	.05534	.05407	.05272	.05129	.04981	.04827	.04671
21	10	13	.05353	.05262	.05161	.05052	.04934	.04810	.04680	.04545
21	10	14	.05082	.05016	.04939	.04853	.04759	.04657	.04548	.04434
21	10	15	.04834	.04790	.04736	.04672	.04600	.04519	.04431	.04337
21	10	16	.04602	.04580	.04547	.04505	.04453	.04393	.04325	.04250
21	10	17	.04382	.04380	.04369	.04347	.04316	.04277	.04229	.04173
21	10	18	.04167	.04187	.04196	.04196	.04186	.04167	.04140	.04105
21	10	19	.03951	.03993	.04024	.04046	.04059	.04063	.04058	.04045
21	10	20	.03721	.03788	.03845	.03892	.03931	.03961	.03982	.03994
21	10	21	.03443	.03542	.03634	.03716	.03791	.03856	.03913	.03960

Covariances of Inverse Gaussian Order Statistics
Shape Parameter k

n	i	j	0.8	0.9	1.0	1.1	1.2	1.3	1.4	1.5
21	11	12	.06382	.06275	.06156	.06026	.05887	.05740	.05586	.05426
21	11	13	.06046	.05968	.05878	.05776	.05665	.05544	.05416	.05281
21	11	14	.05742	.05690	.05626	.05551	.05465	.05369	.05265	.05154
21	11	15	.05463	.05435	.05396	.05345	.05283	.05211	.05130	.05041
21	11	16	.05202	.05198	.05181	.05154	.05115	.05067	.05008	.04942
21	11	17	.04953	.04972	.04979	.04974	.04959	.04933	.04897	.04853
21	11	18	.04712	.04753	.04783	.04801	.04810	.04807	.04795	.04774
21	11	19	.04468	.04533	.04587	.04631	.04665	.04688	.04701	.04704
21	11	20	.04209	.04301	.04383	.04456	.04518	.04570	.04613	.04645
21	11	21	.03894	.04023	.04143	.04254	.04357	.04450	.04533	.04606
21	12	13	.06862	.06801	.06725	.06636	.06534	.06421	.06298	.06166
21	12	14	.06519	.06486	.06439	.06378	.06305	.06220	.06124	.06018
21	12	15	.06203	.06197	.06177	.06143	.06096	.06038	.05968	.05888
21	12	16	.05908	.05927	.05932	.05925	.05904	.05871	.05827	.05772
21	12	17	.05627	.05671	.05701	.05719	.05724	.05717	.05699	.05669
21	12	18	.05353	.05422	.05478	.05521	.05553	.05572	.05580	.05577
21	12	19	.05078	.05172	.05255	.05326	.05386	.05434	.05471	.05497
21	12	20	.04784	.04908	.05022	.05125	.05217	.05299	.05369	.05429
21	12	21	.04427	.04591	.04747	.04894	.05032	.05159	.05277	.05383
21	13	14	.07453	.07446	.07422	.07382	.07326	.07257	.07174	.07079
21	13	15	.07094	.07116	.07121	.07111	.07086	.07046	.06993	.06927
21	13	16	.06759	.06808	.06841	.06860	.06864	.06853	.06829	.06792
21	13	17	.06439	.06515	.06576	.06623	.06656	.06675	.06680	.06672
21	13	18	.06127	.06230	.06320	.06396	.06458	.06507	.06542	.06565
21	13	19	.05812	.05944	.06064	.06171	.06265	.06346	.06415	.06470
21	13	20	.05477	.05642	.05796	.05939	.06070	.06189	.06296	.06391
21	13	21	.05070	.05279	.05480	.05672	.05855	.06027	.06188	.06338
21	14	15	.08194	.08253	.08294	.08316	.08321	.08309	.08280	.08236
21	14	16	.07809	.07898	.07970	.08025	.08062	.08083	.08088	.08077
21	14	17	.07441	.07560	.07663	.07750	.07820	.07874	.07912	.07935
21	14	18	.07082	.07232	.07366	.07485	.07589	.07678	.07751	.07809
21	14	19	.06721	.06902	.07070	.07224	.07364	.07490	.07601	.07698
21	14	20	.06335	.06552	.06758	.06953	.07136	.07305	.07462	.07605
21	14	21	.05865	.06132	.06392	.06643	.06884	.07115	.07335	.07543
21	15	16	.09146	.09291	.09417	.09522	.09608	.09674	.09721	.09749
21	15	17	.08719	.08897	.09057	.09199	.09322	.09426	.09513	.09580
21	15	18	.08302	.08513	.08709	.08887	.09049	.09194	.09321	.09430
21	15	19	.07880	.08127	.08360	.08579	.08783	.08971	.09142	.09298
21	15	20	.07430	.07718	.07995	.08260	.08513	.08752	.08976	.09186
21	15	21	.06881	.07226	.07563	.07894	.08215	.08526	.08826	.09113
21	16	17	.10415	.10677	.10919	.11141	.11341	.11520	.11677	.11813
21	16	18	.09921	.10221	.10503	.10767	.11013	.11239	.11445	.11631
21	16	19	.09421	.09761	.10087	.10398	.10692	.10970	.11229	.11471
21	16	20	.08887	.09273	.09649	.10014	.10366	.10705	.11028	.11336
21	16	21	.08235	.08685	.09132	.09573	.10007	.10431	.10845	.11247
21	17	18	.12193	.12624	.13038	.13432	.13806	.14158	.14488	.14794
21	17	19	.11586	.12063	.12528	.12977	.13410	.13825	.14220	.14595
21	17	20	.10935	.11466	.11990	.12505	.13007	.13496	.13970	.14427
21	17	21	.10138	.10746	.11354	.11960	.12562	.13157	.13743	.14319
21	18	19	.14872	.15574	.16265	.16942	.17604	.18247	.18871	.19472
21	18	20	.14048	.14815	.15578	.16336	.17085	.17823	.18548	.19257
21	18	21	.13038	.13896	.14764	.15636	.16511	.17385	.18255	.19119
21	19	20	.19405	.20605	.21813	.23026	.24239	.25449	.26652	.27845
21	19	21	.18037	.19355	.20699	.22066	.23449	.24846	.26252	.27663
21	20	21	.28836	.31226	.33694	.36232	.38833	.41489	.44195	.46943

Covariances of Inverse Gaussian Order Statistics
Shape Parameter k

n	i	j	0.8	0.9	1.0	1.1	1.2	1.3	1.4	1.5
22	1	2	.04670	.04061	.03527	.03060	.02652	.02298	.01990	.01724
22	1	3	.03467	.03049	.02676	.02346	.02055	.01799	.01573	.01376
22	1	4	.02832	.02510	.02221	.01963	.01732	.01527	.01346	.01185
22	1	5	.02432	.02169	.01932	.01717	.01525	.01353	.01199	.01063
22	1	6	.02151	.01930	.01728	.01545	.01379	.01230	.01096	.00976
22	1	7	.01942	.01750	.01575	.01415	.01269	.01138	.01018	.00911
22	1	8	.01778	.01609	.01455	.01313	.01183	.01065	.00958	.00861
22	1	9	.01644	.01495	.01357	.01230	.01113	.01006	.00908	.00820
22	1	10	.01533	.01400	.01276	.01161	.01054	.00957	.00868	.00786
22	1	11	.01439	.01318	.01206	.01102	.01005	.00915	.00833	.00757
22	1	12	.01356	.01248	.01146	.01050	.00962	.00879	.00803	.00733
22	1	13	.01283	.01185	.01092	.01005	.00924	.00848	.00777	.00712
22	1	14	.01218	.01129	.01045	.00965	.00890	.00820	.00755	.00694
22	1	15	.01159	.01078	.01001	.00928	.00860	.00795	.00734	.00678
22	1	16	.01104	.01031	.00961	.00895	.00832	.00772	.00716	.00663
22	1	17	.01052	.00987	.00924	.00864	.00806	.00751	.00699	.00651
22	1	18	.01003	.00945	.00889	.00834	.00782	.00732	.00684	.00639
22	1	19	.00955	.00904	.00854	.00806	.00759	.00714	.00670	.00629
22	1	20	.00906	.00863	.00820	.00778	.00736	.00696	.00657	.00620
22	1	21	.00854	.00819	.00784	.00749	.00714	.00679	.00645	.00612
22	1	22	.00791	.00767	.00741	.00715	.00688	.00661	.00634	.00608
22	2	3	.04611	.04113	.03661	.03254	.02888	.02561	.02268	.02007
22	2	4	.03775	.03394	.03045	.02728	.02440	.02179	.01944	.01733
22	2	5	.03245	.02936	.02652	.02390	.02151	.01933	.01735	.01556
22	2	6	.02874	.02615	.02374	.02151	.01947	.01759	.01587	.01431
22	2	7	.02595	.02373	.02165	.01972	.01793	.01627	.01476	.01336
22	2	8	.02377	.02183	.02001	.01830	.01672	.01524	.01388	.01263
22	2	9	.02200	.02029	.01867	.01715	.01573	.01440	.01317	.01203
22	2	10	.02052	.01900	.01756	.01619	.01491	.01371	.01258	.01154
22	2	11	.01926	.01790	.01660	.01537	.01421	.01311	.01208	.01112
22	2	12	.01816	.01694	.01578	.01466	.01360	.01260	.01165	.01077
22	2	13	.01718	.01610	.01504	.01403	.01307	.01215	.01128	.01046
22	2	14	.01631	.01534	.01439	.01347	.01259	.01175	.01095	.01019
22	2	15	.01552	.01464	.01379	.01296	.01216	.01139	.01066	.00995
22	2	16	.01478	.01401	.01324	.01250	.01177	.01107	.01039	.00974
22	2	17	.01409	.01341	.01273	.01206	.01141	.01077	.01015	.00956
22	2	18	.01344	.01284	.01224	.01165	.01107	.01049	.00993	.00939
22	2	19	.01279	.01228	.01177	.01125	.01074	.01023	.00973	.00924
22	2	20	.01214	.01173	.01130	.01086	.01042	.00998	.00954	.00911
22	2	21	.01145	.01114	.01080	.01046	.01010	.00974	.00937	.00900
22	2	22	.01060	.01042	.01022	.00999	.00975	.00948	.00921	.00893
22	3	4	.04564	.04140	.03748	.03388	.03056	.02754	.02478	.02227
22	3	5	.03927	.03586	.03267	.02971	.02698	.02445	.02214	.02001
22	3	6	.03480	.03195	.02927	.02677	.02443	.02226	.02026	.01841
22	3	7	.03145	.02902	.02671	.02454	.02251	.02061	.01885	.01721
22	3	8	.02882	.02671	.02470	.02279	.02100	.01931	.01774	.01627
22	3	9	.02667	.02483	.02305	.02137	.01977	.01826	.01683	.01550
22	3	10	.02489	.02325	.02168	.02018	.01874	.01738	.01609	.01487
22	3	11	.02336	.02191	.02051	.01916	.01786	.01663	.01545	.01434
22	3	12	.02203	.02074	.01949	.01827	.01710	.01598	.01490	.01388
22	3	13	.02085	.01971	.01859	.01749	.01643	.01541	.01443	.01349
22	3	14	.01980	.01878	.01778	.01680	.01584	.01491	.01401	.01314
22	3	15	.01883	.01794	.01704	.01616	.01530	.01445	.01363	.01284
22	3	16	.01794	.01716	.01637	.01558	.01481	.01404	.01330	.01257
22	3	17	.01711	.01643	.01574	.01504	.01435	.01366	.01299	.01233

Covariances of Inverse Gaussian Order Statistics
Shape Parameter k

n	i	j	0.8	0.9	1.0	1.1	1.2	1.3	1.4	1.5
22	3	18	.01631	.01573	.01513	.01453	.01392	.01331	.01271	.01211
22	3	19	.01553	.01505	.01455	.01404	.01351	.01298	.01245	.01192
22	3	20	.01474	.01437	.01397	.01355	.01311	.01267	.01221	.01175
22	3	21	.01390	.01364	.01336	.01304	.01271	.01236	.01199	.01161
22	3	22	.01287	.01277	.01264	.01246	.01226	.01204	.01179	.01152
22	4	5	.04577	.04209	.03862	.03536	.03233	.02950	.02688	.02447
22	4	6	.04059	.03753	.03462	.03188	.02930	.02688	.02462	.02253
22	4	7	.03670	.03410	.03161	.02925	.02701	.02490	.02292	.02106
22	4	8	.03364	.03139	.02924	.02717	.02520	.02334	.02157	.01992
22	4	9	.03115	.02919	.02730	.02548	.02373	.02207	.02048	.01899
22	4	10	.02907	.02735	.02568	.02406	.02250	.02101	.01958	.01822
22	4	11	.02729	.02578	.02430	.02285	.02145	.02010	.01881	.01757
22	4	12	.02574	.02441	.02309	.02180	.02054	.01932	.01814	.01701
22	4	13	.02437	.02319	.02203	.02087	.01974	.01864	.01757	.01653
22	4	14	.02314	.02210	.02107	.02004	.01903	.01803	.01706	.01611
22	4	15	.02201	.02111	.02020	.01929	.01838	.01748	.01660	.01574
22	4	16	.02098	.02020	.01940	.01860	.01779	.01699	.01619	.01541
22	4	17	.02000	.01934	.01865	.01795	.01724	.01653	.01582	.01511
22	4	18	.01907	.01852	.01794	.01734	.01673	.01611	.01548	.01485
22	4	19	.01816	.01772	.01725	.01676	.01624	.01571	.01517	.01462
22	4	20	.01724	.01692	.01656	.01618	.01576	.01533	.01488	.01441
22	4	21	.01625	.01607	.01584	.01557	.01528	.01495	.01461	.01424
22	4	22	.01506	.01504	.01498	.01488	.01474	.01457	.01436	.01412
22	5	6	.04642	.04317	.04007	.03711	.03430	.03165	.02916	.02683
22	5	7	.04199	.03925	.03660	.03406	.03164	.02933	.02715	.02510
22	5	8	.03851	.03615	.03386	.03165	.02953	.02750	.02557	.02374
22	5	9	.03567	.03362	.03163	.02969	.02782	.02601	.02429	.02264
22	5	10	.03329	.03151	.02976	.02805	.02638	.02477	.02322	.02172
22	5	11	.03126	.02970	.02816	.02664	.02516	.02371	.02231	.02095
22	5	12	.02949	.02813	.02677	.02542	.02409	.02279	.02152	.02029
22	5	13	.02792	.02673	.02554	.02434	.02316	.02199	.02084	.01972
22	5	14	.02652	.02548	.02443	.02338	.02232	.02127	.02024	.01922
22	5	15	.02523	.02434	.02343	.02250	.02157	.02063	.01970	.01878
22	5	16	.02404	.02329	.02250	.02169	.02087	.02005	.01921	.01839
22	5	17	.02293	.02230	.02163	.02094	.02023	.01951	.01878	.01804
22	5	18	.02186	.02136	.02081	.02024	.01963	.01901	.01837	.01773
22	5	19	.02082	.02044	.02001	.01955	.01906	.01854	.01800	.01745
22	5	20	.01977	.01951	.01921	.01887	.01850	.01809	.01766	.01720
22	5	21	.01864	.01853	.01837	.01817	.01793	.01765	.01734	.01700
22	5	22	.01727	.01735	.01738	.01737	.01730	.01719	.01704	.01686
22	6	7	.04751	.04463	.04184	.03914	.03654	.03405	.03168	.02943
22	6	8	.04358	.04112	.03872	.03639	.03412	.03194	.02985	.02785
22	6	9	.04038	.03826	.03618	.03414	.03215	.03022	.02836	.02656
22	6	10	.03770	.03587	.03405	.03226	.03050	.02878	.02711	.02550
22	6	11	.03541	.03382	.03223	.03065	.02909	.02756	.02606	.02460
22	6	12	.03341	.03203	.03064	.02925	.02786	.02649	.02515	.02383
22	6	13	.03164	.03045	.02923	.02801	.02678	.02556	.02435	.02316
22	6	14	.03005	.02902	.02797	.02690	.02582	.02473	.02365	.02257
22	6	15	.02860	.02773	.02683	.02590	.02495	.02399	.02302	.02205
22	6	16	.02725	.02653	.02577	.02497	.02415	.02331	.02246	.02160
22	6	17	.02599	.02541	.02478	.02411	.02341	.02269	.02194	.02119
22	6	18	.02478	.02433	.02384	.02330	.02272	.02211	.02148	.02082
22	6	19	.02360	.02329	.02292	.02251	.02206	.02156	.02104	.02050
22	6	20	.02241	.02224	.02201	.02173	.02141	.02104	.02064	.02021
22	6	21	.02113	.02112	.02105	.02092	.02075	.02053	.02027	.01997

Covariances of Inverse Gaussian Order Statistics
Shape Parameter k

n	i	j	0.8	0.9	1.0	1.1	1.2	1.3	1.4	1.5
22	6	22	.01958	.01978	.01992	.02000	.02002	.02000	.01993	.01981
22	7	8	.04899	.04645	.04394	.04149	.03909	.03676	.03451	.03235
22	7	9	.04541	.04323	.04107	.03894	.03684	.03479	.03280	.03086
22	7	10	.04240	.04054	.03866	.03680	.03496	.03314	.03136	.02963
22	7	11	.03983	.03823	.03660	.03497	.03335	.03174	.03015	.02859
22	7	12	.03759	.03621	.03480	.03338	.03195	.03052	.02910	.02770
22	7	13	.03561	.03443	.03321	.03197	.03071	.02945	.02818	.02693
22	7	14	.03382	.03282	.03178	.03071	.02961	.02850	.02737	.02625
22	7	15	.03219	.03136	.03048	.02957	.02862	.02764	.02665	.02565
22	7	16	.03068	.03001	.02928	.02851	.02770	.02686	.02600	.02512
22	7	17	.02926	.02874	.02816	.02753	.02686	.02615	.02541	.02464
22	7	18	.02791	.02753	.02709	.02660	.02606	.02548	.02487	.02422
22	7	19	.02658	.02635	.02606	.02571	.02531	.02486	.02437	.02384
22	7	20	.02524	.02516	.02502	.02482	.02456	.02426	.02390	.02351
22	7	21	.02380	.02390	.02393	.02390	.02381	.02367	.02347	.02323
22	7	22	.02205	.02238	.02264	.02284	.02298	.02306	.02308	.02304
22	8	9	.05086	.04864	.04641	.04420	.04200	.03984	.03772	.03565
22	8	10	.04751	.04562	.04370	.04178	.03986	.03796	.03608	.03424
22	8	11	.04465	.04303	.04138	.03971	.03804	.03636	.03469	.03304
22	8	12	.04214	.04077	.03936	.03791	.03645	.03497	.03349	.03201
22	8	13	.03992	.03877	.03757	.03632	.03505	.03375	.03244	.03112
22	8	14	.03793	.03697	.03595	.03489	.03379	.03266	.03151	.03034
22	8	15	.03610	.03533	.03449	.03360	.03266	.03168	.03068	.02965
22	8	16	.03441	.03381	.03313	.03240	.03162	.03080	.02993	.02904
22	8	17	.03283	.03238	.03187	.03129	.03066	.02998	.02926	.02850
22	8	18	.03131	.03102	.03066	.03024	.02975	.02922	.02863	.02801
22	8	19	.02982	.02969	.02949	.02922	.02889	.02850	.02806	.02757
22	8	20	.02832	.02835	.02832	.02821	.02805	.02781	.02753	.02719
22	8	21	.02671	.02693	.02709	.02717	.02719	.02714	.02703	.02687
22	8	22	.02475	.02522	.02563	.02597	.02624	.02644	.02658	.02665
22	9	10	.05315	.05125	.04930	.04733	.04534	.04336	.04138	.03943
22	9	11	.04996	.04835	.04669	.04500	.04328	.04154	.03980	.03806
22	9	12	.04717	.04582	.04442	.04297	.04148	.03996	.03843	.03688
22	9	13	.04469	.04358	.04240	.04117	.03989	.03857	.03723	.03586
22	9	14	.04246	.04156	.04059	.03956	.03847	.03733	.03617	.03497
22	9	15	.04043	.03972	.03894	.03809	.03718	.03622	.03522	.03418
22	9	16	.03854	.03802	.03742	.03674	.03600	.03521	.03436	.03348
22	9	17	.03677	.03642	.03599	.03548	.03491	.03428	.03359	.03285
22	9	18	.03507	.03489	.03463	.03429	.03388	.03341	.03288	.03229
22	9	19	.03341	.03340	.03331	.03314	.03290	.03259	.03222	.03179
22	9	20	.03173	.03190	.03199	.03200	.03194	.03181	.03161	.03135
22	9	21	.02993	.03030	.03060	.03082	.03097	.03104	.03104	.03098
22	9	22	.02773	.02838	.02896	.02946	.02989	.03024	.03053	.03073
22	10	11	.05591	.05433	.05268	.05097	.04921	.04743	.04562	.04380
22	10	12	.05280	.05150	.05012	.04868	.04718	.04564	.04406	.04246
22	10	13	.05004	.04899	.04786	.04665	.04538	.04406	.04269	.04129
22	10	14	.04755	.04673	.04582	.04483	.04377	.04265	.04148	.04027
22	10	15	.04528	.04467	.04396	.04318	.04232	.04139	.04040	.03936
22	10	16	.04318	.04276	.04225	.04165	.04098	.04023	.03943	.03856
22	10	17	.04119	.04096	.04064	.04023	.03974	.03917	.03854	.03784
22	10	18	.03930	.03925	.03911	.03889	.03858	.03819	.03773	.03720
22	10	19	.03744	.03758	.03763	.03759	.03746	.03726	.03698	.03663
22	10	20	.03556	.03589	.03614	.03630	.03637	.03636	.03628	.03612
22	10	21	.03354	.03410	.03457	.03496	.03526	.03549	.03563	.03569
22	10	22	.03109	.03194	.03272	.03342	.03404	.03458	.03504	.03541

Covariances of Inverse Gaussian Order Statistics
Shape Parameter k

n	i	j	0.8	0.9	1.0	1.1	1.2	1.3	1.4	1.5
22	11	12	.05921	.05798	.05665	.05524	.05374	.05219	.05058	.04894
22	11	13	.05613	.05517	.05411	.05295	.05171	.05039	.04902	.04760
22	11	14	.05335	.05264	.05181	.05089	.04988	.04880	.04764	.04643
22	11	15	.05081	.05032	.04972	.04902	.04823	.04735	.04641	.04539
22	11	16	.04846	.04818	.04779	.04730	.04671	.04604	.04529	.04447
22	11	17	.04624	.04616	.04598	.04569	.04531	.04483	.04428	.04364
22	11	18	.04412	.04424	.04425	.04417	.04398	.04371	.04335	.04291
22	11	19	.04204	.04236	.04258	.04269	.04272	.04265	.04249	.04225
22	11	20	.03993	.04046	.04090	.04123	.04148	.04163	.04169	.04167
22	11	21	.03767	.03844	.03913	.03972	.04022	.04063	.04095	.04118
22	11	22	.03492	.03601	.03704	.03797	.03883	.03959	.04027	.04086
22	12	13	.06318	.06234	.06137	.06029	.05911	.05783	.05647	.05505
22	12	14	.06007	.05949	.05879	.05796	.05703	.05601	.05489	.05370
22	12	15	.05722	.05689	.05642	.05584	.05515	.05436	.05348	.05251
22	12	16	.05458	.05447	.05424	.05389	.05343	.05286	.05220	.05145
22	12	17	.05209	.05220	.05219	.05207	.05183	.05148	.05104	.05050
22	12	18	.04971	.05003	.05024	.05034	.05032	.05020	.04997	.04965
22	12	19	.04737	.04791	.04834	.04867	.04888	.04898	.04899	.04889
22	12	20	.04500	.04577	.04644	.04701	.04746	.04782	.04807	.04823
22	12	21	.04246	.04350	.04444	.04528	.04603	.04667	.04722	.04766
22	12	22	.03936	.04075	.04207	.04330	.04444	.04549	.04644	.04730
22	13	14	.06799	.06759	.06705	.06637	.06556	.06462	.06358	.06244
22	13	15	.06478	.06465	.06437	.06396	.06341	.06274	.06195	.06106
22	13	16	.06180	.06192	.06189	.06173	.06144	.06102	.06049	.05984
22	13	17	.05900	.05935	.05957	.05965	.05961	.05944	.05915	.05875
22	13	18	.05631	.05689	.05735	.05768	.05788	.05796	.05792	.05777
22	13	19	.05367	.05449	.05519	.05577	.05623	.05657	.05679	.05689
22	13	20	.05099	.05207	.05303	.05388	.05461	.05523	.05573	.05612
22	13	21	.04813	.04949	.05075	.05191	.05296	.05391	.05474	.05547
22	13	22	.04462	.04637	.04805	.04964	.05114	.05255	.05385	.05505
22	14	15	.07389	.07403	.07400	.07381	.07346	.07297	.07234	.07157
22	14	16	.07051	.07092	.07117	.07126	.07119	.07098	.07064	.07015
22	14	17	.06733	.06799	.06851	.06887	.06909	.06916	.06909	.06888
22	14	18	.06427	.06519	.06597	.06661	.06710	.06745	.06766	.06774
22	14	19	.06127	.06245	.06350	.06442	.06519	.06584	.06635	.06672
22	14	20	.05823	.05969	.06102	.06224	.06333	.06429	.06512	.06583
22	14	21	.05496	.05674	.05841	.05997	.06143	.06276	.06398	.06507
22	14	22	.05097	.05318	.05531	.05736	.05932	.06118	.06294	.06458
22	15	16	.08127	.08207	.08269	.08313	.08339	.08347	.08339	.08315
22	15	17	.07763	.07871	.07962	.08037	.08094	.08134	.08158	.08166
22	15	18	.07412	.07549	.07669	.07774	.07863	.07935	.07991	.08032
22	15	19	.07068	.07233	.07384	.07520	.07641	.07747	.07837	.07913
22	15	20	.06719	.06914	.07097	.07267	.07424	.07566	.07694	.07807
22	15	21	.06344	.06574	.06795	.07004	.07202	.07387	.07559	.07719
22	15	22	.05884	.06163	.06436	.06701	.06956	.07202	.07438	.07661
22	16	17	.09076	.09242	.09388	.09515	.09623	.09711	.09780	.09829
22	16	18	.08670	.08867	.09046	.09207	.09351	.09476	.09582	.09671
22	16	19	.08270	.08498	.08712	.08909	.09089	.09253	.09400	.09529
22	16	20	.07863	.08126	.08376	.08612	.08833	.09039	.09229	.09403
22	16	21	.07427	.07729	.08021	.08302	.08571	.08827	.09070	.09298
22	16	22	.06891	.07248	.07599	.07944	.08281	.08608	.08925	.09230
22	17	18	.10341	.10623	.10886	.11130	.11353	.11555	.11736	.11895
22	17	19	.09868	.10186	.10488	.10773	.11039	.11287	.11515	.11724
22	17	20	.09387	.09743	.10087	.10417	.10731	.11029	.11309	.11572
22	17	21	.08869	.09270	.09663	.10045	.10416	.10773	.11116	.11444

Covariances of Inverse Gaussian Order Statistics
Shape Parameter k

n	i	j	0.8	0.9	1.0	1.1	1.2	1.3	1.4	1.5
22	17	22	.08232	.08697	.09159	.09616	.10066	.10509	.10941	.11363
22	18	19	.12112	.12563	.12999	.13416	.13814	.14191	.14546	.14879
22	18	20	.11527	.12024	.12508	.12979	.13434	.13872	.14291	.14691
22	18	21	.10897	.11446	.11988	.12522	.13045	.13555	.14052	.14533
22	18	22	.10122	.10744	.11368	.11992	.12612	.13227	.13835	.14434
22	19	20	.14781	.15504	.16217	.16919	.17607	.18278	.18930	.19562
22	19	21	.13985	.14770	.15554	.16334	.17107	.17870	.18622	.19359
22	19	22	.13001	.13876	.14761	.15654	.16550	.17447	.18343	.19233
22	20	21	.19300	.20521	.21754	.22994	.24236	.25478	.26714	.27942
22	20	22	.17970	.19306	.20672	.22062	.23471	.24896	.26333	.27776
22	21	22	.28710	.31124	.33619	.36188	.38823	.41518	.44265	.47059
23	1	2	.04545	.03946	.03422	.02963	.02565	.02218	.01918	.01659
23	1	3	.03371	.02959	.02593	.02269	.01984	.01733	.01514	.01321
23	1	4	.02752	.02435	.02150	.01896	.01670	.01470	.01293	.01137
23	1	5	.02361	.02102	.01869	.01658	.01470	.01301	.01151	.01018
23	1	6	.02089	.01870	.01671	.01491	.01328	.01182	.01051	.00935
23	1	7	.01885	.01695	.01522	.01365	.01222	.01093	.00976	.00872
23	1	8	.01726	.01559	.01406	.01266	.01138	.01022	.00917	.00823
23	1	9	.01596	.01448	.01311	.01186	.01070	.00965	.00870	.00783
23	1	10	.01489	.01356	.01233	.01119	.01014	.00918	.00830	.00750
23	1	11	.01397	.01277	.01166	.01062	.00966	.00878	.00797	.00723
23	1	12	.01318	.01209	.01107	.01013	.00925	.00843	.00768	.00699
23	1	13	.01248	.01149	.01056	.00969	.00888	.00813	.00743	.00679
23	1	14	.01185	.01095	.01010	.00930	.00856	.00786	.00721	.00661
23	1	15	.01128	.01047	.00969	.00895	.00826	.00762	.00702	.00646
23	1	16	.01076	.01002	.00931	.00864	.00800	.00740	.00684	.00632
23	1	17	.01028	.00960	.00896	.00834	.00775	.00720	.00668	.00619
23	1	18	.00982	.00921	.00862	.00806	.00753	.00702	.00653	.00608
23	1	19	.00937	.00883	.00831	.00780	.00731	.00685	.00640	.00598
23	1	20	.00894	.00847	.00800	.00755	.00711	.00668	.00628	.00589
23	1	21	.00850	.00809	.00769	.00729	.00690	.00653	.00616	.00581
23	1	22	.00803	.00769	.00736	.00703	.00670	.00637	.00605	.00575
23	1	23	.00744	.00721	.00697	.00672	.00647	.00621	.00596	.00570
23	2	3	.04467	.03977	.03534	.03135	.02777	.02458	.02173	.01920
23	2	4	.03655	.03280	.02937	.02625	.02343	.02089	.01860	.01655
23	2	5	.03141	.02836	.02555	.02299	.02064	.01852	.01659	.01484
23	2	6	.02780	.02524	.02287	.02068	.01867	.01683	.01516	.01363
23	2	7	.02511	.02290	.02085	.01895	.01719	.01557	.01408	.01273
23	2	8	.02299	.02107	.01926	.01758	.01602	.01457	.01324	.01202
23	2	9	.02128	.01958	.01798	.01647	.01507	.01377	.01256	.01144
23	2	10	.01985	.01834	.01690	.01555	.01428	.01309	.01199	.01097
23	2	11	.01864	.01728	.01598	.01476	.01361	.01252	.01151	.01057
23	2	12	.01758	.01636	.01519	.01408	.01302	.01203	.01110	.01022
23	2	13	.01665	.01555	.01449	.01348	.01251	.01160	.01074	.00993
23	2	14	.01581	.01482	.01386	.01294	.01206	.01122	.01042	.00967
23	2	15	.01506	.01416	.01329	.01245	.01165	.01087	.01014	.00944
23	2	16	.01436	.01356	.01277	.01201	.01127	.01056	.00989	.00924
23	2	17	.01371	.01300	.01229	.01160	.01093	.01028	.00966	.00906
23	2	18	.01310	.01247	.01184	.01122	.01061	.01002	.00945	.00889
23	2	19	.01251	.01196	.01140	.01085	.01031	.00977	.00925	.00875
23	2	20	.01194	.01146	.01098	.01050	.01002	.00954	.00907	.00862
23	2	21	.01135	.01096	.01056	.01015	.00973	.00932	.00891	.00850
23	2	22	.01071	.01042	.01011	.00978	.00944	.00910	.00875	.00841
23	2	23	.00994	.00977	.00957	.00936	.00912	.00887	.00861	.00834
23	3	4	.04405	.03988	.03603	.03249	.02926	.02630	.02362	.02119

Covariances of Inverse Gaussian Order Statistics
Shape Parameter k

n	i	j	0.8	0.9	1.0	1.1	1.2	1.3	1.4	1.5
23	3	5	.03789	.03452	.03139	.02848	.02580	.02334	.02108	.01902
23	3	6	.03357	.03075	.02811	.02564	.02335	.02123	.01928	.01748
23	3	7	.03033	.02791	.02564	.02350	.02151	.01965	.01792	.01633
23	3	8	.02779	.02569	.02370	.02182	.02005	.01840	.01685	.01542
23	3	9	.02572	.02388	.02212	.02045	.01887	.01738	.01599	.01469
23	3	10	.02400	.02237	.02080	.01931	.01788	.01654	.01527	.01408
23	3	11	.02254	.02108	.01968	.01833	.01704	.01582	.01466	.01357
23	3	12	.02126	.01996	.01870	.01749	.01632	.01520	.01414	.01313
23	3	13	.02014	.01898	.01784	.01674	.01568	.01466	.01368	.01275
23	3	14	.01913	.01809	.01707	.01608	.01511	.01418	.01328	.01242
23	3	15	.01822	.01729	.01637	.01548	.01460	.01374	.01292	.01213
23	3	16	.01738	.01656	.01574	.01493	.01413	.01335	.01260	.01187
23	3	17	.01660	.01587	.01514	.01442	.01370	.01300	.01231	.01164
23	3	18	.01586	.01522	.01458	.01394	.01330	.01266	.01204	.01143
23	3	19	.01515	.01460	.01405	.01349	.01292	.01236	.01179	.01124
23	3	20	.01445	.01400	.01353	.01305	.01256	.01206	.01157	.01107
23	3	21	.01373	.01338	.01301	.01261	.01220	.01178	.01136	.01093
23	3	22	.01297	.01273	.01245	.01216	.01184	.01150	.01116	.01080
23	3	23	.01203	.01193	.01180	.01163	.01144	.01122	.01098	.01072
23	4	5	.04402	.04039	.03697	.03378	.03081	.02806	.02551	.02317
23	4	6	.03903	.03600	.03313	.03044	.02791	.02554	.02335	.02131
23	4	7	.03528	.03270	.03024	.02791	.02571	.02365	.02171	.01991
23	4	8	.03233	.03010	.02796	.02593	.02398	.02215	.02043	.01881
23	4	9	.02994	.02799	.02611	.02430	.02258	.02094	.01938	.01792
23	4	10	.02795	.02623	.02456	.02295	.02140	.01992	.01852	.01718
23	4	11	.02624	.02472	.02323	.02179	.02040	.01906	.01778	.01656
23	4	12	.02476	.02341	.02208	.02079	.01953	.01832	.01715	.01603
23	4	13	.02346	.02226	.02107	.01991	.01877	.01766	.01660	.01557
23	4	14	.02229	.02122	.02016	.01912	.01809	.01709	.01611	.01517
23	4	15	.02123	.02029	.01934	.01841	.01748	.01657	.01568	.01481
23	4	16	.02025	.01942	.01859	.01775	.01692	.01610	.01529	.01449
23	4	17	.01934	.01862	.01789	.01715	.01641	.01567	.01493	.01421
23	4	18	.01848	.01787	.01723	.01658	.01593	.01527	.01461	.01396
23	4	19	.01765	.01714	.01660	.01605	.01548	.01490	.01431	.01373
23	4	20	.01684	.01643	.01599	.01552	.01504	.01455	.01404	.01353
23	4	21	.01601	.01571	.01537	.01500	.01462	.01421	.01378	.01335
23	4	22	.01512	.01494	.01472	.01446	.01418	.01387	.01354	.01319
23	4	23	.01403	.01401	.01394	.01384	.01370	.01353	.01332	.01310
23	5	6	.04450	.04128	.03822	.03531	.03256	.02997	.02755	.02528
23	5	7	.04024	.03751	.03489	.03239	.03001	.02776	.02563	.02363
23	5	8	.03690	.03455	.03228	.03009	.02800	.02601	.02412	.02233
23	5	9	.03418	.03213	.03014	.02822	.02637	.02459	.02290	.02128
23	5	10	.03191	.03011	.02836	.02665	.02500	.02341	.02188	.02041
23	5	11	.02997	.02839	.02684	.02532	.02384	.02240	.02101	.01968
23	5	12	.02828	.02689	.02552	.02416	.02283	.02153	.02027	.01905
23	5	13	.02679	.02557	.02435	.02313	.02194	.02076	.01962	.01850
23	5	14	.02546	.02439	.02330	.02222	.02115	.02008	.01904	.01803
23	5	15	.02425	.02331	.02236	.02139	.02043	.01948	.01853	.01760
23	5	16	.02314	.02232	.02149	.02064	.01978	.01892	.01807	.01723
23	5	17	.02210	.02140	.02068	.01994	.01918	.01842	.01766	.01689
23	5	18	.02112	.02053	.01992	.01928	.01862	.01795	.01727	.01659
23	5	19	.02017	.01970	.01919	.01866	.01810	.01752	.01692	.01632
23	5	20	.01924	.01888	.01848	.01805	.01759	.01710	.01660	.01608
23	5	21	.01830	.01805	.01777	.01745	.01709	.01671	.01630	.01587
23	5	22	.01728	.01717	.01702	.01682	.01658	.01631	.01601	.01569

Covariances of Inverse Gaussian Order Statistics
Shape Parameter k

n	i	j	0.8	0.9	1.0	1.1	1.2	1.3	1.4	1.5
23	5	23	.01603	.01610	.01612	.01609	.01602	.01591	.01576	.01557
23	6	7	.04537	.04251	.03975	.03709	.03453	.03210	.02979	.02760
23	6	8	.04162	.03917	.03678	.03447	.03224	.03009	.02804	.02609
23	6	9	.03856	.03644	.03436	.03233	.03036	.02846	.02663	.02487
23	6	10	.03601	.03416	.03234	.03054	.02879	.02709	.02545	.02386
23	6	11	.03383	.03221	.03061	.02902	.02746	.02593	.02445	.02301
23	6	12	.03193	.03052	.02910	.02769	.02630	.02492	.02358	.02227
23	6	13	.03026	.02902	.02777	.02652	.02528	.02404	.02283	.02164
23	6	14	.02875	.02768	.02658	.02548	.02437	.02326	.02216	.02108
23	6	15	.02739	.02646	.02551	.02453	.02355	.02256	.02157	.02059
23	6	16	.02613	.02534	.02452	.02367	.02280	.02192	.02104	.02015
23	6	17	.02496	.02430	.02360	.02287	.02211	.02134	.02055	.01976
23	6	18	.02386	.02331	.02273	.02211	.02147	.02080	.02011	.01941
23	6	19	.02279	.02237	.02190	.02140	.02086	.02029	.01970	.01910
23	6	20	.02174	.02144	.02110	.02071	.02028	.01982	.01933	.01882
23	6	21	.02067	.02050	.02028	.02001	.01970	.01936	.01898	.01857
23	6	22	.01953	.01950	.01942	.01929	.01912	.01890	.01865	.01836
23	6	23	.01812	.01829	.01840	.01846	.01847	.01843	.01835	.01822
23	7	8	.04661	.04407	.04158	.03914	.03678	.03449	.03229	.03018
23	7	9	.04321	.04102	.03885	.03673	.03465	.03263	.03067	.02878
23	7	10	.04035	.03846	.03657	.03471	.03287	.03107	.02931	.02761
23	7	11	.03792	.03627	.03462	.03298	.03135	.02974	.02816	.02663
23	7	12	.03580	.03437	.03293	.03148	.03003	.02859	.02717	.02578
23	7	13	.03392	.03269	.03143	.03015	.02887	.02758	.02631	.02505
23	7	14	.03224	.03118	.03009	.02897	.02783	.02669	.02555	.02441
23	7	15	.03072	.02981	.02887	.02790	.02690	.02589	.02486	.02384
23	7	16	.02931	.02855	.02775	.02691	.02605	.02516	.02425	.02334
23	7	17	.02800	.02738	.02671	.02601	.02526	.02449	.02370	.02289
23	7	18	.02676	.02628	.02574	.02515	.02453	.02387	.02319	.02248
23	7	19	.02557	.02521	.02480	.02434	.02384	.02329	.02272	.02212
23	7	20	.02439	.02417	.02389	.02355	.02317	.02275	.02229	.02179
23	7	21	.02320	.02311	.02297	.02277	.02252	.02222	.02188	.02151
23	7	22	.02191	.02198	.02200	.02195	.02185	.02170	.02150	.02126
23	7	23	.02033	.02062	.02084	.02100	.02111	.02116	.02116	.02111
23	8	9	.04820	.04596	.04372	.04151	.03933	.03719	.03510	.03308
23	8	10	.04504	.04311	.04117	.03923	.03731	.03542	.03356	.03175
23	8	11	.04233	.04066	.03898	.03729	.03560	.03392	.03225	.03062
23	8	12	.03997	.03854	.03708	.03560	.03410	.03261	.03113	.02966
23	8	13	.03788	.03666	.03540	.03410	.03279	.03147	.03014	.02882
23	8	14	.03601	.03497	.03389	.03277	.03162	.03045	.02927	.02808
23	8	15	.03431	.03344	.03252	.03156	.03056	.02953	.02849	.02743
23	8	16	.03274	.03203	.03127	.03045	.02960	.02871	.02779	.02685
23	8	17	.03128	.03072	.03010	.02943	.02871	.02795	.02715	.02634
23	8	18	.02990	.02948	.02900	.02846	.02788	.02724	.02657	.02587
23	8	19	.02857	.02829	.02795	.02755	.02709	.02659	.02604	.02546
23	8	20	.02726	.02712	.02692	.02666	.02634	.02596	.02554	.02508
23	8	21	.02592	.02594	.02589	.02577	.02559	.02536	.02508	.02475
23	8	22	.02449	.02467	.02479	.02485	.02484	.02477	.02465	.02448
23	8	23	.02272	.02314	.02349	.02378	.02400	.02416	.02426	.02430
23	9	10	.05015	.04820	.04622	.04423	.04223	.04025	.03830	.03637
23	9	11	.04715	.04548	.04378	.04204	.04030	.03855	.03681	.03509
23	9	12	.04453	.04311	.04165	.04014	.03862	.03708	.03553	.03399
23	9	13	.04221	.04102	.03976	.03847	.03714	.03578	.03441	.03303
23	9	14	.04013	.03914	.03808	.03697	.03581	.03463	.03342	.03219
23	9	15	.03824	.03743	.03655	.03561	.03462	.03359	.03253	.03145

Covariances of Inverse Gaussian Order Statistics
Shape Parameter k

n	i	j	0.8	0.9	1.0	1.1	1.2	1.3	1.4	1.5
23	9	16	.03650	.03586	.03514	.03436	.03353	.03265	.03174	.03079
23	9	17	.03488	.03439	.03383	.03321	.03252	.03179	.03101	.03020
23	9	18	.03334	.03301	.03260	.03212	.03159	.03099	.03035	.02967
23	9	19	.03186	.03167	.03142	.03109	.03070	.03025	.02974	.02919
23	9	20	.03040	.03037	.03026	.03009	.02985	.02954	.02918	.02877
23	9	21	.02891	.02904	.02910	.02909	.02901	.02886	.02865	.02839
23	9	22	.02731	.02763	.02787	.02805	.02815	.02819	.02816	.02807
23	9	23	.02535	.02592	.02641	.02684	.02720	.02749	.02771	.02787
23	10	11	.05250	.05084	.04912	.04736	.04557	.04377	.04195	.04014
23	10	12	.04959	.04820	.04674	.04523	.04368	.04210	.04050	.03889
23	10	13	.04702	.04587	.04464	.04335	.04201	.04063	.03923	.03780
23	10	14	.04471	.04377	.04275	.04166	.04052	.03933	.03810	.03685
23	10	15	.04261	.04187	.04104	.04014	.03918	.03816	.03710	.03600
23	10	16	.04068	.04011	.03946	.03874	.03795	.03710	.03619	.03525
23	10	17	.03887	.03848	.03800	.03744	.03681	.03612	.03537	.03458
23	10	18	.03716	.03693	.03662	.03622	.03575	.03522	.03462	.03397
23	10	19	.03551	.03545	.03529	.03506	.03475	.03437	.03393	.03343
23	10	20	.03389	.03399	.03400	.03393	.03379	.03357	.03329	.03294
23	10	21	.03224	.03251	.03270	.03281	.03284	.03280	.03269	.03251
23	10	22	.03046	.03093	.03132	.03164	.03187	.03204	.03213	.03215
23	10	23	.02827	.02901	.02968	.03028	.03080	.03125	.03162	.03192
23	11	12	.05529	.05394	.05251	.05100	.04944	.04783	.04618	.04452
23	11	13	.05243	.05134	.05015	.04889	.04756	.04617	.04474	.04328
23	11	14	.04987	.04900	.04804	.04700	.04588	.04470	.04347	.04219
23	11	15	.04754	.04688	.04612	.04528	.04436	.04337	.04233	.04123
23	11	16	.04538	.04492	.04436	.04371	.04298	.04217	.04130	.04037
23	11	17	.04337	.04309	.04272	.04225	.04170	.04107	.04037	.03960
23	11	18	.04147	.04137	.04117	.04088	.04050	.04004	.03951	.03891
23	11	19	.03964	.03971	.03968	.03957	.03937	.03909	.03873	.03829
23	11	20	.03783	.03808	.03823	.03830	.03828	.03818	.03800	.03774
23	11	21	.03598	.03642	.03677	.03704	.03721	.03730	.03732	.03725
23	11	22	.03400	.03465	.03523	.03571	.03612	.03644	.03668	.03684
23	11	23	.03156	.03251	.03339	.03418	.03490	.03554	.03610	.03657
23	12	13	.05861	.05760	.05648	.05526	.05396	.05258	.05114	.04965
23	12	14	.05576	.05499	.05411	.05313	.05206	.05091	.04969	.04841
23	12	15	.05316	.05261	.05196	.05120	.05035	.04941	.04839	.04731
23	12	16	.05076	.05043	.04998	.04943	.04878	.04804	.04722	.04633
23	12	17	.04852	.04838	.04814	.04779	.04734	.04679	.04616	.04546
23	12	18	.04640	.04645	.04640	.04624	.04598	.04563	.04519	.04467
23	12	19	.04435	.04459	.04473	.04477	.04470	.04455	.04430	.04396
23	12	20	.04233	.04277	.04310	.04334	.04347	.04352	.04347	.04333
23	12	21	.04027	.04091	.04146	.04191	.04226	.04252	.04269	.04277
23	12	22	.03806	.03893	.03972	.04042	.04102	.04154	.04196	.04230
23	12	23	.03533	.03653	.03765	.03869	.03965	.04052	.04130	.04200
23	13	14	.06259	.06195	.06119	.06030	.05931	.05821	.05702	.05576
23	13	15	.05968	.05929	.05877	.05812	.05737	.05650	.05554	.05450
23	13	16	.05700	.05684	.05654	.05612	.05559	.05495	.05421	.05338
23	13	17	.05450	.05454	.05447	.05427	.05395	.05353	.05300	.05238
23	13	18	.05212	.05237	.05251	.05252	.05242	.05221	.05189	.05148
23	13	19	.04983	.05028	.05063	.05085	.05096	.05097	.05087	.05067
23	13	20	.04756	.04823	.04879	.04923	.04957	.04979	.04992	.04994
23	13	21	.04526	.04615	.04693	.04761	.04819	.04866	.04903	.04930
23	13	22	.04277	.04392	.04497	.04592	.04678	.04754	.04820	.04876
23	13	23	.03971	.04121	.04263	.04397	.04522	.04638	.04744	.04841
23	14	15	.06739	.06720	.06685	.06637	.06574	.06499	.06413	.06315

Covariances of Inverse Gaussian Order Statistics
Shape Parameter k

n	i	j	0.8	0.9	1.0	1.1	1.2	1.3	1.4	1.5
23	14	16	.06438	.06443	.06433	.06409	.06372	.06322	.06260	.06187
23	14	17	.06156	.06184	.06198	.06198	.06185	.06159	.06121	.06071
23	14	18	.05889	.05939	.05976	.06000	.06010	.06008	.05994	.05967
23	14	19	.05631	.05703	.05763	.05810	.05844	.05866	.05876	.05874
23	14	20	.05376	.05471	.05555	.05626	.05685	.05732	.05767	.05791
23	14	21	.05116	.05236	.05344	.05442	.05528	.05602	.05665	.05717
23	14	22	.04836	.04983	.05121	.05249	.05367	.05473	.05569	.05654
23	14	23	.04491	.04677	.04855	.05026	.05188	.05340	.05483	.05615
23	15	16	.07328	.07362	.07378	.07379	.07363	.07333	.07288	.07229
23	15	17	.07009	.07068	.07110	.07137	.07148	.07145	.07127	.07096
23	15	18	.06706	.06789	.06857	.06910	.06948	.06971	.06980	.06975
23	15	19	.06413	.06521	.06614	.06693	.06757	.06808	.06844	.06867
23	15	20	.06124	.06256	.06375	.06481	.06574	.06653	.06718	.06770
23	15	21	.05830	.05988	.06135	.06270	.06393	.06503	.06600	.06685
23	15	22	.05511	.05700	.05880	.06049	.06207	.06354	.06489	.06612
23	15	23	.05119	.05351	.05576	.05793	.06001	.06200	.06389	.06567
23	16	17	.08065	.08164	.08245	.08308	.08353	.08381	.08393	.08388
23	16	18	.07718	.07844	.07953	.08045	.08120	.08179	.08221	.08247
23	16	19	.07383	.07536	.07673	.07794	.07899	.07989	.08063	.08120
23	16	20	.07052	.07232	.07398	.07549	.07686	.07808	.07915	.08007
23	16	21	.06714	.06923	.07120	.07305	.07476	.07634	.07777	.07907
23	16	22	.06349	.06592	.06826	.07049	.07261	.07460	.07648	.07822
23	16	23	.05899	.06189	.06474	.06752	.07021	.07281	.07530	.07769
23	17	18	.09011	.09195	.09361	.09508	.09635	.09743	.09833	.09903
23	17	19	.08622	.08836	.09033	.09213	.09375	.09519	.09645	.09753
23	17	20	.08238	.08482	.08712	.08926	.09125	.09306	.09471	.09619
23	17	21	.07845	.08123	.08387	.08639	.08877	.09100	.09308	.09500
23	17	22	.07421	.07736	.08042	.08338	.08623	.08895	.09154	.09399
23	17	23	.06897	.07266	.07630	.07989	.08340	.08683	.09015	.09337
23	18	19	.10270	.10571	.10854	.11118	.11362	.11586	.11789	.11971
23	18	20	.09816	.10152	.10472	.10775	.11061	.11329	.11579	.11808
23	18	21	.09352	.09725	.10085	.10432	.10764	.11081	.11382	.11665
23	18	22	.08850	.09265	.09673	.10072	.10459	.10835	.11197	.11544
23	18	23	.08228	.08705	.09181	.09653	.10119	.10578	.11029	.11469
23	19	20	.12035	.12505	.12961	.13399	.13820	.14220	.14600	.14957
23	19	21	.11471	.11985	.12488	.12978	.13454	.13914	.14356	.14779
23	19	22	.10861	.11425	.11984	.12535	.13078	.13609	.14127	.14630
23	19	23	.10104	.10740	.11379	.12019	.12657	.13291	.13919	.14539
23	20	21	.14695	.15437	.16172	.16896	.17608	.18304	.18984	.19644
23	20	22	.13923	.14726	.15529	.16330	.17125	.17912	.18689	.19453
23	20	23	.12965	.13855	.14757	.15668	.16585	.17504	.18422	.19338
23	21	22	.19201	.20442	.21697	.22962	.24232	.25502	.26770	.28031
23	21	23	.17905	.19259	.20644	.22056	.23490	.24941	.26406	.27881
23	22	23	.28591	.31027	.33547	.36145	.38812	.41543	.44329	.47166
24	1	2	.04430	.03841	.03325	.02875	.02484	.02145	.01852	.01600
24	1	3	.03282	.02876	.02516	.02198	.01919	.01674	.01459	.01272
24	1	4	.02677	.02365	.02085	.01835	.01614	.01418	.01245	.01093
24	1	5	.02297	.02041	.01811	.01604	.01419	.01254	.01108	.00978
24	1	6	.02031	.01814	.01618	.01441	.01282	.01139	.01011	.00897
24	1	7	.01832	.01645	.01474	.01319	.01178	.01052	.00938	.00836
24	1	8	.01677	.01512	.01361	.01223	.01097	.00984	.00881	.00788
24	1	9	.01552	.01405	.01269	.01145	.01032	.00928	.00835	.00750
24	1	10	.01448	.01315	.01193	.01080	.00977	.00883	.00797	.00718
24	1	11	.01359	.01239	.01128	.01025	.00931	.00844	.00764	.00692
24	1	12	.01282	.01173	.01072	.00978	.00891	.00810	.00736	.00669

Covariances of Inverse Gaussian Order Statistics
Shape Parameter k

n	i	j	0.8	0.9	1.0	1.1	1.2	1.3	1.4	1.5
24	1	13	.01214	.01115	.01022	.00936	.00855	.00781	.00712	.00649
24	1	14	.01154	.01064	.00978	.00899	.00824	.00755	.00691	.00632
24	1	15	.01100	.01017	.00939	.00865	.00796	.00732	.00672	.00616
24	1	16	.01050	.00974	.00902	.00834	.00771	.00711	.00655	.00603
24	1	17	.01004	.00935	.00869	.00806	.00747	.00692	.00640	.00591
24	1	18	.00960	.00898	.00838	.00780	.00726	.00674	.00626	.00580
24	1	19	.00919	.00862	.00808	.00755	.00705	.00658	.00613	.00570
24	1	20	.00879	.00829	.00779	.00732	.00686	.00642	.00601	.00561
24	1	21	.00840	.00795	.00751	.00709	.00668	.00628	.00590	.00553
24	1	22	.00799	.00761	.00723	.00686	.00649	.00614	.00579	.00546
24	1	23	.00756	.00725	.00693	.00662	.00630	.00600	.00570	.00541
24	1	24	.00702	.00680	.00657	.00634	.00610	.00585	.00561	.00537
24	2	3	.04336	.03853	.03417	.03026	.02676	.02364	.02086	.01840
24	2	4	.03545	.03175	.02837	.02532	.02256	.02007	.01784	.01585
24	2	5	.03044	.02744	.02467	.02215	.01986	.01777	.01589	.01420
24	2	6	.02694	.02441	.02207	.01992	.01795	.01615	.01451	.01303
24	2	7	.02433	.02214	.02012	.01824	.01651	.01493	.01348	.01215
24	2	8	.02228	.02037	.01858	.01692	.01538	.01397	.01266	.01147
24	2	9	.02062	.01893	.01734	.01585	.01447	.01319	.01200	.01091
24	2	10	.01924	.01773	.01630	.01496	.01371	.01254	.01146	.01045
24	2	11	.01806	.01670	.01542	.01420	.01306	.01199	.01099	.01007
24	2	12	.01704	.01582	.01465	.01354	.01250	.01151	.01059	.00974
24	2	13	.01615	.01504	.01398	.01297	.01201	.01110	.01025	.00945
24	2	14	.01535	.01434	.01338	.01245	.01157	.01073	.00994	.00920
24	2	15	.01462	.01371	.01283	.01199	.01118	.01040	.00967	.00898
24	2	16	.01396	.01314	.01234	.01156	.01082	.01011	.00943	.00878
24	2	17	.01335	.01261	.01188	.01118	.01049	.00984	.00921	.00861
24	2	18	.01277	.01211	.01145	.01081	.01019	.00959	.00900	.00845
24	2	19	.01223	.01164	.01105	.01047	.00991	.00935	.00882	.00831
24	2	20	.01170	.01118	.01066	.01014	.00964	.00914	.00865	.00818
24	2	21	.01117	.01073	.01028	.00983	.00937	.00893	.00849	.00806
24	2	22	.01064	.01027	.00989	.00951	.00912	.00873	.00834	.00796
24	2	23	.01006	.00978	.00948	.00917	.00885	.00853	.00820	.00788
24	2	24	.00935	.00918	.00899	.00879	.00856	.00832	.00808	.00782
24	3	4	.04260	.03849	.03471	.03124	.02807	.02519	.02257	.02021
24	3	5	.03662	.03330	.03021	.02736	.02474	.02233	.02013	.01812
24	3	6	.03243	.02965	.02704	.02462	.02237	.02030	.01839	.01665
24	3	7	.02930	.02691	.02466	.02256	.02059	.01877	.01709	.01553
24	3	8	.02684	.02476	.02279	.02093	.01919	.01757	.01606	.01466
24	3	9	.02485	.02301	.02127	.01962	.01806	.01660	.01523	.01396
24	3	10	.02319	.02156	.02000	.01852	.01711	.01578	.01454	.01337
24	3	11	.02178	.02032	.01892	.01758	.01630	.01509	.01395	.01288
24	3	12	.02055	.01925	.01798	.01677	.01561	.01450	.01345	.01246
24	3	13	.01948	.01830	.01716	.01605	.01499	.01398	.01301	.01209
24	3	14	.01851	.01746	.01642	.01542	.01445	.01352	.01263	.01177
24	3	15	.01764	.01669	.01576	.01485	.01396	.01310	.01228	.01149
24	3	16	.01685	.01599	.01515	.01432	.01352	.01273	.01197	.01124
24	3	17	.01611	.01535	.01459	.01384	.01311	.01239	.01169	.01102
24	3	18	.01541	.01474	.01407	.01340	.01273	.01208	.01144	.01082
24	3	19	.01475	.01417	.01357	.01297	.01238	.01179	.01120	.01063
24	3	20	.01412	.01361	.01309	.01257	.01204	.01151	.01099	.01047
24	3	21	.01348	.01306	.01262	.01217	.01171	.01125	.01079	.01032
24	3	22	.01284	.01251	.01215	.01178	.01139	.01100	.01060	.01019
24	3	23	.01214	.01191	.01165	.01137	.01107	.01075	.01042	.01008
24	3	24	.01128	.01118	.01105	.01089	.01070	.01049	.01026	.01002

Covariances of Inverse Gaussian Order Statistics
Shape Parameter k

n	i	j	0.8	0.9	1.0	1.1	1.2	1.3	1.4	1.5
24	4	5	.04244	.03885	.03549	.03236	.02945	.02676	.02428	.02200
24	4	6	.03761	.03461	.03179	.02913	.02665	.02434	.02220	.02022
24	4	7	.03399	.03143	.02900	.02670	.02454	.02252	.02063	.01888
24	4	8	.03115	.02893	.02681	.02479	.02288	.02109	.01940	.01782
24	4	9	.02885	.02690	.02503	.02324	.02153	.01992	.01840	.01697
24	4	10	.02693	.02520	.02354	.02194	.02041	.01895	.01757	.01626
24	4	11	.02529	.02376	.02227	.02083	.01945	.01813	.01686	.01567
24	4	12	.02387	.02251	.02117	.01987	.01862	.01741	.01626	.01516
24	4	13	.02262	.02140	.02020	.01903	.01789	.01679	.01573	.01471
24	4	14	.02151	.02042	.01934	.01828	.01724	.01624	.01526	.01433
24	4	15	.02050	.01953	.01856	.01760	.01666	.01574	.01485	.01398
24	4	16	.01957	.01871	.01785	.01698	.01613	.01530	.01448	.01368
24	4	17	.01872	.01795	.01719	.01642	.01565	.01489	.01414	.01341
24	4	18	.01791	.01725	.01657	.01588	.01520	.01451	.01383	.01316
24	4	19	.01715	.01658	.01599	.01538	.01477	.01416	.01355	.01294
24	4	20	.01640	.01593	.01542	.01491	.01437	.01383	.01329	.01274
24	4	21	.01567	.01529	.01487	.01444	.01399	.01352	.01305	.01257
24	4	22	.01492	.01464	.01432	.01397	.01360	.01322	.01282	.01241
24	4	23	.01411	.01394	.01373	.01348	.01321	.01292	.01261	.01228
24	4	24	.01311	.01309	.01302	.01291	.01278	.01261	.01241	.01219
24	5	6	.04275	.03957	.03655	.03369	.03100	.02847	.02610	.02390
24	5	7	.03866	.03595	.03336	.03090	.02856	.02635	.02427	.02233
24	5	8	.03544	.03310	.03085	.02869	.02664	.02468	.02283	.02109
24	5	9	.03283	.03079	.02881	.02690	.02507	.02332	.02166	.02008
24	5	10	.03065	.02885	.02710	.02540	.02376	.02219	.02068	.01925
24	5	11	.02879	.02721	.02565	.02413	.02265	.02123	.01986	.01855
24	5	12	.02718	.02577	.02438	.02302	.02169	.02040	.01915	.01795
24	5	13	.02576	.02451	.02327	.02205	.02084	.01967	.01853	.01742
24	5	14	.02450	.02339	.02228	.02118	.02009	.01902	.01798	.01697
24	5	15	.02335	.02237	.02138	.02040	.01941	.01845	.01749	.01656
24	5	16	.02230	.02144	.02056	.01968	.01880	.01792	.01706	.01620
24	5	17	.02132	.02057	.01980	.01902	.01823	.01745	.01666	.01588
24	5	18	.02041	.01976	.01910	.01841	.01771	.01701	.01630	.01559
24	5	19	.01954	.01899	.01842	.01783	.01722	.01660	.01597	.01533
24	5	20	.01869	.01825	.01778	.01728	.01675	.01621	.01566	.01510
24	5	21	.01786	.01752	.01714	.01674	.01630	.01585	.01537	.01489
24	5	22	.01701	.01677	.01650	.01619	.01586	.01549	.01511	.01470
24	5	23	.01608	.01597	.01582	.01563	.01540	.01514	.01486	.01454
24	5	24	.01495	.01500	.01501	.01497	.01489	.01478	.01463	.01445
24	6	7	.04346	.04062	.03788	.03526	.03275	.03037	.02811	.02598
24	6	8	.03986	.03741	.03504	.03276	.03056	.02845	.02645	.02455
24	6	9	.03693	.03481	.03273	.03072	.02877	.02689	.02510	.02338
24	6	10	.03449	.03263	.03080	.02901	.02728	.02559	.02397	.02242
24	6	11	.03240	.03077	.02915	.02756	.02600	.02449	.02302	.02161
24	6	12	.03060	.02916	.02772	.02630	.02490	.02353	.02220	.02091
24	6	13	.02900	.02773	.02646	.02519	.02393	.02270	.02148	.02030
24	6	14	.02758	.02646	.02533	.02420	.02307	.02195	.02085	.01977
24	6	15	.02629	.02531	.02432	.02331	.02230	.02129	.02029	.01930
24	6	16	.02511	.02426	.02339	.02249	.02159	.02069	.01978	.01888
24	6	17	.02401	.02328	.02252	.02174	.02095	.02014	.01932	.01851
24	6	18	.02298	.02237	.02172	.02104	.02035	.01963	.01891	.01817
24	6	19	.02200	.02150	.02096	.02038	.01978	.01916	.01852	.01787
24	6	20	.02105	.02066	.02022	.01975	.01925	.01872	.01817	.01760
24	6	21	.02012	.01983	.01950	.01913	.01873	.01830	.01784	.01735
24	6	22	.01916	.01899	.01877	.01852	.01822	.01789	.01753	.01714

Covariances of Inverse Gaussian Order Statistics
Shape Parameter k

n	i	j	0.8	0.9	1.0	1.1	1.2	1.3	1.4	1.5
24	6	23	.01812	.01808	.01800	.01787	.01770	.01748	.01723	.01696
24	6	24	.01684	.01698	.01708	.01712	.01711	.01706	.01697	.01684
24	7	8	.04449	.04196	.03948	.03707	.03474	.03249	.03034	.02828
24	7	9	.04124	.03904	.03689	.03477	.03271	.03072	.02880	.02695
24	7	10	.03852	.03661	.03472	.03285	.03102	.02924	.02751	.02584
24	7	11	.03620	.03453	.03286	.03121	.02958	.02798	.02642	.02491
24	7	12	.03419	.03273	.03126	.02979	.02833	.02690	.02548	.02411
24	7	13	.03241	.03113	.02984	.02854	.02723	.02594	.02467	.02341
24	7	14	.03082	.02971	.02857	.02742	.02626	.02510	.02394	.02280
24	7	15	.02939	.02842	.02743	.02641	.02538	.02434	.02330	.02226
24	7	16	.02807	.02724	.02638	.02549	.02458	.02365	.02272	.02178
24	7	17	.02684	.02615	.02541	.02464	.02384	.02302	.02219	.02135
24	7	18	.02570	.02512	.02450	.02385	.02316	.02245	.02171	.02097
24	7	19	.02460	.02415	.02365	.02310	.02252	.02191	.02127	.02062
24	7	20	.02354	.02321	.02282	.02238	.02191	.02140	.02087	.02030
24	7	21	.02250	.02228	.02201	.02169	.02132	.02092	.02049	.02002
24	7	22	.02142	.02133	.02119	.02099	.02074	.02046	.02013	.01977
24	7	23	.02026	.02032	.02031	.02026	.02015	.02000	.01980	.01956
24	7	24	.01883	.01908	.01927	.01941	.01949	.01952	.01950	.01943
24	8	9	.04585	.04359	.04135	.03915	.03698	.03487	.03283	.03085
24	8	10	.04284	.04088	.03893	.03699	.03508	.03320	.03137	.02959
24	8	11	.04027	.03857	.03686	.03515	.03346	.03178	.03013	.02852
24	8	12	.03803	.03656	.03506	.03356	.03205	.03055	.02907	.02761
24	8	13	.03606	.03478	.03348	.03215	.03081	.02947	.02814	.02682
24	8	14	.03430	.03320	.03206	.03089	.02971	.02851	.02732	.02612
24	8	15	.03270	.03176	.03078	.02976	.02872	.02766	.02658	.02550
24	8	16	.03124	.03045	.02961	.02873	.02781	.02688	.02592	.02496
24	8	17	.02988	.02923	.02852	.02777	.02698	.02617	.02533	.02447
24	8	18	.02860	.02808	.02751	.02688	.02621	.02551	.02478	.02402
24	8	19	.02739	.02699	.02654	.02604	.02549	.02490	.02428	.02363
24	8	20	.02621	.02594	.02562	.02523	.02480	.02433	.02382	.02327
24	8	21	.02505	.02491	.02471	.02445	.02414	.02378	.02338	.02295
24	8	22	.02385	.02385	.02379	.02366	.02348	.02326	.02298	.02266
24	8	23	.02257	.02272	.02281	.02284	.02281	.02273	.02260	.02242
24	8	24	.02097	.02134	.02164	.02188	.02206	.02219	.02226	.02227
24	9	10	.04752	.04553	.04353	.04152	.03953	.03757	.03563	.03374
24	9	11	.04468	.04296	.04122	.03947	.03771	.03596	.03423	.03253
24	9	12	.04221	.04073	.03922	.03768	.03613	.03458	.03303	.03149
24	9	13	.04003	.03876	.03745	.03611	.03474	.03336	.03198	.03059
24	9	14	.03807	.03700	.03587	.03470	.03350	.03228	.03104	.02980
24	9	15	.03631	.03540	.03444	.03343	.03239	.03131	.03021	.02910
24	9	16	.03468	.03394	.03313	.03227	.03137	.03043	.02947	.02848
24	9	17	.03318	.03258	.03192	.03120	.03044	.02963	.02879	.02792
24	9	18	.03177	.03131	.03079	.03021	.02957	.02889	.02817	.02742
24	9	19	.03042	.03010	.02971	.02926	.02876	.02820	.02760	.02697
24	9	20	.02911	.02893	.02868	.02836	.02798	.02756	.02708	.02656
24	9	21	.02782	.02777	.02766	.02748	.02724	.02694	.02659	.02619
24	9	22	.02650	.02660	.02663	.02660	.02650	.02634	.02613	.02587
24	9	23	.02507	.02534	.02554	.02567	.02574	.02575	.02570	.02559
24	9	24	.02330	.02380	.02423	.02459	.02490	.02513	.02531	.02542
24	10	11	.04952	.04780	.04604	.04425	.04244	.04063	.03882	.03702
24	10	12	.04679	.04533	.04381	.04225	.04067	.03907	.03746	.03585
24	10	13	.04438	.04314	.04184	.04050	.03911	.03770	.03627	.03483
24	10	14	.04223	.04119	.04008	.03893	.03772	.03648	.03522	.03393
24	10	15	.04027	.03942	.03849	.03751	.03647	.03539	.03428	.03314

Covariances of Inverse Gaussian Order Statistics
Shape Parameter k

n	i	j	0.8	0.9	1.0	1.1	1.2	1.3	1.4	1.5
24	10	16	.03848	.03779	.03703	.03621	.03533	.03440	.03343	.03243
24	10	17	.03681	.03628	.03568	.03501	.03428	.03350	.03267	.03180
24	10	18	.03525	.03487	.03442	.03390	.03331	.03266	.03197	.03123
24	10	19	.03376	.03353	.03322	.03284	.03239	.03189	.03133	.03072
24	10	20	.03231	.03222	.03206	.03183	.03153	.03116	.03073	.03025
24	10	21	.03088	.03094	.03093	.03084	.03069	.03046	.03018	.02984
24	10	22	.02941	.02963	.02978	.02985	.02986	.02979	.02966	.02947
24	10	23	.02783	.02823	.02856	.02882	.02900	.02912	.02917	.02916
24	10	24	.02586	.02651	.02710	.02761	.02805	.02843	.02873	.02897
24	11	12	.05190	.05046	.04895	.04738	.04577	.04413	.04247	.04079
24	11	13	.04924	.04804	.04676	.04542	.04403	.04259	.04113	.03964
24	11	14	.04685	.04587	.04480	.04367	.04247	.04123	.03994	.03862
24	11	15	.04469	.04390	.04303	.04208	.04107	.04000	.03888	.03773
24	11	16	.04271	.04210	.04140	.04063	.03979	.03888	.03793	.03693
24	11	17	.04086	.04042	.03990	.03929	.03861	.03787	.03706	.03621
24	11	18	.03913	.03885	.03849	.03804	.03752	.03693	.03627	.03556
24	11	19	.03748	.03736	.03715	.03686	.03649	.03605	.03555	.03498
24	11	20	.03587	.03591	.03586	.03573	.03552	.03523	.03487	.03445
24	11	21	.03429	.03448	.03459	.03462	.03457	.03445	.03425	.03398
24	11	22	.03266	.03303	.03331	.03351	.03364	.03369	.03366	.03356
24	11	23	.03090	.03146	.03195	.03235	.03268	.03293	.03311	.03321
24	11	24	.02873	.02956	.03032	.03100	.03161	.03215	.03261	.03299
24	12	13	.05472	.05357	.05234	.05102	.04963	.04819	.04669	.04516
24	12	14	.05208	.05116	.05015	.04906	.04788	.04665	.04535	.04402
24	12	15	.04968	.04898	.04817	.04728	.04631	.04526	.04416	.04300
24	12	16	.04748	.04697	.04636	.04566	.04487	.04401	.04308	.04209
24	12	17	.04544	.04511	.04468	.04416	.04355	.04286	.04210	.04128
24	12	18	.04352	.04336	.04311	.04276	.04232	.04180	.04121	.04054
24	12	19	.04168	.04170	.04161	.04144	.04117	.04082	.04039	.03988
24	12	20	.03991	.04009	.04017	.04017	.04007	.03989	.03962	.03929
24	12	21	.03814	.03850	.03876	.03893	.03901	.03900	.03891	.03875
24	12	22	.03634	.03688	.03732	.03768	.03796	.03815	.03825	.03827
24	12	23	.03438	.03513	.03580	.03638	.03688	.03729	.03762	.03788
24	12	24	.03197	.03301	.03397	.03486	.03568	.03641	.03706	.03763
24	13	14	.05806	.05724	.05631	.05527	.05414	.05293	.05164	.05030
24	13	15	.05540	.05480	.05410	.05328	.05237	.05137	.05029	.04914
24	13	16	.05295	.05257	.05207	.05146	.05075	.04995	.04907	.04811
24	13	17	.05068	.05049	.05019	.04978	.04926	.04866	.04796	.04718
24	13	18	.04854	.04854	.04843	.04820	.04788	.04746	.04695	.04635
24	13	19	.04650	.04668	.04675	.04672	.04658	.04634	.04601	.04560
24	13	20	.04453	.04489	.04514	.04529	.04534	.04529	.04515	.04492
24	13	21	.04256	.04311	.04355	.04390	.04414	.04429	.04434	.04431
24	13	22	.04055	.04130	.04195	.04250	.04296	.04332	.04359	.04377
24	13	23	.03838	.03935	.04024	.04103	.04174	.04236	.04288	.04331
24	13	24	.03568	.03697	.03819	.03932	.04038	.04135	.04224	.04303
24	14	15	.06204	.06159	.06101	.06030	.05948	.05855	.05752	.05641
24	14	16	.05931	.05909	.05873	.05825	.05765	.05694	.05613	.05523
24	14	17	.05678	.05677	.05662	.05636	.05597	.05548	.05487	.05417
24	14	18	.05439	.05458	.05464	.05458	.05441	.05412	.05372	.05322
24	14	19	.05211	.05250	.05276	.05291	.05294	.05285	.05266	.05237
24	14	20	.04990	.05048	.05095	.05130	.05153	.05166	.05168	.05159
24	14	21	.04771	.04849	.04916	.04972	.05018	.05052	.05076	.05089
24	14	22	.04547	.04646	.04735	.04815	.04883	.04942	.04990	.05028
24	14	23	.04303	.04427	.04543	.04649	.04745	.04832	.04909	.04976
24	14	24	.04002	.04160	.04312	.04456	.04591	.04718	.04836	.04944

Covariances of Inverse Gaussian Order Statistics
Shape Parameter k

n	i	j	0.8	0.9	1.0	1.1	1.2	1.3	1.4	1.5
24	15	16	.06684	.06682	.06666	.06635	.06590	.06532	.06462	.06380
24	15	17	.06399	.06421	.06428	.06420	.06399	.06365	.06318	.06259
24	15	18	.06132	.06175	.06204	.06219	.06221	.06210	.06186	.06150
24	15	19	.05876	.05940	.05991	.06029	.06054	.06065	.06065	.06052
24	15	20	.05628	.05713	.05786	.05846	.05894	.05929	.05952	.05963
24	15	21	.05381	.05489	.05584	.05668	.05740	.05799	.05847	.05883
24	15	22	.05129	.05259	.05380	.05489	.05587	.05673	.05748	.05812
24	15	23	.04855	.05013	.05161	.05300	.05429	.05548	.05656	.05753
24	15	24	.04515	.04711	.04900	.05081	.05254	.05417	.05572	.05716
24	16	17	.07271	.07322	.07357	.07375	.07377	.07364	.07337	.07295
24	16	18	.06969	.07043	.07102	.07146	.07174	.07186	.07185	.07169
24	16	19	.06679	.06777	.06860	.06928	.06982	.07020	.07045	.07055
24	16	20	.06398	.06519	.06626	.06720	.06799	.06864	.06915	.06952
24	16	21	.06119	.06264	.06396	.06515	.06622	.06714	.06794	.06860
24	16	22	.05833	.06003	.06163	.06310	.06446	.06569	.06680	.06778
24	16	23	.05522	.05723	.05914	.06095	.06265	.06425	.06573	.06709
24	16	24	.05137	.05379	.05615	.05843	.06063	.06274	.06476	.06667
24	17	18	.08006	.08122	.08221	.08302	.08366	.08412	.08441	.08454
24	17	19	.07675	.07817	.07943	.08051	.08144	.08219	.08278	.08321
24	17	20	.07354	.07521	.07674	.07810	.07932	.08037	.08127	.08201
24	17	21	.07035	.07228	.07408	.07575	.07726	.07864	.07986	.08093
24	17	22	.06707	.06929	.07139	.07337	.07523	.07695	.07853	.07997
24	17	23	.06351	.06606	.06852	.07088	.07313	.07527	.07728	.07917
24	17	24	.05910	.06211	.06507	.06797	.07079	.07352	.07615	.07868
24	18	19	.08948	.09150	.09334	.09499	.09645	.09773	.09881	.09971
24	18	20	.08576	.08806	.09020	.09217	.09396	.09558	.09703	.09829
24	18	21	.08206	.08465	.08710	.08941	.09155	.09354	.09536	.09701
24	18	22	.07826	.08117	.08396	.08663	.08916	.09155	.09379	.09588
24	18	23	.07413	.07741	.08060	.08370	.08669	.08957	.09231	.09493
24	18	24	.06900	.07280	.07657	.08028	.08393	.08750	.09097	.09435
24	19	20	.10203	.10522	.10823	.11105	.11369	.11613	.11837	.12040
24	19	21	.09767	.10118	.10455	.10776	.11081	.11368	.11636	.11886
24	19	22	.09318	.09705	.10081	.10444	.10794	.11129	.11447	.11750
24	19	23	.08830	.09259	.09681	.10094	.10498	.10890	.11270	.11635
24	19	24	.08222	.08711	.09199	.09685	.10166	.10641	.11108	.11566
24	20	21	.11962	.12450	.12924	.13382	.13823	.14246	.14648	.15029
24	20	22	.11418	.11947	.12467	.12976	.13471	.13951	.14415	.14860
24	20	23	.10824	.11403	.11978	.12546	.13107	.13657	.14195	.14719
24	20	24	.10085	.10734	.11388	.12043	.12698	.13349	.13996	.14635
24	21	22	.14613	.15373	.16127	.16873	.17607	.18328	.19033	.19719
24	21	23	.13865	.14683	.15504	.16324	.17140	.17950	.18750	.19539
24	21	24	.12929	.13833	.14751	.15680	.16615	.17555	.18495	.19434
24	22	23	.19106	.20367	.21643	.22931	.24226	.25524	.26821	.28114
24	22	24	.17842	.19212	.20616	.22049	.23505	.24982	.26474	.27978
24	23	24	.28477	.30934	.33478	.36103	.38800	.41565	.44388	.47265
25	1	2	.04324	.03742	.03235	.02793	.02410	.02078	.01792	.01545
25	1	3	.03199	.02799	.02445	.02133	.01859	.01619	.01409	.01226
25	1	4	.02608	.02300	.02024	.01779	.01562	.01370	.01201	.01053
25	1	5	.02237	.01984	.01757	.01554	.01372	.01211	.01068	.00941
25	1	6	.01977	.01763	.01570	.01395	.01239	.01099	.00973	.00862
25	1	7	.01784	.01598	.01429	.01276	.01138	.01014	.00903	.00803
25	1	8	.01633	.01469	.01319	.01183	.01060	.00948	.00847	.00757
25	1	9	.01511	.01365	.01230	.01108	.00996	.00895	.00803	.00720
25	1	10	.01409	.01278	.01156	.01045	.00943	.00850	.00766	.00689
25	1	11	.01323	.01204	.01094	.00992	.00898	.00812	.00734	.00663

Covariances of Inverse Gaussian Order Statistics
Shape Parameter k

n	i	j	0.8	0.9	1.0	1.1	1.2	1.3	1.4	1.5
25	1	12	.01249	.01140	.01039	.00946	.00859	.00780	.00707	.00641
25	1	13	.01183	.01084	.00991	.00905	.00825	.00752	.00684	.00622
25	1	14	.01125	.01034	.00949	.00869	.00795	.00727	.00663	.00605
25	1	15	.01072	.00989	.00910	.00837	.00768	.00704	.00645	.00590
25	1	16	.01025	.00948	.00876	.00807	.00744	.00684	.00628	.00577
25	1	17	.00980	.00910	.00844	.00781	.00721	.00666	.00614	.00565
25	1	18	.00939	.00875	.00814	.00756	.00700	.00649	.00600	.00555
25	1	19	.00900	.00842	.00786	.00732	.00681	.00633	.00588	.00545
25	1	20	.00863	.00810	.00759	.00710	.00663	.00618	.00576	.00536
25	1	21	.00827	.00779	.00733	.00689	.00646	.00605	.00566	.00528
25	1	22	.00791	.00749	.00708	.00668	.00629	.00591	.00556	.00521
25	1	23	.00754	.00718	.00682	.00647	.00612	.00579	.00546	.00515
25	1	24	.00714	.00684	.00655	.00625	.00595	.00566	.00538	.00510
25	1	25	.00664	.00643	.00621	.00599	.00576	.00553	.00530	.00507
25	2	3	.04214	.03738	.03310	.02926	.02584	.02278	.02008	.01768
25	2	4	.03443	.03078	.02746	.02446	.02175	.01932	.01715	.01520
25	2	5	.02956	.02659	.02387	.02139	.01914	.01710	.01526	.01361
25	2	6	.02615	.02365	.02134	.01922	.01729	.01552	.01393	.01248
25	2	7	.02361	.02145	.01944	.01760	.01590	.01434	.01292	.01163
25	2	8	.02162	.01972	.01796	.01632	.01481	.01341	.01213	.01097
25	2	9	.02001	.01833	.01675	.01528	.01392	.01266	.01150	.01043
25	2	10	.01867	.01716	.01575	.01442	.01318	.01203	.01097	.00999
25	2	11	.01753	.01618	.01489	.01369	.01256	.01150	.01052	.00961
25	2	12	.01655	.01532	.01415	.01305	.01202	.01104	.01014	.00929
25	2	13	.01568	.01457	.01351	.01250	.01154	.01064	.00980	.00902
25	2	14	.01491	.01390	.01293	.01200	.01112	.01029	.00951	.00877
25	2	15	.01422	.01330	.01241	.01156	.01074	.00997	.00925	.00856
25	2	16	.01358	.01275	.01193	.01115	.01040	.00969	.00901	.00837
25	2	17	.01300	.01224	.01150	.01078	.01009	.00943	.00880	.00820
25	2	18	.01246	.01177	.01109	.01044	.00980	.00919	.00860	.00805
25	2	19	.01194	.01132	.01071	.01011	.00953	.00897	.00843	.00791
25	2	20	.01145	.01090	.01035	.00981	.00928	.00876	.00826	.00778
25	2	21	.01097	.01048	.01000	.00951	.00904	.00857	.00811	.00767
25	2	22	.01049	.01007	.00965	.00922	.00880	.00838	.00797	.00757
25	2	23	.01000	.00966	.00930	.00894	.00857	.00820	.00784	.00748
25	2	24	.00947	.00921	.00892	.00863	.00833	.00802	.00771	.00740
25	2	25	.00881	.00866	.00847	.00828	.00806	.00783	.00760	.00735
25	3	4	.04127	.03722	.03350	.03009	.02699	.02418	.02163	.01933
25	3	5	.03546	.03218	.02914	.02634	.02377	.02141	.01927	.01732
25	3	6	.03140	.02864	.02607	.02369	.02148	.01945	.01759	.01589
25	3	7	.02836	.02599	.02377	.02169	.01977	.01798	.01633	.01482
25	3	8	.02598	.02391	.02196	.02013	.01842	.01682	.01534	.01398
25	3	9	.02405	.02222	.02049	.01885	.01732	.01588	.01454	.01330
25	3	10	.02244	.02082	.01927	.01779	.01641	.01510	.01388	.01273
25	3	11	.02108	.01962	.01822	.01689	.01563	.01444	.01331	.01226
25	3	12	.01990	.01859	.01732	.01611	.01496	.01386	.01283	.01185
25	3	13	.01886	.01768	.01653	.01543	.01437	.01336	.01241	.01150
25	3	14	.01794	.01687	.01583	.01482	.01385	.01292	.01203	.01119
25	3	15	.01711	.01614	.01519	.01427	.01338	.01252	.01170	.01092
25	3	16	.01634	.01547	.01461	.01377	.01296	.01217	.01141	.01068
25	3	17	.01564	.01486	.01408	.01331	.01257	.01184	.01114	.01046
25	3	18	.01499	.01428	.01358	.01289	.01221	.01154	.01089	.01027
25	3	19	.01437	.01374	.01312	.01249	.01187	.01126	.01067	.01009
25	3	20	.01378	.01323	.01267	.01211	.01156	.01101	.01046	.00993
25	3	21	.01320	.01273	.01224	.01175	.01126	.01076	.01027	.00979

Covariances of Inverse Gaussian Order Statistics
Shape Parameter k

n	i	j	0.8	0.9	1.0	1.1	1.2	1.3	1.4	1.5
25	3	22	.01263	.01223	.01182	.01140	.01096	.01053	.01009	.00966
25	3	23	.01204	.01173	.01139	.01104	.01067	.01030	.00992	.00954
25	3	24	.01140	.01118	.01093	.01066	.01038	.01007	.00976	.00944
25	3	25	.01061	.01051	.01038	.01022	.01004	.00984	.00962	.00939
25	4	5	.04098	.03744	.03414	.03106	.02821	.02558	.02317	.02095
25	4	6	.03631	.03335	.03056	.02795	.02552	.02326	.02117	.01924
25	4	7	.03281	.03027	.02787	.02561	.02349	.02151	.01966	.01795
25	4	8	.03007	.02786	.02576	.02377	.02189	.02013	.01847	.01693
25	4	9	.02784	.02590	.02404	.02227	.02059	.01901	.01751	.01612
25	4	10	.02599	.02427	.02261	.02102	.01951	.01807	.01672	.01544
25	4	11	.02442	.02288	.02139	.01996	.01859	.01728	.01604	.01486
25	4	12	.02305	.02168	.02034	.01904	.01780	.01660	.01546	.01437
25	4	13	.02185	.02062	.01941	.01824	.01710	.01600	.01495	.01395
25	4	14	.02078	.01968	.01858	.01752	.01648	.01547	.01450	.01358
25	4	15	.01982	.01882	.01784	.01687	.01592	.01500	.01411	.01325
25	4	16	.01894	.01805	.01716	.01628	.01542	.01457	.01375	.01295
25	4	17	.01813	.01733	.01654	.01574	.01496	.01418	.01343	.01269
25	4	18	.01737	.01667	.01596	.01524	.01453	.01383	.01313	.01245
25	4	19	.01665	.01604	.01541	.01477	.01413	.01349	.01286	.01224
25	4	20	.01597	.01544	.01489	.01433	.01376	.01318	.01261	.01205
25	4	21	.01530	.01485	.01438	.01390	.01340	.01289	.01238	.01187
25	4	22	.01464	.01428	.01389	.01348	.01305	.01261	.01217	.01172
25	4	23	.01396	.01368	.01338	.01305	.01271	.01234	.01196	.01158
25	4	24	.01322	.01305	.01284	.01261	.01235	.01207	.01177	.01146
25	4	25	.01230	.01227	.01220	.01209	.01196	.01179	.01160	.01139
25	5	6	.04117	.03803	.03504	.03223	.02959	.02712	.02481	.02267
25	5	7	.03722	.03454	.03198	.02955	.02725	.02509	.02306	.02116
25	5	8	.03412	.03179	.02956	.02743	.02540	.02348	.02167	.01997
25	5	9	.03160	.02957	.02760	.02571	.02390	.02218	.02055	.01901
25	5	10	.02951	.02771	.02596	.02427	.02265	.02110	.01962	.01821
25	5	11	.02772	.02613	.02457	.02305	.02159	.02018	.01883	.01754
25	5	12	.02618	.02476	.02336	.02199	.02067	.01938	.01815	.01696
25	5	13	.02482	.02355	.02230	.02106	.01986	.01869	.01755	.01646
25	5	14	.02361	.02248	.02135	.02024	.01914	.01807	.01703	.01603
25	5	15	.02252	.02151	.02050	.01949	.01850	.01752	.01657	.01564
25	5	16	.02152	.02062	.01972	.01881	.01791	.01702	.01615	.01529
25	5	17	.02060	.01981	.01900	.01819	.01738	.01657	.01577	.01498
25	5	18	.01974	.01905	.01833	.01761	.01688	.01615	.01542	.01471
25	5	19	.01893	.01833	.01771	.01707	.01642	.01577	.01511	.01445
25	5	20	.01815	.01764	.01711	.01655	.01599	.01540	.01482	.01423
25	5	21	.01739	.01698	.01653	.01606	.01557	.01506	.01455	.01402
25	5	22	.01664	.01632	.01596	.01558	.01517	.01474	.01429	.01384
25	5	23	.01586	.01564	.01538	.01509	.01477	.01442	.01405	.01367
25	5	24	.01502	.01491	.01476	.01458	.01436	.01411	.01383	.01353
25	5	25	.01398	.01402	.01402	.01398	.01389	.01378	.01363	.01345
25	6	7	.04172	.03891	.03620	.03362	.03115	.02882	.02662	.02454
25	6	8	.03826	.03583	.03348	.03122	.02905	.02699	.02503	.02317
25	6	9	.03545	.03333	.03127	.02927	.02734	.02550	.02374	.02206
25	6	10	.03311	.03124	.02942	.02764	.02592	.02426	.02267	.02114
25	6	11	.03111	.02947	.02784	.02625	.02470	.02320	.02176	.02036
25	6	12	.02938	.02792	.02648	.02505	.02365	.02229	.02097	.01970
25	6	13	.02786	.02657	.02527	.02399	.02273	.02150	.02029	.01912
25	6	14	.02651	.02536	.02420	.02305	.02191	.02079	.01969	.01861
25	6	15	.02528	.02427	.02324	.02221	.02118	.02016	.01915	.01816
25	6	16	.02416	.02327	.02236	.02144	.02051	.01958	.01867	.01776

Covariances of Inverse Gaussian Order Statistics
Shape Parameter k

n	i	j	0.8	0.9	1.0	1.1	1.2	1.3	1.4	1.5
25	6	17	.02313	.02235	.02155	.02073	.01990	.01906	.01823	.01740
25	6	18	.02217	.02149	.02079	.02007	.01933	.01859	.01783	.01708
25	6	19	.02125	.02068	.02008	.01945	.01880	.01814	.01747	.01679
25	6	20	.02038	.01991	.01940	.01887	.01831	.01773	.01713	.01653
25	6	21	.01953	.01916	.01875	.01831	.01783	.01734	.01682	.01629
25	6	22	.01869	.01842	.01810	.01775	.01737	.01696	.01653	.01607
25	6	23	.01782	.01766	.01745	.01720	.01691	.01660	.01625	.01588
25	6	24	.01688	.01683	.01675	.01661	.01644	.01624	.01599	.01572
25	6	25	.01571	.01583	.01590	.01593	.01592	.01586	.01576	.01563
25	7	8	.04259	.04006	.03760	.03522	.03292	.03071	.02861	.02660
25	7	9	.03947	.03728	.03512	.03303	.03099	.02903	.02714	.02534
25	7	10	.03687	.03495	.03305	.03120	.02938	.02762	.02592	.02428
25	7	11	.03466	.03297	.03129	.02963	.02801	.02642	.02488	.02339
25	7	12	.03274	.03125	.02976	.02828	.02682	.02539	.02399	.02263
25	7	13	.03105	.02973	.02841	.02709	.02578	.02449	.02321	.02197
25	7	14	.02954	.02838	.02721	.02603	.02485	.02368	.02252	.02139
25	7	15	.02817	.02716	.02613	.02508	.02402	.02296	.02191	.02087
25	7	16	.02693	.02605	.02514	.02421	.02327	.02231	.02136	.02041
25	7	17	.02578	.02502	.02423	.02341	.02257	.02172	.02086	.02000
25	7	18	.02471	.02406	.02338	.02267	.02193	.02118	.02041	.01963
25	7	19	.02369	.02316	.02259	.02198	.02134	.02067	.01999	.01930
25	7	20	.02272	.02230	.02183	.02132	.02077	.02020	.01961	.01900
25	7	21	.02178	.02146	.02109	.02068	.02023	.01976	.01925	.01873
25	7	22	.02084	.02062	.02036	.02006	.01971	.01933	.01892	.01848
25	7	23	.01987	.01977	.01963	.01943	.01919	.01892	.01860	.01826
25	7	24	.01882	.01885	.01884	.01877	.01866	.01851	.01831	.01808
25	7	25	.01751	.01773	.01789	.01800	.01806	.01808	.01804	.01797
25	8	9	.04374	.04148	.03924	.03705	.03491	.03283	.03083	.02889
25	8	10	.04087	.03890	.03694	.03501	.03311	.03125	.02944	.02770
25	8	11	.03842	.03670	.03498	.03326	.03157	.02990	.02827	.02669
25	8	12	.03630	.03479	.03327	.03175	.03023	.02874	.02726	.02582
25	8	13	.03443	.03311	.03177	.03042	.02906	.02772	.02638	.02507
25	8	14	.03276	.03161	.03043	.02923	.02802	.02681	.02560	.02441
25	8	15	.03125	.03025	.02922	.02816	.02709	.02600	.02491	.02382
25	8	16	.02988	.02902	.02812	.02719	.02624	.02527	.02428	.02330
25	8	17	.02860	.02788	.02710	.02630	.02546	.02460	.02372	.02283
25	8	18	.02741	.02681	.02616	.02546	.02474	.02398	.02321	.02241
25	8	19	.02629	.02580	.02527	.02469	.02407	.02341	.02273	.02203
25	8	20	.02522	.02484	.02442	.02395	.02343	.02288	.02230	.02169
25	8	21	.02417	.02391	.02360	.02323	.02282	.02238	.02189	.02138
25	8	22	.02312	.02298	.02278	.02253	.02224	.02189	.02151	.02110
25	8	23	.02205	.02204	.02196	.02183	.02165	.02143	.02116	.02085
25	8	24	.02089	.02101	.02108	.02109	.02105	.02096	.02082	.02064
25	8	25	.01944	.01976	.02002	.02023	.02038	.02047	.02052	.02051
25	9	10	.04518	.04316	.04115	.03915	.03716	.03522	.03331	.03145
25	9	11	.04248	.04073	.03897	.03720	.03544	.03370	.03199	.03031
25	9	12	.04014	.03862	.03708	.03552	.03395	.03239	.03085	.02933
25	9	13	.03808	.03676	.03541	.03403	.03264	.03125	.02986	.02848
25	9	14	.03624	.03510	.03392	.03271	.03148	.03023	.02898	.02773
25	9	15	.03457	.03360	.03258	.03152	.03043	.02932	.02820	.02707
25	9	16	.03305	.03223	.03135	.03043	.02948	.02849	.02749	.02648
25	9	17	.03165	.03096	.03022	.02943	.02860	.02774	.02686	.02595
25	9	18	.03033	.02978	.02917	.02850	.02780	.02705	.02627	.02547
25	9	19	.02909	.02866	.02818	.02763	.02704	.02641	.02574	.02504
25	9	20	.02790	.02760	.02723	.02681	.02633	.02581	.02525	.02465

Covariances of Inverse Gaussian Order Statistics
Shape Parameter k

n	i	j	0.8	0.9	1.0	1.1	1.2	1.3	1.4	1.5
25	9	21	.02675	.02656	.02632	.02601	.02565	.02524	.02479	.02430
25	9	22	.02559	.02553	.02541	.02523	.02499	.02470	.02436	.02398
25	9	23	.02441	.02448	.02449	.02444	.02434	.02417	.02396	.02370
25	9	24	.02312	.02335	.02351	.02362	.02366	.02365	.02358	.02346
25	9	25	.02152	.02196	.02234	.02265	.02291	.02310	.02324	.02332
25	10	11	.04690	.04514	.04334	.04153	.03972	.03790	.03611	.03434
25	10	12	.04433	.04281	.04125	.03966	.03805	.03644	.03483	.03323
25	10	13	.04206	.04075	.03940	.03801	.03659	.03515	.03371	.03227
25	10	14	.04003	.03891	.03775	.03653	.03529	.03401	.03272	.03143
25	10	15	.03820	.03726	.03626	.03521	.03411	.03299	.03184	.03068
25	10	16	.03652	.03574	.03489	.03400	.03305	.03207	.03105	.03001
25	10	17	.03497	.03434	.03364	.03288	.03207	.03122	.03033	.02942
25	10	18	.03352	.03303	.03247	.03185	.03117	.03045	.02968	.02888
25	10	19	.03216	.03180	.03137	.03088	.03033	.02973	.02908	.02839
25	10	20	.03084	.03062	.03032	.02996	.02953	.02905	.02852	.02795
25	10	21	.02957	.02947	.02930	.02907	.02877	.02842	.02801	.02755
25	10	22	.02829	.02833	.02830	.02820	.02803	.02781	.02753	.02719
25	10	23	.02699	.02717	.02728	.02732	.02730	.02722	.02707	.02688
25	10	24	.02556	.02591	.02619	.02640	.02654	.02663	.02665	.02661
25	10	25	.02379	.02437	.02488	.02532	.02570	.02601	.02626	.02644
25	11	12	.04895	.04744	.04587	.04426	.04262	.04096	.03929	.03762
25	11	13	.04645	.04517	.04382	.04243	.04099	.03952	.03803	.03654
25	11	14	.04422	.04314	.04199	.04079	.03953	.03824	.03692	.03559
25	11	15	.04220	.04131	.04034	.03931	.03823	.03710	.03594	.03474
25	11	16	.04035	.03963	.03883	.03796	.03704	.03606	.03505	.03399
25	11	17	.03865	.03808	.03744	.03672	.03595	.03512	.03424	.03332
25	11	18	.03705	.03663	.03614	.03557	.03494	.03425	.03350	.03271
25	11	19	.03554	.03527	.03492	.03449	.03400	.03344	.03283	.03216
25	11	20	.03409	.03396	.03375	.03346	.03311	.03268	.03220	.03167
25	11	21	.03268	.03269	.03262	.03247	.03226	.03197	.03162	.03122
25	11	22	.03128	.03143	.03150	.03150	.03143	.03129	.03108	.03081
25	11	23	.02983	.03014	.03037	.03053	.03061	.03062	.03057	.03045
25	11	24	.02826	.02875	.02916	.02950	.02976	.02996	.03009	.03015
25	11	25	.02631	.02704	.02770	.02829	.02882	.02927	.02965	.02996
25	12	13	.05136	.05011	.04879	.04740	.04595	.04446	.04293	.04139
25	12	14	.04890	.04787	.04676	.04557	.04433	.04303	.04169	.04032
25	12	15	.04667	.04584	.04492	.04393	.04287	.04175	.04058	.03937
25	12	16	.04464	.04399	.04325	.04243	.04154	.04059	.03958	.03852
25	12	17	.04275	.04227	.04170	.04105	.04032	.03953	.03867	.03776
25	12	18	.04099	.04067	.04026	.03977	.03919	.03855	.03784	.03707
25	12	19	.03933	.03916	.03890	.03856	.03814	.03764	.03708	.03645
25	12	20	.03773	.03771	.03761	.03741	.03714	.03680	.03638	.03589
25	12	21	.03617	.03630	.03635	.03631	.03619	.03599	.03572	.03538
25	12	22	.03462	.03491	.03511	.03523	.03527	.03523	.03511	.03493
25	12	23	.03302	.03348	.03385	.03414	.03435	.03448	.03454	.03452
25	12	24	.03129	.03193	.03250	.03299	.03340	.03373	.03399	.03418
25	12	25	.02913	.03003	.03087	.03164	.03234	.03296	.03350	.03397
25	13	14	.05419	.05323	.05217	.05103	.04980	.04851	.04715	.04576
25	13	15	.05173	.05099	.05014	.04919	.04817	.04707	.04590	.04469
25	13	16	.04948	.04893	.04827	.04752	.04668	.04576	.04478	.04373
25	13	17	.04740	.04703	.04655	.04598	.04532	.04457	.04376	.04287
25	13	18	.04546	.04525	.04495	.04455	.04406	.04348	.04282	.04210
25	13	19	.04361	.04358	.04344	.04320	.04287	.04246	.04197	.04140
25	13	20	.04184	.04197	.04199	.04192	.04176	.04151	.04117	.04076
25	13	21	.04012	.04041	.04059	.04069	.04069	.04060	.04044	.04019

Covariances of Inverse Gaussian Order Statistics
Shape Parameter k

n	i	j	0.8	0.9	1.0	1.1	1.2	1.3	1.4	1.5
25	13	22	.03840	.03885	.03921	.03948	.03965	.03974	.03975	.03967
25	13	23	.03664	.03726	.03781	.03826	.03862	.03890	.03910	.03921
25	13	24	.03471	.03555	.03630	.03697	.03756	.03806	.03848	.03882
25	13	25	.03232	.03344	.03449	.03547	.03637	.03719	.03793	.03859
25	14	15	.05754	.05690	.05614	.05527	.05430	.05324	.05210	.05089
25	14	16	.05504	.05461	.05406	.05340	.05263	.05177	.05083	.04981
25	14	17	.05274	.05250	.05214	.05168	.05110	.05043	.04967	.04883
25	14	18	.05058	.05052	.05035	.05007	.04969	.04920	.04862	.04796
25	14	19	.04853	.04866	.04867	.04856	.04836	.04805	.04765	.04716
25	14	20	.04657	.04687	.04705	.04713	.04710	.04698	.04676	.04644
25	14	21	.04466	.04513	.04549	.04575	.04590	.04596	.04592	.04579
25	14	22	.04275	.04340	.04394	.04439	.04474	.04499	.04514	.04520
25	14	23	.04079	.04163	.04237	.04302	.04358	.04404	.04441	.04468
25	14	24	.03865	.03971	.04069	.04158	.04238	.04309	.04372	.04425
25	14	25	.03599	.03736	.03866	.03989	.04104	.04211	.04309	.04398
25	15	16	.06152	.06124	.06083	.06029	.05963	.05885	.05797	.05700
25	15	17	.05895	.05888	.05868	.05836	.05790	.05734	.05667	.05590
25	15	18	.05655	.05668	.05668	.05655	.05631	.05594	.05547	.05490
25	15	19	.05427	.05459	.05479	.05486	.05481	.05465	.05437	.05400
25	15	20	.05208	.05259	.05298	.05324	.05339	.05343	.05335	.05318
25	15	21	.04995	.05064	.05122	.05169	.05204	.05228	.05241	.05243
25	15	22	.04782	.04871	.04949	.05016	.05072	.05117	.05152	.05176
25	15	23	.04563	.04673	.04772	.04862	.04941	.05010	.05069	.05117
25	15	24	.04324	.04458	.04583	.04699	.04806	.04903	.04990	.05067
25	15	25	.04027	.04195	.04356	.04509	.04654	.04791	.04919	.05037
25	16	17	.06631	.06647	.06647	.06632	.06604	.06562	.06507	.06440
25	16	18	.06362	.06399	.06421	.06429	.06422	.06403	.06371	.06326
25	16	19	.06107	.06164	.06207	.06237	.06253	.06255	.06245	.06223
25	16	20	.05862	.05939	.06003	.06054	.06092	.06117	.06129	.06129
25	16	21	.05622	.05720	.05805	.05878	.05938	.05986	.06021	.06044
25	16	22	.05384	.05502	.05609	.05705	.05788	.05860	.05920	.05968
25	16	23	.05138	.05279	.05410	.05530	.05640	.05738	.05824	.05900
25	16	24	.04870	.05037	.05196	.05346	.05486	.05615	.05734	.05843
25	16	25	.04536	.04741	.04939	.05130	.05313	.05488	.05653	.05808
25	17	18	.07218	.07285	.07336	.07371	.07389	.07393	.07381	.07355
25	17	19	.06929	.07019	.07093	.07152	.07195	.07224	.07237	.07236
25	17	20	.06653	.06764	.06861	.06944	.07012	.07065	.07103	.07128
25	17	21	.06382	.06516	.06636	.06743	.06835	.06914	.06979	.07030
25	17	22	.06112	.06269	.06413	.06545	.06664	.06770	.06862	.06942
25	17	23	.05834	.06015	.06186	.06346	.06494	.06629	.06753	.06863
25	17	24	.05531	.05741	.05943	.06135	.06317	.06489	.06649	.06798
25	17	25	.05152	.05404	.05649	.05888	.06119	.06342	.06555	.06758
25	18	19	.07950	.08082	.08197	.08295	.08376	.08439	.08486	.08515
25	18	20	.07634	.07790	.07931	.08056	.08164	.08255	.08330	.08389
25	18	21	.07325	.07506	.07672	.07824	.07960	.08081	.08186	.08275
25	18	22	.07017	.07223	.07416	.07596	.07762	.07913	.08050	.08172
25	18	23	.06699	.06932	.07155	.07366	.07564	.07750	.07922	.08081
25	18	24	.06352	.06617	.06875	.07123	.07360	.07587	.07802	.08004
25	18	25	.05919	.06230	.06536	.06837	.07131	.07416	.07692	.07959
25	19	20	.08889	.09107	.09307	.09489	.09653	.09799	.09925	.10033
25	19	21	.08532	.08777	.09006	.09219	.09415	.09593	.09755	.09898
25	19	22	.08175	.08448	.08707	.08952	.09182	.09397	.09595	.09777
25	19	23	.07807	.08110	.08402	.08683	.08950	.09204	.09444	.09669
25	19	24	.07404	.07743	.08075	.08398	.08711	.09012	.09302	.09578
25	19	25	.06902	.07292	.07680	.08063	.08441	.08811	.09173	.09525

Covariances of Inverse Gaussian Order Statistics
Shape Parameter k

n	i	j	0.8	0.9	1.0	1.1	1.2	1.3	1.4	1.5
25	20	21	.10140	.10474	.10792	.11093	.11375	.11638	.11881	.12104
25	20	22	.09719	.10085	.10437	.10775	.11097	.11402	.11689	.11957
25	20	23	.09285	.09685	.10075	.10454	.10820	.11171	.11508	.11828
25	20	24	.08809	.09250	.09686	.10114	.10532	.10940	.11336	.11719
25	20	25	.08214	.08714	.09215	.09714	.10209	.10699	.11182	.11656
25	21	22	.11892	.12397	.12888	.13365	.13826	.14268	.14692	.15095
25	21	23	.11366	.11910	.12446	.12972	.13485	.13985	.14468	.14935
25	21	24	.10789	.11381	.11970	.12555	.13132	.13700	.14257	.14801
25	21	25	.10066	.10727	.11394	.12063	.12733	.13402	.14066	.14724
25	22	23	.14536	.15312	.16085	.16850	.17606	.18349	.19077	.19789
25	22	24	.13808	.14642	.15479	.16317	.17153	.17984	.18806	.19619
25	22	25	.12894	.13811	.14744	.15689	.16642	.17601	.18562	.19523
25	23	24	.19017	.20294	.21590	.22900	.24219	.25543	.26868	.28191
25	23	25	.17782	.19167	.20588	.22040	.23518	.25018	.26536	.28067
25	24	25	.28370	.30846	.33412	.36062	.38788	.41584	.44442	.47357

Covariances of Inverse Gaussian Order Statistics
Shape Parameter k

n	i	j	1.6	1.7	1.8	1.9	2.0	2.1	2.2	2.3
2	1	2	.27318	.26858	.26392	.25922	.25450	.24978	.24506	.24036
3	1	2	.14713	.14043	.13399	.12780	.12187	.11619	.11077	.10558
3	1	3	.13278	.12960	.12641	.12321	.12002	.11684	.11368	.11057
3	2	3	.34692	.34758	.34787	.34781	.34741	.34670	.34570	.34444
4	1	2	.09837	.09217	.08634	.08085	.07570	.07087	.06634	.06211
4	1	3	.08685	.08283	.07895	.07521	.07161	.06816	.06486	.06169
4	1	4	.08112	.07885	.07657	.07430	.07204	.06981	.06760	.06543
4	2	3	.18921	.18449	.17972	.17493	.17014	.16536	.16061	.15591
4	2	4	.17778	.17656	.17513	.17353	.17175	.16983	.16777	.16559
4	3	4	.38538	.39040	.39495	.39903	.40266	.40585	.40861	.41096
5	1	2	.07339	.06787	.06275	.05800	.05361	.04955	.04579	.04233
5	1	3	.06387	.05999	.05632	.05284	.04956	.04647	.04357	.04083
5	1	4	.05878	.05601	.05334	.05076	.04829	.04591	.04363	.04144
5	1	5	.05596	.05423	.05251	.05080	.04910	.04743	.04579	.04418
5	2	3	.12623	.12092	.11573	.11068	.10578	.10103	.09644	.09201
5	2	4	.11673	.11342	.11009	.10676	.10343	.10013	.09685	.09362
5	2	5	.11145	.11009	.10861	.10703	.10534	.10357	.10174	.09984
5	3	4	.21290	.21030	.20749	.20449	.20133	.19802	.19459	.19106
5	3	5	.20403	.20480	.20528	.20549	.20542	.20511	.20457	.20381
5	4	5	.40959	.41808	.42609	.43360	.44062	.44716	.45322	.45881
6	1	2	.05844	.05351	.04899	.04484	.04104	.03756	.03439	.03149
6	1	3	.05032	.04673	.04337	.04024	.03733	.03462	.03210	.02976
6	1	4	.04588	.04314	.04054	.03808	.03575	.03355	.03147	.02952
6	1	5	.04316	.04111	.03912	.03720	.03535	.03358	.03189	.03026
6	1	6	.04159	.04022	.03886	.03751	.03618	.03487	.03358	.03233
6	2	3	.09347	.08839	.08352	.07885	.07440	.07016	.06613	.06230
6	2	4	.08560	.08195	.07837	.07489	.07150	.06821	.06503	.06197
6	2	5	.08074	.07827	.07578	.07330	.07084	.06839	.06598	.06360
6	2	6	.07792	.07669	.07537	.07398	.07253	.07104	.06951	.06794
6	3	4	.14258	.13848	.13435	.13021	.12608	.12197	.11789	.11386
6	3	5	.13489	.13263	.13024	.12775	.12516	.12250	.11977	.11700
6	3	6	.13043	.13017	.12971	.12908	.12827	.12731	.12622	.12499
6	4	5	.22863	.22791	.22689	.22558	.22401	.22220	.22017	.21794
6	4	6	.22167	.22420	.22640	.22828	.22984	.23110	.23207	.23276
6	5	6	.42646	.43776	.44860	.45897	.46885	.47825	.48715	.49557
7	1	2	.04859	.04414	.04009	.03641	.03307	.03004	.02729	.02480
7	1	3	.04146	.03816	.03511	.03230	.02970	.02731	.02510	.02308
7	1	4	.03754	.03495	.03252	.03024	.02811	.02612	.02427	.02254
7	1	5	.03508	.03301	.03104	.02917	.02740	.02573	.02415	.02266
7	1	6	.03346	.03185	.03029	.02879	.02734	.02595	.02463	.02336
7	1	7	.03251	.03139	.03027	.02917	.02808	.02702	.02598	.02496
7	2	3	.07376	.06906	.06460	.06038	.05641	.05267	.04916	.04587
7	2	4	.06705	.06348	.06003	.05673	.05357	.05054	.04766	.04492
7	2	5	.06280	.06009	.05743	.05484	.05232	.04987	.04751	.04522
7	2	6	.05999	.05805	.05611	.05417	.05225	.05035	.04848	.04664
7	2	7	.05834	.05725	.05611	.05492	.05369	.05243	.05114	.04985
7	3	4	.10554	.10123	.09699	.09284	.08878	.08482	.08098	.07726
7	3	5	.09910	.09606	.09300	.08993	.08688	.08384	.08084	.07789
7	3	6	.09482	.09294	.09097	.08894	.08685	.08472	.08255	.08037
7	3	7	.09232	.09175	.09105	.09022	.08928	.08824	.08710	.08588
7	4	5	.15383	.15089	.14783	.14465	.14139	.13805	.13465	.13121
7	4	6	.14750	.14628	.14486	.14328	.14153	.13963	.13761	.13547
7	4	7	.14380	.14458	.14513	.14545	.14556	.14548	.14520	.14474
7	5	6	.24002	.24092	.24147	.24168	.24158	.24117	.24048	.23952
7	5	7	.23449	.23854	.24225	.24561	.24864	.25135	.25373	.25579

Covariances of Inverse Gaussian Order Statistics
Shape Parameter k

n	i	j	1.6	1.7	1.8	1.9	2.0	2.1	2.2	2.3
7	6	7	.43899	.45262	.46584	.47861	.49093	.50279	.51417	.52507
8	1	2	.04163	.03757	.03391	.03060	.02762	.02493	.02251	.02033
8	1	3	.03526	.03223	.02944	.02689	.02455	.02241	.02047	.01869
8	1	4	.03175	.02932	.02706	.02496	.02302	.02123	.01957	.01804
8	1	5	.02952	.02752	.02565	.02388	.02223	.02069	.01925	.01790
8	1	6	.02800	.02636	.02480	.02331	.02191	.02058	.01932	.01813
8	1	7	.02696	.02564	.02438	.02315	.02198	.02085	.01977	.01875
8	1	8	.02635	.02540	.02446	.02354	.02263	.02174	.02088	.02003
8	2	3	.06074	.05641	.05234	.04853	.04498	.04167	.03858	.03572
8	2	4	.05489	.05150	.04827	.04521	.04232	.03958	.03700	.03458
8	2	5	.05114	.04844	.04584	.04334	.04094	.03864	.03645	.03436
8	2	6	.04858	.04645	.04438	.04235	.04038	.03846	.03662	.03483
8	2	7	.04681	.04523	.04365	.04208	.04053	.03900	.03749	.03602
8	2	8	.04578	.04482	.04383	.04280	.04174	.04067	.03958	.03849
8	3	4	.08309	.07891	.07486	.07095	.06719	.06357	.06010	.05679
8	3	5	.07759	.07438	.07123	.06814	.06511	.06216	.05930	.05652
8	3	6	.07381	.07143	.06904	.06666	.06429	.06194	.05962	.05733
8	3	7	.07119	.06961	.06797	.06629	.06457	.06283	.06107	.05930
8	3	8	.06967	.06902	.06828	.06744	.06652	.06553	.06447	.06336
8	4	5	.11406	.11053	.10698	.10343	.09988	.09636	.09287	.08943
8	4	6	.10869	.10632	.10386	.10133	.09874	.09611	.09346	.09079
8	4	7	.10496	.10372	.10235	.10085	.09925	.09755	.09578	.09393
8	4	8	.10281	.10292	.10200	.10264	.10227	.10175	.10111	.10035
8	5	6	.16220	.16032	.15824	.15599	.15358	.15102	.14835	.14557
8	5	7	.15689	.15663	.15614	.15543	.15451	.15340	.15211	.15065
8	5	8	.15384	.15557	.15704	.15827	.15926	.16002	.16056	.16089
8	6	7	.24871	.25102	.25295	.25452	.25574	.25662	.25718	.25742
8	6	8	.24429	.24966	.25468	.25937	.26372	.26773	.27140	.27474
8	7	8	.44873	.46433	.47957	.49442	.50885	.52285	.53640	.54948
9	1	2	.03648	.03274	.02939	.02638	.02368	.02127	.01910	.01717
9	1	3	.03070	.02789	.02532	.02299	.02087	.01894	.01720	.01561
9	1	4	.02751	.02523	.02314	.02120	.01943	.01780	.01630	.01493
9	1	5	.02547	.02357	.02180	.02016	.01863	.01721	.01589	.01468
9	1	6	.02406	.02246	.02095	.01953	.01820	.01696	.01579	.01470
9	1	7	.02306	.02171	.02043	.01921	.01806	.01696	.01593	.01496
9	1	8	.02235	.02125	.02019	.01917	.01819	.01725	.01635	.01549
9	1	9	.02194	.02112	.02032	.01953	.01875	.01800	.01726	.01655
9	2	3	.05157	.04756	.04383	.04037	.03717	.03421	.03147	.02895
9	2	4	.04636	.04318	.04018	.03736	.03471	.03223	.02992	.02776
9	2	5	.04301	.04041	.03793	.03557	.03333	.03122	.02922	.02733
9	2	6	.04069	.03855	.03649	.03451	.03261	.03079	.02905	.02739
9	2	7	.03902	.03730	.03561	.03397	.03237	.03082	.02932	.02788
9	2	8	.03784	.03652	.03521	.03390	.03261	.03134	.03009	.02888
9	2	9	.03717	.03632	.03545	.03455	.03363	.03271	.03178	.03084
9	3	4	.06819	.06424	.06045	.05683	.05339	.05011	.04701	.04407
9	3	5	.06339	.06024	.05718	.05421	.05136	.04861	.04597	.04344
9	3	6	.06005	.05754	.05508	.05266	.05029	.04799	.04575	.04358
9	3	7	.05764	.05572	.05379	.05186	.04995	.04806	.04620	.04437
9	3	8	.05593	.05459	.05320	.05178	.05034	.04889	.04743	.04596
9	3	9	.05496	.05431	.05358	.05279	.05193	.05103	.05007	.04909
9	4	5	.08979	.08617	.08260	.07909	.07565	.07228	.06901	.06583
9	4	6	.08518	.08243	.07967	.07691	.07416	.07144	.06874	.06608
9	4	7	.08185	.07989	.07788	.07582	.07372	.07160	.06946	.06732
9	4	8	.07948	.07833	.07707	.07574	.07432	.07285	.07132	.06974
9	4	9	.07814	.07796	.07765	.07722	.07668	.07604	.07530	.07447

Covariances of Inverse Gaussian Order Statistics
Shape Parameter k

n	i	j	1.6	1.7	1.8	1.9	2.0	2.1	2.2	2.3
9	5	6	.12053	.11774	.11486	.11191	.10891	.10586	.10279	.09971
9	5	7	.11597	.11425	.11240	.11042	.10835	.10618	.10394	.10163
9	5	8	.11272	.11210	.11132	.11038	.10929	.10808	.10674	.10530
9	5	9	.11089	.11164	.11220	.11258	.11278	.11281	.11269	.11241
9	6	7	.16873	.16780	.16663	.16523	.16363	.16183	.15987	.15775
9	6	8	.16422	.16483	.16519	.16530	.16516	.16481	.16424	.16347
9	6	9	.16171	.16429	.16660	.16866	.17047	.17203	.17336	.17444
9	7	8	.25560	.25914	.26230	.26508	.26749	.26955	.27125	.27261
9	7	9	.25206	.25858	.26477	.27064	.27616	.28135	.28620	.29071
9	8	9	.45656	.47386	.49085	.50750	.52377	.53966	.55513	.57016
10	1	2	.03251	.02904	.02595	.02318	.02072	.01852	.01657	.01483
10	1	3	.02721	.02459	.02221	.02006	.01812	.01637	.01479	.01336
10	1	4	.02428	.02214	.02019	.01840	.01677	.01528	.01392	.01268
10	1	5	.02240	.02060	.01894	.01740	.01598	.01468	.01348	.01237
10	1	6	.02110	.01955	.01812	.01677	.01552	.01436	.01329	.01229
10	1	7	.02015	.01882	.01757	.01639	.01529	.01425	.01328	.01238
10	1	8	.01944	.01831	.01723	.01621	.01524	.01432	.01345	.01263
10	1	9	.01894	.01800	.01709	.01622	.01539	.01459	.01382	.01309
10	1	10	.01866	.01795	.01725	.01656	.01588	.01523	.01459	.01397
10	2	3	.04478	.04106	.03763	.03446	.03155	.02887	.02642	.02417
10	2	4	.04008	.03710	.03431	.03170	.02928	.02702	.02493	.02300
10	2	5	.03705	.03458	.03224	.03003	.02795	.02600	.02418	.02247
10	2	6	.03494	.03286	.03087	.02898	.02718	.02547	.02386	.02234
10	2	7	.03339	.03165	.02996	.02834	.02678	.02529	.02387	.02251
10	2	8	.03225	.03081	.02940	.02803	.02670	.02541	.02417	.02297
10	2	9	.03142	.03030	.02918	.02807	.02697	.02589	.02484	.02381
10	2	10	.03096	.03021	.02944	.02865	.02785	.02704	.02623	.02542
10	3	4	.05767	.05395	.05043	.04709	.04394	.04097	.03817	.03555
10	3	5	.05341	.05037	.04746	.04468	.04201	.03948	.03707	.03478
10	3	6	.05042	.04792	.04550	.04316	.04089	.03871	.03661	.03460
10	3	7	.04823	.04620	.04420	.04224	.04032	.03845	.03664	.03488
10	3	8	.04660	.04500	.04340	.04180	.04022	.03866	.03712	.03561
10	3	9	.04543	.04426	.04307	.04186	.04063	.03939	.03815	.03692
10	3	10	.04478	.04416	.04347	.04274	.04195	.04113	.04028	.03940
10	4	5	.07361	.07008	.06664	.06330	.06007	.05695	.05394	.05105
10	4	6	.06958	.06676	.06397	.06122	.05852	.05589	.05333	.05083
10	4	7	.06662	.06441	.06218	.05996	.05775	.05556	.05340	.05127
10	4	8	.06442	.06278	.06109	.05937	.05763	.05587	.05411	.05235
10	4	9	.06283	.06178	.06066	.05947	.05823	.05695	.05563	.05428
10	4	10	.06195	.06165	.06123	.06073	.06013	.05946	.05872	.05792
10	5	6	.09496	.09189	.08880	.08571	.08263	.07957	.07655	.07357
10	5	7	.09102	.08875	.08641	.08402	.08160	.07916	.07670	.07424
10	5	8	.08808	.08656	.08495	.08325	.08147	.07964	.07775	.07583
10	5	9	.08596	.08523	.08439	.08342	.08235	.08119	.07994	.07862
10	5	10	.08480	.08508	.08521	.08520	.08505	.08478	.08439	.08388
10	6	7	.12567	.12356	.12132	.11895	.11648	.11392	.11128	.10859
10	6	8	.12173	.12063	.11936	.11793	.11637	.11467	.11286	.11095
10	6	9	.11889	.11886	.11864	.11824	.11767	.11694	.11606	.11504
10	6	10	.11736	.11870	.11984	.12079	.12154	.12210	.12249	.12271
10	7	8	.17401	.17392	.17357	.17296	.17211	.17104	.16976	.16829
10	7	9	.17013	.17153	.17266	.17352	.17413	.17448	.17461	.17450
10	7	10	.16807	.17141	.17449	.17731	.17988	.18219	.18424	.18605
10	8	9	.26124	.26586	.27010	.27397	.27746	.28058	.28334	.28574
10	8	10	.25840	.26593	.27316	.28008	.28667	.29293	.29885	.30443
10	9	10	.46301	.48180	.50032	.51856	.53647	.55403	.57121	.58800

Covariances of Inverse Gaussian Order Statistics
Shape Parameter k

n	i	j	1.6	1.7	1.8	1.9	2.0	2.1	2.2	2.3
11	1	2	.02937	.02613	.02325	.02069	.01842	.01640	.01461	.01303
11	1	3	.02445	.02200	.01979	.01779	.01600	.01439	.01295	.01165
11	1	4	.02174	.01973	.01790	.01624	.01473	.01336	.01211	.01099
11	1	5	.02000	.01830	.01673	.01529	.01397	.01276	.01166	.01065
11	1	6	.01878	.01731	.01594	.01467	.01350	.01242	.01143	.01051
11	1	7	.01789	.01660	.01540	.01428	.01323	.01225	.01135	.01051
11	1	8	.01722	.01609	.01503	.01403	.01310	.01222	.01139	.01062
11	1	9	.01670	.01573	.01481	.01393	.01310	.01231	.01156	.01086
11	1	10	.01633	.01552	.01473	.01398	.01325	.01256	.01189	.01126
11	1	11	.01614	.01551	.01489	.01428	.01369	.01311	.01255	.01201
11	2	3	.03957	.03611	.03292	.03001	.02734	.02490	.02267	.02065
11	2	4	.03529	.03248	.02988	.02746	.02523	.02317	.02127	.01952
11	2	5	.03252	.03017	.02796	.02590	.02397	.02217	.02050	.01895
11	2	6	.03058	.02857	.02668	.02488	.02319	.02160	.02011	.01872
11	2	7	.02915	.02743	.02579	.02422	.02273	.02132	.01998	.01872
11	2	8	.02806	.02660	.02519	.02383	.02252	.02127	.02007	.01893
11	2	9	.02724	.02602	.02482	.02366	.02252	.02143	.02037	.01935
11	2	10	.02664	.02567	.02470	.02374	.02279	.02186	.02096	.02007
11	2	11	.02633	.02565	.02496	.02426	.02355	.02283	.02212	.02141
11	3	4	.04988	.04639	.04310	.04002	.03713	.03442	.03189	.02954
11	3	5	.04604	.04315	.04041	.03780	.03533	.03299	.03079	.02872
11	3	6	.04334	.04092	.03859	.03635	.03421	.03217	.03023	.02838
11	3	7	.04134	.03931	.03733	.03541	.03355	.03177	.03005	.02840
11	3	8	.03983	.03814	.03648	.03484	.03325	.03170	.03019	.02873
11	3	9	.03868	.03732	.03596	.03460	.03327	.03194	.03065	.02938
11	3	10	.03784	.03682	.03579	.03473	.03367	.03260	.03153	.03047
11	3	11	.03740	.03681	.03618	.03550	.03479	.03404	.03328	.03250
11	4	5	.06215	.05877	.05551	.05237	.04936	.04648	.04374	.04112
11	4	6	.05857	.05578	.05306	.05042	.04785	.04537	.04298	.04068
11	4	7	.05592	.05363	.05137	.04915	.04696	.04483	.04275	.04073
11	4	8	.05391	.05207	.05022	.04838	.04655	.04475	.04296	.04121
11	4	9	.05237	.05097	.04953	.04807	.04659	.04511	.04362	.04214
11	4	10	.05126	.05031	.04931	.04825	.04716	.04604	.04488	.04371
11	4	11	.05067	.05030	.04985	.04933	.04873	.04808	.04737	.04661
11	5	6	.07785	.07474	.07167	.06863	.06565	.06273	.05988	.05711
11	5	7	.07440	.07193	.06944	.06695	.06448	.06202	.05960	.05721
11	5	8	.07177	.06988	.06793	.06595	.06395	.06193	.05992	.05790
11	5	9	.06977	.06843	.06702	.06555	.06402	.06245	.06085	.05922
11	5	10	.06831	.06757	.06674	.06582	.06482	.06375	.06261	.06143
11	5	11	.06755	.06758	.06749	.06729	.06698	.06657	.06608	.06550
11	6	7	.09913	.09657	.09395	.09127	.08856	.08583	.08309	.08035
11	6	8	.09571	.09389	.09197	.08997	.08789	.08575	.08357	.08135
11	6	9	.09310	.09201	.09079	.08946	.08803	.08650	.08490	.08323
11	6	10	.09119	.09089	.09044	.08985	.08914	.08831	.08737	.08633
11	6	11	.09021	.09093	.09148	.09187	.09212	.09222	.09219	.09202
11	7	8	.12988	.12840	.12673	.12492	.12296	.12087	.11868	.11639
11	7	9	.12644	.12591	.12519	.12428	.12321	.12197	.12060	.11910
11	7	10	.12394	.12445	.12476	.12488	.12480	.12454	.12412	.12353
11	7	11	.12267	.12455	.12624	.12771	.12898	.13006	.13094	.13165
11	8	9	.17838	.17906	.17945	.17956	.17942	.17902	.17839	.17754
11	8	10	.17501	.17712	.17896	.18052	.18181	.18284	.18361	.18415
11	8	11	.17333	.17736	.18114	.18466	.18792	.19092	.19366	.19615
11	9	10	.26594	.27152	.27674	.28159	.28606	.29016	.29389	.29726
11	9	11	.26368	.27212	.28028	.28814	.29569	.30292	.30982	.31640
11	10	11	.46845	.48854	.50843	.52808	.54745	.56652	.58524	.60361

Covariances of Inverse Gaussian Order Statistics
Shape Parameter k

n	i	j	1.6	1.7	1.8	1.9	2.0	2.1	2.2	2.3
12	1	2	.02681	.02377	.02107	.01869	.01658	.01471	.01307	.01161
12	1	3	.02223	.01992	.01785	.01599	.01432	.01284	.01150	.01032
12	1	4	.01970	.01780	.01608	.01453	.01312	.01185	.01070	.00967
12	1	5	.01807	.01646	.01498	.01363	.01240	.01127	.01025	.00933
12	1	6	.01693	.01553	.01423	.01303	.01193	.01092	.01000	.00915
12	1	7	.01609	.01485	.01370	.01263	.01164	.01072	.00988	.00910
12	1	8	.01545	.01435	.01332	.01236	.01147	.01063	.00986	.00914
12	1	9	.01495	.01398	.01307	.01221	.01140	.01064	.00992	.00926
12	1	10	.01457	.01372	.01292	.01215	.01143	.01074	.01009	.00948
12	1	11	.01429	.01357	.01288	.01221	.01158	.01097	.01038	.00983
12	1	12	.01415	.01359	.01303	.01249	.01197	.01145	.01096	.01048
12	2	3	.03546	.03222	.02925	.02654	.02408	.02184	.01981	.01796
12	2	4	.03152	.02888	.02644	.02419	.02212	.02022	.01848	.01688
12	2	5	.02896	.02673	.02465	.02272	.02093	.01926	.01772	.01630
12	2	6	.02717	.02525	.02344	.02175	.02016	.01868	.01730	.01602
12	2	7	.02584	.02417	.02259	.02109	.01968	.01835	.01710	.01593
12	2	8	.02482	.02337	.02198	.02066	.01940	.01820	.01707	.01600
12	2	9	.02403	.02278	.02157	.02041	.01929	.01822	.01719	.01622
12	2	10	.02342	.02236	.02133	.02032	.01934	.01839	.01748	.01660
12	2	11	.02298	.02212	.02126	.02042	.01959	.01878	.01799	.01722
12	2	12	.02275	.02214	.02152	.02089	.02025	.01962	.01898	.01835
12	3	4	.04390	.04062	.03755	.03469	.03202	.02954	.02724	.02510
12	3	5	.04041	.03767	.03508	.03264	.03034	.02818	.02616	.02427
12	3	6	.03794	.03561	.03339	.03127	.02926	.02736	.02556	.02387
12	3	7	.03611	.03411	.03219	.03035	.02858	.02689	.02528	.02375
12	3	8	.03471	.03300	.03134	.02974	.02818	.02668	.02524	.02386
12	3	9	.03362	.03218	.03076	.02938	.02802	.02671	.02543	.02419
12	3	10	.03277	.03160	.03042	.02926	.02810	.02697	.02585	.02476
12	3	11	.03216	.03126	.03034	.02941	.02848	.02754	.02661	.02569
12	3	12	.03185	.03130	.03071	.03009	.02944	.02877	.02808	.02737
12	4	5	.05365	.05043	.04735	.04442	.04162	.03897	.03646	.03409
12	4	6	.05043	.04773	.04512	.04260	.04018	.03787	.03565	.03354
12	4	7	.04803	.04576	.04353	.04137	.03927	.03724	.03528	.03339
12	4	8	.04619	.04429	.04241	.04056	.03874	.03697	.03524	.03356
12	4	9	.04476	.04320	.04164	.04008	.03853	.03701	.03550	.03403
12	4	10	.04365	.04243	.04119	.03992	.03865	.03737	.03610	.03484
12	4	11	.04284	.04198	.04108	.04014	.03917	.03818	.03717	.03614
12	4	12	.04244	.04205	.04159	.04107	.04050	.03987	.03921	.03851
12	5	6	.06569	.06265	.05966	.05675	.05393	.05119	.04854	.04599
12	5	7	.06263	.06011	.05761	.05515	.05274	.05037	.04806	.04581
12	5	8	.06027	.05822	.05616	.05410	.05205	.05003	.04803	.04606
12	5	9	.05842	.05681	.05516	.05349	.05180	.05010	.04840	.04671
12	5	10	.05700	.05582	.05458	.05329	.05196	.05060	.04922	.04783
12	5	11	.05595	.05524	.05445	.05359	.05267	.05170	.05068	.04962
12	5	12	.05544	.05534	.05513	.05483	.05445	.05399	.05346	.05286
12	6	7	.08131	.07860	.07588	.07315	.07044	.06774	.06508	.06246
12	6	8	.07831	.07618	.07401	.07180	.06956	.06732	.06507	.06283
12	6	9	.07596	.07438	.07273	.07101	.06924	.06743	.06559	.06373
12	6	10	.07413	.07311	.07198	.07077	.06949	.06813	.06672	.06526
12	6	11	.07280	.07238	.07184	.07119	.07044	.06961	.06869	.06770
12	6	12	.07215	.07252	.07275	.07285	.07283	.07270	.07246	.07212
12	7	8	.10259	.10051	.09832	.09605	.09370	.09130	.08885	.08638
12	7	9	.09958	.09820	.09668	.09505	.09332	.09150	.08960	.08764
12	7	10	.09724	.09656	.09573	.09476	.09367	.09247	.09116	.08975
12	7	11	.09554	.09563	.09557	.09535	.09498	.09448	.09385	.09311

Covariances of Inverse Gaussian Order Statistics
Shape Parameter k

n	i	j	1.6	1.7	1.8	1.9	2.0	2.1	2.2	2.3
12	7	12	.09472	.09584	.09679	.09758	.09820	.09867	.09899	.09916
12	8	9	.13341	.13249	.13137	.13007	.12860	.12698	.12521	.12332
12	8	10	.13038	.13037	.13016	.12974	.12914	.12836	.12741	.12631
12	8	11	.12816	.12918	.12999	.13058	.13098	.13118	.13119	.13103
12	8	12	.12712	.12951	.13168	.13366	.13542	.13698	.13834	.13950
12	9	10	.18207	.18343	.18450	.18529	.18579	.18603	.18602	.18576
12	9	11	.17912	.18188	.18436	.18656	.18849	.19015	.19154	.19268
12	9	12	.17777	.18242	.18683	.19099	.19489	.19854	.20192	.20505
12	10	11	.26993	.27638	.28248	.28821	.29358	.29858	.30321	.30747
12	10	12	.26816	.27741	.28641	.29512	.30355	.31167	.31947	.32696
12	11	12	.47310	.49436	.51547	.53639	.55708	.57751	.59763	.61744
13	1	2	.02470	.02183	.01929	.01705	.01508	.01335	.01182	.01047
13	1	3	.02040	.01821	.01626	.01452	.01296	.01158	.01034	.00925
13	1	4	.01802	.01623	.01460	.01314	.01183	.01064	.00958	.00862
13	1	5	.01649	.01496	.01356	.01229	.01113	.01008	.00914	.00828
13	1	6	.01542	.01408	.01284	.01171	.01068	.00973	.00887	.00809
13	1	7	.01463	.01343	.01233	.01132	.01038	.00952	.00873	.00800
13	1	8	.01402	.01295	.01196	.01104	.01019	.00940	.00867	.00799
13	1	9	.01354	.01259	.01169	.01086	.01008	.00935	.00868	.00805
13	1	10	.01316	.01231	.01151	.01076	.01005	.00938	.00876	.00817
13	1	11	.01286	.01212	.01141	.01073	.01009	.00948	.00891	.00837
13	1	12	.01265	.01201	.01139	.01080	.01023	.00969	.00918	.00868
13	1	13	.01255	.01204	.01154	.01106	.01058	.01012	.00968	.00925
13	2	3	.03214	.02908	.02630	.02378	.02149	.01942	.01755	.01586
13	2	4	.02848	.02598	.02369	.02158	.01966	.01790	.01629	.01483
13	2	5	.02611	.02399	.02202	.02021	.01853	.01698	.01556	.01425
13	2	6	.02443	.02260	.02088	.01928	.01779	.01641	.01513	.01394
13	2	7	.02319	.02158	.02007	.01864	.01731	.01606	.01489	.01380
13	2	8	.02224	.02082	.01947	.01820	.01699	.01586	.01479	.01379
13	2	9	.02149	.02024	.01905	.01791	.01682	.01579	.01481	.01389
13	2	10	.02089	.01980	.01875	.01774	.01677	.01584	.01495	.01410
13	2	11	.02043	.01950	.01859	.01770	.01684	.01601	.01521	.01444
13	2	12	.02009	.01932	.01857	.01782	.01709	.01637	.01567	.01499
13	2	13	.01993	.01938	.01881	.01824	.01767	.01709	.01652	.01596
13	3	4	.03918	.03609	.03322	.03055	.02807	.02579	.02367	.02173
13	3	5	.03597	.03337	.03093	.02864	.02650	.02450	.02264	.02091
13	3	6	.03370	.03147	.02936	.02736	.02547	.02369	.02203	.02047
13	3	7	.03201	.03008	.02823	.02646	.02479	.02320	.02169	.02027
13	3	8	.03071	.02903	.02740	.02584	.02435	.02292	.02156	.02026
13	3	9	.02968	.02823	.02681	.02544	.02411	.02283	.02159	.02041
13	3	10	.02887	.02763	.02641	.02521	.02404	.02290	.02180	.02073
13	3	11	.02824	.02720	.02618	.02516	.02415	.02316	.02218	.02124
13	3	12	.02777	.02697	.02615	.02533	.02450	.02367	.02285	.02203
13	3	13	.02756	.02704	.02650	.02593	.02533	.02472	.02409	.02346
13	4	5	.04712	.04407	.04116	.03841	.03581	.03336	.03106	.02890
13	4	6	.04419	.04160	.03910	.03672	.03445	.03229	.03024	.02830
13	4	7	.04200	.03978	.03763	.03555	.03355	.03163	.02980	.02805
13	4	8	.04032	.03841	.03655	.03473	.03297	.03127	.02962	.02804
13	4	9	.03898	.03737	.03577	.03420	.03265	.03115	.02968	.02826
13	4	10	.03793	.03658	.03524	.03390	.03257	.03125	.02996	.02870
13	4	11	.03710	.03603	.03494	.03383	.03272	.03161	.03050	.02940
13	4	12	.03650	.03572	.03491	.03406	.03319	.03231	.03141	.03050
13	4	13	.03622	.03582	.03537	.03487	.03433	.03374	.03312	.03248
13	5	6	.05666	.05371	.05086	.04810	.04544	.04288	.04043	.03809
13	5	7	.05390	.05141	.04897	.04659	.04428	.04203	.03986	.03777

Covariances of Inverse Gaussian Order Statistics
Shape Parameter k

n	i	j	1.6	1.7	1.8	1.9	2.0	2.1	2.2	2.3
13	5	8	.05176	.04966	.04759	.04554	.04353	.04157	.03964	.03778
13	5	9	.05007	.04834	.04659	.04485	.04313	.04142	.03973	.03808
13	5	10	.04873	.04734	.04591	.04447	.04302	.04157	.04011	.03867
13	5	11	.04768	.04663	.04553	.04440	.04323	.04204	.04084	.03962
13	5	12	.04691	.04624	.04550	.04471	.04386	.04298	.04206	.04111
13	5	13	.04656	.04638	.04611	.04577	.04536	.04488	.04435	.04376
13	6	7	.06861	.06589	.06318	.06051	.05788	.05531	.05279	.05034
13	6	8	.06594	.06369	.06144	.05919	.05694	.05472	.05253	.05037
13	6	9	.06382	.06202	.06018	.05831	.05643	.05454	.05266	.05078
13	6	10	.06214	.06076	.05932	.05783	.05631	.05475	.05318	.05159
13	6	11	.06082	.05987	.05884	.05775	.05659	.05539	.05414	.05286
13	6	12	.05985	.05938	.05881	.05816	.05743	.05662	.05576	.05484
13	6	13	.05942	.05957	.05961	.05955	.05938	.05913	.05879	.05837
13	7	8	.08421	.08188	.07949	.07707	.07463	.07217	.06971	.06727
13	7	9	.08156	.07977	.07790	.07597	.07399	.07196	.06991	.06784
13	7	10	.07944	.07818	.07682	.07537	.07385	.07226	.07061	.06892
13	7	11	.07778	.07706	.07622	.07528	.07423	.07311	.07190	.07063
13	7	12	.07657	.07645	.07620	.07582	.07534	.07474	.07405	.07327
13	7	13	.07603	.07671	.07724	.07764	.07791	.07805	.07807	.07798
13	8	9	.10552	.10388	.10210	.10021	.09822	.09614	.09399	.09178
13	8	10	.10285	.10186	.10073	.09946	.09807	.09656	.09496	.09327
13	8	11	.10074	.10044	.09998	.09937	.09861	.09772	.09670	.09558
13	8	12	.09921	.09968	.09998	.10011	.10009	.09991	.09960	.09915
13	8	13	.09854	.10004	.10137	.10252	.10351	.10433	.10499	.10550
13	9	10	.13643	.13603	.13541	.13459	.13358	.13239	.13104	.12954
13	9	11	.13372	.13420	.13447	.13451	.13436	.13400	.13347	.13276
13	9	12	.13175	.13324	.13451	.13556	.13639	.13703	.13747	.13771
13	9	13	.13091	.13376	.13640	.13883	.14105	.14307	.14488	.14649
13	10	11	.18524	.18723	.18892	.19032	.19143	.19226	.19283	.19313
13	10	12	.18264	.18599	.18906	.19186	.19438	.19662	.19859	.20030
13	10	13	.18157	.18679	.19178	.19652	.20102	.20526	.20925	.21298
13	11	12	.27338	.28061	.28751	.29405	.30024	.30606	.31152	.31662
13	11	13	.27201	.28200	.29176	.30126	.31048	.31942	.32805	.33638
13	12	13	.47714	.49945	.52166	.54373	.56562	.58728	.60869	.62981
14	1	2	.02293	.02020	.01780	.01569	.01384	.01221	.01079	.00953
14	1	3	.01886	.01679	.01494	.01330	.01184	.01054	.00939	.00837
14	1	4	.01662	.01491	.01338	.01200	.01076	.00965	.00866	.00777
14	1	5	.01518	.01371	.01238	.01118	.01010	.00911	.00823	.00743
14	1	6	.01416	.01288	.01170	.01063	.00966	.00877	.00796	.00723
14	1	7	.01341	.01227	.01121	.01024	.00936	.00855	.00780	.00713
14	1	8	.01283	.01180	.01085	.00997	.00916	.00841	.00772	.00709
14	1	9	.01237	.01144	.01058	.00978	.00903	.00834	.00770	.00711
14	1	10	.01201	.01117	.01038	.00965	.00896	.00832	.00773	.00717
14	1	11	.01171	.01096	.01025	.00958	.00895	.00836	.00781	.00729
14	1	12	.01148	.01081	.01018	.00957	.00900	.00846	.00795	.00747
14	1	13	.01131	.01073	.01018	.00965	.00914	.00865	.00819	.00775
14	1	14	.01124	.01077	.01032	.00988	.00945	.00904	.00864	.00825
14	2	3	.02941	.02651	.02390	.02153	.01939	.01746	.01573	.01417
14	2	4	.02598	.02361	.02145	.01947	.01767	.01603	.01454	.01319
14	2	5	.02376	.02175	.01989	.01818	.01660	.01516	.01384	.01263
14	2	6	.02220	.02044	.01881	.01730	.01590	.01460	.01340	.01230
14	2	7	.02103	.01949	.01803	.01668	.01541	.01424	.01314	.01213
14	2	8	.02013	.01876	.01746	.01624	.01509	.01401	.01301	.01207
14	2	9	.01942	.01820	.01703	.01593	.01488	.01390	.01297	.01210
14	2	10	.01885	.01776	.01672	.01572	.01478	.01388	.01302	.01221

Covariances of Inverse Gaussian Order Statistics
Shape Parameter k

n	i	j	1.6	1.7	1.8	1.9	2.0	2.1	2.2	2.3
14	2	11	.01839	.01743	.01651	.01562	.01476	.01394	.01316	.01241
14	2	12	.01803	.01720	.01640	.01561	.01485	.01411	.01341	.01272
14	2	13	.01777	.01708	.01640	.01573	.01507	.01443	.01381	.01320
14	2	14	.01765	.01715	.01663	.01611	.01559	.01507	.01456	.01405
14	3	4	.03538	.03246	.02975	.02725	.02494	.02282	.02087	.01908
14	3	5	.03240	.02993	.02762	.02547	.02347	.02161	.01989	.01829
14	3	6	.03030	.02816	.02615	.02426	.02249	.02083	.01928	.01784
14	3	7	.02873	.02686	.02509	.02340	.02182	.02032	.01891	.01759
14	3	8	.02751	.02587	.02430	.02279	.02137	.02001	.01873	.01751
14	3	9	.02655	.02510	.02371	.02237	.02108	.01985	.01868	.01756
14	3	10	.02578	.02451	.02328	.02209	.02093	.01982	.01875	.01773
14	3	11	.02515	.02406	.02299	.02194	.02092	.01992	.01895	.01802
14	3	12	.02466	.02375	.02284	.02193	.02104	.02017	.01931	.01847
14	3	13	.02430	.02358	.02284	.02210	.02136	.02062	.01989	.01916
14	3	14	.02415	.02367	.02317	.02264	.02209	.02153	.02096	.02039
14	4	5	.04196	.03906	.03633	.03375	.03132	.02905	.02692	.02494
14	4	6	.03928	.03679	.03442	.03217	.03004	.02802	.02612	.02433
14	4	7	.03726	.03511	.03304	.03106	.02916	.02736	.02564	.02402
14	4	8	.03570	.03383	.03201	.03026	.02857	.02695	.02539	.02391
14	4	9	.03447	.03284	.03125	.02970	.02820	.02674	.02533	.02398
14	4	10	.03347	.03208	.03069	.02934	.02800	.02671	.02544	.02421
14	4	11	.03267	.03149	.03032	.02915	.02799	.02684	.02572	.02462
14	4	12	.03204	.03109	.03012	.02914	.02816	.02717	.02620	.02524
14	4	13	.03158	.03087	.03013	.02937	.02859	.02779	.02699	.02618
14	4	14	.03138	.03100	.03056	.03008	.02957	.02902	.02845	.02785
14	5	6	.04971	.04688	.04416	.04155	.03905	.03667	.03440	.03225
14	5	7	.04719	.04477	.04242	.04014	.03794	.03582	.03379	.03185
14	5	8	.04524	.04316	.04112	.03913	.03718	.03530	.03348	.03172
14	5	9	.04369	.04192	.04015	.03842	.03671	.03504	.03341	.03182
14	5	10	.04245	.04095	.03945	.03795	.03647	.03500	.03356	.03214
14	5	11	.04144	.04022	.03897	.03772	.03645	.03519	.03393	.03268
14	5	12	.04065	.03970	.03872	.03771	.03668	.03562	.03456	.03350
14	5	13	.04007	.03943	.03875	.03801	.03724	.03643	.03560	.03475
14	5	14	.03983	.03960	.03930	.03894	.03852	.03804	.03752	.03696
14	6	7	.05916	.05648	.05385	.05127	.04877	.04633	.04397	.04169
14	6	8	.05675	.05448	.05223	.05001	.04782	.04568	.04358	.04154
14	6	9	.05483	.05293	.05102	.04912	.04723	.04535	.04350	.04169
14	6	10	.05328	.05173	.05014	.04854	.04693	.04531	.04370	.04210
14	6	11	.05203	.05081	.04955	.04825	.04691	.04556	.04419	.04281
14	6	12	.05105	.05017	.04924	.04825	.04721	.04613	.04502	.04389
14	6	13	.05033	.04984	.04927	.04864	.04794	.04718	.04638	.04553
14	6	14	.05003	.05005	.04998	.04982	.04958	.04927	.04888	.04843
14	7	8	.07108	.06866	.06623	.06379	.06137	.05897	.05660	.05426
14	7	9	.06872	.06674	.06473	.06269	.06063	.05857	.05651	.05447
14	7	10	.06680	.06525	.06364	.06197	.06027	.05854	.05678	.05502
14	7	11	.06526	.06412	.06290	.06161	.06026	.05886	.05743	.05596
14	7	12	.06404	.06332	.06251	.06162	.06065	.05961	.05851	.05736
14	7	13	.06315	.06292	.06257	.06213	.06159	.06097	.06027	.05950
14	7	14	.06279	.06319	.06348	.06364	.06370	.06366	.06352	.06328
14	8	9	.08669	.08471	.08264	.08052	.07834	.07612	.07388	.07162
14	8	10	.08432	.08285	.08128	.07963	.07790	.07610	.07425	.07236
14	8	11	.08241	.08145	.08037	.07918	.07790	.07654	.07510	.07360
14	8	12	.08089	.08046	.07990	.07921	.07842	.07752	.07653	.07545
14	8	13	.07979	.07996	.07998	.07988	.07964	.07929	.07883	.07826
14	8	14	.07935	.08032	.08115	.08183	.08238	.08279	.08307	.08322

Covariances of Inverse Gaussian Order Statistics
Shape Parameter k

n	i	j	1.6	1.7	1.8	1.9	2.0	2.1	2.2	2.3
14	9	10	.10804	.10681	.10542	.10389	.10224	.10048	.09862	.09668
14	9	11	.10564	.10504	.10427	.10335	.10228	.10109	.09977	.09835
14	9	12	.10374	.10380	.10369	.10341	.10298	.10240	.10168	.10083
14	9	13	.10237	.10319	.10383	.10430	.10460	.10474	.10473	.10458
14	9	14	.10183	.10368	.10536	.10686	.10819	.10935	.11035	.11118
14	10	11	.13903	.13911	.13896	.13859	.13801	.13724	.13629	.13518
14	10	12	.13661	.13754	.13824	.13872	.13899	.13905	.13891	.13859
14	10	13	.13485	.13677	.13847	.13994	.14120	.14225	.14309	.14373
14	10	14	.13419	.13746	.14053	.14339	.14604	.14849	.15073	.15276
14	11	12	.18799	.19055	.19281	.19478	.19645	.19784	.19896	.19980
14	11	13	.18569	.18958	.19320	.19655	.19962	.20241	.20493	.20718
14	11	14	.18487	.19061	.19613	.20141	.20646	.21125	.21580	.22009
14	12	13	.27638	.28433	.29196	.29925	.30619	.31278	.31901	.32488
14	12	14	.27538	.28603	.29648	.30669	.31665	.32634	.33575	.34486
14	13	14	.48068	.50395	.52716	.55028	.57326	.59606	.61864	.64098
15	1	2	.02142	.01882	.01654	.01454	.01279	.01126	.00992	.00875
15	1	3	.01756	.01558	.01383	.01227	.01090	.00968	.00860	.00764
15	1	4	.01543	.01380	.01234	.01104	.00987	.00883	.00790	.00707
15	1	5	.01407	.01267	.01140	.01026	.00923	.00831	.00748	.00673
15	1	6	.01310	.01187	.01075	.00973	.00881	.00797	.00722	.00653
15	1	7	.01239	.01129	.01028	.00936	.00851	.00775	.00705	.00642
15	1	8	.01183	.01084	.00992	.00908	.00831	.00760	.00695	.00636
15	1	9	.01140	.01049	.00966	.00888	.00817	.00751	.00691	.00635
15	1	10	.01104	.01022	.00946	.00874	.00808	.00747	.00690	.00638
15	1	11	.01075	.01001	.00931	.00865	.00804	.00747	.00694	.00645
15	1	12	.01052	.00985	.00921	.00861	.00805	.00752	.00702	.00656
15	1	13	.01033	.00973	.00916	.00862	.00810	.00762	.00716	.00673
15	1	14	.01020	.00968	.00917	.00869	.00823	.00779	.00737	.00698
15	1	15	.01014	.00972	.00931	.00891	.00852	.00814	.00777	.00742
15	2	3	.02711	.02437	.02189	.01966	.01766	.01586	.01424	.01279
15	2	4	.02390	.02164	.01959	.01773	.01604	.01450	.01311	.01186
15	2	5	.02181	.01989	.01812	.01651	.01502	.01367	.01244	.01131
15	2	6	.02034	.01866	.01710	.01567	.01434	.01313	.01201	.01098
15	2	7	.01924	.01775	.01636	.01507	.01387	.01276	.01174	.01079
15	2	8	.01839	.01706	.01581	.01464	.01355	.01253	.01158	.01070
15	2	9	.01772	.01652	.01539	.01432	.01332	.01238	.01151	.01068
15	2	10	.01717	.01610	.01507	.01410	.01319	.01232	.01150	.01074
15	2	11	.01672	.01576	.01484	.01396	.01312	.01233	.01157	.01085
15	2	12	.01636	.01551	.01469	.01389	.01313	.01240	.01171	.01104
15	2	13	.01607	.01533	.01461	.01391	.01322	.01257	.01193	.01132
15	2	14	.01587	.01525	.01463	.01403	.01343	.01285	.01229	.01175
15	2	15	.01579	.01532	.01485	.01437	.01390	.01342	.01295	.01249
15	3	4	.03224	.02947	.02692	.02457	.02241	.02043	.01862	.01697
15	3	5	.02947	.02712	.02493	.02290	.02102	.01928	.01768	.01621
15	3	6	.02751	.02546	.02355	.02176	.02009	.01853	.01709	.01575
15	3	7	.02604	.02424	.02254	.02094	.01944	.01803	.01671	.01548
15	3	8	.02490	.02330	.02179	.02035	.01899	.01770	.01649	.01535
15	3	9	.02399	.02257	.02122	.01992	.01868	.01750	.01639	.01534
15	3	10	.02326	.02200	.02078	.01961	.01849	.01742	.01639	.01542
15	3	11	.02266	.02155	.02047	.01942	.01840	.01743	.01649	.01559
15	3	12	.02217	.02121	.02026	.01933	.01842	.01754	.01668	.01586
15	3	13	.02179	.02097	.02015	.01934	.01855	.01777	.01700	.01626
15	3	14	.02151	.02085	.02018	.01951	.01884	.01818	.01752	.01687
15	3	15	.02140	.02095	.02048	.02000	.01949	.01898	.01846	.01793
15	4	5	.03780	.03504	.03245	.03003	.02776	.02564	.02367	.02184

Covariances of Inverse Gaussian Order Statistics
Shape Parameter k

n	i	j	1.6	1.7	1.8	1.9	2.0	2.1	2.2	2.3
15	4	6	.03531	.03294	.03068	.02855	.02655	.02466	.02289	.02124
15	4	7	.03345	.03137	.02939	.02750	.02570	.02401	.02240	.02089
15	4	8	.03200	.03017	.02842	.02673	.02512	.02358	.02211	.02073
15	4	9	.03085	.02924	.02768	.02617	.02472	.02332	.02198	.02071
15	4	10	.02991	.02850	.02712	.02578	.02447	.02321	.02199	.02082
15	4	11	.02915	.02792	.02672	.02553	.02436	.02323	.02212	.02105
15	4	12	.02853	.02749	.02644	.02541	.02438	.02338	.02239	.02142
15	4	13	.02803	.02718	.02631	.02543	.02456	.02368	.02282	.02196
15	4	14	.02768	.02703	.02635	.02566	.02495	.02423	.02351	.02278
15	4	15	.02754	.02716	.02675	.02629	.02581	.02530	.02476	.02422
15	5	6	.04421	.04151	.03892	.03645	.03411	.03188	.02978	.02779
15	5	7	.04191	.03956	.03730	.03513	.03304	.03105	.02916	.02735
15	5	8	.04011	.03807	.03608	.03416	.03230	.03051	.02879	.02715
15	5	9	.03868	.03690	.03516	.03346	.03180	.03019	.02863	.02713
15	5	10	.03752	.03598	.03446	.03296	.03149	.03005	.02864	.02728
15	5	11	.03657	.03526	.03395	.03265	.03135	.03007	.02882	.02758
15	5	12	.03580	.03471	.03361	.03250	.03139	.03027	.02916	.02807
15	5	13	.03518	.03433	.03345	.03254	.03161	.03067	.02973	.02878
15	5	14	.03474	.03414	.03350	.03283	.03212	.03138	.03063	.02985
15	5	15	.03457	.03432	.03401	.03364	.03322	.03276	.03226	.03173
15	6	7	.05187	.04927	.04673	.04427	.04189	.03960	.03740	.03528
15	6	8	.04968	.04743	.04523	.04307	.04097	.03893	.03695	.03503
15	6	9	.04792	.04600	.04409	.04220	.04035	.03853	.03675	.03502
15	6	10	.04650	.04487	.04323	.04159	.03997	.03836	.03677	.03521
15	6	11	.04533	.04398	.04260	.04120	.03980	.03840	.03700	.03561
15	6	12	.04438	.04330	.04218	.04102	.03984	.03865	.03745	.03624
15	6	13	.04363	.04283	.04197	.04107	.04014	.03917	.03818	.03716
15	6	14	.04308	.04260	.04205	.04144	.04078	.04007	.03933	.03855
15	6	15	.04288	.04282	.04268	.04247	.04218	.04184	.04143	.04097
15	7	8	.06129	.05886	.05645	.05407	.05172	.04941	.04716	.04496
15	7	9	.05915	.05710	.05505	.05299	.05095	.04892	.04692	.04496
15	7	10	.05741	.05572	.05399	.05224	.05048	.04872	.04696	.04522
15	7	11	.05599	.05463	.05321	.05176	.05028	.04878	.04726	.04574
15	7	12	.05483	.05380	.05270	.05155	.05035	.04911	.04784	.04655
15	7	13	.05391	.05322	.05245	.05162	.05072	.04977	.04877	.04773
15	7	14	.05324	.05294	.05255	.05208	.05154	.05092	.05024	.04951
15	7	15	.05300	.05323	.05335	.05337	.05331	.05316	.05292	.05262
15	8	9	.07321	.07108	.06890	.06670	.06448	.06226	.06004	.05784
15	8	10	.07109	.06938	.06760	.06578	.06391	.06202	.06011	.05819
15	8	11	.06936	.06804	.06665	.06519	.06367	.06211	.06050	.05887
15	8	12	.06794	.06703	.06602	.06493	.06377	.06254	.06125	.05991
15	8	13	.06681	.06632	.06572	.06503	.06425	.06338	.06244	.06144
15	8	14	.06600	.06599	.06586	.06562	.06528	.06485	.06433	.06372
15	8	15	.06571	.06635	.06686	.06725	.06753	.06770	.06775	.06771
15	9	10	.08884	.08719	.08543	.08358	.08166	.07968	.07765	.07558
15	9	11	.08671	.08554	.08426	.08287	.08138	.07981	.07818	.07648
15	9	12	.08496	.08429	.08348	.08256	.08152	.08038	.07915	.07784
15	9	13	.08358	.08342	.08313	.08270	.08214	.08148	.08070	.07983
15	9	14	.08258	.08302	.08331	.08346	.08348	.08337	.08313	.08278
15	9	15	.08223	.08348	.08458	.08554	.08635	.08702	.08755	.08796
15	10	11	.11024	.10938	.10835	.10717	.10585	.10439	.10282	.10115
15	10	12	.10807	.10782	.10740	.10680	.10605	.10516	.10412	.10297
15	10	13	.10635	.10675	.10697	.10701	.10688	.10660	.10617	.10559
15	10	14	.10511	.10626	.10722	.10801	.10863	.10908	.10936	.10950
15	10	15	.10469	.10687	.10888	.11071	.11236	.11385	.11517	.11631

Covariances of Inverse Gaussian Order Statistics
Shape Parameter k

n	i	j	1.6	1.7	1.8	1.9	2.0	2.1	2.2	2.3
15	11	12	.14132	.14183	.14211	.14217	.14200	.14163	.14106	.14031
15	11	13	.13913	.14048	.14159	.14248	.14314	.14360	.14384	.14388
15	11	14	.13756	.13988	.14197	.14385	.14550	.14694	.14817	.14919
15	11	15	.13705	.14072	.14418	.14744	.15050	.15335	.15599	.15843
15	12	13	.19041	.19350	.19629	.19878	.20098	.20289	.20452	.20587
15	12	14	.18836	.19276	.19688	.20074	.20432	.20762	.21066	.21342
15	12	15	.18776	.19398	.19999	.20577	.21133	.21664	.22171	.22653
15	13	14	.27904	.28764	.29593	.30391	.31155	.31885	.32580	.33240
15	13	15	.27834	.28961	.30069	.31156	.32220	.33259	.34271	.35254
15	14	15	.48383	.50796	.53209	.55617	.58016	.60401	.62767	.65113
16	1	2	.02011	.01763	.01545	.01356	.01190	.01045	.00919	.00809
16	1	3	.01644	.01455	.01288	.01140	.01010	.00894	.00793	.00703
16	1	4	.01441	.01286	.01146	.01022	.00912	.00813	.00726	.00648
16	1	5	.01311	.01177	.01057	.00948	.00851	.00763	.00685	.00615
16	1	6	.01220	.01101	.00994	.00897	.00810	.00731	.00659	.00595
16	1	7	.01151	.01045	.00949	.00861	.00781	.00708	.00642	.00583
16	1	8	.01099	.01003	.00914	.00834	.00760	.00693	.00632	.00576
16	1	9	.01056	.00969	.00888	.00814	.00746	.00683	.00626	.00573
16	1	10	.01022	.00942	.00868	.00799	.00736	.00677	.00623	.00574
16	1	11	.00994	.00921	.00853	.00789	.00730	.00675	.00624	.00577
16	1	12	.00971	.00904	.00842	.00783	.00728	.00677	.00629	.00584
16	1	13	.00952	.00891	.00834	.00780	.00729	.00682	.00637	.00595
16	1	14	.00937	.00883	.00831	.00782	.00735	.00691	.00649	.00610
16	1	15	.00926	.00879	.00833	.00789	.00747	.00707	.00669	.00633
16	1	16	.00922	.00884	.00846	.00809	.00773	.00738	.00704	.00672
16	2	3	.02517	.02256	.02021	.01810	.01621	.01451	.01300	.01165
16	2	4	.02213	.01998	.01803	.01627	.01467	.01323	.01193	.01076
16	2	5	.02016	.01832	.01664	.01511	.01371	.01244	.01128	.01023
16	2	6	.01877	.01716	.01568	.01431	.01306	.01191	.01086	.00990
16	2	7	.01773	.01630	.01497	.01374	.01260	.01155	.01059	.00970
16	2	8	.01693	.01564	.01444	.01332	.01227	.01131	.01041	.00959
16	2	9	.01628	.01512	.01403	.01300	.01204	.01115	.01032	.00954
16	2	10	.01576	.01471	.01371	.01277	.01189	.01106	.01028	.00955
16	2	11	.01533	.01438	.01347	.01261	.01180	.01103	.01030	.00962
16	2	12	.01498	.01412	.01330	.01251	.01176	.01105	.01037	.00974
16	2	13	.01469	.01392	.01318	.01247	.01179	.01113	.01051	.00991
16	2	14	.01446	.01378	.01313	.01249	.01188	.01129	.01071	.01017
16	2	15	.01429	.01372	.01316	.01261	.01207	.01155	.01104	.01054
16	2	16	.01423	.01380	.01337	.01293	.01249	.01205	.01162	.01120
16	3	4	.02962	.02699	.02457	.02235	.02032	.01847	.01678	.01525
16	3	5	.02702	.02478	.02270	.02078	.01901	.01738	.01589	.01451
16	3	6	.02518	.02323	.02140	.01970	.01812	.01666	.01531	.01406
16	3	7	.02380	.02207	.02045	.01892	.01750	.01617	.01493	.01378
16	3	8	.02273	.02119	.01973	.01835	.01705	.01583	.01469	.01362
16	3	9	.02188	.02049	.01917	.01792	.01673	.01561	.01456	.01357
16	3	10	.02118	.01994	.01875	.01761	.01652	.01549	.01451	.01359
16	3	11	.02061	.01950	.01842	.01739	.01640	.01545	.01454	.01368
16	3	12	.02013	.01915	.01818	.01725	.01635	.01548	.01465	.01385
16	3	13	.01975	.01888	.01803	.01720	.01638	.01560	.01483	.01410
16	3	14	.01944	.01870	.01796	.01723	.01652	.01581	.01513	.01446
16	3	15	.01921	.01861	.01801	.01740	.01679	.01618	.01558	.01499
16	3	16	.01914	.01872	.01828	.01783	.01736	.01689	.01641	.01593
16	4	5	.03438	.03175	.02929	.02700	.02487	.02289	.02106	.01937
16	4	6	.03206	.02978	.02764	.02562	.02373	.02196	.02031	.01877
16	4	7	.03032	.02832	.02642	.02462	.02292	.02132	.01982	.01841

Covariances of Inverse Gaussian Order Statistics
Shape Parameter k

n	i	j	1.6	1.7	1.8	1.9	2.0	2.1	2.2	2.3
16	4	8	.02897	.02720	.02550	.02388	.02235	.02089	.01951	.01821
16	4	9	.02789	.02631	.02479	.02333	.02194	.02061	.01934	.01813
16	4	10	.02701	.02561	.02425	.02293	.02166	.02045	.01928	.01816
16	4	11	.02628	.02505	.02383	.02265	.02150	.02039	.01932	.01829
16	4	12	.02569	.02460	.02353	.02248	.02144	.02044	.01946	.01851
16	4	13	.02519	.02426	.02333	.02241	.02149	.02059	.01971	.01885
16	4	14	.02480	.02403	.02324	.02245	.02166	.02088	.02010	.01933
16	4	15	.02452	.02392	.02330	.02267	.02202	.02137	.02071	.02005
16	4	16	.02443	.02406	.02366	.02324	.02278	.02230	.02181	.02130
16	5	6	.03977	.03718	.03472	.03239	.03018	.02810	.02614	.02430
16	5	7	.03764	.03538	.03321	.03114	.02917	.02730	.02552	.02384
16	5	8	.03598	.03399	.03207	.03022	.02845	.02675	.02513	.02359
16	5	9	.03465	.03290	.03119	.02954	.02794	.02640	.02492	.02350
16	5	10	.03356	.03202	.03051	.02903	.02760	.02620	.02485	.02354
16	5	11	.03267	.03133	.03000	.02868	.02739	.02613	.02491	.02371
16	5	12	.03193	.03078	.02962	.02847	.02732	.02620	.02509	.02400
16	5	13	.03132	.03035	.02937	.02838	.02739	.02640	.02541	.02444
16	5	14	.03084	.03006	.02926	.02844	.02761	.02676	.02592	.02507
16	5	15	.03049	.02994	.02934	.02872	.02806	.02739	.02670	.02599
16	5	16	.03038	.03011	.02980	.02943	.02903	.02859	.02811	.02761
16	6	7	.04611	.04359	.04116	.03881	.03656	.03441	.03235	.03038
16	6	8	.04409	.04190	.03976	.03768	.03567	.03373	.03187	.03007
16	6	9	.04248	.04056	.03868	.03684	.03504	.03330	.03160	.02997
16	6	10	.04116	.03950	.03785	.03622	.03462	.03305	.03152	.03003
16	6	11	.04007	.03865	.03722	.03579	.03438	.03298	.03160	.03024
16	6	12	.03918	.03797	.03676	.03553	.03429	.03306	.03183	.03062
16	6	13	.03844	.03746	.03645	.03542	.03437	.03331	.03224	.03118
16	6	14	.03785	.03711	.03632	.03550	.03465	.03378	.03288	.03197
16	6	15	.03742	.03695	.03642	.03585	.03523	.03457	.03388	.03316
16	6	16	.03729	.03717	.03699	.03674	.03644	.03608	.03567	.03522
16	7	8	.05373	.05134	.04899	.04668	.04443	.04224	.04011	.03805
16	7	9	.05178	.04972	.04768	.04565	.04366	.04170	.03979	.03793
16	7	10	.05020	.04844	.04667	.04490	.04315	.04141	.03969	.03801
16	7	11	.04888	.04740	.04590	.04438	.04285	.04132	.03980	.03829
16	7	12	.04780	.04659	.04534	.04405	.04275	.04143	.04010	.03877
16	7	13	.04690	.04596	.04497	.04393	.04285	.04175	.04062	.03948
16	7	14	.04619	.04553	.04481	.04404	.04321	.04234	.04143	.04049
16	7	15	.04568	.04535	.04494	.04446	.04392	.04333	.04268	.04198
16	7	16	.04552	.04563	.04564	.04558	.04543	.04522	.04494	.04459
16	8	9	.06313	.06095	.05875	.05655	.05437	.05220	.05006	.04796
16	8	10	.06122	.05939	.05752	.05564	.05374	.05184	.04995	.04808
16	8	11	.05964	.05814	.05659	.05500	.05338	.05174	.05009	.04844
16	8	12	.05833	.05715	.05591	.05461	.05327	.05189	.05048	.04905
16	8	13	.05725	.05640	.05547	.05447	.05341	.05230	.05114	.04995
16	8	14	.05639	.05588	.05528	.05460	.05385	.05303	.05216	.05123
16	8	15	.05577	.05566	.05544	.05514	.05474	.05427	.05373	.05312
16	8	16	.05558	.05600	.05631	.05652	.05663	.05664	.05657	.05641
16	9	10	.07507	.07320	.07128	.06930	.06729	.06525	.06319	.06112
16	9	11	.07316	.07169	.07014	.06853	.06685	.06513	.06338	.06159
16	9	12	.07157	.07049	.06932	.06806	.06672	.06533	.06387	.06238
16	9	13	.07026	.06958	.06878	.06789	.06691	.06585	.06472	.06352
16	9	14	.06922	.06895	.06856	.06807	.06747	.06678	.06600	.06515
16	9	15	.06848	.06869	.06877	.06874	.06859	.06834	.06799	.06755
16	9	16	.06825	.06912	.06985	.07047	.07095	.07132	.07158	.07173
16	10	11	.09073	.08938	.08791	.08634	.08467	.08292	.08110	.07922

Covariances of Inverse Gaussian Order Statistics
Shape Parameter k

n	i	j	1.6	1.7	1.8	1.9	2.0	2.1	2.2	2.3
16	10	12	.08879	.08791	.08690	.08577	.08452	.08318	.08175	.08024
16	10	13	.08720	.08680	.08625	.08557	.08477	.08386	.08284	.08172
16	10	14	.08593	.08603	.08599	.08581	.08549	.08505	.08449	.08382
16	10	15	.08502	.08572	.08626	.08666	.08692	.08704	.08703	.08690
16	10	16	.08476	.08627	.08763	.08885	.08991	.09084	.09162	.09226
16	11	12	.11218	.11167	.11098	.11013	.10911	.10795	.10666	.10525
16	11	13	.11021	.11029	.11019	.10990	.10945	.10885	.10809	.10720
16	11	14	.10864	.10935	.10988	.11023	.11040	.11041	.11025	.10994
16	11	15	.10752	.10898	.11025	.11134	.11226	.11300	.11357	.11398
16	11	16	.10721	.10969	.11201	.11415	.11612	.11791	.11953	.12098
16	12	13	.14334	.14426	.14494	.14539	.14561	.14562	.14541	.14502
16	12	14	.14135	.14309	.14459	.14586	.14690	.14772	.14833	.14873
16	12	15	.13995	.14263	.14510	.14735	.14938	.15119	.15279	.15417
16	12	16	.13958	.14361	.14744	.15108	.15451	.15775	.16077	.16359
16	13	14	.19255	.19613	.19941	.20239	.20508	.20749	.20960	.21143
16	13	15	.19074	.19559	.20018	.20451	.20857	.21236	.21588	.21912
16	13	16	.19032	.19698	.20344	.20969	.21572	.22152	.22709	.23240
16	14	15	.28140	.29060	.29952	.30813	.31642	.32438	.33200	.33928
16	14	16	.28097	.29280	.30447	.31595	.32722	.33825	.34904	.35956
16	15	16	.48664	.51157	.53655	.56152	.58643	.61125	.63592	.66042
17	1	2	.01897	.01659	.01452	.01271	.01113	.00976	.00856	.00752
17	1	3	.01546	.01365	.01206	.01065	.00941	.00832	.00736	.00651
17	1	4	.01353	.01204	.01071	.00952	.00847	.00754	.00671	.00598
17	1	5	.01229	.01100	.00985	.00881	.00789	.00706	.00632	.00566
17	1	6	.01141	.01028	.00925	.00832	.00749	.00674	.00606	.00546
17	1	7	.01076	.00974	.00881	.00797	.00721	.00652	.00589	.00533
17	1	8	.01025	.00933	.00848	.00771	.00700	.00636	.00578	.00525
17	1	9	.00985	.00900	.00822	.00751	.00686	.00626	.00571	.00521
17	1	10	.00952	.00874	.00802	.00736	.00675	.00619	.00568	.00520
17	1	11	.00925	.00853	.00787	.00725	.00668	.00615	.00567	.00522
17	1	12	.00902	.00836	.00775	.00717	.00664	.00615	.00569	.00526
17	1	13	.00883	.00823	.00766	.00713	.00663	.00617	.00573	.00533
17	1	14	.00868	.00813	.00761	.00711	.00665	.00622	.00581	.00543
17	1	15	.00855	.00806	.00758	.00713	.00671	.00631	.00593	.00557
17	1	16	.00847	.00803	.00761	.00721	.00682	.00646	.00611	.00578
17	1	17	.00844	.00808	.00773	.00739	.00706	.00674	.00643	.00613
17	2	3	.02350	.02100	.01877	.01676	.01497	.01338	.01195	.01068
17	2	4	.02061	.01856	.01671	.01503	.01352	.01216	.01093	.00983
17	2	5	.01875	.01699	.01538	.01393	.01260	.01140	.01031	.00932
17	2	6	.01743	.01588	.01446	.01316	.01197	.01089	.00990	.00900
17	2	7	.01644	.01506	.01379	.01261	.01153	.01054	.00963	.00879
17	2	8	.01568	.01443	.01328	.01220	.01121	.01029	.00945	.00867
17	2	9	.01506	.01394	.01288	.01189	.01098	.01012	.00933	.00860
17	2	10	.01457	.01354	.01257	.01166	.01081	.01002	.00928	.00859
17	2	11	.01415	.01321	.01233	.01149	.01070	.00996	.00927	.00862
17	2	12	.01381	.01296	.01214	.01137	.01064	.00995	.00930	.00869
17	2	13	.01352	.01275	.01201	.01130	.01063	.00998	.00938	.00880
17	2	14	.01328	.01259	.01192	.01128	.01066	.01007	.00950	.00896
17	2	15	.01310	.01248	.01189	.01131	.01075	.01021	.00969	.00919
17	2	16	.01296	.01244	.01193	.01142	.01093	.01045	.00999	.00953
17	2	17	.01292	.01252	.01212	.01171	.01131	.01091	.01051	.01012
17	3	4	.02740	.02489	.02259	.02049	.01857	.01683	.01525	.01382
17	3	5	.02495	.02281	.02083	.01901	.01733	.01580	.01440	.01311
17	3	6	.02322	.02134	.01960	.01798	.01649	.01511	.01384	.01267
17	3	7	.02192	.02025	.01869	.01724	.01588	.01463	.01346	.01238

Covariances of Inverse Gaussian Order Statistics
Shape Parameter k

n	i	j	1.6	1.7	1.8	1.9	2.0	2.1	2.2	2.3
17	3	8	.02090	.01941	.01800	.01668	.01545	.01429	.01322	.01221
17	3	9	.02009	.01875	.01747	.01627	.01513	.01406	.01306	.01213
17	3	10	.01943	.01822	.01705	.01595	.01491	.01392	.01299	.01211
17	3	11	.01889	.01779	.01673	.01572	.01476	.01384	.01297	.01215
17	3	12	.01843	.01744	.01648	.01556	.01467	.01383	.01302	.01225
17	3	13	.01805	.01716	.01630	.01546	.01465	.01388	.01313	.01241
17	3	14	.01774	.01695	.01618	.01543	.01470	.01399	.01330	.01264
17	3	15	.01748	.01681	.01614	.01548	.01483	.01419	.01357	.01297
17	3	16	.01731	.01675	.01620	.01564	.01508	.01453	.01398	.01344
17	3	17	.01726	.01686	.01645	.01603	.01560	.01516	.01472	.01427
17	4	5	.03151	.02901	.02667	.02451	.02250	.02064	.01893	.01735
17	4	6	.02934	.02716	.02512	.02320	.02141	.01975	.01820	.01677
17	4	7	.02772	.02579	.02397	.02225	.02064	.01913	.01772	.01640
17	4	8	.02645	.02473	.02310	.02155	.02008	.01870	.01740	.01618
17	4	9	.02543	.02389	.02242	.02101	.01968	.01841	.01720	.01607
17	4	10	.02460	.02322	.02189	.02061	.01939	.01822	.01711	.01605
17	4	11	.02391	.02268	.02148	.02032	.01920	.01812	.01709	.01611
17	4	12	.02334	.02224	.02116	.02011	.01909	.01811	.01715	.01624
17	4	13	.02286	.02189	.02093	.01999	.01907	.01817	.01730	.01645
17	4	14	.02246	.02162	.02078	.01995	.01913	.01832	.01753	.01676
17	4	15	.02215	.02144	.02073	.02001	.01930	.01859	.01788	.01719
17	4	16	.02192	.02137	.02080	.02022	.01962	.01902	.01842	.01782
17	4	17	.02186	.02151	.02113	.02073	.02030	.01985	.01939	.01892
17	5	6	.03612	.03364	.03129	.02908	.02700	.02505	.02322	.02151
17	5	7	.03413	.03196	.02988	.02791	.02604	.02428	.02261	.02105
17	5	8	.03259	.03066	.02881	.02703	.02535	.02374	.02221	.02077
17	5	9	.03134	.02963	.02797	.02637	.02484	.02337	.02197	.02063
17	5	10	.03033	.02880	.02732	.02587	.02448	.02314	.02185	.02061
17	5	11	.02949	.02813	.02681	.02551	.02424	.02302	.02183	.02068
17	5	12	.02878	.02759	.02641	.02525	.02411	.02300	.02191	.02086
17	5	13	.02820	.02716	.02613	.02510	.02409	.02308	.02210	.02113
17	5	14	.02771	.02684	.02595	.02506	.02417	.02328	.02240	.02153
17	5	15	.02732	.02661	.02588	.02514	.02438	.02361	.02285	.02208
17	5	16	.02705	.02653	.02597	.02539	.02479	.02417	.02354	.02289
17	5	17	.02697	.02670	.02639	.02603	.02564	.02522	.02477	.02430
17	6	7	.04144	.03902	.03669	.03446	.03233	.03030	.02837	.02654
17	6	8	.03958	.03745	.03538	.03339	.03147	.02964	.02788	.02620
17	6	9	.03808	.03620	.03437	.03258	.03085	.02918	.02758	.02603
17	6	10	.03686	.03520	.03357	.03197	.03042	.02890	.02743	.02601
17	6	11	.03585	.03439	.03295	.03153	.03013	.02875	.02741	.02611
17	6	12	.03500	.03374	.03248	.03122	.02997	.02874	.02752	.02633
17	6	13	.03429	.03322	.03213	.03104	.02994	.02884	.02775	.02668
17	6	14	.03370	.03282	.03191	.03098	.03004	.02909	.02813	.02718
17	6	15	.03323	.03255	.03183	.03108	.03030	.02951	.02870	.02788
17	6	16	.03290	.03245	.03194	.03140	.03082	.03020	.02956	.02890
17	6	17	.03281	.03267	.03246	.03219	.03188	.03152	.03111	.03068
17	7	8	.04774	.04541	.04313	.04091	.03876	.03669	.03468	.03276
17	7	9	.04595	.04392	.04191	.03994	.03801	.03614	.03432	.03256
17	7	10	.04449	.04271	.04095	.03920	.03748	.03579	.03414	.03253
17	7	11	.04328	.04174	.04020	.03866	.03713	.03562	.03413	.03266
17	7	12	.04226	.04095	.03963	.03829	.03694	.03560	.03427	.03294
17	7	13	.04141	.04033	.03921	.03807	.03691	.03574	.03456	.03338
17	7	14	.04071	.03985	.03895	.03801	.03704	.03604	.03503	.03401
17	7	15	.04015	.03953	.03885	.03813	.03737	.03656	.03574	.03488
17	7	16	.03975	.03940	.03899	.03852	.03800	.03743	.03681	.03616

Covariances of Inverse Gaussian Order Statistics
Shape Parameter k

n	i	j	1.6	1.7	1.8	1.9	2.0	2.1	2.2	2.3
17	7	17	.03965	.03967	.03962	.03950	.03931	.03905	.03874	.03838
17	8	9	.05535	.05316	.05099	.04884	.04671	.04463	.04259	.04060
17	8	10	.05361	.05173	.04984	.04795	.04608	.04422	.04238	.04058
17	8	11	.05216	.05056	.04894	.04730	.04566	.04401	.04237	.04075
17	8	12	.05095	.04962	.04825	.04685	.04543	.04399	.04255	.04110
17	8	13	.04993	.04887	.04775	.04659	.04539	.04416	.04292	.04165
17	8	14	.04909	.04830	.04744	.04652	.04555	.04455	.04350	.04243
17	8	15	.04842	.04791	.04733	.04667	.04596	.04519	.04438	.04352
17	8	16	.04795	.04776	.04750	.04716	.04674	.04626	.04572	.04512
17	8	17	.04783	.04809	.04827	.04835	.04835	.04827	.04811	.04788
17	9	10	.06476	.06280	.06081	.05879	.05677	.05474	.05273	.05073
17	9	11	.06303	.06140	.05973	.05801	.05626	.05450	.05272	.05095
17	9	12	.06158	.06027	.05890	.05747	.05600	.05449	.05295	.05139
17	9	13	.06037	.05937	.05830	.05716	.05596	.05471	.05341	.05208
17	9	14	.05936	.05869	.05792	.05708	.05616	.05518	.05415	.05306
17	9	15	.05856	.05823	.05780	.05727	.05667	.05599	.05524	.05443
17	9	16	.05799	.05806	.05801	.05787	.05763	.05731	.05690	.05642
17	9	17	.05786	.05846	.05895	.05934	.05962	.05980	.05988	.05987
17	10	11	.07671	.07510	.07341	.07165	.06983	.06797	.06608	.06415
17	10	12	.07497	.07374	.07241	.07100	.06952	.06797	.06637	.06473
17	10	13	.07352	.07266	.07169	.07063	.06948	.06825	.06696	.06560
17	10	14	.07231	.07183	.07124	.07054	.06974	.06886	.06788	.06684
17	10	15	.07134	.07128	.07109	.07079	.07038	.06986	.06925	.06855
17	10	16	.07066	.07108	.07137	.07153	.07158	.07151	.07134	.07106
17	10	17	.07050	.07158	.07253	.07335	.07404	.07461	.07506	.07540
17	11	12	.09240	.09134	.09015	.08883	.08740	.08588	.08427	.08258
17	11	13	.09064	.09003	.08927	.08838	.08737	.08625	.08503	.08371
17	11	14	.08917	.08903	.08873	.08829	.08772	.08702	.08621	.08529
17	11	15	.08800	.08836	.08856	.08861	.08852	.08830	.08795	.08748
17	11	16	.08718	.08812	.08891	.08955	.09004	.09038	.09060	.09067
17	11	17	.08699	.08875	.09036	.09182	.09313	.09430	.09532	.09619
17	12	13	.11391	.11372	.11335	.11280	.11208	.11121	.11019	.10903
17	12	14	.11210	.11250	.11269	.11271	.11255	.11222	.11173	.11109
17	12	15	.11067	.11168	.11250	.11314	.11359	.11388	.11399	.11394
17	12	16	.10966	.11140	.11296	.11434	.11554	.11656	.11741	.11809
17	12	17	.10944	.11222	.11482	.11725	.11951	.12160	.12351	.12525
17	13	14	.14514	.14644	.14750	.14832	.14890	.14927	.14941	.14935
17	13	15	.14333	.14543	.14729	.14892	.15031	.15149	.15244	.15318
17	13	16	.14207	.14510	.14792	.15052	.15290	.15507	.15701	.15874
17	13	17	.14183	.14619	.15037	.15436	.15815	.16174	.16513	.16832
17	14	15	.19447	.19849	.20223	.20567	.20883	.21169	.21426	.21656
17	14	16	.19285	.19813	.20316	.20793	.21244	.21669	.22066	.22436
17	14	17	.19260	.19967	.20655	.21324	.21972	.22597	.23200	.23778
17	15	16	.28352	.29328	.30276	.31196	.32086	.32944	.33770	.34561
17	15	17	.28332	.29568	.30789	.31994	.33179	.34343	.35484	.36599
17	16	17	.48917	.51484	.54060	.56639	.59217	.61788	.64349	.66896
18	1	2	.01797	.01568	.01369	.01196	.01046	.00915	.00801	.00703
18	1	3	.01461	.01287	.01134	.00999	.00881	.00777	.00686	.00606
18	1	4	.01276	.01132	.01005	.00892	.00791	.00703	.00624	.00555
18	1	5	.01157	.01033	.00922	.00823	.00735	.00656	.00586	.00524
18	1	6	.01073	.00963	.00865	.00776	.00697	.00625	.00561	.00504
18	1	7	.01010	.00912	.00823	.00742	.00669	.00603	.00544	.00491
18	1	8	.00962	.00872	.00790	.00716	.00649	.00588	.00533	.00483
18	1	9	.00923	.00841	.00766	.00697	.00634	.00577	.00525	.00478
18	1	10	.00891	.00815	.00746	.00682	.00623	.00570	.00521	.00476

Covariances of Inverse Gaussian Order Statistics
Shape Parameter k

n	i	j	1.6	1.7	1.8	1.9	2.0	2.1	2.2	2.3
18	1	11	.00865	.00795	.00730	.00671	.00616	.00565	.00519	.00476
18	1	12	.00843	.00778	.00718	.00662	.00611	.00563	.00519	.00478
18	1	13	.00824	.00764	.00709	.00657	.00608	.00563	.00521	.00483
18	1	14	.00809	.00754	.00702	.00653	.00608	.00566	.00526	.00489
18	1	15	.00796	.00745	.00698	.00653	.00610	.00571	.00533	.00499
18	1	16	.00785	.00740	.00696	.00655	.00616	.00579	.00544	.00511
18	1	17	.00778	.00738	.00699	.00662	.00627	.00593	.00561	.00530
18	1	18	.00776	.00743	.00711	.00679	.00648	.00618	.00590	.00562
18	2	3	.02204	.01966	.01752	.01561	.01392	.01240	.01106	.00986
18	2	4	.01930	.01733	.01556	.01396	.01253	.01124	.01008	.00905
18	2	5	.01752	.01584	.01430	.01291	.01165	.01051	.00948	.00855
18	2	6	.01627	.01478	.01342	.01218	.01105	.01002	.00909	.00824
18	2	7	.01533	.01400	.01278	.01165	.01062	.00968	.00882	.00803
18	2	8	.01460	.01340	.01228	.01126	.01031	.00943	.00863	.00790
18	2	9	.01402	.01292	.01190	.01095	.01007	.00926	.00851	.00782
18	2	10	.01354	.01254	.01160	.01072	.00990	.00915	.00844	.00779
18	2	11	.01314	.01222	.01136	.01055	.00978	.00907	.00841	.00779
18	2	12	.01281	.01197	.01117	.01042	.00971	.00904	.00841	.00783
18	2	13	.01253	.01176	.01102	.01033	.00967	.00904	.00846	.00790
18	2	14	.01229	.01159	.01092	.01028	.00967	.00908	.00853	.00801
18	2	15	.01210	.01147	.01085	.01027	.00970	.00917	.00865	.00816
18	2	16	.01194	.01138	.01084	.01031	.00979	.00930	.00883	.00837
18	2	17	.01183	.01135	.01000	.01042	.00996	.00952	.00909	.00869
18	2	18	.01181	.01143	.01106	.01068	.01030	.00993	.00957	.00921
18	3	4	.02550	.02309	.02090	.01891	.01710	.01545	.01396	.01262
18	3	5	.02318	.02112	.01923	.01750	.01592	.01447	.01315	.01194
18	3	6	.02154	.01973	.01807	.01653	.01511	.01380	.01261	.01151
18	3	7	.02031	.01870	.01721	.01582	.01453	.01334	.01224	.01122
18	3	8	.01935	.01790	.01655	.01528	.01410	.01301	.01199	.01104
18	3	9	.01858	.01727	.01604	.01488	.01379	.01277	.01182	.01094
18	3	10	.01795	.01676	.01563	.01456	.01356	.01261	.01172	.01089
18	3	11	.01743	.01634	.01531	.01433	.01340	.01252	.01168	.01090
18	3	12	.01699	.01600	.01506	.01415	.01329	.01247	.01169	.01096
18	3	13	.01662	.01573	.01486	.01403	.01324	.01248	.01175	.01106
18	3	14	.01631	.01550	.01472	.01397	.01324	.01253	.01186	.01121
18	3	15	.01605	.01534	.01464	.01396	.01329	.01265	.01202	.01142
18	3	16	.01584	.01522	.01461	.01401	.01341	.01283	.01227	.01172
18	3	17	.01570	.01519	.01467	.01416	.01364	.01314	.01264	.01214
18	3	18	.01567	.01530	.01491	.01452	.01411	.01371	.01329	.01288
18	4	5	.02909	.02669	.02447	.02241	.02051	.01876	.01715	.01568
18	4	6	.02705	.02495	.02300	.02118	.01948	.01791	.01646	.01512
18	4	7	.02551	.02366	.02191	.02028	.01875	.01732	.01599	.01475
18	4	8	.02432	.02266	.02108	.01960	.01820	.01689	.01566	.01452
18	4	9	.02336	.02186	.02044	.01908	.01780	.01659	.01545	.01438
18	4	10	.02257	.02122	.01993	.01869	.01751	.01639	.01533	.01433
18	4	11	.02192	.02070	.01952	.01839	.01730	.01626	.01528	.01434
18	4	12	.02137	.02027	.01920	.01817	.01717	.01621	.01529	.01441
18	4	13	.02091	.01992	.01896	.01802	.01710	.01622	.01537	.01455
18	4	14	.02052	.01964	.01878	.01793	.01710	.01629	.01551	.01475
18	4	15	.02020	.01943	.01867	.01792	.01717	.01644	.01572	.01503
18	4	16	.01994	.01929	.01864	.01798	.01733	.01668	.01604	.01541
18	4	17	.01976	.01924	.01872	.01818	.01763	.01708	.01653	.01598
18	4	18	.01972	.01938	.01902	.01864	.01824	.01782	.01739	.01695
18	5	6	.03306	.03069	.02845	.02635	.02438	.02254	.02083	.01923
18	5	7	.03121	.02911	.02712	.02524	.02347	.02180	.02024	.01877

Covariances of Inverse Gaussian Order Statistics
Shape Parameter k

n	i	j	1.6	1.7	1.8	1.9	2.0	2.1	2.2	2.3
18	5	8	.02976	.02789	.02611	.02441	.02280	.02128	.01984	.01848
18	5	9	.02859	.02692	.02531	.02377	.02230	.02090	.01957	.01831
18	5	10	.02763	.02614	.02468	.02329	.02194	.02065	.01942	.01825
18	5	11	.02684	.02550	.02419	.02291	.02168	.02050	.01936	.01826
18	5	12	.02617	.02497	.02379	.02264	.02152	.02043	.01938	.01836
18	5	13	.02561	.02455	.02349	.02246	.02144	.02044	.01947	.01853
18	5	14	.02514	.02421	.02328	.02235	.02144	.02054	.01965	.01879
18	5	15	.02474	.02395	.02314	.02234	.02153	.02072	.01993	.01914
18	5	16	.02443	.02378	.02311	.02242	.02173	.02103	.02033	.01964
18	5	17	.02421	.02372	.02320	.02266	.02210	.02153	.02095	.02035
18	5	18	.02416	.02389	.02358	.02324	.02286	.02246	.02204	.02159
18	6	7	.03760	.03527	.03304	.03091	.02889	.02698	.02517	.02346
18	6	8	.03587	.03380	.03181	.02990	.02808	.02634	.02468	.02311
18	6	9	.03447	.03264	.03085	.02913	.02747	.02588	.02436	.02290
18	6	10	.03333	.03169	.03010	.02854	.02703	.02558	.02417	.02282
18	6	11	.03238	.03093	.02949	.02809	.02672	.02539	.02410	.02285
18	6	12	.03158	.03030	.02902	.02776	.02652	.02531	.02412	.02297
18	6	13	.03091	.02978	.02866	.02754	.02642	.02533	.02425	.02319
18	6	14	.03034	.02937	.02840	.02741	.02643	.02544	.02447	.02351
18	6	15	.02987	.02906	.02823	.02739	.02654	.02568	.02481	.02395
18	6	16	.02949	.02886	.02819	.02750	.02679	.02606	.02532	.02457
18	6	17	.02922	.02879	.02831	.02779	.02725	.02668	.02608	.02547
18	6	18	.02917	.02900	.02878	.02850	.02819	.02783	.02744	.02702
18	7	8	.04289	.04063	.03843	.03630	.03426	.03229	.03040	.02860
18	7	9	.04124	.03924	.03728	.03538	.03353	.03174	.03002	.02836
18	7	10	.03988	.03812	.03638	.03467	.03300	.03137	.02979	.02827
18	7	11	.03875	.03720	.03566	.03413	.03262	.03115	.02971	.02830
18	7	12	.03781	.03645	.03509	.03373	.03239	.03105	.02974	.02845
18	7	13	.03700	.03584	.03466	.03346	.03227	.03108	.02989	.02872
18	7	14	.03633	.03535	.03434	.03332	.03227	.03123	.03017	.02912
18	7	15	.03577	.03498	.03415	.03329	.03241	.03151	.03060	.02968
18	7	16	.03532	.03473	.03410	.03343	.03272	.03198	.03122	.03044
18	7	17	.03500	.03465	.03424	.03379	.03328	.03274	.03216	.03155
18	7	18	.03494	.03491	.03481	.03465	.03443	.03415	.03383	.03347
18	8	9	.04917	.04702	.04489	.04280	.04076	.03877	.03683	.03496
18	8	10	.04757	.04569	.04381	.04195	.04012	.03832	.03656	.03485
18	8	11	.04624	.04460	.04295	.04131	.03968	.03806	.03646	.03490
18	8	12	.04511	.04371	.04228	.04084	.03939	.03795	.03651	.03509
18	8	13	.04416	.04298	.04176	.04052	.03925	.03798	.03670	.03543
18	8	14	.04336	.04240	.04139	.04034	.03926	.03816	.03705	.03592
18	8	15	.04270	.04196	.04116	.04032	.03944	.03852	.03757	.03660
18	8	16	.04217	.04167	.04110	.04048	.03981	.03909	.03834	.03755
18	8	17	.04179	.04157	.04128	.04092	.04050	.04002	.03949	.03892
18	8	18	.04172	.04188	.04196	.04196	.04189	.04175	.04154	.04127
18	9	10	.05678	.05479	.05279	.05079	.04880	.04682	.04488	.04297
18	9	11	.05520	.05350	.05177	.05002	.04826	.04651	.04476	.04303
18	9	12	.05388	.05244	.05097	.04946	.04793	.04638	.04483	.04328
18	9	13	.05275	.05158	.05035	.04908	.04777	.04643	.04507	.04370
18	9	14	.05181	.05089	.04991	.04887	.04778	.04666	.04549	.04431
18	9	15	.05102	.05037	.04964	.04885	.04799	.04709	.04614	.04515
18	9	16	.05039	.05002	.04957	.04905	.04845	.04779	.04708	.04631
18	9	17	.04995	.04991	.04979	.04958	.04929	.04893	.04849	.04800
18	9	18	.04986	.05029	.05061	.05084	.05098	.05104	.05101	.05090
18	10	11	.06620	.06446	.06266	.06083	.05896	.05708	.05519	.05329
18	10	12	.06463	.06320	.06170	.06015	.05856	.05693	.05527	.05360

Covariances of Inverse Gaussian Order Statistics
Shape Parameter k

n	i	j	1.6	1.7	1.8	1.9	2.0	2.1	2.2	2.3
18	10	13	.06329	.06217	.06097	.05970	.05837	.05700	.05558	.05413
18	10	14	.06217	.06135	.06045	.05946	.05840	.05728	.05611	.05489
18	10	15	.06123	.06073	.06013	.05944	.05867	.05782	.05691	.05593
18	10	16	.06048	.06032	.06005	.05969	.05923	.05869	.05806	.05737
18	10	17	.05996	.06020	.06032	.06034	.06025	.06008	.05981	.05946
18	10	18	.05987	.06065	.06132	.06188	.06232	.06267	.06291	.06305
18	11	12	.07817	.07680	.07534	.07379	.07216	.07048	.06874	.06697
18	11	13	.07658	.07558	.07446	.07325	.07195	.07057	.06913	.06763
18	11	14	.07524	.07460	.07383	.07296	.07199	.07093	.06980	.06858
18	11	15	.07412	.07385	.07346	.07295	.07233	.07161	.07079	.06989
18	11	16	.07323	.07336	.07337	.07326	.07303	.07268	.07224	.07169
18	11	17	.07261	.07322	.07370	.07406	.07429	.07440	.07440	.07430
18	11	18	.07250	.07378	.07493	.07595	.07684	.07761	.07825	.07877
18	12	13	.09390	.09311	.09217	.09110	.08991	.08860	.08720	.08570
18	12	14	.09228	.09192	.09141	.09076	.08998	.08907	.08805	.08692
18	12	15	.09093	.09103	.09097	.09076	.09040	.08992	.08931	.08858
18	12	16	.08985	.09044	.09087	.09115	.09129	.09128	.09113	.09086
18	12	17	.08910	.09028	.09130	.09216	.09287	.09344	.09386	.09415
18	12	18	.08899	.09098	.09282	.09452	.09606	.09746	.09870	.09981
18	13	14	.11545	.11558	.11551	.11525	.11481	.11420	.11344	.11253
18	13	15	.11380	.11448	.11497	.11526	.11537	.11531	.11508	.11469
18	13	16	.11248	.11377	.11487	.11578	.11651	.11706	.11743	.11763
18	13	17	.11157	.11359	.11542	.11707	.11854	.11983	.12094	.12189
18	13	18	.11144	.11448	.11736	.12007	.12261	.12497	.12716	.12918
18	14	15	.14676	.14841	.14982	.15099	.15192	.15263	.15311	.15337
18	14	16	.14511	.14754	.14974	.15170	.15344	.15495	.15624	.15730
18	14	17	.14397	.14733	.15048	.15341	.15612	.15862	.16090	.16296
18	14	18	.14385	.14852	.15302	.15734	.16147	.16540	.16914	.17267
18	15	16	.19620	.20064	.20480	.20868	.21226	.21556	.21857	.22130
18	15	17	.19475	.20043	.20587	.21106	.21599	.22066	.22507	.22921
18	15	18	.19465	.20210	.20938	.21647	.22337	.23005	.23651	.24274
18	16	17	.28544	.29571	.30573	.31548	.32494	.33411	.34295	.35147
18	16	18	.28545	.29829	.31100	.32358	.33598	.34818	.36018	.37193
18	17	18	.49147	.51783	.54431	.57086	.59744	.62400	.65048	.67686
19	1	2	.01708	.01488	.01297	.01131	.00987	.00862	.00754	.00660
19	1	3	.01385	.01218	.01071	.00942	.00829	.00730	.00643	.00567
19	1	4	.01207	.01069	.00947	.00838	.00743	.00658	.00583	.00518
19	1	5	.01093	.00974	.00867	.00773	.00688	.00613	.00547	.00488
19	1	6	.01013	.00907	.00812	.00727	.00651	.00583	.00522	.00468
19	1	7	.00953	.00857	.00771	.00694	.00624	.00562	.00505	.00455
19	1	8	.00906	.00819	.00740	.00669	.00605	.00546	.00494	.00446
19	1	9	.00868	.00789	.00716	.00650	.00590	.00535	.00486	.00441
19	1	10	.00838	.00764	.00697	.00635	.00579	.00527	.00481	.00438
19	1	11	.00812	.00744	.00681	.00624	.00571	.00522	.00478	.00437
19	1	12	.00791	.00728	.00669	.00615	.00565	.00519	.00477	.00438
19	1	13	.00773	.00714	.00659	.00608	.00561	.00518	.00478	.00441
19	1	14	.00757	.00703	.00652	.00604	.00560	.00519	.00481	.00445
19	1	15	.00744	.00694	.00646	.00602	.00560	.00521	.00485	.00451
19	1	16	.00734	.00687	.00643	.00602	.00563	.00526	.00492	.00460
19	1	17	.00725	.00683	.00643	.00604	.00568	.00534	.00502	.00472
19	1	18	.00719	.00681	.00646	.00611	.00578	.00547	.00517	.00489
19	1	19	.00718	.00687	.00656	.00627	.00598	.00571	.00544	.00519
19	2	3	.02077	.01848	.01644	.01462	.01300	.01156	.01028	.00915
19	2	4	.01815	.01626	.01457	.01304	.01167	.01045	.00935	.00838
19	2	5	.01646	.01483	.01336	.01203	.01083	.00975	.00878	.00790

Covariances of Inverse Gaussian Order Statistics
Shape Parameter k

n	i	j	1.6	1.7	1.8	1.9	2.0	2.1	2.2	2.3
19	2	6	.01526	.01383	.01252	.01134	.01026	.00928	.00839	.00759
19	2	7	.01436	.01308	.01190	.01083	.00984	.00894	.00812	.00738
19	2	8	.01366	.01250	.01143	.01044	.00953	.00870	.00794	.00724
19	2	9	.01311	.01204	.01106	.01015	.00930	.00853	.00781	.00716
19	2	10	.01265	.01167	.01076	.00992	.00913	.00840	.00773	.00711
19	2	11	.01227	.01137	.01053	.00974	.00900	.00832	.00769	.00710
19	2	12	.01194	.01112	.01034	.00960	.00892	.00827	.00767	.00711
19	2	13	.01167	.01091	.01019	.00950	.00886	.00826	.00769	.00716
19	2	14	.01144	.01074	.01007	.00944	.00884	.00827	.00773	.00723
19	2	15	.01124	.01060	.00999	.00940	.00884	.00831	.00781	.00733
19	2	16	.01108	.01050	.00994	.00940	.00889	.00839	.00792	.00747
19	2	17	.01095	.01043	.00993	.00944	.00897	.00852	.00808	.00767
19	2	18	.01086	.01042	.00998	.00955	.00913	.00872	.00833	.00795
19	2	19	.01085	.01050	.01015	.00979	.00944	.00910	.00876	.00842
19	3	4	.02384	.02154	.01945	.01755	.01583	.01427	.01287	.01160
19	3	5	.02164	.01967	.01786	.01621	.01470	.01333	.01208	.01095
19	3	6	.02008	.01835	.01675	.01528	.01393	.01269	.01156	.01053
19	3	7	.01891	.01737	.01593	.01460	.01338	.01224	.01120	.01025
19	3	8	.01800	.01661	.01530	.01409	.01296	.01192	.01095	.01006
19	3	9	.01727	.01600	.01481	.01369	.01265	.01168	.01078	.00994
19	3	10	.01667	.01551	.01442	.01339	.01242	.01151	.01067	.00988
19	3	11	.01617	.01511	.01410	.01315	.01225	.01140	.01061	.00986
19	3	12	.01575	.01478	.01385	.01297	.01213	.01134	.01059	.00989
19	3	13	.01539	.01450	.01365	.01284	.01206	.01132	.01061	.00995
19	3	14	.01509	.01428	.01350	.01275	.01203	.01134	.01068	.01005
19	3	15	.01483	.01410	.01339	.01270	.01204	.01140	.01078	.01019
19	3	16	.01462	.01396	.01333	.01270	.01209	.01151	.01094	.01039
19	3	17	.01444	.01388	.01331	.01276	.01221	.01168	.01116	.01066
19	3	18	.01433	.01385	.01338	.01290	.01243	.01196	.01149	.01104
19	3	19	.01431	.01396	.01360	.01323	.01285	.01247	.01209	.01171
19	4	5	.02701	.02471	.02259	.02063	.01883	.01718	.01566	.01428
19	4	6	.02508	.02307	.02120	.01946	.01785	.01637	.01500	.01374
19	4	7	.02363	.02184	.02017	.01860	.01715	.01579	.01454	.01337
19	4	8	.02250	.02089	.01938	.01796	.01662	.01538	.01421	.01313
19	4	9	.02159	.02014	.01876	.01746	.01623	.01508	.01400	.01298
19	4	10	.02084	.01952	.01827	.01707	.01594	.01486	.01385	.01290
19	4	11	.02022	.01902	.01787	.01677	.01572	.01472	.01378	.01288
19	4	12	.01970	.01861	.01755	.01654	.01557	.01464	.01376	.01292
19	4	13	.01926	.01827	.01730	.01637	.01548	.01461	.01379	.01300
19	4	14	.01888	.01798	.01711	.01626	.01544	.01464	.01387	.01313
19	4	15	.01856	.01776	.01698	.01620	.01545	.01472	.01401	.01332
19	4	16	.01829	.01759	.01689	.01621	.01553	.01486	.01421	.01357
19	4	17	.01808	.01748	.01688	.01628	.01568	.01508	.01450	.01392
19	4	18	.01793	.01745	.01696	.01646	.01595	.01544	.01493	.01443
19	4	19	.01791	.01759	.01724	.01688	.01650	.01611	.01570	.01529
19	5	6	.03047	.02819	.02606	.02406	.02219	.02046	.01884	.01735
19	5	7	.02873	.02671	.02480	.02301	.02133	.01975	.01827	.01690
19	5	8	.02736	.02556	.02384	.02222	.02068	.01924	.01788	.01660
19	5	9	.02626	.02464	.02309	.02161	.02020	.01886	.01760	.01641
19	5	10	.02536	.02390	.02248	.02113	.01984	.01860	.01743	.01632
19	5	11	.02461	.02329	.02200	.02076	.01957	.01843	.01734	.01629
19	5	12	.02398	.02278	.02161	.02048	.01938	.01833	.01731	.01634
19	5	13	.02344	.02237	.02131	.02028	.01927	.01829	.01735	.01644
19	5	14	.02298	.02202	.02107	.02014	.01922	.01833	.01745	.01661
19	5	15	.02260	.02175	.02091	.02007	.01924	.01843	.01763	.01685

Covariances of Inverse Gaussian Order Statistics
Shape Parameter k

n	i	j	1.6	1.7	1.8	1.9	2.0	2.1	2.2	2.3
19	5	16	.02227	.02154	.02081	.02007	.01934	.01860	.01788	.01717
19	5	17	.02201	.02141	.02079	.02016	.01953	.01889	.01825	.01761
19	5	18	.02183	.02137	.02089	.02039	.01987	.01933	.01879	.01825
19	5	19	.02181	.02154	.02124	.02091	.02055	.02017	.01976	.01935
19	6	7	.03439	.03214	.03000	.02798	.02606	.02425	.02255	.02096
19	6	8	.03277	.03077	.02885	.02702	.02528	.02363	.02207	.02059
19	6	9	.03146	.02967	.02795	.02629	.02470	.02318	.02174	.02037
19	6	10	.03039	.02878	.02722	.02572	.02426	.02287	.02153	.02025
19	6	11	.02950	.02805	.02664	.02527	.02394	.02265	.02142	.02023
19	6	12	.02874	.02745	.02618	.02493	.02372	.02253	.02139	.02028
19	6	13	.02810	.02695	.02581	.02469	.02358	.02250	.02144	.02041
19	6	14	.02755	.02654	.02553	.02452	.02352	.02254	.02157	.02062
19	6	15	.02709	.02622	.02533	.02444	.02355	.02266	.02178	.02092
19	6	16	.02671	.02597	.02521	.02444	.02366	.02288	.02210	.02132
19	6	17	.02640	.02581	.02519	.02455	.02390	.02323	.02255	.02187
19	6	18	.02618	.02577	.02531	.02483	.02431	.02378	.02323	.02266
19	6	19	.02615	.02597	.02574	.02546	.02515	.02480	.02442	.02402
19	7	8	.03889	.03670	.03458	.03255	.03060	.02874	.02696	.02527
19	7	9	.03736	.03540	.03351	.03167	.02990	.02820	.02656	.02500
19	7	10	.03609	.03435	.03265	.03099	.02938	.02782	.02631	.02486
19	7	11	.03504	.03349	.03196	.03046	.02899	.02756	.02618	.02483
19	7	12	.03415	.03277	.03141	.03006	.02873	.02742	.02615	.02490
19	7	13	.03339	.03218	.03097	.02976	.02856	.02738	.02621	.02507
19	7	14	.03275	.03170	.03064	.02957	.02850	.02743	.02637	.02532
19	7	15	.03220	.03131	.03040	.02947	.02853	.02758	.02663	.02569
19	7	16	.03174	.03102	.03026	.02947	.02867	.02785	.02702	.02618
19	7	17	.03138	.03083	.03024	.02961	.02895	.02827	.02757	.02686
19	7	18	.03113	.03078	.03038	.02994	.02946	.02894	.02840	.02783
19	7	19	.03109	.03102	.03089	.03071	.03047	.03019	.02986	.02950
19	8	9	.04417	.04206	.03999	.03797	.03601	.03411	.03228	.03052
19	8	10	.04269	.04082	.03897	.03716	.03539	.03366	.03198	.03035
19	8	11	.04145	.03980	.03816	.03654	.03494	.03336	.03182	.03032
19	8	12	.04040	.03896	.03751	.03606	.03462	.03319	.03179	.03041
19	8	13	.03951	.03826	.03699	.03571	.03442	.03314	.03187	.03061
19	8	14	.03876	.03769	.03659	.03548	.03435	.03321	.03207	.03093
19	8	15	.03811	.03723	.03631	.03536	.03439	.03339	.03239	.03138
19	8	16	.03758	.03689	.03615	.03537	.03456	.03372	.03286	.03198
19	8	17	.03715	.03666	.03612	.03554	.03490	.03423	.03353	.03280
19	8	18	.03685	.03661	.03630	.03593	.03551	.03505	.03454	.03399
19	8	19	.03682	.03690	.03691	.03686	.03673	.03655	.03631	.03603
19	9	10	.05045	.04846	.04648	.04452	.04258	.04068	.03882	.03700
19	9	11	.04900	.04726	.04552	.04377	.04204	.04033	.03863	.03697
19	9	12	.04777	.04627	.04475	.04321	.04167	.04013	.03860	.03708
19	9	13	.04673	.04545	.04414	.04280	.04144	.04007	.03870	.03733
19	9	14	.04584	.04478	.04367	.04253	.04135	.04015	.03894	.03772
19	9	15	.04509	.04424	.04334	.04239	.04140	.04038	.03933	.03827
19	9	16	.04445	.04383	.04315	.04241	.04161	.04078	.03991	.03900
19	9	17	.04395	.04357	.04312	.04260	.04203	.04140	.04072	.04001
19	9	18	.04361	.04351	.04333	.04308	.04276	.04238	.04194	.04145
19	9	19	.04356	.04386	.04407	.04419	.04423	.04420	.04410	.04393
19	10	11	.05806	.05625	.05442	.05256	.05070	.04885	.04700	.04517
19	10	12	.05662	.05509	.05351	.05190	.05026	.04862	.04696	.04531
19	10	13	.05540	.05412	.05279	.05141	.05000	.04855	.04709	.04562
19	10	14	.05435	.05333	.05224	.05109	.04989	.04866	.04739	.04610
19	10	15	.05347	.05269	.05185	.05093	.04996	.04894	.04787	.04677

Covariances of Inverse Gaussian Order Statistics
Shape Parameter k

n	i	j	1.6	1.7	1.8	1.9	2.0	2.1	2.2	2.3
19	10	16	.05273	.05222	.05162	.05096	.05022	.04942	.04857	.04767
19	10	17	.05213	.05191	.05160	.05120	.05072	.05018	.04956	.04889
19	10	18	.05173	.05184	.05185	.05177	.05161	.05137	.05105	.05066
19	10	19	.05168	.05226	.05273	.05310	.05338	.05357	.05367	.05369
19	11	12	.06749	.06595	.06435	.06268	.06097	.05923	.05747	.05568
19	11	13	.06605	.06481	.06349	.06211	.06066	.05917	.05763	.05607
19	11	14	.06482	.06388	.06284	.06173	.06055	.05930	.05800	.05666
19	11	15	.06377	.06313	.06238	.06155	.06063	.05965	.05859	.05748
19	11	16	.06290	.06256	.06212	.06158	.06095	.06024	.05945	.05859
19	11	17	.06220	.06220	.06209	.06188	.06156	.06116	.06067	.06010
19	11	18	.06172	.06212	.06240	.06258	.06264	.06261	.06248	.06227
19	11	19	.06167	.06263	.06346	.06418	.06479	.06529	.06569	.06598
19	12	13	.07949	.07835	.07709	.07574	.07431	.07280	.07122	.06959
19	12	14	.07803	.07723	.07632	.07530	.07418	.07297	.07168	.07033
19	12	15	.07678	.07634	.07577	.07509	.07429	.07340	.07242	.07135
19	12	16	.07574	.07567	.07546	.07514	.07469	.07414	.07348	.07273
19	12	17	.07491	.07524	.07544	.07550	.07545	.07527	.07499	.07460
19	12	18	.07435	.07515	.07582	.07636	.07677	.07706	.07723	.07729
19	12	19	.07430	.07577	.07711	.07832	.07940	.08036	.08118	.08188
19	13	14	.09525	.09471	.09402	.09318	.09221	.09112	.08991	.08860
19	13	15	.09375	.09364	.09336	.09294	.09237	.09167	.09084	.08990
19	13	16	.09251	.09283	.09300	.09301	.09287	.09259	.09218	.09164
19	13	17	.09151	.09232	.09298	.09347	.09382	.09401	.09407	.09399
19	13	18	.09083	.09222	.09346	.09454	.09547	.09625	.09688	.09737
19	13	19	.09078	.09299	.09506	.09697	.09874	.10036	.10182	.10315
19	14	15	.11685	.11726	.11747	.11749	.11732	.11697	.11646	.11580
19	14	16	.11533	.11628	.11704	.11760	.11797	.11816	.11818	.11803
19	14	17	.11411	.11567	.11703	.11820	.11919	.11999	.12061	.12106
19	14	18	.11329	.11556	.11765	.11956	.12129	.12284	.12420	.12540
19	14	19	.11325	.11654	.11967	.12264	.12544	.12807	.13053	.13281
19	15	16	.14822	.15021	.15195	.15345	.15471	.15573	.15653	.15711
19	15	17	.14671	.14946	.15197	.15426	.15632	.15814	.15975	.16113
19	15	18	.14569	.14935	.15281	.15605	.15908	.16190	.16449	.16687
19	15	19	.14567	.15064	.15544	.16007	.16451	.16877	.17283	.17669
19	16	17	.19777	.20260	.20716	.21143	.21543	.21914	.22256	.22570
19	16	18	.19648	.20252	.20834	.21392	.21925	.22433	.22915	.23370
19	16	19	.19651	.20431	.21196	.21943	.22672	.23380	.24068	.24733
19	17	18	.28718	.29793	.30845	.31872	.32871	.33842	.34782	.35691
19	17	19	.28738	.30066	.31385	.32692	.33983	.35257	.36511	.37743
19	18	19	.49357	.52056	.54772	.57499	.60232	.62966	.65697	.68420
20	1	2	.01629	.01416	.01232	.01073	.00935	.00815	.00712	.00622
20	1	3	.01318	.01156	.01015	.00891	.00783	.00688	.00605	.00533
20	1	4	.01147	.01013	.00895	.00791	.00700	.00619	.00548	.00485
20	1	5	.01037	.00922	.00819	.00728	.00647	.00576	.00512	.00456
20	1	6	.00959	.00857	.00766	.00684	.00611	.00546	.00488	.00437
20	1	7	.00901	.00809	.00727	.00652	.00585	.00525	.00472	.00424
20	1	8	.00856	.00772	.00696	.00628	.00566	.00510	.00460	.00415
20	1	9	.00820	.00743	.00673	.00609	.00551	.00499	.00452	.00409
20	1	10	.00791	.00719	.00654	.00594	.00540	.00491	.00446	.00405
20	1	11	.00766	.00700	.00639	.00583	.00532	.00485	.00442	.00404
20	1	12	.00745	.00683	.00626	.00574	.00526	.00481	.00441	.00404
20	1	13	.00728	.00670	.00616	.00567	.00521	.00479	.00441	.00405
20	1	14	.00713	.00659	.00608	.00562	.00519	.00479	.00442	.00408
20	1	15	.00700	.00649	.00602	.00559	.00518	.00480	.00445	.00412
20	1	16	.00689	.00642	.00598	.00557	.00519	.00483	.00450	.00418

Covariances of Inverse Gaussian Order Statistics
Shape Parameter k

n	i	j	1.6	1.7	1.8	1.9	2.0	2.1	2.2	2.3
20	1	17	.00679	.00636	.00596	.00558	.00522	.00488	.00456	.00427
20	1	18	.00672	.00633	.00596	.00560	.00527	.00495	.00466	.00437
20	1	19	.00667	.00632	.00599	.00567	.00536	.00507	.00479	.00453
20	1	20	.00667	.00637	.00609	.00581	.00555	.00529	.00504	.00480
20	2	3	.01965	.01744	.01548	.01374	.01219	.01082	.00961	.00854
20	2	4	.01714	.01532	.01369	.01223	.01093	.00976	.00872	.00779
20	2	5	.01552	.01396	.01254	.01127	.01012	.00909	.00816	.00733
20	2	6	.01437	.01299	.01174	.01060	.00957	.00863	.00779	.00703
20	2	7	.01351	.01227	.01114	.01011	.00916	.00831	.00753	.00682
20	2	8	.01284	.01172	.01068	.00973	.00886	.00807	.00734	.00668
20	2	9	.01231	.01128	.01032	.00945	.00864	.00790	.00721	.00659
20	2	10	.01187	.01092	.01004	.00922	.00847	.00777	.00713	.00654
20	2	11	.01150	.01062	.00980	.00904	.00834	.00768	.00707	.00651
20	2	12	.01119	.01038	.00962	.00890	.00824	.00762	.00705	.00651
20	2	13	.01093	.01017	.00947	.00880	.00817	.00759	.00704	.00653
20	2	14	.01070	.01000	.00935	.00872	.00814	.00758	.00707	.00658
20	2	15	.01051	.00986	.00925	.00867	.00812	.00760	.00711	.00665
20	2	16	.01034	.00975	.00919	.00865	.00814	.00765	.00719	.00675
20	2	17	.01020	.00967	.00915	.00866	.00818	.00773	.00729	.00688
20	2	18	.01009	.00962	.00915	.00870	.00826	.00784	.00744	.00706
20	2	19	.01002	.00960	.00920	.00880	.00841	.00803	.00767	.00731
20	2	20	.01001	.00968	.00935	.00903	.00870	.00838	.00806	.00775
20	3	4	.02240	.02019	.01818	.01637	.01473	.01325	.01192	.01073
20	3	5	.02030	.01841	.01667	.01509	.01366	.01235	.01117	.01010
20	3	6	.01882	.01715	.01561	.01421	.01292	.01174	.01067	.00969
20	3	7	.01770	.01621	.01483	.01356	.01238	.01131	.01032	.00941
20	3	8	.01683	.01548	.01423	.01306	.01198	.01099	.01007	.00922
20	3	9	.01613	.01490	.01375	.01268	.01168	.01075	.00989	.00910
20	3	10	.01556	.01443	.01337	.01238	.01145	.01058	.00977	.00902
20	3	11	.01508	.01404	.01306	.01214	.01127	.01046	.00970	.00899
20	3	12	.01468	.01372	.01282	.01196	.01115	.01038	.00966	.00899
20	3	13	.01433	.01345	.01261	.01182	.01106	.01034	.00966	.00902
20	3	14	.01404	.01323	.01246	.01171	.01101	.01033	.00969	.00909
20	3	15	.01378	.01305	.01233	.01165	.01099	.01036	.00976	.00919
20	3	16	.01357	.01290	.01225	.01162	.01101	.01042	.00986	.00932
20	3	17	.01339	.01279	.01220	.01163	.01107	.01053	.01001	.00950
20	3	18	.01324	.01272	.01220	.01169	.01118	.01069	.01021	.00975
20	3	19	.01315	.01271	.01226	.01182	.01138	.01094	.01052	.01010
20	3	20	.01314	.01281	.01247	.01212	.01177	.01141	.01106	.01070
20	4	5	.02520	.02300	.02097	.01910	.01739	.01583	.01439	.01309
20	4	6	.02338	.02144	.01965	.01799	.01646	.01505	.01375	.01256
20	4	7	.02200	.02028	.01867	.01717	.01578	.01450	.01331	.01221
20	4	8	.02093	.01938	.01792	.01655	.01528	.01409	.01299	.01197
20	4	9	.02006	.01866	.01732	.01607	.01490	.01379	.01277	.01181
20	4	10	.01936	.01807	.01685	.01569	.01460	.01358	.01261	.01171
20	4	11	.01876	.01759	.01646	.01540	.01438	.01343	.01252	.01167
20	4	12	.01826	.01719	.01615	.01517	.01422	.01333	.01248	.01167
20	4	13	.01784	.01685	.01590	.01499	.01411	.01328	.01248	.01172
20	4	14	.01747	.01657	.01570	.01486	.01405	.01327	.01252	.01180
20	4	15	.01716	.01635	.01555	.01478	.01403	.01330	.01260	.01193
20	4	16	.01689	.01616	.01544	.01474	.01405	.01338	.01273	.01210
20	4	17	.01667	.01602	.01538	.01475	.01413	.01352	.01292	.01234
20	4	18	.01649	.01594	.01538	.01483	.01427	.01373	.01319	.01266
20	4	19	.01637	.01592	.01546	.01500	.01453	.01405	.01358	.01311
20	4	20	.01636	.01605	.01573	.01538	.01502	.01465	.01428	.01389

Covariances of Inverse Gaussian Order Statistics
Shape Parameter k

n	i	j	1.6	1.7	1.8	1.9	2.0	2.1	2.2	2.3
20	5	6	.02825	.02606	.02402	.02211	.02034	.01870	.01717	.01577
20	5	7	.02660	.02466	.02283	.02112	.01951	.01802	.01662	.01533
20	5	8	.02532	.02357	.02192	.02036	.01890	.01752	.01623	.01503
20	5	9	.02428	.02270	.02120	.01977	.01843	.01715	.01596	.01483
20	5	10	.02343	.02199	.02062	.01931	.01807	.01689	.01577	.01471
20	5	11	.02271	.02141	.02016	.01895	.01780	.01670	.01566	.01466
20	5	12	.02211	.02092	.01978	.01867	.01760	.01658	.01560	.01467
20	5	13	.02160	.02052	.01947	.01845	.01747	.01652	.01560	.01473
20	5	14	.02116	.02018	.01923	.01830	.01739	.01651	.01566	.01483
20	5	15	.02078	.01991	.01904	.01820	.01736	.01655	.01576	.01499
20	5	16	.02046	.01968	.01891	.01815	.01740	.01665	.01593	.01521
20	5	17	.02019	.01952	.01884	.01817	.01749	.01682	.01616	.01551
20	5	18	.01997	.01941	.01884	.01826	.01767	.01708	.01649	.01591
20	5	19	.01983	.01939	.01894	.01847	.01798	.01749	.01699	.01648
20	5	20	.01982	.01956	.01926	.01894	.01860	.01823	.01785	.01746
20	6	7	.03166	.02950	.02745	.02552	.02370	.02198	.02038	.01888
20	6	8	.03014	.02821	.02636	.02461	.02295	.02139	.01991	.01852
20	6	9	.02891	.02717	.02550	.02391	.02239	.02094	.01957	.01828
20	6	10	.02790	.02633	.02482	.02336	.02196	.02062	.01935	.01814
20	6	11	.02706	.02564	.02426	.02292	.02164	.02040	.01921	.01808
20	6	12	.02635	.02506	.02381	.02258	.02140	.02025	.01915	.01809
20	6	13	.02574	.02458	.02344	.02233	.02124	.02018	.01915	.01816
20	6	14	.02521	.02418	.02315	.02214	.02114	.02017	.01922	.01829
20	6	15	.02477	.02385	.02293	.02202	.02111	.02022	.01935	.01849
20	6	16	.02438	.02358	.02278	.02196	.02115	.02035	.01955	.01876
20	6	17	.02407	.02339	.02269	.02198	.02127	.02055	.01984	.01913
20	6	18	.02381	.02326	.02269	.02210	.02149	.02087	.02025	.01962
20	6	19	.02364	.02324	.02281	.02235	.02187	.02137	.02085	.02032
20	6	20	.02363	.02343	.02320	.02292	.02262	.02228	.02192	.02153
20	7	8	.03554	.03342	.03139	.02944	.02758	.02582	.02414	.02255
20	7	9	.03411	.03221	.03038	.02861	.02691	.02529	.02374	.02227
20	7	10	.03293	.03122	.02956	.02796	.02640	.02491	.02347	.02210
20	7	11	.03194	.03040	.02890	.02744	.02602	.02464	.02331	.02203
20	7	12	.03110	.02973	.02837	.02704	.02574	.02447	.02324	.02204
20	7	13	.03039	.02916	.02794	.02673	.02554	.02438	.02324	.02213
20	7	14	.02977	.02868	.02760	.02651	.02543	.02437	.02332	.02230
20	7	15	.02925	.02830	.02733	.02637	.02540	.02444	.02348	.02254
20	7	16	.02880	.02798	.02715	.02630	.02545	.02459	.02373	.02287
20	7	17	.02842	.02775	.02705	.02633	.02559	.02484	.02408	.02332
20	7	18	.02812	.02760	.02705	.02646	.02585	.02522	.02457	.02391
20	7	19	.02792	.02758	.02719	.02677	.02631	.02582	.02531	.02477
20	7	20	.02791	.02781	.02766	.02746	.02721	.02692	.02660	.02624
20	8	9	.04004	.03798	.03598	.03404	.03216	.03035	.02862	.02695
20	8	10	.03866	.03682	.03502	.03327	.03156	.02990	.02830	.02676
20	8	11	.03751	.03587	.03425	.03266	.03110	.02958	.02811	.02668
20	8	12	.03653	.03507	.03362	.03218	.03077	.02938	.02802	.02669
20	8	13	.03569	.03441	.03311	.03182	.03054	.02928	.02803	.02680
20	8	14	.03498	.03385	.03271	.03156	.03041	.02927	.02813	.02700
20	8	15	.03436	.03340	.03240	.03140	.03038	.02935	.02832	.02730
20	8	16	.03384	.03303	.03219	.03132	.03043	.02953	.02862	.02770
20	8	17	.03340	.03276	.03207	.03135	.03061	.02983	.02904	.02824
20	8	18	.03305	.03259	.03207	.03151	.03092	.03029	.02964	.02897
20	8	19	.03281	.03256	.03224	.03188	.03147	.03102	.03053	.03001
20	8	20	.03280	.03283	.03280	.03270	.03255	.03234	.03208	.03178
20	9	10	.04531	.04334	.04140	.03949	.03762	.03580	.03403	.03231

Covariances of Inverse Gaussian Order Statistics
Shape Parameter k

n	i	j	1.6	1.7	1.8	1.9	2.0	2.1	2.2	2.3
20	9	11	.04397	.04223	.04050	.03878	.03709	.03543	.03380	.03221
20	9	12	.04283	.04130	.03976	.03823	.03670	.03519	.03370	.03224
20	9	13	.04186	.04052	.03917	.03780	.03643	.03507	.03371	.03237
20	9	14	.04102	.03987	.03870	.03750	.03628	.03506	.03384	.03262
20	9	15	.04031	.03934	.03834	.03730	.03624	.03516	.03407	.03298
20	9	16	.03970	.03892	.03809	.03722	.03631	.03538	.03443	.03346
20	9	17	.03919	.03860	.03795	.03726	.03652	.03575	.03494	.03411
20	9	18	.03878	.03840	.03795	.03745	.03690	.03630	.03566	.03499
20	9	19	.03850	.03836	.03815	.03788	.03755	.03716	.03673	.03624
20	9	20	.03849	.03869	.03881	.03886	.03884	.03875	.03860	.03839
20	10	11	.05159	.04976	.04792	.04609	.04426	.04245	.04067	.03891
20	10	12	.05027	.04868	.04706	.04543	.04380	.04217	.04055	.03895
20	10	13	.04914	.04777	.04637	.04494	.04349	.04203	.04057	.03912
20	10	14	.04816	.04701	.04581	.04458	.04331	.04203	.04073	.03942
20	10	15	.04733	.04639	.04539	.04435	.04327	.04215	.04101	.03985
20	10	16	.04662	.04589	.04510	.04425	.04336	.04242	.04145	.04044
20	10	17	.04602	.04552	.04494	.04430	.04361	.04286	.04206	.04123
20	10	18	.04555	.04529	.04495	.04454	.04406	.04352	.04293	.04229
20	10	19	.04523	.04525	.04519	.04505	.04484	.04456	.04421	.04380
20	10	20	.04521	.04564	.04597	.04621	.04637	.04646	.04646	.04639
20	11	12	.05921	.05758	.05590	.05419	.05246	.05072	.04897	.04723
20	11	13	.05789	.05652	.05509	.05361	.05210	.05056	.04901	.04744
20	11	14	.05676	.05563	.05444	.05319	.05190	.05056	.04919	.04780
20	11	15	.05578	.05490	.05395	.05293	.05185	.05072	.04954	.04833
20	11	16	.05495	.05432	.05361	.05282	.05196	.05104	.05007	.04905
20	11	17	.05426	.05388	.05342	.05288	.05226	.05157	.05081	.05001
20	11	18	.05370	.05361	.05343	.05316	.05280	.05237	.05186	.05129
20	11	19	.05333	.05357	.05372	.05378	.05374	.05361	.05341	.05313
20	11	20	.05332	.05404	.05465	.05516	.05558	.05590	.05612	.05626
20	12	13	.06866	.06731	.06588	.06439	.06283	.06123	.05959	.05792
20	12	14	.06733	.06627	.06512	.06389	.06260	.06124	.05982	.05837
20	12	15	.06619	.06542	.06454	.06358	.06254	.06143	.06025	.05902
20	12	16	.06521	.06473	.06414	.06346	.06268	.06183	.06090	.05990
20	12	17	.06439	.06422	.06393	.06354	.06305	.06247	.06181	.06107
20	12	18	.06374	.06390	.06394	.06388	.06371	.06344	.06308	.06263
20	12	19	.06330	.06386	.06430	.06462	.06484	.06495	.06496	.06487
20	12	20	.06330	.06441	.06541	.06629	.06706	.06771	.06826	.06870
20	13	14	.08068	.07975	.07870	.07754	.07629	.07494	.07352	.07204
20	13	15	.07933	.07874	.07802	.07718	.07623	.07519	.07406	.07284
20	13	16	.07817	.07792	.07754	.07704	.07641	.07568	.07485	.07393
20	13	17	.07720	.07732	.07729	.07714	.07686	.07647	.07597	.07537
20	13	18	.07643	.07694	.07732	.07756	.07767	.07766	.07754	.07730
20	13	19	.07592	.07690	.07775	.07847	.07905	.07951	.07985	.08006
20	13	20	.07592	.07757	.07910	.08049	.08176	.08289	.08389	.08477
20	14	15	.09647	.09617	.09571	.09510	.09434	.09345	.09244	.09132
20	14	16	.09509	.09520	.09515	.09493	.09457	.09407	.09344	.09268
20	14	17	.09393	.09447	.09485	.09507	.09514	.09506	.09484	.09449
20	14	18	.09301	.09403	.09490	.09560	.09615	.09654	.09679	.09691
20	14	19	.09239	.09399	.09544	.09672	.09786	.09884	.09967	.10036
20	14	20	.09240	.09482	.09709	.09922	.10120	.10303	.10471	.10624
20	15	16	.11812	.11880	.11927	.11955	.11964	.11955	.11928	.11884
20	15	17	.11671	.11792	.11893	.11975	.12037	.12081	.12107	.12115
20	15	18	.11559	.11739	.11900	.12042	.12165	.12270	.12356	.12425
20	15	19	.11485	.11736	.11969	.12185	.12382	.12561	.12723	.12866
20	15	20	.11488	.11841	.12178	.12500	.12805	.13093	.13364	.13618

Covariances of Inverse Gaussian Order Statistics
Shape Parameter k

n	i	j	1.6	1.7	1.8	1.9	2.0	2.1	2.2	2.3
20	16	17	.14955	.15185	.15390	.15571	.15728	.15862	.15972	.16060
20	16	18	.14816	.15121	.15402	.15661	.15897	.16110	.16301	.16469
20	16	19	.14725	.15120	.15494	.15848	.16181	.16493	.16783	.17051
20	16	20	.14732	.15257	.15765	.16258	.16732	.17188	.17625	.18042
20	17	18	.19920	.20439	.20932	.21398	.21836	.22246	.22628	.22981
20	17	19	.19804	.20444	.21062	.21656	.22227	.22773	.23293	.23788
20	17	20	.19820	.20633	.21433	.22216	.22981	.23728	.24454	.25159
20	18	19	.28877	.29997	.31096	.32171	.33220	.34243	.35236	.36199
20	18	20	.28914	.30284	.31648	.33000	.34340	.35664	.36970	.38255
20	19	20	.49550	.52309	.55087	.57881	.60684	.63492	.66300	.69104
21	1	2	.01558	.01352	.01175	.01021	.00888	.00773	.00674	.00588
21	1	3	.01257	.01101	.00965	.00846	.00742	.00651	.00571	.00502
21	1	4	.01092	.00963	.00850	.00750	.00662	.00584	.00516	.00456
21	1	5	.00986	.00875	.00776	.00689	.00611	.00542	.00482	.00428
21	1	6	.00912	.00813	.00725	.00646	.00576	.00514	.00458	.00409
21	1	7	.00856	.00767	.00687	.00615	.00551	.00493	.00442	.00396
21	1	8	.00812	.00731	.00657	.00591	.00532	.00478	.00430	.00387
21	1	9	.00778	.00702	.00634	.00573	.00517	.00467	.00422	.00381
21	1	10	.00749	.00679	.00616	.00558	.00506	.00459	.00416	.00377
21	1	11	.00725	.00660	.00601	.00547	.00498	.00453	.00412	.00375
21	1	12	.00705	.00644	.00589	.00538	.00491	.00449	.00410	.00374
21	1	13	.00688	.00631	.00579	.00531	.00486	.00446	.00409	.00375
21	1	14	.00673	.00620	.00571	.00525	.00483	.00445	.00409	.00376
21	1	15	.00660	.00610	.00564	.00521	.00482	.00445	.00411	.00379
21	1	16	.00649	.00603	.00559	.00519	.00481	.00446	.00414	.00384
21	1	17	.00640	.00596	.00556	.00518	.00482	.00449	.00418	.00389
21	1	18	.00632	.00592	.00554	.00519	.00485	.00454	.00425	.00397
21	1	19	.00625	.00589	.00554	.00521	.00490	.00461	.00433	.00407
21	1	20	.00621	.00589	.00557	.00528	.00499	.00472	.00446	.00422
21	1	21	.00621	.00594	.00567	.00541	.00516	.00492	.00469	.00447
21	2	3	.01865	.01652	.01464	.01296	.01149	.01018	.00902	.00800
21	2	4	.01624	.01449	.01292	.01152	.01027	.00916	.00816	.00728
21	2	5	.01469	.01318	.01182	.01059	.00950	.00851	.00763	.00684
21	2	6	.01358	.01225	.01104	.00995	.00896	.00807	.00727	.00654
21	2	7	.01276	.01156	.01047	.00948	.00857	.00775	.00701	.00634
21	2	8	.01212	.01103	.01003	.00912	.00828	.00752	.00683	.00620
21	2	9	.01160	.01060	.00968	.00883	.00806	.00735	.00670	.00610
21	2	10	.01118	.01026	.00940	.00861	.00789	.00722	.00660	.00604
21	2	11	.01082	.00997	.00917	.00844	.00775	.00712	.00654	.00601
21	2	12	.01052	.00973	.00899	.00830	.00766	.00706	.00651	.00599
21	2	13	.01027	.00953	.00884	.00819	.00758	.00702	.00649	.00600
21	2	14	.01005	.00936	.00871	.00811	.00753	.00700	.00650	.00603
21	2	15	.00986	.00922	.00862	.00805	.00751	.00700	.00653	.00608
21	2	16	.00970	.00911	.00854	.00801	.00750	.00702	.00657	.00615
21	2	17	.00956	.00901	.00849	.00799	.00752	.00707	.00664	.00624
21	2	18	.00944	.00894	.00846	.00801	.00757	.00714	.00674	.00636
21	2	19	.00934	.00890	.00847	.00805	.00765	.00726	.00688	.00653
21	2	20	.00928	.00890	.00852	.00814	.00778	.00743	.00709	.00676
21	2	21	.00928	.00897	.00866	.00835	.00805	.00774	.00745	.00716
21	3	4	.02113	.01900	.01707	.01534	.01377	.01236	.01110	.00997
21	3	5	.01912	.01730	.01563	.01412	.01275	.01151	.01038	.00937
21	3	6	.01770	.01609	.01462	.01327	.01204	.01092	.00990	.00897
21	3	7	.01664	.01520	.01387	.01264	.01152	.01049	.00955	.00869
21	3	8	.01581	.01450	.01329	.01217	.01113	.01018	.00931	.00850
21	3	9	.01514	.01394	.01283	.01180	.01084	.00995	.00913	.00837

Covariances of Inverse Gaussian Order Statistics
Shape Parameter k

n	i	j	1.6	1.7	1.8	1.9	2.0	2.1	2.2	2.3
21	3	10	.01459	.01349	.01246	.01150	.01061	.00978	.00901	.00829
21	3	11	.01413	.01312	.01216	.01127	.01043	.00965	.00892	.00824
21	3	12	.01374	.01280	.01192	.01108	.01030	.00956	.00887	.00823
21	3	13	.01341	.01254	.01172	.01094	.01020	.00951	.00886	.00824
21	3	14	.01312	.01232	.01156	.01083	.01014	.00948	.00887	.00828
21	3	15	.01287	.01214	.01143	.01075	.01010	.00949	.00890	.00835
21	3	16	.01266	.01198	.01133	.01070	.01010	.00952	.00897	.00844
21	3	17	.01248	.01186	.01126	.01068	.01012	.00958	.00907	.00857
21	3	18	.01233	.01177	.01123	.01070	.01018	.00968	.00920	.00874
21	3	19	.01220	.01172	.01123	.01076	.01029	.00983	.00939	.00896
21	3	20	.01212	.01171	.01130	.01088	.01047	.01007	.00967	.00928
21	3	21	.01212	.01181	.01149	.01116	.01083	.01050	.01016	.00983
21	4	5	.02363	.02151	.01956	.01778	.01615	.01466	.01330	.01207
21	4	6	.02189	.02003	.01831	.01672	.01526	.01392	.01269	.01156
21	4	7	.02058	.01892	.01737	.01594	.01461	.01338	.01225	.01121
21	4	8	.01956	.01806	.01665	.01534	.01412	.01299	.01194	.01097
21	4	9	.01874	.01737	.01608	.01488	.01375	.01270	.01172	.01081
21	4	10	.01806	.01681	.01563	.01451	.01346	.01248	.01156	.01070
21	4	11	.01750	.01635	.01525	.01422	.01324	.01232	.01145	.01064
21	4	12	.01702	.01596	.01495	.01399	.01307	.01221	.01139	.01063
21	4	13	.01661	.01563	.01470	.01381	.01295	.01214	.01137	.01064
21	4	14	.01625	.01536	.01450	.01367	.01287	.01211	.01139	.01070
21	4	15	.01595	.01513	.01434	.01357	.01283	.01212	.01143	.01078
21	4	16	.01569	.01494	.01422	.01351	.01282	.01216	.01152	.01090
21	4	17	.01546	.01479	.01413	.01349	.01285	.01224	.01164	.01107
21	4	18	.01527	.01468	.01409	.01350	.01293	.01237	.01182	.01128
21	4	19	.01512	.01461	.01409	.01358	.01307	.01256	.01206	.01157
21	4	20	.01502	.01460	.01417	.01374	.01330	.01286	.01242	.01199
21	4	21	.01502	.01473	.01442	.01409	.01375	.01341	.01305	.01269
21	5	6	.02633	.02422	.02226	.02044	.01875	.01719	.01575	.01443
21	5	7	.02477	.02289	.02114	.01950	.01797	.01654	.01522	.01400
21	5	8	.02355	.02186	.02027	.01877	.01737	.01606	.01484	.01370
21	5	9	.02256	.02103	.01958	.01821	.01692	.01570	.01456	.01350
21	5	10	.02175	.02036	.01903	.01776	.01657	.01544	.01437	.01337
21	5	11	.02108	.01980	.01858	.01741	.01630	.01524	.01424	.01330
21	5	12	.02050	.01933	.01821	.01713	.01609	.01511	.01417	.01328
21	5	13	.02001	.01894	.01791	.01691	.01595	.01502	.01414	.01330
21	5	14	.01959	.01861	.01766	.01674	.01585	.01499	.01416	.01337
21	5	15	.01922	.01834	.01747	.01662	.01580	.01499	.01422	.01347
21	5	16	.01891	.01811	.01732	.01655	.01579	.01505	.01433	.01363
21	5	17	.01864	.01793	.01722	.01652	.01583	.01515	.01448	.01383
21	5	18	.01841	.01779	.01717	.01654	.01592	.01531	.01470	.01410
21	5	19	.01823	.01771	.01718	.01664	.01609	.01555	.01500	.01446
21	5	20	.01811	.01770	.01727	.01683	.01638	.01592	.01545	.01498
21	5	21	.01811	.01786	.01757	.01727	.01694	.01659	.01623	.01586
21	6	7	.02933	.02724	.02528	.02343	.02169	.02007	.01855	.01714
21	6	8	.02789	.02602	.02425	.02257	.02098	.01949	.01809	.01678
21	6	9	.02673	.02504	.02343	.02190	.02044	.01906	.01776	.01653
21	6	10	.02578	.02425	.02277	.02136	.02002	.01874	.01753	.01638
21	6	11	.02498	.02358	.02224	.02094	.01970	.01851	.01737	.01629
21	6	12	.02430	.02303	.02180	.02061	.01945	.01835	.01728	.01627
21	6	13	.02372	.02257	.02144	.02034	.01928	.01825	.01725	.01630
21	6	14	.02322	.02218	.02115	.02014	.01916	.01820	.01728	.01638
21	6	15	.02279	.02185	.02092	.02000	.01910	.01821	.01735	.01651
21	6	16	.02242	.02158	.02074	.01991	.01909	.01828	.01748	.01670

Covariances of Inverse Gaussian Order Statistics
Shape Parameter k

n	i	j	1.6	1.7	1.8	1.9	2.0	2.1	2.2	2.3
21	6	17	.02210	.02137	.02062	.01988	.01914	.01840	.01767	.01695
21	6	18	.02183	.02120	.02056	.01991	.01925	.01859	.01794	.01728
21	6	19	.02162	.02111	.02057	.02002	.01946	.01888	.01831	.01773
21	6	20	.02148	.02110	.02069	.02026	.01980	.01933	.01885	.01836
21	6	21	.02148	.02129	.02105	.02078	.02048	.02015	.01981	.01944
21	7	8	.03271	.03066	.02870	.02683	.02506	.02338	.02180	.02030
21	7	9	.03136	.02951	.02774	.02604	.02442	.02287	.02140	.02001
21	7	10	.03025	.02858	.02697	.02542	.02392	.02249	.02113	.01983
21	7	11	.02932	.02781	.02634	.02492	.02354	.02222	.02094	.01973
21	7	12	.02853	.02716	.02582	.02452	.02325	.02202	.02084	.01970
21	7	13	.02785	.02662	.02540	.02421	.02304	.02191	.02080	.01974
21	7	14	.02727	.02616	.02506	.02397	.02290	.02186	.02083	.01984
21	7	15	.02676	.02578	.02479	.02381	.02283	.02187	.02092	.02000
21	7	16	.02633	.02546	.02458	.02370	.02282	.02195	.02108	.02023
21	7	17	.02595	.02521	.02444	.02366	.02288	.02209	.02131	.02053
21	7	18	.02564	.02502	.02437	.02370	.02302	.02233	.02163	.02093
21	7	19	.02539	.02490	.02438	.02383	.02326	.02268	.02208	.02147
21	7	20	.02523	.02489	.02452	.02412	.02368	.02322	.02273	.02223
21	7	21	.02523	.02511	.02495	.02474	.02449	.02420	.02388	.02354
21	8	9	.03658	.03458	.03264	.03077	.02898	.02726	.02561	.02405
21	8	10	.03529	.03349	.03174	.03004	.02840	.02681	.02529	.02383
21	8	11	.03421	.03259	.03101	.02946	.02795	.02649	.02507	.02371
21	8	12	.03329	.03184	.03040	.02899	.02761	.02626	.02495	.02368
21	8	13	.03251	.03121	.02991	.02863	.02736	.02612	.02491	.02373
21	8	14	.03183	.03067	.02951	.02835	.02720	.02607	.02495	.02385
21	8	15	.03125	.03023	.02919	.02816	.02712	.02608	.02506	.02404
21	8	16	.03074	.02986	.02895	.02803	.02711	.02618	.02524	.02432
21	8	17	.03031	.02956	.02879	.02799	.02718	.02635	.02552	.02469
21	8	18	.02994	.02934	.02870	.02804	.02734	.02663	.02590	.02517
21	8	19	.02966	.02921	.02872	.02819	.02763	.02705	.02644	.02581
21	8	20	.02946	.02920	.02889	.02853	.02813	.02769	.02723	.02673
21	8	21	.02947	.02946	.02939	.02926	.02909	.02887	.02860	.02830
21	9	10	.04106	.03913	.03724	.03539	.03359	.03185	.03016	.02853
21	9	11	.03982	.03809	.03639	.03471	.03307	.03147	.02991	.02840
21	9	12	.03876	.03722	.03569	.03417	.03268	.03121	.02977	.02836
21	9	13	.03785	.03648	.03511	.03375	.03239	.03105	.02972	.02842
21	9	14	.03706	.03586	.03465	.03343	.03220	.03098	.02977	.02857
21	9	15	.03639	.03535	.03428	.03320	.03210	.03100	.02990	.02881
21	9	16	.03580	.03492	.03400	.03306	.03209	.03111	.03013	.02914
21	9	17	.03530	.03457	.03381	.03301	.03218	.03133	.03046	.02957
21	9	18	.03488	.03432	.03371	.03306	.03238	.03166	.03092	.03015
21	9	19	.03455	.03417	.03373	.03325	.03272	.03215	.03155	.03092
21	9	20	.03432	.03416	.03393	.03364	.03330	.03292	.03249	.03203
21	9	21	.03433	.03446	.03452	.03451	.03444	.03431	.03413	.03390
21	10	11	.04633	.04451	.04269	.04089	.03911	.03736	.03565	.03398
21	10	12	.04511	.04350	.04188	.04026	.03865	.03706	.03549	.03395
21	10	13	.04406	.04265	.04121	.03977	.03832	.03687	.03544	.03402
21	10	14	.04315	.04193	.04067	.03939	.03810	.03680	.03550	.03420
21	10	15	.04237	.04133	.04024	.03913	.03798	.03683	.03566	.03448
21	10	16	.04169	.04083	.03992	.03896	.03798	.03696	.03593	.03488
21	10	17	.04111	.04043	.03969	.03891	.03808	.03722	.03632	.03541
21	10	18	.04063	.04014	.03958	.03897	.03831	.03761	.03687	.03610
21	10	19	.04024	.03996	.03961	.03919	.03872	.03820	.03763	.03702
21	10	20	.03998	.03995	.03984	.03966	.03942	.03911	.03875	.03834
21	10	21	.04000	.04031	.04053	.04069	.04076	.04077	.04071	.04058

Covariances of Inverse Gaussian Order Statistics
Shape Parameter k

n	i	j	1.6	1.7	1.8	1.9	2.0	2.1	2.2	2.3
21	11	12	.05262	.05094	.04924	.04753	.04581	.04410	.04239	.04071
21	11	13	.05141	.04996	.04847	.04696	.04542	.04388	.04234	.04080
21	11	14	.05036	.04912	.04784	.04652	.04517	.04380	.04241	.04102
21	11	15	.04945	.04842	.04734	.04621	.04504	.04384	.04261	.04136
21	11	16	.04867	.04785	.04696	.04602	.04503	.04400	.04293	.04184
21	11	17	.04800	.04739	.04670	.04596	.04516	.04430	.04341	.04247
21	11	18	.04743	.04704	.04658	.04604	.04544	.04478	.04406	.04330
21	11	19	.04698	.04684	.04661	.04630	.04592	.04548	.04497	.04441
21	11	20	.04669	.04683	.04688	.04685	.04675	.04656	.04631	.04599
21	11	21	.04670	.04725	.04770	.04807	.04834	.04853	.04864	.04868
21	12	13	.06026	.05879	.05727	.05570	.05409	.05247	.05082	.04916
21	12	14	.05904	.05782	.05653	.05519	.05380	.05237	.05091	.04943
21	12	15	.05798	.05701	.05595	.05483	.05365	.05242	.05115	.04985
21	12	16	.05707	.05633	.05551	.05461	.05365	.05262	.05154	.05042
21	12	17	.05629	.05580	.05521	.05454	.05380	.05299	.05211	.05119
21	12	18	.05564	.05540	.05506	.05464	.05414	.05355	.05290	.05219
21	12	19	.05512	.05516	.05511	.05496	.05472	.05440	.05399	.05352
21	12	20	.05477	.05515	.05543	.05561	.05570	.05569	.05560	.05542
21	12	21	.05480	.05565	.05640	.05705	.05760	.05804	.05840	.05866
21	13	14	.06972	.06856	.06730	.06596	.06456	.06309	.06158	.06002
21	13	15	.06849	.06760	.06662	.06554	.06439	.06316	.06187	.06053
21	13	16	.06742	.06682	.06610	.06529	.06439	.06341	.06235	.06123
21	13	17	.06651	.06619	.06575	.06521	.06458	.06385	.06304	.06216
21	13	18	.06574	.06572	.06558	.06534	.06498	.06454	.06400	.06338
21	13	19	.06514	.06545	.06564	.06572	.06569	.06555	.06532	.06500
21	13	20	.06474	.06544	.06603	.06650	.06686	.06711	.06726	.06730
21	13	21	.06477	.06604	.06719	.06822	.06914	.06995	.07064	.07122
21	14	15	.08176	.08104	.08018	.07921	.07812	.07695	.07568	.07433
21	14	16	.08051	.08011	.07957	.07891	.07814	.07725	.07627	.07520
21	14	17	.07943	.07937	.07916	.07883	.07837	.07780	.07712	.07635
21	14	18	.07853	.07882	.07897	.07899	.07887	.07864	.07829	.07784
21	14	19	.07781	.07850	.07904	.07945	.07973	.07988	.07991	.07982
21	14	20	.07734	.07850	.07952	.08040	.08116	.08178	.08228	.08265
21	14	21	.07739	.07922	.08092	.08249	.08392	.08523	.08641	.08746
21	15	16	.09759	.09751	.09727	.09687	.09632	.09563	.09480	.09386
21	15	17	.09630	.09663	.09678	.09678	.09662	.09631	.09587	.09529
21	15	18	.09522	.09598	.09656	.09698	.09724	.09736	.09732	.09715
21	15	19	.09437	.09560	.09666	.09756	.09830	.09889	.09933	.09963
21	15	20	.09381	.09561	.09725	.09874	.10007	.10124	.10227	.10315
21	15	21	.09388	.09649	.09896	.10129	.10347	.10551	.10739	.10913
21	16	17	.11928	.12021	.12094	.12146	.12180	.12194	.12190	.12170
21	16	18	.11797	.11942	.12068	.12173	.12260	.12327	.12376	.12407
21	16	19	.11693	.11897	.12082	.12247	.12394	.12522	.12631	.12723
21	16	20	.11627	.11901	.12157	.12396	.12617	.12819	.13004	.13171
21	16	21	.11636	.12012	.12372	.12717	.13045	.13358	.13653	.13932
21	17	18	.15077	.15336	.15570	.15781	.15967	.16130	.16270	.16386
21	17	19	.14949	.15281	.15591	.15879	.16144	.16386	.16605	.16802
21	17	20	.14867	.15289	.15691	.16073	.16434	.16774	.17093	.17391
21	17	21	.14883	.15433	.15969	.16489	.16991	.17476	.17943	.18390
21	18	19	.20051	.20605	.21133	.21635	.22109	.22556	.22975	.23366
21	18	20	.19948	.20620	.21271	.21901	.22507	.23089	.23647	.24179
21	18	21	.19974	.20819	.21651	.22467	.23268	.24050	.24814	.25556
21	19	20	.29023	.30185	.31328	.32448	.33545	.34616	.35660	.36674
21	19	21	.29075	.30485	.31890	.33286	.34671	.36042	.37397	.38733
21	20	21	.49728	.52542	.55380	.58237	.61106	.63984	.66865	.69744

Covariances of Inverse Gaussian Order Statistics
Shape Parameter k

n	i	j	1.6	1.7	1.8	1.9	2.0	2.1	2.2	2.3
22	1	2	.01493	.01294	.01123	.00974	.00846	.00736	.00641	.00558
22	1	3	.01203	.01052	.00920	.00805	.00705	.00618	.00542	.00475
22	1	4	.01043	.00919	.00809	.00712	.00628	.00553	.00488	.00431
22	1	5	.00941	.00833	.00738	.00653	.00579	.00513	.00454	.00403
22	1	6	.00869	.00773	.00688	.00612	.00545	.00485	.00432	.00385
22	1	7	.00815	.00729	.00651	.00582	.00520	.00465	.00416	.00372
22	1	8	.00773	.00694	.00623	.00559	.00502	.00450	.00404	.00363
22	1	9	.00739	.00666	.00600	.00541	.00487	.00439	.00396	.00357
22	1	10	.00711	.00644	.00582	.00527	.00476	.00431	.00389	.00352
22	1	11	.00688	.00625	.00568	.00515	.00468	.00424	.00385	.00350
22	1	12	.00669	.00610	.00556	.00506	.00461	.00420	.00382	.00348
22	1	13	.00652	.00597	.00546	.00499	.00456	.00417	.00381	.00348
22	1	14	.00637	.00585	.00537	.00493	.00452	.00415	.00381	.00349
22	1	15	.00625	.00576	.00531	.00489	.00450	.00414	.00381	.00351
22	1	16	.00614	.00568	.00525	.00486	.00449	.00415	.00383	.00354
22	1	17	.00605	.00562	.00521	.00484	.00449	.00416	.00386	.00358
22	1	18	.00597	.00556	.00519	.00483	.00450	.00419	.00391	.00364
22	1	19	.00590	.00553	.00517	.00484	.00453	.00424	.00397	.00371
22	1	20	.00584	.00550	.00518	.00487	.00458	.00431	.00405	.00380
22	1	21	.00581	.00550	.00521	.00493	.00466	.00441	.00417	.00394
22	1	22	.00581	.00555	.00530	.00506	.00482	.00459	.00438	.00417
22	2	3	.01776	.01570	.01388	.01228	.01086	.00960	.00850	.00753
22	2	4	.01544	.01375	.01223	.01089	.00969	.00862	.00767	.00683
22	2	5	.01394	.01248	.01117	.01000	.00894	.00800	.00716	.00640
22	2	6	.01288	.01159	.01043	.00938	.00843	.00757	.00680	.00612
22	2	7	.01209	.01093	.00988	.00892	.00805	.00726	.00656	.00592
22	2	8	.01147	.01042	.00945	.00857	.00777	.00704	.00637	.00577
22	2	9	.01098	.01000	.00911	.00830	.00755	.00687	.00624	.00568
22	2	10	.01057	.00967	.00884	.00808	.00738	.00674	.00615	.00561
22	2	11	.01022	.00939	.00862	.00791	.00725	.00664	.00608	.00557
22	2	12	.00994	.00916	.00844	.00777	.00714	.00657	.00604	.00555
22	2	13	.00969	.00896	.00829	.00766	.00707	.00652	.00602	.00555
22	2	14	.00947	.00880	.00816	.00757	.00701	.00649	.00601	.00556
22	2	15	.00929	.00866	.00806	.00750	.00698	.00648	.00602	.00559
22	2	16	.00913	.00854	.00798	.00746	.00696	.00649	.00605	.00564
22	2	17	.00899	.00844	.00792	.00743	.00696	.00652	.00610	.00571
22	2	18	.00887	.00836	.00788	.00742	.00698	.00656	.00617	.00579
22	2	19	.00877	.00831	.00786	.00743	.00703	.00663	.00626	.00591
22	2	20	.00869	.00827	.00787	.00748	.00710	.00674	.00639	.00606
22	2	21	.00863	.00827	.00792	.00757	.00723	.00690	.00658	.00627
22	2	22	.00864	.00835	.00806	.00776	.00748	.00719	.00691	.00664
22	3	4	.02000	.01794	.01609	.01443	.01293	.01159	.01038	.00930
22	3	5	.01808	.01631	.01471	.01326	.01195	.01076	.00969	.00873
22	3	6	.01672	.01516	.01374	.01244	.01126	.01019	.00922	.00834
22	3	7	.01570	.01430	.01302	.01184	.01077	.00978	.00889	.00807
22	3	8	.01490	.01363	.01246	.01138	.01039	.00948	.00865	.00788
22	3	9	.01426	.01310	.01202	.01102	.01010	.00925	.00847	.00775
22	3	10	.01373	.01266	.01167	.01074	.00988	.00908	.00834	.00766
22	3	11	.01329	.01230	.01137	.01051	.00970	.00895	.00825	.00760
22	3	12	.01291	.01200	.01114	.01032	.00957	.00886	.00819	.00758
22	3	13	.01259	.01174	.01094	.01018	.00946	.00879	.00816	.00758
22	3	14	.01232	.01153	.01078	.01006	.00939	.00876	.00816	.00760
22	3	15	.01208	.01134	.01064	.00998	.00934	.00874	.00818	.00764
22	3	16	.01187	.01119	.01054	.00991	.00932	.00875	.00822	.00771
22	3	17	.01169	.01106	.01046	.00988	.00932	.00879	.00828	.00780

Covariances of Inverse Gaussian Order Statistics
Shape Parameter k

n	i	j	1.6	1.7	1.8	1.9	2.0	2.1	2.2	2.3
22	3	18	.01153	.01096	.01041	.00987	.00935	.00885	.00837	.00792
22	3	19	.01140	.01088	.01038	.00989	.00941	.00895	.00850	.00807
22	3	20	.01130	.01084	.01039	.00995	.00951	.00909	.00868	.00828
22	3	21	.01123	.01084	.01045	.01007	.00968	.00930	.00893	.00857
22	3	22	.01124	.01094	.01064	.01033	.01001	.00970	.00938	.00907
22	4	5	.02224	.02020	.01833	.01662	.01506	.01364	.01236	.01119
22	4	6	.02058	.01879	.01713	.01561	.01421	.01293	.01176	.01070
22	4	7	.01934	.01773	.01624	.01486	.01359	.01242	.01134	.01036
22	4	8	.01836	.01690	.01555	.01429	.01312	.01204	.01104	.01012
22	4	9	.01757	.01625	.01500	.01384	.01276	.01175	.01081	.00995
22	4	10	.01693	.01571	.01456	.01348	.01248	.01153	.01065	.00984
22	4	11	.01638	.01526	.01420	.01320	.01226	.01137	.01054	.00977
22	4	12	.01593	.01489	.01390	.01297	.01209	.01125	.01047	.00973
22	4	13	.01553	.01457	.01366	.01279	.01196	.01117	.01043	.00973
22	4	14	.01519	.01431	.01346	.01264	.01187	.01113	.01042	.00976
22	4	15	.01490	.01408	.01329	.01253	.01181	.01111	.01045	.00982
22	4	16	.01464	.01389	.01316	.01246	.01178	.01113	.01050	.00990
22	4	17	.01442	.01373	.01306	.01241	.01178	.01117	.01058	.01002
22	4	18	.01423	.01361	.01300	.01240	.01182	.01125	.01070	.01017
22	4	19	.01407	.01351	.01297	.01243	.01189	.01137	.01086	.01037
22	4	20	.01394	.01346	.01298	.01250	.01202	.01155	.01109	.01063
22	4	21	.01386	.01346	.01306	.01265	.01224	.01183	.01142	.01101
22	4	22	.01386	.01358	.01329	.01298	.01266	.01233	.01199	.01165
22	5	6	.02465	.02262	.02074	.01899	.01738	.01590	.01453	.01328
22	5	7	.02317	.02136	.01967	.01809	.01663	.01527	.01402	.01286
22	5	8	.02201	.02037	.01884	.01740	.01606	.01481	.01365	.01257
22	5	9	.02107	.01958	.01818	.01686	.01562	.01446	.01337	.01236
22	5	10	.02030	.01894	.01765	.01643	.01528	.01419	.01318	.01222
22	5	11	.01965	.01840	.01722	.01608	.01501	.01400	.01304	.01214
22	5	12	.01910	.01796	.01686	.01581	.01481	.01385	.01295	.01210
22	5	13	.01863	.01758	.01656	.01559	.01465	.01376	.01291	.01210
22	5	14	.01823	.01726	.01632	.01541	.01454	.01370	.01290	.01213
22	5	15	.01787	.01699	.01612	.01528	.01447	.01368	.01293	.01220
22	5	16	.01757	.01676	.01596	.01519	.01443	.01370	.01299	.01231
22	5	17	.01730	.01657	.01585	.01513	.01444	.01376	.01310	.01245
22	5	18	.01707	.01642	.01577	.01512	.01448	.01386	.01324	.01264
22	5	19	.01688	.01631	.01573	.01515	.01457	.01401	.01344	.01289
22	5	20	.01673	.01624	.01574	.01524	.01473	.01423	.01372	.01322
22	5	21	.01663	.01624	.01584	.01542	.01500	.01456	.01413	.01369
22	5	22	.01664	.01639	.01612	.01582	.01551	.01518	.01484	.01449
22	6	7	.02730	.02529	.02340	.02163	.01998	.01843	.01700	.01566
22	6	8	.02594	.02414	.02243	.02082	.01930	.01788	.01655	.01531
22	6	9	.02485	.02321	.02165	.02017	.01878	.01746	.01622	.01506
22	6	10	.02394	.02245	.02102	.01966	.01837	.01714	.01599	.01489
22	6	11	.02318	.02182	.02051	.01925	.01805	.01691	.01582	.01479
22	6	12	.02254	.02129	.02008	.01892	.01780	.01674	.01572	.01475
22	6	13	.02199	.02084	.01973	.01866	.01762	.01662	.01566	.01475
22	6	14	.02151	.02047	.01944	.01845	.01749	.01655	.01565	.01479
22	6	15	.02109	.02014	.01921	.01829	.01740	.01653	.01569	.01488
22	6	16	.02073	.01988	.01902	.01819	.01736	.01656	.01577	.01501
22	6	17	.02042	.01965	.01888	.01812	.01737	.01662	.01589	.01518
22	6	18	.02015	.01947	.01879	.01810	.01742	.01674	.01607	.01541
22	6	19	.01993	.01934	.01875	.01814	.01753	.01692	.01632	.01572
22	6	20	.01975	.01927	.01877	.01825	.01772	.01719	.01666	.01612
22	6	21	.01963	.01927	.01888	.01847	.01804	.01760	.01715	.01669

Covariances of Inverse Gaussian Order Statistics
Shape Parameter k

n	i	j	1.6	1.7	1.8	1.9	2.0	2.1	2.2	2.3
22	6	22	.01965	.01945	.01921	.01895	.01866	.01834	.01801	.01766
22	7	8	.03027	.02829	.02641	.02462	.02292	.02133	.01983	.01841
22	7	9	.02900	.02721	.02550	.02386	.02231	.02083	.01944	.01812
22	7	10	.02795	.02633	.02476	.02326	.02183	.02046	.01916	.01792
22	7	11	.02707	.02559	.02416	.02278	.02145	.02018	.01896	.01780
22	7	12	.02632	.02498	.02367	.02239	.02116	.01998	.01884	.01775
22	7	13	.02568	.02446	.02326	.02208	.02095	.01984	.01878	.01775
22	7	14	.02513	.02401	.02292	.02184	.02079	.01976	.01877	.01781
22	7	15	.02464	.02364	.02264	.02166	.02069	.01974	.01881	.01791
22	7	16	.02422	.02333	.02243	.02153	.02064	.01977	.01891	.01807
22	7	17	.02386	.02307	.02226	.02146	.02065	.01985	.01906	.01828
22	7	18	.02355	.02286	.02215	.02144	.02072	.01999	.01927	.01856
22	7	19	.02329	.02270	.02210	.02148	.02085	.02021	.01957	.01892
22	7	20	.02308	.02262	.02212	.02161	.02108	.02053	.01997	.01941
22	7	21	.02294	.02262	.02226	.02187	.02146	.02102	.02056	.02009
22	7	22	.02296	.02283	.02265	.02244	.02219	.02191	.02159	.02126
22	8	9	.03364	.03170	.02983	.02803	.02632	.02468	.02312	.02163
22	8	10	.03243	.03068	.02898	.02734	.02575	.02424	.02279	.02140
22	8	11	.03142	.02983	.02828	.02677	.02532	.02391	.02256	.02126
22	8	12	.03055	.02911	.02770	.02632	.02498	.02367	.02241	.02120
22	8	13	.02981	.02851	.02723	.02596	.02472	.02352	.02234	.02120
22	8	14	.02917	.02800	.02683	.02568	.02454	.02342	.02233	.02127
22	8	15	.02861	.02756	.02651	.02546	.02442	.02340	.02239	.02140
22	8	16	.02813	.02720	.02626	.02532	.02437	.02343	.02250	.02158
22	8	17	.02771	.02690	.02607	.02523	.02438	.02353	.02268	.02184
22	8	18	.02735	.02666	.02594	.02521	.02446	.02370	.02294	.02217
22	8	19	.02704	.02648	.02588	.02526	.02462	.02396	.02329	.02261
22	8	20	.02680	.02638	.02591	.02541	.02489	.02434	.02377	.02318
22	8	21	.02665	.02638	.02607	.02572	.02533	.02492	.02447	.02400
22	8	22	.02667	.02663	.02653	.02639	.02620	.02597	.02570	.02540
22	9	10	.03751	.03562	.03378	.03200	.03027	.02860	.02699	.02545
22	9	11	.03634	.03464	.03297	.03134	.02976	.02821	.02672	.02528
22	9	12	.03535	.03382	.03231	.03082	.02936	.02794	.02655	.02521
22	9	13	.03449	.03312	.03176	.03040	.02907	.02775	.02647	.02522
22	9	14	.03376	.03253	.03130	.03007	.02886	.02765	.02646	.02530
22	9	15	.03311	.03203	.03093	.02983	.02872	.02762	.02653	.02545
22	9	16	.03256	.03161	.03064	.02965	.02866	.02766	.02666	.02567
22	9	17	.03207	.03126	.03042	.02955	.02867	.02778	.02688	.02598
22	9	18	.03166	.03098	.03027	.02953	.02876	.02798	.02718	.02638
22	9	19	.03131	.03078	.03020	.02959	.02895	.02829	.02760	.02689
22	9	20	.03103	.03066	.03024	.02977	.02927	.02873	.02817	.02758
22	9	21	.03085	.03066	.03042	.03013	.02980	.02942	.02900	.02855
22	9	22	.03087	.03095	.03096	.03091	.03081	.03066	.03045	.03021
22	10	11	.04199	.04019	.03840	.03665	.03492	.03324	.03160	.03001
22	10	12	.04085	.03924	.03763	.03604	.03447	.03292	.03141	.02993
22	10	13	.03987	.03844	.03700	.03556	.03412	.03271	.03131	.02994
22	10	14	.03902	.03776	.03647	.03518	.03388	.03259	.03131	.03004
22	10	15	.03829	.03718	.03604	.03489	.03372	.03255	.03139	.03022
22	10	16	.03765	.03669	.03570	.03469	.03365	.03261	.03155	.03049
22	10	17	.03709	.03629	.03545	.03457	.03367	.03275	.03180	.03085
22	10	18	.03661	.03597	.03528	.03455	.03378	.03299	.03216	.03132
22	10	19	.03621	.03573	.03520	.03463	.03400	.03335	.03265	.03194
22	10	20	.03589	.03560	.03524	.03483	.03438	.03387	.03333	.03275
22	10	21	.03568	.03561	.03546	.03526	.03499	.03468	.03432	.03391
22	10	22	.03571	.03594	.03609	.03617	.03619	.03614	.03603	.03587

Covariances of Inverse Gaussian Order Statistics
Shape Parameter k

n	i	j	1.6	1.7	1.8	1.9	2.0	2.1	2.2	2.3
22	11	12	.04726	.04557	.04388	.04218	.04049	.03882	.03718	.03556
22	11	13	.04614	.04465	.04314	.04162	.04010	.03858	.03707	.03558
22	11	14	.04517	.04387	.04253	.04118	.03981	.03844	.03706	.03570
22	11	15	.04432	.04320	.04204	.04085	.03963	.03840	.03716	.03591
22	11	16	.04358	.04264	.04165	.04062	.03956	.03847	.03736	.03623
22	11	17	.04294	.04218	.04136	.04049	.03958	.03863	.03766	.03666
22	11	18	.04239	.04181	.04116	.04046	.03971	.03892	.03809	.03723
22	11	19	.04193	.04153	.04107	.04055	.03997	.03934	.03867	.03795
22	11	20	.04156	.04138	.04112	.04080	.04041	.03997	.03947	.03892
22	11	21	.04133	.04139	.04138	.04129	.04114	.04092	.04063	.04030
22	11	22	.04136	.04178	.04211	.04236	.04254	.04264	.04267	.04263
22	12	13	.05356	.05203	.05046	.04887	.04726	.04564	.04401	.04240
22	12	14	.05244	.05112	.04976	.04836	.04693	.04548	.04401	.04254
22	12	15	.05146	.05035	.04919	.04798	.04672	.04544	.04413	.04281
22	12	16	.05061	.04971	.04874	.04771	.04663	.04552	.04437	.04319
22	12	17	.04987	.04917	.04840	.04756	.04666	.04572	.04473	.04370
22	12	18	.04924	.04874	.04817	.04753	.04682	.04605	.04524	.04437
22	12	19	.04871	.04843	.04807	.04764	.04713	.04656	.04593	.04524
22	12	20	.04828	.04825	.04813	.04793	.04765	.04730	.04688	.04640
22	12	21	.04801	.04827	.04843	.04851	.04851	.04842	.04826	.04803
22	12	22	.04806	.04872	.04929	.04977	.05016	.05046	.05067	.05081
22	13	14	.06121	.05990	.05852	.05709	.05561	.05409	.05255	.05098
22	13	15	.06008	.05901	.05786	.05665	.05537	.05405	.05269	.05130
22	13	16	.05910	.05826	.05734	.05634	.05527	.05415	.05298	.05176
22	13	17	.05824	.05764	.05694	.05616	.05531	.05439	.05341	.05238
22	13	18	.05751	.05714	.05668	.05613	.05550	.05480	.05402	.05319
22	13	19	.05689	.05678	.05657	.05627	.05587	.05540	.05485	.05423
22	13	20	.05640	.05657	.05664	.05661	.05649	.05628	.05598	.05561
22	13	21	.05609	.05660	.05700	.05730	.05751	.05762	.05764	.05757
22	13	22	.05614	.05713	.05801	.05879	.05946	.06004	.06051	.06089
22	14	15	.07069	.06970	.06861	.06742	.06616	.06483	.06344	.06199
22	14	16	.06955	.06882	.06799	.06707	.06605	.06495	.06379	.06256
22	14	17	.06855	.06810	.06753	.06687	.06610	.06525	.06431	.06331
22	14	18	.06769	.06752	.06723	.06683	.06633	.06574	.06505	.06428
22	14	19	.06697	.06710	.06710	.06700	.06678	.06646	.06604	.06554
22	14	20	.06641	.06686	.06720	.06741	.06752	.06752	.06741	.06721
22	14	21	.06604	.06689	.06762	.06824	.06873	.06912	.06940	.06957
22	14	22	.06611	.06753	.06882	.07001	.07107	.07202	.07286	.07358
22	15	16	.08276	.08222	.08155	.08075	.07984	.07882	.07770	.07650
22	15	17	.08159	.08137	.08101	.08052	.07991	.07918	.07835	.07742
22	15	18	.08058	.08069	.08066	.08049	.08019	.07978	.07925	.07861
22	15	19	.07973	.08019	.08051	.08069	.08074	.08066	.08046	.08015
22	15	20	.07907	.07992	.08063	.08120	.08164	.08194	.08213	.08219
22	15	21	.07864	.07996	.08115	.08219	.08311	.08389	.08454	.08507
22	15	22	.07873	.08072	.08259	.08432	.08593	.08740	.08875	.08996
22	16	17	.09861	.09875	.09871	.09851	.09816	.09766	.09702	.09625
22	16	18	.09741	.09794	.09830	.09849	.09852	.09840	.09814	.09774
22	16	19	.09641	.09735	.09813	.09875	.09920	.09950	.09965	.09965
22	16	20	.09561	.09703	.09828	.09937	.10031	.10108	.10170	.10218
22	16	21	.09511	.09709	.09892	.10060	.10212	.10348	.10469	.10575
22	16	22	.09523	.09802	.10069	.10320	.10558	.10781	.10989	.11182
22	17	18	.12034	.12151	.12248	.12324	.12381	.12418	.12437	.12438
22	17	19	.11912	.12081	.12229	.12358	.12467	.12557	.12628	.12681
22	17	20	.11817	.12042	.12249	.12438	.12607	.12757	.12889	.13002
22	17	21	.11756	.12052	.12330	.12591	.12834	.13059	.13267	.13456

Covariances of Inverse Gaussian Order Statistics
Shape Parameter k

n	i	j	1.6	1.7	1.8	1.9	2.0	2.1	2.2	2.3
22	17	22	.11773	.12169	.12551	.12918	.13269	.13604	.13923	.14225
22	18	19	.15189	.15475	.15738	.15976	.16191	.16382	.16549	.16693
22	18	20	.15071	.15430	.15766	.16081	.16373	.16643	.16890	.17115
22	18	21	.14998	.15444	.15872	.16281	.16669	.17037	.17384	.17709
22	18	22	.15021	.15596	.16157	.16703	.17232	.17745	.18239	.18715
22	19	20	.20171	.20757	.21319	.21854	.22364	.22846	.23300	.23727
22	19	21	.20080	.20783	.21466	.22128	.22768	.23385	.23977	.24545
22	19	22	.20116	.20990	.21852	.22701	.23535	.24351	.25149	.25928
22	20	21	.29158	.30360	.31544	.32707	.33849	.34966	.36057	.37120
22	20	22	.29224	.30671	.32115	.33552	.34980	.36396	.37797	.39181
22	21	22	.49892	.52759	.55653	.58569	.61501	.64444	.67393	.70345
23	1	2	.01435	.01242	.01075	.00932	.00809	.00702	.00610	.00531
23	1	3	.01153	.01007	.00879	.00768	.00672	.00588	.00515	.00451
23	1	4	.00999	.00878	.00772	.00679	.00597	.00525	.00463	.00408
23	1	5	.00900	.00796	.00703	.00622	.00550	.00486	.00430	.00381
23	1	6	.00830	.00737	.00655	.00582	.00517	.00459	.00408	.00363
23	1	7	.00778	.00694	.00619	.00552	.00493	.00440	.00392	.00350
23	1	8	.00737	.00661	.00592	.00530	.00475	.00425	.00381	.00341
23	1	9	.00705	.00634	.00570	.00512	.00461	.00414	.00372	.00335
23	1	10	.00678	.00612	.00552	.00498	.00450	.00406	.00366	.00331
23	1	11	.00655	.00594	.00538	.00487	.00441	.00399	.00362	.00328
23	1	12	.00636	.00579	.00526	.00478	.00434	.00395	.00359	.00326
23	1	13	.00620	.00566	.00516	.00471	.00429	.00391	.00357	.00325
23	1	14	.00606	.00555	.00508	.00465	.00425	.00389	.00356	.00326
23	1	15	.00594	.00545	.00501	.00460	.00422	.00388	.00356	.00327
23	1	16	.00583	.00537	.00495	.00457	.00421	.00387	.00357	.00329
23	1	17	.00573	.00531	.00491	.00454	.00420	.00388	.00359	.00332
23	1	18	.00565	.00525	.00488	.00453	.00420	.00390	.00362	.00336
23	1	19	.00558	.00521	.00486	.00453	.00422	.00393	.00366	.00341
23	1	20	.00552	.00518	.00485	.00454	.00425	.00397	.00372	.00348
23	1	21	.00548	.00516	.00485	.00457	.00429	.00404	.00379	.00356
23	1	22	.00545	.00516	.00488	.00462	.00437	.00413	.00390	.00369
23	1	23	.00545	.00521	.00497	.00474	.00452	.00430	.00410	.00390
23	2	3	.01695	.01496	.01321	.01166	.01029	.00909	.00803	.00710
23	2	4	.01472	.01308	.01162	.01032	.00917	.00815	.00724	.00644
23	2	5	.01327	.01186	.01060	.00946	.00845	.00755	.00674	.00602
23	2	6	.01225	.01101	.00988	.00886	.00795	.00713	.00640	.00574
23	2	7	.01149	.01037	.00935	.00842	.00759	.00683	.00615	.00554
23	2	8	.01089	.00987	.00893	.00808	.00731	.00661	.00598	.00540
23	2	9	.01041	.00947	.00861	.00782	.00710	.00644	.00584	.00530
23	2	10	.01002	.00915	.00834	.00761	.00693	.00631	.00575	.00523
23	2	11	.00969	.00888	.00813	.00744	.00680	.00621	.00568	.00519
23	2	12	.00941	.00865	.00795	.00730	.00670	.00614	.00563	.00516
23	2	13	.00917	.00846	.00780	.00718	.00662	.00609	.00560	.00515
23	2	14	.00896	.00830	.00768	.00710	.00656	.00605	.00559	.00516
23	2	15	.00878	.00816	.00757	.00703	.00651	.00603	.00559	.00518
23	2	16	.00862	.00804	.00749	.00697	.00649	.00603	.00561	.00521
23	2	17	.00848	.00794	.00742	.00694	.00648	.00604	.00564	.00526
23	2	18	.00836	.00786	.00737	.00692	.00648	.00607	.00568	.00532
23	2	19	.00826	.00779	.00734	.00691	.00650	.00612	.00575	.00540
23	2	20	.00817	.00774	.00733	.00693	.00655	.00618	.00584	.00551
23	2	21	.00810	.00772	.00734	.00697	.00662	.00628	.00596	.00565
23	2	22	.00806	.00772	.00739	.00706	.00674	.00643	.00613	.00585
23	2	23	.00807	.00779	.00752	.00724	.00697	.00670	.00644	.00618
23	3	4	.01899	.01700	.01522	.01362	.01218	.01090	.00975	.00872

Covariances of Inverse Gaussian Order Statistics
Shape Parameter k

n	i	j	1.6	1.7	1.8	1.9	2.0	2.1	2.2	2.3
23	3	5	.01714	.01544	.01389	.01250	.01124	.01010	.00908	.00816
23	3	6	.01584	.01433	.01296	.01171	.01058	.00956	.00863	.00779
23	3	7	.01486	.01351	.01227	.01114	.01010	.00916	.00830	.00753
23	3	8	.01409	.01286	.01173	.01069	.00974	.00886	.00807	.00734
23	3	9	.01347	.01235	.01131	.01034	.00946	.00864	.00789	.00720
23	3	10	.01297	.01193	.01096	.01006	.00923	.00847	.00776	.00711
23	3	11	.01254	.01158	.01068	.00984	.00906	.00834	.00767	.00705
23	3	12	.01218	.01129	.01044	.00966	.00892	.00824	.00760	.00701
23	3	13	.01187	.01104	.01025	.00951	.00882	.00817	.00756	.00700
23	3	14	.01160	.01083	.01009	.00939	.00874	.00812	.00755	.00701
23	3	15	.01137	.01064	.00996	.00930	.00868	.00810	.00755	.00703
23	3	16	.01117	.01049	.00985	.00923	.00865	.00810	.00757	.00708
23	3	17	.01099	.01036	.00976	.00918	.00863	.00811	.00762	.00714
23	3	18	.01083	.01025	.00970	.00916	.00864	.00815	.00768	.00723
23	3	19	.01070	.01017	.00965	.00915	.00867	.00821	.00777	.00734
23	3	20	.01059	.01011	.00964	.00918	.00873	.00830	.00789	.00749
23	3	21	.01050	.01007	.00965	.00923	.00883	.00843	.00805	.00768
23	3	22	.01044	.01008	.00971	.00935	.00899	.00863	.00829	.00795
23	3	23	.01045	.01017	.00988	.00959	.00929	.00900	.00870	.00841
23	4	5	.02101	.01904	.01724	.01560	.01411	.01275	.01153	.01042
23	4	6	.01942	.01769	.01609	.01463	.01329	.01207	.01096	.00994
23	4	7	.01823	.01668	.01524	.01391	.01270	.01158	.01055	.00961
23	4	8	.01730	.01589	.01458	.01336	.01224	.01121	.01025	.00938
23	4	9	.01654	.01526	.01405	.01293	.01189	.01092	.01003	.00921
23	4	10	.01592	.01474	.01363	.01259	.01161	.01071	.00987	.00909
23	4	11	.01540	.01431	.01328	.01231	.01140	.01055	.00975	.00901
23	4	12	.01496	.01395	.01299	.01208	.01123	.01042	.00967	.00897
23	4	13	.01458	.01364	.01275	.01190	.01109	.01034	.00962	.00895
23	4	14	.01426	.01338	.01255	.01175	.01099	.01028	.00960	.00896
23	4	15	.01397	.01316	.01238	.01164	.01093	.01025	.00960	.00900
23	4	16	.01372	.01297	.01225	.01155	.01088	.01024	.00963	.00905
23	4	17	.01350	.01281	.01214	.01149	.01087	.01026	.00969	.00914
23	4	18	.01331	.01268	.01206	.01146	.01088	.01031	.00977	.00925
23	4	19	.01315	.01257	.01201	.01145	.01091	.01039	.00988	.00939
23	4	20	.01301	.01250	.01199	.01148	.01099	.01050	.01003	.00957
23	4	21	.01290	.01245	.01200	.01156	.01111	.01067	.01024	.00982
23	4	22	.01283	.01246	.01208	.01170	.01131	.01093	.01054	.01016
23	4	23	.01285	.01258	.01230	.01200	.01170	.01138	.01107	.01075
23	5	6	.02317	.02121	.01940	.01773	.01619	.01477	.01348	.01229
23	5	7	.02176	.02001	.01838	.01687	.01547	.01417	.01298	.01188
23	5	8	.02065	.01907	.01759	.01621	.01492	.01372	.01262	.01159
23	5	9	.01976	.01831	.01696	.01569	.01449	.01338	.01235	.01139
23	5	10	.01902	.01770	.01645	.01527	.01416	.01312	.01215	.01124
23	5	11	.01840	.01719	.01603	.01493	.01390	.01292	.01201	.01115
23	5	12	.01788	.01675	.01568	.01466	.01369	.01277	.01191	.01109
23	5	13	.01743	.01639	.01539	.01444	.01353	.01267	.01185	.01107
23	5	14	.01704	.01608	.01515	.01426	.01341	.01260	.01182	.01109
23	5	15	.01670	.01581	.01495	.01413	.01333	.01256	.01183	.01113
23	5	16	.01640	.01558	.01479	.01402	.01328	.01256	.01187	.01120
23	5	17	.01614	.01539	.01466	.01395	.01326	.01258	.01193	.01131
23	5	18	.01591	.01524	.01457	.01391	.01327	.01264	.01203	.01144
23	5	19	.01572	.01511	.01451	.01391	.01332	.01274	.01217	.01162
23	5	20	.01555	.01502	.01448	.01394	.01341	.01288	.01236	.01185
23	5	21	.01542	.01497	.01450	.01403	.01356	.01308	.01261	.01215
23	5	22	.01534	.01498	.01459	.01420	.01380	.01339	.01299	.01258

Covariances of Inverse Gaussian Order Statistics
Shape Parameter k

n	i	j	1.6	1.7	1.8	1.9	2.0	2.1	2.2	2.3
23	5	23	.01536	.01512	.01485	.01457	.01427	.01396	.01363	.01330
23	6	7	.02554	.02360	.02178	.02008	.01850	.01703	.01566	.01440
23	6	8	.02424	.02250	.02085	.01930	.01785	.01649	.01523	.01405
23	6	9	.02320	.02161	.02011	.01868	.01734	.01608	.01490	.01380
23	6	10	.02234	.02089	.01950	.01819	.01695	.01577	.01467	.01363
23	6	11	.02162	.02029	.01901	.01779	.01663	.01554	.01450	.01352
23	6	12	.02101	.01978	.01860	.01747	.01639	.01536	.01438	.01345
23	6	13	.02048	.01935	.01826	.01721	.01620	.01523	.01431	.01343
23	6	14	.02002	.01898	.01798	.01700	.01606	.01515	.01428	.01345
23	6	15	.01962	.01867	.01774	.01684	.01596	.01511	.01429	.01350
23	6	16	.01927	.01841	.01755	.01671	.01590	.01510	.01433	.01359
23	6	17	.01897	.01818	.01740	.01663	.01587	.01513	.01441	.01371
23	6	18	.01870	.01800	.01729	.01658	.01589	.01520	.01453	.01388
23	6	19	.01848	.01785	.01721	.01658	.01595	.01532	.01470	.01409
23	6	20	.01828	.01774	.01718	.01662	.01606	.01549	.01493	.01437
23	6	21	.01813	.01768	.01721	.01673	.01623	.01574	.01524	.01474
23	6	22	.01804	.01769	.01732	.01693	.01653	.01611	.01569	.01525
23	6	23	.01806	.01786	.01763	.01737	.01709	.01679	.01647	.01613
23	7	8	.02817	.02625	.02443	.02272	.02110	.01957	.01815	.01681
23	7	9	.02696	.02523	.02357	.02200	.02051	.01910	.01777	.01652
23	7	10	.02597	.02439	.02287	.02142	.02004	.01873	.01749	.01632
23	7	11	.02513	.02369	.02229	.02096	.01967	.01845	.01729	.01618
23	7	12	.02442	.02310	.02182	.02058	.01939	.01824	.01715	.01611
23	7	13	.02381	.02260	.02142	.02027	.01916	.01810	.01707	.01608
23	7	14	.02328	.02217	.02109	.02003	.01900	.01800	.01703	.01610
23	7	15	.02282	.02181	.02081	.01984	.01888	.01795	.01704	.01617
23	7	16	.02242	.02150	.02059	.01969	.01881	.01794	.01710	.01627
23	7	17	.02207	.02124	.02042	.01959	.01878	.01798	.01719	.01642
23	7	18	.02176	.02102	.02028	.01954	.01880	.01807	.01734	.01662
23	7	19	.02149	.02085	.02020	.01954	.01887	.01820	.01754	.01688
23	7	20	.02127	.02073	.02016	.01959	.01900	.01841	.01781	.01721
23	7	21	.02110	.02066	.02019	.01971	.01921	.01870	.01818	.01765
23	7	22	.02098	.02067	.02033	.01995	.01956	.01914	.01871	.01827
23	7	23	.02101	.02087	.02069	.02047	.02022	.01995	.01964	.01932
23	8	9	.03112	.02924	.02743	.02571	.02406	.02250	.02101	.01961
23	8	10	.02998	.02827	.02662	.02504	.02352	.02207	.02069	.01937
23	8	11	.02902	.02747	.02596	.02450	.02309	.02174	.02045	.01922
23	8	12	.02821	.02679	.02541	.02406	.02276	.02150	.02029	.01913
23	8	13	.02751	.02621	.02495	.02371	.02250	.02133	.02020	.01910
23	8	14	.02690	.02572	.02456	.02342	.02231	.02122	.02016	.01913
23	8	15	.02637	.02530	.02425	.02320	.02217	.02116	.02017	.01921
23	8	16	.02590	.02495	.02399	.02303	.02209	.02115	.02023	.01933
23	8	17	.02550	.02465	.02378	.02292	.02206	.02120	.02035	.01951
23	8	18	.02514	.02440	.02363	.02286	.02208	.02130	.02052	.01975
23	8	19	.02484	.02420	.02353	.02285	.02216	.02146	.02076	.02005
23	8	20	.02458	.02405	.02349	.02291	.02231	.02170	.02108	.02045
23	8	21	.02438	.02397	.02353	.02306	.02256	.02205	.02151	.02097
23	8	22	.02425	.02399	.02368	.02334	.02297	.02257	.02215	.02170
23	8	23	.02428	.02422	.02411	.02395	.02375	.02352	.02325	.02295
23	9	10	.03449	.03265	.03086	.02914	.02748	.02588	.02435	.02289
23	9	11	.03339	.03172	.03010	.02852	.02698	.02550	.02408	.02271
23	9	12	.03246	.03095	.02946	.02801	.02660	.02522	.02389	.02261
23	9	13	.03165	.03029	.02893	.02760	.02630	.02502	.02378	.02258
23	9	14	.03096	.02972	.02849	.02727	.02607	.02489	.02374	.02261
23	9	15	.03035	.02924	.02813	.02702	.02591	.02483	.02375	.02270

Covariances of Inverse Gaussian Order Statistics
Shape Parameter k

n	i	j	1.6	1.7	1.8	1.9	2.0	2.1	2.2	2.3
23	9	16	.02982	.02883	.02783	.02682	.02582	.02482	.02383	.02285
23	9	17	.02935	.02848	.02759	.02669	.02579	.02488	.02397	.02306
23	9	18	.02895	.02820	.02742	.02662	.02581	.02499	.02417	.02335
23	9	19	.02860	.02797	.02731	.02662	.02591	.02519	.02445	.02371
23	9	20	.02830	.02780	.02726	.02669	.02609	.02547	.02483	.02417
23	9	21	.02807	.02771	.02730	.02686	.02638	.02587	.02534	.02479
23	9	22	.02793	.02773	.02748	.02719	.02686	.02649	.02609	.02565
23	9	23	.02796	.02799	.02797	.02789	.02777	.02760	.02738	.02713
23	10	11	.03835	.03657	.03483	.03313	.03146	.02984	.02828	.02677
23	10	12	.03728	.03568	.03410	.03254	.03101	.02952	.02806	.02665
23	10	13	.03637	.03493	.03349	.03207	.03067	.02929	.02794	.02662
23	10	14	.03557	.03428	.03299	.03169	.03041	.02914	.02789	.02666
23	10	15	.03487	.03373	.03257	.03140	.03023	.02906	.02791	.02677
23	10	16	.03427	.03326	.03222	.03118	.03012	.02906	.02800	.02695
23	10	17	.03374	.03286	.03195	.03103	.03008	.02913	.02816	.02720
23	10	18	.03327	.03253	.03175	.03095	.03012	.02927	.02840	.02753
23	10	19	.03287	.03227	.03162	.03094	.03023	.02949	.02873	.02796
23	10	20	.03254	.03208	.03157	.03102	.03044	.02982	.02917	.02850
23	10	21	.03227	.03197	.03162	.03122	.03078	.03029	.02978	.02923
23	10	22	.03210	.03200	.03183	.03161	.03133	.03101	.03065	.03025
23	10	23	.03215	.03230	.03240	.03243	.03240	.03231	.03218	.03199
23	11	12	.04284	.04115	.03947	.03781	.03616	.03454	.03296	.03141
23	11	13	.04179	.04028	.03877	.03726	.03570	.03420	.03281	.03138
23	11	14	.04088	.03954	.03819	.03683	.03547	.03411	.03276	.03143
23	11	15	.04009	.03891	.03771	.03649	.03526	.03402	.03279	.03156
23	11	16	.03939	.03837	.03732	.03624	.03513	.03402	.03290	.03177
23	11	17	.03878	.03792	.03701	.03606	.03509	.03410	.03309	.03207
23	11	18	.03825	.03754	.03678	.03597	.03513	.03426	.03337	.03246
23	11	19	.03780	.03724	.03663	.03597	.03527	.03453	.03376	.03296
23	11	20	.03741	.03702	.03657	.03606	.03551	.03491	.03428	.03361
23	11	21	.03711	.03690	.03663	.03630	.03591	.03547	.03499	.03446
23	11	22	.03692	.03693	.03687	.03674	.03656	.03631	.03601	.03567
23	11	23	.03697	.03729	.03753	.03770	.03780	.03783	.03780	.03771
23	12	13	.04812	.04655	.04497	.04338	.04178	.04019	.03861	.03705
23	12	14	.04708	.04570	.04430	.04288	.04144	.03999	.03855	.03711
23	12	15	.04617	.04498	.04375	.04249	.04120	.03990	.03859	.03727
23	12	16	.04537	.04436	.04330	.04219	.04106	.03990	.03872	.03752
23	12	17	.04468	.04384	.04294	.04200	.04101	.03999	.03894	.03788
23	12	18	.04407	.04341	.04268	.04189	.04106	.04019	.03928	.03834
23	12	19	.04355	.04306	.04251	.04189	.04122	.04050	.03973	.03893
23	12	20	.04311	.04281	.04244	.04200	.04151	.04095	.04035	.03970
23	12	21	.04276	.04268	.04251	.04228	.04197	.04160	.04118	.04070
23	12	22	.04255	.04271	.04279	.04280	.04273	.04259	.04239	.04213
23	12	23	.04260	.04312	.04356	.04391	.04418	.04437	.04449	.04454
23	13	14	.05442	.05303	.05159	.05011	.04860	.04708	.04554	.04399
23	13	15	.05338	.05219	.05095	.04966	.04833	.04697	.04558	.04419
23	13	16	.05247	.05148	.05043	.04932	.04817	.04697	.04574	.04449
23	13	17	.05167	.05088	.05002	.04910	.04811	.04708	.04601	.04491
23	13	18	.05097	.05038	.04972	.04898	.04818	.04732	.04641	.04546
23	13	19	.05037	.04999	.04952	.04898	.04836	.04768	.04695	.04616
23	13	20	.04987	.04970	.04945	.04911	.04870	.04822	.04767	.04707
23	13	21	.04947	.04955	.04953	.04943	.04925	.04899	.04866	.04826
23	13	22	.04922	.04959	.04986	.05004	.05014	.05015	.05009	.04995
23	13	23	.04929	.05007	.05075	.05134	.05184	.05225	.05257	.05280
23	14	15	.06208	.06092	.05969	.05839	.05703	.05562	.05418	.05270

Covariances of Inverse Gaussian Order Statistics
Shape Parameter k

n	i	j	1.6	1.7	1.8	1.9	2.0	2.1	2.2	2.3
23	14	16	.06103	.06010	.05909	.05800	.05684	.05563	.05437	.05307
23	14	17	.06011	.05941	.05861	.05774	.05679	.05577	.05469	.05357
23	14	18	.05930	.05883	.05826	.05760	.05686	.05605	.05517	.05423
23	14	19	.05861	.05837	.05804	.05761	.05709	.05649	.05581	.05507
23	14	20	.05803	.05804	.05796	.05777	.05749	.05712	.05667	.05615
23	14	21	.05757	.05787	.05806	.05815	.05814	.05803	.05784	.05757
23	14	22	.05729	.05792	.05844	.05887	.05919	.05941	.05954	.05958
23	14	23	.05737	.05848	.05949	.06040	.06119	.06189	.06249	.06298
23	15	16	.07158	.07075	.06982	.06878	.06766	.06646	.06519	.06386
23	15	17	.07051	.06995	.06927	.06848	.06760	.06663	.06559	.06447
23	15	18	.06958	.06928	.06886	.06833	.06770	.06697	.06615	.06526
23	15	19	.06877	.06875	.06860	.06834	.06797	.06750	.06693	.06627
23	15	20	.06810	.06836	.06851	.06854	.06845	.06826	.06796	.06757
23	15	21	.06757	.06816	.06863	.06899	.06922	.06935	.06937	.06928
23	15	22	.06724	.06822	.06909	.06984	.07048	.07099	.07140	.07170
23	15	23	.06734	.06889	.07033	.07166	.07286	.07395	.07493	.07579
23	16	17	.08367	.08331	.08282	.08219	.08144	.08057	.07960	.07854
23	16	18	.08257	.08253	.08234	.08201	.08156	.08099	.08030	.07951
23	16	19	.08163	.08191	.08204	.08203	.08189	.08163	.08124	.08074
23	16	20	.08084	.08146	.08194	.08228	.08248	.08255	.08250	.08232
23	16	21	.08022	.08123	.08209	.08282	.08341	.08387	.08420	.08441
23	16	22	.07983	.08131	.08265	.08385	.08492	.08586	.08666	.08734
23	16	23	.07996	.08211	.08413	.08603	.08780	.08943	.09094	.09231
23	17	18	.09955	.09989	.10005	.10005	.09988	.09957	.09911	.09851
23	17	19	.09843	.09915	.09970	.10008	.10030	.10036	.10027	.10004
23	17	20	.09749	.09862	.09959	.10039	.10102	.10150	.10182	.10200
23	17	21	.09675	.09835	.09979	.10106	.10217	.10313	.10393	.10457
23	17	22	.09630	.09846	.10047	.10232	.10402	.10557	.10696	.10820
23	17	23	.09647	.09944	.10228	.10498	.10754	.10995	.11222	.11434
23	18	19	.12132	.12272	.12391	.12490	.12569	.12628	.12669	.12691
23	18	20	.12018	.12209	.12379	.12530	.12661	.12772	.12865	.12939
23	18	21	.11930	.12177	.12405	.12614	.12805	.12977	.13130	.13265
23	18	22	.11876	.12192	.12491	.12773	.13037	.13284	.13513	.13724
23	18	23	.11898	.12314	.12716	.13104	.13477	.13834	.14175	.14500
23	19	20	.15292	.15605	.15893	.16158	.16400	.16617	.16811	.16982
23	19	21	.15183	.15567	.15929	.16269	.16588	.16884	.17157	.17409
23	19	22	.15118	.15588	.16041	.16475	.16889	.17283	.17656	.18008
23	19	23	.15148	.15746	.16331	.16902	.17457	.17996	.18517	.19020
23	20	21	.20283	.20899	.21492	.22060	.22602	.23118	.23606	.24067
23	20	22	.20201	.20933	.21647	.22340	.23012	.23662	.24288	.24890
23	20	23	.20247	.21149	.22040	.22919	.23783	.24632	.25464	.26277
23	21	22	.29283	.30522	.31745	.32949	.34133	.35294	.36430	.37539
23	21	23	.29361	.30843	.32324	.33800	.35269	.36727	.38172	.39602
23	22	23	.50045	.52961	.55908	.58879	.61871	.64876	.67891	.70910
24	1	2	.01382	.01194	.01033	.00894	.00774	.00672	.00583	.00507
24	1	3	.01108	.00966	.00843	.00735	.00642	.00561	.00490	.00429
24	1	4	.00959	.00841	.00739	.00648	.00569	.00500	.00440	.00387
24	1	5	.00863	.00761	.00672	.00593	.00523	.00462	.00409	.00361
24	1	6	.00795	.00705	.00625	.00554	.00491	.00436	.00387	.00344
24	1	7	.00745	.00663	.00590	.00526	.00468	.00417	.00372	.00331
24	1	8	.00705	.00630	.00564	.00504	.00450	.00403	.00360	.00322
24	1	9	.00673	.00604	.00542	.00487	.00437	.00392	.00352	.00316
24	1	10	.00647	.00583	.00525	.00473	.00426	.00383	.00345	.00311
24	1	11	.00626	.00565	.00511	.00462	.00417	.00377	.00341	.00308
24	1	12	.00607	.00551	.00499	.00453	.00410	.00372	.00337	.00306

Covariances of Inverse Gaussian Order Statistics
Shape Parameter k

n	i	j	1.6	1.7	1.8	1.9	2.0	2.1	2.2	2.3
24	1	13	.00591	.00538	.00490	.00445	.00405	.00368	.00335	.00305
24	1	14	.00577	.00527	.00481	.00439	.00401	.00366	.00334	.00305
24	1	15	.00565	.00518	.00474	.00434	.00398	.00364	.00333	.00305
24	1	16	.00555	.00510	.00469	.00431	.00396	.00363	.00334	.00307
24	1	17	.00545	.00503	.00464	.00428	.00394	.00364	.00335	.00309
24	1	18	.00537	.00498	.00460	.00426	.00394	.00364	.00337	.00312
24	1	19	.00530	.00493	.00458	.00425	.00395	.00366	.00340	.00316
24	1	20	.00524	.00489	.00456	.00425	.00396	.00369	.00344	.00320
24	1	21	.00519	.00486	.00455	.00426	.00399	.00373	.00349	.00327
24	1	22	.00515	.00485	.00456	.00429	.00403	.00379	.00357	.00335
24	1	23	.00512	.00485	.00459	.00434	.00411	.00388	.00367	.00347
24	1	24	.00513	.00490	.00468	.00446	.00425	.00405	.00385	.00367
24	2	3	.01622	.01430	.01260	.01110	.00979	.00863	.00762	.00673
24	2	4	.01406	.01248	.01107	.00981	.00870	.00772	.00685	.00608
24	2	5	.01267	.01130	.01008	.00899	.00801	.00714	.00637	.00568
24	2	6	.01169	.01048	.00939	.00841	.00753	.00674	.00603	.00540
24	2	7	.01095	.00986	.00887	.00798	.00717	.00645	.00580	.00521
24	2	8	.01037	.00938	.00847	.00765	.00690	.00623	.00562	.00507
24	2	9	.00991	.00899	.00815	.00739	.00670	.00606	.00549	.00497
24	2	10	.00953	.00868	.00790	.00718	.00653	.00594	.00539	.00490
24	2	11	.00921	.00842	.00769	.00702	.00640	.00584	.00532	.00485
24	2	12	.00894	.00820	.00751	.00688	.00630	.00576	.00527	.00482
24	2	13	.00870	.00801	.00737	.00677	.00622	.00571	.00524	.00480
24	2	14	.00850	.00785	.00724	.00668	.00615	.00567	.00522	.00480
24	2	15	.00833	.00771	.00714	.00660	.00610	.00564	.00521	.00481
24	2	16	.00817	.00760	.00705	.00655	.00607	.00563	.00522	.00483
24	2	17	.00804	.00749	.00698	.00650	.00605	.00563	.00524	.00487
24	2	18	.00792	.00741	.00693	.00648	.00605	.00565	.00527	.00491
24	2	19	.00781	.00734	.00689	.00646	.00606	.00567	.00531	.00497
24	2	20	.00772	.00728	.00686	.00646	.00608	.00572	.00538	.00505
24	2	21	.00765	.00724	.00685	.00648	.00612	.00578	.00546	.00515
24	2	22	.00759	.00722	.00687	.00652	.00619	.00588	.00557	.00528
24	2	23	.00755	.00723	.00691	.00661	.00631	.00602	.00574	.00546
24	2	24	.00756	.00730	.00704	.00678	.00652	.00627	.00602	.00578
24	3	4	.01808	.01616	.01444	.01290	.01152	.01028	.00918	.00820
24	3	5	.01630	.01466	.01316	.01182	.01061	.00952	.00854	.00767
24	3	6	.01505	.01359	.01227	.01106	.00998	.00899	.00810	.00730
24	3	7	.01410	.01280	.01160	.01051	.00951	.00861	.00779	.00704
24	3	8	.01337	.01218	.01108	.01008	.00916	.00832	.00756	.00686
24	3	9	.01277	.01168	.01067	.00974	.00888	.00810	.00738	.00673
24	3	10	.01229	.01127	.01034	.00947	.00867	.00793	.00725	.00663
24	3	11	.01188	.01094	.01006	.00925	.00850	.00780	.00716	.00656
24	3	12	.01153	.01065	.00983	.00907	.00836	.00770	.00709	.00652
24	3	13	.01123	.01041	.00964	.00892	.00825	.00762	.00704	.00650
24	3	14	.01097	.01020	.00948	.00880	.00817	.00757	.00702	.00650
24	3	15	.01074	.01003	.00935	.00871	.00811	.00754	.00701	.00651
24	3	16	.01054	.00987	.00924	.00863	.00806	.00752	.00702	.00654
24	3	17	.01037	.00974	.00915	.00858	.00804	.00753	.00704	.00659
24	3	18	.01021	.00963	.00908	.00854	.00803	.00755	.00709	.00665
24	3	19	.01008	.00954	.00902	.00852	.00804	.00758	.00715	.00673
24	3	20	.00996	.00947	.00899	.00852	.00808	.00764	.00723	.00684
24	3	21	.00987	.00942	.00898	.00855	.00813	.00773	.00734	.00697
24	3	22	.00979	.00939	.00899	.00861	.00823	.00785	.00749	.00715
24	3	23	.00974	.00940	.00905	.00871	.00837	.00804	.00771	.00740
24	3	24	.00976	.00949	.00922	.00894	.00866	.00838	.00810	.00782

Covariances of Inverse Gaussian Order Statistics
Shape Parameter k

n	i	j	1.6	1.7	1.8	1.9	2.0	2.1	2.2	2.3
24	4	5	.01991	.01801	.01627	.01470	.01327	.01197	.01080	.00974
24	4	6	.01839	.01671	.01517	.01377	.01248	.01131	.01025	.00928
24	4	7	.01725	.01574	.01435	.01308	.01191	.01083	.00985	.00896
24	4	8	.01635	.01498	.01372	.01255	.01147	.01047	.00956	.00873
24	4	9	.01563	.01438	.01321	.01213	.01113	.01020	.00934	.00856
24	4	10	.01503	.01388	.01280	.01179	.01086	.00999	.00918	.00844
24	4	11	.01453	.01347	.01246	.01152	.01064	.00982	.00906	.00835
24	4	12	.01411	.01312	.01218	.01130	.01047	.00970	.00898	.00830
24	4	13	.01374	.01282	.01195	.01112	.01034	.00961	.00892	.00827
24	4	14	.01343	.01257	.01175	.01097	.01024	.00954	.00889	.00827
24	4	15	.01315	.01235	.01158	.01085	.01016	.00950	.00888	.00829
24	4	16	.01291	.01216	.01145	.01076	.01011	.00948	.00889	.00833
24	4	17	.01269	.01200	.01133	.01069	.01007	.00948	.00892	.00839
24	4	18	.01251	.01187	.01125	.01065	.01007	.00951	.00898	.00847
24	4	19	.01234	.01176	.01118	.01062	.01008	.00956	.00905	.00857
24	4	20	.01220	.01167	.01114	.01063	.01012	.00963	.00916	.00870
24	4	21	.01208	.01160	.01113	.01066	.01019	.00974	.00930	.00887
24	4	22	.01199	.01157	.01115	.01073	.01031	.00990	.00949	.00910
24	4	23	.01193	.01158	.01122	.01086	.01050	.01013	.00977	.00942
24	4	24	.01195	.01170	.01142	.01114	.01085	.01056	.01026	.00995
24	5	6	.02186	.01997	.01822	.01662	.01514	.01379	.01255	.01142
24	5	7	.02051	.01882	.01725	.01579	.01445	.01321	.01207	.01103
24	5	8	.01945	.01792	.01649	.01516	.01392	.01278	.01172	.01074
24	5	9	.01860	.01720	.01588	.01466	.01351	.01244	.01145	.01054
24	5	10	.01789	.01660	.01539	.01425	.01319	.01219	.01126	.01039
24	5	11	.01730	.01611	.01499	.01393	.01293	.01199	.01111	.01029
24	5	12	.01680	.01570	.01465	.01366	.01272	.01184	.01101	.01022
24	5	13	.01636	.01534	.01437	.01344	.01256	.01173	.01094	.01019
24	5	14	.01599	.01504	.01413	.01326	.01243	.01165	.01090	.01019
24	5	15	.01566	.01478	.01394	.01312	.01234	.01160	.01089	.01021
24	5	16	.01537	.01456	.01377	.01301	.01228	.01158	.01090	.01026
24	5	17	.01512	.01437	.01364	.01293	.01224	.01158	.01094	.01033
24	5	18	.01490	.01421	.01353	.01287	.01223	.01161	.01101	.01043
24	5	19	.01470	.01407	.01345	.01285	.01225	.01167	.01111	.01056
24	5	20	.01453	.01397	.01341	.01285	.01230	.01176	.01124	.01072
24	5	21	.01439	.01389	.01339	.01289	.01239	.01190	.01141	.01094
24	5	22	.01428	.01385	.01341	.01297	.01253	.01209	.01165	.01121
24	5	23	.01421	.01387	.01351	.01313	.01276	.01237	.01199	.01160
24	5	24	.01424	.01400	.01375	.01348	.01319	.01289	.01258	.01226
24	6	7	.02398	.02211	.02035	.01872	.01721	.01580	.01450	.01330
24	6	8	.02275	.02106	.01947	.01798	.01658	.01529	.01408	.01296
24	6	9	.02176	.02021	.01876	.01738	.01610	.01489	.01377	.01272
24	6	10	.02094	.01952	.01818	.01691	.01571	.01459	.01353	.01254
24	6	11	.02025	.01895	.01770	.01653	.01541	.01435	.01336	.01242
24	6	12	.01966	.01846	.01731	.01621	.01517	.01417	.01323	.01234
24	6	13	.01916	.01805	.01698	.01595	.01497	.01404	.01315	.01231
24	6	14	.01872	.01769	.01670	.01574	.01483	.01395	.01310	.01230
24	6	15	.01833	.01739	.01647	.01558	.01472	.01389	.01309	.01233
24	6	16	.01800	.01713	.01628	.01545	.01464	.01386	.01311	.01239
24	6	17	.01770	.01690	.01612	.01535	.01460	.01387	.01316	.01248
24	6	18	.01744	.01672	.01599	.01528	.01459	.01391	.01324	.01260
24	6	19	.01722	.01656	.01590	.01525	.01461	.01398	.01336	.01275
24	6	20	.01702	.01644	.01585	.01526	.01467	.01409	.01351	.01295
24	6	21	.01686	.01635	.01583	.01530	.01477	.01425	.01372	.01321
24	6	22	.01673	.01630	.01586	.01540	.01494	.01448	.01401	.01354

Covariances of Inverse Gaussian Order Statistics
Shape Parameter k

n	i	j	1.6	1.7	1.8	1.9	2.0	2.1	2.2	2.3
24	6	23	.01665	.01632	.01596	.01559	.01521	.01482	.01442	.01401
24	6	24	.01667	.01648	.01625	.01600	.01573	.01544	.01513	.01481
24	7	8	.02632	.02447	.02272	.02107	.01952	.01806	.01671	.01544
24	7	9	.02518	.02350	.02190	.02038	.01895	.01760	.01634	.01515
24	7	10	.02424	.02270	.02123	.01983	.01850	.01724	.01606	.01494
24	7	11	.02344	.02203	.02068	.01938	.01814	.01697	.01585	.01480
24	7	12	.02277	.02147	.02022	.01901	.01786	.01676	.01571	.01471
24	7	13	.02218	.02099	.01983	.01871	.01764	.01660	.01561	.01467
24	7	14	.02168	.02058	.01951	.01847	.01746	.01649	.01556	.01466
24	7	15	.02124	.02023	.01924	.01827	.01733	.01642	.01555	.01470
24	7	16	.02085	.01993	.01901	.01812	.01725	.01639	.01557	.01477
24	7	17	.02051	.01967	.01883	.01801	.01720	.01640	.01563	.01487
24	7	18	.02021	.01945	.01869	.01793	.01718	.01645	.01572	.01502
24	7	19	.01995	.01927	.01858	.01790	.01721	.01653	.01586	.01520
24	7	20	.01972	.01912	.01852	.01790	.01728	.01666	.01605	.01544
24	7	21	.01953	.01902	.01849	.01795	.01741	.01685	.01630	.01574
24	7	22	.01938	.01897	.01853	.01807	.01760	.01712	.01663	.01614
24	7	23	.01929	.01899	.01865	.01830	.01792	.01753	.01712	.01670
24	7	24	.01932	.01917	.01899	.01877	.01853	.01826	.01797	.01765
24	8	9	.02894	.02711	.02537	.02371	.02213	.02063	.01922	.01789
24	8	10	.02786	.02620	.02460	.02307	.02161	.02022	.01890	.01765
24	8	11	.02695	.02543	.02396	.02255	.02119	.01990	.01866	.01749
24	8	12	.02618	.02479	.02344	.02213	.02087	.01965	.01849	.01738
24	8	13	.02551	.02424	.02299	.02178	.02061	.01947	.01838	.01733
24	8	14	.02494	.02377	.02262	.02150	.02041	.01935	.01832	.01733
24	8	15	.02443	.02336	.02231	.02127	.02026	.01927	.01830	.01737
24	8	16	.02399	.02301	.02205	.02110	.02016	.01923	.01833	.01745
24	8	17	.02359	.02272	.02184	.02096	.02010	.01924	.01840	.01758
24	8	18	.02325	.02247	.02167	.02088	.02008	.01930	.01852	.01775
24	8	19	.02295	.02226	.02155	.02084	.02012	.01940	.01868	.01797
24	8	20	.02269	.02209	.02147	.02084	.02020	.01955	.01890	.01825
24	8	21	.02247	.02197	.02145	.02091	.02034	.01977	.01919	.01861
24	8	22	.02230	.02191	.02149	.02104	.02058	.02009	.01959	.01908
24	8	23	.02220	.02194	.02164	.02131	.02095	.02057	.02016	.01974
24	8	24	.02224	.02215	.02203	.02186	.02166	.02142	.02116	.02087
24	9	10	.03189	.03010	.02838	.02671	.02511	.02358	.02213	.02074
24	9	11	.03086	.02923	.02765	.02611	.02464	.02321	.02185	.02055
24	9	12	.02998	.02849	.02704	.02563	.02426	.02293	.02165	.02043
24	9	13	.02922	.02786	.02653	.02523	.02396	.02272	.02153	.02037
24	9	14	.02856	.02733	.02611	.02490	.02373	.02258	.02146	.02037
24	9	15	.02798	.02686	.02575	.02464	.02356	.02249	.02144	.02042
24	9	16	.02748	.02647	.02545	.02444	.02344	.02245	.02147	.02052
24	9	17	.02703	.02613	.02521	.02429	.02337	.02246	.02156	.02066
24	9	18	.02664	.02584	.02502	.02419	.02336	.02252	.02169	.02086
24	9	19	.02630	.02560	.02488	.02414	.02340	.02264	.02188	.02112
24	9	20	.02600	.02541	.02479	.02415	.02349	.02282	.02214	.02145
24	9	21	.02575	.02527	.02476	.02423	.02366	.02308	.02248	.02187
24	9	22	.02556	.02520	.02481	.02439	.02393	.02345	.02295	.02243
24	9	23	.02544	.02523	.02498	.02469	.02437	.02401	.02362	.02321
24	9	24	.02548	.02548	.02543	.02533	.02519	.02501	.02478	.02453
24	10	11	.03525	.03352	.03182	.03016	.02856	.02701	.02551	.02407
24	10	12	.03425	.03268	.03112	.02960	.02812	.02668	.02528	.02394
24	10	13	.03339	.03196	.03054	.02915	.02778	.02644	.02514	.02387
24	10	14	.03264	.03135	.03006	.02878	.02751	.02627	.02506	.02387
24	10	15	.03198	.03082	.02965	.02848	.02732	.02617	.02504	.02393

Covariances of Inverse Gaussian Order Statistics
Shape Parameter k

n	i	j	1.6	1.7	1.8	1.9	2.0	2.1	2.2	2.3
24	10	16	.03141	.03036	.02931	.02825	.02718	.02613	.02508	.02405
24	10	17	.03090	.02998	.02903	.02807	.02711	.02614	.02518	.02422
24	10	18	.03045	.02965	.02881	.02796	.02709	.02622	.02534	.02445
24	10	19	.03006	.02937	.02865	.02791	.02714	.02635	.02556	.02476
24	10	20	.02973	.02916	.02855	.02792	.02725	.02656	.02586	.02514
24	10	21	.02944	.02900	.02852	.02800	.02745	.02687	.02626	.02564
24	10	22	.02922	.02892	.02858	.02819	.02776	.02730	.02681	.02629
24	10	23	.02909	.02896	.02877	.02854	.02826	.02795	.02759	.02720
24	10	24	.02914	.02924	.02929	.02928	.02922	.02911	.02895	.02875
24	11	12	.03912	.03745	.03580	.03417	.03258	.03102	.02949	.02802
24	11	13	.03814	.03664	.03514	.03365	.03218	.03074	.02932	.02794
24	11	14	.03729	.03594	.03458	.03323	.03188	.03055	.02923	.02795
24	11	15	.03654	.03533	.03411	.03288	.03165	.03043	.02922	.02802
24	11	16	.03589	.03482	.03373	.03262	.03150	.03038	.02926	.02815
24	11	17	.03531	.03437	.03341	.03242	.03142	.03040	.02938	.02836
24	11	18	.03480	.03400	.03316	.03229	.03140	.03049	.02957	.02863
24	11	19	.03436	.03369	.03298	.03223	.03145	.03065	.02983	.02899
24	11	20	.03398	.03344	.03286	.03224	.03159	.03090	.03018	.02944
24	11	21	.03365	.03327	.03283	.03234	.03181	.03125	.03065	.03002
24	11	22	.03340	.03318	.03290	.03256	.03218	.03175	.03128	.03078
24	11	23	.03325	.03322	.03312	.03297	.03276	.03250	.03220	.03185
24	11	24	.03330	.03355	.03372	.03382	.03387	.03385	.03378	.03366
24	12	13	.04361	.04204	.04046	.03889	.03732	.03577	.03424	.03274
24	12	14	.04264	.04124	.03982	.03840	.03697	.03555	.03414	.03274
24	12	15	.04179	.04056	.03929	.03801	.03671	.03541	.03412	.03283
24	12	16	.04105	.03997	.03885	.03770	.03654	.03536	.03418	.03299
24	12	17	.04039	.03946	.03849	.03748	.03644	.03539	.03431	.03323
24	12	18	.03982	.03903	.03820	.03733	.03643	.03549	.03453	.03355
24	12	19	.03931	.03868	.03800	.03726	.03649	.03568	.03484	.03397
24	12	20	.03888	.03840	.03787	.03728	.03664	.03596	.03525	.03450
24	12	21	.03851	.03820	.03783	.03740	.03691	.03637	.03580	.03518
24	12	22	.03822	.03810	.03790	.03765	.03733	.03696	.03654	.03607
24	12	23	.03805	.03814	.03817	.03812	.03801	.03784	.03760	.03732
24	12	24	.03811	.03852	.03885	.03911	.03929	.03941	.03945	.03944
24	13	14	.04890	.04746	.04599	.04449	.04299	.04147	.03996	.03846
24	13	15	.04793	.04667	.04538	.04405	.04269	.04132	.03994	.03856
24	13	16	.04708	.04600	.04487	.04370	.04249	.04126	.04002	.03876
24	13	17	.04634	.04542	.04446	.04344	.04238	.04129	.04018	.03904
24	13	18	.04568	.04494	.04413	.04327	.04237	.04142	.04043	.03942
24	13	19	.04510	.04453	.04389	.04320	.04244	.04164	.04080	.03991
24	13	20	.04461	.04421	.04375	.04322	.04262	.04198	.04128	.04054
24	13	21	.04419	.04398	.04370	.04335	.04293	.04245	.04192	.04133
24	13	22	.04386	.04387	.04379	.04365	.04342	.04314	.04279	.04238
24	13	23	.04366	.04392	.04410	.04419	.04421	.04416	.04404	.04385
24	13	24	.04374	.04436	.04489	.04534	.04571	.04599	.04620	.04633
24	14	15	.05521	.05395	.05263	.05127	.04986	.04843	.04697	.04551
24	14	16	.05424	.05318	.05205	.05087	.04964	.04837	.04706	.04574
24	14	17	.05339	.05252	.05158	.05057	.04951	.04841	.04726	.04607
24	14	18	.05263	.05196	.05121	.05038	.04949	.04855	.04756	.04653
24	14	19	.05198	.05150	.05093	.05029	.04959	.04881	.04799	.04711
24	14	20	.05141	.05113	.05077	.05032	.04980	.04921	.04855	.04784
24	14	21	.05093	.05087	.05072	.05048	.05016	.04977	.04931	.04878
24	14	22	.05056	.05074	.05083	.05082	.05074	.05057	.05033	.05001
24	14	23	.05033	.05080	.05118	.05146	.05166	.05177	.05180	.05175
24	14	24	.05042	.05131	.05210	.05280	.05341	.05392	.05434	.05468

Covariances of Inverse Gaussian Order Statistics
Shape Parameter k

n	i	j	1.6	1.7	1.8	1.9	2.0	2.1	2.2	2.3
24	15	16	.06288	.06187	.06077	.05960	.05836	.05706	.05571	.05433
24	15	17	.06190	.06111	.06023	.05926	.05822	.05711	.05595	.05473
24	15	18	.06104	.06047	.05980	.05904	.05820	.05729	.05631	.05527
24	15	19	.06028	.05993	.05948	.05894	.05831	.05760	.05682	.05596
24	15	20	.05963	.05951	.05929	.05897	.05856	.05807	.05749	.05684
24	15	21	.05908	.05921	.05924	.05917	.05900	.05873	.05838	.05795
24	15	22	.05865	.05906	.05937	.05957	.05967	.05968	.05959	.05942
24	15	23	.05839	.05914	.05978	.06032	.06076	.06109	.06133	.06147
24	15	24	.05850	.05973	.06086	.06189	.06281	.06362	.06434	.06495
24	16	17	.07240	.07173	.07094	.07005	.06907	.06799	.06684	.06562
24	16	18	.07140	.07098	.07045	.06980	.06905	.06821	.06728	.06627
24	16	19	.07052	.07037	.07009	.06969	.06919	.06859	.06789	.06710
24	16	20	.06976	.06988	.06987	.06974	.06949	.06915	.06870	.06815
24	16	21	.06913	.06953	.06981	.06997	.07001	.06994	.06976	.06949
24	16	22	.06863	.06936	.06996	.07045	.07081	.07106	.07120	.07124
24	16	23	.06833	.06945	.07046	.07134	.07210	.07275	.07328	.07370
24	16	24	.06847	.07016	.07173	.07319	.07453	.07576	.07687	.07786
24	17	18	.08451	.08433	.08400	.08353	.08294	.08222	.08140	.08047
24	17	19	.08348	.08360	.08358	.08341	.08311	.08268	.08213	.08148
24	17	20	.08260	.08303	.08332	.08347	.08348	.08336	.08311	.08275
24	17	21	.08185	.08263	.08326	.08375	.08410	.08432	.08441	.08437
24	17	22	.08127	.08243	.08345	.08433	.08507	.08568	.08615	.08650
24	17	23	.08093	.08255	.08404	.08540	.08662	.08770	.08866	.08948
24	17	24	.08109	.08339	.08557	.08762	.08954	.09133	.09299	.09452
24	18	19	.10042	.10095	.10130	.10148	.10150	.10136	.10107	.10064
24	18	20	.09937	.10028	.10101	.10157	.10196	.10220	.10228	.10222
24	18	21	.09849	.09980	.10094	.10192	.10273	.10338	.10387	.10422
24	18	22	.09781	.09957	.10118	.10263	.10392	.10504	.10601	.10683
24	18	23	.09740	.09973	.10190	.10393	.10581	.10753	.10909	.11051
24	18	24	.09761	.10074	.10375	.10663	.10937	.11196	.11441	.11671
24	19	20	.12222	.12384	.12525	.12645	.12745	.12826	.12887	.12930
24	19	21	.12116	.12327	.12519	.12690	.12842	.12974	.13088	.13182
24	19	22	.12034	.12301	.12550	.12779	.12991	.13183	.13357	.13513
24	19	23	.11986	.12321	.12640	.12942	.13227	.13494	.13744	.13976
24	19	24	.12013	.12448	.12870	.13278	.13672	.14050	.14412	.14758
24	20	21	.15388	.15725	.16038	.16329	.16596	.16839	.17059	.17255
24	20	22	.15287	.15694	.16080	.16446	.16789	.17110	.17409	.17686
24	20	23	.15229	.15722	.16198	.16656	.17094	.17513	.17912	.18290
24	20	24	.15266	.15886	.16494	.17088	.17667	.18231	.18778	.19307
24	21	22	.20386	.21031	.21654	.22252	.22826	.23373	.23894	.24388
24	21	23	.20314	.21073	.21815	.22538	.23241	.23922	.24580	.25214
24	21	24	.20368	.21296	.22215	.23122	.24017	.24896	.25760	.26606
24	22	23	.29399	.30673	.31933	.33176	.34400	.35603	.36782	.37935
24	22	24	.29489	.31004	.32520	.34033	.35540	.37038	.38525	.39998
24	23	24	.50187	.53150	.56147	.59171	.62219	.65283	.68360	.71444
25	1	2	.01333	.01150	.00994	.00859	.00743	.00644	.00559	.00485
25	1	3	.01067	.00929	.00809	.00705	.00615	.00536	.00469	.00410
25	1	4	.00922	.00808	.00708	.00621	.00544	.00478	.00420	.00369
25	1	5	.00829	.00730	.00643	.00567	.00500	.00441	.00389	.00343
25	1	6	.00763	.00676	.00598	.00529	.00469	.00415	.00368	.00326
25	1	7	.00714	.00635	.00564	.00502	.00446	.00397	.00353	.00314
25	1	8	.00676	.00603	.00538	.00480	.00429	.00383	.00342	.00305
25	1	9	.00645	.00578	.00518	.00463	.00415	.00372	.00333	.00299
25	1	10	.00620	.00557	.00501	.00450	.00404	.00364	.00327	.00294
25	1	11	.00598	.00540	.00487	.00439	.00396	.00357	.00322	.00290

Covariances of Inverse Gaussian Order Statistics
Shape Parameter k

n	i	j	1.6	1.7	1.8	1.9	2.0	2.1	2.2	2.3
25	1	12	.00580	.00525	.00475	.00430	.00389	.00352	.00318	.00288
25	1	13	.00565	.00513	.00466	.00423	.00384	.00348	.00316	.00287
25	1	14	.00551	.00502	.00458	.00417	.00379	.00345	.00314	.00286
25	1	15	.00540	.00493	.00451	.00412	.00376	.00343	.00314	.00286
25	1	16	.00529	.00485	.00445	.00408	.00373	.00342	.00314	.00287
25	1	17	.00520	.00479	.00440	.00405	.00372	.00342	.00314	.00289
25	1	18	.00512	.00473	.00436	.00402	.00371	.00342	.00315	.00291
25	1	19	.00505	.00468	.00433	.00401	.00371	.00343	.00318	.00294
25	1	20	.00499	.00464	.00431	.00400	.00372	.00345	.00320	.00297
25	1	21	.00493	.00460	.00429	.00400	.00373	.00348	.00324	.00302
25	1	22	.00489	.00458	.00429	.00402	.00376	.00352	.00329	.00308
25	1	23	.00485	.00457	.00430	.00404	.00380	.00358	.00336	.00316
25	1	24	.00483	.00458	.00433	.00410	.00387	.00366	.00346	.00327
25	1	25	.00484	.00462	.00441	.00420	.00400	.00381	.00363	.00345
25	2	3	.01556	.01369	.01205	.01060	.00933	.00822	.00724	.00639
25	2	4	.01347	.01193	.01057	.00936	.00828	.00734	.00650	.00576
25	2	5	.01212	.01080	.00961	.00855	.00761	.00678	.00603	.00537
25	2	6	.01117	.01000	.00894	.00799	.00715	.00639	.00571	.00511
25	2	7	.01046	.00940	.00844	.00758	.00680	.00610	.00548	.00492
25	2	8	.00990	.00893	.00806	.00726	.00654	.00589	.00531	.00478
25	2	9	.00945	.00856	.00775	.00701	.00634	.00573	.00518	.00468
25	2	10	.00908	.00826	.00750	.00681	.00618	.00560	.00508	.00461
25	2	11	.00877	.00800	.00729	.00664	.00605	.00550	.00501	.00455
25	2	12	.00851	.00779	.00712	.00651	.00594	.00542	.00495	.00452
25	2	13	.00829	.00761	.00698	.00640	.00586	.00537	.00491	.00450
25	2	14	.00809	.00745	.00685	.00630	.00579	.00532	.00489	.00449
25	2	15	.00792	.00731	.00675	.00623	.00574	.00529	.00488	.00449
25	2	16	.00777	.00720	.00667	.00617	.00571	.00528	.00488	.00450
25	2	17	.00763	.00710	.00659	.00612	.00568	.00527	.00489	.00453
25	2	18	.00751	.00701	.00654	.00609	.00567	.00527	.00491	.00456
25	2	19	.00741	.00694	.00649	.00607	.00567	.00529	.00494	.00461
25	2	20	.00732	.00688	.00646	.00606	.00568	.00532	.00498	.00466
25	2	21	.00724	.00683	.00644	.00606	.00570	.00536	.00504	.00474
25	2	22	.00718	.00680	.00643	.00608	.00575	.00543	.00512	.00483
25	2	23	.00712	.00678	.00645	.00612	.00581	.00551	.00523	.00495
25	2	24	.00709	.00679	.00649	.00620	.00592	.00564	.00538	.00512
25	2	25	.00711	.00686	.00661	.00636	.00612	.00588	.00564	.00541
25	3	4	.01726	.01540	.01374	.01225	.01092	.00974	.00868	.00774
25	3	5	.01555	.01395	.01251	.01121	.01004	.00900	.00806	.00722
25	3	6	.01434	.01292	.01164	.01048	.00943	.00849	.00764	.00687
25	3	7	.01343	.01216	.01100	.00994	.00898	.00812	.00733	.00662
25	3	8	.01272	.01156	.01050	.00953	.00864	.00784	.00710	.00644
25	3	9	.01214	.01108	.01010	.00920	.00837	.00762	.00693	.00630
25	3	10	.01167	.01069	.00978	.00894	.00816	.00745	.00680	.00620
25	3	11	.01128	.01036	.00951	.00872	.00799	.00732	.00670	.00613
25	3	12	.01094	.01008	.00929	.00855	.00786	.00722	.00663	.00609
25	3	13	.01065	.00985	.00910	.00840	.00775	.00714	.00658	.00606
25	3	14	.01040	.00965	.00894	.00828	.00766	.00709	.00655	.00605
25	3	15	.01018	.00947	.00881	.00818	.00760	.00705	.00653	.00605
25	3	16	.00998	.00932	.00870	.00811	.00755	.00702	.00653	.00607
25	3	17	.00981	.00919	.00860	.00804	.00752	.00702	.00655	.00610
25	3	18	.00966	.00908	.00853	.00800	.00750	.00702	.00657	.00615
25	3	19	.00953	.00899	.00847	.00797	.00750	.00705	.00662	.00621
25	3	20	.00941	.00891	.00843	.00796	.00751	.00708	.00668	.00629
25	3	21	.00931	.00885	.00840	.00796	.00754	.00714	.00676	.00639

Covariances of Inverse Gaussian Order Statistics
Shape Parameter k

n	i	j	1.6	1.7	1.8	1.9	2.0	2.1	2.2	2.3
25	3	22	.00923	.00881	.00839	.00799	.00760	.00722	.00686	.00651
25	3	23	.00916	.00878	.00841	.00805	.00769	.00734	.00700	.00668
25	3	24	.00912	.00880	.00847	.00815	.00783	.00751	.00721	.00691
25	3	25	.00914	.00888	.00862	.00836	.00809	.00783	.00756	.00730
25	4	5	.01893	.01708	.01541	.01389	.01251	.01127	.01015	.00914
25	4	6	.01746	.01584	.01435	.01300	.01176	.01064	.00962	.00870
25	4	7	.01636	.01490	.01356	.01233	.01121	.01018	.00924	.00838
25	4	8	.01550	.01418	.01295	.01182	.01078	.00983	.00895	.00816
25	4	9	.01481	.01359	.01246	.01142	.01045	.00956	.00874	.00799
25	4	10	.01424	.01311	.01207	.01109	.01019	.00935	.00858	.00786
25	4	11	.01376	.01271	.01174	.01083	.00998	.00919	.00846	.00778
25	4	12	.01335	.01238	.01146	.01061	.00981	.00906	.00837	.00772
25	4	13	.01299	.01209	.01123	.01043	.00967	.00896	.00830	.00768
25	4	14	.01269	.01184	.01104	.01028	.00957	.00889	.00826	.00767
25	4	15	.01242	.01163	.01088	.01016	.00948	.00885	.00824	.00768
25	4	16	.01218	.01145	.01074	.01007	.00942	.00882	.00824	.00770
25	4	17	.01198	.01129	.01063	.00999	.00938	.00881	.00826	.00774
25	4	18	.01179	.01115	.01053	.00994	.00936	.00882	.00829	.00780
25	4	19	.01163	.01104	.01046	.00990	.00936	.00884	.00835	.00788
25	4	20	.01149	.01094	.01041	.00989	.00938	.00889	.00842	.00797
25	4	21	.01137	.01087	.01037	.00989	.00942	.00897	.00853	.00810
25	4	22	.01126	.01081	.01037	.00992	.00949	.00907	.00866	.00826
25	4	23	.01118	.01079	.01039	.00999	.00960	.00922	.00884	.00847
25	4	24	.01113	.01080	.01046	.01012	.00978	.00943	.00909	.00876
25	4	25	.01116	.01091	.01065	.01038	.01011	.00982	.00954	.00925
25	5	6	.02069	.01886	.01718	.01563	.01421	.01292	.01174	.01066
25	5	7	.01940	.01776	.01624	.01484	.01355	.01236	.01127	.01028
25	5	8	.01838	.01689	.01551	.01423	.01304	.01194	.01093	.01000
25	5	9	.01756	.01620	.01493	.01374	.01264	.01162	.01067	.00980
25	5	10	.01689	.01563	.01446	.01336	.01233	.01137	.01047	.00965
25	5	11	.01632	.01516	.01407	.01304	.01207	.01117	.01033	.00954
25	5	12	.01583	.01476	.01374	.01278	.01187	.01102	.01022	.00947
25	5	13	.01542	.01442	.01347	.01256	.01171	.01090	.01014	.00943
25	5	14	.01506	.01413	.01324	.01239	.01158	.01082	.01009	.00941
25	5	15	.01474	.01387	.01304	.01224	.01148	.01076	.01007	.00942
25	5	16	.01446	.01365	.01288	.01213	.01141	.01072	.01007	.00945
25	5	17	.01421	.01347	.01274	.01204	.01136	.01071	.01009	.00950
25	5	18	.01400	.01330	.01263	.01197	.01134	.01072	.01013	.00957
25	5	19	.01381	.01317	.01254	.01193	.01133	.01076	.01020	.00967
25	5	20	.01364	.01305	.01248	.01191	.01136	.01082	.01029	.00979
25	5	21	.01349	.01296	.01244	.01192	.01141	.01091	.01042	.00994
25	5	22	.01337	.01290	.01243	.01196	.01149	.01103	.01058	.01013
25	5	23	.01328	.01287	.01246	.01204	.01163	.01121	.01080	.01039
25	5	24	.01322	.01289	.01255	.01219	.01184	.01148	.01111	.01075
25	5	25	.01325	.01302	.01277	.01251	.01224	.01195	.01166	.01136
25	6	7	.02260	.02079	.01910	.01753	.01607	.01473	.01349	.01235
25	6	8	.02143	.01978	.01825	.01681	.01548	.01423	.01308	.01202
25	6	9	.02048	.01898	.01757	.01624	.01501	.01385	.01277	.01177
25	6	10	.01969	.01832	.01701	.01579	.01463	.01355	.01254	.01160
25	6	11	.01903	.01776	.01656	.01541	.01434	.01332	.01237	.01147
25	6	12	.01847	.01730	.01618	.01511	.01410	.01314	.01224	.01139
25	6	13	.01799	.01690	.01585	.01486	.01391	.01300	.01215	.01134
25	6	14	.01757	.01656	.01558	.01465	.01375	.01290	.01209	.01132
25	6	15	.01720	.01626	.01535	.01448	.01364	.01283	.01206	.01133
25	6	16	.01687	.01601	.01516	.01434	.01355	.01279	.01206	.01136

Covariances of Inverse Gaussian Order Statistics
Shape Parameter k

n	i	j	1.6	1.7	1.8	1.9	2.0	2.1	2.2	2.3
25	6	17	.01659	.01579	.01500	.01424	.01350	.01278	.01209	.01142
25	6	18	.01634	.01560	.01487	.01416	.01347	.01279	.01214	.01151
25	6	19	.01611	.01544	.01477	.01411	.01347	.01283	.01222	.01162
25	6	20	.01592	.01531	.01470	.01409	.01349	.01291	.01233	.01177
25	6	21	.01575	.01520	.01465	.01410	.01355	.01301	.01248	.01196
25	6	22	.01561	.01513	.01464	.01415	.01365	.01316	.01267	.01219
25	6	23	.01550	.01509	.01467	.01425	.01381	.01337	.01293	.01250
25	6	24	.01543	.01511	.01478	.01443	.01406	.01369	.01331	.01293
25	6	25	.01546	.01527	.01504	.01480	.01454	.01426	.01396	.01366
25	7	8	.02470	.02291	.02122	.01963	.01814	.01675	.01545	.01425
25	7	9	.02361	.02198	.02043	.01897	.01759	.01630	.01509	.01397
25	7	10	.02271	.02122	.01979	.01844	.01716	.01596	.01482	.01376
25	7	11	.02196	.02058	.01926	.01801	.01681	.01568	.01462	.01361
25	7	12	.02131	.02004	.01882	.01765	.01653	.01547	.01447	.01351
25	7	13	.02076	.01958	.01845	.01736	.01631	.01531	.01436	.01345
25	7	14	.02027	.01919	.01813	.01712	.01614	.01519	.01429	.01343
25	7	15	.01985	.01885	.01787	.01692	.01600	.01511	.01426	.01344
25	7	16	.01948	.01855	.01765	.01676	.01590	.01507	.01426	.01349
25	7	17	.01915	.01830	.01746	.01664	.01584	.01505	.01429	.01356
25	7	18	.01886	.01808	.01731	.01655	.01580	.01507	.01436	.01366
25	7	19	.01860	.01789	.01719	.01649	.01580	.01512	.01445	.01380
25	7	20	.01837	.01774	.01711	.01647	.01583	.01520	.01458	.01397
25	7	21	.01818	.01762	.01705	.01648	.01590	.01533	.01476	.01419
25	7	22	.01802	.01754	.01704	.01654	.01602	.01550	.01499	.01447
25	7	23	.01789	.01749	.01708	.01665	.01621	.01575	.01530	.01483
25	7	24	.01781	.01752	.01720	.01686	.01650	.01613	.01574	.01535
25	7	25	.01785	.01770	.01751	.01730	.01706	.01680	.01651	.01621
25	8	9	.02703	.02526	.02357	.02197	.02046	.01903	.01768	.01642
25	8	10	.02601	.02439	.02284	.02136	.01996	.01863	.01737	.01618
25	8	11	.02515	.02366	.02223	.02087	.01956	.01831	.01713	.01601
25	8	12	.02441	.02305	.02173	.02046	.01924	.01807	.01695	.01589
25	8	13	.02378	.02252	.02130	.02012	.01898	.01788	.01683	.01583
25	8	14	.02323	.02207	.02094	.01984	.01877	.01774	.01675	.01580
25	8	15	.02274	.02168	.02063	.01961	.01862	.01765	.01672	.01582
25	8	16	.02232	.02134	.02038	.01943	.01850	.01760	.01672	.01587
25	8	17	.02194	.02105	.02017	.01929	.01843	.01758	.01676	.01595
25	8	18	.02161	.02080	.01999	.01919	.01839	.01760	.01683	.01607
25	8	19	.02132	.02059	.01986	.01912	.01839	.01766	.01694	.01623
25	8	20	.02106	.02041	.01976	.01909	.01843	.01776	.01710	.01644
25	8	21	.02084	.02028	.01970	.01911	.01851	.01791	.01730	.01670
25	8	22	.02065	.02018	.01968	.01917	.01865	.01811	.01757	.01702
25	8	23	.02051	.02013	.01973	.01931	.01886	.01840	.01793	.01745
25	8	24	.02042	.02016	.01987	.01955	.01921	.01884	.01846	.01806
25	8	25	.02046	.02037	.02023	.02006	.01985	.01962	.01936	.01907
25	9	10	.02965	.02791	.02623	.02462	.02309	.02162	.02023	.01891
25	9	11	.02867	.02708	.02554	.02405	.02263	.02126	.01996	.01872
25	9	12	.02783	.02638	.02496	.02358	.02226	.02098	.01976	.01858
25	9	13	.02712	.02578	.02447	.02320	.02196	.02077	.01962	.01851
25	9	14	.02649	.02526	.02406	.02288	.02173	.02061	.01953	.01848
25	9	15	.02594	.02482	.02371	.02262	.02155	.02050	.01949	.01850
25	9	16	.02546	.02443	.02342	.02241	.02142	.02044	.01949	.01856
25	9	17	.02503	.02410	.02317	.02225	.02133	.02042	.01953	.01866
25	9	18	.02465	.02382	.02298	.02213	.02129	.02045	.01962	.01880
25	9	19	.02432	.02358	.02282	.02206	.02129	.02052	.01975	.01899
25	9	20	.02403	.02338	.02271	.02202	.02133	.02063	.01993	.01923

Covariances of Inverse Gaussian Order Statistics
Shape Parameter k

n	i	j	1.6	1.7	1.8	1.9	2.0	2.1	2.2	2.3
25	9	21	.02377	.02322	.02264	.02204	.02143	.02080	.02017	.01953
25	9	22	.02356	.02311	.02263	.02212	.02159	.02104	.02048	.01991
25	9	23	.02340	.02305	.02268	.02227	.02184	.02138	.02091	.02042
25	9	24	.02330	.02309	.02284	.02255	.02223	.02189	.02151	.02112
25	9	25	.02335	.02332	.02325	.02314	.02298	.02279	.02257	.02231
25	10	11	.03260	.03090	.02925	.02764	.02610	.02461	.02318	.02181
25	10	12	.03165	.03010	.02859	.02711	.02567	.02428	.02294	.02166
25	10	13	.03084	.02942	.02803	.02667	.02533	.02404	.02278	.02157
25	10	14	.03013	.02884	.02756	.02630	.02507	.02386	.02268	.02154
25	10	15	.02951	.02833	.02716	.02601	.02486	.02374	.02264	.02156
25	10	16	.02896	.02790	.02683	.02577	.02471	.02367	.02264	.02163
25	10	17	.02848	.02752	.02655	.02558	.02461	.02365	.02269	.02175
25	10	18	.02805	.02720	.02633	.02545	.02456	.02368	.02279	.02192
25	10	19	.02767	.02692	.02615	.02536	.02456	.02376	.02295	.02214
25	10	20	.02734	.02670	.02602	.02533	.02462	.02389	.02316	.02242
25	10	21	.02705	.02652	.02595	.02535	.02473	.02409	.02343	.02277
25	10	22	.02681	.02639	.02593	.02544	.02491	.02437	.02380	.02321
25	10	23	.02663	.02633	.02599	.02561	.02520	.02476	.02429	.02380
25	10	24	.02651	.02637	.02618	.02594	.02566	.02535	.02500	.02462
25	10	25	.02657	.02664	.02665	.02661	.02653	.02639	.02622	.02601
25	11	12	.03596	.03432	.03270	.03112	.02957	.02806	.02661	.02519
25	11	13	.03504	.03355	.03207	.03061	.02918	.02778	.02642	.02510
25	11	14	.03424	.03288	.03154	.03020	.02888	.02758	.02631	.02506
25	11	15	.03353	.03231	.03109	.02986	.02864	.02744	.02626	.02509
25	11	16	.03291	.03182	.03071	.02959	.02847	.02736	.02626	.02517
25	11	17	.03237	.03139	.03039	.02938	.02836	.02734	.02632	.02531
25	11	18	.03188	.03102	.03013	.02923	.02831	.02738	.02644	.02551
25	11	19	.03146	.03071	.02993	.02913	.02831	.02747	.02662	.02576
25	11	20	.03108	.03045	.02979	.02909	.02837	.02762	.02686	.02609
25	11	21	.03076	.03025	.02970	.02912	.02850	.02785	.02718	.02650
25	11	22	.03048	.03011	.02968	.02922	.02871	.02817	.02761	.02702
25	11	23	.03027	.03004	.02975	.02942	.02904	.02863	.02818	.02770
25	11	24	.03015	.03008	.02997	.02979	.02957	.02931	.02900	.02865
25	11	25	.03021	.03039	.03051	.03057	.03057	.03052	.03042	.03027
25	12	13	.03983	.03826	.03670	.03515	.03362	.03212	.03064	.02920
25	12	14	.03892	.03751	.03609	.03468	.03327	.03188	.03051	.02917
25	12	15	.03812	.03686	.03558	.03430	.03301	.03173	.03046	.02920
25	12	16	.03742	.03630	.03515	.03399	.03281	.03164	.03046	.02930
25	12	17	.03681	.03581	.03479	.03375	.03269	.03161	.03054	.02946
25	12	18	.03626	.03540	.03450	.03357	.03262	.03166	.03068	.02969
25	12	19	.03577	.03504	.03427	.03346	.03263	.03176	.03088	.02999
25	12	20	.03535	.03475	.03411	.03342	.03270	.03194	.03117	.03037
25	12	21	.03498	.03452	.03401	.03345	.03285	.03221	.03154	.03084
25	12	22	.03467	.03436	.03399	.03357	.03310	.03258	.03203	.03145
25	12	23	.03443	.03428	.03407	.03380	.03348	.03311	.03269	.03224
25	12	24	.03429	.03434	.03431	.03423	.03409	.03389	.03364	.03335
25	12	25	.03437	.03469	.03494	.03512	.03524	.03529	.03529	.03523
25	13	14	.04432	.04286	.04138	.03989	.03840	.03692	.03545	.03399
25	13	15	.04342	.04212	.04080	.03945	.03810	.03674	.03538	.03404
25	13	16	.04263	.04149	.04031	.03910	.03788	.03664	.03540	.03415
25	13	17	.04193	.04094	.03990	.03883	.03773	.03662	.03548	.03434
25	13	18	.04131	.04046	.03957	.03863	.03766	.03667	.03565	.03461
25	13	19	.04076	.04006	.03931	.03851	.03767	.03679	.03589	.03496
25	13	20	.04028	.03973	.03912	.03846	.03775	.03700	.03622	.03540
25	13	21	.03986	.03947	.03901	.03850	.03793	.03731	.03665	.03596

Covariances of Inverse Gaussian Order Statistics
Shape Parameter k

n	i	j	1.6	1.7	1.8	1.9	2.0	2.1	2.2	2.3
25	13	22	.03951	.03929	.03899	.03863	.03821	.03774	.03722	.03666
25	13	23	.03924	.03920	.03908	.03890	.03866	.03835	.03799	.03758
25	13	24	.03908	.03926	.03936	.03940	.03936	.03926	.03910	.03888
25	13	25	.03917	.03966	.04008	.04042	.04069	.04088	.04100	.04106
25	14	15	.04962	.04829	.04693	.04554	.04412	.04268	.04124	.03980
25	14	16	.04872	.04757	.04637	.04514	.04387	.04257	.04126	.03994
25	14	17	.04792	.04694	.04591	.04483	.04370	.04255	.04136	.04016
25	14	18	.04722	.04640	.04553	.04460	.04362	.04261	.04156	.04048
25	14	19	.04659	.04595	.04523	.04446	.04363	.04276	.04184	.04089
25	14	20	.04605	.04557	.04502	.04441	.04373	.04300	.04222	.04140
25	14	21	.04557	.04527	.04490	.04445	.04393	.04336	.04273	.04205
25	14	22	.04518	.04507	.04487	.04461	.04427	.04386	.04340	.04288
25	14	23	.04487	.04497	.04498	.04492	.04478	.04457	.04429	.04396
25	14	24	.04469	.04504	.04531	.04549	.04560	.04562	.04558	.04547
25	14	25	.04479	.04550	.04613	.04667	.04713	.04751	.04780	.04802
25	15	16	.05594	.05481	.05361	.05235	.05105	.04971	.04833	.04694
25	15	17	.05503	.05409	.05308	.05200	.05086	.04968	.04846	.04721
25	15	18	.05423	.05348	.05264	.05174	.05078	.04975	.04868	.04758
25	15	19	.05352	.05296	.05231	.05158	.05079	.04993	.04902	.04806
25	15	20	.05290	.05252	.05206	.05152	.05090	.05022	.04947	.04867
25	15	21	.05236	.05219	.05192	.05157	.05114	.05064	.05007	.04943
25	15	22	.05191	.05195	.05190	.05176	.05153	.05123	.05085	.05040
25	15	23	.05155	.05184	.05203	.05212	.05213	.05205	.05190	.05167
25	15	24	.05135	.05192	.05240	.05279	.05308	.05328	.05340	.05344
25	15	25	.05146	.05246	.05336	.05416	.05487	.05548	.05600	.05644
25	16	17	.06363	.06275	.06178	.06073	.05960	.05841	.05717	.05587
25	16	18	.06271	.06204	.06128	.06044	.05951	.05850	.05744	.05631
25	16	19	.06189	.06145	.06090	.06025	.05952	.05871	.05783	.05689
25	16	20	.06118	.06095	.06062	.06019	.05967	.05906	.05837	.05761
25	16	21	.06056	.06056	.06046	.06025	.05995	.05955	.05907	.05851
25	16	22	.06004	.06029	.06043	.06047	.06040	.06025	.06000	.05966
25	16	23	.05964	.06017	.06059	.06090	.06111	.06122	.06123	.06116
25	16	24	.05940	.06027	.06102	.06168	.06222	.06266	.06301	.06326
25	16	25	.05954	.06089	.06214	.06328	.06431	.06524	.06607	.06680
25	17	18	.07316	.07264	.07200	.07124	.07039	.06944	.06841	.06730
25	17	19	.07222	.07195	.07155	.07104	.07042	.06969	.06888	.06798
25	17	20	.07139	.07137	.07123	.07097	.07059	.07010	.06952	.06885
25	17	21	.07068	.07093	.07105	.07104	.07093	.07070	.07036	.06993
25	17	22	.07008	.07061	.07102	.07130	.07147	.07152	.07146	.07130
25	17	23	.06962	.07047	.07120	.07181	.07230	.07267	.07293	.07308
25	17	24	.06935	.07059	.07172	.07273	.07362	.07439	.07505	.07559
25	17	25	.06951	.07133	.07303	.07462	.07609	.07745	.07869	.07981
25	18	19	.08529	.08527	.08510	.08479	.08435	.08378	.08309	.08229
25	18	20	.08433	.08460	.08473	.08471	.08456	.08428	.08387	.08334
25	18	21	.08349	.08408	.08452	.08481	.08497	.08499	.08488	.08465
25	18	22	.08279	.08372	.08450	.08513	.08562	.08598	.08621	.08631
25	18	23	.08225	.08356	.08472	.08574	.08662	.08737	.08799	.08847
25	18	24	.08194	.08371	.08534	.08684	.08820	.08943	.09053	.09149
25	18	25	.08214	.08458	.08690	.08909	.09116	.09311	.09492	.09660
25	19	20	.10122	.10194	.10247	.10283	.10302	.10305	.10293	.10266
25	19	21	.10024	.10132	.10223	.10296	.10353	.10393	.10418	.10427
25	19	22	.09941	.10090	.10221	.10335	.10433	.10515	.10581	.10631
25	19	23	.09878	.10071	.10249	.10410	.10555	.10685	.10798	.10896
25	19	24	.09841	.10090	.10324	.10544	.10748	.10937	.11110	.11268
25	19	25	.09866	.10196	.10513	.10817	.11108	.11385	.11647	.11895

Covariances of Inverse Gaussian Order Statistics
Shape Parameter k

n	i	j	1.6	1.7	1.8	1.9	2.0	2.1	2.2	2.3
25	20	21	.12306	.12488	.12650	.12791	.12911	.13012	.13093	.13156
25	20	22	.12207	.12438	.12649	.12840	.13012	.13165	.13298	.13413
25	20	23	.12131	.12417	.12685	.12934	.13165	.13378	.13571	.13747
25	20	24	.12088	.12441	.12779	.13101	.13405	.13692	.13962	.14214
25	20	25	.12120	.12573	.13013	.13441	.13854	.14252	.14635	.15001
25	21	22	.15477	.15837	.16174	.16489	.16780	.17048	.17293	.17514
25	21	23	.15384	.15813	.16222	.16611	.16978	.17323	.17647	.17948
25	21	24	.15332	.15847	.16345	.16825	.17287	.17730	.18153	.18556
25	21	25	.15375	.16016	.16645	.17262	.17864	.18452	.19023	.19578
25	22	23	.20482	.21155	.21806	.22433	.23037	.23615	.24167	.24692
25	22	24	.20419	.21204	.21973	.22724	.23456	.24167	.24856	.25521
25	22	25	.20481	.21433	.22378	.23313	.24236	.25145	.26039	.26916
25	23	24	.29508	.30815	.32110	.33390	.34652	.35894	.37114	.38310
25	23	25	.29608	.31154	.32703	.34251	.35795	.37331	.38858	.40373
25	24	25	.50321	.53328	.56372	.59447	.62547	.65668	.68804	.71950

Covariances of Inverse Gaussian Order Statistics
Shape Parameter k

n	i	j	2.4	2.5
2	1	2	.23569	.23105
3	1	2	.10064	.09592
3	1	3	.10749	.10446
3	2	3	.34292	.34118
4	1	2	.05815	.05444
4	1	3	.05867	.05579
4	1	4	.06330	.06122
4	2	3	.15127	.14669
4	2	4	.16331	.16094
4	3	4	.41292	.41450
5	1	2	.03913	.03619
5	1	3	.03827	.03586
5	1	4	.03936	.03737
5	1	5	.04261	.04107
5	2	3	.08774	.08365
5	2	4	.09044	.08732
5	2	5	.09790	.09593
5	3	4	.18744	.18376
5	3	5	.20284	.20169
5	4	5	.46393	.46861
6	1	2	.02883	.02642
6	1	3	.02759	.02558
6	1	4	.02768	.02595
6	1	5	.02871	.02724
6	1	6	.03111	.02992
6	2	3	.05867	.05524
6	2	4	.05901	.05617
6	2	5	.06127	.05899
6	2	6	.06636	.06476
6	3	4	.10990	.10600
6	3	5	.11420	.11138
6	3	6	.12365	.12220
6	4	5	.21552	.21294
6	4	6	.23318	.23335
6	5	6	.50351	.51096
7	1	2	.02255	.02050
7	1	3	.02122	.01951
7	1	4	.02093	.01944
7	1	5	.02126	.01994
7	1	6	.02215	.02099
7	1	7	.02398	.02302
7	2	3	.04278	.03990
7	2	4	.04232	.03986
7	2	5	.04303	.04091
7	2	6	.04484	.04308
7	2	7	.04854	.04724
7	3	4	.07366	.07019
7	3	5	.07498	.07213
7	3	6	.07817	.07597
7	3	7	.08460	.08325
7	4	5	.12776	.12429
7	4	6	.13324	.13092
7	4	7	.14412	.14335
7	5	6	.23831	.23687
7	5	7	.25755	.25902
7	6	7	.53549	.54541
8	1	2	.01837	.01661
8	1	3	.01707	.01559
8	1	4	.01663	.01533
8	1	5	.01665	.01548
8	1	6	.01702	.01596
8	1	7	.01776	.01683
8	1	8	.01922	.01843
8	2	3	.03306	.03059
8	2	4	.03230	.03016
8	2	5	.03238	.03050
8	2	6	.03312	.03147
8	2	7	.03458	.03317
8	2	8	.03740	.03631
8	3	4	.05363	.05062
8	3	5	.05384	.05125
8	3	6	.05509	.05290
8	3	7	.05754	.05578
8	3	8	.06221	.06102
8	4	5	.08605	.08274
8	4	6	.08811	.08544
8	4	7	.09203	.09008
8	4	8	.09947	.09850
8	5	6	.14269	.13975
8	5	7	.14905	.14732
8	5	8	.16102	.16096
8	6	7	.25737	.25703
8	6	8	.27776	.28045
8	7	8	.56210	.57424
9	1	2	.01544	.01389
9	1	3	.01418	.01288
9	1	4	.01368	.01253
9	1	5	.01356	.01252
9	1	6	.01368	.01274
9	1	7	.01404	.01318
9	1	8	.01467	.01390
9	1	9	.01585	.01519
9	2	3	.02662	.02448
9	2	4	.02575	.02388
9	2	5	.02555	.02388
9	2	6	.02581	.02432
9	2	7	.02649	.02516
9	2	8	.02769	.02653
9	2	9	.02991	.02899
9	3	4	.04129	.03867
9	3	5	.04103	.03873
9	3	6	.04148	.03946
9	3	7	.04258	.04083
9	3	8	.04451	.04306
9	3	9	.04807	.04703
9	4	5	.06274	.05976
9	4	6	.06348	.06093
9	4	7	.06519	.06306
9	4	8	.06814	.06650
9	4	9	.07357	.07260

n	i	j	2.4	2.5
9	5	6	.09664	.09357
9	5	7	.09927	.09688
9	5	8	.10376	.10214
9	5	9	.11200	.11145
9	6	7	.15549	.15311
9	6	8	.16252	.16139
9	6	9	.17531	.17596
9	7	8	.27365	.27437
9	7	9	.29489	.29873
9	8	9	.58475	.59888
10	1	2	.01327	.01189
10	1	3	.01208	.01092
10	1	4	.01156	.01053
10	1	5	.01136	.01043
10	1	6	.01136	.01051
10	1	7	.01153	.01074
10	1	8	.01186	.01113
10	1	9	.01240	.01174
10	1	10	.01338	.01280
10	2	3	.02211	.02022
10	2	4	.02121	.01955
10	2	5	.02087	.01939
10	2	6	.02090	.01955
10	2	7	.02121	.01999
10	2	8	.02182	.02071
10	2	9	.02281	.02184
10	2	10	.02461	.02382
10	3	4	.03310	.03080
10	3	5	.03262	.03057
10	3	6	.03268	.03085
10	3	7	.03319	.03155
10	3	8	.03414	.03270
10	3	9	.03569	.03448
10	3	10	.03850	.03759
10	4	5	.04828	.04564
10	4	6	.04842	.04609
10	4	7	.04918	.04715
10	4	8	.05060	.04886
10	4	9	.05290	.05152
10	4	10	.05706	.05615
10	5	6	.07065	.06778
10	5	7	.07180	.06937
10	5	8	.07388	.07190
10	5	9	.07724	.07580
10	5	10	.08328	.08259
10	6	7	.10586	.10310
10	6	8	.10895	.10688
10	6	9	.11390	.11265
10	6	10	.12277	.12267
10	7	8	.16664	.16483
10	7	9	.17419	.17367
10	7	10	.18762	.18896
10	8	9	.28780	.28952
10	8	10	.30968	.31459
10	9	10	.60436	.62030

n	i	j	2.4	2.5
11	1	2	.01162	.01038
11	1	3	.01049	.00945
11	1	4	.00997	.00904
11	1	5	.00973	.00889
11	1	6	.00967	.00889
11	1	7	.00973	.00901
11	1	8	.00990	.00923
11	1	9	.01020	.00957
11	1	10	.01066	.01009
11	1	11	.01149	.01099
11	2	3	.01880	.01712
11	2	4	.01791	.01643
11	2	5	.01751	.01617
11	2	6	.01741	.01619
11	2	7	.01753	.01640
11	2	8	.01784	.01681
11	2	9	.01837	.01744
11	2	10	.01921	.01838
11	2	11	.02070	.02001
11	3	4	.02734	.02531
11	3	5	.02677	.02494
11	3	6	.02664	.02498
11	3	7	.02683	.02533
11	3	8	.02732	.02596
11	3	9	.02814	.02693
11	3	10	.02942	.02838
11	3	11	.03170	.03089
11	4	5	.03864	.03629
11	4	6	.03848	.03637
11	4	7	.03878	.03689
11	4	8	.03949	.03782
11	4	9	.04068	.03923
11	4	10	.04253	.04134
11	4	11	.04582	.04499
11	5	6	.05442	.05182
11	5	7	.05487	.05258
11	5	8	.05590	.05392
11	5	9	.05758	.05593
11	5	10	.06020	.05893
11	5	11	.06484	.06411
11	6	7	.07763	.07493
11	6	8	.07911	.07685
11	6	9	.08150	.07972
11	6	10	.08520	.08399
11	6	11	.09174	.09134
11	7	8	.11402	.11159
11	7	9	.11747	.11575
11	7	10	.12280	.12192
11	7	11	.13217	.13252
11	8	9	.17649	.17524
11	8	10	.18444	.18452
11	8	11	.19840	.20039
11	9	10	.30026	.30292
11	9	11	.32263	.32854
11	10	11	.62159	.63916

n	i	j	2.4	2.5
12	1	2	.01033	.00919
12	1	3	.00925	.00831
12	1	4	.00874	.00790
12	1	5	.00848	.00772
12	1	6	.00838	.00767
12	1	7	.00838	.00772
12	1	8	.00847	.00785
12	1	9	.00864	.00805
12	1	10	.00890	.00836
12	1	11	.00930	.00880
12	1	12	.01002	.00957
12	2	3	.01629	.01478
12	2	4	.01542	.01409
12	2	5	.01499	.01378
12	2	6	.01482	.01371
12	2	7	.01483	.01380
12	2	8	.01499	.01403
12	2	9	.01529	.01440
12	2	10	.01575	.01495
12	2	11	.01647	.01575
12	2	12	.01773	.01712
12	3	4	.02313	.02130
12	3	5	.02251	.02087
12	3	6	.02227	.02077
12	3	7	.02230	.02092
12	3	8	.02254	.02128
12	3	9	.02299	.02184
12	3	10	.02370	.02266
12	3	11	.02477	.02388
12	3	12	.02666	.02595
12	4	5	.03185	.02974
12	4	6	.03153	.02963
12	4	7	.03158	.02985
12	4	8	.03194	.03037
12	4	9	.03258	.03118
12	4	10	.03359	.03235
12	4	11	.03511	.03408
12	4	12	.03778	.03702
12	5	6	.04354	.04119
12	5	7	.04363	.04152
12	5	8	.04413	.04225
12	5	9	.04504	.04338
12	5	10	.04642	.04501
12	5	11	.04853	.04741
12	5	12	.05221	.05150
12	6	7	.05989	.05738
12	6	8	.06060	.05840
12	6	9	.06185	.05997
12	6	10	.06376	.06223
12	6	11	.06665	.06553
12	6	12	.07168	.07117
12	7	8	.08388	.08137
12	7	9	.08563	.08358
12	7	10	.08827	.08672
12	7	11	.09226	.09130
12	7	12	.09920	.09912
12	8	9	.12132	.11922
12	8	10	.12507	.12370
12	8	11	.13070	.13021
12	8	12	.14048	.14127
12	9	10	.18528	.18457
12	9	11	.19356	.19421
12	9	12	.20793	.21055
12	10	11	.31137	.31490
12	10	12	.33412	.34094
12	11	12	.63689	.65596
13	1	2	.00929	.00824
13	1	3	.00827	.00740
13	1	4	.00777	.00700
13	1	5	.00750	.00680
13	1	6	.00737	.00672
13	1	7	.00734	.00673
13	1	8	.00737	.00680
13	1	9	.00747	.00693
13	1	10	.00763	.00712
13	1	11	.00786	.00738
13	1	12	.00822	.00777
13	1	13	.00884	.00844
13	2	3	.01433	.01295
13	2	4	.01349	.01228
13	2	5	.01305	.01195
13	2	6	.01284	.01183
13	2	7	.01278	.01184
13	2	8	.01285	.01197
13	2	9	.01302	.01219
13	2	10	.01329	.01253
13	2	11	.01371	.01300
13	2	12	.01433	.01369
13	2	13	.01540	.01486
13	3	4	.01993	.01829
13	3	5	.01931	.01782
13	3	6	.01901	.01765
13	3	7	.01893	.01767
13	3	8	.01903	.01786
13	3	9	.01928	.01821
13	3	10	.01970	.01871
13	3	11	.02031	.01941
13	3	12	.02123	.02044
13	3	13	.02282	.02218
13	4	5	.02687	.02497
13	4	6	.02647	.02474
13	4	7	.02638	.02480
13	4	8	.02652	.02507
13	4	9	.02688	.02555
13	4	10	.02746	.02626
13	4	11	.02832	.02725
13	4	12	.02959	.02869
13	4	13	.03181	.03112
13	5	6	.03586	.03374
13	5	7	.03576	.03383

n	i	j	2.4	2.5
13	5	8	.03596	.03421
13	5	9	.03646	.03487
13	5	10	.03725	.03584
13	5	11	.03841	.03719
13	5	12	.04014	.03915
13	5	13	.04313	.04246
13	6	7	.04796	.04565
13	6	8	.04825	.04618
13	6	9	.04892	.04709
13	6	10	.04999	.04840
13	6	11	.05155	.05022
13	6	12	.05387	.05286
13	6	13	.05788	.05732
13	7	8	.06484	.06244
13	7	9	.06576	.06368
13	7	10	.06720	.06545
13	7	11	.06930	.06791
13	7	12	.07241	.07147
13	7	13	.07778	.07748
13	8	9	.08953	.08723
13	8	10	.09150	.08966
13	8	11	.09435	.09303
13	8	12	.09858	.09789
13	8	13	.10586	.10609
13	9	10	.12791	.12615
13	9	11	.13190	.13000
13	9	12	.13778	.13767
13	9	13	.14790	.14913
13	10	11	.19320	.19302
13	10	12	.20175	.20295
13	10	13	.21645	.21967
13	11	12	.32134	.32571
13	11	13	.34439	.35207
13	12	13	.65061	.67107
14	1	2	.00843	.00747
14	1	3	.00747	.00666
14	1	4	.00698	.00627
14	1	5	.00671	.00607
14	1	6	.00657	.00597
14	1	7	.00651	.00595
14	1	8	.00651	.00598
14	1	9	.00656	.00606
14	1	10	.00666	.00618
14	1	11	.00680	.00635
14	1	12	.00702	.00659
14	1	13	.00733	.00694
14	1	14	.00788	.00752
14	2	3	.01277	.01150
14	2	4	.01196	.01085
14	2	5	.01152	.01051
14	2	6	.01129	.01036
14	2	7	.01119	.01032
14	2	8	.01119	.01038
14	2	9	.01128	.01051
14	2	10	.01145	.01073

n	i	j	2.4	2.5
14	2	11	.01170	.01103
14	2	12	.01207	.01145
14	2	13	.01261	.01204
14	2	14	.01354	.01305
14	3	4	.01744	.01594
14	3	5	.01682	.01547
14	3	6	.01649	.01525
14	3	7	.01636	.01520
14	3	8	.01637	.01529
14	3	9	.01650	.01549
14	3	10	.01675	.01581
14	3	11	.01712	.01626
14	3	12	.01766	.01687
14	3	13	.01845	.01775
14	3	14	.01981	.01923
14	4	5	.02309	.02137
14	4	6	.02265	.02108
14	4	7	.02248	.02103
14	4	8	.02250	.02116
14	4	9	.02268	.02144
14	4	10	.02303	.02188
14	4	11	.02355	.02250
14	4	12	.02428	.02335
14	4	13	.02537	.02456
14	4	14	.02724	.02661
14	5	6	.03031	.02839
14	5	7	.03000	.02823
14	5	8	.03003	.02841
14	5	9	.03029	.02880
14	5	10	.03075	.02940
14	5	11	.03144	.03023
14	5	12	.03243	.03136
14	5	13	.03388	.03299
14	5	14	.03637	.03574
14	6	7	.03950	.03739
14	6	8	.03956	.03764
14	6	9	.03991	.03817
14	6	10	.04052	.03896
14	6	11	.04144	.04006
14	6	12	.04274	.04157
14	6	13	.04464	.04373
14	6	14	.04792	.04736
14	7	8	.05197	.04974
14	7	9	.05244	.05044
14	7	10	.05326	.05150
14	7	11	.05446	.05295
14	7	12	.05617	.05494
14	7	13	.05867	.05779
14	7	14	.06297	.06257
14	8	9	.06935	.06709
14	8	10	.07044	.06850
14	8	11	.07204	.07044
14	8	12	.07430	.07308
14	8	13	.07760	.07685
14	8	14	.08327	.08319

n	i	j	2.4	2.5
14	9	10	.09467	.09260
14	9	11	.09683	.09522
14	9	12	.09986	.09878
14	9	13	.10428	.10386
14	9	14	.11186	.11239
14	10	11	.13390	.13248
14	10	12	.13809	.13742
14	10	13	.14418	.14444
14	10	14	.15459	.15623
14	11	12	.20038	.20072
14	11	13	.20916	.21088
14	11	14	.22413	.22791
14	12	13	.33039	.33553
14	12	14	.35366	.36215
14	13	14	.66303	.68477
15	1	2	.00772	.00682
15	1	3	.00680	.00605
15	1	4	.00633	.00567
15	1	5	.00607	.00546
15	1	6	.00592	.00536
15	1	7	.00584	.00532
15	1	8	.00582	.00532
15	1	9	.00584	.00537
15	1	10	.00589	.00545
15	1	11	.00599	.00556
15	1	12	.00613	.00572
15	1	13	.00632	.00594
15	1	14	.00660	.00624
15	1	15	.00708	.00676
15	2	3	.01149	.01033
15	2	4	.01072	.00970
15	2	5	.01029	.00936
15	2	6	.01004	.00918
15	2	7	.00992	.00911
15	2	8	.00988	.00913
15	2	9	.00992	.00920
15	2	10	.01002	.00934
15	2	11	.01018	.00954
15	2	12	.01041	.00981
15	2	13	.01074	.01018
15	2	14	.01122	.01071
15	2	15	.01203	.01159
15	3	4	.01546	.01408
15	3	5	.01485	.01361
15	3	6	.01451	.01336
15	3	7	.01433	.01327
15	3	8	.01429	.01329
15	3	9	.01434	.01341
15	3	10	.01449	.01361
15	3	11	.01472	.01390
15	3	12	.01506	.01430
15	3	13	.01554	.01484
15	3	14	.01623	.01560
15	3	15	.01740	.01688
15	4	5	.02015	.01857

n	i	j	2.4	2.5
15	4	6	.01969	.01825
15	4	7	.01947	.01813
15	4	8	.01941	.01817
15	4	9	.01949	.01833
15	4	10	.01969	.01861
15	4	11	.02001	.01901
15	4	12	.02047	.01955
15	4	13	.02112	.02029
15	4	14	.02205	.02133
15	4	15	.02365	.02308
15	5	6	.02592	.02417
15	5	7	.02564	.02402
15	5	8	.02558	.02408
15	5	9	.02568	.02430
15	5	10	.02595	.02467
15	5	11	.02638	.02520
15	5	12	.02699	.02592
15	5	13	.02784	.02690
15	5	14	.02907	.02828
15	5	15	.03117	.03059
15	6	7	.03326	.03133
15	6	8	.03319	.03142
15	6	9	.03334	.03171
15	6	10	.03369	.03220
15	6	11	.03425	.03290
15	6	12	.03504	.03384
15	6	13	.03614	.03511
15	6	14	.03774	.03691
15	6	15	.04047	.03992
15	7	8	.04283	.04076
15	7	9	.04303	.04115
15	7	10	.04349	.04179
15	7	11	.04422	.04270
15	7	12	.04524	.04392
15	7	13	.04666	.04557
15	7	14	.04872	.04790
15	7	15	.05224	.05180
15	8	9	.05566	.05351
15	8	10	.05627	.05436
15	8	11	.05721	.05554
15	8	12	.05854	.05713
15	8	13	.06038	.05927
15	8	14	.06304	.06229
15	8	15	.06757	.06735
15	9	10	.07348	.07137
15	9	11	.07472	.07293
15	9	12	.07646	.07502
15	9	13	.07886	.07782
15	9	14	.08233	.08177
15	9	15	.08823	.08839
15	10	11	.09938	.09753
15	10	12	.10170	.10032
15	10	13	.10488	.10405
15	10	14	.10948	.10932
15	10	15	.11730	.11813

n	i	j	2.4	2.5
15	11	12	.13938	.13830
15	11	13	.14374	.14342
15	11	14	.15001	.15064
15	11	15	.16066	.16269
15	12	13	.20695	.20777
15	12	14	.21591	.21814
15	12	15	.23110	.23542
15	13	14	.33863	.34451
15	13	15	.36209	.37133
15	14	15	.67433	.69726
16	1	2	.00712	.00628
16	1	3	.00624	.00554
16	1	4	.00579	.00517
16	1	5	.00553	.00497
16	1	6	.00537	.00485
16	1	7	.00529	.00480
16	1	8	.00525	.00479
16	1	9	.00525	.00481
16	1	10	.00528	.00486
16	1	11	.00534	.00494
16	1	12	.00543	.00505
16	1	13	.00556	.00519
16	1	14	.00573	.00538
16	1	15	.00598	.00566
16	1	16	.00641	.00612
16	2	3	.01044	.00936
16	2	4	.00970	.00875
16	2	5	.00927	.00841
16	2	6	.00902	.00822
16	2	7	.00888	.00814
16	2	8	.00882	.00812
16	2	9	.00882	.00816
16	2	10	.00888	.00824
16	2	11	.00898	.00838
16	2	12	.00913	.00856
16	2	13	.00935	.00881
16	2	14	.00964	.00914
16	2	15	.01006	.00960
16	2	16	.01079	.01038
16	3	4	.01385	.01258
16	3	5	.01326	.01211
16	3	6	.01291	.01185
16	3	7	.01271	.01172
16	3	8	.01263	.01170
16	3	9	.01263	.01176
16	3	10	.01271	.01189
16	3	11	.01286	.01208
16	3	12	.01308	.01235
16	3	13	.01339	.01271
16	3	14	.01381	.01318
16	3	15	.01441	.01385
16	3	16	.01545	.01497
16	4	5	.01780	.01636
16	4	6	.01734	.01601
16	4	7	.01709	.01586

n	i	j	2.4	2.5
16	4	8	.01698	.01583
16	4	9	.01699	.01591
16	4	10	.01710	.01609
16	4	11	.01730	.01635
16	4	12	.01760	.01672
16	4	13	.01801	.01720
16	4	14	.01858	.01784
16	4	15	.01939	.01874
16	4	16	.02078	.02025
16	5	6	.02258	.02097
16	5	7	.02226	.02077
16	5	8	.02213	.02074
16	5	9	.02215	.02085
16	5	10	.02229	.02109
16	5	11	.02255	.02144
16	5	12	.02294	.02191
16	5	13	.02348	.02254
16	5	14	.02422	.02338
16	5	15	.02528	.02457
16	5	16	.02709	.02654
16	6	7	.02852	.02675
16	6	8	.02836	.02672
16	6	9	.02839	.02687
16	6	10	.02858	.02717
16	6	11	.02892	.02763
16	6	12	.02942	.02824
16	6	13	.03011	.02906
16	6	14	.03106	.03014
16	6	15	.03242	.03166
16	6	16	.03473	.03420
16	7	8	.03607	.03416
16	7	9	.03611	.03436
16	7	10	.03636	.03475
16	7	11	.03680	.03533
16	7	12	.03744	.03612
16	7	13	.03832	.03716
16	7	14	.03952	.03854
16	7	15	.04125	.04049
16	7	16	.04419	.04373
16	8	9	.04590	.04389
16	8	10	.04622	.04439
16	8	11	.04679	.04514
16	8	12	.04761	.04615
16	8	13	.04873	.04748
16	8	14	.05025	.04924
16	8	15	.05245	.05172
16	8	16	.05617	.05586
16	9	10	.05906	.05701
16	9	11	.05979	.05798
16	9	12	.06084	.05928
16	9	13	.06228	.06098
16	9	14	.06423	.06324
16	9	15	.06703	.06642
16	9	16	.07177	.07172
16	10	11	.07730	.07534

n	i	j	2.4	2.5
16	10	12	.07867	.07703
16	10	13	.08052	.07925
16	10	14	.08304	.08217
16	10	15	.08665	.08629
16	10	16	.09277	.09315
16	11	12	.10372	.10210
16	11	13	.10617	.10503
16	11	14	.10949	.10891
16	11	15	.11423	.11434
16	11	16	.12226	.12338
16	12	13	.14443	.14367
16	12	14	.14893	.14894
16	12	15	.15535	.15632
16	12	16	.16620	.16861
16	13	14	.21299	.21428
16	13	15	.22210	.22480
16	13	16	.23747	.24229
16	14	15	.34620	.35277
16	14	16	.36980	.37974
16	15	16	.68470	.70873
17	1	2	.00661	.00582
17	1	3	.00577	.00511
17	1	4	.00533	.00475
17	1	5	.00507	.00455
17	1	6	.00492	.00443
17	1	7	.00482	.00437
17	1	8	.00478	.00434
17	1	9	.00476	.00435
17	1	10	.00477	.00438
17	1	11	.00481	.00443
17	1	12	.00487	.00451
17	1	13	.00496	.00461
17	1	14	.00507	.00474
17	1	15	.00523	.00492
17	1	16	.00546	.00516
17	1	17	.00585	.00557
17	2	3	.00955	.00854
17	2	4	.00884	.00796
17	2	5	.00843	.00763
17	2	6	.00818	.00743
17	2	7	.00803	.00733
17	2	8	.00795	.00729
17	2	9	.00793	.00730
17	2	10	.00795	.00735
17	2	11	.00801	.00744
17	2	12	.00811	.00757
17	2	13	.00826	.00775
17	2	14	.00845	.00797
17	2	15	.00872	.00826
17	2	16	.00910	.00868
17	2	17	.00974	.00937
17	3	4	.01252	.01134
17	3	5	.01194	.01088
17	3	6	.01160	.01061
17	3	7	.01139	.01047
17	3	8	.01128	.01042
17	3	9	.01125	.01043
17	3	10	.01128	.01051
17	3	11	.01137	.01064
17	3	12	.01152	.01083
17	3	13	.01172	.01107
17	3	14	.01200	.01139
17	3	15	.01238	.01181
17	3	16	.01292	.01240
17	3	17	.01383	.01339
17	4	5	.01590	.01456
17	4	6	.01544	.01421
17	4	7	.01517	.01403
17	4	8	.01504	.01396
17	4	9	.01500	.01399
17	4	10	.01504	.01409
17	4	11	.01517	.01427
17	4	12	.01536	.01452
17	4	13	.01564	.01485
17	4	14	.01601	.01528
17	4	15	.01651	.01584
17	4	16	.01722	.01663
17	4	17	.01844	.01795
17	5	6	.01991	.01843
17	5	7	.01958	.01820
17	5	8	.01940	.01812
17	5	9	.01936	.01816
17	5	10	.01942	.01829
17	5	11	.01958	.01852
17	5	12	.01984	.01885
17	5	13	.02019	.01928
17	5	14	.02067	.01983
17	5	15	.02132	.02056
17	5	16	.02224	.02159
17	5	17	.02381	.02330
17	6	7	.02481	.02318
17	6	8	.02461	.02309
17	6	9	.02456	.02314
17	6	10	.02464	.02332
17	6	11	.02484	.02362
17	6	12	.02517	.02403
17	6	13	.02562	.02458
17	6	14	.02623	.02529
17	6	15	.02705	.02622
17	6	16	.02822	.02753
17	6	17	.03020	.02971
17	7	8	.03092	.02915
17	7	9	.03086	.02923
17	7	10	.03097	.02946
17	7	11	.03123	.02983
17	7	12	.03164	.03036
17	7	13	.03221	.03105
17	7	14	.03298	.03194
17	7	15	.03401	.03312
17	7	16	.03548	.03477

n	i	j	2.4	2.5
17	7	17	.03797	.03752
17	8	9	.03867	.03679
17	8	10	.03882	.03709
17	8	11	.03915	.03757
17	8	12	.03966	.03823
17	8	13	.04038	.03910
17	8	14	.04134	.04023
17	8	15	.04263	.04171
17	8	16	.04448	.04379
17	8	17	.04759	.04724
17	9	10	.04875	.04680
17	9	11	.04917	.04741
17	9	12	.04983	.04826
17	9	13	.05073	.04935
17	9	14	.05194	.05078
17	9	15	.05356	.05264
17	9	16	.05587	.05526
17	9	17	.05978	.05960
17	10	11	.06222	.06027
17	10	12	.06305	.06135
17	10	13	.06420	.06275
17	10	14	.06572	.06455
17	10	15	.06778	.06692
17	10	16	.07070	.07024
17	10	17	.07563	.07575
17	11	12	.08084	.07903
17	11	13	.08231	.08084
17	11	14	.08427	.08316
17	11	15	.08690	.08621
17	11	16	.09063	.09046
17	11	17	.09693	.09754
17	12	13	.10775	.10635
17	12	14	.11031	.10940
17	12	15	.11374	.11340
17	12	16	.11861	.11897
17	12	17	.12682	.12822
17	13	14	.14910	.14865
17	13	15	.15372	.15405
17	13	16	.16026	.16158
17	13	17	.17129	.17406
17	14	15	.21857	.22030
17	14	16	.22780	.23096
17	14	17	.24333	.24863
17	15	16	.35319	.36041
17	15	17	.37689	.38750
17	16	17	.69425	.71932
18	1	2	.00617	.00542
18	1	3	.00536	.00474
18	1	4	.00494	.00439
18	1	5	.00468	.00419
18	1	6	.00453	.00407
18	1	7	.00443	.00400
18	1	8	.00438	.00397
18	1	9	.00435	.00396
18	1	10	.00435	.00398

n	i	j	2.4	2.5
18	1	11	.00437	.00401
18	1	12	.00441	.00407
18	1	13	.00447	.00414
18	1	14	.00455	.00423
18	1	15	.00466	.00436
18	1	16	.00481	.00452
18	1	17	.00501	.00474
18	1	18	.00536	.00511
18	2	3	.00880	.00785
18	2	4	.00812	.00729
18	2	5	.00772	.00696
18	2	6	.00747	.00677
18	2	7	.00731	.00666
18	2	8	.00722	.00660
18	2	9	.00718	.00660
18	2	10	.00718	.00662
18	2	11	.00722	.00668
18	2	12	.00728	.00677
18	2	13	.00738	.00689
18	2	14	.00752	.00705
18	2	15	.00770	.00726
18	2	16	.00794	.00752
18	2	17	.00828	.00789
18	2	18	.00886	.00851
18	3	4	.01140	.01031
18	3	5	.01085	.00986
18	3	6	.01050	.00959
18	3	7	.01029	.00943
18	3	8	.01017	.00936
18	3	9	.01011	.00935
18	3	10	.01011	.00939
18	3	11	.01016	.00947
18	3	12	.01026	.00960
18	3	13	.01040	.00978
18	3	14	.01059	.01000
18	3	15	.01084	.01029
18	3	16	.01118	.01067
18	3	17	.01166	.01119
18	3	18	.01247	.01207
18	4	5	.01433	.01309
18	4	6	.01388	.01274
18	4	7	.01360	.01254
18	4	8	.01344	.01245
18	4	9	.01338	.01243
18	4	10	.01338	.01249
18	4	11	.01345	.01260
18	4	12	.01357	.01277
18	4	13	.01376	.01301
18	4	14	.01401	.01330
18	4	15	.01435	.01369
18	4	16	.01479	.01419
18	4	17	.01543	.01489
18	4	18	.01650	.01605
18	5	6	.01775	.01637
18	5	7	.01740	.01613

n	i	j	2.4	2.5
18	5	8	.01721	.01601
18	5	9	.01712	.01600
18	5	10	.01713	.01607
18	5	11	.01722	.01622
18	5	12	.01738	.01644
18	5	13	.01762	.01674
18	5	14	.01794	.01713
18	5	15	.01837	.01762
18	5	16	.01895	.01826
18	5	17	.01976	.01916
18	5	18	.02113	.02066
18	6	7	.02186	.02035
18	6	8	.02162	.02021
18	6	9	.02152	.02020
18	6	10	.02153	.02030
18	6	11	.02165	.02049
18	6	12	.02185	.02077
18	6	13	.02215	.02115
18	6	14	.02256	.02163
18	6	15	.02310	.02226
18	6	16	.02382	.02307
18	6	17	.02484	.02421
18	6	18	.02657	.02609
18	7	8	.02689	.02525
18	7	9	.02677	.02525
18	7	10	.02679	.02537
18	7	11	.02693	.02561
18	7	12	.02719	.02597
18	7	13	.02757	.02644
18	7	14	.02808	.02705
18	7	15	.02875	.02782
18	7	16	.02965	.02884
18	7	17	.03092	.03026
18	7	18	.03306	.03262
18	8	9	.03315	.03140
18	8	10	.03318	.03156
18	8	11	.03336	.03187
18	8	12	.03369	.03231
18	8	13	.03416	.03290
18	8	14	.03479	.03366
18	8	15	.03562	.03462
18	8	16	.03673	.03589
18	8	17	.03830	.03765
18	8	18	.04095	.04058
18	9	10	.04109	.03926
18	9	11	.04133	.03965
18	9	12	.04173	.04020
18	9	13	.04232	.04094
18	9	14	.04310	.04188
18	9	15	.04413	.04308
18	9	16	.04550	.04466
18	9	17	.04745	.04685
18	9	18	.05073	.05048
18	10	11	.05141	.04954
18	10	12	.05192	.05024

n	i	j	2.4	2.5
18	10	13	.05265	.05116
18	10	14	.05363	.05234
18	10	15	.05491	.05384
18	10	16	.05662	.05580
18	10	17	.05903	.05853
18	10	18	.06310	.06306
18	11	12	.06516	.06333
18	11	13	.06608	.06450
18	11	14	.06731	.06598
18	11	15	.06892	.06788
18	11	16	.07106	.07034
18	11	17	.07409	.07378
18	11	18	.07918	.07947
18	12	13	.08413	.08249
18	12	14	.08570	.08439
18	12	15	.08775	.08681
18	12	16	.09046	.08996
18	12	17	.09431	.09434
18	12	18	.10077	.10159
18	13	14	.11149	.11032
18	13	15	.11415	.11347
18	13	16	.11768	.11757
18	13	17	.12266	.12327
18	13	18	.13103	.13270
18	14	15	.15343	.15330
18	14	16	.15816	.15881
18	14	17	.16481	.16645
18	14	18	.17599	.17911
18	15	16	.22374	.22590
18	15	17	.23308	.23667
18	15	18	.24874	.25449
18	16	17	.35966	.36750
18	16	18	.38344	.39468
18	17	18	.70309	.72914
19	1	2	.00578	.00507
19	1	3	.00500	.00442
19	1	4	.00459	.00408
19	1	5	.00435	.00388
19	1	6	.00419	.00376
19	1	7	.00410	.00369
19	1	8	.00403	.00365
19	1	9	.00400	.00363
19	1	10	.00399	.00364
19	1	11	.00400	.00366
19	1	12	.00402	.00370
19	1	13	.00407	.00375
19	1	14	.00412	.00382
19	1	15	.00420	.00391
19	1	16	.00430	.00402
19	1	17	.00443	.00417
19	1	18	.00462	.00437
19	1	19	.00494	.00471
19	2	3	.00815	.00726
19	2	4	.00750	.00672
19	2	5	.00711	.00640

n	i	j	2.4	2.5
19	2	6	.00686	.00621
19	2	7	.00670	.00609
19	2	8	.00661	.00603
19	2	9	.00655	.00600
19	2	10	.00654	.00601
19	2	11	.00655	.00605
19	2	12	.00659	.00611
19	2	13	.00666	.00620
19	2	14	.00676	.00631
19	2	15	.00688	.00646
19	2	16	.00705	.00664
19	2	17	.00727	.00689
19	2	18	.00758	.00722
19	2	19	.00810	.00778
19	3	4	.01046	.00943
19	3	5	.00992	.00899
19	3	6	.00959	.00873
19	3	7	.00937	.00856
19	3	8	.00924	.00848
19	3	9	.00916	.00845
19	3	10	.00914	.00846
19	3	11	.00916	.00851
19	3	12	.00922	.00860
19	3	13	.00932	.00873
19	3	14	.00945	.00889
19	3	15	.00963	.00909
19	3	16	.00986	.00935
19	3	17	.01017	.00969
19	3	18	.01060	.01017
19	3	19	.01133	.01095
19	4	5	.01301	.01186
19	4	6	.01258	.01151
19	4	7	.01230	.01130
19	4	8	.01213	.01119
19	4	9	.01204	.01115
19	4	10	.01201	.01117
19	4	11	.01204	.01124
19	4	12	.01212	.01136
19	4	13	.01224	.01152
19	4	14	.01242	.01174
19	4	15	.01265	.01201
19	4	16	.01295	.01236
19	4	17	.01336	.01280
19	4	18	.01392	.01343
19	4	19	.01488	.01446
19	5	6	.01596	.01468
19	5	7	.01562	.01443
19	5	8	.01540	.01429
19	5	9	.01529	.01424
19	5	10	.01526	.01427
19	5	11	.01530	.01436
19	5	12	.01540	.01451
19	5	13	.01556	.01472
19	5	14	.01579	.01500
19	5	15	.01608	.01534
19	5	16	.01647	.01579
19	5	17	.01698	.01636
19	5	18	.01770	.01716
19	5	19	.01892	.01847
19	6	7	.01946	.01806
19	6	8	.01920	.01789
19	6	9	.01907	.01784
19	6	10	.01903	.01788
19	6	11	.01908	.01799
19	6	12	.01921	.01818
19	6	13	.01941	.01845
19	6	14	.01969	.01879
19	6	15	.02006	.01923
19	6	16	.02054	.01978
19	6	17	.02118	.02050
19	6	18	.02208	.02149
19	6	19	.02359	.02315
19	7	8	.02367	.02215
19	7	9	.02351	.02209
19	7	10	.02347	.02214
19	7	11	.02354	.02229
19	7	12	.02370	.02253
19	7	13	.02395	.02286
19	7	14	.02429	.02328
19	7	15	.02475	.02382
19	7	16	.02534	.02451
19	7	17	.02613	.02540
19	7	18	.02724	.02663
19	7	19	.02910	.02867
19	8	9	.02882	.02720
19	8	10	.02878	.02727
19	8	11	.02887	.02745
19	8	12	.02906	.02775
19	8	13	.02937	.02816
19	8	14	.02980	.02868
19	8	15	.03036	.02935
19	8	16	.03109	.03019
19	8	17	.03205	.03128
19	8	18	.03341	.03280
19	8	19	.03569	.03531
19	9	10	.03524	.03352
19	9	11	.03534	.03375
19	9	12	.03559	.03412
19	9	13	.03597	.03462
19	9	14	.03650	.03527
19	9	15	.03718	.03609
19	9	16	.03807	.03712
19	9	17	.03925	.03847
19	9	18	.04091	.04033
19	9	19	.04370	.04342
19	10	11	.04336	.04159
19	10	12	.04367	.04204
19	10	13	.04414	.04267
19	10	14	.04479	.04347
19	10	15	.04563	.04447

n	i	j	2.4	2.5
19	10	16	.04673	.04575
19	10	17	.04817	.04740
19	10	18	.05021	.04969
19	10	19	.05362	.05349
19	11	12	.05389	.05211
19	11	13	.05448	.05288
19	11	14	.05528	.05387
19	11	15	.05632	.05512
19	11	16	.05767	.05670
19	11	17	.05946	.05875
19	11	18	.06196	.06158
19	11	19	.06617	.06627
19	12	13	.06791	.06620
19	12	14	.06891	.06744
19	12	15	.07021	.06901
19	12	16	.07189	.07098
19	12	17	.07411	.07353
19	12	18	.07723	.07708
19	12	19	.08247	.08293
19	13	14	.08721	.08573
19	13	15	.08886	.08772
19	13	16	.09098	.09022
19	13	17	.09378	.09345
19	13	18	.09772	.09794
19	13	19	.10432	.10536
19	14	15	.11498	.11403
19	14	16	.11773	.11727
19	14	17	.12134	.12147
19	14	18	.12642	.12727
19	14	19	.13492	.13686
19	15	16	.15748	.15764
19	15	17	.16229	.16324
19	15	18	.16904	.17099
19	15	19	.18034	.18379
19	16	17	.22856	.23113
19	16	18	.23798	.24200
19	16	19	.25375	.25993
19	17	18	.36568	.37411
19	17	19	.38952	.40136
19	18	19	.71131	.73828
20	1	2	.00544	.00477
20	1	3	.00469	.00414
20	1	4	.00430	.00381
20	1	5	.00406	.00362
20	1	6	.00391	.00350
20	1	7	.00380	.00342
20	1	8	.00374	.00338
20	1	9	.00370	.00335
20	1	10	.00368	.00335
20	1	11	.00368	.00336
20	1	12	.00370	.00339
20	1	13	.00373	.00343
20	1	14	.00377	.00348
20	1	15	.00382	.00354
20	1	16	.00389	.00363

n	i	j	2.4	2.5
20	1	17	.00399	.00373
20	1	18	.00411	.00386
20	1	19	.00428	.00405
20	1	20	.00458	.00436
20	2	3	.00759	.00675
20	2	4	.00697	.00623
20	2	5	.00659	.00592
20	2	6	.00634	.00572
20	2	7	.00618	.00560
20	2	8	.00608	.00553
20	2	9	.00602	.00550
20	2	10	.00599	.00549
20	2	11	.00599	.00551
20	2	12	.00601	.00556
20	2	13	.00606	.00562
20	2	14	.00613	.00570
20	2	15	.00622	.00581
20	2	16	.00634	.00595
20	2	17	.00649	.00612
20	2	18	.00669	.00634
20	2	19	.00697	.00664
20	2	20	.00744	.00715
20	3	4	.00965	.00868
20	3	5	.00913	.00826
20	3	6	.00880	.00799
20	3	7	.00858	.00783
20	3	8	.00844	.00773
20	3	9	.00836	.00769
20	3	10	.00833	.00768
20	3	11	.00833	.00771
20	3	12	.00836	.00777
20	3	13	.00842	.00786
20	3	14	.00852	.00798
20	3	15	.00864	.00813
20	3	16	.00881	.00832
20	3	17	.00902	.00855
20	3	18	.00930	.00886
20	3	19	.00969	.00929
20	3	20	.01035	.01000
20	4	5	.01190	.01082
20	4	6	.01147	.01048
20	4	7	.01120	.01026
20	4	8	.01102	.01014
20	4	9	.01091	.01008
20	4	10	.01087	.01008
20	4	11	.01087	.01012
20	4	12	.01091	.01020
20	4	13	.01100	.01031
20	4	14	.01112	.01047
20	4	15	.01128	.01067
20	4	16	.01150	.01091
20	4	17	.01177	.01123
20	4	18	.01214	.01163
20	4	19	.01265	.01219
20	4	20	.01350	.01312

n	i	j	2.4	2.5
20	5	6	.01447	.01328
20	5	7	.01413	.01301
20	5	8	.01391	.01286
20	5	9	.01378	.01279
20	5	10	.01372	.01279
20	5	11	.01372	.01284
20	5	12	.01378	.01294
20	5	13	.01389	.01309
20	5	14	.01404	.01328
20	5	15	.01425	.01354
20	5	16	.01452	.01385
20	5	17	.01487	.01425
20	5	18	.01533	.01476
20	5	19	.01597	.01547
20	5	20	.01706	.01664
20	6	7	.01748	.01617
20	6	8	.01721	.01599
20	6	9	.01706	.01591
20	6	10	.01699	.01590
20	6	11	.01700	.01597
20	6	12	.01707	.01609
20	6	13	.01720	.01628
20	6	14	.01739	.01653
20	6	15	.01765	.01684
20	6	16	.01799	.01723
20	6	17	.01842	.01773
20	6	18	.01899	.01836
20	6	19	.01979	.01925
20	6	20	.02112	.02070
20	7	8	.02106	.01964
20	7	9	.02087	.01954
20	7	10	.02079	.01954
20	7	11	.02080	.01963
20	7	12	.02089	.01978
20	7	13	.02106	.02001
20	7	14	.02129	.02032
20	7	15	.02161	.02070
20	7	16	.02202	.02119
20	7	17	.02255	.02179
20	7	18	.02325	.02257
20	7	19	.02422	.02366
20	7	20	.02586	.02545
20	8	9	.02537	.02385
20	8	10	.02528	.02386
20	8	11	.02529	.02396
20	8	12	.02540	.02416
20	8	13	.02561	.02444
20	8	14	.02590	.02481
20	8	15	.02628	.02528
20	8	16	.02678	.02587
20	8	17	.02743	.02661
20	8	18	.02827	.02756
20	8	19	.02946	.02889
20	8	20	.03145	.03107
20	9	10	.03064	.02904
20	9	11	.03067	.02917
20	9	12	.03081	.02941
20	9	13	.03105	.02975
20	9	14	.03141	.03021
20	9	15	.03188	.03078
20	9	16	.03248	.03150
20	9	17	.03326	.03240
20	9	18	.03429	.03356
20	9	19	.03572	.03517
20	9	20	.03813	.03782
20	10	11	.03720	.03552
20	10	12	.03737	.03582
20	10	13	.03767	.03624
20	10	14	.03811	.03680
20	10	15	.03868	.03750
20	10	16	.03942	.03837
20	10	17	.04036	.03947
20	10	18	.04160	.04088
20	10	19	.04334	.04283
20	10	20	.04626	.04606
20	11	12	.04550	.04378
20	11	13	.04587	.04430
20	11	14	.04640	.04498
20	11	15	.04710	.04584
20	11	16	.04800	.04691
20	11	17	.04915	.04824
20	11	18	.05066	.04997
20	11	19	.05277	.05235
20	11	20	.05632	.05629
20	12	13	.05623	.05453
20	12	14	.05688	.05537
20	12	15	.05774	.05643
20	12	16	.05885	.05774
20	12	17	.06026	.05938
20	12	18	.06210	.06150
20	12	19	.06470	.06443
20	12	20	.06903	.06927
20	13	14	.07049	.06890
20	13	15	.07156	.07022
20	13	16	.07293	.07185
20	13	17	.07467	.07389
20	13	18	.07696	.07652
20	13	19	.08016	.08016
20	13	20	.08552	.08616
20	14	15	.09009	.08877
20	14	16	.09181	.09084
20	14	17	.09401	.09341
20	14	18	.09688	.09673
20	14	19	.10090	.10131
20	14	20	.10763	.10887
20	15	16	.11825	.11751
20	15	17	.12107	.12084
20	15	18	.12476	.12511
20	15	19	.12992	.13101
20	15	20	.13854	.14074

n	i	j	2.4	2.5
20	16	17	.16126	.16171
20	16	18	.16615	.16740
20	16	19	.17298	.17524
20	16	20	.18439	.18816
20	17	18	.23306	.23603
20	17	19	.24256	.24698
20	17	20	.25842	.26501
20	18	19	.37130	.38029
20	18	20	.39519	.40759
20	19	20	.71899	.74682
21	1	2	.00514	.00450
21	1	3	.00442	.00389
21	1	4	.00404	.00357
21	1	5	.00380	.00338
21	1	6	.00365	.00326
21	1	7	.00355	.00319
21	1	8	.00348	.00314
21	1	9	.00344	.00311
21	1	10	.00342	.00310
21	1	11	.00341	.00311
21	1	12	.00342	.00312
21	1	13	.00343	.00315
21	1	14	.00346	.00319
21	1	15	.00350	.00324
21	1	16	.00356	.00330
21	1	17	.00363	.00338
21	1	18	.00371	.00347
21	1	19	.00383	.00360
21	1	20	.00399	.00377
21	1	21	.00425	.00405
21	2	3	.00710	.00630
21	2	4	.00650	.00580
21	2	5	.00613	.00550
21	2	6	.00589	.00531
21	2	7	.00573	.00518
21	2	8	.00563	.00511
21	2	9	.00556	.00507
21	2	10	.00552	.00505
21	2	11	.00551	.00506
21	2	12	.00552	.00509
21	2	13	.00555	.00513
21	2	14	.00560	.00519
21	2	15	.00566	.00527
21	2	16	.00575	.00537
21	2	17	.00586	.00550
21	2	18	.00600	.00566
21	2	19	.00618	.00586
21	2	20	.00644	.00614
21	2	21	.00687	.00660
21	3	4	.00895	.00804
21	3	5	.00845	.00763
21	3	6	.00813	.00737
21	3	7	.00791	.00720
21	3	8	.00777	.00710
21	3	9	.00768	.00704

n	i	j	2.4	2.5
21	3	10	.00763	.00702
21	3	11	.00762	.00703
21	3	12	.00763	.00707
21	3	13	.00767	.00713
21	3	14	.00773	.00722
21	3	15	.00783	.00733
21	3	16	.00795	.00747
21	3	17	.00810	.00765
21	3	18	.00829	.00786
21	3	19	.00855	.00814
21	3	20	.00890	.00853
21	3	21	.00950	.00917
21	4	5	.01095	.00993
21	4	6	.01054	.00960
21	4	7	.01026	.00938
21	4	8	.01008	.00925
21	4	9	.00996	.00918
21	4	10	.00990	.00916
21	4	11	.00988	.00917
21	4	12	.00990	.00922
21	4	13	.00995	.00931
21	4	14	.01004	.00942
21	4	15	.01016	.00957
21	4	16	.01031	.00975
21	4	17	.01051	.00998
21	4	18	.01076	.01026
21	4	19	.01109	.01063
21	4	20	.01156	.01113
21	4	21	.01233	.01197
21	5	6	.01321	.01209
21	5	7	.01287	.01182
21	5	8	.01264	.01166
21	5	9	.01250	.01157
21	5	10	.01243	.01155
21	5	11	.01241	.01157
21	5	12	.01243	.01163
21	5	13	.01250	.01174
21	5	14	.01261	.01188
21	5	15	.01276	.01207
21	5	16	.01295	.01230
21	5	17	.01320	.01258
21	5	18	.01351	.01294
21	5	19	.01393	.01340
21	5	20	.01451	.01404
21	5	21	.01548	.01509
21	6	7	.01582	.01460
21	6	8	.01555	.01441
21	6	9	.01538	.01430
21	6	10	.01529	.01427
21	6	11	.01527	.01430
21	6	12	.01530	.01438
21	6	13	.01538	.01451
21	6	14	.01552	.01469
21	6	15	.01570	.01492
21	6	16	.01594	.01520

n	i	j	2.4	2.5
21	6	17	.01625	.01556
21	6	18	.01664	.01600
21	6	19	.01715	.01657
21	6	20	.01786	.01736
21	6	21	.01905	.01866
21	7	8	.01890	.01758
21	7	9	.01870	.01745
21	7	10	.01859	.01742
21	7	11	.01856	.01745
21	7	12	.01860	.01755
21	7	13	.01871	.01771
21	7	14	.01887	.01793
21	7	15	.01909	.01821
21	7	16	.01939	.01856
21	7	17	.01976	.01900
21	7	18	.02023	.01954
21	7	19	.02085	.02023
21	7	20	.02172	.02119
21	7	21	.02317	.02277
21	8	9	.02255	.02114
21	8	10	.02243	.02110
21	8	11	.02240	.02115
21	8	12	.02245	.02127
21	8	13	.02258	.02146
21	8	14	.02278	.02173
21	8	15	.02305	.02207
21	8	16	.02340	.02250
21	8	17	.02385	.02302
21	8	18	.02442	.02367
21	8	19	.02517	.02452
21	8	20	.02621	.02568
21	8	21	.02796	.02759
21	9	10	.02696	.02546
21	9	11	.02693	.02552
21	9	12	.02700	.02567
21	9	13	.02715	.02591
21	9	14	.02739	.02623
21	9	15	.02772	.02665
21	9	16	.02814	.02716
21	9	17	.02868	.02779
21	9	18	.02937	.02858
21	9	19	.03027	.02960
21	9	20	.03153	.03100
21	9	21	.03363	.03331
21	10	11	.03236	.03078
21	10	12	.03244	.03096
21	10	13	.03262	.03125
21	10	14	.03292	.03164
21	10	15	.03331	.03214
21	10	16	.03382	.03276
21	10	17	.03447	.03352
21	10	18	.03530	.03448
21	10	19	.03638	.03570
21	10	20	.03788	.03739
21	10	21	.04040	.04017

n	i	j	2.4	2.5
21	11	12	.03905	.03742
21	11	13	.03928	.03777
21	11	14	.03963	.03824
21	11	15	.04011	.03885
21	11	16	.04073	.03959
21	11	17	.04151	.04052
21	11	18	.04250	.04167
21	11	19	.04380	.04314
21	11	20	.04561	.04518
21	11	21	.04864	.04854
21	12	13	.04751	.04585
21	12	14	.04794	.04643
21	12	15	.04852	.04717
21	12	16	.04927	.04808
21	12	17	.05021	.04920
21	12	18	.05141	.05059
21	12	19	.05298	.05238
21	12	20	.05517	.05485
21	12	21	.05883	.05892
21	13	14	.05843	.05682
21	13	15	.05914	.05772
21	13	16	.06006	.05883
21	13	17	.06121	.06020
21	13	18	.06268	.06191
21	13	19	.06458	.06409
21	13	20	.06725	.06711
21	13	21	.07170	.07207
21	14	15	.07292	.07145
21	14	16	.07405	.07283
21	14	17	.07548	.07452
21	14	18	.07728	.07663
21	14	19	.07963	.07933
21	14	20	.08291	.08305
21	14	21	.08838	.08918
21	15	16	.09280	.09165
21	15	17	.09459	.09378
21	15	18	.09685	.09642
21	15	19	.09978	.09980
21	15	20	.10388	.10447
21	15	21	.11071	.11215
21	16	17	.12132	.12080
21	16	18	.12421	.12419
21	16	19	.12797	.12854
21	16	20	.13320	.13452
21	16	21	.14193	.14437
21	17	18	.16481	.16554
21	17	19	.16977	.17130
21	17	20	.17667	.17922
21	17	21	.18818	.19225
21	18	19	.23729	.24063
21	18	20	.24685	.25164
21	18	21	.26278	.26977
21	19	20	.37657	.38610
21	19	21	.40049	.41342
21	20	21	.72618	.75483

n	i	j	2.4	2.5
22	1	2	.00487	.00426
22	1	3	.00418	.00367
22	1	4	.00380	.00336
22	1	5	.00358	.00318
22	1	6	.00343	.00306
22	1	7	.00333	.00298
22	1	8	.00326	.00293
22	1	9	.00321	.00290
22	1	10	.00319	.00289
22	1	11	.00318	.00288
22	1	12	.00317	.00289
22	1	13	.00318	.00291
22	1	14	.00320	.00294
22	1	15	.00323	.00298
22	1	16	.00327	.00303
22	1	17	.00332	.00308
22	1	18	.00339	.00316
22	1	19	.00347	.00324
22	1	20	.00357	.00336
22	1	21	.00372	.00352
22	1	22	.00397	.00378
22	2	3	.00667	.00591
22	2	4	.00609	.00543
22	2	5	.00573	.00513
22	2	6	.00550	.00494
22	2	7	.00534	.00482
22	2	8	.00523	.00474
22	2	9	.00516	.00469
22	2	10	.00512	.00467
22	2	11	.00510	.00467
22	2	12	.00510	.00468
22	2	13	.00511	.00471
22	2	14	.00515	.00476
22	2	15	.00519	.00482
22	2	16	.00526	.00490
22	2	17	.00534	.00499
22	2	18	.00544	.00511
22	2	19	.00557	.00525
22	2	20	.00574	.00544
22	2	21	.00598	.00569
22	2	22	.00637	.00612
22	3	4	.00834	.00748
22	3	5	.00786	.00708
22	3	6	.00754	.00682
22	3	7	.00733	.00665
22	3	8	.00718	.00655
22	3	9	.00709	.00648
22	3	10	.00703	.00645
22	3	11	.00701	.00645
22	3	12	.00700	.00647
22	3	13	.00703	.00652
22	3	14	.00707	.00658
22	3	15	.00714	.00666
22	3	16	.00723	.00677
22	3	17	.00734	.00690
22	3	18	.00748	.00706
22	3	19	.00766	.00726
22	3	20	.00789	.00752
22	3	21	.00822	.00787
22	3	22	.00876	.00846
22	4	5	.01013	.00917
22	4	6	.00973	.00884
22	4	7	.00945	.00863
22	4	8	.00927	.00849
22	4	9	.00915	.00841
22	4	10	.00908	.00837
22	4	11	.00904	.00837
22	4	12	.00904	.00840
22	4	13	.00907	.00846
22	4	14	.00913	.00854
22	4	15	.00922	.00865
22	4	16	.00933	.00879
22	4	17	.00948	.00896
22	4	18	.00966	.00917
22	4	19	.00989	.00943
22	4	20	.01019	.00976
22	4	21	.01061	.01022
22	4	22	.01131	.01097
22	5	6	.01213	.01108
22	5	7	.01180	.01081
22	5	8	.01157	.01064
22	5	9	.01142	.01055
22	5	10	.01133	.01050
22	5	11	.01129	.01050
22	5	12	.01129	.01054
22	5	13	.01133	.01061
22	5	14	.01141	.01071
22	5	15	.01151	.01085
22	5	16	.01165	.01102
22	5	17	.01184	.01124
22	5	18	.01206	.01150
22	5	19	.01235	.01183
22	5	20	.01273	.01224
22	5	21	.01325	.01282
22	5	22	.01413	.01376
22	6	7	.01442	.01328
22	6	8	.01415	.01308
22	6	9	.01397	.01296
22	6	10	.01387	.01290
22	6	11	.01382	.01290
22	6	12	.01382	.01295
22	6	13	.01387	.01304
22	6	14	.01396	.01317
22	6	15	.01409	.01334
22	6	16	.01427	.01355
22	6	17	.01449	.01381
22	6	18	.01477	.01414
22	6	19	.01512	.01454
22	6	20	.01558	.01505
22	6	21	.01622	.01575

n	i	j	2.4	2.5
22	6	22	.01729	.01692
22	7	8	.01709	.01586
22	7	9	.01688	.01571
22	7	10	.01676	.01565
22	7	11	.01670	.01565
22	7	12	.01671	.01571
22	7	13	.01677	.01582
22	7	14	.01687	.01598
22	7	15	.01703	.01619
22	7	16	.01724	.01645
22	7	17	.01751	.01676
22	7	18	.01785	.01715
22	7	19	.01828	.01764
22	7	20	.01883	.01826
22	7	21	.01961	.01912
22	7	22	.02090	.02053
22	8	9	.02023	.01891
22	8	10	.02009	.01883
22	8	11	.02002	.01884
22	8	12	.02003	.01891
22	8	13	.02010	.01904
22	8	14	.02024	.01924
22	8	15	.02043	.01948
22	8	16	.02068	.01980
22	8	17	.02100	.02018
22	8	18	.02141	.02065
22	8	19	.02192	.02123
22	8	20	.02259	.02198
22	8	21	.02352	.02301
22	8	22	.02507	.02471
22	9	10	.02397	.02256
22	9	11	.02390	.02257
22	9	12	.02391	.02266
22	9	13	.02400	.02282
22	9	14	.02416	.02305
22	9	15	.02439	.02335
22	9	16	.02469	.02372
22	9	17	.02508	.02418
22	9	18	.02556	.02475
22	9	19	.02617	.02545
22	9	20	.02697	.02634
22	9	21	.02808	.02757
22	9	22	.02992	.02960
22	10	11	.02847	.02699
22	10	12	.02849	.02710
22	10	13	.02860	.02729
22	10	14	.02879	.02757
22	10	15	.02907	.02793
22	10	16	.02943	.02838
22	10	17	.02989	.02893
22	10	18	.03047	.02960
22	10	19	.03120	.03044
22	10	20	.03214	.03151
22	10	21	.03346	.03298
22	10	22	.03566	.03540

n	i	j	2.4	2.5
22	11	12	.03398	.03243
22	11	13	.03411	.03267
22	11	14	.03434	.03300
22	11	15	.03467	.03343
22	11	16	.03510	.03397
22	11	17	.03565	.03463
22	11	18	.03634	.03543
22	11	19	.03721	.03643
22	11	20	.03834	.03771
22	11	21	.03991	.03947
22	11	22	.04253	.04237
22	12	13	.04080	.03921
22	12	14	.04108	.03962
22	12	15	.04147	.04014
22	12	16	.04199	.04078
22	12	17	.04265	.04157
22	12	18	.04347	.04254
22	12	19	.04451	.04374
22	12	20	.04586	.04527
22	12	21	.04774	.04738
22	12	22	.05087	.05086
22	13	14	.04941	.04782
22	13	15	.04989	.04845
22	13	16	.05051	.04924
22	13	17	.05131	.05019
22	13	18	.05230	.05136
22	13	19	.05354	.05280
22	13	20	.05516	.05465
22	13	21	.05742	.05719
22	13	22	.06118	.06138
22	14	15	.06051	.05899
22	14	16	.06127	.05994
22	14	17	.06224	.06111
22	14	18	.06344	.06253
22	14	19	.06495	.06428
22	14	20	.06691	.06653
22	14	21	.06964	.06962
22	14	22	.07419	.07470
22	15	16	.07521	.07386
22	15	17	.07640	.07530
22	15	18	.07787	.07705
22	15	19	.07973	.07921
22	15	20	.08213	.08197
22	15	21	.08548	.08576
22	15	22	.09105	.09201
22	16	17	.09536	.09436
22	16	18	.09720	.09655
22	16	19	.09951	.09925
22	16	20	.10251	.10270
22	16	21	.10667	.10744
22	16	22	.11360	.11524
22	17	18	.12422	.12390
22	17	19	.12717	.12735
22	17	20	.13098	.13176
22	17	21	.13628	.13781

n	i	j	2.4	2.5
22	17	22	.14509	.14777
22	18	19	.16815	.16915
22	18	20	.17317	.17497
22	18	21	.18013	.18296
22	18	22	.19172	.19609
22	19	20	.24126	.24497
22	19	21	.25087	.25604
22	19	22	.26686	.27423
22	20	21	.38153	.39156
22	20	22	.40546	.41890
22	21	22	.73294	.76236
23	1	2	.00463	.00404
23	1	3	.00396	.00348
23	1	4	.00360	.00318
23	1	5	.00338	.00299
23	1	6	.00323	.00288
23	1	7	.00313	.00280
23	1	8	.00306	.00275
23	1	9	.00301	.00271
23	1	10	.00298	.00270
23	1	11	.00297	.00269
23	1	12	.00296	.00269
23	1	13	.00297	.00271
23	1	14	.00298	.00273
23	1	15	.00300	.00275
23	1	16	.00303	.00279
23	1	17	.00307	.00284
23	1	18	.00312	.00289
23	1	19	.00318	.00296
23	1	20	.00325	.00304
23	1	21	.00335	.00315
23	1	22	.00349	.00330
23	1	23	.00372	.00354
23	2	3	.00628	.00556
23	2	4	.00572	.00509
23	2	5	.00538	.00481
23	2	6	.00515	.00462
23	2	7	.00499	.00450
23	2	8	.00488	.00442
23	2	9	.00481	.00436
23	2	10	.00476	.00434
23	2	11	.00474	.00433
23	2	12	.00473	.00433
23	2	13	.00474	.00435
23	2	14	.00476	.00439
23	2	15	.00479	.00443
23	2	16	.00484	.00449
23	2	17	.00490	.00457
23	2	18	.00498	.00465
23	2	19	.00507	.00476
23	2	20	.00519	.00490
23	2	21	.00535	.00507
23	2	22	.00557	.00530
23	2	23	.00593	.00569
23	3	4	.00780	.00698

n	i	j	2.4	2.5
23	3	5	.00734	.00660
23	3	6	.00703	.00635
23	3	7	.00682	.00618
23	3	8	.00667	.00607
23	3	9	.00658	.00600
23	3	10	.00651	.00596
23	3	11	.00648	.00595
23	3	12	.00647	.00596
23	3	13	.00648	.00599
23	3	14	.00650	.00603
23	3	15	.00655	.00610
23	3	16	.00662	.00618
23	3	17	.00670	.00628
23	3	18	.00681	.00640
23	3	19	.00694	.00655
23	3	20	.00710	.00674
23	3	21	.00732	.00697
23	3	22	.00762	.00729
23	3	23	.00811	.00783
23	4	5	.00941	.00850
23	4	6	.00902	.00819
23	4	7	.00876	.00797
23	4	8	.00857	.00783
23	4	9	.00845	.00775
23	4	10	.00837	.00770
23	4	11	.00832	.00768
23	4	12	.00831	.00770
23	4	13	.00832	.00773
23	4	14	.00836	.00779
23	4	15	.00842	.00788
23	4	16	.00850	.00798
23	4	17	.00861	.00811
23	4	18	.00875	.00827
23	4	19	.00892	.00846
23	4	20	.00913	.00870
23	4	21	.00940	.00900
23	4	22	.00979	.00942
23	4	23	.01043	.01011
23	5	6	.01120	.01020
23	5	7	.01087	.00994
23	5	8	.01065	.00977
23	5	9	.01049	.00966
23	5	10	.01039	.00961
23	5	11	.01034	.00959
23	5	12	.01033	.00961
23	5	13	.01034	.00965
23	5	14	.01039	.00973
23	5	15	.01046	.00983
23	5	16	.01057	.00996
23	5	17	.01070	.01013
23	5	18	.01087	.01032
23	5	19	.01108	.01056
23	5	20	.01135	.01086
23	5	21	.01169	.01124
23	5	22	.01217	.01176

n	i	j	2.4	2.5
23	5	23	.01296	.01262
23	6	7	.01323	.01215
23	6	8	.01296	.01194
23	6	9	.01277	.01181
23	6	10	.01266	.01174
23	6	11	.01259	.01172
23	6	12	.01257	.01175
23	6	13	.01260	.01180
23	6	14	.01265	.01190
23	6	15	.01275	.01202
23	6	16	.01287	.01218
23	6	17	.01304	.01238
23	6	18	.01324	.01262
23	6	19	.01350	.01292
23	6	20	.01382	.01328
23	6	21	.01424	.01374
23	6	22	.01482	.01438
23	6	23	.01579	.01543
23	7	8	.01556	.01440
23	7	9	.01535	.01425
23	7	10	.01521	.01417
23	7	11	.01514	.01415
23	7	12	.01512	.01417
23	7	13	.01514	.01425
23	7	14	.01521	.01436
23	7	15	.01532	.01451
23	7	16	.01548	.01471
23	7	17	.01567	.01494
23	7	18	.01592	.01524
23	7	19	.01623	.01559
23	7	20	.01662	.01603
23	7	21	.01712	.01659
23	7	22	.01782	.01736
23	7	23	.01898	.01862
23	8	9	.01829	.01704
23	8	10	.01813	.01695
23	8	11	.01804	.01692
23	8	12	.01802	.01696
23	8	13	.01805	.01705
23	8	14	.01814	.01718
23	8	15	.01827	.01736
23	8	16	.01845	.01760
23	8	17	.01869	.01788
23	8	18	.01898	.01823
23	8	19	.01935	.01866
23	8	20	.01981	.01918
23	8	21	.02041	.01985
23	8	22	.02124	.02077
23	8	23	.02263	.02228
23	9	10	.02149	.02017
23	9	11	.02140	.02014
23	9	12	.02137	.02019
23	9	13	.02141	.02029
23	9	14	.02151	.02045
23	9	15	.02167	.02067
23	9	16	.02189	.02095
23	9	17	.02217	.02129
23	9	18	.02252	.02171
23	9	19	.02296	.02221
23	9	20	.02351	.02283
23	9	21	.02421	.02363
23	9	22	.02520	.02472
23	9	23	.02684	.02652
23	10	11	.02531	.02391
23	10	12	.02529	.02397
23	10	13	.02534	.02409
23	10	14	.02546	.02429
23	10	15	.02565	.02455
23	10	16	.02591	.02488
23	10	17	.02624	.02528
23	10	18	.02665	.02578
23	10	19	.02717	.02638
23	10	20	.02782	.02712
23	10	21	.02866	.02806
23	10	22	.02982	.02936
23	10	23	.03176	.03149
23	11	12	.02991	.02844
23	11	13	.02997	.02860
23	11	14	.03012	.02883
23	11	15	.03034	.02914
23	11	16	.03065	.02953
23	11	17	.03104	.03002
23	11	18	.03153	.03060
23	11	19	.03215	.03131
23	11	20	.03291	.03219
23	11	21	.03390	.03331
23	11	22	.03528	.03485
23	11	23	.03757	.03738
23	12	13	.03551	.03400
23	12	14	.03569	.03428
23	12	15	.03596	.03465
23	12	16	.03632	.03512
23	12	17	.03679	.03570
23	12	18	.03737	.03639
23	12	19	.03810	.03724
23	12	20	.03900	.03828
23	12	21	.04018	.03961
23	12	22	.04181	.04144
23	12	23	.04452	.04444
23	13	14	.04245	.04092
23	13	15	.04278	.04137
23	13	16	.04322	.04193
23	13	17	.04377	.04262
23	13	18	.04447	.04345
23	13	19	.04533	.04446
23	13	20	.04641	.04570
23	13	21	.04780	.04728
23	13	22	.04974	.04946
23	13	23	.05296	.05304
23	14	15	.05121	.04969

n	i	j	2.4	2.5
23	14	16	.05173	.05037
23	14	17	.05240	.05119
23	14	18	.05323	.05219
23	14	19	.05426	.05340
23	14	20	.05555	.05489
23	14	21	.05722	.05680
23	14	22	.05953	.05941
23	14	23	.06339	.06370
23	15	16	.06248	.06105
23	15	17	.06329	.06205
23	15	18	.06429	.06326
23	15	19	.06554	.06473
23	15	20	.06709	.06653
23	15	21	.06910	.06883
23	15	22	.07189	.07199
23	15	23	.07654	.07718
23	16	17	.07738	.07615
23	16	18	.07862	.07764
23	16	19	.08014	.07944
23	16	20	.08204	.08165
23	16	21	.08449	.08446
23	16	22	.08789	.08833
23	16	23	.09356	.09468
23	17	18	.09778	.09694
23	17	19	.09967	.09918
23	17	20	.10203	.10193
23	17	21	.10508	.10543
23	17	22	.10930	.11024
23	17	23	.11631	.11814
23	18	19	.12696	.12683
23	18	20	.12995	.13034
23	18	21	.13382	.13481
23	18	22	.13917	.14092
23	18	23	.14807	.15097
23	19	20	.17131	.17257
23	19	21	.17637	.17844
23	19	22	.18339	.18649
23	19	23	.19505	.19970
23	20	21	.24501	.24906
23	20	22	.25467	.26018
23	20	23	.27070	.27843
23	21	22	.38620	.39672
23	21	23	.41014	.42407
23	22	23	.73930	.76946
24	1	2	.00442	.00385
24	1	3	.00376	.00330
24	1	4	.00341	.00301
24	1	5	.00320	.00283
24	1	6	.00305	.00272
24	1	7	.00295	.00264
24	1	8	.00288	.00258
24	1	9	.00284	.00255
24	1	10	.00280	.00253
24	1	11	.00278	.00252
24	1	12	.00277	.00252

n	i	j	2.4	2.5
24	1	13	.00277	.00253
24	1	14	.00278	.00254
24	1	15	.00280	.00256
24	1	16	.00282	.00259
24	1	17	.00285	.00262
24	1	18	.00288	.00267
24	1	19	.00293	.00272
24	1	20	.00299	.00278
24	1	21	.00306	.00286
24	1	22	.00315	.00296
24	1	23	.00328	.00310
24	1	24	.00349	.00332
24	2	3	.00594	.00525
24	2	4	.00540	.00480
24	2	5	.00506	.00452
24	2	6	.00484	.00434
24	2	7	.00469	.00421
24	2	8	.00458	.00413
24	2	9	.00450	.00408
24	2	10	.00445	.00405
24	2	11	.00442	.00403
24	2	12	.00441	.00403
24	2	13	.00441	.00404
24	2	14	.00442	.00406
24	2	15	.00444	.00410
24	2	16	.00448	.00414
24	2	17	.00452	.00420
24	2	18	.00458	.00427
24	2	19	.00465	.00435
24	2	20	.00474	.00445
24	2	21	.00486	.00458
24	2	22	.00500	.00474
24	2	23	.00521	.00496
24	2	24	.00554	.00531
24	3	4	.00733	.00655
24	3	5	.00688	.00618
24	3	6	.00658	.00593
24	3	7	.00637	.00576
24	3	8	.00623	.00565
24	3	9	.00613	.00558
24	3	10	.00606	.00554
24	3	11	.00602	.00552
24	3	12	.00600	.00552
24	3	13	.00600	.00553
24	3	14	.00601	.00556
24	3	15	.00605	.00561
24	3	16	.00609	.00567
24	3	17	.00616	.00575
24	3	18	.00624	.00585
24	3	19	.00634	.00596
24	3	20	.00646	.00610
24	3	21	.00661	.00627
24	3	22	.00681	.00649
24	3	23	.00709	.00678
24	3	24	.00754	.00727

n	i	j	2.4	2.5
24	4	5	.00878	.00792
24	4	6	.00841	.00761
24	4	7	.00814	.00740
24	4	8	.00796	.00726
24	4	9	.00783	.00717
24	4	10	.00775	.00711
24	4	11	.00770	.00709
24	4	12	.00767	.00709
24	4	13	.00767	.00711
24	4	14	.00769	.00715
24	4	15	.00774	.00722
24	4	16	.00780	.00730
24	4	17	.00788	.00740
24	4	18	.00798	.00752
24	4	19	.00811	.00766
24	4	20	.00826	.00784
24	4	21	.00846	.00806
24	4	22	.00871	.00834
24	4	23	.00907	.00872
24	4	24	.00965	.00935
24	5	6	.01039	.00945
24	5	7	.01007	.00919
24	5	8	.00984	.00902
24	5	9	.00969	.00890
24	5	10	.00958	.00884
24	5	11	.00952	.00881
24	5	12	.00949	.00881
24	5	13	.00949	.00884
24	5	14	.00952	.00889
24	5	15	.00957	.00897
24	5	16	.00965	.00907
24	5	17	.00975	.00919
24	5	18	.00988	.00934
24	5	19	.01003	.00952
24	5	20	.01023	.00975
24	5	21	.01047	.01002
24	5	22	.01078	.01036
24	5	23	.01122	.01084
24	5	24	.01194	.01162
24	6	7	.01219	.01118
24	6	8	.01193	.01097
24	6	9	.01174	.01083
24	6	10	.01162	.01075
24	6	11	.01154	.01072
24	6	12	.01151	.01072
24	6	13	.01151	.01076
24	6	14	.01154	.01082
24	6	15	.01161	.01091
24	6	16	.01170	.01104
24	6	17	.01182	.01119
24	6	18	.01197	.01137
24	6	19	.01216	.01159
24	6	20	.01240	.01186
24	6	21	.01270	.01219
24	6	22	.01308	.01262

n	i	j	2.4	2.5
24	6	23	.01360	.01319
24	6	24	.01448	.01414
24	7	8	.01426	.01316
24	7	9	.01404	.01300
24	7	10	.01389	.01291
24	7	11	.01381	.01287
24	7	12	.01377	.01287
24	7	13	.01377	.01292
24	7	14	.01381	.01299
24	7	15	.01389	.01311
24	7	16	.01400	.01325
24	7	17	.01414	.01344
24	7	18	.01433	.01366
24	7	19	.01456	.01392
24	7	20	.01484	.01425
24	7	21	.01519	.01464
24	7	22	.01565	.01515
24	7	23	.01628	.01584
24	7	24	.01733	.01699
24	8	9	.01664	.01547
24	8	10	.01647	.01536
24	8	11	.01637	.01531
24	8	12	.01633	.01532
24	8	13	.01633	.01537
24	8	14	.01638	.01547
24	8	15	.01647	.01560
24	8	16	.01660	.01578
24	8	17	.01678	.01599
24	8	18	.01699	.01626
24	8	19	.01727	.01658
24	8	20	.01760	.01696
24	8	21	.01802	.01743
24	8	22	.01856	.01803
24	8	23	.01931	.01886
24	8	24	.02055	.02022
24	9	10	.01942	.01817
24	9	11	.01930	.01812
24	9	12	.01925	.01813
24	9	13	.01926	.01819
24	9	14	.01932	.01831
24	9	15	.01943	.01847
24	9	16	.01958	.01867
24	9	17	.01979	.01893
24	9	18	.02005	.01924
24	9	19	.02037	.01962
24	9	20	.02076	.02007
24	9	21	.02126	.02063
24	9	22	.02189	.02135
24	9	23	.02277	.02232
24	9	24	.02424	.02393
24	10	11	.02269	.02138
24	10	12	.02264	.02139
24	10	13	.02265	.02146
24	10	14	.02272	.02160
24	10	15	.02285	.02179

n	i	j	2.4	2.5
24	10	16	.02303	.02203
24	10	17	.02327	.02234
24	10	18	.02358	.02271
24	10	19	.02395	.02315
24	10	20	.02442	.02369
24	10	21	.02500	.02435
24	10	22	.02575	.02519
24	10	23	.02678	.02634
24	10	24	.02851	.02823
24	11	12	.02659	.02520
24	11	13	.02660	.02530
24	11	14	.02669	.02546
24	11	15	.02684	.02568
24	11	16	.02705	.02597
24	11	17	.02734	.02633
24	11	18	.02770	.02676
24	11	19	.02814	.02729
24	11	20	.02869	.02792
24	11	21	.02937	.02870
24	11	22	.03025	.02969
24	11	23	.03146	.03104
24	11	24	.03349	.03327
24	12	13	.03127	.02983
24	12	14	.03137	.03002
24	12	15	.03155	.03029
24	12	16	.03181	.03063
24	12	17	.03214	.03106
24	12	18	.03257	.03157
24	12	19	.03309	.03219
24	12	20	.03373	.03293
24	12	21	.03453	.03385
24	12	22	.03556	.03501
24	12	23	.03699	.03661
24	12	24	.03937	.03924
24	13	14	.03697	.03550
24	13	15	.03719	.03582
24	13	16	.03749	.03623
24	13	17	.03789	.03673
24	13	18	.03839	.03734
24	13	19	.03900	.03807
24	13	20	.03976	.03895
24	13	21	.04070	.04003
24	13	22	.04192	.04141
24	13	23	.04360	.04329
24	13	24	.04640	.04639
24	14	15	.04403	.04255
24	14	16	.04439	.04304
24	14	17	.04487	.04364
24	14	18	.04546	.04436
24	14	19	.04619	.04523
24	14	20	.04708	.04627
24	14	21	.04819	.04755
24	14	22	.04963	.04919
24	14	23	.05162	.05143
24	14	24	.05493	.05510

n	i	j	2.4	2.5
24	15	16	.05291	.05147
24	15	17	.05348	.05219
24	15	18	.05419	.05306
24	15	19	.05505	.05409
24	15	20	.05612	.05534
24	15	21	.05745	.05687
24	15	22	.05916	.05882
24	15	23	.06153	.06150
24	15	24	.06546	.06588
24	16	17	.06434	.06301
24	16	18	.06519	.06406
24	16	19	.06624	.06531
24	16	20	.06752	.06681
24	16	21	.06912	.06866
24	16	22	.07117	.07101
24	16	23	.07402	.07423
24	16	24	.07874	.07951
24	17	18	.07944	.07833
24	17	19	.08072	.07986
24	17	20	.08228	.08170
24	17	21	.08422	.08396
24	17	22	.08672	.08682
24	17	23	.09018	.09075
24	17	24	.09592	.09719
24	18	19	.10007	.09938
24	18	20	.10201	.10167
24	18	21	.10441	.10446
24	18	22	.10750	.10802
24	18	23	.11177	.11289
24	18	24	.11887	.12087
24	19	20	.12954	.12962
24	19	21	.13259	.13317
24	19	22	.13650	.13769
24	19	23	.14190	.14386
24	19	24	.15087	.15400
24	20	21	.17429	.17580
24	20	22	.17940	.18173
24	20	23	.18646	.18982
24	20	24	.19818	.20310
24	21	22	.24855	.25294
24	21	23	.25825	.26410
24	21	24	.27433	.28239
24	22	23	.39062	.40159
24	22	24	.41456	.42894
24	23	24	.74532	.77618
25	1	2	.00422	.00368
25	1	3	.00359	.00314
25	1	4	.00325	.00286
25	1	5	.00304	.00269
25	1	6	.00289	.00257
25	1	7	.00280	.00249
25	1	8	.00273	.00244
25	1	9	.00268	.00240
25	1	10	.00264	.00238
25	1	11	.00262	.00237

n	i	j	2.4	2.5
25	1	12	.00261	.00236
25	1	13	.00260	.00237
25	1	14	.00261	.00238
25	1	15	.00262	.00239
25	1	16	.00263	.00241
25	1	17	.00265	.00244
25	1	18	.00268	.00248
25	1	19	.00272	.00252
25	1	20	.00276	.00257
25	1	21	.00282	.00262
25	1	22	.00288	.00270
25	1	23	.00297	.00279
25	1	24	.00309	.00292
25	1	25	.00329	.00313
25	2	3	.00563	.00498
25	2	4	.00511	.00454
25	2	5	.00478	.00426
25	2	6	.00457	.00409
25	2	7	.00441	.00396
25	2	8	.00431	.00388
25	2	9	.00423	.00382
25	2	10	.00418	.00379
25	2	11	.00414	.00377
25	2	12	.00412	.00376
25	2	13	.00412	.00377
25	2	14	.00412	.00378
25	2	15	.00414	.00381
25	2	16	.00416	.00384
25	2	17	.00420	.00389
25	2	18	.00424	.00394
25	2	19	.00430	.00401
25	2	20	.00437	.00409
25	2	21	.00445	.00418
25	2	22	.00456	.00430
25	2	23	.00469	.00444
25	2	24	.00488	.00465
25	2	25	.00519	.00498
25	3	4	.00691	.00616
25	3	5	.00647	.00580
25	3	6	.00618	.00556
25	3	7	.00598	.00540
25	3	8	.00583	.00528
25	3	9	.00573	.00521
25	3	10	.00566	.00516
25	3	11	.00561	.00513
25	3	12	.00559	.00513
25	3	13	.00558	.00513
25	3	14	.00559	.00516
25	3	15	.00561	.00519
25	3	16	.00564	.00524
25	3	17	.00569	.00530
25	3	18	.00575	.00537
25	3	19	.00583	.00546
25	3	20	.00592	.00557
25	3	21	.00603	.00570
25	3	22	.00618	.00586
25	3	23	.00636	.00606
25	3	24	.00661	.00633
25	3	25	.00704	.00678
25	4	5	.00823	.00741
25	4	6	.00786	.00711
25	4	7	.00761	.00690
25	4	8	.00743	.00676
25	4	9	.00730	.00666
25	4	10	.00721	.00660
25	4	11	.00715	.00657
25	4	12	.00712	.00656
25	4	13	.00711	.00657
25	4	14	.00712	.00660
25	4	15	.00714	.00665
25	4	16	.00719	.00671
25	4	17	.00725	.00678
25	4	18	.00733	.00688
25	4	19	.00742	.00699
25	4	20	.00754	.00713
25	4	21	.00769	.00730
25	4	22	.00787	.00750
25	4	23	.00811	.00775
25	4	24	.00843	.00811
25	4	25	.00897	.00868
25	5	6	.00968	.00879
25	5	7	.00937	.00853
25	5	8	.00914	.00836
25	5	9	.00899	.00824
25	5	10	.00888	.00817
25	5	11	.00881	.00813
25	5	12	.00877	.00812
25	5	13	.00876	.00813
25	5	14	.00877	.00817
25	5	15	.00880	.00822
25	5	16	.00886	.00830
25	5	17	.00893	.00840
25	5	18	.00903	.00851
25	5	19	.00915	.00866
25	5	20	.00930	.00883
25	5	21	.00948	.00903
25	5	22	.00970	.00928
25	5	23	.00999	.00960
25	5	24	.01039	.01003
25	5	25	.01105	.01075
25	6	7	.01130	.01033
25	6	8	.01103	.01013
25	6	9	.01085	.00999
25	6	10	.01072	.00990
25	6	11	.01063	.00985
25	6	12	.01059	.00984
25	6	13	.01057	.00986
25	6	14	.01059	.00990
25	6	15	.01063	.00997
25	6	16	.01070	.01006

n	i	j	2.4	2.5
25	6	17	.01079	.01018
25	6	18	.01090	.01032
25	6	19	.01105	.01049
25	6	20	.01123	.01070
25	6	21	.01144	.01094
25	6	22	.01171	.01125
25	6	23	.01206	.01163
25	6	24	.01254	.01216
25	6	25	.01335	.01303
25	7	8	.01313	.01210
25	7	9	.01291	.01193
25	7	10	.01276	.01183
25	7	11	.01266	.01178
25	7	12	.01261	.01176
25	7	13	.01260	.01178
25	7	14	.01261	.01184
25	7	15	.01266	.01192
25	7	16	.01274	.01203
25	7	17	.01285	.01217
25	7	18	.01299	.01234
25	7	19	.01316	.01254
25	7	20	.01337	.01279
25	7	21	.01363	.01308
25	7	22	.01396	.01345
25	7	23	.01437	.01391
25	7	24	.01494	.01454
25	7	25	.01590	.01557
25	8	9	.01524	.01413
25	8	10	.01506	.01401
25	8	11	.01495	.01395
25	8	12	.01488	.01393
25	8	13	.01487	.01396
25	8	14	.01489	.01402
25	8	15	.01495	.01412
25	8	16	.01504	.01425
25	8	17	.01517	.01442
25	8	18	.01534	.01462
25	8	19	.01554	.01486
25	8	20	.01579	.01515
25	8	21	.01610	.01550
25	8	22	.01648	.01593
25	8	23	.01697	.01648
25	8	24	.01764	.01722
25	8	25	.01877	.01845
25	9	10	.01767	.01649
25	9	11	.01754	.01642
25	9	12	.01747	.01640
25	9	13	.01745	.01643
25	9	14	.01748	.01651
25	9	15	.01754	.01662
25	9	16	.01766	.01678
25	9	17	.01781	.01698
25	9	18	.01800	.01721
25	9	19	.01824	.01750
25	9	20	.01853	.01784

n	i	j	2.4	2.5
25	9	21	.01889	.01825
25	9	22	.01934	.01876
25	9	23	.01991	.01940
25	9	24	.02071	.02028
25	9	25	.02203	.02172
25	10	11	.02050	.01926
25	10	12	.02042	.01924
25	10	13	.02040	.01928
25	10	14	.02043	.01937
25	10	15	.02052	.01950
25	10	16	.02065	.01969
25	10	17	.02082	.01992
25	10	18	.02105	.02020
25	10	19	.02133	.02053
25	10	20	.02168	.02094
25	10	21	.02210	.02142
25	10	22	.02262	.02201
25	10	23	.02329	.02276
25	10	24	.02422	.02379
25	10	25	.02576	.02548
25	11	12	.02384	.02253
25	11	13	.02381	.02258
25	11	14	.02386	.02268
25	11	15	.02395	.02284
25	11	16	.02411	.02306
25	11	17	.02431	.02333
25	11	18	.02458	.02366
25	11	19	.02491	.02405
25	11	20	.02531	.02452
25	11	21	.02580	.02509
25	11	22	.02641	.02578
25	11	23	.02719	.02666
25	11	24	.02827	.02787
25	11	25	.03007	.02984
25	12	13	.02780	.02644
25	12	14	.02785	.02656
25	12	15	.02797	.02675
25	12	16	.02814	.02701
25	12	17	.02839	.02732
25	12	18	.02870	.02771
25	12	19	.02908	.02817
25	12	20	.02955	.02872
25	12	21	.03012	.02938
25	12	22	.03083	.03020
25	12	23	.03175	.03123
25	12	24	.03301	.03264
25	12	25	.03511	.03495
25	13	14	.03256	.03116
25	13	15	.03270	.03138
25	13	16	.03291	.03168
25	13	17	.03320	.03206
25	13	18	.03356	.03251
25	13	19	.03401	.03305
25	13	20	.03456	.03370
25	13	21	.03523	.03447

n	i	j	2.4	2.5
25	13	22	.03606	.03542
25	13	23	.03713	.03663
25	13	24	.03860	.03828
25	13	25	.04106	.04099
25	14	15	.03836	.03694
25	14	16	.03861	.03729
25	14	17	.03895	.03773
25	14	18	.03938	.03826
25	14	19	.03990	.03890
25	14	20	.04055	.03966
25	14	21	.04133	.04057
25	14	22	.04231	.04169
25	14	23	.04356	.04311
25	14	24	.04529	.04505
25	14	25	.04817	.04824
25	15	16	.04553	.04411
25	15	17	.04593	.04463
25	15	18	.04643	.04526
25	15	19	.04706	.04602
25	15	20	.04781	.04692
25	15	21	.04874	.04800
25	15	22	.04989	.04932
25	15	23	.05136	.05099
25	15	24	.05340	.05329
25	15	25	.05679	.05705
25	16	17	.05454	.05317
25	16	18	.05514	.05393
25	16	19	.05588	.05483
25	16	20	.05678	.05590
25	16	21	.05788	.05718
25	16	22	.05924	.05875
25	16	23	.06099	.06075
25	16	24	.06341	.06348
25	16	25	.06742	.06795
25	17	18	.06612	.06488
25	17	19	.06701	.06596
25	17	20	.06809	.06725
25	17	21	.06941	.06879
25	17	22	.07104	.07068
25	17	23	.07313	.07308
25	17	24	.07602	.07635
25	17	25	.08082	.08172
25	18	19	.08140	.08040
25	18	20	.08271	.08197
25	18	21	.08431	.08385
25	18	22	.08629	.08615
25	18	23	.08883	.08906
25	18	24	.09233	.09304
25	18	25	.09815	.09957
25	19	20	.10225	.10171
25	19	21	.10422	.10403
25	19	22	.10666	.10687
25	19	23	.10979	.11047
25	19	24	.11411	.11539
25	19	25	.12128	.12346

n	i	j	2.4	2.5
25	20	21	.13200	.13227
25	20	22	.13508	.13586
25	20	23	.13903	.14042
25	20	24	.14448	.14665
25	20	25	.15352	.15685
25	21	22	.17712	.17888
25	21	23	.18227	.18484
25	21	24	.18937	.19298
25	21	25	.20114	.20633
25	22	23	.25191	.25662
25	22	24	.26163	.26781
25	22	25	.27775	.28614
25	23	24	.39480	.40622
25	23	25	.41873	.43356
25	24	25	.75102	.78255

Table 4. Coefficients a_i for the BLUE of the location parameter μ for $n = 2(1)25$ and the censoring number $s = 0(1)\left[\frac{n+1}{2}\right]$

Coefficients a_i for the BLUE of μ
Shape parameter k

n	s	i	0.0	0.1	0.2	0.3	0.4	0.5	0.6
2	0	1	.50000	.50000	.50000	.50000	.50000	.50000	.50000
2	0	2	.50000	.50000	.50000	.50000	.50000	.50000	.50000
3	0	1	.33333	.33318	.33273	.33195	.33085	.32943	.32768
3	0	2	.33333	.33363	.33451	.33597	.33799	.34056	.34365
3	0	3	.33333	.33319	.33276	.33208	.33116	.33001	.32868
3	1	1	.00000	-.01103	-.02242	-.03415	-.04622	-.05863	-.07136
3	1	2	1.00000	1.01103	1.02242	1.03415	1.04622	1.05863	1.07136
4	0	1	.25000	.24980	.24916	.24807	.24649	.24439	.24177
4	0	2	.25000	.25021	.25085	.25197	.25361	.25579	.25854
4	0	3	.25000	.25019	.25074	.25159	.25270	.25404	.25557
4	0	4	.25000	.24981	.24925	.24837	.24720	.24577	.24412
4	1	1	.11607	.10860	.10058	.09201	.08286	.07311	.06277
4	1	2	.24084	.24537	.25057	.25645	.26302	.27029	.27825
4	1	3	.64310	.64603	.64885	.65154	.65412	.65660	.65898
5	0	1	.20000	.19978	.19907	.19783	.19603	.19360	.19052
5	0	2	.20000	.20014	.20060	.20142	.20267	.20439	.20663
5	0	3	.20000	.20017	.20069	.20154	.20272	.20423	.20605
5	0	4	.20000	.20012	.20045	.20094	.20155	.20222	.20294
5	0	5	.20000	.19979	.19920	.19826	.19704	.19556	.19386
5	1	1	.12516	.11978	.11379	.10714	.09980	.09173	.08291
5	1	2	.18305	.18486	.18720	.19013	.19367	.19786	.20274
5	1	3	.21472	.21715	.21993	.22304	.22646	.23017	.23415
5	1	4	.47708	.47821	.47907	.47969	.48008	.48024	.48020
5	2	1	-.06377	-.07804	-.09317	-.10921	-.12616	-.14407	-.16293
5	2	2	.14983	.15682	.16458	.17315	.18252	.19272	.20376
5	2	3	.91395	.92122	.92859	.93606	.94364	.95134	.95917
6	0	1	.16667	.16644	.16570	.16441	.16249	.15990	.15657
6	0	2	.16667	.16676	.16708	.16768	.16862	.16996	.17175
6	0	3	.16667	.16681	.16724	.16798	.16904	.17044	.17218
6	0	4	.16667	.16680	.16719	.16781	.16863	.16965	.17084
6	0	5	.16667	.16674	.16694	.16720	.16750	.16779	.16806
6	0	6	.16667	.16645	.16586	.16492	.16371	.16226	.16061
6	1	1	.11829	.11420	.10948	.10406	.09791	.09095	.08315
6	1	2	.15097	.15169	.15281	.15438	.15645	.15908	.16230
6	1	3	.16803	.16951	.17135	.17355	.17611	.17905	.18235
6	1	4	.18280	.18417	.18574	.18749	.18940	.19144	.19359
6	1	5	.37991	.38042	.38062	.38051	.38013	.37949	.37861
6	2	1	.01848	.00843	-.00247	-.01426	-.02700	-.04072	-.05547
6	2	2	.12261	.12532	.12864	.13263	.13731	.14272	.14891
6	2	3	.17615	.18029	.18479	.18967	.19490	.20048	.20640
6	2	4	.68276	.68596	.68903	.69197	.69480	.69752	.70016
6	3	1	-.21592	-.23643	-.25817	-.28118	-.30549	-.33115	-.35817
6	3	2	.06485	.07329	.08261	.09283	.10396	.11603	.12904
6	3	3	1.15107	1.16314	1.17556	1.18835	1.20153	1.21512	1.22913
7	0	1	.14286	.14263	.14189	.14057	.13861	.13594	.13247
7	0	2	.14286	.14292	.14314	.14357	.14427	.14530	.14672
7	0	3	.14286	.14297	.14332	.14394	.14486	.14609	.14767
7	0	4	.14286	.14298	.14334	.14395	.14478	.14585	.14714
7	0	5	.14286	.14296	.14325	.14369	.14427	.14495	.14571
7	0	6	.14286	.14290	.14300	.14313	.14323	.14329	.14328
7	0	7	.14286	.14265	.14206	.14115	.13998	.13858	.13701
7	1	1	.10882	.10558	.10171	.09715	.09181	.08565	.07858
7	1	2	.12955	.12977	.13028	.13113	.13239	.13411	.13633
7	1	3	.13997	.14087	.14208	.14363	.14554	.14782	.15049
7	1	4	.14874	.14978	.15108	.15261	.15437	.15634	.15853
7	1	5	.15705	.15786	.15879	.15983	.16094	.16211	.16331

Coefficients a_i for the BLUE of μ
Shape parameter k

n	s	i	0.0	0.1	0.2	0.3	0.4	0.5	0.6
7	1	6	.31587	.31614	.31606	.31565	.31495	.31397	.31276
7	2	1	.04655	.03897	.03056	.02128	.01106	-.00017	-.01245
7	2	2	.10721	.10822	.10968	.11165	.11419	.11733	.12114
7	2	3	.13748	.13991	.14272	.14593	.14954	.15356	.15800
7	2	4	.16260	.16508	.16777	.17066	.17374	.17698	.18039
7	2	5	.54616	.54783	.54926	.55048	.55148	.55229	.55292
7	3	1	-.07380	-.08811	-.10351	-.12007	-.13783	-.15686	-.17719
7	3	2	.06772	.07089	.07475	.07932	.08466	.09079	.09776
7	3	3	.13752	.14294	.14877	.15501	.16165	.16870	.17615
7	3	4	.86856	.87427	.87999	.88574	.89153	.89736	.90327
7	4	1	-.34745	-.37370	-.40155	-.43105	-.46225	-.49520	-.52994
7	4	2	-.01346	-.00419	.00605	.01727	.02949	.04273	.05700
7	4	3	1.36090	1.37789	1.39550	1.41378	1.43276	1.45247	1.47294
8	0	1	.12500	.12477	.12404	.12273	.12076	.11805	.11452
8	0	2	.12500	.12504	.12518	.12549	.12600	.12679	.12792
8	0	3	.12500	.12509	.12537	.12589	.12667	.12774	.12913
8	0	4	.12500	.12511	.12543	.12598	.12676	.12779	.12905
8	0	5	.12500	.12510	.12540	.12588	.12653	.12733	.12828
8	0	6	.12500	.12508	.12529	.12561	.12600	.12645	.12692
8	0	7	.12500	.12502	.12506	.12508	.12506	.12497	.12479
8	0	8	.12500	.12480	.12422	.12335	.12222	.12089	.11939
8	1	1	.09967	.09702	.09376	.08981	.08508	.07950	.07299
8	1	2	.11386	.11384	.11402	.11447	.11523	.11637	.11794
8	1	3	.12080	.12134	.12215	.12327	.12470	.12649	.12866
8	1	4	.12649	.12724	.12824	.12949	.13099	.13274	.13474
8	1	5	.13176	.13250	.13341	.13448	.13570	.13705	.13852
8	1	6	.13699	.13748	.13805	.13867	.13933	.13999	.14065
8	1	7	.27043	.27058	.27037	.26983	.26898	.26786	.26651
8	2	1	.05692	.05094	.04417	.03655	.02800	.01845	.00783
8	2	2	.09621	.09643	.09699	.09794	.09934	.10123	.10368
8	2	3	.11532	.11677	.11856	.12071	.12326	.12621	.12958
8	2	4	.13090	.13269	.13473	.13702	.13956	.14235	.14538
8	2	5	.14512	.14668	.14836	.15017	.15207	.15406	.15612
8	2	6	.45552	.45650	.45719	.45761	.45777	.45770	.45741
8	3	1	-.01672	-.02749	-.03927	-.05214	-.06614	-.08136	-.09786
8	3	2	.06765	.06873	.07032	.07246	.07521	.07861	.08272
8	3	3	.10841	.11153	.11506	.11903	.12343	.12829	.13360
8	3	4	.14132	.14474	.14838	.15225	.15633	.16062	.16510
8	3	5	.69935	.70249	.70550	.70839	.71117	.71384	.71644
8	4	1	-.15491	-.17312	-.19268	-.21365	-.23610	-.26009	-.28568
8	4	2	.01760	.02096	.02505	.02991	.03558	.04210	.04950
8	4	3	.10013	.10652	.11336	.12065	.12840	.13660	.14526
8	4	4	1.03717	1.04563	1.05426	1.06308	1.07212	1.08139	1.09092
9	0	1	.11111	.11089	.11017	.10887	.10691	.10420	.10065
9	0	2	.11111	.11113	.11122	.11142	.11180	.11239	.11327
9	0	3	.11111	.11118	.11141	.11184	.11249	.11342	.11464
9	0	4	.11111	.11120	.11148	.11197	.11268	.11363	.11483
9	0	5	.11111	.11121	.11149	.11195	.11260	.11343	.11443
9	0	6	.11111	.11120	.11144	.11181	.11231	.11292	.11361
9	0	7	.11111	.11117	.11132	.11154	.11181	.11208	.11236
9	0	8	.11111	.11112	.11111	.11107	.11096	.11076	.11047
9	0	9	.11111	.11091	.11036	.10952	.10844	.10717	.10575
9	1	1	.09149	.08927	.08646	.08296	.07870	.07358	.06751
9	1	2	.10175	.10160	.10159	.10179	.10223	.10298	.10409
9	1	3	.10665	.10698	.10753	.10833	.10943	.11085	.11262
9	1	4	.11060	.11114	.11190	.11290	.11415	.11567	.11745

Coefficients a_i for the BLUE of μ
Shape parameter k

n	s	i	0.0	0.1	0.2	0.3	0.4	0.5	0.6
9	1	5	.11418	.11477	.11556	.11652	.11767	.11899	.12048
9	1	6	.11766	.11818	.11883	.11959	.12044	.12138	.12238
9	1	7	.12119	.12149	.12184	.12221	.12259	.12294	.12327
9	1	8	.23648	.23657	.23630	.23569	.23478	.23361	.23220
9	2	1	.06024	.05537	.04975	.04330	.03595	.02760	.01818
9	2	2	.08756	.08740	.08748	.08786	.08859	.08973	.09133
9	2	3	.10056	.10142	.10258	.10405	.10589	.10810	.11071
9	2	4	.11100	.11226	.11377	.11554	.11759	.11991	.12250
9	2	5	.12039	.12167	.12314	.12478	.12657	.12853	.13062
9	2	6	.12937	.13038	.13148	.13264	.13386	.13511	.13639
9	2	7	.39087	.39150	.39181	.39182	.39155	.39103	.39026
9	3	1	.01040	.00190	-.00753	-.01799	-.02954	-.04226	-.05624
9	3	2	.06597	.06609	.06659	.06751	.06890	.07083	.07334
9	3	3	.09230	.09414	.09635	.09895	.10197	.10543	.10934
9	3	4	.11332	.11571	.11836	.12129	.12449	.12796	.13170
9	3	5	.13204	.13428	.13666	.13917	.14180	.14453	.14738
9	3	6	.58596	.58788	.58958	.59107	.59238	.59352	.59449
9	4	1	-.07313	-.08684	-.10174	-.11793	-.13546	-.15443	-.17489
9	4	2	.03156	.03258	.03414	.03629	.03908	.04255	.04676
9	4	3	.08087	.08451	.08858	.09312	.09812	.10362	.10961
9	4	4	.11995	.12415	.12860	.13329	.13823	.14340	.14881
9	4	5	.84076	.84561	.85043	.85523	.86003	.86486	.86972
9	5	1	-.22718	-.24899	-.27240	-.29749	-.32432	-.35298	-.38352
9	5	2	-.02842	-.02506	-.02094	-.01600	-.01022	-.00355	.00403
9	5	3	.06444	.07156	.07917	.08729	.09590	.10501	.11463
9	5	4	1.19116	1.20249	1.21416	1.22620	1.23864	1.25152	1.26486
10	0	1	.10000	.09978	.09908	.09780	.09586	.09316	.08961
10	0	2	.10000	.10001	.10005	.10018	.10044	.10087	.10155
10	0	3	.10000	.10005	.10024	.10059	.10114	.10193	.10300
10	0	4	.10000	.10008	.10032	.10075	.10139	.10225	.10336
10	0	5	.10000	.10008	.10034	.10078	.10140	.10220	.10320
10	0	6	.10000	.10008	.10032	.10071	.10125	.10191	.10269
10	0	7	.10000	.10007	.10026	.10056	.10094	.10139	.10189
10	0	8	.10000	.10004	.10015	.10030	.10046	.10062	.10075
10	0	9	.10000	.09999	.09996	.09986	.09969	.09943	.09907
10	0	10	.10000	.09981	.09928	.09847	.09743	.09623	.09489
10	1	1	.08433	.08244	.07997	.07683	.07293	.06818	.06247
10	1	2	.09206	.09184	.09173	.09176	.09198	.09246	.09324
10	1	3	.09568	.09587	.09624	.09683	.09768	.09882	.10027
10	1	4	.09856	.09894	.09953	.10033	.10138	.10268	.10425
10	1	5	.10112	.10159	.10224	.10308	.10411	.10534	.10676
10	1	6	.10357	.10403	.10464	.10539	.10627	.10727	.10838
10	1	7	.10601	.10638	.10685	.10740	.10800	.10865	.10933
10	1	8	.10853	.10871	.10892	.10912	.10932	.10947	.10959
10	1	9	.21014	.21020	.20990	.20926	.20833	.20714	.20573
10	2	1	.06045	.05639	.05161	.04604	.03958	.03214	.02362
10	2	2	.08045	.08008	.07990	.07994	.08026	.08091	.08195
10	2	3	.08978	.09028	.09102	.09205	.09338	.09507	.09712
10	2	4	.09718	.09806	.09918	.10056	.10220	.10412	.10632
10	2	5	.10374	.10475	.10595	.10735	.10894	.11071	.11267
10	2	6	.10995	.11088	.11195	.11313	.11443	.11581	.11729
10	2	7	.11606	.11672	.11745	.11822	.11901	.11981	.12061
10	2	8	.34239	.34283	.34293	.34272	.34221	.34143	.34041
10	3	1	.02443	.01750	.00970	.00093	-.00889	-.01985	-.03203
10	3	2	.06356	.06320	.06312	.06337	.06399	.06504	.06657
10	3	3	.08178	.08286	.08425	.08600	.08811	.09064	.09359

Coefficients a_i for the BLUE of μ
Shape parameter k

n	s	i	0.0	0.1	0.2	0.3	0.4	0.5	0.6
10	3	4	.09617	.09783	.09976	.10197	.10446	.10725	.11034
10	3	5	.10886	.11063	.11258	.11472	.11703	.11952	.12217
10	3	6	.12074	.12225	.12386	.12554	.12730	.12911	.13098
10	3	7	.50447	.50574	.50673	.50748	.50799	.50829	.50838
10	4	1	-.03158	-.04241	-.05432	-.06742	-.08178	-.09749	-.11462
10	4	2	.03829	.03824	.03859	.03939	.04069	.04254	.04499
10	4	3	.07072	.07284	.07535	.07828	.08165	.08549	.08980
10	4	4	.09620	.09908	.10224	.10570	.10944	.11349	.11783
10	4	5	.11852	.12136	.12435	.12749	.13077	.13418	.13772
10	4	6	.70785	.71089	.71379	.71656	.71923	.72180	.72429
10	5	1	-.12396	-.14038	-.15818	-.17747	-.19831	-.22080	-.24500
10	5	2	-.00163	-.00077	.00066	.00271	.00542	.00883	.01300
10	5	3	.05485	.05886	.06333	.06830	.07376	.07975	.08626
10	5	4	.09896	.10381	.10894	.11433	.11999	.12592	.13212
10	5	5	.97177	.97847	.98525	.99213	.99914	1.00629	1.01363
11	0	1	.09091	.09070	.09001	.08875	.08684	.08417	.08064
11	0	2	.09091	.09090	.09092	.09098	.09114	.09145	.09196
11	0	3	.09091	.09095	.09110	.09138	.09185	.09252	.09345
11	0	4	.09091	.09097	.09118	.09156	.09213	.09291	.09392
11	0	5	.09091	.09098	.09122	.09162	.09220	.09296	.09392
11	0	6	.09091	.09099	.09122	.09160	.09213	.09280	.09362
11	0	7	.09091	.09098	.09118	.09151	.09194	.09247	.09309
11	0	8	.09091	.09097	.09112	.09135	.09164	.09197	.09233
11	0	9	.09091	.09094	.09102	.09111	.09120	.09126	.09129
11	0	10	.09091	.09089	.09083	.09070	.09049	.09018	.08977
11	0	11	.09091	.09072	.09021	.08943	.08845	.08730	.08602
11	1	1	.07809	.07645	.07425	.07140	.06780	.06335	.05794
11	1	2	.08411	.08386	.08367	.08359	.08366	.08394	.08446
11	1	3	.08688	.08697	.08722	.08765	.08831	.08923	.09044
11	1	4	.08906	.08933	.08978	.09043	.09130	.09242	.09380
11	1	5	.09097	.09133	.09186	.09259	.09350	.09462	.09593
11	1	6	.09276	.09315	.09369	.09438	.09523	.09621	.09733
11	1	7	.09452	.09488	.09536	.09594	.09662	.09738	.09821
11	1	8	.09632	.09659	.09693	.09733	.09775	.09820	.09865
11	1	9	.09818	.09829	.09840	.09850	.09857	.09859	.09857
11	1	10	.18910	.18915	.18883	.18818	.18725	.18607	.18467
11	2	1	.05923	.05578	.05164	.04674	.04097	.03424	.02644
11	2	2	.07444	.07398	.07364	.07348	.07353	.07385	.07448
11	2	3	.08143	.08169	.08216	.08287	.08386	.08515	.08679
11	2	4	.08690	.08752	.08835	.08942	.09074	.09234	.09422
11	2	5	.09170	.09247	.09344	.09461	.09598	.09755	.09934
11	2	6	.09618	.09697	.09792	.09901	.10024	.10160	.10309
11	2	7	.10054	.10123	.10202	.10289	.10383	.10483	.10588
11	2	8	.10492	.10537	.10585	.10636	.10687	.10737	.10786
11	2	9	.30466	.30499	.30497	.30463	.30398	.30307	.30191
11	3	1	.03195	.02617	.01956	.01204	.00350	-.00613	-.01697
11	3	2	.06086	.06027	.05988	.05973	.05988	.06038	.06127
11	3	3	.07413	.07473	.07560	.07678	.07829	.08016	.08243
11	3	4	.08449	.08565	.08706	.08873	.09069	.09295	.09551
11	3	5	.09355	.09491	.09647	.09824	.10021	.10238	.10475
11	3	6	.10195	.10327	.10472	.10630	.10799	.10980	.11171
11	3	7	.11006	.11111	.11222	.11338	.11458	.11581	.11706
11	3	8	.44302	.44390	.44449	.44481	.44485	.44466	.44424
11	4	1	-.00824	-.01707	-.02691	-.03785	-.04999	-.06341	-.07820
11	4	2	.04147	.04090	.04062	.04068	.04113	.04203	.04342
11	4	3	.06423	.06545	.06701	.06894	.07126	.07401	.07721

Coefficients a_i for the BLUE of μ
Shape parameter k

n	s	i	0.0	0.1	0.2	0.3	0.4	0.5	0.6
11	4	4	.08196	.08394	.08621	.08878	.09165	.09483	.09833
11	4	5	.09738	.09956	.10195	.10452	.10729	.11025	.11339
11	4	6	.11159	.11356	.11565	.11782	.12007	.12241	.12481
11	4	7	.61161	.61365	.61548	.61712	.61859	.61989	.62104
11	5	1	-.06979	-.08276	-.09699	-.11256	-.12956	-.14808	-.16820
11	5	2	.01277	.01249	.01264	.01325	.01438	.01607	.01838
11	5	3	.05043	.05273	.05546	.05862	.06225	.06637	.07099
11	5	4	.07966	.08294	.08652	.09041	.09462	.09915	.10399
11	5	5	.10494	.10830	.11184	.11554	.11939	.12340	.12756
11	5	6	.82199	.82629	.83054	.83474	.83891	.84309	.84728
12	0	1	.08333	.08313	.08245	.08123	.07934	.07671	.07320
12	0	2	.08333	.08332	.08330	.08332	.08341	.08361	.08398
12	0	3	.08333	.08336	.08348	.08371	.08410	.08468	.08548
12	0	4	.08333	.08339	.08357	.08390	.08440	.08510	.08603
12	0	5	.08333	.08340	.08361	.08397	.08451	.08522	.08613
12	0	6	.08333	.08340	.08362	.08398	.08449	.08515	.08597
12	0	7	.08333	.08340	.08360	.08393	.08438	.08495	.08561
12	0	8	.08333	.08339	.08357	.08384	.08419	.08461	.08509
12	0	9	.08333	.08338	.08350	.08368	.08390	.08414	.08438
12	0	10	.08333	.08336	.08340	.08345	.08348	.08348	.08343
12	0	11	.08333	.08331	.08323	.08308	.08283	.08249	.08205
12	0	12	.08333	.08315	.08266	.08191	.08097	.07987	.07866
12	1	1	.07264	.07121	.06923	.06661	.06325	.05905	.05390
12	1	2	.07745	.07718	.07695	.07680	.07676	.07688	.07721
12	1	3	.07963	.07966	.07982	.08013	.08064	.08138	.08239
12	1	4	.08133	.08152	.08187	.08240	.08313	.08409	.08531
12	1	5	.08280	.08308	.08352	.08414	.08494	.08595	.08716
12	1	6	.08416	.08448	.08495	.08558	.08636	.08730	.08839
12	1	7	.08548	.08580	.08625	.08682	.08751	.08830	.08919
12	1	8	.08681	.08709	.08747	.08792	.08845	.08903	.08965
12	1	9	.08818	.08837	.08862	.08890	.08920	.08950	.08979
12	1	10	.08960	.08966	.08970	.08972	.08971	.08964	.08951
12	1	11	.17191	.17195	.17162	.17098	.17006	.16889	.16752
12	2	1	.05736	.05438	.05075	.04637	.04115	.03499	.02777
12	2	2	.06928	.06877	.06835	.06805	.06792	.06801	.06836
12	2	3	.07468	.07479	.07507	.07556	.07629	.07729	.07861
12	2	4	.07887	.07930	.07992	.08076	.08183	.08316	.08477
12	2	5	.08251	.08310	.08388	.08485	.08602	.08741	.08901
12	2	6	.08586	.08652	.08733	.08829	.08942	.09069	.09212
12	2	7	.08910	.08972	.09047	.09132	.09228	.09333	.09447
12	2	8	.09231	.09282	.09341	.09406	.09475	.09547	.09622
12	2	9	.09557	.09587	.09620	.09653	.09685	.09715	.09742
12	2	10	.27446	.27472	.27463	.27421	.27349	.27250	.27126
12	3	1	.03597	.03105	.02535	.01878	.01124	.00262	-.00717
12	3	2	.05812	.05742	.05685	.05647	.05632	.05645	.05691
12	3	3	.06815	.06846	.06898	.06977	.07085	.07225	.07402
12	3	4	.07593	.07673	.07776	.07905	.08060	.08243	.08457
12	3	5	.08265	.08369	.08493	.08638	.08804	.08991	.09201
12	3	6	.08884	.08993	.09118	.09257	.09412	.09581	.09764
12	3	7	.09476	.09575	.09685	.09803	.09929	.10063	.10203
12	3	8	.10059	.10133	.10212	.10293	.10376	.10460	.10544
12	3	9	.39498	.39564	.39598	.39603	.39579	.39530	.39457
12	4	1	.00572	-.00167	-.00999	-.01935	-.02983	-.04155	-.05460
12	4	2	.04275	.04191	.04129	.04093	.04087	.04118	.04189
12	4	3	.05949	.06016	.06111	.06238	.06401	.06601	.06842
12	4	4	.07244	.07381	.07545	.07737	.07959	.08212	.08497

Coefficients a_i for the BLUE of μ
Shape parameter k

n	s	i	0.0	0.1	0.2	0.3	0.4	0.5	0.6
12	4	5	.08361	.08527	.08714	.08923	.09153	.09405	.09679
12	4	6	.09384	.09550	.09730	.09924	.10130	.10350	.10581
12	4	7	.10356	.10498	.10646	.10800	.10960	.11123	.11290
12	4	8	.53860	.54004	.54123	.54219	.54293	.54346	.54381
12	5	1	-.03821	-.04882	-.06057	-.07355	-.08787	-.10361	-.12087
12	5	2	.02103	.02020	.01969	.01952	.01977	.02046	.02166
12	5	3	.04775	.04906	.05073	.05278	.05524	.05815	.06152
12	5	4	.06836	.07060	.07315	.07600	.07918	.08269	.08653
12	5	5	.08608	.08863	.09138	.09434	.09751	.10088	.10446
12	5	6	.10222	.10462	.10713	.10975	.11247	.11527	.11816
12	5	7	.71278	.71571	.71850	.72116	.72370	.72615	.72853
12	6	1	-.10480	-.11979	-.13618	-.15407	-.17356	-.19473	-.21769
12	6	2	-.01092	-.01145	-.01155	-.01117	-.01027	-.00878	-.00668
12	6	3	.03129	.03372	.03659	.03992	.04373	.04805	.05289
12	6	4	.06374	.06734	.07127	.07553	.08012	.08505	.09033
12	6	5	.09151	.09534	.09935	.10354	.10791	.11245	.11716
12	6	6	.92918	.93484	.94052	.94626	.95206	.95797	.96399
13	0	1	.07692	.07672	.07607	.07486	.07301	.07041	.06695
13	0	2	.07692	.07690	.07687	.07684	.07686	.07698	.07722
13	0	3	.07692	.07695	.07704	.07722	.07754	.07803	.07873
13	0	4	.07692	.07697	.07712	.07741	.07785	.07848	.07932
13	0	5	.07692	.07698	.07717	.07750	.07799	.07865	.07951
13	0	6	.07692	.07699	.07718	.07752	.07801	.07865	.07944
13	0	7	.07692	.07699	.07718	.07750	.07795	.07852	.07921
13	0	8	.07692	.07698	.07716	.07744	.07782	.07829	.07884
13	0	9	.07692	.07698	.07712	.07734	.07763	.07796	.07833
13	0	10	.07692	.07696	.07706	.07720	.07735	.07752	.07767
13	0	11	.07692	.07694	.07696	.07698	.07696	.07690	.07679
13	0	12	.07692	.07690	.07680	.07663	.07636	.07600	.07554
13	0	13	.07692	.07675	.07627	.07555	.07465	.07360	.07245
13	1	1	.06787	.06660	.06479	.06237	.05922	.05523	.05031
13	1	2	.07179	.07152	.07126	.07105	.07092	.07093	.07110
13	1	3	.07354	.07353	.07362	.07385	.07424	.07484	.07568
13	1	4	.07490	.07503	.07530	.07573	.07635	.07718	.07825
13	1	5	.07606	.07628	.07664	.07717	.07788	.07878	.07989
13	1	6	.07713	.07739	.07780	.07835	.07907	.07994	.08098
13	1	7	.07815	.07842	.07883	.07937	.08003	.08081	.08171
13	1	8	.07916	.07942	.07979	.08026	.08082	.08145	.08216
13	1	9	.08018	.08041	.08070	.08106	.08147	.08191	.08237
13	1	10	.08125	.08140	.08158	.08178	.08198	.08216	.08233
13	1	11	.08237	.08239	.08239	.08236	.08228	.08215	.08195
13	1	12	.15759	.15763	.15730	.15666	.15576	.15462	.15328
13	2	1	.05523	.05262	.04939	.04544	.04067	.03497	.02823
13	2	2	.06479	.06427	.06380	.06342	.06316	.06308	.06321
13	2	3	.06908	.06908	.06923	.06956	.07009	.07087	.07194
13	2	4	.07237	.07267	.07313	.07379	.07467	.07578	.07716
13	2	5	.07521	.07566	.07629	.07710	.07810	.07932	.08075
13	2	6	.07780	.07833	.07902	.07986	.08087	.08204	.08338
13	2	7	.08027	.08082	.08149	.08229	.08320	.08424	.08538
13	2	8	.08270	.08320	.08379	.08447	.08522	.08603	.08690
13	2	9	.08515	.08554	.08598	.08646	.08697	.08750	.08803
13	2	10	.08766	.08786	.08807	.08828	.08846	.08862	.08874
13	2	11	.24973	.24995	.24981	.24935	.24858	.24755	.24627
13	3	1	.03799	.03375	.02876	.02294	.01618	.00838	-.00058
13	3	2	.05546	.05470	.05404	.05352	.05317	.05305	.05320
13	3	3	.06328	.06338	.06367	.06419	.06496	.06601	.06740

Coefficients a_i for the BLUE of μ
Shape parameter k

n	s	i	0.0	0.1	0.2	0.3	0.4	0.5	0.6
13	3	4	.06929	.06984	.07060	.07159	.07282	.07433	.07612
13	3	5	.07445	.07524	.07622	.07741	.07881	.08042	.08227
13	3	6	.07916	.08005	.08109	.08231	.08368	.08522	.08691
13	3	7	.08363	.08450	.08549	.08660	.08782	.08914	.09056
13	3	8	.08800	.08875	.08958	.09048	.09143	.09243	.09346
13	3	9	.09235	.09288	.09344	.09401	.09459	.09516	.09572
13	3	10	.35639	.35690	.35709	.35696	.35655	.35586	.35494
13	4	1	.01439	.00810	.00093	-.00721	-.01643	-.02684	-.03854
13	4	2	.04296	.04200	.04119	.04058	.04021	.04013	.04038
13	4	3	.05573	.05604	.05659	.05742	.05855	.06002	.06186
13	4	4	.06553	.06648	.06768	.06913	.07086	.07289	.07524
13	4	5	.07394	.07520	.07667	.07836	.08027	.08241	.08478
13	4	6	.08158	.08293	.03445	.08612	.08795	.08994	.09207
13	4	7	.08880	.09008	.09145	.09293	.09449	.09614	.09787
13	4	8	.09581	.09684	.09792	.09904	.10018	.10135	.10253
13	4	9	.48125	.48232	.48311	.48364	.48392	.48397	.48381
13	5	1	-.01853	-.02742	-.03735	-.04843	-.06077	-.07445	-.08960
13	5	2	.02591	.02482	.02395	.02335	.02306	.02314	.02364
13	5	3	.04575	.04644	.04743	.04876	.05045	.05253	.05504
13	5	4	.06095	.06250	.06432	.06644	.06888	.07163	.07473
13	5	5	.07395	.07587	.07801	.08037	.08297	.08579	.08885
13	5	6	.08573	.08770	.08981	.09208	.09448	.09702	.09969
13	5	7	.09680	.09856	.10039	.10229	.10425	.10627	.10833
13	5	8	.62944	.63154	.63343	.63514	.63668	.63807	.63932
13	6	1	-.06588	-.07815	-.09169	-.10659	-.12297	-.14091	-.16053
13	6	2	.00199	.00090	.00012	-.00029	-.00029	.00017	.00114
13	6	3	.03222	.03357	.03530	.03743	.03998	.04299	.04649
13	6	4	.05534	.05779	.06056	.06365	.06708	.07086	.07500
13	6	5	.07504	.07790	.08097	.08427	.08779	.09153	.09549
13	6	6	.09281	.09560	.09851	.10153	.10466	.10789	.11123
13	6	7	.80847	.81239	.81623	.82001	.82374	.82746	.83118
13	7	1	-.13709	-.15397	-.17240	-.19248	-.21432	-.23801	-.26366
13	7	2	-.03302	-.03384	-.03421	-.03410	-.03345	-.03222	-.03037
13	7	3	.01320	.01570	.01866	.02209	.02603	.03049	.03548
13	7	4	.04844	.05231	.05653	.06109	.06600	.07128	.07692
13	7	5	.07836	.08259	.08701	.09164	.09646	.10147	.10667
13	7	6	1.03012	1.03721	1.04441	1.05176	1.05928	1.06700	1.07496
14	0	1	.07143	.07123	.07059	.06942	.06760	.06504	.06161
14	0	2	.07143	.07140	.07135	.07129	.07126	.07130	.07144
14	0	3	.07143	.07144	.07151	.07166	.07192	.07234	.07294
14	0	4	.07143	.07147	.07160	.07185	.07224	.07280	.07357
14	0	5	.07143	.07148	.07164	.07194	.07239	.07300	.07380
14	0	6	.07143	.07149	.07167	.07198	.07244	.07305	.07381
14	0	7	.07143	.07149	.07167	.07198	.07242	.07298	.07367
14	0	8	.07143	.07149	.07166	.07195	.07234	.07282	.07341
14	0	9	.07143	.07148	.07164	.07188	.07220	.07259	.07303
14	0	10	.07143	.07147	.07160	.07178	.07201	.07227	.07255
14	0	11	.07143	.07146	.07153	.07164	.07175	.07185	.07193
14	0	12	.07143	.07144	.07144	.07143	.07138	.07128	.07113
14	0	13	.07143	.07140	.07129	.07111	.07082	.07045	.06998
14	0	14	.07143	.07126	.07080	.07011	.06924	.06824	.06714
14	1	1	.06366	.06252	.06086	.05860	.05563	.05183	.04710
14	1	2	.06691	.06664	.06636	.06611	.06592	.06583	.06588
14	1	3	.06835	.06831	.06835	.06850	.06880	.06929	.06998
14	1	4	.06945	.06954	.06975	.07010	.07062	.07135	.07229
14	1	5	.07039	.07056	.07086	.07131	.07194	.07275	.07376

Coefficients a_i for the BLUE of μ
Shape parameter k

n	s	i	0.0	0.1	0.2	0.3	0.4	0.5	0.6
14	1	6	.07125	.07146	.07181	.07230	.07295	.07376	.07474
14	1	7	.07206	.07229	.07265	.07315	.07377	.07452	.07541
14	1	8	.07285	.07308	.07343	.07389	.07444	.07510	.07584
14	1	9	.07364	.07386	.07416	.07455	.07500	.07552	.07607
14	1	10	.07446	.07463	.07487	.07515	.07546	.07579	.07613
14	1	11	.07531	.07542	.07554	.07568	.07580	.07591	.07599
14	1	12	.07620	.07620	.07617	.07609	.07597	.07579	.07555
14	1	13	.14549	.14552	.14520	.14457	.14368	.14257	.14126
14	2	1	.05302	.05072	.04781	.04421	.03982	.03451	.02817
14	2	2	.06085	.06033	.05983	.05940	.05905	.05885	.05882
14	2	3	.06432	.06426	.06432	.06452	.06491	.06551	.06637
14	2	4	.06697	.06717	.06751	.06803	.06875	.06969	.07088
14	2	5	.06924	.06958	.07009	.07076	.07163	.07269	.07397
14	2	6	.07129	.07172	.07230	.07304	.07393	.07500	.07624
14	2	7	.07323	.07370	.07429	.07501	.07587	.07685	.07796
14	2	8	.07512	.07557	.07613	.07679	.07754	.07838	.07929
14	2	9	.07701	.07740	.07787	.07841	.07900	.07964	.08031
14	2	10	.07893	.07922	.07955	.07991	.08028	.08066	.08103
14	2	11	.08090	.08103	.08116	.08127	.08137	.08142	.08144
14	2	12	.22911	.22930	.22914	.22864	.22785	.22680	.22551
14	3	1	.03884	.03512	.03071	.02549	.01936	.01222	.00395
14	3	2	.05293	.05216	.05145	.05084	.05036	.05006	.04999
14	3	3	.05917	.05914	.05927	.05959	.06013	.06092	.06201
14	3	4	.06394	.06431	.06487	.06563	.06662	.06786	.06938
14	3	5	.06801	.06861	.06939	.07036	.07154	.07293	.07456
14	3	6	.07169	.07240	.07328	.07432	.07553	.07691	.07847
14	3	7	.07516	.07590	.07678	.07778	.07891	.08017	.08154
14	3	8	.07852	.07922	.08001	.08090	.08186	.08290	.08401
14	3	9	.08185	.08243	.08307	.08376	.08448	.08523	.08600
14	3	10	.08521	.08559	.08599	.08639	.08679	.08717	.08753
14	3	11	.32469	.32512	.32520	.32496	.32443	.32362	.32257
14	4	1	.01991	.01447	.00820	.00101	-.00721	-.01658	-.02720
14	4	2	.04256	.04155	.04064	.03988	.03931	.03897	.03890
14	4	3	.05258	.05266	.05293	.05345	.05423	.05531	.05673
14	4	4	.06023	.06087	.06174	.06284	.06421	.06585	.06780
14	4	5	.06674	.06769	.06885	.07022	.07180	.07362	.07568
14	4	6	.07262	.07372	.07497	.07640	.07800	.07977	.08171
14	4	7	.07815	.07925	.08047	.08181	.08327	.08484	.08652
14	4	8	.08349	.08447	.08554	.08667	.08787	.08913	.09043
14	4	9	.08874	.08950	.09029	.09111	.09194	.09278	.09362
14	4	10	.43500	.43583	.43636	.43660	.43658	.43631	.43581
14	5	1	-.00568	-.01326	-.02182	-.03145	-.04226	-.05437	-.06788
14	5	2	.02880	.02758	.02653	.02567	.02506	.02475	.02477
14	5	3	.04404	.04434	.04489	.04572	.04687	.04838	.05026
14	5	4	.05566	.05672	.05804	.05963	.06151	.06370	.06621
14	5	5	.06553	.06698	.06865	.07054	.07267	.07504	.07765
14	5	6	.07444	.07602	.07778	.07970	.08178	.08403	.08645
14	5	7	.08277	.08430	.08594	.08769	.08953	.09147	.09350
14	5	8	.09076	.09207	.09343	.09484	.09628	.09776	.09926
14	5	9	.56368	.56524	.56657	.56766	.56855	.56925	.56977
14	6	1	-.04105	-.05134	-.06279	-.07551	-.08959	-.10515	-.12230
14	6	2	.01018	.00882	.00769	.00683	.00630	.00613	.00638
14	6	3	.03279	.03347	.03447	.03582	.03754	.03967	.04224
14	6	4	.04999	.05167	.05364	.05593	.05853	.06147	.06477
14	6	5	.06459	.06673	.06909	.07170	.07454	.07763	.08097
14	6	6	.07771	.07995	.08235	.08490	.08761	.09046	.09347

Coefficients a_i for the BLUE of μ
Shape parameter k

n	s	i	0.0	0.1	0.2	0.3	0.4	0.5	0.6
14	6	7	.08992	.09200	.09415	.09638	.09868	.10105	.10348
14	6	8	.71588	.71872	.72140	.72395	.72639	.72873	.73099
14	7	1	-.09154	-.10538	-.12061	-.13733	-.15566	-.17570	-.19755
14	7	2	-.01582	-.01719	-.01824	-.01893	-.01919	-.01899	-.01829
14	7	3	.01753	.01890	.02065	.02281	.02541	.02849	.03205
14	7	4	.04286	.04549	.04843	.05172	.05537	.05937	.06374
14	7	5	.06430	.06742	.07078	.07437	.07820	.08226	.08656
14	7	6	.08349	.08661	.08987	.09326	.09677	.10040	.10414
14	7	7	.89918	.90416	.90912	.91410	.91911	.92418	.92934
15	0	1	.06667	.06648	.06585	.06470	.06292	.06039	.05702
15	0	2	.06667	.06664	.06657	.06648	.06641	.06638	.06644
15	0	3	.06667	.06668	.06672	.06684	.06705	.06740	.06791
15	0	4	.06667	.06670	.06681	.06702	.06737	.06788	.06857
15	0	5	.06667	.06671	.06686	.06712	.06753	.06810	.06884
15	0	6	.06667	.06672	.06688	.06717	.06760	.06818	.06891
15	0	7	.06667	.06672	.06689	.06719	.06760	.06815	.06883
15	0	8	.06667	.06672	.06689	.06717	.06756	.06805	.06865
15	0	9	.06667	.06672	.06688	.06713	.06746	.06788	.06837
15	0	10	.06667	.06671	.06685	.06706	.06733	.06765	.06801
15	0	11	.06667	.06670	.06681	.06696	.06714	.06734	.06755
15	0	12	.06667	.06669	.06675	.06682	.06689	.06695	.06697
15	0	13	.06667	.06667	.06666	.06663	.06655	.06642	.06623
15	0	14	.06667	.06663	.06652	.06632	.06603	.06564	.06516
15	0	15	.06667	.06650	.06606	.06539	.06456	.06360	.06255
15	1	1	.05992	.05889	.05736	.05524	.05242	.04879	.04423
15	1	2	.06266	.06239	.06211	.06183	.06159	.06142	.06137
15	1	3	.06386	.06379	.06380	.06390	.06412	.06451	.06509
15	1	4	.06477	.06482	.06498	.06527	.06572	.06634	.06718
15	1	5	.06554	.06567	.06592	.06631	.06686	.06759	.06851
15	1	6	.06624	.06642	.06672	.06716	.06775	.06849	.06941
15	1	7	.06690	.06710	.06742	.06787	.06845	.06917	.07002
15	1	8	.06753	.06774	.06806	.06849	.06904	.06968	.07043
15	1	9	.06816	.06836	.06866	.06905	.06952	.07007	.07068
15	1	10	.06880	.06898	.06923	.06955	.06992	.07033	.07078
15	1	11	.06946	.06960	.06979	.07001	.07025	.07049	.07073
15	1	12	.07016	.07023	.07032	.07041	.07047	.07052	.07052
15	1	13	.07088	.07086	.07080	.07070	.07055	.07033	.07005
15	1	14	.13511	.13514	.13483	.13421	.13334	.13226	.13099
15	2	1	.05084	.04879	.04616	.04285	.03877	.03379	.02780
15	2	2	.05736	.05685	.05634	.05587	.05546	.05516	.05501
15	2	3	.06022	.06012	.06010	.06022	.06049	.06095	.06164
15	2	4	.06240	.06252	.06277	.06318	.06377	.06457	.06560
15	2	5	.06424	.06450	.06491	.06548	.06622	.06716	.06830
15	2	6	.06591	.06625	.06674	.06738	.06817	.06914	.07028
15	2	7	.06746	.06785	.06837	.06902	.06981	.07073	.07179
15	2	8	.06896	.06936	.06987	.07049	.07121	.07204	.07296
15	2	9	.07045	.07082	.07128	.07183	.07244	.07313	.07387
15	2	10	.07195	.07226	.07264	.07307	.07354	.07403	.07455
15	2	11	.07348	.07371	.07396	.07422	.07450	.07476	.07501
15	2	12	.07507	.07515	.07522	.07527	.07529	.07527	.07521
15	2	13	.21164	.21182	.21164	.21113	.21033	.20927	.20798
15	3	1	.03896	.03568	.03173	.02700	.02141	.01482	.00712
15	3	2	.05055	.04979	.04906	.04839	.04783	.04740	.04716
15	3	3	.05564	.05552	.05553	.05570	.05607	.05666	.05751
15	3	4	.05950	.05974	.06014	.06073	.06152	.06255	.06384
15	3	5	.06277	.06323	.06384	.06465	.06564	.06685	.06828

Coefficients a_i for the BLUE of μ
Shape parameter k

n	s	i	0.0	0.1	0.2	0.3	0.4	0.5	0.6
15	3	6	.06572	.06629	.06702	.06791	.06897	.07021	.07163
15	3	7	.06847	.06910	.06986	.07076	.07179	.07296	.07426
15	3	8	.07112	.07174	.07247	.07331	.07424	.07527	.07638
15	3	9	.07373	.07429	.07494	.07565	.07642	.07724	.07811
15	3	10	.07635	.07680	.07729	.07782	.07837	.07893	.07950
15	3	11	.07900	.07927	.07955	.07983	.08010	.08034	.08056
15	3	12	.29820	.29856	.29857	.29826	.29765	.29677	.29564
15	4	1	.02344	.01868	.01314	.00671	-.00070	-.00922	-.01896
15	4	2	.04180	.04078	.03983	.03899	.03828	.03776	.03746
15	4	3	.04985	.04977	.04986	.05015	.05067	.05146	.05255
15	4	4	.05596	.05639	.05701	.05785	.05893	.06027	.06190
15	4	5	.06113	.06185	.06276	.06387	.06520	.06675	.06854
15	4	6	.06578	.06665	.06770	.06891	.07030	.07187	.07362
15	4	7	.07011	.07104	.07210	.07330	.07463	.07608	.07767
15	4	8	.07428	.07517	.07615	.07724	.07841	.07966	.08099
15	4	9	.07835	.07913	.07996	.08084	.08177	.08274	.08373
15	4	10	.08241	.08298	.08357	.08418	.08478	.08539	.08598
15	4	11	.39689	.39756	.39791	.39796	.39773	.39725	.39652
15	5	1	.00298	-.00358	-.01105	-.01954	-.02916	-.04002	-.05222
15	5	2	.03045	.02919	.02804	.02704	.02623	.02566	.02536
15	5	3	.04249	.04253	.04277	.04327	.04404	.04512	.04654
15	5	4	.05162	.05233	.05328	.05448	.05594	.05770	.05977
15	5	5	.05933	.06043	.06173	.06325	.06501	.06700	.06924
15	5	6	.06626	.06753	.06897	.07059	.07239	.07437	.07653
15	5	7	.07270	.07400	.07543	.07699	.07867	.08047	.08239
15	5	8	.07886	.08006	.08135	.08272	.08415	.08565	.08721
15	5	9	.08487	.08586	.08688	.08794	.08902	.09011	.09122
15	5	10	.51044	.51165	.51258	.51327	.51372	.51395	.51397
15	6	1	-.02443	-.03322	-.04309	-.05413	-.06646	-.08019	-.09542
15	6	2	.01552	.01404	.01273	.01163	.01077	.01021	.00999
15	6	3	.03300	.03326	.03379	.03461	.03576	.03727	.03917
15	6	4	.04624	.04739	.04880	.05050	.05250	.05482	.05747
15	6	5	.05743	.05904	.06087	.06294	.06525	.06782	.07064
15	6	6	.06744	.06923	.07119	.07333	.07565	.07814	.08081
15	6	7	.07672	.07849	.08038	.08237	.08448	.08669	.08900
15	6	8	.08557	.08714	.08877	.09045	.09217	.09394	.09574
15	6	9	.64251	.64463	.64655	.64830	.64988	.65130	.65261
15	7	1	-.06206	-.07368	-.08656	-.10082	-.11657	-.13391	-.15296
15	7	2	-.00459	-.00622	-.00762	-.00874	-.00954	-.00997	-.00998
15	7	3	.02052	.02118	.02216	.02350	.02523	.02737	.02996
15	7	4	.03951	.04130	.04339	.04579	.04854	.05163	.05509
15	7	5	.05552	.05785	.06041	.06323	.06629	.06961	.07320
15	7	6	.06981	.07230	.07495	.07777	.08075	.08389	.08719
15	7	7	.08302	.08539	.08784	.09037	.09299	.09569	.09845
15	7	8	.79826	.80190	.80544	.80890	.81231	.81569	.81905
15	8	1	-.11547	-.13079	-.14761	-.16606	-.18625	-.20828	-.23226
15	8	2	-.03256	-.03422	-.03556	-.03652	-.03707	-.03716	-.03674
15	8	3	.00361	.00495	.00669	.00886	.01148	.01458	.01819
15	8	4	.03091	.03366	.03676	.04020	.04402	.04821	.05278
15	8	5	.05387	.05722	.06082	.06467	.06877	.07312	.07772
15	8	6	.07430	.07773	.08131	.08502	.08887	.09286	.09697
15	8	7	.98535	.99144	.99759	1.00382	1.01017	1.01667	1.02334
16	0	1	.06250	.06232	.06171	.06058	.05883	.05635	.05301
16	0	2	.06250	.06247	.06239	.06228	.06216	.06208	.06206
16	0	3	.06250	.06251	.06254	.06262	.06279	.06308	.06352
16	0	4	.06250	.06253	.06262	.06281	.06311	.06356	.06418

Coefficients a_i for the BLUE of μ
Shape parameter k

n	s	i	0.0	0.1	0.2	0.3	0.4	0.5	0.6
16	0	5	.06250	.06254	.06267	.06291	.06328	.06380	.06449
16	0	6	.06250	.06255	.06270	.06296	.06336	.06390	.06460
16	0	7	.06250	.06255	.06271	.06299	.06338	.06391	.06457
16	0	8	.06250	.06255	.06271	.06298	.06336	.06385	.06445
16	0	9	.06250	.06255	.06270	.06295	.06330	.06373	.06424
16	0	10	.06250	.06255	.06269	.06291	.06320	.06356	.06397
16	0	11	.06250	.06254	.06266	.06284	.06306	.06333	.06362
16	0	12	.06250	.06253	.06262	.06274	.06288	.06304	.06318
16	0	13	.06250	.06252	.06256	.06261	.06265	.06266	.06265
16	0	14	.06250	.06250	.06248	.06242	.06232	.06217	.06195
16	0	15	.06250	.06246	.06235	.06214	.06184	.06145	.06097
16	0	16	.06250	.06234	.06191	.06127	.06047	.05954	.05854
16	1	1	.05659	.05565	.05423	.05223	.04954	.04606	.04166
16	1	2	.05892	.05866	.05837	.05807	.05780	.05757	.05743
16	1	3	.05993	.05986	.05983	.05989	.06005	.06035	.06083
16	1	4	.06070	.06073	.06085	.06109	.06146	.06201	.06274
16	1	5	.06135	.06144	.06165	.06199	.06247	.06313	.06397
16	1	6	.06193	.06207	.06233	.06272	.06325	.06394	.06480
16	1	7	.06247	.06264	.06292	.06333	.06387	.06455	.06537
16	1	8	.06299	.06317	.06346	.06387	.06438	.06501	.06575
16	1	9	.06349	.06368	.06396	.06434	.06481	.06536	.06600
16	1	10	.06401	.06418	.06443	.06477	.06516	.06562	.06612
16	1	11	.06453	.06468	.06489	.06515	.06545	.06578	.06613
16	1	12	.06507	.06518	.06533	.06550	.06568	.06586	.06603
16	1	13	.06565	.06570	.06576	.06581	.06583	.06582	.06578
16	1	14	.06625	.06621	.06614	.06602	.06584	.06560	.06529
16	1	15	.12613	.12615	.12584	.12524	.12439	.12334	.12211
16	2	1	.04875	.04690	.04450	.04144	.03762	.03293	.02724
16	2	2	.05425	.05375	.05325	.05275	.05230	.05192	.05167
16	2	3	.05665	.05651	.05645	.05649	.05667	.05702	.05757
16	2	4	.05846	.05852	.05871	.05903	.05951	.06019	.06108
16	2	5	.05999	.06018	.06051	.06098	.06162	.06245	.06348
16	2	6	.06135	.06163	.06204	.06260	.06331	.06418	.06523
16	2	7	.06263	.06295	.06340	.06399	.06470	.06556	.06657
16	2	8	.06384	.06419	.06465	.06522	.06590	.06670	.06761
16	2	9	.06504	.06538	.06581	.06634	.06696	.06765	.06842
16	2	10	.06623	.06654	.06693	.06738	.06789	.06844	.06904
16	2	11	.06745	.06770	.06801	.06835	.06872	.06910	.06950
16	2	12	.06870	.06887	.06906	.06925	.06945	.06962	.06978
16	2	13	.07000	.07004	.07007	.07007	.07004	.06996	.06984
16	2	14	.19666	.19683	.19663	.19612	.19531	.19426	.19298
16	3	1	.03865	.03572	.03215	.02784	.02268	.01656	.00935
16	3	2	.04833	.04758	.04685	.04615	.04553	.04501	.04464
16	3	3	.05254	.05237	.05229	.05236	.05259	.05302	.05369
16	3	4	.05573	.05587	.05615	.05661	.05725	.05810	.05920
16	3	5	.05841	.05875	.05924	.05990	.06075	.06180	.06306
16	3	6	.06081	.06127	.06188	.06264	.06357	.06468	.06597
16	3	7	.06304	.06357	.06422	.06502	.06596	.06703	.06826
16	3	8	.06517	.06572	.06637	.06714	.06802	.06901	.07010
16	3	9	.06726	.06778	.06839	.06908	.06986	.07070	.07161
16	3	10	.06933	.06979	.07031	.07089	.07151	.07216	.07285
16	3	11	.07143	.07178	.07217	.07258	.07299	.07342	.07384
16	3	12	.07358	.07377	.07396	.07415	.07431	.07446	.07457
16	3	13	.27571	.27603	.27599	.27563	.27498	.27405	.27287
16	4	1	.02569	.02148	.01653	.01073	.00398	-.00384	-.01284
16	4	2	.04084	.03984	.03888	.03798	.03719	.03655	.03609

Coefficients a_i for the BLUE of μ
Shape parameter k

n	s	i	0.0	0.1	0.2	0.3	0.4	0.5	0.6
16	4	3	.04743	.04725	.04721	.04733	.04766	.04823	.04906
16	4	4	.05241	.05268	.05313	.05376	.05462	.05572	.05708
16	4	5	.05660	.05715	.05787	.05877	.05988	.06121	.06277
16	4	6	.06035	.06105	.06191	.06294	.06415	.06554	.06712
16	4	7	.06383	.06460	.06552	.06657	.06777	.06910	.07058
16	4	8	.06714	.06793	.06882	.06982	.07092	.07212	.07343
16	4	9	.07038	.07110	.07191	.07279	.07373	.07474	.07579
16	4	10	.07358	.07419	.07485	.07554	.07626	.07701	.07777
16	4	11	.07680	.07722	.07766	.07811	.07855	.07898	.07939
16	4	12	.36495	.36550	.36573	.36565	.36528	.36465	.36377
16	5	1	.00896	.00321	-.00340	-.01097	-.01962	-.02946	-.04060
16	5	2	.03132	.03006	.02887	.02779	.02685	.02610	.02558
16	5	3	.04104	.04090	.04094	.04119	.04168	.04244	.04351
16	5	4	.04837	.04884	.04952	.05042	.05156	.05298	.05469
16	5	5	.05454	.05537	.05639	.05762	.05907	.06075	.06269
16	5	6	.06005	.06107	.06226	.06362	.06517	.06691	.06884
16	5	7	.06516	.06625	.06748	.06885	.07035	.07200	.07378
16	5	8	.07001	.07108	.07225	.07352	.07488	.07633	.07786
16	5	9	.07473	.07569	.07671	.07778	.07891	.08008	.08129
16	5	10	.07938	.08014	.08092	.08172	.08253	.08335	.08417
16	5	11	.46643	.46739	.46807	.46847	.46861	.46852	.46820
16	6	1	-.01290	-.02052	-.02914	-.03887	-.04981	-.06209	-.07582
16	6	2	.01907	.01756	.01616	.01491	.01386	.01301	.01250
16	6	3	.03295	.03293	.03313	.03358	.03432	.03538	.03679
16	6	4	.04341	.04418	.04519	.04645	.04800	.04984	.05201
16	6	5	.05221	.05342	.05485	.05650	.05840	.06054	.06294
16	6	6	.06006	.06149	.06309	.06488	.06686	.06903	.07139
16	6	7	.06731	.06879	.07041	.07216	.07405	.07606	.07820
16	6	8	.07419	.07559	.07709	.07867	.08032	.08205	.08385
16	6	9	.08084	.08204	.08330	.08458	.08590	.08723	.08859
16	6	10	.58288	.58452	.58593	.58712	.58811	.58891	.58955
16	7	1	-.04200	-.05194	-.06304	-.07542	-.08920	-.10447	-.12136
16	7	2	.00302	.00128	-.00029	-.00165	-.00276	-.00358	-.00406
16	7	3	.02255	.02275	.02324	.02402	.02514	.02663	.02852
16	7	4	.03725	.03846	.03995	.04173	.04382	.04624	.04901
16	7	5	.04960	.05135	.05332	.05555	.05802	.06076	.06376
16	7	6	.06059	.06256	.06472	.06706	.06958	.07229	.07519
16	7	7	.07072	.07271	.07481	.07704	.07938	.08184	.08441
16	7	8	.08030	.08212	.08399	.08593	.08792	.08996	.09205
16	7	9	.71796	.72070	.72330	.72575	.72809	.73032	.73249
16	8	1	-.08172	-.09460	-.10885	-.12458	-.14191	-.16096	-.18183
16	8	2	-.01851	-.02041	-.02209	-.02348	-.02455	-.02525	-.02554
16	8	3	.00888	.00949	.01043	.01175	.01345	.01558	.01816
16	8	4	.02947	.03134	.03352	.03603	.03888	.04210	.04569
16	8	5	.04675	.04923	.05197	.05496	.05822	.06175	.06555
16	8	6	.06207	.06478	.06766	.07071	.07394	.07734	.08091
16	8	7	.07616	.07879	.08151	.08433	.08724	.09024	.09332
16	8	8	.87690	.88139	.88584	.89028	.89472	.89920	.90374
17	0	1	.05882	.05864	.05805	.05694	.05523	.05279	.04950
17	0	2	.05882	.05879	.05870	.05857	.05842	.05829	.05821
17	0	3	.05882	.05882	.05884	.05890	.05903	.05927	.05964
17	0	4	.05882	.05884	.05892	.05908	.05935	.05975	.06031
17	0	5	.05882	.05886	.05897	.05919	.05952	.06000	.06065
17	0	6	.05882	.05887	.05900	.05925	.05962	.06012	.06078
17	0	7	.05882	.05887	.05902	.05928	.05965	.06016	.06080
17	0	8	.05882	.05887	.05902	.05928	.05965	.06013	.06072

Coefficients a_i for the BLUE of μ
Shape parameter k

n	s	i	0.0	0.1	0.2	0.3	0.4	0.5	0.6
17	0	9	.05882	.05887	.05902	.05927	.05961	.06004	.06057
17	0	10	.05882	.05887	.05901	.05923	.05954	.05991	.06036
17	0	11	.05882	.05887	.05899	.05918	.05944	.05974	.06008
17	0	12	.05882	.05886	.05896	.05911	.05930	.05952	.05975
17	0	13	.05882	.05885	.05892	.05902	.05913	.05924	.05934
17	0	14	.05882	.05884	.05887	.05889	.05891	.05889	.05884
17	0	15	.05882	.05882	.05879	.05872	.05860	.05843	.05819
17	0	16	.05882	.05878	.05866	.05845	.05815	.05775	.05727
17	0	17	.05882	.05867	.05825	.05763	.05686	.05597	.05500
17	1	1	.05360	.05274	.05141	.04952	.04695	.04360	.03933
17	1	2	.05561	.05536	.05507	.05475	.05444	.05417	.05395
17	1	3	.05647	.05639	.05634	.05636	.05647	.05670	.05709
17	1	4	.05712	.05714	.05722	.05742	.05773	.05821	.05886
17	1	5	.05767	.05775	.05792	.05821	.05864	.05923	.05999
17	1	6	.05816	.05828	.05850	.05885	.05933	.05996	.06076
17	1	7	.05861	.05876	.05901	.05938	.05988	.06052	.06130
17	1	8	.05904	.05920	.05947	.05984	.06034	.06094	.06166
17	1	9	.05946	.05963	.05989	.06025	.06071	.06126	.06190
17	1	10	.05988	.06004	.06029	.06062	.06103	.06150	.06204
17	1	11	.06030	.06045	.06067	.06095	.06129	.06167	.06208
17	1	12	.06074	.06086	.06104	.06126	.06150	.06177	.06204
17	1	13	.06120	.06128	.06140	.06153	.06166	.06178	.06189
17	1	14	.06168	.06171	.06174	.06176	.06175	.06170	.06161
17	1	15	.06218	.06213	.06205	.06191	.06171	.06145	.06113
17	1	16	.11827	.11829	.11799	.11740	.11657	.11554	.11434
17	2	1	.04676	.04508	.04287	.04003	.03644	.03200	.02657
17	2	2	.05146	.05098	.05048	.04997	.04948	.04905	.04871
17	2	3	.05349	.05333	.05323	.05322	.05333	.05358	.05402
17	2	4	.05502	.05504	.05517	.05542	.05582	.05639	.05717
17	2	5	.05630	.05645	.05671	.05711	.05766	.05839	.05932
17	2	6	.05744	.05767	.05801	.05850	.05913	.05992	.06089
17	2	7	.05850	.05877	.05917	.05969	.06034	.06114	.06208
17	2	8	.05951	.05980	.06021	.06074	.06138	.06214	.06301
17	2	9	.06049	.06079	.06119	.06169	.06229	.06297	.06374
17	2	10	.06146	.06175	.06212	.06257	.06309	.06368	.06432
17	2	11	.06243	.06269	.06302	.06339	.06381	.06427	.06475
17	2	12	.06344	.06364	.06389	.06416	.06445	.06475	.06505
17	2	13	.06448	.06460	.06474	.06488	.06501	.06512	.06521
17	2	14	.06556	.06557	.06556	.06553	.06545	.06534	.06517
17	2	15	.18367	.18383	.18362	.18311	.18231	.18126	.17999
17	3	1	.03806	.03543	.03219	.02823	.02344	.01772	.01092
17	3	2	.04626	.04554	.04481	.04410	.04344	.04285	.04239
17	3	3	.04980	.04959	.04945	.04944	.04956	.04987	.05038
17	3	4	.05247	.05254	.05273	.05307	.05359	.05431	.05525
17	3	5	.05470	.05495	.05534	.05589	.05661	.05752	.05865
17	3	6	.05669	.05706	.05757	.05822	.05904	.06003	.06120
17	3	7	.05853	.05897	.05953	.06024	.06108	.06207	.06321
17	3	8	.06028	.06074	.06133	.06203	.06285	.06378	.06483
17	3	9	.06197	.06244	.06301	.06366	.06441	.06525	.06616
17	3	10	.06365	.06409	.06460	.06518	.06582	.06652	.06726
17	3	11	.06533	.06571	.06613	.06660	.06710	.06762	.06816
17	3	12	.06705	.06732	.06762	.06794	.06825	.06857	.06887
17	3	13	.06881	.06894	.06907	.06918	.06928	.06934	.06937
17	3	14	.25640	.25668	.25661	.25622	.25553	.25457	.25336
17	4	1	.02708	.02332	.01886	.01359	.00740	.00017	-.00822
17	4	2	.03978	.03881	.03785	.03693	.03609	.03536	.03477

Coefficients a_i for the BLUE of μ
Shape parameter k

n	s	i	0.0	0.1	0.2	0.3	0.4	0.5	0.6
17	4	3	.04527	.04501	.04488	.04488	.04506	.04545	.04608
17	4	4	.04939	.04954	.04985	.05033	.05101	.05191	.05306
17	4	5	.05285	.05325	.05382	.05456	.05549	.05663	.05800
17	4	6	.05592	.05648	.05720	.05807	.05912	.06035	.06177
17	4	7	.05876	.05941	.06019	.06112	.06219	.06340	.06477
17	4	8	.06145	.06213	.06292	.06383	.06486	.06599	.06724
17	4	9	.06406	.06472	.06547	.06631	.06723	.06823	.06930
17	4	10	.06664	.06723	.06789	.06861	.06938	.07019	.07103
17	4	11	.06921	.06969	.07021	.07076	.07133	.07191	.07249
17	4	12	.07181	.07213	.07246	.07278	.07310	.07340	.07368
17	4	13	.33778	.33826	.33840	.33822	.33776	.33702	.33603
17	5	1	.01315	.00806	.00216	-.00466	-.01251	-.02151	-.03178
17	5	2	.03167	.03044	.02924	.02812	.02711	.02623	.02555
17	5	3	.03966	.03941	.03930	.03937	.03965	.04017	.04097
17	5	4	.04567	.04596	.04643	.04710	.04800	.04915	.05057
17	5	5	.05070	.05132	.05212	.05311	.05432	.05575	.05742
17	5	6	.05517	.05599	.05696	.05811	.05945	.06097	.06269
17	5	7	.05930	.06021	.06126	.06245	.06379	.06528	.06692
17	5	8	.06320	.06413	.06517	.06632	.06759	.06895	.07043
17	5	9	.06698	.06786	.06882	.06986	.07096	.07214	.07338
17	5	10	.07069	.07145	.07227	.07312	.07402	.07494	.07588
17	5	11	.07437	.07496	.07556	.07617	.07679	.07740	.07800
17	5	12	.42944	.43023	.43072	.43091	.43084	.43052	.42996
17	6	1	-.00469	-.01137	-.01900	-.02766	-.03749	-.04859	-.06109
17	6	2	.02143	.01993	.01850	.01719	.01601	.01503	.01426
17	6	3	.03269	.03248	.03246	.03265	.03309	.03382	.03485
17	6	4	.04114	.04164	.04235	.04329	.04448	.04596	.04773
17	6	5	.04822	.04913	.05024	.05157	.05313	.05493	.05698
17	6	6	.05451	.05564	.05696	.05846	.06015	.06204	.06412
17	6	7	.06029	.06153	.06291	.06444	.06611	.06792	.06988
17	6	8	.06576	.06699	.06833	.06977	.07131	.07294	.07467
17	6	9	.07103	.07216	.07336	.07462	.07594	.07730	.07872
17	6	10	.07619	.07713	.07810	.07910	.08011	.08113	.08216
17	6	11	.53343	.53474	.53579	.53659	.53716	.53753	.53770
17	7	1	-.02784	-.03647	-.04618	-.05709	-.06930	-.08293	-.09812
17	7	2	.00831	.00656	.00491	.00342	.00212	.00105	.00026
17	7	3	.02389	.02380	.02395	.02435	.02504	.02606	.02743
17	7	4	.03558	.03638	.03743	.03874	.04035	.04226	.04450
17	7	5	.04536	.04666	.04819	.04996	.05197	.05425	.05679
17	7	6	.05402	.05559	.05734	.05929	.06142	.06376	.06630
17	7	7	.06199	.06364	.06544	.06737	.06944	.07165	.07399
17	7	8	.06950	.07110	.07279	.07457	.07643	.07837	.08039
17	7	9	.07672	.07813	.07960	.08110	.08264	.08421	.08580
17	7	10	.65248	.65461	.65654	.65830	.65989	.66133	.66266
17	8	1	-.05851	-.06954	-.08183	-.09549	-.11064	-.12739	-.14587
17	8	2	-.00879	-.01079	-.01262	-.01424	-.01561	-.01670	-.01744
17	8	3	.01261	.01275	.01317	.01391	.01499	.01644	.01830
17	8	4	.02864	.02990	.03144	.03328	.03544	.03794	.04079
17	8	5	.04205	.04390	.04600	.04836	.05097	.05386	.05702
17	8	6	.05391	.05605	.05837	.06089	.06360	.06650	.06960
17	8	7	.06479	.06697	.06929	.07172	.07429	.07697	.07978
17	8	8	.07502	.07706	.07917	.08135	.08359	.08589	.08824
17	8	9	.79028	.79369	.79700	.80022	.80338	.80649	.80958
17	9	1	-.10021	-.11428	-.12983	-.14696	-.16581	-.18648	-.20910
17	9	2	-.03167	-.03385	-.03579	-.03747	-.03881	-.03980	-.04038
17	9	3	-.00221	-.00166	-.00076	.00050	.00217	.00426	.00682

Coefficients a_i for the BLUE of μ
Shape parameter k

n	s	i	0.0	0.1	0.2	0.3	0.4	0.5	0.6
17	9	4	.01985	.02177	.02402	.02661	.02955	.03286	.03656
17	9	5	.03826	.04088	.04376	.04691	.05034	.05404	.05803
17	9	6	.05451	.05741	.06049	.06376	.06721	.07085	.07467
17	9	7	.06938	.07225	.07522	.07830	.08148	.08475	.08812
17	9	8	.95208	.95747	.96289	.96834	.97388	.97951	.98528
18	0	1	.05556	.05538	.05480	.05372	.05203	.04963	.04639
18	0	2	.05556	.05552	.05542	.05527	.05510	.05493	.05478
18	0	3	.05556	.05555	.05556	.05560	.05570	.05588	.05619
18	0	4	.05556	.05557	.05564	.05577	.05601	.05636	.05687
18	0	5	.05556	.05558	.05569	.05588	.05618	.05662	.05722
18	0	6	.05556	.05559	.05572	.05594	.05628	.05676	.05738
18	0	7	.05556	.05560	.05573	.05598	.05633	.05681	.05743
18	0	8	.05556	.05560	.05574	.05599	.05634	.05680	.05738
18	0	9	.05556	.05560	.05574	.05598	.05632	.05675	.05728
18	0	10	.05556	.05560	.05574	.05596	.05627	.05665	.05712
18	0	11	.05556	.05560	.05572	.05593	.05619	.05652	.05690
18	0	12	.05556	.05559	.05570	.05587	.05609	.05635	.05664
18	0	13	.05556	.05559	.05568	.05580	.05596	.05614	.05632
18	0	14	.05556	.05558	.05564	.05571	.05580	.05587	.05593
18	0	15	.05556	.05557	.05558	.05559	.05558	.05554	.05546
18	0	16	.05556	.05555	.05551	.05543	.05530	.05510	.05485
18	0	17	.05556	.05552	.05539	.05517	.05487	.05447	.05399
18	0	18	.05556	.05541	.05500	.05440	.05366	.05280	.05187
18	1	1	.05090	.05011	.04886	.04706	.04460	.04137	.03723
18	1	2	.05265	.05241	.05212	.05179	.05146	.05114	.05087
18	1	3	.05340	.05330	.05324	.05323	.05329	.05347	.05379
18	1	4	.05396	.05395	.05402	.05417	.05444	.05485	.05543
18	1	5	.05443	.05448	.05462	.05487	.05525	.05578	.05648
18	1	6	.05484	.05494	.05513	.05544	.05588	.05646	.05720
18	1	7	.05523	.05535	.05557	.05591	.05637	.05697	.05770
18	1	8	.05559	.05573	.05597	.05632	.05678	.05736	.05806
18	1	9	.05594	.05609	.05633	.05667	.05712	.05766	.05829
18	1	10	.05629	.05643	.05667	.05699	.05740	.05788	.05844
18	1	11	.05663	.05677	.05699	.05728	.05763	.05804	.05850
18	1	12	.05699	.05712	.05730	.05755	.05783	.05815	.05849
18	1	13	.05736	.05746	.05761	.05779	.05798	.05819	.05840
18	1	14	.05774	.05781	.05790	.05800	.05809	.05817	.05823
18	1	15	.05815	.05817	.05818	.05818	.05814	.05806	.05794
18	1	16	.05858	.05852	.05843	.05827	.05807	.05779	.05746
18	1	17	.11133	.11135	.11105	.11048	.10967	.10867	.10751
18	2	1	.04488	.04335	.04131	.03865	.03527	.03104	.02584
18	2	2	.04894	.04848	.04798	.04747	.04696	.04648	.04607
18	2	3	.05068	.05051	.05038	.05033	.05037	.05055	.05089
18	2	4	.05199	.05198	.05206	.05225	.05258	.05307	.05375
18	2	5	.05308	.05318	.05339	.05373	.05421	.05486	.05569
18	2	6	.05404	.05422	.05451	.05493	.05550	.05622	.05711
18	2	7	.05493	.05516	.05550	.05597	.05656	.05730	.05818
18	2	8	.05578	.05603	.05640	.05688	.05747	.05819	.05903
18	2	9	.05659	.05686	.05723	.05770	.05826	.05893	.05969
18	2	10	.05739	.05766	.05801	.05845	.05897	.05956	.06021
18	2	11	.05819	.05844	.05876	.05915	.05959	.06008	.06062
18	2	12	.05901	.05923	.05950	.05981	.06016	.06053	.06092
18	2	13	.05985	.06002	.06021	.06043	.06066	.06089	.06111
18	2	14	.06072	.06082	.06092	.06101	.06109	.06115	.06118
18	2	15	.06163	.06162	.06159	.06152	.06142	.06127	.06108
18	2	16	.17230	.17245	.17224	.17172	.17093	.16989	.16865

Coefficients a_i for the BLUE of μ
Shape parameter k

n	s	i	0.0	0.1	0.2	0.3	0.4	0.5	0.6
18	3	1	.03732	.03494	.03197	.02831	.02384	.01846	.01203
18	3	2	.04434	.04365	.04293	.04222	.04153	.04089	.04035
18	3	3	.04735	.04711	.04693	.04685	.04690	.04710	.04749
18	3	4	.04961	.04963	.04975	.05001	.05042	.05102	.05182
18	3	5	.05150	.05168	.05199	.05244	.05306	.05385	.05485
18	3	6	.05317	.05346	.05388	.05445	.05517	.05605	.05712
18	3	7	.05470	.05507	.05556	.05618	.05694	.05784	.05890
18	3	8	.05616	.05656	.05708	.05771	.05847	.05934	.06034
18	3	9	.05756	.05797	.05849	.05911	.05982	.06062	.06152
18	3	10	.05894	.05934	.05983	.06039	.06103	.06174	.06250
18	3	11	.06031	.06068	.06111	.06160	.06213	.06271	.06332
18	3	12	.06170	.06201	.06236	.06274	.06314	.06355	.06398
18	3	13	.06313	.06334	.06357	.06381	.06405	.06428	.06449
18	3	14	.06459	.06468	.06476	.06482	.06486	.06486	.06483
18	3	15	.23962	.23988	.23979	.23937	.23866	.23768	.23646
18	4	1	.02789	.02451	.02046	.01563	.00991	.00319	-.00467
18	4	2	.03868	.03774	.03679	.03587	.03499	.03420	.03352
18	4	3	.04330	.04300	.04280	.04271	.04278	.04303	.04349
18	4	4	.04677	.04683	.04703	.04739	.04792	.04866	.04963
18	4	5	.04966	.04996	.05040	.05101	.05179	.05278	.05398
18	4	6	.05222	.05267	.05326	.05400	.05492	.05601	.05729
18	4	7	.05458	.05511	.05578	.05659	.05755	.05866	.05992
18	4	8	.05680	.05738	.05808	.05890	.05984	.06090	.06208
18	4	9	.05894	.05953	.06021	.06100	.06187	.06284	.06389
18	4	10	.06104	.06160	.06223	.06294	.06371	.06454	.06542
18	4	11	.06313	.06362	.06417	.06476	.06538	.06604	.06672
18	4	12	.06524	.06562	.06604	.06647	.06692	.06736	.06781
18	4	13	.06737	.06761	.06785	.06809	.06831	.06851	.06869
18	4	14	.31439	.31481	.31489	.31465	.31411	.31330	.31224
18	5	1	.01610	.01156	.00625	.00006	-.00713	-.01542	-.02494
18	5	2	.03168	.03049	.02931	.02817	.02711	.02615	.02535
18	5	3	.03836	.03802	.03781	.03775	.03787	.03821	.03879
18	5	4	.04335	.04351	.04382	.04432	.04502	.04595	.04714
18	5	5	.04752	.04798	.04861	.04941	.05042	.05164	.05310
18	5	6	.05121	.05186	.05267	.05364	.05479	.05613	.05767
18	5	7	.05460	.05536	.05625	.05730	.05849	.05983	.06134
18	5	8	.05780	.05860	.05952	.06055	.06171	.06298	.06437
18	5	9	.06087	.06166	.06254	.06352	.06458	.06572	.06694
18	5	10	.06388	.06461	.06540	.06626	.06717	.06813	.06913
18	5	11	.06686	.06748	.06814	.06882	.06953	.07026	.07101
18	5	12	.06985	.07030	.07077	.07123	.07170	.07216	.07260
18	5	13	.39790	.39857	.39892	.39897	.39874	.39825	.39752
18	6	1	.00128	-.00464	-.01145	-.01925	-.02815	-.03828	-.04976
18	6	2	.02299	.02152	.02010	.01875	.01751	.01641	.01549
18	6	3	.03228	.03195	.03177	.03178	.03200	.03247	.03321
18	6	4	.03923	.03953	.04002	.04071	.04164	.04282	.04429
18	6	5	.04504	.04572	.04658	.04765	.04894	.05046	.05223
18	6	6	.05017	.05108	.05215	.05341	.05486	.05650	.05835
18	6	7	.05487	.05590	.05708	.05840	.05987	.06151	.06329
18	6	8	.05930	.06037	.06154	.06284	.06425	.06577	.06740
18	6	9	.06356	.06458	.06569	.06688	.06814	.06948	.07089
18	6	10	.06771	.06862	.06959	.07061	.07166	.07275	.07387
18	6	11	.07181	.07255	.07331	.07408	.07487	.07566	.07644
18	6	12	.49176	.49282	.49361	.49413	.49441	.49445	.49429
18	7	1	-.01758	-.02516	-.03374	-.04346	-.05441	-.06672	-.08051
18	7	2	.01206	.01033	.00867	.00711	.00570	.00447	.00347

Coefficients a_i for the BLUE of μ
Shape parameter k

n	s	i	0.0	0.1	0.2	0.3	0.4	0.5	0.6
18	7	3	.02475	.02446	.02437	.02450	.02488	.02555	.02654
18	7	4	.03423	.03474	.03547	.03644	.03767	.03919	.04101
18	7	5	.04214	.04312	.04430	.04572	.04736	.04926	.05142
18	7	6	.04912	.05037	.05180	.05342	.05523	.05725	.05948
18	7	7	.05552	.05689	.05841	.06008	.06190	.06387	.06600
18	7	8	.06154	.06292	.06441	.06601	.06772	.06952	.07143
18	7	9	.06730	.06859	.06996	.07139	.07289	.07444	.07605
18	7	10	.07290	.07402	.07517	.07635	.07755	.07877	.08001
18	7	11	.59802	.59971	.60118	.60244	.60350	.60438	.60510
18	8	1	-.04195	-.05153	-.06228	-.07431	-.08773	-.10267	-.11925
18	8	2	-.00187	-.00387	-.00576	-.00750	-.00905	-.01037	-.01142
18	8	3	.01527	.01511	.01518	.01552	.01615	.01712	.01844
18	8	4	.02807	.02890	.02997	.03132	.03297	.03493	.03723
18	8	5	.03874	.04013	.04175	.04361	.04573	.04811	.05077
18	8	6	.04816	.04985	.05173	.05381	.05609	.05857	.06127
18	8	7	.05677	.05857	.06052	.06262	.06486	.06725	.06978
18	8	8	.06484	.06661	.06848	.07045	.07250	.07464	.07687
18	8	9	.07255	.07416	.07583	.07754	.07929	.08107	.08289
18	8	10	.71942	.72208	.72459	.72696	.72921	.73135	.73342
18	9	1	-.07409	-.08616	-.09957	-.11445	-.13092	-.14909	-.16908
18	9	2	-.01998	-.02224	-.02433	-.02621	-.02785	-.02920	-.03023
18	9	3	.00313	.00320	.00356	.00423	.00525	.00665	.00846
18	9	4	.02039	.02167	.02325	.02513	.02734	.02991	.03283
18	9	5	.03475	.03670	.03890	.04137	.04410	.04712	.05043
18	9	6	.04741	.04968	.05216	.05483	.05771	.06079	.06407
18	9	7	.05896	.06132	.06381	.06645	.06921	.07211	.07513
18	9	8	.06976	.07201	.07434	.07674	.07921	.08175	.08435
18	9	9	.85968	.86381	.86788	.87192	.87594	.87997	.88404
19	0	1	.05263	.05246	.05189	.05083	.04918	.04682	.04362
19	0	2	.05263	.05259	.05249	.05232	.05213	.05192	.05173
19	0	3	.05263	.05263	.05262	.05264	.05271	.05286	.05311
19	0	4	.05263	.05264	.05270	.05281	.05302	.05333	.05379
19	0	5	.05263	.05266	.05275	.05292	.05320	.05360	.05415
19	0	6	.05263	.05267	.05278	.05298	.05330	.05374	.05433
19	0	7	.05263	.05267	.05280	.05302	.05336	.05381	.05440
19	0	8	.05263	.05267	.05281	.05304	.05337	.05382	.05439
19	0	9	.05263	.05268	.05281	.05304	.05337	.05379	.05432
19	0	10	.05263	.05268	.05281	.05303	.05333	.05372	.05419
19	0	11	.05263	.05267	.05280	.05300	.05328	.05362	.05402
19	0	12	.05263	.05267	.05279	.05296	.05320	.05349	.05381
19	0	13	.05263	.05267	.05276	.05291	.05310	.05332	.05356
19	0	14	.05263	.05266	.05274	.05284	.05297	.05311	.05325
19	0	15	.05263	.05265	.05270	.05276	.05282	.05286	.05289
19	0	16	.05263	.05264	.05265	.05264	.05261	.05255	.05244
19	0	17	.05263	.05262	.05257	.05248	.05234	.05214	.05187
19	0	18	.05263	.05259	.05246	.05224	.05194	.05154	.05106
19	0	19	.05263	.05249	.05209	.05151	.05079	.04997	.04907
19	1	1	.04846	.04773	.04655	.04484	.04247	.03934	.03532
19	1	2	.04999	.04976	.04947	.04914	.04878	.04843	.04811
19	1	3	.05064	.05055	.05047	.05043	.05046	.05058	.05083
19	1	4	.05113	.05112	.05116	.05128	.05150	.05186	.05237
19	1	5	.05154	.05157	.05169	.05191	.05225	.05272	.05336
19	1	6	.05190	.05197	.05214	.05241	.05281	.05335	.05404
19	1	7	.05222	.05233	.05252	.05283	.05326	.05382	.05451
19	1	8	.05253	.05265	.05287	.05319	.05363	.05418	.05485
19	1	9	.05283	.05296	.05318	.05351	.05393	.05445	.05508

Coefficients a_i for the BLUE of μ
Shape parameter k

n	s	i	0.0	0.1	0.2	0.3	0.4	0.5	0.6
19	1	10	.05312	.05326	.05348	.05379	.05419	.05467	.05523
19	1	11	.05341	.05354	.05376	.05404	.05440	.05482	.05530
19	1	12	.05371	.05383	.05402	.05427	.05458	.05493	.05532
19	1	13	.05401	.05412	.05428	.05448	.05472	.05499	.05527
19	1	14	.05432	.05441	.05453	.05468	.05483	.05500	.05515
19	1	15	.05465	.05471	.05477	.05484	.05491	.05495	.05496
19	1	16	.05500	.05501	.05501	.05498	.05492	.05481	.05467
19	1	17	.05536	.05531	.05520	.05504	.05482	.05454	.05420
19	1	18	.10516	.10519	.10489	.10433	.10354	.10257	.10144
19	2	1	.04312	.04172	.03982	.03732	.03412	.03008	.02509
19	2	2	.04665	.04621	.04573	.04521	.04468	.04417	.04371
19	2	3	.04816	.04798	.04783	.04774	.04774	.04785	.04811
19	2	4	.04929	.04926	.04930	.04945	.04971	.05013	.05072
19	2	5	.05023	.05030	.05046	.05075	.05117	.05174	.05249
19	2	6	.05105	.05119	.05144	.05181	.05231	.05296	.05378
19	2	7	.05181	.05200	.05230	.05271	.05326	.05394	.05476
19	2	8	.05253	.05275	.05307	.05351	.05406	.05473	.05553
19	2	9	.05321	.05345	.05378	.05422	.05476	.05540	.05614
19	2	10	.05388	.05412	.05446	.05487	.05537	.05596	.05662
19	2	11	.05455	.05478	.05510	.05548	.05593	.05643	.05700
19	2	12	.05522	.05544	.05572	.05605	.05642	.05684	.05728
19	2	13	.05591	.05609	.05632	.05658	.05687	.05717	.05748
19	2	14	.05662	.05676	.05692	.05709	.05726	.05743	.05759
19	2	15	.05737	.05743	.05750	.05756	.05760	.05762	.05760
19	2	16	.05814	.05811	.05806	.05798	.05785	.05768	.05746
19	2	17	.16225	.16240	.16219	.16168	.16089	.15987	.15864
19	3	1	.03648	.03431	.03158	.02817	.02399	.01890	.01279
19	3	2	.04256	.04189	.04119	.04048	.03977	.03910	.03849
19	3	3	.04515	.04489	.04468	.04455	.04453	.04464	.04493
19	3	4	.04708	.04706	.04712	.04731	.04764	.04814	.04883
19	3	5	.04869	.04882	.04906	.04944	.04996	.05066	.05155
19	3	6	.05011	.05034	.05070	.05118	.05182	.05261	.05358
19	3	7	.05141	.05172	.05214	.05268	.05337	.05420	.05518
19	3	8	.05264	.05298	.05344	.05401	.05470	.05552	.05647
19	3	9	.05381	.05418	.05464	.05521	.05588	.05665	.05753
19	3	10	.05496	.05533	.05578	.05632	.05694	.05764	.05841
19	3	11	.05610	.05645	.05687	.05735	.05790	.05850	.05914
19	3	12	.05725	.05756	.05792	.05833	.05877	.05925	.05975
19	3	13	.05841	.05866	.05895	.05926	.05958	.05991	.06024
19	3	14	.05961	.05978	.05995	.06014	.06031	.06047	.06061
19	3	15	.06084	.06089	.06094	.06096	.06095	.06091	.06083
19	3	16	.22491	.22515	.22505	.22462	.22389	.22290	.22168
19	4	1	.02829	.02524	.02154	.01709	.01178	.00548	-.00192
19	4	2	.03756	.03666	.03574	.03482	.03392	.03308	.03233
19	4	3	.04151	.04118	.04092	.04077	.04075	.04089	.04122
19	4	4	.04445	.04445	.04457	.04483	.04525	.04585	.04668
19	4	5	.04691	.04712	.04746	.04796	.04862	.04948	.05054
19	4	6	.04907	.04942	.04991	.05055	.05134	.05231	.05347
19	4	7	.05105	.05149	.05207	.05278	.05363	.05464	.05580
19	4	8	.05291	.05341	.05402	.05476	.05562	.05660	.05771
19	4	9	.05469	.05521	.05583	.05655	.05737	.05829	.05931
19	4	10	.05643	.05694	.05754	.05821	.05896	.05978	.06067
19	4	11	.05816	.05863	.05917	.05976	.06041	.06110	.06183
19	4	12	.05988	.06029	.06074	.06123	.06174	.06227	.06282
19	4	13	.06163	.06194	.06227	.06262	.06296	.06331	.06364
19	4	14	.06342	.06359	.06377	.06393	.06408	.06420	.06430

Coefficients a_i for the BLUE of μ
Shape parameter k

n	s	i	0.0	0.1	0.2	0.3	0.4	0.5	0.6
19	4	15	.29404	.29442	.29445	.29416	.29357	.29271	.29160
19	5	1	.01820	.01411	.00929	.00363	-.00299	-.01068	-.01957
19	5	2	.03147	.03031	.02916	.02802	.02694	.02593	.02504
19	5	3	.03711	.03673	.03645	.03629	.03629	.03647	.03688
19	5	4	.04132	.04138	.04158	.04193	.04248	.04323	.04423
19	5	5	.04483	.04516	.04565	.04631	.04715	.04819	.04946
19	5	6	.04792	.04844	.04910	.04992	.05092	.05210	.05347
19	5	7	.05075	.05137	.05214	.05304	.05410	.05532	.05669
19	5	8	.05340	.05409	.05489	.05582	.05687	.05805	.05935
19	5	9	.05594	.05664	.05744	.05834	.05933	.06042	.06160
19	5	10	.05842	.05909	.05984	.06066	.06156	.06251	.06353
19	5	11	.06087	.06147	.06213	.06284	.06359	.06438	.06519
19	5	12	.06331	.06381	.06434	.06489	.06546	.06604	.06662
19	5	13	.06577	.06612	.06648	.06683	.06718	.06752	.06784
19	5	14	.37070	.37127	.37152	.37147	.37112	.37051	.36966
19	6	1	.00569	.00039	-.00574	-.01281	-.02094	-.03026	-.04087
19	6	2	.02399	.02258	.02118	.01982	.01854	.01737	.01634
19	6	3	.03178	.03136	.03108	.03094	.03100	.03127	.03179
19	6	4	.03758	.03773	.03805	.03855	.03927	.04022	.04143
19	6	5	.04241	.04292	.04359	.04445	.04552	.04681	.04833
19	6	6	.04667	.04739	.04828	.04934	.05058	.05201	.05365
19	6	7	.05056	.05141	.05241	.05355	.05486	.05632	.05794
19	6	8	.05421	.05512	.05615	.05730	.05858	.05999	.06151
19	6	9	.05770	.05861	.05961	.06071	.06190	.06318	.06454
19	6	10	.06109	.06194	.06287	.06385	.06490	.06600	.06715
19	6	11	.06443	.06518	.06596	.06679	.06764	.06852	.06941
19	6	12	.06775	.06834	.06894	.06955	.07016	.07076	.07136
19	6	13	.45615	.45704	.45764	.45795	.45801	.45782	.45740
19	7	1	-.00996	-.01669	-.02436	-.03309	-.04301	-.05422	-.06686
19	7	2	.01474	.01306	.01141	.00983	.00836	.00704	.00588
19	7	3	.02525	.02484	.02458	.02452	.02467	.02507	.02576
19	7	4	.03308	.03338	.03387	.03457	.03551	.03672	.03821
19	7	5	.03959	.04032	.04124	.04237	.04373	.04532	.04717
19	7	6	.04532	.04631	.04748	.04883	.05038	.05213	.05409
19	7	7	.05056	.05169	.05298	.05442	.05601	.05777	.05969
19	7	8	.05546	.05664	.05795	.05938	.06092	.06258	.06437
19	7	9	.06014	.06130	.06254	.06388	.06529	.06678	.06834
19	7	10	.06468	.06574	.06686	.06802	.06923	.07048	.07177
19	7	11	.06914	.07003	.07095	.07188	.07283	.07379	.07475
19	7	12	.55200	.55338	.55451	.55540	.55607	.55654	.55682
19	8	1	-.02978	-.03821	-.04772	-.05844	-.07047	-.08394	-.09898
19	8	2	.00317	.00120	-.00069	-.00248	-.00413	-.00560	-.00685
19	8	3	.01718	.01681	.01664	.01670	.01701	.01762	.01855
19	8	4	.02761	.02812	.02886	.02984	.03110	.03264	.03450
19	8	5	.03628	.03731	.03856	.04004	.04177	.04375	.04600
19	8	6	.04390	.04524	.04677	.04849	.05042	.05256	.05491
19	8	7	.05086	.05234	.05399	.05579	.05774	.05986	.06214
19	8	8	.05736	.05888	.06052	.06226	.06412	.06609	.06817
19	8	9	.06356	.06501	.06653	.06813	.06979	.07152	.07331
19	8	10	.06955	.07084	.07217	.07352	.07491	.07632	.07774
19	8	11	.66032	.66245	.66438	.66614	.66773	.66918	.67051
19	9	1	-.05530	-.06579	-.07753	-.09064	-.10522	-.12141	-.13934
19	9	2	-.01154	-.01379	-.01593	-.01791	-.01971	-.02129	-.02260
19	9	3	.00705	.00680	.00679	.00706	.00763	.00852	.00979
19	9	4	.02088	.02171	.02280	.02417	.02584	.02784	.03018
19	9	5	.03236	.03381	.03550	.03745	.03965	.04213	.04489

Coefficients a_i for the BLUE of μ
Shape parameter k

n	s	i	0.0	0.1	0.2	0.3	0.4	0.5	0.6
19	9	6	.04245	.04425	.04624	.04844	.05085	.05347	.05631
19	9	7	.05164	.05358	.05568	.05792	.06032	.06287	.06558
19	9	8	.06022	.06216	.06419	.06633	.06856	.07089	.07332
19	9	9	.06838	.07018	.07203	.07393	.07588	.07787	.07990
19	9	10	.78386	.78710	.79023	.79325	.79620	.79909	.80196
19	10	1	-.08883	-.10189	-.11638	-.13242	-.15015	-.16968	-.19114
19	10	2	-.03062	-.03313	-.03548	-.03763	-.03954	-.04116	-.04246
19	10	3	-.00591	-.00593	-.00565	-.00505	-.00410	-.00277	-.00102
19	10	4	.01246	.01376	.01535	.01727	.01952	.02212	.02510
19	10	5	.02770	.02972	.03201	.03457	.03742	.04054	.04397
19	10	6	.04107	.04347	.04608	.04889	.05192	.05516	.05862
19	10	7	.05323	.05575	.05842	.06123	.06418	.06727	.07050
19	10	8	.06455	.06700	.06953	.07214	.07482	.07758	.08041
19	10	9	.92636	.93125	.93612	.94101	.94593	.95093	.95604
20	0	1	.05000	.04983	.04928	.04824	.04662	.04429	.04114
20	0	2	.05000	.04996	.04985	.04967	.04946	.04922	.04898
20	0	3	.05000	.04999	.04998	.04998	.05002	.05013	.05034
20	0	4	.05000	.05001	.05005	.05015	.05032	.05060	.05101
20	0	5	.05000	.05002	.05010	.05025	.05050	.05087	.05138
20	0	6	.05000	.05003	.05013	.05032	.05061	.05103	.05158
20	0	7	.05000	.05004	.05015	.05036	.05067	.05111	.05167
20	0	8	.05000	.05004	.05016	.05038	.05070	.05113	.05168
20	0	9	.05000	.05004	.05017	.05039	.05070	.05112	.05164
20	0	10	.05000	.05004	.05017	.05038	.05068	.05107	.05154
20	0	11	.05000	.05004	.05016	.05037	.05064	.05099	.05141
20	0	12	.05000	.05004	.05015	.05034	.05058	.05089	.05124
20	0	13	.05000	.05004	.05014	.05030	.05051	.05075	.05103
20	0	14	.05000	.05003	.05012	.05025	.05041	.05059	.05079
20	0	15	.05000	.05002	.05009	.05018	.05029	.05039	.05050
20	0	16	.05000	.05002	.05005	.05010	.05013	.05015	.05015
20	0	17	.05000	.05000	.05000	.04999	.04994	.04986	.04973
20	0	18	.05000	.04999	.04993	.04984	.04968	.04947	.04919
20	0	19	.05000	.04996	.04983	.04961	.04930	.04891	.04843
20	0	20	.05000	.04986	.04948	.04891	.04822	.04742	.04656
20	1	1	.04624	.04556	.04445	.04280	.04053	.03749	.03358
20	1	2	.04759	.04737	.04708	.04674	.04637	.04599	.04562
20	1	3	.04817	.04807	.04798	.04792	.04791	.04799	.04819
20	1	4	.04859	.04857	.04859	.04868	.04887	.04918	.04963
20	1	5	.04895	.04897	.04907	.04925	.04955	.04998	.05056
20	1	6	.04926	.04932	.04946	.04971	.05007	.05056	.05120
20	1	7	.04954	.04963	.04980	.05008	.05048	.05100	.05166
20	1	8	.04981	.04991	.05011	.05040	.05081	.05133	.05198
20	1	9	.05006	.05018	.05038	.05069	.05109	.05159	.05220
20	1	10	.05031	.05043	.05064	.05094	.05132	.05179	.05235
20	1	11	.05056	.05068	.05088	.05116	.05152	.05194	.05244
20	1	12	.05080	.05092	.05111	.05137	.05168	.05205	.05247
20	1	13	.05105	.05116	.05133	.05155	.05182	.05212	.05244
20	1	14	.05131	.05141	.05155	.05172	.05192	.05214	.05237
20	1	15	.05158	.05166	.05176	.05188	.05200	.05212	.05224
20	1	16	.05187	.05191	.05196	.05201	.05204	.05205	.05204
20	1	17	.05217	.05217	.05216	.05211	.05203	.05191	.05174
20	1	18	.05248	.05242	.05231	.05214	.05192	.05163	.05128
20	1	19	.09965	.09967	.09938	.09883	.09806	.09711	.09602
20	2	1	.04146	.04018	.03841	.03605	.03300	.02914	.02433
20	2	2	.04457	.04415	.04367	.04315	.04262	.04208	.04157
20	2	3	.04589	.04570	.04554	.04542	.04538	.04544	.04562

Coefficients a_i for the BLUE of μ
Shape parameter k

n	s	i	0.0	0.1	0.2	0.3	0.4	0.5	0.6
20	2	4	.04687	.04682	.04684	.04694	.04716	.04751	.04802
20	2	5	.04768	.04773	.04786	.04810	.04846	.04897	.04965
20	2	6	.04840	.04851	.04872	.04904	.04949	.05008	.05083
20	2	7	.04905	.04921	.04947	.04984	.05033	.05096	.05174
20	2	8	.04966	.04985	.05014	.05054	.05105	.05168	.05244
20	2	9	.05025	.05046	.05076	.05116	.05167	.05228	.05300
20	2	10	.05082	.05103	.05134	.05174	.05222	.05279	.05344
20	2	11	.05138	.05160	.05189	.05226	.05271	.05322	.05380
20	2	12	.05194	.05215	.05242	.05276	.05315	.05359	.05407
20	2	13	.05252	.05270	.05294	.05322	.05354	.05389	.05426
20	2	14	.05310	.05326	.05345	.05367	.05390	.05415	.05439
20	2	15	.05371	.05382	.05395	.05408	.05422	.05434	.05444
20	2	16	.05435	.05440	.05444	.05447	.05448	.05446	.05441
20	2	17	.05501	.05497	.05491	.05481	.05466	.05447	.05423
20	2	18	.15332	.15346	.15325	.15274	.15197	.15096	.14976
20	3	1	.03559	.03361	.03108	.02790	.02396	.01914	.01330
20	3	2	.04090	.04026	.03958	.03887	.03815	.03745	.03680
20	3	3	.04315	.04287	.04264	.04248	.04240	.04245	.04264
20	3	4	.04482	.04476	.04479	.04492	.04518	.04559	.04618
20	3	5	.04621	.04629	.04648	.04679	.04724	.04785	.04864
20	3	6	.04743	.04761	.04790	.04832	.04888	.04960	.05048
20	3	7	.04854	.04879	.04915	.04964	.05026	.05101	.05192
20	3	8	.04959	.04988	.05028	.05080	.05144	.05220	.05309
20	3	9	.05058	.05090	.05132	.05185	.05247	.05321	.05405
20	3	10	.05155	.05188	.05230	.05281	.05340	.05408	.05484
20	3	11	.05251	.05283	.05323	.05370	.05424	.05485	.05551
20	3	12	.05346	.05376	.05413	.05454	.05501	.05552	.05607
20	3	13	.05443	.05469	.05500	.05535	.05572	.05611	.05652
20	3	14	.05542	.05563	.05586	.05611	.05637	.05663	.05689
20	3	15	.05644	.05657	.05670	.05684	.05696	.05707	.05715
20	3	16	.05749	.05752	.05753	.05751	.05747	.05740	.05728
20	3	17	.21191	.21214	.21202	.21158	.21085	.20986	.20864
20	4	1	.02842	.02564	.02224	.01811	.01315	.00724	.00023
20	4	2	.03646	.03560	.03470	.03379	.03288	.03201	.03121
20	4	3	.03986	.03951	.03922	.03902	.03892	.03897	.03919
20	4	4	.04239	.04234	.04240	.04257	.04290	.04339	.04409
20	4	5	.04449	.04464	.04491	.04531	.04587	.04661	.04755
20	4	6	.04634	.04662	.04702	.04757	.04826	.04913	.05017
20	4	7	.04802	.04839	.04888	.04951	.05027	.05118	.05225
20	4	8	.04960	.05003	.05056	.05122	.05201	.05292	.05396
20	4	9	.05111	.05156	.05212	.05278	.05354	.05441	.05539
20	4	10	.05257	.05303	.05357	.05421	.05493	.05573	.05660
20	4	11	.05401	.05445	.05496	.05554	.05619	.05689	.05764
20	4	12	.05544	.05584	.05630	.05680	.05735	.05793	.05854
20	4	13	.05689	.05723	.05760	.05800	.05842	.05886	.05930
20	4	14	.05836	.05861	.05888	.05915	.05942	.05968	.05993
20	4	15	.05987	.06000	.06012	.06023	.06032	.06038	.06042
20	4	16	.27617	.27652	.27652	.27619	.27557	.27467	.27353
20	5	1	.01967	.01597	.01157	.00636	.00023	-.00694	-.01528
20	5	2	.03110	.02999	.02887	.02775	.02665	.02561	.02465
20	5	3	.03593	.03551	.03518	.03495	.03485	.03492	.03519
20	5	4	.03952	.03950	.03961	.03986	.04027	.04088	.04172
20	5	5	.04250	.04274	.04312	.04365	.04435	.04525	.04636
20	5	6	.04512	.04553	.04608	.04678	.04764	.04868	.04991
20	5	7	.04751	.04803	.04868	.04947	.05041	.05151	.05276
20	5	8	.04975	.05033	.05103	.05186	.05282	.05390	.05512

Coefficients a_i for the BLUE of μ
Shape parameter k

n	s	i	0.0	0.1	0.2	0.3	0.4	0.5	0.6
20	5	9	.05188	.05249	.05321	.05403	.05495	.05598	.05711
20	5	10	.05395	.05455	.05525	.05602	.05688	.05781	.05882
20	5	11	.05598	.05655	.05719	.05789	.05864	.05945	.06030
20	5	12	.05800	.05850	.05906	.05965	.06027	.06091	.06158
20	5	13	.06003	.06044	.06087	.06132	.06177	.06223	.06269
20	5	14	.06208	.06236	.06263	.06290	.06316	.06340	.06362
20	5	15	.34698	.34749	.34767	.34753	.34710	.34641	.34546
20	6	1	.00897	.00420	-.00136	-.00782	-.01529	-.02391	-.03380
20	6	2	.02460	.02324	.02188	.02053	.01924	.01802	.01691
20	6	3	.03121	.03074	.03037	.03014	.03007	.03018	.03053
20	6	4	.03612	.03616	.03634	.03670	.03724	.03800	.03901
20	6	5	.04019	.04056	.04108	.04177	.04266	.04375	.04508
20	6	6	.04377	.04435	.04507	.04596	.04703	.04829	.04975
20	6	7	.04703	.04774	.04858	.04958	.05073	.05204	.05352
20	6	8	.05008	.05086	.05176	.05279	.05394	.05523	.05665
20	6	9	.05298	.05378	.05469	.05569	.05680	.05800	.05931
20	6	10	.05580	.05658	.05743	.05837	.05938	.06046	.06160
20	6	11	.05856	.05927	.06005	.06087	.06174	.06265	.06360
20	6	12	.06129	.06191	.06255	.06322	.06392	.06462	.06534
20	6	13	.06403	.06450	.06497	.06545	.06593	.06640	.06685
20	6	14	.42536	.42613	.42658	.42674	.42663	.42627	.42567
20	7	1	-.00422	-.01024	-.01715	-.02506	-.03411	-.04440	-.05607
20	7	2	.01667	.01504	.01343	.01186	.01036	.00897	.00772
20	7	3	.02550	.02500	.02463	.02442	.02440	.02460	.02506
20	7	4	.03206	.03220	.03251	.03301	.03372	.03468	.03591
20	7	5	.03750	.03804	.03875	.03965	.04077	.04211	.04370
20	7	6	.04228	.04306	.04402	.04515	.04647	.04800	.04973
20	7	7	.04662	.04756	.04865	.04989	.05129	.05286	.05460
20	7	8	.05068	.05169	.05283	.05410	.05549	.05701	.05866
20	7	9	.05455	.05557	.05669	.05791	.05922	.06062	.06212
20	7	10	.05828	.05926	.06030	.06141	.06259	.06383	.06512
20	7	11	.06194	.06281	.06373	.06468	.06567	.06669	.06773
20	7	12	.06555	.06627	.06701	.06775	.06851	.06926	.07001
20	7	13	.51259	.51374	.51462	.51524	.51561	.51577	.51572
20	8	1	-.02065	-.02813	-.03663	-.04626	-.05715	-.06941	-.08317
20	8	2	.00689	.00499	.00312	.00132	-.00038	-.00193	-.00332
20	8	3	.01854	.01804	.01770	.01756	.01763	.01797	.01859
20	8	4	.02718	.02747	.02795	.02866	.02961	.03083	.03234
20	8	5	.03434	.03511	.03608	.03726	.03867	.04032	.04224
20	8	6	.04063	.04169	.04294	.04437	.04601	.04785	.04991
20	8	7	.04634	.04757	.04896	.05050	.05221	.05409	.05613
20	8	8	.05167	.05297	.05439	.05594	.05761	.05941	.06133
20	8	9	.05674	.05802	.05940	.06086	.06241	.06405	.06576
20	8	10	.06163	.06283	.06408	.06539	.06675	.06815	.06960
20	8	11	.06640	.06744	.06851	.06960	.07070	.07182	.07295
20	8	12	.61026	.61200	.61351	.61481	.61592	.61685	.61763
20	9	1	-.04137	-.05061	-.06101	-.07268	-.08574	-.10033	-.11657
20	9	2	-.00530	-.00749	-.00962	-.01164	-.01353	-.01524	-.01674
20	9	3	.00995	.00949	.00924	.00922	.00946	.00999	.01084
20	9	4	.02126	.02176	.02250	.02349	.02475	.02631	.02819
20	9	5	.03062	.03170	.03300	.03454	.03633	.03838	.04071
20	9	6	.03883	.04025	.04187	.04368	.04571	.04796	.05042
20	9	7	.04629	.04789	.04964	.05156	.05364	.05589	.05831
20	9	8	.05324	.05489	.05665	.05854	.06054	.06266	.06490
20	9	9	.05983	.06143	.06310	.06485	.06667	.06857	.07053
20	9	10	.06618	.06763	.06911	.07064	.07220	.07378	.07540

Coefficients a_i for the BLUE of μ
Shape parameter k

n	s	i	0.0	0.1	0.2	0.3	0.4	0.5	0.6
20	9	11	.72047	.72307	.72550	.72780	.72996	.73203	.73401
20	10	1	-.06796	-.07933	-.09202	-.10615	-.12185	-.13925	-.15846
20	10	2	-.02075	-.02324	-.02562	-.02786	-.02991	-.03174	-.03331
20	10	3	-.00082	-.00115	-.00125	-.00106	-.00057	.00025	.00144
20	10	4	.01396	.01479	.01589	.01727	.01896	.02097	.02334
20	10	5	.02619	.02770	.02945	.03146	.03374	.03630	.03915
20	10	6	.03690	.03879	.04089	.04320	.04572	.04846	.05143
20	10	7	.04662	.04869	.05091	.05330	.05584	.05854	.06141
20	10	8	.05566	.05774	.05993	.06223	.06463	.06714	.06974
20	10	9	.06422	.06619	.06821	.07030	.07243	.07462	.07686
20	10	10	.84597	.84982	.85360	.85732	.86101	.86469	.86840
21	0	1	.04762	.04746	.04691	.04590	.04430	.04201	.03890
21	0	2	.04762	.04758	.04746	.04728	.04704	.04677	.04650
21	0	3	.04762	.04761	.04759	.04757	.04759	.04767	.04783
21	0	4	.04762	.04763	.04766	.04774	.04789	.04814	.04850
21	0	5	.04762	.04764	.04771	.04784	.04807	.04841	.04888
21	0	6	.04762	.04765	.04774	.04791	.04818	.04857	.04909
21	0	7	.04762	.04765	.04776	.04795	.04825	.04866	.04919
21	0	8	.04762	.04766	.04777	.04798	.04828	.04869	.04922
21	0	9	.04762	.04766	.04778	.04799	.04829	.04869	.04920
21	0	10	.04762	.04766	.04778	.04799	.04828	.04866	.04913
21	0	11	.04762	.04766	.04778	.04798	.04825	.04860	.04903
21	0	12	.04762	.04766	.04777	.04796	.04821	.04852	.04889
21	0	13	.04762	.04766	.04776	.04793	.04815	.04842	.04872
21	0	14	.04762	.04765	.04774	.04789	.04807	.04828	.04852
21	0	15	.04762	.04765	.04772	.04784	.04797	.04813	.04829
21	0	16	.04762	.04764	.04770	.04777	.04786	.04794	.04801
21	0	17	.04762	.04763	.04766	.04769	.04771	.04771	.04768
21	0	18	.04762	.04762	.04761	.04758	.04753	.04743	.04728
21	0	19	.04762	.04760	.04755	.04744	.04728	.04706	.04677
21	0	20	.04762	.04758	.04744	.04722	.04692	.04653	.04606
21	0	21	.04762	.04748	.04711	.04656	.04589	.04512	.04429
21	1	1	.04421	.04358	.04252	.04095	.03875	.03580	.03199
21	1	2	.04541	.04520	.04491	.04457	.04419	.04378	.04337
21	1	3	.04592	.04582	.04572	.04565	.04561	.04565	.04579
21	1	4	.04630	.04626	.04627	.04634	.04650	.04676	.04716
21	1	5	.04661	.04662	.04670	.04686	.04712	.04751	.04804
21	1	6	.04688	.04693	.04705	.04727	.04760	.04805	.04865
21	1	7	.04713	.04720	.04736	.04761	.04798	.04846	.04908
21	1	8	.04736	.04745	.04763	.04790	.04828	.04878	.04939
21	1	9	.04758	.04768	.04787	.04816	.04854	.04902	.04961
21	1	10	.04780	.04790	.04810	.04838	.04875	.04921	.04976
21	1	11	.04801	.04812	.04831	.04858	.04893	.04936	.04985
21	1	12	.04822	.04833	.04851	.04876	.04908	.04946	.04989
21	1	13	.04843	.04853	.04870	.04893	.04921	.04953	.04989
21	1	14	.04864	.04874	.04889	.04908	.04931	.04957	.04984
21	1	15	.04887	.04895	.04907	.04922	.04939	.04957	.04975
21	1	16	.04910	.04916	.04925	.04934	.04944	.04953	.04960
21	1	17	.04935	.04938	.04942	.04945	.04946	.04944	.04940
21	1	18	.04962	.04961	.04958	.04952	.04943	.04929	.04910
21	1	19	.04989	.04982	.04971	.04954	.04930	.04901	.04866
21	1	20	.09469	.09471	.09442	.09389	.09313	.09221	.09114
21	2	1	.03992	.03873	.03707	.03484	.03193	.02822	.02358
21	2	2	.04267	.04226	.04180	.04128	.04073	.04018	.03963
21	2	3	.04383	.04364	.04346	.04333	.04325	.04326	.04338
21	2	4	.04469	.04463	.04462	.04469	.04486	.04515	.04560

Coefficients a_i for the BLUE of μ
Shape parameter k

n	s	i	0.0	0.1	0.2	0.3	0.4	0.5	0.6
21	2	5	.04540	.04542	.04553	.04572	.04604	.04649	.04710
21	2	6	.04602	.04611	.04628	.04656	.04697	.04751	.04820
21	2	7	.04659	.04672	.04694	.04728	.04773	.04831	.04903
21	2	8	.04712	.04728	.04754	.04790	.04837	.04897	.04968
21	2	9	.04762	.04781	.04808	.04845	.04893	.04951	.05020
21	2	10	.04811	.04831	.04859	.04896	.04942	.04997	.05061
21	2	11	.04859	.04879	.04907	.04942	.04986	.05036	.05094
21	2	12	.04907	.04926	.04953	.04986	.05025	.05070	.05120
21	2	13	.04955	.04973	.04997	.05027	.05060	.05098	.05139
21	2	14	.05004	.05020	.05041	.05065	.05092	.05122	.05153
21	2	15	.05055	.05068	.05084	.05102	.05121	.05141	.05160
21	2	16	.05108	.05116	.05126	.05137	.05147	.05155	.05161
21	2	17	.05163	.05166	.05168	.05169	.05167	.05162	.05154
21	2	18	.05220	.05215	.05208	.05196	.05180	.05160	.05134
21	2	19	.14533	.14546	.14526	.14475	.14399	.14300	.14182
21	3	1	.03468	.03286	.03051	.02752	.02380	.01921	.01363
21	3	2	.03935	.03875	.03808	.03738	.03666	.03594	.03525
21	3	3	.04132	.04104	.04080	.04060	.04048	.04047	.04059
21	3	4	.04278	.04270	.04269	.04277	.04297	.04331	.04382
21	3	5	.04399	.04404	.04418	.04443	.04482	.04535	.04606
21	3	6	.04505	.04519	.04543	.04580	.04629	.04693	.04774
21	3	7	.04601	.04622	.04653	.04696	.04751	.04821	.04905
21	3	8	.04691	.04716	.04751	.04798	.04856	.04927	.05011
21	3	9	.04776	.04804	.04842	.04890	.04949	.05018	.05098
21	3	10	.04859	.04889	.04927	.04974	.05031	.05096	.05171
21	3	11	.04940	.04970	.05007	.05053	.05105	.05165	.05232
21	3	12	.05021	.05050	.05085	.05126	.05173	.05226	.05283
21	3	13	.05102	.05128	.05160	.05196	.05236	.05279	.05325
21	3	14	.05185	.05207	.05233	.05262	.05294	.05326	.05360
21	3	15	.05270	.05287	.05306	.05326	.05347	.05367	.05386
21	3	16	.05357	.05367	.05377	.05387	.05395	.05401	.05404
21	3	17	.05448	.05448	.05447	.05443	.05436	.05426	.05411
21	3	18	.20033	.20056	.20043	.19998	.19925	.19826	.19705
21	4	1	.02834	.02580	.02266	.01882	.01417	.00858	.00193
21	4	2	.03538	.03456	.03369	.03279	.03188	.03099	.03015
21	4	3	.03834	.03798	.03767	.03742	.03727	.03725	.03737
21	4	4	.04054	.04045	.04045	.04056	.04081	.04121	.04180
21	4	5	.04235	.04245	.04265	.04298	.04345	.04409	.04492
21	4	6	.04395	.04417	.04450	.04496	.04557	.04634	.04729
21	4	7	.04539	.04570	.04612	.04667	.04735	.04818	.04916
21	4	8	.04674	.04711	.04758	.04817	.04888	.04973	.05070
21	4	9	.04803	.04843	.04892	.04953	.05024	.05106	.05199
21	4	10	.04927	.04968	.05018	.05077	.05145	.05222	.05308
21	4	11	.05049	.05089	.05138	.05193	.05256	.05326	.05402
21	4	12	.05170	.05208	.05253	.05303	.05358	.05418	.05483
21	4	13	.05291	.05325	.05364	.05407	.05453	.05501	.05552
21	4	14	.05414	.05442	.05473	.05506	.05541	.05576	.05611
21	4	15	.05539	.05559	.05580	.05601	.05622	.05641	.05659
21	4	16	.05668	.05677	.05685	.05691	.05696	.05697	.05696
21	4	17	.26036	.26068	.26066	.26030	.25966	.25874	.25757
21	5	1	.02070	.01733	.01329	.00847	.00276	-.00396	-.01183
21	5	2	.03063	.02957	.02848	.02738	.02628	.02521	.02421
21	5	3	.03481	.03437	.03400	.03371	.03354	.03352	.03367
21	5	4	.03790	.03783	.03786	.03802	.03833	.03882	.03952
21	5	5	.04046	.04063	.04091	.04134	.04193	.04270	.04367
21	5	6	.04271	.04303	.04348	.04407	.04482	.04575	.04685

Coefficients a_i for the BLUE of μ
Shape parameter k

n	s	i	0.0	0.1	0.2	0.3	0.4	0.5	0.6
21	5	7	.04475	.04518	.04573	.04642	.04726	.04825	.04940
21	5	8	.04665	.04715	.04776	.04850	.04937	.05036	.05150
21	5	9	.04846	.04900	.04964	.05038	.05124	.05220	.05327
21	5	10	.05021	.05075	.05139	.05211	.05292	.05382	.05480
21	5	11	.05192	.05245	.05305	.05372	.05446	.05526	.05612
21	5	12	.05361	.05410	.05465	.05524	.05588	.05656	.05727
21	5	13	.05531	.05573	.05619	.05668	.05720	.05773	.05828
21	5	14	.05702	.05735	.05770	.05806	.05843	.05879	.05915
21	5	15	.05875	.05896	.05917	.05937	.05956	.05973	.05987
21	5	16	.32613	.32659	.32671	.32651	.32602	.32526	.32424
21	6	1	.01144	.00712	.00205	-.00389	-.01081	-.01883	-.02808
21	6	2	.02492	.02362	.02230	.02098	.01968	.01844	.01727
21	6	3	.03060	.03009	.02967	.02936	.02919	.02919	.02939
21	6	4	.03480	.03476	.03484	.03507	.03548	.03609	.03692
21	6	5	.03828	.03853	.03893	.03949	.04022	.04116	.04231
21	6	6	.04132	.04177	.04237	.04312	.04405	.04515	.04645
21	6	7	.04409	.04467	.04539	.04625	.04727	.04844	.04979
21	6	8	.04666	.04733	.04811	.04902	.05007	.05124	.05256
21	6	9	.04911	.04981	.05062	.05153	.05255	.05368	.05491
21	6	10	.05147	.05217	.05296	.05384	.05479	.05583	.05695
21	6	11	.05378	.05445	.05519	.05599	.05684	.05776	.05872
21	6	12	.05607	.05667	.05732	.05801	.05874	.05949	.06028
21	6	13	.05834	.05885	.05938	.05993	.06049	.06107	.06164
21	6	14	.06063	.06100	.06138	.06176	.06213	.06249	.06283
21	6	15	.39848	.39915	.39949	.39954	.39931	.39881	.39807
21	7	1	.00017	-.00525	-.01152	-.01874	-.02705	-.03656	-.04741
21	7	2	.01804	.01648	.01492	.01337	.01187	.01044	.00912
21	7	3	.02556	.02501	.02456	.02424	.02409	.02414	.02441
21	7	4	.03113	.03115	.03132	.03166	.03220	.03296	.03396
21	7	5	.03574	.03612	.03666	.03739	.03831	.03945	.04082
21	7	6	.03977	.04039	.04117	.04212	.04325	.04458	.04611
21	7	7	.04342	.04420	.04512	.04619	.04742	.04882	.05039
21	7	8	.04683	.04769	.04868	.04980	.05105	.05244	.05396
21	7	9	.05006	.05096	.05196	.05306	.05427	.05559	.05700
21	7	10	.05318	.05406	.05502	.05606	.05718	.05838	.05964
21	7	11	.05622	.05704	.05792	.05886	.05984	.06087	.06195
21	7	12	.05922	.05994	.06070	.06149	.06230	.06313	.06397
21	7	13	.06220	.06278	.06337	.06397	.06457	.06517	.06576
21	7	14	.47845	.47943	.48011	.48053	.48069	.48061	.48031
21	8	1	-.01366	-.02036	-.02802	-.03675	-.04667	-.05791	-.07060
21	8	2	.00968	.00785	.00603	.00424	.00253	.00092	-.00056
21	8	3	.01950	.01891	.01845	.01816	.01806	.01819	.01857
21	8	4	.02676	.02688	.02717	.02766	.02838	.02934	.03057
21	8	5	.03277	.03333	.03407	.03501	.03617	.03756	.03920
21	8	6	.03802	.03887	.03988	.04108	.04247	.04407	.04588
21	8	7	.04279	.04381	.04497	.04630	.04779	.04946	.05130
21	8	8	.04722	.04833	.04956	.05093	.05242	.05405	.05582
21	8	9	.05143	.05255	.05378	.05511	.05654	.05806	.05969
21	8	10	.05547	.05656	.05772	.05895	.06025	.06162	.06304
21	8	11	.05941	.06041	.06145	.06253	.06365	.06480	.06598
21	8	12	.06329	.06414	.06500	.06588	.06677	.06767	.06857
21	8	13	.56730	.56874	.56993	.57089	.57164	.57218	.57255
21	9	1	-.03081	-.03903	-.04832	-.05881	-.07063	-.08389	-.09874
21	9	2	-.00059	-.00271	-.00480	-.00682	-.00874	-.01053	-.01215
21	9	3	.01211	.01152	.01110	.01087	.01087	.01113	.01167
21	9	4	.02152	.02179	.02226	.02296	.02390	.02513	.02664

Coefficients a_i for the BLUE of μ
Shape parameter k

n	s	i	0.0	0.1	0.2	0.3	0.4	0.5	0.6
21	9	5	.02928	.03008	.03108	.03230	.03376	.03547	.03744
21	9	6	.03608	.03720	.03851	.04002	.04173	.04366	.04581
21	9	7	.04223	.04355	.04502	.04666	.04847	.05045	.05261
21	9	8	.04795	.04935	.05088	.05254	.05433	.05625	.05830
21	9	9	.05337	.05477	.05627	.05785	.05953	.06130	.06316
21	9	10	.05857	.05990	.06128	.06273	.06423	.06578	.06738
21	9	11	.06363	.06481	.06602	.06725	.06851	.06979	.07108
21	9	12	.66665	.66877	.67070	.67244	.67403	.67547	.67679
21	10	1	-.05239	-.06241	-.07365	-.08624	-.10030	-.11597	-.13336
21	10	2	-.01337	-.01580	-.01815	-.02041	-.02253	-.02448	-.02623
21	10	3	.00302	.00248	.00214	.00203	.00218	.00264	.00341
21	10	4	.01515	.01564	.01637	.01735	.01862	.02018	.02207
21	10	5	.02516	.02628	.02762	.02920	.03105	.03316	.03555
21	10	6	.03391	.03540	.03709	.03899	.04111	.04345	.04601
21	10	7	.04183	.04352	.04538	.04741	.04960	.05197	.05451
21	10	8	.04918	.05095	.05284	.05485	.05699	.05925	.06163
21	10	9	.05613	.05786	.05967	.06157	.06354	.06559	.06771
21	10	10	.06280	.06440	.06604	.06772	.06945	.07120	.07299
21	10	11	.77859	.78169	.78466	.78753	.79030	.79302	.79570
21	11	1	-.08000	-.09221	-.10581	-.12093	-.13770	-.15626	-.17672
21	11	2	-.02955	-.03227	-.03489	-.03737	-.03968	-.04177	-.04361
21	11	3	-.00836	-.00878	-.00897	-.00887	-.00847	-.00773	-.00662
21	11	4	.00731	.00812	.00921	.01059	.01229	.01432	.01670
21	11	5	.02023	.02178	.02358	.02565	.02799	.03062	.03355
21	11	6	.03151	.03348	.03566	.03807	.04069	.04355	.04663
21	11	7	.04171	.04389	.04623	.04874	.05142	.05426	.05727
21	11	8	.05117	.05339	.05572	.05817	.06073	.06339	.06617
21	11	9	.06009	.06222	.06441	.06666	.06898	.07135	.07377
21	11	10	.90589	.91038	.91484	.91929	.92375	.92826	.93285
22	0	1	.04545	.04530	.04476	.04377	.04220	.03994	.03688
22	0	2	.04545	.04541	.04529	.04510	.04485	.04455	.04424
22	0	3	.04545	.04544	.04541	.04539	.04538	.04543	.04555
22	0	4	.04545	.04546	.04548	.04555	.04568	.04589	.04622
22	0	5	.04545	.04547	.04553	.04565	.04586	.04616	.04660
22	0	6	.04545	.04548	.04556	.04572	.04597	.04633	.04682
22	0	7	.04545	.04548	.04558	.04576	.04604	.04643	.04694
22	0	8	.04545	.04549	.04560	.04579	.04608	.04647	.04698
22	0	9	.04545	.04549	.04560	.04580	.04610	.04648	.04698
22	0	10	.04545	.04549	.04561	.04581	.04609	.04647	.04693
22	0	11	.04545	.04549	.04561	.04580	.04607	.04642	.04685
22	0	12	.04545	.04549	.04560	.04579	.04604	.04636	.04674
22	0	13	.04545	.04549	.04560	.04577	.04599	.04627	.04660
22	0	14	.04545	.04549	.04558	.04573	.04593	.04617	.04644
22	0	15	.04545	.04548	.04557	.04569	.04586	.04604	.04624
22	0	16	.04545	.04548	.04555	.04564	.04576	.04589	.04602
22	0	17	.04545	.04547	.04552	.04558	.04565	.04571	.04575
22	0	18	.04545	.04546	.04549	.04550	.04551	.04549	.04544
22	0	19	.04545	.04545	.04544	.04540	.04533	.04522	.04506
22	0	20	.04545	.04544	.04538	.04527	.04510	.04487	.04458
22	0	21	.04545	.04541	.04528	.04506	.04475	.04436	.04390
22	0	22	.04545	.04532	.04496	.04443	.04377	.04303	.04222
22	1	1	.04235	.04176	.04075	.03924	.03711	.03425	.03053
22	1	2	.04343	.04322	.04294	.04259	.04220	.04177	.04133
22	1	3	.04388	.04378	.04367	.04358	.04353	.04353	.04363
22	1	4	.04421	.04417	.04417	.04422	.04435	.04457	.04492
22	1	5	.04449	.04449	.04456	.04469	.04493	.04527	.04576

Coefficients a_i for the BLUE of μ
Shape parameter k

n	s	i	0.0	0.1	0.2	0.3	0.4	0.5	0.6
22	1	6	.04473	.04477	.04487	.04507	.04537	.04578	.04634
22	1	7	.04495	.04501	.04515	.04538	.04571	.04617	.04675
22	1	8	.04515	.04523	.04539	.04564	.04600	.04646	.04705
22	1	9	.04534	.04543	.04561	.04587	.04623	.04670	.04727
22	1	10	.04553	.04563	.04581	.04607	.04643	.04688	.04742
22	1	11	.04571	.04581	.04599	.04625	.04660	.04701	.04751
22	1	12	.04589	.04599	.04617	.04642	.04674	.04712	.04756
22	1	13	.04607	.04617	.04634	.04657	.04685	.04719	.04757
22	1	14	.04626	.04635	.04650	.04670	.04695	.04723	.04754
22	1	15	.04644	.04653	.04666	.04683	.04703	.04724	.04748
22	1	16	.04664	.04671	.04681	.04694	.04708	.04723	.04737
22	1	17	.04685	.04690	.04697	.04704	.04711	.04718	.04722
22	1	18	.04707	.04709	.04711	.04712	.04712	.04708	.04701
22	1	19	.04730	.04729	.04725	.04718	.04707	.04692	.04672
22	1	20	.04753	.04747	.04735	.04717	.04694	.04664	.04629
22	1	21	.09019	.09022	.08994	.08941	.08868	.08777	.08674
22	2	1	.03847	.03737	.03581	.03369	.03090	.02733	.02285
22	2	2	.04092	.04053	.04008	.03957	.03901	.03844	.03786
22	2	3	.04195	.04176	.04157	.04142	.04131	.04128	.04135
22	2	4	.04271	.04263	.04261	.04265	.04278	.04303	.04341
22	2	5	.04333	.04334	.04342	.04358	.04386	.04426	.04481
22	2	6	.04388	.04395	.04409	.04434	.04470	.04519	.04583
22	2	7	.04438	.04449	.04468	.04498	.04539	.04593	.04660
22	2	8	.04484	.04498	.04521	.04554	.04597	.04653	.04721
22	2	9	.04528	.04544	.04569	.04603	.04648	.04703	.04769
22	2	10	.04570	.04587	.04613	.04648	.04692	.04745	.04808
22	2	11	.04612	.04629	.04655	.04689	.04731	.04781	.04839
22	2	12	.04653	.04670	.04696	.04728	.04767	.04812	.04863
22	2	13	.04694	.04711	.04734	.04764	.04799	.04838	.04882
22	2	14	.04735	.04751	.04772	.04798	.04827	.04860	.04895
22	2	15	.04778	.04792	.04809	.04830	.04854	.04878	.04904
22	2	16	.04822	.04833	.04846	.04861	.04877	.04892	.04907
22	2	17	.04868	.04875	.04883	.04890	.04897	.04902	.04905
22	2	18	.04916	.04918	.04918	.04917	.04913	.04906	.04895
22	2	19	.04966	.04960	.04952	.04939	.04922	.04901	.04874
22	2	20	.13813	.13826	.13805	.13756	.13680	.13583	.13467
22	3	1	.03377	.03209	.02989	.02709	.02355	.01917	.01382
22	3	2	.03791	.03733	.03669	.03600	.03528	.03454	.03382
22	3	3	.03965	.03937	.03911	.03889	.03874	.03868	.03873
22	3	4	.04093	.04083	.04079	.04084	.04099	.04127	.04170
22	3	5	.04199	.04201	.04212	.04233	.04265	.04312	.04376
22	3	6	.04292	.04303	.04323	.04354	.04398	.04456	.04530
22	3	7	.04376	.04393	.04420	.04458	.04508	.04571	.04650
22	3	8	.04454	.04475	.04506	.04548	.04602	.04668	.04747
22	3	9	.04528	.04552	.04586	.04630	.04684	.04750	.04826
22	3	10	.04599	.04625	.04660	.04704	.04758	.04821	.04893
22	3	11	.04669	.04696	.04731	.04773	.04824	.04883	.04949
22	3	12	.04738	.04764	.04798	.04838	.04885	.04937	.04996
22	3	13	.04807	.04832	.04863	.04899	.04940	.04986	.05035
22	3	14	.04877	.04899	.04926	.04957	.04992	.05029	.05068
22	3	15	.04948	.04967	.04989	.05013	.05040	.05066	.05094
22	3	16	.05021	.05035	.05051	.05067	.05083	.05099	.05113
22	3	17	.05097	.05104	.05112	.05118	.05123	.05125	.05125
22	3	18	.05175	.05174	.05171	.05165	.05156	.05143	.05127
22	3	19	.18996	.19018	.19004	.18959	.18886	.18788	.18668
22	4	1	.02813	.02579	.02288	.01929	.01491	.00962	.00328

Coefficients a_i for the BLUE of μ
Shape parameter k

n	s	i	0.0	0.1	0.2	0.3	0.4	0.5	0.6
22	4	2	.03434	.03355	.03271	.03183	.03092	.03002	.02914
22	4	3	.03693	.03657	.03624	.03596	.03577	.03568	.03573
22	4	4	.03886	.03874	.03870	.03876	.03894	.03927	.03976
22	4	5	.04044	.04049	.04064	.04091	.04131	.04186	.04260
22	4	6	.04183	.04199	.04227	.04267	.04320	.04389	.04474
22	4	7	.04308	.04333	.04369	.04417	.04479	.04554	.04645
22	4	8	.04425	.04456	.04497	.04550	.04615	.04693	.04784
22	4	9	.04536	.04570	.04615	.04670	.04735	.04812	.04901
22	4	10	.04642	.04679	.04724	.04779	.04843	.04917	.05000
22	4	11	.04746	.04783	.04828	.04881	.04941	.05009	.05085
22	4	12	.04849	.04885	.04928	.04977	.05032	.05092	.05158
22	4	13	.04952	.04985	.05024	.05067	.05115	.05167	.05221
22	4	14	.05055	.05085	.05118	.05154	.05193	.05234	.05276
22	4	15	.05161	.05184	.05210	.05237	.05266	.05294	.05322
22	4	16	.05268	.05284	.05301	.05317	.05333	.05347	.05359
22	4	17	.05379	.05385	.05390	.05393	.05394	.05392	.05386
22	4	18	.24626	.24657	.24652	.24615	.24549	.24455	.24338
22	5	1	.02139	.01830	.01457	.01010	.00476	-.00157	-.00901
22	5	2	.03009	.02909	.02803	.02695	.02586	.02478	.02374
22	5	3	.03374	.03329	.03289	.03257	.03234	.03224	.03229
22	5	4	.03643	.03631	.03629	.03638	.03660	.03699	.03757
22	5	5	.03865	.03875	.03896	.03931	.03980	.04047	.04132
22	5	6	.04060	.04084	.04121	.04172	.04237	.04319	.04418
22	5	7	.04235	.04271	.04318	.04378	.04453	.04542	.04647
22	5	8	.04399	.04441	.04495	.04561	.04639	.04731	.04836
22	5	9	.04554	.04601	.04658	.04726	.04804	.04894	.04996
22	5	10	.04703	.04752	.04810	.04877	.04953	.05038	.05133
22	5	11	.04849	.04897	.04953	.05017	.05088	.05167	.05252
22	5	12	.04992	.05039	.05091	.05150	.05213	.05282	.05356
22	5	13	.05135	.05177	.05224	.05275	.05330	.05387	.05447
22	5	14	.05279	.05315	.05354	.05395	.05438	.05482	.05527
22	5	15	.05425	.05452	.05481	.05510	.05539	.05568	.05595
22	5	16	.05573	.05590	.05605	.05620	.05633	.05644	.05652
22	5	17	.30764	.30807	.30814	.30790	.30736	.30655	.30549
22	6	1	.01330	.00937	.00471	-.00078	-.00721	-.01471	-.02341
22	6	2	.02505	.02381	.02253	.02123	.01994	.01868	.01748
22	6	3	.02996	.02944	.02898	.02861	.02837	.02827	.02835
22	6	4	.03360	.03349	.03349	.03363	.03393	.03441	.03510
22	6	5	.03659	.03676	.03706	.03750	.03811	.03891	.03992
22	6	6	.03921	.03956	.04005	.04069	.04149	.04246	.04362
22	6	7	.04158	.04206	.04267	.04342	.04432	.04538	.04660
22	6	8	.04378	.04435	.04503	.04584	.04678	.04786	.04907
22	6	9	.04587	.04648	.04720	.04802	.04896	.05001	.05118
22	6	10	.04788	.04850	.04922	.05003	.05093	.05191	.05299
22	6	11	.04983	.05044	.05113	.05189	.05272	.05362	.05457
22	6	12	.05176	.05233	.05296	.05365	.05438	.05516	.05597
22	6	13	.05367	.05418	.05473	.05531	.05592	.05655	.05720
22	6	14	.05559	.05601	.05645	.05690	.05736	.05782	.05828
22	6	15	.05753	.05782	.05812	.05842	.05870	.05897	.05923
22	6	16	.37481	.37540	.37566	.37562	.37529	.37469	.37385
22	7	1	.00357	-.00135	-.00707	-.01371	-.02138	-.03022	-.04035
22	7	2	.01902	.01753	.01602	.01450	.01300	.01155	.01019
22	7	3	.02549	.02490	.02439	.02400	.02375	.02367	.02379
22	7	4	.03027	.03020	.03026	.03047	.03086	.03146	.03229
22	7	5	.03421	.03447	.03488	.03546	.03622	.03719	.03837
22	7	6	.03765	.03814	.03878	.03958	.04055	.04171	.04307

Coefficients a_i for the BLUE of μ
Shape parameter k

n	s	i	0.0	0.1	0.2	0.3	0.4	0.5	0.6
22	7	7	.04076	.04140	.04218	.04311	.04420	.04544	.04687
22	7	8	.04366	.04439	.04525	.04624	.04736	.04862	.05003
22	7	9	.04639	.04717	.04806	.04906	.05017	.05139	.05272
22	7	10	.04902	.04981	.05069	.05165	.05271	.05384	.05506
22	7	11	.05158	.05234	.05317	.05407	.05502	.05604	.05711
22	7	12	.05410	.05480	.05554	.05633	.05716	.05803	.05892
22	7	13	.05660	.05720	.05783	.05848	.05915	.05983	.06052
22	7	14	.05909	.05956	.06004	.06052	.06100	.06148	.06194
22	7	15	.44859	.44943	.44997	.45023	.45022	.44996	.44948
22	8	1	-.00823	-.01427	-.02122	-.02919	-.03830	-.04867	-.06044
22	8	2	.01178	.01002	.00826	.00651	.00480	.00316	.00162
22	8	3	.02016	.01951	.01897	.01858	.01834	.01831	.01850
22	8	4	.02634	.02633	.02648	.02680	.02732	.02807	.02908
22	8	5	.03144	.03184	.03240	.03315	.03410	.03527	.03668
22	8	6	.03589	.03655	.03738	.03838	.03957	.04096	.04256
22	8	7	.03992	.04075	.04174	.04288	.04419	.04567	.04732
22	8	8	.04365	.04459	.04566	.04686	.04820	.04968	.05130
22	8	9	.04719	.04817	.04926	.05046	.05176	.05318	.05470
22	8	10	.05058	.05155	.05261	.05375	.05498	.05628	.05766
22	8	11	.05387	.05479	.05578	.05682	.05792	.05906	.06025
22	8	12	.05711	.05793	.05880	.05970	.06063	.06158	.06255
22	8	13	.06031	.06100	.06170	.06242	.06314	.06386	.06458
22	8	14	.53002	.53123	.53218	.53288	.53335	.53359	.53364
22	9	1	-.02266	-.03002	-.03840	-.04791	-.05868	-.07083	-.08450
22	9	2	.00300	.00097	-.00106	-.00305	-.00498	-.00681	-.00851
22	9	3	.01374	.01306	.01252	.01215	.01196	.01201	.01231
22	9	4	.02167	.02176	.02203	.02251	.02321	.02416	.02539
22	9	5	.02820	.02878	.02954	.03051	.03170	.03313	.03481
22	9	6	.03390	.03479	.03586	.03711	.03857	.04023	.04212
22	9	7	.03906	.04014	.04138	.04279	.04436	.04611	.04804
22	9	8	.04383	.04502	.04634	.04780	.04940	.05113	.05300
22	9	9	.04835	.04957	.05090	.05233	.05387	.05551	.05725
22	9	10	.05268	.05387	.05514	.05648	.05790	.05939	.06094
22	9	11	.05687	.05798	.05914	.06034	.06158	.06286	.06418
22	9	12	.06098	.06195	.06294	.06395	.06497	.06601	.06705
22	9	13	.62037	.62213	.62366	.62500	.62614	.62711	.62793
22	10	1	-.04051	-.04942	-.05948	-.07080	-.08351	-.09775	-.11363
22	10	2	-.00775	-.01008	-.01238	-.01462	-.01677	-.01879	-.02065
22	10	3	.00596	.00527	.00476	.00445	.00436	.00453	.00499
22	10	4	.01607	.01631	.01676	.01745	.01839	.01960	.02112
22	10	5	.02440	.02521	.02624	.02749	.02899	.03074	.03276
22	10	6	.03166	.03284	.03421	.03578	.03756	.03956	.04180
22	10	7	.03822	.03961	.04117	.04289	.04480	.04687	.04913
22	10	8	.04430	.04579	.04742	.04918	.05108	.05311	.05528
22	10	9	.05004	.05154	.05315	.05486	.05666	.05856	.06055
22	10	10	.05553	.05697	.05848	.06006	.06169	.06338	.06512
22	10	11	.06084	.06215	.06350	.06487	.06628	.06770	.06915
22	10	12	.72126	.72379	.72617	.72839	.73048	.73247	.73438
22	11	1	-.06290	-.07366	-.08572	-.09919	-.11421	-.13092	-.14943
22	11	2	-.02110	-.02374	-.02632	-.02881	-.03117	-.03336	-.03536
22	11	3	-.00362	-.00426	-.00469	-.00489	-.00482	-.00446	-.00377
22	11	4	.00927	.00974	.01045	.01142	.01268	.01424	.01613
22	11	5	.01988	.02102	.02240	.02402	.02590	.02806	.03051
22	11	6	.02912	.03067	.03243	.03441	.03660	.03902	.04168
22	11	7	.03747	.03925	.04120	.04333	.04563	.04810	.05076
22	11	8	.04519	.04707	.04907	.05120	.05346	.05585	.05837

Coefficients a_i for the BLUE of μ
Shape parameter k

n	s	i	0.0	0.1	0.2	0.3	0.4	0.5	0.6
22	11	9	.05247	.05433	.05627	.05830	.06041	.06261	.06488
22	11	10	.05943	.06117	.06295	.06479	.06667	.06859	.07054
22	11	11	.83479	.83842	.84196	.84543	.84886	.85226	.85568
23	0	1	.04348	.04332	.04280	.04182	.04028	.03806	.03504
23	0	2	.04348	.04344	.04331	.04311	.04284	.04253	.04219
23	0	3	.04348	.04346	.04343	.04339	.04337	.04339	.04347
23	0	4	.04348	.04348	.04350	.04355	.04366	.04384	.04414
23	0	5	.04348	.04349	.04354	.04365	.04383	.04412	.04452
23	0	6	.04348	.04350	.04357	.04372	.04395	.04428	.04474
23	0	7	.04348	.04351	.04360	.04376	.04402	.04439	.04487
23	0	8	.04348	.04351	.04361	.04379	.04407	.04444	.04493
23	0	9	.04348	.04351	.04362	.04381	.04409	.04446	.04494
23	0	10	.04348	.04351	.04362	.04382	.04409	.04446	.04491
23	0	11	.04348	.04351	.04363	.04381	.04408	.04443	.04485
23	0	12	.04348	.04351	.04362	.04380	.04406	.04438	.04477
23	0	13	.04348	.04351	.04362	.04379	.04402	.04431	.04465
23	0	14	.04348	.04351	.04361	.04376	.04397	.04422	.04451
23	0	15	.04348	.04351	.04360	.04373	.04391	.04412	.04435
23	0	16	.04348	.04351	.04358	.04369	.04383	.04399	.04416
23	0	17	.04348	.04350	.04356	.04364	.04374	.04385	.04395
23	0	18	.04348	.04349	.04353	.04358	.04363	.04367	.04369
23	0	19	.04348	.04349	.04350	.04351	.04350	.04346	.04340
23	0	20	.04348	.04348	.04346	.04341	.04333	.04321	.04304
23	0	21	.04348	.04346	.04340	.04328	.04311	.04287	.04258
23	0	22	.04348	.04343	.04330	.04308	.04278	.04239	.04194
23	0	23	.04348	.04335	.04300	.04248	.04184	.04112	.04034
23	1	1	.04064	.04008	.03912	.03766	.03560	.03282	.02918
23	1	2	.04161	.04141	.04113	.04079	.04038	.03994	.03947
23	1	3	.04201	.04191	.04180	.04170	.04162	.04159	.04165
23	1	4	.04231	.04227	.04225	.04229	.04238	.04257	.04288
23	1	5	.04255	.04255	.04260	.04272	.04292	.04324	.04368
23	1	6	.04277	.04280	.04289	.04306	.04333	.04372	.04424
23	1	7	.04296	.04301	.04314	.04335	.04366	.04408	.04463
23	1	8	.04314	.04321	.04335	.04359	.04392	.04436	.04492
23	1	9	.04331	.04339	.04355	.04380	.04414	.04458	.04513
23	1	10	.04347	.04356	.04373	.04398	.04432	.04475	.04528
23	1	11	.04363	.04373	.04390	.04415	.04448	.04489	.04538
23	1	12	.04379	.04389	.04405	.04429	.04461	.04499	.04544
23	1	13	.04395	.04404	.04420	.04443	.04472	.04506	.04546
23	1	14	.04410	.04420	.04435	.04455	.04481	.04511	.04544
23	1	15	.04427	.04435	.04449	.04467	.04488	.04513	.04540
23	1	16	.04443	.04450	.04462	.04477	.04494	.04512	.04532
23	1	17	.04460	.04466	.04475	.04486	.04498	.04509	.04520
23	1	18	.04479	.04483	.04488	.04494	.04499	.04503	.04505
23	1	19	.04498	.04500	.04501	.04501	.04498	.04493	.04484
23	1	20	.04518	.04517	.04512	.04504	.04493	.04476	.04455
23	1	21	.04539	.04532	.04520	.04502	.04479	.04449	.04414
23	1	22	.08611	.08613	.08586	.08534	.08463	.08375	.08273
23	2	1	.03711	.03609	.03461	.03260	.02993	.02648	.02214
23	2	2	.03931	.03894	.03850	.03799	.03743	.03684	.03623
23	2	3	.04022	.04003	.03985	.03968	.03954	.03948	.03950
23	2	4	.04090	.04082	.04078	.04080	.04089	.04110	.04143
23	2	5	.04146	.04145	.04151	.04164	.04188	.04224	.04273
23	2	6	.04194	.04199	.04211	.04233	.04265	.04310	.04369
23	2	7	.04238	.04247	.04264	.04290	.04328	.04378	.04441
23	2	8	.04279	.04290	.04311	.04341	.04381	.04433	.04498

Coefficients a_i for the BLUE of μ
Shape parameter k

n	s	i	0.0	0.1	0.2	0.3	0.4	0.5	0.6
23	2	9	.04317	.04331	.04353	.04385	.04427	.04479	.04543
23	2	10	.04354	.04369	.04393	.04426	.04467	.04518	.04579
23	2	11	.04390	.04406	.04430	.04462	.04503	.04551	.04608
23	2	12	.04425	.04442	.04466	.04497	.04535	.04580	.04631
23	2	13	.04461	.04477	.04500	.04529	.04564	.04604	.04649
23	2	14	.04496	.04512	.04533	.04559	.04590	.04624	.04662
23	2	15	.04533	.04547	.04565	.04588	.04613	.04642	.04672
23	2	16	.04570	.04582	.04597	.04615	.04635	.04655	.04676
23	2	17	.04608	.04618	.04629	.04641	.04654	.04666	.04677
23	2	18	.04649	.04654	.04660	.04666	.04670	.04672	.04672
23	2	19	.04691	.04692	.04691	.04688	.04683	.04674	.04661
23	2	20	.04735	.04729	.04720	.04706	.04689	.04666	.04639
23	2	21	.13161	.13174	.13153	.13104	.13030	.12935	.12821
23	3	1	.03287	.03131	.02926	.02661	.02325	.01905	.01390
23	3	2	.03657	.03601	.03539	.03471	.03399	.03324	.03249
23	3	3	.03811	.03783	.03756	.03733	.03715	.03704	.03704
23	3	4	.03924	.03913	.03907	.03908	.03919	.03941	.03978
23	3	5	.04018	.04018	.04026	.04042	.04070	.04111	.04168
23	3	6	.04100	.04108	.04124	.04151	.04190	.04242	.04310
23	3	7	.04173	.04187	.04211	.04244	.04290	.04348	.04421
23	3	8	.04242	.04260	.04288	.04325	.04375	.04436	.04510
23	3	9	.04306	.04328	.04358	.04398	.04449	.04510	.04583
23	3	10	.04369	.04392	.04424	.04465	.04515	.04575	.04644
23	3	11	.04429	.04453	.04485	.04526	.04574	.04631	.04696
23	3	12	.04489	.04513	.04544	.04583	.04629	.04681	.04739
23	3	13	.04548	.04571	.04601	.04637	.04679	.04725	.04776
23	3	14	.04607	.04629	.04657	.04689	.04725	.04764	.04807
23	3	15	.04668	.04687	.04711	.04738	.04767	.04799	.04832
23	3	16	.04730	.04746	.04765	.04785	.04807	.04829	.04851
23	3	17	.04794	.04805	.04818	.04831	.04844	.04855	.04865
23	3	18	.04860	.04865	.04870	.04874	.04876	.04876	.04872
23	3	19	.04928	.04926	.04921	.04914	.04903	.04889	.04870
23	3	20	.18061	.18082	.18069	.18024	.17951	.17853	.17734
23	4	1	.02783	.02567	.02295	.01958	.01544	.01041	.00436
23	4	2	.03333	.03258	.03177	.03091	.03001	.02910	.02819
23	4	3	.03563	.03526	.03492	.03462	.03439	.03425	.03423
23	4	4	.03732	.03718	.03711	.03713	.03725	.03751	.03793
23	4	5	.03871	.03873	.03884	.03905	.03939	.03987	.04052
23	4	6	.03993	.04005	.04028	.04062	.04109	.04170	.04248
23	4	7	.04103	.04123	.04154	.04196	.04251	.04320	.04404
23	4	8	.04204	.04231	.04267	.04314	.04374	.04445	.04531
23	4	9	.04301	.04331	.04370	.04420	.04481	.04553	.04637
23	4	10	.04393	.04425	.04467	.04518	.04578	.04648	.04727
23	4	11	.04483	.04516	.04558	.04607	.04665	.04731	.04805
23	4	12	.04571	.04604	.04645	.04692	.04746	.04806	.04872
23	4	13	.04659	.04691	.04728	.04772	.04820	.04873	.04930
23	4	14	.04747	.04776	.04810	.04848	.04889	.04933	.04980
23	4	15	.04837	.04862	.04890	.04921	.04954	.04988	.05023
23	4	16	.04928	.04947	.04969	.04992	.05015	.05037	.05059
23	4	17	.05021	.05034	.05047	.05059	.05071	.05080	.05088
23	4	18	.05117	.05121	.05123	.05124	.05121	.05116	.05108
23	4	19	.23362	.23391	.23385	.23347	.23279	.23185	.23067
23	5	1	.02183	.01899	.01554	.01136	.00634	.00037	-.00670
23	5	2	.02952	.02856	.02754	.02648	.02540	.02432	.02326
23	5	3	.03272	.03227	.03185	.03150	.03123	.03106	.03104
23	5	4	.03508	.03493	.03486	.03489	.03505	.03536	.03584

Coefficients a_i for the BLUE of μ
Shape parameter k

n	s	i	0.0	0.1	0.2	0.3	0.4	0.5	0.6
23	5	5	.03703	.03708	.03723	.03750	.03791	.03849	.03924
23	5	6	.03872	.03891	.03921	.03964	.04021	.04094	.04183
23	5	7	.04025	.04054	.04095	.04147	.04214	.04294	.04391
23	5	8	.04167	.04203	.04250	.04309	.04380	.04464	.04562
23	5	9	.04301	.04342	.04393	.04454	.04527	.04611	.04706
23	5	10	.04430	.04473	.04526	.04587	.04659	.04740	.04830
23	5	11	.04555	.04599	.04651	.04711	.04779	.04855	.04938
23	5	12	.04678	.04721	.04771	.04827	.04890	.04958	.05032
23	5	13	.04800	.04840	.04886	.04937	.04992	.05052	.05115
23	5	14	.04922	.04958	.04999	.05042	.05089	.05137	.05188
23	5	15	.05046	.05076	.05109	.05143	.05179	.05215	.05251
23	5	16	.05171	.05193	.05217	.05240	.05263	.05285	.05306
23	5	17	.05299	.05311	.05323	.05333	.05341	.05348	.05351
23	5	18	.29115	.29154	.29159	.29131	.29074	.28989	.28879
23	6	1	.01471	.01110	.00681	.00172	-.00429	-.01134	-.01955
23	6	2	.02502	.02384	.02261	.02134	.02007	.01880	.01757
23	6	3	.02932	.02878	.02830	.02789	.02759	.02741	.02739
23	6	4	.03249	.03233	.03228	.03234	.03254	.03292	.03348
23	6	5	.03509	.03519	.03541	.03576	.03627	.03695	.03783
23	6	6	.03736	.03764	.03803	.03857	.03926	.04012	.04116
23	6	7	.03941	.03981	.04033	.04098	.04178	.04273	.04384
23	6	8	.04132	.04179	.04239	.04311	.04396	.04494	.04607
23	6	9	.04311	.04364	.04420	.04503	.04589	.04686	.04796
23	6	10	.04483	.04539	.04604	.04678	.04762	.04856	.04959
23	6	11	.04650	.04706	.04770	.04842	.04921	.05007	.05101
23	6	12	.04815	.04868	.04929	.04995	.05067	.05144	.05227
23	6	13	.04977	.05027	.05081	.05140	.05203	.05269	.05338
23	6	14	.05140	.05183	.05230	.05279	.05330	.05383	.05437
23	6	15	.05304	.05338	.05375	.05412	.05450	.05487	.05524
23	6	16	.05469	.05493	.05516	.05539	.05561	.05581	.05599
23	6	17	.35379	.35433	.35453	.35442	.35401	.35334	.35241
23	7	1	.00622	.00174	-.00352	-.00965	-.01678	-.02503	-.03454
23	7	2	.01970	.01828	.01682	.01533	.01385	.01240	.01100
23	7	3	.02532	.02470	.02416	.02371	.02339	.02321	.02321
23	7	4	.02946	.02932	.02929	.02941	.02968	.03015	.03082
23	7	5	.03286	.03303	.03334	.03379	.03442	.03524	.03627
23	7	6	.03583	.03621	.03673	.03740	.03824	.03926	.04047
23	7	7	.03851	.03904	.03970	.04050	.04146	.04257	.04386
23	7	8	.04099	.04161	.04236	.04323	.04424	.04539	.04669
23	7	9	.04333	.04401	.04480	.04570	.04671	.04784	.04909
23	7	10	.04558	.04628	.04707	.04796	.04894	.05001	.05118
23	7	11	.04775	.04845	.04922	.05006	.05098	.05196	.05301
23	7	12	.04989	.05055	.05126	.05203	.05285	.05372	.05463
23	7	13	.05200	.05259	.05323	.05390	.05460	.05533	.05607
23	7	14	.05411	.05461	.05514	.05568	.05623	.05679	.05736
23	7	15	.05622	.05660	.05699	.05738	.05776	.05813	.05849
23	7	16	.42225	.42299	.42342	.42355	.42341	.42302	.42238
23	8	1	-.00396	-.00945	-.01580	-.02311	-.03152	-.04115	-.05213
23	8	2	.01337	.01169	.00999	.00828	.00658	.00494	.00336
23	8	3	.02059	.01991	.01932	.01884	.01851	.01835	.01839
23	8	4	.02591	.02581	.02584	.02603	.02640	.02698	.02779
23	8	5	.03028	.03055	.03098	.03157	.03235	.03334	.03455
23	8	6	.03410	.03462	.03529	.03613	.03715	.03836	.03977
23	8	7	.03754	.03823	.03906	.04005	.04119	.04251	.04400
23	8	8	.04072	.04152	.04244	.04350	.04470	.04604	.04753
23	8	9	.04372	.04458	.04554	.04662	.04781	.04912	.05055

Coefficients a_i for the BLUE of μ
Shape parameter k

n	s	i	0.0	0.1	0.2	0.3	0.4	0.5	0.6
23	8	10	.04660	.04747	.04843	.04948	.05062	.05185	.05317
23	8	11	.04939	.05023	.05115	.05214	.05319	.05430	.05547
23	8	12	.05211	.05290	.05374	.05463	.05555	.05652	.05752
23	8	13	.05481	.05550	.05623	.05698	.05775	.05854	.05934
23	8	14	.05749	.05806	.05864	.05922	.05981	.06039	.06097
23	8	15	.49735	.49839	.49916	.49965	.49990	.49992	.49972
23	9	1	-.01627	-.02291	-.03051	-.03919	-.04907	-.06028	-.07295
23	9	2	.00578	.00383	.00187	-.00008	-.00199	-.00384	-.00559
23	9	3	.01497	.01423	.01361	.01312	.01281	.01269	.01280
23	9	4	.02173	.02169	.02181	.02211	.02261	.02335	.02434
23	9	5	.02729	.02770	.02827	.02904	.03001	.03121	.03264
23	9	6	.03213	.03283	.03370	.03475	.03598	.03743	.03908
23	9	7	.03650	.03739	.03844	.03965	.04102	.04257	.04430
23	9	8	.04054	.04155	.04269	.04397	.04539	.04696	.04867
23	9	9	.04435	.04541	.04659	.04787	.04927	.05078	.05241
23	9	10	.04799	.04906	.05021	.05145	.05277	.05418	.05566
23	9	11	.05152	.05254	.05362	.05477	.05597	.05722	.05853
23	9	12	.05497	.05590	.05687	.05788	.05891	.05998	.06107
23	9	13	.05836	.05916	.05998	.06081	.06165	.06249	.06334
23	9	14	.58013	.58161	.58285	.58386	.58466	.58527	.58570
23	10	1	-.03127	-.03926	-.04833	-.05859	-.07018	-.08321	-.09782
23	10	2	-.00339	-.00562	-.00786	-.01006	-.01220	-.01424	-.01617
23	10	3	.00822	.00745	.00682	.00635	.00609	.00604	.00626
23	10	4	.01677	.01683	.01707	.01753	.01821	.01915	.02036
23	10	5	.02379	.02438	.02516	.02615	.02736	.02882	.03053
23	10	6	.02990	.03083	.03194	.03325	.03476	.03648	.03843
23	10	7	.03542	.03656	.03787	.03934	.04099	.04282	.04484
23	10	8	.04051	.04178	.04318	.04472	.04641	.04823	.05021
23	10	9	.04531	.04662	.04804	.04957	.05121	.05296	.05481
23	10	10	.04990	.05119	.05256	.05401	.05554	.05714	.05881
23	10	11	.05433	.05554	.05681	.05813	.05949	.06090	.06234
23	10	12	.05865	.05974	.06085	.06198	.06313	.06430	.06547
23	10	13	.67186	.67397	.67588	.67761	.67919	.68062	.68193
23	11	1	-.04977	-.05935	-.07014	-.08226	-.09584	-.11101	-.12791
23	11	2	-.01460	-.01714	-.01965	-.02211	-.02448	-.02673	-.02882
23	11	3	.00004	-.00074	-.00134	-.00175	-.00192	-.00184	-.00147
23	11	4	.01082	.01103	.01146	.01212	.01305	.01425	.01575
23	11	5	.01967	.02050	.02155	.02282	.02435	.02613	.02819
23	11	6	.02737	.02859	.03001	.03164	.03348	.03555	.03785
23	11	7	.03430	.03576	.03739	.03920	.04118	.04335	.04571
23	11	8	.04071	.04229	.04401	.04587	.04787	.05001	.05229
23	11	9	.04674	.04835	.05007	.05188	.05380	.05582	.05793
23	11	10	.05249	.05405	.05568	.05738	.05914	.06096	.06284
23	11	11	.05804	.05948	.06096	.06247	.06401	.06557	.06717
23	11	12	.77418	.77716	.78001	.78274	.78537	.78794	.79047
23	12	1	-.07293	-.08441	-.09725	-.11158	-.12753	-.14523	-.16482
23	12	2	-.02849	-.03135	-.03416	-.03687	-.03947	-.04191	-.04416
23	12	3	-.01001	-.01074	-.01127	-.01156	-.01159	-.01131	-.01072
23	12	4	.00359	.00404	.00472	.00568	.00692	.00847	.01036
23	12	5	.01476	.01592	.01732	.01898	.02090	.02310	.02559
23	12	6	.02447	.02607	.02789	.02993	.03219	.03469	.03743
23	12	7	.03321	.03507	.03710	.03932	.04171	.04429	.04706
23	12	8	.04128	.04325	.04536	.04760	.04997	.05249	.05513
23	12	9	.04885	.05083	.05289	.05505	.05729	.05963	.06204
23	12	10	.05607	.05795	.05987	.06185	.06388	.06595	.06806
23	12	11	.88920	.89338	.89752	.90162	.90571	.90984	.91402

Coefficients a_i for the BLUE of μ
Shape parameter k

n	s	i	0.0	0.1	0.2	0.3	0.4	0.5	0.6
24	0	1	.04167	.04151	.04100	.04004	.03853	.03634	.03336
24	0	2	.04167	.04162	.04149	.04129	.04101	.04068	.04031
24	0	3	.04167	.04165	.04161	.04156	.04152	.04152	.04157
24	0	4	.04167	.04167	.04168	.04172	.04180	.04197	.04222
24	0	5	.04167	.04168	.04172	.04182	.04198	.04224	.04261
24	0	6	.04167	.04169	.04175	.04188	.04210	.04241	.04284
24	0	7	.04167	.04169	.04177	.04193	.04217	.04252	.04298
24	0	8	.04167	.04170	.04179	.04196	.04222	.04258	.04305
24	0	9	.04167	.04170	.04180	.04198	.04225	.04261	.04307
24	0	10	.04167	.04170	.04181	.04199	.04225	.04261	.04306
24	0	11	.04167	.04170	.04181	.04199	.04225	.04259	.04301
24	0	12	.04167	.04170	.04181	.04198	.04223	.04255	.04294
24	0	13	.04167	.04170	.04180	.04197	.04221	.04250	.04285
24	0	14	.04167	.04170	.04180	.04195	.04217	.04243	.04274
24	0	15	.04167	.04170	.04179	.04193	.04212	.04234	.04260
24	0	16	.04167	.04170	.04178	.04190	.04206	.04224	.04244
24	0	17	.04167	.04169	.04176	.04186	.04198	.04212	.04226
24	0	18	.04167	.04169	.04174	.04181	.04189	.04198	.04205
24	0	19	.04167	.04168	.04171	.04175	.04179	.04181	.04181
24	0	20	.04167	.04167	.04168	.04168	.04166	.04161	.04153
24	0	21	.04167	.04166	.04164	.04159	.04150	.04136	.04119
24	0	22	.04167	.04165	.04158	.04146	.04128	.04105	.04075
24	0	23	.04167	.04162	.04149	.04127	.04097	.04059	.04014
24	0	24	.04167	.04154	.04120	.04070	.04008	.03938	.03862
24	1	1	.03906	.03853	.03761	.03621	.03421	.03149	.02794
24	1	2	.03994	.03974	.03947	.03912	.03871	.03825	.03776
24	1	3	.04030	.04020	.04008	.03997	.03987	.03982	.03984
24	1	4	.04057	.04052	.04050	.04051	.04059	.04075	.04102
24	1	5	.04079	.04078	.04082	.04092	.04110	.04138	.04178
24	1	6	.04098	.04100	.04108	.04123	.04148	.04183	.04231
24	1	7	.04115	.04119	.04130	.04149	.04178	.04218	.04270
24	1	8	.04131	.04137	.04150	.04171	.04203	.04244	.04298
24	1	9	.04146	.04153	.04168	.04191	.04223	.04265	.04318
24	1	10	.04160	.04168	.04184	.04208	.04240	.04282	.04333
24	1	11	.04174	.04183	.04199	.04223	.04254	.04294	.04343
24	1	12	.04188	.04197	.04213	.04236	.04267	.04304	.04349
24	1	13	.04202	.04211	.04226	.04248	.04277	.04312	.04352
24	1	14	.04215	.04224	.04239	.04260	.04286	.04316	.04352
24	1	15	.04229	.04237	.04251	.04270	.04293	.04319	.04349
24	1	16	.04243	.04251	.04263	.04279	.04298	.04320	.04343
24	1	17	.04258	.04265	.04275	.04288	.04302	.04318	.04334
24	1	18	.04273	.04279	.04286	.04295	.04305	.04314	.04322
24	1	19	.04290	.04293	.04297	.04302	.04305	.04307	.04306
24	1	20	.04307	.04308	.04308	.04307	.04303	.04296	.04285
24	1	21	.04325	.04323	.04318	.04309	.04297	.04279	.04258
24	1	22	.04343	.04336	.04324	.04306	.04282	.04253	.04217
24	1	23	.08238	.08240	.08214	.08163	.08093	.08007	.07908
24	2	1	.03584	.03488	.03349	.03156	.02899	.02567	.02145
24	2	2	.03782	.03746	.03703	.03653	.03597	.03537	.03474
24	2	3	.03864	.03845	.03826	.03808	.03793	.03783	.03781
24	2	4	.03924	.03916	.03910	.03910	.03917	.03933	.03962
24	2	5	.03974	.03972	.03976	.03987	.04008	.04039	.04084
24	2	6	.04017	.04021	.04031	.04050	.04079	.04119	.04174
24	2	7	.04056	.04063	.04078	.04102	.04136	.04182	.04242
24	2	8	.04092	.04102	.04120	.04148	.04185	.04234	.04295
24	2	9	.04126	.04138	.04159	.04188	.04227	.04277	.04337

Coefficients a_i for the BLUE of μ
Shape parameter k

n	s	i	0.0	0.1	0.2	0.3	0.4	0.5	0.6
24	2	10	.04159	.04172	.04194	.04224	.04264	.04313	.04371
24	2	11	.04190	.04205	.04227	.04258	.04296	.04343	.04399
24	2	12	.04221	.04236	.04259	.04288	.04325	.04369	.04421
24	2	13	.04252	.04267	.04289	.04317	.04352	.04392	.04438
24	2	14	.04283	.04297	.04318	.04344	.04375	.04411	.04451
24	2	15	.04314	.04328	.04346	.04370	.04397	.04427	.04460
24	2	16	.04346	.04358	.04374	.04394	.04416	.04441	.04466
24	2	17	.04378	.04389	.04402	.04417	.04434	.04451	.04468
24	2	18	.04412	.04420	.04429	.04440	.04450	.04459	.04466
24	2	19	.04448	.04452	.04457	.04461	.04463	.04463	.04460
24	2	20	.04485	.04485	.04483	.04479	.04472	.04462	.04447
24	2	21	.04524	.04518	.04508	.04494	.04476	.04453	.04425
24	2	22	.12568	.12581	.12561	.12512	.12439	.12346	.12234
24	3	1	.03200	.03055	.02861	.02610	.02289	.01887	.01391
24	3	2	.03531	.03478	.03418	.03351	.03279	.03204	.03127
24	3	3	.03669	.03641	.03614	.03589	.03568	.03554	.03549
24	3	4	.03770	.03758	.03750	.03748	.03755	.03773	.03804
24	3	5	.03853	.03851	.03856	.03869	.03893	.03929	.03980
24	3	6	.03925	.03931	.03945	.03968	.04002	.04050	.04112
24	3	7	.03991	.04002	.04022	.04052	.04093	.04147	.04214
24	3	8	.04051	.04066	.04091	.04125	.04170	.04227	.04297
24	3	9	.04108	.04126	.04154	.04190	.04237	.04295	.04365
24	3	10	.04162	.04183	.04212	.04250	.04297	.04354	.04421
24	3	11	.04215	.04237	.04267	.04305	.04351	.04406	.04469
24	3	12	.04267	.04289	.04319	.04356	.04400	.04451	.04509
24	3	13	.04318	.04340	.04369	.04404	.04445	.04491	.04543
24	3	14	.04370	.04391	.04418	.04450	.04486	.04527	.04572
24	3	15	.04422	.04441	.04465	.04493	.04525	.04559	.04596
24	3	16	.04474	.04491	.04512	.04535	.04561	.04587	.04615
24	3	17	.04529	.04542	.04558	.04576	.04594	.04612	.04629
24	3	18	.04585	.04594	.04604	.04615	.04624	.04632	.04639
24	3	19	.04643	.04647	.04650	.04652	.04651	.04648	.04642
24	3	20	.04703	.04700	.04694	.04686	.04673	.04657	.04637
24	3	21	.17214	.17235	.17221	.17176	.17104	.17007	.16890
24	4	1	.02745	.02545	.02291	.01974	.01581	.01102	.00521
24	4	2	.03237	.03165	.03087	.03002	.02913	.02822	.02730
24	4	3	.03441	.03404	.03369	.03338	.03312	.03294	.03287
24	4	4	.03591	.03576	.03566	.03564	.03572	.03592	.03626
24	4	5	.03714	.03713	.03720	.03737	.03765	.03807	.03865
24	4	6	.03822	.03831	.03849	.03878	.03919	.03975	.04045
24	4	7	.03918	.03935	.03961	.03999	.04048	.04111	.04188
24	4	8	.04008	.04030	.04062	.04104	.04158	.04225	.04305
24	4	9	.04092	.04118	.04154	.04199	.04255	.04323	.04402
24	4	10	.04173	.04201	.04239	.04286	.04342	.04408	.04485
24	4	11	.04251	.04281	.04319	.04366	.04421	.04484	.04556
24	4	12	.04328	.04358	.04396	.04441	.04493	.04552	.04617
24	4	13	.04404	.04433	.04469	.04512	.04559	.04612	.04670
24	4	14	.04479	.04507	.04541	.04579	.04621	.04667	.04717
24	4	15	.04556	.04581	.04611	.04644	.04679	.04717	.04757
24	4	16	.04634	.04655	.04679	.04706	.04734	.04762	.04791
24	4	17	.04713	.04730	.04747	.04766	.04785	.04803	.04820
24	4	18	.04795	.04805	.04815	.04824	.04832	.04838	.04841
24	4	19	.04879	.04881	.04881	.04879	.04874	.04867	.04855
24	4	20	.22221	.22250	.22243	.22203	.22135	.22040	.21922
24	5	1	.02209	.01947	.01625	.01234	.00761	.00194	-.00479
24	5	2	.02892	.02801	.02703	.02599	.02493	.02384	.02276

Coefficients a_i for the BLUE of μ
Shape parameter k

n	s	i	0.0	0.1	0.2	0.3	0.4	0.5	0.6
24	5	3	.03176	.03131	.03088	.03050	.03019	.02998	.02989
24	5	4	.03384	.03367	.03356	.03355	.03364	.03388	.03427
24	5	5	.03556	.03557	.03567	.03588	.03622	.03672	.03738
24	5	6	.03705	.03719	.03743	.03779	.03829	.03894	.03975
24	5	7	.03839	.03863	.03897	.03943	.04002	.04075	.04164
24	5	8	.03963	.03994	.04035	.04087	.04151	.04229	.04320
24	5	9	.04080	.04115	.04160	.04216	.04283	.04361	.04451
24	5	10	.04192	.04230	.04277	.04334	.04401	.04477	.04564
24	5	11	.04300	.04340	.04388	.04444	.04508	.04581	.04662
24	5	12	.04406	.04446	.04493	.04546	.04607	.04674	.04748
24	5	13	.04512	.04550	.04594	.04644	.04699	.04759	.04823
24	5	14	.04616	.04652	.04692	.04736	.04784	.04836	.04890
24	5	15	.04722	.04753	.04788	.04825	.04865	.04906	.04949
24	5	16	.04829	.04854	.04882	.04911	.04940	.04970	.05000
24	5	17	.04938	.04956	.04975	.04993	.05012	.05028	.05044
24	5	18	.05049	.05058	.05066	.05073	.05077	.05079	.05079
24	5	19	.27634	.27671	.27673	.27642	.27582	.27495	.27383
24	6	1	.01577	.01245	.00847	.00372	-.00191	-.00855	-.01634
24	6	2	.02489	.02376	.02257	.02134	.02008	.01882	.01758
24	6	3	.02867	.02813	.02763	.02720	.02684	.02660	.02650
24	6	4	.03145	.03127	.03116	.03116	.03129	.03158	.03204
24	6	5	.03374	.03378	.03393	.03421	.03463	.03521	.03598
24	6	6	.03573	.03593	.03626	.03671	.03731	.03807	.03900
24	6	7	.03752	.03784	.03828	.03885	.03956	.04041	.04143
24	6	8	.03917	.03958	.04010	.04073	.04150	.04240	.04344
24	6	9	.04073	.04119	.04176	.04243	.04322	.04412	.04515
24	6	10	.04222	.04271	.04330	.04398	.04476	.04565	.04663
24	6	11	.04366	.04417	.04475	.04542	.04617	.04700	.04791
24	6	12	.04508	.04557	.04614	.04677	.04747	.04823	.04905
24	6	13	.04647	.04694	.04747	.04805	.04868	.04935	.05006
24	6	14	.04786	.04829	.04876	.04927	.04981	.05037	.05096
24	6	15	.04926	.04963	.05002	.05044	.05087	.05131	.05176
24	6	16	.05067	.05096	.05126	.05156	.05187	.05217	.05246
24	6	17	.05210	.05229	.05247	.05265	.05281	.05295	.05307
24	6	18	.33502	.33551	.33566	.33549	.33503	.33430	.33331
24	7	1	.00830	.00419	-.00066	-.00634	-.01299	-.02073	-.02970
24	7	2	.02015	.01880	.01739	.01595	.01449	.01304	.01162
24	7	3	.02507	.02445	.02388	.02339	.02300	.02275	.02265
24	7	4	.02869	.02850	.02841	.02844	.02862	.02897	.02952
24	7	5	.03166	.03175	.03197	.03233	.03284	.03354	.03443
24	7	6	.03424	.03453	.03495	.03552	.03624	.03714	.03822
24	7	7	.03656	.03700	.03756	.03825	.03910	.04010	.04127
24	7	8	.03871	.03924	.03989	.04066	.04157	.04262	.04381
24	7	9	.04074	.04133	.04202	.04283	.04376	.04481	.04597
24	7	10	.04267	.04329	.04401	.04482	.04573	.04674	.04785
24	7	11	.04454	.04517	.04588	.04666	.04753	.04848	.04950
24	7	12	.04637	.04698	.04765	.04839	.04919	.05005	.05096
24	7	13	.04817	.04874	.04936	.05003	.05074	.05148	.05226
24	7	14	.04997	.05048	.05102	.05159	.05218	.05279	.05342
24	7	15	.05176	.05219	.05263	.05308	.05354	.05400	.05446
24	7	16	.05357	.05389	.05420	.05451	.05482	.05511	.05538
24	7	17	.39884	.39950	.39984	.39988	.39963	.39913	.39838
24	8	1	-.00058	-.00559	-.01141	-.01816	-.02597	-.03495	-.04524
24	8	2	.01457	.01297	.01133	.00966	.00800	.00635	.00476
24	8	3	.02085	.02015	.01952	.01899	.01858	.01832	.01824
24	8	4	.02547	.02530	.02524	.02533	.02558	.02602	.02667

Coefficients a_i for the BLUE of μ
Shape parameter k

n	s	i	0.0	0.1	0.2	0.3	0.4	0.5	0.6
24	8	5	.02926	.02943	.02974	.03020	.03084	.03167	.03272
24	8	6	.03256	.03296	.03351	.03421	.03509	.03614	.03740
24	8	7	.03552	.03609	.03680	.03765	.03866	.03983	.04118
24	8	8	.03827	.03894	.03974	.04068	.04175	.04297	.04433
24	8	9	.04084	.04159	.04244	.04341	.04450	.04570	.04703
24	8	10	.04331	.04408	.04495	.04591	.04697	.04812	.04937
24	8	11	.04569	.04646	.04730	.04823	.04922	.05029	.05144
24	8	12	.04801	.04875	.04955	.05040	.05131	.05226	.05327
24	8	13	.05031	.05098	.05170	.05246	.05324	.05406	.05491
24	8	14	.05258	.05317	.05378	.05441	.05506	.05571	.05637
24	8	15	.05485	.05532	.05580	.05628	.05676	.05723	.05768
24	8	16	.46849	.46940	.47001	.47035	.47043	.47026	.46988
24	9	1	-.01119	-.01722	-.02417	-.03213	-.04125	-.05165	-.06345
24	9	2	.00795	.00609	.00420	.00230	.00041	-.00144	-.00321
24	9	3	.01588	.01511	.01443	.01387	.01345	.01321	.01317
24	9	4	.02171	.02157	.02157	.02174	.02208	.02264	.02344
24	9	5	.02650	.02677	.02719	.02779	.02859	.02959	.03082
24	9	6	.03065	.03120	.03190	.03278	.03383	.03509	.03655
24	9	7	.03440	.03513	.03601	.03705	.03825	.03963	.04118
24	9	8	.03785	.03871	.03969	.04082	.04208	.04350	.04506
24	9	9	.04110	.04202	.04306	.04421	.04548	.04687	.04838
24	9	10	.04420	.04515	.04619	.04732	.04854	.04986	.05127
24	9	11	.04720	.04812	.04913	.05020	.05134	.05255	.05382
24	9	12	.05012	.05099	.05192	.05290	.05392	.05498	.05609
24	9	13	.05299	.05378	.05460	.05545	.05632	.05721	.05811
24	9	14	.05584	.05650	.05718	.05787	.05856	.05925	.05993
24	9	15	.54481	.54608	.54709	.54785	.54839	.54871	.54883
24	10	1	-.02396	-.03118	-.03942	-.04879	-.05941	-.07142	-.08496
24	10	2	.00003	-.00211	-.00426	-.00641	-.00852	-.01058	-.01254
24	10	3	.00998	.00915	.00844	.00787	.00746	.00726	.00729
24	10	4	.01729	.01721	.01730	.01757	.01805	.01877	.01974
24	10	5	.02328	.02369	.02427	.02505	.02604	.02725	.02872
24	10	6	.02849	.02922	.03012	.03120	.03248	.03397	.03568
24	10	7	.03318	.03412	.03522	.03648	.03791	.03953	.04133
24	10	8	.03750	.03857	.03978	.04113	.04263	.04427	.04607
24	10	9	.04156	.04270	.04395	.04532	.04680	.04840	.05012
24	10	10	.04544	.04659	.04782	.04915	.05057	.05207	.05366
24	10	11	.04918	.05029	.05146	.05270	.05400	.05536	.05678
24	10	12	.05282	.05385	.05492	.05603	.05717	.05835	.05955
24	10	13	.05639	.05730	.05822	.05916	.06011	.06107	.06203
24	10	14	.62884	.63062	.63217	.63353	.63469	.63569	.63654
24	11	1	-.03950	-.04811	-.05784	-.06883	-.08120	-.09509	-.11062
24	11	2	-.00952	-.01195	-.01438	-.01679	-.01914	-.02140	-.02355
24	11	3	.00291	.00204	.00132	.00077	.00041	.00028	.00041
24	11	4	.01204	.01207	.01228	.01271	.01337	.01428	.01548
24	11	5	.01952	.02012	.02091	.02191	.02314	.02462	.02636
24	11	6	.02602	.02698	.02813	.02948	.03104	.03281	.03481
24	11	7	.03187	.03307	.03443	.03597	.03769	.03959	.04168
24	11	8	.03726	.03859	.04007	.04169	.04346	.04537	.04744
24	11	9	.04232	.04371	.04522	.04684	.04857	.05042	.05238
24	11	10	.04714	.04853	.05000	.05155	.05318	.05489	.05668
24	11	11	.05179	.05310	.05448	.05591	.05738	.05891	.06047
24	11	12	.05630	.05750	.05873	.05998	.06125	.06254	.06385
24	11	13	.72186	.72434	.72666	.72882	.73085	.73277	.73461
24	12	1	-.05864	-.06887	-.08036	-.09325	-.10767	-.12374	-.14161
24	12	2	-.02117	-.02392	-.02664	-.02931	-.03190	-.03438	-.03671

Coefficients a_i for the BLUE of μ
Shape parameter k

n	s	i	0.0	0.1	0.2	0.3	0.4	0.5	0.6
24	12	3	-.00565	-.00652	-.00722	-.00772	-.00799	-.00800	-.00772
24	12	4	.00575	.00593	.00633	.00697	.00787	.00905	.01054
24	12	5	.01509	.01593	.01699	.01829	.01983	.02164	.02373
24	12	6	.02320	.02446	.02592	.02760	.02949	.03162	.03399
24	12	7	.03048	.03200	.03370	.03558	.03764	.03988	.04233
24	12	8	.03720	.03886	.04066	.04261	.04470	.04694	.04932
24	12	9	.04350	.04520	.04701	.04893	.05096	.05309	.05532
24	12	10	.04949	.05116	.05290	.05471	.05659	.05853	.06054
24	12	11	.05525	.05681	.05841	.06005	.06172	.06342	.06515
24	12	12	.82551	.82896	.83230	.83557	.83877	.84195	.84512
25	0	1	.04000	.03985	.03935	.03841	.03692	.03476	.03182
25	0	2	.04000	.03996	.03982	.03961	.03932	.03898	.03858
25	0	3	.04000	.03998	.03994	.03988	.03982	.03980	.03982
25	0	4	.04000	.04000	.04000	.04003	.04010	.04024	.04047
25	0	5	.04000	.04001	.04005	.04013	.04028	.04051	.04085
25	0	6	.04000	.04002	.04008	.04020	.04039	.04068	.04109
25	0	7	.04000	.04002	.04010	.04024	.04047	.04079	.04123
25	0	8	.04000	.04003	.04011	.04027	.04052	.04086	.04131
25	0	9	.04000	.04003	.04012	.04029	.04055	.04090	.04135
25	0	10	.04000	.04003	.04013	.04031	.04056	.04091	.04134
25	0	11	.04000	.04003	.04014	.04031	.04056	.04090	.04131
25	0	12	.04000	.04003	.04014	.04031	.04055	.04087	.04126
25	0	13	.04000	.04003	.04013	.04030	.04053	.04083	.04118
25	0	14	.04000	.04003	.04013	.04029	.04050	.04077	.04109
25	0	15	.04000	.04003	.04012	.04027	.04046	.04070	.04097
25	0	16	.04000	.04003	.04011	.04024	.04041	.04061	.04084
25	0	17	.04000	.04003	.04010	.04021	.04035	.04051	.04069
25	0	18	.04000	.04002	.04008	.04017	.04028	.04039	.04051
25	0	19	.04000	.04002	.04006	.04012	.04019	.04026	.04031
25	0	20	.04000	.04001	.04004	.04007	.04009	.04010	.04008
25	0	21	.04000	.04000	.04001	.04000	.03997	.03990	.03981
25	0	22	.04000	.03999	.03997	.03991	.03981	.03967	.03948
25	0	23	.04000	.03998	.03991	.03979	.03961	.03937	.03907
25	0	24	.04000	.03996	.03982	.03960	.03930	.03893	.03849
25	0	25	.04000	.03988	.03954	.03905	.03845	.03777	.03704
25	1	1	.03760	.03710	.03622	.03486	.03291	.03027	.02679
25	1	2	.03839	.03821	.03794	.03759	.03718	.03670	.03619
25	1	3	.03872	.03862	.03850	.03838	.03827	.03819	.03817
25	1	4	.03896	.03892	.03888	.03889	.03894	.03907	.03930
25	1	5	.03916	.03915	.03918	.03926	.03942	.03967	.04004
25	1	6	.03933	.03935	.03942	.03955	.03978	.04010	.04055
25	1	7	.03949	.03952	.03962	.03979	.04006	.04043	.04092
25	1	8	.03963	.03968	.03980	.04000	.04029	.04068	.04119
25	1	9	.03976	.03983	.03996	.04017	.04048	.04088	.04139
25	1	10	.03989	.03996	.04011	.04033	.04064	.04104	.04154
25	1	11	.04002	.04009	.04024	.04047	.04077	.04116	.04164
25	1	12	.04014	.04022	.04037	.04059	.04089	.04126	.04170
25	1	13	.04026	.04034	.04049	.04071	.04099	.04133	.04174
25	1	14	.04038	.04046	.04060	.04081	.04107	.04138	.04175
25	1	15	.04050	.04058	.04071	.04090	.04114	.04141	.04173
25	1	16	.04062	.04069	.04082	.04099	.04119	.04143	.04168
25	1	17	.04074	.04081	.04092	.04106	.04123	.04142	.04162
25	1	18	.04088	.04093	.04102	.04113	.04126	.04139	.04152
25	1	19	.04101	.04106	.04112	.04120	.04127	.04134	.04140
25	1	20	.04116	.04118	.04122	.04125	.04127	.04127	.04124
25	1	21	.04131	.04131	.04131	.04129	.04124	.04115	.04103

Coefficients a_i for the BLUE of μ
Shape parameter k

n	s	i	0.0	0.1	0.2	0.3	0.4	0.5	0.6
25	1	22	.04147	.04145	.04139	.04130	.04117	.04099	.04076
25	1	23	.04163	.04156	.04144	.04126	.04102	.04073	.04037
25	1	24	.07896	.07898	.07872	.07823	.07754	.07670	.07574
25	2	1	.03465	.03375	.03243	.03058	.02811	.02489	.02079
25	2	2	.03644	.03610	.03568	.03518	.03462	.03401	.03336
25	2	3	.03718	.03699	.03680	.03660	.03644	.03631	.03626
25	2	4	.03772	.03763	.03757	.03755	.03759	.03772	.03796
25	2	5	.03817	.03814	.03816	.03825	.03843	.03871	.03911
25	2	6	.03855	.03857	.03866	.03882	.03908	.03945	.03995
25	2	7	.03890	.03896	.03909	.03930	.03962	.04004	.04060
25	2	8	.03922	.03931	.03947	.03972	.04006	.04052	.04110
25	2	9	.03952	.03963	.03981	.04008	.04045	.04092	.04150
25	2	10	.03981	.03993	.04013	.04041	.04079	.04125	.04182
25	2	11	.04009	.04022	.04043	.04072	.04108	.04154	.04208
25	2	12	.04036	.04050	.04071	.04099	.04135	.04178	.04229
25	2	13	.04063	.04077	.04098	.04125	.04159	.04199	.04245
25	2	14	.04090	.04104	.04124	.04150	.04181	.04217	.04258
25	2	15	.04117	.04130	.04149	.04173	.04201	.04232	.04268
25	2	16	.04144	.04157	.04174	.04194	.04219	.04245	.04274
25	2	17	.04173	.04183	.04198	.04215	.04235	.04256	.04277
25	2	18	.04202	.04211	.04222	.04235	.04249	.04263	.04277
25	2	19	.04232	.04238	.04246	.04254	.04262	.04269	.04274
25	2	20	.04264	.04267	.04270	.04272	.04272	.04270	.04266
25	2	21	.04297	.04296	.04293	.04288	.04280	.04268	.04252
25	2	22	.04331	.04324	.04314	.04300	.04281	.04258	.04229
25	2	23	.12027	.12039	.12019	.11971	.11900	.11808	.11698
25	3	1	.03115	.02979	.02797	.02558	.02252	.01865	.01385
25	3	2	.03413	.03363	.03304	.03239	.03168	.03092	.03013
25	3	3	.03537	.03510	.03482	.03456	.03433	.03416	.03407
25	3	4	.03628	.03615	.03605	.03601	.03605	.03619	.03645
25	3	5	.03702	.03699	.03701	.03712	.03732	.03763	.03809
25	3	6	.03766	.03770	.03781	.03801	.03832	.03875	.03931
25	3	7	.03825	.03834	.03851	.03877	.03915	.03964	.04027
25	3	8	.03878	.03891	.03913	.03944	.03985	.04038	.04103
25	3	9	.03928	.03945	.03969	.04003	.04046	.04101	.04167
25	3	10	.03977	.03995	.04021	.04056	.04101	.04155	.04219
25	3	11	.04023	.04043	.04070	.04106	.04150	.04202	.04264
25	3	12	.04069	.04089	.04117	.04152	.04194	.04244	.04301
25	3	13	.04113	.04134	.04161	.04195	.04235	.04281	.04333
25	3	14	.04158	.04178	.04204	.04236	.04272	.04314	.04360
25	3	15	.04203	.04222	.04246	.04275	.04307	.04343	.04382
25	3	16	.04249	.04266	.04287	.04312	.04340	.04369	.04401
25	3	17	.04295	.04310	.04328	.04348	.04370	.04392	.04415
25	3	18	.04343	.04355	.04368	.04383	.04398	.04412	.04426
25	3	19	.04393	.04400	.04408	.04416	.04423	.04429	.04432
25	3	20	.04444	.04447	.04448	.04448	.04446	.04441	.04432
25	3	21	.04498	.04493	.04487	.04477	.04464	.04447	.04426
25	3	22	.16444	.16464	.16450	.16405	.16334	.16238	.16122
25	4	1	.02703	.02517	.02279	.01979	.01606	.01147	.00590
25	4	2	.03145	.03076	.03000	.02918	.02830	.02738	.02645
25	4	3	.03327	.03291	.03256	.03223	.03194	.03173	.03161
25	4	4	.03461	.03445	.03433	.03428	.03432	.03447	.03475
25	4	5	.03571	.03568	.03572	.03584	.03608	.03644	.03696
25	4	6	.03666	.03673	.03687	.03712	.03748	.03798	.03862
25	4	7	.03752	.03766	.03788	.03821	.03865	.03922	.03994
25	4	8	.03831	.03850	.03878	.03916	.03965	.04027	.04101

Coefficients a_i for the BLUE of μ
Shape parameter k

n	s	i	0.0	0.1	0.2	0.3	0.4	0.5	0.6
25	4	9	.03905	.03928	.03960	.04001	.04053	.04116	.04191
25	4	10	.03976	.04002	.04036	.04079	.04132	.04194	.04267
25	4	11	.04045	.04072	.04107	.04151	.04203	.04263	.04333
25	4	12	.04112	.04140	.04175	.04218	.04268	.04325	.04389
25	4	13	.04178	.04206	.04240	.04281	.04328	.04380	.04438
25	4	14	.04244	.04271	.04303	.04341	.04383	.04430	.04481
25	4	15	.04310	.04335	.04365	.04398	.04436	.04476	.04519
25	4	16	.04377	.04399	.04425	.04454	.04485	.04517	.04551
25	4	17	.04445	.04464	.04484	.04507	.04531	.04555	.04578
25	4	18	.04515	.04529	.04544	.04559	.04574	.04588	.04600
25	4	19	.04587	.04595	.04602	.04609	.04614	.04617	.04617
25	4	20	.04661	.04661	.04660	.04656	.04649	.04640	.04626
25	4	21	.21188	.21215	.21207	.21167	.21098	.21004	.20885
25	5	1	.02221	.01978	.01677	.01309	.00862	.00323	-.00321
25	5	2	.02832	.02744	.02650	.02549	.02444	.02336	.02227
25	5	3	.03084	.03039	.02996	.02956	.02923	.02897	.02882
25	5	4	.03270	.03250	.03237	.03231	.03236	.03253	.03285
25	5	5	.03421	.03419	.03425	.03442	.03470	.03512	.03571
25	5	6	.03553	.03563	.03583	.03614	.03657	.03714	.03788
25	5	7	.03672	.03691	.03720	.03761	.03813	.03880	.03961
25	5	8	.03781	.03807	.03843	.03890	.03948	.04019	.04104
25	5	9	.03884	.03915	.03955	.04005	.04067	.04139	.04224
25	5	10	.03982	.04016	.04059	.04111	.04173	.04245	.04327
25	5	11	.04077	.04112	.04156	.04209	.04269	.04339	.04417
25	5	12	.04169	.04206	.04249	.04300	.04358	.04423	.04495
25	5	13	.04261	.04296	.04338	.04386	.04440	.04500	.04565
25	5	14	.04351	.04385	.04424	.04468	.04517	.04570	.04626
25	5	15	.04442	.04473	.04508	.04547	.04589	.04634	.04680
25	5	16	.04534	.04561	.04591	.04623	.04657	.04692	.04728
25	5	17	.04627	.04649	.04672	.04696	.04721	.04746	.04769
25	5	18	.04722	.04737	.04753	.04767	.04781	.04794	.04804
25	5	19	.04820	.04826	.04832	.04835	.04837	.04836	.04832
25	5	20	.26297	.26332	.26332	.26299	.26237	.26148	.26034
25	6	1	.01656	.01349	.00979	.00535	.00005	-.00624	-.01364
25	6	2	.02468	.02360	.02246	.02126	.02002	.01877	.01751
25	6	3	.02803	.02749	.02698	.02652	.02614	.02584	.02568
25	6	4	.03049	.03028	.03013	.03009	.03015	.03036	.03073
25	6	5	.03251	.03251	.03261	.03282	.03316	.03366	.03433
25	6	6	.03426	.03441	.03467	.03506	.03557	.03625	.03709
25	6	7	.03584	.03610	.03647	.03697	.03760	.03837	.03930
25	6	8	.03729	.03763	.03808	.03865	.03934	.04017	.04113
25	6	9	.03865	.03905	.03955	.04016	.04088	.04172	.04269
25	6	10	.03995	.04039	.04092	.04154	.04227	.04310	.04403
25	6	11	.04121	.04166	.04220	.04282	.04353	.04432	.04520
25	6	12	.04243	.04289	.04342	.04402	.04469	.04543	.04623
25	6	13	.04364	.04408	.04459	.04515	.04577	.04644	.04715
25	6	14	.04484	.04526	.04572	.04623	.04678	.04736	.04797
25	6	15	.04604	.04641	.04682	.04726	.04772	.04821	.04871
25	6	16	.04725	.04756	.04790	.04826	.04862	.04899	.04936
25	6	17	.04847	.04871	.04896	.04922	.04946	.04970	.04993
25	6	18	.04972	.04987	.05001	.05014	.05026	.05035	.05043
25	6	19	.31814	.31859	.31870	.31849	.31799	.31721	.31618
25	7	1	.00993	.00615	.00166	-.00363	-.00986	-.01715	-.02563
25	7	2	.02043	.01914	.01779	.01639	.01496	.01352	.01209
25	7	3	.02477	.02415	.02357	.02305	.02261	.02229	.02211
25	7	4	.02796	.02773	.02759	.02756	.02766	.02791	.02835

Coefficients a_i for the BLUE of μ
Shape parameter k

n	s	i	0.0	0.1	0.2	0.3	0.4	0.5	0.6
25	7	5	.03057	.03060	.03075	.03102	.03144	.03203	.03281
25	7	6	.03283	.03305	.03339	.03387	.03449	.03527	.03624
25	7	7	.03486	.03522	.03569	.03630	.03704	.03794	.03901
25	7	8	.03674	.03719	.03775	.03844	.03925	.04021	.04131
25	7	9	.03850	.03901	.03963	.04036	.04121	.04217	.04327
25	7	10	.04018	.04073	.04138	.04212	.04297	.04391	.04496
25	7	11	.04180	.04237	.04302	.04375	.04457	.04547	.04645
25	7	12	.04338	.04394	.04458	.04528	.04605	.04688	.04777
25	7	13	.04494	.04548	.04607	.04672	.04742	.04816	.04895
25	7	14	.04648	.04698	.04752	.04809	.04870	.04934	.05001
25	7	15	.04802	.04846	.04892	.04941	.04991	.05043	.05096
25	7	16	.04957	.04993	.05030	.05068	.05106	.05144	.05181
25	7	17	.05113	.05139	.05164	.05189	.05213	.05235	.05256
25	7	18	.37789	.37849	.37876	.37872	.37840	.37781	.37697
25	8	1	.00213	-.00246	-.00783	-.01410	-.02137	-.02978	-.03947
25	8	2	.01547	.01395	.01237	.01075	.00912	.00748	.00588
25	8	3	.02099	.02028	.01962	.01905	.01858	.01824	.01806
25	8	4	.02503	.02480	.02467	.02468	.02483	.02515	.02567
25	8	5	.02834	.02843	.02864	.02900	.02952	.03023	.03113
25	8	6	.03121	.03152	.03197	.03255	.03330	.03423	.03534
25	8	7	.03380	.03426	.03486	.03559	.03648	.03753	.03875
25	8	8	.03618	.03675	.03745	.03827	.03923	.04034	.04159
25	8	9	.03841	.03906	.03981	.04068	.04167	.04278	.04402
25	8	10	.04054	.04122	.04200	.04288	.04386	.04494	.04612
25	8	11	.04259	.04329	.04406	.04492	.04586	.04688	.04798
25	8	12	.04459	.04527	.04602	.04683	.04771	.04864	.04963
25	8	13	.04656	.04720	.04790	.04864	.04942	.05025	.05111
25	8	14	.04851	.04909	.04971	.05035	.05103	.05172	.05244
25	8	15	.05045	.05095	.05147	.05200	.05254	.05308	.05363
25	8	16	.05240	.05279	.05318	.05357	.05396	.05434	.05470
25	8	17	.44280	.44360	.44411	.44432	.44426	.44395	.44341
25	9	1	-.00711	-.01262	-.01899	-.02635	-.03480	-.04449	-.05554
25	9	2	.00964	.00787	.00605	.00420	.00235	.00052	-.00127
25	9	3	.01656	.01577	.01505	.01443	.01394	.01360	.01344
25	9	4	.02163	.02142	.02132	.02138	.02160	.02201	.02264
25	9	5	.02578	.02595	.02625	.02672	.02737	.02821	.02927
25	9	6	.02939	.02981	.03038	.03111	.03201	.03310	.03439
25	9	7	.03262	.03323	.03397	.03487	.03592	.03714	.03854
25	9	8	.03561	.03633	.03719	.03818	.03931	.04058	.04201
25	9	9	.03841	.03921	.04012	.04115	.04231	.04358	.04498
25	9	10	.04107	.04191	.04284	.04388	.04501	.04624	.04757
25	9	11	.04364	.04448	.04540	.04640	.04747	.04863	.04985
25	9	12	.04614	.04695	.04783	.04876	.04975	.05079	.05188
25	9	13	.04860	.04935	.05015	.05099	.05186	.05277	.05370
25	9	14	.05102	.05169	.05239	.05311	.05384	.05459	.05534
25	9	15	.05344	.05399	.05456	.05513	.05570	.05626	.05682
25	9	16	.51356	.51467	.51549	.51606	.51638	.51647	.51636
25	10	1	-.01812	-.02468	-.03220	-.04080	-.05060	-.06174	-.07434
25	10	2	.00274	.00070	-.00137	-.00346	-.00554	-.00758	-.00956
25	10	3	.01135	.01049	.00972	.00907	.00857	.00824	.00811
25	10	4	.01766	.01749	.01745	.01758	.01790	.01843	.01920
25	10	5	.02283	.02310	.02353	.02413	.02493	.02595	.02720
25	10	6	.02731	.02788	.02861	.02951	.03060	.03189	.03339
25	10	7	.03134	.03211	.03304	.03412	.03537	.03680	.03842
25	10	8	.03505	.03596	.03700	.03818	.03951	.04099	.04263
25	10	9	.03853	.03951	.04061	.04183	.04318	.04464	.04623

Coefficients a_i for the BLUE of μ
Shape parameter k

n	s	i	0.0	0.1	0.2	0.3	0.4	0.5	0.6
25	10	10	.04184	.04285	.04396	.04517	.04648	.04788	.04938
25	10	11	.04502	.04603	.04711	.04826	.04949	.05079	.05215
25	10	12	.04812	.04908	.05009	.05116	.05227	.05343	.05463
25	10	13	.05116	.05203	.05294	.05388	.05485	.05584	.05685
25	10	14	.05415	.05491	.05569	.05647	.05726	.05805	.05885
25	10	15	.59103	.59254	.59382	.59488	.59572	.59638	.59686
25	11	1	-.03135	-.03912	-.04797	-.05799	-.06934	-.08213	-.09651
25	11	2	-.00550	-.00782	-.01016	-.01250	-.01481	-.01707	-.01925
25	11	3	.00518	.00426	.00345	.00279	.00230	.00201	.00195
25	11	4	.01300	.01289	.01294	.01318	.01363	.01432	.01527
25	11	5	.01940	.01981	.02040	.02118	.02218	.02341	.02488
25	11	6	.02495	.02571	.02664	.02775	.02907	.03060	.03235
25	11	7	.02994	.03092	.03207	.03338	.03487	.03654	.03841
25	11	8	.03452	.03565	.03692	.03834	.03990	.04162	.04349
25	11	9	.03882	.04003	.04135	.04280	.04436	.04604	.04785
25	11	10	.04291	.04414	.04546	.04687	.04837	.04997	.05165
25	11	11	.04684	.04804	.04931	.05064	.05204	.05349	.05501
25	11	12	.05066	.05178	.05295	.05416	.05541	.05669	.05800
25	11	13	.05440	.05540	.05643	.05748	.05854	.05961	.06069
25	11	14	.67622	.67831	.68021	.68193	.68349	.68490	.68620
25	12	1	-.04740	-.05659	-.06697	-.07866	-.09179	-.10650	-.12292
25	12	2	-.01541	-.01803	-.02066	-.02327	-.02583	-.02831	-.03068
25	12	3	-.00221	.00316	-.00398	-.00463	-.00508	-.00530	-.00526
25	12	4	.00747	.00746	.00764	.00804	.00867	.00956	.01074
25	12	5	.01539	.01598	.01678	.01779	.01904	.02053	.02229
25	12	6	.02225	.02324	.02442	.02580	.02740	.02922	.03127
25	12	7	.02841	.02965	.03107	.03267	.03444	.03641	.03857
25	12	8	.03407	.03547	.03701	.03871	.04055	.04255	.04470
25	12	9	.03937	.04085	.04243	.04414	.04596	.04790	.04995
25	12	10	.04441	.04589	.04745	.04910	.05083	.05264	.05454
25	12	11	.04926	.05067	.05215	.05368	.05527	.05690	.05859
25	12	12	.05395	.05526	.05659	.05796	.05935	.06076	.06219
25	12	13	.77044	.77332	.77606	.77867	.78119	.78363	.78602
25	13	1	-.06713	-.07799	-.09016	-.10380	-.11902	-.13597	-.15479
25	13	2	-.02749	-.03044	-.03337	-.03625	-.03906	-.04176	-.04432
25	13	3	-.01113	-.01210	-.01290	-.01349	-.01386	-.01396	-.01378
25	13	4	.00085	.00100	.00136	.00197	.00284	.00400	.00547
25	13	5	.01065	.01150	.01257	.01387	.01543	.01726	.01938
25	13	6	.01914	.02043	.02194	.02365	.02560	.02777	.03020
25	13	7	.02675	.02833	.03008	.03202	.03415	.03648	.03899
25	13	8	.03375	.03548	.03736	.03939	.04157	.04390	.04638
25	13	9	.04030	.04209	.04399	.04601	.04814	.05037	.05271
25	13	10	.04651	.04828	.05013	.05205	.05404	.05611	.05824
25	13	11	.05248	.05415	.05587	.05762	.05942	.06125	.06312
25	13	12	.87533	.87926	.88313	.88695	.89075	.89456	.89840

Coefficients a_i for the BLUE of μ
Shape parameter k

n	s	i	0.7	0.8	0.9	1.0	1.1	1.2	1.3
2	0	1	.50000	.50000	.50000	.50000	.50000	.50000	.50000
2	0	2	.50000	.50000	.50000	.50000	.50000	.50000	.50000
3	0	1	.32560	.32321	.32051	.31751	.31422	.31065	.30682
3	0	2	.34723	.35128	.35576	.36065	.36591	.37152	.37745
3	0	3	.32717	.32551	.32373	.32184	.31986	.31783	.31574
3	1	1	-.08441	-.09778	-.11145	-.12541	-.13966	-.15419	-.16898
3	1	2	1.08441	1.09778	1.11145	1.12541	1.13966	1.15419	1.16898
4	0	1	.23860	.23487	.23057	.22570	.22026	.21424	.20764
4	0	2	.26188	.26582	.27036	.27552	.28128	.28765	.29462
4	0	3	.25724	.25903	.26090	.26284	.26483	.26684	.26886
4	0	4	.24228	.24029	.23816	.23594	.23363	.23127	.22887
4	1	1	.05182	.04026	.02808	.01528	.00185	-.01220	-.02687
4	1	2	.28691	.29626	.30629	.31700	.32838	.34042	.35310
4	1	3	.66127	.66348	.66563	.66772	.66977	.67178	.67377
5	0	1	.18674	.18224	.17699	.17096	.16413	.15649	.14802
5	0	2	.20943	.21283	.21686	.22154	.22691	.23297	.23974
5	0	3	.20816	.21056	.21323	.21615	.21932	.22270	.22629
5	0	4	.20367	.20438	.20505	.20568	.20626	.20677	.20721
5	0	5	.19200	.18999	.18787	.18566	.18339	.18108	.17874
5	1	1	.07330	.06287	.05160	.03947	.02646	.01256	-.00225
5	1	2	.20833	.21466	.22175	.22962	.23827	.24773	.25801
5	1	3	.23839	.24287	.24758	.25250	.25761	.26291	.26838
5	1	4	.47998	.47960	.47907	.47841	.47765	.47679	.47586
5	2	1	-.18279	-.20364	-.22550	-.24839	-.27232	-.29729	-.32333
5	2	2	.21565	.22838	.24198	.25645	.27178	.28800	.30509
5	2	3	.96714	.97525	.98352	.99194	1.00053	1.00930	1.01824
6	0	1	.15245	.14749	.14164	.13487	.12713	.11839	.10863
6	0	2	.17404	.17689	.18034	.18443	.18920	.19468	.20090
6	0	3	.17427	.17671	.17950	.18264	.18613	.18996	.19412
6	0	4	.17217	.17363	.17520	.17686	.17860	.18040	.18227
6	0	5	.16826	.16840	.16846	.16844	.16832	.16812	.16783
6	0	6	.15881	.15688	.15485	.15276	.15062	.14845	.14626
6	1	1	.07445	.06480	.05417	.04251	.02979	.01598	.00104
6	1	2	.16617	.17073	.17601	.18206	.18889	.19655	.20505
6	1	3	.18602	.19006	.19446	.19921	.20431	.20976	.21554
6	1	4	.19584	.19816	.20055	.20299	.20548	.20799	.21054
6	1	5	.37752	.37625	.37481	.37323	.37152	.36972	.36784
6	2	1	-.07127	-.08818	-.10622	-.12542	-.14582	-.16744	-.19030
6	2	2	.15590	.16372	.17240	.18197	.19245	.20385	.21620
6	2	3	.21266	.21925	.22616	.23339	.24092	.24875	.25688
6	2	4	.70271	.70521	.70765	.71006	.71245	.71483	.71721
6	3	1	-.38660	-.41645	-.44776	-.48054	-.51482	-.55062	-.58795
6	3	2	.14302	.15797	.17391	.19085	.20880	.22779	.24783
6	3	3	1.24358	1.25849	1.27385	1.28970	1.30602	1.32283	1.34012
7	0	1	.12816	.12292	.11670	.10944	.10109	.09160	.08092
7	0	2	.14859	.15096	.15388	.15740	.16156	.16642	.17200
7	0	3	.14960	.15190	.15459	.15767	.16115	.16504	.16933
7	0	4	.14864	.15035	.15225	.15434	.15660	.15902	.16159
7	0	5	.14654	.14741	.14830	.14922	.15013	.15104	.15194
7	0	6	.14318	.14300	.14272	.14234	.14186	.14129	.14064
7	0	7	.13529	.13347	.13156	.12960	.12761	.12559	.12357
7	1	1	.07055	.06149	.05134	.04005	.02757	.01384	-.00117
7	1	2	.13912	.14253	.14659	.15137	.15689	.16320	.17033
7	1	3	.15356	.15704	.16093	.16525	.17000	.17517	.18077
7	1	4	.16092	.16349	.16624	.16915	.17223	.17545	.17881
7	1	5	.16453	.16576	.16698	.16819	.16938	.17054	.17168

Coefficients a_i for the BLUE of μ
Shape parameter k

n	s	i	0.7	0.8	0.9	1.0	1.1	1.2	1.3
7	1	6	.31133	.30970	.30791	.30599	.30394	.30180	.29958
7	2	1	-.02585	-.04041	-.05620	-.07325	-.09162	-.11134	-.13247
7	2	2	.12565	.13090	.13695	.14382	.15155	.16017	.16973
7	2	3	.16286	.16815	.17386	.18000	.18658	.19358	.20101
7	2	4	.18394	.18763	.19145	.19539	.19945	.20361	.20787
7	2	5	.55340	.55373	.55394	.55404	.55405	.55399	.55387
7	3	1	-.19887	-.22196	-.24648	-.27249	-.30001	-.32909	-.35976
7	3	2	.10560	.11434	.12401	.13463	.14625	.15888	.17254
7	3	3	.18401	.19226	.20092	.20997	.21942	.22926	.23950
7	3	4	.90926	.91535	.92156	.92788	.93434	.94095	.94772
7	4	1	-.56652	-.60497	-.64533	-.68763	-.73191	-.77820	-.82652
7	4	2	.07232	.08869	.10615	.12471	.14439	.16521	.18720
7	4	3	1.49420	1.51627	1.53917	1.56292	1.58752	1.61298	1.63932
8	0	1	.11009	.10469	.09823	.09065	.08189	.07187	.06054
8	0	2	.12942	.13138	.13384	.13684	.14046	.14472	.14968
8	0	3	.13087	.13298	.13548	.13840	.14173	.14551	.14973
8	0	4	.13056	.13232	.13432	.13656	.13903	.14173	.14466
8	0	5	.12935	.13054	.13183	.13322	.13468	.13621	.13781
8	0	6	.12741	.12790	.12837	.12881	.12922	.12960	.12993
8	0	7	.12451	.12413	.12365	.12308	.12240	.12164	.12080
8	0	8	.11777	.11606	.11428	.11245	.11059	.10872	.10685
8	1	1	.06548	.05688	.04712	.03614	.02387	.01024	-.00480
8	1	2	.11999	.12259	.12578	.12962	.13414	.13940	.14545
8	1	3	.13122	.13419	.13760	.14146	.14578	.15057	.15583
8	1	4	.13699	.13949	.14224	.14522	.14844	.15189	.15556
8	1	5	.14009	.14176	.14351	.14533	.14721	.14915	.15114
8	1	6	.14129	.14190	.14248	.14301	.14349	.14392	.14431
8	1	7	.26494	.26318	.26127	.25922	.25707	.25482	.25251
8	2	1	-.00394	-.01692	-.03118	-.04679	-.06380	-.08227	-.10226
8	2	2	.10673	.11044	.11485	.12000	.12595	.13273	.14039
8	2	3	.13340	.13767	.14240	.14761	.15330	.15948	.16616
8	2	4	.14864	.15214	.15586	.15981	.16396	.16833	.17290
8	2	5	.15824	.16042	.16264	.16489	.16718	.16949	.17182
8	2	6	.45692	.45626	.45543	.45448	.45341	.45224	.45100
8	3	1	-.11569	-.13493	-.15562	-.17783	-.20160	-.22699	-.25406
8	3	2	.08757	.09320	.09967	.10700	.11524	.12443	.13459
8	3	3	.13938	.14564	.15237	.15960	.16731	.17553	.18424
8	3	4	.16978	.17464	.17968	.18489	.19028	.19584	.20155
8	3	5	.71897	.72145	.72390	.72633	.72876	.73120	.73367
8	4	1	-.31292	-.34187	-.37257	-.40507	-.43942	-.47567	-.51386
8	4	2	.05782	.06708	.07732	.08857	.10085	.11421	.12868
8	4	3	.15438	.16396	.17401	.18451	.19549	.20694	.21887
8	4	4	1.10073	1.11083	1.12124	1.13199	1.14308	1.15451	1.16631
9	0	1	.09616	.09066	.08406	.07627	.06722	.05683	.04503
9	0	2	.11448	.11609	.11814	.12071	.12382	.12755	.13193
9	0	3	.11619	.11809	.12039	.12309	.12623	.12982	.13387
9	0	4	.11628	.11799	.11998	.12224	.12477	.12758	.13066
9	0	5	.11559	.11692	.11839	.12001	.12177	.12366	.12566
9	0	6	.11437	.11519	.11606	.11697	.11790	.11885	.11981
9	0	7	.11262	.11285	.11303	.11317	.11326	.11330	.11328
9	0	8	.11008	.10959	.10899	.10830	.10752	.10666	.10573
9	0	9	.10423	.10261	.10094	.09924	.09751	.09577	.09405
9	1	1	.06041	.05219	.04277	.03205	.01998	.00646	-.00857
9	1	2	.10562	.10763	.11015	.11326	.11701	.12143	.12658
9	1	3	.11477	.11732	.12030	.12373	.12763	.13202	.13691
9	1	4	.11950	.12183	.12444	.12734	.13052	.13398	.13772

Coefficients a_i for the BLUE of μ
Shape parameter k

n	s	i	0.7	0.8	0.9	1.0	1.1	1.2	1.3
9	1	5	.12213	.12393	.12587	.12794	.13014	.13247	.13491
9	1	6	.12342	.12451	.12563	.12676	.12791	.12906	.13022
9	1	7	.12356	.12380	.12398	.12411	.12418	.12419	.12414
9	1	8	.23059	.22880	.22686	.22480	.22263	.22039	.21809
9	2	1	.00760	-.00423	-.01737	-.03192	-.04793	-.06549	-.08466
9	2	2	.09344	.09612	.09943	.10340	.10809	.11354	.11981
9	2	3	.11375	.11724	.12120	.12565	.13059	.13605	.14204
9	2	4	.12538	.12853	.13197	.13569	.13968	.14396	.14851
9	2	5	.13285	.13521	.13769	.14028	.14298	.14577	.14867
9	2	6	.13769	.13900	.14031	.14161	.14290	.14418	.14544
9	2	7	.38928	.38812	.38678	.38530	.38369	.38198	.38019
9	3	1	-.07154	-.08825	-.10642	-.12614	-.14748	-.17048	-.19523
9	3	2	.07648	.08030	.08485	.09017	.09631	.10331	.11122
9	3	3	.11372	.11860	.12397	.12986	.13628	.14324	.15074
9	3	4	.13570	.13998	.14452	.14932	.15439	.15971	.16529
9	3	5	.15031	.15333	.15643	.15961	.16286	.16617	.16955
9	3	6	.59533	.59604	.59665	.59718	.59764	.59805	.59843
9	4	1	-.19692	-.22058	-.24594	-.27305	-.30199	-.33280	-.36556
9	4	2	.05173	.05753	.06418	.07172	.08020	.08966	.10013
9	4	3	.11610	.12312	.13065	.13873	.14734	.15650	.16622
9	4	4	.15444	.16031	.16639	.17270	.17922	.18597	.19293
9	4	5	.87464	.87963	.88471	.88991	.89522	.90068	.90628
9	5	1	-.41601	-.45051	-.48708	-.52578	-.56666	-.60977	-.65517
9	5	2	.01256	.02207	.03259	.04415	.05678	.07052	.08541
9	5	3	.12475	.13539	.14654	.15821	.17041	.18315	.19643
9	5	4	1.27870	1.29306	1.30796	1.32342	1.33946	1.35610	1.37333
10	0	1	.08511	.07956	.07288	.06495	.05571	.04506	.03292
10	0	2	.10251	.10382	.10554	.10771	.11039	.11363	.11748
10	0	3	.10437	.10608	.10817	.11066	.11358	.11694	.12078
10	0	4	.10472	.10636	.10828	.11048	.11299	.11580	.11892
10	0	5	.10438	.10574	.10729	.10902	.11093	.11301	.11525
10	0	6	.10359	.10458	.10567	.10684	.10809	.10940	.11077
10	0	7	.10243	.10298	.10355	.10412	.10469	.10524	.10578
10	0	8	.10085	.10090	.10089	.10083	.10070	.10052	.10028
10	0	9	.09860	.09804	.09738	.09662	.09579	.09488	.09390
10	0	10	.09344	.09193	.09036	.08876	.08715	.08553	.08393
10	1	1	.05572	.04782	.03868	.02821	.01632	.00292	-.01207
10	1	2	.09437	.09592	.09793	.10046	.10357	.10731	.11171
10	1	3	.10209	.10428	.10690	.10995	.11346	.11747	.12197
10	1	4	.10609	.10823	.11067	.11341	.11647	.11984	.12353
10	1	5	.10836	.11016	.11214	.11429	.11662	.11913	.12179
10	1	6	.10959	.11089	.11226	.11372	.11523	.11681	.11843
10	1	7	.11002	.11072	.11141	.11210	.11277	.11342	.11405
10	1	8	.10965	.10965	.10959	.10947	.10928	.10904	.10874
10	1	9	.20411	.20234	.20042	.19839	.19627	.19408	.19183
10	2	1	.01395	.00301	-.00927	-.02299	-.03823	-.05508	-.07360
10	2	2	.08342	.08538	.08789	.09099	.09474	.09919	.10439
10	2	3	.09958	.10247	.10582	.10964	.11396	.11881	.12419
10	2	4	.10883	.11164	.11475	.11818	.12193	.12600	.13039
10	2	5	.11482	.11714	.11963	.12229	.12511	.12810	.13123
10	2	6	.11884	.12046	.12214	.12387	.12564	.12746	.12932
10	2	7	.12140	.12217	.12292	.12364	.12433	.12499	.12562
10	2	8	.33917	.33774	.33614	.33438	.33251	.33052	.32846
10	3	1	-.04553	-.06042	-.07680	-.09474	-.11433	-.13564	-.15875
10	3	2	.06864	.07129	.07458	.07855	.08326	.08875	.09507
10	3	3	.09699	.10087	.10524	.11013	.11555	.12153	.12807

Coefficients a_i for the BLUE of μ
Shape parameter k

n	s	i	0.7	0.8	0.9	1.0	1.1	1.2	1.3
10	3	4	.11373	.11744	.12145	.12578	.13042	.13538	.14065
10	3	5	.12498	.12795	.13107	.13433	.13774	.14128	.14495
10	3	6	.13288	.13482	.13679	.13879	.14080	.14284	.14489
10	3	7	.50830	.50805	.50767	.50716	.50656	.50588	.50513
10	4	1	-.13327	-.15351	-.17541	-.19905	-.22451	-.25186	-.28115
10	4	2	.04809	.05189	.05643	.06176	.06793	.07497	.08293
10	4	3	.09462	.09995	.10582	.11224	.11922	.12677	.13492
10	4	4	.12246	.12740	.13263	.13816	.14399	.15012	.15656
10	4	5	.14138	.14515	.14904	.15304	.15714	.16134	.16565
10	4	6	.72672	.72912	.73149	.73386	.73624	.73865	.74110
10	5	1	-.27099	-.29885	-.32865	-.36044	-.39431	-.43031	-.46852
10	5	2	.01795	.02374	.03040	.03798	.04650	.05602	.06657
10	5	3	.09331	.10092	.10908	.11782	.12713	.13704	.14756
10	5	4	.13857	.14529	.15227	.15951	.16701	.17478	.18281
10	5	5	1.02115	1.02890	1.03689	1.04514	1.05366	1.06247	1.07158
11	0	1	.07614	.07058	.06385	.05584	.04647	.03564	.02325
11	0	2	.09272	.09378	.09519	.09702	.09931	.10212	.10549
11	0	3	.09466	.09620	.09809	.10036	.10306	.10619	.10980
11	0	4	.09519	.09673	.09855	.10068	.10311	.10587	.10895
11	0	5	.09507	.09643	.09799	.09976	.10173	.10390	.10628
11	0	6	.09457	.09565	.09686	.09818	.09962	.10116	.10280
11	0	7	.09377	.09452	.09531	.09616	.09703	.09793	.09886
11	0	8	.09269	.09305	.09340	.09373	.09404	.09432	.09457
11	0	9	.09127	.09119	.09105	.09084	.09058	.09025	.08986
11	0	10	.08926	.08865	.08795	.08717	.08631	.08538	.08439
11	0	11	.08466	.08323	.08175	.08026	.07875	.07724	.07574
11	1	1	.05148	.04386	.03497	.02472	.01301	-.00027	-.01519
11	1	2	.08529	.08648	.08808	.09014	.09273	.09588	.09966
11	1	3	.09197	.09387	.09617	.09888	.10205	.10570	.10985
11	1	4	.09546	.09740	.09966	.10223	.10513	.10837	.11194
11	1	5	.09746	.09919	.10113	.10328	.10564	.10821	.11097
11	1	6	.09859	.09997	.10146	.10308	.10479	.10661	.10853
11	1	7	.09910	.10004	.10102	.10204	.10308	.10414	.10522
11	1	8	.09909	.09952	.09993	.10031	.10066	.10098	.10126
11	1	9	.09848	.09832	.09811	.09782	.09748	.09707	.09661
11	1	10	.18308	.18134	.17947	.17749	.17543	.17331	.17114
11	2	1	.01748	.00725	-.00434	-.01740	-.03201	-.04827	-.06627
11	2	2	.07549	.07693	.07884	.08128	.08431	.08797	.09231
11	2	3	.08880	.09122	.09407	.09738	.10118	.10549	.11033
11	2	4	.09639	.09888	.10169	.10484	.10832	.11215	.11633
11	2	5	.10133	.10352	.10593	.10854	.11135	.11437	.11759
11	2	6	.10469	.10641	.10823	.11016	.11218	.11428	.11647
11	2	7	.10697	.10809	.10923	.11039	.11156	.11274	.11393
11	2	8	.10832	.10874	.10913	.10949	.10980	.11007	.11031
11	2	9	.30053	.29896	.29721	.29532	.29331	.29120	.28901
11	3	1	-.02911	-.04264	-.05766	-.07426	-.09253	-.11255	-.13442
11	3	2	.06262	.06447	.06688	.06989	.07355	.07793	.08305
11	3	3	.08511	.08825	.09187	.09598	.10062	.10580	.11155
11	3	4	.09838	.10159	.10512	.10900	.11322	.11779	.12272
11	3	5	.10733	.11011	.11308	.11625	.11961	.12316	.12690
11	3	6	.11372	.11582	.11801	.12027	.12261	.12501	.12749
11	3	7	.11832	.11958	.12085	.12211	.12337	.12462	.12586
11	3	8	.44362	.44282	.44186	.44076	.43955	.43824	.43685
11	4	1	-.09446	-.11227	-.13173	-.15292	-.17593	-.20083	-.22771
11	4	2	.04536	.04789	.05107	.05495	.05956	.06497	.07120
11	4	3	.08088	.08505	.08973	.09496	.10074	.10710	.11405

Coefficients a_i for the BLUE of μ
Shape parameter k

n	s	i	0.7	0.8	0.9	1.0	1.1	1.2	1.3
11	4	4	.10216	.10632	.11082	.11565	.12084	.12637	.13225
11	4	5	.11672	.12022	.12390	.12776	.13179	.13598	.14034
11	4	6	.12727	.12979	.13237	.13499	.13766	.14038	.14314
11	4	7	.62207	.62300	.62384	.62461	.62534	.62604	.62672
11	5	1	-.19002	-.21361	-.23906	-.26645	-.29585	-.32734	-.36100
11	5	2	.02135	.02503	.02946	.03469	.04075	.04770	.05557
11	5	3	.07613	.08182	.08806	.09488	.10229	.11030	.11894
11	5	4	.10916	.11465	.12048	.12663	.13311	.13993	.14709
11	5	5	.13187	.13631	.14090	.14562	.15048	.15547	.16061
11	5	6	.85151	.85580	.86017	.86464	.86922	.87394	.87880
12	0	1	.06873	.06318	.05643	.04838	.03893	.02797	.01541
12	0	2	.08455	.08540	.08656	.08809	.09004	.09246	.09540
12	0	3	.08655	.08792	.08962	.09170	.09418	.09709	.10046
12	0	4	.08720	.08863	.09035	.09238	.09472	.09738	.10039
12	0	5	.08724	.08856	.09010	.09186	.09384	.09605	.09849
12	0	6	.08693	.08805	.08931	.09072	.09227	.09395	.09576
12	0	7	.08638	.08723	.08817	.08918	.09026	.09140	.09260
12	0	8	.08561	.08616	.08674	.08733	.08794	.08854	.08915
12	0	9	.08461	.08483	.08502	.08517	.08530	.08538	.08543
12	0	10	.08332	.08315	.08291	.08261	.08225	.08182	.08134
12	0	11	.08151	.08088	.08016	.07936	.07849	.07756	.07658
12	0	12	.07737	.07601	.07462	.07321	.07180	.07038	.06898
12	1	1	.04770	.04032	.03166	.02161	.01007	-.00308	-.01791
12	1	2	.07779	.07869	.07995	.08163	.08378	.08644	.08967
12	1	3	.08370	.08535	.08737	.08979	.09265	.09597	.09978
12	1	4	.08679	.08856	.09064	.09303	.09576	.09884	.10228
12	1	5	.08858	.09023	.09210	.09420	.09653	.09909	.10188
12	1	6	.08963	.09102	.09256	.09424	.09606	.09802	.10010
12	1	7	.09017	.09123	.09237	.09357	.09484	.09617	.09754
12	1	8	.09030	.09098	.09167	.09237	.09307	.09377	.09446
12	1	9	.09005	.09029	.09050	.09066	.09079	.09088	.09092
12	1	10	.08932	.08906	.08874	.08836	.08791	.08740	.08684
12	1	11	.16597	.16428	.16246	.16054	.15855	.15651	.15442
12	2	1	.01939	.00975	-.00127	-.01378	-.02787	-.04364	-.06120
12	2	2	.06902	.07006	.07151	.07343	.07587	.07889	.08253
12	2	3	.08027	.08230	.08475	.08763	.09099	.09484	.09922
12	2	4	.08667	.08888	.09141	.09428	.09750	.10108	.10504
12	2	5	.09083	.09288	.09515	.09766	.10040	.10337	.10658
12	2	6	.09369	.09541	.09726	.09926	.10138	.10363	.10600
12	2	7	.09568	.09696	.09831	.09971	.10116	.10265	.10419
12	2	8	.09699	.09776	.09853	.09930	.10006	.10080	.10154
12	2	9	.09765	.09785	.09800	.09811	.09817	.09819	.09816
12	2	10	.26981	.26816	.26635	.26440	.26234	.26017	.25794
12	3	1	-.01825	-.03071	-.04465	-.06019	-.07741	-.09642	-.11731
12	3	2	.05775	.05902	.06078	.06308	.06595	.06947	.07367
12	3	3	.07617	.07874	.08176	.08526	.08927	.09381	.09889
12	3	4	.08702	.08979	.09291	.09639	.10022	.10442	.10901
12	3	5	.09434	.09688	.09966	.10266	.10589	.10935	.11304
12	3	6	.09960	.10170	.10392	.10626	.10872	.11130	.11398
12	3	7	.10348	.10499	.10654	.10813	.10975	.11140	.11309
12	3	8	.10627	.10710	.10790	.10869	.10945	.11019	.11091
12	3	9	.39362	.39248	.39118	.38973	.38815	.38648	.38472
12	4	1	-.06907	-.08508	-.10271	-.12206	-.14323	-.16630	-.19137
12	4	2	.04306	.04474	.04698	.04983	.05334	.05756	.06252
12	4	3	.07127	.07459	.07840	.08273	.08760	.09303	.09905
12	4	4	.08816	.09169	.09558	.09982	.10444	.10943	.11480

Coefficients a_i for the BLUE of μ
Shape parameter k

n	s	i	0.7	0.8	0.9	1.0	1.1	1.2	1.3
12	4	5	.09975	.10293	.10632	.10993	.11376	.11781	.12207
12	4	6	.10823	.11077	.11341	.11615	.11899	.12193	.12496
12	4	7	.11460	.11632	.11807	.11983	.12160	.12339	.12519
12	4	8	.54400	.54404	.54395	.54377	.54350	.54316	.54278
12	5	1	-.13976	-.16035	-.18275	-.20705	-.23332	-.26166	-.29215
12	5	2	.02342	.02577	.02877	.03247	.03691	.04213	.04819
12	5	3	.06539	.06977	.07468	.08015	.08620	.09285	.10011
12	5	4	.09073	.09527	.10018	.10545	.11109	.11710	.12350
12	5	5	.10824	.11222	.11640	.12078	.12536	.13013	.13510
12	5	6	.12114	.12419	.12731	.13051	.13378	.13711	.14052
12	5	7	.73085	.73314	.73541	.73769	.73999	.74234	.74473
12	6	1	-.24251	-.26930	-.29813	-.32910	-.36227	-.39774	-.43558
12	6	2	-.00391	-.00043	.00381	.00884	.01471	.02146	.02914
12	6	3	.05827	.06422	.07075	.07787	.08560	.09397	.10297
12	6	4	.09595	.10192	.10824	.11492	.12196	.12937	.13714
12	6	5	.12203	.12708	.13228	.13765	.14319	.14889	.15475
12	6	6	.97017	.97651	.98306	.98982	.99681	1.00406	1.01157
13	0	1	.06251	.05697	.05023	.04217	.03268	.02164	.00896
13	0	2	.07765	.07831	.07924	.08051	.08216	.08423	.08679
13	0	3	.07967	.08088	.08242	.08431	.08658	.08928	.09242
13	0	4	.08040	.08173	.08335	.08526	.08749	.09006	.09297
13	0	5	.08056	.08183	.08333	.08505	.08701	.08922	.09167
13	0	6	.00040	.00152	.00200	.00121	.08585	.08761	.08951
13	0	7	.08001	.08093	.08195	.08307	.08428	.08558	.08697
13	0	8	.07945	.08013	.08085	.08162	.08243	.08327	.08414
13	0	9	.07872	.07913	.07953	.07994	.08034	.08073	.08110
13	0	10	.07780	.07791	.07797	.07800	.07799	.07793	.07783
13	0	11	.07662	.07639	.07608	.07572	.07529	.07480	.07425
13	0	12	.07499	.07434	.07361	.07281	.07195	.07103	.07006
13	0	13	.07122	.06994	.06862	.06729	.06596	.06463	.06331
13	1	1	.04432	.03716	.02871	.01885	.00747	-.00554	-.02028
13	1	2	.07149	.07215	.07313	.07448	.07625	.07849	.08126
13	1	3	.07680	.07823	.08000	.08217	.08475	.08777	.09127
13	1	4	.07958	.08119	.08309	.08532	.08789	.09080	.09408
13	1	5	.08121	.08276	.08454	.08656	.08883	.09135	.09412
13	1	6	.08218	.08355	.08509	.08679	.08865	.09067	.09285
13	1	7	.08272	.08384	.08506	.08637	.08778	.08928	.09085
13	1	8	.08292	.08374	.08461	.08551	.08645	.08741	.08839
13	1	9	.08285	.08332	.08380	.08427	.08472	.08516	.08558
13	1	10	.08247	.08257	.08263	.08264	.08261	.08254	.08242
13	1	11	.08169	.08136	.08097	.08051	.08000	.07943	.07881
13	1	12	.15177	.15013	.14837	.14652	.14461	.14264	.14064
13	2	1	.02034	.01118	.00064	-.01139	-.02503	-.04038	-.05755
13	2	2	.06362	.06433	.06542	.06692	.06889	.07138	.07443
13	2	3	.07331	.07504	.07715	.07967	.08265	.08610	.09006
13	2	4	.07882	.08079	.08307	.08570	.08867	.09201	.09573
13	2	5	.08241	.08430	.08644	.08882	.09145	.09434	.09748
13	2	6	.08488	.08655	.08838	.09037	.09252	.09483	.09729
13	2	7	.08662	.08797	.08941	.09094	.09255	.09423	.09600
13	2	8	.08782	.08878	.08978	.09080	.09184	.09290	.09398
13	2	9	.08856	.08908	.08959	.09008	.09055	.09099	.09142
13	2	10	.08882	.08886	.08885	.08878	.08868	.08852	.08833
13	2	11	.24479	.24312	.24128	.23932	.23724	.23507	.23283
13	3	1	-.01080	-.02239	-.03547	-.05014	-.06651	-.08469	-.10477
13	3	2	.05368	.05452	.05579	.05752	.05979	.06262	.06607
13	3	3	.06913	.07126	.07381	.07682	.08030	.08430	.08883

Coefficients a_i for the BLUE of μ
Shape parameter k

n	s	i	0.7	0.8	0.9	1.0	1.1	1.2	1.3
13	3	4	.07822	.08064	.08341	.08652	.09001	.09387	.09813
13	3	5	.08435	.08667	.08924	.09205	.09512	.09844	.10202
13	3	6	.08877	.09079	.09297	.09529	.09777	.10040	.10318
13	3	7	.09207	.09366	.09534	.09709	.09891	.10079	.10274
13	3	8	.09453	.09561	.09672	.09784	.09897	.10011	.10125
13	3	9	.09626	.09678	.09727	.09774	.09817	.09857	.09895
13	3	10	.35379	.35245	.35093	.34927	.34748	.34558	.34360
13	4	1	-.05163	-.06622	-.08243	-.10034	-.12007	-.14171	-.16536
13	4	2	.04102	.04210	.04366	.04575	.04843	.05175	.05574
13	4	3	.06411	.06679	.06993	.07356	.07771	.08240	.08766
13	4	4	.07791	.08093	.08431	.08805	.09218	.09669	.10161
13	4	5	.08739	.09025	.09334	.09669	.10028	.10413	.10822
13	4	6	.09436	.09680	.09937	.10209	.10495	.10795	.11107
13	4	7	.09966	.10153	.10345	.10543	.10747	.10956	.11169
13	4	8	.10372	.10491	.10610	.10728	.10847	.10964	.11081
13	4	9	.48345	.48293	.48227	.48147	.48057	.47959	.47855
13	5	1	-.10630	-.12467	-.14480	-.16679	-.19073	-.21673	-.24486
13	5	2	.02459	.02606	.02809	.03073	.03402	.03801	.04276
13	5	3	.05801	.06145	.06539	.06987	.07490	.08051	.08672
13	5	4	.07817	.08197	.08615	.09071	.09565	.10099	.10674
13	5	5	.09214	.09567	.09943	.10344	.10768	.11216	.11688
13	5	6	.10250	.10543	.10849	.11166	.11496	.11837	.12190
13	5	7	.11044	.11258	.11477	.11699	.11924	.12152	.12384
13	5	8	.64046	.64151	.64248	.64340	.64429	.64516	.64603
13	6	1	-.18191	-.20516	-.23037	-.25763	-.28703	-.31867	-.35263
13	6	2	.00266	.00478	.00756	.01102	.01522	.02020	.02600
13	6	3	.05049	.05501	.06009	.06574	.07198	.07883	.08632
13	6	4	.07950	.08437	.08961	.09525	.10127	.10770	.11453
13	6	5	.09967	.10407	.10870	.11354	.11860	.12389	.12939
13	6	6	.11467	.11820	.12183	.12555	.12936	.13326	.13725
13	6	7	.83493	.83872	.84259	.84654	.85061	.85480	.85914
13	7	1	-.29135	-.32119	-.35325	-.38764	-.42443	-.46371	-.50556
13	7	2	-.02785	-.02463	-.02065	-.01588	-.01027	-.00380	.00360
13	7	3	.04104	.04718	.05391	.06126	.06923	.07785	.08714
13	7	4	.08292	.08929	.09604	.10317	.11069	.11860	.12690
13	7	5	.11205	.11763	.12339	.12933	.13547	.14180	.14832
13	7	6	1.08319	1.09171	1.10056	1.10975	1.11931	1.12925	1.13960
14	0	1	.05721	.05171	.04499	.03694	.02743	.01635	.00359
14	0	2	.07174	.07223	.07297	.07401	.07538	.07715	.07936
14	0	3	.07376	.07484	.07622	.07794	.08002	.08251	.08542
14	0	4	.07455	.07579	.07730	.07910	.08122	.08366	.08646
14	0	5	.07480	.07601	.07745	.07913	.08105	.08322	.08565
14	0	6	.07475	.07585	.07712	.07858	.08021	.08201	.08400
14	0	7	.07449	.07543	.07649	.07768	.07897	.08038	.08190
14	0	8	.07407	.07482	.07564	.07653	.07748	.07849	.07955
14	0	9	.07352	.07405	.07461	.07519	.07579	.07640	.07702
14	0	10	.07284	.07312	.07340	.07366	.07391	.07413	.07432
14	0	11	.07199	.07201	.07198	.07191	.07180	.07164	.07144
14	0	12	.07091	.07062	.07027	.06986	.06938	.06885	.06826
14	0	13	.06941	.06876	.06804	.06724	.06639	.06548	.06454
14	0	14	.06597	.06475	.06350	.06224	.06098	.05973	.05849
14	1	1	.04131	.03435	.02609	.01640	.00518	-.00769	-.02233
14	1	2	.06611	.06658	.06732	.06839	.06984	.07171	.07406
14	1	3	.07094	.07218	.07375	.07568	.07801	.08077	.08398
14	1	4	.07348	.07494	.07669	.07876	.08116	.08391	.08702
14	1	5	.07498	.07643	.07812	.08005	.08224	.08470	.08742

Coefficients a_i for the BLUE of μ
Shape parameter k

n	s	i	0.7	0.8	0.9	1.0	1.1	1.2	1.3
14	1	6	.07589	.07722	.07872	.08041	.08227	.08432	.08655
14	1	7	.07642	.07755	.07881	.08018	.08167	.08327	.08498
14	1	8	.07666	.07756	.07853	.07956	.08065	.08180	.08299
14	1	9	.07667	.07730	.07796	.07863	.07931	.08000	.08069
14	1	10	.07647	.07680	.07711	.07741	.07768	.07792	.07814
14	1	11	.07602	.07602	.07597	.07588	.07573	.07555	.07531
14	1	12	.07524	.07486	.07442	.07392	.07336	.07275	.07210
14	1	13	.13980	.13821	.13651	.13473	.13289	.13100	.12908
14	2	1	.02069	.01195	.00183	-.00980	-.02304	-.03802	-.05484
14	2	2	.05901	.05948	.06026	.06142	.06299	.06503	.06759
14	2	3	.06751	.06898	.07081	.07303	.07568	.07878	.08237
14	2	4	.07234	.07409	.07616	.07856	.08130	.08442	.08791
14	2	5	.07548	.07723	.07922	.08147	.08398	.08676	.08981
14	2	6	.07765	.07925	.08102	.08297	.08511	.08742	.08992
14	2	7	.07920	.08055	.08202	.08361	.08530	.08710	.08900
14	2	8	.08028	.08134	.08246	.08364	.08486	.08613	.08745
14	2	9	.08101	.08172	.08246	.08320	.08394	.08468	.08543
14	2	10	.08139	.08173	.08205	.08234	.08260	.08282	.08302
14	2	11	.08141	.08133	.08120	.08102	.08080	.08053	.08022
14	2	12	.22401	.22234	.22051	.21855	.21649	.21434	.21212
14	3	1	-.00556	-.01644	-.02880	-.04275	-.05840	-.07588	-.09528
14	3	2	.05019	.05070	.05158	.05288	.05464	.05692	.05977
14	3	3	.06342	.06519	.06736	.06995	.07300	.07654	.08059
14	3	4	.07118	.07331	.07577	.07857	.08175	.08530	.08925
14	3	5	.07642	.07853	.08089	.08352	.08641	.08958	.09302
14	3	6	.08020	.08211	.08420	.08646	.08891	.09153	.09432
14	3	7	.08303	.08463	.08635	.08816	.09008	.09209	.09419
14	3	8	.08518	.08640	.08767	.08899	.09034	.09173	.09315
14	3	9	.08678	.08756	.08835	.08914	.08991	.09069	.09145
14	3	10	.08787	.08817	.08844	.08868	.08888	.08905	.08918
14	3	11	.32130	.31983	.31819	.31640	.31448	.31246	.31035
14	4	1	-.03918	-.05264	-.06770	-.08445	-.10301	-.12350	-.14600
14	4	2	.03916	.03979	.04084	.04237	.04441	.04702	.05024
14	4	3	.05851	.06069	.06331	.06638	.06995	.07404	.07867
14	4	4	.07006	.07266	.07561	.07893	.08263	.08673	.09123
14	4	5	.07799	.08055	.08337	.08645	.08980	.09342	.09732
14	4	6	.08383	.08612	.08858	.09121	.09400	.09697	.10010
14	4	7	.08830	.09018	.09216	.09423	.09638	.09862	.10095
14	4	8	.09178	.09316	.09458	.09603	.09750	.09900	.10052
14	4	9	.09445	.09527	.09608	.09688	.09766	.09842	.09916
14	4	10	.43510	.43422	.43317	.43198	.43068	.42928	.42780
14	5	1	-.08290	-.09954	-.11790	-.13809	-.16023	-.18440	-.21071
14	5	2	.02519	.02604	.02737	.02923	.03167	.03474	.03849
14	5	3	.05256	.05531	.05852	.06223	.06648	.07127	.07665
14	5	4	.06907	.07229	.07588	.07985	.08421	.08898	.09417
14	5	5	.08052	.08364	.08702	.09066	.09457	.09875	.10320
14	5	6	.08903	.09177	.09467	.09773	.10095	.10432	.10784
14	5	7	.09561	.09781	.10008	.10242	.10484	.10732	.10987
14	5	8	.10078	.10232	.10387	.10543	.10700	.10859	.11018
14	5	9	.57014	.57037	.57050	.57054	.57050	.57042	.57031
14	6	1	-.14113	-.16175	-.18427	-.20878	-.23539	-.26420	-.29529
14	6	2	.00710	.00832	.01010	.01248	.01551	.01924	.02370
14	6	3	.04527	.04878	.05282	.05739	.06253	.06826	.07460
14	6	4	.06842	.07245	.07687	.08169	.08690	.09254	.09860
14	6	5	.08456	.08840	.09249	.09684	.10144	.10631	.11143
14	6	6	.09661	.09991	.10334	.10691	.11062	.11446	.11845

Coefficients a_i for the BLUE of μ
Shape parameter k

n	s	i	0.7	0.8	0.9	1.0	1.1	1.2	1.3
14	6	7	.10597	.10851	.11110	.11374	.11644	.11918	.12197
14	6	8	.73320	.73538	.73755	.73974	.74195	.74422	.74655
14	7	1	-.22132	-.24710	-.27500	-.30510	-.33751	-.37232	-.40962
14	7	2	-.01703	-.01518	-.01268	-.00950	-.00559	-.00091	.00459
14	7	3	.03614	.04076	.04595	.05171	.05808	.06508	.07272
14	7	4	.06850	.07364	.07918	.08512	.09147	.09824	.10544
14	7	5	.09110	.09587	.10089	.10614	.11164	.11737	.12336
14	7	6	.10800	.11197	.11605	.12025	.12455	.12897	.13349
14	7	7	.93461	.94003	.94561	.95138	.95736	.96356	.97001
15	0	1	.05266	.04720	.04051	.03248	.02297	.01187	-.00094
15	0	2	.06662	.06697	.06754	.06837	.06950	.07099	.07289
15	0	3	.06863	.06959	.07083	.07239	.07429	.07658	.07929
15	0	4	.06947	.07061	.07202	.07371	.07571	.07804	.08072
15	0	5	.06978	.07094	.07231	.07393	.07579	.07792	.08031
15	0	6	.06981	.07088	.07214	.07358	.07521	.07703	.07904
15	0	7	.06964	.07059	.07168	.07290	.07425	.07573	.07733
15	0	8	.06934	.07013	.07102	.07199	.07304	.07416	.07536
15	0	9	.06892	.06954	.07020	.07090	.07164	.07242	.07323
15	0	10	.06840	.06881	.06923	.06966	.07010	.07053	.07095
15	0	11	.06775	.06795	.06812	.06827	.06839	.06849	.06855
15	0	12	.06697	.06692	.06682	.06668	.06649	.06626	.06598
15	0	13	.06597	.06565	.06527	.06482	.06431	.06375	.06314
15	0	14	.06460	.06395	.06323	.06244	.06161	.06072	.05980
15	0	15	.06143	.06027	.05908	.05789	.05669	.05551	.05434
15	1	1	.03862	.03184	.02375	.01423	.00316	-.00958	-.02411
15	1	2	.06147	.06176	.06230	.06313	.06430	.06585	.06784
15	1	3	.06590	.06698	.06837	.07010	.07220	.07471	.07766
15	1	4	.06825	.06957	.07118	.07310	.07534	.07793	.08089
15	1	5	.06964	.07100	.07259	.07443	.07654	.07891	.08156
15	1	6	.07050	.07177	.07323	.07488	.07673	.07877	.08101
15	1	7	.07101	.07214	.07341	.07480	.07634	.07800	.07979
15	1	8	.07128	.07222	.07325	.07436	.07556	.07682	.07816
15	1	9	.07135	.07207	.07284	.07365	.07449	.07536	.07626
15	1	10	.07124	.07172	.07221	.07270	.07319	.07368	.07415
15	1	11	.07097	.07118	.07137	.07153	.07167	.07177	.07184
15	1	12	.07049	.07041	.07028	.07010	.06988	.06961	.06929
15	1	13	.06971	.06930	.06883	.06830	.06771	.06708	.06641
15	1	14	.12957	.12803	.12639	.12468	.12290	.12109	.11925
15	2	1	.02068	.01230	.00253	-.00874	-.02163	-.03628	-.05279
15	2	2	.05504	.05530	.05584	.05670	.05794	.05960	.06173
15	2	3	.06260	.06385	.06543	.06739	.06975	.07254	.07580
15	2	4	.06689	.06845	.07032	.07252	.07505	.07796	.08124
15	2	5	.06967	.07128	.07314	.07526	.07764	.08031	.08325
15	2	6	.07161	.07312	.07482	.07672	.07881	.08110	.08358
15	2	7	.07299	.07432	.07579	.07739	.07912	.08098	.08296
15	2	8	.07398	.07508	.07626	.07753	.07887	.08028	.08175
15	2	9	.07466	.07549	.07636	.07727	.07820	.07915	.08013
15	2	10	.07508	.07561	.07614	.07667	.07719	.07770	.07819
15	2	11	.07524	.07544	.07561	.07575	.07586	.07593	.07597
15	2	12	.07510	.07494	.07473	.07447	.07416	.07381	.07342
15	2	13	.20649	.20482	.20301	.20108	.19904	.19693	.19475
15	3	1	-.00181	-.01208	-.02382	-.03716	-.05221	-.06908	-.08790
15	3	2	.04714	.04740	.04798	.04892	.05028	.05209	.05442
15	3	3	.05866	.06014	.06199	.06424	.06692	.07007	.07371
15	3	4	.06541	.06728	.06947	.07201	.07490	.07818	.08185
15	3	5	.06995	.07187	.07404	.07649	.07921	.08221	.08550

Coefficients a_i for the BLUE of μ
Shape parameter k

n	s	i	0.7	0.8	0.9	1.0	1.1	1.2	1.3
15	3	6	.07323	.07503	.07701	.07919	.08156	.08414	.08690
15	3	7	.07570	.07727	.07897	.08079	.08274	.08481	.08700
15	3	8	.07758	.07886	.08022	.08164	.08313	.08467	.08628
15	3	9	.07902	.07996	.08093	.08192	.08293	.08395	.08499
15	3	10	.08007	.08063	.08119	.08173	.08225	.08276	.08324
15	3	11	.08074	.08089	.08101	.08108	.08112	.08112	.08109
15	3	12	.29429	.29274	.29103	.28916	.28717	.28508	.28291
15	4	1	-.03004	-.04257	-.05667	-.07247	-.09007	-.10959	-.13115
15	4	2	.03744	.03774	.03840	.03947	.04101	.04305	.04565
15	4	3	.05397	.05576	.05795	.06058	.06367	.06725	.07136
15	4	4	.06382	.06608	.06867	.07163	.07497	.07870	.08284
15	4	5	.07058	.07288	.07544	.07828	.08140	.08480	.08850
15	4	6	.07556	.07769	.08001	.08252	.08522	.08811	.09120
15	4	7	.07938	.08122	.08318	.08526	.08745	.08976	.09217
15	4	8	.08239	.08385	.08538	.08697	.08861	.09030	.09204
15	4	9	.08475	.08578	.08683	.08789	.08896	.09004	.09112
15	4	10	.08656	.08712	.08765	.08817	.08866	.08913	.08957
15	4	11	.39558	.39445	.39315	.39170	.39013	.38846	.38671
15	5	1	-.06590	-.08115	-.09810	-.11686	-.13753	-.16024	-.18508
15	5	2	.02539	.02578	.02660	.02788	.02968	.03203	.03499
15	5	3	.04834	.05054	.05319	.05630	.05992	.06406	.06875
15	5	4	.06216	.06490	.06801	.07150	.07538	.07966	.08437
15	5	5	.07174	.07451	.07755	.08086	.08446	.08835	.09253
15	5	6	.07888	.08141	.08413	.08703	.09011	.09338	.09683
15	5	7	.08442	.08657	.08882	.09118	.09364	.09621	.09887
15	5	8	.08882	.09048	.09219	.09394	.09574	.09756	.09943
15	5	9	.09232	.09343	.09454	.09565	.09675	.09785	.09894
15	5	10	.51382	.51352	.51307	.51252	.51187	.51115	.51037
15	6	1	-.11228	-.13088	-.15131	-.17370	-.19814	-.22475	-.25362
15	6	2	.01016	.01075	.01183	.01344	.01561	.01840	.02186
15	6	3	.04149	.04426	.04751	.05127	.05556	.06042	.06586
15	6	4	.06048	.06386	.06762	.07178	.07634	.08133	.08674
15	6	5	.07373	.07709	.08071	.08462	.08881	.09328	.09804
15	6	6	.08365	.08667	.08986	.09323	.09677	.10048	.10436
15	6	7	.09140	.09390	.09649	.09917	.10193	.10478	.10771
15	6	8	.09758	.09944	.10133	.10324	.10517	.10713	.10912
15	6	9	.65380	.65491	.65596	.65697	.65795	.65893	.65993
15	7	1	-.17383	-.19661	-.22142	-.24836	-.27753	-.30904	-.34297
15	7	2	-.00953	-.00858	-.00707	-.00498	-.00224	.00118	.00532
15	7	3	.03302	.03657	.04065	.04528	.05048	.05628	.06270
15	7	4	.05892	.06314	.06776	.07279	.07824	.08412	.09044
15	7	5	.07704	.08115	.08553	.09019	.09511	.10032	.10580
15	7	6	.09065	.09427	.09804	.10197	.10606	.11029	.11469
15	7	7	.10129	.10419	.10716	.11020	.11330	.11646	.11969
15	7	8	.82243	.82586	.82934	.83291	.83659	.84039	.84433
15	8	1	-.25831	-.28652	-.31699	-.34983	-.38513	-.42299	-.46350
15	8	2	-.03578	-.03422	-.03203	-.02916	-.02558	-.02123	-.01609
15	8	3	.02231	.02699	.03224	.03808	.04453	.05162	.05936
15	8	4	.05775	.06312	.06891	.07511	.08173	.08880	.09630
15	8	5	.08257	.08768	.09304	.09866	.10454	.11068	.11709
15	8	6	.10122	.10559	.11009	.11471	.11947	.12435	.12937
15	8	7	1.03023	1.03735	1.04474	1.05243	1.06043	1.06878	1.07747
16	0	1	.04870	.04329	.03664	.02864	.01915	.00805	-.00478
16	0	2	.06215	.06237	.06279	.06343	.06435	.06559	.06720
16	0	3	.06414	.06499	.06610	.06751	.06925	.07136	.07386
16	0	4	.06501	.06606	.06737	.06896	.07085	.07306	.07562

Coefficients a_i for the BLUE of μ
Shape parameter k

n	s	i	0.7	0.8	0.9	1.0	1.1	1.2	1.3
16	0	5	.06538	.06646	.06778	.06933	.07113	.07320	.07554
16	0	6	.06546	.06650	.06773	.06914	.07076	.07257	.07459
16	0	7	.06537	.06632	.06740	.06863	.07001	.07153	.07320
16	0	8	.06516	.06597	.06689	.06791	.06903	.07024	.07154
16	0	9	.06484	.06550	.06623	.06702	.06787	.06877	.06972
16	0	10	.06443	.06492	.06545	.06601	.06658	.06718	.06778
16	0	11	.06392	.06424	.06456	.06487	.06517	.06546	.06574
16	0	12	.06332	.06344	.06353	.06359	.06362	.06361	.06358
16	0	13	.06259	.06249	.06234	.06214	.06189	.06160	.06127
16	0	14	.06167	.06132	.06091	.06044	.05991	.05934	.05871
16	0	15	.06040	.05976	.05905	.05827	.05745	.05659	.05570
16	0	16	.05747	.05636	.05523	.05410	.05296	.05184	.05073
16	1	1	.03621	.02959	.02166	.01230	.00138	-.01123	-.02565
16	1	2	.05741	.05757	.05793	.05855	.05947	.06074	.06240
16	1	3	.06152	.06246	.06368	.06523	.06713	.06941	.07212
16	1	4	.06370	.06491	.06639	.06816	.07026	.07270	.07549
16	1	5	.06501	.06627	.06777	.06952	.07154	.07382	.07639
16	1	6	.06583	.06704	.06845	.07005	.07186	.07388	.07611
16	1	7	.06633	.06743	.06868	.07008	.07163	.07333	.07517
16	1	8	.06660	.06756	.06862	.06978	.07104	.07239	.07384
16	1	9	.06671	.06749	.06833	.06923	.07019	.07120	.07225
16	1	10	.06667	.06725	.06786	.06849	.06914	.06980	.07047
16	1	11	.06649	.06685	.06721	.06756	.06790	.06822	.06853
16	1	12	.06618	.06630	.06639	.06645	.06648	.06647	.06643
16	1	13	.06568	.06554	.06536	.06512	.06483	.06450	.06413
16	1	14	.06492	.06449	.06400	.06345	.06285	.06221	.06153
16	1	15	.12073	.11924	.11766	.11601	.11430	.11256	.11079
16	2	1	.02043	.01236	.00292	-.00803	-.02061	-.03496	-.05118
16	2	2	.05156	.05165	.05199	.05261	.05356	.05489	.05665
16	2	3	.05836	.05943	.06081	.06253	.06464	.06716	.07012
16	2	4	.06222	.06362	.06532	.06733	.06968	.07238	.07546
16	2	5	.06473	.06621	.06794	.06992	.07219	.07473	.07756
16	2	6	.06647	.06789	.06952	.07134	.07338	.07562	.07807
16	2	7	.06772	.06901	.07045	.07204	.07378	.07566	.07769
16	2	8	.06862	.06973	.07095	.07226	.07366	.07516	.07674
16	2	9	.06925	.07015	.07111	.07212	.07318	.07428	.07543
16	2	10	.06967	.07033	.07101	.07170	.07241	.07312	.07384
16	2	11	.06989	.07028	.07066	.07103	.07138	.07171	.07202
16	2	12	.06991	.07001	.07007	.07010	.07009	.07005	.06997
16	2	13	.06967	.06945	.06918	.06886	.06850	.06809	.06765
16	2	14	.19150	.18986	.18808	.18618	.18418	.18211	.17998
16	3	1	.00093	-.00883	-.02005	-.03285	-.04737	-.06373	-.08204
16	3	2	.04446	.04451	.04484	.04549	.04650	.04793	.04983
16	3	3	.05462	.05586	.05745	.05940	.06177	.06457	.06785
16	3	4	.06056	.06222	.06418	.06648	.06913	.07215	.07556
16	3	5	.06456	.06631	.06831	.07058	.07314	.07598	.07912
16	3	6	.06745	.06913	.07100	.07308	.07537	.07788	.08059
16	3	7	.06962	.07114	.07279	.07460	.07654	.07863	.08086
16	3	8	.07129	.07258	.07397	.07544	.07700	.07865	.08038
16	3	9	.07258	.07360	.07468	.07580	.07696	.07815	.07938
16	3	10	.07355	.07428	.07502	.07576	.07651	.07726	.07800
16	3	11	.07425	.07464	.07502	.07537	.07571	.07602	.07630
16	3	12	.07464	.07468	.07468	.07463	.07455	.07444	.07429
16	3	13	.27147	.26988	.26813	.26622	.26420	.26207	.25987
16	4	1	-.02316	-.03491	-.04822	-.06321	-.08000	-.09871	-.11945
16	4	2	.03585	.03589	.03624	.03695	.03807	.03965	.04173

Coefficients a_i for the BLUE of μ
Shape parameter k

n	s	i	0.7	0.8	0.9	1.0	1.1	1.2	1.3
16	4	3	.05020	.05167	.05352	.05577	.05846	.06162	.06527
16	4	4	.05873	.06069	.06299	.06564	.06865	.07206	.07587
16	4	5	.06458	.06664	.06898	.07159	.07449	.07769	.08119
16	4	6	.06889	.07086	.07303	.07541	.07800	.08080	.08381
16	4	7	.07220	.07397	.07587	.07792	.08010	.08241	.08487
16	4	8	.07482	.07630	.07787	.07952	.08125	.08305	.08493
16	4	9	.07690	.07805	.07924	.08046	.08171	.08300	.08430
16	4	10	.07854	.07932	.08009	.08087	.08164	.08240	.08316
16	4	11	.07978	.08015	.08049	.08080	.08109	.08135	.08158
16	4	12	.36267	.36137	.35990	.35828	.35653	.35468	.35275
16	5	1	-.05318	-.06730	-.08309	-.10066	-.12014	-.14163	-.16526
16	5	2	.02532	.02539	.02581	.02664	.02792	.02970	.03203
16	5	3	.04492	.04671	.04890	.05153	.05463	.05823	.06236
16	5	4	.05671	.05907	.06178	.06486	.06832	.07219	.07648
16	5	5	.06488	.06734	.07007	.07309	.07641	.08002	.08394
16	5	6	.07096	.07329	.07581	.07854	.08148	.08462	.08796
16	5	7	.07570	.07775	.07994	.08225	.08470	.08727	.08997
16	5	8	.07948	.08117	.08294	.08478	.08668	.08865	.09068
16	5	9	.08253	.08380	.08510	.08642	.08776	.08911	.09049
16	5	10	.08498	.08578	.08657	.08735	.08811	.08886	.08960
16	5	11	.46769	.46701	.46617	.46521	.46413	.46297	.46175
16	6	1	-.09111	-.10808	-.12685	-.14754	-.17025	-.19509	-.22219
16	6	2	.01228	.01243	.01300	.01402	.01554	.01762	.02028
16	6	3	.03858	.04078	.04343	.04655	.05018	.05433	.05905
16	6	4	.05451	.05737	.06060	.06422	.06824	.07268	.07756
16	6	5	.06561	.06856	.07179	.07531	.07912	.08324	.08766
16	6	6	.07394	.07669	.07964	.08278	.08612	.08967	.09341
16	6	7	.08047	.08286	.08536	.08799	.09074	.09360	.09657
16	6	8	.08571	.08763	.08961	.09164	.09373	.09587	.09806
16	6	9	.08996	.09134	.09273	.09413	.09554	.09695	.09837
16	6	10	.59005	.59042	.59070	.59089	.59104	.59114	.59123
16	7	1	-.13999	-.16046	-.18289	-.20738	-.23405	-.26300	-.29434
16	7	2	-.00415	-.00382	-.00302	-.00170	.00018	.00267	.00581
16	7	3	.03084	.03361	.03687	.04064	.04495	.04983	.05529
16	7	4	.05214	.05565	.05955	.06386	.06859	.07374	.07935
16	7	5	.06705	.07061	.07445	.07859	.08302	.08775	.09279
16	7	6	.07827	.08154	.08499	.08864	.09246	.09648	.10069
16	7	7	.08708	.08986	.09275	.09573	.09882	.10200	.10529
16	7	8	.09418	.09634	.09855	.10079	.10306	.10537	.10772
16	7	9	.73460	.73667	.73875	.74084	.74296	.74515	.74740
16	8	1	-.20464	-.22950	-.25651	-.28578	-.31742	-.35153	-.38820
16	8	2	-.02538	-.02471	-.02351	-.02172	-.01931	-.01622	-.01243
16	8	3	.02122	.02479	.02888	.03353	.03876	.04458	.05103
16	8	4	.04967	.05404	.05883	.06403	.06967	.07576	.08229
16	8	5	.06962	.07397	.07861	.08353	.08874	.09424	.10004
16	8	6	.08466	.08857	.09266	.09691	.10134	.10593	.11070
16	8	7	.09648	.09972	.10304	.10644	.10991	.11347	.11710
16	8	8	.90837	.91312	.91801	.92307	.92832	.93378	.93946
17	0	1	.04524	.03987	.03328	.02531	.01585	.00477	-.00806
17	0	2	.05820	.05832	.05860	.05908	.05981	.06082	.06217
17	0	3	.06018	.06093	.06192	.06319	.06478	.06671	.06903
17	0	4	.06107	.06204	.06326	.06474	.06652	.06862	.07106
17	0	5	.06147	.06250	.06375	.06523	.06697	.06897	.07124
17	0	6	.06161	.06261	.06380	.06518	.06677	.06856	.07057
17	0	7	.06158	.06250	.06358	.06481	.06620	.06774	.06944
17	0	8	.06143	.06225	.06319	.06424	.06540	.06667	.06805

Coefficients a_i for the BLUE of μ
Shape parameter k

n	s	i	0.7	0.8	0.9	1.0	1.1	1.2	1.3
17	0	9	.06118	.06188	.06266	.06351	.06444	.06543	.06649
17	0	10	.06086	.06142	.06202	.06267	.06335	.06407	.06482
17	0	11	.06046	.06086	.06129	.06172	.06216	.06261	.06305
17	0	12	.05999	.06022	.06045	.06067	.06087	.06105	.06120
17	0	13	.05942	.05948	.05950	.05949	.05945	.05937	.05925
17	0	14	.05874	.05860	.05841	.05816	.05787	.05754	.05716
17	0	15	.05789	.05752	.05709	.05660	.05607	.05548	.05485
17	0	16	.05671	.05607	.05537	.05462	.05381	.05298	.05211
17	0	17	.05398	.05292	.05185	.05076	.04968	.04861	.04756
17	1	1	.03403	.02756	.01979	.01058	-.00020	-.01268	-.02698
17	1	2	.05384	.05387	.05409	.05452	.05523	.05624	.05762
17	1	3	.05768	.05849	.05957	.06095	.06266	.06474	.06723
17	1	4	.05972	.06082	.06218	.06382	.06578	.06807	.07071
17	1	5	.06095	.06213	.06354	.06520	.06712	.06932	.07180
17	1	6	.06173	.06289	.06424	.06579	.06756	.06953	.07173
17	1	7	.06222	.06329	.06452	.06591	.06746	.06916	.07102
17	1	8	.06250	.06346	.06453	.06572	.06701	.06842	.06994
17	1	9	.06263	.06344	.06433	.06530	.06633	.06743	.06860
17	1	10	.06264	.06328	.06397	.06470	.06547	.06627	.06709
17	1	11	.06253	.06299	.06347	.06396	.06445	.06495	.06544
17	1	12	.06231	.06258	.06283	.06307	.06329	.06349	.06366
17	1	13	.06197	.06202	.06204	.06202	.06197	.06188	.06175
17	1	14	.06148	.06129	.06106	.06078	.06045	.06007	.05966
17	1	15	.06074	.06030	.05979	.05923	.05863	.05798	.05729
17	1	16	.11301	.11157	.11004	.10845	.10681	.10513	.10344
17	2	1	.02003	.01225	.00309	-.00757	-.01987	-.03394	-.04990
17	2	2	.04849	.04845	.04860	.04901	.04972	.05076	.05219
17	2	3	.05468	.05558	.05678	.05830	.06018	.06246	.06515
17	2	4	.05818	.05944	.06098	.06282	.06499	.06751	.07040
17	2	5	.06045	.06182	.06343	.06529	.06743	.06986	.07258
17	2	6	.06204	.06338	.06492	.06667	.06864	.07083	.07323
17	2	7	.06318	.06442	.06583	.06739	.06911	.07099	.07303
17	2	8	.06400	.06511	.06633	.06766	.06911	.07066	.07231
17	2	9	.06460	.06553	.06654	.06762	.06876	.06997	.07124
17	2	10	.06501	.06574	.06651	.06732	.06816	.06903	.06991
17	2	11	.06525	.06577	.06629	.06682	.06735	.06787	.06839
17	2	12	.06534	.06562	.06588	.06612	.06633	.06652	.06669
17	2	13	.06526	.06528	.06526	.06520	.06511	.06497	.06481
17	2	14	.06496	.06470	.06438	.06402	.06362	.06317	.06269
17	2	15	.17854	.17692	.17517	.17331	.17136	.16934	.16726
17	3	1	.00294	-.00637	-.01713	-.02947	-.04353	-.05942	-.07728
17	3	2	.04207	.04195	.04208	.04248	.04321	.04431	.04582
17	3	3	.05114	.05218	.05353	.05524	.05733	.05983	.06279
17	3	4	.05643	.05790	.05966	.06175	.06418	.06697	.07015
17	3	5	.05999	.06158	.06343	.06554	.06794	.07063	.07362
17	3	6	.06256	.06413	.06589	.06788	.07007	.07250	.07514
17	3	7	.06450	.06595	.06755	.06931	.07123	.07331	.07554
17	3	8	.06599	.06727	.06865	.07015	.07174	.07345	.07525
17	3	9	.06715	.06822	.06935	.07054	.07180	.07311	.07448
17	3	10	.06805	.06887	.06972	.07060	.07151	.07243	.07337
17	3	11	.06871	.06927	.06982	.07037	.07092	.07146	.07199
17	3	12	.06916	.06942	.06966	.06988	.07007	.07023	.07037
17	3	13	.06936	.06932	.06923	.06910	.06893	.06873	.06849
17	3	14	.25194	.25033	.24855	.24664	.24460	.24247	.24026
17	4	1	-.01789	-.02898	-.04161	-.05590	-.07199	-.09000	-.11005
17	4	2	.03437	.03420	.03431	.03473	.03551	.03669	.03833

Coefficients a_i for the BLUE of μ
Shape parameter k

n	s	i	0.7	0.8	0.9	1.0	1.1	1.2	1.3
17	4	3	.04698	.04820	.04976	.05170	.05405	.05684	.06011
17	4	4	.05448	.05620	.05823	.06061	.06335	.06647	.06999
17	4	5	.05961	.06147	.06360	.06601	.06871	.07170	.07501
17	4	6	.06339	.06521	.06724	.06949	.07196	.07465	.07756
17	4	7	.06630	.06797	.06980	.07179	.07393	.07623	.07868
17	4	8	.06859	.07006	.07162	.07329	.07506	.07692	.07888
17	4	9	.07044	.07164	.07290	.07422	.07559	.07701	.07847
17	4	10	.07191	.07282	.07374	.07469	.07565	.07662	.07761
17	4	11	.07307	.07365	.07422	.07478	.07533	.07586	.07638
17	4	12	.07393	.07415	.07435	.07451	.07464	.07475	.07482
17	4	13	.33482	.33341	.33183	.33010	.32823	.32627	.32421
17	5	1	-.04344	-.05661	-.07143	-.08801	-.10648	-.12695	-.14956
17	5	2	.02508	.02489	.02501	.02548	.02635	.02767	.02948
17	5	3	.04208	.04353	.04535	.04758	.05026	.05340	.05705
17	5	4	.05229	.05432	.05670	.05943	.06255	.06606	.06998
17	5	5	.05935	.06154	.06401	.06677	.06983	.07319	.07687
17	5	6	.06462	.06675	.06909	.07165	.07442	.07742	.08065
17	5	7	.06872	.07066	.07275	.07499	.07738	.07992	.08261
17	5	8	.07200	.07367	.07544	.07730	.07925	.08129	.08341
17	5	9	.07467	.07602	.07741	.07885	.08033	.08186	.08342
17	5	10	.07685	.07782	.07881	.07981	.08081	.08182	.08283
17	5	11	.07859	.07917	.07972	.08026	.08078	.08128	.08176
17	5	12	.42920	.42825	.42713	.42500	.42161	.42305	.42151
17	6	1	-.07510	-.09075	-.10816	-.12745	-.14873	-.17213	-.19776
17	6	2	.01377	.01359	.01376	.01433	.01535	.01686	.01889
17	6	3	.03623	.03799	.04017	.04278	.04586	.04945	.05357
17	6	4	.04983	.05227	.05506	.05824	.06181	.06579	.07021
17	6	5	.05930	.06190	.06479	.06797	.07145	.07525	.07936
17	6	6	.06641	.06891	.07163	.07455	.07769	.08105	.08463
17	6	7	.07199	.07424	.07663	.07917	.08185	.08466	.08762
17	6	8	.07649	.07839	.08038	.08245	.08460	.08683	.08913
17	6	9	.08017	.08167	.08319	.08476	.08635	.08797	.08962
17	6	10	.08319	.08423	.08526	.08628	.08731	.08833	.08934
17	6	11	.53770	.53756	.53728	.53691	.53646	.53594	.53539
17	7	1	-.11496	-.13359	-.15412	-.17666	-.20133	-.22824	-.25750
17	7	2	-.00021	-.00032	-.00003	.00072	.00195	.00373	.00608
17	7	3	.02920	.03138	.03400	.03711	.04072	.04486	.04957
17	7	4	.04708	.05003	.05336	.05709	.06123	.06580	.07081
17	7	5	.05960	.06271	.06610	.06980	.07381	.07812	.08276
17	7	6	.06904	.07199	.07514	.07850	.08208	.08586	.08986
17	7	7	.07647	.07908	.08183	.08470	.08770	.09084	.09410
17	7	8	.08248	.08465	.08688	.08918	.09154	.09397	.09646
17	7	9	.08742	.08906	.09072	.09239	.09408	.09579	.09752
17	7	10	.66388	.66502	.66611	.66717	.66821	.66926	.67034
17	8	1	-.16619	-.18846	-.21280	-.23932	-.26814	-.29935	-.33307
17	8	2	-.01781	-.01776	-.01725	-.01623	-.01466	-.01250	-.00970
17	8	3	.02059	.02334	.02658	.03034	.03463	.03950	.04496
17	8	4	.04402	.04764	.05166	.05609	.06095	.06625	.07201
17	8	5	.06047	.06422	.06825	.07259	.07724	.08220	.08748
17	8	6	.07290	.07640	.08009	.08398	.08808	.09238	.09687
17	8	7	.08270	.08574	.08890	.09216	.09555	.09904	.10265
17	8	8	.09064	.09309	.09559	.09813	.10073	.10337	.10606
17	8	9	.81267	.81580	.81899	.82225	.82562	.82911	.83275
17	9	1	-.23377	-.26062	-.28975	-.32127	-.35529	-.39190	-.43122
17	9	2	-.04051	-.04016	-.03927	-.03781	-.03574	-.03301	-.02959
17	9	3	.00985	.01340	.01748	.02212	.02734	.03316	.03961

Coefficients a_i for the BLUE of μ
Shape parameter k

n	s	i	0.7	0.8	0.9	1.0	1.1	1.2	1.3
17	9	4	.04065	.04516	.05008	.05543	.06123	.06748	.07419
17	9	5	.06231	.06687	.07173	.07689	.08235	.08812	.09420
17	9	6	.07867	.08286	.08722	.09177	.09651	.10143	.10654
17	9	7	.09158	.09513	.09877	.10251	.10633	.11025	.11426
17	9	8	.99122	.99736	1.00374	1.01036	1.01727	1.02448	1.03201
18	0	1	.04218	.03687	.03032	.02240	.01298	.00193	-.01089
18	0	2	.05471	.05473	.05488	.05522	.05577	.05658	.05769
18	0	3	.05666	.05732	.05820	.05935	.06079	.06256	.06470
18	0	4	.05756	.05845	.05958	.06097	.06265	.06463	.06695
18	0	5	.05799	.05896	.06015	.06156	.06323	.06516	.06737
18	0	6	.05817	.05913	.06028	.06162	.06317	.06494	.06692
18	0	7	.05818	.05909	.06015	.06137	.06275	.06430	.06602
18	0	8	.05808	.05891	.05985	.06092	.06211	.06342	.06484
18	0	9	.05790	.05862	.05943	.06032	.06130	.06236	.06351
18	0	10	.05765	.05824	.05890	.05962	.06038	.06120	.06206
18	0	11	.05733	.05779	.05829	.05882	.05937	.05994	.06052
18	0	12	.05695	.05727	.05760	.05794	.05827	.05860	.05891
18	0	13	.05650	.05667	.05682	.05696	.05708	.05717	.05723
18	0	14	.05597	.05597	.05594	.05588	.05577	.05563	.05546
18	0	15	.05533	.05515	.05493	.05465	.05433	.05396	.05356
18	0	16	.05453	.05415	.05371	.05322	.05267	.05208	.05145
18	0	17	.05343	.05281	.05212	.05138	.05060	.04978	.04894
18	0	18	.05089	.04988	.04885	.04781	.04678	.04576	.04476
18	1	1	.03207	.02573	.01810	.00903	-.00161	-.01396	-.02814
18	1	2	.05067	.05060	.05068	.05095	.05146	.05225	.05337
18	1	3	.05428	.05498	.05593	.05715	.05870	.06059	.06288
18	1	4	.05620	.05720	.05845	.05997	.06180	.06395	.06645
18	1	5	.05737	.05847	.05979	.06136	.06319	.06530	.06769
18	1	6	.05811	.05921	.06051	.06200	.06371	.06565	.06780
18	1	7	.05859	.05963	.06083	.06219	.06372	.06542	.06729
18	1	8	.05888	.05982	.06089	.06209	.06340	.06484	.06640
18	1	9	.05902	.05985	.06077	.06177	.06286	.06403	.06528
18	1	10	.05906	.05975	.06049	.06130	.06215	.06304	.06398
18	1	11	.05900	.05953	.06010	.06069	.06130	.06192	.06256
18	1	12	.05885	.05921	.05959	.05996	.06033	.06069	.06104
18	1	13	.05860	.05879	.05896	.05911	.05923	.05933	.05941
18	1	14	.05825	.05825	.05820	.05812	.05801	.05785	.05766
18	1	15	.05777	.05755	.05728	.05696	.05660	.05620	.05575
18	1	16	.05706	.05661	.05609	.05553	.05492	.05427	.05359
18	1	17	.10622	.10482	.10334	.10181	.10022	.09861	.09699
18	2	1	.01955	.01202	.00312	-.00728	-.01932	-.03313	-.04885
18	2	2	.04577	.04560	.04561	.04583	.04632	.04711	.04825
18	2	3	.05143	.05220	.05324	.05458	.05626	.05831	.06077
18	2	4	.05464	.05577	.05717	.05886	.06087	.06322	.06593
18	2	5	.05672	.05798	.05948	.06123	.06325	.06556	.06817
18	2	6	.05817	.05944	.06090	.06258	.06447	.06659	.06894
18	2	7	.05922	.06042	.06178	.06330	.06500	.06686	.06890
18	2	8	.05999	.06107	.06228	.06362	.06508	.06665	.06835
18	2	9	.06055	.06149	.06252	.06364	.06484	.06612	.06747
18	2	10	.06094	.06172	.06256	.06344	.06437	.06535	.06636
18	2	11	.06119	.06179	.06241	.06306	.06372	.06439	.06508
18	2	12	.06131	.06172	.06212	.06251	.06290	.06328	.06364
18	2	13	.06131	.06150	.06166	.06180	.06192	.06200	.06206
18	2	14	.06117	.06112	.06104	.06092	.06076	.06056	.06033
18	2	15	.06083	.06054	.06019	.05980	.05937	.05889	.05838
18	2	16	.16722	.16563	.16392	.16210	.16019	.15822	.15620

Coefficients a_i for the BLUE of μ
Shape parameter k

n	s	i	0.7	0.8	0.9	1.0	1.1	1.2	1.3
18	3	1	.00442	-.00449	-.01485	-.02678	-.04042	-.05590	-.07335
18	3	2	.03993	.03967	.03962	.03981	.04029	.04111	.04230
18	3	3	.04810	.04897	.05013	.05161	.05346	.05570	.05837
18	3	4	.05286	.05416	.05575	.05765	.05988	.06246	.06542
18	3	5	.05606	.05751	.05922	.06119	.06343	.06598	.06882
18	3	6	.05838	.05983	.06150	.06338	.06548	.06782	.07038
18	3	7	.06012	.06149	.06304	.06474	.06662	.06867	.07089
18	3	8	.06146	.06271	.06407	.06556	.06717	.06890	.07074
18	3	9	.06251	.06359	.06475	.06599	.06730	.06869	.07015
18	3	10	.06333	.06421	.06514	.06611	.06712	.06817	.06925
18	3	11	.06395	.06462	.06529	.06599	.06669	.06740	.06811
18	3	12	.06440	.06483	.06524	.06565	.06604	.06641	.06677
18	3	13	.06468	.06485	.06498	.06509	.06517	.06522	.06524
18	3	14	.06476	.06465	.06449	.06430	.06407	.06380	.06350
18	3	15	.23503	.23341	.23164	.22972	.22769	.22557	.22337
18	4	1	-.01380	-.02431	-.03635	-.05004	-.06552	-.08291	-.10235
18	4	2	.03299	.03266	.03256	.03274	.03323	.03409	.03535
18	4	3	.04421	.04521	.04652	.04819	.05025	.05273	.05565
18	4	4	.05086	.05237	.05419	.05633	.05882	.06168	.06494
18	4	5	.05541	.05710	.05904	.06126	.06378	.06659	.06972
18	4	6	.05876	.06045	.06234	.06446	.06681	.06939	.07220
18	4	7	.06134	.06293	.06468	.06659	.06868	.07093	.07336
18	4	8	.06338	.06481	.06635	.06801	.06978	.07167	.07367
18	4	9	.06503	.06624	.06754	.06890	.07034	.07185	.07342
18	4	10	.06635	.06733	.06835	.06940	.07049	.07161	.07276
18	4	11	.06742	.06813	.06885	.06958	.07032	.07105	.07178
18	4	12	.06825	.06867	.06908	.06947	.06985	.07021	.07054
18	4	13	.06883	.06895	.06903	.06908	.06910	.06909	.06905
18	4	14	.31096	.30947	.30782	.30601	.30407	.30203	.29990
18	5	1	-.03583	-.04820	-.06219	-.07792	-.09553	-.11513	-.13686
18	5	2	.02473	.02433	.02421	.02439	.02492	.02586	.02723
18	5	3	.03965	.04083	.04235	.04425	.04657	.04933	.05256
18	5	4	.04861	.05037	.05247	.05491	.05771	.06091	.06451
18	5	5	.05480	.05676	.05900	.06152	.06434	.06747	.07092
18	5	6	.05941	.06136	.06353	.06593	.06855	.07141	.07450
18	5	7	.06300	.06483	.06681	.06897	.07129	.07377	.07642
18	5	8	.06588	.06750	.06924	.07109	.07304	.07511	.07728
18	5	9	.06823	.06960	.07104	.07254	.07411	.07573	.07742
18	5	10	.07017	.07125	.07235	.07349	.07465	.07583	.07703
18	5	11	.07175	.07251	.07326	.07401	.07476	.07550	.07624
18	5	12	.07302	.07342	.07380	.07416	.07450	.07481	.07510
18	5	13	.39658	.39544	.39413	.39268	.39110	.38942	.38766
18	6	1	-.06271	-.07725	-.09352	-.11164	-.13174	-.15392	-.17833
18	6	2	.01480	.01436	.01424	.01446	.01507	.01611	.01764
18	6	3	.03428	.03568	.03747	.03967	.04231	.04542	.04904
18	6	4	.04605	.04814	.05058	.05338	.05657	.06016	.06417
18	6	5	.05426	.05656	.05915	.06203	.06522	.06872	.07255
18	6	6	.06041	.06268	.06518	.06789	.07084	.07402	.07744
18	6	7	.06524	.06734	.06961	.07203	.07462	.07736	.08026
18	6	8	.06914	.07099	.07294	.07500	.07716	.07941	.08177
18	6	9	.07236	.07389	.07548	.07712	.07882	.08056	.08236
18	6	10	.07502	.07619	.07738	.07859	.07981	.08104	.08229
18	6	11	.07722	.07799	.07876	.07951	.08025	.08098	.08169
18	6	12	.49393	.49341	.49275	.49197	.49108	.49013	.48911
18	7	1	-.09591	-.11303	-.13200	-.15294	-.17598	-.20122	-.22878
18	7	2	.00273	.00229	.00221	.00251	.00325	.00446	.00620

Coefficients a_i for the BLUE of μ
Shape parameter k

n	s	i	0.7	0.8	0.9	1.0	1.1	1.2	1.3
18	7	3	.02788	.02960	.03173	.03430	.03735	.04090	.04499
18	7	4	.04316	.04566	.04853	.05178	.05543	.05950	.06401
18	7	5	.05386	.05658	.05959	.06291	.06654	.07049	.07477
18	7	6	.06192	.06458	.06746	.07056	.07389	.07745	.08124
18	7	7	.06828	.07071	.07329	.07603	.07892	.08196	.08515
18	7	8	.07344	.07554	.07774	.08003	.08240	.08487	.08742
18	7	9	.07770	.07941	.08115	.08294	.08477	.08664	.08855
18	7	10	.08125	.08250	.08376	.08503	.08629	.08756	.08884
18	7	11	.60569	.60616	.60654	.60686	.60712	.60737	.60760
18	8	1	-.13759	-.15781	-.18003	-.20436	-.23092	-.25982	-.29119
18	8	2	-.01216	-.01254	-.01252	-.01207	-.01113	-.00967	-.00765
18	8	3	.02016	.02230	.02489	.02795	.03153	.03564	.04031
18	8	4	.03987	.04290	.04630	.05012	.05435	.05902	.06413
18	8	5	.05372	.05696	.06050	.06435	.06852	.07301	.07783
18	8	6	.06418	.06731	.07065	.07421	.07799	.08200	.08623
18	8	7	.07245	.07527	.07823	.08133	.08457	.08796	.09148
18	8	8	.07918	.08157	.08404	.08659	.08921	.09190	.09468
18	8	9	.08475	.08663	.08854	.09047	.09243	.09442	.09644
18	8	10	.73543	.73742	.73940	.74140	.74344	.74555	.74773
18	9	1	-.19103	-.21504	-.24122	-.26970	-.30059	-.33399	-.37001
18	9	2	-.03088	-.03112	-.03091	-.03020	-.02896	-.02714	-.02470
18	9	3	.01071	.01341	.01662	.02033	.02460	.02943	.03485
18	9	4	.03613	.03983	.04394	.04847	.05344	.05886	.06474
18	9	5	.05402	.05792	.06213	.06665	.07148	.07664	.08213
18	9	6	.06757	.07127	.07518	.07930	.08364	.08819	.09295
18	9	7	.07829	.08157	.08497	.08850	.09216	.09594	.09985
18	9	8	.08700	.08972	.09250	.09533	.09822	.10117	.10418
18	9	9	.88819	.89243	.89680	.90132	.90602	.91091	.91602
19	0	1	.03946	.03420	.02771	.01984	.01047	-.00054	-.01333
19	0	2	.05158	.05151	.05156	.05176	.05215	.05278	.05367
19	0	3	.05351	.05408	.05487	.05589	.05720	.05883	.06080
19	0	4	.05442	.05524	.05628	.05758	.05916	.06103	.06323
19	0	5	.05487	.05578	.05690	.05826	.05985	.06171	.06385
19	0	6	.05508	.05600	.05710	.05841	.05992	.06165	.06360
19	0	7	.05513	.05601	.05705	.05825	.05962	.06117	.06289
19	0	8	.05508	.05589	.05683	.05790	.05910	.06044	.06190
19	0	9	.05494	.05567	.05649	.05741	.05843	.05954	.06075
19	0	10	.05474	.05536	.05606	.05682	.05765	.05854	.05948
19	0	11	.05448	.05499	.05555	.05615	.05678	.05745	.05814
19	0	12	.05417	.05456	.05497	.05540	.05584	.05628	.05673
19	0	13	.05381	.05406	.05432	.05457	.05482	.05505	.05526
19	0	14	.05338	.05350	.05359	.05367	.05371	.05373	.05373
19	0	15	.05288	.05285	.05278	.05266	.05252	.05233	.05211
19	0	16	.05229	.05208	.05183	.05153	.05119	.05080	.05037
19	0	17	.05154	.05115	.05070	.05020	.04965	.04906	.04844
19	0	18	.05051	.04990	.04922	.04850	.04774	.04695	.04613
19	0	19	.04813	.04716	.04617	.04518	.04419	.04322	.04226
19	1	1	.03028	.02408	.01658	.00765	-.00286	-.01508	-.02915
19	1	2	.04784	.04767	.04763	.04776	.04810	.04869	.04957
19	1	3	.05124	.05185	.05268	.05377	.05516	.05688	.05897
19	1	4	.05307	.05398	.05512	.05654	.05824	.06026	.06262
19	1	5	.05418	.05520	.05645	.05793	.05967	.06169	.06399
19	1	6	.05489	.05594	.05717	.05861	.06027	.06214	.06426
19	1	7	.05536	.05636	.05752	.05886	.06036	.06205	.06391
19	1	8	.05565	.05657	.05763	.05882	.06015	.06160	.06319
19	1	9	.05581	.05664	.05756	.05859	.05972	.06093	.06224

Coefficients a_i for the BLUE of μ
Shape parameter k

n	s	i	0.7	0.8	0.9	1.0	1.1	1.2	1.3
19	1	10	.05587	.05658	.05736	.05821	.05913	.06010	.06113
19	1	11	.05584	.05642	.05705	.05771	.05841	.05914	.05990
19	1	12	.05573	.05617	.05663	.05711	.05759	.05808	.05857
19	1	13	.05555	.05584	.05613	.05641	.05667	.05692	.05716
19	1	14	.05530	.05542	.05552	.05560	.05565	.05567	.05566
19	1	15	.05494	.05489	.05480	.05467	.05450	.05430	.05406
19	1	16	.05447	.05422	.05393	.05359	.05320	.05277	.05231
19	1	17	.05379	.05333	.05282	.05225	.05165	.05100	.05033
19	1	18	.10019	.09884	.09741	.09592	.09440	.09284	.09128
19	2	1	.01901	.01171	.00305	-.00711	-.01892	-.03249	-.04797
19	2	2	.04332	.04305	.04293	.04300	.04330	.04387	.04475
19	2	3	.04855	.04920	.05010	.05128	.05278	.05463	.05687
19	2	4	.05151	.05253	.05380	.05536	.05722	.05941	.06196
19	2	5	.05343	.05459	.05599	.05763	.05955	.06174	.06423
19	2	6	.05478	.05596	.05735	.05895	.06077	.06282	.06511
19	2	7	.05575	.05689	.05820	.05968	.06134	.06318	.06519
19	2	8	.05646	.05752	.05871	.06003	.06149	.06307	.06480
19	2	9	.05698	.05793	.05897	.06011	.06134	.06267	.06408
19	2	10	.05736	.05816	.05904	.05998	.06098	.06204	.06315
19	2	11	.05761	.05826	.05896	.05969	.06045	.06124	.06205
19	2	12	.05775	.05824	.05874	.05926	.05977	.06029	.06082
19	2	13	.05779	.05810	.05840	.05869	.05897	.05923	.05947
19	2	14	.05773	.05785	.05794	.05800	.05803	.05803	.05800
19	2	15	.05754	.05745	.05732	.05714	.05693	.05668	.05640
19	2	16	.05719	.05687	.05650	.05609	.05563	.05514	.05462
19	2	17	.15724	.15569	.15401	.15223	.15038	.14846	.14649
19	3	1	.00552	-.00305	-.01304	-.02461	-.03787	-.05298	-.07006
19	3	2	.03799	.03761	.03742	.03743	.03770	.03826	.03917
19	3	3	.04541	.04613	.04712	.04842	.05005	.05206	.05448
19	3	4	.04974	.05090	.05233	.05406	.05611	.05850	.06126
19	3	5	.05264	.05397	.05554	.05737	.05949	.06189	.06460
19	3	6	.05474	.05610	.05766	.05944	.06146	.06370	.06619
19	3	7	.05632	.05763	.05910	.06075	.06258	.06459	.06678
19	3	8	.05754	.05875	.06008	.06155	.06315	.06489	.06675
19	3	9	.05850	.05957	.06073	.06199	.06334	.06478	.06631
19	3	10	.05925	.06016	.06113	.06216	.06325	.06439	.06558
19	3	11	.05983	.06056	.06132	.06211	.06293	.06377	.06462
19	3	12	.06027	.06080	.06134	.06188	.06242	.06296	.06350
19	3	13	.06057	.06088	.06119	.06147	.06174	.06199	.06222
19	3	14	.06072	.06081	.06087	.06089	.06088	.06085	.06078
19	3	15	.06071	.06055	.06035	.06011	.05983	.05951	.05917
19	3	16	.22025	.21863	.21686	.21496	.21295	.21084	.20868
19	4	1	-.01056	-.02058	-.03210	-.04526	-.06021	-.07706	-.09595
19	4	2	.03170	.03124	.03097	.03094	.03120	.03177	.03271
19	4	3	.04177	.04259	.04370	.04514	.04695	.04914	.05177
19	4	4	.04774	.04907	.05069	.05263	.05490	.05753	.06055
19	4	5	.05182	.05334	.05512	.05718	.05952	.06216	.06512
19	4	6	.05482	.05637	.05814	.06014	.06237	.06483	.06754
19	4	7	.05713	.05862	.06029	.06213	.06415	.06634	.06873
19	4	8	.05895	.06033	.06183	.06346	.06522	.06711	.06912
19	4	9	.06043	.06164	.06294	.06432	.06580	.06736	.06900
19	4	10	.06163	.06264	.06371	.06484	.06601	.06723	.06850
19	4	11	.06260	.06340	.06422	.06507	.06593	.06682	.06771
19	4	12	.06337	.06394	.06450	.06506	.06561	.06616	.06670
19	4	13	.06397	.06427	.06456	.06482	.06506	.06528	.06548
19	4	14	.06437	.06440	.06440	.06436	.06429	.06419	.06406

Coefficients a_i for the BLUE of μ
Shape parameter k

n	s	i	0.7	0.8	0.9	1.0	1.1	1.2	1.3
19	4	15	.29027	.28874	.28703	.28518	.28320	.28111	.27895
19	5	1	-.02979	-.04147	-.05474	-.06975	-.08661	-.10545	-.12642
19	5	2	.02429	.02374	.02342	.02336	.02362	.02422	.02523
19	5	3	.03754	.03849	.03976	.04138	.04339	.04582	.04870
19	5	4	.04548	.04702	.04888	.05106	.05360	.05651	.05982
19	5	5	.05097	.05273	.05476	.05708	.05969	.06261	.06585
19	5	6	.05505	.05684	.05885	.06110	.06357	.06629	.06925
19	5	7	.05823	.05994	.06182	.06388	.06612	.06853	.07112
19	5	8	.06079	.06234	.06403	.06584	.06778	.06984	.07202
19	5	9	.06287	.06423	.06568	.06721	.06881	.07050	.07226
19	5	10	.06460	.06573	.06690	.06812	.06938	.07068	.07202
19	5	11	.06603	.06690	.06778	.06867	.06958	.07050	.07143
19	5	12	.06721	.06778	.06835	.06892	.06947	.07001	.07053
19	5	13	.06813	.06840	.06865	.06887	.06907	.06923	.06938
19	5	14	.36858	.36731	.36586	.36426	.36253	.36070	.35879
19	6	1	-.05292	-.06654	-.08184	-.09897	-.11805	-.13920	-.16255
19	6	2	.01550	.01487	.01451	.01444	.01472	.01539	.01649
19	6	3	.03260	.03372	.03519	.03704	.03931	.04202	.04521
19	6	4	.04293	.04473	.04686	.04935	.05220	.05545	.05911
19	6	5	.05011	.05216	.05449	.05711	.06004	.06329	.06686
19	6	6	.05550	.05757	.05987	.06240	.06516	.06817	.07142
19	6	7	.05974	.06170	.06383	.06614	.06862	.07127	.07410
19	6	8	.06316	.06493	.06682	.06884	.07097	.07322	.07558
19	6	9	.06599	.06751	.06911	.07079	.07254	.07436	.07624
19	6	10	.06835	.06958	.07086	.07218	.07352	.07490	.07631
19	6	11	.07032	.07124	.07217	.07310	.07404	.07498	.07592
19	6	12	.07195	.07253	.07309	.07363	.07416	.07467	.07516
19	6	13	.45679	.45599	.45504	.45396	.45276	.45148	.45014
19	7	1	-.08105	-.09693	-.11460	-.13421	-.15587	-.17971	-.20585
19	7	2	.00495	.00427	.00390	.00386	.00420	.00496	.00619
19	7	3	.02677	.02812	.02986	.03200	.03459	.03765	.04121
19	7	4	.04002	.04215	.04463	.04748	.05072	.05438	.05845
19	7	5	.04928	.05168	.05437	.05736	.06067	.06430	.06826
19	7	6	.05627	.05867	.06130	.06416	.06726	.07060	.07419
19	7	7	.06178	.06403	.06646	.06905	.07181	.07474	.07784
19	7	8	.06626	.06827	.07040	.07264	.07498	.07744	.08001
19	7	9	.06998	.07168	.07346	.07529	.07719	.07915	.08116
19	7	10	.07309	.07445	.07583	.07723	.07866	.08011	.08158
19	7	11	.07571	.07667	.07763	.07859	.07954	.08048	.08143
19	7	12	.55694	.55692	.55678	.55656	.55626	.55591	.55553
19	8	1	-.11571	-.13425	-.15473	-.17727	-.20199	-.22901	-.25844
19	8	2	-.00784	-.00854	-.00889	-.00886	-.00842	-.00750	-.00609
19	8	3	.01982	.02149	.02357	.02609	.02908	.03258	.03661
19	8	4	.03669	.03924	.04215	.04546	.04918	.05331	.05789
19	8	5	.04853	.05136	.05448	.05792	.06168	.06577	.07020
19	8	6	.05749	.06029	.06332	.06658	.07007	.07381	.07779
19	8	7	.06457	.06717	.06993	.07285	.07594	.07918	.08259
19	8	8	.07035	.07264	.07503	.07752	.08011	.08280	.08558
19	8	9	.07516	.07706	.07901	.08101	.08307	.08517	.08732
19	8	10	.07919	.08065	.08212	.08361	.08511	.08662	.08815
19	8	11	.67174	.67289	.67400	.67508	.67616	.67726	.67839
19	9	1	-.15911	-.18086	-.20471	-.23077	-.25916	-.28999	-.32339
19	9	2	-.02360	-.02426	-.02453	-.02438	-.02375	-.02262	-.02093
19	9	3	.01145	.01353	.01606	.01907	.02259	.02664	.03126
19	9	4	.03287	.03595	.03941	.04329	.04759	.05234	.05754
19	9	5	.04795	.05131	.05498	.05897	.06328	.06792	.07291

Coefficients a_i for the BLUE of μ
Shape parameter k

n	s	i	0.7	0.8	0.9	1.0	1.1	1.2	1.3
19	9	6	.05938	.06266	.06618	.06992	.07389	.07810	.08254
19	9	7	.06844	.07144	.07460	.07791	.08137	.08499	.08875
19	9	8	.07583	.07843	.08112	.08390	.08676	.08970	.09274
19	9	9	.08198	.08409	.08623	.08841	.09063	.09288	.09517
19	9	10	.80483	.80772	.81066	.81368	.81680	.82004	.82341
19	10	1	-.21465	-.24033	-.26830	-.29867	-.33155	-.36707	-.40531
19	10	2	-.04341	-.04394	-.04404	-.04366	-.04275	-.04129	-.03923
19	10	3	.00116	.00381	.00696	.01062	.01482	.01960	.02496
19	10	4	.02846	.03222	.03640	.04101	.04606	.05157	.05754
19	10	5	.04770	.05173	.05609	.06076	.06576	.07110	.07677
19	10	6	.06229	.06618	.07029	.07461	.07917	.08395	.08895
19	10	7	.07386	.07736	.08100	.08477	.08868	.09273	.09691
19	10	8	.08331	.08627	.08930	.09240	.09557	.09881	.10212
19	10	9	.96128	.96669	.97231	.97815	.98424	.99061	.99728
20	0	1	.03703	.03182	.02539	.01758	.00826	-.00271	-.01547
20	0	2	.04877	.04863	.04858	.04866	.04890	.04936	.05006
20	0	3	.05067	.05117	.05186	.05278	.05397	.05546	.05727
20	0	4	.05158	.05234	.05331	.05452	.05600	.05777	.05986
20	0	5	.05206	.05291	.05398	.05526	.05679	.05858	.06064
20	0	6	.05229	.05317	.05423	.05549	.05696	.05864	.06056
20	0	7	.05237	.05323	.05424	.05542	.05677	.05830	.06001
20	0	8	.05235	.05315	.05409	.05515	.05636	.05770	.05919
20	0	9	.05226	.05298	.05382	.05476	.05580	.05695	.05819
20	0	10	.05210	.05274	.05346	.05426	.05513	.05608	.05709
20	0	11	.05189	.05244	.05303	.05368	.05438	.05512	.05591
20	0	12	.05164	.05207	.05254	.05304	.05356	.05411	.05466
20	0	13	.05134	.05166	.05199	.05234	.05268	.05303	.05337
20	0	14	.05099	.05119	.05138	.05157	.05173	.05189	.05202
20	0	15	.05059	.05066	.05070	.05072	.05071	.05068	.05061
20	0	16	.05012	.05005	.04994	.04979	.04961	.04938	.04912
20	0	17	.04956	.04933	.04906	.04874	.04837	.04797	.04752
20	0	18	.04885	.04846	.04801	.04750	.04696	.04637	.04575
20	0	19	.04789	.04729	.04663	.04592	.04518	.04441	.04362
20	0	20	.04565	.04472	.04377	.04282	.04187	.04094	.04003
20	1	1	.02866	.02258	.01521	.00640	-.00398	-.01608	-.03002
20	1	2	.04530	.04505	.04490	.04490	.04509	.04549	.04616
20	1	3	.04852	.04903	.04976	.05073	.05198	.05354	.05546
20	1	4	.05026	.05108	.05214	.05344	.05503	.05693	.05916
20	1	5	.05132	.05227	.05344	.05485	.05650	.05843	.06064
20	1	6	.05201	.05299	.05417	.05555	.05715	.05898	.06104
20	1	7	.05246	.05342	.05455	.05585	.05733	.05898	.06083
20	1	8	.05275	.05366	.05470	.05587	.05719	.05866	.06026
20	1	9	.05292	.05374	.05467	.05571	.05686	.05811	.05946
20	1	10	.05300	.05372	.05453	.05541	.05637	.05740	.05850
20	1	11	.05299	.05361	.05428	.05500	.05577	.05658	.05743
20	1	12	.05292	.05342	.05394	.05449	.05507	.05566	.05627
20	1	13	.05279	.05315	.05353	.05390	.05428	.05466	.05503
20	1	14	.05260	.05282	.05303	.05323	.05341	.05358	.05373
20	1	15	.05233	.05240	.05245	.05247	.05246	.05241	.05234
20	1	16	.05198	.05189	.05176	.05160	.05139	.05114	.05086
20	1	17	.05152	.05125	.05094	.05058	.05017	.04973	.04925
20	1	18	.05087	.05041	.04989	.04933	.04873	.04809	.04743
20	1	19	.09480	.09349	.09211	.09067	.08920	.08771	.08620
20	2	1	.01845	.01136	.00291	-.00703	-.01861	-.03197	-.04723
20	2	2	.04112	.04076	.04053	.04046	.04058	.04095	.04160
20	2	3	.04597	.04652	.04729	.04833	.04967	.05134	.05337

Coefficients a_i for the BLUE of μ
Shape parameter k

n	s	i	0.7	0.8	0.9	1.0	1.1	1.2	1.3
20	2	4	.04872	.04964	.05080	.05223	.05396	.05600	.05839
20	2	5	.05051	.05158	.05288	.05442	.05623	.05832	.06070
20	2	6	.05176	.05287	.05418	.05571	.05746	.05944	.06166
20	2	7	.05266	.05375	.05501	.05645	.05806	.05986	.06186
20	2	8	.05333	.05436	.05552	.05682	.05827	.05985	.06158
20	2	9	.05382	.05476	.05580	.05694	.05819	.05955	.06100
20	2	10	.05418	.05500	.05590	.05687	.05792	.05903	.06021
20	2	11	.05443	.05512	.05586	.05665	.05749	.05836	.05927
20	2	12	.05458	.05513	.05571	.05631	.05693	.05756	.05821
20	2	13	.05465	.05505	.05545	.05585	.05625	.05665	.05704
20	2	14	.05463	.05487	.05509	.05529	.05547	.05564	.05578
20	2	15	.05453	.05459	.05461	.05461	.05458	.05451	.05442
20	2	16	.05431	.05418	.05401	.05380	.05355	.05326	.05294
20	2	17	.05394	.05361	.05322	.05280	.05233	.05183	.05129
20	2	18	.14838	.14687	.14523	.14349	.14168	.13981	.13791
20	3	1	.00633	-.00192	-.01160	-.02283	-.03576	-.05053	-.06727
20	3	2	.03622	.03575	.03543	.03529	.03537	.03572	.03637
20	3	3	.04302	.04361	.04446	.04558	.04703	.04883	.05101
20	3	4	.04698	.04801	.04930	.05088	.05277	.05499	.05757
20	3	5	.04963	.05085	.05230	.05401	.05599	.05827	.06084
20	3	6	.05155	.05281	.05429	.05598	.05789	.06005	.06245
20	3	7	.05299	.05423	.05564	.05723	.05900	.06096	.06312
20	3	8	.05411	.05527	.05657	.05801	.05960	.06132	.06319
20	3	9	.05499	.05604	.05720	.05847	.05983	.06130	.06286
20	3	10	.05568	.05660	.05760	.05867	.05980	.06100	.06227
20	3	11	.05623	.05700	.05781	.05867	.05957	.06050	.06147
20	3	12	.05664	.05725	.05787	.05851	.05917	.05984	.06051
20	3	13	.05694	.05737	.05779	.05821	.05862	.05903	.05942
20	3	14	.05713	.05736	.05757	.05776	.05793	.05808	.05821
20	3	15	.05720	.05722	.05722	.05717	.05710	.05700	.05686
20	3	16	.05712	.05693	.05669	.05641	.05609	.05574	.05536
20	3	17	.20721	.20561	.20385	.20197	.19998	.19791	.19577
20	4	1	-.00799	-.01756	-.02863	-.04132	-.05578	-.07215	-.09056
20	4	2	.03050	.02993	.02952	.02932	.02936	.02969	.03035
20	4	3	.03962	.04028	.04121	.04245	.04403	.04599	.04835
20	4	4	.04501	.04619	.04764	.04939	.05147	.05389	.05669
20	4	5	.04870	.05008	.05171	.05362	.05580	.05829	.06108
20	4	6	.05140	.05284	.05450	.05638	.05849	.06085	.06345
20	4	7	.05349	.05489	.05648	.05824	.06019	.06233	.06465
20	4	8	.05514	.05645	.05791	.05950	.06123	.06311	.06512
20	4	9	.05647	.05766	.05895	.06034	.06182	.06341	.06509
20	4	10	.05756	.05859	.05969	.06085	.06208	.06337	.06472
20	4	11	.05845	.05930	.06019	.06111	.06207	.06306	.06408
20	4	12	.05917	.05982	.06049	.06117	.06185	.06255	.06325
20	4	13	.05974	.06018	.06061	.06103	.06144	.06185	.06223
20	4	14	.06016	.06037	.06056	.06072	.06086	.06097	.06106
20	4	15	.06042	.06039	.06032	.06022	.06008	.05992	.05972
20	4	16	.27217	.27061	.26887	.26699	.26499	.26288	.26070
20	5	1	-.02493	-.03601	-.04866	-.06302	-.07923	-.09741	-.11770
20	5	2	.02381	.02313	.02265	.02239	.02241	.02275	.02343
20	5	3	.03569	.03644	.03750	.03888	.04063	.04277	.04534
20	5	4	.04279	.04414	.04578	.04774	.05005	.05271	.05577
20	5	5	.04770	.04928	.05113	.05326	.05568	.05841	.06146
20	5	6	.05135	.05299	.05486	.05696	.05929	.06188	.06471
20	5	7	.05419	.05579	.05757	.05953	.06168	.06401	.06654
20	5	8	.05647	.05796	.05959	.06135	.06325	.06529	.06746

Coefficients a_i for the BLUE of μ
Shape parameter k

n	s	i	0.7	0.8	0.9	1.0	1.1	1.2	1.3
20	5	9	.05834	.05967	.06110	.06263	.06425	.06597	.06777
20	5	10	.05989	.06104	.06224	.06351	.06484	.06622	.06765
20	5	11	.06119	.06212	.06308	.06407	.06509	.06613	.06720
20	5	12	.06226	.06295	.06366	.06436	.06507	.06578	.06649
20	5	13	.06314	.06358	.06401	.06442	.06482	.06520	.06556
20	5	14	.06382	.06399	.06413	.06425	.06433	.06439	.06443
20	5	15	.34429	.34292	.34137	.33967	.33784	.33590	.33388
20	6	1	-.04508	-.05789	-.07236	-.08863	-.10683	-.12709	-.14953
20	6	2	.01595	.01518	.01462	.01433	.01434	.01469	.01542
20	6	3	.03113	.03201	.03322	.03479	.03674	.03911	.04192
20	6	4	.04028	.04184	.04372	.04593	.04850	.05145	.05480
20	6	5	.04665	.04848	.05058	.05297	.05567	.05868	.06202
20	6	6	.05142	.05331	.05542	.05777	.06036	.06320	.06630
20	6	7	.05516	.05699	.05899	.06118	.06355	.06611	.06886
20	6	8	.05820	.05988	.06170	.06365	.06574	.06796	.07031
20	6	9	.06071	.06220	.06378	.06546	.06723	.06908	.07101
20	6	10	.06280	.06407	.06539	.06676	.06819	.06966	.07118
20	6	11	.06457	.06558	.06661	.06767	.06874	.06983	.07094
20	6	12	.06606	.06679	.06751	.06823	.06895	.06966	.07037
20	6	13	.06729	.06771	.06812	.06850	.06886	.06920	.06953
20	6	14	.42486	.42386	.42270	.42139	.41997	.41845	.41686
20	7	1	-.06925	-.08406	-.10064	-.11911	-.13961	-.16226	-.18719
20	7	2	.00664	.00578	.00518	.00486	.00488	.00528	.00609
20	7	3	.02580	.02686	.02827	.03006	.03227	.03491	.03804
20	7	4	.03742	.03925	.04141	.04393	.04682	.05011	.05381
20	7	5	.04555	.04767	.05008	.05279	.05581	.05915	.06283
20	7	6	.05167	.05385	.05625	.05890	.06179	.06493	.06832
20	7	7	.05651	.05859	.06086	.06330	.06593	.06874	.07173
20	7	8	.06044	.06235	.06439	.06655	.06884	.07126	.07381
20	7	9	.06371	.06538	.06714	.06898	.07089	.07289	.07497
20	7	10	.06646	.06785	.06929	.07077	.07230	.07386	.07547
20	7	11	.06879	.06987	.07096	.07207	.07319	.07432	.07546
20	7	12	.07076	.07150	.07223	.07295	.07366	.07436	.07505
20	7	13	.51550	.51511	.51459	.51396	.51323	.51245	.51161
20	8	1	-.09856	-.11571	-.13474	-.15578	-.17897	-.20441	-.23224
20	8	2	-.00449	-.00542	-.00605	-.00636	-.00630	-.00582	-.00490
20	8	3	.01953	.02082	.02249	.02457	.02709	.03009	.03358
20	8	4	.03416	.03632	.03883	.04172	.04500	.04870	.05282
20	8	5	.04443	.04691	.04968	.05277	.05617	.05991	.06399
20	8	6	.05220	.05471	.05746	.06046	.06369	.06718	.07092
20	8	7	.05835	.06074	.06331	.06605	.06897	.07207	.07535
20	8	8	.06337	.06553	.06782	.07023	.07275	.07540	.07816
20	8	9	.06756	.06943	.07137	.07338	.07547	.07763	.07985
20	8	10	.07109	.07262	.07417	.07577	.07739	.07905	.08073
20	8	11	.07409	.07523	.07638	.07753	.07868	.07983	.08099
20	8	12	.61828	.61882	.61928	.61968	.62004	.62039	.62074
20	9	1	-.13458	-.15449	-.17643	-.20051	-.22687	-.25561	-.28686
20	9	2	-.01799	-.01894	-.01957	-.01983	-.01968	-.01907	-.01798
20	9	3	.01205	.01365	.01566	.01811	.02104	.02447	.02843
20	9	4	.03041	.03299	.03594	.03929	.04305	.04723	.05187
20	9	5	.04333	.04624	.04947	.05301	.05688	.06108	.06563
20	9	6	.05312	.05605	.05921	.06261	.06626	.07016	.07431
20	9	7	.06089	.06364	.06656	.06965	.07292	.07635	.07996
20	9	8	.06725	.06970	.07227	.07495	.07774	.08064	.08364
20	9	9	.07255	.07464	.07678	.07899	.08126	.08358	.08596
20	9	10	.07703	.07869	.08037	.08207	.08379	.08553	.08729

Coefficients a_i for the BLUE of μ
Shape parameter k

n	s	i	0.7	0.8	0.9	1.0	1.1	1.2	1.3
20	9	11	.73594	.73784	.73973	.74165	.74361	.74564	.74775
20	10	1	-.17962	-.20285	-.22827	-.25601	-.28617	-.31887	-.35423
20	10	2	-.03459	-.03553	-.03609	-.03624	-.03594	-.03514	-.03381
20	10	3	.00303	.00504	.00750	.01044	.01389	.01787	.02241
20	10	4	.02607	.02918	.03269	.03661	.04097	.04577	.05103
20	10	5	.04231	.04577	.04955	.05365	.05809	.06287	.06799
20	10	6	.05463	.05807	.06173	.06564	.06978	.07417	.07881
20	10	7	.06443	.06761	.07096	.07446	.07812	.08195	.08594
20	10	8	.07245	.07525	.07814	.08113	.08422	.08740	.09068
20	10	9	.07914	.08146	.08383	.08625	.08870	.09120	.09375
20	10	10	.87216	.87601	.87997	.88407	.88833	.89278	.89744
21	0	1	.03484	.02969	.02331	.01556	.00630	-.00462	-.01733
21	0	2	.04624	.04602	.04588	.04585	.04597	.04627	.04678
21	0	3	.04811	.04853	.04914	.04997	.05104	.05239	.05406
21	0	4	.04902	.04971	.05061	.05174	.05313	.05480	.05677
21	0	5	.04951	.05031	.05131	.05254	.05400	.05571	.05770
21	0	6	.04976	.05060	.05162	.05283	.05425	.05590	.05777
21	0	7	.04987	.05069	.05168	.05283	.05416	.05567	.05736
21	0	8	.04988	.05066	.05158	.05264	.05384	.05519	.05668
21	0	9	.04981	.05054	.05137	.05232	.05337	.05454	.05583
21	0	10	.04969	.05034	.05108	.05190	.05281	.05379	.05486
21	0	11	.04953	.05009	.05072	.05141	.05216	.05296	.05381
21	0	12	.04932	.04979	.05030	.05086	.05144	.05206	.05271
21	0	13	.04907	.04944	.04983	.05025	.05067	.05111	.05156
21	0	14	.04878	.04904	.04931	.04958	.04985	.05011	.05036
21	0	15	.04845	.04860	.04874	.04886	.04897	.04905	.04912
21	0	16	.04806	.04810	.04810	.04808	.04802	.04794	.04782
21	0	17	.04762	.04752	.04738	.04721	.04699	.04674	.04645
21	0	18	.04709	.04685	.04656	.04623	.04585	.04543	.04498
21	0	19	.04643	.04603	.04558	.04507	.04453	.04395	.04333
21	0	20	.04552	.04493	.04428	.04360	.04287	.04213	.04136
21	0	21	.04341	.04251	.04160	.04069	.03978	.03889	.03801
21	1	1	.02717	.02121	.01396	.00528	-.00498	-.01696	-.03079
21	1	2	.04300	.04267	.04244	.04232	.04237	.04260	.04308
21	1	3	.04606	.04650	.04713	.04798	.04910	.05052	.05227
21	1	4	.04772	.04847	.04944	.05065	.05213	.05391	.05601
21	1	5	.04874	.04963	.05073	.05205	.05363	.05547	.05759
21	1	6	.04941	.05034	.05146	.05279	.05433	.05609	.05810
21	1	7	.04985	.05077	.05186	.05312	.05456	.05619	.05801
21	1	8	.05014	.05102	.05204	.05320	.05451	.05596	.05757
21	1	9	.05031	.05113	.05205	.05309	.05425	.05552	.05690
21	1	10	.05040	.05113	.05195	.05286	.05385	.05492	.05607
21	1	11	.05042	.05106	.05176	.05252	.05334	.05421	.05513
21	1	12	.05038	.05091	.05148	.05209	.05274	.05341	.05411
21	1	13	.05028	.05070	.05114	.05159	.05206	.05254	.05302
21	1	14	.05013	.05043	.05073	.05102	.05131	.05160	.05187
21	1	15	.04993	.05009	.05024	.05038	.05049	.05059	.05066
21	1	16	.04966	.04969	.04969	.04966	.04960	.04950	.04938
21	1	17	.04932	.04920	.04904	.04884	.04860	.04832	.04801
21	1	18	.04887	.04859	.04826	.04788	.04747	.04701	.04652
21	1	19	.04825	.04779	.04727	.04671	.04612	.04549	.04484
21	1	20	.08996	.08869	.08735	.08596	.08454	.08310	.08164
21	2	1	.01788	.01098	.00273	-.00701	-.01839	-.03153	-.04659
21	2	2	.03912	.03869	.03835	.03816	.03814	.03833	.03877
21	2	3	.04365	.04410	.04477	.04568	.04687	.04837	.05022
21	2	4	.04622	.04705	.04810	.04942	.05102	.05293	.05517

Coefficients a_i for the BLUE of μ
Shape parameter k

n	s	i	0.7	0.8	0.9	1.0	1.1	1.2	1.3
21	2	5	.04789	.04888	.05009	.05154	.05325	.05523	.05751
21	2	6	.04906	.05010	.05135	.05280	.05448	.05639	.05854
21	2	7	.04991	.05095	.05215	.05354	.05511	.05687	.05883
21	2	8	.05054	.05153	.05266	.05394	.05536	.05694	.05866
21	2	9	.05100	.05192	.05295	.05409	.05535	.05672	.05820
21	2	10	.05135	.05217	.05308	.05407	.05514	.05630	.05753
21	2	11	.05159	.05230	.05308	.05391	.05479	.05573	.05672
21	2	12	.05175	.05234	.05297	.05363	.05433	.05505	.05579
21	2	13	.05183	.05229	.05277	.05326	.05376	.05427	.05478
21	2	14	.05184	.05216	.05248	.05280	.05310	.05340	.05368
21	2	15	.05179	.05196	.05211	.05224	.05235	.05244	.05250
21	2	16	.05165	.05166	.05164	.05158	.05150	.05138	.05123
21	2	17	.05142	.05125	.05105	.05081	.05053	.05021	.04986
21	2	18	.05104	.05069	.05030	.04986	.04939	.04888	.04834
21	2	19	.14047	.13898	.13738	.13569	.13393	.13211	.13025
21	3	1	.00693	-.00105	-.01043	-.02136	-.03398	-.04844	-.06488
21	3	2	.03461	.03406	.03363	.03335	.03327	.03342	.03385
21	3	3	.04087	.04136	.04207	.04305	.04432	.04593	.04790
21	3	4	.04452	.04544	.04661	.04805	.04979	.05185	.05426
21	3	5	.04696	.04808	.04942	.05102	.05288	.05503	.05748
21	3	6	.04873	.04990	.05129	.05289	.05472	.05679	.05911
21	3	7	.05006	.05123	.05257	.05410	.05581	.05772	.05983
21	3	8	.05109	.05220	.05346	.05487	.05642	.05812	.05998
21	3	9	.05190	.05293	.05407	.05532	.05669	.05817	.05976
21	3	10	.05254	.05346	.05446	.05555	.05671	.05796	.05928
21	3	11	.05305	.05384	.05469	.05559	.05655	.05756	.05861
21	3	12	.05344	.05409	.05478	.05549	.05623	.05700	.05779
21	3	13	.05373	.05423	.05474	.05526	.05579	.05632	.05685
21	3	14	.05393	.05427	.05460	.05492	.05523	.05552	.05580
21	3	15	.05404	.05420	.05434	.05446	.05455	.05461	.05466
21	3	16	.05405	.05402	.05396	.05387	.05374	.05358	.05340
21	3	17	.05393	.05370	.05343	.05313	.05278	.05241	.05200
21	3	18	.19563	.19404	.19231	.19045	.18849	.18645	.18435
21	4	1	-.00591	-.01509	-.02575	-.03803	-.05206	-.06799	-.08596
21	4	2	.02937	.02871	.02818	.02783	.02770	.02782	.02823
21	4	3	.03769	.03822	.03900	.04006	.04145	.04318	.04530
21	4	4	.04260	.04364	.04494	.04653	.04843	.05067	.05327
21	4	5	.04596	.04721	.04872	.05048	.05252	.05486	.05750
21	4	6	.04842	.04975	.05130	.05307	.05508	.05732	.05983
21	4	7	.05031	.05164	.05314	.05482	.05670	.05877	.06104
21	4	8	.05182	.05307	.05447	.05602	.05772	.05957	.06156
21	4	9	.05303	.05418	.05545	.05683	.05832	.05991	.06162
21	4	10	.05402	.05505	.05615	.05734	.05860	.05994	.06135
21	4	11	.05484	.05571	.05664	.05762	.05865	.05972	.06083
21	4	12	.05550	.05621	.05695	.05772	.05850	.05931	.06013
21	4	13	.05604	.05657	.05711	.05765	.05820	.05874	.05928
21	4	14	.05646	.05680	.05712	.05744	.05774	.05802	.05829
21	4	15	.05675	.05689	.05699	.05708	.05713	.05716	.05717
21	4	16	.05691	.05683	.05671	.05656	.05637	.05615	.05591
21	4	17	.25619	.25461	.25286	.25097	.24896	.24685	.24466
21	5	1	-.02097	-.03152	-.04363	-.05742	-.07305	-.09064	-.11033
21	5	2	.02330	.02252	.02190	.02148	.02130	.02139	.02180
21	5	3	.03403	.03463	.03550	.03667	.03819	.04008	.04237
21	5	4	.04044	.04162	.04308	.04485	.04694	.04939	.05221
21	5	5	.04487	.04630	.04799	.04995	.05220	.05475	.05762
21	5	6	.04816	.04967	.05140	.05337	.05557	.05803	.06074

Coefficients a_i for the BLUE of μ
Shape parameter k

n	s	i	0.7	0.8	0.9	1.0	1.1	1.2	1.3
21	5	7	.05072	.05221	.05389	.05576	.05781	.06007	.06252
21	5	8	.05277	.05419	.05575	.05746	.05931	.06131	.06346
21	5	9	.05446	.05575	.05715	.05866	.06028	.06201	.06383
21	5	10	.05586	.05700	.05822	.05951	.06087	.06230	.06380
21	5	11	.05703	.05800	.05901	.06006	.06116	.06229	.06346
21	5	12	.05802	.05878	.05957	.06038	.06121	.06205	.06290
21	5	13	.05883	.05939	.05994	.06050	.06105	.06160	.06214
21	5	14	.05949	.05982	.06013	.06043	.06071	.06096	.06120
21	5	15	.05999	.06008	.06014	.06018	.06018	.06016	.06011
21	5	16	.32300	.32156	.31994	.31817	.31627	.31426	.31217
21	6	1	-.03870	-.05081	-.06456	-.08009	-.09752	-.11699	-.13864
21	6	2	.01623	.01533	.01463	.01414	.01392	.01401	.01443
21	6	3	.02981	.03051	.03150	.03281	.03449	.03656	.03906
21	6	4	.03800	.03936	.04101	.04299	.04531	.04800	.05108
21	6	5	.04370	.04533	.04724	.04943	.05192	.05472	.05784
21	6	6	.04796	.04968	.05163	.05382	.05626	.05894	.06189
21	6	7	.05130	.05300	.05488	.05695	.05922	.06167	.06433
21	6	8	.05401	.05561	.05735	.05923	.06126	.06344	.06576
21	6	9	.05625	.05770	.05926	.06091	.06267	.06453	.06649
21	6	10	.05814	.05940	.06073	.06214	.06360	.06513	.06673
21	6	11	.05973	.06078	.06187	.06300	.06417	.06537	.06660
21	6	12	.06108	.06190	.06273	.06358	.06444	.06530	.06617
21	6	13	.06221	.06278	.06335	.06390	.06445	.06499	.06551
21	6	14	.06315	.06345	.06373	.06399	.06423	.06444	.06464
21	6	15	.39712	.39596	.39464	.39317	.39158	.38989	.38812
21	7	1	-.05972	-.07362	-.08926	-.10675	-.12625	-.14787	-.17176
21	7	2	.00794	.00694	.00615	.00561	.00537	.00546	.00592
21	7	3	.02493	.02575	.02689	.02839	.03027	.03257	.03531
21	7	4	.03524	.03681	.03870	.04093	.04352	.04649	.04987
21	7	5	.04244	.04432	.04649	.04895	.05172	.05481	.05823
21	7	6	.04786	.04983	.05204	.05449	.05718	.06013	.06335
21	7	7	.05214	.05407	.05618	.05849	.06098	.06367	.06656
21	7	8	.05562	.05743	.05937	.06145	.06368	.06604	.06855
21	7	9	.05852	.06014	.06186	.06368	.06559	.06760	.06970
21	7	10	.06097	.06237	.06383	.06535	.06692	.06856	.07025
21	7	11	.06305	.06420	.06537	.06657	.06781	.06906	.07034
21	7	12	.06483	.06570	.06657	.06744	.06832	.06920	.07008
21	7	13	.06634	.06690	.06746	.06800	.06852	.06903	.06953
21	7	14	.47982	.47916	.47835	.47741	.47637	.47525	.47407
21	8	1	-.08486	-.10083	-.11864	-.13842	-.16030	-.18440	-.21086
21	8	2	-.00186	-.00296	-.00381	-.00438	-.00463	-.00451	-.00399
21	8	3	.01924	.02023	.02157	.02329	.02542	.02799	.03104
21	8	4	.03209	.03393	.03611	.03864	.04155	.04487	.04860
21	8	5	.04110	.04328	.04576	.04854	.05164	.05506	.05883
21	8	6	.04792	.05018	.05269	.05544	.05844	.06170	.06522
21	8	7	.05331	.05551	.05790	.06047	.06323	.06618	.06933
21	8	8	.05772	.05976	.06193	.06424	.06668	.06926	.07197
21	8	9	.06140	.06322	.06512	.06711	.06919	.07136	.07361
21	8	10	.06452	.06606	.06765	.06930	.07099	.07273	.07452
21	8	11	.06719	.06842	.06967	.07095	.07224	.07355	.07487
21	8	12	.06947	.07037	.07126	.07215	.07304	.07392	.07479
21	8	13	.57276	.57284	.57281	.57269	.57252	.57230	.57206
21	9	1	-.11528	-.13367	-.15401	-.17645	-.20111	-.22810	-.25756
21	9	2	-.01357	-.01475	-.01564	-.01622	-.01643	-.01626	-.01564
21	9	3	.01253	.01375	.01534	.01735	.01980	.02271	.02612
21	9	4	.02848	.03065	.03318	.03609	.03940	.04312	.04728

Coefficients a_i for the BLUE of μ
Shape parameter k

n	s	i	0.7	0.8	0.9	1.0	1.1	1.2	1.3
21	9	5	.03969	.04224	.04509	.04825	.05174	.05557	.05974
21	9	6	.04820	.05081	.05367	.05678	.06014	.06376	.06764
21	9	7	.05495	.05746	.06016	.06305	.06611	.06937	.07282
21	9	8	.06048	.06278	.06522	.06778	.07047	.07329	.07623
21	9	9	.06510	.06713	.06923	.07141	.07368	.07602	.07844
21	9	10	.06903	.07072	.07245	.07422	.07603	.07787	.07976
21	9	11	.07238	.07370	.07502	.07636	.07770	.07905	.08041
21	9	12	.67802	.67918	.68029	.68138	.68248	.68360	.68476
21	10	1	-.15261	-.17385	-.19719	-.22277	-.25071	-.28112	-.31413
21	10	2	-.02774	-.02896	-.02986	-.03041	-.03056	-.03028	-.02952
21	10	3	.00454	.00606	.00799	.01037	.01322	.01656	.02043
21	10	4	.02431	.02690	.02988	.03325	.03704	.04126	.04593
21	10	5	.03824	.04123	.04453	.04816	.05213	.05644	.06110
21	10	6	.04882	.05186	.05514	.05868	.06246	.06651	.07081
21	10	7	.05723	.06012	.06319	.06644	.06987	.07348	.07727
21	10	8	.06413	.06675	.06948	.07234	.07531	.07840	.08161
21	10	9	.06991	.07217	.07450	.07689	.07936	.08189	.08449
21	10	10	.07481	.07665	.07853	.08043	.08236	.08431	.08630
21	10	11	.79837	.80107	.80381	.80662	.80953	.81256	.81573
21	11	1	-.19922	-.22388	-.25082	-.28017	-.31204	-.34656	-.38383
21	11	2	-.04516	-.04638	-.04724	-.04770	-.04773	-.04727	-.04630
21	11	3	-.00511	-.00318	-.00080	.00206	.00542	.00931	.01376
21	11	4	.01945	.02258	.02612	.03008	.03447	.03931	.04462
21	11	5	.03678	.04033	.04420	.04841	.05295	.05785	.06310
21	11	6	.04996	.05352	.05732	.06138	.06568	.07023	.07504
21	11	7	.06045	.06379	.06730	.07099	.07484	.07886	.08306
21	11	8	.06905	.07203	.07512	.07831	.08161	.08501	.08852
21	11	9	.07625	.07878	.08136	.08399	.08667	.08941	.09220
21	11	10	.93756	.94241	.94743	.95267	.95813	.96385	.96985
22	0	1	.03287	.02777	.02145	.01376	.00456	-.00631	-.01897
22	0	2	.04393	.04366	.04344	.04331	.04330	.04346	.04381
22	0	3	.04578	.04614	.04667	.04740	.04837	.04960	.05113
22	0	4	.04669	.04732	.04815	.04920	.05050	.05208	.05395
22	0	5	.04718	.04794	.04889	.05005	.05144	.05309	.05500
22	0	6	.04745	.04825	.04923	.05040	.05177	.05337	.05520
22	0	7	.04758	.04838	.04933	.05046	.05176	.05324	.05492
22	0	8	.04762	.04838	.04928	.05033	.05152	.05286	.05436
22	0	9	.04758	.04829	.04912	.05007	.05114	.05232	.05362
22	0	10	.04749	.04814	.04889	.04972	.05065	.05167	.05278
22	0	11	.04736	.04793	.04858	.04930	.05009	.05094	.05185
22	0	12	.04718	.04768	.04823	.04883	.04947	.05015	.05087
22	0	13	.04698	.04738	.04783	.04830	.04879	.04931	.04984
22	0	14	.04673	.04705	.04738	.04772	.04807	.04842	.04877
22	0	15	.04646	.04667	.04689	.04710	.04730	.04749	.04766
22	0	16	.04614	.04625	.04635	.04642	.04648	.04651	.04651
22	0	17	.04578	.04578	.04575	.04568	.04559	.04547	.04531
22	0	18	.04536	.04523	.04507	.04487	.04463	.04436	.04405
22	0	19	.04486	.04460	.04430	.04396	.04357	.04314	.04268
22	0	20	.04423	.04383	.04338	.04288	.04234	.04176	.04115
22	0	21	.04338	.04280	.04216	.04149	.04079	.04006	.03932
22	0	22	.04138	.04051	.03964	.03876	.03789	.03703	.03619
22	1	1	.02581	.01995	.01282	.00425	-.00588	-.01774	-.03145
22	1	2	.04090	.04052	.04020	.03998	.03990	.03999	.04028
22	1	3	.04384	.04420	.04474	.04549	.04649	.04778	.04937
22	1	4	.04542	.04611	.04699	.04812	.04950	.05117	.05315
22	1	5	.04641	.04723	.04826	.04951	.05101	.05277	.05480

Coefficients a_i for the BLUE of μ
Shape parameter k

n	s	i	0.7	0.8	0.9	1.0	1.1	1.2	1.3
22	1	6	.04705	.04793	.04900	.05027	.05175	.05346	.05541
22	1	7	.04748	.04837	.04941	.05064	.05204	.05364	.05543
22	1	8	.04777	.04862	.04962	.05076	.05205	.05349	.05509
22	1	9	.04795	.04875	.04967	.05070	.05186	.05313	.05453
22	1	10	.04805	.04878	.04960	.05052	.05153	.05263	.05382
22	1	11	.04809	.04873	.04945	.05024	.05110	.05202	.05300
22	1	12	.04806	.04862	.04923	.04988	.05058	.05132	.05209
22	1	13	.04799	.04845	.04894	.04946	.04999	.05055	.05113
22	1	14	.04788	.04823	.04859	.04897	.04935	.04973	.05011
22	1	15	.04771	.04795	.04819	.04842	.04864	.04884	.04904
22	1	16	.04750	.04762	.04773	.04781	.04787	.04790	.04791
22	1	17	.04724	.04723	.04719	.04712	.04702	.04689	.04672
22	1	18	.04690	.04676	.04657	.04635	.04609	.04579	.04546
22	1	19	.04647	.04618	.04584	.04545	.04503	.04457	.04407
22	1	20	.04588	.04542	.04491	.04436	.04377	.04315	.04251
22	1	21	.08559	.08436	.08306	.08171	.08034	.07894	.07754
22	2	1	.01731	.01058	.00251	-.00704	-.01821	-.03117	-.04603
22	2	2	.03730	.03680	.03638	.03608	.03592	.03595	.03620
22	2	3	.04155	.04192	.04248	.04328	.04433	.04568	.04736
22	2	4	.04396	.04471	.04567	.04688	.04836	.05015	.05225
22	2	5	.04553	.04645	.04758	.04894	.05055	.05244	.05461
22	2	6	.04663	.04761	.04879	.05017	.05178	.05362	.05570
22	2	7	.04743	.04842	.04958	.05091	.05244	.05415	.05607
22	2	8	.04802	.04898	.05008	.05132	.05272	.05428	.05600
22	2	9	.04847	.04936	.05038	.05151	.05276	.05414	.05563
22	2	10	.04880	.04961	.05052	.05152	.05262	.05380	.05507
22	2	11	.04904	.04976	.05055	.05141	.05234	.05332	.05437
22	2	12	.04920	.04981	.05048	.05119	.05195	.05274	.05356
22	2	13	.04929	.04980	.05033	.05089	.05146	.05206	.05267
22	2	14	.04932	.04971	.05010	.05050	.05090	.05130	.05170
22	2	15	.04930	.04955	.04980	.05004	.05026	.05047	.05067
22	2	16	.04921	.04932	.04942	.04949	.04954	.04957	.04957
22	2	17	.04905	.04902	.04896	.04886	.04874	.04858	.04839
22	2	18	.04880	.04862	.04839	.04812	.04782	.04748	.04711
22	2	19	.04843	.04807	.04767	.04723	.04675	.04624	.04570
22	2	20	.13335	.13190	.13034	.12869	.12697	.12520	.12340
22	3	1	.00735	-.00036	-.00948	-.02014	-.03248	-.04665	-.06279
22	3	2	.03313	.03251	.03198	.03159	.03137	.03135	.03156
22	3	3	.03894	.03932	.03992	.04076	.04189	.04332	.04510
22	3	4	.04231	.04313	.04419	.04551	.04711	.04902	.05127
22	3	5	.04457	.04560	.04684	.04833	.05008	.05211	.05444
22	3	6	.04621	.04731	.04861	.05013	.05187	.05385	.05609
22	3	7	.04744	.04854	.04983	.05129	.05295	.05480	.05686
22	3	8	.04839	.04946	.05068	.05204	.05356	.05524	.05708
22	3	9	.04915	.05014	.05126	.05250	.05386	.05534	.05695
22	3	10	.04974	.05065	.05165	.05275	.05393	.05520	.05656
22	3	11	.05022	.05102	.05189	.05283	.05382	.05488	.05599
22	3	12	.05059	.05127	.05200	.05277	.05358	.05442	.05529
22	3	13	.05087	.05143	.05200	.05260	.05321	.05384	.05447
22	3	14	.05108	.05149	.05191	.05233	.05274	.05316	.05357
22	3	15	.05121	.05147	.05172	.05196	.05218	.05238	.05257
22	3	16	.05126	.05136	.05144	.05149	.05152	.05152	.05149
22	3	17	.05122	.05115	.05105	.05091	.05074	.05054	.05031
22	3	18	.05106	.05081	.05052	.05020	.04983	.04944	.04902
22	3	19	.18528	.18371	.18199	.18016	.17823	.17623	.17416
22	4	1	-.00423	-.01306	-.02336	-.03525	-.04890	-.06443	-.08200

Coefficients a_i for the BLUE of μ
Shape parameter k

n	s	i	0.7	0.8	0.9	1.0	1.1	1.2	1.3
22	4	2	.02832	.02758	.02695	.02647	.02618	.02611	.02630
22	4	3	.03595	.03636	.03701	.03792	.03913	.04067	.04257
22	4	4	.04046	.04137	.04254	.04398	.04573	.04779	.05021
22	4	5	.04353	.04467	.04606	.04769	.04960	.05180	.05431
22	4	6	.04578	.04702	.04847	.05013	.05204	.05418	.05659
22	4	7	.04752	.04876	.05018	.05180	.05360	.05560	.05781
22	4	8	.04889	.05009	.05144	.05294	.05459	.05641	.05838
22	4	9	.05001	.05112	.05236	.05372	.05519	.05679	.05850
22	4	10	.05092	.05193	.05303	.05422	.05550	.05686	.05831
22	4	11	.05167	.05255	.05351	.05452	.05559	.05672	.05790
22	4	12	.05228	.05303	.05382	.05465	.05550	.05639	.05731
22	4	13	.05279	.05338	.05400	.05463	.05527	.05592	.05658
22	4	14	.05319	.05362	.05405	.05448	.05491	.05533	.05574
22	4	15	.05349	.05375	.05399	.05422	.05442	.05461	.05478
22	4	16	.05369	.05376	.05381	.05383	.05382	.05378	.05372
22	4	17	.05377	.05365	.05349	.05330	.05308	.05282	.05254
22	4	18	.24198	.24040	.23865	.23676	.23475	.23264	.23046
22	5	1	-.01771	-.02780	-.03941	-.05271	-.06781	-.08487	-.10403
22	5	2	.02278	.02191	.02118	.02062	.02026	.02015	.02032
22	5	3	.03254	.03300	.03371	.03471	.03602	.03769	.03973
22	5	4	.03837	.03940	.04070	.04230	.04420	.04645	.04906
22	5	5	.04239	.04368	.04523	.04703	.04913	.05152	.05422
22	5	6	.04537	.04677	.04838	.05022	.05230	.05463	.05722
22	5	7	.04770	.04909	.05068	.05245	.05442	.05659	.05897
22	5	8	.04956	.05091	.05240	.05405	.05585	.05781	.05992
22	5	9	.05109	.05234	.05371	.05519	.05679	.05851	.06034
22	5	10	.05236	.05349	.05470	.05600	.05738	.05884	.06038
22	5	11	.05343	.05441	.05545	.05654	.05769	.05889	.06014
22	5	12	.05433	.05515	.05600	.05688	.05778	.05872	.05968
22	5	13	.05509	.05572	.05637	.05703	.05770	.05837	.05905
22	5	14	.05571	.05616	.05660	.05703	.05745	.05787	.05827
22	5	15	.05621	.05645	.05667	.05688	.05706	.05721	.05735
22	5	16	.05658	.05660	.05660	.05657	.05651	.05642	.05630
22	5	17	.30420	.30271	.30104	.29922	.29727	.29522	.29308
22	6	1	-.03344	-.04495	-.05806	-.07293	-.08968	-.10846	-.12940
22	6	2	.01637	.01538	.01455	.01391	.01349	.01335	.01351
22	6	3	.02863	.02916	.02996	.03107	.03251	.03433	.03654
22	6	4	.03602	.03720	.03866	.04043	.04253	.04498	.04782
22	6	5	.04115	.04262	.04435	.04636	.04866	.05127	.05420
22	6	6	.04499	.04656	.04837	.05041	.05269	.05523	.05804
22	6	7	.04800	.04958	.05134	.05330	.05546	.05782	.06038
22	6	8	.05044	.05195	.05361	.05542	.05738	.05950	.06178
22	6	9	.05245	.05385	.05536	.05698	.05872	.06057	.06253
22	6	10	.05415	.05540	.05672	.05813	.05962	.06118	.06282
22	6	11	.05559	.05666	.05779	.05896	.06019	.06146	.06277
22	6	12	.05682	.05770	.05860	.05954	.06049	.06147	.06246
22	6	13	.05786	.05853	.05921	.05989	.06057	.06125	.06194
22	6	14	.05874	.05918	.05962	.06004	.06045	.06085	.06123
22	6	15	.05946	.05966	.05985	.06001	.06014	.06026	.06035
22	6	16	.37278	.37152	.37008	.36849	.36677	.36495	.36305
22	7	1	-.05191	-.06503	-.07984	-.09649	-.11511	-.13584	-.15880
22	7	2	.00893	.00782	.00689	.00617	.00571	.00554	.00570
22	7	3	.02415	.02477	.02568	.02692	.02853	.03052	.03294
22	7	4	.03336	.03472	.03638	.03836	.04069	.04338	.04647
22	7	5	.03980	.04148	.04343	.04567	.04822	.05108	.05427
22	7	6	.04464	.04644	.04847	.05073	.05325	.05603	.05907

Coefficients a_i for the BLUE of μ
Shape parameter k

n	s	i	0.7	0.8	0.9	1.0	1.1	1.2	1.3
22	7	7	.04847	.05025	.05223	.05440	.05676	.05933	.06211
22	7	8	.05158	.05328	.05512	.05712	.05926	.06156	.06402
22	7	9	.05417	.05573	.05739	.05917	.06106	.06305	.06515
22	7	10	.05636	.05774	.05919	.06072	.06232	.06399	.06574
22	7	11	.05823	.05940	.06062	.06188	.06318	.06453	.06591
22	7	12	.05984	.06078	.06174	.06272	.06372	.06473	.06576
22	7	13	.06121	.06191	.06260	.06329	.06398	.06467	.06535
22	7	14	.06238	.06281	.06323	.06362	.06400	.06436	.06471
22	7	15	.44879	.44791	.44688	.44571	.44443	.44305	.44162
22	8	1	-.07374	-.08871	-.10547	-.12416	-.14491	-.16786	-.19314
22	8	2	.00023	-.00100	-.00202	-.00280	-.00330	-.00348	-.00330
22	8	3	.01896	.01970	.02076	.02218	.02398	.02620	.02886
22	8	4	.03035	.03192	.03382	.03605	.03865	.04164	.04503
22	8	5	.03834	.04027	.04249	.04500	.04783	.05098	.05447
22	8	6	.04438	.04642	.04871	.05125	.05403	.05708	.06040
22	8	7	.04916	.05118	.05340	.05581	.05842	.06123	.06424
22	8	8	.05307	.05498	.05704	.05924	.06159	.06409	.06674
22	8	9	.05633	.05807	.05991	.06186	.06391	.06606	.06832
22	8	10	.05911	.06063	.06222	.06388	.06560	.06738	.06923
22	8	11	.06148	.06276	.06407	.06542	.06680	.06821	.06965
22	8	12	.06353	.06453	.06555	.06657	.06760	.06864	.06969
22	8	13	.06530	.06600	.06670	.06739	.06807	.06874	.06940
22	8	14	.53352	.53324	.53284	.53233	.53174	.53109	.53041
22	9	1	-.09981	-.11691	-.13591	-.15695	-.18017	-.20568	-.23362
22	9	2	-.01005	-.01138	-.01248	-.01331	-.01382	-.01399	-.01377
22	9	3	.01290	.01381	.01507	.01671	.01876	.02124	.02420
22	9	4	.02691	.02875	.03093	.03348	.03640	.03973	.04347
22	9	5	.03676	.03899	.04153	.04437	.04753	.05103	.05487
22	9	6	.04423	.04658	.04917	.05201	.05511	.05848	.06212
22	9	7	.05016	.05246	.05495	.05764	.06053	.06361	.06690
22	9	8	.05502	.05717	.05947	.06191	.06449	.06722	.07009
22	9	9	.05909	.06103	.06307	.06520	.06743	.06976	.07218
22	9	10	.06255	.06423	.06597	.06776	.06961	.07152	.07348
22	9	11	.06553	.06691	.06831	.06974	.07120	.07268	.07418
22	9	12	.06810	.06914	.07020	.07125	.07230	.07336	.07441
22	9	13	.62862	.62922	.62973	.63019	.63063	.63106	.63150
22	10	1	-.13130	-.15088	-.17249	-.19628	-.22237	-.25087	-.28191
22	10	2	-.02231	-.02373	-.02489	-.02574	-.02625	-.02637	-.02607
22	10	3	.00577	.00690	.00842	.01034	.01270	.01552	.01884
22	10	4	.02295	.02513	.02767	.03059	.03391	.03765	.04182
22	10	5	.03506	.03767	.04058	.04381	.04737	.05128	.05553
22	10	6	.04426	.04697	.04993	.05314	.05661	.06034	.06435
22	10	7	.05158	.05421	.05703	.06004	.06325	.06665	.07025
22	10	8	.05759	.06003	.06260	.06531	.06815	.07113	.07425
22	10	9	.06263	.06480	.06705	.06940	.07183	.07434	.07694
22	10	10	.06692	.06876	.07066	.07260	.07458	.07661	.07869
22	10	11	.07061	.07209	.07358	.07509	.07662	.07816	.07972
22	10	12	.73623	.73805	.73987	.74171	.74360	.74555	.74760
22	11	1	-.16987	-.19239	-.21710	-.24412	-.27359	-.30563	-.34034
22	11	2	-.03712	-.03862	-.03980	-.04064	-.04110	-.04114	-.04072
22	11	3	-.00272	-.00129	.00055	.00284	.00559	.00884	.01261
22	11	4	.01837	.02097	.02396	.02735	.03115	.03539	.04008
22	11	5	.03326	.03631	.03969	.04340	.04744	.05184	.05659
22	11	6	.04458	.04773	.05112	.05477	.05868	.06286	.06730
22	11	7	.05360	.05662	.05983	.06322	.06680	.07057	.07453
22	11	8	.06101	.06378	.06668	.06969	.07284	.07611	.07951

Coefficients a_i for the BLUE of μ
Shape parameter k

n	s	i	0.7	0.8	0.9	1.0	1.1	1.2	1.3
22	11	9	.06723	.06966	.07216	.07474	.07739	.08012	.08292
22	11	10	.07253	.07456	.07661	.07870	.08083	.08299	.08519
22	11	11	.85913	.86266	.86629	.87005	.87396	.87805	.88234
23	0	1	.03108	.02604	.01978	.01215	.00300	-.00780	-.02041
23	0	2	.04184	.04150	.04121	.04099	.04088	.04090	.04109
23	0	3	.04365	.04395	.04441	.04506	.04592	.04704	.04844
23	0	4	.04456	.04514	.04590	.04688	.04810	.04958	.05136
23	0	5	.04506	.04577	.04667	.04777	.04910	.05067	.05252
23	0	6	.04535	.04610	.04704	.04816	.04949	.05104	.05282
23	0	7	.04549	.04626	.04718	.04827	.04954	.05100	.05265
23	0	8	.04555	.04629	.04717	.04820	.04938	.05071	.05220
23	0	9	.04553	.04623	.04706	.04800	.04907	.05026	.05157
23	0	10	.04547	.04612	.04686	.04771	.04865	.04969	.05083
23	0	11	.04536	.04595	.04661	.04735	.04817	.04905	.05001
23	0	12	.04522	.04573	.04631	.04694	.04762	.04835	.04913
23	0	13	.04504	.04548	.04596	.04647	.04702	.04760	.04821
23	0	14	.04484	.04519	.04557	.04597	.04638	.04681	.04725
23	0	15	.04461	.04487	.04515	.04543	.04571	.04598	.04625
23	0	16	.04434	.04451	.04468	.04484	.04498	.04511	.04523
23	0	17	.04404	.04411	.04417	.04420	.04421	.04420	.04416
23	0	18	.04369	.04366	.04360	.04351	.04339	.04323	.04305
23	0	19	.04329	.04315	.04297	.04275	.04249	.04220	.04187
23	0	20	.04282	.04256	.04225	.04189	.04150	.04106	.04060
23	0	21	.04223	.04183	.04138	.04088	.04034	.03977	.03918
23	0	22	.04142	.04085	.04023	.03958	.03889	.03819	.03746
23	0	23	.03953	.03869	.03785	.03700	.03616	.03534	.03453
23	1	1	.02456	.01881	.01178	.00333	-.00669	-.01843	-.03203
23	1	2	.03900	.03855	.03816	.03785	.03765	.03760	.03773
23	1	3	.04181	.04210	.04256	.04322	.04411	.04527	.04672
23	1	4	.04333	.04395	.04476	.04580	.04709	.04865	.05052
23	1	5	.04428	.04505	.04601	.04720	.04862	.05029	.05224
23	1	6	.04490	.04574	.04675	.04797	.04939	.05104	.05293
23	1	7	.04533	.04617	.04718	.04836	.04973	.05129	.05305
23	1	8	.04561	.04644	.04740	.04852	.04979	.05122	.05281
23	1	9	.04580	.04658	.04748	.04851	.04966	.05094	.05234
23	1	10	.04591	.04663	.04745	.04837	.04939	.05051	.05173
23	1	11	.04595	.04660	.04734	.04814	.04903	.04998	.05101
23	1	12	.04595	.04652	.04715	.04784	.04858	.04937	.05021
23	1	13	.04590	.04639	.04691	.04748	.04807	.04870	.04935
23	1	14	.04581	.04620	.04662	.04706	.04751	.04797	.04844
23	1	15	.04568	.04598	.04628	.04658	.04689	.04719	.04748
23	1	16	.04551	.04570	.04589	.04606	.04621	.04636	.04648
23	1	17	.04530	.04538	.04544	.04547	.04549	.04547	.04543
23	1	18	.04504	.04500	.04493	.04483	.04469	.04452	.04433
23	1	19	.04471	.04455	.04434	.04410	.04381	.04350	.04315
23	1	20	.04430	.04399	.04364	.04325	.04282	.04236	.04186
23	1	21	.04373	.04327	.04276	.04222	.04164	.04103	.04040
23	1	22	.08162	.08042	.07916	.07786	.07653	.07518	.07382
23	2	1	.01675	.01018	.00228	-.00709	-.01808	-.03085	-.04553
23	2	2	.03564	.03508	.03458	.03418	.03390	.03379	.03387
23	2	3	.03964	.03994	.04041	.04110	.04203	.04324	.04476
23	2	4	.04191	.04258	.04346	.04457	.04595	.04761	.04959
23	2	5	.04340	.04424	.04530	.04657	.04810	.04989	.05197
23	2	6	.04443	.04536	.04647	.04779	.04932	.05109	.05311
23	2	7	.04519	.04613	.04724	.04852	.05000	.05167	.05355
23	2	8	.04575	.04667	.04773	.04895	.05032	.05186	.05355

Coefficients a_i for the BLUE of μ
Shape parameter k

n	s	i	0.7	0.8	0.9	1.0	1.1	1.2	1.3
23	2	9	.04618	.04705	.04804	.04916	.05040	.05178	.05328
23	2	10	.04649	.04730	.04820	.04921	.05031	.05151	.05281
23	2	11	.04673	.04745	.04825	.04913	.05008	.05111	.05220
23	2	12	.04689	.04752	.04821	.04896	.04976	.05060	.05149
23	2	13	.04699	.04752	.04810	.04871	.04935	.05001	.05070
23	2	14	.04703	.04746	.04792	.04838	.04886	.04935	.04985
23	2	15	.04703	.04735	.04767	.04799	.04831	.04862	.04893
23	2	16	.04697	.04717	.04736	.04754	.04769	.04784	.04796
23	2	17	.04686	.04694	.04699	.04701	.04701	.04698	.04693
23	2	18	.04669	.04663	.04653	.04640	.04624	.04605	.04583
23	2	19	.04644	.04623	.04599	.04570	.04538	.04502	.04464
23	2	20	.04607	.04571	.04530	.04485	.04437	.04386	.04332
23	2	21	.12692	.12550	.12397	.12236	.12069	.11897	.11721
23	3	1	.00766	.00017	-.00870	-.01911	-.03119	-.04509	-.06097
23	3	2	.03176	.03108	.03048	.02998	.02963	.02945	.02949
23	3	3	.03717	.03747	.03797	.03869	.03968	.04096	.04256
23	3	4	.04032	.04105	.04200	.04321	.04469	.04647	.04857
23	3	5	.04242	.04336	.04452	.04591	.04756	.04948	.05169
23	3	6	.04394	.04497	.04619	.04763	.04930	.05120	.05335
23	3	7	.04509	.04613	.04736	.04876	.05036	.05216	.05416
23	3	8	.04598	.04700	.04817	.04950	.05098	.05263	.05444
23	3	9	.04668	.04765	.04874	.04996	.05131	.05278	.05438
23	3	10	.04724	.04813	.04913	.05022	.05141	.05270	.05408
23	3	11	.04769	.04849	.04937	.05032	.05135	.05244	.05360
23	3	12	.04804	.04874	.04949	.05030	.05115	.05205	.05299
23	3	13	.04831	.04890	.04952	.05018	.05086	.05156	.05228
23	3	14	.04851	.04898	.04946	.04996	.05046	.05097	.05149
23	3	15	.04865	.04899	.04933	.04966	.04999	.05031	.05062
23	3	16	.04872	.04892	.04911	.04928	.04943	.04956	.04967
23	3	17	.04873	.04878	.04881	.04881	.04879	.04874	.04866
23	3	18	.04865	.04855	.04842	.04825	.04805	.04782	.04756
23	3	19	.04848	.04821	.04790	.04756	.04719	.04678	.04634
23	3	20	.17596	.17441	.17272	.17092	.16903	.16706	.16503
23	4	1	-.00286	-.01137	-.02134	-.03289	-.04618	-.06135	-.07855
23	4	2	.02733	.02652	.02581	.02522	.02479	.02455	.02454
23	4	3	.03437	.03468	.03521	.03599	.03704	.03840	.04011
23	4	4	.03853	.03934	.04039	.04170	.04330	.04521	.04746
23	4	5	.04136	.04240	.04368	.04520	.04699	.04906	.05143
23	4	6	.04344	.04458	.04594	.04751	.04932	.05137	.05367
23	4	7	.04503	.04620	.04755	.04909	.05082	.05276	.05490
23	4	8	.04630	.04744	.04873	.05018	.05179	.05357	.05550
23	4	9	.04733	.04840	.04961	.05094	.05239	.05397	.05568
23	4	10	.04817	.04916	.05025	.05144	.05272	.05410	.05557
23	4	11	.04886	.04975	.05071	.05174	.05284	.05401	.05524
23	4	12	.04943	.05020	.05102	.05189	.05280	.05375	.05474
23	4	13	.04990	.05054	.05121	.05191	.05263	.05336	.05412
23	4	14	.05028	.05078	.05129	.05181	.05234	.05286	.05339
23	4	15	.05058	.05093	.05128	.05161	.05194	.05226	.05257
23	4	16	.05080	.05099	.05116	.05132	.05145	.05156	.05165
23	4	17	.05093	.05095	.05095	.05091	.05085	.05076	.05064
23	4	18	.05096	.05080	.05061	.05039	.05013	.04984	.04953
23	4	19	.22927	.22769	.22594	.22405	.22205	.21996	.21780
23	5	1	-.01500	-.02467	-.03586	-.04870	-.06333	-.07992	-.09858
23	5	2	.02224	.02131	.02048	.01980	.01930	.01900	.01896
23	5	3	.03118	.03153	.03210	.03294	.03408	.03554	.03736
23	5	4	.03652	.03743	.03859	.04002	.04176	.04383	.04625

Coefficients a_i for the BLUE of μ
Shape parameter k

n	s	i	0.7	0.8	0.9	1.0	1.1	1.2	1.3
23	5	5	.04019	.04137	.04278	.04445	.04640	.04864	.05119
23	5	6	.04292	.04421	.04571	.04744	.04940	.05162	.05409
23	5	7	.04504	.04635	.04784	.04953	.05142	.05351	.05581
23	5	8	.04675	.04802	.04945	.05103	.05278	.05469	.05677
23	5	9	.04814	.04934	.05067	.05212	.05369	.05539	.05722
23	5	10	.04931	.05041	.05160	.05289	.05427	.05575	.05732
23	5	11	.05029	.05126	.05231	.05343	.05460	.05585	.05715
23	5	12	.05111	.05195	.05284	.05377	.05474	.05574	.05678
23	5	13	.05181	.05250	.05321	.05395	.05470	.05547	.05625
23	5	14	.05240	.05292	.05345	.05399	.05453	.05506	.05559
23	5	15	.05287	.05323	.05357	.05390	.05422	.05453	.05482
23	5	16	.05325	.05342	.05357	.05369	.05379	.05387	.05393
23	5	17	.05351	.05349	.05344	.05335	.05324	.05310	.05293
23	5	18	.28747	.28594	.28424	.28239	.28041	.27833	.27617
23	6	1	-.02907	-.04004	-.05259	-.06687	-.08302	-.10118	-.12149
23	6	2	.01641	.01535	.01441	.01364	.01306	.01271	.01264
23	6	3	.02756	.02795	.02859	.02951	.03075	.03233	.03430
23	6	4	.03427	.03529	.03659	.03817	.04008	.04232	.04493
23	6	5	.03892	.04024	.04182	.04366	.04580	.04823	.05098
23	6	6	.04240	.04385	.04552	.04742	.04957	.05198	.05464
23	6	7	.04513	.04660	.04826	.05011	.05217	.05443	.05690
23	6	8	.04734	.04877	.05035	.05208	.05398	.05604	.05827
23	6	9	.04917	.05051	.05197	.05355	.05525	.05708	.05903
23	6	10	.05071	.05193	.05324	.05464	.05613	.05770	.05937
23	6	11	.05202	.05309	.05423	.05543	.05669	.05801	.05939
23	6	12	.05314	.05405	.05500	.05599	.05702	.05807	.05916
23	6	13	.05409	.05483	.05559	.05636	.05714	.05793	.05874
23	6	14	.05491	.05546	.05600	.05655	.05709	.05763	.05816
23	6	15	.05559	.05594	.05627	.05658	.05688	.05716	.05743
23	6	16	.05615	.05628	.05639	.05647	.05652	.05656	.05657
23	6	17	.35127	.34992	.34839	.34671	.34490	.34298	.34098
23	7	1	-.04544	-.05787	-.07196	-.08787	-.10572	-.12565	-.14781
23	7	2	.00969	.00849	.00745	.00658	.00594	.00555	.00545
23	7	3	.02342	.02387	.02460	.02562	.02699	.02872	.03085
23	7	4	.03173	.03290	.03436	.03612	.03822	.04067	.04350
23	7	5	.03753	.03903	.04080	.04284	.04519	.04784	.05082
23	7	6	.04189	.04352	.04539	.04750	.04985	.05247	.05535
23	7	7	.04533	.04698	.04883	.05087	.05312	.05557	.05824
23	7	8	.04813	.04973	.05148	.05338	.05545	.05768	.06007
23	7	9	.05046	.05195	.05356	.05529	.05714	.05910	.06119
23	7	10	.05244	.05378	.05522	.05673	.05834	.06003	.06180
23	7	11	.05413	.05530	.05654	.05783	.05917	.06057	.06202
23	7	12	.05558	.05657	.05759	.05864	.05972	.06082	.06195
23	7	13	.05684	.05762	.05841	.05921	.06002	.06083	.06165
23	7	14	.05792	.05847	.05903	.05957	.06010	.06063	.06115
23	7	15	.05883	.05915	.05946	.05974	.06000	.06025	.06047
23	7	16	.42153	.42050	.41929	.41794	.41648	.41492	.41328
23	8	1	-.06460	-.07869	-.09453	-.11228	-.13206	-.15400	-.17825
23	8	2	.00190	.00057	-.00059	-.00153	-.00224	-.00266	-.00277
23	8	3	.01867	.01921	.02004	.02121	.02273	.02464	.02696
23	8	4	.02886	.03021	.03186	.03384	.03616	.03886	.04195
23	8	5	.03600	.03772	.03971	.04199	.04458	.04749	.05074
23	8	6	.04140	.04326	.04535	.04769	.05028	.05314	.05627
23	8	7	.04567	.04754	.04960	.05186	.05432	.05700	.05988
23	8	8	.04917	.05096	.05290	.05500	.05726	.05967	.06225
23	8	9	.05209	.05375	.05552	.05741	.05942	.06154	.06377

Coefficients a_i for the BLUE of μ
Shape parameter k

n	s	i	0.7	0.8	0.9	1.0	1.1	1.2	1.3
23	8	10	.05457	.05606	.05762	.05927	.06100	.06280	.06469
23	8	11	.05670	.05799	.05932	.06071	.06214	.06362	.06515
23	8	12	.05855	.05961	.06069	.06180	.06293	.06408	.06526
23	8	13	.06015	.06096	.06178	.06261	.06343	.06425	.06508
23	8	14	.06153	.06209	.06263	.06316	.06367	.06417	.06467
23	8	15	.49933	.49878	.49809	.49727	.49637	.49539	.49436
23	9	1	-.08721	-.10320	-.12105	-.14089	-.16287	-.18711	-.21374
23	9	2	-.00721	-.00866	-.00992	-.01094	-.01170	-.01214	-.01225
23	9	3	.01317	.01383	.01481	.01615	.01786	.01999	.02256
23	9	4	.02560	.02717	.02906	.03129	.03389	.03687	.04026
23	9	5	.03434	.03631	.03857	.04113	.04401	.04721	.05076
23	9	6	.04096	.04308	.04543	.04805	.05092	.05405	.05746
23	9	7	.04622	.04833	.05064	.05315	.05586	.05878	.06191
23	9	8	.05053	.05254	.05470	.05702	.05949	.06212	.06490
23	9	9	.05414	.05599	.05795	.06002	.06220	.06449	.06689
23	9	10	.05722	.05886	.06058	.06237	.06422	.06615	.06815
23	9	11	.05988	.06128	.06272	.06420	.06572	.06728	.06888
23	9	12	.06218	.06331	.06446	.06562	.06679	.06799	.06919
23	9	13	.06418	.06502	.06585	.06669	.06751	.06834	.06916
23	9	14	.58598	.58614	.58619	.58617	.58609	.58598	.58587
23	10	1	-.11416	-.13233	-.15249	-.17477	-.19928	-.22616	-.25555
23	10	2	-.01794	-.01951	-.02086	-.02195	-.02273	-.02318	-.02326
23	10	3	.00677	.00759	.00876	.01031	.01227	.01466	.01752
23	10	4	.02188	.02371	.02589	.02843	.03136	.03469	.03843
23	10	5	.03252	.03480	.03738	.04027	.04348	.04704	.05094
23	10	6	.04061	.04303	.04570	.04863	.05182	.05528	.05902
23	10	7	.04704	.04944	.05203	.05483	.05783	.06103	.06445
23	10	8	.05233	.05459	.05701	.05957	.06228	.06514	.06816
23	10	9	.05677	.05883	.06100	.06326	.06564	.06811	.07069
23	10	10	.06055	.06236	.06424	.06617	.06817	.07024	.07236
23	10	11	.06382	.06534	.06689	.06847	.07008	.07172	.07340
23	10	12	.06666	.06785	.06906	.07026	.07148	.07270	.07393
23	10	13	.68315	.68430	.68541	.68650	.68760	.68873	.68991
23	11	1	-.14666	-.16739	-.19024	-.21534	-.24281	-.27277	-.30536
23	11	2	-.03073	-.03241	-.03383	-.03496	-.03575	-.03618	-.03620
23	11	3	-.00077	.00027	.00169	.00352	.00578	.00850	.01172
23	11	4	.01758	.01976	.02229	.02521	.02854	.03228	.03646
23	11	5	.03054	.03319	.03615	.03944	.04306	.04703	.05136
23	11	6	.04040	.04319	.04623	.04953	.05310	.05694	.06106
23	11	7	.04825	.05099	.05392	.05705	.06038	.06391	.06765
23	11	8	.05471	.05727	.05998	.06282	.06581	.06894	.07221
23	11	9	.06014	.06245	.06485	.06734	.06993	.07261	.07538
23	11	10	.06478	.06677	.06882	.07092	.07307	.07527	.07753
23	11	11	.06878	.07042	.07208	.07375	.07545	.07717	.07892
23	11	12	.79297	.79550	.79807	.80070	.80343	.80628	.80927
23	12	1	-.18642	-.21018	-.23621	-.26464	-.29560	-.32921	-.36559
23	12	2	-.04618	-.04794	-.04941	-.05054	-.05131	-.05167	-.05160
23	12	3	-.00976	-.00842	-.00667	-.00449	-.00184	.00131	.00497
23	12	4	.01259	.01519	.01818	.02157	.02538	.02963	.03433
23	12	5	.02839	.03150	.03494	.03871	.04282	.04729	.05212
23	12	6	.04042	.04365	.04715	.05090	.05492	.05922	.06378
23	12	7	.05001	.05315	.05649	.06001	.06373	.06765	.07177
23	12	8	.05791	.06081	.06386	.06703	.07033	.07377	.07735
23	12	9	.06455	.06713	.06979	.07254	.07537	.07828	.08127
23	12	10	.07021	.07241	.07464	.07691	.07923	.08158	.08398
23	12	11	.91829	.92269	.92725	.93200	.93696	.94217	.94763

Coefficients a_i for the BLUE of μ
Shape parameter k

n	s	i	0.7	0.8	0.9	1.0	1.1	1.2	1.3
24	0	1	.02945	.02446	.01826	.01069	.00161	-.00914	-.02168
24	0	2	.03992	.03953	.03917	.03887	.03865	.03856	.03861
24	0	3	.04170	.04195	.04234	.04291	.04368	.04469	.04597
24	0	4	.04260	.04313	.04384	.04475	.04589	.04729	.04896
24	0	5	.04311	.04378	.04462	.04567	.04694	.04845	.05022
24	0	6	.04341	.04413	.04502	.04610	.04739	.04889	.05062
24	0	7	.04357	.04431	.04520	.04626	.04750	.04892	.05055
24	0	8	.04364	.04436	.04522	.04623	.04739	.04871	.05019
24	0	9	.04365	.04434	.04515	.04608	.04714	.04834	.04966
24	0	10	.04360	.04425	.04499	.04584	.04679	.04785	.04901
24	0	11	.04352	.04411	.04478	.04554	.04637	.04729	.04828
24	0	12	.04340	.04393	.04452	.04518	.04589	.04666	.04749
24	0	13	.04326	.04372	.04422	.04477	.04536	.04599	.04666
24	0	14	.04308	.04347	.04389	.04433	.04480	.04529	.04579
24	0	15	.04288	.04319	.04351	.04385	.04419	.04455	.04490
24	0	16	.04266	.04288	.04311	.04333	.04356	.04377	.04397
24	0	17	.04240	.04254	.04267	.04278	.04288	.04296	.04302
24	0	18	.04212	.04216	.04218	.04218	.04216	.04210	.04203
24	0	19	.04179	.04173	.04165	.04153	.04138	.04120	.04099
24	0	20	.04141	.04125	.04105	.04082	.04054	.04023	.03989
24	0	21	.04096	.04069	.04037	.04001	.03961	.03917	.03871
24	0	22	.04040	.04000	.03955	.03906	.03853	.03796	.03738
24	0	23	.03963	.03907	.03847	.03783	.03716	.03647	.03577
24	0	24	.03783	.03702	.03621	.03539	.03459	.03379	.03301
24	1	1	.02340	.01775	.01082	.00248	-.00742	-.01905	-.03253
24	1	2	.03725	.03675	.03629	.03590	.03560	.03542	.03541
24	1	3	.03995	.04018	.04057	.04114	.04193	.04297	.04429
24	1	4	.04142	.04198	.04272	.04368	.04488	.04635	.04811
24	1	5	.04233	.04305	.04396	.04507	.04642	.04802	.04989
24	1	6	.04294	.04373	.04469	.04586	.04723	.04882	.05065
24	1	7	.04335	.04416	.04513	.04628	.04760	.04912	.05084
24	1	8	.04364	.04443	.04537	.04646	.04771	.04912	.05069
24	1	9	.04382	.04459	.04547	.04649	.04763	.04890	.05031
24	1	10	.04394	.04465	.04547	.04639	.04742	.04855	.04978
24	1	11	.04400	.04465	.04539	.04621	.04711	.04809	.04915
24	1	12	.04401	.04459	.04524	.04595	.04673	.04756	.04844
24	1	13	.04398	.04449	.04504	.04564	.04628	.04696	.04767
24	1	14	.04391	.04434	.04479	.04528	.04578	.04631	.04685
24	1	15	.04381	.04414	.04450	.04486	.04524	.04562	.04600
24	1	16	.04367	.04392	.04416	.04441	.04465	.04488	.04510
24	1	17	.04350	.04365	.04378	.04390	.04401	.04410	.04416
24	1	18	.04328	.04333	.04335	.04335	.04332	.04327	.04319
24	1	19	.04303	.04296	.04287	.04274	.04257	.04238	.04216
24	1	20	.04271	.04253	.04231	.04205	.04175	.04142	.04106
24	1	21	.04231	.04200	.04164	.04125	.04081	.04035	.03985
24	1	22	.04177	.04131	.04081	.04027	.03970	.03911	.03849
24	1	23	.07800	.07683	.07561	.07435	.07306	.07175	.07044
24	2	1	.01620	.00978	.00205	-.00716	-.01799	-.03058	-.04508
24	2	2	.03411	.03350	.03293	.03244	.03206	.03181	.03173
24	2	3	.03790	.03812	.03851	.03910	.03992	.04101	.04238
24	2	4	.04005	.04065	.04145	.04247	.04375	.04530	.04715
24	2	5	.04145	.04223	.04322	.04442	.04586	.04756	.04954
24	2	6	.04243	.04330	.04435	.04561	.04708	.04878	.05073
24	2	7	.04315	.04404	.04510	.04634	.04777	.04940	.05123
24	2	8	.04369	.04457	.04560	.04678	.04812	.04963	.05130
24	2	9	.04409	.04494	.04591	.04701	.04824	.04960	.05110

Coefficients a_i for the BLUE of μ
Shape parameter k

n	s	i	0.7	0.8	0.9	1.0	1.1	1.2	1.3
24	2	10	.04440	.04519	.04608	.04708	.04819	.04940	.05071
24	2	11	.04462	.04535	.04615	.04704	.04801	.04906	.05019
24	2	12	.04479	.04543	.04614	.04691	.04774	.04863	.04957
24	2	13	.04489	.04545	.04606	.04670	.04739	.04811	.04887
24	2	14	.04495	.04541	.04591	.04643	.04697	.04753	.04810
24	2	15	.04496	.04533	.04571	.04610	.04649	.04689	.04729
24	2	16	.04492	.04519	.04545	.04571	.04596	.04620	.04642
24	2	17	.04485	.04500	.04514	.04526	.04536	.04545	.04551
24	2	18	.04472	.04476	.04477	.04475	.04471	.04464	.04455
24	2	19	.04454	.04445	.04433	.04417	.04398	.04376	.04352
24	2	20	.04429	.04407	.04380	.04350	.04317	.04280	.04240
24	2	21	.04392	.04356	.04315	.04270	.04222	.04171	.04117
24	2	22	.12107	.11969	.11820	.11663	.11499	.11332	.11161
24	3	1	.00786	.00059	-.00806	-.01823	-.03007	-.04373	-.05936
24	3	2	.03050	.02977	.02909	.02851	.02804	.02772	.02759
24	3	3	.03557	.03579	.03619	.03681	.03767	.03880	.04025
24	3	4	.03851	.03916	.04002	.04112	.04249	.04414	.04611
24	3	5	.04047	.04134	.04241	.04371	.04526	.04708	.04919
24	3	6	.04190	.04285	.04401	.04537	.04696	.04878	.05085
24	3	7	.04297	.04396	.04512	.04647	.04801	.04975	.05170
24	3	8	.04380	.04478	.04590	.04719	.04864	.05025	.05203
24	3	9	.04446	.04539	.04646	.04765	.04898	.05044	.05203
24	3	10	.04498	.04586	.04684	.04792	.04911	.05041	.05180
24	3	11	.04540	.04620	.04708	.04805	.04909	.05021	.05140
24	3	12	.04574	.04645	.04722	.04805	.04894	.04988	.05088
24	3	13	.04600	.04661	.04727	.04797	.04870	.04946	.05026
24	3	14	.04620	.04671	.04724	.04779	.04837	.04895	.04955
24	3	15	.04634	.04674	.04714	.04755	.04796	.04837	.04879
24	3	16	.04643	.04670	.04697	.04723	.04749	.04773	.04795
24	3	17	.04646	.04661	.04674	.04685	.04694	.04701	.04706
24	3	18	.04643	.04644	.04643	.04639	.04632	.04623	.04611
24	3	19	.04633	.04620	.04604	.04584	.04562	.04536	.04507
24	3	20	.04613	.04585	.04554	.04518	.04480	.04438	.04394
24	3	21	.16753	.16601	.16435	.16258	.16072	.15878	.15680
24	4	1	-.00173	-.00996	-.01962	-.03086	-.04383	-.05867	-.07553
24	4	2	.02639	.02553	.02474	.02406	.02351	.02312	.02294
24	4	3	.03293	.03315	.03358	.03423	.03514	.03634	.03787
24	4	4	.03678	.03750	.03844	.03963	.04110	.04287	.04497
24	4	5	.03941	.04036	.04154	.04295	.04463	.04658	.04883
24	4	6	.04133	.04240	.04366	.04515	.04686	.04882	.05103
24	4	7	.04281	.04391	.04519	.04666	.04832	.05019	.05227
24	4	8	.04398	.04507	.04631	.04771	.04927	.05100	.05290
24	4	9	.04493	.04597	.04714	.04843	.04987	.05143	.05313
24	4	10	.04571	.04668	.04775	.04893	.05021	.05159	.05307
24	4	11	.04635	.04724	.04820	.04924	.05035	.05155	.05281
24	4	12	.04689	.04767	.04851	.04940	.05035	.05135	.05240
24	4	13	.04733	.04800	.04871	.04945	.05022	.05103	.05186
24	4	14	.04769	.04824	.04881	.04939	.04999	.05060	.05123
24	4	15	.04798	.04840	.04882	.04925	.04967	.05009	.05051
24	4	16	.04820	.04848	.04875	.04901	.04926	.04949	.04971
24	4	17	.04835	.04849	.04860	.04869	.04876	.04881	.04884
24	4	18	.04842	.04841	.04836	.04828	.04818	.04805	.04789
24	4	19	.04841	.04823	.04801	.04777	.04749	.04718	.04684
24	4	20	.21783	.21625	.21451	.21264	.21065	.20858	.20644
24	5	1	-.01274	-.02203	-.03282	-.04525	-.05947	-.07562	-.09384
24	5	2	.02171	.02072	.01982	.01903	.01839	.01794	.01771

Coefficients a_i for the BLUE of μ
Shape parameter k

n	s	i	0.7	0.8	0.9	1.0	1.1	1.2	1.3
24	5	3	.02995	.03019	.03064	.03134	.03232	.03360	.03523
24	5	4	.03486	.03565	.03669	.03799	.03957	.04148	.04372
24	5	5	.03823	.03930	.04060	.04214	.04396	.04606	.04847
24	5	6	.04074	.04193	.04333	.04495	.04681	.04892	.05128
24	5	7	.04269	.04391	.04532	.04693	.04873	.05074	.05296
24	5	8	.04425	.04546	.04682	.04835	.05004	.05190	.05393
24	5	9	.04554	.04669	.04797	.04938	.05093	.05260	.05441
24	5	10	.04661	.04768	.04885	.05012	.05150	.05298	.05456
24	5	11	.04751	.04848	.04952	.05065	.05184	.05311	.05445
24	5	12	.04827	.04912	.05003	.05099	.05200	.05306	.05416
24	5	13	.04892	.04964	.05040	.05119	.05201	.05285	.05372
24	5	14	.04946	.05005	.05065	.05126	.05188	.05252	.05316
24	5	15	.04992	.05035	.05079	.05122	.05165	.05207	.05249
24	5	16	.05029	.05056	.05083	.05108	.05131	.05153	.05173
24	5	17	.05057	.05068	.05076	.05083	.05087	.05088	.05087
24	5	18	.05075	.05069	.05059	.05047	.05031	.05013	.04992
24	5	19	.27248	.27093	.26921	.26735	.26535	.26325	.26108
24	6	1	-.02540	-.03589	-.04794	-.06169	-.07730	-.09490	-.11464
24	6	2	.01638	.01525	.01423	.01334	.01262	.01210	.01182
24	6	3	.02657	.02684	.02734	.02810	.02916	.03054	.03228
24	6	4	.03270	.03359	.03474	.03617	.03790	.03995	.04236
24	6	5	.03695	.03815	.03958	.04128	.04326	.04553	.04812
24	6	6	.04013	.04146	.04301	.04479	.04681	.04909	.05162
24	6	7	.04262	.04399	.04554	.04730	.04926	.05142	.05381
24	6	8	.04463	.04598	.04748	.04915	.05098	.05298	.05515
24	6	9	.04630	.04758	.04899	.05052	.05219	.05399	.05592
24	6	10	.04771	.04889	.05017	.05155	.05303	.05461	.05629
24	6	11	.04890	.04997	.05110	.05231	.05359	.05494	.05636
24	6	12	.04993	.05085	.05183	.05286	.05393	.05505	.05620
24	6	13	.05081	.05159	.05239	.05323	.05409	.05497	.05587
24	6	14	.05156	.05218	.05281	.05344	.05409	.05473	.05538
24	6	15	.05221	.05265	.05309	.05352	.05395	.05437	.05477
24	6	16	.05274	.05300	.05325	.05348	.05368	.05387	.05405
24	6	17	.05317	.05324	.05328	.05330	.05329	.05326	.05321
24	6	18	.33210	.33069	.32909	.32735	.32547	.32349	.32142
24	7	1	-.04002	-.05184	-.06530	-.08054	-.09771	-.11694	-.13837
24	7	2	.01027	.00900	.00786	.00687	.00607	.00549	.00517
24	7	3	.02274	.02305	.02361	.02446	.02561	.02711	.02899
24	7	4	.03028	.03130	.03258	.03415	.03605	.03828	.04089
24	7	5	.03554	.03689	.03849	.04037	.04253	.04500	.04779
24	7	6	.03950	.04099	.04271	.04467	.04688	.04935	.05208
24	7	7	.04262	.04415	.04587	.04780	.04993	.05228	.05484
24	7	8	.04515	.04665	.04831	.05013	.05212	.05428	.05661
24	7	9	.04727	.04869	.05023	.05191	.05371	.05564	.05770
24	7	10	.04906	.05036	.05176	.05326	.05486	.05654	.05832
24	7	11	.05059	.05176	.05299	.05429	.05566	.05710	.05859
24	7	12	.05192	.05292	.05397	.05507	.05620	.05738	.05859
24	7	13	.05306	.05390	.05475	.05563	.05652	.05744	.05837
24	7	14	.05406	.05470	.05535	.05601	.05666	.05731	.05797
24	7	15	.05491	.05536	.05579	.05622	.05663	.05703	.05742
24	7	16	.05563	.05587	.05608	.05627	.05644	.05659	.05673
24	7	17	.39741	.39625	.39491	.39343	.39182	.39011	.38832
24	8	1	-.05698	-.07030	-.08535	-.10227	-.12119	-.14225	-.16558
24	8	2	.00324	.00184	.00058	-.00051	-.00139	-.00203	-.00238
24	8	3	.01837	.01874	.01939	.02033	.02161	.02326	.02529
24	8	4	.02756	.02872	.03016	.03192	.03401	.03645	.03927

Coefficients a_i for the BLUE of μ
Shape parameter k

n	s	i	0.7	0.8	0.9	1.0	1.1	1.2	1.3
24	8	5	.03400	.03553	.03732	.03940	.04178	.04447	.04749
24	8	6	.03886	.04055	.04247	.04463	.04705	.04973	.05269
24	8	7	.04271	.04443	.04634	.04846	.05079	.05334	.05610
24	8	8	.04585	.04753	.04936	.05136	.05352	.05585	.05835
24	8	9	.04848	.05006	.05176	.05359	.05554	.05762	.05982
24	8	10	.05072	.05216	.05369	.05531	.05703	.05883	.06073
24	8	11	.05264	.05392	.05526	.05666	.05812	.05964	.06121
24	8	12	.05431	.05540	.05653	.05769	.05889	.06012	.06138
24	8	13	.05577	.05665	.05755	.05847	.05940	.06034	.06129
24	8	14	.05704	.05770	.05836	.05903	.05968	.06034	.06099
24	8	15	.05813	.05856	.05898	.05938	.05977	.06014	.06050
24	8	16	.46929	.46852	.46760	.46655	.46539	.46415	.46285
24	9	1	-.07680	-.09183	-.10868	-.12749	-.14839	-.17152	-.19701
24	9	2	-.00489	-.00643	-.00782	-.00900	-.00995	-.01063	-.01101
24	9	3	.01336	.01382	.01457	.01565	.01708	.01889	.02113
24	9	4	.02449	.02582	.02746	.02943	.03174	.03443	.03751
24	9	5	.03230	.03405	.03607	.03838	.04101	.04396	.04725
24	9	6	.03822	.04013	.04229	.04469	.04735	.05028	.05349
24	9	7	.04293	.04486	.04700	.04934	.05190	.05466	.05765
24	9	8	.04678	.04866	.05069	.05289	.05525	.05777	.06046
24	9	9	.05001	.05177	.05364	.05564	.05775	.05999	.06235
24	9	10	.05277	.05436	.05604	.05780	.05964	.06157	.06359
24	9	11	.05515	.05654	.05799	.05950	.06105	.06266	.06432
24	9	12	.05722	.05840	.05960	.06083	.06209	.06337	.06468
24	9	13	.05903	.05996	.06090	.06185	.06281	.06377	.06473
24	9	14	.06061	.06129	.06195	.06261	.06326	.06390	.06453
24	9	15	.54879	.54861	.54830	.54789	.54742	.54689	.54633
24	10	1	-.10015	-.11713	-.13604	-.15701	-.18018	-.20567	-.23362
24	10	2	-.01438	-.01606	-.01755	-.01882	-.01984	-.02056	-.02095
24	10	3	.00757	.00815	.00904	.01028	.01190	.01393	.01640
24	10	4	.02099	.02254	.02442	.02664	.02923	.03220	.03559
24	10	5	.03044	.03244	.03473	.03733	.04024	.04349	.04709
24	10	6	.03761	.03979	.04221	.04489	.04783	.05105	.05454
24	10	7	.04332	.04551	.04790	.05050	.05331	.05633	.05957
24	10	8	.04801	.05012	.05238	.05480	.05738	.06012	.06303
24	10	9	.05196	.05391	.05598	.05816	.06046	.06288	.06541
24	10	10	.05533	.05708	.05891	.06082	.06281	.06488	.06702
24	10	11	.05824	.05976	.06132	.06293	.06459	.06630	.06804
24	10	12	.06078	.06204	.06331	.06461	.06592	.06726	.06862
24	10	13	.06300	.06397	.06494	.06591	.06688	.06785	.06882
24	10	14	.63726	.63789	.63845	.63896	.63945	.63994	.64044
24	11	1	-.12794	-.14717	-.16845	-.19191	-.21769	-.24590	-.27669
24	11	2	-.02555	-.02736	-.02896	-.03031	-.03137	-.03211	-.03249
24	11	3	.00082	.00156	.00264	.00409	.00595	.00824	.01099
24	11	4	.01698	.01880	.02097	.02350	.02642	.02974	.03348
24	11	5	.02838	.03069	.03331	.03624	.03950	.04310	.04706
24	11	6	.03705	.03954	.04228	.04529	.04856	.05210	.05593
24	11	7	.04397	.04645	.04914	.05203	.05513	.05845	.06198
24	11	8	.04966	.05202	.05455	.05722	.06006	.06305	.06619
24	11	9	.05444	.05662	.05891	.06130	.06381	.06642	.06914
24	11	10	.05854	.06047	.06247	.06455	.06669	.06890	.07118
24	11	11	.06208	.06373	.06542	.06714	.06890	.07070	.07254
24	11	12	.06517	.06651	.06785	.06921	.07058	.07196	.07335
24	11	13	.73638	.73813	.73988	.74165	.74346	.74535	.74733
24	12	1	-.16141	-.18326	-.20731	-.23368	-.26249	-.29388	-.32797
24	12	2	-.03886	-.04079	-.04248	-.04388	-.04497	-.04570	-.04605

Coefficients a_i for the BLUE of μ
Shape parameter k

n	s	i	0.7	0.8	0.9	1.0	1.1	1.2	1.3
24	12	3	-.00712	-.00617	-.00485	-.00312	-.00096	.00165	.00475
24	12	4	.01235	.01452	.01704	.01995	.02327	.02700	.03118
24	12	5	.02612	.02881	.03181	.03515	.03882	.04284	.04723
24	12	6	.03660	.03946	.04258	.04597	.04963	.05356	.05778
24	12	7	.04496	.04780	.05083	.05407	.05752	.06117	.06504
24	12	8	.05185	.05453	.05735	.06033	.06345	.06672	.07014
24	12	9	.05765	.06009	.06262	.06526	.06799	.07082	.07376
24	12	10	.06262	.06475	.06694	.06919	.07150	.07387	.07630
24	12	11	.06691	.06870	.07051	.07235	.07422	.07611	.07803
24	12	12	.84832	.85158	.85494	.85841	.86204	.86583	.86981
25	0	1	.02796	.02302	.01688	.00937	.00035	-.01033	-.02281
25	0	2	.03815	.03772	.03730	.03692	.03661	.03640	.03632
25	0	3	.03991	.04011	.04044	.04093	.04162	.04253	.04369
25	0	4	.04081	.04129	.04194	.04278	.04385	.04516	.04675
25	0	5	.04132	.04194	.04274	.04373	.04494	.04638	.04808
25	0	6	.04162	.04231	.04316	.04420	.04544	.04689	.04857
25	0	7	.04180	.04250	.04337	.04439	.04560	.04699	.04859
25	0	8	.04188	.04258	.04342	.04441	.04555	.04685	.04832
25	0	9	.04191	.04258	.04338	.04430	.04536	.04654	.04787
25	0	10	.04188	.04252	.04326	.04411	.04506	.04613	.04730
25	0	11	.04182	.04241	.04308	.04385	.04470	.04563	.04665
25	0	12	.04172	.04226	.04286	.04353	.04427	.04508	.04595
25	0	13	.04160	.04207	.04260	.04318	.04381	.04448	.04520
25	0	14	.04145	.04186	.04231	.04279	.04330	.04384	.04441
25	0	15	.04128	.04162	.04198	.04236	.04276	.04318	.04360
25	0	16	.04109	.04135	.04163	.04191	.04219	.04248	.04276
25	0	17	.04087	.04106	.04124	.04142	.04159	.04175	.04190
25	0	18	.04063	.04073	.04082	.04090	.04095	.04099	.04101
25	0	19	.04035	.04037	.04036	.04033	.04028	.04019	.04008
25	0	20	.04004	.03997	.03986	.03972	.03955	.03935	.03911
25	0	21	.03968	.03951	.03930	.03905	.03876	.03844	.03809
25	0	22	.03925	.03897	.03865	.03828	.03788	.03744	.03698
25	0	23	.03872	.03832	.03787	.03738	.03686	.03631	.03573
25	0	24	.03799	.03744	.03685	.03623	.03558	.03491	.03423
25	0	25	.03628	.03549	.03470	.03392	.03314	.03237	.03162
25	1	1	.02234	.01677	.00995	.00171	-.00809	-.01960	-.03297
25	1	2	.03565	.03510	.03458	.03411	.03371	.03342	.03327
25	1	3	.03824	.03842	.03874	.03923	.03993	.04086	.04205
25	1	4	.03966	.04016	.04085	.04173	.04285	.04422	.04588
25	1	5	.04054	.04121	.04206	.04312	.04440	.04592	.04771
25	1	6	.04114	.04188	.04280	.04391	.04523	.04677	.04854
25	1	7	.04154	.04232	.04325	.04435	.04564	.04712	.04880
25	1	8	.04182	.04259	.04350	.04456	.04579	.04717	.04872
25	1	9	.04201	.04276	.04362	.04462	.04575	.04702	.04842
25	1	10	.04213	.04284	.04364	.04456	.04559	.04672	.04797
25	1	11	.04220	.04285	.04359	.04441	.04533	.04633	.04742
25	1	12	.04222	.04281	.04347	.04420	.04500	.04586	.04678
25	1	13	.04221	.04273	.04330	.04393	.04461	.04533	.04609
25	1	14	.04215	.04261	.04309	.04362	.04417	.04475	.04536
25	1	15	.04207	.04245	.04284	.04326	.04369	.04413	.04458
25	1	16	.04196	.04225	.04255	.04286	.04316	.04347	.04378
25	1	17	.04182	.04202	.04222	.04242	.04260	.04278	.04294
25	1	18	.04165	.04176	.04186	.04194	.04200	.04204	.04206
25	1	19	.04144	.04145	.04144	.04141	.04135	.04126	.04114
25	1	20	.04118	.04110	.04098	.04083	.04064	.04043	.04018
25	1	21	.04088	.04068	.04045	.04017	.03986	.03952	.03915

Coefficients a_i for the BLUE of μ
Shape parameter k

n	s	i	0.7	0.8	0.9	1.0	1.1	1.2	1.3
25	1	22	.04049	.04017	.03982	.03942	.03898	.03852	.03803
25	1	23	.03997	.03952	.03903	.03850	.03794	.03735	.03675
25	1	24	.07468	.07355	.07236	.07114	.06988	.06862	.06735
25	2	1	.01567	.00939	.00180	-.00725	-.01791	-.03033	-.04467
25	2	2	.03270	.03204	.03142	.03084	.03036	.02999	.02978
25	2	3	.03630	.03646	.03678	.03728	.03799	.03896	.04020
25	2	4	.03834	.03887	.03960	.04054	.04172	.04317	.04491
25	2	5	.03967	.04039	.04131	.04244	.04380	.04542	.04731
25	2	6	.04060	.04142	.04242	.04361	.04502	.04666	.04853
25	2	7	.04129	.04214	.04315	.04434	.04572	.04730	.04908
25	2	8	.04180	.04265	.04364	.04479	.04610	.04757	.04922
25	2	9	.04219	.04301	.04396	.04503	.04625	.04760	.04909
25	2	10	.04249	.04326	.04414	.04513	.04623	.04745	.04877
25	2	11	.04271	.04342	.04423	.04512	.04610	.04717	.04832
25	2	12	.04287	.04352	.04423	.04502	.04588	.04679	.04777
25	2	13	.04297	.04355	.04418	.04485	.04558	.04634	.04715
25	2	14	.04304	.04353	.04406	.04462	.04521	.04583	.04647
25	2	15	.04306	.04347	.04389	.04434	.04479	.04526	.04574
25	2	16	.04304	.04336	.04368	.04400	.04433	.04465	.04496
25	2	17	.04299	.04321	.04342	.04362	.04381	.04399	.04415
25	2	18	.04290	.04301	.04311	.04318	.04324	.04328	.04329
25	2	19	.04276	.04277	.04275	.04270	.04262	.04251	.04238
25	2	20	.04258	.04247	.04232	.04214	.04193	.04169	.04142
25	2	21	.04232	.04209	.04181	.04150	.04116	.04078	.04038
25	2	22	.04197	.04160	.04119	.04074	.04026	.03975	.03922
25	2	23	.11574	.11438	.11293	.11140	.10981	.10818	.10652
25	3	1	.00799	.00091	-.00753	-.01749	-.02910	-.04253	-.05792
25	3	2	.02934	.02856	.02782	.02715	.02657	.02613	.02585
25	3	3	.03409	.03424	.03456	.03508	.03583	.03683	.03813
25	3	4	.03685	.03743	.03821	.03922	.04048	.04201	.04385
25	3	5	.03870	.03949	.04049	.04171	.04317	.04489	.04689
25	3	6	.04003	.04093	.04202	.04331	.04482	.04657	.04856
25	3	7	.04104	.04198	.04308	.04438	.04586	.04754	.04944
25	3	8	.04182	.04276	.04384	.04509	.04649	.04807	.04982
25	3	9	.04244	.04335	.04438	.04555	.04685	.04829	.04988
25	3	10	.04294	.04379	.04475	.04583	.04701	.04830	.04971
25	3	11	.04334	.04413	.04500	.04597	.04702	.04815	.04937
25	3	12	.04365	.04437	.04515	.04600	.04691	.04789	.04892
25	3	13	.04391	.04454	.04521	.04594	.04671	.04752	.04838
25	3	14	.04410	.04464	.04521	.04581	.04643	.04708	.04775
25	3	15	.04424	.04468	.04514	.04561	.04609	.04658	.04707
25	3	16	.04434	.04467	.04501	.04534	.04568	.04601	.04633
25	3	17	.04438	.04460	.04482	.04502	.04521	.04538	.04554
25	3	18	.04438	.04449	.04457	.04464	.04468	.04470	.04470
25	3	19	.04433	.04431	.04426	.04419	.04408	.04395	.04380
25	3	20	.04421	.04406	.04388	.04366	.04341	.04313	.04283
25	3	21	.04400	.04372	.04339	.04303	.04263	.04221	.04176
25	3	22	.15987	.15837	.15674	.15500	.15317	.15128	.14934
25	4	1	-.00080	-.00877	-.01815	-.02911	-.04177	-.05631	-.07285
25	4	2	.02552	.02461	.02376	.02298	.02232	.02180	.02146
25	4	3	.03161	.03176	.03209	.03263	.03341	.03447	.03583
25	4	4	.03520	.03583	.03667	.03776	.03910	.04074	.04270
25	4	5	.03764	.03851	.03960	.04092	.04249	.04433	.04646
25	4	6	.03943	.04042	.04161	.04301	.04464	.04651	.04863
25	4	7	.04081	.04184	.04305	.04445	.04605	.04785	.04987
25	4	8	.04190	.04293	.04412	.04546	.04698	.04867	.05053

Coefficients a_i for the BLUE of μ
Shape parameter k

n	s	i	0.7	0.8	0.9	1.0	1.1	1.2	1.3
25	4	9	.04278	.04378	.04491	.04617	.04757	.04911	.05080
25	4	10	.04351	.04445	.04550	.04666	.04793	.04930	.05079
25	4	11	.04411	.04497	.04593	.04697	.04809	.04930	.05059
25	4	12	.04461	.04539	.04624	.04715	.04812	.04916	.05025
25	4	13	.04502	.04571	.04644	.04722	.04803	.04889	.04978
25	4	14	.04536	.04594	.04655	.04719	.04785	.04853	.04923
25	4	15	.04564	.04611	.04659	.04708	.04758	.04809	.04860
25	4	16	.04586	.04620	.04655	.04690	.04724	.04757	.04790
25	4	17	.04601	.04624	.04645	.04664	.04682	.04699	.04713
25	4	18	.04611	.04620	.04627	.04631	.04633	.04633	.04630
25	4	19	.04615	.04609	.04601	.04590	.04576	.04560	.04540
25	4	20	.04610	.04590	.04566	.04539	.04510	.04477	.04442
25	4	21	.20747	.20590	.20417	.20232	.20036	.19830	.19619
25	5	1	-.01083	-.01979	-.03022	-.04228	-.05611	-.07186	-.08968
25	5	2	.02119	.02015	.01918	.01830	.01755	.01696	.01655
25	5	3	.02881	.02896	.02931	.02989	.03072	.03184	.03329
25	5	4	.03335	.03405	.03497	.03614	.03760	.03935	.04143
25	5	5	.03647	.03744	.03863	.04007	.04176	.04374	.04601
25	5	6	.03879	.03989	.04119	.04272	.04448	.04648	.04874
25	5	7	.04059	.04174	.04307	.04459	.04631	.04825	.05040
25	5	8	.04203	.04318	.04448	.04594	.04757	.04938	.05136
25	5	9	.04322	.04432	.04556	.04693	.04844	.05008	.05187
25	5	10	.04420	.04524	.04639	.04764	.04901	.05048	.05206
25	5	11	.04504	.04599	.04703	.04815	.04935	.05064	.05200
25	5	12	.04574	.04660	.04751	.04849	.04953	.05063	.05177
25	5	13	.04634	.04709	.04787	.04870	.04957	.05047	.05141
25	5	14	.04685	.04748	.04813	.04880	.04949	.05020	.05092
25	5	15	.04728	.04778	.04828	.04879	.04931	.04983	.05035
25	5	16	.04764	.04800	.04835	.04870	.04904	.04937	.04969
25	5	17	.04792	.04813	.04833	.04851	.04867	.04882	.04895
25	5	18	.04813	.04819	.04823	.04824	.04823	.04819	.04813
25	5	19	.04825	.04816	.04803	.04787	.04768	.04746	.04722
25	5	20	.25898	.25742	.25569	.25381	.25181	.24971	.24753
25	6	1	-.02230	-.03235	-.04395	-.05723	-.07235	-.08945	-.10867
25	6	2	.01629	.01511	.01401	.01303	.01218	.01152	.01106
25	6	3	.02566	.02583	.02621	.02683	.02772	.02892	.03046
25	6	4	.03129	.03207	.03309	.03437	.03594	.03783	.04006
25	6	5	.03519	.03627	.03759	.03916	.04099	.04312	.04555
25	6	6	.03811	.03934	.04077	.04244	.04435	.04650	.04892
25	6	7	.04040	.04167	.04314	.04480	.04666	.04874	.05104
25	6	8	.04225	.04351	.04494	.04654	.04830	.05024	.05235
25	6	9	.04378	.04500	.04635	.04784	.04946	.05123	.05313
25	6	10	.04506	.04621	.04746	.04881	.05028	.05185	.05352
25	6	11	.04616	.04721	.04833	.04954	.05083	.05220	.05364
25	6	12	.04710	.04803	.04902	.05007	.05117	.05233	.05354
25	6	13	.04791	.04872	.04956	.05044	.05135	.05230	.05327
25	6	14	.04861	.04928	.04996	.05067	.05139	.05212	.05287
25	6	15	.04921	.04973	.05025	.05078	.05130	.05183	.05235
25	6	16	.04972	.05008	.05044	.05078	.05111	.05143	.05173
25	6	17	.05014	.05034	.05051	.05067	.05081	.05092	.05102
25	6	18	.05047	.05049	.05049	.05045	.05040	.05031	.05021
25	6	19	.31492	.31346	.31182	.31003	.30810	.30607	.30395
25	7	1	-.03544	-.04672	-.05961	-.07426	-.09081	-.10941	-.13019
25	7	2	.01071	.00939	.00817	.00707	.00614	.00539	.00488
25	7	3	.02211	.02230	.02272	.02340	.02437	.02567	.02732
25	7	4	.02899	.02987	.03100	.03240	.03412	.03616	.03855

Coefficients a_i for the BLUE of μ
Shape parameter k

n	s	i	0.7	0.8	0.9	1.0	1.1	1.2	1.3
25	7	5	.03379	.03500	.03646	.03818	.04018	.04248	.04509
25	7	6	.03740	.03877	.04036	.04219	.04426	.04659	.04918
25	7	7	.04024	.04167	.04328	.04510	.04712	.04936	.05182
25	7	8	.04256	.04397	.04554	.04728	.04919	.05127	.05353
25	7	9	.04449	.04584	.04732	.04894	.05069	.05258	.05461
25	7	10	.04612	.04738	.04874	.05021	.05178	.05346	.05524
25	7	11	.04752	.04866	.04988	.05119	.05256	.05401	.05554
25	7	12	.04873	.04974	.05081	.05193	.05310	.05432	.05558
25	7	13	.04978	.05064	.05154	.05247	.05343	.05442	.05543
25	7	14	.05070	.05140	.05212	.05285	.05360	.05435	.05511
25	7	15	.05149	.05202	.05256	.05309	.05362	.05414	.05466
25	7	16	.05217	.05253	.05287	.05319	.05350	.05380	.05408
25	7	17	.05274	.05291	.05305	.05317	.05326	.05334	.05339
25	7	18	.37591	.37465	.37321	.37162	.36990	.36808	.36618
25	8	1	-.05057	-.06322	-.07756	-.09374	-.11190	-.13217	-.15470
25	8	2	.00433	.00287	.00152	.00032	-.00071	-.00152	-.00209
25	8	3	.01808	.01830	.01878	.01954	.02061	.02202	.02380
25	8	4	.02642	.02741	.02867	.03023	.03211	.03433	.03691
25	8	5	.03225	.03362	.03524	.03714	.03933	.04182	.04464
25	8	6	.03666	.03820	.03996	.04197	.04423	.04675	.04955
25	8	7	.04015	.04174	.04352	.04552	.04772	.05014	.05278
25	8	8	.04300	.04457	.04630	.04820	.05027	.05251	.05493
25	8	9	.04538	.04688	.04850	.05027	.05216	.05419	.05635
25	8	10	.04741	.04880	.05029	.05188	.05357	.05536	.05725
25	8	11	.04916	.05041	.05174	.05313	.05461	.05615	.05775
25	8	12	.05068	.05177	.05292	.05411	.05535	.05664	.05796
25	8	13	.05200	.05293	.05388	.05486	.05586	.05689	.05793
25	8	14	.05316	.05390	.05465	.05541	.05617	.05694	.05771
25	8	15	.05418	.05472	.05526	.05579	.05631	.05682	.05733
25	8	16	.05505	.05539	.05570	.05600	.05628	.05655	.05680
25	8	17	.44266	.44172	.44062	.43939	.43804	.43660	.43509
25	9	1	-.06810	-.08229	-.09827	-.11617	-.13612	-.15828	-.18276
25	9	2	-.00298	-.00460	-.00607	-.00739	-.00850	-.00938	-.01000
25	9	3	.01348	.01377	.01433	.01519	.01638	.01792	.01986
25	9	4	.02352	.02465	.02608	.02781	.02988	.03231	.03512
25	9	5	.03056	.03211	.03392	.03603	.03843	.04115	.04420
25	9	6	.03590	.03763	.03959	.04181	.04428	.04703	.05005
25	9	7	.04013	.04191	.04389	.04608	.04849	.05111	.05396
25	9	8	.04360	.04535	.04726	.04934	.05159	.05402	.05662
25	9	9	.04651	.04817	.04996	.05188	.05393	.05611	.05842
25	9	10	.04900	.05053	.05216	.05388	.05570	.05761	.05962
25	9	11	.05115	.05252	.05395	.05546	.05703	.05866	.06036
25	9	12	.05302	.05421	.05544	.05671	.05802	.05937	.06075
25	9	13	.05467	.05565	.05666	.05768	.05872	.05979	.06086
25	9	14	.05611	.05688	.05765	.05842	.05919	.05997	.06075
25	9	15	.05737	.05791	.05844	.05895	.05946	.05995	.06043
25	9	16	.51606	.51561	.51501	.51431	.51352	.51266	.51176
25	10	1	-.08854	-.10449	-.12231	-.14216	-.16415	-.18844	-.21514
25	10	2	-.01144	-.01321	-.01481	-.01623	-.01743	-.01837	-.01902
25	10	3	.00822	.00859	.00925	.01024	.01157	.01329	.01542
25	10	4	.02023	.02155	.02317	.02512	.02742	.03009	.03315
25	10	5	.02869	.03046	.03250	.03484	.03750	.04048	.04380
25	10	6	.03511	.03707	.03927	.04173	.04445	.04745	.05072
25	10	7	.04022	.04223	.04443	.04685	.04948	.05233	.05541
25	10	8	.04442	.04637	.04849	.05078	.05323	.05586	.05865
25	10	9	.04795	.04979	.05176	.05386	.05608	.05843	.06090

Coefficients a_i for the BLUE of μ
Shape parameter k

n	s	i	0.7	0.8	0.9	1.0	1.1	1.2	1.3
25	10	10	.05097	.05265	.05443	.05630	.05826	.06031	.06244
25	10	11	.05358	.05508	.05663	.05825	.05993	.06166	.06345
25	10	12	.05587	.05715	.05846	.05981	.06119	.06260	.06404
25	10	13	.05788	.05892	.05997	.06104	.06212	.06321	.06431
25	10	14	.05964	.06043	.06121	.06199	.06277	.06354	.06431
25	10	15	.59720	.59741	.59754	.59759	.59759	.59757	.59755
25	11	1	-.11260	-.13055	-.15048	-.17255	-.19687	-.22358	-.25282
25	11	2	-.02130	-.02321	-.02495	-.02647	-.02774	-.02873	-.02941
25	11	3	.00214	.00262	.00342	.00457	.00609	.00802	.01037
25	11	4	.01650	.01803	.01989	.02209	.02466	.02762	.03099
25	11	5	.02663	.02865	.03097	.03359	.03654	.03983	.04346
25	11	6	.03433	.03655	.03903	.04177	.04478	.04806	.05163
25	11	7	.04047	.04273	.04520	.04788	.05077	.05389	.05722
25	11	8	.04552	.04771	.05007	.05258	.05527	.05812	.06114
25	11	9	.04978	.05183	.05400	.05629	.05871	.06124	.06390
25	11	10	.05342	.05528	.05722	.05925	.06136	.06356	.06583
25	11	11	.05658	.05821	.05990	.06163	.06342	.06525	.06714
25	11	12	.05935	.06072	.06212	.06354	.06499	.06647	.06797
25	11	13	.06178	.06287	.06397	.06507	.06618	.06729	.06841
25	11	14	.68740	.68854	.68964	.69073	.69183	.69297	.69416
25	12	1	-.14119	-.16145	-.18382	-.20844	-.23544	-.26495	-.29710
25	12	2	-.03291	-.03496	-.03681	-.03842	-.03975	-.04078	-.04147
25	12	3	-.00494	-.00430	-.00332	-.00197	-.00021	.00197	.00460
25	12	4	.01221	.01402	.01617	.01868	.02158	.02488	.02861
25	12	5	.02434	.02667	.02932	.03228	.03558	.03922	.04322
25	12	6	.03357	.03611	.03892	.04198	.04533	.04895	.05286
25	12	7	.04093	.04350	.04627	.04925	.05245	.05587	.05950
25	12	8	.04701	.04947	.05210	.05488	.05782	.06093	.06420
25	12	9	.05212	.05441	.05681	.05933	.06196	.06470	.06756
25	12	10	.05651	.05856	.06069	.06289	.06517	.06752	.06995
25	12	11	.06032	.06209	.06391	.06577	.06767	.06962	.07161
25	12	12	.06364	.06511	.06660	.06810	.06961	.07114	.07270
25	12	13	.78839	.79076	.79318	.79567	.79825	.80094	.80377
25	13	1	-.17560	-.19854	-.22375	-.25135	-.28148	-.31425	-.34980
25	13	2	-.04671	-.04890	-.05085	-.05254	-.05392	-.05496	-.05564
25	13	3	-.01328	-.01243	-.01121	-.00959	-.00754	-.00504	-.00206
25	13	4	.00726	.00940	.01191	.01481	.01810	.02182	.02598
25	13	5	.02179	.02451	.02755	.03092	.03464	.03871	.04314
25	13	6	.03287	.03580	.03899	.04245	.04619	.05020	.05451
25	13	7	.04172	.04464	.04777	.05111	.05466	.05842	.06241
25	13	8	.04901	.05180	.05473	.05783	.06107	.06447	.06803
25	13	9	.05516	.05772	.06038	.06315	.06602	.06900	.07209
25	13	10	.06044	.06271	.06504	.06743	.06989	.07242	.07501
25	13	11	.06501	.06694	.06891	.07090	.07292	.07498	.07707
25	13	12	.90233	.90636	.91054	.91489	.91945	.92422	.92925

Coefficients a_i for the BLUE of μ
Shape parameter k

n	s	i	1.4	1.5	1.6	1.7	1.8	1.9	2.0
2	0	1	.50000	.50000	.50000	.50000	.50000	.50000	.50000
2	0	2	.50000	.50000	.50000	.50000	.50000	.50000	.50000
3	0	1	.30273	.29840	.29384	.28907	.28410	.27893	.27358
3	0	2	.38366	.39013	.39683	.40375	.41086	.41815	.42558
3	0	3	.31362	.31147	.30932	.30718	.30504	.30292	.30083
3	1	1	-.18404	-.19935	-.21491	-.23071	-.24675	-.26301	-.27949
3	1	2	1.18404	1.19935	1.21491	1.23071	1.24675	1.26301	1.27949
4	0	1	.20048	.19276	.18448	.17566	.16630	.15641	.14600
4	0	2	.30218	.31031	.31900	.32824	.33802	.34832	.35912
4	0	3	.27089	.27291	.27492	.27691	.27888	.28082	.28275
4	0	4	.22645	.22402	.22160	.21919	.21681	.21445	.21214
4	1	1	-.04216	-.05806	-.07458	-.09171	-.10944	-.12778	-.14671
4	1	2	.36641	.38034	.39488	.41002	.42574	.44204	.45890
4	1	3	.67575	.67772	.67970	.68169	.68370	.68574	.68781
5	0	1	.13871	.12856	.11756	.10570	.09299	.07942	.06499
5	0	2	.24722	.25544	.26439	.27407	.28450	.29566	.30756
5	0	3	.23008	.23404	.23818	.24246	.24690	.25146	.25615
5	0	4	.20759	.20790	.20815	.20833	.20847	.20854	.20858
5	0	5	.17640	.17405	.17173	.16942	.16715	.16492	.16272
5	1	1	-.01798	-.03464	-.05225	-.07080	-.09031	-.11078	-.13221
5	1	2	.26910	.28101	.29375	.30733	.32175	.33700	.35310
5	1	3	.27401	.27979	.28571	.29176	.29795	.30425	.31066
5	1	4	.47487	.47384	.47278	.47170	.47062	.46953	.46846
5	2	1	-.35043	-.37860	-.40785	-.43820	-.46964	-.50218	-.53583
5	2	2	.32306	.34193	.36169	.38235	.40392	.42640	.44980
5	2	3	1.02736	1.03667	1.04616	1.05585	1.06572	1.07578	1.08603
6	0	1	.09781	.08592	.07293	.05882	.04359	.02721	.00968
6	0	2	.20788	.21564	.22421	.23361	.24385	.25494	.26690
6	0	3	.19861	.20341	.20852	.21394	.21964	.22563	.23190
6	0	4	.18417	.18611	.18809	.19008	.19209	.19412	.19616
6	0	5	.16746	.16701	.16650	.16592	.16529	.16461	.16389
6	0	6	.14408	.14190	.13975	.13763	.13554	.13348	.13148
6	1	1	-.01505	-.03232	-.05079	-.07049	-.09143	-.11363	-.13712
6	1	2	.21441	.22467	.23584	.24793	.26097	.27497	.28995
6	1	3	.22165	.22808	.23482	.24187	.24921	.25685	.26478
6	1	4	.21310	.21568	.21826	.22086	.22347	.22608	.22869
6	1	5	.36589	.36390	.36187	.35983	.35778	.35573	.35370
6	2	1	-.21443	-.23986	-.26661	-.29469	-.32414	-.35496	-.38718
6	2	2	.22952	.24383	.25914	.27547	.29284	.31127	.33077
6	2	3	.26530	.27400	.28299	.29225	.30179	.31159	.32167
6	2	4	.71961	.72203	.72448	.72697	.72951	.73210	.73474
6	3	1	-.62684	-.66731	-.70937	-.75304	-.79834	-.84529	-.89390
6	3	2	.26893	.29112	.31442	.33883	.36440	.39112	.41904
6	3	3	1.35791	1.37619	1.39495	1.41421	1.43395	1.45416	1.47486
7	0	1	.06902	.05586	.04140	.02561	.00846	-.01007	-.03001
7	0	2	.17835	.18548	.19345	.20227	.21197	.22257	.23410
7	0	3	.17403	.17914	.18466	.19057	.19689	.20361	.21072
7	0	4	.16431	.16717	.17015	.17325	.17646	.17978	.18319
7	0	5	.15283	.15369	.15453	.15534	.15613	.15690	.15764
7	0	6	.13990	.13910	.13824	.13732	.13635	.13535	.13432
7	0	7	.12155	.11956	.11758	.11564	.11373	.11186	.11004
7	1	1	-.01752	-.03524	-.05437	-.07494	-.09699	-.12055	-.14565
7	1	2	.17832	.18721	.19701	.20777	.21950	.23224	.24602
7	1	3	.18680	.19326	.20015	.20746	.21520	.22335	.23193
7	1	4	.18230	.18591	.18965	.19349	.19743	.20148	.20563
7	1	5	.17279	.17386	.17491	.17593	.17692	.17788	.17882

Coefficients a_i for the BLUE of μ
Shape parameter k

n	s	i	1.4	1.5	1.6	1.7	1.8	1.9	2.0
7	1	6	.29731	.29499	.29265	.29030	.28794	.28559	.28326
7	2	1	-.15504	-.17908	-.20464	-.23175	-.26044	-.29076	-.32272
7	2	2	.18024	.19174	.20425	.21782	.23245	.24818	.26504
7	2	3	.20886	.21715	.22586	.23499	.24454	.25452	.26492
7	2	4	.21222	.21667	.22121	.22583	.23054	.23533	.24020
7	2	5	.55371	.55352	.55332	.55312	.55291	.55272	.55256
7	3	1	-.39206	-.42602	-.46167	-.49905	-.53819	-.57911	-.62186
7	3	2	.18728	.20311	.22007	.23817	.25745	.27793	.29965
7	3	3	.25014	.26117	.27260	.28443	.29667	.30931	.32236
7	3	4	.95464	.96174	.96900	.97645	.98407	.99187	.99985
7	4	1	-.87692	-.92941	-.98402	-1.04079	-1.09974	-1.16090	-1.22430
7	4	2	.21039	.23480	.26047	.28742	.31569	.34531	.37631
7	4	3	1.66652	1.69460	1.72355	1.75337	1.78405	1.81559	1.84799
8	0	1	.04785	.03375	.01818	.00112	-.01750	-.03769	-.05951
8	0	2	.15538	.16185	.16914	.17728	.18630	.19624	.20713
8	0	3	.15441	.15955	.16516	.17124	.17779	.18482	.19233
8	0	4	.14781	.15117	.15473	.15850	.16245	.16660	.17093
8	0	5	.13945	.14115	.14288	.14465	.14646	.14829	.15014
8	0	6	.13022	.13046	.13067	.13083	.13095	.13104	.13110
8	0	7	.11989	.11892	.11789	.11682	.11571	.11458	.11342
8	0	8	.10500	.10316	.10135	.09957	.09783	.09613	.09446
8	1	1	-.02131	-.03935	-.05896	-.08019	-.10309	-.12770	-.15406
8	1	2	.15232	.16004	.16868	.17824	.18878	.20033	.21292
8	1	3	.16158	.16783	.17457	.18181	.18956	.19782	.20658
8	1	4	.15945	.16355	.16786	.17236	.17706	.18196	.18704
8	1	5	.15316	.15523	.15732	.15945	.16160	.16378	.16597
8	1	6	.14465	.14494	.14519	.14540	.14557	.14571	.14582
8	1	7	.25015	.24776	.24535	.24293	.24051	.23811	.23573
8	2	1	-.12383	-.14701	-.17188	-.19846	-.22682	-.25699	-.28902
8	2	2	.14896	.15848	.16899	.18052	.19312	.20680	.22162
8	2	3	.17334	.18103	.18923	.19795	.20719	.21695	.22724
8	2	4	.17766	.18262	.18777	.19311	.19863	.20433	.21021
8	2	5	.17417	.17654	.17892	.18131	.18373	.18615	.18859
8	2	6	.44969	.44834	.44696	.44557	.44416	.44276	.44136
8	3	1	-.28284	-.31339	-.34575	-.37997	-.41610	-.45417	-.49424
8	3	2	.14577	.15801	.17132	.18576	.20134	.21812	.23612
8	3	3	.19346	.20318	.21342	.22418	.23546	.24727	.25962
8	3	4	.20743	.21347	.21967	.22603	.23255	.23922	.24605
8	3	5	.73618	.73873	.74134	.74401	.74675	.74956	.75245
8	4	1	-.55403	-.59623	-.64049	-.68687	-.73540	-.78612	-.83908
8	4	2	.14427	.16104	.17901	.19821	.21868	.24045	.26357
8	4	3	.23128	.24418	.25758	.27148	.28590	.30084	.31630
8	4	4	1.17847	1.19100	1.20391	1.21718	1.23082	1.24483	1.25921
9	0	1	.03176	.01695	.00055	-.01749	-.03724	-.05874	-.08203
9	0	2	.13701	.14283	.14945	.15689	.16520	.17441	.18456
9	0	3	.13840	.14343	.14896	.15501	.16158	.16867	.17631
9	0	4	.13401	.13763	.14152	.14567	.15008	.15474	.15966
9	0	5	.12778	.13001	.13233	.13475	.13727	.13986	.14254
9	0	6	.12078	.12175	.12271	.12368	.12464	.12560	.12654
9	0	7	.11321	.11308	.11291	.11270	.11244	.11215	.11182
9	0	8	.10473	.10368	.10259	.10146	.10030	.09913	.09794
9	0	9	.09233	.09064	.08897	.08734	.08574	.08418	.08266
9	1	1	-.02518	-.04343	-.06339	-.08511	-.10866	-.13408	-.16143
9	1	2	.13251	.13926	.14687	.15539	.16485	.17529	.18676
9	1	3	.14231	.14824	.15470	.16172	.16929	.17742	.18613
9	1	4	.14174	.14604	.15061	.15545	.16056	.16593	.17157

Coefficients a_i for the BLUE of μ
Shape parameter k

n	s	i	1.4	1.5	1.6	1.7	1.8	1.9	2.0
9	1	5	.13745	.14010	.14284	.14567	.14859	.15159	.15468
9	1	6	.13137	.13252	.13366	.13480	.13593	.13705	.13816
9	1	7	.12405	.12390	.12371	.12348	.12321	.12291	.12258
9	1	8	.21575	.21338	.21100	.20862	.20624	.20388	.20155
9	2	1	-.10550	-.12809	-.15248	-.17873	-.20690	-.23704	-.26922
9	2	2	.12693	.13496	.14392	.15386	.16483	.17686	.18999
9	2	3	.14857	.15565	.16330	.17151	.18030	.18968	.19966
9	2	4	.15333	.15842	.16378	.16940	.17529	.18144	.18786
9	2	5	.15165	.15471	.15786	.16108	.16439	.16776	.17121
9	2	6	.14669	.14793	.14914	.15035	.15154	.15272	.15389
9	2	7	.37833	.37642	.37448	.37252	.37055	.36858	.36662
9	3	1	-.22178	-.25019	-.28052	-.31283	-.34718	-.38361	-.42219
9	3	2	.12007	.12991	.14078	.15271	.16575	.17994	.19531
9	3	3	.15881	.16744	.17665	.18644	.19684	.20783	.21944
9	3	4	.17112	.17720	.18354	.19013	.19697	.20405	.21139
9	3	5	.17299	.17650	.18006	.18368	.18737	.19111	.19491
9	3	6	.59879	.59914	.59949	.59986	.60026	.60068	.60114
9	4	1	-.40030	-.43710	-.47600	-.51705	-.56031	-.60584	-.65368
9	4	2	.11165	.12425	.13799	.15288	.16899	.18633	.20496
9	4	3	.17651	.18737	.19882	.21086	.22350	.23676	.25064
9	4	4	.20012	.20752	.21515	.22300	.23107	.23938	.24791
9	4	5	.91203	.91795	.92405	.93031	.93675	.94337	.95017
9	5	1	-.70291	-.75303	-.80560	-.86066	-.91826	-.97844	-1.04126
9	5	2	.10147	.11875	.13727	.15709	.17824	.20076	.22468
9	5	3	.21027	.22468	.23967	.25524	.27142	.28822	.30565
9	5	4	1.39116	1.40961	1.42866	1.44832	1.46859	1.48946	1.51093
10	0	1	.01922	.00388	-.01316	-.03197	-.05260	-.07512	-.09959
10	0	2	.12199	.12721	.13318	.13994	.14754	.15602	.16541
10	0	3	.12511	.12995	.13531	.14121	.14766	.15468	.16227
10	0	4	.12234	.12608	.13012	.13448	.13914	.14412	.14940
10	0	5	.11766	.12022	.12294	.12580	.12880	.13194	.13521
10	0	6	.11220	.11367	.11518	.11674	.11833	.11995	.12160
10	0	7	.10630	.10680	.10728	.10773	.10817	.10858	.10898
10	0	8	.09998	.09963	.09923	.09879	.09831	.09780	.09725
10	0	9	.09287	.09180	.09069	.08955	.08839	.08722	.08604
10	0	10	.08234	.08077	.07923	.07773	.07625	.07482	.07342
10	1	1	-.02872	-.04711	-.06732	-.08941	-.11344	-.13949	-.16761
10	1	2	.11685	.12275	.12947	.13705	.14554	.15498	.16541
10	1	3	.12701	.13259	.13872	.14543	.15273	.16062	.16913
10	1	4	.12754	.13188	.13654	.14152	.14682	.15245	.15839
10	1	5	.12462	.12761	.13074	.13402	.13745	.14101	.14471
10	1	6	.12011	.12182	.12358	.12537	.12719	.12904	.13092
10	1	7	.11466	.11524	.11579	.11633	.11684	.11732	.11779
10	1	8	.10838	.10798	.10754	.10705	.10654	.10599	.10542
10	1	9	.18955	.18725	.18494	.18264	.18035	.17808	.17584
10	2	1	-.09389	-.11601	-.14004	-.16604	-.19408	-.22423	-.25656
10	2	2	.11038	.11721	.12493	.13358	.14320	.15384	.16554
10	2	3	.13012	.13663	.14372	.15141	.15971	.16865	.17822
10	2	4	.13511	.14015	.14551	.15120	.15722	.16356	.17023
10	2	5	.13452	.13795	.14152	.14523	.14908	.15306	.15717
10	2	6	.13120	.13312	.13507	.13704	.13904	.14106	.14310
10	2	7	.12622	.12679	.12734	.12785	.12835	.12882	.12927
10	2	8	.32633	.32415	.32195	.31972	.31748	.31525	.31302
10	3	1	-.18373	-.21065	-.23958	-.27058	-.30374	-.33910	-.37674
10	3	2	.10225	.11036	.11943	.12950	.14062	.15283	.16617
10	3	3	.13518	.14290	.15122	.16016	.16974	.17997	.19086

Coefficients a_i for the BLUE of μ
Shape parameter k

n	s	i	1.4	1.5	1.6	1.7	1.8	1.9	2.0
10	3	4	.14624	.15215	.15838	.16493	.17180	.17900	.18651
10	3	5	.14876	.15269	.15675	.16093	.16523	.16966	.17420
10	3	6	.14695	.14903	.15113	.15324	.15536	.15750	.15966
10	3	7	.50434	.50351	.50267	.50182	.50097	.50014	.49933
10	4	1	-.31248	-.34589	-.38145	-.41924	-.45931	-.50173	-.54656
10	4	2	.09186	.10179	.11277	.12484	.13804	.15241	.16800
10	4	3	.14366	.15302	.16301	.17364	.18492	.19687	.20950
10	4	4	.16329	.17032	.17766	.18531	.19326	.20153	.21010
10	4	5	.17006	.17457	.17919	.18390	.18872	.19365	.19868
10	4	6	.74361	.74618	.74883	.75155	.75437	.75728	.76028
10	5	1	-.50898	-.55177	-.59695	-.64457	-.69470	-.74739	-.80271
10	5	2	.07818	.09091	.10478	.11985	.13615	.15372	.17261
10	5	3	.15869	.17045	.18285	.19590	.20963	.22403	.23913
10	5	4	.19112	.19970	.20855	.21769	.22711	.23683	.24684
10	5	5	1.08099	1.09072	1.10077	1.11113	1.12181	1.13282	1.14413
11	0	1	.00923	-.00650	-.02403	-.04343	-.06476	-.08810	-.11350
11	0	2	.10949	.11414	.11951	.12563	.13255	.14032	.14896
11	0	3	.11390	.11851	.12365	.12934	.13561	.14245	.14989
11	0	4	.11237	.11612	.12022	.12466	.12946	.13460	.14009
11	0	5	.10886	.11163	.11459	.11774	.12108	.12460	.12830
11	0	6	.10453	.10635	.10826	.11024	.11230	.11443	.11663
11	0	7	.09980	.10076	.10173	.10270	.10368	.10467	.10565
11	0	8	.09479	.09497	.09513	.09525	.09534	.09541	.09545
11	0	9	.08942	.08893	.08839	.08781	.08720	.08656	.08589
11	0	10	.08336	.08229	.08118	.08006	.07892	.07778	.07663
11	0	11	.07426	.07281	.07138	.06998	.06862	.06729	.06600
11	1	1	-.03184	-.05032	-.07069	-.09304	-.11745	-.14398	-.17270
11	1	2	.10411	.10928	.11522	.12197	.12958	.13810	.14757
11	1	3	.11452	.11974	.12553	.13190	.13887	.14647	.15469
11	1	4	.11587	.12015	.12479	.12979	.13516	.14089	.14699
11	1	5	.11395	.11712	.12048	.12404	.12779	.13172	.13584
11	1	6	.11053	.11262	.11479	.11704	.11936	.12175	.12421
11	1	7	.10631	.10740	.10851	.10962	.11073	.11185	.11296
11	1	8	.10151	.10172	.10190	.10205	.10217	.10227	.10234
11	1	9	.09611	.09555	.09496	.09433	.09367	.09299	.09229
11	1	10	.16894	.16673	.16452	.16231	.16012	.15795	.15581
11	2	1	-.08610	-.10784	-.13156	-.15735	-.18529	-.21544	-.24789
11	2	2	.09738	.10324	.10993	.11749	.12597	.13542	.14588
11	2	3	.11573	.12171	.12828	.13546	.14328	.15174	.16086
11	2	4	.12087	.12577	.13103	.13666	.14266	.14904	.15580
11	2	5	.12100	.12461	.12840	.13239	.13656	.14092	.14546
11	2	6	.11874	.12109	.12350	.12599	.12854	.13116	.13384
11	2	7	.11511	.11630	.11749	.11867	.11986	.12105	.12223
11	2	8	.11050	.11066	.11079	.11089	.11096	.11101	.11104
11	2	9	.28676	.28446	.28214	.27979	.27745	.27510	.27278
11	3	1	-.15821	-.18401	-.21190	-.24194	-.27423	-.30883	-.34581
11	3	2	.08898	.09576	.10343	.11204	.12164	.13226	.14397
11	3	3	.11788	.12482	.13237	.14056	.14941	.15893	.16914
11	3	4	.12801	.13366	.13967	.14606	.15282	.15995	.16746
11	3	5	.13082	.13493	.13921	.14368	.14832	.15314	.15814
11	3	6	.13002	.13262	.13528	.13800	.14077	.14360	.14649
11	3	7	.12709	.12831	.12953	.13074	.13194	.13314	.13433
11	3	8	.43541	.43392	.43240	.43087	.42934	.42780	.42628
11	4	1	-.25664	-.28770	-.32097	-.35653	-.39443	-.43477	-.47760
11	4	2	.07832	.08636	.09536	.10538	.11646	.12864	.14197
11	4	3	.12161	.12981	.13864	.14814	.15832	.16919	.18077

Coefficients a_i for the BLUE of μ
Shape parameter k

n	s	i	1.4	1.5	1.6	1.7	1.8	1.9	2.0
11	4	4	.13849	.14509	.15204	.15937	.16706	.17512	.18356
11	4	5	.14487	.14957	.15442	.15945	.16463	.16998	.17549
11	4	6	.14594	.14879	.15168	.15462	.15760	.16062	.16369
11	4	7	.62740	.62810	.62882	.62957	.63036	.63121	.63211
11	5	1	-.39689	-.43510	-.47569	-.51874	-.56430	-.61246	-.66327
11	5	2	.06442	.07427	.08518	.09719	.11034	.12468	.14024
11	5	3	.12821	.13813	.14872	.15999	.17197	.18465	.19807
11	5	4	.15458	.16243	.17062	.17916	.18807	.19734	.20698
11	5	5	.16588	.17129	.17684	.18253	.18837	.19435	.20048
11	5	6	.88381	.88899	.89433	.89986	.90556	.91144	.91750
12	0	1	.00116	-.01488	-.03278	-.05264	-.07452	-.09851	-.12467
12	0	2	.09892	.10306	.10787	.11339	.11967	.12676	.13469
12	0	3	.10431	.10868	.11358	.11903	.12506	.13168	.13890
12	0	4	.10375	.10747	.11155	.11600	.12083	.12604	.13163
12	0	5	.10115	.10404	.10715	.11049	.11405	.11783	.12183
12	0	6	.09770	.09977	.10195	.10424	.10665	.10916	.11178
12	0	7	.09385	.09514	.09648	.09785	.09926	.10070	.10217
12	0	8	.08975	.09034	.09093	.09150	.09207	.09262	.09317
12	0	9	.08543	.08540	.08534	.08524	.08510	.08495	.08476
12	0	10	.08081	.08023	.07961	.07895	.07826	.07754	.07681
12	0	11	.07556	.07451	.07343	.07233	.07123	.07012	.06901
12	0	12	.06760	.06624	.06491	.06361	.06234	.06111	.05991
12	1	1	-.03454	-.05306	-.07354	-.09608	-.12076	-.14766	-.17685
12	1	2	.09353	.09806	.10330	.10931	.11613	.12381	.13240
12	1	3	.10411	.10898	.11442	.12044	.12707	.13433	.14223
12	1	4	.10608	.11026	.11481	.11976	.12510	.13084	.13698
12	1	5	.10491	.10817	.11166	.11538	.11933	.12350	.12791
12	1	6	.10231	.10465	.10710	.10967	.11234	.11513	.11802
12	1	7	.09897	.10043	.10194	.10348	.10505	.10666	.10829
12	1	8	.09515	.09582	.09648	.09714	.09778	.09841	.09903
12	1	9	.09093	.09090	.09083	.09074	.09061	.09045	.09027
12	1	10	.08624	.08559	.08491	.08420	.08347	.08271	.08194
12	1	11	.15232	.15020	.14808	.14597	.14388	.14182	.13978
12	2	1	-.08063	-.10203	-.12549	-.15109	-.17892	-.20906	-.24159
12	2	2	.08685	.09189	.09770	.10433	.11183	.12024	.12961
12	2	3	.10414	.10963	.11572	.12242	.12975	.13774	.14641
12	2	4	.10936	.11408	.11918	.12469	.13059	.13691	.14364
12	2	5	.11002	.11368	.11758	.12171	.12607	.13065	.13547
12	2	6	.10849	.11110	.11382	.11665	.11959	.12263	.12577
12	2	7	.10577	.10738	.10902	.11069	.11239	.11412	.11588
12	2	8	.10226	.10296	.10365	.10432	.10498	.10563	.10627
12	2	9	.09810	.09800	.09787	.09770	.09751	.09730	.09707
12	2	10	.25564	.25331	.25094	.24857	.24620	.24383	.24148
12	3	1	-.14016	-.16508	-.19214	-.22143	-.25304	-.28705	-.32353
12	3	2	.07860	.08432	.09086	.09828	.10663	.11594	.12627
12	3	3	.10456	.11083	.11771	.12524	.13342	.14229	.15185
12	3	4	.11397	.11933	.12509	.13125	.13782	.14481	.15222
12	3	5	.11695	.12109	.12545	.13004	.13485	.13989	.14515
12	3	6	.11677	.11967	.12267	.12577	.12897	.13227	.13566
12	3	7	.11480	.11653	.11828	.12006	.12186	.12369	.12553
12	3	8	.11161	.11229	.11295	.11359	.11422	.11484	.11544
12	3	9	.38289	.38102	.37912	.37719	.37526	.37333	.37141
12	4	1	-.21852	-.24784	-.27941	-.31332	-.34965	-.38847	-.42988
12	4	2	.06829	.07491	.08242	.09087	.10031	.11077	.12232
12	4	3	.10567	.11292	.12082	.12938	.13863	.14858	.15926
12	4	4	.12056	.12672	.13328	.14024	.14761	.15541	.16362

Coefficients a_i for the BLUE of μ
Shape parameter k

n	s	i	1.4	1.5	1.6	1.7	1.8	1.9	2.0
12	4	5	.12655	.13124	.13615	.14127	.14660	.15216	.15792
12	4	6	.12808	.13129	.13458	.13797	.14144	.14500	.14865
12	4	7	.12700	.12882	.13066	.13251	.13437	.13625	.13815
12	4	8	.54237	.54194	.54151	.54109	.54069	.54031	.53996
12	5	1	-.32488	-.35992	-.39735	-.43726	-.47973	-.52482	-.57262
12	5	2	.05513	.06299	.07181	.08165	.09255	.10454	.11768
12	5	3	.10801	.11656	.12578	.13570	.14632	.15768	.16978
12	5	4	.13028	.13746	.14503	.15300	.16139	.17019	.17941
12	5	5	.14026	.14563	.15119	.15695	.16292	.16908	.17546
12	5	6	.14400	.14754	.15116	.15485	.15861	.16245	.16636
12	5	7	.74720	.74975	.75238	.75511	.75795	.76089	.76394
12	6	1	-.47586	-.51868	-.56410	-.61219	-.66304	-.71671	-.77328
12	6	2	.03778	.04743	.05814	.06995	.08289	.09702	.11237
12	6	3	.11265	.12300	.13405	.14582	.15832	.17157	.18560
12	6	4	.14529	.15382	.16274	.17206	.18178	.19190	.20245
12	6	5	.16079	.16700	.17339	.17995	.18670	.19363	.20076
12	6	6	1.01935	1.02742	1.03577	1.04442	1.05335	1.06258	1.07210
13	0	1	-.00546	-.02173	-.03992	-.06014	-.08245	-.10696	-.13372
13	0	2	.08988	.09355	.09784	.10281	.10849	.11494	.12219
13	0	3	.09603	.10015	.10479	.10999	.11575	.12211	.12909
13	0	4	.09623	.09987	.10389	.10830	.11311	.11831	.12393
13	0	5	.09436	.09731	.10051	.10396	.10766	.11161	.11581
13	0	6	.09161	.09384	.09621	.09873	.10140	.10420	.10713
13	0	7	.08844	.08998	.09159	.09327	.09501	.09681	.09867
13	0	8	.08503	.08593	.08686	.08779	.08874	.08970	.09067
13	0	9	.08145	.08178	.08210	.08239	.08267	.08293	.08317
13	0	10	.07769	.07751	.07729	.07704	.07676	.07645	.07612
13	0	11	.07366	.07302	.07235	.07165	.07091	.07016	.06939
13	0	12	.06906	.06803	.06699	.06593	.06486	.06379	.06273
13	0	13	.06202	.06074	.05950	.05828	.05710	.05595	.05483
13	1	1	-.03686	-.05538	-.07592	-.09859	-.12347	-.15065	-.18021
13	1	2	.08459	.08855	.09318	.09852	.10463	.11155	.11932
13	1	3	.09528	.09982	.10492	.11060	.11688	.12379	.13134
13	1	4	.09773	.10177	.10621	.11106	.11632	.12200	.12811
13	1	5	.09715	.10044	.10398	.10778	.11185	.11617	.12075
13	1	6	.09519	.09768	.10032	.10311	.10604	.10911	.11232
13	1	7	.09251	.09423	.09603	.09789	.09981	.10180	.10384
13	1	8	.08940	.09042	.09145	.09249	.09355	.09461	.09568
13	1	9	.08598	.08635	.08671	.08705	.08737	.08766	.08795
13	1	10	.08226	.08207	.08183	.08157	.08128	.08096	.08062
13	1	11	.07815	.07745	.07672	.07597	.07519	.07440	.07359
13	1	12	.13862	.13659	.13457	.13256	.13057	.12861	.12668
13	2	1	-.07663	-.09774	-.12095	-.14638	-.17410	-.20420	-.23678
13	2	2	.07811	.08245	.08751	.09334	.09997	.10747	.11587
13	2	3	.09455	.09961	.10525	.11149	.11837	.12590	.13411
13	2	4	.09984	.10435	.10927	.11461	.12037	.12657	.13321
13	2	5	.10088	.10454	.10847	.11265	.11710	.12181	.12678
13	2	6	.09991	.10267	.10558	.10863	.11183	.11517	.11864
13	2	7	.09783	.09973	.10169	.10372	.10580	.10794	.11014
13	2	8	.09507	.09617	.09728	.09839	.09951	.10064	.10178
13	2	9	.09181	.09219	.09254	.09286	.09317	.09347	.09374
13	2	10	.08809	.08782	.08752	.08719	.08683	.08645	.08605
13	2	11	.23054	.22821	.22586	.22350	.22114	.21880	.21647
13	3	1	-.12686	-.15104	-.17743	-.20610	-.23716	-.27068	-.30676
13	3	2	.07020	.07505	.08066	.08709	.09438	.10257	.11173
13	3	3	.09392	.09961	.10590	.11283	.12042	.12868	.13765

Coefficients a_i for the BLUE of μ
Shape parameter k

n	s	i	1.4	1.5	1.6	1.7	1.8	1.9	2.0
13	3	4	.10278	.10785	.11333	.11924	.12559	.13238	.13962
13	3	5	.10585	.10995	.11430	.11891	.12378	.12892	.13432
13	3	6	.10610	.10916	.11236	.11570	.11917	.12278	.12652
13	3	7	.10475	.10682	.10894	.11112	.11335	.11563	.11796
13	3	8	.10240	.10355	.10470	.10585	.10701	.10817	.10933
13	3	9	.09929	.09961	.09991	.10018	.10044	.10067	.10090
13	3	10	.34156	.33946	.33734	.33519	.33303	.33088	.32874
13	4	1	-.19112	-.21908	-.24933	-.28197	-.31708	-.35474	-.39505
13	4	2	.06047	.06597	.07229	.07949	.08760	.09668	.10676
13	4	3	.09351	.09998	.10709	.11486	.12331	.13246	.14234
13	4	4	.10693	.11267	.11883	.12543	.13247	.13996	.14790
13	4	5	.11257	.11716	.12202	.12712	.13249	.13811	.14399
13	4	6	.11433	.11772	.12124	.12489	.12866	.13256	.13659
13	4	7	.11388	.11611	.11838	.12070	.12306	.12547	.12792
13	4	8	.11198	.11314	.11430	.11545	.11660	.11775	.11891
13	4	9	.47745	.47633	.47518	.47403	.47289	.47176	.47065
13	5	1	-.27524	-.30794	-.34305	-.38067	-.42087	-.46374	-.50936
13	5	2	.04829	.05467	.06193	.07013	.07931	.08951	.10077
13	5	3	.09355	.10103	.10917	.11800	.12753	.13780	.14882
13	5	4	.11290	.11948	.12648	.13393	.14182	.15016	.15896
13	5	5	.12185	.12705	.13250	.13820	.14414	.15033	.15678
13	5	6	.12555	.12931	.13318	.13717	.14127	.14549	.14983
13	5	7	.12618	.12856	.13097	.13342	.13590	.13841	.14097
13	5	8	.64693	.64785	.64881	.64982	.65090	.65203	.65324
13	6	1	-.38900	-.42786	-.46930	-.51340	-.56025	-.60991	-.66248
13	6	2	.03268	.04027	.04882	.05838	.06898	.08067	.09351
13	6	3	.09446	.10327	.11278	.12300	.13395	.14566	.15815
13	6	4	.12177	.12944	.13753	.14605	.15502	.16444	.17433
13	6	5	.13512	.14108	.14727	.15369	.16034	.16723	.17437
13	6	6	.14134	.14552	.14980	.15418	.15865	.16324	.16792
13	6	7	.86363	.86828	.87311	.87811	.88330	.88866	.89421
13	7	1	-.55007	-.59732	-.64738	-.70034	-.75627	-.81525	-.87736
13	7	2	.01195	.02131	.03171	.04320	.05582	.06961	.08462
13	7	3	.09711	.10778	.11918	.13132	.14422	.15790	.17239
13	7	4	.13561	.14473	.15427	.16425	.17466	.18552	.19684
13	7	5	.15504	.16197	.16910	.17645	.18401	.19179	.19981
13	7	6	1.15035	1.16152	1.17311	1.18513	1.19756	1.21042	1.22369
14	0	1	-.01096	-.02739	-.04580	-.06629	-.08895	-.11387	-.14113
14	0	2	.08206	.08530	.08912	.09357	.09870	.10454	.11115
14	0	3	.08880	.09267	.09706	.10199	.10749	.11358	.12028
14	0	4	.08962	.09316	.09709	.10142	.10616	.11132	.11690
14	0	5	.08835	.09131	.09454	.09804	.10182	.10588	.11021
14	0	6	.08616	.08849	.09100	.09368	.09652	.09953	.10271
14	0	7	.08353	.08525	.08707	.08898	.09098	.09307	.09524
14	0	8	.08066	.08180	.08299	.08422	.08548	.08677	.08809
14	0	9	.07764	.07826	.07888	.07950	.08012	.08073	.08134
14	0	10	.07449	.07463	.07475	.07485	.07492	.07497	.07500
14	0	11	.07119	.07090	.07058	.07022	.06984	.06943	.06899
14	0	12	.06764	.06697	.06626	.06553	.06478	.06401	.06323
14	0	13	.06356	.06257	.06155	.06053	.05950	.05848	.05746
14	0	14	.05727	.05607	.05490	.05376	.05266	.05158	.05053
14	1	1	-.03885	-.05734	-.07791	-.10066	-.12568	-.15307	-.18291
14	1	2	.07694	.08039	.08447	.08921	.09467	.10089	.10792
14	1	3	.08769	.09192	.09669	.10203	.10797	.11452	.12173
14	1	4	.09052	.09441	.09871	.10342	.10857	.11415	.12018
14	1	5	.09041	.09368	.09723	.10106	.10517	.10958	.11427

Coefficients a_i for the BLUE of μ
Shape parameter k

n	s	i	1.4	1.5	1.6	1.7	1.8	1.9	2.0
14	1	6	.08895	.09154	.09430	.09723	.10034	.10362	.10706
14	1	7	.08679	.08870	.09071	.09281	.09500	.09728	.09964
14	1	8	.08423	.08550	.08682	.08817	.08955	.09097	.09241
14	1	9	.08139	.08208	.08278	.08346	.08415	.08483	.08551
14	1	10	.07833	.07850	.07864	.07875	.07884	.07891	.07896
14	1	11	.07504	.07473	.07438	.07401	.07360	.07318	.07273
14	1	12	.07140	.07068	.06992	.06915	.06835	.06755	.06673
14	1	13	.12715	.12521	.12328	.12137	.11947	.11761	.11577
14	2	1	-.07362	-.09445	-.11744	-.14269	-.17029	-.20034	-.23294
14	2	2	.07072	.07446	.07887	.08398	.08986	.09654	.10407
14	2	3	.08648	.09113	.09636	.10218	.10862	.11572	.12348
14	2	4	.09180	.09610	.10082	.10597	.11156	.11761	.12411
14	2	5	.09315	.09676	.10065	.10484	.10931	.11407	.11912
14	2	6	.09259	.09544	.09846	.10166	.10503	.10857	.11228
14	2	7	.09100	.09310	.09529	.09756	.09993	.10238	.10491
14	2	8	.08880	.09018	.09160	.09305	.09453	.09604	.09758
14	2	9	.08617	.08690	.08763	.08835	.08906	.08978	.09048
14	2	10	.08319	.08333	.08345	.08354	.08361	.08366	.08369
14	2	11	.07987	.07948	.07906	.07862	.07816	.07767	.07717
14	2	12	.20986	.20756	.20525	.20293	.20062	.19832	.19604
14	3	1	-.11672	-.14029	-.16610	-.19425	-.22483	-.25794	-.29367
14	3	2	.06322	.06734	.07216	.07774	.08413	.09136	.09949
14	3	3	.08519	.09037	.09614	.10253	.10957	.11728	.12569
14	3	4	.09361	.09839	.10360	.10925	.11535	.12192	.12896
14	3	5	.09674	.10074	.10503	.10960	.11447	.11963	.12508
14	3	6	.09728	.10042	.10373	.10721	.11086	.11467	.11865
14	3	7	.09639	.09868	.10105	.10350	.10604	.10865	.11135
14	3	8	.09461	.09609	.09760	.09913	.10069	.10227	.10388
14	3	9	.09220	.09294	.09368	.09440	.09512	.09582	.09653
14	3	10	.08929	.08936	.08941	.08944	.08945	.08943	.08941
14	3	11	.30818	.30596	.30371	.30144	.29917	.29690	.29464
14	4	1	-.17064	-.19750	-.22668	-.25829	-.29241	-.32915	-.36859
14	4	2	.05413	.05873	.06408	.07024	.07726	.08517	.09402
14	4	3	.08388	.08969	.09612	.10321	.11096	.11941	.12858
14	4	4	.09616	.10151	.10730	.11355	.12025	.12742	.13508
14	4	5	.10150	.10596	.11070	.11573	.12104	.12665	.13256
14	4	6	.10340	.10686	.11049	.11428	.11823	.12234	.12661
14	4	7	.10335	.10583	.10839	.11102	.11373	.11651	.11936
14	4	8	.10206	.10362	.10519	.10679	.10841	.11004	.11170
14	4	9	.09989	.10060	.10130	.10199	.10267	.10334	.10401
14	4	10	.42627	.42470	.42310	.42149	.41987	.41827	.41668
14	5	1	-.23926	-.27014	-.30346	-.33930	-.37776	-.41893	-.46290
14	5	2	.04295	.04818	.05423	.06113	.06893	.07769	.08744
14	5	3	.08262	.08923	.09649	.10442	.11305	.12240	.13249
14	5	4	.09979	.10584	.11235	.11930	.12673	.13463	.14302
14	5	5	.10793	.11293	.11820	.12376	.12961	.13574	.14216
14	5	6	.11152	.11535	.11933	.12347	.12776	.13220	.13680
14	5	7	.11249	.11517	.11792	.12074	.12362	.12656	.12958
14	5	8	.11178	.11339	.11501	.11664	.11829	.11995	.12162
14	5	9	.57018	.57006	.56994	.56985	.56979	.56976	.56979
14	6	1	-.32876	-.36471	-.40323	-.44441	-.48833	-.53509	-.58477
14	6	2	.02895	.03503	.04198	.04986	.05869	.06854	.07944
14	6	3	.08157	.08920	.09751	.10652	.11626	.12674	.13800
14	6	4	.10509	.11202	.11941	.12726	.13558	.14438	.15367
14	6	5	.11682	.12248	.12841	.13461	.14109	.14785	.15489
14	6	6	.12257	.12682	.13122	.13576	.14043	.14526	.15022

Coefficients a_i for the BLUE of μ
Shape parameter k

n	s	i	1.4	1.5	1.6	1.7	1.8	1.9	2.0
14	6	7	.12481	.12770	.13064	.13364	.13669	.13980	.14297
14	6	8	.74895	.75145	.75405	.75676	.75958	.76252	.76559
14	7	1	-.44949	-.49203	-.53733	-.58546	-.63652	-.69059	-.74775
14	7	2	.01095	.01821	.02641	.03560	.04583	.05714	.06956
14	7	3	.08103	.09003	.09974	.11017	.12136	.13333	.14609
14	7	4	.11307	.12115	.12968	.13867	.14814	.15809	.16853
14	7	5	.12959	.13607	.14282	.14982	.15708	.16462	.17243
14	7	6	.13814	.14290	.14778	.15279	.15792	.16318	.16858
14	7	7	.97671	.98367	.99090	.99841	1.00618	1.01424	1.02256
15	0	1	-.01556	-.03211	-.05069	-.07139	-.09432	-.11957	-.14722
15	0	2	.07524	.07808	.08146	.08544	.09005	.09534	.10134
15	0	3	.08244	.08607	.09021	.09487	.10010	.10591	.11232
15	0	4	.08377	.08719	.09101	.09523	.09988	.10495	.11046
15	0	5	.08298	.08592	.08915	.09268	.09649	.10060	.10501
15	0	6	.08125	.08365	.08624	.08903	.09200	.09517	.09853
15	0	7	.07906	.08092	.08289	.08498	.08718	.08949	.09191
15	0	8	.07663	.07797	.07936	.08081	.08232	.08388	.08549
15	0	9	.07406	.07491	.07577	.07666	.07756	.07847	.07940
15	0	10	.07137	.07178	.07217	.07256	.07293	.07329	.07364
15	0	11	.06858	.06858	.06856	.06851	.06843	.06833	.06821
15	0	12	.06565	.06529	.06489	.06447	.06401	.06353	.06304
15	0	13	.06249	.06181	.06109	.06035	.05959	.05882	.05803
15	0	14	.05886	.05789	.05691	.05593	.05494	.05396	.05299
15	0	15	.05319	.05206	.05096	.04988	.04884	.04783	.04684
15	1	1	-.04054	-.05899	-.07956	-.10235	-.12747	-.15500	-.18505
15	1	2	.07031	.07331	.07688	.08109	.08596	.09154	.09789
15	1	3	.08109	.08502	.08948	.09449	.10010	.10631	.11316
15	1	4	.08422	.08795	.09210	.09666	.10167	.10712	.11304
15	1	5	.08449	.08772	.09124	.09506	.09918	.10361	.10835
15	1	6	.08345	.08608	.08891	.09194	.09516	.09859	.10220
15	1	7	.08171	.08374	.08590	.08818	.09057	.09307	.09569
15	1	8	.07957	.08104	.08257	.08416	.08580	.08750	.08925
15	1	9	.07718	.07812	.07908	.08005	.08104	.08204	.08305
15	1	10	.07461	.07507	.07551	.07593	.07635	.07676	.07715
15	1	11	.07188	.07188	.07186	.07182	.07175	.07166	.07154
15	1	12	.06894	.06855	.06812	.06767	.06720	.06670	.06619
15	1	13	.06570	.06496	.06419	.06341	.06261	.06181	.06099
15	1	14	.11740	.11555	.11371	.11188	.11008	.10830	.10655
15	2	1	-.07127	-.09185	-.11462	-.13970	-.16718	-.19716	-.22974
15	2	2	.06438	.06760	.07143	.07592	.08111	.08707	.09381
15	2	3	.07956	.08385	.08869	.09412	.10016	.10683	.11417
15	2	4	.08491	.08900	.09352	.09847	.10387	.10974	.11608
15	2	5	.08649	.09003	.09386	.09801	.10246	.10722	.11230
15	2	6	.08627	.08915	.09224	.09552	.09899	.10267	.10654
15	2	7	.08507	.08729	.08964	.09210	.09467	.09735	.10014
15	2	8	.08329	.08488	.08653	.08824	.09000	.09181	.09367
15	2	9	.08112	.08213	.08315	.08418	.08523	.08628	.08735
15	2	10	.07868	.07914	.07960	.08003	.08046	.08087	.08128
15	2	11	.07598	.07595	.07590	.07582	.07572	.07560	.07546
15	2	12	.07299	.07253	.07204	.07152	.07099	.07044	.06988
15	2	13	.19254	.19029	.18803	.18577	.18351	.18127	.17905
15	3	1	-.10878	-.13182	-.15713	-.18482	-.21498	-.24773	-.28315
15	3	2	.05731	.06081	.06496	.06980	.07540	.08179	.08901
15	3	3	.07787	.08259	.08789	.09380	.10034	.10754	.11543
15	3	4	.08593	.09044	.09538	.10077	.10662	.11295	.11977
15	3	5	.08909	.09298	.09717	.10167	.10649	.11162	.11707

Coefficients a_i for the BLUE of μ
Shape parameter k

n	s	i	1.4	1.5	1.6	1.7	1.8	1.9	2.0
15	3	6	.08987	.09303	.09638	.09993	.10368	.10763	.11177
15	3	7	.08931	.09173	.09427	.09691	.09967	.10253	.10550
15	3	8	.08795	.08966	.09143	.09325	.09511	.09702	.09898
15	3	9	.08604	.08709	.08816	.08923	.09031	.09140	.09250
15	3	10	.08371	.08417	.08460	.08502	.08543	.08582	.08621
15	3	11	.08103	.08093	.08081	.08067	.08051	.08033	.08013
15	3	12	.28067	.27839	.27608	.27375	.27142	.26909	.26678
15	4	1	-.15484	-.18078	-.20908	-.23983	-.27314	-.30910	-.34782
15	4	2	.04885	.05270	.05725	.06254	.06862	.07554	.08333
15	4	3	.07602	.08126	.08711	.09359	.10073	.10855	.11708
15	4	4	.08740	.09239	.09784	.10374	.11012	.11698	.12434
15	4	5	.09249	.09679	.10139	.10630	.11153	.11707	.12294
15	4	6	.09448	.09794	.10161	.10546	.10950	.11374	.11817
15	4	7	.09470	.09733	.10007	.10291	.10585	.10890	.11205
15	4	8	.09382	.09565	.09753	.09945	.10141	.10341	.10545
15	4	9	.09220	.09328	.09437	.09546	.09656	.09766	.09877
15	4	10	.09000	.09040	.09079	.09116	.09151	.09185	.09219
15	4	11	.38489	.38303	.38114	.37923	.37732	.37541	.37351
15	5	1	-.21216	-.24159	-.27347	-.30789	-.34496	-.38477	-.42742
15	5	2	.03860	.04291	.04797	.05382	.06050	.06807	.07656
15	5	3	.07402	.07990	.08642	.09359	.10145	.11001	.11930
15	5	4	.08951	.09510	.10115	.10766	.11466	.12215	.13014
15	5	5	.09700	.10178	.10686	.11225	.11795	.12396	.13030
15	5	6	.10047	.10429	.10829	.11248	.11686	.12142	.12617
15	5	7	.10163	.10449	.10744	.11049	.11363	.11687	.12021
15	5	8	.10133	.10327	.10524	.10725	.10930	.11138	.11349
15	5	9	.10003	.10111	.10219	.10327	.10436	.10544	.10653
15	5	10	.50956	.50873	.50790	.50707	.50626	.50547	.50472
15	6	1	-.28486	-.31856	-.35483	-.39375	-.43544	-.47997	-.52745
15	6	2	.02602	.03094	.03665	.04321	.05065	.05903	.06839
15	6	3	.07191	.07859	.08594	.09397	.10271	.11219	.12241
15	6	4	.09260	.09892	.10570	.11296	.12070	.12895	.13770
15	6	5	.10310	.10845	.11410	.12006	.12632	.13290	.13979
15	6	6	.10841	.11264	.11705	.12162	.12638	.13131	.13642
15	6	7	.11073	.11383	.11702	.12029	.12364	.12708	.13062
15	6	8	.11112	.11315	.11521	.11729	.11940	.12154	.12371
15	6	9	.66096	.66204	.66317	.66436	.66563	.66697	.66840
15	7	1	-.37943	-.41852	-.46032	-.50492	-.55243	-.60294	-.65652
15	7	2	.01024	.01597	.02256	.03006	.03850	.04793	.05839
15	7	3	.06976	.07748	.08590	.09503	.10489	.11551	.12692
15	7	4	.09721	.10444	.11214	.12032	.12900	.13819	.14789
15	7	5	.11157	.11763	.12398	.13062	.13757	.14483	.15240
15	7	6	.11924	.12395	.12882	.13385	.13905	.14442	.14995
15	7	7	.12299	.12636	.12979	.13330	.13688	.14054	.14428
15	7	8	.84843	.85269	.85712	.86174	.86653	.87151	.87668
15	8	1	-.50675	-.55284	-.60185	-.65388	-.70901	-.76733	-.82892
15	8	2	-.01010	-.00323	.00457	.01335	.02314	.03398	.04592
15	8	3	.06778	.07689	.08673	.09731	.10865	.12079	.13373
15	8	4	.10427	.11269	.12159	.13098	.14087	.15126	.16217
15	8	5	.12377	.13072	.13796	.14548	.15329	.16140	.16982
15	8	6	.13452	.13981	.14525	.15083	.15656	.16245	.16849
15	8	7	1.08652	1.09595	1.10575	1.11594	1.12651	1.13746	1.14878
16	0	1	-.01945	-.03608	-.05478	-.07565	-.09878	-.12429	-.15226
16	0	2	.06923	.07171	.07470	.07823	.08236	.08713	.09257
16	0	3	.07680	.08020	.08409	.08850	.09345	.09898	.10510
16	0	4	.07854	.08184	.08554	.08964	.09417	.09914	.10455

Coefficients a_i for the BLUE of μ
Shape parameter k

n	s	i	1.4	1.5	1.6	1.7	1.8	1.9	2.0
16	0	5	.07816	.08107	.08427	.08778	.09160	.09572	.10017
16	0	6	.07682	.07925	.08190	.08475	.08782	.09109	.09458
16	0	7	.07500	.07695	.07903	.08126	.08361	.08610	.08871
16	0	8	.07293	.07441	.07596	.07760	.07931	.08109	.08294
16	0	9	.07071	.07175	.07281	.07392	.07505	.07622	.07741
16	0	10	.06840	.06902	.06964	.07027	.07090	.07153	.07216
16	0	11	.06600	.06624	.06646	.06667	.06686	.06704	.06720
16	0	12	.06351	.06340	.06327	.06311	.06293	.06272	.06249
16	0	13	.06089	.06047	.06003	.05955	.05905	.05853	.05799
16	0	14	.05805	.05736	.05664	.05590	.05514	.05437	.05360
16	0	15	.05478	.05385	.05290	.05196	.05101	.05007	.04914
16	0	16	.04964	.04857	.04753	.04652	.04553	.04458	.04365
16	1	1	-.04199	-.06038	-.08092	-.10373	-.12890	-.15654	-.18674
16	1	2	.06451	.06710	.07023	.07394	.07827	.08327	.08898
16	1	3	.07529	.07894	.08311	.08782	.09310	.09898	.10548
16	1	4	.07867	.08224	.08622	.09063	.09548	.10079	.10657
16	1	5	.07926	.08242	.08589	.08967	.09377	.09819	.10294
16	1	6	.07854	.08120	.08406	.08715	.09045	.09396	.09770
16	1	7	.07715	.07928	.08154	.08394	.08648	.08916	.09197
16	1	8	.07537	.07698	.07867	.08045	.08230	.08422	.08621
16	1	9	.07334	.07447	.07564	.07684	.07807	.07933	.08062
16	1	10	.07115	.07183	.07252	.07320	.07389	.07458	.07528
16	1	11	.06882	.06909	.06934	.06958	.06979	.07000	.07018
16	1	12	.06636	.06625	.06611	.06595	.06577	.06556	.06533
16	1	13	.06372	.06327	.06280	.06230	.06177	.06123	.06067
16	1	14	.06081	.06007	.05930	.05852	.05773	.05693	.05613
16	1	15	.10902	.10725	.10549	.10375	.10202	.10033	.09867
16	2	1	-.06940	-.08974	-.11231	-.13721	-.16456	-.19445	-.22699
16	2	2	.05888	.06163	.06495	.06888	.07347	.07876	.08480
16	2	3	.07357	.07752	.08201	.08707	.09273	.09901	.10593
16	2	4	.07893	.08282	.08713	.09188	.09708	.10276	.10892
16	2	5	.08070	.08414	.08790	.09197	.09638	.10111	.10618
16	2	6	.08075	.08363	.08674	.09006	.09360	.09736	.10134
16	2	7	.07986	.08216	.08461	.08720	.08992	.09278	.09577
16	2	8	.07841	.08016	.08198	.08389	.08586	.08791	.09002
16	2	9	.07661	.07783	.07908	.08036	.08167	.08300	.08437
16	2	10	.07456	.07528	.07601	.07673	.07745	.07818	.07890
16	2	11	.07231	.07258	.07283	.07306	.07327	.07347	.07365
16	2	12	.06985	.06971	.06954	.06934	.06912	.06888	.06863
16	2	13	.06716	.06665	.06611	.06556	.06498	.06439	.06378
16	2	14	.17782	.17563	.17343	.17122	.16903	.16686	.16471
16	3	1	-.10242	-.12499	-.14985	-.17713	-.20692	-.23933	-.27447
16	3	2	.05223	.05519	.05875	.06295	.06785	.07349	.07992
16	3	3	.07163	.07594	.08082	.08629	.09237	.09910	.10650
16	3	4	.07939	.08363	.08832	.09345	.09906	.10515	.11173
16	3	5	.08257	.08633	.09041	.09481	.09954	.10461	.11002
16	3	6	.08352	.08666	.09003	.09361	.09740	.10142	.10566
16	3	7	.08323	.08573	.08837	.09115	.09406	.09710	.10027
16	3	8	.08218	.08407	.08602	.08805	.09014	.09231	.09454
16	3	9	.08065	.08194	.08326	.08460	.08597	.08737	.08879
16	3	10	.07875	.07949	.08024	.08097	.08171	.08244	.08318
16	3	11	.07657	.07681	.07703	.07723	.07742	.07759	.07774
16	3	12	.07410	.07389	.07365	.07339	.07311	.07281	.07250
16	3	13	.25761	.25530	.25297	.25063	.24828	.24594	.24361
16	4	1	-.14235	-.16751	-.19505	-.22507	-.25767	-.29298	-.33107
16	4	2	.04435	.04757	.05143	.05598	.06126	.06732	.07419

Coefficients a_i for the BLUE of μ
Shape parameter k

n	s	i	1.4	1.5	1.6	1.7	1.8	1.9	2.0
16	4	3	.06946	.07421	.07954	.08548	.09207	.09933	.10727
16	4	4	.08011	.08477	.08989	.09548	.10154	.10810	.11516
16	4	5	.08500	.08912	.09357	.09834	.10345	.10889	.11468
16	4	6	.08703	.09046	.09411	.09798	.10206	.10636	.11088
16	4	7	.08745	.09017	.09301	.09599	.09909	.10233	.10569
16	4	8	.08688	.08889	.09098	.09313	.09534	.09762	.09996
16	4	9	.08564	.08699	.08837	.08977	.09119	.09263	.09410
16	4	10	.08391	.08465	.08539	.08612	.08685	.08757	.08829
16	4	11	.08178	.08197	.08213	.08227	.08240	.08252	.08262
16	4	12	.35075	.34870	.34662	.34452	.34242	.34032	.33823
16	5	1	-.19113	-.21936	-.25004	-.28329	-.31922	-.35791	-.39948
16	5	2	.03495	.03851	.04275	.04771	.05346	.06001	.06743
16	5	3	.06704	.07231	.07819	.08471	.09189	.09976	.10835
16	5	4	.08121	.08638	.09202	.09813	.10472	.11182	.11944
16	5	5	.08817	.09272	.09759	.10279	.10833	.11420	.12042
16	5	6	.09152	.09528	.09925	.10344	.10784	.11245	.11727
16	5	7	.09279	.09574	.09881	.10200	.10532	.10875	.11231
16	5	8	.09277	.09493	.09714	.09941	.10174	.10413	.10658
16	5	9	.09188	.09328	.09470	.09614	.09760	.09907	.10056
16	5	10	.09033	.09104	.09175	.09244	.09313	.09382	.09450
16	5	11	.46047	.45917	.45784	.45652	.45519	.45389	.45261
16	6	1	-.25163	-.28353	-.31800	-.35514	-.39504	-.43781	-.48354
16	6	2	.02359	.02758	.03230	.03780	.04411	.05129	.05937
16	6	3	.06435	.07026	.07681	.08403	.09193	.10055	.10990
16	6	4	.08288	.08866	.09491	.10165	.10888	.11663	.12489
16	6	5	.09240	.09745	.10282	.10853	.11456	.12094	.12765
16	6	6	.09735	.10149	.10584	.11039	.11514	.12011	.12528
16	6	7	.09966	.10286	.10618	.10961	.11315	.11681	.12058
16	6	8	.10030	.10259	.10493	.10732	.10976	.11226	.11480
16	6	9	.09979	.10123	.10267	.10413	.10559	.10707	.10857
16	6	10	.59131	.59141	.59154	.59170	.59190	.59216	.59248
16	7	1	-.32816	-.36458	-.40370	-.44560	-.49039	-.53817	-.58903
16	7	2	.00964	.01420	.01955	.02573	.03277	.04073	.04964
16	7	3	.06138	.06810	.07549	.08357	.09236	.10190	.11220
16	7	4	.08541	.09194	.09895	.10645	.11446	.12299	.13204
16	7	5	.09814	.10380	.10977	.11608	.12271	.12968	.13699
16	7	6	.10508	.10967	.11445	.11943	.12461	.12998	.13556
16	7	7	.10867	.11216	.11574	.11943	.12323	.12713	.13114
16	7	8	.11011	.11253	.11499	.11749	.12004	.12263	.12527
16	7	9	.74975	.75219	.75474	.75742	.76021	.76314	.76619
16	8	1	-.42754	-.46965	-.51462	-.56254	-.61351	-.66763	-.72498
16	8	2	-.00788	-.00253	.00366	.01073	.01873	.02770	.03767
16	8	3	.05813	.06591	.07438	.08356	.09350	.10419	.11568
16	8	4	.08929	.09677	.10474	.11320	.12218	.13169	.14173
16	8	5	.10615	.11256	.11928	.12632	.13369	.14139	.14942
16	8	6	.11564	.12076	.12606	.13155	.13722	.14308	.14914
16	8	7	.12082	.12463	.12852	.13250	.13658	.14075	.14502
16	8	8	.94538	.95155	.95798	.96467	.97162	.97883	.98631
17	0	1	-.02276	-.03944	-.05822	-.07921	-.10251	-.12822	-.15643
17	0	2	.06390	.06605	.06867	.07180	.07548	.07976	.08468
17	0	3	.07176	.07494	.07859	.08275	.08744	.09269	.09853
17	0	4	.07385	.07702	.08059	.08457	.08897	.09381	.09911
17	0	5	.07381	.07667	.07983	.08330	.08710	.09121	.09566
17	0	6	.07280	.07524	.07791	.08081	.08393	.08727	.09085
17	0	7	.07130	.07331	.07547	.07779	.08026	.08288	.08565
17	0	8	.06953	.07111	.07279	.07457	.07644	.07840	.08045

Coefficients a_i for the BLUE of μ
Shape parameter k

n	s	i	1.4	1.5	1.6	1.7	1.8	1.9	2.0
17	0	9	.06761	.06878	.07001	.07129	.07262	.07400	.07542
17	0	10	.06559	.06638	.06720	.06803	.06888	.06974	.07061
17	0	11	.06350	.06394	.06438	.06480	.06523	.06565	.06606
17	0	12	.06134	.06146	.06155	.06162	.06168	.06171	.06173
17	0	13	.05910	.05892	.05871	.05847	.05820	.05792	.05761
17	0	14	.05674	.05629	.05581	.05530	.05477	.05422	.05365
17	0	15	.05419	.05349	.05278	.05204	.05129	.05053	.04977
17	0	16	.05122	.05032	.04941	.04850	.04759	.04669	.04580
17	0	17	.04652	.04551	.04453	.04357	.04264	.04173	.04085
17	1	1	-.04323	-.06155	-.08204	-.10483	-.13003	-.15773	-.18805
17	1	2	.05939	.06162	.06434	.06759	.07143	.07590	.08103
17	1	3	.07015	.07355	.07744	.08185	.08683	.09238	.09855
17	1	4	.07373	.07714	.08096	.08521	.08990	.09505	.10067
17	1	5	.07458	.07767	.08107	.08479	.08885	.09323	.09797
17	1	6	.07415	.07680	.07968	.08279	.08613	.08970	.09352
17	1	7	.07305	.07523	.07757	.08006	.08271	.08551	.08847
17	1	8	.07155	.07327	.07509	.07701	.07902	.08112	.08332
17	1	9	.06982	.07110	.07244	.07382	.07525	.07673	.07826
17	1	10	.06793	.06880	.06969	.07059	.07151	.07244	.07339
17	1	11	.06593	.06641	.06689	.06736	.06783	.06828	.06874
17	1	12	.06382	.06395	.06406	.06414	.06421	.06427	.06430
17	1	13	.06159	.06140	.06118	.06093	.06066	.06037	.06006
17	1	14	.05921	.05873	.05821	.05768	.05712	.05655	.05597
17	1	15	.05658	.05584	.05508	.05431	.05353	.05275	.05196
17	1	16	.10174	.10004	.09836	.09669	.09505	.09343	.09184
17	2	1	-.06787	-.08798	-.11035	-.13508	-.16229	-.19208	-.22457
17	2	2	.05405	.05639	.05925	.06268	.06672	.07142	.07681
17	2	3	.06831	.07196	.07613	.08084	.08614	.09204	.09858
17	2	4	.07368	.07737	.08148	.08603	.09103	.09651	.10248
17	2	5	.07560	.07894	.08261	.08660	.09093	.09560	.10062
17	2	6	.07587	.07874	.08183	.08517	.08874	.09254	.09659
17	2	7	.07523	.07759	.08011	.08278	.08561	.08859	.09173
17	2	8	.07407	.07592	.07788	.07992	.08207	.08430	.08662
17	2	9	.07256	.07393	.07536	.07684	.07836	.07992	.08153
17	2	10	.07082	.07174	.07268	.07364	.07461	.07559	.07659
17	2	11	.06890	.06941	.06991	.07040	.07088	.07135	.07182
17	2	12	.06684	.06696	.06705	.06713	.06719	.06723	.06726
17	2	13	.06461	.06438	.06412	.06383	.06353	.06321	.06287
17	2	14	.06217	.06163	.06106	.06047	.05987	.05925	.05862
17	2	15	.16516	.16303	.16089	.15875	.15663	.15452	.15244
17	3	1	-.09722	-.11937	-.14383	-.17074	-.20018	-.23229	-.26715
17	3	2	.04780	.05029	.05333	.05697	.06126	.06623	.07193
17	3	3	.06623	.07018	.07467	.07974	.08540	.09169	.09864
17	3	4	.07373	.07773	.08217	.08706	.09242	.09827	.10462
17	3	5	.07693	.08055	.08451	.08880	.09343	.09841	.10375
17	3	6	.07802	.08112	.08446	.08804	.09185	.09590	.10020
17	3	7	.07794	.08048	.08319	.08605	.08906	.09223	.09555
17	3	8	.07715	.07914	.08123	.08341	.08568	.08804	.09049
17	3	9	.07590	.07736	.07888	.08043	.08204	.08368	.08537
17	3	10	.07433	.07530	.07628	.07727	.07828	.07929	.08032
17	3	11	.07251	.07302	.07352	.07400	.07448	.07496	.07542
17	3	12	.07048	.07056	.07062	.07067	.07069	.07070	.07069
17	3	13	.06822	.06792	.06760	.06726	.06690	.06652	.06613
17	3	14	.23800	.23570	.23337	.23104	.22870	.22637	.22405
17	4	1	-.13225	-.15674	-.18362	-.21300	-.24499	-.27972	-.31727
17	4	2	.04047	.04315	.04642	.05032	.05490	.06020	.06626

Coefficients a_i for the BLUE of μ
Shape parameter k

n	s	i	1.4	1.5	1.6	1.7	1.8	1.9	2.0
17	4	3	.06388	.06819	.07307	.07853	.08463	.09137	.09878
17	4	4	.07392	.07829	.08311	.08840	.09416	.10042	.10719
17	4	5	.07864	.08259	.08687	.09150	.09647	.10180	.10748
17	4	6	.08070	.08408	.08769	.09153	.09561	.09993	.10449
17	4	7	.08128	.08403	.08694	.09000	.09321	.09657	.10008
17	4	8	.08093	.08307	.08530	.08762	.09002	.09251	.09508
17	4	9	.07998	.08153	.08313	.08476	.08644	.08815	.08990
17	4	10	.07860	.07960	.08061	.08163	.08265	.08369	.08473
17	4	11	.07688	.07737	.07785	.07832	.07877	.07922	.07966
17	4	12	.07487	.07489	.07490	.07488	.07484	.07480	.07473
17	4	13	.32210	.31993	.31773	.31552	.31330	.31108	.30888
17	5	1	-.17441	-.20161	-.23129	-.26355	-.29851	-.33626	-.37691
17	5	2	.03182	.03474	.03829	.04251	.04745	.05314	.05963
17	5	3	.06123	.06597	.07130	.07725	.08384	.09110	.09906
17	5	4	.07433	.07913	.08439	.09013	.09636	.10310	.11035
17	5	5	.08087	.08519	.08986	.09487	.10022	.10594	.11202
17	5	6	.08410	.08778	.09169	.09583	.10020	.10482	.10967
17	5	7	.08545	.08843	.09156	.09483	.09825	.10182	.10553
17	5	8	.08562	.08792	.09029	.09275	.09528	.09790	.10059
17	5	9	.08502	.08665	.08832	.09002	.09176	.09353	.09534
17	5	10	.08384	.08486	.08588	.08690	.08793	.08897	.09001
17	5	11	.08222	.08267	.08310	.08352	.08393	.08433	.08473
17	5	12	.41991	.41827	.41661	.41494	.41328	.41162	.40998
17	6	1	-.22573	-.25616	-.28915	-.32481	-.36325	-.40457	-.44887
17	6	2	.02151	.02475	.02865	.03327	.03864	.04481	.05181
17	6	3	.05824	.06351	.06938	.07590	.08309	.09097	.09957
17	6	4	.07506	.08038	.08616	.09244	.09921	.10650	.11432
17	6	5	.08380	.08857	.09367	.09913	.10493	.11109	.11761
17	6	6	.08844	.09246	.09672	.10120	.10591	.11086	.11604
17	6	7	.09072	.09395	.09732	.10084	.10449	.10828	.11222
17	6	8	.09151	.09396	.09649	.09908	.10176	.10450	.10733
17	6	9	.09130	.09300	.09474	.09650	.09829	.10011	.10196
17	6	10	.09035	.09136	.09237	.09338	.09439	.09540	.09642
17	6	11	.53481	.53423	.53365	.53308	.53255	.53205	.53160
17	7	1	-.28923	-.32352	-.36049	-.40024	-.44288	-.48850	-.53720
17	7	2	.00906	.01271	.01707	.02219	.02810	.03485	.04249
17	7	3	.05486	.06077	.06732	.07453	.08244	.09107	.10043
17	7	4	.07627	.08221	.08863	.09554	.10297	.11092	.11941
17	7	5	.08772	.09301	.09865	.10462	.11095	.11763	.12468
17	7	6	.09408	.09851	.10316	.10804	.11314	.11847	.12403
17	7	7	.09748	.10100	.10465	.10843	.11234	.11638	.12056
17	7	8	.09902	.10163	.10432	.10706	.10988	.11276	.11571
17	7	9	.09926	.10103	.10281	.10462	.10645	.10830	.11018
17	7	10	.67147	.67264	.67389	.67521	.67662	.67812	.67971
17	8	1	-.36940	-.40845	-.45031	-.49509	-.54288	-.59379	-.64791
17	8	2	-.00623	-.00203	.00292	.00868	.01529	.02278	.03121
17	8	3	.05104	.05776	.06515	.07323	.08204	.09158	.10190
17	8	4	.07823	.08494	.09213	.09984	.10806	.11682	.12613
17	8	5	.09308	.09901	.10527	.11188	.11884	.12615	.13383
17	8	6	.10158	.10649	.11162	.11695	.12251	.12829	.13429
17	8	7	.10637	.11021	.11417	.11825	.12245	.12678	.13123
17	8	8	.10880	.11159	.11443	.11733	.12029	.12330	.12638
17	8	9	.83653	.84048	.84461	.84892	.85341	.85809	.86296
17	9	1	-.47335	-.51838	-.56641	-.61753	-.67186	-.72947	-.79046
17	9	2	-.02543	-.02050	-.01474	-.00813	-.00061	.00786	.01730
17	9	3	.04671	.05449	.06297	.07217	.08212	.09284	.10435

Coefficients a_i for the BLUE of μ
Shape parameter k

n	s	i	1.4	1.5	1.6	1.7	1.8	1.9	2.0
17	9	4	.08139	.08907	.09725	.10595	.11518	.12495	.13527
17	9	5	.10060	.10732	.11438	.12177	.12951	.13760	.14604
17	9	6	.11184	.11733	.12302	.12892	.13502	.14134	.14787
17	9	7	.11838	.12259	.12690	.13133	.13586	.14051	.14527
17	9	8	1.03987	1.04807	1.05662	1.06552	1.07478	1.08439	1.09434
18	0	1	-.02559	-.04230	-.06114	-.08221	-.10563	-.13149	-.15991
18	0	2	.05915	.06100	.06328	.06603	.06930	.07313	.07755
18	0	3	.06724	.07021	.07363	.07755	.08199	.08697	.09252
18	0	4	.06962	.07266	.07609	.07994	.08420	.08891	.09408
18	0	5	.06986	.07266	.07577	.07919	.08294	.08703	.09145
18	0	6	.06913	.07157	.07425	.07716	.08031	.08370	.08734
18	0	7	.06790	.06995	.07217	.07456	.07711	.07984	.08272
18	0	8	.06639	.06805	.06983	.07172	.07372	.07582	.07803
18	0	9	.06472	.06601	.06737	.06879	.07028	.07183	.07344
18	0	10	.06295	.06389	.06486	.06586	.06689	.06795	.06904
18	0	11	.06112	.06172	.06233	.06295	.06358	.06421	.06484
18	0	12	.05922	.05952	.05981	.06009	.06035	.06060	.06085
18	0	13	.05727	.05729	.05728	.05725	.05720	.05713	.05705
18	0	14	.05525	.05501	.05473	.05444	.05411	.05377	.05341
18	0	15	.05311	.05263	.05213	.05160	.05105	.05048	.04990
18	0	16	.05079	.05010	.04939	.04866	.04793	.04718	.04644
18	0	17	.04808	.04721	.04634	.04546	.04459	.04372	.04287
18	0	18	.04377	.04281	.04188	.04096	.04008	.03922	.03839
18	1	1	-.04429	-.06252	-.08296	-.10572	-.13091	-.15865	-.18903
18	1	2	.05485	.05674	.05909	.06193	.06532	.06929	.07388
18	1	3	.06557	.06872	.07235	.07649	.08117	.08642	.09226
18	1	4	.06931	.07257	.07623	.08031	.08484	.08982	.09528
18	1	5	.07038	.07339	.07671	.08036	.08435	.08869	.09337
18	1	6	.07019	.07282	.07569	.07880	.08216	.08576	.08962
18	1	7	.06933	.07154	.07393	.07649	.07921	.08211	.08518
18	1	8	.06808	.06988	.07179	.07382	.07595	.07820	.08056
18	1	9	.06660	.06800	.06946	.07099	.07259	.07425	.07597
18	1	10	.06496	.06598	.06703	.06811	.06922	.07035	.07152
18	1	11	.06321	.06387	.06454	.06521	.06589	.06657	.06725
18	1	12	.06138	.06171	.06202	.06233	.06262	.06290	.06317
18	1	13	.05946	.05948	.05948	.05946	.05941	.05935	.05928
18	1	14	.05743	.05718	.05689	.05659	.05625	.05590	.05554
18	1	15	.05527	.05476	.05423	.05367	.05310	.05251	.05191
18	1	16	.05288	.05215	.05141	.05065	.04988	.04912	.04835
18	1	17	.09535	.09373	.09211	.09052	.08895	.08740	.08589
18	2	1	-.06659	-.08649	-.10866	-.13322	-.16028	-.18997	-.22238
18	2	2	.04979	.05176	.05421	.05718	.06072	.06487	.06968
18	2	3	.06366	.06703	.07090	.07529	.08025	.08580	.09196
18	2	4	.06903	.07252	.07644	.08080	.08560	.09089	.09665
18	2	5	.07108	.07431	.07787	.08177	.08601	.09061	.09557
18	2	6	.07153	.07436	.07743	.08075	.08432	.08815	.09223
18	2	7	.07111	.07349	.07604	.07877	.08167	.08475	.08799
18	2	8	.07017	.07210	.07415	.07630	.07857	.08095	.08344
18	2	9	.06890	.07040	.07196	.07359	.07528	.07704	.07885
18	2	10	.06741	.06850	.06961	.07076	.07193	.07313	.07435
18	2	11	.06577	.06646	.06717	.06787	.06858	.06929	.07001
18	2	12	.06399	.06433	.06465	.06496	.06526	.06555	.06583
18	2	13	.06209	.06210	.06208	.06204	.06199	.06191	.06182
18	2	14	.06006	.05976	.05944	.05910	.05873	.05835	.05795
18	2	15	.05784	.05728	.05669	.05609	.05547	.05484	.05421
18	2	16	.15415	.15208	.15001	.14794	.14588	.14385	.14184

Coefficients a_i for the BLUE of μ
Shape parameter k

n	s	i	1.4	1.5	1.6	1.7	1.8	1.9	2.0
18	3	1	-.09290	-.11466	-.13876	-.16532	-.19445	-.22627	-.26087
18	3	2	.04391	.04598	.04856	.05170	.05543	.05980	.06485
18	3	3	.06150	.06512	.06927	.07396	.07924	.08513	.09165
18	3	4	.06878	.07256	.07676	.08142	.08654	.09215	.09826
18	3	5	.07199	.07548	.07930	.08347	.08799	.09287	.09812
18	3	6	.07319	.07624	.07954	.08309	.08689	.09095	.09526
18	3	7	.07328	.07585	.07858	.08149	.08457	.08782	.09125
18	3	8	.07270	.07477	.07696	.07925	.08165	.08416	.08678
18	3	9	.07168	.07328	.07494	.07666	.07844	.08029	.08219
18	3	10	.07037	.07151	.07269	.07388	.07511	.07636	.07763
18	3	11	.06883	.06956	.07028	.07100	.07173	.07246	.07319
18	3	12	.06712	.06745	.06776	.06806	.06835	.06862	.06889
18	3	13	.06524	.06520	.06515	.06507	.06498	.06487	.06474
18	3	14	.06317	.06281	.06243	.06203	.06161	.06118	.06073
18	3	15	.22113	.21885	.21654	.21423	.21191	.20961	.20733
18	4	1	-.12395	-.14784	-.17413	-.20295	-.23440	-.26861	-.30567
18	4	2	.03707	.03928	.04203	.04537	.04932	.05395	.05930
18	4	3	.05906	.06299	.06745	.07249	.07813	.08440	.09133
18	4	4	.06860	.07270	.07723	.08224	.08772	.09369	.10018
18	4	5	.07317	.07695	.08107	.08554	.09037	.09556	.10113
18	4	6	.07526	.07855	.08210	.08590	.08995	.09426	.09882
18	4	7	.07595	.07871	.08164	.08474	.08802	.09146	.09507
18	4	8	.07578	.07800	.08033	.08276	.08530	.08795	.09070
18	4	9	.07505	.07675	.07850	.08032	.08219	.08411	.08609
18	4	10	.07393	.07513	.07635	.07760	.07886	.08015	.08146
18	4	11	.07252	.07325	.07398	.07471	.07544	.07617	.07690
18	4	12	.07086	.07116	.07145	.07172	.07197	.07222	.07246
18	4	13	.06898	.06889	.06877	.06864	.06849	.06832	.06814
18	4	14	.29771	.29548	.29321	.29093	.28864	.28636	.28409
18	5	1	-.16083	-.18716	-.21598	-.24739	-.28150	-.31843	-.35829
18	5	2	.02909	.03147	.03443	.03801	.04224	.04718	.05286
18	5	3	.05631	.06059	.06543	.07087	.07694	.08365	.09104
18	5	4	.06853	.07299	.07792	.08331	.08920	.09560	.10251
18	5	5	.07470	.07882	.08328	.08810	.09327	.09882	.10474
18	5	6	.07784	.08141	.08524	.08931	.09364	.09823	.10307
18	5	7	.07923	.08221	.08536	.08867	.09215	.09580	.09962
18	5	8	.07955	.08193	.08441	.08699	.08967	.09246	.09534
18	5	9	.07916	.08095	.08280	.08470	.08665	.08866	.09072
18	5	10	.07826	.07950	.08076	.08204	.08334	.08465	.08599
18	5	11	.07697	.07770	.07842	.07914	.07985	.08057	.08128
18	5	12	.07537	.07562	.07585	.07607	.07627	.07647	.07665
18	5	13	.38584	.38397	.38208	.38017	.37827	.37636	.37448
18	6	1	-.20506	-.23424	-.26599	-.30041	-.33762	-.37773	-.42084
18	6	2	.01968	.02230	.02552	.02939	.03395	.03926	.04534
18	6	3	.05318	.05789	.06319	.06911	.07567	.08290	.09083
18	6	4	.06862	.07352	.07889	.08475	.09111	.09799	.10540
18	6	5	.07672	.08122	.08607	.09128	.09686	.10280	.10913
18	6	6	.08109	.08499	.08913	.09352	.09816	.10305	.10820
18	6	7	.08333	.08655	.08993	.09348	.09719	.10106	.10509
18	6	8	.08422	.08676	.08941	.09214	.09498	.09790	.10093
18	6	9	.08421	.08610	.08804	.09002	.09206	.09414	.09627
18	6	10	.08355	.08483	.08611	.08741	.08873	.09006	.09141
18	6	11	.08240	.08310	.08379	.08448	.08516	.08584	.08651
18	6	12	.48806	.48699	.48590	.48483	.48377	.48273	.48173
18	7	1	-.25878	-.29134	-.32655	-.36454	-.40541	-.44927	-.49621
18	7	2	.00849	.01140	.01495	.01919	.02416	.02992	.03648

Coefficients a_i for the BLUE of μ
Shape parameter k

n	s	i	1.4	1.5	1.6	1.7	1.8	1.9	2.0
18	7	3	.04963	.05486	.06070	.06719	.07434	.08219	.09076
18	7	4	.06897	.07439	.08030	.08670	.09362	.10106	.10904
18	7	5	.07939	.08435	.08966	.09533	.10137	.10777	.11456
18	7	6	.08526	.08953	.09403	.09878	.10378	.10902	.11452
18	7	7	.08850	.09199	.09565	.09945	.10341	.10753	.11181
18	7	8	.09006	.09279	.09561	.09851	.10150	.10458	.10775
18	7	9	.09050	.09249	.09451	.09658	.09869	.10084	.10303
18	7	10	.09012	.09141	.09271	.09401	.09533	.09666	.09800
18	7	11	.60785	.60813	.60844	.60880	.60922	.60970	.61025
18	8	1	-.32512	-.36173	-.40112	-.44340	-.48868	-.53705	-.58862
18	8	2	-.00501	-.00173	.00225	.00696	.01245	.01875	.02591
18	8	3	.04557	.05144	.05796	.06514	.07301	.08160	.09094
18	8	4	.06971	.07577	.08232	.08938	.09696	.10508	.11375
18	8	5	.08299	.08850	.09436	.10057	.10715	.11411	.12144
18	8	6	.09070	.09539	.10032	.10549	.11090	.11655	.12246
18	8	7	.09515	.09895	.10291	.10701	.11125	.11565	.12020
18	8	8	.09752	.10045	.10345	.10653	.10969	.11293	.11626
18	8	9	.09848	.10056	.10266	.10480	.10698	.10919	.11143
18	8	10	.75001	.75239	.75490	.75753	.76029	.76318	.76622
18	9	1	-.40876	-.45034	-.49487	-.54243	-.59313	-.64707	-.70435
18	9	2	-.02160	-.01781	-.01327	-.00795	-.00181	.00520	.01311
18	9	3	.04090	.04758	.05494	.06300	.07177	.08128	.09157
18	9	4	.07109	.07794	.08529	.09316	.10156	.11050	.12001
18	9	5	.08796	.09413	.10064	.10752	.11476	.12238	.13038
18	9	6	.09794	.10315	.10858	.11425	.12015	.12629	.13267
18	9	7	.10388	.10805	.11235	.11679	.12137	.12608	.13094
18	9	8	.10725	.11038	.11358	.11685	.12019	.12361	.12710
18	9	9	.92135	.92692	.93274	.93882	.94515	.95173	.95857
19	0	1	-.02803	-.04475	-.06361	-.08474	-.10824	-.13423	-.16280
19	0	2	.05489	.05645	.05842	.06082	.06371	.06711	.07108
19	0	3	.06316	.06592	.06914	.07282	.07701	.08172	.08700
19	0	4	.06578	.06869	.07199	.07570	.07983	.08440	.08942
19	0	5	.06627	.06900	.07204	.07541	.07910	.08314	.08752
19	0	6	.06578	.06820	.07087	.07378	.07694	.08035	.08402
19	0	7	.06478	.06686	.06911	.07154	.07416	.07696	.07994
19	0	8	.06349	.06522	.06706	.06904	.07114	.07336	.07570
19	0	9	.06204	.06342	.06488	.06642	.06804	.06974	.07151
19	0	10	.06048	.06153	.06263	.06377	.06496	.06619	.06746
19	0	11	.05886	.05960	.06036	.06114	.06194	.06275	.06358
19	0	12	.05718	.05764	.05809	.05854	.05899	.05943	.05988
19	0	13	.05546	.05565	.05582	.05597	.05611	.05624	.05635
19	0	14	.05369	.05363	.05354	.05343	.05330	.05315	.05299
19	0	15	.05185	.05156	.05124	.05090	.05054	.05015	.04975
19	0	16	.04990	.04941	.04888	.04834	.04778	.04720	.04661
19	0	17	.04778	.04710	.04640	.04569	.04496	.04423	.04351
19	0	18	.04530	.04446	.04362	.04277	.04193	.04110	.04029
19	0	19	.04132	.04041	.03952	.03865	.03781	.03699	.03620
19	1	1	-.04519	-.06333	-.08370	-.10641	-.13158	-.15932	-.18975
19	1	2	.05078	.05237	.05438	.05684	.05981	.06333	.06742
19	1	3	.06146	.06438	.06777	.07164	.07604	.08100	.08653
19	1	4	.06534	.06844	.07194	.07586	.08022	.08504	.09033
19	1	5	.06659	.06951	.07274	.07632	.08023	.08450	.08912
19	1	6	.06661	.06920	.07204	.07514	.07849	.08211	.08599
19	1	7	.06595	.06818	.07059	.07318	.07596	.07893	.08208
19	1	8	.06491	.06676	.06874	.07085	.07309	.07545	.07793
19	1	9	.06364	.06512	.06669	.06834	.07007	.07188	.07376

Coefficients a_i for the BLUE of μ
Shape parameter k

n	s	i	1.4	1.5	1.6	1.7	1.8	1.9	2.0
19	1	10	.06221	.06334	.06453	.06575	.06702	.06833	.06968
19	1	11	.06068	.06148	.06230	.06314	.06400	.06487	.06576
19	1	12	.05907	.05956	.06005	.06054	.06103	.06151	.06199
19	1	13	.05738	.05759	.05778	.05795	.05811	.05826	.05839
19	1	14	.05563	.05557	.05548	.05538	.05525	.05510	.05494
19	1	15	.05378	.05348	.05315	.05279	.05242	.05202	.05162
19	1	16	.05181	.05129	.05074	.05017	.04959	.04899	.04839
19	1	17	.04963	.04891	.04818	.04744	.04669	.04594	.04519
19	1	18	.08971	.08815	.08660	.08507	.08357	.08209	.08064
19	2	1	-.06550	-.08519	-.10717	-.13156	-.15848	-.18804	-.22036
19	2	2	.04599	.04762	.04970	.05226	.05535	.05900	.06327
19	2	3	.05952	.06263	.06622	.07032	.07496	.08017	.08598
19	2	4	.06488	.06819	.07192	.07608	.08070	.08578	.09135
19	2	5	.06703	.07016	.07361	.07740	.08155	.08606	.09093
19	2	6	.06764	.07042	.07345	.07675	.08030	.08412	.08821
19	2	7	.06740	.06978	.07236	.07511	.07806	.08120	.08452
19	2	8	.06665	.06863	.07074	.07298	.07535	.07784	.08046
19	2	9	.06558	.06717	.06884	.07059	.07242	.07433	.07631
19	2	10	.06431	.06551	.06677	.06807	.06941	.07079	.07221
19	2	11	.06288	.06373	.06460	.06548	.06638	.06730	.06822
19	2	12	.06134	.06185	.06237	.06288	.06338	.06388	.06438
19	2	13	.05969	.05990	.06009	.06026	.06042	.06056	.06069
19	2	14	.05795	.05787	.05776	.05764	.05749	.05732	.05715
19	2	15	.05609	.05574	.05538	.05499	.05458	.05415	.05372
19	2	16	.05406	.05349	.05289	.05228	.05165	.05102	.05038
19	2	17	.14450	.14250	.14049	.13848	.13649	.13452	.13258
19	3	1	-.08925	-.11066	-.13442	-.16066	-.18949	-.22103	-.25540
19	3	2	.04045	.04216	.04433	.04701	.05024	.05407	.05853
19	3	3	.05733	.06065	.06447	.06883	.07375	.07925	.08538
19	3	4	.06441	.06797	.07196	.07639	.08129	.08667	.09254
19	3	5	.06762	.07098	.07467	.07871	.08311	.08788	.09302
19	3	6	.06892	.07191	.07515	.07865	.08242	.08646	.09077
19	3	7	.06915	.07171	.07446	.07739	.08051	.08382	.08732
19	3	8	.06874	.07086	.07311	.07549	.07799	.08061	.08336
19	3	9	.06792	.06961	.07138	.07323	.07515	.07716	.07923
19	3	10	.06681	.06809	.06941	.07078	.07218	.07362	.07510
19	3	11	.06550	.06639	.06730	.06822	.06915	.07009	.07105
19	3	12	.06403	.06456	.06508	.06560	.06611	.06662	.06712
19	3	13	.06242	.06262	.06279	.06294	.06308	.06321	.06333
19	3	14	.06068	.06056	.06042	.06026	.06007	.05988	.05966
19	3	15	.05879	.05839	.05796	.05752	.05706	.05659	.05611
19	3	16	.20646	.20420	.20193	.19965	.19738	.19511	.19287
19	4	1	-.11701	-.14036	-.16614	-.19445	-.22541	-.25915	-.29577
19	4	2	.03406	.03586	.03816	.04099	.04440	.04843	.05312
19	4	3	.05486	.05843	.06253	.06718	.07241	.07825	.08473
19	4	4	.06397	.06780	.07208	.07682	.08203	.08774	.09395
19	4	5	.06840	.07202	.07598	.08029	.08497	.09003	.09546
19	4	6	.07050	.07372	.07719	.08092	.08493	.08920	.09375
19	4	7	.07129	.07404	.07697	.08009	.08339	.08689	.09057
19	4	8	.07127	.07354	.07593	.07844	.08108	.08384	.08673
19	4	9	.07072	.07252	.07439	.07634	.07836	.08045	.08262
19	4	10	.06980	.07115	.07254	.07396	.07542	.07692	.07845
19	4	11	.06862	.06955	.07048	.07142	.07238	.07334	.07432
19	4	12	.06723	.06776	.06828	.06879	.06929	.06979	.07028
19	4	13	.06566	.06582	.06596	.06608	.06619	.06628	.06636
19	4	14	.06391	.06373	.06353	.06331	.06307	.06282	.06256

Coefficients a_i for the BLUE of μ
Shape parameter k

n	s	i	1.4	1.5	1.6	1.7	1.8	1.9	2.0
19	4	15	.27672	.27445	.27214	.26982	.26750	.26518	.26288
19	5	1	-.14962	-.17519	-.20325	-.23391	-.26729	-.30350	-.34267
19	5	2	.02667	.02859	.03104	.03405	.03767	.04194	.04690
19	5	3	.05206	.05594	.06036	.06535	.07094	.07716	.08403
19	5	4	.06355	.06771	.07233	.07741	.08298	.08905	.09565
19	5	5	.06942	.07334	.07760	.08223	.08722	.09259	.09835
19	5	6	.07247	.07594	.07966	.08366	.08792	.09245	.09726
19	5	7	.07390	.07685	.07999	.08331	.08682	.09051	.09439
19	5	8	.07433	.07675	.07930	.08196	.08475	.08765	.09067
19	5	9	.07409	.07600	.07798	.08003	.08214	.08433	.08659
19	5	10	.07340	.07481	.07625	.07773	.07924	.08079	.08236
19	5	11	.07236	.07330	.07425	.07521	.07618	.07715	.07813
19	5	12	.07105	.07156	.07205	.07254	.07302	.07350	.07397
19	5	13	.06950	.06961	.06969	.06976	.06982	.06986	.06989
19	5	14	.35681	.35479	.35274	.35067	.34860	.34653	.34448
19	6	1	-.18823	-.21635	-.24703	-.28039	-.31655	-.35562	-.39770
19	6	2	.01805	.02013	.02277	.02600	.02988	.03443	.03971
19	6	3	.04890	.05313	.05793	.06332	.06932	.07598	.08332
19	6	4	.06320	.06774	.07274	.07822	.08421	.09070	.09773
19	6	5	.07077	.07503	.07964	.08462	.08997	.09570	.10182
19	6	6	.07492	.07868	.08270	.08698	.09153	.09635	.10144
19	6	7	.07711	.08029	.08366	.08720	.09093	.09484	.09893
19	6	8	.07806	.08066	.08337	.08620	.08914	.09219	.09536
19	6	9	.07820	.08022	.08230	.08445	.08666	.08894	.09129
19	6	10	.07775	.07922	.08072	.08224	.08380	.08538	.08699
19	6	11	.07687	.07782	.07877	.07973	.08069	.08166	.08264
19	6	12	.07564	.07611	.07656	.07701	.07745	.07788	.07831
19	6	13	.44874	.44731	.44587	.44442	.44298	.44156	.44016
19	7	1	-.23441	-.26551	-.29926	-.33577	-.37516	-.41754	-.46302
19	7	2	.00792	.01021	.01308	.01659	.02078	.02567	.03133
19	7	3	.04531	.04996	.05521	.06107	.06757	.07474	.08261
19	7	4	.06297	.06795	.07341	.07936	.08582	.09280	.10032
19	7	5	.07256	.07722	.08223	.08761	.09337	.09952	.10606
19	7	6	.07803	.08212	.08647	.09108	.09595	.10110	.10652
19	7	7	.08111	.08455	.08817	.09196	.09593	.10008	.10441
19	7	8	.08268	.08546	.08835	.09135	.09446	.09767	.10100
19	7	9	.08324	.08538	.08757	.08982	.09213	.09450	.09693
19	7	10	.08308	.08459	.08613	.08770	.08929	.09090	.09254
19	7	11	.08237	.08331	.08424	.08518	.08612	.08707	.08802
19	7	12	.55514	.55475	.55438	.55404	.55374	.55349	.55328
19	8	1	-.29041	-.32503	-.36240	-.40264	-.44586	-.49217	-.54166
19	8	2	-.00412	-.00157	.00161	.00546	.01002	.01534	.02145
19	8	3	.04120	.04637	.05216	.05858	.06567	.07346	.08196
19	8	4	.06293	.06844	.07444	.08094	.08797	.09553	.10363
19	8	5	.07497	.08009	.08558	.09144	.09767	.10430	.11132
19	8	6	.08202	.08649	.09123	.09621	.10146	.10698	.11277
19	8	7	.08617	.08990	.09381	.09788	.10213	.10654	.11114
19	8	8	.08847	.09146	.09454	.09773	.10101	.10440	.10790
19	8	9	.08953	.09178	.09408	.09643	.09884	.10130	.10381
19	8	10	.08968	.09124	.09281	.09440	.09601	.09764	.09929
19	8	11	.67957	.68082	.68215	.68356	.68507	.68668	.68840
19	9	1	-.35946	-.39830	-.44004	-.48477	-.53261	-.58365	-.63800
19	9	2	-.01866	-.01575	-.01218	-.00789	-.00285	.00299	.00965
19	9	3	.03646	.04227	.04873	.05585	.06366	.07219	.08146
19	9	4	.06321	.06936	.07602	.08319	.09089	.09914	.10795
19	9	5	.07824	.08393	.08998	.09641	.10321	.11040	.11799

Coefficients a_i for the BLUE of μ
Shape parameter k

n	s	i	1.4	1.5	1.6	1.7	1.8	1.9	2.0
19	9	6	.08723	.09216	.09734	.10277	.10846	.11441	.12063
19	9	7	.09267	.09675	.10098	.10538	.10994	.11466	.11955
19	9	8	.09586	.09907	.10237	.10576	.10924	.11282	.11650
19	9	9	.09750	.09987	.10228	.10473	.10723	.10978	.11237
19	9	10	.82695	.83065	.83452	.83858	.84283	.84726	.85189
19	10	1	-.44641	-.49045	-.53755	-.58781	-.64133	-.69821	-.75855
19	10	2	-.03652	-.03314	-.02904	-.02418	-.01852	-.01202	-.00465
19	10	3	.03095	.03758	.04488	.05287	.06157	.07103	.08125
19	10	4	.06401	.07096	.07844	.08643	.09497	.10407	.11373
19	10	5	.08280	.08917	.09592	.10303	.11052	.11840	.12668
19	10	6	.09419	.09967	.10539	.11135	.11756	.12403	.13076
19	10	7	.10124	.10571	.11033	.11510	.12002	.12510	.13034
19	10	8	.10550	.10896	.11249	.11611	.11981	.12360	.12748
19	10	9	1.00425	1.01154	1.01916	1.02711	1.03539	1.04400	1.05295
20	0	1	-.03014	-.04685	-.06572	-.08688	-.11044	-.13651	-.16520
20	0	2	.05104	.05235	.05403	.05611	.05863	.06164	.06518
20	0	3	.05945	.06203	.06504	.06850	.07245	.07691	.08192
20	0	4	.06228	.06507	.06824	.07181	.07579	.08022	.08509
20	0	5	.06299	.06565	.06861	.07191	.07554	.07951	.08384
20	0	6	.06271	.06510	.06774	.07064	.07379	.07721	.08089
20	0	7	.06191	.06399	.06626	.06873	.07138	.07423	.07728
20	0	8	.06081	.06257	.06448	.06652	.06869	.07101	.07345
20	0	9	.05954	.06099	.06253	.06417	.06590	.06772	.06962
20	0	10	.05817	.05931	.06052	.06178	.06310	.06447	.06589
20	0	11	.05673	.05758	.05847	.05938	.06033	.06130	.06230
20	0	12	.05524	.05582	.05641	.05701	.05762	.05824	.05886
20	0	13	.05370	.05403	.05436	.05467	.05498	.05528	.05558
20	0	14	.05213	.05223	.05230	.05236	.05240	.05243	.05244
20	0	15	.05051	.05039	.05024	.05007	.04988	.04967	.04944
20	0	16	.04883	.04851	.04815	.04778	.04738	.04697	.04654
20	0	17	.04705	.04654	.04601	.04546	.04489	.04431	.04372
20	0	18	.04510	.04443	.04374	.04304	.04233	.04162	.04091
20	0	19	.04281	.04200	.04119	.04038	.03957	.03877	.03799
20	0	20	.03913	.03826	.03741	.03658	.03578	.03500	.03425
20	1	1	-.04596	-.06400	-.08429	-.10694	-.13207	-.15980	-.19024
20	1	2	.04713	.04844	.05013	.05225	.05484	.05792	.06156
20	1	3	.05775	.06046	.06361	.06724	.07137	.07604	.08127
20	1	4	.06173	.06469	.06803	.07180	.07599	.08064	.08575
20	1	5	.06315	.06597	.06912	.07260	.07643	.08062	.08517
20	1	6	.06334	.06589	.06870	.07176	.07510	.07870	.08259
20	1	7	.06286	.06509	.06751	.07013	.07294	.07595	.07916
20	1	8	.06200	.06389	.06592	.06809	.07040	.07285	.07543
20	1	9	.06091	.06246	.06411	.06585	.06769	.06962	.07164
20	1	10	.05966	.06089	.06218	.06352	.06493	.06638	.06790
20	1	11	.05831	.05923	.06018	.06116	.06217	.06321	.06427
20	1	12	.05689	.05752	.05816	.05880	.05946	.06012	.06078
20	1	13	.05540	.05576	.05611	.05645	.05679	.05712	.05744
20	1	14	.05385	.05396	.05405	.05412	.05417	.05421	.05424
20	1	15	.05224	.05212	.05197	.05179	.05160	.05139	.05116
20	1	16	.05055	.05021	.04985	.04946	.04905	.04863	.04819
20	1	17	.04874	.04821	.04765	.04708	.04649	.04589	.04529
20	1	18	.04674	.04604	.04532	.04459	.04386	.04313	.04241
20	1	19	.08469	.08319	.08170	.08023	.07879	.07737	.07598
20	2	1	-.06455	-.08404	-.10583	-.13005	-.15683	-.18626	-.21848
20	2	2	.04258	.04391	.04565	.04784	.05051	.05371	.05747
20	2	3	.05580	.05867	.06200	.06582	.07016	.07506	.08054

Coefficients a_i for the BLUE of μ
Shape parameter k

n	s	i	1.4	1.5	1.6	1.7	1.8	1.9	2.0
20	2	4	.06114	.06429	.06784	.07181	.07624	.08113	.08650
20	2	5	.06339	.06640	.06975	.07343	.07748	.08189	.08667
20	2	6	.06413	.06685	.06984	.07309	.07661	.08041	.08449
20	2	7	.06404	.06642	.06899	.07176	.07473	.07791	.08128
20	2	8	.06345	.06547	.06763	.06992	.07236	.07494	.07766
20	2	9	.06256	.06421	.06596	.06781	.06975	.07178	.07390
20	2	10	.06146	.06277	.06413	.06555	.06703	.06857	.07015
20	2	11	.06021	.06119	.06220	.06323	.06429	.06538	.06649
20	2	12	.05886	.05953	.06020	.06088	.06156	.06225	.06295
20	2	13	.05742	.05780	.05816	.05852	.05886	.05920	.05954
20	2	14	.05590	.05600	.05609	.05615	.05620	.05623	.05626
20	2	15	.05430	.05415	.05398	.05378	.05357	.05334	.05309
20	2	16	.05259	.05221	.05181	.05139	.05095	.05050	.05003
20	2	17	.05073	.05015	.04955	.04893	.04831	.04768	.04704
20	2	18	.13597	.13402	.13208	.13014	.12821	.12631	.12443
20	3	1	-.08612	-.10721	-.13066	-.15660	-.18515	-.21643	-.25055
20	3	2	.03736	.03874	.04054	.04281	.04559	.04893	.05285
20	3	3	.05361	.05666	.06019	.06423	.06881	.07397	.07972
20	3	4	.06052	.06388	.06766	.07188	.07655	.08171	.08735
20	3	5	.06373	.06695	.07051	.07443	.07870	.08335	.08838
20	3	6	.06511	.06802	.07120	.07465	.07837	.08237	.08666
20	3	7	.06546	.06800	.07074	.07368	.07681	.08015	.08370
20	3	8	.06519	.06734	.06963	.07206	.07463	.07734	.08019
20	3	9	.06453	.06629	.06814	.07009	.07213	.07425	.07647
20	3	10	.06359	.06498	.06642	.06791	.06946	.07106	.07272
20	3	11	.06247	.06349	.06455	.06563	.06673	.06786	.06901
20	3	12	.06120	.06189	.06258	.06328	.06398	.06469	.06540
20	3	13	.05981	.06018	.06055	.06090	.06125	.06158	.06191
20	3	14	.05831	.05839	.05845	.05850	.05853	.05854	.05854
20	3	15	.05670	.05651	.05630	.05607	.05582	.05556	.05528
20	3	16	.05495	.05452	.05407	.05360	.05311	.05262	.05211
20	3	17	.19358	.19137	.18914	.18690	.18467	.18245	.18025
20	4	1	-.11113	-.13401	-.15931	-.18716	-.21767	-.25098	-.28719
20	4	2	.03138	.03282	.03470	.03709	.04000	.04349	.04760
20	4	3	.05114	.05440	.05817	.06246	.06732	.07276	.07882
20	4	4	.05988	.06348	.06752	.07201	.07696	.08241	.08836
20	4	5	.06420	.06766	.07146	.07562	.08015	.08506	.09036
20	4	6	.06631	.06944	.07283	.07649	.08043	.08465	.08916
20	4	7	.06718	.06989	.07281	.07592	.07924	.08276	.08648
20	4	8	.06727	.06957	.07200	.07457	.07727	.08012	.08310
20	4	9	.06687	.06874	.07070	.07275	.07489	.07712	.07943
20	4	10	.06612	.06758	.06910	.07066	.07228	.07394	.07566
20	4	11	.06513	.06620	.06730	.06842	.06956	.07072	.07190
20	4	12	.06395	.06465	.06536	.06607	.06678	.06750	.06822
20	4	13	.06261	.06297	.06332	.06366	.06400	.06432	.06463
20	4	14	.06113	.06117	.06120	.06121	.06120	.06118	.06115
20	4	15	.05950	.05926	.05899	.05871	.05841	.05810	.05777
20	4	16	.25845	.25616	.25385	.25151	.24918	.24685	.24454
20	5	1	-.14023	-.16512	-.19251	-.22250	-.25523	-.29081	-.32935
20	5	2	.02451	.02603	.02803	.03054	.03362	.03729	.04161
20	5	3	.04836	.05188	.05591	.06050	.06566	.07143	.07783
20	5	4	.05923	.06311	.06744	.07223	.07751	.08328	.08957
20	5	5	.06484	.06856	.07264	.07708	.08189	.08709	.09269
20	5	6	.06780	.07116	.07479	.07869	.08287	.08733	.09208
20	5	7	.06926	.07217	.07528	.07859	.08209	.08580	.08971
20	5	8	.06977	.07222	.07480	.07752	.08037	.08336	.08649

Coefficients a_i for the BLUE of μ
Shape parameter k

n	s	i	1.4	1.5	1.6	1.7	1.8	1.9	2.0
20	5	9	.06967	.07165	.07372	.07588	.07812	.08045	.08286
20	5	10	.06914	.07067	.07226	.07389	.07557	.07729	.07907
20	5	11	.06829	.06940	.07053	.07168	.07285	.07404	.07525
20	5	12	.06720	.06791	.06862	.06933	.07004	.07076	.07147
20	5	13	.06592	.06625	.06658	.06689	.06719	.06748	.06777
20	5	14	.06444	.06443	.06441	.06436	.06431	.06424	.06416
20	5	15	.33180	.32967	.32751	.32533	.32315	.32097	.31881
20	6	1	-.17429	-.20149	-.23125	-.26369	-.29893	-.33708	-.37827
20	6	2	.01658	.01821	.02034	.02302	.02629	.03018	.03475
20	6	3	.04522	.04903	.05338	.05830	.06382	.06997	.07676
20	6	4	.05857	.06278	.06745	.07259	.07823	.08437	.09104
20	6	5	.06570	.06972	.07411	.07886	.08400	.08952	.09544
20	6	6	.06966	.07328	.07717	.08133	.08578	.09051	.09553
20	6	7	.07179	.07492	.07825	.08177	.08548	.08940	.09352
20	6	8	.07279	.07541	.07815	.08103	.08404	.08718	.09046
20	6	9	.07304	.07514	.07732	.07959	.08194	.08437	.08688
20	6	10	.07275	.07436	.07601	.07771	.07945	.08123	.08306
20	6	11	.07207	.07321	.07436	.07554	.07673	.07793	.07915
20	6	12	.07108	.07178	.07248	.07317	.07387	.07456	.07526
20	6	13	.06984	.07013	.07041	.07067	.07093	.07118	.07142
20	6	14	.41521	.41353	.41182	.41010	.40839	.40669	.40501
20	7	1	-.21452	-.24437	-.27687	-.31212	-.35026	-.39138	-.43560
20	7	2	.00736	.00912	.01142	.01430	.01780	.02196	.02682
20	7	3	.04166	.04583	.05055	.05586	.06179	.06837	.07562
20	7	4	.05795	.06254	.06760	.07314	.07919	.08576	.09286
20	7	5	.06685	.07123	.07597	.08108	.08658	.09247	.09877
20	7	6	.07198	.07590	.08009	.08455	.08930	.09433	.09964
20	7	7	.07492	.07829	.08185	.08561	.08956	.09370	.09805
20	7	8	.07648	.07928	.08220	.08525	.08843	.09174	.09517
20	7	9	.07712	.07935	.08166	.08404	.08649	.08903	.09164
20	7	10	.07711	.07879	.08051	.08227	.08406	.08590	.08777
20	7	11	.07661	.07777	.07895	.08013	.08133	.08254	.08377
20	7	12	.07573	.07641	.07708	.07774	.07841	.07907	.07973
20	7	13	.51075	.50987	.50900	.50814	.50731	.50651	.50575
20	8	1	-.26256	-.29551	-.33120	-.36974	-.41124	-.45582	-.50359
20	8	2	-.00348	-.00153	.00098	.00412	.00790	.01238	.01758
20	8	3	.03760	.04219	.04735	.05313	.05955	.06664	.07442
20	8	4	.05739	.06243	.06795	.07396	.08050	.08756	.09517
20	8	5	.06841	.07320	.07835	.08388	.08980	.09612	.10284
20	8	6	.07492	.07919	.08372	.08853	.09362	.09898	.10464
20	8	7	.07881	.08246	.08629	.09031	.09452	.09892	.10352
20	8	8	.08104	.08404	.08716	.09040	.09376	.09724	.10085
20	8	9	.08214	.08450	.08693	.08943	.09200	.09465	.09736
20	8	10	.08245	.08420	.08598	.08779	.08963	.09151	.09343
20	8	11	.08215	.08332	.08450	.08568	.08687	.08807	.08929
20	8	12	.62112	.62153	.62198	.62250	.62309	.62375	.62449
20	9	1	-.32073	-.35734	-.39681	-.43924	-.48473	-.53341	-.58538
20	9	2	-.01636	-.01418	-.01138	-.00793	-.00379	.00107	.00670
20	9	3	.03294	.03804	.04374	.05008	.05709	.06479	.07320
20	9	4	.05696	.06254	.06861	.07518	.08229	.08993	.09814
20	9	5	.07054	.07581	.08145	.08747	.09388	.10069	.10791
20	9	6	.07872	.08339	.08832	.09352	.09900	.10476	.11081
20	9	7	.08374	.08770	.09184	.09616	.10066	.10536	.11024
20	9	8	.08675	.08998	.09331	.09676	.10033	.10401	.10780
20	9	9	.08840	.09090	.09346	.09609	.09877	.10153	.10435
20	9	10	.08907	.09088	.09271	.09457	.09645	.09837	.10032

Coefficients a_i for the BLUE of μ
Shape parameter k

n	s	i	1.4	1.5	1.6	1.7	1.8	1.9	2.0
20	9	11	.74997	.75229	.75474	.75733	.76005	.76292	.76593
20	10	1	-.39237	-.43339	-.47740	-.52451	-.57483	-.62846	-.68550
20	10	2	-.03191	-.02939	-.02623	-.02238	-.01779	-.01245	-.00629
20	10	3	.02753	.03327	.03964	.04667	.05439	.06283	.07200
20	10	4	.05677	.06300	.06973	.07699	.08479	.09313	.10205
20	10	5	.07348	.07933	.08555	.09216	.09915	.10655	.11436
20	10	6	.08369	.08884	.09424	.09992	.10586	.11208	.11858
20	10	7	.09009	.09442	.09891	.10358	.10842	.11345	.11865
20	10	8	.09405	.09753	.10111	.10479	.10857	.11247	.11648
20	10	9	.09635	.09899	.10169	.10444	.10724	.11011	.11304
20	10	10	.90231	.90742	.91276	.91835	.92419	.93028	.93663
21	0	1	-.03197	-.04866	-.06753	-.08870	-.11229	-.13842	-.16719
21	0	2	.04756	.04863	.05004	.05182	.05401	.05665	.05978
21	0	3	.05608	.05848	.06129	.06454	.06826	.07248	.07722
21	0	4	.05908	.06175	.06478	.06822	.07206	.07634	.08106
21	0	5	.05998	.06256	.06545	.06867	.07222	.07613	.08039
21	0	6	.05988	.06223	.06484	.06771	.07085	.07425	.07794
21	0	7	.05925	.06133	.06361	.06609	.06877	.07166	.07475
21	0	8	.05832	.06011	.06205	.06414	.06638	.06876	.07130
21	0	9	.05722	.05872	.06033	.06204	.06385	.06577	.06779
21	0	10	.05600	.05722	.05851	.05987	.06130	.06279	.06435
21	0	11	.05472	.05566	.05666	.05769	.05876	.05987	.06102
21	0	12	.05338	.05407	.05479	.05552	.05627	.05703	.05781
21	0	13	.05201	.05246	.05292	.05338	.05383	.05429	.05475
21	0	14	.05060	.05083	.05105	.05126	.05146	.05164	.05182
21	0	15	.04916	.04918	.04918	.04917	.04913	.04908	.04901
21	0	16	.04768	.04750	.04731	.04709	.04685	.04659	.04632
21	0	17	.04613	.04578	.04540	.04501	.04459	.04415	.04371
21	0	18	.04449	.04398	.04344	.04289	.04232	.04174	.04115
21	0	19	.04269	.04203	.04136	.04068	.03999	.03929	.03860
21	0	20	.04058	.03980	.03901	.03823	.03745	.03669	.03593
21	0	21	.03715	.03632	.03551	.03472	.03395	.03321	.03249
21	1	1	-.04661	-.06455	-.08476	-.10734	-.13241	-.16011	-.19054
21	1	2	.04382	.04488	.04629	.04808	.05031	.05301	.05621
21	1	3	.05438	.05689	.05983	.06322	.06710	.07150	.07644
21	1	4	.05846	.06127	.06446	.06807	.07210	.07658	.08153
21	1	5	.06001	.06274	.06579	.06919	.07293	.07703	.08150
21	1	6	.06035	.06285	.06561	.06864	.07195	.07553	.07940
21	1	7	.06003	.06224	.06466	.06729	.07012	.07316	.07641
21	1	8	.05933	.06124	.06330	.06551	.06787	.07039	.07306
21	1	9	.05839	.05999	.06170	.06352	.06544	.06747	.06960
21	1	10	.05730	.05860	.05997	.06142	.06293	.06451	.06616
21	1	11	.05610	.05712	.05818	.05928	.06042	.06159	.06281
21	1	12	.05483	.05558	.05634	.05712	.05792	.05874	.05957
21	1	13	.05351	.05400	.05449	.05498	.05547	.05597	.05646
21	1	14	.05213	.05239	.05262	.05285	.05307	.05328	.05347
21	1	15	.05071	.05074	.05075	.05074	.05071	.05067	.05061
21	1	16	.04923	.04905	.04885	.04863	.04838	.04812	.04785
21	1	17	.04767	.04731	.04692	.04650	.04607	.04563	.04517
21	1	18	.04601	.04547	.04491	.04433	.04375	.04315	.04255
21	1	19	.04416	.04347	.04277	.04206	.04135	.04064	.03994
21	1	20	.08019	.07875	.07731	.07590	.07452	.07316	.07183
21	2	1	-.06371	-.08301	-.10462	-.12867	-.15529	-.18460	-.21670
21	2	2	.03950	.04056	.04200	.04384	.04613	.04890	.05221
21	2	3	.05245	.05509	.05818	.06174	.06580	.07040	.07556
21	2	4	.05777	.06075	.06413	.06793	.07216	.07686	.08204

Coefficients a_i for the BLUE of μ
Shape parameter k

n	s	i	1.4	1.5	1.6	1.7	1.8	1.9	2.0
21	2	5	.06009	.06299	.06623	.06981	.07374	.07805	.08273
21	2	6	.06094	.06360	.06653	.06973	.07321	.07698	.08103
21	2	7	.06098	.06334	.06590	.06868	.07166	.07485	.07826
21	2	8	.06054	.06257	.06476	.06709	.06958	.07223	.07502
21	2	9	.05979	.06150	.06331	.06523	.06725	.06938	.07161
21	2	10	.05884	.06023	.06168	.06321	.06480	.06646	.06818
21	2	11	.05775	.05883	.05995	.06111	.06230	.06354	.06481
21	2	12	.05656	.05735	.05815	.05897	.05981	.06066	.06153
21	2	13	.05529	.05580	.05632	.05683	.05734	.05785	.05837
21	2	14	.05395	.05421	.05445	.05469	.05491	.05512	.05532
21	2	15	.05254	.05256	.05256	.05254	.05250	.05245	.05239
21	2	16	.05106	.05086	.05063	.05039	.05012	.04984	.04955
21	2	17	.04948	.04908	.04866	.04821	.04775	.04728	.04680
21	2	18	.04777	.04719	.04659	.04598	.04536	.04473	.04410
21	2	19	.12838	.12649	.12460	.12272	.12086	.11902	.11720
21	3	1	-.08342	-.10420	-.12736	-.15302	-.18130	-.21232	-.24621
21	3	2	.03458	.03566	.03713	.03903	.04140	.04428	.04771
21	3	3	.05027	.05307	.05633	.06008	.06436	.06918	.07458
21	3	4	.05703	.06020	.06378	.06780	.07226	.07720	.08262
21	3	5	.06024	.06333	.06676	.07055	.07469	.07922	.08413
21	3	6	.06168	.06452	.06762	.07101	.07467	.07863	.08288
21	3	7	.06213	.06464	.06736	.07029	.07343	.07678	.08035
21	3	8	.06199	.06415	.06646	.06893	.07155	.07432	.07724
21	3	9	.06146	.06327	.06518	.06720	.06933	.07156	.07389
21	3	10	.06067	.06213	.06367	.06527	.06694	.06867	.07047
21	3	11	.05970	.06083	.06201	.06322	.06447	.06575	.06707
21	3	12	.05859	.05941	.06025	.06110	.06197	.06285	.06374
21	3	13	.05738	.05790	.05843	.05895	.05948	.06000	.06052
21	3	14	.05607	.05632	.05656	.05679	.05700	.05721	.05740
21	3	15	.05467	.05467	.05465	.05460	.05455	.05447	.05439
21	3	16	.05318	.05294	.05268	.05240	.05210	.05179	.05147
21	3	17	.05157	.05112	.05064	.05016	.04965	.04914	.04862
21	3	18	.18220	.18002	.17784	.17564	.17346	.17129	.16914
21	4	1	-.10609	-.12853	-.15340	-.18083	-.21093	-.24384	-.27967
21	4	2	.02897	.03008	.03161	.03358	.03605	.03906	.04263
21	4	3	.04783	.05081	.05427	.05824	.06275	.06782	.07349
21	4	4	.05625	.05964	.06345	.06770	.07242	.07761	.08331
21	4	5	.06047	.06377	.06743	.07144	.07582	.08058	.08574
21	4	6	.06259	.06562	.06892	.07250	.07637	.08053	.08499
21	4	7	.06351	.06619	.06908	.07217	.07548	.07900	.08274
21	4	8	.06371	.06601	.06846	.07106	.07381	.07672	.07978
21	4	9	.06343	.06535	.06737	.06949	.07172	.07405	.07649
21	4	10	.06282	.06437	.06598	.06766	.06940	.07120	.07307
21	4	11	.06198	.06317	.06440	.06566	.06695	.06828	.06964
21	4	12	.06097	.06182	.06268	.06356	.06445	.06535	.06626
21	4	13	.05981	.06035	.06088	.06140	.06192	.06244	.06296
21	4	14	.05854	.05878	.05900	.05920	.05940	.05958	.05976
21	4	15	.05715	.05711	.05705	.05697	.05688	.05677	.05665
21	4	16	.05564	.05535	.05503	.05470	.05436	.05400	.05363
21	4	17	.24241	.24013	.23781	.23548	.23316	.23084	.22853
21	5	1	-.13226	-.15654	-.18333	-.21273	-.24487	-.27987	-.31785
21	5	2	.02257	.02373	.02533	.02740	.02999	.03314	.03688
21	5	3	.04510	.04829	.05198	.05619	.06097	.06632	.07229
21	5	4	.05543	.05906	.06312	.06765	.07264	.07814	.08414
21	5	5	.06081	.06436	.06825	.07252	.07716	.08218	.08761
21	5	6	.06371	.06695	.07047	.07427	.07836	.08275	.08743

Coefficients a_i for the BLUE of μ
Shape parameter k

n	s	i	1.4	1.5	1.6	1.7	1.8	1.9	2.0
21	5	7	.06518	.06804	.07111	.07438	.07787	.08158	.08550
21	5	8	.06576	.06821	.07081	.07355	.07645	.07950	.08270
21	5	9	.06576	.06780	.06993	.07217	.07450	.07693	.07947
21	5	10	.06536	.06699	.06868	.07043	.07225	.07412	.07605
21	5	11	.06467	.06591	.06718	.06849	.06982	.07119	.07259
21	5	12	.06376	.06463	.06551	.06641	.06731	.06822	.06914
21	5	13	.06267	.06320	.06372	.06423	.06475	.06525	.06576
21	5	14	.06142	.06163	.06182	.06199	.06216	.06231	.06246
21	5	15	.06003	.05994	.05983	.05970	.05956	.05940	.05924
21	5	16	.31002	.30781	.30558	.30333	.30108	.29883	.29660
21	6	1	-.16259	-.18897	-.21792	-.24954	-.28397	-.32133	-.36172
21	6	2	.01524	.01647	.01816	.02035	.02309	.02640	.03033
21	6	3	.04202	.04546	.04941	.05391	.05899	.06467	.07098
21	6	4	.05456	.05848	.06284	.06767	.07298	.07880	.08514
21	6	5	.06131	.06512	.06929	.07384	.07876	.08408	.08981
21	6	6	.06510	.06858	.07234	.07639	.08072	.08535	.09029
21	6	7	.06719	.07025	.07352	.07700	.08069	.08459	.08871
21	6	8	.06822	.07083	.07359	.07649	.07954	.08274	.08609
21	6	9	.06855	.07070	.07295	.07530	.07775	.08029	.08293
21	6	10	.06838	.07009	.07186	.07369	.07557	.07752	.07951
21	6	11	.06785	.06914	.07046	.07180	.07317	.07457	.07600
21	6	12	.06705	.06794	.06883	.06973	.07064	.07155	.07248
21	6	13	.06603	.06654	.06704	.06753	.06802	.06851	.06899
21	6	14	.06481	.06497	.06511	.06524	.06536	.06546	.06556
21	6	15	.38629	.38441	.38251	.38060	.37869	.37679	.37491
21	7	1	-.19802	-.22680	-.25821	-.29237	-.32941	-.36944	-.41257
21	7	2	.00679	.00811	.00992	.01226	.01516	.01868	.02284
21	7	3	.03854	.04227	.04653	.05137	.05679	.06284	.06954
21	7	4	.05367	.05791	.06261	.06779	.07347	.07966	.08638
21	7	5	.06199	.06611	.07060	.07547	.08072	.08637	.09243
21	7	6	.06683	.07058	.07462	.07894	.08355	.08845	.09366
21	7	7	.06965	.07294	.07643	.08013	.08404	.08816	.09249
21	7	8	.07119	.07398	.07691	.07999	.08320	.08656	.09007
21	7	9	.07189	.07418	.07655	.07902	.08159	.08424	.08699
21	7	10	.07199	.07379	.07564	.07754	.07949	.08150	.08356
21	7	11	.07165	.07298	.07433	.07571	.07711	.07854	.07999
21	7	12	.07097	.07186	.07275	.07364	.07454	.07545	.07636
21	7	13	.07001	.07049	.07095	.07141	.07186	.07230	.07275
21	7	14	.47285	.47161	.47036	.46912	.46789	.46669	.46552
21	8	1	-.23979	-.27133	-.30558	-.34267	-.38271	-.42582	-.47211
21	8	2	-.00303	-.00159	.00037	.00289	.00601	.00976	.01419
21	8	3	.03458	.03866	.04329	.04852	.05435	.06083	.06797
21	8	4	.05277	.05739	.06249	.06807	.07417	.08079	.08795
21	8	5	.06295	.06743	.07228	.07751	.08314	.08916	.09560
21	8	6	.06901	.07308	.07742	.08205	.08697	.09218	.09770
21	8	7	.07268	.07622	.07996	.08391	.08806	.09242	.09699
21	8	8	.07482	.07781	.08093	.08419	.08758	.09112	.09480
21	8	9	.07595	.07837	.08088	.08348	.08617	.08894	.09180
21	8	10	.07635	.07823	.08016	.08214	.08416	.08624	.08836
21	8	11	.07622	.07758	.07897	.08037	.08179	.08323	.08470
21	8	12	.07567	.07654	.07742	.07829	.07917	.08006	.08095
21	8	13	.57183	.57160	.57141	.57125	.57115	.57109	.57110
21	9	1	-.28961	-.32435	-.36192	-.40243	-.44598	-.49269	-.54268
21	9	2	-.01456	-.01296	-.01080	-.00806	-.00468	-.00064	.00412
21	9	3	.03006	.03456	.03963	.04531	.05163	.05861	.06628
21	9	4	.05188	.05696	.06252	.06858	.07516	.08228	.08995

Coefficients a_i for the BLUE of μ
Shape parameter k

n	s	i	1.4	1.5	1.6	1.7	1.8	1.9	2.0
21	9	5	.06427	.06917	.07444	.08010	.08616	.09262	.09949
21	9	6	.07179	.07621	.08091	.08590	.09117	.09674	.10260
21	9	7	.07645	.08029	.08431	.08854	.09297	.09760	.10244
21	9	8	.07931	.08251	.08584	.08931	.09290	.09664	.10051
21	9	9	.08093	.08351	.08616	.08889	.09171	.09460	.09758
21	9	10	.08169	.08365	.08566	.08771	.08980	.09193	.09411
21	9	11	.08179	.08317	.08457	.08599	.08742	.08887	.09035
21	9	12	.68598	.68728	.68867	.69015	.69174	.69343	.69524
21	10	1	-.34985	-.38840	-.42990	-.47444	-.52215	-.57314	-.62750
21	10	2	-.02825	-.02644	-.02403	-.02099	-.01730	-.01289	-.00775
21	10	3	.02485	.02986	.03546	.04170	.04860	.05619	.06448
21	10	4	.05107	.05668	.06280	.06943	.07659	.08430	.09257
21	10	5	.06612	.07151	.07728	.08344	.09000	.09697	.10436
21	10	6	.07538	.08022	.08534	.09074	.09642	.10240	.10867
21	10	7	.08124	.08541	.08976	.09431	.09905	.10399	.10914
21	10	8	.08494	.08838	.09195	.09565	.09947	.10342	.10750
21	10	9	.08715	.08989	.09269	.09557	.09852	.10155	.10465
21	10	10	.08831	.09035	.09243	.09455	.09670	.09889	.10113
21	10	11	.81905	.82254	.82620	.83005	.83409	.83832	.84275
21	11	1	-.42398	-.46710	-.51332	-.56274	-.61547	-.67161	-.73126
21	11	2	-.04478	-.04266	-.03992	-.03652	-.03241	-.02756	-.02193
21	11	3	.01879	.02442	.03069	.03762	.04523	.05355	.06260
21	11	4	.05040	.05668	.06347	.07079	.07866	.08708	.09607
21	11	5	.06871	.07470	.08108	.08784	.09501	.10258	.11058
21	11	6	.08011	.08545	.09106	.09695	.10312	.10958	.11633
21	11	7	.08743	.09198	.09671	.10163	.10674	.11204	.11754
21	11	8	.09213	.09586	.09969	.10365	.10772	.11191	.11623
21	11	9	.09505	.09795	.10092	.10395	.10705	.11022	.11346
21	11	10	.97613	.98272	.98962	.99683	1.00436	1.01221	1.02038
22	0	1	-.03356	-.05022	-.06907	-.09024	-.11384	-.14001	-.16883
22	0	2	.04439	.04524	.04640	.04790	.04978	.05207	.05482
22	0	3	.05300	.05522	.05785	.06089	.06440	.06838	.07287
22	0	4	.05615	.05869	.06160	.06490	.06860	.07273	.07730
22	0	5	.05720	.05970	.06252	.06565	.06913	.07296	.07715
22	0	6	.05726	.05958	.06215	.06498	.06809	.07147	.07514
22	0	7	.05679	.05886	.06114	.06362	.06631	.06922	.07234
22	0	8	.05601	.05782	.05978	.06190	.06419	.06663	.06922
22	0	9	.05505	.05659	.05824	.06002	.06190	.06390	.06602
22	0	10	.05397	.05525	.05661	.05805	.05957	.06117	.06284
22	0	11	.05282	.05385	.05493	.05606	.05724	.05847	.05974
22	0	12	.05162	.05241	.05322	.05407	.05494	.05583	.05675
22	0	13	.05039	.05095	.05152	.05210	.05269	.05328	.05389
22	0	14	.04912	.04947	.04981	.05015	.05048	.05081	.05114
22	0	15	.04783	.04797	.04811	.04823	.04833	.04843	.04851
22	0	16	.04650	.04646	.04640	.04632	.04623	.04611	.04598
22	0	17	.04513	.04492	.04468	.04443	.04415	.04386	.04355
22	0	18	.04370	.04333	.04294	.04252	.04209	.04164	.04118
22	0	19	.04219	.04167	.04114	.04058	.04002	.03944	.03886
22	0	20	.04052	.03988	.03922	.03855	.03788	.03720	.03653
22	0	21	.03856	.03780	.03705	.03629	.03554	.03481	.03408
22	0	22	.03537	.03456	.03379	.03303	.03230	.03159	.03090
22	1	1	-.04716	-.06500	-.08511	-.10761	-.13263	-.16028	-.19068
22	1	2	.04082	.04164	.04279	.04429	.04619	.04852	.05132
22	1	3	.05131	.05364	.05637	.05954	.06318	.06732	.07199
22	1	4	.05546	.05813	.06119	.06464	.06851	.07283	.07760
22	1	5	.05713	.05977	.06273	.06603	.06968	.07368	.07806

Coefficients a_i for the BLUE of μ
Shape parameter k

n	s	i	1.4	1.5	1.6	1.7	1.8	1.9	2.0
22	1	6	.05760	.06005	.06277	.06575	.06902	.07256	.07640
22	1	7	.05742	.05962	.06202	.06464	.06748	.07053	.07380
22	1	8	.05685	.05877	.06085	.06310	.06550	.06807	.07079
22	1	9	.05605	.05769	.05945	.06132	.06331	.06542	.06764
22	1	10	.05509	.05646	.05790	.05943	.06103	.06272	.06448
22	1	11	.05403	.05513	.05628	.05748	.05873	.06003	.06137
22	1	12	.05290	.05374	.05461	.05551	.05644	.05739	.05836
22	1	13	.05172	.05232	.05293	.05355	.05418	.05481	.05546
22	1	14	.05049	.05086	.05123	.05160	.05196	.05231	.05267
22	1	15	.04921	.04938	.04952	.04966	.04978	.04989	.04999
22	1	16	.04790	.04786	.04781	.04773	.04764	.04753	.04741
22	1	17	.04653	.04631	.04607	.04581	.04552	.04523	.04491
22	1	18	.04510	.04471	.04430	.04387	.04342	.04296	.04249
22	1	19	.04356	.04301	.04246	.04188	.04130	.04071	.04011
22	1	20	.04185	.04117	.04048	.03979	.03910	.03841	.03773
22	1	21	.07614	.07475	.07337	.07201	.07068	.06937	.06809
22	2	1	-.06295	-.08207	-.10350	-.12739	-.15386	-.18303	-.21501
22	2	2	.03672	.03753	.03868	.04020	.04214	.04453	.04741
22	2	3	.04940	.05184	.05470	.05801	.06182	.06613	.07099
22	2	4	.05470	.05753	.06074	.06437	.06843	.07294	.07792
22	2	5	.05709	.05988	.06300	.06647	.07030	.07450	.07908
22	2	6	.05804	.06063	.06350	.06664	.07007	.07379	.07780
22	2	7	.05819	.06052	.06306	.06582	.06880	.07200	.07542
22	2	8	.05787	.05991	.06211	.06447	.06699	.06968	.07254
22	2	9	.05725	.05899	.06084	.06282	.06491	.06712	.06944
22	2	10	.05643	.05787	.05940	.06101	.06269	.06446	.06630
22	2	11	.05548	.05663	.05784	.05911	.06042	.06178	.06318
22	2	12	.05442	.05531	.05622	.05717	.05813	.05913	.06014
22	2	13	.05329	.05392	.05456	.05521	.05587	.05653	.05720
22	2	14	.05210	.05249	.05287	.05325	.05363	.05400	.05436
22	2	15	.05085	.05101	.05116	.05130	.05142	.05153	.05162
22	2	16	.04954	.04950	.04943	.04934	.04924	.04912	.04898
22	2	17	.04817	.04793	.04766	.04738	.04708	.04676	.04643
22	2	18	.04671	.04629	.04585	.04539	.04492	.04443	.04394
22	2	19	.04513	.04455	.04395	.04335	.04273	.04211	.04149
22	2	20	.12157	.11974	.11791	.11609	.11429	.11251	.11075
22	3	1	-.08105	-.10155	-.12443	-.14982	-.17784	-.20863	-.24229
22	3	2	.03206	.03287	.03404	.03560	.03760	.04006	.04303
22	3	3	.04726	.04983	.05284	.05632	.06030	.06482	.06989
22	3	4	.05388	.05688	.06027	.06409	.06835	.07308	.07829
22	3	5	.05708	.06005	.06335	.06701	.07103	.07543	.08022
22	3	6	.05858	.06134	.06437	.06768	.07128	.07518	.07938
22	3	7	.05912	.06159	.06428	.06719	.07031	.07367	.07725
22	3	8	.05908	.06124	.06356	.06605	.06870	.07151	.07449
22	3	9	.05867	.06051	.06246	.06454	.06673	.06904	.07147
22	3	10	.05800	.05953	.06113	.06282	.06458	.06642	.06834
22	3	11	.05716	.05838	.05965	.06097	.06234	.06375	.06521
22	3	12	.05619	.05712	.05808	.05906	.06006	.06109	.06214
22	3	13	.05512	.05578	.05644	.05711	.05778	.05847	.05915
22	3	14	.05397	.05437	.05476	.05514	.05552	.05589	.05626
22	3	15	.05274	.05290	.05304	.05316	.05327	.05337	.05346
22	3	16	.05144	.05137	.05128	.05117	.05104	.05090	.05075
22	3	17	.05006	.04978	.04948	.04916	.04883	.04848	.04812
22	3	18	.04857	.04810	.04761	.04711	.04660	.04608	.04555
22	3	19	.17206	.16993	.16779	.16564	.16351	.16139	.15929
22	4	1	-.10173	-.12377	-.14824	-.17527	-.20500	-.23754	-.27301

Coefficients a_i for the BLUE of μ
Shape parameter k

n	s	i	1.4	1.5	1.6	1.7	1.8	1.9	2.0
22	4	2	.02679	.02761	.02881	.03042	.03249	.03504	.03813
22	4	3	.04487	.04759	.05077	.05444	.05862	.06336	.06866
22	4	4	.05300	.05618	.05978	.06381	.06830	.07326	.07872
22	4	5	.05713	.06029	.06379	.06765	.07189	.07651	.08152
22	4	6	.05925	.06218	.06540	.06890	.07269	.07678	.08117
22	4	7	.06023	.06286	.06570	.06877	.07206	.07557	.07931
22	4	8	.06051	.06280	.06525	.06787	.07065	.07360	.07671
22	4	9	.06033	.06228	.06434	.06652	.06882	.07123	.07375
22	4	10	.05984	.06145	.06314	.06490	.06674	.06866	.07065
22	4	11	.05913	.06041	.06174	.06311	.06454	.06600	.06751
22	4	12	.05825	.05922	.06021	.06123	.06227	.06332	.06440
22	4	13	.05725	.05792	.05860	.05928	.05997	.06066	.06136
22	4	14	.05614	.05653	.05692	.05730	.05767	.05804	.05840
22	4	15	.05493	.05507	.05519	.05529	.05538	.05546	.05552
22	4	16	.05363	.05353	.05340	.05325	.05309	.05292	.05273
22	4	17	.05223	.05190	.05155	.05118	.05080	.05041	.05002
22	4	18	.22822	.22595	.22365	.22134	.21903	.21673	.21444
22	5	1	-.12541	-.14915	-.17539	-.20425	-.23586	-.27033	-.30780
22	5	2	.02081	.02165	.02290	.02458	.02673	.02939	.03261
22	5	3	.04219	.04509	.04847	.05235	.05676	.06174	.06731
22	5	4	.05205	.05545	.05928	.06355	.06829	.07351	.07924
22	5	5	.05725	.06062	.06434	.06844	.07291	.07776	.08302
22	5	6	.06008	.06321	.06662	.07032	.07432	.07862	.08322
22	5	7	.06156	.06436	.06738	.07061	.07407	.07776	.08167
22	5	8	.06220	.06463	.06723	.06999	.07291	.07599	.07924
22	5	9	.06229	.06435	.06653	.06882	.07122	.07374	.07637
22	5	10	.06200	.06369	.06546	.06731	.06923	.07122	.07328
22	5	11	.06143	.06277	.06416	.06558	.06706	.06857	.07012
22	5	12	.06066	.06166	.06268	.06373	.06479	.06587	.06697
22	5	13	.05973	.06041	.06109	.06178	.06247	.06317	.06387
22	5	14	.05866	.05904	.05941	.05978	.06013	.06048	.06083
22	5	15	.05747	.05757	.05766	.05773	.05778	.05783	.05786
22	5	16	.05616	.05600	.05583	.05563	.05542	.05521	.05498
22	5	17	.29088	.28863	.28636	.28406	.28177	.27948	.27720
22	6	1	-.15264	-.17830	-.20652	-.23742	-.27113	-.30776	-.34744
22	6	2	.01401	.01489	.01619	.01795	.02021	.02300	.02637
22	6	3	.03919	.04230	.04590	.05002	.05470	.05996	.06582
22	6	4	.05104	.05469	.05877	.06332	.06834	.07385	.07988
22	6	5	.05746	.06107	.06505	.06939	.07412	.07925	.08478
22	6	6	.06111	.06446	.06809	.07202	.07624	.08076	.08560
22	6	7	.06316	.06615	.06935	.07278	.07643	.08030	.08440
22	6	8	.06421	.06680	.06955	.07246	.07553	.07876	.08216
22	6	9	.06460	.06679	.06908	.07149	.07401	.07663	.07937
22	6	10	.06453	.06632	.06818	.07010	.07210	.07417	.07630
22	6	11	.06413	.06553	.06697	.06845	.06997	.07153	.07312
22	6	12	.06348	.06451	.06555	.06662	.06770	.06880	.06992
22	6	13	.06262	.06330	.06399	.06467	.06536	.06604	.06673
22	6	14	.06160	.06195	.06230	.06264	.06296	.06329	.06360
22	6	15	.06042	.06047	.06051	.06053	.06054	.06054	.06054
22	6	16	.36108	.35907	.35703	.35498	.35292	.35088	.34885
22	7	1	-.18414	-.21197	-.24243	-.27563	-.31171	-.35078	-.39296
22	7	2	.00623	.00717	.00854	.01041	.01279	.01574	.01928
22	7	3	.03581	.03916	.04303	.04743	.05241	.05798	.06418
22	7	4	.04997	.05390	.05828	.06313	.06847	.07431	.08068
22	7	5	.05780	.06169	.06594	.07057	.07560	.08102	.08685
22	7	6	.06239	.06598	.06987	.07404	.07851	.08329	.08838

Coefficients a_i for the BLUE of μ
Shape parameter k

n	s	i	1.4	1.5	1.6	1.7	1.8	1.9	2.0
22	7	7	.06510	.06830	.07171	.07534	.07920	.08327	.08758
22	7	8	.06662	.06939	.07230	.07538	.07861	.08200	.08555
22	7	9	.06736	.06968	.07210	.07463	.07726	.08000	.08285
22	7	10	.06755	.06942	.07137	.07338	.07545	.07760	.07981
22	7	11	.06732	.06878	.07027	.07180	.07336	.07496	.07660
22	7	12	.06680	.06785	.06891	.06999	.07109	.07220	.07332
22	7	13	.06602	.06669	.06736	.06803	.06870	.06937	.07005
22	7	14	.06504	.06536	.06566	.06596	.06624	.06653	.06680
22	7	15	.44013	.43861	.43708	.43554	.43402	.43251	.43103
22	8	1	-.22087	-.25118	-.28419	-.32003	-.35881	-.40065	-.44567
22	8	2	-.00272	-.00172	-.00023	.00176	.00429	.00742	.01117
22	8	3	.03199	.03563	.03980	.04453	.04985	.05579	.06237
22	8	4	.04884	.05310	.05782	.06302	.06872	.07494	.08169
22	8	5	.05831	.06252	.06709	.07205	.07740	.08316	.08933
22	8	6	.06399	.06787	.07203	.07649	.08124	.08630	.09167
22	8	7	.06747	.07090	.07455	.07841	.08249	.08680	.09133
22	8	8	.06954	.07249	.07559	.07884	.08225	.08581	.08953
22	8	9	.07067	.07313	.07569	.07835	.08111	.08397	.08694
22	8	10	.07115	.07312	.07515	.07725	.07941	.08163	.08391
22	8	11	.07113	.07264	.07418	.07575	.07735	.07899	.08066
22	8	12	.07074	.07181	.07288	.07397	.07506	.07617	.07730
22	8	13	.07005	.07070	.07134	.07198	.07262	.07326	.07390
22	8	14	.52970	.52900	.52831	.52764	.52700	.52641	.52587
22	9	1	-.26411	-.29728	-.33323	-.37210	-.41400	-.45905	-.50735
22	9	2	-.01312	-.01201	-.01040	-.00825	-.00552	-.00218	.00181
22	9	3	.02765	.03163	.03617	.04128	.04700	.05336	.06038
22	9	4	.04766	.05230	.05742	.06303	.06915	.07580	.08299
22	9	5	.05907	.06363	.06857	.07391	.07964	.08578	.09234
22	9	6	.06603	.07022	.07470	.07948	.08455	.08993	.09562
22	9	7	.07039	.07409	.07800	.08212	.08646	.09102	.09580
22	9	8	.07310	.07626	.07956	.08301	.08661	.09037	.09427
22	9	9	.07469	.07730	.08001	.08281	.08570	.08870	.09179
22	9	10	.07549	.07757	.07969	.08188	.08412	.08642	.08877
22	9	11	.07571	.07727	.07885	.08046	.08209	.08376	.08546
22	9	12	.07548	.07654	.07761	.07869	.07978	.08088	.08199
22	9	13	.63197	.63248	.63306	.63370	.63442	.63523	.63612
22	10	1	-.31562	-.35212	-.39151	-.43393	-.47947	-.52826	-.58039
22	10	2	-.02532	-.02407	-.02229	-.01994	-.01698	-.01337	-.00909
22	10	3	.02268	.02707	.03204	.03761	.04381	.05066	.05820
22	10	4	.04645	.05155	.05713	.06322	.06983	.07699	.08469
22	10	5	.06015	.06514	.07052	.07629	.08246	.08904	.09605
22	10	6	.06863	.07319	.07804	.08319	.08863	.09438	.10044
22	10	7	.07405	.07805	.08225	.08667	.09130	.09614	.10121
22	10	8	.07750	.08089	.08442	.08810	.09191	.09588	.09999
22	10	9	.07963	.08241	.08527	.08822	.09126	.09440	.09763
22	10	10	.08081	.08299	.08521	.08748	.08980	.09218	.09461
22	10	11	.08129	.08289	.08450	.08614	.08780	.08949	.09121
22	10	12	.74975	.75201	.75441	.75695	.75963	.76247	.76545
22	11	1	-.37786	-.41830	-.46176	-.50837	-.55823	-.61146	-.66815
22	11	2	-.03981	-.03837	-.03636	-.03375	-.03050	-.02657	-.02193
22	11	3	.01693	.02183	.02732	.03344	.04022	.04767	.05583
22	11	4	.04524	.05089	.05704	.06370	.07090	.07864	.08696
22	11	5	.06171	.06721	.07309	.07938	.08607	.09318	.10071
22	11	6	.07202	.07701	.08229	.08787	.09374	.09991	.10639
22	11	7	.07869	.08304	.08759	.09234	.09731	.10248	.10787
22	11	8	.08303	.08669	.09048	.09440	.09846	.10266	.10700

Coefficients a_i for the BLUE of μ
Shape parameter k

n	s	i	1.4	1.5	1.6	1.7	1.8	1.9	2.0
22	11	9	.08579	.08875	.09178	.09490	.09810	.10139	.10477
22	11	10	.08742	.08969	.09201	.09437	.09677	.09922	.10173
22	11	11	.88684	.89156	.89652	.90173	.90717	.91287	.91881
23	0	1	-.03495	-.05157	-.07039	-.09154	-.11514	-.14132	-.17018
23	0	2	.04150	.04214	.04306	.04430	.04589	.04787	.05026
23	0	3	.05016	.05223	.05468	.05753	.06082	.06458	.06883
23	0	4	.05344	.05587	.05865	.06182	.06538	.06937	.07379
23	0	5	.05464	.05706	.05979	.06285	.06625	.06999	.07410
23	0	6	.05484	.05711	.05964	.06243	.06550	.06885	.07250
23	0	7	.05450	.05655	.05882	.06130	.06399	.06691	.07005
23	0	8	.05385	.05567	.05765	.05979	.06211	.06459	.06724
23	0	9	.05302	.05459	.05628	.05810	.06004	.06211	.06430
23	0	10	.05206	.05339	.05481	.05631	.05791	.05959	.06136
23	0	11	.05103	.05212	.05328	.05449	.05576	.05710	.05848
23	0	12	.04995	.05082	.05172	.05266	.05364	.05465	.05569
23	0	13	.04884	.04949	.05016	.05084	.05155	.05227	.05301
23	0	14	.04769	.04814	.04859	.04905	.04950	.04996	.05043
23	0	15	.04652	.04678	.04703	.04727	.04750	.04773	.04795
23	0	16	.04532	.04540	.04547	.04552	.04555	.04557	.04558
23	0	17	.04410	.04401	.04390	.04377	.04363	.04347	.04329
23	0	18	.04283	.04259	.04233	.04204	.04173	.04141	.04108
23	0	19	.04151	.04113	.04072	.04029	.03985	.03939	.03892
23	0	20	.04011	.03959	.03906	.03851	.03794	.03737	.03680
23	0	21	.03856	.03793	.03728	.03663	.03597	.03532	.03466
23	0	22	.03673	.03600	.03527	.03454	.03382	.03311	.03241
23	0	23	.03374	.03297	.03222	.03150	.03080	.03011	.02946
23	1	1	-.04763	-.06537	-.08538	-.10779	-.13273	-.16032	-.19068
23	1	2	.03809	.03869	.03959	.04082	.04241	.04440	.04683
23	1	3	.04851	.05065	.05319	.05615	.05956	.06346	.06786
23	1	4	.05271	.05525	.05817	.06147	.06519	.06934	.07394
23	1	5	.05448	.05703	.05990	.06310	.06665	.07057	.07485
23	1	6	.05507	.05746	.06012	.06306	.06628	.06978	.07359
23	1	7	.05501	.05718	.05957	.06217	.06500	.06805	.07134
23	1	8	.05456	.05648	.05857	.06083	.06326	.06587	.06864
23	1	9	.05388	.05554	.05733	.05925	.06130	.06347	.06577
23	1	10	.05304	.05445	.05595	.05754	.05922	.06099	.06285
23	1	11	.05210	.05326	.05448	.05577	.05711	.05851	.05997
23	1	12	.05109	.05201	.05297	.05396	.05500	.05606	.05717
23	1	13	.05002	.05071	.05143	.05216	.05291	.05367	.05445
23	1	14	.04891	.04939	.04988	.05036	.05085	.05134	.05184
23	1	15	.04777	.04805	.04832	.04858	.04883	.04908	.04932
23	1	16	.04659	.04668	.04675	.04681	.04685	.04688	.04690
23	1	17	.04537	.04528	.04517	.04505	.04490	.04474	.04457
23	1	18	.04410	.04385	.04358	.04328	.04297	.04264	.04230
23	1	19	.04277	.04237	.04195	.04150	.04105	.04058	.04010
23	1	20	.04134	.04080	.04025	.03968	.03910	.03851	.03793
23	1	21	.03975	.03909	.03843	.03775	.03708	.03641	.03575
23	1	22	.07247	.07113	.06980	.06849	.06721	.06595	.06472
23	2	1	-.06227	-.08121	-.10247	-.12619	-.15250	-.18153	-.21339
23	2	2	.03418	.03477	.03565	.03689	.03850	.04053	.04301
23	2	3	.04663	.04887	.05152	.05460	.05816	.06221	.06678
23	2	4	.05190	.05458	.05764	.06110	.06498	.06931	.07411
23	2	5	.05434	.05703	.06004	.06340	.06712	.07121	.07568
23	2	6	.05537	.05790	.06070	.06379	.06716	.07082	.07479
23	2	7	.05563	.05792	.06044	.06318	.06614	.06933	.07276
23	2	8	.05542	.05745	.05965	.06202	.06457	.06729	.07019

Coefficients a_i for the BLUE of μ
Shape parameter k

n	s	i	1.4	1.5	1.6	1.7	1.8	1.9	2.0
23	2	9	.05491	.05667	.05855	.06057	.06271	.06498	.06738
23	2	10	.05420	.05569	.05727	.05894	.06070	.06255	.06449
23	2	11	.05336	.05458	.05587	.05722	.05863	.06009	.06161
23	2	12	.05242	.05339	.05440	.05545	.05653	.05764	.05879
23	2	13	.05141	.05214	.05289	.05366	.05444	.05524	.05605
23	2	14	.05035	.05085	.05135	.05186	.05237	.05288	.05339
23	2	15	.04923	.04952	.04980	.05007	.05033	.05058	.05083
23	2	16	.04806	.04815	.04822	.04828	.04832	.04835	.04836
23	2	17	.04685	.04675	.04663	.04649	.04633	.04616	.04597
23	2	18	.04558	.04530	.04501	.04469	.04436	.04401	.04366
23	2	19	.04423	.04379	.04334	.04287	.04239	.04190	.04140
23	2	20	.04276	.04218	.04159	.04099	.04039	.03978	.03917
23	2	21	.11544	.11367	.11189	.11013	.10838	.10666	.10496
23	3	1	-.07896	-.09919	-.12181	-.14694	-.17472	-.20527	-.23871
23	3	2	.02977	.03034	.03123	.03248	.03413	.03621	.03877
23	3	3	.04453	.04688	.04966	.05289	.05660	.06082	.06558
23	3	4	.05102	.05385	.05707	.06070	.06477	.06929	.07429
23	3	5	.05421	.05706	.06024	.06377	.06767	.07194	.07660
23	3	6	.05576	.05843	.06139	.06463	.06816	.07200	.07614
23	3	7	.05637	.05880	.06145	.06433	.06744	.07078	.07435
23	3	8	.05642	.05857	.06090	.06339	.06606	.06890	.07192
23	3	9	.05611	.05797	.05996	.06207	.06432	.06669	.06918
23	3	10	.05556	.05713	.05879	.06054	.06238	.06431	.06633
23	3	11	.05483	.05612	.05747	.05887	.06034	.06187	.06345
23	3	12	.05397	.05499	.05605	.05714	.05826	.05942	.06060
23	3	13	.05303	.05379	.05457	.05536	.05617	.05699	.05783
23	3	14	.05200	.05252	.05304	.05356	.05408	.05461	.05513
23	3	15	.05092	.05121	.05149	.05176	.05202	.05227	.05252
23	3	16	.04977	.04984	.04990	.04994	.04997	.04999	.04999
23	3	17	.04856	.04843	.04828	.04812	.04794	.04775	.04754
23	3	18	.04727	.04696	.04663	.04628	.04592	.04554	.04516
23	3	19	.04589	.04541	.04491	.04440	.04389	.04336	.04283
23	3	20	.16297	.16089	.15879	.15670	.15461	.15255	.15050
23	4	1	-.09791	-.11958	-.14368	-.17035	-.19972	-.23191	-.26705
23	4	2	.02480	.02536	.02627	.02755	.02925	.03139	.03403
23	4	3	.04219	.04467	.04760	.05099	.05488	.05929	.06426
23	4	4	.05007	.05306	.05646	.06029	.06456	.06930	.07452
23	4	5	.05412	.05713	.06050	.06422	.06831	.07278	.07765
23	4	6	.05624	.05908	.06220	.06561	.06932	.07333	.07765
23	4	7	.05726	.05984	.06264	.06566	.06892	.07241	.07614
23	4	8	.05761	.05989	.06234	.06496	.06775	.07072	.07387
23	4	9	.05752	.05949	.06158	.06380	.06614	.06861	.07121
23	4	10	.05713	.05879	.06053	.06236	.06428	.06629	.06839
23	4	11	.05653	.05788	.05930	.06076	.06229	.06387	.06551
23	4	12	.05577	.05684	.05793	.05906	.06022	.06142	.06264
23	4	13	.05489	.05568	.05648	.05730	.05813	.05897	.05982
23	4	14	.05392	.05445	.05497	.05550	.05602	.05655	.05707
23	4	15	.05286	.05314	.05341	.05367	.05392	.05417	.05440
23	4	16	.05172	.05177	.05181	.05183	.05183	.05183	.05181
23	4	17	.05051	.05034	.05016	.04997	.04975	.04953	.04929
23	4	18	.04920	.04884	.04846	.04807	.04767	.04726	.04684
23	4	19	.21558	.21332	.21105	.20876	.20647	.20420	.20195
23	5	1	-.11947	-.14272	-.16846	-.19683	-.22794	-.26194	-.29893
23	5	2	.01920	.01977	.02069	.02201	.02377	.02600	.02874
23	5	3	.03958	.04222	.04531	.04888	.05297	.05760	.06280
23	5	4	.04904	.05222	.05583	.05986	.06436	.06933	.07480

Coefficients a_i for the BLUE of μ
Shape parameter k

n	s	i	1.4	1.5	1.6	1.7	1.8	1.9	2.0
23	5	5	.05406	.05727	.06083	.06476	.06907	.07376	.07886
23	5	6	.05684	.05986	.06316	.06676	.07066	.07486	.07938
23	5	7	.05832	.06106	.06402	.06721	.07063	.07428	.07818
23	5	8	.05901	.06142	.06401	.06676	.06969	.07279	.07607
23	5	9	.05917	.06125	.06345	.06578	.06823	.07081	.07351
23	5	10	.05897	.06071	.06254	.06446	.06646	.06855	.07072
23	5	11	.05851	.05993	.06140	.06293	.06451	.06615	.06783
23	5	12	.05785	.05896	.06010	.06126	.06246	.06369	.06494
23	5	13	.05705	.05786	.05868	.05951	.06036	.06121	.06208
23	5	14	.05613	.05665	.05718	.05770	.05823	.05875	.05927
23	5	15	.05510	.05536	.05561	.05586	.05609	.05631	.05653
23	5	16	.05397	.05399	.05399	.05397	.05395	.05391	.05386
23	5	17	.05274	.05253	.05230	.05206	.05180	.05153	.05126
23	5	18	.27394	.27167	.26937	.26705	.26473	.26242	.26012
23	6	1	-.14408	-.16910	-.19666	-.22691	-.25996	-.29595	-.33499
23	6	2	.01287	.01345	.01440	.01578	.01761	.01993	.02279
23	6	3	.03667	.03948	.04277	.04655	.05086	.05573	.06119
23	6	4	.04793	.05133	.05516	.05944	.06418	.06942	.07515
23	6	5	.05406	.05749	.06127	.06543	.06997	.07491	.08025
23	6	6	.05758	.06080	.06431	.06811	.07222	.07664	.08137
23	6	7	.05959	.06251	.06564	.06901	.07260	.07643	.08051
23	6	8	.06066	.06322	.06595	.06886	.07193	.07517	.07860
23	6	9	.06111	.06331	.06562	.06807	.07063	.07332	.07613
23	6	10	.06112	.06295	.06487	.06687	.06896	.07112	.07337
23	6	11	.06082	.06230	.06383	.06542	.06706	.06875	.07049
23	6	12	.06028	.06142	.06260	.06380	.06502	.06628	.06756
23	6	13	.05956	.06038	.06121	.06205	.06291	.06377	.06464
23	6	14	.05868	.05920	.05972	.06023	.06074	.06125	.06176
23	6	15	.05768	.05792	.05814	.05835	.05856	.05875	.05894
23	6	16	.05656	.05653	.05648	.05642	.05635	.05627	.05618
23	6	17	.33892	.33681	.33467	.33252	.33036	.32821	.32608
23	7	1	-.17232	-.19931	-.22893	-.26128	-.29651	-.33472	-.37605
23	7	2	.00568	.00628	.00728	.00873	.01065	.01308	.01607
23	7	3	.03341	.03643	.03993	.04395	.04852	.05366	.05941
23	7	4	.04673	.05038	.05447	.05902	.06405	.06957	.07561
23	7	5	.05414	.05781	.06185	.06626	.07107	.07627	.08189
23	7	6	.05851	.06196	.06569	.06972	.07406	.07871	.08368
23	7	7	.06113	.06423	.06756	.07112	.07491	.07893	.08320
23	7	8	.06263	.06535	.06824	.07130	.07453	.07793	.08150
23	7	9	.06340	.06572	.06817	.07073	.07342	.07622	.07914
23	7	10	.06365	.06558	.06759	.06968	.07185	.07410	.07643
23	7	11	.06352	.06507	.06667	.06831	.07000	.07174	.07353
23	7	12	.06311	.06429	.06550	.06673	.06798	.06925	.07056
23	7	13	.06248	.06331	.06415	.06499	.06584	.06670	.06757
23	7	14	.06166	.06216	.06266	.06315	.06364	.06412	.06461
23	7	15	.06068	.06088	.06106	.06123	.06139	.06154	.06169
23	7	16	.41159	.40986	.40811	.40636	.40461	.40288	.40117
23	8	1	-.20493	-.23416	-.26609	-.30083	-.33851	-.37924	-.42315
23	8	2	-.00253	-.00189	-.00083	.00070	.00273	.00530	.00844
23	8	3	.02974	.03300	.03676	.04105	.04592	.05137	.05744
23	8	4	.04545	.04938	.05377	.05862	.06397	.06982	.07620
23	8	5	.05433	.05828	.06260	.06730	.07240	.07791	.08383
23	8	6	.05968	.06338	.06736	.07165	.07625	.08116	.08638
23	8	7	.06298	.06631	.06985	.07362	.07763	.08186	.08634
23	8	8	.06499	.06789	.07095	.07418	.07758	.08115	.08489
23	8	9	.06612	.06859	.07117	.07386	.07667	.07960	.08265

Coefficients a_i for the BLUE of μ
Shape parameter k

n	s	i	1.4	1.5	1.6	1.7	1.8	1.9	2.0
23	8	10	.06664	.06868	.07078	.07297	.07522	.07756	.07997
23	8	11	.06672	.06833	.06999	.07169	.07343	.07522	.07705
23	8	12	.06645	.06766	.06889	.07014	.07141	.07271	.07403
23	8	13	.06591	.06674	.06757	.06841	.06925	.07010	.07096
23	8	14	.06515	.06562	.06608	.06654	.06699	.06745	.06789
23	8	15	.49330	.49222	.49114	.49008	.48904	.48804	.48707
23	9	1	-.24290	-.27469	-.30926	-.34673	-.38720	-.43081	-.47766
23	9	2	-.01197	-.01128	-.01013	-.00850	-.00633	-.00360	-.00028
23	9	3	.02559	.02913	.03319	.03781	.04301	.04882	.05526
23	9	4	.04408	.04834	.05307	.05828	.06399	.07021	.07698
23	9	5	.05466	.05893	.06358	.06861	.07405	.07990	.08617
23	9	6	.06116	.06514	.06941	.07399	.07887	.08407	.08959
23	9	7	.06526	.06882	.07261	.07662	.08086	.08534	.09005
23	9	8	.06784	.07094	.07419	.07761	.08120	.08495	.08887
23	9	9	.06939	.07201	.07474	.07757	.08052	.08358	.08676
23	9	10	.07022	.07236	.07457	.07685	.07920	.08162	.08412
23	9	11	.07052	.07220	.07391	.07567	.07746	.07930	.08118
23	9	12	.07041	.07165	.07290	.07416	.07545	.07675	.07808
23	9	13	.06998	.07079	.07161	.07243	.07325	.07408	.07491
23	9	14	.58575	.58567	.58562	.58561	.58567	.58579	.58598
23	10	1	-.28755	-.32230	-.35992	-.40052	-.44422	-.49114	-.54140
23	10	2	-.02293	-.02216	-.02090	-.01913	-.01680	-.01387	-.01033
23	10	3	.02087	.02474	.02916	.03415	.03975	.04598	.05286
23	10	4	.04262	.04727	.05240	.05801	.06415	.07081	.07801
23	10	5	.05521	.05985	.06487	.07029	.07611	.08235	.08901
23	10	6	.06304	.06735	.07195	.07686	.08207	.08760	.09345
23	10	7	.06807	.07191	.07597	.08025	.08476	.08950	.09447
23	10	8	.07132	.07464	.07811	.08174	.08553	.08949	.09360
23	10	9	.07336	.07615	.07903	.08202	.08512	.08832	.09164
23	10	10	.07455	.07680	.07912	.08149	.08394	.08645	.08903
23	10	11	.07510	.07684	.07861	.08042	.08226	.08414	.08605
23	10	12	.07517	.07642	.07768	.07895	.08024	.08154	.08287
23	10	13	.69116	.69250	.69392	.69545	.69709	.69885	.70073
23	11	1	-.34069	-.37888	-.42006	-.46433	-.51181	-.56261	-.61684
23	11	2	-.03579	-.03489	-.03349	-.03154	-.02900	-.02584	-.02203
23	11	3	.01545	.01973	.02457	.03002	.03608	.04280	.05019
23	11	4	.04109	.04619	.05179	.05789	.06452	.07169	.07941
23	11	5	.05605	.06113	.06659	.07245	.07872	.08541	.09253
23	11	6	.06546	.07015	.07514	.08042	.08601	.09192	.09815
23	11	7	.07160	.07575	.08013	.08472	.08953	.09457	.09984
23	11	8	.07563	.07920	.08292	.08678	.09081	.09498	.09932
23	11	9	.07825	.08121	.08427	.08743	.09069	.09405	.09752
23	11	10	.07984	.08221	.08464	.08712	.08966	.09227	.09494
23	11	11	.08069	.08248	.08430	.08615	.08803	.08995	.09190
23	11	12	.81241	.81572	.81921	.82288	.82675	.83081	.83506
23	12	1	-.40485	-.44711	-.49250	-.54111	-.59306	-.64847	-.70742
23	12	2	-.05105	-.05000	-.04839	-.04621	-.04341	-.03996	-.03582
23	12	3	.00917	.01394	.01931	.02530	.03194	.03924	.04724
23	12	4	.03950	.04516	.05132	.05800	.06521	.07298	.08132
23	12	5	.05733	.06292	.06890	.07528	.08208	.08931	.09697
23	12	6	.06863	.07377	.07920	.08493	.09097	.09732	.10399
23	12	7	.07608	.08061	.08534	.09029	.09545	.10084	.10646
23	12	8	.08106	.08491	.08890	.09304	.09732	.10175	.10634
23	12	9	.08434	.08751	.09076	.09410	.09754	.10107	.10471
23	12	10	.08642	.08891	.09145	.09404	.09668	.09939	.10215
23	12	11	.95337	.95939	.96572	.97234	.97928	.98652	.99407

Coefficients a_i for the BLUE of μ
Shape parameter k

n	s	i	1.4	1.5	1.6	1.7	1.8	1.9	2.0
24	0	1	-.03617	-.05274	-.07152	-.09264	-.11623	-.14240	-.17128
24	0	2	.03884	.03930	.04001	.04100	.04232	.04399	.04605
24	0	3	.04756	.04947	.05175	.05442	.05751	.06105	.06507
24	0	4	.05094	.05326	.05592	.05895	.06238	.06622	.07049
24	0	5	.05226	.05461	.05726	.06023	.06354	.06720	.07123
24	0	6	.05259	.05481	.05729	.06004	.06307	.06638	.06999
24	0	7	.05237	.05440	.05665	.05911	.06180	.06472	.06786
24	0	8	.05184	.05365	.05564	.05780	.06013	.06264	.06533
24	0	9	.05112	.05270	.05443	.05628	.05827	.06039	.06265
24	0	10	.05027	.05163	.05310	.05466	.05632	.05808	.05993
24	0	11	.04935	.05049	.05171	.05299	.05434	.05576	.05725
24	0	12	.04837	.04930	.05028	.05130	.05237	.05348	.05464
24	0	13	.04736	.04809	.04884	.04962	.05043	.05126	.05212
24	0	14	.04632	.04685	.04740	.04796	.04853	.04910	.04969
24	0	15	.04525	.04561	.04596	.04631	.04666	.04701	.04736
24	0	16	.04417	.04435	.04453	.04469	.04484	.04498	.04511
24	0	17	.04306	.04308	.04309	.04308	.04305	.04301	.04296
24	0	18	.04192	.04180	.04165	.04148	.04129	.04109	.04088
24	0	19	.04075	.04048	.04019	.03988	.03956	.03921	.03886
24	0	20	.03952	.03913	.03871	.03827	.03782	.03736	.03689
24	0	21	.03821	.03770	.03717	.03662	.03607	.03551	.03494
24	0	22	.03677	.03615	.03552	.03489	.03425	.03361	.03297
24	0	23	.03507	.03435	.03364	.03294	.03225	.03156	.03089
24	0	24	.03225	.03151	.03080	.03010	.02942	.02877	.02814
24	1	1	-.04802	-.06565	-.08556	-.10788	-.13274	-.16026	-.19056
24	1	2	.03558	.03599	.03666	.03763	.03894	.04061	.04269
24	1	3	.04593	.04791	.05026	.05302	.05622	.05988	.06404
24	1	4	.05018	.05260	.05537	.05854	.06210	.06609	.07053
24	1	5	.05204	.05450	.05727	.06038	.06384	.06765	.07184
24	1	6	.05273	.05506	.05767	.06055	.06371	.06717	.07093
24	1	7	.05277	.05491	.05727	.05986	.06267	.06572	.06900
24	1	8	.05243	.05435	.05644	.05871	.06115	.06378	.06659
24	1	9	.05186	.05354	.05535	.05730	.05939	.06161	.06397
24	1	10	.05112	.05257	.05411	.05575	.05750	.05934	.06128
24	1	11	.05029	.05150	.05278	.05413	.05556	.05705	.05861
24	1	12	.04938	.05036	.05140	.05248	.05361	.05478	.05599
24	1	13	.04841	.04919	.04999	.05082	.05167	.05255	.05345
24	1	14	.04741	.04798	.04857	.04916	.04976	.05037	.05099
24	1	15	.04638	.04676	.04713	.04751	.04789	.04826	.04863
24	1	16	.04531	.04551	.04570	.04588	.04604	.04620	.04635
24	1	17	.04421	.04424	.04426	.04425	.04423	.04420	.04415
24	1	18	.04308	.04295	.04280	.04263	.04245	.04224	.04203
24	1	19	.04190	.04163	.04133	.04101	.04068	.04033	.03997
24	1	20	.04067	.04025	.03982	.03937	.03891	.03843	.03795
24	1	21	.03934	.03880	.03825	.03769	.03712	.03654	.03596
24	1	22	.03786	.03721	.03656	.03591	.03525	.03460	.03396
24	1	23	.06913	.06784	.06655	.06529	.06406	.06285	.06166
24	2	1	-.06164	-.08040	-.10150	-.12506	-.15122	-.18010	-.21183
24	2	2	.03186	.03224	.03289	.03385	.03516	.03686	.03897
24	2	3	.04409	.04615	.04860	.05147	.05479	.05859	.06289
24	2	4	.04934	.05187	.05478	.05808	.06180	.06595	.07056
24	2	5	.05182	.05440	.05731	.06056	.06417	.06815	.07250
24	2	6	.05292	.05538	.05812	.06114	.06444	.06805	.07197
24	2	7	.05327	.05552	.05801	.06071	.06366	.06683	.07025
24	2	8	.05315	.05517	.05737	.05974	.06230	.06504	.06797
24	2	9	.05274	.05451	.05642	.05846	.06064	.06296	.06542

Coefficients a_i for the BLUE of μ
Shape parameter k

n	s	i	1.4	1.5	1.6	1.7	1.8	1.9	2.0
24	2	10	.05213	.05365	.05528	.05700	.05882	.06075	.06277
24	2	11	.05139	.05267	.05402	.05544	.05692	.05848	.06010
24	2	12	.05056	.05160	.05268	.05382	.05499	.05621	.05748
24	2	13	.04965	.05047	.05131	.05217	.05306	.05398	.05491
24	2	14	.04869	.04929	.04990	.05052	.05115	.05178	.05243
24	2	15	.04769	.04808	.04848	.04887	.04926	.04964	.05002
24	2	16	.04664	.04685	.04704	.04722	.04739	.04755	.04771
24	2	17	.04556	.04558	.04559	.04558	.04555	.04551	.04546
24	2	18	.04442	.04428	.04412	.04393	.04373	.04352	.04329
24	2	19	.04324	.04294	.04262	.04228	.04193	.04156	.04118
24	2	20	.04198	.04154	.04108	.04060	.04012	.03962	.03912
24	2	21	.04062	.04005	.03946	.03887	.03828	.03768	.03708
24	2	22	.10989	.10817	.10644	.10473	.10304	.10137	.09973
24	3	1	-.07709	-.09707	-.11944	-.14433	-.17187	-.20219	-.23541
24	3	2	.02768	.02803	.02867	.02963	.03096	.03269	.03486
24	3	3	.04203	.04419	.04675	.04975	.05320	.05715	.06162
24	3	4	.04841	.05108	.05413	.05759	.06147	.06581	.07060
24	3	5	.05159	.05432	.05738	.06079	.06456	.06871	.07325
24	3	6	.05318	.05577	.05865	.06182	.06528	.06904	.07312
24	3	7	.05386	.05624	.05885	.06169	.06477	.06808	.07164
24	3	8	.05399	.05612	.05843	.06093	.06360	.06646	.06950
24	3	9	.05377	.05563	.05764	.05978	.06206	.06448	.06703
24	3	10	.05330	.05491	.05661	.05841	.06032	.06232	.06442
24	3	11	.05267	.05402	.05543	.05691	.05846	.06008	.06176
24	3	12	.05192	.05301	.05415	.05533	.05655	.05782	.05912
24	3	13	.05108	.05193	.05281	.05370	.05463	.05557	.05654
24	3	14	.05017	.05079	.05142	.05206	.05270	.05336	.05402
24	3	15	.04919	.04960	.05000	.05040	.05079	.05119	.05157
24	3	16	.04817	.04837	.04856	.04874	.04890	.04906	.04921
24	3	17	.04709	.04710	.04710	.04707	.04703	.04698	.04692
24	3	18	.04596	.04579	.04560	.04540	.04518	.04494	.04469
24	3	19	.04476	.04443	.04407	.04371	.04332	.04293	.04253
24	3	20	.04347	.04299	.04249	.04198	.04146	.04093	.04040
24	3	21	.15478	.15275	.15070	.14865	.14662	.14460	.14261
24	4	1	-.09455	-.11587	-.13962	-.16595	-.19499	-.22685	-.26167
24	4	2	.02299	.02332	.02395	.02493	.02629	.02806	.03029
24	4	3	.03975	.04202	.04471	.04784	.05146	.05557	.06023
24	4	4	.04741	.05023	.05344	.05707	.06114	.06566	.07066
24	4	5	.05138	.05427	.05749	.06107	.06503	.06936	.07409
24	4	6	.05350	.05625	.05928	.06260	.06623	.07016	.07441
24	4	7	.05457	.05709	.05983	.06282	.06603	.06950	.07320
24	4	8	.05498	.05723	.05966	.06227	.06507	.06805	.07122
24	4	9	.05496	.05693	.05904	.06128	.06366	.06618	.06883
24	4	10	.05466	.05635	.05813	.06002	.06200	.06408	.06626
24	4	11	.05415	.05556	.05704	.05858	.06019	.06187	.06361
24	4	12	.05349	.05463	.05582	.05704	.05831	.05962	.06096
24	4	13	.05272	.05360	.05451	.05544	.05639	.05736	.05835
24	4	14	.05186	.05250	.05314	.05380	.05445	.05512	.05579
24	4	15	.05092	.05133	.05173	.05213	.05252	.05291	.05330
24	4	16	.04992	.05011	.05028	.05045	.05060	.05075	.05088
24	4	17	.04885	.04883	.04880	.04875	.04869	.04862	.04853
24	4	18	.04771	.04750	.04728	.04704	.04679	.04652	.04625
24	4	19	.04648	.04611	.04571	.04530	.04488	.04445	.04402
24	4	20	.20424	.20201	.19976	.19751	.19525	.19301	.19079
24	5	1	-.11428	-.13708	-.16236	-.19027	-.22093	-.25448	-.29103
24	5	2	.01773	.01804	.01868	.01968	.02107	.02290	.02521

Coefficients a_i for the BLUE of μ
Shape parameter k

n	s	i	1.4	1.5	1.6	1.7	1.8	1.9	2.0
24	5	3	.03722	.03962	.04245	.04574	.04953	.05383	.05869
24	5	4	.04632	.04931	.05270	.05652	.06079	.06553	.07075
24	5	5	.05119	.05425	.05766	.06143	.06557	.07011	.07504
24	5	6	.05392	.05683	.06003	.06352	.06732	.07144	.07587
24	5	7	.05541	.05808	.06098	.06411	.06748	.07110	.07496
24	5	8	.05613	.05852	.06108	.06382	.06674	.06985	.07315
24	5	9	.05636	.05844	.06065	.06300	.06548	.06811	.07086
24	5	10	.05623	.05801	.05988	.06185	.06391	.06608	.06833
24	5	11	.05586	.05734	.05888	.06049	.06216	.06390	.06570
24	5	12	.05531	.05649	.05772	.05899	.06030	.06165	.06303
24	5	13	.05461	.05552	.05646	.05741	.05838	.05937	.06038
24	5	14	.05380	.05445	.05511	.05577	.05643	.05710	.05778
24	5	15	.05290	.05330	.05370	.05409	.05448	.05486	.05523
24	5	16	.05191	.05208	.05224	.05238	.05252	.05264	.05275
24	5	17	.05085	.05080	.05073	.05065	.05056	.05045	.05034
24	5	18	.04969	.04944	.04918	.04889	.04860	.04829	.04798
24	5	19	.25884	.25655	.25424	.25192	.24959	.24727	.24497
24	6	1	-.13666	-.16109	-.18806	-.21771	-.25017	-.28557	-.32402
24	6	2	.01182	.01212	.01277	.01380	.01524	.01714	.01953
24	6	3	.03441	.03695	.03995	.04342	.04740	.05191	.05699
24	6	4	.04515	.04832	.05192	.05595	.06044	.06541	.07088
24	6	5	.05103	.05428	.05789	.06187	.06623	.07099	.07615
24	6	6	.05444	.05753	.06092	.06460	.06860	.07290	.07753
24	6	7	.05641	.05925	.06231	.06561	.06915	.07293	.07696
24	6	8	.05750	.06002	.06272	.06560	.06866	.07191	.07534
24	6	9	.05799	.06018	.06251	.06497	.06757	.07030	.07316
24	6	10	.05806	.05993	.06189	.06395	.06610	.06834	.07068
24	6	11	.05784	.05939	.06100	.06267	.06440	.06620	.06805
24	6	12	.05740	.05864	.05991	.06122	.06256	.06395	.06537
24	6	13	.05679	.05773	.05868	.05966	.06065	.06165	.06268
24	6	14	.05604	.05669	.05735	.05802	.05868	.05935	.06002
24	6	15	.05517	.05556	.05595	.05632	.05669	.05705	.05742
24	6	16	.05420	.05434	.05447	.05458	.05469	.05478	.05486
24	6	17	.05313	.05304	.05293	.05281	.05267	.05252	.05237
24	6	18	.31929	.31711	.31490	.31268	.31045	.30823	.30602
24	7	1	-.16214	-.18839	-.21725	-.24884	-.28330	-.32075	-.36131
24	7	2	.00515	.00545	.00612	.00718	.00869	.01067	.01316
24	7	3	.03127	.03399	.03717	.04084	.04504	.04979	.05512
24	7	4	.04387	.04727	.05109	.05536	.06010	.06533	.07106
24	7	5	.05091	.05438	.05822	.06243	.06703	.07202	.07744
24	7	6	.05510	.05839	.06199	.06588	.07009	.07461	.07945
24	7	7	.05762	.06063	.06388	.06736	.07108	.07504	.07925
24	7	8	.05911	.06178	.06464	.06767	.07088	.07427	.07785
24	7	9	.05990	.06222	.06467	.06725	.06996	.07281	.07579
24	7	10	.06020	.06216	.06422	.06637	.06861	.07094	.07336
24	7	11	.06015	.06176	.06344	.06518	.06697	.06882	.07074
24	7	12	.05983	.06111	.06243	.06378	.06516	.06658	.06803
24	7	13	.05931	.06027	.06124	.06223	.06324	.06426	.06529
24	7	14	.05862	.05928	.05993	.06059	.06125	.06191	.06257
24	7	15	.05780	.05816	.05852	.05887	.05922	.05955	.05989
24	7	16	.05684	.05694	.05702	.05709	.05715	.05721	.05725
24	7	17	.38648	.38459	.38268	.38076	.37885	.37694	.37506
24	8	1	-.19133	-.21963	-.25059	-.28436	-.32106	-.36081	-.40373
24	8	2	-.00242	-.00211	-.00141	-.00029	.00129	.00337	.00597
24	8	3	.02776	.03067	.03407	.03798	.04243	.04744	.05305
24	8	4	.04249	.04613	.05022	.05476	.05978	.06530	.07133

Coefficients a_i for the BLUE of μ
Shape parameter k

n	s	i	1.4	1.5	1.6	1.7	1.8	1.9	2.0
24	8	5	.05085	.05457	.05866	.06313	.06799	.07327	.07895
24	8	6	.05593	.05945	.06327	.06740	.07184	.07660	.08169
24	8	7	.05908	.06229	.06574	.06941	.07333	.07749	.08190
24	8	8	.06102	.06387	.06689	.07008	.07345	.07701	.08075
24	8	9	.06215	.06461	.06719	.06990	.07274	.07571	.07881
24	8	10	.06271	.06478	.06693	.06918	.07151	.07393	.07644
24	8	11	.06285	.06454	.06628	.06808	.06994	.07185	.07382
24	8	12	.06267	.06400	.06535	.06674	.06815	.06960	.07108
24	8	13	.06225	.06323	.06421	.06521	.06622	.06725	.06829
24	8	14	.06163	.06228	.06292	.06356	.06420	.06484	.06549
24	8	15	.06084	.06117	.06149	.06181	.06212	.06242	.06272
24	8	16	.46151	.46015	.45878	.45741	.45606	.45474	.45345
24	9	1	-.22500	-.25560	-.28896	-.32519	-.36442	-.40677	-.45236
24	9	2	-.01105	-.01072	-.00997	-.00878	-.00711	-.00492	-.00218
24	9	3	.02380	.02695	.03060	.03478	.03951	.04483	.05077
24	9	4	.04100	.04493	.04931	.05415	.05949	.06534	.07171
24	9	5	.05088	.05489	.05926	.06402	.06919	.07476	.08076
24	9	6	.05698	.06076	.06484	.06923	.07393	.07896	.08431
24	9	7	.06085	.06429	.06795	.07185	.07599	.08037	.08500
24	9	8	.06332	.06635	.06955	.07293	.07648	.08021	.08412
24	9	9	.06484	.06744	.07017	.07302	.07599	.07910	.08232
24	9	10	.06568	.06786	.07012	.07247	.07490	.07741	.08001
24	9	11	.06604	.06780	.06961	.07148	.07340	.07536	.07739
24	9	12	.06602	.06738	.06877	.07019	.07163	.07310	.07460
24	9	13	.06571	.06669	.06768	.06868	.06969	.07071	.07174
24	9	14	.06515	.06578	.06639	.06701	.06762	.06824	.06886
24	9	15	.54577	.54521	.54467	.54417	.54371	.54330	.54294
24	10	1	-.26416	-.29740	-.33349	-.37253	-.41465	-.45997	-.50860
24	10	2	-.02097	-.02060	-.01979	-.01851	-.01672	-.01439	-.01149
24	10	3	.01933	.02275	.02670	.03119	.03626	.04193	.04823
24	10	4	.03939	.04365	.04837	.05357	.05927	.06550	.07226
24	10	5	.05104	.05537	.06007	.06517	.07068	.07660	.08295
24	10	6	.05832	.06240	.06677	.07146	.07645	.08178	.08743
24	10	7	.06303	.06672	.07063	.07478	.07916	.08379	.08866
24	10	8	.06610	.06934	.07274	.07631	.08006	.08398	.08808
24	10	9	.06806	.07083	.07371	.07671	.07984	.08308	.08645
24	10	10	.06924	.07153	.07391	.07636	.07889	.08150	.08419
24	10	11	.06984	.07167	.07356	.07549	.07746	.07949	.08156
24	10	12	.06999	.07139	.07281	.07425	.07572	.07721	.07873
24	10	13	.06980	.07078	.07176	.07275	.07375	.07476	.07578
24	10	14	.64099	.64159	.64226	.64301	.64384	.64476	.64578
24	11	1	-.31017	-.34646	-.38569	-.42798	-.47344	-.52219	-.57434
24	11	2	-.03248	-.03205	-.03115	-.02975	-.02782	-.02532	-.02223
24	11	3	.01423	.01798	.02228	.02714	.03260	.03868	.04540
24	11	4	.03766	.04230	.04742	.05304	.05916	.06582	.07303
24	11	5	.05138	.05608	.06117	.06665	.07255	.07887	.08562
24	11	6	.06005	.06446	.06917	.07419	.07953	.08519	.09118
24	11	7	.06573	.06970	.07390	.07833	.08299	.08789	.09304
24	11	8	.06950	.07297	.07660	.08040	.08436	.08850	.09281
24	11	9	.07198	.07492	.07798	.08114	.08443	.08783	.09135
24	11	10	.07353	.07595	.07845	.08101	.08365	.08636	.08915
24	11	11	.07441	.07632	.07827	.08026	.08230	.08438	.08651
24	11	12	.07476	.07618	.07762	.07909	.08057	.08207	.08360
24	11	13	.74942	.75164	.75399	.75648	.75912	.76192	.76487
24	12	1	-.36488	-.40472	-.44763	-.49370	-.54307	-.59583	-.65210
24	12	2	-.04598	-.04545	-.04442	-.04287	-.04076	-.03806	-.03472

Coefficients a_i for the BLUE of μ
Shape parameter k

n	s	i	1.4	1.5	1.6	1.7	1.8	1.9	2.0
24	12	3	.00837	.01252	.01724	.02254	.02846	.03503	.04225
24	12	4	.03580	.04091	.04650	.05260	.05922	.06639	.07411
24	12	5	.05198	.05712	.06266	.06860	.07495	.08173	.08895
24	12	6	.06229	.06709	.07219	.07761	.08334	.08939	.09577
24	12	7	.06912	.07342	.07794	.08269	.08767	.09289	.09835
24	12	8	.07372	.07745	.08133	.08538	.08960	.09397	.09852
24	12	9	.07680	.07994	.08318	.08653	.09000	.09357	.09726
24	12	10	.07879	.08135	.08396	.08665	.08940	.09222	.09512
24	12	11	.07998	.08197	.08399	.08604	.08813	.09027	.09244
24	12	12	.87400	.87842	.88306	.88794	.89306	.89843	.90404
25	0	1	-.03724	-.05375	-.07248	-.09357	-.11713	-.14328	-.17216
25	0	2	.03641	.03668	.03719	.03795	.03901	.04040	.04215
25	0	3	.04515	.04692	.04903	.05153	.05443	.05776	.06156
25	0	4	.04863	.05083	.05338	.05628	.05958	.06328	.06740
25	0	5	.05006	.05232	.05489	.05778	.06101	.06458	.06851
25	0	6	.05049	.05266	.05509	.05779	.06077	.06404	.06761
25	0	7	.05038	.05239	.05461	.05705	.05973	.06263	.06578
25	0	8	.04995	.05176	.05375	.05591	.05826	.06079	.06350
25	0	9	.04933	.05093	.05267	.05456	.05658	.05874	.06105
25	0	10	.04858	.04998	.05148	.05308	.05479	.05661	.05854
25	0	11	.04776	.04894	.05021	.05155	.05297	.05447	.05604
25	0	12	.04687	.04786	.04890	.05000	.05114	.05234	.05359
25	0	13	.04595	.04675	.04758	.04844	.04934	.05026	.05122
25	0	14	.04500	.04561	.04625	.04689	.04756	.04824	.04894
25	0	15	.04403	.04447	.04492	.04536	.04582	.04627	.04673
25	0	16	.04304	.04332	.04359	.04385	.04411	.04437	.04461
25	0	17	.04203	.04215	.04226	.04236	.04244	.04251	.04258
25	0	18	.04100	.04098	.04094	.04088	.04080	.04071	.04061
25	0	19	.03995	.03979	.03961	.03941	.03919	.03895	.03871
25	0	20	.03885	.03857	.03826	.03793	.03759	.03723	.03686
25	0	21	.03771	.03731	.03688	.03644	.03599	.03552	.03505
25	0	22	.03649	.03598	.03545	.03491	.03436	.03381	.03326
25	0	23	.03514	.03453	.03392	.03330	.03267	.03205	.03144
25	0	24	.03354	.03285	.03216	.03148	.03081	.03015	.02950
25	0	25	.03089	.03018	.02949	.02882	.02817	.02754	.02693
25	1	1	-.04835	-.06587	-.08567	-.10790	-.13267	-.16011	-.19034
25	1	2	.03329	.03351	.03397	.03470	.03574	.03711	.03886
25	1	3	.04355	.04537	.04756	.05013	.05312	.05656	.06047
25	1	4	.04784	.05014	.05279	.05581	.05923	.06307	.06734
25	1	5	.04978	.05215	.05483	.05784	.06120	.06492	.06900
25	1	6	.05056	.05283	.05538	.05820	.06131	.06472	.06842
25	1	7	.05069	.05280	.05513	.05769	.06048	.06351	.06678
25	1	8	.05045	.05236	.05444	.05671	.05916	.06180	.06463
25	1	9	.04997	.05166	.05349	.05546	.05758	.05984	.06224
25	1	10	.04933	.05080	.05238	.05406	.05586	.05776	.05977
25	1	11	.04858	.04984	.05117	.05258	.05407	.05564	.05728
25	1	12	.04777	.04881	.04991	.05106	.05227	.05353	.05484
25	1	13	.04690	.04774	.04862	.04953	.05047	.05145	.05245
25	1	14	.04599	.04664	.04730	.04799	.04869	.04941	.05015
25	1	15	.04504	.04551	.04599	.04646	.04694	.04743	.04792
25	1	16	.04408	.04437	.04466	.04495	.04522	.04550	.04577
25	1	17	.04308	.04321	.04333	.04344	.04354	.04362	.04369
25	1	18	.04206	.04204	.04200	.04195	.04187	.04179	.04169
25	1	19	.04100	.04084	.04066	.04045	.04023	.04000	.03975
25	1	20	.03991	.03961	.03929	.03896	.03861	.03824	.03787
25	1	21	.03875	.03833	.03790	.03744	.03697	.03650	.03602

Coefficients a_i for the BLUE of μ
Shape parameter k

n	s	i	1.4	1.5	1.6	1.7	1.8	1.9	2.0
25	1	22	.03751	.03698	.03643	.03588	.03532	.03475	.03418
25	1	23	.03613	.03550	.03486	.03423	.03359	.03296	.03233
25	1	24	.06609	.06483	.06359	.06238	.06118	.06001	.05887
25	2	1	-.06106	-.07965	-.10058	-.12398	-.14998	-.17872	-.21031
25	2	2	.02974	.02992	.03035	.03106	.03209	.03348	.03525
25	2	3	.04175	.04364	.04591	.04858	.05167	.05523	.05928
25	2	4	.04697	.04937	.05214	.05529	.05884	.06283	.06726
25	2	5	.04949	.05197	.05478	.05793	.06143	.06529	.06954
25	2	6	.05066	.05305	.05572	.05867	.06191	.06546	.06931
25	2	7	.05108	.05330	.05574	.05842	.06133	.06449	.06789
25	2	8	.05105	.05305	.05524	.05761	.06017	.06292	.06586
25	2	9	.05072	.05250	.05442	.05649	.05870	.06105	.06355
25	2	10	.05021	.05175	.05341	.05517	.05704	.05903	.06112
25	2	11	.04956	.05087	.05227	.05375	.05530	.05694	.05865
25	2	12	.04881	.04991	.05106	.05227	.05353	.05484	.05620
25	2	13	.04800	.04888	.04980	.05075	.05174	.05275	.05380
25	2	14	.04713	.04781	.04851	.04923	.04996	.05071	.05147
25	2	15	.04622	.04671	.04720	.04770	.04820	.04870	.04921
25	2	16	.04528	.04558	.04588	.04618	.04647	.04675	.04703
25	2	17	.04430	.04443	.04455	.04466	.04476	.04484	.04492
25	2	18	.04328	.04326	.04321	.04315	.04307	.04298	.04287
25	2	19	.04223	.04205	.04185	.04163	.04140	.04115	.04089
25	2	20	.04112	.04080	.04046	.04011	.03974	.03936	.03896
25	2	21	.03995	.03950	.03904	.03856	.03807	.03757	.03707
25	2	22	.03867	.03811	.03754	.03695	.03637	.03578	.03519
25	2	23	.10484	.10317	.10149	.09984	.09820	.09658	.09499
25	3	1	-.07541	-.09516	-.11729	-.14194	-.16925	-.19935	-.23235
25	3	2	.02576	.02591	.02631	.02702	.02805	.02946	.03126
25	3	3	.03975	.04172	.04408	.04686	.05008	.05377	.05796
25	3	4	.04602	.04854	.05144	.05473	.05843	.06258	.06718
25	3	5	.04919	.05181	.05475	.05804	.06169	.06571	.07013
25	3	6	.05081	.05333	.05613	.05921	.06260	.06629	.07029
25	3	7	.05155	.05388	.05645	.05925	.06229	.06557	.06911
25	3	8	.05175	.05386	.05615	.05864	.06131	.06417	.06723
25	3	9	.05160	.05347	.05549	.05764	.05995	.06240	.06500
25	3	10	.05122	.05285	.05458	.05642	.05838	.06044	.06260
25	3	11	.05068	.05206	.05352	.05507	.05668	.05838	.06015
25	3	12	.05001	.05116	.05237	.05362	.05493	.05629	.05770
25	3	13	.04926	.05019	.05115	.05214	.05316	.05421	.05529
25	3	14	.04845	.04915	.04988	.05062	.05138	.05215	.05293
25	3	15	.04757	.04808	.04858	.04909	.04961	.05012	.05064
25	3	16	.04665	.04696	.04727	.04756	.04785	.04814	.04842
25	3	17	.04569	.04581	.04593	.04603	.04612	.04620	.04626
25	3	18	.04468	.04463	.04457	.04449	.04440	.04430	.04418
25	3	19	.04362	.04341	.04319	.04295	.04270	.04243	.04215
25	3	20	.04250	.04214	.04178	.04139	.04099	.04059	.04017
25	3	21	.04129	.04081	.04031	.03980	.03928	.03875	.03823
25	3	22	.14737	.14537	.14337	.14138	.13939	.13742	.13548
25	4	1	-.09155	-.11255	-.13599	-.16200	-.19071	-.22226	-.25677
25	4	2	.02133	.02144	.02183	.02253	.02358	.02501	.02685
25	4	3	.03753	.03960	.04207	.04497	.04832	.05216	.05651
25	4	4	.04499	.04764	.05068	.05412	.05799	.06231	.06710
25	4	5	.04889	.05165	.05474	.05819	.06201	.06620	.07079
25	4	6	.05101	.05366	.05661	.05984	.06338	.06722	.07139
25	4	7	.05210	.05457	.05726	.06019	.06337	.06679	.07046
25	4	8	.05257	.05479	.05720	.05980	.06259	.06557	.06874

Coefficients a_i for the BLUE of μ
Shape parameter k

n	s	i	1.4	1.5	1.6	1.7	1.8	1.9	2.0
25	4	9	.05262	.05459	.05670	.05896	.06136	.06391	.06661
25	4	10	.05239	.05410	.05592	.05784	.05987	.06201	.06426
25	4	11	.05197	.05342	.05495	.05655	.05823	.05999	.06182
25	4	12	.05139	.05260	.05385	.05516	.05651	.05792	.05937
25	4	13	.05071	.05168	.05267	.05369	.05474	.05583	.05694
25	4	14	.04995	.05068	.05143	.05219	.05296	.05375	.05455
25	4	15	.04911	.04963	.05014	.05066	.05118	.05170	.05222
25	4	16	.04821	.04852	.04882	.04912	.04940	.04968	.04996
25	4	17	.04726	.04738	.04748	.04757	.04764	.04770	.04776
25	4	18	.04626	.04619	.04610	.04600	.04589	.04576	.04562
25	4	19	.04519	.04495	.04470	.04442	.04414	.04384	.04354
25	4	20	.04404	.04365	.04324	.04282	.04239	.04195	.04150
25	4	21	.19402	.19182	.18960	.18738	.18516	.18296	.18077
25	5	1	-.10970	-.13208	-.15694	-.18443	-.21467	-.24780	-.28393
25	5	2	.01638	.01646	.01683	.01754	.01861	.02007	.02197
25	5	3	.03508	.03726	.03985	.04288	.04639	.05039	.05492
25	5	4	.04386	.04667	.04986	.05348	.05753	.06204	.06703
25	5	5	.04860	.05151	.05477	.05839	.06238	.06676	.07154
25	5	6	.05127	.05408	.05718	.06057	.06427	.06829	.07262
25	5	7	.05277	.05537	.05820	.06128	.06460	.06817	.07199
25	5	8	.05353	.05587	.05841	.06113	.06404	.06714	.07044
25	5	9	.05380	.05588	.05809	.06045	.06296	.06561	.06840
25	5	10	.05374	.05554	.05744	.05945	.06156	.06378	.06611
25	5	11	.05345	.05497	.05656	.05824	.05998	.06180	.06369
25	5	12	.05298	.05423	.05553	.05689	.05829	.05974	.06123
25	5	13	.05237	.05337	.05440	.05545	.05653	.05764	.05878
25	5	14	.05166	.05241	.05318	.05396	.05475	.05555	.05636
25	5	15	.05087	.05139	.05191	.05243	.05295	.05347	.05399
25	5	16	.05000	.05030	.05059	.05087	.05114	.05141	.05167
25	5	17	.04906	.04915	.04923	.04929	.04934	.04939	.04942
25	5	18	.04805	.04795	.04783	.04770	.04755	.04739	.04722
25	5	19	.04696	.04668	.04639	.04607	.04575	.04542	.04508
25	5	20	.24529	.24301	.24070	.23838	.23605	.23374	.23144
25	6	1	-.13016	-.15405	-.18049	-.20960	-.24151	-.27636	-.31427
25	6	2	.01084	.01089	.01126	.01198	.01308	.01459	.01655
25	6	3	.03237	.03467	.03740	.04059	.04426	.04844	.05317
25	6	4	.04264	.04562	.04899	.05280	.05705	.06177	.06698
25	6	5	.04831	.05140	.05484	.05865	.06284	.06742	.07241
25	6	6	.05161	.05459	.05786	.06143	.06530	.06950	.07402
25	6	7	.05356	.05631	.05930	.06253	.06600	.06973	.07371
25	6	8	.05465	.05713	.05979	.06265	.06569	.06893	.07236
25	6	9	.05517	.05736	.05969	.06216	.06477	.06753	.07043
25	6	10	.05531	.05719	.05918	.06128	.06348	.06578	.06819
25	6	11	.05516	.05675	.05841	.06015	.06196	.06384	.06580
25	6	12	.05480	.05610	.05745	.05885	.06030	.06179	.06333
25	6	13	.05427	.05531	.05636	.05745	.05855	.05969	.06085
25	6	14	.05363	.05439	.05517	.05596	.05676	.05757	.05839
25	6	15	.05287	.05339	.05391	.05442	.05494	.05545	.05597
25	6	16	.05203	.05231	.05258	.05285	.05310	.05335	.05360
25	6	17	.05110	.05116	.05121	.05124	.05126	.05128	.05128
25	6	18	.05008	.04994	.04978	.04960	.04942	.04922	.04901
25	6	19	.30177	.29955	.29729	.29502	.29275	.29048	.28823
25	7	1	-.15331	-.17888	-.20705	-.23796	-.27172	-.30847	-.34833
25	7	2	.00462	.00466	.00503	.00577	.00690	.00846	.01050
25	7	3	.02935	.03180	.03468	.03805	.04191	.04630	.05125
25	7	4	.04132	.04448	.04806	.05208	.05655	.06150	.06695

Coefficients a_i for the BLUE of μ
Shape parameter k

n	s	i	1.4	1.5	1.6	1.7	1.8	1.9	2.0
25	7	5	.04804	.05132	.05497	.05899	.06339	.06819	.07341
25	7	6	.05206	.05522	.05867	.06243	.06651	.07090	.07563
25	7	7	.05451	.05743	.06058	.06398	.06762	.07152	.07567
25	7	8	.05597	.05859	.06140	.06440	.06758	.07096	.07453
25	7	9	.05678	.05908	.06153	.06411	.06684	.06971	.07273
25	7	10	.05712	.05910	.06119	.06338	.06567	.06806	.07055
25	7	11	.05713	.05880	.06053	.06234	.06422	.06616	.06817
25	7	12	.05689	.05825	.05965	.06110	.06259	.06412	.06570
25	7	13	.05647	.05753	.05861	.05972	.06085	.06201	.06319
25	7	14	.05588	.05666	.05745	.05824	.05905	.05986	.06068
25	7	15	.05517	.05568	.05619	.05670	.05720	.05770	.05820
25	7	16	.05435	.05461	.05486	.05509	.05532	.05555	.05576
25	7	17	.05343	.05345	.05346	.05345	.05343	.05340	.05337
25	7	18	.36422	.36221	.36018	.35814	.35609	.35406	.35204
25	8	1	-.17962	-.20707	-.23718	-.27008	-.30590	-.34477	-.38680
25	8	2	-.00238	-.00235	-.00197	-.00121	-.00003	.00160	.00372
25	8	3	.02599	.02860	.03167	.03523	.03931	.04393	.04912
25	8	4	.03988	.04326	.04707	.05132	.05604	.06125	.06697
25	8	5	.04780	.05130	.05518	.05943	.06407	.06912	.07459
25	8	6	.05262	.05599	.05966	.06363	.06792	.07253	.07748
25	8	7	.05565	.05875	.06209	.06567	.06950	.07359	.07792
25	8	8	.05753	.06031	.06328	.06643	.06977	.07330	.07703
25	8	9	.05865	.06109	.06366	.06638	.06923	.07222	.07536
25	8	10	.05924	.06132	.06351	.06580	.06818	.07066	.07324
25	8	11	.05943	.06117	.06298	.06486	.06680	.06881	.07088
25	8	12	.05933	.06074	.06219	.06368	.06521	.06678	.06839
25	8	13	.05900	.06009	.06120	.06233	.06348	.06465	.06584
25	8	14	.05849	.05927	.06006	.06086	.06166	.06246	.06328
25	8	15	.05783	.05832	.05881	.05929	.05977	.06025	.06073
25	8	16	.05703	.05725	.05746	.05766	.05785	.05804	.05822
25	8	17	.43353	.43194	.43034	.42874	.42714	.42557	.42403
25	9	1	-.20972	-.23928	-.27156	-.30671	-.34484	-.38607	-.43054
25	9	2	-.01031	-.01029	-.00989	-.00909	-.00785	-.00614	-.00392
25	9	3	.02222	.02503	.02831	.03210	.03642	.04130	.04677
25	9	4	.03832	.04195	.04601	.05053	.05554	.06104	.06705
25	9	5	.04760	.05136	.05549	.06000	.06491	.07023	.07598
25	9	6	.05335	.05695	.06085	.06506	.06959	.07444	.07963
25	9	7	.05703	.06034	.06388	.06767	.07170	.07599	.08053
25	9	8	.05939	.06235	.06549	.06881	.07231	.07601	.07990
25	9	9	.06087	.06345	.06616	.06901	.07200	.07512	.07838
25	9	10	.06172	.06392	.06622	.06860	.07109	.07366	.07634
25	9	11	.06212	.06394	.06583	.06778	.06979	.07186	.07400
25	9	12	.06217	.06363	.06513	.06666	.06823	.06984	.07149
25	9	13	.06196	.06307	.06420	.06534	.06650	.06769	.06889
25	9	14	.06152	.06230	.06308	.06387	.06466	.06546	.06626
25	9	15	.06091	.06137	.06183	.06228	.06273	.06318	.06363
25	9	16	.51084	.50991	.50898	.50808	.50722	.50639	.50561
25	10	1	-.24440	-.27634	-.31109	-.34877	-.38951	-.43343	-.48065
25	10	2	-.01935	-.01932	-.01890	-.01805	-.01673	-.01493	-.01259
25	10	3	.01799	.02102	.02455	.02860	.03320	.03838	.04416
25	10	4	.03662	.04053	.04489	.04972	.05504	.06087	.06723
25	10	5	.04748	.05152	.05594	.06075	.06596	.07159	.07765
25	10	6	.05428	.05814	.06231	.06679	.07158	.07671	.08217
25	10	7	.05871	.06225	.06602	.07004	.07430	.07881	.08358
25	10	8	.06162	.06477	.06810	.07161	.07530	.07918	.08325
25	10	9	.06351	.06625	.06911	.07211	.07524	.07850	.08190

Coefficients a_i for the BLUE of μ
Shape parameter k

n	s	i	1.4	1.5	1.6	1.7	1.8	1.9	2.0
25	10	10	.06467	.06699	.06939	.07189	.07448	.07716	.07994
25	10	11	.06530	.06720	.06917	.07119	.07327	.07541	.07761
25	10	12	.06552	.06703	.06857	.07014	.07175	.07339	.07507
25	10	13	.06543	.06655	.06769	.06885	.07002	.07121	.07241
25	10	14	.06507	.06584	.06660	.06737	.06814	.06892	.06970
25	10	15	.59754	.59756	.59764	.59776	.59796	.59823	.59858
25	11	1	-.28470	-.31936	-.35692	-.39750	-.44123	-.48822	-.53858
25	11	2	-.02974	-.02969	-.02922	-.02830	-.02689	-.02496	-.02249
25	11	3	.01319	.01649	.02031	.02467	.02960	.03512	.04126
25	11	4	.03478	.03902	.04372	.04890	.05459	.06079	.06754
25	11	5	.04746	.05182	.05658	.06173	.06729	.07327	.07968
25	11	6	.05549	.05965	.06411	.06889	.07399	.07942	.08519
25	11	7	.06079	.06458	.06861	.07289	.07740	.08217	.08719
25	11	8	.06433	.06769	.07123	.07495	.07885	.08293	.08720
25	11	9	.06668	.06959	.07262	.07578	.07907	.08248	.08603
25	11	10	.06819	.07064	.07317	.07578	.07848	.08127	.08415
25	11	11	.06908	.07107	.07312	.07521	.07736	.07957	.08183
25	11	12	.06950	.07105	.07263	.07424	.07588	.07755	.07925
25	11	13	.06953	.07067	.07181	.07297	.07414	.07532	.07652
25	11	14	.69542	.69677	.69823	.69979	.70147	.70328	.70522
25	12	1	-.33201	-.36981	-.41061	-.45454	-.50172	-.55224	-.60624
25	12	2	-.04179	-.04169	-.04115	-.04013	-.03861	-.03654	-.03389
25	12	3	.00772	.01135	.01551	.02023	.02554	.03146	.03802
25	12	4	.03277	.03740	.04250	.04810	.05421	.06085	.06803
25	12	5	.04759	.05234	.05748	.06302	.06898	.07536	.08218
25	12	6	.05706	.06156	.06637	.07150	.07694	.08272	.08884
25	12	7	.06336	.06746	.07178	.07634	.08115	.08620	.09151
25	12	8	.06764	.07125	.07503	.07898	.08311	.08742	.09191
25	12	9	.07054	.07363	.07685	.08019	.08365	.08723	.09095
25	12	10	.07245	.07503	.07769	.08043	.08325	.08616	.08915
25	12	11	.07364	.07572	.07784	.08001	.08223	.08451	.08684
25	12	12	.07427	.07586	.07747	.07911	.08078	.08248	.08420
25	12	13	.80676	.80991	.81325	.81677	.82048	.82439	.82850
25	13	1	-.38824	-.42970	-.47429	-.52212	-.57332	-.62800	-.68625
25	13	2	-.05590	-.05574	-.05510	-.05395	-.05227	-.05002	-.04717
25	13	3	.00143	.00545	.01002	.01518	.02094	.02734	.03440
25	13	4	.03059	.03568	.04126	.04734	.05395	.06110	.06881
25	13	5	.04795	.05314	.05873	.06474	.07116	.07801	.08531
25	13	6	.05911	.06402	.06923	.07476	.08061	.08679	.09331
25	13	7	.06661	.07105	.07571	.08061	.08575	.09113	.09676
25	13	8	.07176	.07564	.07969	.08391	.08829	.09286	.09761
25	13	9	.07529	.07859	.08201	.08555	.08920	.09298	.09687
25	13	10	.07767	.08040	.08320	.08607	.08903	.09206	.09517
25	13	11	.07920	.08136	.08357	.08582	.08811	.09045	.09284
25	13	12	.93454	.94011	.94596	.95211	.95856	.96530	.97235

Coefficients a_i for the BLUE of μ
Shape parameter k

n	s	i	2.1	2.2	2.3	2.4	2.5
2	0	1	.50000	.50000	.50000	.50000	.50000
2	0	2	.50000	.50000	.50000	.50000	.50000
3	0	1	.26807	.26239	.25657	.25060	.24450
3	0	2	.43316	.44086	.44867	.45657	.46456
3	0	3	.29877	.29675	.29477	.29283	.29093
3	1	1	-.29619	-.31310	-.33021	-.34753	-.36504
3	1	2	1.29619	1.31310	1.33021	1.34753	1.36504
4	0	1	.13508	.12366	.11176	.09937	.08651
4	0	2	.37041	.38218	.39442	.40711	.42024
4	0	3	.28464	.28652	.28837	.29019	.29199
4	0	4	.20986	.20764	.20546	.20333	.20126
4	1	1	-.16624	-.18636	-.20706	-.22834	-.25021
4	1	2	.47633	.49430	.51281	.53185	.55143
4	1	3	.68992	.69206	.69425	.69649	.69878
5	0	1	.04971	.03356	.01656	-.00129	-.02000
5	0	2	.32020	.33358	.34769	.36254	.37812
5	0	3	.26095	.26587	.27088	.27599	.28119
5	0	4	.20857	.20852	.20844	.20834	.20821
5	0	5	.16057	.15847	.15642	.15442	.15247
5	1	1	-.15462	-.17801	-.20238	-.22774	-.25409
5	1	2	.37004	.38783	.40647	.42597	.44632
5	1	3	.31718	.32381	.33054	.33737	.34430
5	1	4	.46740	.46637	.46536	.46439	.46346
5	2	1	-.57060	-.60649	-.64351	-.68166	-.72096
5	2	2	.47413	.49939	.52560	.55276	.58088
5	2	3	1.09647	1.10709	1.11791	1.12890	1.14008
6	0	1	-.00903	-.02891	-.04998	-.07226	-.09574
6	0	2	.27973	.29347	.30810	.32366	.34014
6	0	3	.23843	.24522	.25227	.25956	.26710
6	0	4	.19821	.20027	.20233	.20439	.20646
6	0	5	.16314	.16236	.16156	.16074	.15990
6	0	6	.12951	.12760	.12573	.12391	.12214
6	1	1	-.16190	-.18801	-.21544	-.24423	-.27438
6	1	2	.30592	.32291	.34091	.35996	.38007
6	1	3	.27299	.28147	.29022	.29925	.30853
6	1	4	.23131	.23394	.23657	.23921	.24185
6	1	5	.35168	.34969	.34773	.34581	.34392
6	2	1	-.42082	-.45590	-.49244	-.53045	-.56996
6	2	2	.35137	.37307	.39590	.41988	.44502
6	2	3	.33201	.34261	.35348	.36462	.37601
6	2	4	.73745	.74022	.74306	.74596	.74893
6	3	1	-.94419	-.99618	-1.04988	-1.10532	-1.16250
6	3	2	.44817	.47852	.51014	.54303	.57722
6	3	3	1.49602	1.51765	1.53975	1.56229	1.58528
7	0	1	-.05138	-.07421	-.09852	-.12433	-.15166
7	0	2	.24659	.26005	.27450	.28997	.30648
7	0	3	.21821	.22610	.23437	.24302	.25205
7	0	4	.18670	.19030	.19398	.19775	.20159
7	0	5	.15836	.15906	.15974	.16040	.16104
7	0	6	.13326	.13219	.13110	.13001	.12892
7	0	7	.10826	.10652	.10483	.10318	.10158
7	1	1	-.17232	-.20059	-.23049	-.26205	-.29530
7	1	2	.26085	.27677	.29380	.31196	.33129
7	1	3	.24092	.25033	.26015	.27039	.28104
7	1	4	.20986	.21419	.21860	.22310	.22768
7	1	5	.17973	.18063	.18151	.18238	.18323

Coefficients a_i for the BLUE of μ
Shape parameter k

n	s	i	2.1	2.2	2.3	2.4	2.5
7	1	6	.28095	.27868	.27643	.27423	.27206
7	2	1	-.35636	-.39172	-.42883	-.46771	-.50840
7	2	2	.28306	.30225	.32266	.34430	.36721
7	2	3	.27575	.28699	.29866	.31076	.32328
7	2	4	.24514	.25017	.25528	.26046	.26572
7	2	5	.55241	.55230	.55223	.55219	.55219
7	3	1	-.66646	-.71294	-.76134	-.81168	-.86399
7	3	2	.32262	.34689	.37247	.39941	.42771
7	3	3	.33582	.34970	.36399	.37871	.39386
7	3	4	1.00801	1.01635	1.02487	1.03356	1.04242
7	4	1	-1.28995	-1.35790	-1.42816	-1.50075	-1.57572
7	4	2	.40873	.44260	.47795	.51482	.55324
7	4	3	1.88122	1.91530	1.95021	1.98594	2.02248
8	0	1	-.08298	-.10815	-.13503	-.16368	-.19411
8	0	2	.21899	.23186	.24577	.26075	.27681
8	0	3	.20032	.20879	.21774	.22718	.23710
8	0	4	.17543	.18011	.18496	.18997	.19514
8	0	5	.15202	.15393	.15585	.15779	.15975
8	0	6	.13112	.13112	.13110	.13105	.13099
8	0	7	.11225	.11108	.10989	.10871	.10754
8	0	8	.09284	.09126	.08973	.08824	.08679
8	1	1	-.18222	-.21221	-.24408	-.27787	-.31361
8	1	2	.22659	.24136	.25727	.27436	.29265
8	1	3	.21587	.22567	.23599	.24684	.25822
8	1	4	.19230	.19774	.20336	.20915	.21512
8	1	5	.16819	.17043	.17269	.17497	.17727
8	1	6	.14590	.14596	.14600	.14603	.14604
8	1	7	.23337	.23105	.22877	.22652	.22432
8	2	1	-.32295	-.35883	-.39670	-.43660	-.47857
8	2	2	.23759	.25477	.27317	.29284	.31381
8	2	3	.23806	.24942	.26133	.27377	.28677
8	2	4	.21626	.22248	.22888	.23545	.24219
8	2	5	.19105	.19352	.19601	.19851	.20104
8	2	6	.43999	.43863	.43731	.43601	.43476
8	3	1	-.53634	-.58051	-.62681	-.67527	-.72594
8	3	2	.25537	.27592	.29779	.32102	.34565
8	3	3	.27250	.28593	.29990	.31444	.32954
8	3	4	.25304	.26019	.26750	.27497	.28260
8	3	5	.75542	.75848	.76161	.76484	.76814
8	4	1	-.89430	-.95184	-1.01173	-1.07402	-1.13873
8	4	2	.28805	.31395	.34130	.37013	.40048
8	4	3	.33230	.34885	.36595	.38362	.40186
8	4	4	1.27395	1.28904	1.30448	1.32027	1.33640
9	0	1	-.10717	-.13420	-.16315	-.19409	-.22703
9	0	2	.19569	.20783	.22101	.23527	.25064
9	0	3	.18449	.19321	.20250	.21234	.22276
9	0	4	.16483	.17024	.17589	.18179	.18793
9	0	5	.14530	.14813	.15103	.15400	.15703
9	0	6	.12749	.12842	.12935	.13028	.13120
9	0	7	.11147	.11109	.11069	.11027	.10984
9	0	8	.09674	.09555	.09435	.09316	.09198
9	0	9	.08118	.07974	.07834	.07698	.07566
9	1	1	-.19076	-.22212	-.25557	-.29114	-.32890
9	1	2	.19929	.21292	.22769	.24363	.26079
9	1	3	.19542	.20531	.21578	.22687	.23856
9	1	4	.17746	.18362	.19003	.19670	.20362

Coefficients a_i for the BLUE of μ
Shape parameter k

n	s	i	2.1	2.2	2.3	2.4	2.5
9	1	5	.15784	.16108	.16439	.16777	.17123
9	1	6	.13926	.14036	.14145	.14254	.14363
9	1	7	.12222	.12185	.12146	.12106	.12065
9	1	8	.19925	.19698	.19475	.19257	.19042
9	2	1	-.30349	-.33989	-.37848	-.41932	-.46246
9	2	2	.20426	.21971	.23639	.25431	.27354
9	2	3	.21024	.22144	.23326	.24572	.25881
9	2	4	.19453	.20146	.20865	.21610	.22380
9	2	5	.17473	.17832	.18198	.18570	.18950
9	2	6	.15505	.15621	.15736	.15851	.15965
9	2	7	.36467	.36275	.36085	.35899	.35716
9	3	1	-.46296	-.50599	-.55133	-.59902	-.64912
9	3	2	.21190	.22975	.24891	.26941	.29129
9	3	3	.23167	.24453	.25803	.27219	.28700
9	3	4	.21898	.22682	.23491	.24325	.25185
9	3	5	.19877	.20269	.20668	.21073	.21484
9	3	6	.60165	.60220	.60279	.60344	.60414
9	4	1	-.70389	-.75652	-.81161	-.86922	-.92940
9	4	2	.22490	.24621	.26891	.29305	.31867
9	4	3	.26516	.28031	.29613	.31261	.32976
9	4	4	.25668	.26569	.27493	.28442	.29415
9	4	5	.95715	.96431	.97164	.97914	.98681
9	5	1	-1.10676	-1.17499	-1.24599	-1.31982	-1.39651
9	5	2	.25006	.27692	.30532	.33529	.36686
9	5	3	.32371	.34244	.36184	.38191	.40269
9	5	4	1.53298	1.55562	1.57884	1.60262	1.62696
10	0	1	-.12606	-.15458	-.18521	-.21800	-.25299
10	0	2	.17576	.18711	.19948	.21292	.22747
10	0	3	.17045	.17922	.18860	.19860	.20922
10	0	4	.15500	.16089	.16710	.17361	.18043
10	0	5	.13862	.14215	.14580	.14957	.15347
10	0	6	.12328	.12499	.12672	.12848	.13026
10	0	7	.10935	.10971	.11005	.11038	.11070
10	0	8	.09669	.09610	.09549	.09488	.09425
10	0	9	.08486	.08368	.08251	.08135	.08020
10	0	10	.07206	.07074	.06945	.06821	.06700
10	1	1	-.19787	-.23032	-.26503	-.30205	-.34144
10	1	2	.17687	.18940	.20304	.21785	.23384
10	1	3	.17827	.18804	.19845	.20953	.22128
10	1	4	.16467	.17126	.17818	.18542	.19299
10	1	5	.14854	.15250	.15660	.16082	.16516
10	1	6	.13283	.13476	.13672	.13871	.14072
10	1	7	.11824	.11868	.11910	.11951	.11991
10	1	8	.10483	.10423	.10362	.10299	.10236
10	1	9	.17362	.17145	.16931	.16722	.16516
10	2	1	-.29112	-.32799	-.36721	-.40886	-.45299
10	2	2	.17834	.19229	.20743	.22379	.24142
10	2	3	.18844	.19933	.21090	.22315	.23611
10	2	4	.17723	.18456	.19221	.20020	.20852
10	2	5	.16141	.16578	.17027	.17489	.17964
10	2	6	.14517	.14726	.14938	.15152	.15368
10	2	7	.12971	.13013	.13055	.13095	.13135
10	2	8	.31082	.30864	.30648	.30436	.30228
10	3	1	-.41672	-.45910	-.50394	-.55131	-.60126
10	3	2	.18069	.19643	.21343	.23172	.25136
10	3	3	.20243	.21468	.22763	.24129	.25568

Coefficients a_i for the BLUE of μ
Shape parameter k

n	s	i	2.1	2.2	2.3	2.4	2.5
10	3	4	.19435	.20252	.21102	.21985	.22902
10	3	5	.17887	.18365	.18856	.19358	.19873
10	3	6	.16184	.16403	.16625	.16849	.17075
10	3	7	.49854	.49778	.49706	.49637	.49573
10	4	1	-.59385	-.64368	-.69610	-.75117	-.80895
10	4	2	.18484	.20298	.22247	.24333	.26562
10	4	3	.22282	.23685	.25159	.26707	.28330
10	4	4	.21900	.22822	.23775	.24762	.25782
10	4	5	.20381	.20906	.21441	.21988	.22546
10	4	6	.76338	.76658	.76987	.77327	.77676
10	5	1	-.86071	-.92145	-.98498	-1.05136	-1.12065
10	5	2	.19285	.21449	.23757	.26214	.28822
10	5	3	.25494	.27148	.28875	.30678	.32557
10	5	4	.25715	.26778	.27871	.28996	.30154
10	5	5	1.15576	1.16770	1.17994	1.19248	1.20531
11	0	1	-.14104	-.17078	-.20277	-.23707	-.27374
11	0	2	.15854	.16908	.18062	.19320	.20687
11	0	3	.15795	.16663	.17595	.18593	.19657
11	0	4	.14594	.15215	.15871	.16563	.17291
11	0	5	.13217	.13622	.14044	.14484	.14940
11	0	6	.11889	.12122	.12360	.12605	.12856
11	0	7	.10665	.10764	.10864	.10964	.11065
11	0	8	.09547	.09547	.09546	.09543	.09538
11	0	9	.08521	.08451	.08379	.08307	.08234
11	0	10	.07548	.07434	.07321	.07210	.07100
11	0	11	.06474	.06352	.06234	.06119	.06007
11	1	1	-.20369	-.23702	-.27274	-.31092	-.35164
11	1	2	.15803	.16952	.18209	.19579	.21064
11	1	3	.16357	.17312	.18334	.19426	.20590
11	1	4	.15346	.16031	.16753	.17513	.18312
11	1	5	.14015	.14463	.14930	.15414	.15916
11	1	6	.12673	.12932	.13197	.13469	.13746
11	1	7	.11409	.11521	.11634	.11747	.11861
11	1	8	.10240	.10243	.10245	.10245	.10245
11	1	9	.09157	.09085	.09012	.08938	.08864
11	1	10	.15370	.15163	.14960	.14761	.14566
11	2	1	-.28270	-.31994	-.35969	-.40201	-.44697
11	2	2	.15739	.17000	.18376	.19870	.21487
11	2	3	.17066	.18116	.19237	.20431	.21699
11	2	4	.16293	.17045	.17836	.18665	.19535
11	2	5	.15019	.15509	.16018	.16544	.17088
11	2	6	.13659	.13939	.14226	.14518	.14817
11	2	7	.12342	.12461	.12580	.12699	.12819
11	2	8	.11105	.11105	.11103	.11100	.11097
11	2	9	.27047	.26819	.26594	.26372	.26154
11	3	1	-.38525	-.42722	-.47178	-.51902	-.56899
11	3	2	.15679	.17077	.18597	.20241	.22015
11	3	3	.18005	.19168	.20405	.21717	.23106
11	3	4	.17536	.18365	.19232	.20139	.21086
11	3	5	.16331	.16866	.17418	.17988	.18576
11	3	6	.14943	.15242	.15548	.15859	.16175
11	3	7	.13553	.13673	.13793	.13913	.14034
11	3	8	.42478	.42331	.42186	.42045	.41907
11	4	1	-.52300	-.57104	-.62178	-.67530	-.73166
11	4	2	.15649	.17223	.18926	.20761	.22731
11	4	3	.19308	.20613	.21993	.23452	.24989

Coefficients a_i for the BLUE of μ
Shape parameter k

n	s	i	2.1	2.2	2.3	2.4	2.5
11	4	4	.19239	.20160	.21120	.22119	.23158
11	4	5	.18117	.18702	.19304	.19922	.20558
11	4	6	.16681	.16998	.17320	.17647	.17979
11	4	7	.63306	.63408	.63515	.63629	.63750
11	5	1	-.71681	-.77314	-.83233	-.89444	-.95953
11	5	2	.15707	.17522	.19472	.21563	.23797
11	5	3	.21224	.22717	.24288	.25939	.27671
11	5	4	.21699	.22738	.23816	.24933	.26090
11	5	5	.20677	.21321	.21981	.22657	.23349
11	5	6	.92374	.93016	.93676	.94353	.95047
12	0	1	-.15308	-.18380	-.21690	-.25244	-.29050
12	0	2	.14351	.15326	.16397	.17570	.18848
12	0	3	.14676	.15526	.16442	.17426	.18479
12	0	4	.13762	.14400	.15077	.15795	.16554
12	0	5	.12605	.13048	.13512	.13998	.14506
12	0	6	.11450	.11732	.12023	.12324	.12634
12	0	7	.10367	.10520	.10676	.10834	.10995
12	0	8	.09371	.09423	.09475	.09527	.09578
12	0	9	.08455	.08433	.08408	.08383	.08356
12	0	10	.07606	.07530	.07452	.07375	.07297
12	0	11	.06791	.06682	.06574	.06467	.06362
12	0	12	.05875	.05761	.05652	.05545	.05442
12	1	1	-.20842	-.24244	-.27898	-.31811	-.35990
12	1	2	.14193	.15245	.16401	.17665	.19040
12	1	3	.15080	.16005	.17000	.18067	.19207
12	1	4	.14353	.15050	.15788	.16569	.17393
12	1	5	.13254	.13739	.14247	.14777	.15329
12	1	6	.12102	.12411	.12731	.13061	.13400
12	1	7	.10996	.11165	.11337	.11512	.11690
12	1	8	.09964	.10024	.10084	.10143	.10201
12	1	9	.09008	.08986	.08963	.08939	.08913
12	1	10	.08116	.08037	.07958	.07879	.07800
12	1	11	.13778	.13581	.13389	.13200	.13016
12	2	1	-.27659	-.31414	-.35431	-.39717	-.44281
12	2	2	.13998	.15140	.16391	.17756	.19239
12	2	3	.15576	.16583	.17663	.18818	.20050
12	2	4	.15080	.15838	.16639	.17483	.18372
12	2	5	.14051	.14577	.15127	.15699	.16294
12	2	6	.12901	.13236	.13580	.13934	.14298
12	2	7	.11766	.11947	.12131	.12317	.12506
12	2	8	.10690	.10752	.10813	.10874	.10934
12	2	9	.09682	.09655	.09628	.09600	.09571
12	2	10	.23915	.23685	.23459	.23235	.23016
12	3	1	-.36257	-.40425	-.44864	-.49581	-.54585
12	3	2	.13766	.15015	.16380	.17864	.19471
12	3	3	.16214	.17316	.18494	.19749	.21084
12	3	4	.16005	.16832	.17703	.18618	.19578
12	3	5	.15063	.15634	.16227	.16844	.17483
12	3	6	.13914	.14273	.14641	.15018	.15405
12	3	7	.12740	.12929	.13120	.13314	.13510
12	3	8	.11603	.11662	.11720	.11778	.11836
12	3	9	.36951	.36764	.36579	.36397	.36219
12	4	1	-.47393	-.52072	-.57032	-.62279	-.67822
12	4	2	.13498	.14881	.16385	.18015	.19774
12	4	3	.17067	.18285	.19581	.20957	.22415
12	4	4	.17227	.18135	.19087	.20084	.21126

Coefficients a_i for the BLUE of μ
Shape parameter k

n	s	i	2.1	2.2	2.3	2.4	2.5
12	4	5	.16391	.17012	.17654	.18319	.19007
12	4	6	.15238	.15621	.16012	.16412	.16821
12	4	7	.14006	.14200	.14396	.14593	.14794
12	4	8	.53965	.53939	.53917	.53899	.53886
12	5	1	-.62320	-.67664	-.73300	-.79236	-.85480
12	5	2	.13201	.14758	.16442	.18259	.20211
12	5	3	.18264	.19629	.21075	.22602	.24214
12	5	4	.18906	.19914	.20967	.22065	.23208
12	5	5	.18204	.18883	.19583	.20305	.21049
12	5	6	.17035	.17442	.17858	.18281	.18714
12	5	7	.76710	.77037	.77375	.77724	.78084
12	6	1	-.83281	-.89539	-.96107	-1.02992	-1.10202
12	6	2	.12900	.14694	.16624	.18695	.20909
12	6	3	.20042	.21604	.23250	.24980	.26797
12	6	4	.21342	.22482	.23666	.24895	.26169
12	6	5	.20808	.21559	.22332	.23124	.23938
12	6	6	1.08190	1.09199	1.10235	1.11299	1.12389
13	0	1	-.16283	-.19436	-.22837	-.26494	-.30414
13	0	2	.13028	.13927	.14919	.16007	.17197
13	0	3	.13670	.14496	.15390	.16352	.17385
13	0	4	.12997	.13643	.14331	.15063	.15840
13	0	5	.12026	.12497	.12992	.13513	.14059
13	0	6	.11020	.11341	.11674	.12020	.12379
13	0	7	.10059	.10256	.10458	.10666	.10879
13	0	8	.09165	.09263	.09362	.09463	.09564
13	0	9	.08339	.08361	.08381	.08400	.08418
13	0	10	.07576	.07539	.07500	.07460	.07419
13	0	11	.06861	.06782	.06702	.06623	.06543
13	0	12	.06167	.06063	.05960	.05858	.05759
13	0	13	.05375	.05269	.05167	.05068	.04972
13	1	1	-.21224	-.24681	-.28400	-.32390	-.36657
13	1	2	.12800	.13762	.14822	.15986	.17257
13	1	3	.13957	.14848	.15811	.16846	.17957
13	1	4	.13466	.14165	.14909	.15699	.16536
13	1	5	.12560	.13070	.13607	.14171	.14760
13	1	6	.11567	.11916	.12278	.12654	.13043
13	1	7	.10594	.10810	.11031	.11258	.11490
13	1	8	.09677	.09786	.09895	.10006	.10118
13	1	9	.08822	.08847	.08872	.08895	.08918
13	1	10	.08026	.07988	.07950	.07910	.07870
13	1	11	.07278	.07197	.07115	.07034	.06953
13	1	12	.12478	.12291	.12109	.11931	.11756
13	2	1	-.27192	-.30969	-.35019	-.39350	-.43968
13	2	2	.12522	.13556	.14694	.15941	.17300
13	2	3	.14301	.15264	.16300	.17412	.18602
13	2	4	.14030	.14785	.15586	.16435	.17331
13	2	5	.13202	.13752	.14330	.14934	.15565
13	2	6	.12225	.12600	.12988	.13390	.13806
13	2	7	.11239	.11469	.11705	.11946	.12192
13	2	8	.10292	.10407	.10522	.10639	.10757
13	2	9	.09400	.09425	.09449	.09472	.09494
13	2	10	.08564	.08522	.08479	.08435	.08391
13	2	11	.21417	.21190	.20966	.20745	.20529
13	3	1	-.34548	-.38693	-.43118	-.47832	-.52844
13	3	2	.12188	.13308	.14537	.15879	.17339
13	3	3	.14734	.15777	.16897	.18095	.19374

Coefficients a_i for the BLUE of μ
Shape parameter k

n	s	i	2.1	2.2	2.3	2.4	2.5
13	3	4	.14732	.15549	.16413	.17325	.18286
13	3	5	.13998	.14591	.15211	.15858	.16532
13	3	6	.13040	.13441	.13855	.14282	.14723
13	3	7	.12034	.12277	.12526	.12779	.13037
13	3	8	.11050	.11167	.11285	.11404	.11524
13	3	9	.10111	.10131	.10150	.10169	.10187
13	3	10	.32662	.32452	.32245	.32041	.31840
13	4	1	-.43809	-.48395	-.53270	-.58442	-.63920
13	4	2	.11789	.13013	.14351	.15808	.17387
13	4	3	.15297	.16437	.17655	.18954	.20336
13	4	4	.15631	.16519	.17456	.18441	.19476
13	4	5	.15013	.15654	.16321	.17015	.17736
13	4	6	.14074	.14502	.14943	.15397	.15864
13	4	7	.13042	.13296	.13555	.13819	.14087
13	4	8	.12007	.12123	.12240	.12357	.12476
13	4	9	.46956	.46851	.46750	.46652	.46558
13	5	1	-.55782	-.60920	-.66358	-.72103	-.78163
13	5	2	.11315	.12669	.14142	.15741	.17468
13	5	3	.16060	.17318	.18657	.20079	.21586
13	5	4	.16823	.17797	.18820	.19892	.21014
13	5	5	.16348	.17044	.17766	.18515	.19290
13	5	6	.15429	.15887	.16356	.16838	.17333
13	5	7	.14356	.14619	.14887	.15159	.15435
13	5	8	.65451	.65586	.65729	.65879	.66037
13	6	1	-.71803	-.77664	-.83839	-.90334	-.97158
13	6	2	.10752	.12275	.13926	.15707	.17623
13	6	3	.17143	.18553	.20046	.21625	.23291
13	6	4	.18468	.19551	.20682	.21863	.23094
13	6	5	.18175	.18938	.19726	.20541	.21381
13	6	6	.17272	.17762	.18264	.18778	.19303
13	6	7	.89994	.90586	.91195	.91821	.92465
13	7	1	-.94266	-1.01124	-1.08317	-1.15851	-1.23733
13	7	2	.10090	.11848	.13740	.15772	.17947
13	7	3	.18770	.20385	.22087	.23878	.25759
13	7	4	.20863	.22089	.23365	.24690	.26065
13	7	5	.20806	.21654	.22528	.23426	.24350
13	7	6	1.23738	1.25147	1.26597	1.28085	1.29613
14	0	1	-.17081	-.20299	-.23775	-.27517	-.31532
14	0	2	.11857	.12683	.13597	.14605	.15709
14	0	3	.12761	.13560	.14426	.15362	.16370
14	0	4	.12292	.12939	.13631	.14369	.15154
14	0	5	.11483	.11972	.12489	.13035	.13610
14	0	6	.10605	.10956	.11322	.11704	.12103
14	0	7	.09749	.09983	.10224	.10473	.10730
14	0	8	.08943	.09081	.09222	.09365	.09510
14	0	9	.08195	.08255	.08316	.08376	.08436
14	0	10	.07501	.07500	.07499	.07496	.07492
14	0	11	.06854	.06808	.06760	.06711	.06661
14	0	12	.06244	.06164	.06084	.06004	.05925
14	0	13	.05645	.05546	.05448	.05351	.05256
14	0	14	.04951	.04853	.04757	.04665	.04575
14	1	1	-.21529	-.25030	-.28802	-.32854	-.37193
14	1	2	.11580	.12458	.13430	.14500	.15672
14	1	3	.12960	.13816	.14743	.15743	.16819
14	1	4	.12666	.13362	.14104	.14896	.15736
14	1	5	.11925	.12452	.13009	.13595	.14212

Coefficients a_i for the BLUE of μ
Shape parameter k

n	s	i	2.1	2.2	2.3	2.4	2.5
14	1	6	.11068	.11446	.11842	.12253	.12682
14	1	7	.10209	.10462	.10723	.10993	.11270
14	1	8	.09388	.09538	.09691	.09847	.10006
14	1	9	.08618	.08685	.08753	.08820	.08887
14	1	10	.07900	.07902	.07902	.07902	.07901
14	1	11	.07227	.07180	.07131	.07082	.07032
14	1	12	.06591	.06509	.06427	.06346	.06265
14	1	13	.11397	.11220	.11047	.10878	.10713
14	2	1	-.26816	-.30611	-.34686	-.39051	-.43714
14	2	2	.11250	.12187	.13222	.14360	.15606
14	2	3	.13194	.14111	.15102	.16170	.17316
14	2	4	.13109	.13854	.14649	.15493	.16389
14	2	5	.12448	.13013	.13608	.14234	.14890
14	2	6	.11615	.12020	.12442	.12881	.13336
14	2	7	.10753	.11023	.11301	.11587	.11881
14	2	8	.09914	.10073	.10235	.10400	.10567
14	2	9	.09118	.09188	.09258	.09328	.09398
14	2	10	.08370	.08370	.08369	.08367	.08365
14	2	11	.07666	.07614	.07561	.07508	.07455
14	2	12	.19379	.19157	.18938	.18723	.18512
14	3	1	-.33211	-.37336	-.41749	-.46461	-.51478
14	3	2	.10856	.11862	.12970	.14187	.15515
14	3	3	.13482	.14468	.15532	.16674	.17897
14	3	4	.13648	.14449	.15300	.16202	.17156
14	3	5	.13084	.13689	.14325	.14991	.15689
14	3	6	.12280	.12712	.13160	.13625	.14107
14	3	7	.11413	.11698	.11991	.12292	.12601
14	3	8	.10551	.10716	.10885	.11055	.11228
14	3	9	.09723	.09792	.09862	.09931	.10000
14	3	10	.08937	.08931	.08926	.08919	.08912
14	3	11	.29240	.29018	.28799	.28584	.28372
14	4	1	-.41082	-.45594	-.50403	-.55517	-.60945
14	4	2	.10386	.11474	.12669	.13977	.15401
14	4	3	.13849	.14917	.16063	.17291	.18602
14	4	4	.14322	.15186	.16102	.17069	.18089
14	4	5	.13876	.14526	.15206	.15917	.16660
14	4	6	.13105	.13565	.14042	.14535	.15044
14	4	7	.12229	.12530	.12837	.13153	.13476
14	4	8	.11337	.11507	.11679	.11854	.12031
14	4	9	.10467	.10532	.10598	.10664	.10730
14	4	10	.41511	.41357	.41206	.41058	.40914
14	5	1	-.50975	-.55958	-.61247	-.66850	-.72776
14	5	2	.09823	.11011	.12311	.13729	.15268
14	5	3	.14335	.15500	.16746	.18075	.19490
14	5	4	.15190	.16129	.17120	.18163	.19259
14	5	5	.14888	.15589	.16321	.17083	.17876
14	5	6	.14155	.14647	.15154	.15677	.16216
14	5	7	.13266	.13580	.13902	.14231	.14567
14	5	8	.12332	.12503	.12677	.12852	.13030
14	5	9	.56986	.56999	.57017	.57041	.57070
14	6	1	-.63745	-.69323	-.75217	-.81437	-.87991
14	6	2	.09144	.10457	.11890	.13444	.15126
14	6	3	.15004	.16289	.17658	.19113	.20655
14	6	4	.16346	.17376	.18458	.19593	.20781
14	6	5	.16222	.16985	.17777	.18599	.19452
14	6	6	.15533	.16060	.16601	.17158	.17730

Coefficients a_i for the BLUE of μ
Shape parameter k

n	s	i	2.1	2.2	2.3	2.4	2.5
14	6	7	.14619	.14948	.15284	.15626	.15975
14	6	8	.76877	.77207	.77550	.77905	.78271
14	7	1	-.80809	-.87167	-.93859	-1.00891	-1.08272
14	7	2	.08315	.09795	.11399	.13132	.14999
14	7	3	.15966	.17408	.18936	.20552	.22258
14	7	4	.17947	.19093	.20291	.21543	.22849
14	7	5	.18053	.18891	.19758	.20654	.21581
14	7	6	.17411	.17979	.18560	.19157	.19768
14	7	7	1.03116	1.04002	1.04915	1.05854	1.06818
15	0	1	-.17736	-.21008	-.24546	-.28359	-.32453
15	0	2	.10811	.11568	.12410	.13340	.14363
15	0	3	.11936	.12706	.13543	.14449	.15426
15	0	4	.11643	.12286	.12975	.13713	.14500
15	0	5	.10972	.11474	.12007	.12570	.13165
15	0	6	.10207	.10581	.10973	.11384	.11814
15	0	7	.09444	.09707	.09980	.10264	.10558
15	0	8	.08716	.08886	.09062	.09242	.09427
15	0	9	.08034	.08128	.08224	.08322	.08420
15	0	10	.07399	.07432	.07465	.07497	.07528
15	0	11	.06807	.06792	.06775	.06757	.06738
15	0	12	.06252	.06200	.06146	.06092	.06037
15	0	13	.05724	.05645	.05566	.05487	.05408
15	0	14	.05202	.05107	.05014	.04922	.04832
15	0	15	.04589	.04496	.04407	.04320	.04236
15	1	1	-.21771	-.25307	-.29121	-.33223	-.37621
15	1	2	.10504	.11303	.12192	.13174	.14253
15	1	3	.12068	.12887	.13777	.14740	.15779
15	1	4	.11942	.12629	.13365	.14151	.14988
15	1	5	.11341	.11878	.12448	.13050	.13684
15	1	6	.10601	.11002	.11423	.11862	.12322
15	1	7	.09841	.10124	.10418	.10723	.11038
15	1	8	.09104	.09289	.09478	.09673	.09872
15	1	9	.08407	.08511	.08615	.08721	.08828
15	1	10	.07754	.07791	.07828	.07865	.07901
15	1	11	.07142	.07127	.07112	.07095	.07078
15	1	12	.06566	.06512	.06458	.06402	.06347
15	1	13	.06018	.05936	.05855	.05775	.05695
15	1	14	.10484	.10316	.10152	.09992	.09836
15	2	1	-.26502	-.30308	-.34403	-.38794	-.43491
15	2	2	.10141	.10989	.11930	.12969	.14109
15	2	3	.12219	.13092	.14039	.15061	.16161
15	2	4	.12290	.13023	.13806	.14641	.15528
15	2	5	.11771	.12344	.12949	.13588	.14261
15	2	6	.11061	.11488	.11935	.12401	.12888
15	2	7	.10305	.10606	.10917	.11240	.11573
15	2	8	.09558	.09753	.09954	.10159	.10368
15	2	9	.08843	.08951	.09061	.09172	.09285
15	2	10	.08167	.08206	.08244	.08281	.08318
15	2	11	.07531	.07514	.07497	.07478	.07459
15	2	12	.06930	.06872	.06814	.06755	.06696
15	2	13	.17686	.17470	.17258	.17050	.16845
15	3	1	-.32134	-.36240	-.40641	-.45348	-.50369
15	3	2	.09712	.10616	.11617	.12720	.13929
15	3	3	.12403	.13336	.14345	.15432	.16599
15	3	4	.12708	.13490	.14324	.15211	.16152
15	3	5	.12285	.12896	.13540	.14217	.14929

Coefficients a_i for the BLUE of μ
Shape parameter k

n	s	i	2.1	2.2	2.3	2.4	2.5
15	3	6	.11610	.12064	.12537	.13030	.13544
15	3	7	.10858	.11176	.11505	.11844	.12194
15	3	8	.10098	.10303	.10512	.10726	.10944
15	3	9	.09361	.09473	.09585	.09699	.09814
15	3	10	.08658	.08695	.08731	.08767	.08802
15	3	11	.07992	.07970	.07947	.07924	.07901
15	3	12	.26449	.26222	.25998	.25778	.25561
15	4	1	-.38938	-.43389	-.48143	-.53210	-.58598
15	4	2	.09206	.10175	.11246	.12422	.13709
15	4	3	.12634	.13636	.14716	.15876	.17120
15	4	4	.13220	.14058	.14949	.15895	.16895
15	4	5	.12913	.13565	.14251	.14970	.15725
15	4	6	.12279	.12761	.13262	.13784	.14325
15	4	7	.11530	.11865	.12211	.12566	.12932
15	4	8	.10753	.10966	.11183	.11404	.11630
15	4	9	.09988	.10101	.10214	.10328	.10443
15	4	10	.09251	.09283	.09314	.09346	.09377
15	4	11	.37164	.36979	.36797	.36618	.36443
15	5	1	-.47300	-.52161	-.57333	-.62826	-.68648
15	5	2	.08602	.09650	.10803	.12067	.13445
15	5	3	.12935	.14018	.15182	.16428	.17759
15	5	4	.13864	.14768	.15725	.16736	.17804
15	5	5	.13697	.14396	.15129	.15696	.16697
15	5	6	.13111	.13624	.14156	.14708	.15279
15	5	7	.12364	.12717	.13079	.13452	.13834
15	5	8	.11564	.11784	.12007	.12234	.12465
15	5	9	.10763	.10873	.10984	.11096	.11209
15	5	10	.50400	.50332	.50268	.50209	.50155
15	6	1	-.57796	-.63159	-.68844	-.74858	-.81211
15	6	2	.07876	.09020	.10275	.11645	.13133
15	6	3	.13342	.14523	.15787	.17136	.18571
15	6	4	.14698	.15678	.16714	.17804	.18952
15	6	5	.14701	.15456	.16243	.17065	.17921
15	6	6	.14172	.14720	.15287	.15872	.16477
15	6	7	.13424	.13795	.14175	.14565	.14964
15	6	8	.12592	.12816	.13043	.13275	.13510
15	6	9	.66991	.67151	.67320	.67497	.67683
15	7	1	-.71327	-.77327	-.83661	-.90337	-.97363
15	7	2	.06993	.08259	.09641	.11144	.12770
15	7	3	.13913	.15217	.16606	.18082	.19647
15	7	4	.15811	.16888	.18019	.19207	.20451
15	7	5	.16029	.16850	.17704	.18591	.19513
15	7	6	.15566	.16155	.16762	.17387	.18031
15	7	7	.14810	.15201	.15600	.16008	.16425
15	7	8	.88203	.88757	.89329	.89918	.90525
15	8	1	-.89386	-.96225	-1.03415	-1.10964	-1.18881
15	8	2	.05901	.07328	.08878	.10554	.12360
15	8	3	.14751	.16214	.17765	.19407	.21140
15	8	4	.17362	.18561	.19815	.21127	.22496
15	8	5	.17854	.18759	.19696	.20666	.21669
15	8	6	.17470	.18108	.18763	.19436	.20127
15	8	7	1.16048	1.17254	1.18497	1.19775	1.21089
16	0	1	-.18278	-.21594	-.25183	-.29053	-.33213
16	0	2	.09873	.10566	.11338	.12195	.13139
16	0	3	.11185	.11924	.12729	.13604	.14549
16	0	4	.11043	.11678	.12361	.13093	.13876

Coefficients a_i for the BLUE of μ
Shape parameter k

n	s	i	2.1	2.2	2.3	2.4	2.5
16	0	5	.10494	.11003	.11545	.12120	.12729
16	0	6	.09827	.10218	.10630	.11064	.11518
16	0	7	.09146	.09433	.09733	.10045	.10370
16	0	8	.08486	.08685	.08891	.09103	.09321
16	0	9	.07863	.07988	.08115	.08245	.08378
16	0	10	.07280	.07343	.07407	.07471	.07535
16	0	11	.06735	.06748	.06761	.06773	.06784
16	0	12	.06225	.06199	.06171	.06143	.06114
16	0	13	.05743	.05687	.05630	.05572	.05514
16	0	14	.05282	.05204	.05126	.05049	.04972
16	0	15	.04822	.04731	.04642	.04555	.04470
16	0	16	.04275	.04188	.04103	.04022	.03942
16	1	1	-.21961	-.25524	-.29372	-.33514	-.37959
16	1	2	.09546	.10273	.11084	.11984	.12975
16	1	3	.11264	.12047	.12900	.13824	.14824
16	1	4	.11282	.11957	.12682	.13459	.14289
16	1	5	.10803	.11345	.11922	.12533	.13179
16	1	6	.10165	.10582	.11022	.11483	.11966
16	1	7	.09491	.09798	.10118	.10452	.10798
16	1	8	.08828	.09041	.09261	.09488	.09722
16	1	9	.08194	.08329	.08466	.08606	.08749
16	1	10	.07597	.07666	.07736	.07806	.07877
16	1	11	.07036	.07053	.07068	.07083	.07097
16	1	12	.06509	.06483	.06457	.06429	.06401
16	1	13	.06010	.05953	.05895	.05836	.05777
16	1	14	.05533	.05453	.05374	.05295	.05217
16	1	15	.09704	.09544	.09388	.09236	.09088
16	2	1	-.26229	-.30042	-.34150	-.38562	-.43286
16	2	2	.09163	.09930	.10785	.11732	.12775
16	2	3	.11353	.12183	.13085	.14062	.15117
16	2	4	.11557	.12273	.13041	.13862	.14738
16	2	5	.11158	.11734	.12345	.12991	.13673
16	2	6	.10554	.10997	.11462	.11949	.12459
16	2	7	.09889	.10214	.10553	.10905	.11270
16	2	8	.09221	.09447	.09679	.09918	.10164
16	2	9	.08576	.08718	.08862	.09009	.09159
16	2	10	.07962	.08035	.08107	.08180	.08254
16	2	11	.07383	.07399	.07414	.07429	.07443
16	2	12	.06836	.06808	.06778	.06749	.06718
16	2	13	.06317	.06256	.06194	.06132	.06070
16	2	14	.16259	.16050	.15844	.15643	.15445
16	3	1	-.31242	-.35330	-.39720	-.44421	-.49442
16	3	2	.08717	.09529	.10433	.11433	.12533
16	3	3	.11460	.12342	.13299	.14333	.15446
16	3	4	.11882	.12643	.13457	.14325	.15250
16	3	5	.11578	.12189	.12835	.13518	.14237
16	3	6	.11012	.11481	.11972	.12486	.13022
16	3	7	.10357	.10701	.11057	.11427	.11809
16	3	8	.09684	.09920	.10164	.10414	.10670
16	3	9	.09024	.09171	.09321	.09473	.09628
16	3	10	.08391	.08464	.08538	.08611	.08685
16	3	11	.07789	.07802	.07815	.07827	.07839
16	3	12	.07218	.07185	.07151	.07116	.07082
16	3	13	.24131	.23903	.23679	.23458	.23240
16	4	1	-.37206	-.41604	-.46312	-.51337	-.56691
16	4	2	.08194	.09059	.10019	.11080	.12243

Coefficients a_i for the BLUE of μ
Shape parameter k

n	s	i	2.1	2.2	2.3	2.4	2.5
16	4	3	.11594	.12535	.13553	.14651	.15830
16	4	4	.12274	.13085	.13950	.14871	.15848
16	4	5	.12082	.12731	.13416	.14138	.14897
16	4	6	.11562	.12058	.12577	.13118	.13682
16	4	7	.10917	.11279	.11654	.12041	.12441
16	4	8	.10237	.10483	.10737	.10996	.11262
16	4	9	.09559	.09709	.09863	.10018	.10176
16	4	10	.08901	.08973	.09045	.09118	.09190
16	4	11	.08271	.08279	.08287	.08295	.08302
16	4	12	.33616	.33412	.33211	.33013	.32818
16	5	1	-.44401	-.49162	-.54239	-.59642	-.65380
16	5	2	.07576	.08503	.09530	.10660	.11898
16	5	3	.11768	.12778	.13867	.15037	.16291
16	5	4	.12758	.13626	.14549	.15529	.16566
16	5	5	.12699	.13392	.14120	.14886	.15689
16	5	6	.12232	.12758	.13307	.13878	.14472
16	5	7	.11600	.11980	.12373	.12779	.13197
16	5	8	.10908	.11165	.11427	.11696	.11971
16	5	9	.10207	.10360	.10515	.10672	.10832
16	5	10	.09518	.09586	.09655	.09723	.09793
16	5	11	.45135	.45014	.44895	.44781	.44671
16	6	1	-.53233	-.58428	-.63947	-.69801	-.75997
16	6	2	.06840	.07843	.08949	.10162	.11487
16	6	3	.12002	.13093	.14265	.15521	.16863
16	6	4	.13369	.14304	.15295	.16343	.17450
16	6	5	.13472	.14214	.14993	.15808	.16661
16	6	6	.13067	.13628	.14210	.14814	.15441
16	6	7	.12447	.12848	.13261	.13686	.14124
16	6	8	.11740	.12006	.12277	.12554	.12837
16	6	9	.11008	.11161	.11316	.11473	.11632
16	6	10	.59286	.59330	.59381	.59438	.59503
16	7	1	-.64306	-.70035	-.76100	-.82508	-.89269
16	7	2	.05954	.07049	.08252	.09567	.10998
16	7	3	.12320	.13510	.14791	.16151	.17598
16	7	4	.14164	.15180	.16252	.17382	.18571
16	7	5	.14465	.15266	.16103	.16977	.17888
16	7	6	.14135	.14734	.15355	.15997	.16662
16	7	7	.13526	.13949	.14384	.14830	.15288
16	7	8	.12796	.13070	.13349	.13634	.13924
16	7	9	.76938	.77269	.77614	.77971	.78342
16	8	1	-.78565	-.84973	-.91730	-.98845	-1.06326
16	8	2	.04870	.06082	.07408	.08851	.10415
16	8	3	.12799	.14113	.15513	.17002	.18581
16	8	4	.15232	.16347	.17520	.18751	.20042
16	8	5	.15780	.16653	.17562	.18507	.19489
16	8	6	.15540	.16185	.16852	.17539	.18248
16	8	7	.14940	.15387	.15846	.16316	.16797
16	8	8	.99405	1.00205	1.01030	1.01880	1.02754
17	0	1	-.18726	-.22078	-.25709	-.29627	-.33842
17	0	2	.09027	.09659	.10366	.11152	.12022
17	0	3	.10497	.11205	.11979	.12821	.13733
17	0	4	.10487	.11111	.11785	.12508	.13284
17	0	5	.10044	.10557	.11104	.11686	.12304
17	0	6	.09466	.09870	.10297	.10747	.11221
17	0	7	.08857	.09163	.09485	.09821	.10171
17	0	8	.08259	.08481	.08712	.08951	.09199

Coefficients a_i for the BLUE of μ
Shape parameter k

n	s	i	2.1	2.2	2.3	2.4	2.5
17	0	9	.07688	.07839	.07993	.08152	.08315
17	0	10	.07150	.07241	.07332	.07425	.07520
17	0	11	.06646	.06687	.06727	.06766	.06806
17	0	12	.06174	.06173	.06171	.06169	.06165
17	0	13	.05729	.05696	.05661	.05626	.05590
17	0	14	.05307	.05249	.05190	.05130	.05071
17	0	15	.04900	.04824	.04748	.04673	.04598
17	0	16	.04492	.04406	.04321	.04238	.04156
17	0	17	.04001	.03918	.03839	.03761	.03687
17	1	1	-.22107	-.25691	-.29565	-.33740	-.38223
17	1	2	.08688	.09347	.10086	.10909	.11819
17	1	3	.10536	.11282	.12098	.12984	.13945
17	1	4	.10678	.11339	.12050	.12815	.13633
17	1	5	.10305	.10848	.11427	.12043	.12697
17	1	6	.09757	.10186	.10639	.11116	.11618
17	1	7	.09158	.09485	.09826	.10183	.10555
17	1	8	.08560	.08797	.09043	.09298	.09561
17	1	9	.07983	.08144	.08310	.08480	.08654
17	1	10	.07435	.07532	.07631	.07731	.07833
17	1	11	.06919	.06963	.07007	.07051	.07095
17	1	12	.06432	.06433	.06433	.06433	.06431
17	1	13	.05974	.05940	.05906	.05870	.05834
17	1	14	.05538	.05478	.05418	.05357	.05297
17	1	15	.05118	.05040	.04962	.04886	.04810
17	1	16	.09029	.08877	.08729	.08584	.08443
17	2	1	-.25985	-.29802	-.33919	-.38345	-.43090
17	2	2	.08295	.08988	.09763	.10625	.11577
17	2	3	.10577	.11365	.12225	.13157	.14166
17	2	4	.10895	.11593	.12343	.13148	.14009
17	2	5	.10601	.11175	.11786	.12435	.13122
17	2	6	.10088	.10541	.11019	.11522	.12050
17	2	7	.09502	.09847	.10207	.10582	.10973
17	2	8	.08904	.09154	.09413	.09680	.09957
17	2	9	.08319	.08489	.08663	.08841	.09023
17	2	10	.07759	.07861	.07965	.08070	.08176
17	2	11	.07229	.07275	.07320	.07365	.07411
17	2	12	.06727	.06727	.06726	.06725	.06722
17	2	13	.06252	.06216	.06179	.06141	.06103
17	2	14	.05799	.05736	.05673	.05609	.05546
17	2	15	.15039	.14837	.14639	.14445	.14254
17	3	1	-.30488	-.34558	-.38935	-.43628	-.48648
17	3	2	.07841	.08571	.09386	.10292	.11293
17	3	3	.10627	.11460	.12367	.13350	.14411
17	3	4	.11147	.11886	.12678	.13526	.14431
17	3	5	.10945	.11552	.12197	.12880	.13602
17	3	6	.10474	.10952	.11455	.11983	.12537
17	3	7	.09902	.10265	.10643	.11036	.11445
17	3	8	.09303	.09565	.09836	.10116	.10404
17	3	9	.08709	.08886	.09067	.09252	.09441
17	3	10	.08136	.08241	.08347	.08454	.08563
17	3	11	.07588	.07633	.07679	.07724	.07768
17	3	12	.07067	.07065	.07061	.07057	.07052
17	3	13	.06573	.06532	.06491	.06449	.06407
17	3	14	.22176	.21950	.21727	.21508	.21292
17	4	1	-.35775	-.40127	-.44793	-.49782	-.55105
17	4	2	.07314	.08086	.08948	.09904	.10957

Coefficients a_i for the BLUE of μ
Shape parameter k

n	s	i	2.1	2.2	2.3	2.4	2.5
17	4	3	.10690	.11575	.12535	.13573	.14692
17	4	4	.11448	.12232	.13070	.13965	.14917
17	4	5	.11354	.11997	.12677	.13397	.14155
17	4	6	.10929	.11434	.11964	.12519	.13100
17	4	7	.10374	.10755	.11152	.11564	.11991
17	4	8	.09774	.10048	.10330	.10621	.10921
17	4	9	.09169	.09352	.09538	.09729	.09924
17	4	10	.08578	.08684	.08792	.08901	.09010
17	4	11	.08010	.08053	.08096	.08138	.08181
17	4	12	.07466	.07458	.07450	.07441	.07432
17	4	13	.30669	.30453	.30240	.30031	.29824
17	5	1	-.42056	-.46733	-.51730	-.57057	-.62724
17	5	2	.06697	.07520	.08435	.09447	.10561
17	5	3	.10774	.11718	.12738	.13839	.15022
17	5	4	.11814	.12648	.13538	.14485	.15491
17	5	5	.11846	.12529	.13249	.14009	.14808
17	5	6	.11477	.12011	.12570	.13154	.13763
17	5	7	.10939	.11341	.11756	.12187	.12634
17	5	8	.10337	.10623	.10917	.11219	.11529
17	5	9	.09718	.09906	.10097	.10292	.10492
17	5	10	.09106	.09212	.09319	.09427	.09536
17	5	11	.08512	.08550	.08589	.08627	.08665
17	5	12	.40836	.40678	.40522	.40371	.40223
17	6	1	-.49626	-.54684	-.60069	-.65792	-.71862
17	6	2	.05970	.06852	.07830	.08909	.10093
17	6	3	.10891	.11902	.12994	.14167	.15424
17	6	4	.12268	.13160	.14109	.15116	.16183
17	6	5	.12451	.13178	.13944	.14749	.15593
17	6	6	.12146	.12713	.13304	.13919	.14560
17	6	7	.11629	.12051	.12488	.12940	.13406
17	6	8	.11022	.11320	.11625	.11938	.12259
17	6	9	.10384	.10576	.10771	.10970	.11172
17	6	10	.09745	.09848	.09953	.10058	.10165
17	6	11	.53119	.53083	.53052	.53027	.53007
17	7	1	-.58908	-.64424	-.70277	-.76477	-.83031
17	7	2	.05105	.06058	.07111	.08269	.09536
17	7	3	.11057	.12150	.13325	.14583	.15928
17	7	4	.12845	.13806	.14824	.15902	.17040
17	7	5	.13210	.13990	.14808	.15665	.16562
17	7	6	.12982	.13585	.14213	.14865	.15541
17	7	7	.12488	.12933	.13392	.13866	.14354
17	7	8	.11873	.12182	.12498	.12822	.13154
17	7	9	.11209	.11403	.11600	.11801	.12005
17	7	10	.68139	.68318	.68506	.68704	.68911
17	8	1	-.70534	-.76616	-.83047	-.89835	-.96989
17	8	2	.04060	.05101	.06247	.07503	.08871
17	8	3	.11301	.12493	.13769	.15132	.16584
17	8	4	.13599	.14643	.15745	.16907	.18131
17	8	5	.14187	.15030	.15910	.16830	.17789
17	8	6	.14052	.14698	.15368	.16062	.16780
17	8	7	.13581	.14052	.14537	.15036	.15549
17	8	8	.12952	.13273	.13601	.13936	.14279
17	8	9	.86801	.87326	.87869	.88429	.89008
17	9	1	-.85492	-.92294	-.99460	-1.06999	-1.14918
17	9	2	.02777	.03931	.05195	.06573	.08069
17	9	3	.11669	.12987	.14391	.15885	.17470

Coefficients a_i for the BLUE of μ
Shape parameter k

n	s	i	2.1	2.2	2.3	2.4	2.5
17	9	4	.14617	.15764	.16971	.18238	.19568
17	9	5	.15486	.16405	.17362	.18358	.19394
17	9	6	.15462	.16160	.16881	.17626	.18394
17	9	7	.15016	.15518	.16032	.16559	.17100
17	9	8	1.10465	1.11530	1.12628	1.13759	1.14923
18	0	1	-.19097	-.22478	-.26143	-.30102	-.34362
18	0	2	.08261	.08835	.09480	.10201	.11000
18	0	3	.09866	.10543	.11285	.12094	.12971
18	0	4	.09971	.10583	.11244	.11956	.12721
18	0	5	.09623	.10135	.10684	.11270	.11892
18	0	6	.09122	.09535	.09973	.10436	.10925
18	0	7	.08578	.08900	.09238	.09594	.09966
18	0	8	.08035	.08277	.08529	.08792	.09064
18	0	9	.07511	.07684	.07863	.08047	.08237
18	0	10	.07015	.07129	.07245	.07364	.07486
18	0	11	.06548	.06612	.06676	.06742	.06807
18	0	12	.06108	.06131	.06153	.06174	.06195
18	0	13	.05695	.05683	.05671	.05657	.05643
18	0	14	.05304	.05265	.05225	.05185	.05144
18	0	15	.04931	.04871	.04811	.04751	.04691
18	0	16	.04569	.04494	.04420	.04347	.04275
18	0	17	.04203	.04121	.04040	.03960	.03883
18	0	18	.03759	.03681	.03605	.03532	.03461
18	1	1	-.22217	-.25816	-.29711	-.33911	-.38425
18	1	2	.07914	.08511	.09183	.09934	.10766
18	1	3	.09873	.10584	.11363	.12211	.13132
18	1	4	.10123	.10767	.11464	.12213	.13017
18	1	5	.09842	.10384	.10963	.11580	.12235
18	1	6	.09373	.09810	.10273	.10762	.11278
18	1	7	.08842	.09184	.09542	.09918	.10311
18	1	8	.08302	.08559	.08826	.09104	.09393
18	1	9	.07775	.07959	.08149	.08345	.08546
18	1	10	.07271	.07393	.07517	.07645	.07774
18	1	11	.06794	.06863	.06933	.07004	.07075
18	1	12	.06344	.06369	.06394	.06419	.06443
18	1	13	.05919	.05908	.05897	.05884	.05871
18	1	14	.05516	.05477	.05437	.05396	.05355
18	1	15	.05131	.05070	.05009	.04948	.04887
18	1	16	.04758	.04682	.04607	.04533	.04460
18	1	17	.08440	.08296	.08154	.08016	.07882
18	2	1	-.25762	-.29580	-.33702	-.38138	-.42897
18	2	2	.07518	.08142	.08843	.09627	.10496
18	2	3	.09877	.10625	.11443	.12332	.13297
18	2	4	.10292	.10971	.11703	.12490	.13333
18	2	5	.10089	.10660	.11269	.11916	.12604
18	2	6	.09657	.10117	.10604	.11118	.11658
18	2	7	.09141	.09500	.09877	.10271	.10683
18	2	8	.08604	.08874	.09155	.09446	.09749
18	2	9	.08073	.08266	.08465	.08670	.08881
18	2	10	.07560	.07688	.07818	.07951	.08087
18	2	11	.07072	.07145	.07217	.07291	.07365
18	2	12	.06610	.06636	.06661	.06686	.06711
18	2	13	.06171	.06160	.06147	.06134	.06120
18	2	14	.05755	.05713	.05671	.05628	.05585
18	2	15	.05357	.05292	.05228	.05165	.05101
18	2	16	.13986	.13791	.13600	.13412	.13229

Coefficients a_i for the BLUE of μ
Shape parameter k

n	s	i	2.1	2.2	2.3	2.4	2.5
18	3	1	-.29838	-.33890	-.38253	-.42937	-.47952
18	3	2	.07063	.07718	.08453	.09273	.10181
18	3	3	.09883	.10671	.11531	.12465	.13475
18	3	4	.10489	.11204	.11974	.12799	.13683
18	3	5	.10374	.10975	.11615	.12295	.13015
18	3	6	.09984	.10468	.10979	.11517	.12083
18	3	7	.09485	.09862	.10257	.10669	.11099
18	3	8	.08951	.09234	.09527	.09832	.10147
18	3	9	.08415	.08616	.08824	.09037	.09256
18	3	10	.07893	.08025	.08160	.08297	.08436
18	3	11	.07392	.07466	.07540	.07614	.07689
18	3	12	.06915	.06940	.06964	.06988	.07012
18	3	13	.06461	.06446	.06431	.06414	.06398
18	3	14	.06028	.05982	.05936	.05889	.05843
18	3	15	.20507	.20284	.20064	.19848	.19636
18	4	1	-.34570	-.38880	-.43508	-.48464	-.53758
18	4	2	.06539	.07229	.08002	.08863	.09816
18	4	3	.09894	.10727	.11633	.12615	.13676
18	4	4	.10719	.11475	.12286	.13154	.14081
18	4	5	.10708	.11341	.12015	.12728	.13483
18	4	6	.10365	.10875	.11411	.11975	.12567
18	4	7	.09886	.10282	.10695	.11126	.11575
18	4	8	.09355	.09651	.09957	.10274	.10601
18	4	9	.08813	.09022	.09237	.09457	.09683
18	4	10	.08279	.08415	.08553	.08693	.08835
18	4	11	.07763	.07837	.07911	.07985	.08060
18	4	12	.07268	.07291	.07312	.07333	.07354
18	4	13	.06796	.06776	.06756	.06736	.06715
18	4	14	.28184	.27961	.27742	.27525	.27313
18	5	1	-.40119	-.44722	-.49650	-.54911	-.60517
18	5	2	.05933	.06662	.07479	.08386	.09389
18	5	3	.09914	.10797	.11755	.12792	.13909
18	5	4	.10996	.11797	.12654	.13569	.14543
18	5	5	.11104	.11774	.12484	.13235	.14027
18	5	6	.10817	.11355	.11919	.12510	.13129
18	5	7	.10360	.10776	.11209	.11659	.12126
18	5	8	.09833	.10141	.10460	.10789	.11128
18	5	9	.09283	.09499	.09720	.09947	.10180
18	5	10	.08734	.08872	.09012	.09154	.09298
18	5	11	.08200	.08271	.08343	.08416	.08489
18	5	12	.07683	.07701	.07718	.07735	.07751
18	5	13	.37262	.37078	.36898	.36721	.36548
18	6	1	-.46705	-.51647	-.56921	-.62535	-.68500
18	6	2	.05225	.06001	.06868	.07829	.08889
18	6	3	.09949	.10889	.11907	.13006	.14187
18	6	4	.11335	.12186	.13095	.14063	.15091
18	6	5	.11584	.12294	.13045	.13837	.14671
18	6	6	.11362	.11929	.12524	.13146	.13795
18	6	7	.10930	.11366	.11820	.12291	.12779
18	6	8	.10405	.10726	.11058	.11399	.11751
18	6	9	.09844	.10067	.10295	.10527	.10765
18	6	10	.09278	.09416	.09557	.09699	.09844
18	6	11	.08719	.08787	.08856	.08925	.08995
18	6	12	.48076	.47984	.47896	.47812	.47733
18	7	1	-.54635	-.59977	-.65659	-.71689	-.78077
18	7	2	.04391	.05223	.06150	.07174	.08299

Coefficients a_i for the BLUE of μ
Shape parameter k

n	s	i	2.1	2.2	2.3	2.4	2.5
18	7	3	.10007	.11016	.12104	.13274	.14529
18	7	4	.11758	.12668	.13638	.14667	.15757
18	7	5	.12174	.12932	.13729	.14568	.15449
18	7	6	.12028	.12630	.13259	.13915	.14598
18	7	7	.11625	.12085	.12561	.13055	.13565
18	7	8	.11101	.11437	.11782	.12136	.12500
18	7	9	.10527	.10755	.10988	.11226	.11469
18	7	10	.09936	.10074	.10213	.10355	.10499
18	7	11	.61087	.61157	.61234	.61319	.61412
18	8	1	-.64349	-.70175	-.76349	-.82882	-.89781
18	8	2	.03397	.04297	.05295	.06394	.07599
18	8	3	.10105	.11195	.12366	.13622	.14965
18	8	4	.12298	.13279	.14320	.15421	.16584
18	8	5	.12917	.13729	.14582	.15476	.16412
18	8	6	.12862	.13504	.14172	.14866	.15588
18	8	7	.12491	.12977	.13479	.13998	.14533
18	8	8	.11967	.12317	.12676	.13044	.13421
18	8	9	.11372	.11606	.11843	.12085	.12332
18	8	10	.76940	.77271	.77616	.77975	.78347
18	9	1	-.76506	-.82929	-.89714	-.96869	-1.04402
18	9	2	.02197	.03180	.04266	.05458	.06759
18	9	3	.10265	.11455	.12729	.14089	.15539
18	9	4	.13009	.14075	.15202	.16390	.17642
18	9	5	.13876	.14754	.15673	.16633	.17635
18	9	6	.13930	.14619	.15333	.16074	.16841
18	9	7	.13595	.14112	.14643	.15190	.15754
18	9	8	.13067	.13433	.13807	.14189	.14581
18	9	9	.96567	.97301	.98061	.98844	.99652
19	0	1	-.19405	-.22810	-.26503	-.30493	-.34791
19	0	2	.07564	.08084	.08671	.09329	.10062
19	0	3	.09285	.09932	.10642	.11417	.12260
19	0	4	.09491	.10089	.10736	.11435	.12187
19	0	5	.09226	.09737	.10285	.10870	.11494
19	0	6	.08795	.09214	.09660	.10133	.10633
19	0	7	.08310	.08644	.08996	.09367	.09756
19	0	8	.07816	.08075	.08345	.08627	.08921
19	0	9	.07336	.07528	.07727	.07933	.08146
19	0	10	.06877	.07011	.07149	.07292	.07437
19	0	11	.06442	.06527	.06614	.06703	.06793
19	0	12	.06032	.06076	.06120	.06164	.06208
19	0	13	.05646	.05655	.05663	.05671	.05678
19	0	14	.05281	.05261	.05241	.05220	.05198
19	0	15	.04934	.04892	.04848	.04805	.04761
19	0	16	.04602	.04542	.04482	.04421	.04361
19	0	17	.04278	.04205	.04134	.04063	.03993
19	0	18	.03948	.03869	.03792	.03716	.03642
19	0	19	.03544	.03470	.03398	.03329	.03261
19	1	1	-.22296	-.25907	-.29817	-.34036	-.38574
19	1	2	.07214	.07753	.08362	.09045	.09805
19	1	3	.09266	.09943	.10686	.11497	.12378
19	1	4	.09610	.10238	.10918	.11651	.12438
19	1	5	.09412	.09949	.10525	.11140	.11795
19	1	6	.09013	.09455	.09925	.10422	.10948
19	1	7	.08542	.08895	.09267	.09658	.10067
19	1	8	.08054	.08327	.08612	.08910	.09220
19	1	9	.07572	.07775	.07986	.08204	.08430

Coefficients a_i for the BLUE of μ
Shape parameter k

n	s	i	2.1	2.2	2.3	2.4	2.5
19	1	10	.07108	.07251	.07398	.07549	.07703
19	1	11	.06666	.06757	.06850	.06945	.07041
19	1	12	.06247	.06295	.06343	.06391	.06439
19	1	13	.05851	.05863	.05873	.05883	.05892
19	1	14	.05476	.05457	.05438	.05417	.05396
19	1	15	.05120	.05077	.05033	.04989	.04945
19	1	16	.04778	.04716	.04655	.04594	.04533
19	1	17	.04445	.04371	.04298	.04226	.04155
19	1	18	.07922	.07784	.07649	.07517	.07389
19	2	1	-.25555	-.29371	-.33494	-.37935	-.42705
19	2	2	.06818	.07379	.08012	.08722	.09514
19	2	3	.09242	.09951	.10728	.11576	.12497
19	2	4	.09742	.10401	.11113	.11880	.12703
19	2	5	.09619	.10183	.10787	.11431	.12116
19	2	6	.09257	.09721	.10214	.10734	.11284
19	2	7	.08803	.09174	.09563	.09972	.10401
19	2	8	.08320	.08607	.08906	.09217	.09541
19	2	9	.07837	.08050	.08270	.08499	.08734
19	2	10	.07367	.07517	.07670	.07828	.07989
19	2	11	.06917	.07012	.07109	.07208	.07308
19	2	12	.06488	.06538	.06587	.06637	.06687
19	2	13	.06082	.06093	.06103	.06113	.06122
19	2	14	.05695	.05675	.05654	.05632	.05610
19	2	15	.05327	.05282	.05236	.05190	.05144
19	2	16	.04974	.04910	.04846	.04783	.04720
19	2	17	.13067	.12879	.12695	.12514	.12337
19	3	1	-.29269	-.33303	-.37651	-.42324	-.47333
19	3	2	.06367	.06953	.07614	.08355	.09179
19	3	3	.09215	.09959	.10774	.11661	.12623
19	3	4	.09893	.10585	.11332	.12134	.12995
19	3	5	.09855	.10448	.11080	.11754	.12469
19	3	6	.09536	.10023	.10538	.11082	.11656
19	3	7	.09101	.09489	.09896	.10323	.10770
19	3	8	.08623	.08923	.09235	.09560	.09897
19	3	9	.08138	.08360	.08590	.08827	.09071
19	3	10	.07662	.07817	.07976	.08139	.08306
19	3	11	.07202	.07300	.07400	.07501	.07604
19	3	12	.06762	.06812	.06862	.06912	.06962
19	3	13	.06343	.06353	.06362	.06370	.06378
19	3	14	.05944	.05921	.05897	.05872	.05847
19	3	15	.05562	.05513	.05464	.05414	.05365
19	3	16	.19065	.18846	.18631	.18419	.18211
19	4	1	-.33538	-.37810	-.42403	-.47328	-.52594
19	4	2	.05852	.06466	.07158	.07933	.08794
19	4	3	.09187	.09970	.10825	.11755	.12761
19	4	4	.10069	.10797	.11581	.12422	.13322
19	4	5	.10129	.10752	.11415	.12120	.12868
19	4	6	.09857	.10368	.10908	.11477	.12075
19	4	7	.09444	.09850	.10276	.10721	.11186
19	4	8	.08973	.09286	.09611	.09949	.10299
19	4	9	.08486	.08717	.08955	.09200	.09453
19	4	10	.08001	.08162	.08326	.08493	.08664
19	4	11	.07531	.07630	.07732	.07834	.07938
19	4	12	.07077	.07126	.07175	.07224	.07273
19	4	13	.06644	.06650	.06656	.06662	.06667
19	4	14	.06229	.06201	.06173	.06144	.06115

Coefficients a_i for the BLUE of μ
Shape parameter k

n	s	i	2.1	2.2	2.3	2.4	2.5
19	4	15	.26060	.25834	.25612	.25393	.25177
19	5	1	-.38489	-.43028	-.47894	-.53098	-.58649
19	5	2	.05260	.05906	.06634	.07448	.08350
19	5	3	.09159	.09986	.10887	.11864	.12920
19	5	4	.10278	.11046	.11871	.12754	.13697
19	5	5	.10451	.11107	.11804	.12543	.13325
19	5	6	.10234	.10771	.11337	.11932	.12557
19	5	7	.09845	.10271	.10716	.11180	.11664
19	5	8	.09382	.09708	.10046	.10397	.10760
19	5	9	.08891	.09130	.09377	.09630	.09890
19	5	10	.08397	.08561	.08729	.08900	.09075
19	5	11	.07913	.08013	.08115	.08217	.08321
19	5	12	.07443	.07490	.07536	.07583	.07630
19	5	13	.06992	.06994	.06995	.06996	.06998
19	5	14	.34245	.34045	.33847	.33653	.33463
19	6	1	-.44290	-.49134	-.54311	-.59832	-.65707
19	6	2	.04575	.05260	.06028	.06885	.07834
19	6	3	.09135	.10012	.10964	.11995	.13106
19	6	4	.10530	.11343	.12214	.13144	.14135
19	6	5	.10834	.11527	.12262	.13039	.13859
19	6	6	.10682	.11247	.11842	.12465	.13119
19	6	7	.10321	.10767	.11233	.11718	.12222
19	6	8	.09865	.10205	.10557	.10921	.11297
19	6	9	.09369	.09617	.09871	.10132	.10399
19	6	10	.08864	.09031	.09202	.09376	.09553
19	6	11	.08362	.08462	.08562	.08664	.08767
19	6	12	.07873	.07916	.07958	.08000	.08043
19	6	13	.43880	.43747	.43618	.43493	.43373
19	7	1	-.51169	-.56367	-.61906	-.67796	-.74046
19	7	2	.03778	.04506	.05322	.06229	.07232
19	7	3	.09120	.10054	.11066	.12158	.13332
19	7	4	.10840	.11705	.12629	.13613	.14658
19	7	5	.11299	.12034	.12811	.13631	.14494
19	7	6	.11221	.11819	.12445	.13100	.13785
19	7	7	.10892	.11362	.11850	.12357	.12883
19	7	8	.10444	.10799	.11166	.11544	.11933
19	7	9	.09942	.10197	.10459	.10727	.11002
19	7	10	.09421	.09590	.09763	.09939	.10118
19	7	11	.08898	.08995	.09093	.09192	.09293
19	7	12	.55313	.55305	.55302	.55306	.55316
19	8	1	-.59445	-.65063	-.71030	-.77356	-.84050
19	8	2	.02838	.03619	.04491	.05458	.06523
19	8	3	.09121	.10123	.11204	.12367	.13614
19	8	4	.11231	.12156	.13141	.14187	.15295
19	8	5	.11874	.12657	.13483	.14351	.15263
19	8	6	.11883	.12517	.13180	.13871	.14592
19	8	7	.11591	.12085	.12599	.13131	.13681
19	8	8	.11150	.11521	.11903	.12296	.12701
19	8	9	.10639	.10902	.11171	.11446	.11728
19	8	10	.10097	.10268	.10441	.10618	.10797
19	8	11	.69022	.69214	.69418	.69631	.69856
19	9	1	-.69575	-.75701	-.82186	-.89040	-.96272
19	9	2	.01719	.02564	.03503	.04541	.05680
19	9	3	.09149	.10232	.11397	.12645	.13980
19	9	4	.11733	.12731	.13788	.14908	.16091
19	9	5	.12598	.13439	.14322	.15248	.16218

Coefficients a_i for the BLUE of μ
Shape parameter k

n	s	i	2.1	2.2	2.3	2.4	2.5
19	9	6	.12712	.13388	.14093	.14826	.15589
19	9	7	.12462	.12986	.13527	.14087	.14666
19	9	8	.12028	.12416	.12815	.13225	.13645
19	9	9	.11502	.11773	.12049	.12331	.12619
19	9	10	.85671	.86171	.86691	.87228	.87784
19	10	1	-.82244	-.88998	-.96125	-1.03634	-1.11535
19	10	2	.00364	.01288	.02311	.03437	.04668
19	10	3	.09226	.10409	.11676	.13029	.14471
19	10	4	.12398	.13483	.14630	.15839	.17113
19	10	5	.13537	.14446	.15399	.16394	.17433
19	10	6	.13776	.14503	.15258	.16041	.16853
19	10	7	.13575	.14132	.14707	.15300	.15910
19	10	8	.13146	.13553	.13971	.14398	.14837
19	10	9	1.06222	1.07182	1.08174	1.09197	1.10250
20	0	1	-.19660	-.23083	-.26799	-.30816	-.35144
20	0	2	.06927	.07396	.07928	.08528	.09197
20	0	3	.08749	.09365	.10043	.10786	.11595
20	0	4	.09043	.09626	.10259	.10943	.11680
20	0	5	.08854	.09360	.09904	.10488	.11110
20	0	6	.08485	.08908	.09359	.09838	.10346
20	0	7	.08052	.08396	.08759	.09142	.09546
20	0	8	.07604	.07875	.08161	.08459	.08771
20	0	9	.07162	.07370	.07587	.07812	.08046
20	0	10	.06737	.06889	.07047	.07210	.07377
20	0	11	.06332	.06437	.06544	.06653	.06765
20	0	12	.05949	.06012	.06076	.06141	.06207
20	0	13	.05587	.05615	.05643	.05671	.05698
20	0	14	.05244	.05243	.05241	.05239	.05235
20	0	15	.04920	.04895	.04868	.04841	.04814
20	0	16	.04610	.04566	.04520	.04474	.04428
20	0	17	.04313	.04253	.04193	.04133	.04073
20	0	18	.04020	.03950	.03881	.03812	.03744
20	0	19	.03722	.03646	.03572	.03500	.03429
20	0	20	.03352	.03281	.03213	.03147	.03083
20	1	1	-.22350	-.25968	-.29889	-.34123	-.38680
20	1	2	.06578	.07062	.07612	.08231	.08924
20	1	3	.08709	.09353	.10060	.10835	.11678
20	1	4	.09136	.09746	.10408	.11123	.11893
20	1	5	.09010	.09542	.10112	.10723	.11375
20	1	6	.08675	.09119	.09592	.10095	.10627
20	1	7	.08258	.08619	.09001	.09404	.09827
20	1	8	.07816	.08102	.08402	.08716	.09044
20	1	9	.07374	.07594	.07823	.08060	.08306
20	1	10	.06946	.07108	.07274	.07446	.07623
20	1	11	.06536	.06647	.06761	.06877	.06995
20	1	12	.06146	.06214	.06282	.06352	.06422
20	1	13	.05776	.05807	.05838	.05868	.05899
20	1	14	.05425	.05426	.05425	.05424	.05422
20	1	15	.05092	.05067	.05041	.05014	.04987
20	1	16	.04774	.04729	.04683	.04636	.04589
20	1	17	.04468	.04407	.04346	.04285	.04225
20	1	18	.04168	.04097	.04026	.03957	.03889
20	1	19	.07463	.07330	.07201	.07076	.06953
20	2	1	-.25359	-.29171	-.33293	-.37737	-.42512
20	2	2	.06185	.06686	.07257	.07899	.08617
20	2	3	.08662	.09335	.10073	.10880	.11759

Coefficients a_i for the BLUE of μ
Shape parameter k

n	s	i	2.1	2.2	2.3	2.4	2.5
20	2	4	.09237	.09875	.10567	.11314	.12116
20	2	5	.09184	.09740	.10337	.10975	.11655
20	2	6	.08885	.09351	.09846	.10370	.10926
20	2	7	.08486	.08865	.09264	.09685	.10126
20	2	8	.08052	.08351	.08665	.08993	.09335
20	2	9	.07611	.07841	.08080	.08327	.08584
20	2	10	.07179	.07348	.07522	.07701	.07885
20	2	11	.06763	.06879	.06997	.07118	.07242
20	2	12	.06365	.06436	.06507	.06579	.06652
20	2	13	.05986	.06019	.06050	.06082	.06113
20	2	14	.05626	.05626	.05625	.05624	.05622
20	2	15	.05284	.05257	.05230	.05202	.05173
20	2	16	.04956	.04908	.04860	.04812	.04763
20	2	17	.04640	.04577	.04514	.04451	.04389
20	2	18	.12259	.12077	.11900	.11725	.11555
20	3	1	-.28763	-.32778	-.37111	-.41772	-.46772
20	3	2	.05740	.06262	.06856	.07524	.08270
20	3	3	.08610	.09313	.10084	.10927	.11842
20	3	4	.09351	.10020	.10743	.11522	.12359
20	3	5	.09380	.09963	.10586	.11252	.11960
20	3	6	.09123	.09611	.10127	.10675	.11253
20	3	7	.08745	.09141	.09558	.09995	.10455
20	3	8	.08318	.08631	.08958	.09299	.09655
20	3	9	.07878	.08117	.08366	.08623	.08889
20	3	10	.07442	.07617	.07798	.07983	.08174
20	3	11	.07019	.07139	.07261	.07386	.07513
20	3	12	.06612	.06684	.06757	.06831	.06905
20	3	13	.06224	.06256	.06287	.06318	.06350
20	3	14	.05853	.05851	.05849	.05846	.05842
20	3	15	.05499	.05470	.05440	.05409	.05378
20	3	16	.05160	.05109	.05058	.05006	.04955
20	3	17	.17808	.17594	.17383	.17176	.16973
20	4	1	-.32642	-.36878	-.41438	-.46333	-.51572
20	4	2	.05236	.05782	.06401	.07097	.07874
20	4	3	.08552	.09290	.10097	.10977	.11932
20	4	4	.09483	.10185	.10942	.11756	.12628
20	4	5	.09606	.10216	.10869	.11563	.12302
20	4	6	.09396	.09906	.10446	.11017	.11619
20	4	7	.09041	.09454	.09889	.10345	.10822
20	4	8	.08623	.08949	.09289	.09644	.10012
20	4	9	.08183	.08432	.08690	.08956	.09231
20	4	10	.07742	.07924	.08110	.08301	.08497
20	4	11	.07311	.07434	.07559	.07686	.07816
20	4	12	.06894	.06967	.07040	.07114	.07189
20	4	13	.06494	.06524	.06554	.06584	.06614
20	4	14	.06111	.06106	.06101	.06094	.06088
20	4	15	.05744	.05711	.05676	.05642	.05608
20	4	16	.24225	.23999	.23776	.23557	.23341
20	5	1	-.37097	-.41578	-.46389	-.51541	-.57043
20	5	2	.04661	.05233	.05881	.06609	.07420
20	5	3	.08490	.09265	.10113	.11034	.12032
20	5	4	.09639	.10376	.11170	.12022	.12934
20	5	5	.09868	.10509	.11192	.11919	.12689
20	5	6	.09713	.10247	.10812	.11407	.12034
20	5	7	.09383	.09816	.10269	.10744	.11240
20	5	8	.08975	.09315	.09669	.10037	.10418

Coefficients a_i for the BLUE of μ
Shape parameter k

n	s	i	2.1	2.2	2.3	2.4	2.5
20	5	9	.08535	.08794	.09060	.09335	.09619
20	5	10	.08089	.08275	.08467	.08663	.08864
20	5	11	.07648	.07773	.07900	.08029	.08160
20	5	12	.07219	.07291	.07363	.07437	.07511
20	5	13	.06805	.06832	.06860	.06887	.06914
20	5	14	.06407	.06397	.06387	.06377	.06367
20	5	15	.31666	.31455	.31246	.31041	.30840
20	6	1	-.42260	-.47017	-.52111	-.57550	-.63347
20	6	2	.04003	.04605	.05286	.06050	.06899
20	6	3	.08424	.09243	.10135	.11103	.12149
20	6	4	.09825	.10603	.11438	.12332	.13287
20	6	5	.10177	.10852	.11570	.12330	.13136
20	6	6	.10084	.10645	.11237	.11859	.12513
20	6	7	.09784	.10236	.10710	.11205	.11721
20	6	8	.09387	.09741	.10109	.10491	.10886
20	6	9	.08947	.09214	.09490	.09774	.10066
20	6	10	.08493	.08684	.08880	.09081	.09286
20	6	11	.08040	.08166	.08294	.08424	.08556
20	6	12	.07596	.07666	.07737	.07808	.07880
20	6	13	.07165	.07189	.07212	.07235	.07258
20	6	14	.40336	.40174	.40016	.39861	.39711
20	7	1	-.48303	-.53378	-.58795	-.64565	-.70697
20	7	2	.03242	.03880	.04599	.05403	.06296
20	7	3	.08357	.09225	.10168	.11188	.12289
20	7	4	.10052	.10875	.11757	.12698	.13701
20	7	5	.10547	.11260	.12016	.12815	.13659
20	7	6	.10525	.11116	.11738	.12390	.13073
20	7	7	.10259	.10734	.11229	.11746	.12283
20	7	8	.09874	.10243	.10626	.11023	.11433
20	7	9	.09433	.09709	.09994	.10287	.10587
20	7	10	.08969	.09165	.09365	.09569	.09778
20	7	11	.08501	.08627	.08755	.08885	.09017
20	7	12	.08039	.08106	.08174	.08242	.08310
20	7	13	.50504	.50438	.50376	.50321	.50270
20	8	1	-.55464	-.60909	-.66704	-.72859	-.79383
20	8	2	.02356	.03035	.03798	.04649	.05592
20	8	3	.08292	.09217	.10218	.11300	.12463
20	8	4	.10334	.11209	.12144	.13139	.14197
20	8	5	.10997	.11753	.12552	.13395	.14283
20	8	6	.11059	.11683	.12338	.13023	.13740
20	8	7	.10831	.11330	.11850	.12390	.12951
20	8	8	.10458	.10845	.11244	.11656	.12081
20	8	9	.10015	.10301	.10595	.10896	.11206
20	8	10	.09539	.09738	.09942	.10149	.10361
20	8	11	.09052	.09177	.09304	.09432	.09563
20	8	12	.62531	.62622	.62721	.62828	.62944
20	9	1	-.64073	-.69958	-.76201	-.82812	-.89802
20	9	2	.01314	.02041	.02857	.03764	.04765
20	9	3	.08235	.09226	.10297	.11449	.12685
20	9	4	.10691	.11628	.12625	.13683	.14805
20	9	5	.11554	.12360	.13209	.14103	.15043
20	9	6	.11714	.12377	.13070	.13794	.14549
20	9	7	.11531	.12058	.12605	.13172	.13760
20	9	8	.11172	.11576	.11993	.12422	.12865
20	9	9	.10723	.11019	.11323	.11633	.11952
20	9	10	.10231	.10433	.10639	.10849	.11063

Coefficients a_i for the BLUE of μ
Shape parameter k

n	s	i	2.1	2.2	2.3	2.4	2.5
20	9	11	.76908	.77239	.77584	.77943	.78315
20	10	1	-.74605	-.81021	-.87807	-.94973	-1.02527
20	10	2	.00070	.00858	.01737	.02710	.03783
20	10	3	.08194	.09266	.10420	.11658	.12981
20	10	4	.11155	.12164	.13236	.14369	.15567
20	10	5	.12259	.13124	.14033	.14987	.15986
20	10	6	.12536	.13244	.13982	.14751	.15550
20	10	7	.12405	.12964	.13542	.14140	.14758
20	10	8	.12060	.12485	.12921	.13369	.13831
20	10	9	.11603	.11909	.12222	.12542	.12869
20	10	10	.94322	.95006	.95714	.96446	.97201
21	0	1	-.19871	-.23308	-.27041	-.31080	-.35433
21	0	2	.06343	.06764	.07245	.07789	.08398
21	0	3	.08252	.08839	.09486	.10196	.10972
21	0	4	.08625	.09193	.09809	.10478	.11199
21	0	5	.08502	.09003	.09543	.10122	.10741
21	0	6	.08190	.08615	.09068	.09552	.10065
21	0	7	.07805	.08156	.08528	.08921	.09335
21	0	8	.07398	.07680	.07978	.08291	.08618
21	0	9	.06991	.07213	.07445	.07687	.07939
21	0	10	.06597	.06765	.06940	.07121	.07308
21	0	11	.06220	.06341	.06466	.06595	.06726
21	0	12	.05861	.05942	.06024	.06108	.06193
21	0	13	.05521	.05567	.05613	.05659	.05706
21	0	14	.05199	.05215	.05231	.05246	.05260
21	0	15	.04893	.04885	.04875	.04864	.04853
21	0	16	.04603	.04573	.04543	.04511	.04479
21	0	17	.04325	.04278	.04231	.04184	.04137
21	0	18	.04056	.03997	.03938	.03878	.03820
21	0	19	.03791	.03723	.03656	.03589	.03524
21	0	20	.03519	.03447	.03376	.03307	.03239
21	0	21	.03179	.03112	.03047	.02984	.02923
21	1	1	-.22381	-.26004	-.29932	-.34177	-.38748
21	1	2	.05996	.06429	.06924	.07484	.08113
21	1	3	.08196	.08807	.09481	.10219	.11025
21	1	4	.08695	.09287	.09931	.10628	.11380
21	1	5	.08635	.09159	.09723	.10328	.10974
21	1	6	.08355	.08800	.09275	.09781	.10317
21	1	7	.07987	.08355	.08745	.09157	.09590
21	1	8	.07587	.07885	.08197	.08525	.08868
21	1	9	.07183	.07417	.07660	.07914	.08179
21	1	10	.06787	.06965	.07149	.07339	.07535
21	1	11	.06406	.06534	.06666	.06802	.06941
21	1	12	.06041	.06128	.06215	.06304	.06395
21	1	13	.05695	.05745	.05794	.05844	.05894
21	1	14	.05366	.05385	.05402	.05419	.05436
21	1	15	.05054	.05046	.05037	.05027	.05017
21	1	16	.04756	.04726	.04695	.04664	.04633
21	1	17	.04470	.04423	.04376	.04328	.04280
21	1	18	.04195	.04134	.04074	.04014	.03954
21	1	19	.03924	.03854	.03786	.03719	.03653
21	1	20	.07053	.06926	.06802	.06682	.06564
21	2	1	-.25173	-.28978	-.33097	-.37540	-.42318
21	2	2	.05608	.06056	.06567	.07146	.07797
21	2	3	.08131	.08768	.09469	.10237	.11075
21	2	4	.08771	.09389	.10060	.10785	.11567

Coefficients a_i for the BLUE of μ
Shape parameter k

n	s	i	2.1	2.2	2.3	2.4	2.5
21	2	5	.08780	.09328	.09916	.10546	.11219
21	2	6	.08538	.09003	.09498	.10025	.10583
21	2	7	.08188	.08573	.08979	.09408	.09860
21	2	8	.07797	.08107	.08433	.08775	.09132
21	2	9	.07395	.07639	.07893	.08158	.08432
21	2	10	.06997	.07182	.07374	.07572	.07777
21	2	11	.06611	.06745	.06883	.07024	.07169
21	2	12	.06241	.06331	.06422	.06514	.06608
21	2	13	.05888	.05939	.05991	.06042	.06094
21	2	14	.05551	.05570	.05588	.05606	.05623
21	2	15	.05231	.05222	.05212	.05202	.05191
21	2	16	.04925	.04893	.04861	.04829	.04796
21	2	17	.04631	.04582	.04532	.04482	.04433
21	2	18	.04347	.04284	.04222	.04160	.04099
21	2	19	.11542	.11367	.11196	.11028	.10864
21	3	1	-.28308	-.32305	-.36621	-.41269	-.46258
21	3	2	.05172	.05636	.06167	.06767	.07441
21	3	3	.08059	.08723	.09454	.10253	.11124
21	3	4	.08855	.09501	.10201	.10956	.11769
21	3	5	.08944	.09515	.10129	.10784	.11484
21	3	6	.08742	.09228	.09744	.10292	.10872
21	3	7	.08414	.08815	.09239	.09685	.10154
21	3	8	.08032	.08356	.08695	.09049	.09420
21	3	9	.07633	.07886	.08150	.08425	.08710
21	3	10	.07233	.07425	.07624	.07829	.08040
21	3	11	.06842	.06981	.07123	.07268	.07417
21	3	12	.06465	.06557	.06651	.06746	.06842
21	3	13	.06104	.06156	.06208	.06260	.06313
21	3	14	.05759	.05777	.05794	.05811	.05828
21	3	15	.05429	.05418	.05407	.05395	.05383
21	3	16	.05113	.05079	.05045	.05010	.04975
21	3	17	.04810	.04758	.04705	.04653	.04601
21	3	18	.16702	.16493	.16288	.16086	.15888
21	4	1	-.31854	-.36057	-.40585	-.45451	-.50664
21	4	2	.04681	.05164	.05715	.06339	.07039
21	4	3	.07979	.08673	.09436	.10269	.11174
21	4	4	.08953	.09628	.10358	.11145	.11991
21	4	5	.09130	.09727	.10367	.11050	.11778
21	4	6	.08975	.09482	.10020	.10590	.11193
21	4	7	.08670	.09089	.09529	.09993	.10480
21	4	8	.08299	.08636	.08988	.09356	.09740
21	4	9	.07902	.08165	.08439	.08723	.09017
21	4	10	.07500	.07699	.07904	.08115	.08333
21	4	11	.07103	.07246	.07392	.07541	.07694
21	4	12	.06718	.06812	.06907	.07004	.07102
21	4	13	.06348	.06400	.06452	.06504	.06556
21	4	14	.05993	.06009	.06025	.06040	.06055
21	4	15	.05652	.05639	.05624	.05610	.05595
21	4	16	.05326	.05288	.05250	.05211	.05173
21	4	17	.22625	.22400	.22178	.21960	.21745
21	5	1	-.35893	-.40321	-.45081	-.50185	-.55641
21	5	2	.04125	.04629	.05205	.05855	.06583
21	5	3	.07891	.08619	.09416	.10286	.11230
21	5	4	.09067	.09774	.10538	.11359	.12241
21	5	5	.09344	.09970	.10638	.11351	.12108
21	5	6	.09242	.09772	.10333	.10927	.11553

Coefficients a_i for the BLUE of μ
Shape parameter k

n	s	i	2.1	2.2	2.3	2.4	2.5
21	5	7	.08964	.09401	.09860	.10341	.10846
21	5	8	.08605	.08955	.09321	.09702	.10100
21	5	9	.08210	.08484	.08767	.09060	.09364
21	5	10	.07805	.08010	.08221	.08439	.08662
21	5	11	.07402	.07548	.07697	.07849	.08005
21	5	12	.07008	.07102	.07198	.07295	.07394
21	5	13	.06627	.06677	.06728	.06778	.06830
21	5	14	.06259	.06273	.06285	.06298	.06310
21	5	15	.05906	.05889	.05870	.05852	.05833
21	5	16	.29438	.29220	.29004	.28792	.28583
21	6	1	-.40527	-.45208	-.50227	-.55594	-.61320
21	6	2	.03493	.04021	.04624	.05303	.06062
21	6	3	.07795	.08560	.09396	.10307	.11293
21	6	4	.09202	.09945	.10746	.11605	.12525
21	6	5	.09595	.10251	.10951	.11695	.12484
21	6	6	.09552	.10107	.10694	.11313	.11965
21	6	7	.09305	.09761	.10239	.10740	.11264
21	6	8	.08958	.09323	.09703	.10099	.10510
21	6	9	.08566	.08850	.09143	.09446	.09759
21	6	10	.08157	.08368	.08586	.08809	.09038
21	6	11	.07746	.07894	.08046	.08201	.08359
21	6	12	.07341	.07435	.07531	.07628	.07726
21	6	13	.06947	.06995	.07043	.07092	.07140
21	6	14	.06566	.06575	.06583	.06592	.06600
21	6	15	.37305	.37122	.36943	.36767	.36595
21	7	1	-.45893	-.50861	-.56172	-.61838	-.67869
21	7	2	.02769	.03325	.03958	.04670	.05466
21	7	3	.07691	.08499	.09379	.10335	.11369
21	7	4	.09365	.10149	.10991	.11892	.12855
21	7	5	.09891	.10582	.11316	.12095	.12920
21	7	6	.09917	.10500	.11114	.11761	.12440
21	7	7	.09704	.10181	.10681	.11203	.11748
21	7	8	.09372	.09752	.10148	.10558	.10984
21	7	9	.08983	.09276	.09579	.09892	.10214
21	7	10	.08568	.08785	.09008	.09236	.09470
21	7	11	.08147	.08297	.08451	.08607	.08767
21	7	12	.07728	.07822	.07916	.08012	.08109
21	7	13	.07319	.07363	.07408	.07452	.07498
21	7	14	.46438	.46329	.46225	.46125	.46030
21	8	1	-.52170	-.57468	-.63117	-.69126	-.75506
21	8	2	.01933	.02523	.03190	.03940	.04776
21	8	3	.07581	.08437	.09368	.10375	.11463
21	8	4	.09567	.10396	.11284	.12234	.13245
21	8	5	.10246	.10976	.11749	.12568	.13432
21	8	6	.10352	.10965	.11610	.12288	.12999
21	8	7	.10178	.10679	.11201	.11746	.12314
21	8	8	.09862	.10259	.10671	.11097	.11538
21	8	9	.09475	.09779	.10092	.10415	.10747
21	8	10	.09053	.09276	.09504	.09737	.09976
21	8	11	.08619	.08771	.08926	.09083	.09243
21	8	12	.08185	.08275	.08367	.08460	.08555
21	8	13	.57118	.57133	.57154	.57183	.57218
21	9	1	-.59603	-.65287	-.71330	-.77740	-.84529
21	9	2	.00961	.01588	.02297	.03091	.03973
21	9	3	.07467	.08379	.09368	.10435	.11584
21	9	4	.09819	.10701	.11644	.12647	.13714

Coefficients a_i for the BLUE of μ
Shape parameter k

n	s	i	2.1	2.2	2.3	2.4	2.5
21	9	5	.10680	.11453	.12271	.13135	.14045
21	9	6	.10878	.11526	.12206	.12919	.13664
21	9	7	.10749	.11276	.11824	.12395	.12988
21	9	8	.10451	.10866	.11296	.11739	.12198
21	9	9	.10065	.10380	.10704	.11038	.11380
21	9	10	.09634	.09862	.10095	.10333	.10576
21	9	11	.09184	.09336	.09490	.09647	.09807
21	9	12	.69716	.69920	.70135	.70361	.70599
21	10	1	-.68534	-.74677	-.81187	-.88076	-.95350
21	10	2	-.00184	.00489	.01247	.02093	.03030
21	10	3	.07351	.08330	.09387	.10525	.11746
21	10	4	.10142	.11087	.12092	.13160	.14292
21	10	5	.11218	.12043	.12914	.13829	.14792
21	10	6	.11525	.12214	.12935	.13687	.14473
21	10	7	.11449	.12006	.12583	.13183	.13805
21	10	8	.11171	.11606	.12055	.12518	.12996
21	10	9	.10784	.11111	.11447	.11791	.12144
21	10	10	.10340	.10573	.10810	.11052	.11300
21	10	11	.84737	.85218	.85718	.86237	.86773
21	11	1	-.79453	-.86151	-.93229	-1.00697	-1.08563
21	11	2	-.01549	-.00820	-.00003	.00905	.01908
21	11	3	.07241	.08300	.09440	.10663	.11971
21	11	4	.10565	.11584	.12665	.13810	.15019
21	11	5	.11901	.12788	.13720	.14698	.15723
21	11	6	.12339	.13076	.13843	.14643	.15476
21	11	7	.12324	.12915	.13527	.14160	.14816
21	11	8	.12067	.12525	.12996	.13482	.13981
21	11	9	.11678	.12017	.12365	.12721	.13086
21	11	10	1.02886	1.03765	1.04675	1.05615	1.06583
22	0	1	-.20043	-.23492	-.27239	-.31294	-.35668
22	0	2	.05806	.06183	.06614	.07105	.07659
22	0	3	.07790	.08349	.08966	.09645	.10387
22	0	4	.08234	.08785	.09386	.10037	.10741
22	0	5	.08171	.08665	.09198	.09771	.10386
22	0	6	.07910	.08335	.08789	.09274	.09790
22	0	7	.07568	.07924	.08302	.08703	.09126
22	0	8	.07198	.07490	.07798	.08122	.08462
22	0	9	.06824	.07058	.07303	.07559	.07826
22	0	10	.06459	.06641	.06830	.07027	.07231
22	0	11	.06106	.06243	.06384	.06529	.06679
22	0	12	.05770	.05866	.05965	.06067	.06170
22	0	13	.05450	.05512	.05575	.05638	.05703
22	0	14	.05146	.05178	.05210	.05242	.05274
22	0	15	.04858	.04865	.04870	.04876	.04880
22	0	16	.04584	.04569	.04553	.04536	.04519
22	0	17	.04323	.04290	.04256	.04221	.04186
22	0	18	.04071	.04024	.03976	.03928	.03880
22	0	19	.03828	.03769	.03711	.03652	.03595
22	0	20	.03586	.03520	.03455	.03391	.03327
22	0	21	.03337	.03267	.03199	.03133	.03068
22	0	22	.03024	.02959	.02897	.02837	.02778
22	1	1	-.22394	-.26019	-.29952	-.34204	-.38785
22	1	2	.05463	.05848	.06291	.06796	.07365
22	1	3	.07721	.08301	.08942	.09646	.10416
22	1	4	.08285	.08859	.09484	.10162	.10895
22	1	5	.08283	.08798	.09354	.09952	.10592

Coefficients a_i for the BLUE of μ
Shape parameter k

n	s	i	2.1	2.2	2.3	2.4	2.5
22	1	6	.08054	.08498	.08973	.09479	.10017
22	1	7	.07730	.08103	.08498	.08916	.09358
22	1	8	.07369	.07674	.07997	.08335	.08691
22	1	9	.06998	.07243	.07500	.07768	.08047
22	1	10	.06632	.06823	.07022	.07228	.07442
22	1	11	.06276	.06420	.06569	.06722	.06879
22	1	12	.05936	.06038	.06143	.06250	.06359
22	1	13	.05611	.05677	.05744	.05811	.05880
22	1	14	.05302	.05336	.05371	.05405	.05440
22	1	15	.05007	.05015	.05023	.05029	.05036
22	1	16	.04727	.04712	.04697	.04681	.04664
22	1	17	.04459	.04426	.04392	.04357	.04322
22	1	18	.04201	.04153	.04105	.04056	.04007
22	1	19	.03951	.03892	.03833	.03774	.03715
22	1	20	.03705	.03638	.03572	.03507	.03443
22	1	21	.06685	.06563	.06444	.06329	.06216
22	2	1	-.24994	-.28791	-.32904	-.37345	-.42122
22	2	2	.05081	.05478	.05935	.06455	.07042
22	2	3	.07642	.08245	.08911	.09641	.10439
22	2	4	.08339	.08937	.09587	.10292	.11052
22	2	5	.08405	.08942	.09521	.10142	.10806
22	2	6	.08212	.08675	.09169	.09696	.10255
22	2	7	.07908	.08296	.08707	.09143	.09602
22	2	8	.07556	.07874	.08210	.08562	.08931
22	2	9	.07188	.07444	.07711	.07990	.08280
22	2	10	.06821	.07021	.07228	.07442	.07664
22	2	11	.06464	.06613	.06768	.06926	.07090
22	2	12	.06118	.06225	.06333	.06444	.06558
22	2	13	.05788	.05856	.05926	.05996	.06067
22	2	14	.05472	.05508	.05544	.05580	.05615
22	2	15	.05171	.05179	.05187	.05194	.05200
22	2	16	.04884	.04869	.04852	.04835	.04818
22	2	17	.04609	.04574	.04539	.04503	.04467
22	2	18	.04344	.04294	.04244	.04193	.04143
22	2	19	.04087	.04026	.03965	.03904	.03844
22	2	20	.10903	.10734	.10569	.10407	.10249
22	3	1	-.27895	-.31872	-.36172	-.40805	-.45783
22	3	2	.04655	.05065	.05538	.06075	.06681
22	3	3	.07555	.08182	.08874	.09632	.10460
22	3	4	.08400	.09022	.09699	.10431	.11220
22	3	5	.08541	.09100	.09702	.10347	.11037
22	3	6	.08389	.08871	.09385	.09931	.10511
22	3	7	.08105	.08510	.08938	.09389	.09865
22	3	8	.07764	.08096	.08444	.08809	.09192
22	3	9	.07401	.07666	.07943	.08232	.08533
22	3	10	.07033	.07240	.07455	.07676	.07906
22	3	11	.06672	.06827	.06987	.07151	.07319
22	3	12	.06322	.06431	.06543	.06657	.06774
22	3	13	.05985	.06055	.06126	.06198	.06270
22	3	14	.05662	.05698	.05734	.05770	.05806
22	3	15	.05354	.05361	.05367	.05373	.05379
22	3	16	.05058	.05041	.05023	.05004	.04986
22	3	17	.04775	.04738	.04700	.04662	.04624
22	3	18	.04502	.04449	.04396	.04343	.04291
22	3	19	.15722	.15519	.15318	.15122	.14929
22	4	1	-.31154	-.35324	-.39822	-.44660	-.49847

Coefficients a_i for the BLUE of μ
Shape parameter k

n	s	i	2.1	2.2	2.3	2.4	2.5
22	4	2	.04178	.04603	.05093	.05650	.06278
22	4	3	.07458	.08112	.08832	.09620	.10479
22	4	4	.08469	.09118	.09823	.10584	.11403
22	4	5	.08694	.09278	.09904	.10575	.11290
22	4	6	.08588	.09090	.09625	.10192	.10794
22	4	7	.08328	.08749	.09194	.09663	.10156
22	4	8	.07999	.08343	.08705	.09084	.09479
22	4	9	.07639	.07915	.08202	.08501	.08811
22	4	10	.07272	.07486	.07707	.07936	.08172
22	4	11	.06906	.07066	.07230	.07399	.07572
22	4	12	.06550	.06663	.06777	.06894	.07012
22	4	13	.06206	.06277	.06349	.06421	.06495
22	4	14	.05875	.05911	.05946	.05981	.06016
22	4	15	.05558	.05563	.05568	.05572	.05576
22	4	16	.05254	.05234	.05213	.05192	.05170
22	4	17	.04961	.04920	.04879	.04838	.04797
22	4	18	.21218	.20995	.20776	.20559	.20347
22	5	1	-.34838	-.39217	-.43931	-.48990	-.54404
22	5	2	.03641	.04084	.04593	.05171	.05823
22	5	3	.07350	.08034	.08785	.09606	.10499
22	5	4	.08549	.09228	.09963	.10755	.11606
22	5	5	.08870	.09479	.10132	.10830	.11573
22	5	6	.08814	.09338	.09895	.10485	.11109
22	5	7	.08582	.09020	.09483	.09969	.10480
22	5	8	.08266	.08624	.08999	.09391	.09800
22	5	9	.07911	.08196	.08493	.08802	.09122
22	5	10	.07542	.07763	.07991	.08227	.08470
22	5	11	.07172	.07336	.07505	.07677	.07854
22	5	12	.06809	.06923	.07040	.07158	.07279
22	5	13	.06457	.06528	.06599	.06672	.06745
22	5	14	.06117	.06150	.06184	.06217	.06251
22	5	15	.05789	.05791	.05793	.05794	.05795
22	5	16	.05474	.05450	.05425	.05401	.05376
22	5	17	.27494	.27272	.27052	.26836	.26623
22	6	1	-.39029	-.43642	-.48594	-.53896	-.59558
22	6	2	.03035	.03497	.04028	.04630	.05308
22	6	3	.07232	.07948	.08734	.09590	.10521
22	6	4	.08644	.09355	.10123	.10949	.11836
22	6	5	.09073	.09711	.10393	.11119	.11892
22	6	6	.09075	.09623	.10203	.10816	.11464
22	6	7	.08874	.09331	.09811	.10316	.10846
22	6	8	.08572	.08944	.09333	.09739	.10163
22	6	9	.08222	.08517	.08825	.09143	.09473
22	6	10	.07851	.08079	.08314	.08556	.08805
22	6	11	.07476	.07644	.07816	.07992	.08172
22	6	12	.07105	.07221	.07338	.07457	.07579
22	6	13	.06743	.06813	.06883	.06955	.07027
22	6	14	.06391	.06423	.06453	.06484	.06515
22	6	15	.06052	.06050	.06048	.06045	.06043
22	6	16	.34684	.34487	.34292	.34101	.33914
22	7	1	-.43836	-.48710	-.53929	-.59503	-.65444
22	7	2	.02346	.02830	.03385	.04015	.04722
22	7	3	.07103	.07856	.08679	.09576	.10548
22	7	4	.08759	.09506	.10311	.11174	.12099
22	7	5	.09311	.09980	.10693	.11452	.12257
22	7	6	.09379	.09952	.10558	.11198	.11871

Coefficients a_i for the BLUE of μ
Shape parameter k

n	s	i	2.1	2.2	2.3	2.4	2.5
22	7	7	.09212	.09689	.10190	.10715	.11264
22	7	8	.08926	.09314	.09718	.10139	.10577
22	7	9	.08581	.08888	.09205	.09534	.09874
22	7	10	.08209	.08443	.08685	.08933	.09189
22	7	11	.07827	.07998	.08173	.08352	.08536
22	7	12	.07446	.07562	.07680	.07800	.07921
22	7	13	.07072	.07140	.07209	.07278	.07348
22	7	14	.06707	.06735	.06762	.06789	.06816
22	7	15	.42959	.42818	.42681	.42549	.42421
22	8	1	-.49399	-.54570	-.60092	-.65976	-.72231
22	8	2	.01557	.02068	.02651	.03311	.04051
22	8	3	.06962	.07756	.08623	.09564	.10582
22	8	4	.08899	.09687	.10533	.11439	.12407
22	8	5	.09594	.10297	.11046	.11840	.12682
22	8	6	.09737	.10338	.10973	.11641	.12343
22	8	7	.09609	.10108	.10631	.11178	.11749
22	8	8	.09341	.09745	.10165	.10602	.11056
22	8	9	.09001	.09319	.09648	.09987	.10338
22	8	10	.08626	.08867	.09115	.09370	.09632
22	8	11	.08236	.08410	.08588	.08770	.08955
22	8	12	.07843	.07959	.08076	.08195	.08316
22	8	13	.07455	.07520	.07585	.07652	.07719
22	8	14	.52538	.52495	.52458	.52427	.52402
22	9	1	-.55902	-.61416	-.67289	-.73529	-.80148
22	9	2	.00649	.01189	.01804	.02499	.03277
22	9	3	.06809	.07651	.08567	.09560	.10631
22	9	4	.09075	.09908	.10801	.11755	.12772
22	9	5	.09934	.10677	.11466	.12300	.13182
22	9	6	.10164	.10797	.11464	.12164	.12898
22	9	7	.10080	.10604	.11152	.11723	.12318
22	9	8	.09833	.10255	.10693	.11147	.11618
22	9	9	.09499	.09829	.10169	.10520	.10881
22	9	10	.09119	.09367	.09622	.09883	.10150
22	9	11	.08719	.08895	.09075	.09258	.09446
22	9	12	.08311	.08425	.08541	.08658	.08778
22	9	13	.63711	.63819	.63936	.64062	.64198
22	10	1	-.63599	-.69514	-.75797	-.82455	-.89499
22	10	2	-.00409	.00166	.00820	.01555	.02376
22	10	3	.06645	.07543	.08516	.09568	.10700
22	10	4	.09297	.10184	.11131	.12140	.13213
22	10	5	.10350	.11139	.11973	.12854	.13782
22	10	6	.10681	.11351	.12054	.12791	.13561
22	10	7	.10650	.11201	.11776	.12375	.12997
22	10	8	.10425	.10867	.11324	.11798	.12287
22	10	9	.10096	.10438	.10791	.11154	.11527
22	10	10	.09710	.09965	.10225	.10492	.10766
22	10	11	.09295	.09473	.09654	.09839	.10028
22	10	12	.76859	.77187	.77531	.77890	.78263
22	11	1	-.72841	-.79234	-.86004	-.93161	-1.00713
22	11	2	-.01654	-.01036	-.00338	.00446	.01318
22	11	3	.06471	.07435	.08476	.09598	.10801
22	11	4	.09586	.10535	.11546	.12621	.13759
22	11	5	.10868	.11711	.12599	.13533	.14516
22	11	6	.11319	.12032	.12777	.13556	.14368
22	11	7	.11348	.11932	.12538	.13167	.13821
22	11	8	.11149	.11613	.12092	.12587	.13098

Coefficients a_i for the BLUE of μ
Shape parameter k

n	s	i	2.1	2.2	2.3	2.4	2.5
22	11	9	.10824	.11180	.11546	.11922	.12309
22	11	10	.10429	.10690	.10957	.11230	.11509
22	11	11	.92500	.93143	.93810	.94501	.95214
23	0	1	-.20184	-.23640	-.27398	-.31466	-.35856
23	0	2	.05311	.05645	.06031	.06472	.06972
23	0	3	.07360	.07891	.08480	.09127	.09837
23	0	4	.07867	.08402	.08986	.09620	.10307
23	0	5	.07858	.08345	.08870	.09437	.10044
23	0	6	.07643	.08067	.08521	.09006	.09524
23	0	7	.07341	.07701	.08084	.08490	.08920
23	0	8	.07006	.07305	.07621	.07954	.08305
23	0	9	.06662	.06905	.07161	.07430	.07710
23	0	10	.06322	.06516	.06718	.06929	.07148
23	0	11	.05993	.06143	.06298	.06458	.06624
23	0	12	.05677	.05788	.05902	.06019	.06139
23	0	13	.05376	.05452	.05530	.05610	.05691
23	0	14	.05089	.05136	.05183	.05230	.05278
23	0	15	.04817	.04837	.04858	.04878	.04898
23	0	16	.04557	.04556	.04554	.04551	.04548
23	0	17	.04310	.04290	.04269	.04248	.04226
23	0	18	.04073	.04037	.04001	.03965	.03928
23	0	19	.03845	.03797	.03748	.03700	.03651
23	0	20	.03622	.03565	.03507	.03450	.03394
23	0	21	.03402	.03338	.03274	.03212	.03151
23	0	22	.03172	.03106	.03040	.02977	.02915
23	0	23	.02882	.02820	.02761	.02703	.02647
23	1	1	-.22392	-.26016	-.29951	-.34207	-.38795
23	1	2	.04973	.05313	.05708	.06160	.06672
23	1	3	.07280	.07831	.08440	.09111	.09845
23	1	4	.07902	.08458	.09065	.09723	.10436
23	1	5	.07952	.08459	.09006	.09595	.10226
23	1	6	.07769	.08211	.08684	.09190	.09728
23	1	7	.07485	.07861	.08260	.08683	.09131
23	1	8	.07159	.07471	.07802	.08149	.08515
23	1	9	.06819	.07074	.07341	.07622	.07914
23	1	10	.06480	.06683	.06895	.07115	.07344
23	1	11	.06149	.06306	.06469	.06637	.06811
23	1	12	.05830	.05947	.06066	.06189	.06316
23	1	13	.05525	.05606	.05688	.05772	.05858
23	1	14	.05233	.05283	.05333	.05384	.05435
23	1	15	.04956	.04978	.05001	.05023	.05045
23	1	16	.04691	.04691	.04690	.04688	.04686
23	1	17	.04438	.04418	.04398	.04377	.04355
23	1	18	.04195	.04160	.04123	.04086	.04049
23	1	19	.03962	.03913	.03864	.03815	.03766
23	1	20	.03734	.03675	.03617	.03560	.03502
23	1	21	.03509	.03444	.03380	.03317	.03256
23	1	22	.06352	.06235	.06121	.06010	.05902
23	2	1	-.24820	-.28608	-.32715	-.37150	-.41925
23	2	2	.04598	.04948	.05353	.05818	.06345
23	2	3	.07191	.07762	.08393	.09087	.09847
23	2	4	.07938	.08516	.09146	.09829	.10568
23	2	5	.08054	.08581	.09149	.09760	.10414
23	2	6	.07907	.08366	.08858	.09382	.09940
23	2	7	.07642	.08033	.08448	.08887	.09352
23	2	8	.07326	.07651	.07994	.08355	.08735

Coefficients a_i for the BLUE of μ
Shape parameter k

n	s	i	2.1	2.2	2.3	2.4	2.5
23	2	9	.06991	.07256	.07534	.07825	.08128
23	2	10	.06652	.06864	.07084	.07313	.07550
23	2	11	.06319	.06483	.06652	.06827	.07007
23	2	12	.05997	.06118	.06243	.06370	.06501
23	2	13	.05687	.05771	.05857	.05944	.06033
23	2	14	.05391	.05442	.05495	.05547	.05600
23	2	15	.05107	.05131	.05154	.05177	.05200
23	2	16	.04837	.04836	.04835	.04834	.04832
23	2	17	.04578	.04557	.04536	.04514	.04492
23	2	18	.04329	.04292	.04254	.04216	.04178
23	2	19	.04089	.04039	.03988	.03937	.03887
23	2	20	.03856	.03795	.03736	.03676	.03618
23	2	21	.10330	.10167	.10007	.09851	.09698
23	3	1	-.27516	-.31474	-.35756	-.40375	-.45339
23	3	2	.04183	.04543	.04961	.05440	.05983
23	3	3	.07091	.07684	.08339	.09058	.09845
23	3	4	.07979	.08579	.09233	.09942	.10706
23	3	5	.08167	.08714	.09304	.09938	.10616
23	3	6	.08059	.08537	.09047	.09591	.10169
23	3	7	.07816	.08222	.08653	.09108	.09589
23	3	8	.07512	.07849	.08205	.08579	.08972
23	3	9	.07181	.07456	.07744	.08045	.08359
23	3	10	.06843	.07062	.07290	.07527	.07772
23	3	11	.06508	.06678	.06852	.07033	.07218
23	3	12	.06182	.06307	.06436	.06567	.06702
23	3	13	.05868	.05954	.06042	.06131	.06222
23	3	14	.05565	.05618	.05671	.05724	.05778
23	3	15	.05276	.05299	.05322	.05345	.05368
23	3	16	.04998	.04997	.04995	.04992	.04989
23	3	17	.04732	.04710	.04687	.04663	.04639
23	3	18	.04477	.04437	.04397	.04357	.04317
23	3	19	.04230	.04177	.04124	.04071	.04019
23	3	20	.14848	.14650	.14455	.14264	.14076
23	4	1	-.30525	-.34664	-.39133	-.43943	-.49106
23	4	2	.03720	.04092	.04524	.05019	.05581
23	4	3	.06981	.07598	.08277	.09024	.09839
23	4	4	.08025	.08650	.09329	.10064	.10857
23	4	5	.08293	.08863	.09476	.10133	.10835
23	4	6	.08230	.08726	.09256	.09820	.10418
23	4	7	.08011	.08433	.08879	.09351	.09849
23	4	8	.07719	.08070	.08438	.08825	.09231
23	4	9	.07394	.07679	.07977	.08288	.08612
23	4	10	.07057	.07283	.07519	.07763	.08015
23	4	11	.06720	.06894	.07074	.07259	.07450
23	4	12	.06389	.06518	.06649	.06784	.06922
23	4	13	.06069	.06157	.06246	.06337	.06429
23	4	14	.05760	.05813	.05866	.05919	.05973
23	4	15	.05463	.05486	.05508	.05530	.05551
23	4	16	.05178	.05175	.05171	.05166	.05161
23	4	17	.04905	.04879	.04854	.04828	.04801
23	4	18	.04641	.04598	.04555	.04512	.04469
23	4	19	.19972	.19751	.19535	.19321	.19112
23	5	1	-.33904	-.38239	-.42909	-.47926	-.53300
23	5	2	.03202	.03588	.04036	.04549	.05130
23	5	3	.06860	.07502	.08209	.08984	.09830
23	5	4	.08078	.08730	.09437	.10200	.11022

Coefficients a_i for the BLUE of μ
Shape parameter k

n	s	i	2.1	2.2	2.3	2.4	2.5
23	5	5	.08437	.09030	.09668	.10350	.11078
23	5	6	.08423	.08940	.09490	.10075	.10694
23	5	7	.08231	.08670	.09133	.09622	.10136
23	5	8	.07953	.08317	.08699	.09099	.09518
23	5	9	.07634	.07929	.08237	.08558	.08892
23	5	10	.07297	.07531	.07774	.08025	.08285
23	5	11	.06958	.07137	.07322	.07512	.07707
23	5	12	.06622	.06753	.06888	.07025	.07165
23	5	13	.06295	.06384	.06475	.06566	.06659
23	5	14	.05979	.06031	.06084	.06136	.06190
23	5	15	.05674	.05694	.05715	.05735	.05755
23	5	16	.05380	.05374	.05367	.05360	.05352
23	5	17	.05097	.05068	.05039	.05010	.04981
23	5	18	.25784	.25560	.25338	.25120	.24905
23	6	1	-.37720	-.42271	-.47161	-.52403	-.58008
23	6	2	.02620	.03022	.03488	.04020	.04623
23	6	3	.06726	.07396	.08134	.08941	.09819
23	6	4	.08142	.08822	.09559	.10353	.11207
23	6	5	.08602	.09221	.09885	.10594	.11350
23	6	6	.08643	.09182	.09755	.10362	.11004
23	6	7	.08482	.08938	.09420	.09926	.10459
23	6	8	.08219	.08597	.08993	.09407	.09840
23	6	9	.07906	.08212	.08531	.08862	.09206
23	6	10	.07571	.07812	.08062	.08320	.08587
23	6	11	.07228	.07412	.07601	.07795	.07995
23	6	12	.06886	.07020	.07156	.07296	.07438
23	6	13	.06552	.06641	.06732	.06823	.06917
23	6	14	.06227	.06277	.06329	.06380	06432
23	6	15	.05912	.05930	.05947	.05965	.05982
23	6	16	.05608	.05598	.05587	.05576	.05566
23	6	17	.32397	.32189	.31984	.31782	.31585
23	7	1	-.42060	-.46850	-.51986	-.57478	-.63338
23	7	2	.01964	.02384	.02869	.03424	.04051
23	7	3	.06578	.07281	.08052	.08893	.09808
23	7	4	.08218	.08931	.09701	.10529	.11417
23	7	5	.08793	.09441	.10134	.10873	.11658
23	7	6	.08897	.09460	.10057	.10688	.11355
23	7	7	.08771	.09246	.09746	.10272	.10824
23	7	8	.08525	.08918	.09328	.09757	.10204
23	7	9	.08219	.08536	.08865	.09207	.09562
23	7	10	.07884	.08132	.08390	.08655	.08928
23	7	11	.07537	.07725	.07919	.08117	.08321
23	7	12	.07188	.07324	.07462	.07603	.07746
23	7	13	.06845	.06934	.07024	.07115	.07207
23	7	14	.06509	.06558	.06606	.06656	.06705
23	7	15	.06183	.06197	.06211	.06225	.06239
23	7	16	.39948	.39784	.39623	.39466	.39313
23	8	1	-.47035	-.52095	-.57506	-.63280	-.69426
23	8	2	.01220	.01660	.02167	.02746	.03400
23	8	3	.06415	.07154	.07963	.08843	.09799
23	8	4	.08312	.09061	.09867	.10733	.11660
23	8	5	.09019	.09698	.10423	.11194	.12012
23	8	6	.09194	.09783	.10406	.11064	.11757
23	8	7	.09106	.09602	.10123	.10670	.11242
23	8	8	.08880	.09288	.09715	.10159	.10622
23	8	9	.08581	.08909	.09250	.09603	.09968

Coefficients a_i for the BLUE of μ
Shape parameter k

n	s	i	2.1	2.2	2.3	2.4	2.5
23	8	10	.08245	.08502	.08766	.09039	.09319
23	8	11	.07893	.08086	.08284	.08486	.08693
23	8	12	.07537	.07673	.07812	.07954	.08098
23	8	13	.07182	.07270	.07358	.07448	.07539
23	8	14	.06834	.06879	.06925	.06971	.07017
23	8	15	.48615	.48528	.48446	.48370	.48299
23	9	1	-.52787	-.58155	-.63881	-.69975	-.76448
23	9	2	.00369	.00832	.01365	.01972	.02656
23	9	3	.06237	.07017	.07867	.08792	.09793
23	9	4	.08430	.09219	.10066	.10974	.11944
23	9	5	.09287	.10002	.10763	.11570	.12425
23	9	6	.09544	.10162	.10815	.11502	.12225
23	9	7	.09500	.10019	.10563	.11133	.11728
23	9	8	.09295	.09721	.10165	.10626	.11105
23	9	9	.09005	.09345	.09698	.10063	.10439
23	9	10	.08668	.08933	.09205	.09484	.09772
23	9	11	.08310	.08506	.08707	.08913	.09123
23	9	12	.07943	.08080	.08219	.08361	.08505
23	9	13	.07575	.07660	.07746	.07834	.07923
23	9	14	.58625	.58659	.58702	.58752	.58809
23	10	1	-.59509	-.65233	-.71322	-.77787	-.84637
23	10	2	-.00612	-.00122	.00441	.01080	.01798
23	10	3	.06042	.06868	.07768	.08743	.09795
23	10	4	.08579	.09414	.10309	.11265	.12284
23	10	5	.09612	.10367	.11168	.12016	.12912
23	10	6	.09963	.10614	.11299	.12020	.12776
23	10	7	.09968	.10514	.11084	.11679	.12299
23	10	8	.09789	.10234	.10696	.11176	.11674
23	10	9	.09506	.09860	.10226	.10603	.10992
23	10	10	.09168	.09441	.09720	.10007	.10302
23	10	11	.08801	.09001	.09205	.09414	.09628
23	10	12	.08421	.08557	.08696	.08838	.08982
23	10	13	.70273	.70486	.70710	.70947	.71195
23	11	1	-.67460	-.73601	-.80117	-.87016	-.94308
23	11	2	-.01753	-.01231	-.00633	.00043	.00801
23	11	3	.05828	.06710	.07666	.08699	.09812
23	11	4	.08771	.09660	.10610	.11622	.12697
23	11	5	.10010	.10812	.11660	.12555	.13499
23	11	6	.10471	.11160	.11883	.12642	.13435
23	11	7	.10535	.11109	.11708	.12332	.12981
23	11	8	.10382	.10849	.11332	.11833	.12351
23	11	9	.10109	.10477	.10857	.11248	.11650
23	11	10	.09768	.10048	.10336	.10630	.10932
23	11	11	.09389	.09592	.09799	.10010	.10226
23	11	12	.83952	.84416	.84900	.85402	.85922
23	12	1	-.77004	-.83641	-.90663	-.98081	-1.05902
23	12	2	-.03097	-.02536	-.01897	-.01176	-.00371
23	12	3	.05597	.06543	.07567	.08669	.09853
23	12	4	.09024	.09977	.10991	.12068	.13211
23	12	5	.10508	.11364	.12267	.13218	.14218
23	12	6	.11099	.11832	.12599	.13402	.14239
23	12	7	.11230	.11839	.12471	.13128	.13810
23	12	8	.11109	.11600	.12108	.12632	.13173
23	12	9	.10844	.11229	.11624	.12030	.12447
23	12	10	.10497	.10786	.11082	.11385	.11695
23	12	11	1.00192	1.01007	1.01851	1.02725	1.03627

Coefficients a_i for the BLUE of μ
Shape parameter k

n	s	i	2.1	2.2	2.3	2.4	2.5
24	0	1	-.20297	-.23759	-.27524	-.31602	-.36004
24	0	2	.04853	.05147	.05490	.05884	.06333
24	0	3	.06959	.07463	.08023	.08641	.09319
24	0	4	.07522	.08041	.08608	.09225	.09894
24	0	5	.07562	.08040	.08558	.09116	.09716
24	0	6	.07390	.07811	.08263	.08748	.09264
24	0	7	.07124	.07486	.07872	.08282	.08717
24	0	8	.06820	.07125	.07447	.07789	.08148
24	0	9	.06503	.06755	.07021	.07299	.07591
24	0	10	.06188	.06392	.06606	.06829	.07061
24	0	11	.05880	.06041	.06209	.06383	.06564
24	0	12	.05583	.05707	.05834	.05965	.06101
24	0	13	.05299	.05389	.05481	.05575	.05671
24	0	14	.05028	.05088	.05149	.05211	.05274
24	0	15	.04770	.04804	.04839	.04873	.04907
24	0	16	.04524	.04536	.04547	.04558	.04569
24	0	17	.04289	.04282	.04274	.04265	.04256
24	0	18	.04065	.04041	.04017	.03992	.03966
24	0	19	.03849	.03812	.03774	.03736	.03697
24	0	20	.03641	.03593	.03544	.03496	.03448
24	0	21	.03437	.03381	.03325	.03269	.03214
24	0	22	.03235	.03172	.03111	.03051	.02992
24	0	23	.03023	.02959	.02896	.02835	.02775
24	0	24	.02753	.02694	.02637	.02581	.02528
24	1	1	-.22376	-.25998	-.29932	-.34190	-.38782
24	1	2	.04520	.04819	.05168	.05570	.06030
24	1	3	.06871	.07392	.07971	.08610	.09310
24	1	4	.07543	.08081	.08669	.09309	.10002
24	1	5	.07641	.08138	.08675	.09255	.09877
24	1	6	.07500	.07938	.08409	.08912	.09449
24	1	7	.07252	.07629	.08031	.08457	.08910
24	1	8	.06958	.07276	.07612	.07967	.08341
24	1	9	.06646	.06909	.07186	.07476	.07781
24	1	10	.06332	.06545	.06769	.07001	.07244
24	1	11	.06023	.06192	.06368	.06550	.06738
24	1	12	.05724	.05854	.05987	.06125	.06267
24	1	13	.05437	.05532	.05628	.05727	.05828
24	1	14	.05162	.05226	.05290	.05356	.05423
24	1	15	.04900	.04936	.04973	.05010	.05047
24	1	16	.04649	.04662	.04675	.04688	.04700
24	1	17	.04410	.04403	.04396	.04388	.04379
24	1	18	.04180	.04157	.04132	.04107	.04082
24	1	19	.03960	.03922	.03884	.03845	.03807
24	1	20	.03747	.03698	.03649	.03600	.03551
24	1	21	.03539	.03481	.03424	.03368	.03312
24	1	22	.03332	.03269	.03207	.03147	.03087
24	1	23	.06051	.05938	.05828	.05722	.05618
24	2	1	-.24652	-.28430	-.32527	-.36956	-.41726
24	2	2	.04154	.04459	.04817	.05230	.05701
24	2	3	.06773	.07312	.07911	.08571	.09294
24	2	4	.07565	.08123	.08732	.09394	.10111
24	2	5	.07725	.08241	.08798	.09398	.10042
24	2	6	.07619	.08074	.08562	.09084	.09639
24	2	7	.07392	.07783	.08199	.08642	.09110
24	2	8	.07108	.07438	.07787	.08155	.08542
24	2	9	.06801	.07075	.07362	.07663	.07978

Coefficients a_i for the BLUE of μ
Shape parameter k

n	s	i	2.1	2.2	2.3	2.4	2.5
24	2	10	.06489	.06710	.06942	.07183	.07434
24	2	11	.06179	.06355	.06537	.06725	.06921
24	2	12	.05878	.06012	.06151	.06293	.06440
24	2	13	.05587	.05685	.05785	.05888	.05993
24	2	14	.05308	.05374	.05441	.05509	.05578
24	2	15	.05041	.05079	.05117	.05155	.05193
24	2	16	.04785	.04799	.04812	.04825	.04838
24	2	17	.04540	.04533	.04526	.04517	.04509
24	2	18	.04306	.04281	.04256	.04230	.04204
24	2	19	.04080	.04041	.04001	.03962	.03922
24	2	20	.03861	.03811	.03760	.03710	.03660
24	2	21	.03648	.03589	.03531	.03473	.03416
24	2	22	.09812	.09655	.09500	.09349	.09202
24	3	1	-.27166	-.31104	-.35369	-.39971	-.44921
24	3	2	.03749	.04063	.04430	.04855	.05340
24	3	3	.06664	.07223	.07843	.08525	.09273
24	3	4	.07589	.08168	.08799	.09484	.10226
24	3	5	.07818	.08354	.08931	.09552	.10218
24	3	6	.07751	.08223	.08729	.09269	.09843
24	3	7	.07545	.07951	.08383	.08840	.09324
24	3	8	.07273	.07616	.07977	.08358	.08758
24	3	9	.06972	.07256	.07553	.07864	.08189
24	3	10	.06661	.06891	.07130	.07380	.07639
24	3	11	.06351	.06533	.06721	.06915	.07116
24	3	12	.06047	.06186	.06329	.06475	.06626
24	3	13	.05752	.05853	.05956	.06062	.06169
24	3	14	.05469	.05536	.05605	.05674	.05745
24	3	15	.05196	.05235	.05273	.05312	.05351
24	3	16	.04935	.04948	.04961	.04974	.04986
24	3	17	.04684	.04676	.04667	.04658	.04648
24	3	18	.04444	.04417	.04390	.04363	.04335
24	3	19	.04212	.04171	.04129	.04087	.04046
24	3	20	.03987	.03934	.03882	.03829	.03777
24	3	21	.14064	.13871	.13681	.13495	.13313
24	4	1	-.29956	-.34066	-.38506	-.43289	-.48426
24	4	2	.03300	.03624	.04002	.04440	.04940
24	4	3	.06544	.07125	.07767	.08473	.09246
24	4	4	.07616	.08217	.08872	.09582	.10348
24	4	5	.07922	.08478	.09077	.09720	.10409
24	4	6	.07898	.08388	.08912	.09470	.10064
24	4	7	.07716	.08137	.08584	.09057	.09557
24	4	8	.07457	.07812	.08186	.08579	.08992
24	4	9	.07163	.07456	.07763	.08084	.08419
24	4	10	.06854	.07091	.07338	.07595	.07861
24	4	11	.06542	.06729	.06923	.07123	.07330
24	4	12	.06235	.06378	.06524	.06675	.06829
24	4	13	.05936	.06039	.06145	.06252	.06362
24	4	14	.05647	.05715	.05785	.05855	.05926
24	4	15	.05368	.05407	.05445	.05484	.05522
24	4	16	.05101	.05113	.05125	.05136	.05147
24	4	17	.04844	.04834	.04823	.04812	.04800
24	4	18	.04596	.04567	.04538	.04508	.04479
24	4	19	.04358	.04313	.04269	.04225	.04181
24	4	20	.18859	.18643	.18429	.18220	.18014
24	5	1	-.33071	-.37363	-.41993	-.46969	-.52305
24	5	2	.02801	.03135	.03527	.03979	.04495

Coefficients a_i for the BLUE of μ
Shape parameter k

n	s	i	2.1	2.2	2.3	2.4	2.5
24	5	3	.06412	.07015	.07682	.08414	.09214
24	5	4	.07648	.08273	.08953	.09688	.10482
24	5	5	.08040	.08617	.09239	.09905	.10618
24	5	6	.08062	.08572	.09115	.09693	.10307
24	5	7	.07907	.08345	.08808	.09297	.09814
24	5	8	.07664	.08031	.08418	.08824	.09251
24	5	9	.07376	.07679	.07996	.08328	.08673
24	5	10	.07069	.07314	.07568	.07833	.08107
24	5	11	.06756	.06949	.07148	.07353	.07565
24	5	12	.06445	.06592	.06741	.06895	.07053
24	5	13	.06141	.06246	.06353	.06462	.06574
24	5	14	.05846	.05915	.05985	.06055	.06127
24	5	15	.05561	.05598	.05636	.05673	.05711
24	5	16	.05286	.05296	.05306	.05316	.05325
24	5	17	.05022	.05009	.04995	.04981	.04967
24	5	18	.04766	.04734	.04701	.04669	.04636
24	5	19	.24269	.24044	.23822	.23603	.23388
24	6	1	-.36565	-.41058	-.45892	-.51079	-.56630
24	6	2	.02244	.02590	.02996	.03464	.03998
24	6	3	.06266	.06895	.07588	.08348	.09178
24	6	4	.07686	.08337	.09044	.09808	.10630
24	6	5	.08173	.08775	.09421	.10112	.10850
24	6	6	.08249	.08779	.09343	.09943	.10578
24	6	7	.08124	.08578	.09058	.09565	.10099
24	6	8	.07897	.08278	.08678	.09098	.09538
24	6	9	.07616	.07930	.08258	.08599	.08955
24	6	10	.07311	.07564	.07826	.08098	.08380
24	6	11	.06997	.07195	.07399	.07609	.07826
24	6	12	.06682	.06832	.06985	.07141	.07302
24	6	13	.06372	.06479	.06587	.06697	.06809
24	6	14	.06070	.06139	.06208	.06278	.06349
24	6	15	.05777	.05813	.05849	.05885	.05921
24	6	16	.05494	.05502	.05509	.05516	.05522
24	6	17	.05221	.05204	.05187	.05170	.05153
24	6	18	.30384	.30169	.29956	.29748	.29543
24	7	1	-.40510	-.45224	-.50285	-.55703	-.61489
24	7	2	.01619	.01979	.02401	.02887	.03440
24	7	3	.06106	.06762	.07485	.08276	.09138
24	7	4	.07732	.08413	.09149	.09943	.10797
24	7	5	.08328	.08955	.09627	.10346	.11111
24	7	6	.08463	.09015	.09602	.10224	.10882
24	7	7	.08372	.08844	.09342	.09867	.10419
24	7	8	.08162	.08558	.08972	.09407	.09861
24	7	9	.07890	.08215	.08553	.08906	.09272
24	7	10	.07587	.07848	.08118	.08397	.08686
24	7	11	.07271	.07474	.07683	.07898	.08120
24	7	12	.06951	.07103	.07259	.07418	.07581
24	7	13	.06634	.06741	.06850	.06961	.07074
24	7	14	.06324	.06392	.06460	.06529	.06599
24	7	15	.06022	.06055	.06089	.06122	.06156
24	7	16	.05729	.05733	.05736	.05740	.05743
24	7	17	.37320	.37137	.36958	.36783	.36612
24	8	1	-.44993	-.49955	-.55268	-.60943	-.66992
24	8	2	.00914	.01291	.01730	.02236	.02811
24	8	3	.05929	.06616	.07372	.08197	.09094
24	8	4	.07790	.08502	.09272	.10100	.10988

Coefficients a_i for the BLUE of μ
Shape parameter k

n	s	i	2.1	2.2	2.3	2.4	2.5
24	8	5	.08507	.09164	.09865	.10613	.11409
24	8	6	.08711	.09287	.09898	.10545	.11227
24	8	7	.08657	.09149	.09667	.10212	.10783
24	8	8	.08467	.08878	.09308	.09758	.10227
24	8	9	.08204	.08541	.08890	.09254	.09631
24	8	10	.07903	.08172	.08450	.08737	.09033
24	8	11	.07584	.07792	.08006	.08226	.08452
24	8	12	.07259	.07413	.07571	.07732	.07897
24	8	13	.06934	.07041	.07150	.07260	.07373
24	8	14	.06614	.06680	.06746	.06813	.06881
24	8	15	.06301	.06331	.06361	.06391	.06421
24	8	16	.45219	.45099	.44982	.44871	.44764
24	9	1	-.50129	-.55370	-.60968	-.66934	-.73279
24	9	2	.00115	.00510	.00969	.01498	.02097
24	9	3	.05733	.06456	.07248	.08111	.09048
24	9	4	.07863	.08611	.09417	.10283	.11210
24	9	5	.08720	.09408	.10142	.10922	.11751
24	9	6	.09000	.09603	.10241	.10915	.11625
24	9	7	.08988	.09502	.10042	.10608	.11201
24	9	8	.08821	.09249	.09696	.10162	.10647
24	9	9	.08568	.08917	.09279	.09654	.10043
24	9	10	.08269	.08546	.08832	.09127	.09431
24	9	11	.07946	.08159	.08378	.08602	.08833
24	9	12	.07614	.07770	.07929	.08092	.08259
24	9	13	.07279	.07385	.07493	.07603	.07715
24	9	14	.06948	.07010	.07074	.07138	.07203
24	9	15	.54265	.54243	.54227	.54217	.54215
24	10	1	-.56066	-.61625	-.67548	-.73846	-.80529
24	10	2	-.00798	-.00382	.00100	.00654	.01281
24	10	3	.05518	.06281	.07115	.08021	.09002
24	10	4	.07957	.08746	.09593	.10501	.11471
24	10	5	.08974	.09698	.10468	.11286	.12151
24	10	6	.09341	.09975	.10643	.11347	.12087
24	10	7	.09379	.09916	.10480	.11070	.11687
24	10	8	.09236	.09683	.10148	.10632	.11135
24	10	9	.08994	.09356	.09731	.10120	.10521
24	10	10	.08697	.08982	.09277	.09580	.09892
24	10	11	.08369	.08587	.08810	.09039	.09273
24	10	12	.08027	.08185	.08346	.08510	.08677
24	10	13	.07681	.07786	.07892	.08000	.08110
24	10	14	.64691	.64813	.64945	.65087	.65239
24	11	1	-.63000	-.68928	-.75228	-.81910	-.88984
24	11	2	-.01849	-.01410	-.00900	-.00318	.00341
24	11	3	.05280	.06089	.06971	.07927	.08959
24	11	4	.08080	.08915	.09810	.10766	.11785
24	11	5	.09282	.10047	.10859	.11718	.12626
24	11	6	.09751	.10418	.11120	.11859	.12633
24	11	7	.09844	.10408	.10999	.11616	.12259
24	11	8	.09730	.10197	.10682	.11186	.11709
24	11	9	.09499	.09876	.10265	.10667	.11083
24	11	10	.09202	.09498	.09801	.10113	.10434
24	11	11	.08868	.09091	.09319	.09552	.09790
24	11	12	.08515	.08674	.08835	.08999	.09167
24	11	13	.76798	.77125	.77467	.77825	.78197
24	12	1	-.71198	-.77558	-.84300	-.91433	-.98967
24	12	2	-.03073	-.02604	-.02063	-.01447	-.00752

Coefficients a_i for the BLUE of μ
Shape parameter k

n	s	i	2.1	2.2	2.3	2.4	2.5
24	12	3	.05017	.05881	.06818	.07832	.08923
24	12	4	.08241	.09131	.10081	.11093	.12169
24	12	5	.09661	.10474	.11334	.12241	.13198
24	12	6	.10249	.10955	.11697	.12474	.13288
24	12	7	.10406	.11002	.11623	.12270	.12944
24	12	8	.10324	.10814	.11322	.11849	.12394
24	12	9	.10106	.10499	.10904	.11321	.11752
24	12	10	.09809	.10114	.10427	.10748	.11078
24	12	11	.09466	.09693	.09925	.10162	.10404
24	12	12	.90990	.91599	.92232	.92889	.93568
25	0	1	-.20386	-.23851	-.27621	-.31707	-.36118
25	0	2	.04429	.04685	.04986	.05336	.05737
25	0	3	.06583	.07062	.07595	.08184	.08831
25	0	4	.07197	.07700	.08250	.08850	.09501
25	0	5	.07282	.07751	.08260	.08810	.09401
25	0	6	.07148	.07566	.08016	.08498	.09013
25	0	7	.06916	.07279	.07666	.08079	.08517
25	0	8	.06641	.06950	.07278	.07625	.07992
25	0	9	.06349	.06608	.06882	.07169	.07471
25	0	10	.06056	.06269	.06493	.06727	.06971
25	0	11	.05768	.05940	.06119	.06305	.06499
25	0	12	.05489	.05624	.05763	.05908	.06057
25	0	13	.05221	.05323	.05427	.05535	.05645
25	0	14	.04965	.05037	.05111	.05186	.05263
25	0	15	.04720	.04766	.04814	.04861	.04909
25	0	16	.04486	.04510	.04534	.04558	.04582
25	0	17	.04263	.04267	.04271	.04275	.04278
25	0	18	.04050	.04037	.04024	.04010	.03996
25	0	19	.03845	.03818	.03791	.03763	.03735
25	0	20	.03648	.03610	.03570	.03531	.03492
25	0	21	.03457	.03409	.03361	.03312	.03264
25	0	22	.03270	.03214	.03159	.03105	.03051
25	0	23	.03083	.03022	.02963	.02905	.02848
25	0	24	.02887	.02825	.02764	.02705	.02648
25	0	25	.02635	.02578	.02523	.02470	.02418
25	1	1	-.22349	-.25967	-.29899	-.34156	-.38749
25	1	2	.04102	.04361	.04667	.05023	.05432
25	1	3	.06489	.06983	.07532	.08140	.08807
25	1	4	.07207	.07727	.08297	.08918	.09591
25	1	5	.07347	.07834	.08361	.08931	.09543
25	1	6	.07245	.07679	.08145	.08645	.09179
25	1	7	.07030	.07407	.07810	.08239	.08694
25	1	8	.06765	.07087	.07428	.07789	.08170
25	1	9	.06480	.06749	.07033	.07332	.07646
25	1	10	.06188	.06410	.06643	.06887	.07141
25	1	11	.05900	.06080	.06266	.06460	.06662
25	1	12	.05620	.05761	.05906	.06057	.06213
25	1	13	.05349	.05456	.05565	.05678	.05794
25	1	14	.05089	.05166	.05244	.05323	.05404
25	1	15	.04841	.04890	.04940	.04991	.05042
25	1	16	.04603	.04629	.04655	.04681	.04707
25	1	17	.04376	.04382	.04387	.04392	.04396
25	1	18	.04158	.04147	.04134	.04121	.04107
25	1	19	.03949	.03923	.03896	.03868	.03840
25	1	20	.03748	.03709	.03670	.03630	.03590
25	1	21	.03553	.03504	.03455	.03406	.03358

Coefficients a_i for the BLUE of μ
Shape parameter k

n	s	i	2.1	2.2	2.3	2.4	2.5
25	1	22	.03362	.03305	.03250	.03195	.03140
25	1	23	.03171	.03110	.03051	.02992	.02935
25	1	24	.05776	.05667	.05562	.05459	.05359
25	2	1	-.24488	-.28255	-.32342	-.36763	-.41526
25	2	2	.03744	.04008	.04320	.04684	.05103
25	2	3	.06384	.06894	.07461	.08088	.08776
25	2	4	.07216	.07755	.08344	.08985	.09681
25	2	5	.07417	.07921	.08467	.09056	.09689
25	2	6	.07348	.07798	.08281	.08799	.09351
25	2	7	.07154	.07545	.07962	.08405	.08876
25	2	8	.06900	.07234	.07587	.07960	.08354
25	2	9	.06620	.06900	.07194	.07504	.07828
25	2	10	.06331	.06562	.06803	.07055	.07318
25	2	11	.06043	.06229	.06423	.06623	.06832
25	2	12	.05761	.05907	.06058	.06214	.06375
25	2	13	.05488	.05598	.05712	.05828	.05948
25	2	14	.05225	.05304	.05384	.05466	.05550
25	2	15	.04972	.05023	.05075	.05128	.05180
25	2	16	.04730	.04757	.04784	.04811	.04837
25	2	17	.04498	.04504	.04510	.04514	.04519
25	2	18	.04276	.04264	.04251	.04237	.04223
25	2	19	.04063	.04035	.04007	.03978	.03949
25	2	20	.03857	.03816	.03776	.03735	.03694
25	2	21	.03657	.03606	.03556	.03506	.03456
25	2	22	.03461	.03403	.03346	.03290	.03235
25	2	23	.09343	.09191	.09041	.08895	.08753
25	3	1	-.26839	-.30759	-.35005	-.39591	-.44525
25	3	2	.03350	.03620	.03940	.04314	.04744
25	3	3	.06268	.06796	.07382	.08029	.08739
25	3	4	.07226	.07784	.08393	.09056	.09774
25	3	5	.07494	.08016	.08581	.09189	.09842
25	3	6	.07462	.07928	.08428	.08963	.09533
25	3	7	.07290	.07695	.08126	.08584	.09070
25	3	8	.07048	.07393	.07759	.08145	.08551
25	3	9	.06774	.07064	.07368	.07688	.08022
25	3	10	.06488	.06726	.06975	.07236	.07506
25	3	11	.06200	.06392	.06592	.06799	.07014
25	3	12	.05916	.06067	.06222	.06383	.06549
25	3	13	.05640	.05754	.05870	.05990	.06113
25	3	14	.05373	.05454	.05537	.05621	.05706
25	3	15	.05116	.05168	.05221	.05274	.05328
25	3	16	.04869	.04896	.04923	.04950	.04977
25	3	17	.04632	.04638	.04643	.04647	.04651
25	3	18	.04405	.04392	.04377	.04363	.04348
25	3	19	.04186	.04157	.04127	.04097	.04066
25	3	20	.03975	.03933	.03890	.03848	.03805
25	3	21	.03770	.03717	.03665	.03613	.03562
25	3	22	.13357	.13169	.12984	.12803	.12626
25	4	1	-.29437	-.33518	-.37931	-.42687	-.47799
25	4	2	.02915	.03193	.03522	.03906	.04348
25	4	3	.06141	.06688	.07294	.07962	.08695
25	4	4	.07238	.07816	.08447	.09132	.09873
25	4	5	.07578	.08120	.08704	.09333	.10008
25	4	6	.07588	.08071	.08589	.09141	.09729
25	4	7	.07440	.07859	.08305	.08779	.09279
25	4	8	.07212	.07569	.07947	.08345	.08764

Coefficients a_i for the BLUE of μ
Shape parameter k

n	s	i	2.1	2.2	2.3	2.4	2.5
25	4	9	.06945	.07244	.07559	.07888	.08233
25	4	10	.06661	.06908	.07164	.07432	.07711
25	4	11	.06373	.06571	.06777	.06990	.07210
25	4	12	.06087	.06242	.06402	.06567	.06737
25	4	13	.05808	.05924	.06044	.06166	.06292
25	4	14	.05536	.05619	.05703	.05789	.05876
25	4	15	.05275	.05328	.05381	.05435	.05489
25	4	16	.05023	.05049	.05076	.05102	.05129
25	4	17	.04780	.04784	.04788	.04791	.04794
25	4	18	.04547	.04532	.04516	.04499	.04483
25	4	19	.04323	.04291	.04259	.04226	.04194
25	4	20	.04105	.04060	.04015	.03970	.03926
25	4	21	.17861	.17649	.17439	.17233	.17031
25	5	1	-.32321	-.36573	-.41164	-.46103	-.51402
25	5	2	.02434	.02720	.03059	.03455	.03911
25	5	3	.06001	.06568	.07196	.07887	.08645
25	5	4	.07252	.07852	.08505	.09214	.09980
25	5	5	.07673	.08235	.08841	.09491	.10188
25	5	6	.07729	.08230	.08766	.09337	.09944
25	5	7	.07607	.08042	.08503	.08992	.09509
25	5	8	.07394	.07764	.08154	.08565	.08997
25	5	9	.07135	.07444	.07769	.08108	.08463
25	5	10	.06854	.07108	.07373	.07649	.07935
25	5	11	.06566	.06770	.06981	.07200	.07426
25	5	12	.06278	.06437	.06601	.06769	.06943
25	5	13	.05994	.06114	.06236	.06360	.06488
25	5	14	.05718	.05802	.05887	.05974	.06062
25	5	15	.05451	.05504	.05557	.05610	.05665
25	5	16	.05193	.05219	.05244	.05269	.05294
25	5	17	.04945	.04947	.04948	.04949	.04950
25	5	18	.04705	.04687	.04668	.04649	.04630
25	5	19	.04473	.04439	.04403	.04368	.04333
25	5	20	.22917	.22692	.22471	.22253	.22039
25	6	1	-.35537	-.39977	-.44759	-.49894	-.55395
25	6	2	.01899	.02195	.02546	.02955	.03425
25	6	3	.05847	.06436	.07088	.07804	.08588
25	6	4	.07269	.07893	.08571	.09306	.10098
25	6	5	.07781	.08365	.08993	.09667	.10387
25	6	6	.07888	.08408	.08963	.09554	.10182
25	6	7	.07795	.08246	.08724	.09229	.09762
25	6	8	.07599	.07982	.08385	.08809	.09254
25	6	9	.07348	.07668	.08003	.08353	.08718
25	6	10	.07071	.07333	.07605	.07889	.08183
25	6	11	.06782	.06992	.07209	.07433	.07664
25	6	12	.06491	.06654	.06821	.06994	.07170
25	6	13	.06203	.06325	.06448	.06575	.06704
25	6	14	.05922	.06006	.06092	.06179	.06267
25	6	15	.05648	.05700	.05753	.05805	.05859
25	6	16	.05384	.05407	.05431	.05454	.05478
25	6	17	.05128	.05127	.05126	.05124	.05123
25	6	18	.04880	.04859	.04837	.04815	.04793
25	6	19	.28600	.28379	.28162	.27949	.27739
25	7	1	-.39143	-.43788	-.48780	-.54131	-.59850
25	7	2	.01303	.01610	.01974	.02397	.02883
25	7	3	.05678	.06292	.06969	.07713	.08525
25	7	4	.07292	.07942	.08647	.09409	.10230

Coefficients a_i for the BLUE of μ
Shape parameter k

n	s	i	2.1	2.2	2.3	2.4	2.5
25	7	5	.07905	.08512	.09165	.09864	.10609
25	7	6	.08069	.08610	.09186	.09798	.10446
25	7	7	.08009	.08477	.08972	.09495	.10045
25	7	8	.07830	.08228	.08645	.09083	.09542
25	7	9	.07589	.07920	.08265	.08626	.09002
25	7	10	.07315	.07585	.07866	.08157	.08459
25	7	11	.07026	.07241	.07464	.07693	.07930
25	7	12	.06732	.06898	.07069	.07244	.07424
25	7	13	.06439	.06562	.06687	.06815	.06946
25	7	14	.06151	.06235	.06321	.06408	.06496
25	7	15	.05870	.05921	.05972	.06023	.06075
25	7	16	.05598	.05619	.05639	.05660	.05681
25	7	17	.05333	.05329	.05324	.05320	.05315
25	7	18	.35005	.34809	.34617	.34428	.34243
25	8	1	-.43212	-.48085	-.53309	-.58897	-.64858
25	8	2	.00636	.00955	.01332	.01771	.02275
25	8	3	.05491	.06132	.06838	.07612	.08456
25	8	4	.07321	.08000	.08735	.09527	.10379
25	8	5	.08048	.08682	.09361	.10087	.10860
25	8	6	.08276	.08839	.09438	.10073	.10745
25	8	7	.08253	.08739	.09253	.09794	.10364
25	8	8	.08095	.08507	.08939	.09392	.09865
25	8	9	.07863	.08206	.08563	.08934	.09321
25	8	10	.07593	.07871	.08160	.08459	.08769
25	8	11	.07303	.07524	.07752	.07987	.08228
25	8	12	.07005	.07174	.07348	.07527	.07709
25	8	13	.06706	.06830	.06956	.07085	.07217
25	8	14	.06411	.06494	.06579	.06665	.06752
25	8	15	.06121	.06170	.06218	.06267	.06317
25	8	16	.05839	.05857	.05874	.05892	.05910
25	8	17	.42252	.42105	.41962	.41824	.41691
25	9	1	-.47835	-.52962	-.58447	-.64300	-.70531
25	9	2	-.00117	.00216	.00610	.01067	.01590
25	9	3	.05285	.05957	.06695	.07502	.08380
25	9	4	.07360	.08071	.08838	.09664	.10551
25	9	5	.08216	.08879	.09587	.10343	.11146
25	9	6	.08516	.09104	.09728	.10388	.11085
25	9	7	.08534	.09041	.09575	.10137	.10726
25	9	8	.08398	.08827	.09275	.09744	.10233
25	9	9	.08178	.08533	.08902	.09286	.09684
25	9	10	.07912	.08199	.08496	.08804	.09122
25	9	11	.07620	.07847	.08080	.08320	.08567
25	9	12	.07317	.07490	.07667	.07848	.08033
25	9	13	.07011	.07136	.07263	.07392	.07524
25	9	14	.06707	.06789	.06873	.06957	.07043
25	9	15	.06408	.06453	.06499	.06545	.06591
25	9	16	.50488	.50421	.50360	.50304	.50255
25	10	1	-.53128	-.58543	-.64322	-.70475	-.77012
25	10	2	-.00969	-.00620	-.00209	.00268	.00814
25	10	3	.05057	.05763	.06537	.07382	.08298
25	10	4	.07413	.08159	.08963	.09827	.10751
25	10	5	.08415	.09110	.09851	.10640	.11477
25	10	6	.08797	.09413	.10064	.10752	.11477
25	10	7	.08861	.09391	.09947	.10531	.11143
25	10	8	.08751	.09197	.09663	.10149	.10655
25	10	9	.08544	.08911	.09293	.09690	.10101

Coefficients a_i for the BLUE of μ
Shape parameter k

n	s	i	2.1	2.2	2.3	2.4	2.5
25	10	10	.08281	.08577	.08884	.09200	.09526
25	10	11	.07987	.08219	.08458	.08704	.08956
25	10	12	.07678	.07853	.08033	.08216	.08404
25	10	13	.07364	.07488	.07615	.07744	.07876
25	10	14	.07049	.07129	.07210	.07292	.07376
25	10	15	.59901	.59952	.60012	.60081	.60158
25	11	1	-.59243	-.64988	-.71103	-.77599	-.84486
25	11	2	-.01943	-.01575	-.01143	-.00644	-.00073
25	11	3	.04805	.05550	.06365	.07252	.08212
25	11	4	.07483	.08270	.09115	.10021	.10990
25	11	5	.08654	.09386	.10164	.10990	.11865
25	11	6	.09130	.09776	.10459	.11177	.11934
25	11	7	.09247	.09802	.10383	.10992	.11629
25	11	8	.09166	.09631	.10116	.10621	.11146
25	11	9	.08972	.09354	.09750	.10160	.10585
25	11	10	.08712	.09019	.09335	.09660	.09996
25	11	11	.08416	.08654	.08898	.09149	.09407
25	11	12	.08099	.08277	.08458	.08643	.08833
25	11	13	.07774	.07898	.08024	.08152	.08283
25	11	14	.70728	.70947	.71179	.71424	.71681
25	12	1	-.66382	-.72508	-.79013	-.85907	-.93199
25	12	2	-.03064	-.02675	-.02219	-.01693	-.01095
25	12	3	.04525	.05316	.06178	.07113	.08123
25	12	4	.07579	.08412	.09305	.10260	.11277
25	12	5	.08945	.09718	.10539	.11408	.12325
25	12	6	.09530	.10211	.10929	.11683	.12474
25	12	7	.09708	.10290	.10900	.11537	.12201
25	12	8	.09659	.10146	.10652	.11179	.11725
25	12	9	.09479	.09877	.10289	.10714	.11154
25	12	10	.09223	.09540	.09866	.10202	.10547
25	12	11	.08922	.09166	.09416	.09673	.09935
25	12	12	.08596	.08775	.08958	.09145	.09336
25	12	13	.83280	.83731	.84200	.84688	.85195
25	13	1	-.74820	-.81393	-.88356	-.95717	-1.03486
25	13	2	-.04369	-.03954	-.03469	-.02913	-.02281
25	13	3	.04214	.05058	.05975	.06967	.08036
25	13	4	.07710	.08598	.09546	.10557	.11632
25	13	5	.09306	.10127	.10997	.11915	.12882
25	13	6	.10017	.10739	.11497	.12291	.13124
25	13	7	.10265	.10880	.11522	.12191	.12887
25	13	8	.10253	.10765	.11296	.11846	.12416
25	13	9	.10090	.10506	.10934	.11377	.11833
25	13	10	.09836	.10165	.10502	.10848	.11204
25	13	11	.09529	.09779	.10034	.10296	.10564
25	13	12	.97968	.98731	.99523	1.00343	1.01190

Table 5. Coefficients b_i for the BLUE of the scale parameter σ for $n = 2(1)25$ and the censoring number $s = 0(1)\left[\frac{n+1}{2}\right]$

Coefficients b_i for the BLUE of σ
Shape parameter k

n	s	i	0.0	0.1	0.2	0.3	0.4	0.5	0.6
2	0	1	-.88623	-.88653	-.88746	-.88898	-.89111	-.89382	-.89710
2	0	2	.88623	.88653	.88746	.88898	.89111	.89382	.89710
3	0	1	-.59082	-.60555	-.62114	-.63755	-.65478	-.67278	-.69154
3	0	2	.00000	.02859	.05714	.08563	.11402	.14230	.17046
3	0	3	.59082	.57696	.56400	.55193	.54076	.53048	.52108
3	1	1	-1.18164	-1.20161	-1.22307	-1.24603	-1.27052	-1.29656	-1.32417
3	1	2	1.18164	1.20161	1.22307	1.24603	1.27052	1.29656	1.32417
4	0	1	-.45394	-.47183	-.49085	-.51097	-.53219	-.55450	-.57787
4	0	2	-.11018	-.09255	-.07431	-.05548	-.03606	-.01607	.00449
4	0	3	.11018	.12721	.14363	.15944	.17467	.18933	.20345
4	0	4	.45394	.43717	.42153	.40701	.39359	.38124	.36993
4	1	1	-.69713	-.71894	-.74212	-.76671	-.79273	-.82019	-.84912
4	1	2	-.12682	-.10101	-.07479	-.04815	-.02108	.00641	.03435
4	1	3	.82395	.81995	.81691	.81486	.81381	.81378	.81476
5	0	1	-.37238	-.39088	-.41067	-.43176	-.45414	-.47782	-.50281
5	0	2	-.13521	-.12421	-.11249	-.10004	-.08685	-.07290	-.05820
5	0	3	.00000	.01440	.02878	.04314	.05746	.07172	.08594
5	0	4	.13521	.14553	.15518	.16419	.17260	.18046	.18780
5	0	5	.37238	.35516	.33920	.32446	.31093	.29855	.28727
5	1	1	-.51173	-.53308	-.55589	-.58018	-.60599	-.63334	-.66226
5	1	2	-.16678	-.15137	-.13529	-.11852	-.10105	-.08288	-.06397
5	1	3	.02740	.04459	.06156	.07833	.09491	.11132	.12758
5	1	4	.65111	.63987	.62962	.62037	.61213	.60489	.59865
5	2	1	-.76958	-.79778	-.82789	-.85998	-.89411	-.93034	-.96875
5	2	2	-.21212	-.18889	-.16502	-.14048	-.11526	-.08934	-.06270
5	2	3	.98170	.98667	.99290	1.00045	1.00937	1.01969	1.03144
6	0	1	-.31752	-.33581	-.35549	-.37658	-.39911	-.42308	-.44852
6	0	2	-.13856	-.13169	-.12414	-.11587	-.10687	-.09710	-.08655
6	0	3	-.04321	-.03203	-.02059	-.00892	.00299	.01511	.02746
6	0	4	.04321	.05414	.06480	.07520	.08534	.09521	.10484
6	0	5	.13856	.14479	.15041	.15547	.16002	.16411	.16779
6	0	6	.31752	.30060	.28501	.27071	.25764	.24575	.23499
6	1	1	-.40969	-.43014	-.45211	-.47563	-.50075	-.52751	-.55594
6	1	2	-.16846	-.15890	-.14865	-.13770	-.12602	-.11358	-.10037
6	1	3	-.04061	-.02714	-.01353	.00022	.01411	.02815	.04234
6	1	4	.07395	.08551	.09669	.10752	.11801	.12821	.13813
6	1	5	.54481	.53067	.51760	.50560	.49465	.48474	.47585
6	2	1	-.55282	-.57768	-.60434	-.63285	-.66328	-.69570	-.73015
6	2	2	-.20914	-.19569	-.18152	-.16661	-.15093	-.13447	-.11721
6	2	3	-.02897	-.01211	.00475	.02164	.03855	.05552	.07256
6	2	4	.79093	.78548	.78110	.77782	.77566	.77465	.77479
6	3	1	-.82436	-.85807	-.89421	-.93288	-.97419	-1.01824	-1.06512
6	3	2	-.27604	-.25526	-.23370	-.21134	-.18815	-.16412	-.13919
6	3	3	1.10040	1.11334	1.12792	1.14423	1.16235	1.18235	1.20431
7	0	1	-.27781	-.29559	-.31482	-.33555	-.35779	-.38159	-.40696
7	0	2	-.13510	-.13095	-.12620	-.12080	-.11472	-.10791	-.10036
7	0	3	-.06246	-.05389	-.04495	-.03565	-.02599	-.01597	-.00560
7	0	4	.00000	.00963	.01926	.02886	.03844	.04798	.05749
7	0	5	.06246	.07068	.07853	.08604	.09322	.10009	.10665
7	0	6	.13510	.13868	.14174	.14434	.14653	.14836	.14989
7	0	7	.27781	.26145	.24644	.23275	.22031	.20905	.19889
7	1	1	-.34400	-.36348	-.38452	-.40716	-.43146	-.45746	-.48520
7	1	2	-.16098	-.15506	-.14851	-.14130	-.13341	-.12480	-.11545
7	1	3	-.06808	-.05774	-.04711	-.03617	-.02492	-.01336	-.00150
7	1	4	.01144	.02211	.03267	.04314	.05352	.06381	.07403
7	1	5	.09007	.09799	.10551	.11266	.11947	.12597	.13220

Coefficients b_i for the BLUE of σ
Shape parameter k

n	s	i	0.0	0.1	0.2	0.3	0.4	0.5	0.6
7	1	6	.47156	.45619	.44195	.42883	.41680	.40584	.39593
7	2	1	-.43696	-.45961	-.48401	-.51023	-.53833	-.56838	-.60044
7	2	2	-.19433	-.18616	-.17731	-.16777	-.15750	-.14649	-.13469
7	2	3	-.07179	-.05912	-.04621	-.03305	-.01963	-.00595	.00801
7	2	4	.03213	.04418	.05602	.06767	.07916	.09049	.10169
7	2	5	.67095	.66071	.65151	.64337	.63630	.63032	.62542
7	3	1	-.58481	-.61287	-.64304	-.67543	-.71012	-.74720	-.78678
7	3	2	-.24284	-.23117	-.21875	-.20556	-.19157	-.17677	-.16112
7	3	3	-.07174	-.05547	-.03904	-.02244	-.00566	.01133	.02854
7	3	4	.89940	.89950	.90082	.90342	.90735	.91265	.91936
7	4	1	-.86817	-.90670	-.94813	-.99262	-1.04029	-1.09131	-1.14581
7	4	2	-.32690	-.30841	-.28907	-.26885	-.24772	-.22566	-.20262
7	4	3	1.19507	1.21512	1.23721	1.26146	1.28801	1.31697	1.34843
8	0	1	-.24759	-.26476	-.28342	-.30363	-.32541	-.34881	-.37388
8	0	2	-.12945	-.12719	-.12439	-.12103	-.11706	-.11243	-.10711
8	0	3	-.07131	-.06475	-.05779	-.05041	-.04262	-.03440	-.02576
8	0	4	-.02296	-.01482	-.00654	.00187	.01039	.01903	.02778
8	0	5	.02296	.03096	.03880	.04650	.05404	.06143	.06867
8	0	6	.07131	.07747	.08326	.08868	.09377	.09854	.10302
8	0	7	.12945	.13123	.13259	.13357	.13423	.13463	.13482
8	0	8	.24759	.23185	.21749	.20445	.19265	.18201	.17246
8	1	1	-.29776	-.31631	-.33644	-.35820	-.38165	-.40686	-.43387
8	1	2	-.15151	-.14800	-.14394	-.13930	-.13404	-.12812	-.12152
8	1	3	-.07964	-.07172	-.06343	-.05476	-.04572	-.03628	-.02645
8	1	4	-.02000	-.01084	-.00161	.00768	.01705	.02649	.03600
8	1	5	.03636	.04469	.05282	.06075	.06849	.07605	.08345
8	1	6	.09505	.10051	.10560	.11034	.11478	.11894	.12285
8	1	7	.41750	.40166	.38699	.37348	.36109	.34978	.33953
8	2	1	-.36376	-.38473	-.40742	-.43191	-.45827	-.48658	-.51689
8	2	2	-.17876	-.17383	-.16831	-.16217	-.15537	-.14789	-.13969
8	2	3	-.08809	-.07851	-.06858	-.05830	-.04766	-.03665	-.02527
8	2	4	-.01320	-.00277	.00767	.01811	.02856	.03904	.04955
8	2	5	.05698	.06574	.07423	.08247	.09047	.09827	.10588
8	2	6	.58682	.57408	.56241	.55180	.54227	.53381	.52641
8	3	1	-.45862	-.48335	-.51007	-.53885	-.56979	-.60299	-.63852
8	3	2	-.21555	-.20866	-.20112	-.19289	-.18395	-.17427	-.16381
8	3	3	-.09700	-.08509	-.07287	-.06033	-.04745	-.03423	-.02064
8	3	4	.00022	.01239	.02447	.03648	.04843	.06035	.07225
8	3	5	.77095	.76472	.75959	.75559	.75277	.75114	.75072
8	4	1	-.61096	-.64188	-.67523	-.71113	-.74970	-.79106	-.83533
8	4	2	-.27072	-.26066	-.24986	-.23828	-.22590	-.21269	-.19862
8	4	3	-.10612	-.09054	-.07471	-.05860	-.04219	-.02548	-.00842
8	4	4	.98780	.99308	.99979	1.00800	1.01779	1.02922	1.04237
9	0	1	-.22373	-.24028	-.25834	-.27797	-.29921	-.32213	-.34677
9	0	2	-.12327	-.12235	-.12098	-.11912	-.11671	-.11372	-.11011
9	0	3	-.07510	-.07010	-.06469	-.05885	-.05258	-.04586	-.03868
9	0	4	-.03597	-.02918	-.02217	-.01494	-.00750	.00014	.00798
9	0	5	.00000	.00724	.01447	.02169	.02889	.03606	.04320
9	0	6	.03597	.04254	.04889	.05502	.06093	.06664	.07216
9	0	7	.07510	.07971	.08395	.08784	.09141	.09469	.09771
9	0	8	.12327	.12379	.12396	.12384	.12348	.12293	.12224
9	0	9	.22373	.20863	.19491	.18249	.17130	.16125	.15227
9	1	1	-.26325	-.28094	-.30021	-.32114	-.34378	-.36821	-.39448
9	1	2	-.14211	-.14028	-.13798	-.13518	-.13182	-.12788	-.12332
9	1	3	-.08408	-.07801	-.07155	-.06469	-.05741	-.04972	-.04158
9	1	4	-.03700	-.02930	-.02143	-.01339	-.00518	.00320	.01176

Coefficients b_i for the BLUE of σ
Shape parameter k

n	s	i	0.0	0.1	0.2	0.3	0.4	0.5	0.6
9	1	5	.00618	.01395	.02166	.02931	.03689	.04443	.05191
9	1	6	.04916	.05568	.06194	.06797	.07378	.07937	.08478
9	1	7	.09539	.09913	.10252	.10562	.10844	.11103	.11342
9	1	8	.37571	.35978	.34505	.33150	.31909	.30777	.29751
9	2	1	-.31289	-.33250	-.35382	-.37692	-.40188	-.42878	-.45768
9	2	2	-.16465	-.16188	-.15859	-.15477	-.15036	-.14534	-.13968
9	2	3	-.09375	-.08646	-.07878	-.07071	-.06223	-.05335	-.04403
9	2	4	-.03636	-.02760	-.01870	-.00968	-.00051	.00879	.01823
9	2	5	.01604	.02444	.03273	.04091	.04899	.05699	.06490
9	2	6	.06776	.07423	.08041	.08633	.09201	.09747	.10275
9	2	7	.52386	.50976	.49675	.48483	.47399	.46422	.45551
9	3	1	-.37969	-.40212	-.42644	-.45276	-.48116	-.51172	-.54455
9	3	2	-.19359	-.18962	-.18508	-.17995	-.17419	-.16778	-.16068
9	3	3	-.10482	-.09593	-.08667	-.07702	-.06697	-.05651	-.04563
9	3	4	-.03325	-.02311	-.01288	-.00257	.00784	.01834	.02896
9	3	5	.03166	.04086	.04987	.05872	.06742	.07599	.08446
9	3	6	.67969	.66991	.66120	.65358	.64707	.64168	.63743
9	4	1	-.47659	-.50324	-.53210	-.56326	-.59686	-.63299	-.67176
9	4	2	-.23352	-.22781	-.22147	-.21447	-.20677	-.19835	-.18918
9	4	3	-.11808	-.10691	-.09538	-.08347	-.07118	-.05847	-.04535
9	4	4	-.02557	-.01349	-.00140	.01071	.02285	.03504	.04731
9	4	5	.85375	.85145	.85035	.85050	.85196	.85477	.85897
9	5	1	-.63301	-.66652	-.70274	-.74183	-.78394	-.82922	-.87782
9	5	2	-.29442	-.28585	-.27654	-.26647	-.25560	-.24392	-.23137
9	5	3	-.13477	-.11995	-.10479	-.08927	-.07338	-.05710	-.04039
9	5	4	1.06220	1.07231	1.08407	1.09757	1.11293	1.13024	1.14958
10	0	1	-.20438	-.22032	-.23778	-.25682	-.27751	-.29990	-.32405
10	0	2	-.11719	-.11726	-.11695	-.11622	-.11501	-.11329	-.11101
10	0	3	-.07626	-.07247	-.06829	-.06369	-.05866	-.05318	-.04723
10	0	4	-.04358	-.03794	-.03204	-.02587	-.01943	-.01274	-.00578
10	0	5	-.01422	-.00783	-.00136	.00520	.01182	.01851	.02526
10	0	6	.01422	.02051	.02671	.03281	.03880	.04469	.05048
10	0	7	.04358	.04896	.05409	.05896	.06360	.06802	.07223
10	0	8	.07626	.07967	.08273	.08547	.08793	.09012	.09209
10	0	9	.11719	.11680	.11612	.11522	.11415	.11296	.11169
10	0	10	.20438	.18989	.17677	.16494	.15431	.14480	.13633
10	1	1	-.23642	-.25332	-.27181	-.29195	-.31382	-.33750	-.36305
10	1	2	-.13341	-.13279	-.13178	-.13033	-.12840	-.12595	-.12295
10	1	3	-.08508	-.08044	-.07541	-.06999	-.06414	-.05787	-.05115
10	1	4	-.04652	-.04010	-.03345	-.02657	-.01945	-.01210	-.00451
10	1	5	-.01192	-.00497	.00202	.00905	.01612	.02323	.03038
10	1	6	.02151	.02802	.03439	.04064	.04676	.05276	.05865
10	1	7	.05587	.06098	.06582	.07042	.07478	.07894	.08291
10	1	8	.09368	.09616	.09834	.10025	.10195	.10345	.10479
10	1	9	.34230	.32648	.31188	.29847	.28621	.27505	.26493
10	2	1	-.27531	-.29378	-.31394	-.33587	-.35964	-.38536	-.41308
10	2	2	-.15234	-.15106	-.14935	-.14718	-.14450	-.14129	-.13749
10	2	3	-.09469	-.08912	-.08317	-.07681	-.07004	-.06284	-.05520
10	2	4	-.04877	-.04147	-.03396	-.02625	-.01832	-.01018	-.00183
10	2	5	-.00765	-.00006	.00754	.01514	.02274	.03036	.03800
10	2	6	.03190	.03866	.04526	.05168	.05796	.06410	.07012
10	2	7	.07223	.07704	.08157	.08585	.08990	.09376	.09744
10	2	8	.47463	.45978	.44605	.43343	.42190	.41145	.40204
10	3	1	-.32524	-.34593	-.36846	-.39292	-.41940	-.44800	-.47881
10	3	2	-.17575	-.17370	-.17117	-.16814	-.16456	-.16041	-.15565
10	3	3	-.10579	-.09907	-.09197	-.08447	-.07654	-.06818	-.05938

Coefficients b_i for the BLUE of σ
Shape parameter k

n	s	i	0.0	0.1	0.2	0.3	0.4	0.5	0.6
10	3	4	-.05018	-.04178	-.03321	-.02447	-.01553	-.00641	.00292
10	3	5	-.00056	.00782	.01616	.02446	.03273	.04097	.04921
10	3	6	.04686	.05391	.06074	.06738	.07383	.08013	.08629
10	3	7	.61066	.59875	.58791	.57815	.56948	.56190	.55542
10	4	1	-.39305	-.41686	-.44273	-.47078	-.50111	-.53383	-.56904
10	4	2	-.20633	-.20325	-.19963	-.19546	-.19069	-.18529	-.17923
10	4	3	-.11917	-.11094	-.10230	-.09326	-.08379	-.07388	-.06351
10	4	4	-.05013	-.04030	-.03033	-.02022	-.00995	.00048	.01109
10	4	5	.01113	.02053	.02982	.03901	.04813	.05719	.06620
10	4	6	.75755	.75082	.74518	.74070	.73740	.73533	.73449
10	5	1	-.49191	-.52033	-.55116	-.58454	-.62059	-.65945	-.70126
10	5	2	-.24906	-.24445	-.23923	-.23337	-.22685	-.21963	-.21167
10	5	3	-.13615	-.12570	-.11485	-.10357	-.09187	-.07972	-.06710
10	5	4	-.04718	-.03530	-.02334	-.01129	.00087	.01315	.02559
10	5	5	.92430	.92578	.92858	.93277	.93844	.94564	.95445
11	0	1	-.18834	-.20370	-.22058	-.23905	-.25918	-.28104	-.30468
11	0	2	-.11146	-.11227	-.11277	-.11291	-.11263	-.11191	-.11068
11	0	3	-.07599	-.07317	-.06997	-.06638	-.06236	-.05791	-.05300
11	0	4	-.04806	-.04339	-.03843	-.03318	-.02763	-.02179	-.01565
11	0	5	-.02338	-.01780	-.01208	-.00622	-.00022	.00592	.01218
11	0	6	.00000	.00580	.01159	.01737	.02314	.02888	.03460
11	0	7	.02338	.02880	.03408	.03920	.04417	.04900	.05369
11	0	8	.04806	.05246	.05659	.06046	.06410	.06751	.07071
11	0	9	.07599	.07847	.08062	.08248	.08409	.08547	.08667
11	0	10	.11146	.11039	.10909	.10764	.10607	.10442	.10275
11	0	11	.18834	.17442	.16186	.15057	.14046	.13144	.12343
11	1	1	-.21490	-.23108	-.24884	-.26826	-.28942	-.31239	-.33725
11	1	2	-.12555	-.12582	-.12577	-.12535	-.12451	-.12323	-.12145
11	1	3	-.08434	-.08082	-.07693	-.07266	-.06798	-.06287	-.05732
11	1	4	-.05190	-.04655	-.04095	-.03508	-.02893	-.02252	-.01582
11	1	5	-.02325	-.01714	-.01092	-.00458	.00186	.00841	.01507
11	1	6	.00384	.00996	.01603	.02207	.02806	.03401	.03993
11	1	7	.03087	.03630	.04157	.04666	.05160	.05639	.06104
11	1	8	.05927	.06327	.06702	.07051	.07379	.07688	.07978
11	1	9	.09107	.09260	.09387	.09493	.09580	.09651	.09711
11	1	10	.31489	.29929	.28492	.27175	.25973	.24880	.23891
11	2	1	-.24630	-.26380	-.28296	-.30388	-.32663	-.35131	-.37800
11	2	2	-.14165	-.14146	-.14090	-.13995	-.13857	-.13672	-.13435
11	2	3	-.09342	-.08918	-.08457	-.07957	-.07416	-.06832	-.06204
11	2	4	-.05549	-.04942	-.04310	-.03653	-.02971	-.02263	-.01528
11	2	5	-.02204	-.01533	-.00853	-.00166	.00530	.01234	.01947
11	2	6	.00953	.01601	.02241	.02875	.03501	.04122	.04737
11	2	7	.04089	.04635	.05161	.05669	.06160	.06635	.07096
11	2	8	.07359	.07716	.08048	.08356	.08644	.08914	.09170
11	2	9	.43489	.41966	.40556	.39259	.38072	.36993	.36018
11	3	1	-.28524	-.30454	-.32563	-.34860	-.37356	-.40059	-.42980
11	3	2	-.16103	-.16032	-.15921	-.15767	-.15566	-.15316	-.15011
11	3	3	-.10384	-.09876	-.09329	-.08742	-.08113	-.07442	-.06725
11	3	4	-.05893	-.05199	-.04482	-.03742	-.02978	-.02189	-.01374
11	3	5	-.01940	-.01198	-.00451	.00301	.01059	.01823	.02594
11	3	6	.01777	.02467	.03146	.03814	.04473	.05123	.05767
11	3	7	.05447	.05994	.06518	.07021	.07507	.07975	.08430
11	3	8	.55621	.54297	.53081	.51974	.50975	.50084	.49301
11	4	1	-.33570	-.35743	-.38112	-.40690	-.43485	-.46510	-.49774
11	4	2	-.18537	-.18401	-.18221	-.17993	-.17715	-.17383	-.16993
11	4	3	-.11627	-.11011	-.10355	-.09658	-.08918	-.08134	-.07304

Coefficients b_i for the BLUE of σ
Shape parameter k

n	s	i	0.0	0.1	0.2	0.3	0.4	0.5	0.6
11	4	4	-.06211	-.05407	-.04583	-.03737	-.02868	-.01977	-.01061
11	4	5	-.01459	-.00628	.00203	.01036	.01871	.02709	.03552
11	4	6	.02986	.03727	.04451	.05160	.05857	.06543	.07219
11	4	7	.68418	.67464	.66618	.65882	.65259	.64752	.64361
11	5	1	-.40455	-.42965	-.45698	-.48665	-.51880	-.55354	-.59102
11	5	2	-.21749	-.21524	-.21249	-.20921	-.20537	-.20094	-.19587
11	5	3	-.13171	-.12409	-.11605	-.10759	-.09869	-.08933	-.07949
11	5	4	-.06468	-.05518	-.04550	-.03562	-.02555	-.01526	-.00474
11	5	5	-.00613	.00333	.01274	.02212	.03148	.04083	.05021
11	5	6	.82456	.82083	.81828	.81696	.81692	.81823	.82092
12	0	1	-.17480	-.18962	-.20594	-.22386	-.24345	-.26478	-.28793
12	0	2	-.10615	-.10753	-.10865	-.10946	-.10993	-.11000	-.10964
12	0	3	-.07494	-.07289	-.07049	-.06772	-.06455	-.06096	-.05692
12	0	4	-.05063	-.04677	-.04261	-.03815	-.03337	-.02829	-.02289
12	0	5	-.02942	-.02458	-.01955	-.01434	-.00894	-.00338	.00237
12	0	6	-.00966	-.00441	.00090	.00626	.01167	.01712	.02261
12	0	7	.00966	.01485	.01997	.02501	.02998	.03488	.03969
12	0	8	.02942	.03408	.03856	.04285	.04697	.05093	.05473
12	0	9	.05063	.05422	.05753	.06060	.06343	.06605	.06848
12	0	10	.07494	.07667	.07811	.07929	.08024	.08101	.08162
12	0	11	.10615	.10456	.10281	.10095	.09902	.09706	.09511
12	0	12	.17480	.16142	.14937	.13858	.12894	.12037	.11277
12	1	1	-.19722	-.21275	-.22985	-.24860	-.26908	-.29139	-.31560
12	1	2	-.11849	-.11944	-.12013	-.12050	-.12052	-.12015	-.11934
12	1	3	-.08270	-.08008	-.07711	-.07378	-.07005	-.06592	-.06135
12	1	4	-.05484	-.05040	-.04568	-.04068	-.03539	-.02981	-.02392
12	1	5	-.03054	-.02520	-.01971	-.01405	-.00824	-.00228	.00384
12	1	6	-.00792	-.00232	.00331	.00897	.01464	.02035	.02607
12	1	7	.01418	.01951	.02475	.02990	.03495	.03992	.04481
12	1	8	.03671	.04125	.04560	.04976	.05375	.05758	.06126
12	1	9	.06079	.06391	.06678	.06943	.07187	.07413	.07623
12	1	10	.08809	.08890	.08949	.08990	.09016	.09030	.09035
12	1	11	.29195	.27662	.26254	.24966	.23792	.22726	.21764
12	2	1	-.22317	-.23982	-.25811	-.27814	-.30000	-.32377	-.34955
12	2	2	-.13237	-.13297	-.13329	-.13327	-.13288	-.13208	-.13084
12	2	3	-.09112	-.08792	-.08438	-.08046	-.07615	-.07142	-.06627
12	2	4	-.05901	-.05397	-.04866	-.04307	-.03721	-.03106	-.02462
12	2	5	-.03104	-.02517	-.01916	-.01302	-.00674	-.00032	.00625
12	2	6	-.00503	.00096	.00694	.01293	.01892	.02492	.03092
12	2	7	.02031	.02582	.03120	.03647	.04164	.04670	.05168
12	2	8	.04605	.05048	.05469	.05872	.06257	.06626	.06981
12	2	9	.07335	.07598	.07837	.08056	.08257	.08442	.08614
12	2	10	.40203	.38663	.37238	.35928	.34728	.33635	.32647
12	3	1	-.25451	-.27265	-.29255	-.31429	-.33798	-.36372	-.39160
12	3	2	-.14871	-.14896	-.14888	-.14845	-.14761	-.14635	-.14462
12	3	3	-.10067	-.09683	-.09263	-.08804	-.08305	-.07764	-.07179
12	3	4	-.06333	-.05758	-.05158	-.04532	-.03878	-.03196	-.02486
12	3	5	-.03082	-.02433	-.01773	-.01101	-.00418	.00278	.00986
12	3	6	-.00067	.00576	.01216	.01854	.02489	.03123	.03756
12	3	7	.02861	.03430	.03985	.04526	.05054	.05570	.06077
12	3	8	.05818	.06245	.06650	.07034	.07401	.07752	.08090
12	3	9	.51192	.49784	.48486	.47297	.46217	.45244	.44377
12	4	1	-.29372	-.31383	-.33583	-.35983	-.38594	-.41428	-.44494
12	4	2	-.16863	-.16846	-.16793	-.16701	-.16565	-.16383	-.16151
12	4	3	-.11190	-.10727	-.10227	-.09686	-.09104	-.08479	-.07808
12	4	4	-.06785	-.06126	-.05441	-.04731	-.03995	-.03232	-.02440

Coefficients b_i for the BLUE of σ
Shape parameter k

n	s	i	0.0	0.1	0.2	0.3	0.4	0.5	0.6
12	4	5	-.02959	-.02235	-.01502	-.00761	-.00010	.00752	.01524
12	4	6	.00580	.01277	.01966	.02649	.03328	.04003	.04675
12	4	7	.04001	.04591	.05162	.05717	.06257	.06784	.07300
12	4	8	.62587	.61449	.60418	.59496	.58683	.57983	.57394
12	5	1	-.34477	-.36748	-.39229	-.41931	-.44867	-.48049	-.51489
12	5	2	-.19387	-.19316	-.19205	-.19049	-.18846	-.18593	-.18286
12	5	3	-.12554	-.11990	-.11386	-.10741	-.10051	-.09317	-.08536
12	5	4	-.07259	-.06490	-.05698	-.04882	-.04040	-.03171	-.02276
12	5	5	-.02671	-.01853	-.01029	-.00200	.00636	.01480	.02333
12	5	6	.01554	.02315	.03064	.03803	.04535	.05259	.05979
12	5	7	.74794	.74084	.73484	.72999	.72634	.72391	.72274
12	6	1	-.41464	-.44094	-.46962	-.50082	-.53467	-.57133	-.61094
12	6	2	-.22740	-.22593	-.22400	-.22157	-.21861	-.21509	-.21097
12	6	3	-.14282	-.13578	-.12832	-.12042	-.11207	-.10325	-.09393
12	6	4	-.07743	-.06828	-.05890	-.04930	-.03945	-.02935	-.01899
12	6	5	-.02102	-.01158	-.00214	.00731	.01680	.02633	.03593
12	6	6	.88330	.88251	.88298	.88479	.88800	.89268	.89890
13	0	1	-.16321	-.17751	-.19332	-.21072	-.22979	-.25061	-.27325
13	0	2	-.10126	-.10308	-.10469	-.10605	-.10711	-.10783	-.10816
13	0	3	-.07346	-.07204	-.07030	-.06821	-.06574	-.06287	-.05958
13	0	4	-.05200	-.04881	-.04533	-.04155	-.03745	-.03304	-.02829
13	0	5	-.03346	-.02926	-.02485	-.02024	-.01541	-.01038	-.00514
13	0	6	-.01641	-.01170	-.00688	-.00196	.00305	.00815	.01334
13	0	7	.00000	.00484	.00967	.01449	.01930	.02408	.02885
13	0	8	.01641	.02101	.02551	.02989	.03416	.03832	.04238
13	0	9	.03346	.03745	.04125	.04484	.04826	.05150	.05458
13	0	10	.05200	.05490	.05755	.05996	.06215	.06414	.06595
13	0	11	.07346	.07458	.07545	.07610	.07655	.07684	.07700
13	0	12	.10126	.09928	.09719	.09503	.09283	.09065	.08851
13	0	13	.16321	.15032	.13875	.12841	.11921	.11104	.10382
13	1	1	-.18242	-.19735	-.21383	-.23196	-.25183	-.27351	-.29710
13	1	2	-.11216	-.11363	-.11489	-.11589	-.11659	-.11696	-.11694
13	1	3	-.08063	-.07873	-.07651	-.07394	-.07101	-.06769	-.06395
13	1	4	-.05630	-.05261	-.04865	-.04440	-.03985	-.03500	-.02983
13	1	5	-.03529	-.03064	-.02581	-.02079	-.01558	-.01018	-.00459
13	1	6	-.01597	-.01091	-.00577	-.00055	.00474	.01011	.01554
13	1	7	.00260	.00765	.01267	.01766	.02261	.02754	.03243
13	1	8	.02115	.02579	.03029	.03468	.03894	.04309	.04714
13	1	9	.04038	.04417	.04776	.05116	.05438	.05745	.06036
13	1	10	.06118	.06360	.06578	.06775	.06953	.07115	.07263
13	1	11	.08502	.08526	.08533	.08524	.08504	.08474	.08439
13	1	12	.27242	.25740	.24363	.23106	.21962	.20925	.19991
13	2	1	-.20427	-.22017	-.23769	-.25693	-.27798	-.30093	-.32589
13	2	2	-.12425	-.12547	-.12645	-.12715	-.12754	-.12757	-.12722
13	2	3	-.08835	-.08599	-.08330	-.08027	-.07686	-.07305	-.06883
13	2	4	-.06066	-.05647	-.05201	-.04726	-.04223	-.03689	-.03125
13	2	5	-.03676	-.03164	-.02636	-.02090	-.01527	-.00946	-.00347
13	2	6	-.01481	-.00937	-.00387	.00168	.00729	.01295	.01867
13	2	7	.00627	.01156	.01679	.02196	.02709	.03217	.03722
13	2	8	.02728	.03196	.03649	.04088	.04514	.04929	.05333
13	2	9	.04897	.05256	.05594	.05913	.06215	.06502	.06775
13	2	10	.07226	.07415	.07583	.07733	.07868	.07989	.08099
13	2	11	.37432	.35889	.34464	.33153	.31952	.30860	.29870
13	3	1	-.23010	-.24727	-.26615	-.28684	-.30946	-.33409	-.36084
13	3	2	-.13824	-.13920	-.13990	-.14031	-.14038	-.14008	-.13936
13	3	3	-.09704	-.09418	-.09097	-.08741	-.08346	-.07911	-.07434

Coefficients b_i for the BLUE of σ
Shape parameter k

n	s	i	0.0	0.1	0.2	0.3	0.4	0.5	0.6
13	3	4	-.06528	-.06052	-.05550	-.05019	-.04460	-.03871	-.03251
13	3	5	-.03790	-.03225	-.02645	-.02048	-.01436	-.00808	-.00162
13	3	6	-.01277	-.00691	-.00101	.00493	.01090	.01691	.02296
13	3	7	.01130	.01685	.02231	.02770	.03303	.03829	.04350
13	3	8	.03521	.03993	.04448	.04887	.05313	.05726	.06128
13	3	9	.05976	.06310	.06623	.06917	.07194	.07457	.07708
13	3	10	.47505	.46045	.44696	.43457	.42327	.41303	.40386
13	4	1	-.26156	-.28035	-.30098	-.32355	-.34817	-.37496	-.40402
13	4	2	-.15490	-.15559	-.15599	-.15606	-.15577	-.15508	-.15396
13	4	3	-.10710	-.10365	-.09984	-.09565	-.09107	-.08607	-.08063
13	4	4	-.07029	-.06486	-.05916	-.05318	-.04692	-.04037	-.03351
13	4	5	-.03858	-.03230	-.02589	-.01933	-.01263	-.00578	.00123
13	4	6	-.00955	-.00318	.00319	.00957	.01597	.02239	.02883
13	4	7	.01820	.02404	.02977	.03540	.04095	.04641	.05182
13	4	8	.04563	.05037	.05492	.05929	.06352	.06761	.07160
13	4	9	.57815	.56553	.55398	.54350	.53412	.52584	.51865
13	5	1	-.30112	-.32200	-.34488	-.36987	-.39710	-.42669	-.45877
13	5	2	-.17538	-.17574	-.17576	-.17543	-.17469	-.17353	-.17190
13	5	3	-.11910	-.11491	-.11034	-.10538	-.10001	-.09420	-.08794
13	5	4	-.07579	-.06953	-.06300	-.05620	-.04912	-.04174	-.03406
13	5	5	-.03857	-.03152	-.02435	-.01706	-.00965	-.00210	.00559
13	5	6	-.00456	.00240	.00934	.01626	.02317	.03008	.03700
13	5	7	.02781	.03399	.04002	.04592	.05172	.05741	.06303
13	5	8	.68670	.67730	.66897	.66176	.65568	.65077	.64705
13	6	1	-.35277	-.37641	-.40227	-.43047	-.46116	-.49447	-.53055
13	6	2	-.20148	-.20139	-.20092	-.20005	-.19874	-.19696	-.19467
13	6	3	-.13385	-.12870	-.12316	-.11719	-.11079	-.10393	-.09661
13	6	4	-.08192	-.07458	-.06698	-.05911	-.05096	-.04253	-.03379
13	6	5	-.03738	-.02934	-.02122	-.01301	-.00468	.00375	.01232
13	6	6	.00317	.01088	.01852	.02611	.03365	.04117	.04868
13	6	7	.80423	.79954	.79602	.79371	.79268	.79297	.79462
13	7	1	-.42361	-.45103	-.48098	-.51361	-.54907	-.58753	-.62914
13	7	2	-.23631	-.23558	-.23441	-.23278	-.23065	-.22800	-.22480
13	7	3	-.15278	-.14629	-.13939	-.13203	-.12422	-.11592	-.10712
13	7	4	-.08878	-.07997	-.07091	-.06159	-.05200	-.04213	-.03195
13	7	5	-.03408	-.02473	-.01533	-.00587	.00366	.01328	.02300
13	7	6	.93555	.93760	.94101	.94588	.95228	.96030	.97002
14	0	1	-.15316	-.16699	-.18231	-.19922	-.21779	-.23812	-.26028
14	0	2	-.09678	-.09894	-.10094	-.10273	-.10427	-.10552	-.10644
14	0	3	-.07175	-.07085	-.06965	-.06813	-.06626	-.06401	-.06136
14	0	4	-.05258	-.04997	-.04707	-.04387	-.04037	-.03654	-.03237
14	0	5	-.03616	-.03253	-.02867	-.02459	-.02028	-.01574	-.01098
14	0	6	-.02121	-.01699	-.01265	-.00816	-.00355	.00119	.00605
14	0	7	-.00699	-.00254	.00196	.00650	.01106	.01566	.02028
14	0	8	.00699	.01140	.01576	.02006	.02430	.02849	.03261
14	0	9	.02121	.02528	.02922	.03301	.03668	.04022	.04363
14	0	10	.03616	.03958	.04279	.04580	.04862	.05127	.05376
14	0	11	.05258	.05492	.05702	.05889	.06055	.06203	.06335
14	0	12	.07175	.07239	.07280	.07301	.07306	.07298	.07280
14	0	13	.09678	.09450	.09214	.08976	.08737	.08503	.08276
14	0	14	.15316	.14073	.12961	.11969	.11087	.10307	.09619
14	1	1	-.16982	-.18421	-.20013	-.21768	-.23697	-.25806	-.28107
14	1	2	-.10647	-.10835	-.11006	-.11156	-.11281	-.11377	-.11440
14	1	3	-.07836	-.07705	-.07544	-.07352	-.07125	-.06862	-.06559
14	1	4	-.05682	-.05378	-.05046	-.04686	-.04295	-.03874	-.03420
14	1	5	-.03838	-.03435	-.03011	-.02566	-.02100	-.01613	-.01104

Coefficients b_i for the BLUE of σ
Shape parameter k

n	s	i	0.0	0.1	0.2	0.3	0.4	0.5	0.6
14	1	6	-.02160	-.01705	-.01239	-.00761	-.00273	.00227	.00738
14	1	7	-.00565	-.00095	.00376	.00849	.01323	.01799	.02277
14	1	8	.01003	.01455	.01900	.02337	.02768	.03192	.03610
14	1	9	.02595	.02997	.03384	.03757	.04117	.04464	.04799
14	1	10	.04265	.04581	.04877	.05154	.05414	.05659	.05889
14	1	11	.06090	.06274	.06436	.06578	.06704	.06816	.06916
14	1	12	.08199	.08179	.08144	.08097	.08041	.07979	.07913
14	1	13	.25558	.24088	.22743	.21517	.20403	.19396	.18490
14	2	1	-.18851	-.20374	-.22057	-.23909	-.25942	-.28164	-.30585
14	2	2	-.11711	-.11880	-.12029	-.12156	-.12257	-.12328	-.12365
14	2	3	-.08543	-.08374	-.08176	-.07944	-.07678	-.07375	-.07032
14	2	4	-.06117	-.05770	-.05396	-.04993	-.04561	-.04098	-.03604
14	2	5	-.04041	-.03596	-.03132	-.02648	-.02144	-.01620	-.01076
14	2	6	-.02151	-.01661	-.01161	-.00652	-.00133	.00395	.00934
14	2	7	-.00358	.00137	.00632	.01127	.01621	.02116	.02611
14	2	8	.01403	.01867	.02322	.02769	.03207	.03638	.04062
14	2	9	.03186	.03584	.03965	.04332	.04684	.05024	.05353
14	2	10	.05050	.05341	.05611	.05863	.06100	.06321	.06531
14	2	11	.07072	.07203	.07315	.07411	.07494	.07566	.07629
14	2	12	.35060	.33523	.32105	.30802	.29609	.28524	.27541
14	3	1	-.21022	-.22654	-.24454	-.26432	-.28600	-.30966	-.33543
14	3	2	-.12923	-.13074	-.13204	-.13310	-.13387	-.13433	-.13443
14	3	3	-.09331	-.09122	-.08883	-.08609	-.08300	-.07952	-.07561
14	3	4	-.06581	-.06187	-.05766	-.05317	-.04838	-.04329	-.03788
14	3	5	-.04229	-.03739	-.03230	-.02703	-.02156	-.01590	-.01005
14	3	6	-.02091	-.01562	-.01024	-.00479	.00074	.00636	.01206
14	3	7	-.00064	.00459	.00980	.01499	.02017	.02533	.03048
14	3	8	.01923	.02400	.02866	.03323	.03769	.04208	.04638
14	3	9	.03928	.04319	.04693	.05052	.05396	.05728	.06048
14	3	10	.06012	.06272	.06513	.06736	.06944	.07140	.07324
14	3	11	.44379	.42887	.41508	.40240	.39080	.38027	.37079
14	4	1	-.23609	-.25379	-.27326	-.29463	-.31801	-.34351	-.37123
14	4	2	-.14341	-.14474	-.14583	-.14666	-.14718	-.14737	-.14718
14	4	3	-.10232	-.09978	-.09691	-.09369	-.09010	-.08612	-.08172
14	4	4	-.07089	-.06641	-.06166	-.05662	-.05129	-.04565	-.03970
14	4	5	-.04403	-.03859	-.03298	-.02720	-.02124	-.01509	-.00875
14	4	6	-.01963	-.01388	-.00808	-.00221	.00372	.00972	.01579
14	4	7	.00346	.00901	.01452	.01998	.02542	.03082	.03620
14	4	8	.02602	.03093	.03571	.04038	.04493	.04939	.05376
14	4	9	.04869	.05251	.05615	.05963	.06296	.06615	.06924
14	4	10	.53821	.52474	.51234	.50103	.49080	.48165	.47358
14	5	1	-.26775	-.28717	-.30851	-.33188	-.35741	-.38523	-.41544
14	5	2	-.16043	-.16155	-.16241	-.16297	-.16320	-.16306	-.16253
14	5	3	-.11288	-.10980	-.10636	-.10256	-.09837	-.09377	-.08874
14	5	4	-.07654	-.07141	-.06600	-.06031	-.05432	-.04803	-.04142
14	5	5	-.04552	-.03945	-.03322	-.02683	-.02027	-.01353	-.00661
14	5	6	-.01739	-.01111	-.00479	.00157	.00797	.01442	.02094
14	5	7	.00917	.01509	.02094	.02673	.03246	.03814	.04379
14	5	8	.03502	.04008	.04499	.04975	.05439	.05892	.06335
14	5	9	.63633	.62531	.61536	.60650	.59875	.59214	.58666
14	6	1	-.30768	-.32930	-.35301	-.37896	-.40726	-.43805	-.47147
14	6	2	-.18146	-.18231	-.18287	-.18310	-.18296	-.18243	-.18146
14	6	3	-.12559	-.12182	-.11767	-.11314	-.10819	-.10282	-.09700
14	6	4	-.08294	-.07700	-.07077	-.06426	-.05746	-.05034	-.04291
14	6	5	-.04659	-.03973	-.03274	-.02559	-.01829	-.01083	-.00320
14	6	6	-.01370	-.00677	.00017	.00713	.01410	.02111	.02816

Coefficients b_i for the BLUE of σ
Shape parameter k

n	s	i	0.0	0.1	0.2	0.3	0.4	0.5	0.6
14	6	7	.01725	.02361	.02986	.03602	.04209	.04811	.05407
14	6	8	.74070	.73332	.72703	.72190	.71796	.71525	.71381
14	7	1	-.35992	-.38444	-.41128	-.44061	-.47256	-.50729	-.54496
14	7	2	-.20836	-.20885	-.20900	-.20878	-.20815	-.20709	-.20556
14	7	3	-.14137	-.13669	-.13161	-.12611	-.12018	-.11380	-.10695
14	7	4	-.09031	-.08331	-.07602	-.06845	-.06058	-.05241	-.04391
14	7	5	-.04688	-.03902	-.03104	-.02293	-.01468	-.00629	.00226
14	7	6	-.00772	.00004	.00776	.01546	.02316	.03086	.03859
14	7	7	.85456	.85227	.85120	.85142	.85300	.85602	.86052
15	0	1	-.14436	-.15775	-.17261	-.18905	-.20716	-.22701	-.24870
15	0	2	-.09266	-.09510	-.09741	-.09955	-.10149	-.10318	-.10458
15	0	3	-.06994	-.06947	-.06872	-.06768	-.06632	-.06460	-.06250
15	0	4	-.05265	-.05052	-.04813	-.04544	-.04244	-.03913	-.03549
15	0	5	-.03794	-.03480	-.03143	-.02783	-.02399	-.01991	-.01558
15	0	6	-.02466	-.02090	-.01699	-.01292	-.00869	-.00431	.00022
15	0	7	-.01215	-.00808	-.00392	.00030	.00460	.00896	.01338
15	0	8	.00000	.00415	.00829	.01243	.01655	.02065	.02474
15	0	9	.01215	.01615	.02005	.02388	.02761	.03126	.03483
15	0	10	.02466	.02825	.03169	.03498	.03813	.04113	.04401
15	0	11	.03794	.04086	.04357	.04608	.04841	.05057	.05258
15	0	12	.05265	.05452	.05615	.05757	.05880	.05987	.06079
15	0	13	.06994	.07018	.07022	.07008	.06981	.06943	.06898
15	0	14	.09266	.09015	.08760	.08504	.08252	.08007	.07770
15	0	15	.14436	.13236	.12164	.11210	.10365	.09619	.08962
15	1	1	-.15897	-.17285	-.18825	-.20527	-.22401	-.24456	-.26702
15	1	2	-.10135	-.10355	-.10562	-.10752	-.10922	-.11067	-.11184
15	1	3	-.07602	-.07520	-.07411	-.07273	-.07102	-.06897	-.06655
15	1	4	-.05676	-.05425	-.05149	-.04844	-.04510	-.04145	-.03748
15	1	5	-.04037	-.03687	-.03316	-.02922	-.02506	-.02068	-.01606
15	1	6	-.02557	-.02150	-.01729	-.01294	-.00846	-.00383	.00094
15	1	7	-.01165	-.00733	-.00295	.00148	.00596	.01050	.01509
15	1	8	.00187	.00617	.01045	.01470	.01893	.02313	.02730
15	1	9	.01539	.01941	.02334	.02717	.03091	.03456	.03813
15	1	10	.02927	.03276	.03609	.03926	.04229	.04520	.04798
15	1	11	.04399	.04662	.04905	.05131	.05339	.05533	.05714
15	1	12	.06021	.06156	.06272	.06371	.06455	.06527	.06588
15	1	13	.07907	.07852	.07784	.07707	.07623	.07536	.07446
15	1	14	.24088	.22651	.21337	.20142	.19059	.18081	.17202
15	2	1	-.17515	-.18978	-.20598	-.22386	-.24353	-.26507	-.28860
15	2	2	-.11079	-.11284	-.11474	-.11647	-.11798	-.11923	-.12020
15	2	3	-.08250	-.08137	-.07996	-.07825	-.07621	-.07383	-.07107
15	2	4	-.06098	-.05812	-.05499	-.05158	-.04788	-.04387	-.03955
15	2	5	-.04269	-.03883	-.03476	-.03048	-.02598	-.02127	-.01633
15	2	6	-.02617	-.02178	-.01726	-.01262	-.00784	-.00294	.00208
15	2	7	-.01064	-.00607	-.00145	.00320	.00790	.01263	.01741
15	2	8	.00443	.00889	.01331	.01769	.02204	.02634	.03062
15	2	9	.01947	.02354	.02749	.03134	.03509	.03875	.04232
15	2	10	.03488	.03826	.04148	.04454	.04746	.05025	.05293
15	2	11	.05116	.05350	.05566	.05764	.05947	.06117	.06275
15	2	12	.06897	.06981	.07048	.07102	.07144	.07177	.07203
15	2	13	.33002	.31478	.30072	.28782	.27603	.26530	.25559
15	3	1	-.19368	-.20926	-.22649	-.24547	-.26631	-.28913	-.31401
15	3	2	-.12141	-.12333	-.12510	-.12667	-.12800	-.12907	-.12984
15	3	3	-.08965	-.08820	-.08646	-.08441	-.08202	-.07927	-.07615
15	3	4	-.06551	-.06225	-.05873	-.05492	-.05083	-.04643	-.04171
15	3	5	-.04498	-.04073	-.03627	-.03161	-.02674	-.02166	-.01636

Coefficients b_i for the BLUE of σ
Shape parameter k

n	s	i	0.0	0.1	0.2	0.3	0.4	0.5	0.6
15	3	6	-.02647	-.02172	-.01686	-.01188	-.00679	-.00159	.00374
15	3	7	-.00907	-.00421	.00067	.00557	.01050	.01546	.02045
15	3	8	.00779	.01243	.01701	.02154	.02601	.03044	.03483
15	3	9	.02458	.02870	.03268	.03655	.04031	.04397	.04754
15	3	10	.04174	.04500	.04808	.05101	.05380	.05646	.05902
15	3	11	.05977	.06178	.06361	.06528	.06682	.06824	.06957
15	3	12	.41689	.40180	.38784	.37500	.36325	.35257	.34293
15	4	1	-.21538	-.23214	-.25063	-.27098	-.29329	-.31768	-.34426
15	4	2	-.13364	-.13545	-.13708	-.13849	-.13965	-.14053	-.14108
15	4	3	-.09775	-.09593	-.09382	-.09139	-.08860	-.08545	-.08191
15	4	4	-.07046	-.06676	-.06279	-.05854	-.05399	-.04914	-.04397
15	4	5	-.04728	-.04257	-.03768	-.03258	-.02728	-.02178	-.01606
15	4	6	-.02638	-.02123	-.01598	-.01063	-.00518	.00038	.00605
15	4	7	-.00677	-.00160	.00358	.00877	.01396	.01917	.02440
15	4	8	.01220	.01704	.02179	.02648	.03110	.03566	.04017
15	4	9	.03105	.03520	.03921	.04308	.04684	.05049	.05406
15	4	10	.05023	.05332	.05624	.05901	.06163	.06413	.06653
15	4	11	.50419	.49013	.47716	.46527	.45447	.44474	.43607
15	5	1	-.24137	-.25958	-.27964	-.30168	-.32581	-.35216	-.38085
15	5	2	-.14805	-.14974	-.15121	-.15245	-.15342	-.15408	-.15440
15	5	3	-.10709	-.10487	-.10232	-.09943	-.09619	-.09255	-.08852
15	5	4	-.07597	-.07176	-.06726	-.06248	-.05741	-.05202	-.04631
15	5	5	-.04956	-.04433	-.03891	-.03331	-.02750	-.02150	-.01529
15	5	6	-.02578	-.02015	-.01445	-.00866	-.00279	.00318	.00925
15	5	7	-.00349	.00205	.00757	.01308	.01858	.02408	.02959
15	5	8	.01803	.02308	.02803	.03288	.03766	.04236	.04701
15	5	9	.03932	.04350	.04751	.05138	.05512	.05875	.06229
15	5	10	.59397	.58180	.57070	.56068	.55175	.54393	.53722
15	6	1	-.27327	-.29329	-.31531	-.33946	-.36587	-.39467	-.42600
15	6	2	-.16543	-.16696	-.16826	-.16929	-.17002	-.17042	-.17046
15	6	3	-.11813	-.11540	-.11232	-.10889	-.10508	-.10086	-.09622
15	6	4	-.08222	-.07738	-.07225	-.06683	-.06111	-.05507	-.04870
15	6	5	-.05178	-.04591	-.03987	-.03365	-.02724	-.02063	-.01382
15	6	6	-.02440	-.01822	-.01198	-.00567	.00071	.00717	.01372
15	6	7	.00119	.00716	.01308	.01896	.02482	.03066	.03650
15	6	8	.02583	.03112	.03628	.04133	.04628	.05114	.05593
15	6	9	.68822	.67889	.67064	.66349	.65750	.65268	.64906
15	7	1	-.31357	-.33589	-.36040	-.38725	-.41657	-.44851	-.48323
15	7	2	-.18698	-.18831	-.18937	-.19014	-.19057	-.19064	-.19032
15	7	3	-.13151	-.12813	-.12438	-.12026	-.11573	-.11078	-.10538
15	7	4	-.08943	-.08380	-.07787	-.07165	-.06511	-.05826	-.05108
15	7	5	-.05382	-.04717	-.04035	-.03336	-.02619	-.01883	-.01128
15	7	6	-.02186	-.01499	-.00808	-.00113	.00587	.01293	.02007
15	7	7	.00794	.01442	.02082	.02715	.03344	.03968	.04590
15	7	8	.78923	.78386	.77964	.77662	.77486	.77441	.77531
15	8	1	-.36638	-.39172	-.41950	-.44988	-.48303	-.51912	-.55829
15	8	2	-.21463	-.21568	-.21642	-.21681	-.21683	-.21646	-.21565
15	8	3	-.14823	-.14399	-.13936	-.13431	-.12884	-.12292	-.11652
15	8	4	-.09794	-.09126	-.08429	-.07701	-.06942	-.06151	-.05326
15	8	5	-.05545	-.04777	-.03995	-.03197	-.02382	-.01550	-.00700
15	8	6	-.01742	-.00968	-.00193	.00583	.01362	.02145	.02933
15	8	7	.90005	.90010	.90144	.90415	.90833	.91406	.92140
16	0	1	-.13659	-.14956	-.16400	-.18000	-.19765	-.21706	-.23830
16	0	2	-.08888	-.09153	-.09409	-.09652	-.09879	-.10085	-.10266
16	0	3	-.06810	-.06798	-.06763	-.06699	-.06606	-.06480	-.06318
16	0	4	-.05237	-.05067	-.04870	-.04645	-.04391	-.04105	-.03787

Coefficients b_i for the BLUE of σ
Shape parameter k

n	s	i	0.0	0.1	0.2	0.3	0.4	0.5	0.6
16	0	5	-.03908	-.03637	-.03344	-.03026	-.02684	-.02317	-.01925
16	0	6	-.02715	-.02381	-.02029	-.01660	-.01274	-.00870	-.00450
16	0	7	-.01601	-.01230	-.00848	-.00456	-.00055	.00356	.00776
16	0	8	-.00529	-.00143	.00248	.00640	.01035	.01432	.01831
16	0	9	.00529	.00913	.01292	.01667	.02037	.02402	.02763
16	0	10	.01601	.01962	.02312	.02651	.02979	.03297	.03605
16	0	11	.02715	.03032	.03333	.03618	.03887	.04142	.04384
16	0	12	.03908	.04157	.04385	.04594	.04785	.04960	.05120
16	0	13	.05237	.05384	.05508	.05613	.05700	.05773	.05834
16	0	14	.06810	.06801	.06774	.06732	.06679	.06618	.06551
16	0	15	.08888	.08619	.08348	.08081	.07819	.07565	.07323
16	0	16	.13659	.12498	.11464	.10546	.09733	.09018	.08389
16	1	1	-.14950	-.16292	-.17784	-.19437	-.21260	-.23264	-.25457
16	1	2	-.09671	-.09916	-.10153	-.10376	-.10582	-.10768	-.10930
16	1	3	-.07371	-.07329	-.07263	-.07171	-.07048	-.06893	-.06703
16	1	4	-.05631	-.05427	-.05198	-.04941	-.04656	-.04341	-.03994
16	1	5	-.04160	-.03857	-.03532	-.03184	-.02814	-.02419	-.02000
16	1	6	-.02840	-.02477	-.02098	-.01703	-.01292	-.00865	-.00422
16	1	7	-.01608	-.01213	-.00809	-.00397	.00024	.00453	.00890
16	1	8	-.00423	-.00019	.00386	.00792	.01200	.01608	.02018
16	1	9	.00747	.01138	.01524	.01905	.02280	.02649	.03014
16	1	10	.01930	.02288	.02635	.02970	.03295	.03609	.03914
16	1	11	.03158	.03460	.03746	.04016	.04272	.04514	.04745
16	1	12	.04470	.04688	.04887	.05069	.05235	.05387	.05528
16	1	13	.05926	.06022	.06100	.06163	.06213	.06252	.06282
16	1	14	.07630	.07546	.07452	.07350	.07245	.07137	.07030
16	1	15	.22794	.21388	.20106	.18941	.17888	.16938	.16085
16	2	1	-.16368	-.17776	-.19339	-.21068	-.22974	-.25067	-.27357
16	2	2	-.10515	-.10748	-.10972	-.11181	-.11372	-.11543	-.11689
16	2	3	-.07965	-.07897	-.07804	-.07684	-.07534	-.07351	-.07133
16	2	4	-.06036	-.05801	-.05540	-.05253	-.04936	-.04590	-.04212
16	2	5	-.04406	-.04071	-.03715	-.03336	-.02936	-.02512	-.02065
16	2	6	-.02944	-.02551	-.02143	-.01721	-.01284	-.00832	-.00364
16	2	7	-.01580	-.01160	-.00732	-.00298	.00144	.00592	.01048
16	2	8	-.00268	.00154	.00576	.00997	.01419	.01840	.02262
16	2	9	.01026	.01427	.01821	.02208	.02589	.02963	.03333
16	2	10	.02333	.02690	.03034	.03366	.03687	.03998	.04299
16	2	11	.03686	.03973	.04244	.04500	.04741	.04970	.05188
16	2	12	.05126	.05313	.05483	.05636	.05776	.05904	.06022
16	2	13	.06713	.06758	.06789	.06808	.06818	.06820	.06818
16	2	14	.31198	.29689	.28300	.27027	.25863	.24806	.23850
16	3	1	-.17970	-.19462	-.21116	-.22942	-.24952	-.27157	-.29568
16	3	2	-.11454	-.11679	-.11892	-.12090	-.12269	-.12426	-.12557
16	3	3	-.08616	-.08522	-.08402	-.08254	-.08074	-.07861	-.07613
16	3	4	-.06469	-.06201	-.05908	-.05586	-.05236	-.04856	-.04445
16	3	5	-.04656	-.04287	-.03897	-.03485	-.03052	-.02595	-.02116
16	3	6	-.03030	-.02605	-.02167	-.01714	-.01248	-.00768	-.00273
16	3	7	-.01514	-.01067	-.00614	-.00155	.00309	.00780	.01257
16	3	8	-.00057	.00384	.00824	.01262	.01699	.02135	.02570
16	3	9	.01378	.01789	.02191	.02586	.02973	.03353	.03727
16	3	10	.02825	.03180	.03521	.03849	.04166	.04472	.04769
16	3	11	.04318	.04589	.04843	.05082	.05307	.05521	.05724
16	3	12	.05899	.06052	.06188	.06311	.06421	.06521	.06613
16	3	13	.39346	.37829	.36427	.35137	.33956	.32881	.31910
16	4	1	-.19820	-.21414	-.23178	-.25123	-.27262	-.29604	-.32163
16	4	2	-.12522	-.12740	-.12944	-.13131	-.13298	-.13441	-.13557

Coefficients b_i for the BLUE of σ
Shape parameter k

n	s	i	0.0	0.1	0.2	0.3	0.4	0.5	0.6
16	4	3	-.09345	-.09223	-.09073	-.08894	-.08683	-.08437	-.08154
16	4	4	-.06943	-.06638	-.06307	-.05949	-.05561	-.05143	-.04693
16	4	5	-.04914	-.04506	-.04078	-.03629	-.03159	-.02666	-.02150
16	4	6	-.03096	-.02635	-.02162	-.01676	-.01177	-.00665	-.00139
16	4	7	-.01402	-.00925	-.00443	.00042	.00533	.01028	.01529
16	4	8	.00224	.00687	.01147	.01603	.02057	.02509	.02959
16	4	9	.01823	.02244	.02656	.03058	.03451	.03837	.04217
16	4	10	.03431	.03782	.04119	.04442	.04754	.05054	.05345
16	4	11	.05083	.05334	.05568	.05787	.05993	.06188	.06374
16	4	12	.47479	.46033	.44697	.43470	.42351	.41340	.40433
16	5	1	-.21996	-.23715	-.25613	-.27704	-.29998	-.32509	-.35248
16	5	2	-.13761	-.13972	-.14167	-.14343	-.14497	-.14625	-.14725
16	5	3	-.10177	-.10023	-.09839	-.09625	-.09377	-.09093	-.08771
16	5	4	-.07468	-.07122	-.06748	-.06347	-.05915	-.05453	-.04959
16	5	5	-.05182	-.04730	-.04259	-.03767	-.03253	-.02718	-.02160
16	5	6	-.03134	-.02633	-.02120	-.01595	-.01059	-.00510	.00052
16	5	7	-.01229	-.00718	-.00204	.00313	.00833	.01356	.01884
16	5	8	.00597	.01084	.01566	.02043	.02516	.02985	.03452
16	5	9	.02389	.02821	.03242	.03651	.04052	.04443	.04827
16	5	10	.04185	.04531	.04861	.05177	.05481	.05773	.06056
16	5	11	.55775	.54474	.53281	.52195	.51218	.50350	.49591
16	6	1	-.24610	-.26481	-.28544	-.30812	-.33298	-.36016	-.38978
16	6	2	-.15226	-.15429	-.15614	-.15778	-.15917	-.16029	-.16110
16	6	3	-.11145	-.10952	-.10729	-.10472	-.10180	-.09851	-.09483
16	6	4	-.08061	-.07665	-.07241	-.06788	-.06305	-.05790	-.05243
16	6	5	-.05460	-.04957	-.04434	-.03891	-.03326	-.02741	-.02133
16	6	6	-.03134	-.02584	-.02025	-.01455	-.00874	-.00282	.00323
16	6	7	-.00972	-.00421	.00130	.00683	.01237	.01793	.02353
16	6	8	.01096	.01610	.02117	.02617	.03111	.03600	.04086
16	6	9	.03119	.03563	.03992	.04409	.04815	.05212	.05601
16	6	10	.64392	.63316	.62347	.61487	.60738	.60104	.59585
16	7	1	-.27824	-.29883	-.32151	-.34640	-.37366	-.40341	-.43581
16	7	2	-.16999	-.17192	-.17364	-.17512	-.17634	-.17725	-.17784
16	7	3	-.12294	-.12054	-.11781	-.11473	-.11128	-.10744	-.10318
16	7	4	-.08742	-.08284	-.07799	-.07283	-.06737	-.06158	-.05546
16	7	5	-.05749	-.05182	-.04596	-.03991	-.03365	-.02718	-.02049
16	7	6	-.03075	-.02468	-.01852	-.01227	-.00593	.00051	.00706
16	7	7	-.00594	.00003	.00599	.01193	.01788	.02383	.02980
16	7	8	.01772	.02317	.02852	.03378	.03896	.04408	.04915
16	7	9	.73504	.72744	.72093	.71556	.71139	.70845	.70678
16	8	1	-.31891	-.34190	-.36717	-.39487	-.42517	-.45821	-.49416
16	8	2	-.19203	-.19382	-.19537	-.19665	-.19762	-.19827	-.19857
16	8	3	-.13693	-.13393	-.13057	-.12684	-.12271	-.11816	-.11318
16	8	4	-.09538	-.09003	-.08440	-.07845	-.07219	-.06560	-.05866
16	8	5	-.06041	-.05395	-.04731	-.04048	-.03346	-.02623	-.01878
16	8	6	-.02923	-.02244	-.01559	-.00867	-.00167	.00541	.01259
16	8	7	-.00037	.00617	.01266	.01912	.02555	.03197	.03839
16	8	8	.83327	.82991	.82774	.82684	.82726	.82908	.83236
17	0	1	-.12966	-.14225	-.15628	-.17187	-.18910	-.20808	-.22889
17	0	2	-.08541	-.08823	-.09099	-.09366	-.09619	-.09856	-.10072
17	0	3	-.06628	-.06646	-.06643	-.06615	-.06559	-.06472	-.06353
17	0	4	-.05187	-.05053	-.04893	-.04707	-.04492	-.04247	-.03970
17	0	5	-.03976	-.03743	-.03488	-.03208	-.02904	-.02575	-.02220
17	0	6	-.02895	-.02598	-.02283	-.01949	-.01596	-.01225	-.00835
17	0	7	-.01893	-.01556	-.01206	-.00844	-.00470	-.00084	.00313
17	0	8	-.00936	-.00577	-.00213	.00158	.00533	.00913	.01298

Coefficients b_i for the BLUE of σ
Shape parameter k

n	s	i	0.0	0.1	0.2	0.3	0.4	0.5	0.6
17	0	9	.00000	.00363	.00726	.01088	.01449	.01808	.02166
17	0	10	.00936	.01288	.01634	.01972	.02304	.02629	.02947
17	0	11	.01893	.02218	.02531	.02831	.03120	.03397	.03663
17	0	12	.02895	.03175	.03437	.03683	.03914	.04130	.04334
17	0	13	.03976	.04187	.04378	.04551	.04706	.04847	.04974
17	0	14	.05187	.05299	.05390	.05464	.05522	.05566	.05600
17	0	15	.06628	.06591	.06538	.06473	.06399	.06319	.06236
17	0	16	.08541	.08257	.07975	.07698	.07429	.07170	.06924
17	0	17	.12966	.11842	.10843	.09958	.09176	.08488	.07885
17	1	1	-.14118	-.15417	-.16864	-.18470	-.20246	-.22201	-.24346
17	1	2	-.09250	-.09515	-.09775	-.10025	-.10262	-.10482	-.10681
17	1	3	-.07146	-.07138	-.07109	-.07055	-.06973	-.06862	-.06718
17	1	4	-.05562	-.05398	-.05210	-.04995	-.04753	-.04481	-.04179
17	1	5	-.04230	-.03967	-.03684	-.03377	-.03047	-.02693	-.02314
17	1	6	-.03041	-.02717	-.02376	-.02018	-.01642	-.01249	-.00839
17	1	7	-.01939	-.01579	-.01207	-.00825	-.00433	-.00029	.00384
17	1	8	-.00887	-.00511	-.00130	.00255	.00644	.01037	.01434
17	1	9	.00141	.00516	.00888	.01259	.01627	.01993	.02357
17	1	10	.01169	.01524	.01872	.02212	.02544	.02870	.03188
17	1	11	.02219	.02538	.02843	.03137	.03418	.03689	.03950
17	1	12	.03317	.03579	.03824	.04053	.04269	.04471	.04662
17	1	13	.04499	.04678	.04839	.04984	.05115	.05232	.05339
17	1	14	.05817	.05879	.05926	.05959	.05981	.05993	.05998
17	1	15	.07368	.07260	.07144	.07024	.06901	.06778	.06657
17	1	16	.21643	.20268	.19017	.17882	.16857	.15934	.15106
17	2	1	-.15370	-.16728	-.18240	-.19915	-.21765	-.23801	-.26032
17	2	2	-.10009	-.10265	-.10515	-.10754	-.10979	-.11187	-.11374
17	2	3	-.07692	-.07662	-.07610	-.07532	-.07427	-.07292	-.07124
17	2	4	-.05947	-.05756	-.05541	-.05299	-.05030	-.04731	-.04402
17	2	5	-.04481	-.04190	-.03879	-.03545	-.03188	-.02808	-.02403
17	2	6	-.03172	-.02821	-.02455	-.02071	-.01672	-.01255	-.00822
17	2	7	-.01960	-.01576	-.01182	-.00779	-.00366	.00056	.00488
17	2	8	-.00803	-.00408	-.00010	.00391	.00795	.01202	.01612
17	2	9	.00328	.00715	.01098	.01478	.01855	.02229	.02600
17	2	10	.01457	.01817	.02167	.02509	.02843	.03169	.03488
17	2	11	.02609	.02922	.03221	.03508	.03783	.04047	.04302
17	2	12	.03811	.04055	.04283	.04496	.04695	.04883	.05060
17	2	13	.05100	.05247	.05378	.05495	.05599	.05693	.05777
17	2	14	.06526	.06540	.06541	.06533	.06516	.06494	.06469
17	2	15	.29601	.28111	.26741	.25485	.24340	.23300	.22361
17	3	1	-.16771	-.18204	-.19796	-.21558	-.23501	-.25637	-.27976
17	3	2	-.10846	-.11097	-.11339	-.11570	-.11786	-.11984	-.12160
17	3	3	-.08286	-.08234	-.08160	-.08059	-.07930	-.07770	-.07576
17	3	4	-.06358	-.06139	-.05896	-.05626	-.05327	-.05000	-.04641
17	3	5	-.04738	-.04418	-.04077	-.03714	-.03329	-.02920	-.02486
17	3	6	-.03294	-.02914	-.02520	-.02109	-.01683	-.01241	-.00783
17	3	7	-.01955	-.01546	-.01129	-.00702	-.00268	.00176	.00628
17	3	8	-.00679	-.00264	.00153	.00571	.00991	.01413	.01837
17	3	9	.00568	.00967	.01362	.01752	.02138	.02521	.02900
17	3	10	.01811	.02174	.02528	.02872	.03207	.03534	.03854
17	3	11	.03076	.03383	.03675	.03954	.04222	.04478	.04726
17	3	12	.04393	.04618	.04827	.05021	.05203	.05373	.05534
17	3	13	.05797	.05910	.06008	.06094	.06169	.06235	.06295
17	3	14	.37283	.35765	.34364	.33074	.31894	.30819	.29848
17	4	1	-.18369	-.19892	-.21581	-.23447	-.25504	-.27761	-.30231
17	4	2	-.11788	-.12035	-.12272	-.12495	-.12703	-.12891	-.13057

Coefficients b_i for the BLUE of σ
Shape parameter k

n	s	i	0.0	0.1	0.2	0.3	0.4	0.5	0.6
17	4	3	-.08946	-.08872	-.08773	-.08647	-.08492	-.08304	-.08083
17	4	4	-.06806	-.06557	-.06282	-.05980	-.05650	-.05290	-.04899
17	4	5	-.05008	-.04655	-.04282	-.03886	-.03468	-.03027	-.02562
17	4	6	-.03406	-.02995	-.02569	-.02129	-.01673	-.01203	-.00716
17	4	7	-.01922	-.01485	-.01041	-.00589	-.00130	.00338	.00813
17	4	8	-.00507	-.00071	.00366	.00804	.01242	.01681	.02121
17	4	9	.00872	.01285	.01692	.02094	.02490	.02882	.03270
17	4	10	.02245	.02613	.02969	.03315	.03651	.03979	.04299
17	4	11	.03640	.03938	.04222	.04492	.04750	.04998	.05236
17	4	12	.05086	.05288	.05474	.05647	.05808	.05958	.06101
17	4	13	.44909	.43436	.42075	.40823	.39679	.38641	.37709
17	5	1	-.20221	-.21852	-.23658	-.25650	-.27843	-.30247	-.32875
17	5	2	-.12867	-.13110	-.13342	-.13559	-.13758	-.13937	-.14092
17	5	3	-.09690	-.09591	-.09466	-.09312	-.09127	-.08909	-.08656
17	5	4	-.07301	-.07017	-.06707	-.06370	-.06003	-.05607	-.05178
17	5	5	-.05293	-.04904	-.04493	-.04061	-.03606	-.03128	-.02627
17	5	6	-.03505	-.03059	-.02598	-.02124	-.01635	-.01131	-.00613
17	5	7	-.01850	-.01382	-.00908	-.00428	.00059	.00553	.01054
17	5	8	-.00275	.00186	.00646	.01104	.01562	.02020	.02479
17	5	9	.01259	.01687	.02108	.02522	.02929	.03331	.03728
17	5	10	.02784	.03155	.03513	.03860	.04196	.04524	.04843
17	5	11	.04327	.04614	.04887	.05145	.05391	.05627	.05854
17	5	12	.52633	.51272	.50018	.48872	.47834	.46904	.46082
17	6	1	-.22406	-.24167	-.26114	-.28259	-.30616	-.33197	-.36016
17	6	2	-.14122	-.14362	-.14589	-.14799	-.14990	-.15158	-.15301
17	6	3	-.10546	-.10417	-.10261	-.10074	-.09855	-.09602	-.09312
17	6	4	-.07856	-.07532	-.07181	-.06802	-.06394	-.05954	-.05482
17	6	5	-.05597	-.05165	-.04711	-.04236	-.03738	-.03218	-.02674
17	6	6	-.03587	-.03099	-.02598	-.02084	-.01557	-.01015	-.00459
17	6	7	-.01728	-.01224	-.00716	-.00203	.00316	.00841	.01371
17	6	8	.00038	.00527	.01012	.01495	.01976	.02455	.02934
17	6	9	.01756	.02201	.02636	.03062	.03481	.03893	.04300
17	6	10	.03458	.03831	.04191	.04537	.04873	.05198	.05516
17	6	11	.60590	.59408	.58332	.57363	.56505	.55758	.55123
17	7	1	-.25037	-.26956	-.29074	-.31404	-.33962	-.36760	-.39812
17	7	2	-.15612	-.15848	-.16069	-.16271	-.16451	-.16608	-.16737
17	7	3	-.11545	-.11381	-.11188	-.10962	-.10702	-.10407	-.10072
17	7	4	-.08488	-.08116	-.07717	-.07288	-.06829	-.06338	-.05814
17	7	5	-.05923	-.05439	-.04934	-.04408	-.03860	-.03289	-.02694
17	7	6	-.03642	-.03105	-.02557	-.01996	-.01423	-.00836	-.00236
17	7	7	-.01536	-.00990	-.00441	.00111	.00667	.01227	.01792
17	7	8	.00463	.00983	.01498	.02008	.02514	.03018	.03520
17	7	9	.02402	.02864	.03314	.03755	.04186	.04609	.05026
17	7	10	.68916	.67987	.67166	.66456	.65860	.65383	.65027
17	8	1	-.28276	-.30391	-.32721	-.35281	-.38088	-.41155	-.44498
17	8	2	-.17419	-.17649	-.17862	-.18054	-.18221	-.18362	-.18474
17	8	3	-.12737	-.12529	-.12289	-.12016	-.11706	-.11357	-.10968
17	8	4	-.09220	-.08789	-.08330	-.07840	-.07319	-.06765	-.06177
17	8	5	-.06272	-.05725	-.05158	-.04570	-.03960	-.03327	-.02671
17	8	6	-.03653	-.03058	-.02452	-.01835	-.01206	-.00566	.00088
17	8	7	-.01240	-.00644	-.00047	.00551	.01151	.01753	.02360
17	8	8	.01046	.01603	.02151	.02693	.03229	.03761	.04290
17	8	9	.77770	.77182	.76707	.76351	.76120	.76018	.76051
17	9	1	-.32379	-.34741	-.37340	-.40192	-.43315	-.46724	-.50438
17	9	2	-.19670	-.19892	-.20093	-.20270	-.20420	-.20540	-.20629
17	9	3	-.14194	-.13930	-.13631	-.13295	-.12921	-.12505	-.12047

Coefficients b_i for the BLUE of σ
Shape parameter k

n	s	i	0.0	0.1	0.2	0.3	0.4	0.5	0.6
17	9	4	-.10085	-.09579	-.09043	-.08476	-.07877	-.07243	-.06575
17	9	5	-.06645	-.06019	-.05374	-.04707	-.04020	-.03310	-.02576
17	9	6	-.03594	-.02925	-.02247	-.01560	-.00863	-.00156	.00564
17	9	7	-.00788	-.00131	.00524	.01178	.01832	.02487	.03144
17	9	8	.87356	.87217	.87204	.87323	.87583	.87992	.88557
18	0	1	-.12345	-.13567	-.14933	-.16453	-.18136	-.19993	-.22033
18	0	2	-.08220	-.08515	-.08808	-.09095	-.09371	-.09634	-.09879
18	0	3	-.06449	-.06493	-.06518	-.06520	-.06496	-.06445	-.06362
18	0	4	-.05122	-.05019	-.04892	-.04739	-.04560	-.04351	-.04111
18	0	5	-.04011	-.03812	-.03590	-.03344	-.03075	-.02780	-.02459
18	0	6	-.03025	-.02761	-.02478	-.02176	-.01855	-.01514	-.01153
18	0	7	-.02115	-.01809	-.01489	-.01155	-.00808	-.00447	-.00073
18	0	8	-.01252	-.00921	-.00581	-.00234	.00121	.00483	.00852
18	0	9	-.00415	-.00073	.00271	.00618	.00966	.01315	.01665
18	0	10	.00415	.00754	.01089	.01421	.01749	.02074	.02394
18	0	11	.01252	.01575	.01889	.02194	.02491	.02779	.03059
18	0	12	.02115	.02407	.02686	.02952	.03206	.03447	.03678
18	0	13	.03025	.03270	.03499	.03711	.03908	.04092	.04262
18	0	14	.04011	.04189	.04348	.04490	.04615	.04726	.04825
18	0	15	.05122	.05204	.05267	.05314	.05347	.05368	.05380
18	0	16	.06449	.06389	.06315	.06231	.06140	.06045	.05948
18	0	17	.08220	.07925	.07634	.07350	.07077	.06815	.06567
18	0	18	.12345	.11256	.10289	.09434	.08680	.08018	.07439
18	1	1	-.13379	-.14638	-.16043	-.17606	-.19338	-.21247	-.23346
18	1	2	-.08866	-.09147	-.09426	-.09698	-.09960	-.10209	-.10441
18	1	3	-.06929	-.06950	-.06952	-.06931	-.06885	-.06811	-.06707
18	1	4	-.05477	-.05348	-.05195	-.05018	-.04813	-.04581	-.04318
18	1	5	-.04262	-.04036	-.03788	-.03519	-.03225	-.02907	-.02564
18	1	6	-.03183	-.02894	-.02588	-.02263	-.01921	-.01560	-.01179
18	1	7	-.02188	-.01860	-.01519	-.01166	-.00801	-.00423	-.00033
18	1	8	-.01245	-.00895	-.00539	-.00177	.00192	.00567	.00949
18	1	9	-.00329	.00025	.00381	.00737	.01095	.01452	.01811
18	1	10	.00577	.00923	.01264	.01600	.01932	.02260	.02583
18	1	11	.01492	.01814	.02126	.02430	.02724	.03010	.03288
18	1	12	.02433	.02716	.02986	.03242	.03487	.03720	.03943
18	1	13	.03425	.03651	.03860	.04055	.04235	.04404	.04561
18	1	14	.04497	.04643	.04772	.04886	.04986	.05075	.05154
18	1	15	.05699	.05734	.05754	.05762	.05760	.05751	.05736
18	1	16	.07121	.06994	.06861	.06725	.06588	.06453	.06322
18	1	17	.20613	.19269	.18047	.16940	.15942	.15045	.14242
18	2	1	-.14494	-.15807	-.17271	-.18896	-.20695	-.22677	-.24855
18	2	2	-.09552	-.09827	-.10098	-.10361	-.10614	-.10854	-.11076
18	2	3	-.07432	-.07434	-.07416	-.07376	-.07310	-.07215	-.07090
18	2	4	-.05842	-.05689	-.05513	-.05311	-.05083	-.04827	-.04541
18	2	5	-.04512	-.04261	-.03989	-.03694	-.03377	-.03036	-.02670
18	2	6	-.03331	-.03018	-.02688	-.02341	-.01976	-.01593	-.01192
18	2	7	-.02242	-.01892	-.01530	-.01157	-.00773	-.00377	.00031
18	2	8	-.01210	-.00842	-.00469	-.00091	.00293	.00683	.01077
18	2	9	-.00209	.00159	.00527	.00894	.01262	.01629	.01996
18	2	10	.00782	.01134	.01482	.01823	.02160	.02492	.02819
18	2	11	.01780	.02102	.02414	.02716	.03009	.03293	.03569
18	2	12	.02807	.03082	.03342	.03589	.03825	.04050	.04265
18	2	13	.03886	.04093	.04284	.04460	.04624	.04777	.04919
18	2	14	.05049	.05163	.05262	.05348	.05422	.05487	.05545
18	2	15	.06343	.06330	.06307	.06275	.06238	.06196	.06152
18	2	16	.28177	.26706	.25356	.24121	.22995	.21973	.21051

Coefficients b_i for the BLUE of σ
Shape parameter k

n	s	i	0.0	0.1	0.2	0.3	0.4	0.5	0.6
18	3	1	-.15731	-.17111	-.18646	-.20350	-.22232	-.24305	-.26579
18	3	2	-.10304	-.10575	-.10841	-.11099	-.11345	-.11577	-.11791
18	3	3	-.07976	-.07960	-.07924	-.07863	-.07777	-.07662	-.07515
18	3	4	-.06231	-.06054	-.05853	-.05627	-.05374	-.05092	-.04781
18	3	5	-.04770	-.04493	-.04195	-.03875	-.03532	-.03166	-.02774
18	3	6	-.03474	-.03135	-.02781	-.02409	-.02020	-.01614	-.01190
18	3	7	-.02280	-.01906	-.01522	-.01128	-.00723	-.00307	.00120
18	3	8	-.01148	-.00761	-.00369	.00027	.00427	.00832	.01241
18	3	9	-.00050	.00332	.00713	.01092	.01471	.01848	.02225
18	3	10	.01034	.01395	.01749	.02096	.02438	.02774	.03105
18	3	11	.02127	.02449	.02759	.03060	.03350	.03632	.03906
18	3	12	.03248	.03513	.03763	.04000	.04226	.04441	.04647
18	3	13	.04422	.04608	.04779	.04936	.05081	.05216	.05342
18	3	14	.05681	.05761	.05828	.05883	.05929	.05968	.06001
18	3	15	.35450	.33938	.32541	.31257	.30082	.29012	.28044
18	4	1	-.17126	-.18586	-.20208	-.22004	-.23987	-.26169	-.28560
18	4	2	-.11142	-.11411	-.11674	-.11927	-.12168	-.12394	-.12600
18	4	3	-.08575	-.08541	-.08485	-.08404	-.08296	-.08159	-.07989
18	4	4	-.06652	-.06449	-.06222	-.05969	-.05689	-.05380	-.05040
18	4	5	-.05042	-.04737	-.04411	-.04062	-.03691	-.03297	-.02878
18	4	6	-.03614	-.03248	-.02865	-.02467	-.02052	-.01620	-.01171
18	4	7	-.02299	-.01900	-.01492	-.01073	-.00645	-.00207	.00241
18	4	8	-.01053	-.00645	-.00233	.00182	.00600	.01022	.01448
18	4	9	.00154	.00552	.00947	.01339	.01730	.02118	.02505
18	4	10	.01346	.01714	.02075	.02429	.02775	.03116	.03450
18	4	11	.02544	.02865	.03174	.03472	.03760	.04039	.04309
18	4	12	.03771	.04024	.04263	.04488	.04702	.04906	.05101
18	4	13	.05051	.05212	.05359	.05494	.05618	.05732	.05839
18	4	14	.42638	.41149	.39772	.38504	.37345	.36292	.35343
18	5	1	-.18724	-.20278	-.22003	-.23911	-.26013	-.28324	-.30854
18	5	2	-.12091	-.12359	-.12620	-.12870	-.13106	-.13325	-.13525
18	5	3	-.09246	-.09192	-.09114	-.09011	-.08879	-.08717	-.08521
18	5	4	-.07115	-.06883	-.06627	-.06345	-.06034	-.05694	-.05323
18	5	5	-.05332	-.04995	-.04637	-.04257	-.03855	-.03428	-.02978
18	5	6	-.03751	-.03353	-.02940	-.02512	-.02067	-.01606	-.01128
18	5	7	-.02295	-.01868	-.01432	-.00988	-.00534	-.00071	.00401
18	5	8	-.00917	-.00486	-.00052	.00384	.00822	.01263	.01707
18	5	9	.00416	.00831	.01241	.01648	.02051	.02452	.02850
18	5	10	.01731	.02108	.02476	.02835	.03187	.03531	.03870
18	5	11	.03050	.03369	.03675	.03970	.04254	.04528	.04795
18	5	12	.04396	.04635	.04860	.05071	.05271	.05461	.05643
18	5	13	.49877	.48472	.47174	.45985	.44904	.43929	.43061
18	6	1	-.20582	-.22249	-.24096	-.26136	-.28381	-.30846	-.33542
18	6	2	-.13181	-.13449	-.13708	-.13955	-.14186	-.14400	-.14593
18	6	3	-.10008	-.09931	-.09829	-.09700	-.09541	-.09350	-.09125
18	6	4	-.07631	-.07367	-.07077	-.06760	-.06415	-.06039	-.05632
18	6	5	-.05643	-.05271	-.04877	-.04461	-.04022	-.03559	-.03072
18	6	6	-.03882	-.03449	-.03001	-.02538	-.02059	-.01565	-.01053
18	6	7	-.02261	-.01802	-.01335	-.00860	-.00378	.00113	.00613
18	6	8	-.00728	-.00270	.00188	.00647	.01108	.01571	.02035
18	6	9	.00753	.01186	.01613	.02035	.02453	.02867	.03278
18	6	10	.02211	.02596	.02971	.03336	.03693	.04042	.04384
18	6	11	.03670	.03985	.04287	.04576	.04854	.05123	.05384
18	6	12	.57283	.56020	.54863	.53814	.52874	.52043	.51322
18	7	1	-.22779	-.24581	-.26574	-.28772	-.31189	-.33839	-.36735
18	7	2	-.14454	-.14722	-.14979	-.15223	-.15449	-.15657	-.15841

Coefficients b_i for the BLUE of σ
Shape parameter k

n	s	i	0.0	0.1	0.2	0.3	0.4	0.5	0.6
18	7	3	-.10885	-.10782	-.10652	-.10492	-.10302	-.10078	-.09818
18	7	4	-.08214	-.07912	-.07583	-.07226	-.06839	-.06422	-.05972
18	7	5	-.05981	-.05566	-.05130	-.04671	-.04190	-.03685	-.03155
18	7	6	-.04004	-.03529	-.03040	-.02537	-.02019	-.01486	-.00936
18	7	7	-.02185	-.01689	-.01186	-.00677	-.00161	.00363	.00894
18	7	8	-.00468	.00020	.00507	.00993	.01479	.01966	.02454
18	7	9	.01188	.01642	.02088	.02527	.02960	.03389	.03814
18	7	10	.02815	.03209	.03591	.03962	.04323	.04676	.05021
18	7	11	.64966	.63909	.62959	.62117	.61388	.60773	.60275
18	8	1	-.25427	-.27392	-.29562	-.31953	-.34579	-.37454	-.40594
18	8	2	-.15967	-.16235	-.16491	-.16730	-.16950	-.17150	-.17325
18	8	3	-.11915	-.11779	-.11614	-.11419	-.11190	-.10926	-.10625
18	8	4	-.08883	-.08534	-.08158	-.07753	-.07317	-.06850	-.06349
18	8	5	-.06350	-.05884	-.05398	-.04888	-.04356	-.03801	-.03220
18	8	6	-.04109	-.03585	-.03048	-.02497	-.01932	-.01353	-.00758
18	8	7	-.02050	-.01510	-.00966	-.00416	.00140	.00702	.01271
18	8	8	-.00109	.00413	.00933	.01450	.01966	.02481	.02996
18	8	9	.01759	.02235	.02702	.03160	.03611	.04056	.04496
18	8	10	.73050	.72271	.71602	.71046	.70609	.70295	.70108
18	9	1	-.28691	-.30857	-.33247	-.35876	-.38760	-.41915	-.45358
18	9	2	-.17806	-.18073	-.18325	-.18558	-.18771	-.18960	-.19122
18	9	3	-.13147	-.12970	-.12763	-.12522	-.12246	-.11932	-.11579
18	9	4	-.09663	-.09257	-.08823	-.08358	-.07862	-.07333	-.06769
18	9	5	-.06755	-.06228	-.05679	-.05108	-.04513	-.03896	-.03253
18	9	6	-.04185	-.03601	-.03005	-.02397	-.01776	-.01140	-.00490
18	9	7	-.01828	-.01235	-.00640	-.00042	.00561	.01169	.01783
18	9	8	.00390	.00954	.01512	.02066	.02616	.03163	.03710
18	9	9	.81684	.81269	.80970	.80795	.80751	.80843	.81078
19	0	1	-.11785	-.12973	-.14303	-.15786	-.17431	-.19249	-.21250
19	0	2	-.07923	-.08229	-.08536	-.08838	-.09134	-.09419	-.09689
19	0	3	-.06277	-.06342	-.06391	-.06419	-.06424	-.06402	-.06352
19	0	4	-.05047	-.04971	-.04872	-.04750	-.04601	-.04424	-.04218
19	0	5	-.04022	-.03852	-.03660	-.03446	-.03207	-.02943	-.02653
19	0	6	-.03116	-.02882	-.02629	-.02356	-.02064	-.01751	-.01418
19	0	7	-.02284	-.02006	-.01714	-.01407	-.01085	-.00748	-.00396
19	0	8	-.01500	-.01194	-.00879	-.00554	-.00219	.00124	.00476
19	0	9	-.00743	-.00423	-.00098	.00231	.00564	.00901	.01241
19	0	10	.00000	.00323	.00646	.00967	.01288	.01608	.01925
19	0	11	.00743	.01058	.01367	.01671	.01969	.02261	.02548
19	0	12	.01500	.01795	.02080	.02355	.02620	.02876	.03123
19	0	13	.02284	.02547	.02796	.03031	.03254	.03464	.03663
19	0	14	.03116	.03332	.03530	.03713	.03881	.04035	.04178
19	0	15	.04022	.04172	.04303	.04417	.04516	.04602	.04677
19	0	16	.05047	.05104	.05143	.05167	.05178	.05180	.05174
19	0	17	.06277	.06196	.06105	.06005	.05900	.05792	.05684
19	0	18	.07923	.07619	.07322	.07034	.06757	.06493	.06245
19	0	19	.11785	.10727	.09791	.08964	.08236	.07598	.07040
19	1	1	-.12718	-.13940	-.15307	-.16829	-.18519	-.20385	-.22440
19	1	2	-.08514	-.08808	-.09102	-.09393	-.09676	-.09949	-.10209
19	1	3	-.06722	-.06767	-.06795	-.06803	-.06788	-.06748	-.06679
19	1	4	-.05383	-.05283	-.05162	-.05017	-.04846	-.04648	-.04421
19	1	5	-.04267	-.04073	-.03858	-.03621	-.03361	-.03076	-.02766
19	1	6	-.03281	-.03024	-.02749	-.02455	-.02143	-.01811	-.01460
19	1	7	-.02375	-.02077	-.01765	-.01440	-.01100	-.00747	-.00380
19	1	8	-.01522	-.01199	-.00867	-.00527	-.00179	.00178	.00543
19	1	9	-.00699	-.00365	-.00028	.00312	.00656	.01002	.01351

Coefficients b_i for the BLUE of σ
Shape parameter k

n	s	i	0.0	0.1	0.2	0.3	0.4	0.5	0.6
19	1	10	.00110	.00442	.00772	.01100	.01427	.01751	.02074
19	1	11	.00918	.01236	.01547	.01852	.02151	.02444	.02732
19	1	12	.01740	.02031	.02312	.02583	.02844	.03096	.03339
19	1	13	.02592	.02843	.03081	.03305	.03517	.03718	.03908
19	1	14	.03494	.03689	.03868	.04032	.04183	.04322	.04450
19	1	15	.04474	.04592	.04693	.04780	.04855	.04919	.04975
19	1	16	.05578	.05589	.05586	.05574	.05552	.05525	.05493
19	1	17	.06888	.06745	.06598	.06450	.06303	.06158	.06018
19	1	18	.19686	.18370	.17176	.16097	.15125	.14253	.13472
19	2	1	-.13719	-.14990	-.16409	-.17988	-.19739	-.21673	-.23799
19	2	2	-.09139	-.09428	-.09716	-.09999	-.10276	-.10541	-.10793
19	2	3	-.07186	-.07216	-.07227	-.07218	-.07186	-.07127	-.07040
19	2	4	-.05728	-.05608	-.05465	-.05299	-.05108	-.04889	-.04641
19	2	5	-.04512	-.04296	-.04059	-.03800	-.03518	-.03213	-.02882
19	2	6	-.03438	-.03159	-.02863	-.02549	-.02216	-.01865	-.01494
19	2	7	-.02452	-.02134	-.01802	-.01458	-.01101	-.00730	-.00347
19	2	8	-.01523	-.01182	-.00834	-.00479	-.00115	.00255	.00633
19	2	9	-.00627	-.00280	.00070	.00422	.00777	.01133	.01492
19	2	10	.00252	.00593	.00932	.01267	.01600	.01931	.02259
19	2	11	.01131	.01452	.01767	.02074	.02374	.02668	.02956
19	2	12	.02024	.02313	.02590	.02856	.03113	.03361	.03600
19	2	13	.02949	.03189	.03415	.03629	.03830	.04021	.04203
19	2	14	.03925	.04099	.04259	.04404	.04538	.04661	.04774
19	2	15	.04982	.05068	.05140	.05200	.05249	.05290	.05325
19	2	16	.06165	.06130	.06086	.06036	.05980	.05923	.05864
19	2	17	.26897	.25447	.24118	.22903	.21797	.20794	.19890
19	3	1	-.14820	-.16151	-.17635	-.19285	-.21111	-.23126	-.25341
19	3	2	-.09818	-.10105	-.10390	-.10669	-.10941	-.11201	-.11447
19	3	3	-.07686	-.07700	-.07696	-.07670	-.07621	-.07545	-.07439
19	3	4	-.06094	-.05953	-.05790	-.05602	-.05389	-.05148	-.04878
19	3	5	-.04767	-.04528	-.04268	-.03986	-.03682	-.03353	-.03000
19	3	6	-.03594	-.03292	-.02973	-.02637	-.02283	-.01910	-.01519
19	3	7	-.02519	-.02178	-.01826	-.01462	-.01086	-.00697	-.00295
19	3	8	-.01505	-.01145	-.00780	-.00407	.00028	.00358	.00751
19	3	9	-.00528	-.00165	.00198	.00563	.00929	.01297	.01666
19	3	10	.00431	.00782	.01129	.01472	.01812	.02149	.02483
19	3	11	.01388	.01713	.02030	.02339	.02641	.02936	.03226
19	3	12	.02359	.02644	.02917	.03180	.03432	.03675	.03910
19	3	13	.03363	.03591	.03806	.04007	.04197	.04377	.04549
19	3	14	.04420	.04572	.04710	.04836	.04950	.05055	.05153
19	3	15	.05559	.05610	.05650	.05680	.05702	.05719	.05730
19	3	16	.33810	.32306	.30917	.29642	.28475	.27414	.26453
19	4	1	-.16050	-.17453	-.19014	-.20748	-.22665	-.24777	-.27097
19	4	2	-.10569	-.10856	-.11139	-.11416	-.11685	-.11941	-.12182
19	4	3	-.08233	-.08232	-.08212	-.08169	-.08102	-.08007	-.07882
19	4	4	-.06489	-.06326	-.06140	-.05930	-.05693	-.05429	-.05134
19	4	5	-.05035	-.04772	-.04487	-.04181	-.03852	-.03498	-.03121
19	4	6	-.03751	-.03425	-.03081	-.02721	-.02343	-.01947	-.01533
19	4	7	-.02574	-.02210	-.01836	-.01450	-.01052	-.00642	-.00221
19	4	8	-.01464	-.01084	-.00699	-.00309	.00088	.00490	.00899
19	4	9	-.00395	-.00017	.00361	.00740	.01119	.01498	.01879
19	4	10	.00652	.01014	.01370	.01722	.02069	.02413	.02753
19	4	11	.01697	.02026	.02346	.02657	.02960	.03257	.03546
19	4	12	.02756	.03037	.03305	.03562	.03809	.04046	.04276
19	4	13	.03847	.04062	.04263	.04451	.04628	.04795	.04955
19	4	14	.04992	.05120	.05234	.05337	.05430	.05515	.05593

Coefficients b_i for the BLUE of σ
Shape parameter k

n	s	i	0.0	0.1	0.2	0.3	0.4	0.5	0.6
19	4	15	.40615	.39117	.37731	.36456	.35288	.34227	.33268
19	5	1	-.17444	-.18931	-.20584	-.22416	-.24439	-.26666	-.29110
19	5	2	-.11411	-.11699	-.11982	-.12258	-.12524	-.12777	-.13014
19	5	3	-.08840	-.08823	-.08786	-.08725	-.08638	-.08523	-.08377
19	5	4	-.06921	-.06734	-.06524	-.06288	-.06026	-.05735	-.05414
19	5	5	-.05322	-.05031	-.04720	-.04386	-.04029	-.03649	-.03243
19	5	6	-.03910	-.03556	-.03186	-.02798	-.02394	-.01972	-.01532
19	5	7	-.02615	-.02226	-.01827	-.01417	-.00996	-.00563	-.00119
19	5	8	-.01396	-.00994	-.00588	-.00177	.00239	.00660	.01086
19	5	9	-.00222	.00173	.00567	.00961	.01354	.01747	.02140
19	5	10	.00927	.01299	.01665	.02026	.02381	.02732	.03079
19	5	11	.02071	.02404	.02726	.03039	.03343	.03640	.03930
19	5	12	.03229	.03504	.03766	.04017	.04256	.04487	.04710
19	5	13	.04418	.04617	.04801	.04973	.05135	.05288	.05433
19	5	14	.47436	.45999	.44670	.43450	.42337	.41331	.40431
19	6	1	-.19045	-.20631	-.22391	-.24339	-.26488	-.28851	-.31440
19	6	2	-.12368	-.12658	-.12942	-.13217	-.13482	-.13732	-.13965
19	6	3	-.09523	-.09488	-.09431	-.09350	-.09241	-.09103	-.08934
19	6	4	-.07400	-.07186	-.06948	-.06684	-.06392	-.06072	-.05720
19	6	5	-.05631	-.05310	-.04968	-.04603	-.04215	-.03803	-.03367
19	6	6	-.04070	-.03685	-.03284	-.02867	-.02433	-.01982	-.01512
19	6	7	-.02639	-.02222	-.01794	-.01357	-.00909	-.00451	.00018
19	6	8	-.01293	-.00866	-.00437	-.00003	.00434	.00876	.01322
19	6	9	.00002	.00416	.00828	.01238	.01647	.02055	.02462
19	6	10	.01268	.01652	.02029	.02399	.02763	.03121	.03475
19	6	11	.02527	.02863	.03187	.03500	.03805	.04102	.04391
19	6	12	.03797	.04065	.04319	.04561	.04792	.05014	.05228
19	6	13	.54373	.53050	.51832	.50722	.49720	.48826	.48041
19	7	1	-.20911	-.22613	-.24500	-.26585	-.28883	-.31406	-.34170
19	7	2	-.13470	-.13763	-.14048	-.14324	-.14587	-.14834	-.15064
19	7	3	-.10301	-.10246	-.10167	-.10062	-.09928	-.09764	-.09567
19	7	4	-.07937	-.07692	-.07422	-.07125	-.06800	-.06444	-.06058
19	7	5	-.05968	-.05611	-.05234	-.04833	-.04410	-.03962	-.03489
19	7	6	-.04230	-.03810	-.03374	-.02923	-.02454	-.01969	-.01466
19	7	7	-.02639	-.02189	-.01730	-.01262	-.00784	-.00296	.00202
19	7	8	-.01143	-.00689	-.00233	.00226	.00688	.01153	.01622
19	7	9	.00293	.00729	.01161	.01589	.02015	.02438	.02861
19	7	10	.01697	.02093	.02481	.02860	.03233	.03600	.03961
19	7	11	.03088	.03426	.03751	.04065	.04368	.04664	.04952
19	7	12	.61521	.60365	.59315	.58373	.57542	.56822	.56216
19	8	1	-.23120	-.24961	-.26999	-.29249	-.31724	-.34441	-.37412
19	8	2	-.14760	-.15056	-.15343	-.15618	-.15880	-.16124	-.16349
19	8	3	-.11201	-.11121	-.11016	-.10883	-.10720	-.10525	-.10295
19	8	4	-.08546	-.08265	-.07958	-.07622	-.07257	-.06861	-.06433
19	8	5	-.06337	-.05939	-.05520	-.05078	-.04612	-.04122	-.03607
19	8	6	-.04389	-.03927	-.03451	-.02959	-.02450	-.01925	-.01383
19	8	7	-.02606	-.02118	-.01622	-.01117	-.00605	-.00083	.00448
19	8	8	-.00931	-.00445	.00042	.00530	.01019	.01511	.02006
19	8	9	.00674	.01134	.01587	.02036	.02481	.02923	.03362
19	8	10	.02240	.02649	.03049	.03439	.03820	.04195	.04564
19	8	11	.68976	.68050	.67230	.66522	.65928	.65452	.65099
19	9	1	-.25785	-.27794	-.30015	-.32464	-.35156	-.38106	-.41330
19	9	2	-.16296	-.16596	-.16884	-.17159	-.17418	-.17659	-.17879
19	9	3	-.12259	-.12150	-.12013	-.11846	-.11647	-.11415	-.11146
19	9	4	-.09250	-.08924	-.08571	-.08188	-.07775	-.07331	-.06852
19	9	5	-.06746	-.06299	-.05829	-.05337	-.04821	-.04280	-.03714

Coefficients b_i for the BLUE of σ
Shape parameter k

n	s	i	0.0	0.1	0.2	0.3	0.4	0.5	0.6
19	9	6	-.04540	-.04029	-.03504	-.02964	-.02408	-.01836	-.01247
19	9	7	-.02524	-.01991	-.01451	-.00904	-.00350	.00212	.00783
19	9	8	-.00632	-.00109	.00414	.00936	.01458	.01981	.02506
19	9	9	.01178	.01664	.02143	.02615	.03082	.03544	.04003
19	9	10	.76855	.76227	.75711	.75312	.75037	.74890	.74877
19	10	1	-.29073	-.31290	-.33737	-.36431	-.39390	-.42629	-.46167
19	10	2	-.18167	-.18469	-.18758	-.19031	-.19287	-.19521	-.19733
19	10	3	-.13529	-.13382	-.13205	-.12996	-.12753	-.12473	-.12155
19	10	4	-.10075	-.09694	-.09284	-.08844	-.08372	-.07866	-.07327
19	10	5	-.07203	-.06695	-.06164	-.05610	-.05032	-.04429	-.03801
19	10	6	-.04676	-.04105	-.03520	-.02921	-.02307	-.01678	-.01032
19	10	7	-.02369	-.01781	-.01188	-.00591	.00013	.00623	.01242
19	10	8	-.00208	.00360	.00925	.01487	.02047	.02607	.03168
19	10	9	.85300	.85055	.84931	.84937	.85079	.85366	.85805
20	0	1	-.11276	-.12433	-.13729	-.15177	-.16786	-.18567	-.20530
20	0	2	-.07648	-.07963	-.08280	-.08596	-.08908	-.09212	-.09504
20	0	3	-.06110	-.06195	-.06264	-.06315	-.06344	-.06350	-.06328
20	0	4	-.04966	-.04913	-.04840	-.04744	-.04623	-.04475	-.04298
20	0	5	-.04015	-.03871	-.03707	-.03519	-.03308	-.03072	-.02811
20	0	6	-.03178	-.02971	-.02745	-.02499	-.02233	-.01946	-.01638
20	0	7	-.02413	-.02161	-.01895	-.01612	-.01314	-.01000	-.00670
20	0	8	-.01695	-.01414	-.01121	-.00818	-.00504	-.00179	.00157
20	0	9	-.01006	-.00707	-.00402	-.00090	.00228	.00551	.00880
20	0	10	-.00334	-.00027	.00281	.00590	.00901	.01213	.01525
20	0	11	.00334	.00637	.00938	.01236	.01531	.01822	.02110
20	0	12	.01006	.01298	.01582	.01860	.02130	.02393	.02649
20	0	13	.01695	.01964	.02223	.02471	.02708	.02935	.03153
20	0	14	.02413	.02649	.02871	.03079	.03274	.03457	.03629
20	0	15	.03178	.03367	.03540	.03696	.03839	.03968	.04086
20	0	16	.04015	.04140	.04246	.04337	.04414	.04479	.04533
20	0	17	.04966	.05000	.05018	.05023	.05017	.05002	.04980
20	0	18	.06110	.06013	.05907	.05794	.05677	.05559	.05442
20	0	19	.07648	.07338	.07036	.06744	.06465	.06201	.05953
20	0	20	.11276	.10249	.09341	.08540	.07836	.07221	.06683
20	1	1	-.12124	-.13311	-.14642	-.16126	-.17776	-.19602	-.21615
20	1	2	-.08191	-.08496	-.08802	-.09108	-.09409	-.09703	-.09986
20	1	3	-.06524	-.06590	-.06642	-.06675	-.06687	-.06675	-.06637
20	1	4	-.05284	-.05210	-.05116	-.05000	-.04859	-.04692	-.04497
20	1	5	-.04253	-.04087	-.03901	-.03694	-.03463	-.03209	-.02929
20	1	6	-.03346	-.03117	-.02871	-.02606	-.02322	-.02017	-.01693
20	1	7	-.02516	-.02246	-.01960	-.01661	-.01346	-.01016	-.00672
20	1	8	-.01738	-.01440	-.01132	-.00814	-.00486	-.00148	.00199
20	1	9	-.00992	-.00679	-.00361	-.00038	.00290	.00623	.00961
20	1	10	-.00263	.00053	.00369	.00687	.01005	.01323	.01641
20	1	11	.00459	.00768	.01074	.01375	.01673	.01967	.02258
20	1	12	.01187	.01479	.01763	.02039	.02308	.02570	.02825
20	1	13	.01932	.02196	.02448	.02690	.02921	.03142	.03355
20	1	14	.02709	.02932	.03141	.03337	.03520	.03693	.03856
20	1	15	.03535	.03703	.03855	.03992	.04118	.04232	.04336
20	1	16	.04437	.04529	.04607	.04671	.04724	.04768	.04804
20	1	17	.05456	.05446	.05425	.05394	.05357	.05314	.05268
20	1	18	.06670	.06514	.06355	.06197	.06041	.05889	.05742
20	1	19	.18845	.17557	.16391	.15338	.14391	.13542	.12783
20	2	1	-.13027	-.14259	-.15638	-.17174	-.18880	-.20767	-.22847
20	2	2	-.08762	-.09062	-.09364	-.09665	-.09960	-.10249	-.10526
20	2	3	-.06954	-.07007	-.07044	-.07062	-.07059	-.07032	-.06978

Coefficients b_i for the BLUE of σ
Shape parameter k

n	s	i	0.0	0.1	0.2	0.3	0.4	0.5	0.6
20	2	4	-.05609	-.05517	-.05405	-.05270	-.05111	-.04925	-.04712
20	2	5	-.04492	-.04306	-.04100	-.03873	-.03623	-.03349	-.03051
20	2	6	-.03508	-.03260	-.02994	-.02710	-.02407	-.02084	-.01742
20	2	7	-.02609	-.02319	-.02016	-.01699	-.01367	-.01021	-.00661
20	2	8	-.01765	-.01450	-.01126	-.00793	-.00451	-.00100	.00261
20	2	9	-.00957	-.00630	-.00299	.00036	.00376	.00719	.01067
20	2	10	-.00167	.00159	.00485	.00811	.01136	.01462	.01787
20	2	11	.00615	.00930	.01240	.01546	.01848	.02145	.02439
20	2	12	.01403	.01695	.01979	.02255	.02523	.02784	.03038
20	2	13	.02209	.02467	.02713	.02949	.03174	.03390	.03597
20	2	14	.03048	.03258	.03454	.03638	.03811	.03972	.04125
20	2	15	.03938	.04084	.04216	.04335	.04443	.04540	.04630
20	2	16	.04906	.04967	.05016	.05053	.05082	.05103	.05119
20	2	17	.05993	.05939	.05879	.05812	.05743	.05671	.05600
20	2	18	.25740	.24312	.23003	.21809	.20723	.19740	.18853
20	3	1	-.14014	-.15301	-.16738	-.18338	-.20114	-.22075	-.24235
20	3	2	-.09379	-.09679	-.09979	-.10276	-.10569	-.10853	-.11126
20	3	3	-.07415	-.07455	-.07478	-.07482	-.07465	-.07422	-.07353
20	3	4	-.05954	-.05844	-.05713	-.05559	-.05381	-.05176	-.04943
20	3	5	-.04740	-.04534	-.04308	-.04060	-.03790	-.03496	-.03178
20	3	6	-.03671	-.03402	-.03116	-.02812	-.02489	-.02147	-.01786
20	3	7	-.02695	-.02385	-.02063	-.01727	-.01378	-.01014	-.00637
20	3	8	-.01778	-.01446	-.01105	-.00756	-.00398	-.00032	.00342
20	3	9	-.00901	-.00559	-.00215	.00133	.00485	.00840	.01199
20	3	10	-.00044	.00293	.00629	.00964	.01297	.01630	.01963
20	3	11	.00804	.01125	.01441	.01752	.02057	.02358	.02654
20	3	12	.01658	.01951	.02235	.02510	.02778	.03037	.03290
20	3	13	.02530	.02783	.03023	.03252	.03471	.03680	.03882
20	3	14	.03436	.03633	.03816	.03987	.04147	.04297	.04439
20	3	15	.04395	.04519	.04630	.04728	.04817	.04897	.04970
20	3	16	.05433	.05461	.05479	.05488	.05490	.05487	.05481
20	3	17	.32332	.30838	.29461	.28197	.27041	.25990	.25038
20	4	1	-.15108	-.16459	-.17966	-.19642	-.21499	-.23549	-.25803
20	4	2	-.10056	-.10357	-.10657	-.10954	-.11245	-.11527	-.11797
20	4	3	-.07916	-.07944	-.07954	-.07943	-.07911	-.07853	-.07767
20	4	4	-.06324	-.06195	-.06045	-.05872	-.05673	-.05448	-.05194
20	4	5	-.05001	-.04774	-.04526	-.04257	-.03965	-.03649	-.03309
20	4	6	-.03837	-.03546	-.03238	-.02912	-.02568	-.02206	-.01824
20	4	7	-.02774	-.02443	-.02100	-.01745	-.01376	-.00994	-.00598
20	4	8	-.01776	-.01425	-.01066	-.00699	-.00325	.00057	.00447
20	4	9	-.00821	-.00464	-.00104	.00258	.00622	.00990	.01360
20	4	10	.00111	.00459	.00806	.01150	.01493	.01834	.02174
20	4	11	.01033	.01361	.01682	.01997	.02306	.02611	.02911
20	4	12	.01960	.02253	.02537	.02811	.03077	.03335	.03587
20	4	13	.02906	.03151	.03384	.03606	.03818	.04020	.04214
20	4	14	.03886	.04067	.04235	.04392	.04538	.04675	.04804
20	4	15	.04919	.05018	.05104	.05180	.05247	.05308	.05362
20	4	16	.38799	.37297	.35909	.34630	.33460	.32395	.31433
20	5	1	-.16337	-.17763	-.19351	-.21116	-.23068	-.25220	-.27586
20	5	2	-.10810	-.11113	-.11414	-.11711	-.12002	-.12283	-.12551
20	5	3	-.08468	-.08483	-.08479	-.08454	-.08405	-.08330	-.08227
20	5	4	-.06727	-.06578	-.06407	-.06212	-.05992	-.05744	-.05467
20	5	5	-.05281	-.05030	-.04759	-.04466	-.04150	-.03810	-.03445
20	5	6	-.04008	-.03693	-.03361	-.03011	-.02644	-.02258	-.01853
20	5	7	-.02846	-.02492	-.02127	-.01749	-.01359	-.00956	-.00539
20	5	8	-.01756	-.01384	-.01005	-.00619	-.00227	.00173	.00580

Coefficients b_i for the BLUE of σ
Shape parameter k

n	s	i	0.0	0.1	0.2	0.3	0.4	0.5	0.6
20	5	9	-.00712	-.00338	.00037	.00414	.00793	.01174	.01558
20	5	10	.00304	.00665	.01023	.01378	.01730	.02080	.02428
20	5	11	.01310	.01644	.01971	.02291	.02604	.02912	.03215
20	5	12	.02319	.02612	.02895	.03168	.03432	.03687	.03936
20	5	13	.03346	.03584	.03808	.04021	.04224	.04418	.04604
20	5	14	.04408	.04572	.04723	.04863	.04993	.05114	.05229
20	5	15	.45255	.43795	.42445	.41204	.40069	.39041	.38118
20	6	1	-.17732	-.19246	-.20930	-.22797	-.24860	-.27134	-.29629
20	6	2	-.11657	-.11964	-.12268	-.12567	-.12857	-.13138	-.13405
20	6	3	-.09084	-.09085	-.09065	-.09024	-.08958	-.08864	-.08742
20	6	4	-.07171	-.07000	-.06806	-.06587	-.06342	-.06068	-.05765
20	6	5	-.05582	-.05305	-.05008	-.04688	-.04345	-.03978	-.03587
20	6	6	-.04184	-.03842	-.03484	-.03108	-.02714	-.02302	-.01871
20	6	7	-.02908	-.02529	-.02138	-.01736	-.01322	-.00896	-.00456
20	6	8	-.01712	-.01317	-.00917	-.00510	-.00097	.00322	.00749
20	6	9	-.00568	-.00175	.00218	.00612	.01006	.01402	.01800
20	6	10	.00546	.00920	.01290	.01656	.02019	.02378	.02735
20	6	11	.01646	.01987	.02320	.02644	.02962	.03273	.03579
20	6	12	.02749	.03041	.03322	.03592	.03853	.04106	.04351
20	6	13	.03868	.04095	.04309	.04512	.04704	.04887	.05064
20	6	14	.51790	.50419	.49156	.48000	.46951	.46010	.45177
20	7	1	-.19339	-.20954	-.22749	-.24737	-.26931	-.29345	-.31993
20	7	2	-.12623	-.12934	-.13241	-.13542	-.13834	-.14115	-.14381
20	7	3	-.09779	-.09764	-.09727	-.09667	-.09581	-.09467	-.09322
20	7	4	-.07665	-.07468	-.07248	-.07002	-.06729	-.06427	-.06094
20	7	5	-.05910	-.05604	-.05276	-.04926	-.04553	-.04155	-.03733
20	7	6	-.04366	-.03994	-.03605	-.03199	-.02776	-.02334	-.01874
20	7	7	-.02958	-.02550	-.02131	-.01701	-.01260	-.00807	-.00342
20	7	8	-.01639	-.01218	-.00793	-.00363	.00073	.00515	.00962
20	7	9	-.00378	.00036	.00448	.00861	.01273	.01686	.02099
20	7	10	.00848	.01237	.01621	.01999	.02372	.02742	.03109
20	7	11	.02058	.02406	.02744	.03073	.03395	.03709	.04018
20	7	12	.03267	.03557	.03835	.04101	.04358	.04606	.04847
20	7	13	.58482	.57250	.56123	.55103	.54193	.53392	.52703
20	8	1	-.21213	-.22948	-.24874	-.27004	-.29352	-.31934	-.34762
20	8	2	-.13738	-.14054	-.14365	-.14669	-.14963	-.15243	-.15509
20	8	3	-.10574	-.10540	-.10483	-.10401	-.10292	-.10154	-.09983
20	8	4	-.08222	-.07996	-.07745	-.07467	-.07161	-.06826	-.06459
20	8	5	-.06270	-.05930	-.05567	-.05182	-.04774	-.04341	-.03882
20	8	6	-.04554	-.04147	-.03723	-.03283	-.02825	-.02349	-.01855
20	8	7	-.02990	-.02549	-.02097	-.01636	-.01163	-.00680	-.00185
20	8	8	-.01526	-.01076	-.00623	-.00165	.00297	.00763	.01235
20	8	9	-.00127	.00310	.00744	.01177	.01609	.02040	.02471
20	8	10	.01230	.01635	.02033	.02424	.02809	.03190	.03567
20	8	11	.02567	.02922	.03265	.03599	.03924	.04241	.04552
20	8	12	.65414	.64372	.63435	.62608	.61892	.61292	.60809
20	9	1	-.23434	-.25313	-.27394	-.29694	-.32226	-.35006	-.38051
20	9	2	-.15044	-.15367	-.15683	-.15989	-.16285	-.16565	-.16830
20	9	3	-.11495	-.11438	-.11358	-.11250	-.11114	-.10947	-.10746
20	9	4	-.08857	-.08596	-.08309	-.07994	-.07649	-.07274	-.06867
20	9	5	-.06669	-.06288	-.05885	-.05459	-.05009	-.04534	-.04033
20	9	6	-.04747	-.04298	-.03834	-.03353	-.02854	-.02339	-.01804
20	9	7	-.02996	-.02515	-.02026	-.01528	-.01019	-.00501	.00029
20	9	8	-.01358	-.00874	-.00389	.00100	.00591	.01087	.01587
20	9	9	.00204	.00667	.01127	.01583	.02037	.02489	.02940
20	9	10	.01718	.02140	.02553	.02958	.03357	.03749	.04138

Coefficients b_i for the BLUE of σ
Shape parameter k

n	s	i	0.0	0.1	0.2	0.3	0.4	0.5	0.6
20	9	11	.72677	.71883	.71197	.70625	.70172	.69841	.69637
20	10	1	-.26116	-.28168	-.30438	-.32942	-.35697	-.38719	-.42025
20	10	2	-.16604	-.16932	-.17253	-.17563	-.17859	-.18140	-.18402
20	10	3	-.12580	-.12497	-.12387	-.12248	-.12078	-.11876	-.11638
20	10	4	-.09593	-.09289	-.08958	-.08597	-.08206	-.07784	-.07328
20	10	5	-.07116	-.06686	-.06234	-.05758	-.05258	-.04732	-.04180
20	10	6	-.04941	-.04443	-.03930	-.03400	-.02854	-.02290	-.01708
20	10	7	-.02962	-.02436	-.01902	-.01360	-.00808	-.00248	.00323
20	10	8	-.01113	-.00590	-.00067	.00458	.00984	.01514	.02046
20	10	9	.00647	.01141	.01629	.02112	.02591	.03067	.03541
20	10	10	.80378	.79901	.79539	.79298	.79186	.79208	.79372
21	0	1	-.10813	-.11939	-.13204	-.14619	-.16194	-.17940	-.19867
21	0	2	-.07392	-.07714	-.08040	-.08368	-.08693	-.09013	-.09324
21	0	3	-.05950	-.06051	-.06138	-.06209	-.06260	-.06289	-.06293
21	0	4	-.04881	-.04849	-.04798	-.04726	-.04629	-.04507	-.04358
21	0	5	-.03996	-.03875	-.03734	-.03571	-.03386	-.03176	-.02940
21	0	6	-.03218	-.03035	-.02833	-.02612	-.02370	-.02108	-.01824
21	0	7	-.02511	-.02283	-.02040	-.01780	-.01504	-.01212	-.00902
21	0	8	-.01849	-.01590	-.01320	-.01037	-.00742	-.00436	-.00117
21	0	9	-.01217	-.00939	-.00653	-.00358	-.00056	.00253	.00569
21	0	10	-.00604	-.00316	-.00023	.00273	.00573	.00875	.01180
21	0	11	.00000	.00291	.00581	.00871	.01160	.01447	.01733
21	0	12	.00604	.00889	.01169	.01444	.01715	.01980	.02241
21	0	13	.01217	.01487	.01749	.02002	.02247	.02484	.02713
21	0	14	.01849	.02095	.02330	.02552	.02764	.02966	.03158
21	0	15	.02511	.02723	.02920	.03104	.03275	.03433	.03581
21	0	16	.03218	.03384	.03533	.03666	.03786	.03894	.03991
21	0	17	.03996	.04098	.04183	.04253	.04310	.04357	.04394
21	0	18	.04881	.04896	.04896	.04884	.04862	.04833	.04799
21	0	19	.05950	.05839	.05720	.05596	.05470	.05344	.05220
21	0	20	.07392	.07077	.06771	.06478	.06198	.05935	.05688
21	0	21	.10813	.09814	.08932	.08155	.07474	.06879	.06360
21	1	1	-.11586	-.12741	-.14037	-.15486	-.17099	-.18886	-.20860
21	1	2	-.07893	-.08206	-.08523	-.08841	-.09157	-.09469	-.09773
21	1	3	-.06336	-.06421	-.06492	-.06547	-.06582	-.06596	-.06586
21	1	4	-.05182	-.05131	-.05061	-.04970	-.04856	-.04717	-.04551
21	1	5	-.04225	-.04085	-.03925	-.03743	-.03540	-.03313	-.03061
21	1	6	-.03386	-.03184	-.02963	-.02724	-.02465	-.02186	-.01887
21	1	7	-.02622	-.02376	-.02116	-.01840	-.01548	-.01241	-.00918
21	1	8	-.01908	-.01633	-.01347	-.01050	-.00742	-.00423	-.00093
21	1	9	-.01226	-.00934	-.00635	-.00329	-.00016	.00303	.00628
21	1	10	-.00564	-.00265	.00037	.00342	.00649	.00959	.01270
21	1	11	.00088	.00386	.00682	.00977	.01270	.01562	.01852
21	1	12	.00740	.01027	.01309	.01586	.01857	.02123	.02385
21	1	13	.01401	.01669	.01927	.02178	.02420	.02654	.02881
21	1	14	.02082	.02320	.02546	.02762	.02966	.03161	.03347
21	1	15	.02794	.02992	.03176	.03346	.03505	.03653	.03791
21	1	16	.03555	.03699	.03827	.03942	.04044	.04137	.04220
21	1	17	.04389	.04460	.04517	.04561	.04595	.04621	.04640
21	1	18	.05335	.05307	.05270	.05224	.05173	.05117	.05060
21	1	19	.06465	.06298	.06130	.05963	.05800	.05642	.05490
21	1	20	.18079	.16819	.15678	.14650	.13727	.12900	.12162
21	2	1	-.12406	-.13602	-.14942	-.16439	-.18103	-.19947	-.21982
21	2	2	-.08417	-.08727	-.09040	-.09354	-.09667	-.09974	-.10273
21	2	3	-.06736	-.06808	-.06867	-.06909	-.06931	-.06932	-.06908
21	2	4	-.05489	-.05422	-.05336	-.05228	-.05098	-.04942	-.04759

Coefficients b_i for the BLUE of σ
Shape parameter k

n	s	i	0.0	0.1	0.2	0.3	0.4	0.5	0.6
21	2	5	-.04456	-.04298	-.04120	-.03921	-.03700	-.03455	-.03186
21	2	6	-.03550	-.03329	-.03091	-.02834	-.02558	-.02263	-.01947
21	2	7	-.02725	-.02462	-.02184	-.01892	-.01585	-.01263	-.00925
21	2	8	-.01954	-.01663	-.01362	-.01051	-.00729	-.00397	-.00054
21	2	9	-.01218	-.00912	-.00600	-.00282	.00041	.00371	.00707
21	2	10	-.00504	-.00194	.00119	.00432	.00748	.01065	.01384
21	2	11	.00200	.00505	.00808	.01108	.01407	.01703	.01997
21	2	12	.00903	.01193	.01478	.01756	.02029	.02297	.02559
21	2	13	.01616	.01882	.02138	.02386	.02625	.02857	.03081
21	2	14	.02349	.02580	.02799	.03007	.03204	.03392	.03572
21	2	15	.03115	.03299	.03469	.03627	.03774	.03911	.04039
21	2	16	.03932	.04054	.04162	.04258	.04343	.04419	.04488
21	2	17	.04824	.04864	.04892	.04911	.04922	.04926	.04926
21	2	18	.05828	.05759	.05684	.05604	.05522	.05440	.05359
21	2	19	.24689	.23282	.21995	.20821	.19755	.18790	.17921
21	3	1	-.13296	-.14542	-.15936	-.17491	-.19219	-.21131	-.23239
21	3	2	-.08980	-.09290	-.09603	-.09915	-.10226	-.10530	-.10826
21	3	3	-.07162	-.07223	-.07270	-.07300	-.07310	-.07298	-.07261
21	3	4	-.05813	-.05730	-.05628	-.05504	-.05357	-.05184	-.04984
21	3	5	-.04696	-.04519	-.04323	-.04106	-.03867	-.03604	-.03318
21	3	6	-.03716	-.03476	-.03220	-.02945	-.02651	-.02338	-.02005
21	3	7	-.02823	-.02542	-.02247	-.01938	-.01614	-.01276	-.00922
21	3	8	-.01990	-.01682	-.01366	-.01039	-.00703	-.00356	.00000
21	3	9	-.01194	-.00874	-.00549	-.00218	.00118	.00459	.00805
21	3	10	-.00422	-.00101	.00222	.00545	.00870	.01195	.01522
21	3	11	.00338	.00650	.00960	.01267	.01571	.01872	.02170
21	3	12	.01097	.01391	.01678	.01958	.02232	.02501	.02765
21	3	13	.01866	.02130	.02384	.02630	.02866	.03095	.03316
21	3	14	.02656	.02879	.03090	.03290	.03480	.03661	.03834
21	3	15	.03480	.03649	.03805	.03950	.04083	.04208	.04325
21	3	16	.04356	.04455	.04542	.04617	.04684	.04743	.04795
21	3	17	.05308	.05315	.05314	.05305	.05291	.05272	.05251
21	3	18	.30993	.29511	.28147	.26895	.25752	.24712	.23772
21	4	1	-.14276	-.15580	-.17038	-.18661	-.20463	-.22455	-.24650
21	4	2	-.09595	-.09907	-.10220	-.10533	-.10843	-.11147	-.11442
21	4	3	-.07622	-.07674	-.07710	-.07728	-.07725	-.07700	-.07648
21	4	4	-.06160	-.06061	-.05942	-.05801	-.05636	-.05445	-.05227
21	4	5	-.04949	-.04753	-.04538	-.04302	-.04043	-.03762	-.03455
21	4	6	-.03886	-.03627	-.03351	-.03056	-.02744	-.02412	-.02060
21	4	7	-.02919	-.02618	-.02305	-.01977	-.01635	-.01279	-.00909
21	4	8	-.02015	-.01690	-.01356	-.01014	-.00661	-.00300	.00071
21	4	9	-.01153	-.00818	-.00478	-.00134	.00215	.00569	.00927
21	4	10	-.00317	.00016	.00350	.00684	.01018	.01353	.01688
21	4	11	.00506	.00826	.01143	.01456	.01766	.02072	.02376
21	4	12	.01327	.01624	.01913	.02196	.02472	.02741	.03006
21	4	13	.02157	.02420	.02671	.02913	.03146	.03372	.03590
21	4	14	.03010	.03224	.03427	.03618	.03799	.03972	.04136
21	4	15	.03897	.04050	.04190	.04319	.04439	.04550	.04654
21	4	16	.04836	.04910	.04973	.05027	.05073	.05112	.05147
21	4	17	.37159	.35658	.34270	.32993	.31824	.30759	.29797
21	5	1	-.15368	-.16739	-.18270	-.19974	-.21862	-.23947	-.26242
21	5	2	-.10273	-.10588	-.10904	-.11219	-.11530	-.11834	-.12128
21	5	3	-.08127	-.08168	-.08193	-.08198	-.08182	-.08143	-.08077
21	5	4	-.06536	-.06420	-.06283	-.06124	-.05940	-.05730	-.05492
21	5	5	-.05218	-.05003	-.04767	-.04510	-.04231	-.03927	-.03600
21	5	6	-.04062	-.03782	-.03485	-.03169	-.02836	-.02483	-.02110

Coefficients b_i for the BLUE of σ
Shape parameter k

n	s	i	0.0	0.1	0.2	0.3	0.4	0.5	0.6
21	5	7	-.03011	-.02690	-.02355	-.02008	-.01647	-.01272	-.00882
21	5	8	-.02028	-.01684	-.01332	-.00971	-.00602	-.00224	.00164
21	5	9	-.01092	-.00740	-.00385	-.00026	.00337	.00704	.01076
21	5	10	-.00184	.00163	.00508	.00853	.01198	.01542	.01887
21	5	11	.00709	.01039	.01363	.01683	.01998	.02310	.02619
21	5	12	.01600	.01900	.02192	.02476	.02753	.03024	.03289
21	5	13	.02499	.02759	.03007	.03245	.03474	.03695	.03909
21	5	14	.03420	.03625	.03817	.03998	.04170	.04332	.04488
21	5	15	.04377	.04511	.04634	.04746	.04849	.04944	.05033
21	5	16	.43294	.41818	.40453	.39197	.38049	.37006	.36067
21	6	1	-.16597	-.18046	-.19662	-.21457	-.23445	-.25638	-.28050
21	6	2	-.11030	-.11350	-.11670	-.11987	-.12300	-.12605	-.12900
21	6	3	-.08686	-.08716	-.08728	-.08720	-.08690	-.08635	-.08553
21	6	4	-.06948	-.06813	-.06657	-.06477	-.06272	-.06040	-.05780
21	6	5	-.05509	-.05271	-.05013	-.04733	-.04430	-.04103	-.03751
21	6	6	-.04247	-.03943	-.03622	-.03283	-.02926	-.02550	-.02155
21	6	7	-.03099	-.02755	-.02398	-.02028	-.01646	-.01249	-.00839
21	6	8	-.02027	-.01662	-.01289	-.00909	-.00521	-.00124	.00281
21	6	9	-.01005	-.00635	-.00263	.00112	.00491	.00872	.01258
21	6	10	-.00016	.00345	.00704	.01061	.01416	.01771	.02125
21	6	11	.00957	.01295	.01628	.01955	.02277	.02594	.02908
21	6	12	.01926	.02229	.02523	.02809	.03087	.03358	.03623
21	6	13	.02903	.03158	.03401	.03635	.03859	.04074	.04283
21	6	14	.03900	.04093	.04273	.04442	.04602	.04753	.04897
21	6	15	.49476	.48071	.46773	.45582	.44499	.43523	.42652
21	7	1	-.17996	-.19536	-.21250	-.23152	-.25255	-.27574	-.30121
21	7	2	-.11885	-.12210	-.12534	-.12855	-.13171	-.13478	-.13774
21	7	3	-.09311	-.09328	-.09327	-.09305	-.09259	-.09187	-.09087
21	7	4	-.07403	-.07248	-.07069	-.06867	-.06639	-.06383	-.06098
21	7	5	-.05824	-.05561	-.05278	-.04972	-.04643	-.04289	-.03911
21	7	6	-.04440	-.04110	-.03763	-.03398	-.03015	-.02612	-.02191
21	7	7	-.03181	-.02811	-.02430	-.02035	-.01628	-.01208	-.00774
21	7	8	-.02006	-.01618	-.01223	-.00820	-.00411	.00006	.00432
21	7	9	-.00887	-.00497	-.00106	.00287	.00683	.01081	.01482
21	7	10	.00196	.00572	.00945	.01315	.01683	.02049	.02414
21	7	11	.01260	.01607	.01948	.02282	.02611	.02934	.03254
21	7	12	.02317	.02623	.02919	.03205	.03484	.03755	.04019
21	7	13	.03382	.03631	.03869	.04096	.04313	.04522	.04724
21	7	14	.55777	.54485	.53298	.52218	.51246	.50383	.49630
21	8	1	-.19608	-.21253	-.23082	-.25108	-.27347	-.29812	-.32517
21	8	2	-.12860	-.13191	-.13521	-.13847	-.14166	-.14476	-.14774
21	8	3	-.10018	-.10021	-.10005	-.09965	-.09901	-.09810	-.09690
21	8	4	-.07912	-.07733	-.07529	-.07301	-.07046	-.06762	-.06448
21	8	5	-.06170	-.05879	-.05566	-.05230	-.04871	-.04487	-.04078
21	8	6	-.04643	-.04283	-.03906	-.03511	-.03098	-.02666	-.02215
21	8	7	-.03255	-.02856	-.02446	-.02023	-.01589	-.01141	-.00680
21	8	8	-.01960	-.01545	-.01125	-.00698	-.00264	.00176	.00624
21	8	9	-.00728	-.00316	.00096	.00509	.00924	.01340	.01759
21	8	10	.00464	.00856	.01244	.01629	.02010	.02389	.02766
21	8	11	.01632	.01990	.02339	.02681	.03016	.03346	.03671
21	8	12	.02792	.03100	.03396	.03683	.03960	.04230	.04494
21	8	13	.62266	.61132	.60103	.59182	.58371	.57673	.57089
21	9	1	-.21491	-.23260	-.25223	-.27396	-.29793	-.32431	-.35323
21	9	2	-.13987	-.14327	-.14663	-.14994	-.15317	-.15630	-.15930
21	9	3	-.10828	-.10815	-.10780	-.10721	-.10636	-.10522	-.10378
21	9	4	-.08488	-.08280	-.08048	-.07789	-.07502	-.07186	-.06840

Coefficients b_i for the BLUE of σ
Shape parameter k

n	s	i	0.0	0.1	0.2	0.3	0.4	0.5	0.6
21	9	5	-.06553	-.06228	-.05881	-.05511	-.05117	-.04698	-.04254
21	9	6	-.04857	-.04462	-.04050	-.03621	-.03173	-.02707	-.02221
21	9	7	-.03316	-.02884	-.02441	-.01986	-.01520	-.01041	-.00549
21	9	8	-.01880	-.01435	-.00985	-.00530	-.00070	.00397	.00871
21	9	9	-.00515	-.00078	.00358	.00794	.01230	.01667	.02106
21	9	10	.00804	.01215	.01620	.02020	.02416	.02808	.03198
21	9	11	.02095	.02463	.02821	.03171	.03513	.03849	.04179
21	9	12	.69015	.68090	.67271	.66562	.65969	.65494	.65141
21	10	1	-.23725	-.25640	-.27764	-.30111	-.32698	-.35541	-.38655
21	10	2	-.15310	-.15658	-.16002	-.16339	-.16667	-.16983	-.17285
21	10	3	-.11770	-.11736	-.11679	-.11596	-.11486	-.11345	-.11173
21	10	4	-.09148	-.08906	-.08639	-.08344	-.08020	-.07666	-.07279
21	10	5	-.06980	-.06615	-.06228	-.05818	-.05383	-.04922	-.04436
21	10	6	-.05081	-.04645	-.04193	-.03723	-.03235	-.02728	-.02202
21	10	7	-.03358	-.02886	-.02405	-.01912	-.01409	-.00894	-.00366
21	10	8	-.01753	-.01273	-.00789	-.00302	.00190	.00688	.01191
21	10	9	-.00229	.00237	.00700	.01162	.01622	.02083	.02544
21	10	10	.01241	.01673	.02097	.02514	.02926	.03334	.03738
21	10	11	.76111	.75451	.74901	.74468	.74157	.73974	.73924
21	11	1	-.26424	-.28516	-.30833	-.33391	-.36207	-.39299	-.42683
21	11	2	-.16891	-.17249	-.17600	-.17943	-.18275	-.18595	-.18899
21	11	3	-.12882	-.12823	-.12739	-.12627	-.12485	-.12312	-.12105
21	11	4	-.09914	-.09631	-.09321	-.08983	-.08614	-.08213	-.07779
21	11	5	-.07462	-.07049	-.06614	-.06154	-.05670	-.05159	-.04622
21	11	6	-.05316	-.04831	-.04329	-.03810	-.03274	-.02719	-.02145
21	11	7	-.03369	-.02851	-.02323	-.01786	-.01239	-.00680	-.00110
21	11	8	-.01558	-.01037	-.00514	.00012	.00541	.01075	.01613
21	11	9	.00159	.00658	.01152	.01644	.02132	.02620	.03107
21	11	10	.83658	.83330	.83121	.83038	.83089	.83282	.83623
22	0	1	-.10388	-.11486	-.12722	-.14105	-.15647	-.17360	-.19252
22	0	2	-.07154	-.07481	-.07815	-.08151	-.08488	-.08822	-.09150
22	0	3	-.05798	-.05912	-.06015	-.06103	-.06174	-.06224	-.06251
22	0	4	-.04795	-.04781	-.04750	-.04698	-.04624	-.04526	-.04401
22	0	5	-.03966	-.03867	-.03747	-.03607	-.03444	-.03258	-.03046
22	0	6	-.03242	-.03080	-.02901	-.02701	-.02482	-.02242	-.01980
22	0	7	-.02584	-.02378	-.02156	-.01918	-.01663	-.01391	-.01101
22	0	8	-.01971	-.01734	-.01483	-.01220	-.00944	-.00655	-.00353
22	0	9	-.01389	-.01130	-.00862	-.00584	-.00298	-.00003	.00300
22	0	10	-.00826	-.00554	-.00277	.00006	.00293	.00585	.00881
22	0	11	-.00274	.00003	.00282	.00562	.00842	.01124	.01406
22	0	12	.00274	.00549	.00822	.01092	.01360	.01625	.01886
22	0	13	.00826	.01092	.01352	.01606	.01854	.02096	.02332
22	0	14	.01389	.01638	.01878	.02109	.02331	.02544	.02749
22	0	15	.01971	.02196	.02408	.02608	.02798	.02976	.03145
22	0	16	.02584	.02774	.02950	.03112	.03261	.03398	.03525
22	0	17	.03242	.03386	.03514	.03627	.03727	.03816	.03895
22	0	18	.03966	.04049	.04115	.04167	.04208	.04238	.04260
22	0	19	.04795	.04792	.04777	.04751	.04716	.04674	.04629
22	0	20	.05798	.05674	.05544	.05411	.05277	.05144	.05014
22	0	21	.07154	.06835	.06527	.06233	.05953	.05690	.05445
22	0	22	.10388	.09416	.08559	.07805	.07145	.06569	.06068
22	1	1	-.11097	-.12222	-.13486	-.14901	-.16478	-.18229	-.20165
22	1	2	-.07617	-.07937	-.08262	-.08591	-.08920	-.09247	-.09568
22	1	3	-.06158	-.06258	-.06346	-.06420	-.06477	-.06514	-.06528
22	1	4	-.05079	-.05048	-.05000	-.04932	-.04841	-.04727	-.04587
22	1	5	-.04188	-.04070	-.03933	-.03775	-.03596	-.03394	-.03167

Coefficients b_i for the BLUE of σ
Shape parameter k

n	s	i	0.0	0.1	0.2	0.3	0.4	0.5	0.6
22	1	6	-.03408	-.03228	-.03031	-.02816	-.02581	-.02325	-.02049
22	1	7	-.02700	-.02477	-.02239	-.01986	-.01716	-.01430	-.01128
22	1	8	-.02041	-.01788	-.01523	-.01246	-.00957	-.00656	-.00343
22	1	9	-.01414	-.01142	-.00861	-.00573	-.00276	.00029	.00342
22	1	10	-.00809	-.00526	-.00239	.00052	.00348	.00647	.00951
22	1	11	-.00216	.00069	.00355	.00641	.00927	.01214	.01500
22	1	12	.00374	.00653	.00930	.01203	.01473	.01740	.02004
22	1	13	.00967	.01233	.01493	.01747	.01994	.02235	.02471
22	1	14	.01572	.01817	.02053	.02279	.02497	.02706	.02908
22	1	15	.02197	.02413	.02616	.02808	.02989	.03160	.03322
22	1	16	.02855	.03030	.03191	.03339	.03476	.03603	.03720
22	1	17	.03560	.03682	.03789	.03883	.03967	.04040	.04106
22	1	18	.04335	.04386	.04425	.04452	.04470	.04481	.04486
22	1	19	.05216	.05173	.05121	.05063	.05000	.04934	.04867
22	1	20	.06272	.06096	.05920	.05746	.05578	.05415	.05260
22	1	21	.17378	.16144	.15028	.14024	.13123	.12317	.11600
22	2	1	-.11845	-.13007	-.14312	-.15771	-.17396	-.19200	-.21192
22	2	2	-.08100	-.08417	-.08740	-.09066	-.09392	-.09715	-.10033
22	2	3	-.06530	-.06620	-.06697	-.06759	-.06804	-.06829	-.06832
22	2	4	-.05369	-.05324	-.05261	-.05178	-.05073	-.04944	-.04789
22	2	5	-.04410	-.04276	-.04123	-.03949	-.03754	-.03536	-.03294
22	2	6	-.03571	-.03375	-.03162	-.02930	-.02679	-.02409	-.02118
22	2	7	-.02810	-.02571	-.02317	-.02048	-.01764	-.01464	-.01148
22	2	8	-.02100	-.01833	-.01553	-.01263	-.00961	-.00647	-.00322
22	2	9	-.01426	-.01141	-.00848	-.00547	-.00239	.00076	.00399
22	2	10	-.00775	-.00482	-.00184	.00116	.00421	.00728	.01039
22	2	11	-.00137	.00156	.00449	.00741	.01034	.01326	.01617
22	2	12	.00496	.00781	.01061	.01338	.01611	.01881	.02147
22	2	13	.01134	.01401	.01661	.01915	.02162	.02402	.02637
22	2	14	.01783	.02025	.02257	.02479	.02693	.02898	.03096
22	2	15	.02454	.02661	.02855	.03039	.03212	.03376	.03531
22	2	16	.03159	.03319	.03466	.03601	.03726	.03841	.03947
22	2	17	.03913	.04013	.04100	.04176	.04242	.04299	.04350
22	2	18	.04738	.04760	.04771	.04773	.04769	.04759	.04745
22	2	19	.05671	.05588	.05500	.05410	.05318	.05227	.05138
22	2	20	.23728	.22343	.21076	.19923	.18876	.17931	.17079
22	3	1	-.12653	-.13860	-.15215	-.16727	-.18411	-.20277	-.22337
22	3	2	-.08617	-.08935	-.09257	-.09582	-.09907	-.10229	-.10545
22	3	3	-.06925	-.07005	-.07073	-.07125	-.07159	-.07173	-.07164
22	3	4	-.05674	-.05615	-.05538	-.05440	-.05320	-.05176	-.05006
22	3	5	-.04641	-.04490	-.04321	-.04131	-.03920	-.03686	-.03428
22	3	6	-.03737	-.03524	-.03294	-.03046	-.02779	-.02492	-.02185
22	3	7	-.02916	-.02661	-.02391	-.02107	-.01807	-.01492	-.01161
22	3	8	-.02153	-.01869	-.01575	-.01270	-.00954	-.00627	-.00289
22	3	9	-.01427	-.01128	-.00821	-.00509	-.00189	.00138	.00471
22	3	10	-.00726	-.00421	-.00113	.00198	.00511	.00828	.01147
22	3	11	-.00039	.00263	.00563	.00863	.01162	.01460	.01757
22	3	12	.00643	.00932	.01217	.01498	.01774	.02046	.02315
22	3	13	.01328	.01597	.01857	.02111	.02357	.02597	.02832
22	3	14	.02026	.02264	.02492	.02710	.02920	.03121	.03315
22	3	15	.02746	.02944	.03129	.03304	.03469	.03624	.03772
22	3	16	.03501	.03646	.03778	.03900	.04011	.04113	.04209
22	3	17	.04306	.04384	.04450	.04506	.04553	.04594	.04629
22	3	18	.05184	.05174	.05157	.05133	.05104	.05072	.05039
22	3	19	.29772	.28305	.26954	.25716	.24586	.23559	.22631
22	4	1	-.13536	-.14797	-.16209	-.17784	-.19536	-.21475	-.23614

Coefficients b_i for the BLUE of σ
Shape parameter k

n	s	i	0.0	0.1	0.2	0.3	0.4	0.5	0.6
22	4	2	-.09177	-.09498	-.09822	-.10148	-.10474	-.10796	-.11112
22	4	3	-.07350	-.07422	-.07480	-.07523	-.07546	-.07549	-.07528
22	4	4	-.05999	-.05926	-.05835	-.05722	-.05587	-.05427	-.05241
22	4	5	-.04884	-.04717	-.04531	-.04324	-.04095	-.03844	-.03568
22	4	6	-.03908	-.03677	-.03430	-.03164	-.02880	-.02576	-.02252
22	4	7	-.03022	-.02749	-.02463	-.02161	-.01845	-.01514	-.01167
22	4	8	-.02198	-.01898	-.01588	-.01268	-.00937	-.00596	-.00244
22	4	9	-.01415	-.01101	-.00781	-.00455	-.00122	.00216	.00561
22	4	10	-.00658	-.00341	-.00022	.00299	.00623	.00948	.01276
22	4	11	.00082	.00393	.00702	.01009	.01314	.01619	.01922
22	4	12	.00817	.01112	.01401	.01686	.01965	.02240	.02512
22	4	13	.01555	.01825	.02086	.02339	.02585	.02824	.03058
22	4	14	.02306	.02540	.02764	.02977	.03182	.03378	.03568
22	4	15	.03080	.03267	.03443	.03608	.03763	.03910	.04049
22	4	16	.03889	.04017	.04133	.04239	.04336	.04424	.04507
22	4	17	.04749	.04802	.04844	.04879	.04906	.04928	.04946
22	4	18	.35669	.34171	.32787	.31514	.30349	.29288	.28328
22	5	1	-.14513	-.15835	-.17314	-.18962	-.20791	-.22815	-.25045
22	5	2	-.09792	-.10116	-.10444	-.10773	-.11100	-.11424	-.11740
22	5	3	-.07813	-.07877	-.07926	-.07958	-.07970	-.07961	-.07928
22	5	4	-.06350	-.06263	-.06156	-.06027	-.05876	-.05699	-.05496
22	5	5	-.05142	-.04958	-.04754	-.04529	-.04282	-.04011	-.03717
22	5	6	-.04086	-.03837	-.03570	-.03286	-.02983	-.02660	-.02317
22	5	7	-.03127	-.02836	-.02531	-.02211	-.01877	-.01528	-.01164
22	5	8	-.02235	-.01918	-.01591	-.01254	-.00907	-.00550	-.00183
22	5	9	-.01388	-.01058	-.00723	-.00383	-.00037	.00315	.00672
22	5	10	-.00570	-.00240	.00092	.00424	.00758	.01094	.01432
22	5	11	.00231	.00551	.00868	.01183	.01496	.01807	.02116
22	5	12	.01024	.01325	.01619	.01907	.02190	.02468	.02742
22	5	13	.01821	.02091	.02352	.02605	.02850	.03088	.03320
22	5	14	.02630	.02859	.03078	.03286	.03485	.03676	.03859
22	5	15	.03463	.03639	.03803	.03957	.04101	.04238	.04367
22	5	16	.04331	.04440	.04538	.04627	.04707	.04780	.04848
22	5	17	.41517	.40032	.38658	.37394	.36236	.35184	.34235
22	6	1	-.15604	-.16997	-.18552	-.20282	-.22202	-.24323	-.26659
22	6	2	-.10473	-.10803	-.11135	-.11467	-.11798	-.12123	-.12442
22	6	3	-.08323	-.08377	-.08416	-.08437	-.08438	-.08417	-.08370
22	6	4	-.06733	-.06630	-.06506	-.06360	-.06191	-.05996	-.05773
22	6	5	-.05421	-.05217	-.04993	-.04748	-.04480	-.04190	-.03874
22	6	6	-.04273	-.04003	-.03716	-.03411	-.03087	-.02744	-.02380
22	6	7	-.03232	-.02920	-.02594	-.02255	-.01901	-.01533	-.01150
22	6	8	-.02263	-.01927	-.01581	-.01226	-.00862	-.00488	-.00103
22	6	9	-.01343	-.00997	-.00646	-.00290	.00071	.00437	.00808
22	6	10	-.00456	-.00112	.00232	.00577	.00923	.01270	.01618
22	6	11	.00412	.00742	.01069	.01392	.01713	.02031	.02347
22	6	12	.01272	.01578	.01876	.02168	.02455	.02736	.03012
22	6	13	.02134	.02404	.02664	.02916	.03160	.03396	.03626
22	6	14	.03008	.03231	.03443	.03644	.03836	.04020	.04198
22	6	15	.03905	.04068	.04219	.04360	.04492	.04616	.04734
22	6	16	.47390	.45958	.44635	.43419	.42310	.41307	.40409
22	7	1	-.16835	-.18308	-.19951	-.21777	-.23800	-.26033	-.28490
22	7	2	-.11235	-.11571	-.11908	-.12246	-.12580	-.12909	-.13231
22	7	3	-.08888	-.08933	-.08961	-.08971	-.08959	-.08924	-.08862
22	7	4	-.07154	-.07033	-.06891	-.06726	-.06537	-.06321	-.06077
22	7	5	-.05722	-.05497	-.05251	-.04984	-.04694	-.04380	-.04041
22	7	6	-.04470	-.04177	-.03867	-.03539	-.03192	-.02826	-.02440

Coefficients b_i for the BLUE of σ
Shape parameter k

n	s	i	0.0	0.1	0.2	0.3	0.4	0.5	0.6
22	7	7	-.03335	-.03000	-.02652	-.02291	-.01916	-.01526	-.01121
22	7	8	-.02279	-.01921	-.01555	-.01180	-.00796	-.00403	.00000
22	7	9	-.01277	-.00912	-.00543	-.00170	.00207	.00589	.00976
22	7	10	-.00311	.00048	.00407	.00765	.01123	.01482	.01841
22	7	11	.00633	.00975	.01311	.01644	.01972	.02298	.02621
22	7	12	.01568	.01879	.02183	.02479	.02768	.03052	.03331
22	7	13	.02504	.02773	.03033	.03282	.03524	.03758	.03985
22	7	14	.03451	.03666	.03870	.04063	.04247	.04423	.04592
22	7	15	.53351	.52011	.50777	.49651	.48631	.47720	.46916
22	8	1	-.18238	-.19804	-.21548	-.23484	-.25627	-.27990	-.30587
22	8	2	-.12096	-.12439	-.12784	-.13127	-.13466	-.13800	-.14124
22	8	3	-.09523	-.09557	-.09573	-.09569	-.09543	-.09492	-.09415
22	8	4	-.07621	-.07480	-.07317	-.07131	-.06919	-.06680	-.06412
22	8	5	-.06051	-.05802	-.05531	-.05239	-.04923	-.04583	-.04218
22	8	6	-.04680	-.04361	-.04025	-.03671	-.03298	-.02906	-.02493
22	8	7	-.03436	-.03076	-.02703	-.02316	-.01916	-.01502	-.01074
22	8	8	-.02280	-.01898	-.01509	-.01111	-.00706	-.00291	.00133
22	8	9	-.01183	-.00797	-.00408	-.00016	.00379	.00779	.01182
22	8	10	-.00126	.00250	.00624	.00997	.01369	.01741	.02112
22	8	11	.00906	.01258	.01606	.01947	.02285	.02618	.02949
22	8	12	.01926	.02242	.02550	.02850	.03143	.03429	.03710
22	8	13	.02945	.03213	.03470	.03717	.03955	.04185	.04409
22	8	14	.59458	.58251	.57148	.56153	.55267	.54492	.53829
22	9	1	-.19857	-.21531	-.23393	-.25457	-.27739	-.30253	-.33014
22	9	2	-.13080	-.13432	-.13785	-.14134	-.14480	-.14818	-.15147
22	9	3	-.10242	-.10263	-.10266	-.10247	-.10204	-.10135	-.10039
22	9	4	-.08145	-.07981	-.07795	-.07583	-.07345	-.07080	-.06784
22	9	5	-.06414	-.06137	-.05838	-.05517	-.05172	-.04802	-.04406
22	9	6	-.04902	-.04554	-.04189	-.03805	-.03402	-.02980	-.02538
22	9	7	-.03532	-.03143	-.02741	-.02326	-.01898	-.01457	-.01001
22	9	8	-.02259	-.01851	-.01435	-.01012	-.00582	-.00143	.00304
22	9	9	-.01052	-.00643	-.00232	.00181	.00597	.01016	.01439
22	9	10	.00109	.00504	.00895	.01284	.01672	.02058	.02443
22	9	11	.01242	.01608	.01966	.02318	.02665	.03007	.03345
22	9	12	.02361	.02683	.02995	.03298	.03593	.03881	.04164
22	9	13	.65773	.64742	.63816	.62999	.62294	.61705	.61234
22	10	1	-.21749	-.23550	-.25549	-.27764	-.30209	-.32901	-.35855
22	10	2	-.14220	-.14582	-.14943	-.15300	-.15652	-.15996	-.16330
22	10	3	-.11068	-.11074	-.11060	-.11023	-.10961	-.10871	-.10753
22	10	4	-.08739	-.08548	-.08334	-.08093	-.07825	-.07528	-.07201
22	10	5	-.06818	-.06508	-.06176	-.05821	-.05442	-.05037	-.04606
22	10	6	-.05140	-.04758	-.04358	-.03939	-.03502	-.03046	-.02569
22	10	7	-.03621	-.03198	-.02763	-.02315	-.01855	-.01382	-.00894
22	10	8	-.02210	-.01770	-.01325	-.00873	-.00414	.00052	.00527
22	10	9	-.00874	-.00438	-.00001	.00436	.00875	.01317	.01761
22	10	10	.00411	.00827	.01237	.01645	.02049	.02451	.02852
22	10	11	.01663	.02042	.02412	.02775	.03132	.03483	.03829
22	10	12	.72365	.71558	.70859	.70273	.69805	.69459	.69240
22	11	1	-.23996	-.25946	-.28110	-.30503	-.33143	-.36046	-.39229
22	11	2	-.15560	-.15933	-.16303	-.16669	-.17028	-.17378	-.17718
22	11	3	-.12029	-.12017	-.11982	-.11924	-.11838	-.11724	-.11579
22	11	4	-.09421	-.09198	-.08950	-.08675	-.08371	-.08037	-.07671
22	11	5	-.07271	-.06923	-.06551	-.06156	-.05736	-.05291	-.04818
22	11	6	-.05395	-.04971	-.04531	-.04072	-.03594	-.03097	-.02580
22	11	7	-.03696	-.03234	-.02760	-.02274	-.01776	-.01265	-.00741
22	11	8	-.02120	-.01645	-.01164	-.00678	-.00186	.00312	.00818

Coefficients b_i for the BLUE of σ
Shape parameter k

n	s	i	0.0	0.1	0.2	0.3	0.4	0.5	0.6
22	11	9	-.00629	-.00163	.00303	.00768	.01234	.01701	.02169
22	11	10	.00803	.01241	.01674	.02101	.02524	.02944	.03362
22	11	11	.79314	.78788	.78375	.78081	.77915	.77881	.77986
23	0	1	-.09998	-.11070	-.12276	-.13630	-.15141	-.16822	-.18681
23	0	2	-.06931	-.07263	-.07602	-.07947	-.08293	-.08639	-.08981
23	0	3	-.05652	-.05778	-.05895	-.05998	-.06086	-.06155	-.06202
23	0	4	-.04708	-.04711	-.04697	-.04664	-.04610	-.04532	-.04430
23	0	5	-.03930	-.03849	-.03749	-.03629	-.03487	-.03322	-.03133
23	0	6	-.03252	-.03110	-.02951	-.02772	-.02573	-.02354	-.02113
23	0	7	-.02638	-.02452	-.02250	-.02032	-.01796	-.01543	-.01272
23	0	8	-.02068	-.01850	-.01619	-.01374	-.01115	-.00843	-.00557
23	0	9	-.01528	-.01287	-.01036	-.00775	-.00505	-.00224	.00066
23	0	10	-.01008	-.00753	-.00491	-.00222	.00053	.00334	.00620
23	0	11	-.00501	-.00238	.00028	.00298	.00569	.00843	.01119
23	0	12	.00000	.00264	.00529	.00792	.01054	.01316	.01576
23	0	13	.00501	.00761	.01016	.01268	.01516	.01759	.01998
23	0	14	.01008	.01256	.01498	.01732	.01959	.02179	.02392
23	0	15	.01528	.01758	.01978	.02189	.02390	.02582	.02765
23	0	16	.02068	.02273	.02465	.02645	.02814	.02972	.03121
23	0	17	.02638	.02808	.02964	.03106	.03236	.03355	.03464
23	0	18	.03252	.03377	.03486	.03581	.03664	.03736	.03799
23	0	19	.03930	.03995	.04044	.04081	.04106	.04123	.04132
23	0	20	.04708	.04691	.04661	.04622	.04576	.04524	.04470
23	0	21	.05652	.05518	.05379	.05238	.05097	.04959	.04824
23	0	22	.06931	.06610	.06300	.06006	.05727	.05465	.05222
23	0	23	.09998	.09051	.08217	.07485	.06844	.06286	.05801
23	1	1	-.10651	-.11746	-.12980	-.14363	-.15907	-.17624	-.19524
23	1	2	-.07361	-.07686	-.08018	-.08356	-.08696	-.09036	-.09372
23	1	3	-.05989	-.06102	-.06206	-.06297	-.06372	-.06429	-.06465
23	1	4	-.04977	-.04964	-.04934	-.04886	-.04818	-.04726	-.04610
23	1	5	-.04143	-.04045	-.03929	-.03793	-.03636	-.03456	-.03253
23	1	6	-.03415	-.03257	-.03081	-.02887	-.02674	-.02440	-.02186
23	1	7	-.02757	-.02555	-.02338	-.02105	-.01856	-.01590	-.01306
23	1	8	-.02145	-.01913	-.01667	-.01410	-.01139	-.00855	-.00558
23	1	9	-.01566	-.01313	-.01050	-.00778	-.00496	-.00206	.00094
23	1	10	-.01009	-.00743	-.00471	-.00193	.00090	.00379	.00673
23	1	11	-.00465	-.00194	.00080	.00356	.00634	.00914	.01195
23	1	12	.00072	.00342	.00611	.00878	.01145	.01409	.01672
23	1	13	.00609	.00871	.01128	.01381	.01630	.01874	.02114
23	1	14	.01152	.01399	.01639	.01871	.02096	.02314	.02526
23	1	15	.01709	.01933	.02148	.02353	.02549	.02736	.02915
23	1	16	.02287	.02481	.02663	.02834	.02994	.03145	.03286
23	1	17	.02897	.03051	.03192	.03321	.03438	.03545	.03644
23	1	18	.03552	.03655	.03744	.03821	.03887	.03944	.03993
23	1	19	.04275	.04310	.04332	.04345	.04349	.04346	.04339
23	1	20	.05100	.05044	.04980	.04910	.04837	.04762	.04688
23	1	21	.06091	.05907	.05724	.05545	.05372	.05206	.05048
23	1	22	.16735	.15525	.14433	.13452	.12572	.11787	.11087
23	2	1	-.11336	-.12466	-.13738	-.15162	-.16750	-.18515	-.20468
23	2	2	-.07808	-.08131	-.08462	-.08797	-.09134	-.09471	-.09806
23	2	3	-.06336	-.06441	-.06534	-.06615	-.06680	-.06726	-.06752
23	2	4	-.05251	-.05225	-.05182	-.05121	-.05039	-.04934	-.04805
23	2	5	-.04356	-.04244	-.04113	-.03963	-.03791	-.03597	-.03380
23	2	6	-.03576	-.03403	-.03212	-.03003	-.02775	-.02528	-.02260
23	2	7	-.02870	-.02653	-.02422	-.02175	-.01912	-.01632	-.01336
23	2	8	-.02214	-.01968	-.01709	-.01438	-.01155	-.00859	-.00551

Coefficients b_i for the BLUE of σ
Shape parameter k

n	s	i	0.0	0.1	0.2	0.3	0.4	0.5	0.6
23	2	9	-.01594	-.01327	-.01052	-.00769	-.00477	-.00176	.00133
23	2	10	-.00996	-.00719	-.00437	-.00150	.00142	.00439	.00741
23	2	11	-.00414	-.00133	.00148	.00431	.00716	.01002	.01289
23	2	12	.00162	.00438	.00712	.00984	.01255	.01523	.01789
23	2	13	.00737	.01002	.01262	.01516	.01766	.02011	.02252
23	2	14	.01319	.01565	.01804	.02034	.02258	.02474	.02684
23	2	15	.01915	.02135	.02344	.02544	.02735	.02917	.03092
23	2	16	.02533	.02718	.02890	.03052	.03204	.03346	.03480
23	2	17	.03184	.03324	.03450	.03565	.03670	.03766	.03854
23	2	18	.03883	.03964	.04033	.04091	.04141	.04182	.04218
23	2	19	.04651	.04656	.04652	.04641	.04623	.04601	.04576
23	2	20	.05520	.05426	.05328	.05228	.05128	.05030	.04934
23	2	21	.22847	.21482	.20237	.19104	.18076	.17149	.16315
23	3	1	-.12072	-.13245	-.14562	-.16035	-.17677	-.19500	-.21515
23	3	2	-.08284	-.08608	-.08939	-.09274	-.09612	-.09948	-.10282
23	3	3	-.06703	-.06800	-.06885	-.06957	-.07013	-.07050	-.07065
23	3	4	-.05538	-.05500	-.05445	-.05371	-.05275	-.05157	-.05014
23	3	5	-.04578	-.04451	-.04305	-.04140	-.03954	-.03746	-.03514
23	3	6	-.03740	-.03551	-.03345	-.03122	-.02879	-.02617	-.02334
23	3	7	-.02982	-.02750	-.02504	-.02242	-.01965	-.01672	-.01362
23	3	8	-.02278	-.02017	-.01745	-.01460	-.01164	-.00856	-.00535
23	3	9	-.01612	-.01333	-.01045	-.00750	-.00446	-.00135	.00185
23	3	10	-.00971	-.00683	-.00390	-.00093	.00209	.00514	.00824
23	3	11	-.00346	-.00057	.00233	.00524	.00815	.01108	.01401
23	3	12	.00271	.00554	.00833	.01110	.01385	.01657	.01927
23	3	13	.00888	.01156	.01418	.01674	.01926	.02172	.02414
23	3	14	.01512	.01757	.01994	.02223	.02445	.02659	.02868
23	3	15	.02150	.02364	.02569	.02763	.02948	.03126	.03296
23	3	16	.02811	.02985	.03148	.03300	.03443	.03576	.03703
23	3	17	.03506	.03629	.03741	.03842	.03933	.04017	.04093
23	3	18	.04250	.04308	.04356	.04395	.04426	.04452	.04472
23	3	19	.05063	.05038	.05007	.04970	.04929	.04886	.04843
23	3	20	.28655	.27202	.25866	.24643	.23527	.22513	.21597
23	4	1	-.12873	-.14094	-.15464	-.16995	-.18699	-.20590	-.22678
23	4	2	-.08797	-.09124	-.09458	-.09795	-.10133	-.10471	-.10805
23	4	3	-.07097	-.07187	-.07264	-.07327	-.07374	-.07401	-.07407
23	4	4	-.05843	-.05793	-.05726	-.05638	-.05529	-.05397	-.05240
23	4	5	-.04811	-.04669	-.04509	-.04328	-.04127	-.03903	-.03656
23	4	6	-.03910	-.03705	-.03483	-.03244	-.02986	-.02708	-.02409
23	4	7	-.03094	-.02847	-.02585	-.02308	-.02015	-.01707	-.01383
23	4	8	-.02338	-.02061	-.01774	-.01476	-.01165	-.00844	-.00510
23	4	9	-.01621	-.01328	-.01028	-.00719	-.00404	-.00081	.00250
23	4	10	-.00932	-.00632	-.00328	-.00020	.00291	.00606	.00925
23	4	11	-.00260	.00038	.00337	.00635	.00934	.01234	.01533
23	4	12	.00403	.00691	.00977	.01259	.01538	.01814	.02088
23	4	13	.01065	.01336	.01600	.01858	.02111	.02358	.02601
23	4	14	.01734	.01978	.02214	.02441	.02660	.02873	.03079
23	4	15	.02417	.02626	.02825	.03013	.03193	.03364	.03529
23	4	16	.03125	.03288	.03441	.03582	.03715	.03839	.03956
23	4	17	.03867	.03974	.04069	.04154	.04231	.04301	.04365
23	4	18	.04658	.04693	.04718	.04736	.04748	.04755	.04759
23	4	19	.34308	.32816	.31439	.30172	.29013	.27958	.27003
23	5	1	-.13753	-.15030	-.16461	-.18058	-.19834	-.21801	-.23973
23	5	2	-.09357	-.09689	-.10026	-.10366	-.10708	-.11047	-.11383
23	5	3	-.07524	-.07606	-.07676	-.07731	-.07768	-.07786	-.07781
23	5	4	-.06172	-.06109	-.06028	-.05927	-.05804	-.05657	-.05485

Coefficients b_i for the BLUE of σ
Shape parameter k

n	s	i	0.0	0.1	0.2	0.3	0.4	0.5	0.6
23	5	5	-.05058	-.04901	-.04725	-.04528	-.04311	-.04070	-.03806
23	5	6	-.04087	-.03865	-.03627	-.03370	-.03095	-.02800	-.02485
23	5	7	-.03208	-.02943	-.02664	-.02371	-.02062	-.01738	-.01398
23	5	8	-.02392	-.02100	-.01797	-.01483	-.01158	-.00821	-.00473
23	5	9	-.01620	-.01312	-.00998	-.00676	-.00347	-.00011	.00332
23	5	10	-.00878	-.00565	-.00249	.00070	.00392	.00717	.01046
23	5	11	-.00154	.00154	.00462	.00769	.01076	.01383	.01689
23	5	12	.00559	.00855	.01146	.01434	.01718	.01998	.02276
23	5	13	.01272	.01545	.01812	.02072	.02326	.02574	.02818
23	5	14	.01991	.02234	.02467	.02692	.02909	.03119	.03322
23	5	15	.02724	.02927	.03119	.03300	.03473	.03638	.03796
23	5	16	.03482	.03634	.03774	.03904	.04025	.04138	.04245
23	5	17	.04275	.04363	.04440	.04508	.04568	.04623	.04673
23	5	18	.39900	.38410	.37031	.35762	.34600	.33542	.32587
23	6	1	-.14729	-.16070	-.17570	-.19242	-.21099	-.23155	-.25422
23	6	2	-.09974	-.10311	-.10653	-.10998	-.11343	-.11686	-.12025
23	6	3	-.07990	-.08066	-.08127	-.08173	-.08201	-.08209	-.08193
23	6	4	-.06528	-.06452	-.06356	-.06241	-.06102	-.05939	-.05750
23	6	5	-.05323	-.05149	-.04956	-.04742	-.04506	-.04248	-.03965
23	6	6	-.04273	-.04033	-.03777	-.03502	-.03208	-.02895	-.02561
23	6	7	-.03323	-.03040	-.02743	-.02431	-.02105	-.01763	-.01405
23	6	8	-.02441	-.02131	-.01811	-.01480	-.01139	-.00787	-.00423
23	6	9	-.01607	-.01283	-.00953	-.00617	-.00274	.00076	.00432
23	6	10	-.00805	-.00478	-.00150	.00181	.00515	.00851	.01191
23	6	11	-.00024	.00295	.00613	.00930	.01245	.01559	.01873
23	6	12	.00747	.01049	.01347	.01640	.01929	.02214	.02496
23	6	13	.01515	.01791	.02060	.02321	.02576	.02826	.03069
23	6	14	.02289	.02530	.02761	.02983	.03196	.03403	.03603
23	6	15	.03078	.03273	.03457	.03630	.03795	.03953	.04103
23	6	16	.03891	.04028	.04154	.04271	.04379	.04480	.04576
23	6	17	.45498	.44047	.42704	.41469	.40341	.39319	.38401
23	7	1	-.15821	-.17234	-.18813	-.20571	-.22522	-.24679	-.27055
23	7	2	-.10658	-.11002	-.11350	-.11701	-.12051	-.12398	-.12740
23	7	3	-.08505	-.08573	-.08626	-.08663	-.08680	-.08676	-.08649
23	7	4	-.06917	-.06826	-.06716	-.06584	-.06428	-.06248	-.06041
23	7	5	-.05610	-.05418	-.05205	-.04972	-.04717	-.04438	-.04135
23	7	6	-.04470	-.04211	-.03934	-.03638	-.03324	-.02990	-.02636
23	7	7	-.03439	-.03136	-.02819	-.02488	-.02142	-.01780	-.01403
23	7	8	-.02483	-.02154	-.01815	-.01466	-.01106	-.00736	-.00356
23	7	9	-.01578	-.01237	-.00890	-.00538	-.00179	.00185	.00556
23	7	10	-.00709	-.00368	-.00025	.00319	.00665	.01014	.01364
23	7	11	.00137	.00468	.00796	.01122	.01446	.01769	.02091
23	7	12	.00971	.01281	.01585	.01883	.02177	.02467	.02754
23	7	13	.01801	.02080	.02351	.02614	.02869	.03119	.03363
23	7	14	.02637	.02875	.03103	.03321	.03531	.03733	.03929
23	7	15	.03487	.03673	.03848	.04012	.04168	.04316	.04458
23	7	16	.51159	.49782	.48512	.47348	.46292	.45343	.44500
23	8	1	-.17055	-.18551	-.20220	-.22076	-.24134	-.26407	-.28909
23	8	2	-.11425	-.11777	-.12133	-.12490	-.12845	-.13198	-.13545
23	8	3	-.09078	-.09137	-.09181	-.09207	-.09213	-.09197	-.09156
23	8	4	-.07347	-.07240	-.07112	-.06961	-.06787	-.06587	-.06359
23	8	5	-.05922	-.05709	-.05476	-.05221	-.04943	-.04642	-.04316
23	8	6	-.04680	-.04398	-.04099	-.03781	-.03444	-.03087	-.02710
23	8	7	-.03557	-.03231	-.02892	-.02538	-.02170	-.01787	-.01389
23	8	8	-.02516	-.02165	-.01805	-.01435	-.01056	-.00667	-.00267
23	8	9	-.01531	-.01170	-.00805	-.00435	-.00059	.00322	.00709

Coefficients b_i for the BLUE of σ
Shape parameter k

n	s	i	0.0	0.1	0.2	0.3	0.4	0.5	0.6
23	8	10	-.00585	-.00228	.00130	.00489	.00849	.01210	.01574
23	8	11	.00335	.00678	.01017	.01354	.01688	.02020	.02351
23	8	12	.01240	.01558	.01869	.02173	.02473	.02767	.03058
23	8	13	.02141	.02423	.02695	.02958	.03214	.03464	.03707
23	8	14	.03047	.03281	.03504	.03717	.03921	.04118	.04309
23	8	15	.56934	.55668	.54507	.53453	.52507	.51670	.50944
23	9	1	-.18463	-.20054	-.21827	-.23797	-.25978	-.28384	-.31031
23	9	2	-.12293	-.12655	-.13019	-.13383	-.13746	-.14105	-.14458
23	9	3	-.09721	-.09771	-.09805	-.09819	-.09812	-.09782	-.09726
23	9	4	-.07826	-.07699	-.07552	-.07381	-.07185	-.06962	-.06712
23	9	5	-.06265	-.06029	-.05771	-.05492	-.05189	-.04862	-.04510
23	9	6	-.04905	-.04597	-.04272	-.03929	-.03566	-.03183	-.02780
23	9	7	-.03676	-.03324	-.02960	-.02581	-.02188	-.01781	-.01358
23	9	8	-.02536	-.02161	-.01778	-.01385	-.00983	-.00572	-.00151
23	9	9	-.01459	-.01077	-.00691	-.00301	.00094	.00494	.00899
23	9	10	-.00426	-.00050	.00325	.00700	.01075	.01451	.01828
23	9	11	.00579	.00936	.01288	.01636	.01980	.02322	.02662
23	9	12	.01567	.01893	.02210	.02521	.02826	.03125	.03420
23	9	13	.02548	.02831	.03104	.03368	.03623	.03872	.04114
23	9	14	.62875	.61759	.60747	.59843	.59049	.58368	.57803
23	10	1	-.20089	-.21790	-.23684	-.25785	-.28109	-.30671	-.33486
23	10	2	-.13287	-.13659	-.14033	-.14406	-.14777	-.15143	-.15502
23	10	3	-.10453	-.10491	-.10512	-.10513	-.10491	-.10445	-.10371
23	10	4	-.08363	-.08216	-.08045	-.07850	-.07629	-.07381	-.07104
23	10	5	-.06644	-.06381	-.06096	-.05788	-.05456	-.05100	-.04718
23	10	6	-.05146	-.04810	-.04455	-.04082	-.03690	-.03278	-.02845
23	10	7	-.03793	-.03413	-.03019	-.02612	-.02191	-.01756	-.01305
23	10	8	-.02539	-.02137	-.01727	-.01308	-.00881	-.00445	.00001
23	10	9	-.01355	-.00949	-.00539	-.00127	.00290	.00710	.01136
23	10	10	-.00220	.00176	.00570	.00963	.01355	.01746	.02139
23	10	11	.00883	.01254	.01620	.01980	.02336	.02689	.03039
23	10	12	.01966	.02300	.02625	.02942	.03251	.03555	.03854
23	10	13	.69041	.68115	.67296	.66587	.65993	.65517	.65163
23	11	1	-.21990	-.23821	-.25856	-.28111	-.30603	-.33347	-.36361
23	11	2	-.14439	-.14823	-.15207	-.15591	-.15970	-.16344	-.16711
23	11	3	-.11293	-.11318	-.11324	-.11309	-.11269	-.11204	-.11110
23	11	4	-.08975	-.08801	-.08604	-.08381	-.08131	-.07853	-.07544
23	11	5	-.07068	-.06773	-.06455	-.06114	-.05749	-.05359	-.04942
23	11	6	-.05407	-.05037	-.04648	-.04241	-.03814	-.03367	-.02900
23	11	7	-.03907	-.03493	-.03066	-.02626	-.02173	-.01705	-.01222
23	11	8	-.02519	-.02085	-.01644	-.01195	-.00739	-.00274	.00200
23	11	9	-.01208	-.00774	-.00338	.00100	.00541	.00986	.01434
23	11	10	.00047	.00466	.00881	.01293	.01704	.02114	.02524
23	11	11	.01265	.01653	.02033	.02406	.02775	.03139	.03499
23	11	12	.75494	.74807	.74230	.73768	.73428	.73214	.73133
23	12	1	-.24249	-.26233	-.28436	-.30874	-.33565	-.36526	-.39776
23	12	2	-.15794	-.16191	-.16588	-.16982	-.17372	-.17755	-.18130
23	12	3	-.12273	-.12282	-.12269	-.12234	-.12173	-.12084	-.11966
23	12	4	-.09679	-.09475	-.09245	-.08989	-.08704	-.08389	-.08044
23	12	5	-.07546	-.07213	-.06857	-.06477	-.06072	-.05641	-.05183
23	12	6	-.05690	-.05279	-.04850	-.04402	-.03934	-.03447	-.02939
23	12	7	-.04014	-.03560	-.03094	-.02615	-.02123	-.01617	-.01097
23	12	8	-.02464	-.01993	-.01516	-.01033	-.00542	-.00044	.00463
23	12	9	-.01002	-.00536	-.00069	.00399	.00868	.01340	.01815
23	12	10	.00396	.00840	.01280	.01715	.02147	.02577	.03007
23	12	11	.82315	.81921	.81644	.81491	.81469	.81586	.81849

Coefficients b_i for the BLUE of σ
Shape parameter k

n	s	i	0.0	0.1	0.2	0.3	0.4	0.5	0.6
24	0	1	-.09638	-.10684	-.11864	-.13190	-.14671	-.16321	-.18149
24	0	2	-.06723	-.07058	-.07402	-.07753	-.08108	-.08464	-.08818
24	0	3	-.05512	-.05649	-.05778	-.05895	-.05997	-.06083	-.06149
24	0	4	-.04621	-.04639	-.04640	-.04624	-.04588	-.04530	-.04448
24	0	5	-.03889	-.03825	-.03742	-.03640	-.03518	-.03372	-.03203
24	0	6	-.03252	-.03128	-.02987	-.02827	-.02647	-.02447	-.02225
24	0	7	-.02677	-.02509	-.02326	-.02125	-.01908	-.01672	-.01419
24	0	8	-.02144	-.01944	-.01730	-.01503	-.01261	-.01005	-.00734
24	0	9	-.01641	-.01418	-.01183	-.00938	-.00682	-.00415	-.00138
24	0	10	-.01159	-.00919	-.00672	-.00417	-.00155	.00115	.00391
24	0	11	-.00690	-.00441	-.00187	.00071	.00333	.00599	.00868
24	0	12	-.00229	.00024	.00278	.00534	.00790	.01046	.01303
24	0	13	.00229	.00481	.00730	.00978	.01222	.01465	.01704
24	0	14	.00690	.00935	.01174	.01408	.01637	.01860	.02079
24	0	15	.01159	.01390	.01614	.01830	.02037	.02238	.02431
24	0	16	.01641	.01854	.02056	.02247	.02429	.02602	.02766
24	0	17	.02144	.02331	.02504	.02666	.02816	.02957	.03087
24	0	18	.02677	.02829	.02967	.03092	.03204	.03306	.03399
24	0	19	.03252	.03359	.03452	.03531	.03599	.03656	.03705
24	0	20	.03889	.03938	.03973	.03995	.04008	.04012	.04010
24	0	21	.04621	.04591	.04550	.04500	.04444	.04383	.04320
24	0	22	.05512	.05369	.05223	.05075	.04929	.04786	.04648
24	0	23	.06723	.06400	.06090	.05795	.05518	.05258	.05017
24	0	24	.09638	.08714	.07902	.07190	.06568	.06027	.05557
24	1	1	-.10241	-.11310	-.12515	-.13867	-.15380	-.17063	-.18929
24	1	2	-.07123	-.07452	-.07790	-.08135	-.08484	-.08835	-.09185
24	1	3	-.05829	-.05954	-.06070	-.06176	-.06267	-.06342	-.06398
24	1	4	-.04876	-.04879	-.04866	-.04837	-.04787	-.04716	-.04621
24	1	5	-.04093	-.04013	-.03916	-.03799	-.03663	-.03504	-.03322
24	1	6	-.03412	-.03272	-.03116	-.02942	-.02748	-.02535	-.02301
24	1	7	-.02796	-.02614	-.02416	-.02202	-.01972	-.01725	-.01460
24	1	8	-.02227	-.02013	-.01786	-.01546	-.01293	-.01026	-.00745
24	1	9	-.01689	-.01453	-.01207	-.00951	-.00685	-.00408	-.00122
24	1	10	-.01174	-.00923	-.00666	-.00402	-.00131	.00147	.00430
24	1	11	-.00673	-.00414	-.00153	.00113	.00381	.00653	.00928
24	1	12	-.00180	.00080	.00340	.00600	.00861	.01121	.01382
24	1	13	.00310	.00565	.00818	.01068	.01315	.01559	.01801
24	1	14	.00803	.01048	.01287	.01521	.01750	.01973	.02191
24	1	15	.01304	.01532	.01753	.01965	.02170	.02368	.02558
24	1	16	.01819	.02024	.02220	.02405	.02581	.02748	.02908
24	1	17	.02355	.02531	.02694	.02846	.02987	.03119	.03242
24	1	18	.02923	.03059	.03182	.03293	.03393	.03484	.03567
24	1	19	.03536	.03621	.03694	.03755	.03806	.03849	.03884
24	1	20	.04213	.04233	.04241	.04240	.04232	.04218	.04201
24	1	21	.04987	.04919	.04845	.04766	.04684	.04602	.04520
24	1	22	.05920	.05729	.05541	.05358	.05181	.05012	.04852
24	1	23	.16141	.14955	.13886	.12926	.12067	.11301	.10619
24	2	1	-.10871	-.11972	-.13212	-.14603	-.16157	-.17885	-.19800
24	2	2	-.07538	-.07866	-.08202	-.08545	-.08893	-.09242	-.09590
24	2	3	-.06154	-.06271	-.06379	-.06475	-.06558	-.06624	-.06671
24	2	4	-.05135	-.05126	-.05102	-.05060	-.04999	-.04916	-.04809
24	2	5	-.04298	-.04205	-.04094	-.03964	-.03815	-.03643	-.03448
24	2	6	-.03569	-.03416	-.03246	-.03058	-.02851	-.02625	-.02378
24	2	7	-.02912	-.02715	-.02504	-.02277	-.02034	-.01774	-.01497
24	2	8	-.02303	-.02076	-.01836	-.01584	-.01319	-.01040	-.00749
24	2	9	-.01728	-.01480	-.01222	-.00955	-.00679	-.00392	-.00097

Coefficients b_i for the BLUE of σ
Shape parameter k

n	s	i	0.0	0.1	0.2	0.3	0.4	0.5	0.6
24	2	10	-.01177	-.00916	-.00649	-.00375	-.00096	.00190	.00482
24	2	11	-.00641	-.00375	-.00105	.00168	.00444	.00722	.01003
24	2	12	-.00115	.00151	.00417	.00683	.00948	.01213	.01478
24	2	13	.00409	.00668	.00924	.01177	.01426	.01673	.01916
24	2	14	.00935	.01181	.01421	.01655	.01883	.02106	.02324
24	2	15	.01469	.01696	.01914	.02123	.02326	.02520	.02709
24	2	16	.02019	.02219	.02408	.02587	.02757	.02919	.03073
24	2	17	.02591	.02756	.02909	.03051	.03184	.03307	.03423
24	2	18	.03196	.03316	.03425	.03522	.03610	.03689	.03761
24	2	19	.03847	.03911	.03963	.04006	.04041	.04069	.04091
24	2	20	.04563	.04555	.04538	.04514	.04485	.04452	.04418
24	2	21	.05377	.05273	.05166	.05059	.04952	.04847	.04745
24	2	22	.22035	.20691	.19466	.18353	.17344	.16434	.15617
24	3	1	-.11545	-.12685	-.13967	-.15404	-.17007	-.18790	-.20763
24	3	2	-.07978	-.08307	-.08644	-.08988	-.09336	-.09685	-.10033
24	3	3	-.06496	-.06606	-.06707	-.06796	-.06871	-.06928	-.06966
24	3	4	-.05406	-.05386	-.05351	-.05298	-.05224	-.05130	-.05011
24	3	5	-.04510	-.04404	-.04280	-.04137	-.03974	-.03789	-.03581
24	3	6	-.03730	-.03563	-.03379	-.03178	-.02958	-.02718	-.02458
24	3	7	-.03026	-.02816	-.02591	-.02351	-.02095	-.01822	-.01532
24	3	8	-.02375	-.02134	-.01882	-.01617	-.01340	-.01049	-.00746
24	3	9	-.01760	-.01499	-.01230	-.00952	-.00664	-.00367	-.00062
24	3	10	-.01170	-.00898	-.00621	-.00338	-.00049	.00246	.00546
24	3	11	-.00598	-.00322	-.00043	.00237	.00520	.00805	.01092
24	3	12	-.00035	.00238	.00511	.00782	.01052	.01322	.01591
24	3	13	.00525	.00789	.01048	.01304	.01556	.01805	.02051
24	3	14	.01087	.01335	.01575	.01810	.02038	.02261	.02479
24	3	15	.01658	.01882	.02098	.02305	.02504	.02696	.02881
24	3	16	.02245	.02438	.02621	.02794	.02958	.03114	.03263
24	3	17	.02855	.03009	.03151	.03284	.03406	.03521	.03628
24	3	18	.03498	.03603	.03696	.03779	.03853	.03920	.03981
24	3	19	.04189	.04231	.04263	.04287	.04304	.04316	.04324
24	3	20	.04946	.04908	.04864	.04816	.04765	.04713	.04661
24	3	21	.27627	.26190	.24869	.23661	.22559	.21560	.20657
24	4	1	-.12275	-.13460	-.14791	-.16281	-.17941	-.19786	-.21826
24	4	2	-.08450	-.08782	-.09123	-.09469	-.09819	-.10169	-.10519
24	4	3	-.06862	-.06966	-.07060	-.07142	-.07209	-.07258	-.07288
24	4	4	-.05693	-.05663	-.05616	-.05551	-.05466	-.05359	-.05228
24	4	5	-.04733	-.04613	-.04476	-.04320	-.04143	-.03944	-.03722
24	4	6	-.03897	-.03715	-.03517	-.03302	-.03067	-.02813	-.02539
24	4	7	-.03142	-.02918	-.02678	-.02424	-.02154	-.01867	-.01564
24	4	8	-.02444	-.02190	-.01924	-.01645	-.01355	-.01052	-.00736
24	4	9	-.01785	-.01512	-.01230	-.00939	-.00640	-.00333	-.00016
24	4	10	-.01154	-.00870	-.00582	-.00288	.00010	.00314	.00623
24	4	11	-.00540	-.00255	.00033	.00321	.00612	.00904	.01198
24	4	12	.00063	.00343	.00622	.00899	.01175	.01449	.01723
24	4	13	.00662	.00930	.01193	.01452	.01707	.01958	.02206
24	4	14	.01263	.01512	.01753	.01988	.02216	.02439	.02656
24	4	15	.01874	.02095	.02308	.02512	.02708	.02896	.03079
24	4	16	.02500	.02687	.02863	.03029	.03187	.03336	.03479
24	4	17	.03150	.03293	.03424	.03546	.03658	.03763	.03861
24	4	18	.03835	.03923	.03999	.04067	.04127	.04180	.04229
24	4	19	.04567	.04586	.04596	.04600	.04598	.04592	.04585
24	4	20	.33060	.31575	.30206	.28948	.27796	.26749	.25801
24	5	1	-.13073	-.14309	-.15695	-.17245	-.18972	-.20887	-.23004
24	5	2	-.08963	-.09300	-.09644	-.09994	-.10347	-.10701	-.11053

Coefficients b_i for the BLUE of σ
Shape parameter k

n	s	i	0.0	0.1	0.2	0.3	0.4	0.5	0.6
24	5	3	-.07256	-.07355	-.07443	-.07517	-.07576	-.07618	-.07638
24	5	4	-.06000	-.05959	-.05901	-.05825	-.05727	-.05607	-.05462
24	5	5	-.04969	-.04836	-.04685	-.04514	-.04322	-.04109	-.03872
24	5	6	-.04071	-.03875	-.03661	-.03430	-.03181	-.02911	-.02622
24	5	7	-.03261	-.03021	-.02766	-.02497	-.02212	-.01910	-.01593
24	5	8	-.02511	-.02242	-.01961	-.01668	-.01364	-.01048	-.00719
24	5	9	-.01803	-.01516	-.01221	-.00918	-.00606	-.00286	.00042
24	5	10	-.01125	-.00830	-.00530	-.00225	.00084	.00398	.00717
24	5	11	-.00467	-.00171	.00125	.00423	.00722	.01022	.01323
24	5	12	.00180	.00468	.00753	.01037	.01318	.01598	.01876
24	5	13	.00822	.01095	.01362	.01624	.01882	.02136	.02386
24	5	14	.01467	.01717	.01958	.02193	.02421	.02643	.02860
24	5	15	.02121	.02339	.02548	.02748	.02941	.03126	.03304
24	5	16	.02790	.02969	.03138	.03296	.03446	.03589	.03724
24	5	17	.03484	.03614	.03733	.03842	.03943	.04037	.04125
24	5	18	.04213	.04282	.04341	.04391	.04435	.04474	.04508
24	5	19	.38420	.36928	.35548	.34278	.33114	.32055	.31098
24	6	1	-.13952	-.15245	-.16695	-.18314	-.20114	-.22111	-.24315
24	6	2	-.09524	-.09867	-.10216	-.10571	-.10928	-.11286	-.11642
24	6	3	-.07685	-.07778	-.07860	-.07927	-.07978	-.08011	-.08023
24	6	4	-.06332	-.06280	-.06210	-.06120	-.06009	-.05875	-.05716
24	6	5	-.05221	-.05074	-.04907	-.04721	-.04514	-.04284	-.04031
24	6	6	-.04254	-.04042	-.03812	-.03565	-.03299	-.03013	-.02706
24	6	7	-.03382	-.03125	-.02854	-.02569	-.02267	-.01950	-.01616
24	6	8	-.02575	-.02289	-.01993	-.01685	-.01365	-.01034	-.00691
24	6	9	-.01813	-.01511	-.01201	-.00884	-.00559	-.00226	.00115
24	6	10	-.01083	-.00774	-.00462	-.00145	.00175	.00500	.00829
24	6	11	-.00375	-.00069	.00238	.00545	.00853	.01161	.01471
24	6	12	.00321	.00616	.00909	.01199	.01486	.01772	.02055
24	6	13	.01011	.01288	.01559	.01825	.02085	.02341	.02594
24	6	14	.01703	.01953	.02195	.02430	.02657	.02878	.03094
24	6	15	.02404	.02619	.02824	.03020	.03208	.03388	.03562
24	6	16	.03121	.03292	.03451	.03601	.03742	.03876	.04004
24	6	17	.03863	.03978	.04083	.04178	.04266	.04348	.04424
24	6	18	.43772	.42307	.40951	.39703	.38562	.37526	.36593
24	7	1	-.14928	-.16287	-.17808	-.19505	-.21390	-.23478	-.25782
24	7	2	-.10142	-.10492	-.10849	-.11209	-.11572	-.11935	-.12295
24	7	3	-.08155	-.08243	-.08317	-.08377	-.08420	-.08444	-.08446
24	7	4	-.06694	-.06629	-.06546	-.06442	-.06316	-.06167	-.05993
24	7	5	-.05493	-.05330	-.05147	-.04944	-.04719	-.04472	-.04201
24	7	6	-.04449	-.04219	-.03971	-.03706	-.03421	-.03117	-.02792
24	7	7	-.03506	-.03232	-.02943	-.02639	-.02320	-.01985	-.01634
24	7	8	-.02635	-.02332	-.02018	-.01693	-.01357	-.01010	-.00650
24	7	9	-.01812	-.01494	-.01169	-.00836	-.00497	-.00150	.00205
24	7	10	-.01025	-.00702	-.00376	-.00046	.00286	.00623	.00963
24	7	11	-.00261	.00057	.00375	.00692	.01009	.01326	.01644
24	7	12	.00489	.00794	.01094	.01391	.01684	.01976	.02264
24	7	13	.01233	.01515	.01790	.02059	.02322	.02581	.02835
24	7	14	.01978	.02229	.02470	.02704	.02930	.03150	.03364
24	7	15	.02732	.02942	.03142	.03332	.03515	.03690	.03859
24	7	16	.03501	.03661	.03810	.03950	.04081	.04206	.04324
24	7	17	.49167	.47761	.46462	.45271	.44186	.43207	.42334
24	8	1	-.16022	-.17456	-.19058	-.20843	-.22825	-.25017	-.27434
24	8	2	-.10831	-.11189	-.11553	-.11921	-.12290	-.12659	-.13025
24	8	3	-.08676	-.08756	-.08824	-.08876	-.08909	-.08923	-.08914
24	8	4	-.07090	-.07012	-.06914	-.06795	-.06653	-.06487	-.06295

Coefficients b_i for the BLUE of σ
Shape parameter k

n	s	i	0.0	0.1	0.2	0.3	0.4	0.5	0.6
24	8	5	-.05789	-.05607	-.05406	-.05184	-.04941	-.04674	-.04382
24	8	6	-.04656	-.04406	-.04139	-.03854	-.03549	-.03225	-.02880
24	8	7	-.03635	-.03340	-.03031	-.02708	-.02369	-.02014	-.01644
24	8	8	-.02690	-.02367	-.02035	-.01692	-.01337	-.00972	-.00595
24	8	9	-.01799	-.01463	-.01120	-.00771	-.00415	-.00053	.00317
24	8	10	-.00946	-.00607	-.00267	.00077	.00423	.00772	.01125
24	8	11	-.00119	.00212	.00541	.00869	.01196	.01523	.01850
24	8	12	.00692	.01005	.01314	.01618	.01918	.02215	.02510
24	8	13	.01496	.01783	.02061	.02333	.02599	.02860	.03116
24	8	14	.02300	.02551	.02791	.03024	.03248	.03466	.03678
24	8	15	.03112	.03317	.03510	.03695	.03870	.04039	.04201
24	8	16	.54650	.53337	.52129	.51028	.50034	.49148	.48371
24	9	1	-.17259	-.18778	-.20473	-.22359	-.24450	-.26762	-.29308
24	9	2	-.11603	-.11971	-.12344	-.12720	-.13097	-.13473	-.13845
24	9	3	-.09256	-.09329	-.09389	-.09431	-.09455	-.09457	-.09436
24	9	4	-.07529	-.07435	-.07320	-.07184	-.07025	-.06840	-.06628
24	9	5	-.06111	-.05910	-.05689	-.05446	-.05180	-.04891	-.04578
24	9	6	-.04878	-.04607	-.04317	-.04009	-.03682	-.03335	-.02967
24	9	7	-.03766	-.03449	-.03118	-.02773	-.02412	-.02036	-.01643
24	9	8	-.02738	-.02394	-.02041	-.01677	-.01302	-.00916	-.00520
24	9	9	-.01769	-.01413	-.01052	-.00684	-.00311	.00070	.00457
24	9	10	-.00842	-.00486	-.00129	.00230	.00591	.00954	.01320
24	9	11	.00057	.00401	.00743	.01083	.01421	.01759	.02096
24	9	12	.00938	.01260	.01577	.01889	.02196	.02500	.02800
24	9	13	.01810	.02100	.02383	.02658	.02926	.03189	.03446
24	9	14	.02680	.02930	.03169	.03398	.03620	.03835	.04044
24	9	15	.60267	.59081	.57999	.57025	.56159	.55404	.54762
24	10	1	-.18673	-.20288	-.22090	-.24092	-.26310	-.28759	-.31454
24	10	2	-.12479	-.12858	-.13241	-.13626	-.14012	-.14396	-.14776
24	10	3	-.09908	-.09974	-.10024	-.10056	-.10068	-.10058	-.10023
24	10	4	-.08019	-.07906	-.07773	-.07618	-.07437	-.07231	-.06997
24	10	5	-.06467	-.06243	-.05998	-.05731	-.05441	-.05127	-.04788
24	10	6	-.05117	-.04821	-.04506	-.04173	-.03820	-.03448	-.03054
24	10	7	-.03901	-.03559	-.03203	-.02832	-.02446	-.02045	-.01629
24	10	8	-.02777	-.02409	-.02031	-.01644	-.01246	-.00839	-.00420
24	10	9	-.01718	-.01340	-.00957	-.00569	-.00175	.00224	.00630
24	10	10	-.00705	-.00330	.00045	.00421	.00798	.01177	.01558
24	10	11	.00276	.00635	.00991	.01344	.01694	.02043	.02391
24	10	12	.01236	.01569	.01895	.02215	.02529	.02839	.03146
24	10	13	.02186	.02481	.02767	.03044	.03315	.03579	.03838
24	10	14	.66066	.65043	.64126	.63317	.62621	.62040	.61577
24	11	1	-.20305	-.22034	-.23959	-.26095	-.28460	-.31068	-.33937
24	11	2	-.13482	-.13873	-.14267	-.14663	-.15059	-.15452	-.15841
24	11	3	-.10651	-.10707	-.10746	-.10766	-.10764	-.10739	-.10689
24	11	4	-.08570	-.08437	-.08282	-.08104	-.07900	-.07669	-.07409
24	11	5	-.06862	-.06611	-.06340	-.06045	-.05727	-.05384	-.05016
24	11	6	-.05377	-.05051	-.04708	-.04345	-.03963	-.03561	-.03137
24	11	7	-.04039	-.03667	-.03282	-.02883	-.02469	-.02040	-.01595
24	11	8	-.02802	-.02407	-.02002	-.01588	-.01164	-.00731	-.00287
24	11	9	-.01638	-.01235	-.00828	-.00417	-.00001	.00421	.00848
24	11	10	-.00526	-.00130	.00265	.00660	.01056	.01452	.01851
24	11	11	.00550	.00926	.01297	.01664	.02028	.02389	.02749
24	11	12	.01603	.01946	.02281	.02610	.02932	.03249	.03562
24	11	13	.72099	.71282	.70571	.69973	.69492	.69133	.68901
24	12	1	-.22216	-.24077	-.26146	-.28440	-.30976	-.33772	-.36844
24	12	2	-.14646	-.15051	-.15458	-.15866	-.16273	-.16676	-.17075

Coefficients b_i for the BLUE of σ
Shape parameter k

n	s	i	0.0	0.1	0.2	0.3	0.4	0.5	0.6
24	12	3	-.11506	-.11550	-.11576	-.11580	-.11563	-.11520	-.11451
24	12	4	-.09198	-.09041	-.08861	-.08655	-.08423	-.08163	-.07873
24	12	5	-.07305	-.07023	-.06720	-.06393	-.06042	-.05665	-.05262
24	12	6	-.05659	-.05300	-.04923	-.04526	-.04110	-.03673	-.03215
24	12	7	-.04177	-.03772	-.03353	-.02921	-.02474	-.02012	-.01534
24	12	8	-.02808	-.02381	-.01945	-.01500	-.01046	-.00583	-.00110
24	12	9	-.01521	-.01089	-.00654	-.00216	.00226	.00672	.01124
24	12	10	-.00291	.00128	.00547	.00963	.01380	.01796	.02213
24	12	11	.00897	.01291	.01679	.02061	.02440	.02815	.03187
24	12	12	.78431	.77864	.77409	.77073	.76861	.76781	.76839
25	0	1	-.09305	-.10327	-.11482	-.12780	-.14234	-.15854	-.17651
25	0	2	-.06528	-.06865	-.07213	-.07569	-.07930	-.08295	-.08661
25	0	3	-.05379	-.05525	-.05664	-.05793	-.05909	-.06010	-.06093
25	0	4	-.04536	-.04566	-.04582	-.04581	-.04561	-.04520	-.04457
25	0	5	-.03844	-.03795	-.03728	-.03643	-.03538	-.03411	-.03260
25	0	6	-.03244	-.03136	-.03012	-.02869	-.02706	-.02524	-.02320
25	0	7	-.02703	-.02552	-.02386	-.02202	-.02002	-.01783	-.01547
25	0	8	-.02204	-.02020	-.01823	-.01612	-.01386	-.01145	-.00889
25	0	9	-.01733	-.01526	-.01307	-.01077	-.00835	-.00582	-.00317
25	0	10	-.01284	-.01060	-.00826	-.00585	-.00335	-.00076	.00190
25	0	11	-.00849	-.00613	-.00372	-.00125	.00127	.00384	.00646
25	0	12	-.00422	-.00181	.00064	.00310	.00559	.00810	.01062
25	0	13	.00000	.00242	.00485	.00726	.00967	.01206	.01445
25	0	14	.00422	.00661	.00896	.01128	.01356	.01580	.01801
25	0	15	.00849	.01078	.01302	.01519	.01731	.01936	.02136
25	0	16	.01284	.01500	.01707	.01905	.02095	.02278	.02453
25	0	17	.01733	.01929	.02115	.02289	.02454	.02609	.02755
25	0	18	.02204	.02374	.02531	.02676	.02809	.02933	.03047
25	0	19	.02703	.02839	.02961	.03070	.03167	.03254	.03332
25	0	20	.03244	.03336	.03413	.03478	.03532	.03576	.03612
25	0	21	.03844	.03879	.03900	.03911	.03912	.03905	.03893
25	0	22	.04536	.04494	.04442	.04382	.04318	.04250	.04180
25	0	23	.05379	.05228	.05075	.04922	.04771	.04625	.04484
25	0	24	.06528	.06203	.05893	.05600	.05324	.05066	.04827
25	0	25	.09305	.08403	.07611	.06919	.06314	.05788	.05332
25	1	1	-.09863	-.10906	-.12084	-.13409	-.14891	-.16543	-.18376
25	1	2	-.06901	-.07233	-.07576	-.07926	-.08283	-.08644	-.09006
25	1	3	-.05676	-.05812	-.05940	-.06059	-.06165	-.06256	-.06330
25	1	4	-.04777	-.04794	-.04797	-.04783	-.04751	-.04699	-.04624
25	1	5	-.04040	-.03976	-.03896	-.03797	-.03679	-.03540	-.03378
25	1	6	-.03399	-.03277	-.03139	-.02983	-.02808	-.02613	-.02398
25	1	7	-.02823	-.02658	-.02478	-.02282	-.02069	-.01839	-.01592
25	1	8	-.02290	-.02094	-.01884	-.01661	-.01423	-.01172	-.00907
25	1	9	-.01789	-.01569	-.01339	-.01098	-.00847	-.00584	-.00311
25	1	10	-.01309	-.01074	-.00831	-.00580	-.00322	-.00056	.00218
25	1	11	-.00845	-.00601	-.00351	-.00097	.00162	.00425	.00693
25	1	12	-.00390	-.00142	.00108	.00361	.00614	.00869	.01126
25	1	13	.00060	.00307	.00553	.00798	.01042	.01284	.01525
25	1	14	.00510	.00750	.00987	.01220	.01449	.01674	.01896
25	1	15	.00964	.01193	.01416	.01632	.01842	.02046	.02244
25	1	16	.01428	.01640	.01843	.02037	.02224	.02402	.02574
25	1	17	.01907	.02095	.02273	.02440	.02598	.02748	.02889
25	1	18	.02407	.02565	.02711	.02846	.02971	.03086	.03193
25	1	19	.02938	.03058	.03164	.03259	.03344	.03420	.03488
25	1	20	.03513	.03583	.03640	.03687	.03725	.03755	.03779
25	1	21	.04149	.04155	.04151	.04139	.04120	.04097	.04070

Coefficients b_i for the BLUE of σ
Shape parameter k

n	s	i	0.0	0.1	0.2	0.3	0.4	0.5	0.6
25	1	22	.04878	.04800	.04717	.04629	.04541	.04452	.04364
25	1	23	.05759	.05562	.05370	.05184	.05004	.04833	.04672
25	1	24	.15591	.14428	.13381	.12442	.11602	.10854	.10190
25	2	1	-.10445	-.11519	-.12729	-.14089	-.15610	-.17304	-.19183
25	2	2	-.07287	-.07618	-.07960	-.08310	-.08665	-.09025	-.09385
25	2	3	-.05981	-.06109	-.06230	-.06341	-.06439	-.06522	-.06588
25	2	4	-.05022	-.05029	-.05021	-.04997	-.04954	-.04890	-.04805
25	2	5	-.04236	-.04160	-.04068	-.03957	-.03827	-.03676	-.03502
25	2	6	-.03553	-.03418	-.03267	-.03099	-.02911	-.02705	-.02478
25	2	7	-.02938	-.02761	-.02568	-.02360	-.02135	-.01894	-.01635
25	2	8	-.02371	-.02162	-.01940	-.01705	-.01457	-.01195	-.00919
25	2	9	-.01836	-.01605	-.01364	-.01113	-.00851	-.00579	-.00297
25	2	10	-.01325	-.01080	-.00827	-.00567	-.00300	-.00026	.00255
25	2	11	-.00830	-.00577	-.00319	-.00057	.00208	.00478	.00752
25	2	12	-.00346	-.00090	.00167	.00424	.00683	.00943	.01204
25	2	13	.00134	.00386	.00636	.00885	.01132	.01377	.01621
25	2	14	.00613	.00856	.01095	.01330	.01560	.01786	.02008
25	2	15	.01097	.01326	.01548	.01763	.01972	.02175	.02372
25	2	16	.01591	.01799	.01999	.02190	.02372	.02548	.02716
25	2	17	.02100	.02282	.02453	.02614	.02765	.02909	.03045
25	2	18	.02632	.02780	.02915	.03040	.03155	.03262	.03361
25	2	19	.03196	.03300	.03392	.03474	.03546	.03611	.03668
25	2	20	.03805	.03854	.03892	.03922	.03943	.03959	.03970
25	2	21	.04477	.04456	.04427	.04393	.04354	.04312	.04269
25	2	22	.05241	.05128	.05014	.04900	.04787	.04676	.04570
25	2	23	.21285	.19961	.18756	.17661	.16671	.15779	.14978
25	3	1	-.11065	-.12175	-.13424	-.14826	-.16394	-.18138	-.20071
25	3	2	-.07695	-.08028	-.08371	-.08722	-.09078	-.09438	-.09799
25	3	3	-.06301	-.06424	-.06538	-.06642	-.06733	-.06809	-.06867
25	3	4	-.05278	-.05275	-.05257	-.05222	-.05169	-.05095	-.04999
25	3	5	-.04439	-.04351	-.04247	-.04125	-.03982	-.03819	-.03633
25	3	6	-.03710	-.03563	-.03399	-.03218	-.03018	-.02799	-.02560
25	3	7	-.03054	-.02864	-.02659	-.02438	-.02201	-.01948	-.01677
25	3	8	-.02448	-.02227	-.01993	-.01747	-.01487	-.01214	-.00927
25	3	9	-.01878	-.01636	-.01383	-.01121	-.00849	-.00567	-.00275
25	3	10	-.01333	-.01077	-.00814	-.00545	-.00269	.00014	.00303
25	3	11	-.00805	-.00543	-.00277	-.00007	.00266	.00543	.00823
25	3	12	-.00289	-.00026	.00237	.00501	.00766	.01031	.01297
25	3	13	.00223	.00480	.00735	.00987	.01238	.01486	.01733
25	3	14	.00734	.00979	.01220	.01456	.01688	.01915	.02139
25	3	15	.01249	.01478	.01699	.01913	.02121	.02323	.02519
25	3	16	.01775	.01980	.02176	.02363	.02542	.02714	.02879
25	3	17	.02317	.02492	.02655	.02809	.02955	.03092	.03222
25	3	18	.02883	.03018	.03143	.03258	.03363	.03461	.03551
25	3	19	.03481	.03569	.03646	.03713	.03772	.03824	.03871
25	3	20	.04125	.04152	.04171	.04181	.04186	.04187	.04184
25	3	21	.04832	.04783	.04729	.04671	.04612	.04551	.04492
25	3	22	.26679	.25257	.23952	.22759	.21672	.20687	.19797
25	4	1	-.11733	-.12884	-.14179	-.15630	-.17250	-.19052	-.21048
25	4	2	-.08131	-.08468	-.08814	-.09167	-.09526	-.09888	-.10251
25	4	3	-.06642	-.06759	-.06868	-.06966	-.07051	-.07119	-.07170
25	4	4	-.05548	-.05536	-.05508	-.05463	-.05399	-.05315	-.05207
25	4	5	-.04652	-.04552	-.04436	-.04301	-.04147	-.03971	-.03772
25	4	6	-.03873	-.03713	-.03536	-.03342	-.03129	-.02897	-.02645
25	4	7	-.03172	-.02968	-.02750	-.02516	-.02267	-.02001	-.01718
25	4	8	-.02525	-.02290	-.02044	-.01785	-.01513	-.01228	-.00930

Coefficients b_i for the BLUE of σ
Shape parameter k

n	s	i	0.0	0.1	0.2	0.3	0.4	0.5	0.6
25	4	9	-.01916	-.01661	-.01397	-.01123	-.00840	-.00547	-.00245
25	4	10	-.01333	-.01066	-.00793	-.00514	-.00228	.00064	.00362
25	4	11	-.00770	-.00498	-.00223	.00055	.00336	.00621	.00908
25	4	12	-.00218	.00052	.00323	.00593	.00864	.01134	.01405
25	4	13	.00328	.00590	.00850	.01107	.01361	.01613	.01863
25	4	14	.00873	.01122	.01365	.01602	.01835	.02064	.02288
25	4	15	.01423	.01651	.01872	.02085	.02291	.02492	.02686
25	4	16	.01984	.02185	.02376	.02559	.02734	.02902	.03063
25	4	17	.02561	.02727	.02883	.03030	.03168	.03298	.03422
25	4	18	.03162	.03285	.03398	.03502	.03597	.03684	.03766
25	4	19	.03796	.03867	.03928	.03980	.04025	.04064	.04098
25	4	20	.04477	.04481	.04478	.04470	.04456	.04440	.04422
25	4	21	.31910	.30435	.29074	.27825	.26683	.25644	.24704
25	5	1	-.12460	-.13658	-.15003	-.16510	-.18191	-.20058	-.22125
25	5	2	-.08603	-.08944	-.09294	-.09652	-.10014	-.10380	-.10746
25	5	3	-.07008	-.07120	-.07224	-.07316	-.07394	-.07456	-.07500
25	5	4	-.05837	-.05815	-.05777	-.05722	-.05647	-.05551	-.05431
25	5	5	-.04877	-.04765	-.04637	-.04489	-.04321	-.04132	-.03920
25	5	6	-.04043	-.03869	-.03679	-.03471	-.03245	-.02999	-.02733
25	5	7	-.03293	-.03075	-.02843	-.02595	-.02332	-.02053	-.01756
25	5	8	-.02600	-.02352	-.02092	-.01820	-.01535	-.01237	-.00927
25	5	9	-.01948	-.01680	-.01404	-.01118	-.00823	-.00519	-.00205
25	5	10	-.01325	-.01046	-.00762	-.00472	-.00176	.00126	.00433
25	5	11	-.00722	-.00440	-.00155	.00131	.00421	.00713	.01008
25	5	12	-.00132	.00147	.00424	.00701	.00978	.01254	.01531
25	5	13	.00452	.00720	.00984	.01245	.01504	.01759	.02012
25	5	14	.01035	.01286	.01531	.01770	.02004	.02234	.02459
25	5	15	.01622	.01849	.02069	.02281	.02486	.02684	.02877
25	5	16	.02220	.02417	.02604	.02782	.02952	.03115	.03272
25	5	17	.02835	.02993	.03141	.03279	.03409	.03531	.03648
25	5	18	.03474	.03585	.03685	.03776	.03859	.03936	.04007
25	5	19	.04148	.04199	.04242	.04278	.04307	.04332	.04353
25	5	20	.37061	.35569	.34190	.32921	.31759	.30701	.29743
25	6	1	-.13256	-.14506	-.15910	-.17480	-.19229	.21170	.23317
25	6	2	-.09115	-.09463	-.09819	-.10181	-.10549	-.10919	-.11289
25	6	3	-.07404	-.07512	-.07610	-.07697	-.07768	-.07823	-.07859
25	6	4	-.06147	-.06115	-.06067	-.06001	-.05914	-.05806	-.05674
25	6	5	-.05117	-.04993	-.04850	-.04689	-.04507	-.04304	-.04077
25	6	6	-.04222	-.04034	-.03829	-.03607	-.03365	-.03104	-.02823
25	6	7	-.03417	-.03185	-.02938	-.02675	-.02397	-.02103	-.01792
25	6	8	-.02674	-.02411	-.02137	-.01851	-.01552	-.01240	-.00916
25	6	9	-.01974	-.01693	-.01403	-.01104	-.00797	-.00480	-.00155
25	6	10	-.01306	-.01015	-.00719	-.00417	-.00111	.00202	.00519
25	6	11	-.00660	-.00367	-.00073	.00224	.00522	.00822	.01125
25	6	12	-.00027	.00259	.00545	.00829	.01112	.01395	.01677
25	6	13	.00598	.00871	.01141	.01407	.01669	.01928	.02184
25	6	14	.01222	.01475	.01722	.01963	.02199	.02429	.02654
25	6	15	.01850	.02077	.02295	.02505	.02708	.02904	.03095
25	6	16	.02489	.02681	.02863	.03036	.03200	.03358	.03510
25	6	17	.03145	.03294	.03432	.03561	.03682	.03795	.03903
25	6	18	.03826	.03922	.04007	.04085	.04155	.04219	.04279
25	6	19	.42191	.40716	.39350	.38093	.36943	.35898	.34955
25	7	1	-.14134	-.15445	-.16914	-.18554	-.20380	-.22405	-.24643
25	7	2	-.09678	-.10033	-.10395	-.10764	-.11138	-.11513	-.11888
25	7	3	-.07836	-.07939	-.08032	-.08113	-.08178	-.08225	-.08253
25	7	4	-.06483	-.06441	-.06381	-.06303	-.06204	-.06082	-.05937

Coefficients b_i for the BLUE of σ
Shape parameter k

n	s	i	0.0	0.1	0.2	0.3	0.4	0.5	0.6
25	7	5	-.05375	-.05236	-.05080	-.04904	-.04707	-.04488	-.04245
25	7	6	-.04412	-.04209	-.03988	-.03749	-.03492	-.03214	-.02917
25	7	7	-.03546	-.03297	-.03034	-.02756	-.02462	-.02151	-.01824
25	7	8	-.02746	-.02468	-.02178	-.01876	-.01562	-.01236	-.00897
25	7	9	-.01994	-.01698	-.01393	-.01080	-.00759	-.00429	-.00091
25	7	10	-.01276	-.00971	-.00662	-.00348	-.00030	.00294	.00623
25	7	11	-.00581	-.00277	.00028	.00335	.00643	.00952	.01264
25	7	12	.00098	.00394	.00688	.00979	.01270	.01559	.01847
25	7	13	.00770	.01049	.01324	.01594	.01861	.02123	.02383
25	7	14	.01439	.01695	.01944	.02187	.02423	.02654	.02880
25	7	15	.02113	.02338	.02554	.02762	.02962	.03156	.03344
25	7	16	.02797	.02983	.03158	.03325	.03484	.03635	.03781
25	7	17	.03498	.03636	.03763	.03881	.03991	.04095	.04194
25	7	18	.47347	.45919	.44598	.43385	.42277	.41276	.40380
25	8	1	-.15112	-.16490	-.18032	-.19753	-.21666	-.23785	-.26125
25	8	2	-.10300	-.10663	-.11033	-.11410	-.11790	-.12172	-.12554
25	8	3	-.08310	-.08409	-.08497	-.08570	-.08628	-.08668	-.08686
25	8	4	-.06850	-.06796	-.06724	-.06633	-.06520	-.06384	-.06224
25	8	5	-.05654	-.05500	-.05328	-.05135	-.04922	-.04685	-.04425
25	8	6	-.04615	-.04394	-.04156	-.03899	-.03624	-.03329	-.03013
25	8	7	-.03680	-.03413	-.03132	-.02836	-.02524	-.02196	-.01852
25	8	8	-.02817	-.02521	-.02214	-.01895	-.01564	-.01222	-.00867
25	8	9	-.02006	-.01693	-.01372	-.01044	-.00708	-.00363	-.00010
25	8	10	-.01231	-.00912	-.00588	-.00261	.00070	.00406	.00747
25	8	11	-.00482	-.00165	.00151	.00469	.00787	.01107	.01428
25	8	12	.00250	.00555	.00858	.01158	.01455	.01751	.02046
25	8	13	.00973	.01259	.01539	.01814	.02084	.02351	.02614
25	8	14	.01693	.01952	.02202	.02446	.02682	.02913	.03140
25	8	15	.02417	.02640	.02853	.03058	.03255	.03446	.03630
25	8	16	.03151	.03330	.03498	.03657	.03808	.03952	.04091
25	8	17	.52571	.51220	.49975	.48836	.47803	.46878	.46061
25	9	1	-.16209	-.17662	-.19288	-.21099	-.23111	-.25338	-.27795
25	9	2	-.10992	-.11365	-.11745	-.12130	-.12518	-.12907	-.13296
25	9	3	-.08836	-.08930	-.09011	-.09078	-.09127	-.09158	-.09167
25	9	4	-.07254	-.07187	-.07101	-.06995	-.06867	-.06716	-.06538
25	9	5	-.05957	-.05786	-.05597	-.05386	-.05154	-.04898	-.04619
25	9	6	-.04832	-.04592	-.04334	-.04058	-.03763	-.03448	-.03112
25	9	7	-.03819	-.03533	-.03232	-.02916	-.02585	-.02237	-.01873
25	9	8	-.02885	-.02569	-.02243	-.01906	-.01557	-.01196	-.00822
25	9	9	-.02006	-.01675	-.01337	-.00992	-.00639	-.00278	.00090
25	9	10	-.01168	-.00832	-.00494	-.00152	.00194	.00544	.00897
25	9	11	-.00358	-.00028	.00302	.00631	.00961	.01291	.01622
25	9	12	.00434	.00749	.01061	.01369	.01675	.01978	.02280
25	9	13	.01215	.01507	.01792	.02072	.02347	.02617	.02883
25	9	14	.01992	.02252	.02504	.02748	.02985	.03216	.03442
25	9	15	.02772	.02992	.03201	.03402	.03595	.03781	.03961
25	9	16	.57902	.56660	.55521	.54489	.53565	.52751	.52047
25	10	1	-.17450	-.18990	-.20710	-.22625	-.24750	-.27100	-.29689
25	10	2	-.11771	-.12154	-.12544	-.12939	-.13336	-.13735	-.14132
25	10	3	-.09423	-.09511	-.09585	-.09644	-.09685	-.09705	-.09704
25	10	4	-.07701	-.07620	-.07519	-.07396	-.07251	-.07081	-.06885
25	10	5	-.06290	-.06100	-.05890	-.05659	-.05406	-.05129	-.04828
25	10	6	-.05065	-.04804	-.04525	-.04227	-.03909	-.03571	-.03213
25	10	7	-.03964	-.03655	-.03333	-.02995	-.02641	-.02272	-.01886
25	10	8	-.02948	-.02611	-.02263	-.01905	-.01535	-.01154	-.00761
25	10	9	-.01993	-.01642	-.01284	-.00920	-.00549	-.00170	.00216

Coefficients b_i for the BLUE of σ
Shape parameter k

n	s	i	0.0	0.1	0.2	0.3	0.4	0.5	0.6
25	10	10	-.01082	-.00729	-.00373	-.00015	.00346	.00711	.01080
25	10	11	-.00202	.00143	.00486	.00828	.01170	.01512	.01854
25	10	12	.00657	.00983	.01305	.01622	.01937	.02248	.02557
25	10	13	.01503	.01802	.02093	.02378	.02657	.02931	.03200
25	10	14	.02345	.02607	.02859	.03103	.03340	.03570	.03795
25	10	15	.63383	.62281	.61283	.60393	.59613	.58946	.58395
25	11	1	-.18869	-.20508	-.22337	-.24371	-.26625	-.29116	-.31858
25	11	2	-.12654	-.13049	-.13451	-.13857	-.14265	-.14673	-.15080
25	11	3	-.10085	-.10166	-.10232	-.10281	-.10312	-.10321	-.10307
25	11	4	-.08201	-.08103	-.07984	-.07843	-.07678	-.07488	-.07270
25	11	5	-.06658	-.06446	-.06213	-.05959	-.05682	-.05380	-.05054
25	11	6	-.05319	-.05033	-.04728	-.04405	-.04063	-.03699	-.03315
25	11	7	-.04114	-.03781	-.03433	-.03070	-.02692	-.02297	-.01887
25	11	8	-.03004	-.02643	-.02271	-.01889	-.01496	-.01092	-.00676
25	11	9	-.01961	-.01587	-.01208	-.00822	-.00430	-.00032	.00374
25	11	10	-.00966	-.00594	-.00219	.00157	.00536	.00918	.01302
25	11	11	-.00007	.00354	.00713	.01069	.01425	.01779	.02134
25	11	12	.00930	.01268	.01600	.01927	.02251	.02570	.02887
25	11	13	.01851	.02156	.02453	.02743	.03026	.03303	.03576
25	11	14	.69057	.68131	.67311	.66601	.66005	.65528	.65173
25	12	1	-.20509	-.22263	-.24218	-.26389	-.28793	-.31447	-.34367
25	12	2	-.13666	-.14075	-.14490	-.14908	-.15328	-.15748	-.16165
25	12	3	-.10839	-.10911	-.10968	-.11006	-.11024	-.11020	-.10992
25	12	4	-.08766	-.08648	-.08508	-.08345	-.08157	-.07943	-.07701
25	12	5	-.07067	-.06830	-.06571	-.06290	-.05985	-.05655	-.05300
25	12	6	-.05595	-.05280	-.04948	-.04596	-.04224	-.03831	-.03417
25	12	7	-.04271	-.03908	-.03531	-.03140	-.02733	-.02310	-.01871
25	12	8	-.03050	-.02661	-.02262	-.01853	-.01433	-.01003	-.00562
25	12	9	-.01905	-.01505	-.01101	-.00691	-.00276	.00146	.00574
25	12	10	-.00813	-.00418	-.00022	.00375	.00773	.01173	.01576
25	12	11	.00240	.00619	.00994	.01366	.01737	.02105	.02473
25	12	12	.01266	.01616	.01961	.02298	.02631	.02960	.03285
25	12	13	.74974	.74265	.73664	.73179	.72813	.72574	.72465
25	13	1	-.22429	-.24318	-.26419	-.28751	-.31331	-.34176	-.37305
25	13	2	-.14841	-.15266	-.15696	-.16128	-.16561	-.16993	-.17423
25	13	3	-.11708	-.11770	-.11814	-.11839	-.11843	-.11823	-.11777
25	13	4	-.09410	-.09269	-.09104	-.08916	-.08701	-.08458	-.08186
25	13	5	-.07529	-.07261	-.06971	-.06658	-.06321	-.05958	-.05568
25	13	6	-.05897	-.05550	-.05184	-.04798	-.04392	-.03965	-.03516
25	13	7	-.04432	-.04035	-.03625	-.03200	-.02760	-.02304	-.01832
25	13	8	-.03081	-.02660	-.02229	-.01789	-.01339	-.00878	-.00407
25	13	9	-.01815	-.01386	-.00953	-.00515	-.00073	.00375	.00829
25	13	10	-.00609	-.00188	.00232	.00652	.01073	.01494	.01917
25	13	11	.00553	.00952	.01347	.01737	.02124	.02508	.02891
25	13	12	.81196	.80749	.80416	.80206	.80124	.80179	.80378

Coefficients b_i for the BLUE of σ
Shape parameter k

n	s	i	0.7	0.8	0.9	1.0	1.1	1.2	1.3
2	0	1	-.90093	-.90529	-.91015	-.91551	-.92133	-.92760	-.93428
2	0	2	.90093	.90529	.91015	.91551	.92133	.92760	.93428
3	0	1	-.71101	-.73118	-.75202	-.77351	-.79563	-.81837	-.84171
3	0	2	.19848	.22636	.25411	.28174	.30926	.33670	.36408
3	0	3	.51253	.50482	.49791	.49178	.48637	.48167	.47762
3	1	1	-1.35334	-1.38408	-1.41641	-1.45031	-1.48579	-1.52284	-1.56146
3	1	2	1.35334	1.38408	1.41641	1.45031	1.48579	1.52284	1.56146
4	0	1	-.60229	-.62775	-.65425	-.68176	-.71031	-.73988	-.77048
4	0	2	.02562	.04730	.06955	.09238	.11578	.13979	.16441
4	0	3	.21705	.23017	.24285	.25512	.26703	.27861	.28990
4	0	4	.35962	.35028	.34184	.33427	.32750	.32149	.31617
4	1	1	-.87953	-.91145	-.94489	-.97989	-1.01646	-1.05464	-1.09444
4	1	2	.06277	.09168	.12113	.15116	.18181	.21313	.24519
4	1	3	.81676	.81977	.82376	.82873	.83466	.84150	.84925
5	0	1	-.52909	-.55667	-.58557	-.61579	-.64736	-.68028	-.71460
5	0	2	-.04274	-.02648	-.00944	.00843	.02713	.04669	.06716
5	0	3	.10010	.11420	.12826	.14229	.15629	.17029	.18430
5	0	4	.19468	.20115	.20724	.21301	.21850	.22375	.22880
5	0	5	.27704	.26781	.25951	.25207	.24544	.23955	.23435
5	1	1	-.69279	-.72494	-.75877	-.79431	-.83160	-.87069	-.91162
5	1	2	-.04433	-.02391	-.00268	.01938	.04234	.06623	.09111
5	1	3	.14371	.15975	.17571	.19163	.20755	.22349	.23948
5	1	4	.59340	.58910	.58575	.58329	.58171	.58097	.58103
5	2	1	-1.00938	-1.05230	-1.09758	-1.14528	-1.19548	-1.24825	-1.30366
5	2	2	-.03528	-.00705	.02206	.05210	.08315	.11529	.14860
5	2	3	1.04466	1.05934	1.07552	1.09318	1.11233	1.13296	1.15506
6	0	1	-.47543	-.50384	-.53377	-.56525	-.59830	-.63298	-.66933
6	0	2	-.07520	-.06302	-.04998	-.03606	-.02123	-.00544	.01134
6	0	3	.04001	.05278	.06576	.07895	.09237	.10603	.11994
6	0	4	.11423	.12339	.13235	.14112	.14973	.15819	.16653
6	0	5	.17112	.17414	.17691	.17947	.18185	.18410	.18625
6	0	6	.22527	.21654	.20873	.20176	.19557	.19009	.18526
6	1	1	-.58608	-.61797	-.65167	-.68722	-.72469	-.76413	-.80561
6	1	2	-.08636	-.07152	-.05581	-.03920	-.02163	-.00305	.01660
6	1	3	.05668	.07120	.08591	.10083	.11598	.13139	.14707
6	1	4	.14780	.15726	.16653	.17564	.18463	.19352	.20234
6	1	5	.46796	.46104	.45505	.44995	.44570	.44227	.43960
6	2	1	-.76670	-.80543	-.84640	-.88968	-.93537	-.98353	-1.03427
6	2	2	-.09910	-.08011	-.06020	-.03930	-.01736	.00569	.02993
6	2	3	.08970	.10697	.12441	.14203	.15990	.17803	.19648
6	2	4	.77610	.77857	.78219	.78695	.79283	.79981	.80786
6	3	1	-1.11496	-1.16785	-1.22391	-1.28325	-1.34600	-1.41227	-1.48219
6	3	2	-.11332	-.08646	-.05854	-.02947	.00084	.03247	.06555
6	3	3	1.22828	1.25431	1.28245	1.31272	1.34516	1.37979	1.41663
7	0	1	-.43394	-.46256	-.49285	-.52485	-.55861	-.59419	-.63165
7	0	2	-.09204	-.08290	-.07293	-.06208	-.05032	-.03760	-.02388
7	0	3	.00514	.01622	.02767	.03949	.05168	.06426	.07725
7	0	4	.06696	.07640	.08581	.09520	.10458	.11396	.12337
7	0	5	.11294	.11898	.12478	.13039	.13581	.14108	.14622
7	0	6	.15117	.15225	.15317	.15397	.15469	.15535	.15599
7	0	7	.18977	.18162	.17435	.16789	.16218	.15714	.15271
7	1	1	-.51475	-.54615	-.57946	-.61474	-.65205	-.69148	-.73311
7	1	2	-.10532	-.09438	-.08258	-.06989	-.05626	-.04163	-.02594
7	1	3	.01069	.02322	.03608	.04931	.06292	.07694	.09139
7	1	4	.08418	.09428	.10434	.11439	.12445	.13452	.14464
7	1	5	.13818	.14395	.14953	.15497	.16027	.16548	.17061

Coefficients b_i for the BLUE of σ
Shape parameter k

n	s	i	0.7	0.8	0.9	1.0	1.1	1.2	1.3
7	1	6	.38702	.37909	.37208	.36596	.36067	.35617	.35241
7	2	1	-.63458	-.67088	-.70940	-.75024	-.79348	-.83922	-.88756
7	2	2	-.12207	-.10860	-.09424	-.07892	-.06260	-.04520	-.02665
7	2	3	.02225	.03681	.05170	.06695	.08260	.09866	.11519
7	2	4	.11280	.12383	.13481	.14577	.15674	.16775	.17882
7	2	5	.62160	.61884	.61713	.61644	.61674	.61801	.62019
7	3	1	-.82893	-.87377	-.92140	-.97192	-1.02546	-1.08213	-1.14207
7	3	2	-.14459	-.12712	-.10866	-.08914	-.06850	-.04665	-.02350
7	3	3	.04601	.06376	.08185	.10029	.11915	.13847	.15830
7	3	4	.92751	.93712	.94821	.96077	.97480	.99031	1.00727
7	4	1	-1.20396	-1.26589	-1.33178	-1.40178	-1.47606	-1.55479	-1.63816
7	4	2	-.17854	-.15337	-.12703	-.09942	-.07044	-.03998	-.00792
7	4	3	1.38250	1.41926	1.45880	1.50119	1.54650	1.59477	1.64608
8	0	1	-.40064	-.42915	-.45944	-.49157	-.52560	-.56160	-.59964
8	0	2	-.10107	-.09427	-.08667	-.07825	-.06894	-.05871	-.04751
8	0	3	-.01669	-.00718	.00277	.01317	.02403	.03538	.04723
8	0	4	.03664	.04559	.05466	.06383	.07312	.08254	.09209
8	0	5	.07576	.08272	.08956	.09629	.10293	.10949	.11598
8	0	6	.10724	.11123	.11500	.11860	.12205	.12536	.12856
8	0	7	.13484	.13474	.13457	.13435	.13411	.13387	.13367
8	0	8	.16392	.15632	.14956	.14358	.13831	.13367	.12961
8	1	1	-.46275	-.49354	-.52633	-.56117	-.59816	-.63736	-.67889
8	1	2	-.11419	-.10611	-.09722	-.08748	-.07684	-.06525	-.05264
8	1	3	-.01621	-.00555	.00554	.01708	.02909	.04159	.05463
8	1	4	.04558	.05526	.06502	.07490	.08489	.09503	.10532
8	1	5	.09071	.09783	.10484	.11176	.11861	.12539	.13215
8	1	6	.12656	.13009	.13347	.13673	.13989	.14298	.14601
8	1	7	.33029	.32202	.31467	.30819	.30252	.29762	.29343
8	2	1	-.54929	-.58384	-.62064	-.65977	-.70132	-.74541	-.79214
8	2	2	-.13073	-.12098	-.11039	-.09891	-.08648	-.07304	-.05852
8	2	3	-.01349	-.00130	.01132	.02439	.03794	.05201	.06663
8	2	4	.06011	.07073	.08143	.09223	.10316	.11422	.12546
8	2	5	.11334	.12066	.12788	.13502	.14210	.14914	.15618
8	2	6	.52005	.51472	.51039	.50703	.50461	.50307	.50239
8	3	1	-.67648	-.71697	-.76010	-.80596	-.85469	-.90640	-.96124
8	3	2	-.15254	-.14042	-.12740	-.11341	-.09839	-.08228	-.06497
8	3	3	-.00668	.00769	.02249	.03776	.05354	.06986	.08677
8	3	4	.08416	.09611	.10812	.12022	.13245	.14482	.15738
8	3	5	.75154	.75359	.75687	.76138	.76710	.77400	.78207
8	4	1	-.88264	-.93313	-.98692	-1.04417	-1.10502	-1.16963	-1.23817
8	4	2	-.18364	-.16771	-.15077	-.13274	-.11354	-.09309	-.07128
8	4	3	.00900	.02683	.04511	.06388	.08320	.10311	.12368
8	4	4	1.05728	1.07401	1.09258	1.11302	1.13536	1.15961	1.18577
9	0	1	-.37317	-.40140	-.43149	-.46352	-.49756	-.53368	-.57197
9	0	2	-.10584	-.10086	-.09515	-.08866	-.08134	-.07315	-.06404
9	0	3	-.03105	-.02293	-.01434	-.00524	.00436	.01450	.02520
9	0	4	.01603	.02427	.03271	.04135	.05021	.05928	.06859
9	0	5	.05031	.05740	.06446	.07152	.07856	.08561	.09268
9	0	6	.07749	.08266	.08767	.09254	.09730	.10195	.10651
9	0	7	.10050	.10309	.10551	.10779	.10994	.11199	.11397
9	0	8	.12146	.12062	.11977	.11892	.11810	.11734	.11664
9	0	9	.14426	.13715	.13086	.12531	.12043	.11615	.11242
9	1	1	-.42266	-.45282	-.48502	-.51936	-.55590	-.59476	-.63604
9	1	2	-.11810	-.11218	-.10551	-.09806	-.08976	-.08057	-.07043
9	1	3	-.03300	-.02396	-.01445	-.00444	.00610	.01717	.02883
9	1	4	.02049	.02940	.03850	.04780	.05731	.06705	.07703

Coefficients b_i for the BLUE of σ
Shape parameter k

n	s	i	0.7	0.8	0.9	1.0	1.1	1.2	1.3
9	1	5	.05936	.06676	.07415	.08153	.08890	.09630	.10373
9	1	6	.09002	.09511	.10007	.10491	.10966	.11434	.11896
9	1	7	.11565	.11773	.11971	.12159	.12342	.12520	.12696
9	1	8	.28825	.27996	.27256	.26602	.26028	.25527	.25096
9	2	1	-.48868	-.52185	-.55728	-.59506	-.63529	-.67810	-.72359
9	2	2	-.13333	-.12625	-.11840	-.10973	-.10019	-.08971	-.07822
9	2	3	-.03428	-.02406	-.01337	-.00217	.00955	.02184	.03474
9	2	4	.02783	.03760	.04754	.05768	.06803	.07861	.08944
9	2	5	.07276	.08057	.08836	.09613	.10391	.11171	.11956
9	2	6	.10786	.11284	.11770	.12248	.12719	.13185	.13648
9	2	7	.44783	.44115	.43545	.43068	.42681	.42379	.42159
9	3	1	-.57972	-.61735	-.65753	-.70039	-.74603	-.79458	-.84620
9	3	2	-.15284	-.14424	-.13482	-.12452	-.11329	-.10106	-.08774
9	3	3	-.03431	-.02252	-.01025	.00254	.01588	.02982	.04439
9	3	4	.03971	.05061	.06167	.07292	.08438	.09608	.10805
9	3	5	.09284	.10117	.10946	.11773	.12602	.13434	.14272
9	3	6	.63432	.63233	.63147	.63171	.63303	.63540	.63879
9	4	1	-.71331	-.75774	-.80519	-.85579	-.90969	-.96704	-1.02802
9	4	2	-.17921	-.16840	-.15669	-.14403	-.13035	-.11556	-.09958
9	4	3	-.03177	-.01773	-.00318	.01192	.02760	.04391	.06091
9	4	4	.05968	.07217	.08482	.09764	.11069	.12398	.13756
9	4	5	.86461	.87169	.88024	.89026	.90176	.91472	.92914
9	5	1	-.92989	-.98560	-1.04512	-1.10862	-1.17629	-1.24833	-1.32493
9	5	2	-.21793	-.20353	-.18812	-.17162	-.15394	-.13500	-.11467
9	5	3	-.02323	-.00557	.01262	.03141	.05084	.07097	.09188
9	5	4	1.17104	1.19470	1.22061	1.24883	1.27939	1.31235	1.34773
10	0	1	-.35003	-.37787	-.40766	-.43945	-.47334	-.50939	-.54772
10	0	2	-.10813	-.10461	-.10042	-.09550	-.08982	-.08332	-.07596
10	0	3	-.04082	-.03391	-.02650	-.01858	-.01011	-.00109	.00851
10	0	4	.00144	.00892	.01666	.02467	.03295	.04152	.05039
10	0	5	.03207	.03894	.04588	.05287	.05994	.06708	.07432
10	0	6	.05617	.06177	.06728	.07273	.07811	.08344	.08873
10	0	7	.07625	.08009	.08378	.08733	.09077	.09410	.09734
10	0	8	.09387	.09548	.09695	.09831	.09959	.10080	.10196
10	0	9	.11037	.10906	.10777	.10653	.10536	.10428	.10328
10	0	10	.12880	.12214	.11626	.11109	.10656	.10260	.09914
10	1	1	-.39055	-.42006	-.45166	-.48544	-.52150	-.55994	-.60086
10	1	2	-.11935	-.11512	-.11021	-.10457	-.09815	-.09091	-.08277
10	1	3	-.04396	-.03630	-.02814	-.01947	-.01025	-.00047	.00992
10	1	4	.00333	.01141	.01974	.02833	.03721	.04637	.05585
10	1	5	.03757	.04481	.05211	.05947	.06690	.07442	.08204
10	1	6	.06444	.07014	.07577	.08133	.08684	.09232	.09778
10	1	7	.08671	.09037	.09390	.09732	.10065	.10391	.10711
10	1	8	.10600	.10711	.10815	.10913	.11008	.11102	.11196
10	1	9	.25582	.24764	.24036	.23390	.22822	.22326	.21897
10	2	1	-.44290	-.47489	-.50917	-.54581	-.58493	-.62666	-.67110
10	2	2	-.13308	-.12802	-.12225	-.11574	-.10842	-.10024	-.09112
10	2	3	-.04710	-.03852	-.02943	-.01983	-.00967	.00108	.01245
10	2	4	.00675	.01557	.02463	.03396	.04356	.05346	.06368
10	2	5	.04566	.05335	.06109	.06889	.07677	.08474	.09282
10	2	6	.07604	.08186	.08761	.09330	.09895	.10458	.11020
10	2	7	.10097	.10438	.10769	.11092	.11409	.11722	.12032
10	2	8	.39366	.38627	.37983	.37431	.36965	.36582	.36276
10	3	1	-.51192	-.54744	-.58547	-.62613	-.66953	-.71582	-.76514
10	3	2	-.15023	-.14413	-.13729	-.12966	-.12119	-.11180	-.10142
10	3	3	-.05011	-.04035	-.03009	-.01928	-.00790	.00409	.01673

Coefficients b_i for the BLUE of σ
Shape parameter k

n	s	i	0.7	0.8	0.9	1.0	1.1	1.2	1.3
10	3	4	.01245	.02221	.03220	.04246	.05299	.06383	.07501
10	3	5	.05746	.06572	.07402	.08237	.09080	.09933	.10797
10	3	6	.09233	.09829	.10417	.11000	.11580	.12160	.12740
10	3	7	.55002	.54571	.54245	.54023	.53902	.53877	.53946
10	4	1	-.60687	-.64743	-.69084	-.73724	-.78677	-.83959	-.89586
10	4	2	-.17247	-.16497	-.15668	-.14755	-.13750	-.12647	-.11438
10	4	3	-.05267	-.04133	-.02947	-.01704	-.00400	.00967	.02405
10	4	4	.02189	.03291	.04415	.05565	.06744	.07954	.09199
10	4	5	.07520	.08420	.09322	.10230	.11145	.12070	.13007
10	4	6	.73492	.73663	.73961	.74387	.74939	.75615	.76413
10	5	1	-.74615	-.79427	-.84578	-.90083	-.95960	-1.02228	-1.08905
10	5	2	-.20295	-.19341	-.18300	-.17166	-.15931	-.14587	-.13126
10	5	3	-.05399	-.04036	-.02617	-.01138	.00405	.02019	.03708
10	5	4	.03819	.05098	.06401	.07729	.09087	.10478	.11907
10	5	5	.96490	.97705	.99094	1.00658	1.02399	1.04318	1.06416
11	0	1	-.33019	-.35761	-.38702	-.41850	-.45213	-.48801	-.52624
11	0	2	-.10893	-.10659	-.10364	-.10003	-.09571	-.09063	-.08474
11	0	3	-.04762	-.04175	-.03538	-.02849	-.02105	-.01304	-.00444
11	0	4	-.00921	-.00247	.00458	.01193	.01961	.02762	.03598
11	0	5	.01856	.02507	.03170	.03845	.04534	.05237	.05955
11	0	6	.04029	.04596	.05162	.05726	.06290	.06854	.07419
11	0	7	.05824	.06268	.06700	.07123	.07537	.07944	.08345
11	0	8	.07373	.07659	.07930	.08189	.08437	.08677	.08908
11	0	9	.08770	.08860	.08941	.09013	.09079	.09142	.09203
11	0	10	.10108	.09945	.09788	.09640	.09501	.09372	.09255
11	0	11	.11634	.11007	.10456	.09972	.09549	.09181	.08860
11	1	1	-.36408	-.39295	-.42395	-.45718	-.49272	-.53069	-.57121
11	1	2	-.11913	-.11624	-.11274	-.10857	-.10369	-.09804	-.09156
11	1	3	-.05131	-.04482	-.03784	-.03033	-.02227	-.01363	-.00439
11	1	4	-.00884	-.00158	.00599	.01386	.02206	.03059	.03948
11	1	5	.02183	.02871	.03571	.04283	.05008	.05748	.06504
11	1	6	.04581	.05167	.05751	.06334	.06918	.07502	.08090
11	1	7	.06557	.06999	.07430	.07854	.08270	.08681	.09088
11	1	8	.08253	.08515	.08765	.09006	.09240	.09468	.09691
11	1	9	.09761	.09804	.09843	.09880	.09916	.09953	.09992
11	1	10	.23001	.22204	.21493	.20863	.20309	.19824	.19403
11	2	1	-.40679	-.43777	-.47103	-.50668	-.54483	-.58560	-.62913
11	2	2	-.13144	-.12794	-.12381	-.11899	-.11344	-.10710	-.09990
11	2	3	-.05530	-.04808	-.04035	-.03209	-.02328	-.01387	-.00383
11	2	4	-.00767	.00024	.00843	.01693	.02575	.03492	.04445
11	2	5	.02670	.03402	.04146	.04901	.05670	.06454	.07254
11	2	6	.05348	.05956	.06562	.07167	.07772	.08380	.08990
11	2	7	.07545	.07984	.08413	.08836	.09252	.09665	.10076
11	2	8	.09412	.09644	.09868	.10085	.10298	.10508	.10716
11	2	9	.35145	.34370	.33688	.33095	.32587	.32159	.31805
11	3	1	-.46128	-.49513	-.53147	-.57040	-.61206	-.65659	-.70412
11	3	2	-.14650	-.14226	-.13736	-.13176	-.12538	-.11818	-.11009
11	3	3	-.05961	-.05148	-.04284	-.03366	-.02390	-.01353	-.00250
11	3	4	-.00534	.00334	.01231	.02159	.03120	.04115	.05148
11	3	5	.03372	.04159	.04956	.05765	.06587	.07424	.08278
11	3	6	.06404	.07038	.07669	.08300	.08931	.09565	.10203
11	3	7	.08872	.09305	.09730	.10149	.10564	.10977	.11389
11	3	8	.48624	.48051	.47580	.47209	.46933	.46749	.46653
11	4	1	-.53290	-.57069	-.61123	-.65466	-.70112	-.75076	-.80375
11	4	2	-.16541	-.16025	-.15438	-.14776	-.14032	-.13201	-.12274
11	4	3	-.06425	-.05496	-.04514	-.03476	-.02377	-.01214	.00017

Coefficients b_i for the BLUE of σ
Shape parameter k

n	s	i	0.7	0.8	0.9	1.0	1.1	1.2	1.3
11	4	4	-.00120	.00848	.01844	.02872	.03933	.05030	.06166
11	4	5	.04401	.05257	.06122	.06998	.07887	.08792	.09714
11	4	6	.07889	.08554	.09216	.09877	.10539	.11204	.11874
11	4	7	.64087	.63932	.63894	.63971	.64163	.64465	.64877
11	5	1	-.63135	-.67468	-.72116	-.77093	-.82416	-.88103	-.94173
11	5	2	-.19015	-.18371	-.17652	-.16851	-.15963	-.14979	-.13892
11	5	3	-.06914	-.05828	-.04686	-.03484	-.02218	-.00885	.00523
11	5	4	.00601	.01703	.02834	.03996	.05192	.06426	.07702
11	5	5	.05961	.06908	.07862	.08827	.09805	.10799	.11812
11	5	6	.82502	.83056	.83757	.84605	.85600	.86741	.88028
12	0	1	-.31295	-.33992	-.36893	-.40005	-.43337	-.46899	-.50703
12	0	2	-.10879	-.10743	-.10551	-.10298	-.09981	-.09593	-.09131
12	0	3	-.05243	-.04746	-.04199	-.03601	-.02948	-.02239	-.01470
12	0	4	-.01716	-.01111	-.00472	.00200	.00908	.01652	.02434
12	0	5	.00828	.01437	.02063	.02706	.03367	.04047	.04747
12	0	6	.02813	.03369	.03929	.04493	.05061	.05635	.06215
12	0	7	.04444	.04912	.05373	.05830	.06282	.06730	.07176
12	0	8	.05838	.06190	.06531	.06860	.07181	.07494	.07800
12	0	9	.07073	.07284	.07482	.07669	.07847	.08017	.08182
12	0	10	.08210	.08249	.08280	.08306	.08329	.08350	.08371
12	0	11	.09320	.09136	.08961	.08797	.08644	.08504	.08376
12	0	12	.10606	.10016	.09497	.09043	.08647	.08302	.08003
12	1	1	-.34179	-.37004	-.40045	-.43311	-.46813	-.50561	-.54569
12	1	2	-.11806	-.11627	-.11391	-.11096	-.10735	-.10303	-.09796
12	1	3	-.05633	-.05084	-.04486	-.03836	-.03132	-.02371	-.01549
12	1	4	-.01772	-.01121	-.00437	.00281	.01034	.01824	.02653
12	1	5	.01012	.01656	.02317	.02994	.03690	.04405	.05140
12	1	6	.03182	.03760	.04342	.04928	.05518	.06115	.06719
12	1	7	.04963	.05438	.05907	.06372	.06833	.07292	.07749
12	1	8	.06481	.06825	.07158	.07482	.07799	.08110	.08416
12	1	9	.07819	.08004	.08179	.08347	.08509	.08666	.08820
12	1	10	.09033	.09028	.09022	.09015	.09010	.09008	.09010
12	1	11	.20898	.20124	.19434	.18824	.18286	.17816	.17407
12	2	1	-.37742	-.40749	-.43985	-.47460	-.51187	-.55179	-.59448
12	2	2	-.12911	-.12685	-.12402	-.12057	-.11646	-.11163	-.10601
12	2	3	-.06066	-.05457	-.04799	-.04089	-.03323	-.02499	-.01612
12	2	4	-.01787	-.01082	-.00344	.00428	.01235	.02079	.02964
12	2	5	.01295	.01981	.02682	.03401	.04137	.04893	.05670
12	2	6	.03694	.04298	.04905	.05516	.06132	.06754	.07384
12	2	7	.05657	.06140	.06618	.07091	.07561	.08030	.08499
12	2	8	.07324	.07656	.07979	.08295	.08605	.08911	.09214
12	2	9	.08776	.08930	.09077	.09220	.09360	.09498	.09636
12	2	10	.31759	.30968	.30269	.29657	.29127	.28675	.28296
12	3	1	-.42173	-.45421	-.48914	-.52666	-.56689	-.60996	-.65604
12	3	2	-.14237	-.13959	-.13621	-.13219	-.12747	-.12201	-.11573
12	3	3	-.06548	-.05869	-.05138	-.04355	-.03514	-.02613	-.01647
12	3	4	-.01746	-.00976	-.00173	.00664	.01537	.02448	.03399
12	3	5	.01708	.02443	.03194	.03962	.04747	.05552	.06378
12	3	6	.04390	.05024	.05661	.06301	.06947	.07599	.08259
12	3	7	.06576	.07067	.07553	.08035	.08515	.08995	.09474
12	3	8	.08417	.08734	.09043	.09348	.09648	.09946	.10242
12	3	9	.43615	.42955	.42394	.41929	.41556	.41272	.41071
12	4	1	-.47805	-.51372	-.55206	-.59323	-.63735	-.68458	-.73510
12	4	2	-.15865	-.15522	-.15116	-.14643	-.14098	-.13473	-.12763
12	4	3	-.07090	-.06323	-.05503	-.04628	-.03693	-.02696	-.01631
12	4	4	-.01619	-.00768	.00116	.01034	.01988	.02982	.04018

Coefficients b_i for the BLUE of σ
Shape parameter k

n	s	i	0.7	0.8	0.9	1.0	1.1	1.2	1.3
12	4	5	.02308	.03105	.03916	.04744	.05590	.06455	.07343
12	4	6	.05346	.06017	.06690	.07366	.08047	.08734	.09431
12	4	7	.07808	.08308	.08803	.09294	.09784	.10274	.10766
12	4	8	.56919	.56554	.56301	.56156	.56118	.56182	.56347
12	5	1	-.55201	-.59197	-.63491	-.68099	-.73037	-.78322	-.83973
12	5	2	-.17921	-.17494	-.17001	-.16436	-.15794	-.15068	-.14251
12	5	3	-.07706	-.06824	-.05888	-.04894	-.03838	-.02715	-.01521
12	5	4	-.01351	-.00396	.00592	.01614	.02675	.03776	.04921
12	5	5	.03196	.04071	.04960	.05864	.06786	.07729	.08695
12	5	6	.06696	.07412	.08129	.08848	.09573	.10305	.11046
12	5	7	.72286	.72428	.72700	.73103	.73635	.74295	.75082
12	6	1	-.65364	-.69960	-.74897	-.80194	-.85869	-.91941	-.98432
12	6	2	-.20623	-.20082	-.19469	-.18778	-.18003	-.17137	-.16172
12	6	3	-.08410	-.07372	-.06277	-.05120	-.03897	-.02603	-.01232
12	6	4	-.00834	.00261	.01389	.02553	.03757	.05003	.06296
12	6	5	.04560	.05539	.06530	.07536	.08561	.09607	.10677
12	6	6	.90671	.91614	.92724	.94003	.95452	.97072	.98863
13	0	1	-.29780	-.32432	-.35290	-.38362	-.41659	-.45191	-.48970
13	0	2	-.10807	-.10752	-.10646	-.10485	-.10265	-.09981	-.09628
13	0	3	-.05585	-.05166	-.04698	-.04180	-.03608	-.02981	-.02295
13	0	4	-.02321	-.01778	-.01201	-.00588	.00062	.00751	.01480
13	0	5	.00031	.00596	.01182	.01789	.02418	.03069	.03744
13	0	6	.01861	.02397	.02940	.03492	.04054	.04624	.05205
13	0	7	.03360	.03832	.04304	.04774	.05243	.05713	.06184
13	0	8	.04633	.05020	.05398	.05768	.06132	.06491	.06845
13	0	9	.05751	.06031	.06299	.06556	.06805	.07047	.07282
13	0	10	.06761	.06914	.07055	.07188	.07313	.07432	.07547
13	0	11	.07707	.07706	.07701	.07693	.07684	.07676	.07669
13	0	12	.08644	.08446	.08259	.08085	.07925	.07777	.07644
13	0	13	.09746	.09187	.08697	.08270	.07897	.07574	.07294
13	1	1	-.32268	-.35034	-.38018	-.41229	-.44678	-.48377	-.52339
13	1	2	-.11650	-.11561	-.11421	-.11226	-.10972	-.10654	-.10266
13	1	3	-.05978	-.05515	-.05004	-.04443	-.03828	-.03158	-.02427
13	1	4	-.02433	-.01850	-.01233	-.00581	.00109	.00838	.01608
13	1	5	.00120	.00718	.01336	.01975	.02636	.03319	.04026
13	1	6	.02105	.02664	.03230	.03805	.04389	.04983	.05588
13	1	7	.03730	.04215	.04698	.05180	.05663	.06146	.06632
13	1	8	.05109	.05495	.05874	.06246	.06613	.06976	.07336
13	1	9	.06315	.06582	.06839	.07088	.07330	.07566	.07798
13	1	10	.07399	.07526	.07645	.07758	.07866	.07972	.08076
13	1	11	.08400	.08360	.08320	.08282	.08248	.08218	.08194
13	1	12	.19151	.18401	.17734	.17144	.16624	.16170	.15774
13	2	1	-.35294	-.38218	-.41372	-.44767	-.48414	-.52326	-.56519
13	2	2	-.12644	-.12519	-.12343	-.12111	-.11819	-.11461	-.11031
13	2	3	-.06418	-.05906	-.05346	-.04735	-.04070	-.03347	-.02563
13	2	4	-.02529	-.01899	-.01236	-.00537	.00200	.00976	.01793
13	2	5	.00271	.00907	.01563	.02239	.02937	.03658	.04403
13	2	6	.02446	.03031	.03624	.04224	.04834	.05454	.06085
13	2	7	.04223	.04721	.05218	.05714	.06210	.06708	.07208
13	2	8	.05727	.06113	.06492	.06865	.07233	.07599	.07962
13	2	9	.07036	.07288	.07532	.07768	.08000	.08228	.08453
13	2	10	.08201	.08297	.08388	.08477	.08564	.08651	.08738
13	2	11	.28981	.28186	.27482	.26863	.26325	.25862	.25469
13	3	1	-.38981	-.42111	-.45486	-.49117	-.53017	-.57201	-.61684
13	3	2	-.13821	-.13657	-.13440	-.13166	-.12829	-.12424	-.11945
13	3	3	-.06912	-.06344	-.05726	-.05056	-.04330	-.03545	-.02698

Coefficients b_i for the BLUE of σ
Shape parameter k

n	s	i	0.7	0.8	0.9	1.0	1.1	1.2	1.3
13	3	4	-.02600	-.01916	-.01198	-.00444	.00348	.01180	.02056
13	3	5	.00501	.01182	.01882	.02602	.03344	.04109	.04899
13	3	6	.02907	.03523	.04146	.04777	.05417	.06067	.06729
13	3	7	.04867	.05381	.05893	.06404	.06916	.07430	.07946
13	3	8	.06520	.06905	.07283	.07655	.08025	.08392	.08758
13	3	9	.07948	.08181	.08407	.08628	.08845	.09061	.09277
13	3	10	.39571	.38857	.38239	.37716	.37281	.36931	.36662
13	4	1	-.43548	-.46943	-.50602	-.54537	-.58763	-.63294	-.68149
13	4	2	-.15236	-.15026	-.14761	-.14437	-.14047	-.13586	-.13048
13	4	3	-.07474	-.06837	-.06149	-.05407	-.04608	-.03748	-.02822
13	4	4	-.02634	-.01884	-.01100	-.00279	.00581	.01482	.02427
13	4	5	.00841	.01576	.02329	.03103	.03898	.04717	.05561
13	4	6	.03532	.04185	.04844	.05511	.06187	.06873	.07572
13	4	7	.05717	.06248	.06777	.07306	.07835	.08366	.08902
13	4	8	.07548	.07929	.08304	.08675	.09043	.09410	.09778
13	4	9	.51255	.50753	.50357	.50065	.49874	.49780	.49780
13	5	1	-.49344	-.53086	-.57115	-.61446	-.66096	-.71081	-.76420
13	5	2	-.16978	-.16711	-.16387	-.15999	-.15543	-.15011	-.14399
13	5	3	-.08121	-.07398	-.06623	-.05791	-.04900	-.03944	-.02920
13	5	4	-.02607	-.01775	-.00907	-.00003	.00941	.01928	.02960
13	5	5	.01344	.02145	.02965	.03804	.04665	.05550	.06462
13	5	6	.04394	.05092	.05796	.06506	.07225	.07955	.08698
13	5	7	.06859	.07410	.07959	.08507	.09056	.09608	.10165
13	5	8	.64453	.64322	.64312	.64422	.64651	.64995	.65453
13	6	1	-.56953	-.61156	-.65680	-.70542	-.75760	-.81351	-.87338
13	6	2	-.19185	-.18845	-.18442	-.17973	-.17429	-.16806	-.16096
13	6	3	-.08878	-.08044	-.07154	-.06205	-.05193	-.04113	-.02961
13	6	4	-.02474	-.01535	-.00561	.00452	.01506	.02604	.03750
13	6	5	.02102	.02988	.03892	.04816	.05761	.06732	.07730
13	6	6	.05619	.06373	.07131	.07896	.08670	.09455	.10253
13	6	7	.79768	.80218	.80814	.81556	.82445	.83481	.84662
13	7	1	-.67409	-.72253	-.77466	-.83067	-.89076	-.95516	-1.02408
13	7	2	-.22100	-.21658	-.21147	-.20564	-.19900	-.19149	-.18304
13	7	3	-.09780	-.08792	-.07746	-.06636	-.05459	-.04209	-.02880
13	7	4	-.02147	-.01064	.00056	.01215	.02418	.03668	.04969
13	7	5	.03285	.04285	.05301	.06337	.07396	.08481	.09595
13	7	6	.98151	.99483	1.01002	1.02714	1.04621	1.06725	1.09028
14	0	1	-.28435	-.31042	-.33857	-.36889	-.40149	-.43647	-.47397
14	0	2	-.10698	-.10711	-.10678	-.10596	-.10459	-.10264	-.10006
14	0	3	-.05829	-.05477	-.05079	-.04631	-.04132	-.03579	-.02968
14	0	4	-.02787	-.02302	-.01781	-.01223	-.00627	.00009	.00687
14	0	5	-.00598	-.00075	.00471	.01041	.01635	.02255	.02901
14	0	6	.01103	.01613	.02135	.02669	.03215	.03775	.04349
14	0	7	.02492	.02958	.03427	.03899	.04374	.04852	.05336
14	0	8	.03668	.04069	.04466	.04859	.05248	.05635	.06020
14	0	9	.04693	.05013	.05323	.05625	.05920	.06208	.06491
14	0	10	.05611	.05833	.06044	.06245	.06438	.06624	.06805
14	0	11	.06454	.06561	.06659	.06750	.06834	.06915	.06992
14	0	12	.07254	.07224	.07192	.07158	.07126	.07096	.07068
14	0	13	.08058	.07851	.07657	.07478	.07312	.07161	.07025
14	0	14	.09014	.08484	.08020	.07616	.07265	.06960	.06697
14	1	1	-.30607	-.33317	-.36245	-.39402	-.42799	-.46449	-.50364
14	1	2	-.11467	-.11452	-.11392	-.11283	-.11120	-.10898	-.10613
14	1	3	-.06214	-.05826	-.05391	-.04907	-.04372	-.03781	-.03132
14	1	4	-.02934	-.02413	-.01857	-.01264	-.00634	.00037	.00751
14	1	5	-.00574	-.00021	.00555	.01154	.01778	.02427	.03103

Coefficients b_i for the BLUE of σ
Shape parameter k

n	s	i	0.7	0.8	0.9	1.0	1.1	1.2	1.3
14	1	6	.01260	.01792	.02337	.02893	.03462	.04044	.04641
14	1	7	.02756	.03237	.03720	.04206	.04695	.05189	.05688
14	1	8	.04022	.04428	.04831	.05230	.05626	.06020	.06413
14	1	9	.05124	.05439	.05746	.06045	.06339	.06628	.06912
14	1	10	.06107	.06314	.06513	.06703	.06887	.07067	.07242
14	1	11	.07005	.07087	.07163	.07235	.07303	.07370	.07436
14	1	12	.07846	.07779	.07715	.07655	.07600	.07551	.07507
14	1	13	.17677	.16951	.16306	.15736	.15235	.14796	.14415
14	2	1	-.33215	-.36064	-.39142	-.42462	-.46035	-.49874	-.53995
14	2	2	-.12365	-.12323	-.12235	-.12097	-.11905	-.11653	-.11335
14	2	3	-.06647	-.06218	-.05742	-.05217	-.04639	-.04006	-.03313
14	2	4	-.03077	-.02517	-.01921	-.01288	-.00617	.00095	.00850
14	2	5	-.00510	.00077	.00687	.01319	.01977	.02660	.03370
14	2	6	.01482	.02041	.02611	.03193	.03787	.04395	.05017
14	2	7	.03107	.03604	.04104	.04606	.05111	.05621	.06137
14	2	8	.04480	.04892	.05301	.05705	.06108	.06510	.06911
14	2	9	.05671	.05981	.06283	.06579	.06870	.07157	.07441
14	2	10	.06730	.06919	.07102	.07279	.07451	.07620	.07788
14	2	11	.07686	.07738	.07787	.07836	.07884	.07933	.07984
14	2	12	.26657	.25867	.25166	.24548	.24008	.23542	.23144
14	3	1	-.36339	-.39367	-.42637	-.46163	-.49956	-.54032	-.58407
14	3	2	-.13415	-.13344	-.13226	-.13056	-.12831	-.12543	-.12189
14	3	3	-.07134	-.06659	-.06136	-.05563	-.04936	-.04252	-.03506
14	3	4	-.03215	-.02608	-.01966	-.01286	-.00568	.00192	.00996
14	3	5	-.00399	.00228	.00877	.01550	.02246	.02969	.03720
14	3	6	.01785	.02374	.02974	.03585	.04208	.04845	.05497
14	3	7	.03564	.04079	.04597	.05117	.05641	.06169	.06704
14	3	8	.05062	.05481	.05895	.06306	.06716	.07124	.07534
14	3	9	.06358	.06660	.06956	.07246	.07532	.07816	.08098
14	3	10	.07500	.07668	.07832	.07991	.08148	.08304	.08460
14	3	11	.36233	.35486	.34834	.34273	.33799	.33408	.33094
14	4	1	-.40130	-.43384	-.46896	-.50680	-.54751	-.59124	-.63816
14	4	2	-.14658	-.14554	-.14401	-.14195	-.13931	-.13603	-.13205
14	4	3	-.07688	-.07158	-.06580	-.05949	-.05264	-.04519	-.03711
14	4	4	-.03342	-.02680	-.01982	-.01247	-.00472	.00345	.01208
14	4	5	-.00222	.00452	.01148	.01867	.02611	.03380	.04179
14	4	6	.02195	.02819	.03453	.04098	.04756	.05427	.06114
14	4	7	.04158	.04695	.05233	.05774	.06318	.06868	.07423
14	4	8	.05807	.06231	.06652	.07069	.07486	.07902	.08319
14	4	9	.07224	.07516	.07802	.08085	.08364	.08642	.08920
14	4	10	.46658	.46063	.45571	.45179	.44883	.44681	.44568
14	5	1	-.44818	-.48358	-.52177	-.56290	-.60713	-.65462	-.70557
14	5	2	-.16157	-.16013	-.15819	-.15569	-.15258	-.14880	-.14429
14	5	3	-.08326	-.07730	-.07083	-.06383	-.05626	-.04807	-.03923
14	5	4	-.03448	-.02719	-.01955	-.01152	-.00308	.00580	.01514
14	5	5	.00049	.00780	.01533	.02308	.03108	.03935	.04791
14	5	6	.02752	.03419	.04094	.04781	.05479	.06192	.06920
14	5	7	.04942	.05504	.06066	.06631	.07199	.07773	.08353
14	5	8	.06772	.07202	.07629	.08053	.08476	.08899	.09325
14	5	9	.58233	.57916	.57712	.57622	.57642	.57771	.58004
14	6	1	-.50766	-.54676	-.58891	-.63430	-.68308	-.73544	-.79159
14	6	2	-.18004	-.17812	-.17566	-.17260	-.16891	-.16450	-.15933
14	6	3	-.09071	-.08392	-.07660	-.06872	-.06025	-.05113	-.04131
14	6	4	-.03514	-.02703	-.01854	-.00966	-.00036	.00940	.01964
14	6	5	.00462	.01263	.02086	.02931	.03802	.04700	.05629
14	6	6	.03527	.04245	.04971	.05708	.06456	.07219	.07999

Coefficients b_i for the BLUE of σ
Shape parameter k

n	s	i	0.7	0.8	0.9	1.0	1.1	1.2	1.3
14	6	7	.05999	.06590	.07182	.07775	.08371	.08974	.09584
14	6	8	.71367	.71484	.71733	.72115	.72629	.73274	.74048
14	7	1	-.58571	-.62972	-.67715	-.72820	-.78304	-.84190	-.90499
14	7	2	-.20353	-.20096	-.19781	-.19403	-.18956	-.18434	-.17829
14	7	3	-.09959	-.09172	-.08328	-.07426	-.06460	-.05425	-.04317
14	7	4	-.03507	-.02587	-.01629	-.00631	.00411	.01501	.02643
14	7	5	.01099	.01990	.02903	.03838	.04800	.05790	.06811
14	7	6	.04635	.05417	.06208	.07008	.07820	.08647	.09491
14	7	7	.86656	.87419	.88344	.89434	.90689	.92111	.93699
15	0	1	-.27232	-.29794	-.32566	-.35557	-.38779	-.42242	-.45959
15	0	2	-.10565	-.10636	-.10666	-.10651	-.10587	-.10469	-.10294
15	0	3	-.06000	-.05708	-.05371	-.04986	-.04552	-.04065	-.03522
15	0	4	-.03150	-.02717	-.02247	-.01740	-.01194	-.00608	.00020
15	0	5	-.01101	-.00619	-.00111	.00422	.00982	.01569	.02184
15	0	6	.00490	.00972	.01470	.01982	.02510	.03054	.03615
15	0	7	.01786	.02240	.02700	.03165	.03637	.04116	.04603
15	0	8	.02881	.03286	.03690	.04092	.04495	.04897	.05301
15	0	9	.03831	.04173	.04508	.04836	.05160	.05480	.05796
15	0	10	.04677	.04941	.05196	.05443	.05682	.05915	.06142
15	0	11	.05445	.05620	.05785	.05941	.06091	.06234	.06372
15	0	12	.06160	.06232	.06295	.06353	.06406	.06457	.06506
15	0	13	.06848	.06795	.06742	.06689	.06639	.06593	.06551
15	0	14	.07545	.07333	.07136	.06953	.06785	.06633	.06496
15	0	15	.08385	.07880	.07440	.07057	.06724	.06437	.06188
15	1	1	-.29147	-.31802	-.34677	-.37782	-.41128	-.44729	-.48598
15	1	2	-.11268	-.11317	-.11325	-.11289	-.11204	-.11066	-.10869
15	1	3	-.06373	-.06049	-.05681	-.05266	-.04800	-.04282	-.03707
15	1	4	-.03317	-.02853	-.02352	-.01815	-.01238	-.00621	.00039
15	1	5	-.01120	-.00611	-.00076	.00483	.01070	.01684	.02327
15	1	6	.00584	.01089	.01608	.02141	.02690	.03256	.03839
15	1	7	.01973	.02442	.02917	.03398	.03885	.04380	.04883
15	1	8	.03145	.03559	.03971	.04382	.04794	.05206	.05620
15	1	9	.04162	.04504	.04840	.05171	.05498	.05821	.06142
15	1	10	.05065	.05322	.05571	.05813	.06049	.06279	.06506
15	1	11	.05883	.06043	.06194	.06339	.06479	.06614	.06747
15	1	12	.06641	.06688	.06730	.06770	.06808	.06845	.06883
15	1	13	.07358	.07272	.07190	.07113	.07043	.06979	.06923
15	1	14	.16414	.15713	.15090	.14539	.14056	.13634	.13267
15	2	1	-.31421	-.34201	-.37210	-.40460	-.43964	-.47735	-.51788
15	2	2	-.12083	-.12110	-.12097	-.12039	-.11931	-.11769	-.11548
15	2	3	-.06792	-.06434	-.06031	-.05581	-.05081	-.04526	-.03914
15	2	4	-.03489	-.02990	-.02455	-.01883	-.01271	-.00619	.00078
15	2	5	-.01116	-.00576	-.00011	.00580	.01196	.01841	.02515
15	2	6	.00724	.01254	.01797	.02355	.02928	.03518	.04125
15	2	7	.02223	.02710	.03202	.03699	.04204	.04715	.05236
15	2	8	.03487	.03910	.04331	.04752	.05172	.05594	.06019
15	2	9	.04581	.04924	.05261	.05593	.05922	.06248	.06572
15	2	10	.05551	.05799	.06041	.06276	.06506	.06732	.06956
15	2	11	.06425	.06566	.06701	.06831	.06958	.07083	.07206
15	2	12	.07225	.07244	.07261	.07279	.07298	.07319	.07342
15	2	13	.24685	.23904	.23210	.22598	.22063	.21599	.21201
15	3	1	-.34109	-.37045	-.40223	-.43654	-.47353	-.51333	-.55611
15	3	2	-.13027	-.13032	-.12996	-.12914	-.12781	-.12593	-.12344
15	3	3	-.07262	-.06866	-.06425	-.05935	-.05394	-.04797	-.04142
15	3	4	-.03666	-.03127	-.02552	-.01940	-.01288	-.00594	.00145
15	3	5	-.01083	-.00507	.00093	.00718	.01370	.02050	.02760

Coefficients b_i for the BLUE of σ
Shape parameter k

n	s	i	0.7	0.8	0.9	1.0	1.1	1.2	1.3
15	3	6	.00919	.01476	.02047	.02633	.03234	.03851	.04486
15	3	7	.02547	.03054	.03565	.04082	.04605	.05136	.05676
15	3	8	.03918	.04351	.04783	.05213	.05644	.06077	.06512
15	3	9	.05103	.05446	.05783	.06116	.06446	.06774	.07101
15	3	10	.06148	.06386	.06617	.06844	.07067	.07287	.07506
15	3	11	.07083	.07202	.07318	.07430	.07542	.07653	.07764
15	3	12	.33429	.32663	.31990	.31407	.30908	.30490	.30147
15	4	1	-.37315	-.40447	-.43834	-.47489	-.51428	-.55666	-.60219
15	4	2	-.14129	-.14110	-.14049	-.13940	-.13779	-.13560	-.13279
15	4	3	-.07795	-.07355	-.06869	-.06333	-.05744	-.05098	-.04392
15	4	4	-.03846	-.03261	-.02640	-.01981	-.01281	-.00539	.00250
15	4	5	-.01012	-.00395	.00246	.00912	.01605	.02327	.03079
15	4	6	.01183	.01773	.02377	.02994	.03627	.04276	.04944
15	4	7	.02966	.03495	.04028	.04567	.05112	.05665	.06226
15	4	8	.04464	.04908	.05350	.05792	.06234	.06678	.07125
15	4	9	.05754	.06096	.06432	.06765	.07096	.07425	.07754
15	4	10	.06884	.07109	.07328	.07544	.07757	.07969	.08180
15	4	11	.42846	.42187	.41629	.41168	.40801	.40523	.40331
15	5	1	-.41199	-.44574	-.48220	-.52155	-.56392	-.60949	-.65844
15	5	2	-.15435	-.15389	-.15298	-.15158	-.14963	-.14710	-.14390
15	5	3	-.08405	-.07913	-.07373	-.06782	-.06136	-.05432	-.04664
15	5	4	-.04026	-.03387	-.02710	-.01995	-.01238	-.00438	.00410
15	5	5	-.00886	-.00220	.00469	.01184	.01926	.02697	.03499
15	5	6	.01542	.02171	.02813	.03468	.04138	.04826	.05532
15	5	7	.03511	.04066	.04625	.05189	.05759	.06337	.06925
15	5	8	.05161	.05617	.06072	.06525	.06980	.07436	.07896
15	5	9	.06574	.06914	.07248	.07580	.07910	.08239	.08570
15	5	10	.53162	.52713	.52374	.52143	.52016	.51992	.52067
15	6	1	-.45999	-.49678	-.53653	-.57938	-.62551	-.67511	-.72836
15	6	2	-.17010	-.16931	-.16805	-.16627	-.16393	-.16096	-.15730
15	6	3	-.09114	-.08558	-.07953	-.07294	-.06579	-.05802	-.04959
15	6	4	-.04200	-.03494	-.02750	-.01966	-.01140	-.00269	.00652
15	6	5	-.00680	.00044	.00793	.01566	.02368	.03199	.04062
15	6	6	.02036	.02712	.03399	.04099	.04815	.05548	.06300
15	6	7	.04233	.04819	.05408	.06002	.06601	.07209	.07827
15	6	8	.06067	.06536	.07004	.07471	.07939	.08410	.08884
15	6	9	.64667	.64550	.64557	.64688	.64940	.65312	.65801
15	7	1	-.52086	-.56157	-.60552	-.65289	-.70387	-.75865	-.81746
15	7	2	-.18957	-.18837	-.18666	-.18440	-.18155	-.17803	-.17379
15	7	3	-.09951	-.09316	-.08628	-.07884	-.07080	-.06212	-.05274
15	7	4	-.04354	-.03565	-.02736	-.01867	-.00953	.00008	.01020
15	7	5	-.00353	.00445	.01267	.02114	.02990	.03896	.04836
15	7	6	.02729	.03460	.04204	.04960	.05732	.06521	.07329
15	7	7	.05212	.05834	.06459	.07088	.07724	.08368	.09022
15	7	8	.77761	.78135	.78653	.79318	.80130	.81088	.82192
15	8	1	-.60074	-.64663	-.69616	-.74952	-.80693	-.86860	-.93478
15	8	2	-.21439	-.21263	-.21032	-.20743	-.20390	-.19965	-.19463
15	8	3	-.10963	-.10222	-.09426	-.08570	-.07650	-.06662	-.05599
15	8	4	-.04465	-.03566	-.02628	-.01646	-.00618	.00459	.01591
15	8	5	.00171	.01063	.01979	.02922	.03893	.04896	.05935
15	8	6	.03728	.04531	.05346	.06173	.07016	.07877	.08759
15	8	7	.93043	.94120	.95377	.96816	.98442	1.00255	1.02257
16	0	1	-.26147	-.28666	-.31395	-.34346	-.37528	-.40955	-.44638
16	0	2	-.10418	-.10538	-.10622	-.10665	-.10665	-.10615	-.10512
16	0	3	-.06119	-.05878	-.05595	-.05267	-.04891	-.04464	-.03982
16	0	4	-.03435	-.03049	-.02626	-.02166	-.01667	-.01127	-.00545

Coefficients b_i for the BLUE of σ
Shape parameter k

n	s	i	0.7	0.8	0.9	1.0	1.1	1.2	1.3
16	0	5	-.01507	-.01063	-.00593	-.00095	.00431	.00985	.01570
16	0	6	-.00012	.00443	.00915	.01404	.01911	.02436	.02981
16	0	7	.01205	.01642	.02089	.02543	.03008	.03481	.03965
16	0	8	.02231	.02632	.03035	.03440	.03847	.04257	.04671
16	0	9	.03119	.03470	.03818	.04162	.04504	.04844	.05182
16	0	10	.03905	.04195	.04478	.04754	.05024	.05290	.05551
16	0	11	.04614	.04834	.05043	.05245	.05439	.05627	.05810
16	0	12	.05268	.05405	.05533	.05653	.05767	.05876	.05982
16	0	13	.05884	.05926	.05962	.05994	.06023	.06050	.06077
16	0	14	.06482	.06411	.06342	.06275	.06212	.06153	.06100
16	0	15	.07094	.06879	.06679	.06495	.06328	.06175	.06038
16	0	16	.07838	.07356	.06937	.06573	.06258	.05985	.05750
16	1	1	-.27851	-.30454	-.33277	-.36332	-.39629	-.43181	-.47004
16	1	2	-.11063	-.11166	-.11232	-.11259	-.11241	-.11175	-.11056
16	1	3	-.06476	-.06209	-.05899	-.05545	-.05142	-.04688	-.04179
16	1	4	-.03614	-.03199	-.02750	-.02263	-.01737	-.01170	-.00560
16	1	5	-.01557	-.01088	-.00593	-.00071	.00479	.01058	.01668
16	1	6	.00037	.00513	.01005	.01514	.02041	.02587	.03152
16	1	7	.01335	.01788	.02250	.02720	.03199	.03689	.04189
16	1	8	.02428	.02840	.03253	.03668	.04085	.04506	.04931
16	1	9	.03374	.03730	.04082	.04431	.04778	.05123	.05468
16	1	10	.04210	.04499	.04780	.05055	.05326	.05592	.05855
16	1	11	.04965	.05175	.05377	.05572	.05761	.05946	.06127
16	1	12	.05657	.05779	.05893	.06001	.06105	.06206	.06305
16	1	13	.06306	.06325	.06341	.06356	.06370	.06385	.06402
16	1	14	.06926	.06825	.06730	.06641	.06559	.06485	.06419
16	1	15	.15322	.14643	.14041	.13510	.13044	.12637	.12284
16	2	1	-.29854	-.32569	-.35514	-.38699	-.42138	-.45845	-.49835
16	2	2	-.11806	-.11892	-.11941	-.11951	-.11916	-.11832	-.11694
16	2	3	-.06877	-.06581	-.06243	-.05858	-.05426	-.04941	-.04401
16	2	4	-.03802	-.03357	-.02877	-.02360	-.01803	-.01205	-.00563
16	2	5	-.01593	-.01096	-.00574	-.00025	.00553	.01160	.01797
16	2	6	.00119	.00617	.01132	.01664	.02214	.02782	.03371
16	2	7	.01511	.01982	.02461	.02948	.03444	.03951	.04468
16	2	8	.02684	.03106	.03530	.03956	.04385	.04817	.05253
16	2	9	.03697	.04057	.04413	.04767	.05119	.05470	.05821
16	2	10	.04592	.04877	.05156	.05430	.05699	.05965	.06229
16	2	11	.05396	.05596	.05789	.05976	.06158	.06337	.06513
16	2	12	.06132	.06234	.06332	.06426	.06518	.06608	.06697
16	2	13	.06812	.06805	.06798	.06792	.06789	.06788	.06792
16	2	14	.22990	.22220	.21537	.20934	.20405	.19946	.19552
16	3	1	-.32195	-.35049	-.38144	-.41490	-.45103	-.48996	-.53187
16	3	2	-.12659	-.12728	-.12760	-.12751	-.12698	-.12594	-.12435
16	3	3	-.07326	-.06998	-.06628	-.06210	-.05744	-.05225	-.04648
16	3	4	-.04001	-.03522	-.03008	-.02456	-.01864	-.01230	-.00552
16	3	5	-.01613	-.01085	-.00531	.00049	.00658	.01297	.01967
16	3	6	.00237	.00762	.01303	.01860	.02435	.03029	.03644
16	3	7	.01740	.02231	.02729	.03235	.03750	.04276	.04813
16	3	8	.03005	.03440	.03876	.04314	.04755	.05199	.05648
16	3	9	.04096	.04461	.04822	.05180	.05537	.05894	.06251
16	3	10	.05058	.05339	.05615	.05886	.06153	.06418	.06681
16	3	11	.05919	.06106	.06287	.06464	.06638	.06809	.06979
16	3	12	.06699	.06781	.06859	.06936	.07012	.07089	.07166
16	3	13	.31038	.30263	.29579	.28983	.28470	.28035	.27673
16	4	1	-.34949	-.37974	-.41252	-.44795	-.48619	-.52738	-.57171
16	4	2	-.13643	-.13695	-.13709	-.13681	-.13606	-.13480	-.13298

Coefficients b_i for the BLUE of σ
Shape parameter k

n	s	i	0.7	0.8	0.9	1.0	1.1	1.2	1.3
16	4	3	-.07832	-.07469	-.07061	-.06606	-.06101	-.05541	-.04922
16	4	4	-.04210	-.03693	-.03139	-.02547	-.01915	-.01240	-.00519
16	4	5	-.01611	-.01047	-.00457	.00159	.00804	.01479	.02186
16	4	6	.00401	.00956	.01527	.02114	.02718	.03342	.03986
16	4	7	.02035	.02548	.03068	.03596	.04133	.04681	.05239
16	4	8	.03408	.03857	.04307	.04758	.05212	.05670	.06133
16	4	9	.04590	.04959	.05325	.05688	.06050	.06412	.06775
16	4	10	.05628	.05904	.06175	.06442	.06706	.06969	.07231
16	4	11	.06552	.06724	.06891	.07055	.07218	.07379	.07541
16	4	12	.39630	.38928	.38325	.37816	.37398	.37068	.36819
16	5	1	-.38229	-.41463	-.44965	-.48748	-.52829	-.57225	-.61953
16	5	2	-.14793	-.14826	-.14819	-.14769	-.14671	-.14520	-.14310
16	5	3	-.08409	-.08003	-.07553	-.07053	-.06502	-.05895	-.05227
16	5	4	-.04431	-.03868	-.03268	-.02630	-.01950	-.01226	-.00455
16	5	5	-.01578	-.00972	-.00341	.00318	.01005	.01723	.02474
16	5	6	.00628	.01218	.01823	.02444	.03083	.03741	.04420
16	5	7	.02417	.02956	.03501	.04054	.04616	.05188	.05772
16	5	8	.03917	.04382	.04847	.05313	.05782	.06255	.06733
16	5	9	.05206	.05579	.05949	.06317	.06684	.07051	.07420
16	5	10	.06331	.06600	.06865	.07126	.07385	.07644	.07903
16	5	11	.48940	.48398	.47961	.47629	.47397	.47263	.47222
16	6	1	-.42198	-.45690	-.49468	-.53548	-.57946	-.62682	-.67774
16	6	2	-.16158	-.16168	-.16137	-.16061	-.15935	-.15754	-.15511
16	6	3	-.09072	-.08618	-.08115	-.07563	-.06957	-.06292	-.05565
16	6	4	-.04662	-.04044	-.03390	-.02695	-.01958	-.01176	-.00345
16	6	5	-.01501	-.00845	-.00164	.00545	.01282	.02052	.02855
16	6	6	.00940	.01571	.02216	.02878	.03558	.04257	.04977
16	6	7	.02916	.03485	.04060	.04642	.05233	.05834	.06447
16	6	8	.04569	.05051	.05533	.06016	.06502	.06992	.07488
16	6	9	.05983	.06360	.06735	.07107	.07479	.07851	.08226
16	6	10	.59183	.58898	.58731	.58680	.58744	.58919	.59204
16	7	1	-.47101	-.50914	-.55039	-.59490	-.64287	-.69450	-.74999
16	7	2	-.17807	-.17790	-.17730	-.17622	-.17462	-.17243	-.16961
16	7	3	-.09849	-.09333	-.08768	-.08150	-.07476	-.06741	-.05941
16	7	4	-.04899	-.04216	-.03494	-.02731	-.01924	-.01070	-.00166
16	7	5	-.01358	-.00641	.00101	.00871	.01670	.02502	.03368
16	7	6	.01374	.02054	.02749	.03459	.04188	.04936	.05706
16	7	7	.03580	.04184	.04793	.05410	.06036	.06672	.07320
16	7	8	.05418	.05920	.06421	.06924	.07429	.07939	.08455
16	7	9	.70641	.70737	.70966	.71329	.71827	.72457	.73218
16	8	1	-.53318	-.57544	-.62111	-.67039	-.72348	-.78059	-.84195
16	8	2	-.19847	-.19796	-.19698	-.19550	-.19346	-.19081	-.18747
16	8	3	-.10773	-.10180	-.09535	-.08835	-.08075	-.07251	-.06359
16	8	4	-.05137	-.04370	-.03563	-.02714	-.01819	-.00875	.00122
16	8	5	-.01110	-.00318	.00500	.01346	.02222	.03132	.04078
16	8	6	.01988	.02730	.03485	.04256	.05045	.05855	.06687
16	8	7	.04483	.05130	.05782	.06441	.07108	.07787	.08478
16	8	8	.83714	.84348	.85141	.86095	.87212	.88492	.89935
17	0	1	-.25163	-.27639	-.30328	-.33238	-.36382	-.39771	-.43419
17	0	2	-.10263	-.10425	-.10556	-.10650	-.10705	-.10715	-.10677
17	0	3	-.06197	-.06003	-.05768	-.05490	-.05166	-.04793	-.04368
17	0	4	-.03661	-.03316	-.02936	-.02519	-.02064	-.01567	-.01029
17	0	5	-.01839	-.01430	-.00995	-.00531	-.00038	.00485	.01040
17	0	6	-.00427	.00000	.00446	.00912	.01397	.01902	.02429
17	0	7	.00721	.01140	.01570	.02012	.02465	.02929	.03407
17	0	8	.01687	.02080	.02478	.02880	.03286	.03698	.04116

Coefficients b_i for the BLUE of σ
Shape parameter k

n	s	i	0.7	0.8	0.9	1.0	1.1	1.2	1.3
17	0	9	.02522	.02876	.03229	.03581	.03933	.04285	.04638
17	0	10	.03258	.03563	.03863	.04158	.04449	.04737	.05022
17	0	11	.03920	.04167	.04407	.04639	.04865	.05086	.05303
17	0	12	.04525	.04707	.04879	.05043	.05201	.05353	.05500
17	0	13	.05089	.05195	.05293	.05383	.05469	.05551	.05629
17	0	14	.05626	.05644	.05659	.05670	.05679	.05688	.05697
17	0	15	.06151	.06066	.05985	.05907	.05834	.05766	.05705
17	0	16	.06693	.06477	.06277	.06093	.05926	.05775	.05639
17	0	17	.07357	.06898	.06497	.06150	.05850	.05591	.05368
17	1	1	-.26690	-.29244	-.32018	-.35024	-.38273	-.41778	-.45555
17	1	2	-.10857	-.11005	-.11122	-.11203	-.11244	-.11242	-.11191
17	1	3	-.06538	-.06321	-.06063	-.05763	-.05416	-.05020	-.04571
17	1	4	-.03844	-.03475	-.03072	-.02631	-.02151	-.01631	-.01067
17	1	5	-.01909	-.01478	-.01020	-.00534	-.00020	.00526	.01103
17	1	6	-.00410	.00037	.00502	.00986	.01489	.02014	.02560
17	1	7	.00809	.01243	.01688	.02145	.02613	.03093	.03586
17	1	8	.01834	.02238	.02646	.03059	.03476	.03900	.04329
17	1	9	.02719	.03080	.03439	.03798	.04156	.04515	.04876
17	1	10	.03500	.03807	.04108	.04405	.04699	.04989	.05279
17	1	11	.04201	.04445	.04681	.04910	.05135	.05355	.05572
17	1	12	.04842	.05013	.05177	.05334	.05486	.05634	.05778
17	1	13	.05437	.05527	.05611	.05690	.05766	.05839	.05911
17	1	14	.05999	.05996	.05991	.05986	.05982	.05979	.05979
17	1	15	.06540	.06429	.06323	.06225	.06135	.06054	.05981
17	1	16	.14367	.13710	.13129	.12616	.12166	.11774	.11433
17	2	1	-.28470	-.31126	-.34010	-.37134	-.40513	-.44159	-.48088
17	2	2	-.11537	-.11672	-.11776	-.11843	-.11872	-.11856	-.11791
17	2	3	-.06920	-.06678	-.06396	-.06070	-.05698	-.05276	-.04801
17	2	4	-.04040	-.03645	-.03215	-.02747	-.02241	-.01693	-.01102
17	2	5	-.01973	-.01517	-.01034	-.00524	.00016	.00586	.01188
17	2	6	-.00371	.00097	.00583	.01088	.01613	.02158	.02726
17	2	7	.00930	.01382	.01844	.02317	.02801	.03298	.03808
17	2	8	.02025	.02441	.02861	.03285	.03715	.04150	.04592
17	2	9	.02969	.03336	.03702	.04067	.04433	.04799	.05167
17	2	10	.03801	.04109	.04411	.04710	.05005	.05299	.05591
17	2	11	.04548	.04786	.05017	.05243	.05465	.05683	.05898
17	2	12	.05227	.05387	.05540	.05689	.05833	.05974	.06113
17	2	13	.05855	.05927	.05995	.06060	.06123	.06186	.06249
17	2	14	.06442	.06414	.06388	.06364	.06343	.06326	.06314
17	2	15	.21516	.20760	.20089	.19496	.18977	.18525	.18136
17	3	1	-.30530	-.33310	-.36329	-.39598	-.43133	-.46947	-.51057
17	3	2	-.12311	-.12434	-.12524	-.12579	-.12593	-.12562	-.12481
17	3	3	-.07346	-.07078	-.06768	-.06415	-.06014	-.05563	-.05057
17	3	4	-.04250	-.03826	-.03366	-.02868	-.02331	-.01753	-.01129
17	3	5	-.02028	-.01545	-.01034	-.00496	.00072	.00670	.01301
17	3	6	-.00308	.00185	.00695	.01223	.01772	.02341	.02932
17	3	7	.01090	.01561	.02042	.02533	.03036	.03551	.04080
17	3	8	.02264	.02694	.03127	.03565	.04007	.04455	.04910
17	3	9	.03277	.03651	.04024	.04397	.04769	.05143	.05519
17	3	10	.04168	.04476	.04779	.05078	.05376	.05671	.05966
17	3	11	.04964	.05196	.05422	.05643	.05860	.06075	.06288
17	3	12	.05687	.05834	.05975	.06112	.06247	.06379	.06511
17	3	13	.06349	.06401	.06450	.06498	.06547	.06597	.06649
17	3	14	.28974	.28196	.27508	.26906	.26386	.25941	.25568
17	4	1	-.32926	-.35857	-.39038	-.42482	-.46203	-.50218	-.54544
17	4	2	-.13197	-.13307	-.13384	-.13424	-.13423	-.13376	-.13278

Coefficients b_i for the BLUE of σ
Shape parameter k

n	s	i	0.7	0.8	0.9	1.0	1.1	1.2	1.3
17	4	3	-.07824	-.07526	-.07186	-.06801	-.06368	-.05883	-.05343
17	4	4	-.04475	-.04018	-.03524	-.02992	-.02420	-.01806	-.01146
17	4	5	-.02073	-.01557	-.01015	-.00445	.00154	.00785	.01449
17	4	6	-.00213	.00306	.00844	.01399	.01975	.02571	.03190
17	4	7	.01296	.01789	.02291	.02803	.03327	.03863	.04413
17	4	8	.02564	.03008	.03456	.03908	.04365	.04827	.05297
17	4	9	.03655	.04037	.04417	.04797	.05178	.05559	.05944
17	4	10	.04612	.04920	.05224	.05524	.05823	.06119	.06416
17	4	11	.05466	.05690	.05909	.06124	.06336	.06546	.06755
17	4	12	.06236	.06367	.06493	.06617	.06740	.06862	.06985
17	4	13	.36878	.36147	.35513	.34971	.34518	.34150	.33861
17	5	1	-.35739	-.38853	-.42229	-.45883	-.49830	-.54086	-.58671
17	5	2	-.14220	-.14317	-.14380	-.14404	-.14386	-.14321	-.14203
17	5	3	-.08364	-.08033	-.07658	-.07237	-.06766	-.06243	-.05662
17	5	4	-.04717	-.04221	-.03688	-.03117	-.02505	-.01849	-.01147
17	5	5	-.02101	-.01549	-.00971	-.00364	.00272	.00941	.01643
17	5	6	-.00078	.00473	.01041	.01628	.02234	.02862	.03512
17	5	7	.01563	.02080	.02606	.03143	.03690	.04250	.04824
17	5	8	.02939	.03400	.03864	.04332	.04805	.05284	.05770
17	5	9	.04121	.04511	.04900	.05288	.05677	.06067	.06460
17	5	10	.05156	.05463	.05766	.06067	.06365	.06663	.06962
17	5	11	.06074	.06288	.06498	.06704	.06909	.07113	.07317
17	5	12	.45366	.44756	.44249	.43842	.43534	.43319	.43194
17	6	1	-.39086	-.42420	-.46034	-.49943	-.54163	-.58712	-.63610
17	6	2	-.15416	-.15498	-.15544	-.15551	-.15514	-.15428	-.15288
17	6	3	-.08982	-.08611	-.08195	-.07732	-.07217	-.06648	-.06019
17	6	4	-.04977	-.04436	-.03857	-.03240	-.02580	-.01876	-.01124
17	6	5	-.02106	-.01512	-.00890	-.00241	.00439	.01151	.01899
17	6	6	.00112	.00700	.01304	.01927	.02570	.03234	.03921
17	6	7	.01909	.02454	.03009	.03573	.04148	.04736	.05337
17	6	8	.03413	.03894	.04377	.04863	.05354	.05851	.06356
17	6	9	.04702	.05102	.05499	.05896	.06294	.06693	.07096
17	6	10	.05827	.06132	.06434	.06733	.07031	.07330	.07629
17	6	11	.54602	.54194	.53899	.53714	.53639	.53669	.53803
17	7	1	-.43134	-.46739	-.50645	-.54866	-.59422	-.64331	-.69614
17	7	2	-.16835	-.16900	-.16928	-.16914	-.16854	-.16743	-.16575
17	7	3	-.09697	-.09278	-.08813	-.08299	-.07732	-.07107	-.06421
17	7	4	-.05256	-.04661	-.04028	-.03355	-.02638	-.01876	-.01063
17	7	5	-.02075	-.01430	-.00758	-.00058	.00674	.01439	.02240
17	7	6	.00379	.01009	.01657	.02323	.03008	.03715	.04446
17	7	7	.02364	.02943	.03529	.04126	.04733	.05354	.05988
17	7	8	.04022	.04524	.05028	.05536	.06048	.06566	.07093
17	7	9	.05438	.05847	.06254	.06660	.07067	.07476	.07890
17	7	10	.64794	.64686	.64704	.64847	.65114	.65505	.66016
17	8	1	-.48133	-.52076	-.56345	-.60957	-.65932	-.71291	-.77056
17	8	2	-.18553	-.18596	-.18600	-.18561	-.18472	-.18331	-.18129
17	8	3	-.10537	-.10060	-.09534	-.08957	-.08325	-.07632	-.06875
17	8	4	-.05554	-.04894	-.04194	-.03452	-.02666	-.01831	-.00945
17	8	5	-.01990	-.01284	-.00550	.00214	.01009	.01838	.02705
17	8	6	.00756	.01438	.02138	.02855	.03593	.04353	.05137
17	8	7	.02972	.03590	.04216	.04851	.05498	.06157	.06831
17	8	8	.04818	.05345	.05874	.06406	.06943	.07486	.08038
17	8	9	.76222	.76536	.76995	.77600	.78353	.79251	.80295
17	9	1	-.54472	-.58847	-.63579	-.68691	-.74203	-.80138	-.86520
17	9	2	-.20682	-.20698	-.20671	-.20598	-.20473	-.20292	-.20047
17	9	3	-.11544	-.10992	-.10390	-.09733	-.09017	-.08238	-.07390

Coefficients b_i for the BLUE of σ
Shape parameter k

n	s	i	0.7	0.8	0.9	1.0	1.1	1.2	1.3
17	9	4	-.05870	-.05127	-.04342	-.03514	-.02639	-.01714	-.00735
17	9	5	-.01819	-.01035	-.00223	.00619	.01494	.02404	.03353
17	9	6	.01297	.02044	.02808	.03590	.04393	.05218	.06068
17	9	7	.03805	.04471	.05145	.05828	.06521	.07228	.07951
17	9	8	.89286	.90182	.91252	.92498	.93924	.95531	.97320
18	0	1	-.24266	-.26701	-.29349	-.32219	-.35325	-.38677	-.42290
18	0	2	-.10103	-.10303	-.10474	-.10613	-.10717	-.10780	-.10799
18	0	3	-.06245	-.06092	-.05901	-.05668	-.05391	-.05067	-.04694
18	0	4	-.03839	-.03533	-.03192	-.02815	-.02399	-.01943	-.01445
18	0	5	-.02111	-.01736	-.01333	-.00901	-.00440	.00053	.00578
18	0	6	-.00773	-.00372	.00049	.00490	.00953	.01437	.01944
18	0	7	.00314	.00714	.01127	.01553	.01993	.02446	.02914
18	0	8	.01228	.01610	.01999	.02394	.02797	.03206	.03624
18	0	9	.02016	.02368	.02722	.03076	.03432	.03791	.04152
18	0	10	.02710	.03023	.03332	.03639	.03943	.04246	.04548
18	0	11	.03332	.03597	.03856	.04110	.04358	.04603	.04844
18	0	12	.03898	.04109	.04312	.04507	.04697	.04881	.05061
18	0	13	.04421	.04571	.04712	.04845	.04973	.05095	.05214
18	0	14	.04913	.04993	.05065	.05132	.05195	.05254	.05312
18	0	15	.05385	.05385	.05381	.05376	.05369	.05364	.05359
18	0	16	.05851	.05756	.05664	.05578	.05497	.05423	.05356
18	0	17	.06335	.06119	.05919	.05737	.05571	.05422	.05288
18	0	18	.06933	.06492	.06110	.05778	.05492	.05245	.05033
18	1	1	-.25643	-.28150	-.30877	-.33835	-.37038	-.40498	-.44230
18	1	2	-.10652	-.10840	-.11000	-.11129	-.11222	-.11276	-.11285
18	1	3	-.06570	-.06397	-.06185	-.05933	-.05636	-.05293	-.04899
18	1	4	-.04024	-.03696	-.03334	-.02935	-.02499	-.02021	-.01501
18	1	5	-.02196	-.01800	-.01377	-.00925	-.00444	.00069	.00614
18	1	6	-.00780	-.00361	.00077	.00536	.01016	.01518	.02043
18	1	7	.00370	.00785	.01212	.01653	.02107	.02575	.03058
18	1	8	.01336	.01730	.02129	.02536	.02949	.03370	.03799
18	1	9	.02169	.02529	.02889	.03251	.03615	.03982	.04351
18	1	10	.02903	.03219	.03531	.03842	.04150	.04458	.04765
18	1	11	.03559	.03824	.04082	.04336	.04585	.04830	.05074
18	1	12	.04157	.04362	.04560	.04752	.04939	.05121	.05300
18	1	13	.04708	.04847	.04979	.05105	.05226	.05344	.05459
18	1	14	.05225	.05289	.05348	.05404	.05457	.05508	.05559
18	1	15	.05717	.05696	.05675	.05655	.05636	.05619	.05606
18	1	16	.06195	.06075	.05962	.05858	.05762	.05675	.05597
18	1	17	.13525	.12889	.12327	.11831	.11397	.11019	.10690
18	2	1	-.27237	-.29837	-.32664	-.35731	-.39052	-.42641	-.46512
18	2	2	-.11277	-.11455	-.11605	-.11723	-.11806	-.11850	-.11849
18	2	3	-.06932	-.06738	-.06506	-.06232	-.05914	-.05548	-.05131
18	2	4	-.04223	-.03872	-.03487	-.03065	-.02604	-.02103	-.01558
18	2	5	-.02278	-.01860	-.01415	-.00941	-.00437	.00098	.00666
18	2	6	-.00773	-.00334	.00124	.00603	.01102	.01624	.02169
18	2	7	.00450	.00882	.01325	.01782	.02252	.02736	.03235
18	2	8	.01478	.01884	.02295	.02714	.03139	.03572	.04014
18	2	9	.02363	.02731	.03099	.03469	.03841	.04215	.04594
18	2	10	.03142	.03461	.03777	.04091	.04404	.04716	.05027
18	2	11	.03838	.04101	.04358	.04611	.04860	.05106	.05351
18	2	12	.04471	.04670	.04862	.05049	.05231	.05410	.05587
18	2	13	.05053	.05180	.05301	.05418	.05531	.05642	.05751
18	2	14	.05596	.05643	.05687	.05729	.05770	.05811	.05853
18	2	15	.06107	.06064	.06023	.05984	.05950	.05921	.05896
18	2	16	.20222	.19481	.18823	.18242	.17733	.17289	.16907

Coefficients b_i for the BLUE of σ
Shape parameter k

n	s	i	0.7	0.8	0.9	1.0	1.1	1.2	1.3
18	3	1	-.29067	-.31779	-.34727	-.37926	-.41388	-.45128	-.49164
18	3	2	-.11984	-.12152	-.12293	-.12401	-.12474	-.12506	-.12494
18	3	3	-.07335	-.07119	-.06863	-.06566	-.06224	-.05833	-.05390
18	3	4	-.04437	-.04061	-.03650	-.03201	-.02714	-.02186	-.01613
18	3	5	-.02358	-.01915	-.01445	-.00946	-.00417	.00144	.00738
18	3	6	-.00748	-.00288	.00193	.00693	.01214	.01758	.02325
18	3	7	.00558	.01008	.01470	.01944	.02431	.02933	.03451
18	3	8	.01656	.02075	.02501	.02933	.03371	.03818	.04273
18	3	9	.02601	.02977	.03354	.03733	.04113	.04496	.04884
18	3	10	.03431	.03754	.04074	.04392	.04708	.05024	.05340
18	3	11	.04173	.04434	.04689	.04941	.05189	.05435	.05679
18	3	12	.04845	.05036	.05221	.05401	.05578	.05753	.05926
18	3	13	.05461	.05574	.05683	.05788	.05891	.05994	.06095
18	3	14	.06030	.06057	.06083	.06109	.06137	.06165	.06197
18	3	15	.27174	.26398	.25710	.25107	.24583	.24134	.23754
18	4	1	-.31173	-.34020	-.37113	-.40468	-.44098	-.48019	-.52248
18	4	2	-.12785	-.12945	-.13076	-.13174	-.13236	-.13257	-.13232
18	4	3	-.07785	-.07544	-.07263	-.06939	-.06570	-.06151	-.05679
18	4	4	-.04669	-.04264	-.03823	-.03346	-.02828	-.02269	-.01665
18	4	5	-.02433	-.01962	-.01464	-.00937	-.00380	.00209	.00833
18	4	6	-.00703	-.00218	.00287	.00812	.01357	.01926	.02518
18	4	7	.00700	.01170	.01652	.02146	.02653	.03175	.03713
18	4	8	.01878	.02313	.02753	.03200	.03653	.04114	.04584
18	4	9	.02891	.03277	.03664	.04051	.04441	.04834	.05231
18	4	10	.03781	.04107	.04430	.04752	.05072	.05392	.05713
18	4	11	.04573	.04831	.05084	.05334	.05580	.05825	.06070
18	4	12	.05289	.05470	.05647	.05820	.05990	.06159	.06327
18	4	13	.05941	.06038	.06132	.06224	.06316	.06407	.06500
18	4	14	.34495	.33745	.33090	.32526	.32049	.31654	.31337
18	5	1	-.33617	-.36624	-.39891	-.43432	-.47261	-.51395	-.55854
18	5	2	-.13702	-.13853	-.13974	-.14062	-.14112	-.14120	-.14081
18	5	3	-.08291	-.08022	-.07712	-.07359	-.06959	-.06508	-.06002
18	5	4	-.04919	-.04482	-.04008	-.03497	-.02945	-.02350	-.01710
18	5	5	-.02501	-.01999	-.01469	-.00910	-.00320	.00302	.00959
18	5	6	-.00632	-.00118	.00414	.00967	.01541	.02138	.02759
18	5	7	.00884	.01377	.01882	.02398	.02928	.03473	.04033
18	5	8	.02155	.02607	.03064	.03527	.03997	.04475	.04961
18	5	9	.03247	.03644	.04041	.04438	.04838	.05241	.05649
18	5	10	.04204	.04534	.04861	.05186	.05510	.05834	.06159
18	5	11	.05054	.05308	.05558	.05804	.06049	.06292	.06535
18	5	12	.05819	.05989	.06155	.06318	.06480	.06641	.06802
18	5	13	.42298	.41639	.41080	.40620	.40254	.39979	.39791
18	6	1	-.36483	-.39684	-.43157	-.46920	-.50987	-.55378	-.60111
18	6	2	-.14761	-.14903	-.15013	-.15089	-.15126	-.15120	-.15065
18	6	3	-.08864	-.08563	-.08220	-.07833	-.07397	-.06909	-.06364
18	6	4	-.05191	-.04716	-.04205	-.03654	-.03063	-.02427	-.01744
18	6	5	-.02559	-.02020	-.01453	-.00857	-.00230	.00430	.01126
18	6	6	-.00525	.00021	.00586	.01171	.01777	.02406	.03060
18	6	7	.01123	.01642	.02173	.02715	.03271	.03841	.04427
18	6	8	.02503	.02974	.03450	.03932	.04420	.04916	.05422
18	6	9	.03687	.04095	.04503	.04912	.05323	.05737	.06156
18	6	10	.04722	.05055	.05385	.05713	.06041	.06369	.06699
18	6	11	.05638	.05886	.06131	.06373	.06614	.06854	.07095
18	6	12	.50712	.50212	.49821	.49537	.49358	.49279	.49299
18	7	1	-.39892	-.43325	-.47048	-.51078	-.55434	-.60133	-.65196
18	7	2	-.16000	-.16131	-.16229	-.16292	-.16314	-.16291	-.16218

Coefficients b_i for the BLUE of σ
Shape parameter k

n	s	i	0.7	0.8	0.9	1.0	1.1	1.2	1.3
18	7	3	-.09521	-.09183	-.08801	-.08373	-.07895	-.07363	-.06772
18	7	4	-.05488	-.04969	-.04412	-.03816	-.03177	-.02493	-.01761
18	7	5	-.02600	-.02018	-.01408	-.00768	-.00097	.00608	.01350
18	7	6	-.00370	.00214	.00816	.01439	.02083	.02750	.03443
18	7	7	.01435	.01985	.02546	.03118	.03704	.04304	.04920
18	7	8	.02944	.03438	.03935	.04438	.04948	.05465	.05992
18	7	9	.04236	.04657	.05077	.05498	.05922	.06348	.06780
18	7	10	.05361	.05697	.06030	.06361	.06693	.07025	.07359
18	7	11	.59895	.59635	.59494	.59473	.59568	.59780	.60103
18	8	1	-.44014	-.47730	-.51758	-.56117	-.60824	-.65901	-.71369
18	8	2	-.17473	-.17590	-.17674	-.17721	-.17725	-.17683	-.17588
18	8	3	-.10284	-.09900	-.09472	-.08995	-.08466	-.07881	-.07235
18	8	4	-.05813	-.05241	-.04630	-.03979	-.03283	-.02541	-.01749
18	8	5	-.02614	-.01981	-.01319	-.00627	.00097	.00856	.01652
18	8	6	-.00146	.00482	.01129	.01797	.02486	.03199	.03937
18	8	7	.01848	.02434	.03030	.03638	.04258	.04894	.05546
18	8	8	.03512	.04031	.04553	.05081	.05615	.06157	.06709
18	8	9	.04932	.05367	.05801	.06236	.06673	.07114	.07560
18	8	10	.70052	.70129	.70340	.70687	.71170	.71786	.72536
18	9	1	-.49104	-.53173	-.57580	-.62347	-.67494	-.73042	-.79016
18	9	2	-.19256	-.19357	-.19423	-.19449	-.19431	-.19364	-.19242
18	9	3	-.11184	-.10745	-.10259	-.09722	-.09130	-.08479	-.07765
18	9	4	-.06170	-.05532	-.04855	-.04135	-.03370	-.02556	-.01690
18	9	5	-.02585	-.01889	-.01164	-.00408	.00380	.01205	.02069
18	9	6	.00176	.00859	.01561	.02282	.03026	.03794	.04589
18	9	7	.02403	.03032	.03671	.04321	.04984	.05663	.06358
18	9	8	.04257	.04806	.05358	.05914	.06478	.07049	.07631
18	9	9	.81462	.81999	.82692	.83544	.84557	.85731	.87066
19	0	1	-.23443	-.25839	-.28447	-.31279	-.34346	-.37661	-.41239
19	0	2	-.09942	-.10174	-.10381	-.10560	-.10706	-.10817	-.10888
19	0	3	-.06270	-.06154	-.06001	-.05809	-.05575	-.05296	-.04969
19	0	4	-.03980	-.03709	-.03404	-.03063	-.02685	-.02266	-.01806
19	0	5	-.02336	-.01992	-.01619	-.01218	-.00786	-.00323	.00173
19	0	6	-.01064	-.00688	-.00292	.00126	.00566	.01030	.01517
19	0	7	-.00030	.00350	.00745	.01155	.01580	.02021	.02478
19	0	8	.00837	.01207	.01585	.01971	.02367	.02771	.03186
19	0	9	.01585	.01932	.02281	.02634	.02991	.03352	.03718
19	0	10	.02242	.02557	.02870	.03183	.03496	.03808	.04122
19	0	11	.02829	.03105	.03377	.03644	.03908	.04169	.04429
19	0	12	.03362	.03593	.03817	.04035	.04247	.04456	.04660
19	0	13	.03852	.04032	.04204	.04368	.04527	.04681	.04831
19	0	14	.04309	.04432	.04546	.04654	.04757	.04855	.04950
19	0	15	.04743	.04801	.04852	.04899	.04943	.04984	.05024
19	0	16	.05161	.05145	.05127	.05108	.05090	.05072	.05057
19	0	17	.05577	.05474	.05375	.05283	.05196	.05117	.05045
19	0	18	.06013	.05798	.05600	.05419	.05256	.05108	.04976
19	0	19	.06554	.06132	.05765	.05448	.05174	.04939	.04736
19	1	1	-.24693	-.27155	-.29836	-.32750	-.35907	-.39323	-.43011
19	1	2	-.10451	-.10673	-.10871	-.11042	-.11181	-.11284	-.11348
19	1	3	-.06578	-.06445	-.06275	-.06066	-.05815	-.05519	-.05174
19	1	4	-.04164	-.03873	-.03549	-.03190	-.02792	-.02355	-.01875
19	1	5	-.02430	-.02067	-.01677	-.01257	-.00807	-.00326	.00189
19	1	6	-.01089	-.00697	-.00284	.00150	.00607	.01087	.01591
19	1	7	.00001	.00395	.00805	.01228	.01667	.02121	.02592
19	1	8	.00915	.01296	.01685	.02082	.02489	.02904	.03330
19	1	9	.01703	.02058	.02415	.02776	.03141	.03511	.03885

Coefficients b_i for the BLUE of σ
Shape parameter k

n	s	i	0.7	0.8	0.9	1.0	1.1	1.2	1.3
19	1	10	.02395	.02715	.03033	.03351	.03668	.03986	.04306
19	1	11	.03013	.03291	.03563	.03833	.04099	.04363	.04626
19	1	12	.03574	.03802	.04024	.04241	.04453	.04661	.04867
19	1	13	.04090	.04263	.04429	.04589	.04744	.04896	.05044
19	1	14	.04570	.04682	.04787	.04887	.04983	.05076	.05167
19	1	15	.05023	.05066	.05105	.05141	.05176	.05210	.05243
19	1	16	.05459	.05424	.05389	.05356	.05325	.05298	.05274
19	1	17	.05884	.05758	.05639	.05530	.05430	.05339	.05257
19	1	18	.12778	.12161	.11617	.11138	.10718	.10353	.10036
19	2	1	-.26130	-.28677	-.31450	-.34464	-.37730	-.41264	-.45081
19	2	2	-.11028	-.11242	-.11432	-.11595	-.11726	-.11822	-.11878
19	2	3	-.06922	-.06770	-.06582	-.06354	-.06085	-.05770	-.05406
19	2	4	-.04362	-.04052	-.03707	-.03326	-.02908	-.02449	-.01948
19	2	5	-.02525	-.02142	-.01731	-.01292	-.00822	-.00320	.00215
19	2	6	-.01104	-.00694	-.00263	.00190	.00664	.01162	.01685
19	2	7	.00050	.00461	.00885	.01324	.01778	.02247	.02734
19	2	8	.01019	.01412	.01813	.02222	.02641	.03069	.03507
19	2	9	.01853	.02216	.02583	.02952	.03326	.03704	.04087
19	2	10	.02585	.02910	.03233	.03556	.03879	.04202	.04528
19	2	11	.03239	.03517	.03791	.04062	.04330	.04597	.04862
19	2	12	.03832	.04057	.04276	.04490	.04701	.04908	.05113
19	2	13	.04375	.04541	.04701	.04855	.05005	.05152	.05297
19	2	14	.04880	.04980	.05075	.05166	.05253	.05339	.05424
19	2	15	.05355	.05381	.05405	.05429	.05452	.05476	.05501
19	2	16	.05805	.05749	.05696	.05646	.05602	.05562	.05528
19	2	17	.19078	.18352	.17708	.17139	.16640	.16206	.15831
19	3	1	-.27767	-.30416	-.33300	-.36433	-.39828	-.43501	-.47468
19	3	2	-.11675	-.11883	-.12066	-.12222	-.12345	-.12433	-.12481
19	3	3	-.07303	-.07132	-.06924	-.06676	-.06386	-.06050	-.05664
19	3	4	-.04577	-.04244	-.03876	-.03473	-.03031	-.02548	-.02022
19	3	5	-.02621	-.02216	-.01783	-.01321	-.00828	-.00304	.00255
19	3	6	-.01108	-.00678	-.00227	.00246	.00740	.01258	.01801
19	3	7	.00120	.00547	.00989	.01444	.01915	.02402	.02905
19	3	8	.01150	.01557	.01971	.02394	.02825	.03266	.03718
19	3	9	.02037	.02410	.02786	.03165	.03548	.03935	.04328
19	3	10	.02815	.03145	.03474	.03802	.04130	.04459	.04790
19	3	11	.03509	.03788	.04063	.04335	.04605	.04873	.05140
19	3	12	.04137	.04358	.04574	.04786	.04994	.05199	.05403
19	3	13	.04712	.04869	.05020	.05168	.05312	.05454	.05594
19	3	14	.05244	.05329	.05412	.05491	.05569	.05647	.05724
19	3	15	.05739	.05747	.05754	.05762	.05772	.05784	.05799
19	3	16	.25589	.24818	.24134	.23533	.23009	.22559	.22177
19	4	1	-.29635	-.32406	-.35421	-.38695	-.42241	-.46077	-.50219
19	4	2	-.12405	-.12607	-.12784	-.12932	-.13048	-.13128	-.13166
19	4	3	-.07725	-.07534	-.07305	-.07035	-.06722	-.06362	-.05952
19	4	4	-.04809	-.04451	-.04058	-.03629	-.03161	-.02652	-.02099
19	4	5	-.02717	-.02287	-.01829	-.01342	-.00825	-.00274	.00310
19	4	6	-.01099	-.00646	-.00173	.00322	.00839	.01380	.01946
19	4	7	.00213	.00660	.01121	.01595	.02084	.02589	.03112
19	4	8	.01314	.01736	.02165	.02603	.03048	.03504	.03970
19	4	9	.02261	.02645	.03031	.03420	.03813	.04210	.04614
19	4	10	.03091	.03427	.03761	.04095	.04429	.04764	.05101
19	4	11	.03831	.04110	.04386	.04659	.04929	.05199	.05469
19	4	12	.04498	.04714	.04926	.05133	.05338	.05541	.05743
19	4	13	.05107	.05253	.05396	.05535	.05671	.05806	.05941
19	4	14	.05667	.05737	.05805	.05871	.05938	.06005	.06073

Coefficients b_i for the BLUE of σ
Shape parameter k

n	s	i	0.7	0.8	0.9	1.0	1.1	1.2	1.3
19	4	15	.32410	.31649	.30981	.30402	.29908	.29494	.29157
19	5	1	-.31782	-.34696	-.37865	-.41305	-.45029	-.49056	-.53404
19	5	2	-.13233	-.13429	-.13599	-.13740	-.13849	-.13920	-.13949
19	5	3	-.08198	-.07983	-.07730	-.07436	-.07097	-.06711	-.06272
19	5	4	-.05061	-.04675	-.04254	-.03796	-.03299	-.02759	-.02174
19	5	5	-.02812	-.02354	-.01868	-.01353	-.00807	-.00228	.00387
19	5	6	-.01073	-.00595	-.00096	.00424	.00966	.01532	.02124
19	5	7	.00337	.00805	.01287	.01782	.02292	.02819	.03363
19	5	8	.01519	.01957	.02403	.02857	.03319	.03790	.04273
19	5	9	.02534	.02929	.03327	.03727	.04131	.04540	.04955
19	5	10	.03423	.03765	.04105	.04445	.04785	.05126	.05469
19	5	11	.04214	.04494	.04770	.05043	.05315	.05586	.05857
19	5	12	.04926	.05136	.05342	.05545	.05746	.05945	.06144
19	5	13	.05572	.05706	.05837	.05966	.06094	.06221	.06349
19	5	14	.39634	.38939	.38343	.37842	.37434	.37115	.36879
19	6	1	-.34270	-.37354	-.40705	-.44341	-.48276	-.52528	-.57118
19	6	2	-.14179	-.14369	-.14533	-.14667	-.14767	-.14829	-.14847
19	6	3	-.08730	-.08489	-.08209	-.07887	-.07519	-.07102	-.06631
19	6	4	-.05336	-.04919	-.04465	-.03974	-.03443	-.02868	-.02247
19	6	5	-.02904	-.02414	-.01897	-.01349	-.00770	-.00158	.00491
19	6	6	-.01024	-.00517	.00010	.00559	.01130	.01725	.02347
19	6	7	.00499	.00991	.01497	.02016	.02550	.03101	.03669
19	6	8	.01774	.02232	.02696	.03168	.03648	.04138	.04639
19	6	9	.02869	.03277	.03687	.04100	.04516	.04937	.05364
19	6	10	.03826	.04174	.04520	.04866	.05212	.05560	.05910
19	6	11	.04675	.04954	.05230	.05503	.05775	.06047	.06319
19	6	12	.05436	.05639	.05838	.06035	.06230	.06424	.06620
19	6	13	.47364	.46795	.46332	.45972	.45713	.45552	.45485
19	7	1	-.37187	-.40472	-.44041	-.47909	-.52095	-.56616	-.61493
19	7	2	-.15272	-.15456	-.15613	-.15739	-.15830	-.15881	-.15888
19	7	3	-.09334	-.09064	-.08752	-.08397	-.07996	-.07543	-.07035
19	7	4	-.05638	-.05184	-.04693	-.04163	-.03592	-.02977	-.02314
19	7	5	-.02990	-.02464	-.01909	-.01324	-.00707	-.00056	.00632
19	7	6	-.00945	-.00405	.00156	.00738	.01342	.01971	.02627
19	7	7	.00711	.01231	.01764	.02311	.02873	.03450	.04046
19	7	8	.02096	.02575	.03060	.03553	.04054	.04564	.05086
19	7	9	.03283	.03705	.04129	.04555	.04985	.05420	.05861
19	7	10	.04318	.04673	.05026	.05378	.05730	.06085	.06442
19	7	11	.05234	.05512	.05786	.06058	.06330	.06602	.06875
19	7	12	.55725	.55349	.55088	.54940	.54905	.54979	.55160
19	8	1	-.40654	-.44181	-.48011	-.52160	-.56646	-.61491	-.66715
19	8	2	-.16552	-.16730	-.16878	-.16995	-.17075	-.17114	-.17107
19	8	3	-.10029	-.09723	-.09375	-.08981	-.08539	-.08044	-.07492
19	8	4	-.05971	-.05473	-.04938	-.04363	-.03745	-.03082	-.02370
19	8	5	-.03065	-.02496	-.01898	-.01269	-.00607	.00090	.00824
19	8	6	-.00823	-.00244	.00355	.00976	.01620	.02289	.02984
19	8	7	.00990	.01543	.02108	.02687	.03280	.03890	.04518
19	8	8	.02505	.03009	.03518	.04035	.04559	.05094	.05640
19	8	9	.03801	.04239	.04678	.05120	.05566	.06016	.06473
19	8	10	.04928	.05289	.05649	.06008	.06367	.06729	.07094
19	8	11	.64869	.64766	.64790	.64942	.65220	.65623	.66149
19	9	1	-.44845	-.48668	-.52815	-.57306	-.62161	-.67400	-.73048
19	9	2	-.18074	-.18243	-.18382	-.18487	-.18554	-.18578	-.18555
19	9	3	-.10838	-.10489	-.10096	-.09656	-.09165	-.08620	-.08014
19	9	4	-.06339	-.05790	-.05201	-.04571	-.03897	-.03176	-.02404
19	9	5	-.03121	-.02500	-.01850	-.01168	-.00453	.00298	.01089

Coefficients b_i for the BLUE of σ
Shape parameter k

n	s	i	0.7	0.8	0.9	1.0	1.1	1.2	1.3
19	9	6	-.00641	-.00015	.00630	.01297	.01988	.02704	.03448
19	9	7	.01363	.01954	.02557	.03173	.03805	.04453	.05119
19	9	8	.03034	.03566	.04104	.04648	.05201	.05763	.06337
19	9	9	.04459	.04916	.05373	.05832	.06295	.06763	.07238
19	9	10	.75002	.75270	.75681	.76238	.76943	.77793	.78790
19	10	1	-.50021	-.54210	-.58752	-.63668	-.68981	-.74712	-.80887
19	10	2	-.19920	-.20077	-.20203	-.20294	-.20344	-.20350	-.20305
19	10	3	-.11796	-.11394	-.10946	-.10448	-.09897	-.09288	-.08616
19	10	4	-.06751	-.06137	-.05482	-.04785	-.04042	-.03249	-.02404
19	10	5	-.03145	-.02461	-.01747	-.01000	-.00219	.00599	.01458
19	10	6	-.00369	.00312	.01014	.01737	.02485	.03259	.04061
19	10	7	.01869	.02506	.03154	.03816	.04493	.05187	.05900
19	10	8	.03731	.04297	.04868	.05445	.06031	.06627	.07235
19	10	9	.86403	.87164	.88094	.89197	.90474	.91927	.93557
20	0	1	-.22686	-.25043	-.27613	-.30407	-.33437	-.36716	-.40258
20	0	2	-.09782	-.10042	-.10281	-.10494	-.10679	-.10832	-.10949
20	0	3	-.06277	-.06194	-.06076	-.05921	-.05726	-.05488	-.05204
20	0	4	-.04091	-.03853	-.03580	-.03273	-.02929	-.02545	-.02121
20	0	5	-.02523	-.02207	-.01864	-.01491	-.01087	-.00652	-.00184
20	0	6	-.01309	-.00959	-.00586	-.00190	.00228	.00671	.01139
20	0	7	-.00324	.00037	.00414	.00808	.01217	.01644	.02088
20	0	8	.00502	.00858	.01224	.01600	.01987	.02384	.02793
20	0	9	.01214	.01553	.01896	.02245	.02600	.02960	.03327
20	0	10	.01838	.02152	.02466	.02781	.03097	.03416	.03736
20	0	11	.02395	.02677	.02955	.03232	.03506	.03779	.04052
20	0	12	.02899	.03143	.03382	.03615	.03845	.04071	.04294
20	0	13	.03362	.03563	.03756	.03944	.04126	.04304	.04478
20	0	14	.03791	.03944	.04089	.04227	.04360	.04489	.04613
20	0	15	.04194	.04294	.04386	.04472	.04554	.04632	.04707
20	0	16	.04579	.04619	.04653	.04684	.04712	.04738	.04764
20	0	17	.04954	.04925	.04895	.04865	.04836	.04809	.04785
20	0	18	.05328	.05218	.05114	.05016	.04926	.04843	.04767
20	0	19	.05722	.05509	.05313	.05135	.04973	.04828	.04699
20	0	20	.06215	.05809	.05457	.05153	.04891	.04665	.04472
20	1	1	-.23825	-.26244	-.28883	-.31753	-.34867	-.38239	-.41884
20	1	2	-.10256	-.10507	-.10739	-.10946	-.11125	-.11273	-.11384
20	1	3	-.06570	-.06471	-.06338	-.06168	-.05958	-.05706	-.05407
20	1	4	-.04272	-.04016	-.03727	-.03402	-.03041	-.02641	-.02199
20	1	5	-.02623	-.02291	-.01930	-.01541	-.01121	-.00669	-.00184
20	1	6	-.01348	-.00981	-.00593	-.00183	.00251	.00709	.01192
20	1	7	-.00312	.00063	.00453	.00859	.01282	.01722	.02180
20	1	8	.00557	.00923	.01300	.01687	.02084	.02493	.02913
20	1	9	.01304	.01651	.02003	.02361	.02723	.03092	.03468
20	1	10	.01960	.02279	.02599	.02920	.03242	.03566	.03893
20	1	11	.02545	.02829	.03111	.03390	.03668	.03945	.04221
20	1	12	.03074	.03318	.03556	.03790	.04020	.04248	.04473
20	1	13	.03559	.03757	.03947	.04133	.04313	.04490	.04664
20	1	14	.04009	.04155	.04294	.04428	.04557	.04682	.04804
20	1	15	.04432	.04520	.04604	.04682	.04757	.04830	.04901
20	1	16	.04833	.04859	.04881	.04901	.04920	.04939	.04959
20	1	17	.05221	.05174	.05129	.05086	.05046	.05010	.04978
20	1	18	.05603	.05471	.05349	.05236	.05133	.05039	.04955
20	1	19	.12109	.11511	.10983	.10520	.10115	.09762	.09456
20	2	1	-.25129	-.27626	-.30349	-.33311	-.36526	-.40007	-.43772
20	2	2	-.10789	-.11035	-.11260	-.11462	-.11635	-.11778	-.11884
20	2	3	-.06895	-.06781	-.06632	-.06447	-.06221	-.05951	-.05635

Coefficients b_i for the BLUE of σ
Shape parameter k

n	s	i	0.7	0.8	0.9	1.0	1.1	1.2	1.3
20	2	4	-.04468	-.04194	-.03886	-.03543	-.03164	-.02744	-.02283
20	2	5	-.02727	-.02376	-.01997	-.01589	-.01151	-.00681	-.00177
20	2	6	-.01380	-.00996	-.00591	-.00164	.00286	.00760	.01261
20	2	7	-.00286	.00103	.00508	.00929	.01366	.01820	.02293
20	2	8	.00631	.01010	.01398	.01797	.02206	.02626	.03058
20	2	9	.01419	.01776	.02137	.02503	.02875	.03253	.03637
20	2	10	.02112	.02437	.02762	.03089	.03418	.03748	.04082
20	2	11	.02729	.03015	.03300	.03582	.03863	.04143	.04424
20	2	12	.03286	.03529	.03767	.04001	.04231	.04459	.04686
20	2	13	.03797	.03990	.04177	.04359	.04537	.04712	.04884
20	2	14	.04270	.04407	.04539	.04667	.04790	.04911	.05030
20	2	15	.04712	.04789	.04862	.04931	.04998	.05064	.05129
20	2	16	.05131	.05140	.05148	.05156	.05164	.05174	.05186
20	2	17	.05531	.05464	.05401	.05343	.05290	.05243	.05202
20	2	18	.18057	.17347	.16717	.16161	.15673	.15248	.14881
20	3	1	-.26604	-.29194	-.32019	-.35090	-.38423	-.42032	-.45934
20	3	2	-.11385	-.11627	-.11847	-.12043	-.12212	-.12349	-.12449
20	3	3	-.07255	-.07124	-.06959	-.06756	-.06512	-.06225	-.05890
20	3	4	-.04680	-.04386	-.04058	-.03695	-.03295	-.02855	-.02372
20	3	5	-.02833	-.02463	-.02064	-.01636	-.01178	-.00687	-.00162
20	3	6	-.01405	-.01003	-.00580	-.00134	.00334	.00827	.01346
20	3	7	-.00246	.00160	.00581	.01017	.01470	.01940	.02429
20	3	8	.00726	.01118	.01520	.01931	.02353	.02786	.03231
20	3	9	.01562	.01928	.02299	.02675	.03056	.03444	.03838
20	3	10	.02294	.02626	.02958	.03291	.03626	.03963	.04303
20	3	11	.02947	.03237	.03524	.03809	.04093	.04377	.04661
20	3	12	.03537	.03778	.04015	.04249	.04479	.04708	.04935
20	3	13	.04076	.04264	.04446	.04624	.04799	.04971	.05141
20	3	14	.04574	.04702	.04826	.04945	.05062	.05177	.05291
20	3	15	.05038	.05101	.05161	.05220	.05277	.05334	.05392
20	3	16	.05473	.05464	.05457	.05450	.05446	.05445	.05448
20	3	17	.24183	.23418	.22740	.22144	.21623	.21174	.20792
20	4	1	-.28274	-.30975	-.33918	-.37118	-.40588	-.44345	-.48407
20	4	2	-.12053	-.12290	-.12507	-.12698	-.12862	-.12993	-.13088
20	4	3	-.07652	-.07504	-.07320	-.07099	-.06836	-.06529	-.06173
20	4	4	-.04910	-.04594	-.04244	-.03859	-.03436	-.02972	-.02465
20	4	5	-.02943	-.02550	-.02129	-.01679	-.01198	-.00685	-.00136
20	4	6	-.01422	-.01000	-.00556	-.00090	.00399	.00912	.01452
20	4	7	-.00189	.00235	.00674	.01128	.01598	.02086	.02592
20	4	8	.00846	.01253	.01669	.02094	.02530	.02977	.03436
20	4	9	.01734	.02112	.02494	.02880	.03272	.03670	.04075
20	4	10	.02513	.02852	.03191	.03531	.03872	.04216	.04564
20	4	11	.03206	.03499	.03789	.04077	.04364	.04651	.04939
20	4	12	.03832	.04072	.04307	.04540	.04769	.04998	.05225
20	4	13	.04402	.04584	.04760	.04934	.05104	.05273	.05440
20	4	14	.04927	.05044	.05158	.05269	.05378	.05487	.05595
20	4	15	.05413	.05461	.05507	.05553	.05600	.05647	.05696
20	4	16	.30570	.29802	.29126	.28538	.28034	.27608	.27255
20	5	1	-.30177	-.33007	-.36089	-.39437	-.43068	-.46998	-.51245
20	5	2	-.12804	-.13038	-.13251	-.13438	-.13597	-.13723	-.13811
20	5	3	-.08093	-.07926	-.07723	-.07481	-.07197	-.06867	-.06488
20	5	4	-.05159	-.04819	-.04445	-.04035	-.03586	-.03096	-.02561
20	5	5	-.03055	-.02638	-.02192	-.01717	-.01211	-.00671	-.00097
20	5	6	-.01428	-.00983	-.00517	-.00028	.00484	.01020	.01584
20	5	7	-.00110	.00334	.00792	.01266	.01755	.02263	.02789
20	5	8	.00995	.01418	.01850	.02292	.02743	.03206	.03681

Coefficients b_i for the BLUE of σ
Shape parameter k

n	s	i	0.7	0.8	0.9	1.0	1.1	1.2	1.3
20	5	9	.01944	.02334	.02727	.03125	.03529	.03938	.04355
20	5	10	.02775	.03122	.03468	.03815	.04164	.04515	.04870
20	5	11	.03514	.03809	.04102	.04393	.04683	.04973	.05265
20	5	12	.04179	.04417	.04651	.04881	.05110	.05337	.05565
20	5	13	.04784	.04958	.05128	.05296	.05461	.05625	.05788
20	5	14	.05338	.05443	.05546	.05646	.05746	.05846	.05947
20	5	15	.37297	.36576	.35953	.35423	.34984	.34631	.34359
20	6	1	-.32360	-.35341	-.38585	-.42108	-.45927	-.50058	-.54521
20	6	2	-.13656	-.13887	-.14096	-.14280	-.14433	-.14553	-.14635
20	6	3	-.08588	-.08399	-.08173	-.07908	-.07600	-.07244	-.06839
20	6	4	-.05431	-.05064	-.04663	-.04224	-.03746	-.03226	-.02661
20	6	5	-.03169	-.02724	-.02251	-.01747	-.01212	-.00644	-.00039
20	6	6	-.01421	-.00950	-.00458	.00056	.00594	.01157	.01747
20	6	7	-.00004	.00462	.00942	.01438	.01950	.02479	.03027
20	6	8	.01182	.01624	.02073	.02532	.03001	.03481	.03974
20	6	9	.02200	.02603	.03010	.03420	.03837	.04259	.04689
20	6	10	.03091	.03445	.03799	.04154	.04511	.04870	.05233
20	6	11	.03881	.04179	.04474	.04768	.05061	.05355	.05649
20	6	12	.04591	.04825	.05056	.05285	.05512	.05738	.05964
20	6	13	.05234	.05399	.05561	.05721	.05880	.06038	.06196
20	6	14	.44450	.43828	.43310	.42892	.42573	.42348	.42214
20	7	1	-.34889	-.38047	-.41482	-.45211	-.49250	-.53618	-.58334
20	7	2	-.14630	-.14858	-.15064	-.15243	-.15392	-.15506	-.15581
20	7	3	-.09145	-.08932	-.08681	-.08389	-.08053	-.07669	-.07232
20	7	4	-.05730	-.05332	-.04899	-.04428	-.03916	-.03362	-.02761
20	7	5	-.03284	-.02807	-.02302	-.01766	-.01198	-.00596	.00043
20	7	6	-.01394	-.00894	-.00373	.00171	.00739	.01332	.01952
20	7	7	.00136	.00628	.01133	.01654	.02190	.02745	.03319
20	7	8	.01417	.01878	.02348	.02827	.03316	.03816	.04329
20	7	9	.02514	.02932	.03353	.03778	.04208	.04645	.05089
20	7	10	.03473	.03836	.04199	.04562	.04928	.05296	.05667
20	7	11	.04322	.04622	.04920	.05216	.05512	.05809	.06107
20	7	12	.05082	.05313	.05540	.05765	.05989	.06213	.06438
20	7	13	.52126	.51661	.51307	.51063	.50926	.50895	.50964
20	8	1	-.37853	-.41221	-.44882	-.48854	-.53155	-.57803	-.62822
20	8	2	-.15756	-.15982	-.16184	-.16358	-.16501	-.16608	-.16675
20	8	3	-.09779	-.09538	-.09257	-.08935	-.08566	-.08148	-.07676
20	8	4	-.06060	-.05626	-.05156	-.04647	-.04097	-.03502	-.02860
20	8	5	-.03397	-.02884	-.02341	-.01768	-.01162	-.00521	.00158
20	8	6	-.01341	-.00807	-.00252	.00326	.00928	.01556	.02211
20	8	7	.00323	.00843	.01377	.01927	.02492	.03076	.03679
20	8	8	.01713	.02198	.02691	.03192	.03703	.04226	.04762
20	8	9	.02904	.03338	.03775	.04216	.04662	.05115	.05575
20	8	10	.03941	.04314	.04686	.05059	.05433	.05810	.06192
20	8	11	.04858	.05160	.05460	.05759	.06057	.06356	.06658
20	8	12	.60446	.60204	.60084	.60084	.60205	.60443	.60797
20	9	1	-.41375	-.44994	-.48927	-.53191	-.57805	-.62791	-.68171
20	9	2	-.17075	-.17297	-.17495	-.17664	-.17800	-.17899	-.17957
20	9	3	-.10510	-.10235	-.09920	-.09560	-.09153	-.08695	-.08181
20	9	4	-.06427	-.05951	-.05437	-.04883	-.04287	-.03644	-.02953
20	9	5	-.03505	-.02948	-.02362	-.01745	-.01094	-.00407	.00319
20	9	6	-.01251	-.00677	-.00083	.00535	.01177	.01846	.02543
20	9	7	.00571	.01125	.01693	.02276	.02876	.03493	.04131
20	9	8	.02092	.02604	.03123	.03650	.04188	.04737	.05299
20	9	9	.03392	.03845	.04301	.04760	.05224	.05695	.06174
20	9	10	.04522	.04905	.05287	.05670	.06054	.06442	.06834

Coefficients b_i for the BLUE of σ
Shape parameter k

n	s	i	0.7	0.8	0.9	1.0	1.1	1.2	1.3
20	9	11	.69564	.69624	.69820	.70152	.70620	.71224	.71963
20	10	1	-.45632	-.49558	-.53820	-.58440	-.63437	-.68834	-.74655
20	10	2	-.18644	-.18863	-.19055	-.19217	-.19345	-.19434	-.19480
20	10	3	-.11363	-.11048	-.10690	-.10286	-.09833	-.09326	-.08760
20	10	4	-.06837	-.06310	-.05744	-.05136	-.04484	-.03784	-.03034
20	10	5	-.03601	-.02993	-.02355	-.01684	-.00979	-.00236	.00547
20	10	6	-.01108	-.00487	.00156	.00821	.01511	.02229	.02976
20	10	7	.00906	.01500	.02108	.02731	.03370	.04028	.04706
20	10	8	.02584	.03127	.03677	.04235	.04803	.05383	.05976
20	10	9	.04015	.04489	.04966	.05446	.05931	.06423	.06924
20	10	10	.79681	.80142	.80758	.81530	.82461	.83552	.84801
21	0	1	-.21985	-.24306	-.26839	-.29597	-.32590	-.35833	-.39339
21	0	2	-.09624	-.09909	-.10175	-.10419	-.10639	-.10830	-.10988
21	0	3	-.06270	-.06217	-.06130	-.06009	-.05849	-.05648	-.05404
21	0	4	-.04179	-.03970	-.03727	-.03451	-.03138	-.02788	-.02396
21	0	5	-.02679	-.02390	-.02073	-.01726	-.01350	-.00941	-.00499
21	0	6	-.01518	-.01190	-.00840	-.00466	-.00069	.00353	.00801
21	0	7	-.00577	-.00234	.00125	.00502	.00896	.01309	.01740
21	0	8	.00213	.00555	.00908	.01272	.01649	.02038	.02440
21	0	9	.00892	.01222	.01558	.01901	.02251	.02609	.02974
21	0	10	.01487	.01797	.02109	.02424	.02741	.03062	.03387
21	0	11	.02018	.02301	.02583	.02865	.03146	.03427	.03708
21	0	12	.02497	.02749	.02997	.03241	.03483	.03722	.03959
21	0	13	.02935	.03151	.03360	.03565	.03765	.03961	.04154
21	0	14	.03341	.03515	.03683	.03845	.04001	.04153	.04301
21	0	15	.03720	.03850	.03972	.04088	.04199	.04306	.04410
21	0	16	.04079	.04159	.04232	.04300	.04364	.04425	.04484
21	0	17	.04423	.04447	.04467	.04484	.04499	.04513	.04527
21	0	18	.04761	.04721	.04681	.04642	.04605	.04570	.04539
21	0	19	.05099	.04984	.04876	.04774	.04681	.04595	.04518
21	0	20	.05458	.05247	.05053	.04878	.04719	.04577	.04450
21	0	21	.05909	.05518	.05180	.04888	.04636	.04420	.04235
21	1	1	-.23030	-.25407	-.28004	-.30832	-.33905	-.37235	-.40838
21	1	2	-.10065	-.10343	-.10604	-.10844	-.11059	-.11246	-.11401
21	1	3	-.06548	-.06481	-.06381	-.06247	-.06074	-.05861	-.05603
21	1	4	-.04356	-.04131	-.03873	-.03582	-.03254	-.02888	-.02481
21	1	5	-.02783	-.02478	-.02146	-.01785	-.01393	-.00969	-.00512
21	1	6	-.01566	-.01224	-.00859	-.00472	-.00060	.00376	.00838
21	1	7	-.00579	-.00224	.00148	.00537	.00944	.01368	.01812
21	1	8	.00249	.00601	.00965	.01340	.01727	.02126	.02539
21	1	9	.00960	.01299	.01643	.01995	.02353	.02719	.03093
21	1	10	.01584	.01900	.02218	.02539	.02863	.03190	.03521
21	1	11	.02140	.02427	.02713	.02998	.03283	.03569	.03855
21	1	12	.02642	.02895	.03144	.03390	.03633	.03875	.04115
21	1	13	.03101	.03314	.03523	.03726	.03926	.04123	.04317
21	1	14	.03525	.03695	.03859	.04018	.04172	.04322	.04469
21	1	15	.03921	.04043	.04160	.04271	.04377	.04481	.04582
21	1	16	.04296	.04365	.04429	.04490	.04548	.04603	.04658
21	1	17	.04654	.04665	.04673	.04679	.04686	.04693	.04701
21	1	18	.05003	.04946	.04892	.04840	.04793	.04750	.04712
21	1	19	.05347	.05212	.05087	.04971	.04866	.04771	.04685
21	1	20	.11507	.10927	.10416	.09967	.09574	.09233	.08938
21	2	1	-.24218	-.26668	-.29344	-.32257	-.35423	-.38854	-.42569
21	2	2	-.10561	-.10834	-.11091	-.11326	-.11538	-.11721	-.11872
21	2	3	-.06857	-.06776	-.06663	-.06514	-.06328	-.06100	-.05828
21	2	4	-.04548	-.04306	-.04032	-.03724	-.03380	-.02997	-.02573

Coefficients b_i for the BLUE of σ
Shape parameter k

n	s	i	0.7	0.8	0.9	1.0	1.1	1.2	1.3
21	2	5	-.02892	-.02571	-.02222	-.01844	-.01435	-.00995	-.00521
21	2	6	-.01611	-.01253	-.00873	-.00470	-.00043	.00408	.00887
21	2	7	-.00572	-.00202	.00183	.00586	.01006	.01444	.01902
21	2	8	.00300	.00664	.01039	.01425	.01824	.02234	.02659
21	2	9	.01048	.01396	.01750	.02110	.02478	.02852	.03236
21	2	10	.01705	.02027	.02352	.02679	.03010	.03344	.03682
21	2	11	.02289	.02580	.02870	.03159	.03448	.03738	.04029
21	2	12	.02817	.03071	.03321	.03568	.03813	.04057	.04300
21	2	13	.03299	.03511	.03717	.03920	.04119	.04315	.04509
21	2	14	.03744	.03909	.04069	.04223	.04374	.04522	.04667
21	2	15	.04159	.04273	.04382	.04486	.04587	.04686	.04783
21	2	16	.04550	.04608	.04662	.04713	.04763	.04812	.04861
21	2	17	.04923	.04918	.04913	.04908	.04904	.04903	.04903
21	2	18	.05281	.05206	.05135	.05070	.05011	.04957	.04910
21	2	19	.17142	.16448	.15831	.15287	.14811	.14395	.14036
21	3	1	-.25555	-.28091	-.30859	-.33873	-.37148	-.40697	-.44539
21	3	2	-.11111	-.11382	-.11635	-.11868	-.12076	-.12255	-.12403
21	3	3	-.07196	-.07101	-.06973	-.06810	-.06609	-.06366	-.06078
21	3	4	-.04756	-.04496	-.04205	-.03879	-.03516	-.03115	-.02672
21	3	5	-.03005	-.02666	-.02299	-.01903	-.01476	-.01018	-.00524
21	3	6	-.01651	-.01277	-.00880	-.00460	-.00016	.00452	.00948
21	3	7	-.00554	-.00169	.00231	.00649	.01083	.01536	.02009
21	3	8	.00367	.00744	.01131	.01530	.01941	.02364	.02801
21	3	9	.01157	.01515	.01879	.02249	.02626	.03011	.03404
21	3	10	.01850	.02180	.02512	.02846	.03183	.03524	.03870
21	3	11	.02467	.02762	.03056	.03349	.03643	.03937	.04232
21	3	12	.03023	.03278	.03529	.03778	.04024	.04270	.04514
21	3	13	.03531	.03740	.03945	.04145	.04343	.04538	.04732
21	3	14	.03999	.04158	.04312	.04462	.04609	.04753	.04896
21	3	15	.04435	.04539	.04639	.04736	.04830	.04923	.05015
21	3	16	.04843	.04888	.04930	.04971	.05011	.05052	.05094
21	3	17	.05229	.05208	.05187	.05169	.05154	.05142	.05134
21	3	18	.22926	.22170	.21500	.20909	.20393	.19948	.19566
21	4	1	-.27059	-.29696	-.32573	-.35704	-.39104	-.42789	-.46776
21	4	2	-.11725	-.11993	-.12244	-.12474	-.12679	-.12855	-.13000
21	4	3	-.07569	-.07460	-.07317	-.07138	-.06920	-.06660	-.06354
21	4	4	-.04980	-.04702	-.04391	-.04046	-.03663	-.03241	-.02777
21	4	5	-.03123	-.02764	-.02378	-.01962	-.01515	-.01036	-.00521
21	4	6	-.01687	-.01294	-.00879	-.00440	.00022	.00510	.01024
21	4	7	-.00523	-.00122	.00295	.00728	.01179	.01649	.02138
21	4	8	.00452	.00843	.01245	.01657	.02081	.02518	.02968
21	4	9	.01290	.01659	.02033	.02414	.02802	.03197	.03601
21	4	10	.02024	.02362	.02701	.03043	.03388	.03736	.04089
21	4	11	.02677	.02976	.03274	.03572	.03870	.04168	.04469
21	4	12	.03265	.03521	.03773	.04022	.04270	.04517	.04763
21	4	13	.03801	.04008	.04209	.04408	.04603	.04797	.04990
21	4	14	.04294	.04447	.04595	.04739	.04881	.05021	.05160
21	4	15	.04752	.04845	.04936	.05024	.05110	.05196	.05282
21	4	16	.05179	.05208	.05237	.05266	.05296	.05327	.05361
21	4	17	.28932	.28163	.27483	.26890	.26379	.25945	.25583
21	5	1	-.28759	-.31513	-.34515	-.37782	-.41327	-.45169	-.49325
21	5	2	-.12411	-.12678	-.12927	-.13154	-.13357	-.13530	-.13671
21	5	3	-.07982	-.07857	-.07697	-.07501	-.07265	-.06986	-.06660
21	5	4	-.05224	-.04925	-.04593	-.04226	-.03821	-.03376	-.02888
21	5	5	-.03246	-.02866	-.02457	-.02019	-.01550	-.01047	-.00510
21	5	6	-.01717	-.01303	-.00867	-.00408	.00075	.00583	.01119

Coefficients b_i for the BLUE of σ
Shape parameter k

n	s	i	0.7	0.8	0.9	1.0	1.1	1.2	1.3
21	5	7	-.00478	-.00059	.00376	.00828	.01297	.01785	.02293
21	5	8	.00560	.00967	.01383	.01811	.02250	.02702	.03167
21	5	9	.01452	.01832	.02219	.02611	.03010	.03417	.03833
21	5	10	.02232	.02578	.02925	.03275	.03628	.03984	.04346
21	5	11	.02925	.03229	.03531	.03833	.04136	.04439	.04744
21	5	12	.03549	.03805	.04057	.04308	.04556	.04804	.05052
21	5	13	.04116	.04319	.04517	.04713	.04906	.05097	.05289
21	5	14	.04637	.04781	.04922	.05060	.05195	.05330	.05465
21	5	15	.05118	.05199	.05278	.05356	.05433	.05510	.05589
21	5	16	.35229	.34490	.33847	.33296	.32834	.32456	.32157
21	6	1	-.30693	-.33582	-.36730	-.40153	-.43867	-.47890	-.52241
21	6	2	-.13182	-.13449	-.13696	-.13922	-.14122	-.14293	-.14430
21	6	3	-.08442	-.08299	-.08121	-.07905	-.07649	-.07349	-.07001
21	6	4	-.05490	-.05168	-.04812	-.04420	-.03990	-.03519	-.03004
21	6	5	-.03374	-.02969	-.02536	-.02073	-.01579	-.01050	-.00486
21	6	6	-.01739	-.01302	-.00843	-.00361	.00146	.00678	.01238
21	6	7	-.00414	.00026	.00481	.00953	.01443	.01951	.02479
21	6	8	.00696	.01119	.01553	.01997	.02453	.02921	.03403
21	6	9	.01648	.02042	.02441	.02846	.03258	.03678	.04106
21	6	10	.02480	.02835	.03192	.03550	.03912	.04277	.04647
21	6	11	.03219	.03527	.03835	.04141	.04448	.04756	.05066
21	6	12	.03883	.04139	.04392	.04642	.04892	.05140	.05390
21	6	13	.04485	.04683	.04877	.05069	.05259	.05448	.05636
21	6	14	.05036	.05171	.05303	.05432	.05561	.05689	.05818
21	6	15	.41887	.41226	.40665	.40204	.39838	.39564	.39377
21	7	1	-.32910	-.35957	-.39275	-.42880	-.46791	-.51024	-.55601
21	7	2	-.14057	-.14323	-.14570	-.14794	-.14992	-.15160	-.15294
21	7	3	-.08957	-.08794	-.08595	-.08358	-.08079	-.07755	-.07381
21	7	4	-.05782	-.05433	-.05050	-.04631	-.04172	-.03672	-.03127
21	7	5	-.03506	-.03074	-.02613	-.02122	-.01599	-.01042	-.00447
21	7	6	-.01749	-.01286	-.00801	-.00293	.00240	.00799	.01386
21	7	7	-.00326	.00137	.00616	.01110	.01623	.02154	.02705
21	7	8	.00865	.01308	.01761	.02224	.02698	.03185	.03687
21	7	9	.01887	.02296	.02709	.03129	.03555	.03989	.04432
21	7	10	.02779	.03144	.03510	.03878	.04249	.04624	.05005
21	7	11	.03570	.03883	.04195	.04506	.04818	.05131	.05447
21	7	12	.04279	.04534	.04787	.05037	.05287	.05536	.05787
21	7	13	.04920	.05112	.05301	.05488	.05673	.05858	.06044
21	7	14	.48986	.48452	.48026	.47706	.47490	.47376	.47358
21	8	1	-.35477	-.38708	-.42225	-.46045	-.50185	-.54666	-.59507
21	8	2	-.15057	-.15324	-.15570	-.15793	-.15989	-.16154	-.16284
21	8	3	-.09538	-.09352	-.09130	-.08868	-.08563	-.08211	-.07809
21	8	4	-.06103	-.05725	-.05311	-.04860	-.04368	-.03834	-.03254
21	8	5	-.03643	-.03179	-.02687	-.02163	-.01607	-.01016	-.00387
21	8	6	-.01743	-.01251	-.00736	-.00198	.00365	.00955	.01573
21	8	7	-.00206	.00283	.00787	.01308	.01847	.02404	.02982
21	8	8	.01080	.01544	.02018	.02502	.02998	.03506	.04029
21	8	9	.02181	.02606	.03036	.03472	.03914	.04364	.04822
21	8	10	.03142	.03518	.03894	.04273	.04655	.05040	.05431
21	8	11	.03992	.04310	.04627	.04943	.05260	.05578	.05899
21	8	12	.04752	.05006	.05258	.05508	.05757	.06006	.06257
21	8	13	.56622	.56271	.56038	.55920	.55917	.56027	.56246
21	9	1	-.38485	-.41934	-.45685	-.49758	-.54171	-.58944	-.64099
21	9	2	-.16215	-.16481	-.16727	-.16949	-.17143	-.17304	-.17430
21	9	3	-.10201	-.09989	-.09739	-.09447	-.09112	-.08728	-.08292
21	9	4	-.06460	-.06047	-.05597	-.05109	-.04579	-.04005	-.03384

Coefficients b_i for the BLUE of σ
Shape parameter k

n	s	i	0.7	0.8	0.9	1.0	1.1	1.2	1.3
21	9	5	-.03782	-.03282	-.02753	-.02191	-.01597	-.00966	-.00297
21	9	6	-.01716	-.01189	-.00640	-.00067	.00531	.01156	.01811
21	9	7	-.00044	.00475	.01009	.01560	.02128	.02716	.03325
21	9	8	.01352	.01841	.02339	.02848	.03368	.03901	.04448
21	9	9	.02546	.02990	.03439	.03892	.04352	.04820	.05297
21	9	10	.03587	.03975	.04363	.04754	.05147	.05544	.05947
21	9	11	.04505	.04829	.05150	.05472	.05793	.06117	.06444
21	9	12	.64913	.64812	.64839	.64996	.65281	.65693	.66230
21	10	1	-.42059	-.45768	-.49801	-.54176	-.58915	-.64039	-.69570
21	10	2	-.17571	-.17838	-.18083	-.18303	-.18494	-.18652	-.18772
21	10	3	-.10966	-.10722	-.10439	-.10113	-.09741	-.09319	-.08842
21	10	4	-.06860	-.06405	-.05912	-.05380	-.04805	-.04184	-.03514
21	10	5	-.03922	-.03379	-.02805	-.02200	-.01560	-.00883	-.00166
21	10	6	-.01656	-.01089	-.00499	.00114	.00754	.01421	.02118
21	10	7	.00174	.00728	.01298	.01884	.02487	.03111	.03755
21	10	8	.01702	.02219	.02746	.03283	.03831	.04392	.04968
21	10	9	.03006	.03472	.03941	.04415	.04896	.05384	.05882
21	10	10	.04140	.04541	.04943	.05346	.05752	.06163	.06579
21	10	11	.74011	.74239	.74612	.75130	.75795	.76606	.77563
21	11	1	-.46379	-.50404	-.54779	-.59522	-.64658	-.70208	-.76198
21	11	2	-.19186	-.19452	-.19696	-.19914	-.20101	-.20254	-.20367
21	11	3	-.11861	-.11579	-.11256	-.10887	-.10471	-.10002	-.09477
21	11	4	-.07310	-.06805	-.06261	-.05675	-.05045	-.04368	-.03639
21	11	5	-.04057	-.03462	-.02836	-.02177	-.01482	-.00750	.00024
21	11	6	-.01550	-.00935	-.00297	.00366	.01054	.01771	.02520
21	11	7	.00472	.01068	.01680	.02307	.02953	.03618	.04306
21	11	8	.02158	.02709	.03269	.03839	.04421	.05015	.05624
21	11	9	.03595	.04084	.04578	.05076	.05581	.06093	.06616
21	11	10	.84119	.84775	.85597	.86588	.87749	.89084	.90591
22	0	1	-.21336	-.23621	-.26118	-.28840	-.31797	-.35005	-.38476
22	0	2	-.09469	-.09775	-.10066	-.10338	-.10589	-.10814	-.11009
22	0	3	-.06252	-.06226	-.06168	-.06077	-.05950	-.05784	-.05575
22	0	4	-.04248	-.04065	-.03850	-.03602	-.03319	-.02999	-.02639
22	0	5	-.02809	-.02545	-.02253	-.01931	-.01580	-.01197	-.00780
22	0	6	-.01697	-.01391	-.01062	-.00709	-.00332	.00071	.00499
22	0	7	-.00794	-.00470	-.00128	.00232	.00611	.01009	.01426
22	0	8	-.00038	.00289	.00629	.00982	.01348	.01727	.02121
22	0	9	.00612	.00931	.01259	.01595	.01939	.02292	.02654
22	0	10	.01181	.01485	.01793	.02105	.02422	.02743	.03069
22	0	11	.01688	.01970	.02253	.02537	.02821	.03107	.03395
22	0	12	.02145	.02401	.02654	.02906	.03156	.03404	.03652
22	0	13	.02562	.02787	.03008	.03224	.03437	.03647	.03855
22	0	14	.02947	.03137	.03322	.03501	.03675	.03845	.04012
22	0	15	.03305	.03458	.03603	.03742	.03876	.04006	.04132
22	0	16	.03643	.03753	.03856	.03953	.04045	.04134	.04220
22	0	17	.03965	.04028	.04085	.04138	.04187	.04234	.04279
22	0	18	.04275	.04286	.04294	.04299	.04303	.04307	.04311
22	0	19	.04581	.04532	.04484	.04438	.04394	.04353	.04316
22	0	20	.04889	.04770	.04658	.04554	.04458	.04371	.04292
22	0	21	.05217	.05008	.04818	.04645	.04489	.04349	.04225
22	0	22	.05632	.05255	.04929	.04648	.04406	.04199	.04022
22	1	1	-.22297	-.24635	-.27192	-.29980	-.33012	-.36301	-.39863
22	1	2	-.09881	-.10182	-.10469	-.10737	-.10985	-.11207	-.11401
22	1	3	-.06517	-.06478	-.06408	-.06306	-.06168	-.05990	-.05771
22	1	4	-.04420	-.04223	-.03994	-.03733	-.03436	-.03102	-.02728
22	1	5	-.02915	-.02637	-.02331	-.01995	-.01630	-.01233	-.00803

Coefficients b_i for the BLUE of σ
Shape parameter k

n	s	i	0.7	0.8	0.9	1.0	1.1	1.2	1.3
22	1	6	-.01752	-.01432	-.01090	-.00724	-.00334	.00081	.00523
22	1	7	-.00808	-.00472	-.00118	.00254	.00644	.01053	.01483
22	1	8	-.00017	.00320	.00671	.01033	.01409	.01799	.02203
22	1	9	.00662	.00990	.01326	.01670	.02023	.02384	.02755
22	1	10	.01257	.01568	.01882	.02201	.02524	.02851	.03185
22	1	11	.01787	.02074	.02361	.02649	.02939	.03229	.03522
22	1	12	.02265	.02523	.02779	.03033	.03285	.03537	.03789
22	1	13	.02701	.02926	.03146	.03363	.03577	.03788	.03998
22	1	14	.03102	.03290	.03473	.03650	.03824	.03994	.04161
22	1	15	.03477	.03624	.03765	.03901	.04032	.04160	.04285
22	1	16	.03829	.03931	.04027	.04119	.04207	.04292	.04375
22	1	17	.04164	.04216	.04265	.04310	.04353	.04395	.04435
22	1	18	.04486	.04484	.04480	.04476	.04472	.04469	.04468
22	1	19	.04801	.04736	.04675	.04617	.04563	.04515	.04471
22	1	20	.05113	.04976	.04849	.04732	.04625	.04529	.04442
22	1	21	.10962	.10399	.09903	.09468	.09088	.08758	.08472
22	2	1	-.23385	-.25791	-.28421	-.31288	-.34407	-.37791	-.41457
22	2	2	-.10342	-.10640	-.10924	-.11190	-.11435	-.11655	-.11847
22	2	3	-.06809	-.06758	-.06677	-.06563	-.06412	-.06222	-.05990
22	2	4	-.04607	-.04395	-.04152	-.03876	-.03564	-.03215	-.02826
22	2	5	-.03027	-.02733	-.02412	-.02062	-.01682	-.01269	-.00823
22	2	6	-.01805	-.01471	-.01115	-.00735	-.00331	.00098	.00555
22	2	7	-.00815	-.00465	-.00099	.00286	.00689	.01111	.01553
22	2	8	.00015	.00364	.00726	.01099	.01486	.01886	.02301
22	2	9	.00729	.01066	.01411	.01764	.02125	.02495	.02875
22	2	10	.01353	.01671	.01992	.02317	.02647	.02982	.03322
22	2	11	.01909	.02200	.02492	.02785	.03079	.03375	.03673
22	2	12	.02410	.02670	.02928	.03185	.03440	.03694	.03949
22	2	13	.02867	.03091	.03312	.03529	.03743	.03956	.04166
22	2	14	.03287	.03473	.03652	.03828	.03999	.04168	.04335
22	2	15	.03679	.03821	.03957	.04088	.04216	.04340	.04463
22	2	16	.04047	.04141	.04230	.04315	.04397	.04477	.04556
22	2	17	.04396	.04437	.04475	.04512	.04547	.04582	.04617
22	2	18	.04730	.04713	.04697	.04681	.04668	.04657	.04648
22	2	19	.05052	.04970	.04893	.04823	.04758	.04700	.04649
22	2	20	.16317	.15637	.15035	.14503	.14037	.13631	.13281
22	3	1	-.24603	-.27088	-.29804	-.32765	-.35984	-.39477	-.43261
22	3	2	-.10853	-.11149	-.11431	-.11695	-.11938	-.12156	-.12346
22	3	3	-.07129	-.07066	-.06973	-.06846	-.06683	-.06480	-.06234
22	3	4	-.04809	-.04582	-.04323	-.04031	-.03703	-.03338	-.02931
22	3	5	-.03144	-.02834	-.02497	-.02131	-.01734	-.01305	-.00841
22	3	6	-.01857	-.01508	-.01136	-.00741	-.00321	.00124	.00596
22	3	7	-.00814	-.00450	-.00070	.00329	.00745	.01181	.01638
22	3	8	.00060	.00422	.00795	.01180	.01579	.01991	.02418
22	3	9	.00811	.01159	.01514	.01876	.02247	.02627	.03016
22	3	10	.01469	.01794	.02123	.02455	.02792	.03134	.03482
22	3	11	.02054	.02350	.02647	.02944	.03243	.03544	.03847
22	3	12	.02580	.02843	.03103	.03362	.03620	.03877	.04135
22	3	13	.03060	.03285	.03505	.03722	.03936	.04149	.04361
22	3	14	.03502	.03684	.03861	.04034	.04203	.04370	.04536
22	3	15	.03913	.04048	.04178	.04304	.04427	.04548	.04668
22	3	16	.04298	.04382	.04462	.04539	.04615	.04689	.04763
22	3	17	.04661	.04689	.04716	.04743	.04769	.04796	.04825
22	3	18	.05006	.04973	.04943	.04915	.04890	.04870	.04854
22	3	19	.21796	.21050	.20388	.19805	.19295	.18853	.18475
22	4	1	-.25966	-.28543	-.31359	-.34426	-.37761	-.41379	-.45298

Coefficients b_i for the BLUE of σ
Shape parameter k

n	s	i	0.7	0.8	0.9	1.0	1.1	1.2	1.3
22	4	2	-.11419	-.11714	-.11995	-.12258	-.12500	-.12717	-.12905
22	4	3	-.07481	-.07405	-.07299	-.07158	-.06981	-.06763	-.06502
22	4	4	-.05027	-.04783	-.04508	-.04198	-.03853	-.03469	-.03044
22	4	5	-.03267	-.02940	-.02585	-.02201	-.01786	-.01338	-.00856
22	4	6	-.01907	-.01541	-.01152	-.00740	-.00303	.00159	.00649
22	4	7	-.00804	-.00425	-.00030	.00384	.00816	.01267	.01739
22	4	8	.00119	.00494	.00880	.01279	.01690	.02115	.02556
22	4	9	.00913	.01271	.01636	.02009	.02391	.02781	.03181
22	4	10	.01607	.01940	.02277	.02617	.02962	.03312	.03668
22	4	11	.02224	.02526	.02828	.03131	.03435	.03741	.04049
22	4	12	.02780	.03045	.03307	.03568	.03829	.04089	.04349
22	4	13	.03285	.03509	.03728	.03945	.04159	.04372	.04584
22	4	14	.03751	.03928	.04101	.04271	.04437	.04602	.04766
22	4	15	.04182	.04309	.04433	.04553	.04671	.04787	.04902
22	4	16	.04584	.04657	.04728	.04797	.04864	.04932	.04999
22	4	17	.04962	.04976	.04990	.05005	.05021	.05040	.05060
22	4	18	.27466	.26697	.26017	.25423	.24908	.24469	.24101
22	5	1	-.27496	-.30179	-.33109	-.36300	-.39768	-.43529	-.47602
22	5	2	-.12048	-.12344	-.12624	-.12887	-.13127	-.13343	-.13530
22	5	3	-.07868	-.07779	-.07658	-.07503	-.07311	-.07077	-.06799
22	5	4	-.05264	-.05002	-.04708	-.04379	-.04014	-.03610	-.03164
22	5	5	-.03397	-.03050	-.02676	-.02272	-.01836	-.01368	-.00865
22	5	6	-.01954	-.01569	-.01161	-.00731	-.00275	.00207	.00716
22	5	7	-.00784	-.00388	.00024	.00454	.00903	.01371	.01860
22	5	8	.00195	.00584	.00985	.01398	.01823	.02263	.02717
22	5	9	.01036	.01406	.01783	.02168	.02560	.02962	.03374
22	5	10	.01771	.02114	.02459	.02808	.03161	.03519	.03884
22	5	11	.02425	.02732	.03040	.03348	.03657	.03969	.04283
22	5	12	.03012	.03280	.03545	.03808	.04071	.04333	.04597
22	5	13	.03547	.03769	.03987	.04203	.04417	.04630	.04842
22	5	14	.04037	.04210	.04379	.04544	.04707	.04869	.05031
22	5	15	.04490	.04610	.04725	.04838	.04950	.05061	.05171
22	5	16	.04912	.04973	.05032	.05091	.05149	.05209	.05269
22	5	17	.33386	.32635	.31978	.31412	.30933	.30535	.30216
22	6	1	-.29223	-.32029	-.35090	-.38423	-.42044	-.45969	-.50218
22	6	2	-.12751	-.13048	-.13329	-.13591	-.13832	-.14047	-.14232
22	6	3	-.08296	-.08193	-.08057	-.07885	-.07676	-.07425	-.07128
22	6	4	-.05522	-.05240	-.04925	-.04575	-.04188	-.03762	-.03293
22	6	5	-.03533	-.03165	-.02769	-.02343	-.01885	-.01394	-.00867
22	6	6	-.01996	-.01591	-.01162	-.00711	-.00234	.00269	.00800
22	6	7	-.00751	-.00337	.00095	.00544	.01011	.01498	.02006
22	6	8	.00291	.00696	.01113	.01542	.01983	.02438	.02909
22	6	9	.01185	.01569	.01958	.02356	.02761	.03175	.03599
22	6	10	.01967	.02319	.02674	.03032	.03395	.03762	.04136
22	6	11	.02661	.02975	.03288	.03602	.03917	.04235	.04555
22	6	12	.03285	.03554	.03822	.04087	.04352	.04618	.04884
22	6	13	.03851	.04071	.04288	.04503	.04716	.04928	.05140
22	6	14	.04369	.04536	.04700	.04860	.05019	.05178	.05336
22	6	15	.04847	.04956	.05062	.05167	.05271	.05375	.05480
22	6	16	.39615	.38924	.38331	.37836	.37434	.37121	.36894
22	7	1	-.31185	-.34132	-.37346	-.40843	-.44639	-.48754	-.53206
22	7	2	-.13542	-.13840	-.14122	-.14385	-.14626	-.14841	-.15025
22	7	3	-.08773	-.08653	-.08500	-.08311	-.08082	-.07811	-.07494
22	7	4	-.05804	-.05499	-.05161	-.04788	-.04377	-.03925	-.03429
22	7	5	-.03676	-.03284	-.02864	-.02413	-.01930	-.01413	-.00860
22	7	6	-.02032	-.01604	-.01152	-.00677	-.00177	.00350	.00905

Coefficients b_i for the BLUE of σ
Shape parameter k

n	s	i	0.7	0.8	0.9	1.0	1.1	1.2	1.3
22	7	7	-.00701	-.00266	.00186	.00656	.01144	.01652	.02182
22	7	8	.00412	.00836	.01270	.01716	.02175	.02648	.03136
22	7	9	.01368	.01765	.02169	.02580	.03000	.03428	.03866
22	7	10	.02202	.02564	.02930	.03298	.03670	.04048	.04432
22	7	11	.02942	.03262	.03581	.03902	.04223	.04547	.04874
22	7	12	.03606	.03877	.04147	.04414	.04682	.04950	.05219
22	7	13	.04207	.04425	.04640	.04853	.05064	.05275	.05486
22	7	14	.04756	.04916	.05073	.05228	.05382	.05535	.05689
22	7	15	.46221	.45632	.45149	.44770	.44492	.44312	.44225
22	8	1	-.33434	-.36545	-.39935	-.43622	-.47623	-.51956	-.56644
22	8	2	-.14438	-.14739	-.15022	-.15287	-.15528	-.15743	-.15927
22	8	3	-.09308	-.09170	-.08997	-.08787	-.08538	-.08244	-.07903
22	8	4	-.06114	-.05784	-.05420	-.05020	-.04580	-.04100	-.03574
22	8	5	-.03826	-.03407	-.02959	-.02480	-.01969	-.01423	-.00840
22	8	6	-.02060	-.01605	-.01128	-.00626	-.00099	.00455	.01038
22	8	7	-.00630	-.00171	.00305	.00798	.01310	.01842	.02395
22	8	8	.00566	.01009	.01463	.01929	.02408	.02901	.03409
22	8	9	.01591	.02004	.02424	.02851	.03285	.03729	.04183
22	8	10	.02485	.02859	.03235	.03615	.03998	.04387	.04782
22	8	11	.03277	.03604	.03930	.04257	.04585	.04915	.05249
22	8	12	.03986	.04260	.04531	.04801	.05070	.05340	.05612
22	8	13	.04628	.04842	.05054	.05264	.05473	.05682	.05892
22	8	14	.53279	.52842	.52519	.52308	.52208	.52215	.52327
22	9	1	-.36037	-.39339	-.42936	-.46844	-.51084	-.55675	-.60638
22	9	2	-.15464	-.15767	-.16053	-.16319	-.16561	-.16776	-.16960
22	9	3	-.09912	-.09753	-.09558	-.09325	-.09051	-.08731	-.08362
22	9	4	-.06458	-.06099	-.05704	-.05273	-.04801	-.04287	-.03727
22	9	5	-.03984	-.03534	-.03054	-.02543	-.01998	-.01419	-.00801
22	9	6	-.02075	-.01590	-.01082	-.00551	.00007	.00592	.01207
22	9	7	-.00530	-.00044	.00458	.00978	.01517	.02076	.02657
22	9	8	.00761	.01227	.01703	.02192	.02693	.03208	.03739
22	9	9	.01866	.02297	.02735	.03179	.03631	.04092	.04564
22	9	10	.02829	.03216	.03605	.03996	.04392	.04793	.05201
22	9	11	.03681	.04015	.04348	.04682	.05017	.05354	.05696
22	9	12	.04442	.04717	.04989	.05261	.05532	.05804	.06079
22	9	13	.60883	.60655	.60550	.60567	.60707	.60967	.61345
22	10	1	-.39087	-.42614	-.46453	-.50624	-.55146	-.60040	-.65329
22	10	2	-.16651	-.16958	-.17246	-.17514	-.17757	-.17972	-.18155
22	10	3	-.10603	-.10419	-.10198	-.09937	-.09634	-.09284	-.08883
22	10	4	-.06841	-.06447	-.06018	-.05550	-.05041	-.04488	-.03888
22	10	5	-.04148	-.03662	-.03145	-.02596	-.02014	-.01394	-.00736
22	10	6	-.02072	-.01552	-.01010	-.00442	.00151	.00772	.01424
22	10	7	-.00393	.00124	.00658	.01209	.01779	.02369	.02982
22	10	8	.01009	.01502	.02004	.02518	.03045	.03586	.04143
22	10	9	.02208	.02660	.03118	.03582	.04054	.04535	.05027
22	10	10	.03252	.03653	.04056	.04461	.04871	.05286	.05707
22	10	11	.04173	.04514	.04855	.05196	.05539	.05884	.06233
22	10	12	.69152	.69198	.69379	.69698	.70154	.70746	.71474
22	11	1	-.42710	-.46506	-.50636	-.55120	-.59979	-.65236	-.70915
22	11	2	-.18043	-.18353	-.18644	-.18914	-.19159	-.19374	-.19555
22	11	3	-.11401	-.11187	-.10935	-.10642	-.10304	-.09918	-.09478
22	11	4	-.07272	-.06837	-.06366	-.05855	-.05301	-.04702	-.04054
22	11	5	-.04318	-.03789	-.03228	-.02635	-.02007	-.01341	-.00635
22	11	6	-.02042	-.01481	-.00898	-.00289	.00346	.01011	.01706
22	11	7	-.00203	.00351	.00920	.01508	.02114	.02742	.03392
22	11	8	.01331	.01854	.02387	.02931	.03487	.04059	.04646

Coefficients b_i for the BLUE of σ
Shape parameter k

n	s	i	0.7	0.8	0.9	1.0	1.1	1.2	1.3
22	11	9	.02641	.03117	.03597	.04084	.04579	.05083	.05598
22	11	10	.03779	.04196	.04614	.05035	.05460	.05891	.06328
22	11	11	.78237	.78636	.79189	.79897	.80763	.81786	.82968
23	0	1	-.20731	-.22982	-.25445	-.28131	-.31054	-.34227	-.37664
23	0	2	-.09317	-.09642	-.09955	-.10252	-.10530	-.10786	-.11016
23	0	3	-.06226	-.06224	-.06192	-.06129	-.06032	-.05897	-.05722
23	0	4	-.04300	-.04142	-.03953	-.03731	-.03476	-.03184	-.02853
23	0	5	-.02918	-.02677	-.02408	-.02110	-.01782	-.01423	-.01031
23	0	6	-.01850	-.01564	-.01255	-.00923	-.00565	-.00182	.00228
23	0	7	-.00983	-.00676	-.00351	-.00007	.00356	.00739	.01143
23	0	8	-.00257	.00056	.00383	.00723	.01078	.01447	.01832
23	0	9	.00366	.00675	.00993	.01321	.01658	.02005	.02363
23	0	10	.00912	.01209	.01512	.01820	.02134	.02453	.02779
23	0	11	.01397	.01677	.01959	.02242	.02528	.02817	.03108
23	0	12	.01835	.02092	.02349	.02604	.02859	.03115	.03371
23	0	13	.02233	.02464	.02692	.02917	.03140	.03360	.03579
23	0	14	.02600	.02801	.02998	.03190	.03378	.03562	.03744
23	0	15	.02941	.03109	.03272	.03428	.03580	.03728	.03873
23	0	16	.03260	.03393	.03518	.03638	.03753	.03864	.03971
23	0	17	.03563	.03656	.03742	.03823	.03899	.03972	.04043
23	0	18	.03854	.03902	.03945	.03985	.04021	.04056	.04089
23	0	19	.04135	.04135	.04131	.04127	.04122	.04117	.04113
23	0	20	.04413	.04357	.04302	.04250	.04200	.04155	.04113
23	0	21	.04695	.04573	.04459	.04353	.04256	.04167	.04087
23	0	22	.04997	.04791	.04603	.04433	.04280	.04143	.04021
23	0	23	.05380	.05016	.04701	.04430	.04198	.03999	.03828
23	1	1	-.21618	-.23919	-.26438	-.29187	-.32180	-.35429	-.38952
23	1	2	-.09703	-.10025	-.10334	-.10629	-.10905	-.11159	-.11389
23	1	3	-.06477	-.06464	-.06422	-.06349	-.06242	-.06098	-.05913
23	1	4	-.04467	-.04296	-.04094	-.03861	-.03593	-.03289	-.02946
23	1	5	-.03025	-.02771	-.02489	-.02179	-.01838	-.01466	-.01061
23	1	6	-.01910	-.01612	-.01291	-.00946	-.00577	-.00182	.00240
23	1	7	-.01006	-.00688	-.00351	.00004	.00378	.00772	.01187
23	1	8	-.00248	.00075	.00411	.00762	.01126	.01505	.01899
23	1	9	.00402	.00720	.01046	.01382	.01727	.02082	.02448
23	1	10	.00972	.01276	.01585	.01899	.02219	.02546	.02879
23	1	11	.01478	.01763	.02049	.02337	.02628	.02922	.03219
23	1	12	.01934	.02194	.02454	.02713	.02971	.03230	.03490
23	1	13	.02349	.02582	.02811	.03037	.03261	.03484	.03705
23	1	14	.02732	.02932	.03128	.03320	.03508	.03693	.03876
23	1	15	.03087	.03252	.03412	.03567	.03717	.03865	.04009
23	1	16	.03420	.03547	.03668	.03784	.03896	.04004	.04110
23	1	17	.03735	.03820	.03899	.03975	.04046	.04116	.04184
23	1	18	.04037	.04075	.04110	.04142	.04172	.04202	.04231
23	1	19	.04328	.04315	.04301	.04288	.04275	.04264	.04255
23	1	20	.04614	.04543	.04475	.04412	.04354	.04301	.04253
23	1	21	.04899	.04760	.04631	.04513	.04406	.04309	.04222
23	1	22	.10468	.09920	.09439	.09017	.08648	.08329	.08052
23	2	1	-.22620	-.24983	-.27570	-.30393	-.33467	-.36805	-.40425
23	2	2	-.10134	-.10453	-.10761	-.11054	-.11328	-.11582	-.11810
23	2	3	-.06755	-.06731	-.06679	-.06595	-.06477	-.06323	-.06127
23	2	4	-.04649	-.04464	-.04250	-.04003	-.03722	-.03404	-.03047
23	2	5	-.03138	-.02870	-.02575	-.02251	-.01896	-.01511	-.01091
23	2	6	-.01970	-.01659	-.01325	-.00967	-.00585	-.00176	.00259
23	2	7	-.01023	-.00693	-.00345	.00022	.00408	.00814	.01241
23	2	8	-.00230	.00104	.00451	.00811	.01186	.01575	.01980

Coefficients b_i for the BLUE of σ
Shape parameter k

n	s	i	0.7	0.8	0.9	1.0	1.1	1.2	1.3
23	2	9	.00451	.00777	.01113	.01457	.01811	.02175	.02550
23	2	10	.01047	.01358	.01674	.01996	.02323	.02656	.02996
23	2	11	.01577	.01867	.02158	.02452	.02748	.03047	.03349
23	2	12	.02054	.02318	.02580	.02842	.03104	.03367	.03630
23	2	13	.02489	.02722	.02952	.03180	.03405	.03630	.03853
23	2	14	.02889	.03088	.03283	.03473	.03661	.03846	.04030
23	2	15	.03260	.03421	.03578	.03730	.03878	.04024	.04167
23	2	16	.03607	.03728	.03844	.03955	.04063	.04168	.04272
23	2	17	.03936	.04012	.04084	.04152	.04219	.04283	.04347
23	2	18	.04249	.04276	.04301	.04325	.04348	.04371	.04395
23	2	19	.04550	.04523	.04498	.04473	.04452	.04433	.04417
23	2	20	.04842	.04755	.04673	.04598	.04529	.04467	.04412
23	2	21	.15568	.14903	.14314	.13795	.13340	.12943	.12600
23	3	1	-.23735	-.26172	-.28839	-.31748	-.34915	-.38355	-.42085
23	3	2	-.10609	-.10928	-.11234	-.11527	-.11801	-.12053	-.12281
23	3	3	-.07058	-.07023	-.06961	-.06866	-.06737	-.06571	-.06364
23	3	4	-.04845	-.04646	-.04418	-.04157	-.03861	-.03529	-.03157
23	3	5	-.03257	-.02975	-.02664	-.02326	-.01957	-.01556	-.01121
23	3	6	-.02030	-.01705	-.01356	-.00984	-.00588	-.00165	.00285
23	3	7	-.01036	-.00692	-.00331	.00049	.00448	.00867	.01307
23	3	8	-.00203	.00143	.00501	.00873	.01259	.01660	.02076
23	3	9	.00513	.00849	.01194	.01548	.01911	.02285	.02669
23	3	10	.01139	.01458	.01781	.02110	.02445	.02785	.03133
23	3	11	.01695	.01990	.02287	.02586	.02887	.03192	.03500
23	3	12	.02196	.02462	.02728	.02993	.03259	.03524	.03792
23	3	13	.02651	.02885	.03117	.03345	.03572	.03798	.04023
23	3	14	.03070	.03268	.03461	.03651	.03838	.04023	.04206
23	3	15	.03459	.03616	.03769	.03918	.04064	.04207	.04349
23	3	16	.03822	.03936	.04046	.04152	.04255	.04356	.04456
23	3	17	.04165	.04231	.04295	.04356	.04415	.04474	.04533
23	3	18	.04490	.04505	.04519	.04533	.04547	.04563	.04581
23	3	19	.04800	.04758	.04719	.04683	.04652	.04624	.04601
23	3	20	.20774	.20039	.19386	.18811	.18307	.17871	.17497
23	4	1	-.24976	-.27498	-.30257	-.33265	-.36539	-.40094	-.43948
23	4	2	-.11133	-.11452	-.11759	-.12051	-.12325	-.12578	-.12806
23	4	3	-.07389	-.07344	-.07270	-.07164	-.07023	-.06844	-.06624
23	4	4	-.05056	-.04843	-.04600	-.04323	-.04012	-.03663	-.03275
23	4	5	-.03383	-.03085	-.02759	-.02404	-.02018	-.01600	-.01148
23	4	6	-.02090	-.01749	-.01385	-.00998	-.00586	-.00147	.00319
23	4	7	-.01042	-.00684	-.00309	.00085	.00498	.00931	.01386
23	4	8	-.00164	.00194	.00565	.00949	.01347	.01760	.02189
23	4	9	.00589	.00936	.01291	.01655	.02029	.02412	.02807
23	4	10	.01248	.01575	.01907	.02244	.02586	.02935	.03291
23	4	11	.01833	.02135	.02437	.02742	.03049	.03359	.03673
23	4	12	.02360	.02630	.02900	.03168	.03437	.03706	.03977
23	4	13	.02839	.03074	.03306	.03536	.03764	.03991	.04218
23	4	14	.03279	.03475	.03667	.03855	.04041	.04225	.04408
23	4	15	.03687	.03840	.03988	.04133	.04276	.04416	.04555
23	4	16	.04067	.04174	.04276	.04376	.04474	.04570	.04666
23	4	17	.04424	.04480	.04534	.04587	.04639	.04691	.04743
23	4	18	.04761	.04763	.04765	.04768	.04773	.04780	.04790
23	4	19	.26145	.25379	.24701	.24107	.23592	.23150	.22778
23	5	1	-.26362	-.28981	-.31844	-.34966	-.38362	-.42048	-.46044
23	5	2	-.11713	-.12033	-.12341	-.12634	-.12908	-.13162	-.13389
23	5	3	-.07752	-.07696	-.07610	-.07491	-.07338	-.07145	-.06911
23	5	4	-.05285	-.05056	-.04796	-.04503	-.04175	-.03808	-.03401

Coefficients b_i for the BLUE of σ
Shape parameter k

n	s	i	0.7	0.8	0.9	1.0	1.1	1.2	1.3
23	5	5	-.03516	-.03200	-.02857	-.02484	-.02081	-.01645	-.01174
23	5	6	-.02149	-.01791	-.01410	-.01006	-.00576	-.00121	.00364
23	5	7	-.01041	-.00668	-.00277	.00133	.00561	.01010	.01481
23	5	8	-.00114	.00258	.00643	.01040	.01452	.01878	.02321
23	5	9	.00682	.01041	.01407	.01782	.02167	.02562	.02968
23	5	10	.01378	.01714	.02055	.02401	.02752	.03109	.03473
23	5	11	.01996	.02304	.02613	.02923	.03237	.03553	.03873
23	5	12	.02552	.02826	.03098	.03370	.03642	.03915	.04190
23	5	13	.03057	.03292	.03525	.03755	.03984	.04213	.04441
23	5	14	.03520	.03713	.03903	.04089	.04273	.04456	.04639
23	5	15	.03948	.04096	.04239	.04379	.04518	.04655	.04791
23	5	16	.04347	.04444	.04539	.04631	.04723	.04813	.04904
23	5	17	.04719	.04763	.04806	.04849	.04892	.04936	.04982
23	5	18	.31732	.30973	.30307	.29730	.29239	.28828	.28493
23	6	1	-.27914	-.30645	-.33628	-.36879	-.40414	-.44250	-.48407
23	6	2	-.12356	-.12679	-.12988	-.13283	-.13559	-.13813	-.14042
23	6	3	-.08152	-.08084	-.07985	-.07853	-.07685	-.07478	-.07227
23	6	4	-.05534	-.05287	-.05010	-.04698	-.04350	-.03964	-.03537
23	6	5	-.03657	-.03322	-.02959	-.02567	-.02144	-.01687	-.01195
23	6	6	-.02206	-.01830	-.01430	-.01007	-.00559	-.00083	.00420
23	6	7	-.01032	-.00641	-.00233	.00194	.00639	.01106	.01594
23	6	8	-.00048	.00339	.00738	.01151	.01577	.02019	.02476
23	6	9	.00796	.01167	.01545	.01933	.02330	.02737	.03155
23	6	10	.01533	.01879	.02229	.02584	.02945	.03311	.03685
23	6	11	.02187	.02502	.02817	.03135	.03454	.03777	.04104
23	6	12	.02775	.03053	.03329	.03604	.03880	.04157	.04436
23	6	13	.03309	.03545	.03778	.04009	.04239	.04468	.04698
23	6	14	.03798	.03988	.04175	.04359	.04541	.04722	.04903
23	6	15	.04248	.04389	.04527	.04662	.04795	.04928	.05060
23	6	16	.04666	.04754	.04839	.04924	.05007	.05091	.05176
23	6	17	.37586	.36872	.36255	.35734	.35304	.34961	.34701
23	7	1	-.29666	-.32524	-.35644	-.39043	-.42737	-.46745	-.51085
23	7	2	-.13076	-.13401	-.13713	-.14010	-.14288	-.14544	-.14773
23	7	3	-.08595	-.08513	-.08400	-.08253	-.08070	-.07846	-.07578
23	7	4	-.05805	-.05539	-.05242	-.04910	-.04541	-.04133	-.03682
23	7	5	-.03806	-.03450	-.03066	-.02652	-.02206	-.01727	-.01211
23	7	6	-.02261	-.01864	-.01444	-.01000	-.00530	-.00033	.00493
23	7	7	-.01010	-.00601	-.00174	.00272	.00737	.01222	.01730
23	7	8	.00036	.00440	.00856	.01285	.01728	.02185	.02660
23	7	9	.00934	.01319	.01711	.02112	.02522	.02943	.03374
23	7	10	.01718	.02075	.02435	.02801	.03171	.03548	.03932
23	7	11	.02413	.02735	.03057	.03382	.03708	.04038	.04372
23	7	12	.03037	.03318	.03598	.03877	.04157	.04437	.04720
23	7	13	.03602	.03838	.04071	.04303	.04533	.04763	.04994
23	7	14	.04119	.04306	.04489	.04670	.04849	.05028	.05208
23	7	15	.04595	.04728	.04858	.04987	.05114	.05242	.05370
23	7	16	.43764	.43133	.42606	.42180	.41852	.41619	.41478
23	8	1	-.31655	-.34659	-.37938	-.41507	-.45384	-.49589	-.54140
23	8	2	-.13885	-.14214	-.14529	-.14829	-.15110	-.15367	-.15598
23	8	3	-.09089	-.08992	-.08863	-.08699	-.08498	-.08255	-.07968
23	8	4	-.06103	-.05815	-.05495	-.05140	-.04747	-.04314	-.03838
23	8	5	-.03964	-.03584	-.03176	-.02737	-.02267	-.01762	-.01220
23	8	6	-.02312	-.01891	-.01448	-.00980	-.00487	.00034	.00585
23	8	7	-.00974	-.00543	-.00095	.00372	.00858	.01365	.01895
23	8	8	.00144	.00566	.01001	.01448	.01909	.02385	.02878
23	8	9	.01103	.01503	.01910	.02326	.02751	.03187	.03633

Coefficients b_i for the BLUE of σ
Shape parameter k

n	s	i	0.7	0.8	0.9	1.0	1.1	1.2	1.3
23	8	10	.01939	.02308	.02680	.03057	.03439	.03827	.04222
23	8	11	.02681	.03010	.03341	.03672	.04007	.04344	.04686
23	8	12	.03345	.03630	.03914	.04196	.04480	.04764	.05051
23	8	13	.03946	.04182	.04415	.04646	.04876	.05107	.05338
23	8	14	.04495	.04676	.04855	.05032	.05208	.05384	.05560
23	8	15	.50328	.49824	.49429	.49144	.48965	.48890	.48916
23	9	1	-.33934	-.37108	-.40569	-.44335	-.48424	-.52856	-.57652
23	9	2	-.14802	-.15136	-.15455	-.15759	-.16042	-.16303	-.16536
23	9	3	-.09642	-.09528	-.09382	-.09199	-.08978	-.08714	-.08404
23	9	4	-.06431	-.06119	-.05773	-.05392	-.04972	-.04511	-.04005
23	9	5	-.04131	-.03725	-.03289	-.02823	-.02324	-.01789	-.01217
23	9	6	-.02356	-.01909	-.01439	-.00945	-.00424	.00124	.00703
23	9	7	-.00919	-.00464	.00008	.00499	.01009	.01541	.02095
23	9	8	.00281	.00725	.01180	.01648	.02129	.02626	.03140
23	9	9	.01310	.01727	.02152	.02584	.03026	.03478	.03941
23	9	10	.02207	.02588	.02973	.03362	.03757	.04157	.04566
23	9	11	.03001	.03339	.03677	.04017	.04360	.04705	.05056
23	9	12	.03711	.04000	.04287	.04574	.04861	.05149	.05441
23	9	13	.04352	.04587	.04819	.05049	.05279	.05510	.05742
23	9	14	.57354	.57024	.56813	.56719	.56743	.56881	.57132
23	10	1	-.36571	-.39942	-.43617	-.47612	-.51949	-.56647	-.61729
23	10	2	-.15852	-.16191	-.16516	-.16823	-.17111	-.17374	-.17610
23	10	3	-.10269	-.10136	-.09968	-.09764	-.09519	-.09231	-.08895
23	10	4	-.06796	-.06455	-.06080	-.05668	-.05217	-.04723	-.04183
23	10	5	-.04309	-.03872	-.03405	-.02906	-.02374	-.01806	-.01200
23	10	6	-.02391	-.01914	-.01414	-.00888	-.00337	.00244	.00854
23	10	7	-.00839	-.00356	.00143	.00662	.01200	.01760	.02343
23	10	8	.00457	.00924	.01403	.01894	.02400	.02920	.03458
23	10	9	.01566	.02003	.02447	.02898	.03359	.03830	.04312
23	10	10	.02533	.02929	.03328	.03731	.04139	.04554	.04976
23	10	11	.03387	.03734	.04081	.04430	.04782	.05136	.05496
23	10	12	.04149	.04442	.04733	.05023	.05314	.05607	.05903
23	10	13	.64935	.64835	.64865	.65024	.65313	.65731	.66276
23	11	1	-.39661	-.43264	-.47189	-.51455	-.56084	-.61095	-.66514
23	11	2	-.17068	-.17413	-.17743	-.18056	-.18347	-.18615	-.18853
23	11	3	-.10986	-.10829	-.10637	-.10407	-.10136	-.09819	-.09453
23	11	4	-.07204	-.06830	-.06421	-.05973	-.05485	-.04953	-.04374
23	11	5	-.04498	-.04024	-.03521	-.02985	-.02414	-.01807	-.01160
23	11	6	-.02411	-.01899	-.01364	-.00803	-.00215	.00402	.01051
23	11	7	-.00724	-.00210	.00322	.00872	.01443	.02035	.02650
23	11	8	.00684	.01178	.01684	.02202	.02735	.03282	.03847
23	11	9	.01887	.02346	.02811	.03284	.03766	.04259	.04763
23	11	10	.02934	.03346	.03761	.04180	.04604	.05035	.05473
23	11	11	.03858	.04215	.04572	.04931	.05292	.05657	.06026
23	11	12	.73188	.73384	.73724	.74209	.74841	.75619	.76543
23	12	1	-.43331	-.47211	-.51436	-.56025	-.61001	-.66388	-.72211
23	12	2	-.18494	-.18846	-.19182	-.19500	-.19797	-.20068	-.20310
23	12	3	-.11816	-.11631	-.11410	-.11149	-.10845	-.10494	-.10091
23	12	4	-.07665	-.07251	-.06801	-.06311	-.05779	-.05201	-.04575
23	12	5	-.04696	-.04180	-.03633	-.03053	-.02437	-.01783	-.01088
23	12	6	-.02409	-.01856	-.01279	-.00676	-.00045	.00615	.01308
23	12	7	-.00562	-.00010	.00559	.01147	.01755	.02385	.03040
23	12	8	.00979	.01505	.02042	.02592	.03156	.03736	.04333
23	12	9	.02294	.02778	.03268	.03766	.04273	.04790	.05320
23	12	10	.03436	.03866	.04299	.04736	.05178	.05626	.06082
23	12	11	.82264	.82837	.83572	.84473	.85542	.86781	.88191

Coefficients b_i for the BLUE of σ
Shape parameter k

n	s	i	0.7	0.8	0.9	1.0	1.1	1.2	1.3
24	0	1	-.20166	-.22384	-.24814	-.27466	-.30355	-.33494	-.36897
24	0	2	-.09168	-.09511	-.09844	-.10163	-.10466	-.10750	-.11011
24	0	3	-.06194	-.06213	-.06205	-.06168	-.06098	-.05992	-.05849
24	0	4	-.04340	-.04204	-.04038	-.03841	-.03611	-.03346	-.03043
24	0	5	-.03010	-.02790	-.02542	-.02267	-.01961	-.01625	-.01255
24	0	6	-.01981	-.01715	-.01425	-.01112	-.00773	-.00409	-.00017
24	0	7	-.01148	-.00858	-.00549	-.00220	.00128	.00496	.00886
24	0	8	-.00449	-.00150	.00163	.00492	.00835	.01194	.01569
24	0	9	.00150	.00448	.00757	.01075	.01405	.01745	.02098
24	0	10	.00674	.00964	.01260	.01563	.01873	.02189	.02514
24	0	11	.01140	.01416	.01695	.01977	.02262	.02552	.02845
24	0	12	.01560	.01817	.02074	.02332	.02590	.02850	.03111
24	0	13	.01942	.02176	.02409	.02639	.02869	.03097	.03325
24	0	14	.02292	.02501	.02706	.02908	.03106	.03302	.03496
24	0	15	.02618	.02798	.02973	.03144	.03310	.03472	.03632
24	0	16	.02922	.03071	.03214	.03351	.03484	.03613	.03739
24	0	17	.03209	.03324	.03432	.03535	.03633	.03728	.03819
24	0	18	.03483	.03560	.03631	.03698	.03760	.03820	.03878
24	0	19	.03746	.03782	.03813	.03841	.03867	.03891	.03915
24	0	20	.04003	.03992	.03980	.03967	.03954	.03942	.03932
24	0	21	.04257	.04195	.04134	.04076	.04023	.03973	.03928
24	0	22	.04516	.04392	.04276	.04168	.04070	.03980	.03900
24	0	23	.04795	.04591	.04406	.04239	.04089	.03955	.03836
24	0	24	.05149	.04797	.04493	.04232	.04008	.03816	.03652
24	1	1	-.20989	-.23254	-.25736	-.28447	-.31402	-.34613	-.38098
24	1	2	-.09531	-.09871	-.10201	-.10519	-.10821	-.11104	-.11365
24	1	3	-.06432	-.06442	-.06426	-.06379	-.06300	-.06187	-.06035
24	1	4	-.04501	-.04354	-.04177	-.03969	-.03728	-.03452	-.03137
24	1	5	-.03116	-.02884	-.02625	-.02338	-.02021	-.01673	-.01292
24	1	6	-.02045	-.01767	-.01466	-.01142	-.00792	-.00416	-.00013
24	1	7	-.01177	-.00876	-.00557	-.00218	.00140	.00519	.00919
24	1	8	-.00450	-.00141	.00182	.00519	.00872	.01240	.01624
24	1	9	.00174	.00481	.00797	.01124	.01461	.01810	.02170
24	1	10	.00720	.01017	.01320	.01629	.01945	.02268	.02599
24	1	11	.01206	.01486	.01770	.02057	.02348	.02642	.02942
24	1	12	.01642	.01903	.02163	.02425	.02687	.02950	.03216
24	1	13	.02039	.02276	.02510	.02743	.02975	.03206	.03436
24	1	14	.02404	.02614	.02819	.03021	.03220	.03417	.03613
24	1	15	.02743	.02922	.03095	.03265	.03430	.03593	.03753
24	1	16	.03059	.03205	.03345	.03480	.03611	.03738	.03863
24	1	17	.03358	.03467	.03571	.03669	.03765	.03857	.03946
24	1	18	.03642	.03711	.03776	.03837	.03895	.03951	.04006
24	1	19	.03915	.03941	.03964	.03985	.04005	.04024	.04044
24	1	20	.04180	.04158	.04136	.04114	.04094	.04076	.04060
24	1	21	.04441	.04364	.04292	.04225	.04162	.04106	.04055
24	1	22	.04702	.04562	.04433	.04314	.04207	.04109	.04022
24	1	23	.10016	.09483	.09015	.08606	.08249	.07938	.07671
24	2	1	-.21914	-.24237	-.26783	-.29564	-.32594	-.35889	-.39464
24	2	2	-.09935	-.10273	-.10602	-.10919	-.11220	-.11504	-.11765
24	2	3	-.06696	-.06696	-.06670	-.06615	-.06527	-.06404	-.06243
24	2	4	-.04677	-.04518	-.04329	-.04109	-.03856	-.03568	-.03241
24	2	5	-.03229	-.02985	-.02713	-.02414	-.02084	-.01724	-.01329
24	2	6	-.02110	-.01820	-.01507	-.01170	-.00808	-.00421	-.00005
24	2	7	-.01203	-.00891	-.00560	-.00211	.00159	.00549	.00961
24	2	8	-.00443	-.00124	.00209	.00556	.00918	.01296	.01691
24	2	9	.00209	.00524	.00849	.01184	.01530	.01887	.02256

Coefficients b_i for the BLUE of σ
Shape parameter k

n	s	i	0.7	0.8	0.9	1.0	1.1	1.2	1.3
24	2	10	.00779	.01083	.01393	.01709	.02032	.02362	.02701
24	2	11	.01286	.01572	.01861	.02154	.02450	.02750	.03055
24	2	12	.01742	.02006	.02271	.02536	.02802	.03069	.03339
24	2	13	.02157	.02395	.02631	.02866	.03100	.03333	.03566
24	2	14	.02538	.02747	.02952	.03155	.03354	.03552	.03749
24	2	15	.02891	.03067	.03239	.03407	.03572	.03734	.03894
24	2	16	.03221	.03362	.03498	.03630	.03758	.03884	.04007
24	2	17	.03531	.03634	.03732	.03826	.03917	.04006	.04093
24	2	18	.03827	.03888	.03945	.04000	.04052	.04103	.04154
24	2	19	.04109	.04125	.04138	.04151	.04164	.04177	.04192
24	2	20	.04382	.04348	.04314	.04283	.04254	.04229	.04207
24	2	21	.04648	.04557	.04471	.04393	.04321	.04256	.04198
24	2	22	.14886	.14236	.13660	.13152	.12707	.12320	.11985
24	3	1	-.22939	-.25331	-.27951	-.30812	-.33930	-.37319	-.40997
24	3	2	-.10378	-.10716	-.11045	-.11362	-.11664	-.11948	-.12210
24	3	3	-.06982	-.06974	-.06939	-.06874	-.06777	-.06644	-.06472
24	3	4	-.04867	-.04695	-.04494	-.04261	-.03995	-.03694	-.03353
24	3	5	-.03349	-.03091	-.02807	-.02493	-.02150	-.01776	-.01367
24	3	6	-.02176	-.01873	-.01547	-.01197	-.00822	-.00421	.00008
24	3	7	-.01226	-.00901	-.00558	-.00197	.00185	.00587	.01011
24	3	8	-.00430	-.00100	.00244	.00602	.00975	.01364	.01769
24	3	9	.00254	.00578	.00913	.01257	.01612	.01978	.02356
24	3	10	.00851	.01163	.01480	.01804	.02134	.02472	.02818
24	3	11	.01382	.01674	.01969	.02267	.02569	.02874	.03185
24	3	12	.01859	.02127	.02396	.02664	.02934	.03206	.03479
24	3	13	.02293	.02533	.02772	.03008	.03244	.03480	.03716
24	3	14	.02692	.02901	.03106	.03308	.03509	.03707	.03905
24	3	15	.03061	.03235	.03405	.03571	.03734	.03895	.04055
24	3	16	.03405	.03542	.03674	.03802	.03927	.04050	.04172
24	3	17	.03729	.03825	.03917	.04006	.04092	.04176	.04259
24	3	18	.04036	.04088	.04137	.04184	.04230	.04276	.04321
24	3	19	.04329	.04333	.04336	.04340	.04344	.04351	.04359
24	3	20	.04609	.04560	.04514	.04472	.04434	.04401	.04372
24	3	21	.19846	.19121	.18478	.17911	.17415	.16985	.16615
24	4	1	-.24075	-.26545	-.29250	-.32204	-.35420	-.38917	-.42710
24	4	2	-.10865	-.11204	-.11534	-.11852	-.12155	-.12440	-.12703
24	4	3	-.07295	-.07277	-.07233	-.07158	-.07050	-.06907	-.06724
24	4	4	-.05071	-.04886	-.04672	-.04426	-.04146	-.03830	-.03474
24	4	5	-.03476	-.03204	-.02905	-.02577	-.02219	-.01829	-.01405
24	4	6	-.02243	-.01926	-.01586	-.01221	-.00832	-.00417	.00027
24	4	7	-.01244	-.00907	-.00551	-.00176	.00219	.00634	.01072
24	4	8	-.00408	-.00066	.00289	.00659	.01044	.01444	.01861
24	4	9	.00310	.00645	.00989	.01343	.01708	.02084	.02472
24	4	10	.00937	.01257	.01583	.01914	.02253	.02598	.02952
24	4	11	.01495	.01793	.02094	.02398	.02706	.03018	.03334
24	4	12	.01995	.02268	.02540	.02813	.03087	.03363	.03640
24	4	13	.02451	.02693	.02933	.03172	.03410	.03648	.03886
24	4	14	.02869	.03077	.03282	.03484	.03685	.03884	.04082
24	4	15	.03255	.03427	.03594	.03758	.03919	.04079	.04237
24	4	16	.03615	.03747	.03874	.03998	.04120	.04239	.04358
24	4	17	.03954	.04042	.04126	.04209	.04289	.04369	.04448
24	4	18	.04273	.04314	.04354	.04393	.04431	.04470	.04510
24	4	19	.04575	.04566	.04558	.04551	.04547	.04545	.04546
24	4	20	.24948	.24187	.23513	.22921	.22406	.21965	.21590
24	5	1	-.25336	-.27896	-.30698	-.33755	-.37084	-.40702	-.44626
24	5	2	-.11401	-.11742	-.12074	-.12394	-.12699	-.12986	-.13250

Coefficients b_i for the BLUE of σ
Shape parameter k

n	s	i	0.7	0.8	0.9	1.0	1.1	1.2	1.3
24	5	3	-.07636	-.07609	-.07554	-.07469	-.07350	-.07195	-.07000
24	5	4	-.05291	-.05093	-.04864	-.04603	-.04308	-.03976	-.03605
24	5	5	-.03610	-.03323	-.03008	-.02664	-.02290	-.01884	-.01443
24	5	6	-.02311	-.01978	-.01622	-.01243	-.00838	-.00406	.00053
24	5	7	-.01258	-.00906	-.00536	-.00147	.00262	.00692	.01145
24	5	8	-.00377	-.00022	.00346	.00729	.01126	.01539	.01969
24	5	9	.00379	.00725	.01080	.01445	.01821	.02207	.02606
24	5	10	.01040	.01369	.01703	.02043	.02390	.02745	.03107
24	5	11	.01627	.01932	.02239	.02550	.02864	.03182	.03506
24	5	12	.02154	.02430	.02707	.02984	.03262	.03542	.03825
24	5	13	.02632	.02877	.03119	.03359	.03599	.03839	.04080
24	5	14	.03071	.03279	.03484	.03686	.03886	.04085	.04284
24	5	15	.03477	.03645	.03810	.03971	.04130	.04288	.04445
24	5	16	.03854	.03980	.04102	.04221	.04338	.04453	.04569
24	5	17	.04208	.04287	.04364	.04439	.04513	.04586	.04661
24	5	18	.04540	.04570	.04599	.04628	.04658	.04690	.04723
24	5	19	.30239	.29476	.28804	.28220	.27720	.27299	.26952
24	6	1	-.26741	-.29403	-.32314	-.35490	-.38947	-.42702	-.46774
24	6	2	-.11993	-.12338	-.12672	-.12995	-.13302	-.13591	-.13858
24	6	3	-.08011	-.07973	-.07908	-.07811	-.07680	-.07513	-.07304
24	6	4	-.05531	-.05317	-.05072	-.04795	-.04483	-.04134	-.03745
24	6	5	-.03753	-.03448	-.03116	-.02755	-.02363	-.01939	-.01479
24	6	6	-.02379	-.02029	-.01657	-.01260	-.00838	-.00389	.00089
24	6	7	-.01266	-.00898	-.00512	-.00107	.00318	.00764	.01232
24	6	8	-.00335	.00034	.00416	.00813	.01224	.01651	.02095
24	6	9	.00464	.00822	.01189	.01566	.01953	.02351	.02762
24	6	10	.01162	.01501	.01844	.02194	.02551	.02914	.03286
24	6	11	.01781	.02094	.02409	.02726	.03047	.03372	.03703
24	6	12	.02337	.02619	.02900	.03181	.03464	.03749	.04036
24	6	13	.02842	.03088	.03332	.03575	.03817	.04059	.04302
24	6	14	.03304	.03511	.03715	.03916	.04116	.04315	.04514
24	6	15	.03731	.03895	.04056	.04214	.04370	.04526	.04680
24	6	16	.04127	.04245	.04361	.04474	.04586	.04697	.04808
24	6	17	.04496	.04565	.04633	.04700	.04766	.04833	.04901
24	6	18	.35763	.35031	.34396	.33855	.33403	.33037	.32751
24	7	1	-.28315	-.31093	-.34129	-.37440	-.41041	-.44952	-.49191
24	7	2	-.12651	-.12999	-.13338	-.13664	-.13974	-.14266	-.14535
24	7	3	-.08423	-.08375	-.08297	-.08188	-.08044	-.07863	-.07640
24	7	4	-.05791	-.05560	-.05298	-.05003	-.04673	-.04305	-.03895
24	7	5	-.03904	-.03581	-.03230	-.02850	-.02438	-.01993	-.01513
24	7	6	-.02447	-.02078	-.01687	-.01272	-.00831	-.00362	.00136
24	7	7	-.01266	-.00881	-.00478	-.00056	.00387	.00850	.01337
24	7	8	-.00279	.00106	.00503	.00915	.01341	.01784	.02243
24	7	9	.00568	.00939	.01319	.01709	.02109	.02520	.02944
24	7	10	.01307	.01657	.02011	.02371	.02737	.03111	.03494
24	7	11	.01963	.02284	.02606	.02931	.03259	.03592	.03930
24	7	12	.02552	.02838	.03124	.03410	.03697	.03986	.04278
24	7	13	.03085	.03333	.03579	.03823	.04067	.04311	.04557
24	7	14	.03573	.03779	.03981	.04181	.04380	.04579	.04777
24	7	15	.04022	.04182	.04338	.04493	.04645	.04798	.04950
24	7	16	.04438	.04549	.04657	.04763	.04869	.04975	.05081
24	7	17	.41566	.40902	.40340	.39876	.39509	.39235	.39048
24	8	1	-.30089	-.32999	-.36178	-.39642	-.43409	-.47497	-.51927
24	8	2	-.13386	-.13739	-.14082	-.14412	-.14727	-.15022	-.15295
24	8	3	-.08880	-.08819	-.08729	-.08606	-.08448	-.08251	-.08011
24	8	4	-.06075	-.05826	-.05545	-.05230	-.04879	-.04489	-.04057

Coefficients b_i for the BLUE of σ
Shape parameter k

n	s	i	0.7	0.8	0.9	1.0	1.1	1.2	1.3
24	8	5	-.04065	-.03722	-.03350	-.02948	-.02514	-.02047	-.01543
24	8	6	-.02513	-.02125	-.01713	-.01276	-.00813	-.00323	.00197
24	8	7	-.01257	-.00852	-.00430	.00012	.00474	.00957	.01464
24	8	8	-.00206	.00196	.00610	.01039	.01483	.01942	.02419
24	8	9	.00695	.01081	.01475	.01879	.02293	.02719	.03156
24	8	10	.01481	.01842	.02208	.02579	.02957	.03342	.03735
24	8	11	.02178	.02507	.02837	.03170	.03507	.03848	.04194
24	8	12	.02803	.03094	.03385	.03676	.03968	.04262	.04559
24	8	13	.03368	.03618	.03865	.04111	.04357	.04603	.04850
24	8	14	.03885	.04088	.04289	.04487	.04685	.04882	.05081
24	8	15	.04359	.04513	.04664	.04813	.04962	.05110	.05259
24	8	16	.47703	.47144	.46693	.46347	.46105	.45964	.45920
24	9	1	-.32104	-.35165	-.38507	-.42147	-.46104	-.50396	-.55046
24	9	2	-.14212	-.14571	-.14920	-.15255	-.15575	-.15874	-.16151
24	9	3	-.09390	-.09315	-.09210	-.09072	-.08897	-.08683	-.08425
24	9	4	-.06388	-.06118	-.05815	-.05477	-.05103	-.04689	-.04232
24	9	5	-.04238	-.03871	-.03475	-.03049	-.02590	-.02097	-.01567
24	9	6	-.02578	-.02166	-.01731	-.01270	-.00784	-.00269	.00276
24	9	7	-.01235	-.00808	-.00364	.00099	.00583	.01088	.01617
24	9	8	-.00111	.00309	.00743	.01191	.01653	.02132	.02628
24	9	9	.00851	.01253	.01663	.02083	.02513	.02954	.03408
24	9	10	.01690	.02064	.02442	.02825	.03216	.03613	.04019
24	9	11	.02433	.02771	.03110	.03452	.03798	.04148	.04502
24	9	12	.03098	.03395	.03691	.03987	.04285	.04584	.04887
24	9	13	.03700	.03951	.04199	.04447	.04694	.04942	.05192
24	9	14	.04248	.04449	.04647	.04843	.05039	.05235	.05432
24	9	15	.54235	.53823	.53526	.53343	.53272	.53312	.53460
24	10	1	-.34412	-.37647	-.41178	-.45021	-.49197	-.53725	-.58628
24	10	2	-.15150	-.15516	-.15871	-.16212	-.16537	-.16842	-.17123
24	10	3	-.09962	-.09872	-.09750	-.09594	-.09401	-.09167	-.08888
24	10	4	-.06734	-.06439	-.06112	-.05749	-.05348	-.04906	-.04421
24	10	5	-.04422	-.04029	-.03606	-.03152	-.02665	-.02142	-.01582
24	10	6	-.02638	-.02200	-.01738	-.01251	-.00737	-.00195	.00379
24	10	7	-.01195	-.00745	-.00276	.00212	.00720	.01251	.01805
24	10	8	.00011	.00453	.00908	.01377	.01861	.02361	.02879
24	10	9	.01043	.01463	.01891	.02329	.02776	.03235	.03707
24	10	10	.01943	.02330	.02723	.03120	.03524	.03935	.04355
24	10	11	.02738	.03086	.03436	.03787	.04142	.04502	.04866
24	10	12	.03450	.03752	.04054	.04355	.04658	.04963	.05272
24	10	13	.04092	.04344	.04594	.04842	.05091	.05341	.05592
24	10	14	.61237	.61019	.60925	.60957	.61112	.61389	.61786
24	11	1	-.37082	-.40520	-.44271	-.48351	-.52782	-.57585	-.62783
24	11	2	-.16223	-.16597	-.16959	-.17308	-.17639	-.17950	-.18237
24	11	3	-.10610	-.10502	-.10361	-.10185	-.09970	-.09713	-.09409
24	11	4	-.07119	-.06797	-.06441	-.06048	-.05616	-.05143	-.04624
24	11	5	-.04620	-.04196	-.03741	-.03256	-.02736	-.02180	-.01585
24	11	6	-.02692	-.02224	-.01731	-.01213	-.00668	-.00093	.00513
24	11	7	-.01133	-.00655	-.00158	.00358	.00895	.01454	.02038
24	11	8	.00168	.00635	.01115	.01608	.02117	.02642	.03185
24	11	9	.01282	.01722	.02171	.02628	.03096	.03575	.04067
24	11	10	.02251	.02655	.03063	.03476	.03895	.04321	.04756
24	11	11	.03107	.03466	.03826	.04189	.04554	.04924	.05300
24	11	12	.03872	.04180	.04487	.04794	.05103	.05414	.05729
24	11	13	.68799	.68832	.69001	.69307	.69751	.70333	.71051
24	12	1	-.40209	-.43887	-.47895	-.52254	-.56986	-.62112	-.67658
24	12	2	-.17467	-.17849	-.18220	-.18576	-.18915	-.19233	-.19526

Coefficients b_i for the BLUE of σ
Shape parameter k

n	s	i	0.7	0.8	0.9	1.0	1.1	1.2	1.3
24	12	3	-.11352	-.11223	-.11059	-.10859	-.10618	-.10334	-.10002
24	12	4	-.07551	-.07197	-.06807	-.06380	-.05912	-.05401	-.04843
24	12	5	-.04831	-.04371	-.03881	-.03358	-.02800	-.02205	-.01569
24	12	6	-.02735	-.02231	-.01703	-.01149	-.00567	.00045	.00689
24	12	7	-.01040	-.00530	.00000	.00548	.01118	.01711	.02328
24	12	8	.00373	.00869	.01377	.01898	.02435	.02988	.03560
24	12	9	.01581	.02045	.02517	.02998	.03488	.03991	.04506
24	12	10	.02632	.03054	.03479	.03910	.04346	.04790	.05243
24	12	11	.03559	.03930	.04301	.04675	.05053	.05435	.05822
24	12	12	.77040	.77390	.77890	.78546	.79357	.80326	.81450
25	0	1	-.19638	-.21824	-.24221	-.26841	-.29697	-.32802	-.36172
25	0	2	-.09024	-.09382	-.09732	-.10072	-.10398	-.10707	-.10996
25	0	3	-.06156	-.06195	-.06209	-.06195	-.06150	-.06072	-.05957
25	0	4	-.04368	-.04253	-.04110	-.03935	-.03729	-.03488	-.03211
25	0	5	-.03086	-.02886	-.02659	-.02404	-.02120	-.01805	-.01457
25	0	6	-.02095	-.01847	-.01575	-.01280	-.00959	-.00613	-.00239
25	0	7	-.01291	-.01017	-.00724	-.00411	-.00078	.00277	.00653
25	0	8	-.00619	-.00333	-.00032	.00284	.00616	.00964	.01329
25	0	9	-.00041	.00246	.00544	.00854	.01176	.01509	.01855
25	0	10	.00464	.00745	.01034	.01331	.01636	.01949	.02270
25	0	11	.00912	.01183	.01457	.01736	.02020	.02309	.02603
25	0	12	.01315	.01570	.01827	.02085	.02345	.02607	.02871
25	0	13	.01682	.01918	.02153	.02387	.02621	.02855	.03089
25	0	14	.02018	.02232	.02443	.02651	.02857	.03062	.03265
25	0	15	.02330	.02519	.02703	.02884	.03061	.03235	.03407
25	0	16	.02621	.02782	.02938	.03090	.03237	.03380	.03520
25	0	17	.02894	.03026	.03152	.03272	.03388	.03500	.03609
25	0	18	.03154	.03253	.03346	.03434	.03518	.03598	.03676
25	0	19	.03402	.03466	.03524	.03578	.03629	.03677	.03724
25	0	20	.03642	.03667	.03688	.03706	.03722	.03738	.03753
25	0	21	.03877	.03859	.03839	.03818	.03799	.03781	.03765
25	0	22	.04111	.04043	.03978	.03916	.03859	.03806	.03758
25	0	23	.04350	.04224	.04107	.03999	.03899	.03809	.03728
25	0	24	.04608	.04407	.04225	.04061	.03914	.03783	.03666
25	0	25	.04937	.04597	.04303	.04051	.03834	.03649	.03491
25	1	1	-.20402	-.22633	-.25080	-.27755	-.30673	-.33847	-.37294
25	1	2	-.09365	-.09721	-.10070	-.10408	-.10734	-.11043	-.11334
25	1	3	-.06383	-.06414	-.06420	-.06398	-.06346	-.06260	-.06138
25	1	4	-.04525	-.04399	-.04245	-.04061	-.03845	-.03594	-.03307
25	1	5	-.03192	-.02980	-.02743	-.02477	-.02183	-.01857	-.01498
25	1	6	-.02161	-.01902	-.01620	-.01314	-.00984	-.00627	-.00242
25	1	7	-.01326	-.01042	-.00739	-.00416	-.00073	.00291	.00677
25	1	8	-.00627	-.00332	-.00022	.00303	.00643	.01000	.01374
25	1	9	-.00026	.00269	.00575	.00892	.01221	.01562	.01916
25	1	10	.00498	.00787	.01082	.01386	.01697	.02016	.02344
25	1	11	.00964	.01240	.01520	.01804	.02093	.02387	.02687
25	1	12	.01383	.01642	.01903	.02165	.02428	.02695	.02964
25	1	13	.01764	.02003	.02240	.02477	.02714	.02950	.03188
25	1	14	.02114	.02329	.02541	.02750	.02958	.03164	.03369
25	1	15	.02437	.02626	.02810	.02990	.03168	.03343	.03516
25	1	16	.02739	.02899	.03053	.03203	.03349	.03492	.03632
25	1	17	.03023	.03151	.03274	.03391	.03505	.03615	.03724
25	1	18	.03293	.03386	.03474	.03558	.03639	.03717	.03792
25	1	19	.03550	.03606	.03658	.03707	.03753	.03798	.03841
25	1	20	.03798	.03814	.03827	.03838	.03849	.03859	.03871
25	1	21	.04040	.04011	.03981	.03953	.03927	.03903	.03882

Coefficients b_i for the BLUE of σ
Shape parameter k

n	s	i	0.7	0.8	0.9	1.0	1.1	1.2	1.3
25	1	22	.04280	.04199	.04123	.04052	.03986	.03927	.03873
25	1	23	.04520	.04380	.04250	.04131	.04024	.03927	.03840
25	1	24	.09602	.09084	.08628	.08230	.07884	.07583	.07323
25	2	1	-.21259	-.23544	-.26051	-.28792	-.31781	-.35034	-.38566
25	2	2	-.09744	-.10099	-.10447	-.10786	-.11111	-.11422	-.11713
25	2	3	-.06633	-.06656	-.06654	-.06624	-.06564	-.06470	-.06340
25	2	4	-.04695	-.04558	-.04394	-.04199	-.03972	-.03710	-.03412
25	2	5	-.03304	-.03082	-.02833	-.02556	-.02249	-.01912	-.01542
25	2	6	-.02230	-.01959	-.01666	-.01349	-.01007	-.00639	-.00243
25	2	7	-.01359	-.01064	-.00750	-.00417	-.00064	.00311	.00708
25	2	8	-.00629	-.00325	-.00006	.00329	.00678	.01045	.01428
25	2	9	-.00003	.00300	.00615	.00940	.01277	.01626	.01988
25	2	10	.00544	.00839	.01142	.01452	.01770	.02096	.02431
25	2	11	.01029	.01311	.01596	.01886	.02181	.02480	.02786
25	2	12	.01466	.01729	.01994	.02260	.02527	.02798	.03071
25	2	13	.01863	.02104	.02344	.02583	.02823	.03062	.03303
25	2	14	.02227	.02443	.02656	.02867	.03075	.03283	.03490
25	2	15	.02564	.02752	.02935	.03115	.03293	.03468	.03641
25	2	16	.02879	.03035	.03187	.03335	.03480	.03622	.03762
25	2	17	.03174	.03297	.03416	.03530	.03641	.03749	.03856
25	2	18	.03454	.03541	.03624	.03703	.03779	.03853	.03926
25	2	19	.03721	.03769	.03813	.03856	.03896	.03936	.03976
25	2	20	.03977	.03983	.03986	.03990	.03994	.03999	.04005
25	2	21	.04226	.04184	.04144	.04107	.04073	.04042	.04015
25	2	22	.04469	.04375	.04286	.04205	.04130	.04063	.04003
25	2	23	.14262	.13626	.13062	.12566	.12131	.11753	.11426
25	3	1	-.22206	-.24555	-.27131	-.29947	-.33017	-.36359	-.39987
25	3	2	-.10159	-.10514	-.10863	-.11203	-.11530	-.11841	-.12134
25	3	3	-.06905	-.06920	-.06910	-.06872	-.06803	-.06701	-.06562
25	3	4	-.04878	-.04730	-.04554	-.04348	-.04110	-.03836	-.03526
25	3	5	-.03424	-.03189	-.02928	-.02638	-.02320	-.01970	-.01587
25	3	6	-.02300	-.02017	-.01712	-.01383	-.01029	-.00649	-.00240
25	3	7	-.01389	-.01083	-.00758	-.00414	-.00049	.00337	.00745
25	3	8	-.00627	-.00312	.00018	.00362	.00722	.01098	.01492
25	3	9	.00028	.00340	.00664	.00998	.01344	.01702	.02073
25	3	10	.00599	.00902	.01213	.01530	.01855	.02189	.02531
25	3	11	.01107	.01395	.01686	.01982	.02282	.02587	.02899
25	3	12	.01563	.01831	.02099	.02370	.02642	.02917	.03195
25	3	13	.01978	.02222	.02464	.02706	.02948	.03191	.03434
25	3	14	.02358	.02575	.02789	.03000	.03210	.03419	.03628
25	3	15	.02710	.02896	.03079	.03259	.03435	.03610	.03784
25	3	16	.03038	.03191	.03341	.03487	.03629	.03770	.03908
25	3	17	.03345	.03464	.03578	.03688	.03796	.03901	.04005
25	3	18	.03636	.03717	.03793	.03867	.03938	.04008	.04077
25	3	19	.03913	.03952	.03989	.04024	.04058	.04093	.04127
25	3	20	.04178	.04172	.04166	.04161	.04158	.04156	.04156
25	3	21	.04434	.04378	.04326	.04279	.04236	.04197	.04164
25	3	22	.18998	.18285	.17652	.17094	.16605	.16181	.15816
25	4	1	-.23250	-.25672	-.28327	-.31228	-.34391	-.37832	-.41569
25	4	2	-.10613	-.10970	-.11321	-.11662	-.11991	-.12304	-.12599
25	4	3	-.07200	-.07208	-.07189	-.07143	-.07065	-.06954	-.06805
25	4	4	-.05074	-.04915	-.04728	-.04510	-.04259	-.03972	-.03648
25	4	5	-.03550	-.03302	-.03028	-.02726	-.02394	-.02030	-.01633
25	4	6	-.02372	-.02076	-.01758	-.01416	-.01049	-.00655	-.00233
25	4	7	-.01417	-.01099	-.00761	-.00405	-.00028	.00370	.00791
25	4	8	-.00618	-.00292	.00048	.00404	.00775	.01162	.01567

Coefficients b_i for the BLUE of σ
Shape parameter k

n	s	i	0.7	0.8	0.9	1.0	1.1	1.2	1.3
25	4	9	.00068	.00390	.00723	.01067	.01422	.01790	.02170
25	4	10	.00667	.00978	.01296	.01621	.01955	.02296	.02647
25	4	11	.01198	.01493	.01790	.02092	.02399	.02710	.03028
25	4	12	.01676	.01948	.02222	.02496	.02773	.03052	.03335
25	4	13	.02110	.02357	.02602	.02847	.03091	.03337	.03584
25	4	14	.02508	.02725	.02940	.03153	.03364	.03574	.03784
25	4	15	.02876	.03061	.03243	.03421	.03598	.03772	.03946
25	4	16	.03218	.03369	.03515	.03658	.03798	.03937	.04074
25	4	17	.03539	.03652	.03761	.03867	.03971	.04072	.04173
25	4	18	.03842	.03915	.03984	.04051	.04117	.04182	.04247
25	4	19	.04129	.04158	.04186	.04213	.04240	.04268	.04298
25	4	20	.04403	.04385	.04368	.04353	.04340	.04331	.04325
25	4	21	.23859	.23103	.22434	.21846	.21334	.20893	.20519
25	5	1	-.24404	-.26909	-.29653	-.32650	-.35918	-.39471	-.43328
25	5	2	-.11110	-.11470	-.11824	-.12167	-.12499	-.12815	-.13113
25	5	3	-.07522	-.07521	-.07494	-.07439	-.07352	-.07230	-.07072
25	5	4	-.05287	-.05115	-.04915	-.04683	-.04419	-.04119	-.03780
25	5	5	-.03684	-.03422	-.03134	-.02817	-.02471	-.02092	-.01680
25	5	6	-.02445	-.02136	-.01804	-.01448	-.01066	-.00658	-.00221
25	5	7	-.01442	-.01110	-.00760	-.00390	.00000	.00412	.00846
25	5	8	-.00602	-.00264	.00088	.00455	.00838	.01238	.01655
25	5	9	.00118	.00451	.00794	.01148	.01514	.01892	.02283
25	5	10	.00747	.01067	.01394	.01728	.02070	.02420	.02779
25	5	11	.01306	.01607	.01911	.02220	.02533	.02851	.03175
25	5	12	.01807	.02084	.02362	.02642	.02923	.03207	.03495
25	5	13	.02263	.02512	.02760	.03007	.03255	.03503	.03753
25	5	14	.02680	.02898	.03113	.03326	.03538	.03750	.03961
25	5	15	.03065	.03249	.03429	.03606	.03781	.03955	.04129
25	5	16	.03423	.03570	.03713	.03852	.03990	.04126	.04261
25	5	17	.03759	.03865	.03968	.04069	.04168	.04266	.04363
25	5	18	.04074	.04138	.04199	.04259	.04319	.04378	.04438
25	5	19	.04372	.04390	.04407	.04425	.04444	.04465	.04488
25	5	20	.28884	.28119	.27445	.26857	.26351	.25923	.25568
25	6	1	-.25683	-.28281	-.31126	-.34233	-.37617	-.41297	-.45291
25	6	2	-.11657	-.12021	-.12378	-.12725	-.13060	-.13380	-.13680
25	6	3	-.07873	-.07863	-.07827	-.07762	-.07666	-.07534	-.07363
25	6	4	-.05516	-.05331	-.05117	-.04871	-.04592	-.04277	-.03923
25	6	5	-.03826	-.03550	-.03246	-.02914	-.02551	-.02156	-.01727
25	6	6	-.02521	-.02196	-.01849	-.01477	-.01080	-.00656	-.00202
25	6	7	-.01464	-.01117	-.00752	-.00368	.00037	.00463	.00912
25	6	8	-.00579	-.00227	.00138	.00518	.00914	.01327	.01757
25	6	9	.00180	.00525	.00879	.01245	.01622	.02010	.02413
25	6	10	.00843	.01173	.01509	.01852	.02203	.02562	.02930
25	6	11	.01431	.01739	.02051	.02367	.02687	.03013	.03344
25	6	12	.01959	.02241	.02524	.02809	.03095	.03384	.03677
25	6	13	.02438	.02690	.02941	.03191	.03442	.03693	.03946
25	6	14	.02876	.03094	.03310	.03524	.03737	.03950	.04162
25	6	15	.03280	.03462	.03640	.03816	.03990	.04163	.04336
25	6	16	.03656	.03798	.03937	.04072	.04207	.04340	.04472
25	6	17	.04007	.04106	.04203	.04297	.04391	.04484	.04577
25	6	18	.04335	.04389	.04442	.04494	.04546	.04599	.04653
25	6	19	.34114	.33370	.32722	.32165	.31697	.31312	.31006
25	7	1	-.27106	-.29810	-.32770	-.35999	-.39517	-.43339	-.47486
25	7	2	-.12262	-.12630	-.12992	-.13343	-.13683	-.14006	-.14310
25	7	3	-.08258	-.08239	-.08194	-.08118	-.08010	-.07867	-.07684
25	7	4	-.05765	-.05566	-.05336	-.05075	-.04780	-.04448	-.04076

Coefficients b_i for the BLUE of σ
Shape parameter k

n	s	i	0.7	0.8	0.9	1.0	1.1	1.2	1.3
25	7	5	-.03978	-.03685	-.03364	-.03015	-.02635	-.02222	-.01774
25	7	6	-.02598	-.02257	-.01892	-.01503	-.01089	-.00647	-.00176
25	7	7	-.01480	-.01118	-.00737	-.00337	.00084	.00526	.00992
25	7	8	-.00545	-.00179	.00201	.00595	.01005	.01432	.01877
25	7	9	.00257	.00614	.00981	.01359	.01748	.02149	.02563
25	7	10	.00957	.01297	.01644	.01997	.02358	.02727	.03105
25	7	11	.01578	.01894	.02214	.02537	.02865	.03198	.03538
25	7	12	.02135	.02423	.02711	.03001	.03293	.03587	.03885
25	7	13	.02640	.02895	.03149	.03402	.03655	.03910	.04166
25	7	14	.03102	.03320	.03536	.03751	.03964	.04178	.04391
25	7	15	.03527	.03706	.03882	.04056	.04228	.04400	.04571
25	7	16	.03921	.04058	.04191	.04323	.04453	.04583	.04712
25	7	17	.04288	.04379	.04469	.04556	.04644	.04731	.04819
25	7	18	.39588	.38898	.38307	.37815	.37416	.37107	.36885
25	8	1	-.28700	-.31524	-.34613	-.37982	-.41649	-.45633	-.49954
25	8	2	-.12933	-.13307	-.13674	-.14031	-.14375	-.14703	-.15012
25	8	3	-.08683	-.08654	-.08597	-.08510	-.08391	-.08234	-.08038
25	8	4	-.06037	-.05821	-.05575	-.05296	-.04983	-.04632	-.04241
25	8	5	-.04140	-.03829	-.03490	-.03121	-.02721	-.02288	-.01819
25	8	6	-.02676	-.02316	-.01933	-.01525	-.01092	-.00630	-.00139
25	8	7	-.01490	-.01111	-.00712	-.00294	.00144	.00605	.01089
25	8	8	-.00498	-.00117	.00278	.00689	.01115	.01557	.02018
25	8	9	.00351	.00722	.01103	.01494	.01896	.02311	.02739
25	8	10	.01093	.01445	.01802	.02166	.02538	.02918	.03307
25	8	11	.01750	.02076	.02404	.02736	.03072	.03413	.03761
25	8	12	.02340	.02634	.02928	.03224	.03521	.03821	.04125
25	8	13	.02874	.03132	.03389	.03645	.03901	.04159	.04418
25	8	14	.03362	.03580	.03796	.04011	.04225	.04439	.04653
25	8	15	.03810	.03986	.04159	.04330	.04500	.04670	.04840
25	8	16	.04225	.04355	.04483	.04609	.04734	.04860	.04986
25	8	17	.45351	.44748	.44251	.43857	.43564	.43368	.43267
25	9	1	-.30496	-.33456	-.36692	-.40220	-.44058	-.48227	-.52745
25	9	2	-.13682	-.14063	-.14436	-.14800	-.15150	-.15484	-.15799
25	9	3	-.09153	-.09113	-.09045	-.08945	-.08812	-.08641	-.08430
25	9	4	-.06334	-.06100	-.05835	-.05538	-.05204	-.04833	-.04420
25	9	5	-.04314	-.03982	-.03622	-.03232	-.02810	-.02355	-.01863
25	9	6	-.02754	-.02374	-.01970	-.01541	-.01086	-.00603	-.00089
25	9	7	-.01492	-.01093	-.00675	-.00238	.00221	.00701	.01206
25	9	8	-.00437	-.00038	.00375	.00803	.01247	.01707	.02186
25	9	9	.00467	.00854	.01249	.01655	.02073	.02502	.02945
25	9	10	.01256	.01620	.01990	.02366	.02750	.03142	.03543
25	9	11	.01955	.02289	.02627	.02968	.03313	.03664	.04020
25	9	12	.02580	.02881	.03181	.03483	.03786	.04092	.04403
25	9	13	.03147	.03408	.03667	.03927	.04186	.04447	.04710
25	9	14	.03663	.03881	.04097	.04312	.04526	.04740	.04955
25	9	15	.04137	.04309	.04479	.04646	.04814	.04981	.05149
25	9	16	.51456	.50977	.50610	.50354	.50207	.50167	.50229
25	10	1	-.32534	-.35651	-.39055	-.42765	-.46799	-.51178	-.55923
25	10	2	-.14526	-.14914	-.15295	-.15666	-.16023	-.16364	-.16685
25	10	3	-.09678	-.09625	-.09544	-.09430	-.09281	-.09095	-.08866
25	10	4	-.06661	-.06407	-.06121	-.05802	-.05446	-.05050	-.04613
25	10	5	-.04500	-.04145	-.03762	-.03348	-.02901	-.02420	-.01902
25	10	6	-.02832	-.02429	-.02002	-.01549	-.01070	-.00562	-.00023
25	10	7	-.01483	-.01062	-.00622	-.00163	.00318	.00821	.01349
25	10	8	-.00355	.00064	.00496	.00944	.01407	.01887	.02386
25	10	9	.00611	.01014	.01426	.01849	.02283	.02729	.03188

Coefficients b_i for the BLUE of σ
Shape parameter k

n	s	i	0.7	0.8	0.9	1.0	1.1	1.2	1.3
25	10	10	.01453	.01830	.02214	.02603	.03000	.03406	.03821
25	10	11	.02197	.02543	.02890	.03241	.03596	.03957	.04323
25	10	12	.02864	.03171	.03478	.03786	.04096	.04409	.04726
25	10	13	.03467	.03731	.03993	.04255	.04518	.04782	.05048
25	10	14	.04015	.04233	.04448	.04661	.04875	.05089	.05305
25	10	15	.57962	.57649	.57456	.57382	.57428	.57590	.57867
25	11	1	-.34869	-.38165	-.41763	-.45683	-.49943	-.54565	-.59571
25	11	2	-.15483	-.15880	-.16270	-.16648	-.17014	-.17362	-.17691
25	11	3	-.10268	-.10201	-.10104	-.09974	-.09808	-.09603	-.09355
25	11	4	-.07024	-.06747	-.06437	-.06092	-.05710	-.05288	-.04822
25	11	5	-.04701	-.04320	-.03909	-.03468	-.02993	-.02483	-.01935
25	11	6	-.02908	-.02479	-.02025	-.01545	-.01038	-.00502	.00065
25	11	7	-.01459	-.01013	-.00549	-.00064	.00442	.00971	.01524
25	11	8	-.00248	.00193	.00648	.01117	.01602	.02105	.02626
25	11	9	.00788	.01210	.01641	.02083	.02535	.03000	.03479
25	11	10	.01691	.02084	.02482	.02887	.03299	.03719	.04149
25	11	11	.02489	.02845	.03204	.03566	.03932	.04303	.04681
25	11	12	.03202	.03516	.03830	.04145	.04462	.04782	.05106
25	11	13	.03845	.04112	.04378	.04643	.04908	.05175	.05445
25	11	14	.64944	.64844	.64874	.65035	.65327	.65749	.66299
25	12	1	-.37570	-.41075	-.44899	-.49062	-.53585	-.58490	-.63801
25	12	2	-.16579	-.16986	-.17385	-.17774	-.18148	-.18506	-.18843
25	12	3	-.10937	-.10853	-.10739	-.10590	-.10404	-.10177	-.09906
25	12	4	-.07428	-.07124	-.06787	-.06413	-.06001	-.05548	-.05050
25	12	5	-.04917	-.04506	-.04065	-.03591	-.03084	-.02541	-.01958
25	12	6	-.02980	-.02520	-.02036	-.01525	-.00987	-.00418	.00182
25	12	7	-.01415	-.00941	-.00448	.00066	.00600	.01159	.01742
25	12	8	-.00108	.00358	.00838	.01333	.01844	.02372	.02919
25	12	9	.01010	.01453	.01906	.02368	.02842	.03328	.03828
25	12	10	.01982	.02393	.02808	.03229	.03658	.04095	.04542
25	12	11	.02841	.03210	.03581	.03955	.04334	.04717	.05107
25	12	12	.03608	.03929	.04251	.04573	.04898	.05225	.05557
25	12	13	.72494	.72662	.72974	.73431	.74034	.74784	.75680
25	13	1	-.40734	-.44484	-.48573	-.53023	-.57855	-.63094	-.68763
25	13	2	-.17849	-.18267	-.18678	-.19077	-.19462	-.19830	-.20176
25	13	3	-.11704	-.11600	-.11464	-.11293	-.11083	-.10831	-.10533
25	13	4	-.07884	-.07548	-.07178	-.06771	-.06324	-.05833	-.05297
25	13	5	-.05151	-.04704	-.04227	-.03717	-.03172	-.02589	-.01966
25	13	6	-.03045	-.02550	-.02029	-.01482	-.00907	-.00301	.00338
25	13	7	-.01343	-.00836	-.00310	.00237	.00805	.01397	.02016
25	13	8	.00076	.00572	.01081	.01605	.02145	.02703	.03280
25	13	9	.01289	.01757	.02234	.02721	.03219	.03730	.04255
25	13	10	.02344	.02773	.03208	.03648	.04096	.04552	.05018
25	13	11	.03273	.03656	.04041	.04429	.04820	.05218	.05622
25	13	12	.80727	.81231	.81895	.82723	.83717	.84878	.86207

Coefficients b_i for the BLUE of σ
Shape parameter k

n	s	i	1.4	1.5	1.6	1.7	1.8	1.9	2.0
2	0	1	-.94136	-.94882	-.95664	-.96480	-.97327	-.98205	-.99111
2	0	2	.94136	.94882	.95664	.96480	.97327	.98205	.99111
3	0	1	-.86563	-.89013	-.91520	-.94083	-.96702	-.99376	-1.02105
3	0	2	.39143	.41876	.44612	.47353	.50100	.52859	.55630
3	0	3	.47420	.47136	.46908	.46731	.46601	.46517	.46476
3	1	1	-1.60165	-1.64340	-1.68671	-1.73158	-1.77800	-1.82597	-1.87549
3	1	2	1.60165	1.64340	1.68671	1.73158	1.77800	1.82597	1.87549
4	0	1	-.80213	-.83483	-.86861	-.90347	-.93945	-.97656	-1.01483
4	0	2	.18968	.21561	.24224	.26961	.29773	.32666	.35641
4	0	3	.30094	.31176	.32241	.33290	.34327	.35354	.36374
4	0	4	.31152	.30746	.30396	.30097	.29845	.29637	.29469
4	1	1	-1.13591	-1.17907	-1.22395	-1.27059	-1.31903	-1.36930	-1.42145
4	1	2	.27803	.31172	.34632	.38189	.41849	.45618	.49503
4	1	3	.85787	.86734	.87763	.88870	.90054	.91312	.92642
5	0	1	-.75035	-.78756	-.82628	-.86655	-.90843	-.95197	-.99723
5	0	2	.08856	.11094	.13433	.15880	.18439	.21116	.23916
5	0	3	.19834	.21243	.22659	.24085	.25522	.26973	.28440
5	0	4	.23368	.23842	.24305	.24760	.25208	.25651	.26092
5	0	5	.22978	.22578	.22230	.21930	.21674	.21456	.21275
5	1	1	-.95446	-.99927	-1.04610	-1.09502	-1.14611	-1.19943	-1.25506
5	1	2	.11705	.14410	.17235	.20185	.23269	.26495	.29870
5	1	3	.25557	.27177	.28813	.30467	.32142	.33841	.35567
5	1	4	.58185	.58339	.58562	.58850	.59200	.59607	.60069
5	2	1	-1.36180	-1.42274	-1.48658	-1.55339	-1.62327	-1.69632	-1.77262
5	2	2	.18317	.21911	.25650	.29545	.33606	.37844	.42271
5	2	3	1.17862	1.20363	1.23008	1.25794	1.28721	1.31787	1.34991
6	0	1	-.70740	-.74725	-.78896	-.83259	-.87822	-.92594	-.97585
6	0	2	.02916	.04808	.06814	.08942	.11198	.13589	.16122
6	0	3	.13412	.14859	.16337	.17848	.19396	.20981	.22608
6	0	4	.17477	.18292	.19102	.19907	.20710	.21512	.22315
6	0	5	.18832	.19033	.19231	.19428	.19624	.19820	.20019
6	0	6	.18103	.17733	.17411	.17133	.16895	.16691	.16520
6	1	1	-.84921	-.89501	-.94310	-.99357	-1.04652	-1.10206	-1.16030
6	1	2	.03738	.05936	.08263	.10725	.13333	.16094	.19019
6	1	3	.16307	.17941	.19613	.21325	.23082	.24885	.26740
6	1	4	.21111	.21986	.22862	.23739	.24621	.25508	.26403
6	1	5	.43764	.43637	.43573	.43567	.43617	.43719	.43868
6	2	1	-1.08769	-1.14388	-1.20296	-1.26503	-1.33022	-1.39865	-1.47044
6	2	2	.05545	.08234	.11069	.14060	.17218	.20555	.24082
6	2	3	.21529	.23449	.25413	.27426	.29491	.31613	.33796
6	2	4	.81695	.82706	.83814	.85017	.86312	.87696	.89166
6	3	1	-1.55589	-1.63351	-1.71518	-1.80106	-1.89128	-1.98600	-2.08538
6	3	2	.10019	.13651	.17463	.21470	.25684	.30121	.34794
6	3	3	1.45570	1.49700	1.54055	1.58636	1.63444	1.68479	1.73744
7	0	1	-.67106	-.71251	-.75607	-.80185	-.84996	-.90049	-.95358
7	0	2	-.00910	.00680	.02388	.04222	.06188	.08295	.10552
7	0	3	.09067	.10454	.11888	.13373	.14911	.16506	.18162
7	0	4	.13280	.14229	.15184	.16148	.17122	.18108	.19108
7	0	5	.15124	.15617	.16103	.16583	.17059	.17532	.18003
7	0	6	.15662	.15725	.15791	.15860	.15933	.16010	.16092
7	0	7	.14883	.14545	.14252	.13999	.13782	.13597	.13441
7	1	1	-.77703	-.82334	-.87215	-.92357	-.97774	-1.03478	-1.09482
7	1	2	-.00912	.00890	.02820	.04888	.07101	.09471	.12007
7	1	3	.10630	.12171	.13766	.15417	.17129	.18906	.20753
7	1	4	.15483	.16510	.17547	.18598	.19664	.20747	.21848
7	1	5	.17568	.18072	.18574	.19075	.19577	.20082	.20589

Coefficients b_i for the BLUE of σ
Shape parameter k

n	s	i	1.4	1.5	1.6	1.7	1.8	1.9	2.0
7	1	6	.34934	.34691	.34508	.34380	.34302	.34272	.34285
7	2	1	-.93861	-.99250	-1.04934	-1.10929	-1.17247	-1.23903	-1.30914
7	2	2	-.00687	.01423	.03674	.06078	.08644	.11384	.14310
7	2	3	.13223	.14980	.16797	.18677	.20625	.22647	.24746
7	2	4	.18999	.20127	.21269	.22429	.23608	.24808	.26033
7	2	5	.62327	.62720	.63194	.63745	.64370	.65065	.65825
7	3	1	-1.20541	-1.27230	-1.34289	-1.41734	-1.49581	-1.57848	-1.66551
7	3	2	.00105	.02712	.05480	.08423	.11554	.14886	.18433
7	3	3	.17868	.19968	.22136	.24376	.26694	.29096	.31589
7	3	4	1.02567	1.04550	1.06674	1.08935	1.11333	1.13866	1.16530
7	4	1	-1.72634	-1.81954	-1.91793	-2.02173	-2.13114	-2.24636	-2.36763
7	4	2	.02588	.06157	.09928	.13918	.18144	.22621	.27367
7	4	3	1.70046	1.75797	1.81865	1.88255	1.94970	2.02016	2.09396
8	0	1	-.63980	-.68220	-.72693	-.77410	-.82385	-.87631	-.93161
8	0	2	-.03527	-.02194	-.00744	.00830	.02535	.04381	.06377
8	0	3	.05960	.07253	.08603	.10016	.11494	.13041	.14661
8	0	4	.10180	.11168	.12175	.13202	.14251	.15325	.16425
8	0	5	.12242	.12883	.13522	.14162	.14802	.15445	.16092
8	0	6	.13168	.13472	.13771	.14065	.14356	.14646	.14934
8	0	7	.13350	.13339	.13333	.13334	.13342	.13356	.13377
8	0	8	.12607	.12299	.12032	.11802	.11605	.11437	.11295
8	1	1	-.72285	-.76934	-.81851	-.87047	-.92538	-.98339	-1.04466
8	1	2	-.03894	-.02409	-.00799	.00944	.02830	.04868	.07069
8	1	3	.06821	.08239	.09721	.11269	.12890	.14587	.16366
8	1	4	.11578	.12644	.13733	.14845	.15984	.17152	.18351
8	1	5	.13888	.14562	.15237	.15915	.16598	.17288	.17985
8	1	6	.14901	.15198	.15495	.15792	.16090	.16391	.16694
8	1	7	.28991	.28699	.28464	.28281	.28146	.28054	.28001
8	2	1	-.84165	-.89406	-.94951	-1.00816	-1.07018	-1.13572	-1.20498
8	2	2	-.04283	-.02590	-.00762	.01209	.03336	.05630	.08102
8	2	3	.08184	.09769	.11422	.13148	.14953	.16842	.18820
8	2	4	.13689	.14854	.16043	.17261	.18508	.19788	.21103
8	2	5	.16323	.17030	.17742	.18461	.19188	.19924	.20671
8	2	6	.50253	.50343	.50506	.50737	.51033	.51389	.51801
8	3	1	-1.01935	-1.08087	-1.14598	-1.21485	-1.28765	-1.36458	-1.44583
8	3	2	-.04639	-.02643	-.00499	.01805	.04282	.06944	.09804
8	3	3	.10432	.12256	.14155	.16135	.18202	.20361	.22619
8	3	4	.17016	.18318	.19648	.21009	.22405	.23838	.25311
8	3	5	.79126	.80157	.81294	.82535	.83877	.85316	.86849
8	4	1	-1.31083	-1.38777	-1.46919	-1.55530	-1.64630	-1.74241	-1.84385
8	4	2	-.04800	-.02314	.00344	.03187	.06229	.09486	.12972
8	4	3	.14497	.16705	.18998	.21383	.23867	.26458	.29162
8	4	4	1.21386	1.24386	1.27578	1.30961	1.34535	1.38298	1.42251
9	0	1	-.61253	-.65546	-.70090	-.74898	-.79982	-.85360	-.91046
9	0	2	-.05394	-.04278	-.03051	-.01705	-.00230	.01381	.03139
9	0	3	.03648	.04837	.06091	.07413	.08808	.10281	.11835
9	0	4	.07814	.08795	.09804	.10844	.11916	.13023	.14166
9	0	5	.09978	.10692	.11411	.12138	.12874	.13620	.14378
9	0	6	.11101	.11545	.11984	.12421	.12856	.13291	.13726
9	0	7	.11588	.11775	.11959	.12140	.12320	.12500	.12679
9	0	8	.11602	.11548	.11502	.11464	.11436	.11415	.11403
9	0	9	.10917	.10634	.10391	.10181	.10002	.09849	.09720
9	1	1	-.67984	-.72631	-.77558	-.82780	-.88312	-.94174	-1.00381
9	1	2	-.05925	-.04698	-.03352	-.01880	-.00271	.01485	.03398
9	1	3	.04109	.05401	.06761	.08195	.09708	.11305	.12991
9	1	4	.08728	.09781	.10866	.11984	.13138	.14331	.15567

Coefficients b_i for the BLUE of σ
Shape parameter k

n	s	i	1.4	1.5	1.6	1.7	1.8	1.9	2.0
9	1	5	.11121	.11875	.12638	.13411	.14195	.14993	.15805
9	1	6	.12354	.12809	.13263	.13717	.14173	.14631	.15092
9	1	7	.12870	.13044	.13220	.13397	.13576	.13759	.13944
9	1	8	.24728	.24418	.24162	.23955	.23793	.23671	.23586
9	2	1	-.77191	-.82319	-.87760	-.93530	-.99646	-1.06128	-1.12996
9	2	2	-.06564	-.05190	-.03690	-.02055	-.00273	.01667	.03776
9	2	3	.04827	.06249	.07745	.09319	.10978	.12728	.14573
9	2	4	.10056	.11198	.12374	.13586	.14838	.16133	.17473
9	2	5	.12748	.13548	.14358	.15181	.16017	.16870	.17739
9	2	6	.14110	.14572	.15036	.15503	.15974	.16450	.16933
9	2	7	.42015	.41942	.41937	.41995	.42111	.42281	.42502
9	3	1	-.90104	-.95924	-1.02099	-1.08647	-1.15588	-1.22941	-1.30729
9	3	2	-.07326	-.05752	-.04042	-.02184	-.00168	.02020	.04392
9	3	3	.05964	.07563	.09240	.11003	.12857	.14810	.16866
9	3	4	.12032	.13291	.14587	.15923	.17301	.18726	.20201
9	3	5	.15119	.15975	.16845	.17728	.18629	.19548	.20487
9	3	6	.64316	.64847	.65469	.66177	.66968	.67838	.68783
9	4	1	-1.09279	-1.16154	-1.23446	-1.31177	-1.39367	-1.48039	-1.57217
9	4	2	-.08231	-.06365	-.04347	-.02165	.00193	.02742	.05496
9	4	3	.07865	.09720	.11661	.13697	.15833	.18077	.20437
9	4	4	.15146	.16573	.18039	.19549	.21107	.22716	.24380
9	4	5	.94499	.96226	.98093	1.00096	1.02235	1.04505	1.06904
9	5	1	-1.40632	-1.49272	-1.58436	-1.68147	-1.78431	-1.89315	-2.00823
9	5	2	-.09286	-.06942	-.04422	-.01713	.01203	.04340	.07715
9	5	3	.11363	.13631	.15998	.18472	.21063	.23777	.26625
9	5	4	1.38555	1.42583	1.46861	1.51388	1.56166	1.61197	1.66483
10	0	1	-.58843	-.63165	-.67749	-.72612	-.77769	-.83236	-.89032
10	0	2	-.06766	-.05837	-.04802	-.03653	-.02382	-.00979	.00564
10	0	3	.01873	.02960	.04115	.05342	.06647	.08034	.09508
10	0	4	.05957	.06909	.07896	.08920	.09985	.11092	.12245
10	0	5	.08165	.08910	.09668	.10440	.11228	.12033	.12858
10	0	6	.09400	.09925	.10451	.10978	.11507	.12040	.12578
10	0	7	.10052	.10364	.10672	.10977	.11279	.11581	.11882
10	0	8	.10309	.10420	.10529	.10638	.10748	.10858	.10970
10	0	9	.10239	.10160	.10090	.10031	.09981	.09941	.09909
10	0	10	.09614	.09355	.09131	.08939	.08775	.08635	.08517
10	1	1	-.64441	-.69071	-.73991	-.79218	-.84770	-.90664	-.96922
10	1	2	-.07367	-.06353	-.05229	-.03985	-.02612	-.01099	.00563
10	1	3	.02095	.03265	.04508	.05828	.07230	.08720	.10304
10	1	4	.06565	.07581	.08635	.09730	.10868	.12053	.13288
10	1	5	.08978	.09765	.10567	.11386	.12223	.13080	.13960
10	1	6	.10323	.10869	.11418	.11970	.12526	.13089	.13659
10	1	7	.11028	.11341	.11653	.11965	.12277	.12590	.12905
10	1	8	.11290	.11387	.11486	.11589	.11694	.11804	.11918
10	1	9	.21528	.21215	.20953	.20737	.20563	.20428	.20326
10	2	1	-.71842	-.76876	-.82229	-.87919	-.93964	-1.00386	-1.07205
10	2	2	-.08101	-.06981	-.05743	-.04379	-.02878	-.01230	.00579
10	2	3	.02448	.03723	.05074	.06506	.08026	.09640	.11354
10	2	4	.07424	.08518	.09652	.10830	.12054	.13329	.14657
10	2	5	.10102	.10937	.11788	.12659	.13549	.14463	.15401
10	2	6	.11584	.12150	.12720	.13295	.13878	.14468	.15068
10	2	7	.12342	.12651	.12961	.13274	.13589	.13908	.14232
10	2	8	.36043	.35878	.35777	.35734	.35746	.35808	.35916
10	3	1	-.81765	-.87351	-.93291	-.99604	-1.06311	-1.13433	-1.20994
10	3	2	-.08998	-.07739	-.06355	-.04835	-.03169	-.01344	.00651
10	3	3	.03007	.04417	.05907	.07485	.09156	.10927	.12805

Coefficients b_i for the BLUE of σ
Shape parameter k

n	s	i	1.4	1.5	1.6	1.7	1.8	1.9	2.0
10	3	4	.08654	.09847	.11082	.12364	.13696	.15081	.16525
10	3	5	.11675	.12568	.13480	.14413	.15368	.16348	.17355
10	3	6	.13323	.13911	.14504	.15106	.15716	.16336	.16967
10	3	7	.54104	.54347	.54671	.55072	.55545	.56086	.56691
10	4	1	-.95576	-1.01948	-1.08722	-1.15918	-1.23560	-1.31670	-1.40274
10	4	2	-.10113	-.08664	-.07079	-.05347	-.03455	-.01392	.00858
10	4	3	.03917	.05510	.07190	.08964	.10838	.12822	.14921
10	4	4	.10482	.11808	.13179	.14600	.16075	.17608	.19203
10	4	5	.13960	.14930	.15921	.16934	.17972	.19038	.20133
10	4	6	.77331	.78364	.79511	.80767	.82130	.83594	.85159
10	5	1	-1.16012	-1.23570	-1.31603	-1.40134	-1.49187	-1.58789	-1.68966
10	5	2	-.11536	-.09807	-.07927	-.05883	-.03661	-.01247	.01374
10	5	3	.05480	.07340	.09296	.11356	.13528	.15820	.18240
10	5	4	.13377	.14893	.16459	.18080	.19760	.21505	.23318
10	5	5	1.08692	1.11145	1.13775	1.16580	1.19560	1.22711	1.26034
11	0	1	-.56694	-.61025	-.65629	-.70523	-.75725	-.81252	-.87124
11	0	2	-.07798	-.07030	-.06161	-.05185	-.04093	-.02876	-.01526
11	0	3	.00478	.01467	.02527	.03661	.04874	.06172	.07560
11	0	4	.04470	.05380	.06331	.07325	.08365	.09454	.10595
11	0	5	.06689	.07440	.08210	.09000	.09813	.10650	.11513
11	0	6	.07988	.08559	.09136	.09719	.10310	.10909	.11518
11	0	7	.08741	.09133	.09524	.09913	.10302	.10692	.11084
11	0	8	.09134	.09355	.09573	.09788	.10001	.10214	.10426
11	0	9	.09262	.09321	.09381	.09442	.09505	.09571	.09638
11	0	10	.09150	.09056	.08973	.08901	.08839	.08787	.08744
11	0	11	.08582	.08342	.08136	.07959	.07808	.07679	.07571
11	1	1	-.61441	-.66045	-.70947	-.76166	-.81720	-.87629	-.93915
11	1	2	-.08420	-.07587	-.06650	-.05601	-.04431	-.03129	-.01686
11	1	3	.00551	.01609	.02741	.03951	.05246	.06630	.08110
11	1	4	.04875	.05842	.06852	.07908	.09014	.10172	.11386
11	1	5	.07277	.08069	.08881	.09717	.10577	.11464	.12379
11	1	6	.08681	.09278	.09881	.10492	.11112	.11744	.12387
11	1	7	.09492	.09895	.10297	.10700	.11104	.11512	.11923
11	1	8	.09911	.10128	.10345	.10562	.10779	.10997	.11216
11	1	9	.10034	.10080	.10129	.10183	.10242	.10305	.10372
11	1	10	.19041	.18732	.18471	.18254	.18077	.17936	.17827
11	2	1	-.67557	-.72507	-.77781	-.83399	-.89379	-.95745	-1.02518
11	2	2	-.09178	-.08265	-.07244	-.06105	-.04838	-.03434	-.01880
11	2	3	.00687	.01830	.03050	.04352	.05743	.07229	.08816
11	2	4	.05438	.06472	.07552	.08681	.09861	.11098	.12394
11	2	5	.08072	.08910	.09771	.10656	.11568	.12508	.13480
11	2	6	.09606	.10229	.10859	.11499	.12149	.12813	.13490
11	2	7	.10485	.10894	.11305	.11718	.12135	.12557	.12983
11	2	8	.10925	.11133	.11344	.11556	.11771	.11989	.12211
11	2	9	.31522	.31303	.31145	.31042	.30990	.30985	.31023
11	3	1	-.75483	-.80890	-.86650	-.92784	-.99314	-1.06263	-1.13654
11	3	2	-.10101	-.09089	-.07961	-.06710	-.05323	-.03789	-.02097
11	3	3	.00923	.02172	.03501	.04918	.06428	.08039	.09758
11	3	4	.06223	.07341	.08506	.09723	.10995	.12326	.13721
11	3	5	.09152	.10046	.10964	.11908	.12881	.13884	.14921
11	3	6	.10847	.11498	.12159	.12831	.13515	.14214	.14928
11	3	7	.11801	.12216	.12634	.13056	.13484	.13918	.14360
11	3	8	.46640	.46706	.46846	.47057	.47333	.47670	.48064
11	4	1	-.86027	-.92051	-.98467	-1.05298	-1.12566	-1.20296	-1.28514
11	4	2	-.11244	-.10101	-.08835	-.07436	-.05893	-.04192	-.02322
11	4	3	.01323	.02709	.04181	.05746	.07410	.09182	.11069

Coefficients b_i for the BLUE of σ
Shape parameter k

n	s	i	1.4	1.5	1.6	1.7	1.8	1.9	2.0
11	4	4	.07346	.08571	.09847	.11177	.12565	.14017	.15536
11	4	5	.10657	.11622	.12612	.13630	.14679	.15761	.16878
11	4	6	.12552	.13239	.13936	.14646	.15370	.16111	.16868
11	4	7	.65393	.66011	.66726	.67535	.68434	.69418	.70484
11	5	1	-1.00645	-1.07542	-1.14885	-1.22698	-1.31007	-1.39838	-1.49217
11	5	2	-.12693	-.11371	-.09915	-.08315	-.06557	-.04629	-.02515
11	5	3	.02010	.03584	.05250	.07017	.08892	.10883	.12998
11	5	4	.09023	.10393	.11818	.13301	.14846	.16460	.18147
11	5	5	.12846	.13905	.14990	.16107	.17256	.18441	.19665
11	5	6	.89459	.91031	.92742	.94589	.96571	.98683	1.00923
12	0	1	-.54761	-.59087	-.63696	-.68606	-.73833	-.79397	-.85320
12	0	2	-.08588	-.07958	-.07235	-.06410	-.05477	-.04426	-.03249
12	0	3	-.00639	.00260	.01228	.02272	.03396	.04606	.05907
12	0	4	.03256	.04120	.05028	.05983	.06989	.08048	.09164
12	0	5	.05467	.06210	.06976	.07768	.08588	.09438	.10319
12	0	6	.06803	.07398	.08004	.08620	.09248	.09890	.10547
12	0	7	.07620	.08065	.08510	.08957	.09407	.09861	.10321
12	0	8	.08101	.08398	.08691	.08983	.09274	.09565	.09856
12	0	9	.08342	.08498	.08652	.08805	.08956	.09107	.09258
12	0	10	.08393	.08417	.08443	.08472	.08503	.08538	.08576
12	0	11	.08261	.08159	.08068	.07989	.07920	.07862	.07813
12	0	12	.07745	.07522	.07330	.07166	.07027	.06908	.06808
12	1	1	-.58851	-.63422	-.68299	-.73499	-.79044	-.84953	-.91250
12	1	2	-.09206	-.08527	-.07751	-.06870	-.05876	-.04759	-.03509
12	1	3	-.00662	.00294	.01323	.02431	.03624	.04906	.06285
12	1	4	.03523	.04437	.05397	.06407	.07470	.08590	.09771
12	1	5	.05897	.06678	.07485	.08319	.09183	.10079	.11010
12	1	6	.07331	.07953	.08585	.09231	.09890	.10565	.11257
12	1	7	.08207	.08665	.09126	.09591	.10060	.10535	.11016
12	1	8	.08719	.09020	.09319	.09618	.09917	.10218	.10521
12	1	9	.08972	.09123	.09273	.09424	.09576	.09729	.09884
12	1	10	.09016	.09027	.09043	.09064	.09091	.09122	.09159
12	1	11	.17054	.16753	.16498	.16284	.16109	.15967	.15855
12	2	1	-.64011	-.68885	-.74087	-.79637	-.85556	-.91867	-.98593
12	2	2	-.09954	-.09214	-.08374	-.07425	-.06357	-.05161	-.03825
12	2	3	-.00659	.00366	.01468	.02652	.03924	.05290	.06759
12	2	4	.03891	.04863	.05884	.06956	.08085	.09274	.10527
12	2	5	.06469	.07294	.08145	.09026	.09938	.10884	.11867
12	2	6	.08023	.08673	.09334	.10010	.10701	.11409	.12136
12	2	7	.08968	.09440	.09916	.10396	.10882	.11376	.11877
12	2	8	.09515	.09816	.10117	.10420	.10724	.11032	.11343
12	2	9	.09774	.09915	.10057	.10201	.10349	.10500	.10655
12	2	10	.27983	.27733	.27541	.27401	.27310	.27262	.27255
12	3	1	-.70528	-.75787	-.81401	-.87390	-.93777	-1.00586	-1.07842
12	3	2	-.10857	-.10043	-.09125	-.08092	-.06935	-.05642	-.04202
12	3	3	-.00612	.00497	.01687	.02962	.04331	.05799	.07374
12	3	4	.04395	.05438	.06532	.07680	.08887	.10157	.11495
12	3	5	.07228	.08104	.09008	.09943	.10912	.11916	.12959
12	3	6	.08929	.09611	.10306	.11015	.11742	.12487	.13252
12	3	7	.09956	.10442	.10932	.11429	.11933	.12445	.12966
12	3	8	.10539	.10838	.11138	.11442	.11749	.12061	.12378
12	3	9	.40949	.40901	.40923	.41011	.41159	.41363	.41620
12	4	1	-.78908	-.84672	-.90822	-.97381	-1.04373	-1.11824	-1.19758
12	4	2	-.11959	-.11053	-.10036	-.08898	-.07628	-.06215	-.04645
12	4	3	-.00494	.00722	.02022	.03413	.04902	.06496	.08203
12	4	4	.05100	.06231	.07416	.08657	.09961	.11331	.12773

Coefficients b_i for the BLUE of σ
Shape parameter k

n	s	i	1.4	1.5	1.6	1.7	1.8	1.9	2.0
12	4	5	.08255	.09194	.10163	.11164	.12201	.13276	.14391
12	4	6	.10138	.10858	.11591	.12342	.13110	.13898	.14707
12	4	7	.11261	.11762	.12268	.12782	.13305	.13837	.14380
12	4	8	.56607	.56959	.57399	.57922	.58524	.59201	.59949
12	5	1	-.90009	-.96451	-1.03323	-1.10649	-1.18453	-1.26763	-1.35606
12	5	2	-.13333	-.12306	-.11161	-.09885	-.08468	-.06897	-.05160
12	5	3	-.00250	.01104	.02548	.04089	.05734	.07492	.09371
12	5	4	.06114	.07359	.08661	.10024	.11452	.12951	.14525
12	5	5	.09686	.10706	.11758	.12843	.13967	.15130	.16338
12	5	6	.11799	.12566	.13348	.14148	.14968	.15809	.16674
12	5	7	.75992	.77022	.78169	.79430	.80800	.82278	.83858
12	6	1	-1.05364	-1.12761	-1.20647	-1.29049	-1.37995	-1.47512	-1.57632
12	6	2	-.15097	-.13904	-.12581	-.11116	-.09497	-.07711	-.05742
12	6	3	.00222	.01766	.03407	.05154	.07013	.08995	.11108
12	6	4	.07641	.09041	.10502	.12028	.13625	.15299	.17054
12	6	5	.11774	.12902	.14064	.15262	.16502	.17785	.19115
12	6	6	1.00825	1.02956	1.05256	1.07721	1.10352	1.13144	1.16097
13	0	1	-.53009	-.57322	-.61925	-.66837	-.72075	-.77660	-.83615
13	0	2	-.09200	-.08691	-.08095	-.07405	-.06612	-.05710	-.04688
13	0	3	-.01547	-.00733	.00151	.01110	.02149	.03274	.04489
13	0	4	.02251	.03066	.03929	.04841	.05807	.06829	.07912
13	0	5	.04443	.05168	.05921	.06704	.07519	.08368	.09254
13	0	6	.05798	.06403	.07022	.07656	.08306	.08974	.09663
13	0	7	.06658	.07134	.07615	.08101	.08594	.09094	.09603
13	0	8	.07196	.07545	.07893	.08241	.08589	.08939	.09292
13	0	9	.07512	.07738	.07961	.08182	.08402	.08621	.08840
13	0	10	.07659	.07768	.07876	.07983	.08090	.08198	.08305
13	0	11	.07664	.07663	.07665	.07671	.07681	.07694	.07712
13	0	12	.07524	.07417	.07322	.07239	.07167	.07105	.07052
13	0	13	.07052	.06844	.06665	.06513	.06383	.06274	.06181
13	1	1	-.56579	-.61113	-.65959	-.71134	-.76661	-.82560	-.88856
13	1	2	-.09801	-.09254	-.08618	-.07884	-.07044	-.06090	-.05011
13	1	3	-.01632	-.00770	.00165	.01178	.02275	.03461	.04743
13	1	4	.02421	.03280	.04188	.05149	.06166	.07242	.08383
13	1	5	.04760	.05520	.06310	.07132	.07988	.08880	.09811
13	1	6	.06205	.06836	.07481	.08144	.08825	.09525	.10247
13	1	7	.07121	.07614	.08112	.08618	.09131	.09653	.10186
13	1	8	.07693	.08050	.08408	.08766	.09126	.09490	.09857
13	1	9	.08027	.08253	.08478	.08703	.08927	.09153	.09379
13	1	10	.08178	.08281	.08385	.08489	.08595	.08703	.08813
13	1	11	.08175	.08162	.08155	.08154	.08159	.08170	.08186
13	1	12	.15433	.15141	.14893	.14685	.14513	.14373	.14261
13	2	1	-.61007	-.65809	-.70942	-.76428	-.82287	-.88545	-.95224
13	2	2	-.10523	-.09930	-.09245	-.08458	-.07562	-.06546	-.05400
13	2	3	-.01713	-.00794	.00202	.01278	.02441	.03697	.05055
13	2	4	.02656	.03566	.04527	.05542	.06616	.07753	.08957
13	2	5	.05175	.05976	.06807	.07671	.08571	.09510	.10490
13	2	6	.06730	.07389	.08064	.08756	.09469	.10202	.10959
13	2	7	.07713	.08223	.08739	.09263	.09796	.10339	.10894
13	2	8	.08325	.08688	.09052	.09419	.09789	.10164	.10543
13	2	9	.08676	.08899	.09123	.09347	.09573	.09801	.10031
13	2	10	.08828	.08919	.09014	.09111	.09212	.09317	.09425
13	2	11	.25141	.24873	.24660	.24498	.24382	.24307	.24270
13	3	1	-.66484	-.71619	-.77108	-.82975	-.89240	-.95930	-1.03070
13	3	2	-.11386	-.10738	-.09993	-.09143	-.08179	-.07089	-.05864
13	3	3	-.01782	-.00794	.00273	.01424	.02666	.04006	.05452

Coefficients b_i for the BLUE of σ
Shape parameter k

n	s	i	1.4	1.5	1.6	1.7	1.8	1.9	2.0
13	3	4	.02977	.03947	.04970	.06050	.07192	.08399	.09676
13	3	5	.05717	.06564	.07443	.08357	.09309	.10300	.11335
13	3	6	.07405	.08096	.08804	.09530	.10278	.11048	.11842
13	3	7	.08468	.08996	.09531	.10075	.10628	.11194	.11771
13	3	8	.09124	.09492	.09863	.10237	.10616	.11000	.11390
13	3	9	.09492	.09709	.09928	.10149	.10373	.10602	.10834
13	3	10	.36469	.36346	.36290	.36295	.36357	.36472	.36635
13	4	1	-.73346	-.78903	-.84844	-.91189	-.97965	-1.05196	-1.12910
13	4	2	-.12425	-.11709	-.10893	-.09966	-.08918	-.07739	-.06418
13	4	3	-.01826	-.00753	.00401	.01644	.02982	.04423	.05975
13	4	4	.03419	.04463	.05562	.06720	.07943	.09234	.10599
13	4	5	.06434	.07337	.08274	.09247	.10259	.11313	.12413
13	4	6	.08284	.09013	.09759	.10526	.11314	.12126	.12964
13	4	7	.09442	.09990	.10546	.11112	.11689	.12278	.12881
13	4	8	.10147	.10519	.10895	.11276	.11663	.12056	.12457
13	4	9	.49869	.50044	.50299	.50630	.51034	.51504	.52038
13	5	1	-.82132	-.88239	-.94764	-1.01731	-1.09166	-1.17095	-1.25548
13	5	2	-.13696	-.12896	-.11989	-.10965	-.09813	-.08522	-.07079
13	5	3	-.01822	-.00644	.00621	.01979	.03438	.05006	.06690
13	5	4	.04043	.05178	.06372	.07628	.08952	.10348	.11821
13	5	5	.07403	.08376	.09384	.10429	.11516	.12648	.13827
13	5	6	.09456	.10230	.11023	.11838	.12675	.13538	.14429
13	5	7	.10728	.11299	.11879	.12470	.13074	.13691	.14324
13	5	8	.66022	.66697	.67475	.68352	.69324	.70387	.71537
13	6	1	-.93742	-1.00586	-1.07894	-1.15693	-1.24010	-1.32875	-1.42317
13	6	2	-.15290	-.14379	-.13353	-.12202	-.10913	-.09475	-.07875
13	6	3	-.01729	-.00413	.00996	.02505	.04122	.05855	.07713
13	6	4	.04948	.06204	.07520	.08903	.10358	.11890	.13504
13	6	5	.08758	.09820	.10919	.12059	.13242	.14472	.15753
13	6	6	.11067	.11899	.12751	.13626	.14526	.15453	.16410
13	6	7	.85987	.87454	.89059	.90801	.92676	.94681	.96813
13	7	1	-1.09779	-1.17653	-1.26058	-1.35023	-1.44577	-1.54752	-1.65580
13	7	2	-.17353	-.16289	-.15099	-.13772	-.12295	-.10654	-.08837
13	7	3	-.01465	.00042	.01650	.03366	.05199	.07159	.09254
13	7	4	.06326	.07744	.09229	.10785	.12418	.14135	.15942
13	7	5	.10742	.11924	.13146	.14412	.15725	.17089	.18507
13	7	6	1.11530	1.14231	1.17132	1.20232	1.23529	1.27024	1.30714
14	0	1	-.51410	-.55704	-.60294	-.65198	-.70437	-.76031	-.82004
14	0	2	-.09678	-.09276	-.08792	-.08221	-.07554	-.06784	-.05903
14	0	3	-.02296	-.01559	-.00753	.00126	.01085	.02128	.03261
14	0	4	.01408	.02175	.02991	.03859	.04782	.05764	.06809
14	0	5	.03574	.04277	.05011	.05778	.06580	.07420	.08300
14	0	6	.04937	.05542	.06164	.06804	.07466	.08149	.08855
14	0	7	.05824	.06320	.06822	.07334	.07855	.08388	.08932
14	0	8	.06404	.06788	.07174	.07562	.07953	.08348	.08748
14	0	9	.06771	.07047	.07321	.07595	.07868	.08141	.08415
14	0	10	.06981	.07154	.07324	.07493	.07660	.07827	.07994
14	0	11	.07068	.07142	.07216	.07290	.07365	.07440	.07517
14	0	12	.07045	.07026	.07011	.07001	.06995	.06994	.06996
14	0	13	.06903	.06793	.06697	.06612	.06539	.06475	.06421
14	0	14	.06470	.06275	.06108	.05966	.05845	.05743	.05657
14	1	1	-.54561	-.59056	-.63866	-.69012	-.74514	-.80395	-.86681
14	1	2	-.10257	-.09825	-.09310	-.08705	-.08001	-.07191	-.06264
14	1	3	-.02421	-.01644	-.00795	.00130	.01138	.02233	.03424
14	1	4	.01509	.02314	.03171	.04081	.05049	.06079	.07176
14	1	5	.03807	.04542	.05310	.06112	.06952	.07832	.08754

Coefficients b_i for the BLUE of σ
Shape parameter k

n	s	i	1.4	1.5	1.6	1.7	1.8	1.9	2.0
14	1	6	.05253	.05883	.06531	.07199	.07889	.08603	.09343
14	1	7	.06194	.06706	.07228	.07759	.08302	.08857	.09425
14	1	8	.06807	.07202	.07600	.08000	.08405	.08816	.09232
14	1	9	.07195	.07475	.07755	.08034	.08315	.08597	.08882
14	1	10	.07415	.07586	.07756	.07926	.08096	.08266	.08438
14	1	11	.07503	.07570	.07639	.07710	.07783	.07858	.07936
14	1	12	.07471	.07441	.07418	.07401	.07391	.07387	.07388
14	1	13	.14086	.13804	.13565	.13363	.13195	.13058	.12948
14	2	1	-.58413	-.63147	-.68215	-.73639	-.79441	-.85645	-.92276
14	2	2	-.10946	-.10479	-.09926	-.09280	-.08533	-.07674	-.06695
14	2	3	-.02556	-.01730	-.00831	.00147	.01210	.02366	.03620
14	2	4	.01651	.02501	.03403	.04362	.05380	.06463	.07616
14	2	5	.04110	.04882	.05687	.06528	.07408	.08330	.09298
14	2	6	.05656	.06313	.06989	.07686	.08407	.09153	.09925
14	2	7	.06660	.07191	.07731	.08283	.08846	.09423	.10015
14	2	8	.07313	.07718	.08126	.08538	.08956	.09379	.09810
14	2	9	.07724	.08006	.08289	.08572	.08858	.09146	.09438
14	2	10	.07954	.08120	.08286	.08453	.08622	.08793	.08966
14	2	11	.08037	.08094	.08154	.08218	.08286	.08357	.08432
14	2	12	.22809	.22532	.22308	.22132	.22000	.21908	.21851
14	3	1	-.63098	-.68124	-.73504	-.79263	-.85422	-.92007	-.99046
14	3	2	-.11761	-.11252	-.10655	-.09961	-.09161	-.08246	-.07206
14	3	3	-.02695	-.01813	-.00855	.00185	.01314	.02539	.03866
14	3	4	.01848	.02749	.03705	.04719	.05796	.06940	.08156
14	3	5	.04501	.05314	.06162	.07048	.07975	.08945	.09962
14	3	6	.06166	.06854	.07562	.08292	.09046	.09827	.10636
14	3	7	.07246	.07796	.08357	.08930	.09516	.10116	.10732
14	3	8	.07945	.08359	.08777	.09201	.09630	.10067	.10512
14	3	9	.08380	.08662	.08946	.09232	.09522	.09815	.10112
14	3	10	.08616	.08774	.08934	.09097	.09263	.09432	.09604
14	3	11	.32853	.32680	.32570	.32519	.32522	.32574	.32671
14	4	1	-.68846	-.74234	-.80001	-.86171	-.92768	-.99820	-1.07353
14	4	2	-.12730	-.12172	-.11521	-.10770	-.09908	-.08926	-.07812
14	4	3	-.02835	-.01885	-.00856	.00258	.01465	.02772	.04186
14	4	4	.02119	.03083	.04102	.05183	.06328	.07544	.08834
14	4	5	.05008	.05871	.06770	.07709	.08689	.09715	.10790
14	4	6	.06818	.07542	.08286	.09054	.09848	.10668	.11519
14	4	7	.07987	.08560	.09145	.09741	.10352	.10978	.11621
14	4	8	.08739	.09163	.09592	.10027	.10469	.10920	.11379
14	4	9	.09199	.09480	.09764	.10051	.10343	.10639	.10941
14	4	10	.44540	.44591	.44719	.44917	.45182	.45509	.45894
14	5	1	-.76016	-.81861	-.88116	-.94804	-1.01953	-1.09588	-1.17741
14	5	2	-.13898	-.13279	-.12563	-.11741	-.10804	-.09739	-.08537
14	5	3	-.02967	-.01934	-.00818	.00387	.01690	.03097	.04617
14	5	4	.02499	.03538	.04635	.05796	.07026	.08328	.09709
14	5	5	.05680	.06603	.07563	.08565	.09610	.10704	.11848
14	5	6	.07666	.08433	.09221	.10034	.10873	.11742	.12641
14	5	7	.08942	.09542	.10153	.10777	.11416	.12072	.12746
14	5	8	.09755	.10189	.10629	.11077	.11533	.11998	.12473
14	5	9	.58339	.58771	.59296	.59910	.60609	.61388	.62243
14	6	1	-.85174	-.91611	-.98496	-1.05854	-1.13714	-1.22103	-1.31054
14	6	2	-.15330	-.14635	-.13837	-.12926	-.11893	-.10725	-.09411
14	6	3	-.03075	-.01937	-.00712	.00609	.02032	.03565	.05219
14	6	4	.03041	.04175	.05370	.06633	.07967	.09378	.10872
14	6	5	.06590	.07588	.08625	.09705	.10832	.12009	.13239
14	6	6	.08797	.09616	.10458	.11326	.12222	.13148	.14107

Coefficients b_i for the BLUE of σ
Shape parameter k

n	s	i	1.4	1.5	1.6	1.7	1.8	1.9	2.0
14	6	7	.10203	.10833	.11476	.12133	.12807	.13498	.14208
14	6	8	.74948	.75971	.77115	.78375	.79748	.81231	.82819
14	7	1	-.97255	-1.04483	-1.12210	-1.20463	-1.29273	-1.38669	-1.48685
14	7	2	-.17132	-.16336	-.15429	-.14402	-.13243	-.11940	-.10479
14	7	3	-.03128	-.01853	-.00484	.00987	.02567	.04267	.06094
14	7	4	.03840	.05097	.06421	.07815	.09285	.10838	.12480
14	7	5	.07868	.08962	.10098	.11280	.12511	.13795	.15137
14	7	6	.10355	.11241	.12152	.13090	.14058	.15058	.16093
14	7	7	.95453	.97371	.99452	1.01694	1.04094	1.06651	1.09361
15	0	1	-.49944	-.54214	-.58785	-.63675	-.68906	-.74500	-.80479
15	0	2	-.10054	-.09746	-.09362	-.08897	-.08343	-.07693	-.06938
15	0	3	-.02919	-.02254	-.01521	-.00715	.00168	.01134	.02189
15	0	4	.00693	.01413	.02183	.03006	.03886	.04826	.05831
15	0	5	.02830	.03507	.04218	.04964	.05749	.06574	.07442
15	0	6	.04193	.04791	.05409	.06049	.06713	.07402	.08118
15	0	7	.05098	.05603	.06118	.06645	.07185	.07739	.08309
15	0	8	.05706	.06114	.06526	.06943	.07365	.07794	.08231
15	0	9	.06110	.06423	.06735	.07048	.07361	.07677	.07996
15	0	10	.06366	.06586	.06804	.07021	.07236	.07452	.07668
15	0	11	.06507	.06639	.06769	.06898	.07026	.07154	.07282
15	0	12	.06554	.06602	.06650	.06699	.06750	.06802	.06856
15	0	13	.06513	.06481	.06454	.06432	.06415	.06403	.06396
15	0	14	.06373	.06263	.06166	.06081	.06008	.05944	.05889
15	0	15	.05975	.05791	.05635	.05501	.05388	.05292	.05212
15	1	1	-.52751	-.57204	-.61977	-.67089	-.72562	-.78421	-.84689
15	1	2	-.10608	-.10277	-.09869	-.09378	-.08795	-.08113	-.07323
15	1	3	-.03071	-.02371	-.01601	-.00757	.00168	.01179	.02283
15	1	4	.00744	.01498	.02304	.03164	.04084	.05067	.06117
15	1	5	.03000	.03707	.04448	.05227	.06046	.06907	.07815
15	1	6	.04440	.05061	.05704	.06370	.07062	.07780	.08527
15	1	7	.05395	.05917	.06451	.06998	.07559	.08136	.08729
15	1	8	.06036	.06456	.06881	.07312	.07749	.08194	.08648
15	1	9	.06462	.06781	.07101	.07422	.07745	.08072	.08402
15	1	10	.06730	.06952	.07173	.07393	.07614	.07835	.08058
15	1	11	.06877	.07006	.07134	.07263	.07392	.07522	.07653
15	1	12	.06923	.06964	.07007	.07053	.07101	.07152	.07206
15	1	13	.06873	.06832	.06797	.06769	.06749	.06734	.06725
15	1	14	.12950	.12678	.12447	.12252	.12090	.11956	.11849
15	2	1	-.56140	-.60810	-.65815	-.71179	-.76924	-.83075	-.89658
15	2	2	-.11261	-.10903	-.10466	-.09944	-.09327	-.08607	-.07776
15	2	3	-.03240	-.02499	-.01687	-.00798	.00175	.01236	.02394
15	2	4	.00821	.01613	.02459	.03362	.04326	.05355	.06455
15	2	5	.03221	.03960	.04736	.05550	.06406	.07306	.08254
15	2	6	.04751	.05399	.06068	.06762	.07482	.08231	.09010
15	2	7	.05766	.06307	.06860	.07427	.08009	.08608	.09225
15	2	8	.06446	.06878	.07315	.07759	.08210	.08670	.09140
15	2	9	.06896	.07220	.07546	.07874	.08206	.08541	.08881
15	2	10	.07178	.07399	.07620	.07842	.08065	.08290	.08517
15	2	11	.07329	.07452	.07576	.07701	.07828	.07957	.08088
15	2	12	.07370	.07401	.07436	.07475	.07518	.07565	.07616
15	2	13	.20864	.20582	.20352	.20169	.20027	.19923	.19854
15	3	1	-.60205	-.65133	-.70416	-.76078	-.82141	-.88633	-.95580
15	3	2	-.12028	-.11638	-.11167	-.10607	-.09951	-.09187	-.08308
15	3	3	-.03423	-.02635	-.01774	-.00833	.00194	.01314	.02534
15	3	4	.00931	.01768	.02660	.03612	.04626	.05708	.06864
15	3	5	.03503	.04280	.05094	.05948	.06845	.07789	.08783

Coefficients b_i for the BLUE of σ
Shape parameter k

n	s	i	1.4	1.5	1.6	1.7	1.8	1.9	2.0
15	3	6	.05141	.05817	.06517	.07242	.07994	.08775	.09589
15	3	7	.06225	.06787	.07361	.07950	.08555	.09177	.09819
15	3	8	.06951	.07395	.07845	.08302	.08768	.09243	.09728
15	3	9	.07428	.07757	.08088	.08422	.08760	.09103	.09452
15	3	10	.07724	.07943	.08162	.08384	.08607	.08834	.09064
15	3	11	.07877	.07991	.08108	.08228	.08351	.08477	.08606
15	3	12	.29874	.29667	.29521	.29430	.29391	.29399	.29449
15	4	1	-.65108	-.70351	-.75971	-.81992	-.88439	-.95338	-1.02718
15	4	2	-.12929	-.12502	-.11991	-.11389	-.10685	-.09870	-.08935
15	4	3	-.03619	-.02777	-.01857	-.00855	.00236	.01424	.02716
15	4	4	.01087	.01977	.02924	.03931	.05005	.06149	.07369
15	4	5	.03865	.04686	.05545	.06446	.07391	.08385	.09430
15	4	6	.05632	.06342	.07076	.07835	.08624	.09443	.10296
15	4	7	.06799	.07383	.07981	.08594	.09224	.09873	.10542
15	4	8	.07576	.08033	.08497	.08969	.09449	.09940	.10443
15	4	9	.08084	.08417	.08753	.09092	.09437	.09787	.10144
15	4	10	.08393	.08607	.08824	.09043	.09266	.09493	.09724
15	4	11	.40220	.40185	.40222	.40325	.40491	.40715	.40992
15	5	1	-.71097	-.76730	-.82766	-.89229	-.96146	-1.03544	-1.11454
15	5	2	-.13999	-.13528	-.12970	-.12316	-.11556	-.10680	-.09679
15	5	3	-.03829	-.02919	-.01930	-.00855	.00313	.01583	.02961
15	5	4	.01308	.02261	.03273	.04348	.05492	.06709	.08005
15	5	5	.04336	.05209	.06122	.07078	.08080	.09132	.10238
15	5	6	.06258	.07007	.07781	.08583	.09414	.10276	.11174
15	5	7	.07523	.08134	.08759	.09401	.10059	.10738	.11437
15	5	8	.08361	.08833	.09311	.09799	.10296	.10805	.11325
15	5	9	.08902	.09238	.09578	.09923	.10274	.10631	.10995
15	5	10	.52236	.52496	.52842	.53269	.53774	.54351	.54997
15	6	1	-.78549	-.84673	-.91230	-.98249	-1.05756	-1.13781	-1.22354
15	6	2	-.15289	-.14764	-.14148	-.13430	-.12602	-.11652	-.10569
15	6	3	-.04046	-.03055	-.01980	-.00815	.00448	.01817	.03299
15	6	4	.01625	.02655	.03747	.04904	.06134	.07440	.08829
15	6	5	.04961	.05897	.06875	.07898	.08969	.10093	.11272
15	6	6	.07073	.07869	.08692	.09543	.10425	.11340	.12291
15	6	7	.08456	.09098	.09756	.10430	.11123	.11836	.12571
15	6	8	.09365	.09852	.10348	.10853	.11369	.11898	.12439
15	6	9	.66405	.67120	.67941	.68866	.69891	.71010	.72221
15	7	1	-.88051	-.94806	-1.02038	-1.09773	-1.18041	-1.26872	-1.36300
15	7	2	-.16874	-.16281	-.15591	-.14793	-.13878	-.12834	-.11649
15	7	3	-.04261	-.03167	-.01984	-.00706	.00677	.02171	.03786
15	7	4	.02087	.03214	.04406	.05668	.07005	.08424	.09930
15	7	5	.05812	.06828	.07887	.08994	.10151	.11363	.12635
15	7	6	.08160	.09016	.09898	.10810	.11755	.12735	.13753
15	7	7	.09688	.10368	.11065	.11779	.12513	.13269	.14048
15	7	8	.83440	.84829	.86357	.88021	.89818	.91744	.93797
15	8	1	-1.00572	-1.08169	-1.16298	-1.24988	-1.34270	-1.44178	-1.54745
15	8	2	-.18875	-.18191	-.17403	-.16500	-.15471	-.14302	-.12983
15	8	3	-.04456	-.03226	-.01901	-.00473	.01066	.02726	.04515
15	8	4	.02782	.04036	.05359	.06757	.08235	.09800	.11458
15	8	5	.07011	.08130	.09296	.10511	.11780	.13107	.14498
15	8	6	.09663	.10594	.11553	.12544	.13570	.14633	.15737
15	8	7	1.04447	1.06827	1.09395	1.12149	1.15089	1.18214	1.21520
16	0	1	-.48593	-.52835	-.57382	-.62254	-.67471	-.73057	-.79035
16	0	2	-.10351	-.10126	-.09832	-.09462	-.09010	-.08468	-.07828
16	0	3	-.03444	-.02843	-.02177	-.01441	-.00628	.00265	.01245
16	0	4	.00082	.00757	.01482	.02261	.03097	.03995	.04958

Coefficients b_i for the BLUE of σ
Shape parameter k

n	s	i	1.4	1.5	1.6	1.7	1.8	1.9	2.0
16	0	5	.02187	.02837	.03522	.04245	.05009	.05815	.06667
16	0	6	.03546	.04132	.04742	.05376	.06036	.06725	.07444
16	0	7	.04461	.04968	.05489	.06024	.06575	.07143	.07730
16	0	8	.05089	.05512	.05942	.06378	.06823	.07277	.07741
16	0	9	.05520	.05859	.06199	.06541	.06886	.07236	.07590
16	0	10	.05809	.06065	.06320	.06574	.06828	.07083	.07340
16	0	11	.05990	.06166	.06340	.06513	.06685	.06856	.07028
16	0	12	.06084	.06184	.06283	.06381	.06479	.06577	.06675
16	0	13	.06104	.06131	.06160	.06191	.06223	.06258	.06294
16	0	14	.06053	.06011	.05975	.05945	.05920	.05900	.05885
16	0	15	.05916	.05807	.05711	.05626	.05553	.05489	.05435
16	0	16	.05548	.05375	.05227	.05102	.04995	.04905	.04830
16	1	1	-.51113	-.55524	-.60258	-.65334	-.70775	-.76605	-.82851
16	1	2	-.10878	-.10636	-.10323	-.09933	-.09459	-.08892	-.08224
16	1	3	-.03612	-.02982	-.02285	-.01515	-.00667	.00265	.01288
16	1	4	.00096	.00801	.01558	.02370	.03241	.04177	.05180
16	1	5	.02310	.02986	.03700	.04452	.05247	.06086	.06974
16	1	6	.03739	.04347	.04980	.05638	.06325	.07041	.07789
16	1	7	.04701	.05225	.05765	.06319	.06891	.07481	.08091
16	1	8	.05361	.05797	.06240	.06691	.07151	.07621	.08103
16	1	9	.05814	.06160	.06509	.06862	.07218	.07579	.07946
16	1	10	.06116	.06376	.06636	.06896	.07156	.07419	.07684
16	1	11	.06305	.06481	.06657	.06831	.07006	.07182	.07358
16	1	12	.06402	.06499	.06596	.06693	.06791	.06890	.06990
16	1	13	.06420	.06441	.06465	.06492	.06522	.06555	.06592
16	1	14	.06361	.06310	.06268	.06232	.06204	.06182	.06166
16	1	15	.11978	.11717	.11494	.11306	.11150	.11020	.10916
16	2	1	-.54124	-.58732	-.63677	-.68983	-.74672	-.80770	-.87304
16	2	2	-.11497	-.11234	-.10899	-.10484	-.09983	-.09387	-.08687
16	2	3	-.03801	-.03137	-.02404	-.01596	-.00708	.00268	.01337
16	2	4	.00125	.00864	.01656	.02505	.03416	.04393	.05440
16	2	5	.02468	.03175	.03919	.04704	.05532	.06407	.07332
16	2	6	.03980	.04613	.05271	.05956	.06669	.07414	.08192
16	2	7	.04998	.05541	.06099	.06674	.07266	.07878	.08511
16	2	8	.05696	.06144	.06600	.07065	.07540	.08027	.08525
16	2	9	.06173	.06527	.06884	.07245	.07611	.07982	.08360
16	2	10	.06491	.06754	.07016	.07280	.07545	.07814	.08085
16	2	11	.06688	.06862	.07036	.07211	.07386	.07563	.07742
16	2	12	.06787	.06877	.06969	.07062	.07158	.07255	.07355
16	2	13	.06799	.06811	.06827	.06847	.06873	.06902	.06936
16	2	14	.19216	.18935	.18702	.18514	.18367	.18256	.18178
16	3	1	-.57692	-.62532	-.67727	-.73300	-.79276	-.85681	-.92543
16	3	2	-.12216	-.11929	-.11568	-.11125	-.10593	-.09963	-.09226
16	3	3	-.04011	-.03307	-.02533	-.01681	-.00746	.00278	.01400
16	3	4	.00174	.00952	.01784	.02676	.03631	.04654	.05751
16	3	5	.02671	.03411	.04190	.05011	.05876	.06790	.07756
16	3	6	.04280	.04940	.05626	.06339	.07083	.07858	.08669
16	3	7	.05362	.05926	.06505	.07101	.07716	.08351	.09008
16	3	8	.06103	.06565	.07036	.07515	.08006	.08508	.09023
16	3	9	.06609	.06970	.07335	.07704	.08079	.08460	.08848
16	3	10	.06944	.07208	.07472	.07739	.08008	.08280	.08557
16	3	11	.07149	.07319	.07490	.07662	.07837	.08013	.08193
16	3	12	.07246	.07328	.07412	.07500	.07591	.07685	.07782
16	3	13	.27380	.27150	.26978	.26859	.26790	.26766	.26782
16	4	1	-.61936	-.67054	-.72546	-.78437	-.84752	-.91519	-.98766
16	4	2	-.13053	-.12738	-.12347	-.11872	-.11305	-.10636	-.09855

Coefficients b_i for the BLUE of σ
Shape parameter k

n	s	i	1.4	1.5	1.6	1.7	1.8	1.9	2.0
16	4	3	-.04241	-.03492	-.02669	-.01767	-.00779	.00303	.01485
16	4	4	.00251	.01073	.01953	.02893	.03899	.04975	.06128
16	4	5	.02928	.03707	.04526	.05389	.06297	.07256	.08268
16	4	6	.04653	.05344	.06062	.06808	.07585	.08395	.09242
16	4	7	.05811	.06397	.07000	.07620	.08259	.08920	.09603
16	4	8	.06602	.07079	.07564	.08060	.08567	.09086	.09619
16	4	9	.07140	.07508	.07880	.08258	.08642	.09033	.09431
16	4	10	.07493	.07756	.08022	.08290	.08562	.08838	.09119
16	4	11	.07703	.07867	.08034	.08203	.08375	.08550	.08729
16	4	12	.36649	.36551	.36522	.36557	.36650	.36798	.36996
16	5	1	-.67033	-.72488	-.78341	-.84615	-.91339	-.98540	-1.06248
16	5	2	-.14035	-.13688	-.13263	-.12749	-.12140	-.11425	-.10595
16	5	3	-.04494	-.03691	-.02812	-.01850	-.00798	.00350	.01603
16	5	4	.00366	.01242	.02176	.03174	.04239	.05378	.06596
16	5	5	.03260	.04085	.04951	.05861	.06820	.07830	.08897
16	5	6	.05122	.05849	.06603	.07387	.08203	.09054	.09942
16	5	7	.06369	.06982	.07610	.08258	.08925	.09615	.10328
16	5	8	.07218	.07711	.08214	.08727	.09252	.09790	.10343
16	5	9	.07792	.08167	.08548	.08934	.09327	.09728	.10138
16	5	10	.08163	.08426	.08692	.08961	.09235	.09514	.09798
16	5	11	.47271	.47406	.47621	.47913	.48276	.48706	.49198
16	6	1	-.73244	-.79114	-.85409	-.92155	-.99380	-1.07113	-1.15386
16	6	2	-.15201	-.14816	-.14349	-.13790	-.13131	-.12361	-.11471
16	6	3	-.04770	-.03902	-.02955	-.01921	-.00794	.00434	.01771
16	6	4	.00538	.01477	.02477	.03543	.04680	.05893	.07189
16	6	5	.03694	.04573	.05495	.06463	.07481	.08553	.09682
16	6	6	.05721	.06490	.07288	.08116	.08978	.09876	.10813
16	6	7	.07074	.07717	.08377	.09056	.09756	.10479	.11227
16	6	8	.07991	.08502	.09024	.09557	.10103	.10662	.11237
16	6	9	.08604	.08988	.09376	.09772	.10175	.10587	.11008
16	6	10	.59594	.60086	.60676	.61359	.62132	.62990	.63929
16	7	1	-.80957	-.87349	-.94199	-1.01536	-1.09389	-1.17789	-1.26768
16	7	2	-.16608	-.16176	-.15657	-.15042	-.14322	-.13485	-.12522
16	7	3	-.05070	-.04122	-.03091	-.01969	-.00749	.00578	.02019
16	7	4	.00793	.01810	.02892	.04042	.05266	.06570	.07960
16	7	5	.04272	.05218	.06208	.07246	.08336	.09483	.10690
16	7	6	.06501	.07322	.08172	.09054	.09971	.10926	.11921
16	7	7	.07982	.08661	.09358	.10075	.10814	.11577	.12366
16	7	8	.08979	.09512	.10056	.10612	.11181	.11766	.12367
16	7	9	.74108	.75124	.76262	.77519	.78891	.80374	.81966
16	8	1	-.90780	-.97842	-1.05407	-1.13505	-1.22166	-1.31424	-1.41312
16	8	2	-.18339	-.17847	-.17263	-.16577	-.15779	-.14858	-.13802
16	8	3	-.05391	-.04341	-.03203	-.01969	-.00631	.00820	.02392
16	8	4	.01177	.02293	.03476	.04733	.06067	.07486	.08996
16	8	5	.05064	.06093	.07168	.08294	.09476	.10716	.12020
16	8	6	.07544	.08429	.09345	.10295	.11281	.12306	.13374
16	8	7	.09184	.09907	.10649	.11413	.12199	.13012	.13851
16	8	8	.91541	.93308	.95234	.97316	.99553	1.01942	1.04480
17	0	1	-.47341	-.51554	-.56075	-.60924	-.66123	-.71694	-.77664
17	0	2	-.10585	-.10435	-.10221	-.09937	-.09576	-.09132	-.08597
17	0	3	-.03887	-.03347	-.02743	-.02071	-.01325	-.00500	.00410
17	0	4	-.00445	.00187	.00869	.01605	.02398	.03253	.04174
17	0	5	.01627	.02249	.02908	.03606	.04347	.05131	.05964
17	0	6	.02977	.03550	.04147	.04772	.05424	.06108	.06825
17	0	7	.03898	.04403	.04925	.05463	.06019	.06595	.07193
17	0	8	.04540	.04972	.05413	.05863	.06323	.06795	.07280

Coefficients b_i for the BLUE of σ
Shape parameter k

n	s	i	1.4	1.5	1.6	1.7	1.8	1.9	2.0
17	0	9	.04992	.05349	.05709	.06073	.06443	.06818	.07200
17	0	10	.05306	.05588	.05871	.06154	.06439	.06726	.07016
17	0	11	.05516	.05726	.05935	.06143	.06351	.06558	.06767
17	0	12	.05645	.05786	.05925	.06064	.06201	.06338	.06475
17	0	13	.05706	.05781	.05856	.05930	.06004	.06079	.06155
17	0	14	.05708	.05720	.05734	.05750	.05768	.05789	.05812
17	0	15	.05650	.05601	.05559	.05522	.05492	.05467	.05446
17	0	16	.05518	.05410	.05315	.05232	.05160	.05097	.05043
17	0	17	.05177	.05013	.04874	.04755	.04655	.04570	.04499
17	1	1	-.49620	-.53989	-.58682	-.63721	-.69127	-.74927	-.81145
17	1	2	-.11087	-.10924	-.10695	-.10396	-.10018	-.09555	-.08999
17	1	3	-.04066	-.03501	-.02870	-.02169	-.01392	-.00534	.00413
17	1	4	-.00458	.00200	.00910	.01675	.02500	.03389	.04347
17	1	5	.01713	.02360	.03044	.03769	.04538	.05353	.06218
17	1	6	.03128	.03721	.04341	.04988	.05665	.06374	.07118
17	1	7	.04093	.04615	.05154	.05711	.06287	.06884	.07504
17	1	8	.04766	.05211	.05665	.06129	.06605	.07094	.07596
17	1	9	.05238	.05604	.05974	.06349	.06730	.07118	.07513
17	1	10	.05567	.05855	.06144	.06434	.06726	.07022	.07322
17	1	11	.05786	.05999	.06211	.06422	.06634	.06847	.07062
17	1	12	.05920	.06060	.06200	.06339	.06478	.06618	.06758
17	1	13	.05983	.06054	.06126	.06199	.06272	.06348	.06424
17	1	14	.05982	.05988	.05997	.06009	.06025	.06045	.06068
17	1	15	.05916	.05860	.05811	.05770	.05736	.05709	.05688
17	1	16	.11139	.10887	.10673	.10492	.10341	.10216	.10114
17	2	1	-.52317	-.56866	-.61754	-.67003	-.72637	-.78683	-.85167
17	2	2	-.11671	-.11492	-.11247	-.10929	-.10531	-.10045	-.09463
17	2	3	-.04268	-.03673	-.03012	-.02278	-.01467	-.00571	.00416
17	2	4	-.00464	.00224	.00965	.01764	.02623	.03549	.04546
17	2	5	.01825	.02498	.03211	.03965	.04764	.05612	.06511
17	2	6	.03316	.03932	.04575	.05246	.05949	.06685	.07457
17	2	7	.04332	.04872	.05430	.06006	.06602	.07220	.07862
17	2	8	.05041	.05499	.05967	.06445	.06936	.07441	.07960
17	2	9	.05538	.05912	.06292	.06676	.07068	.07466	.07874
17	2	10	.05883	.06175	.06469	.06765	.07064	.07366	.07674
17	2	11	.06112	.06325	.06538	.06752	.06966	.07183	.07402
17	2	12	.06251	.06388	.06525	.06663	.06802	.06942	.07083
17	2	13	.06313	.06378	.06445	.06514	.06585	.06658	.06734
17	2	14	.06306	.06303	.06305	.06312	.06324	.06340	.06360
17	2	15	.17804	.17524	.17292	.17103	.16953	.16837	.16753
17	3	1	-.55481	-.60239	-.65352	-.70844	-.76739	-.83064	-.89847
17	3	2	-.12345	-.12148	-.11883	-.11544	-.11122	-.10610	-.10000
17	3	3	-.04493	-.03865	-.03168	-.02398	-.01547	-.00609	.00422
17	3	4	-.00458	.00263	.01040	.01875	.02774	.03741	.04781
17	3	5	.01968	.02671	.03415	.04202	.05035	.05918	.06854
17	3	6	.03548	.04189	.04857	.05555	.06286	.07051	.07853
17	3	7	.04623	.05183	.05761	.06358	.06976	.07616	.08282
17	3	8	.05373	.05845	.06327	.06821	.07328	.07849	.08386
17	3	9	.05898	.06281	.06669	.07064	.07466	.07876	.08295
17	3	10	.06261	.06557	.06855	.07156	.07461	.07770	.08084
17	3	11	.06501	.06713	.06926	.07140	.07357	.07575	.07797
17	3	12	.06643	.06775	.06909	.07044	.07181	.07320	.07461
17	3	13	.06703	.06760	.06819	.06883	.06949	.07019	.07092
17	3	14	.25261	.25015	.24825	.24688	.24597	.24549	.24540
17	4	1	-.59200	-.64206	-.69584	-.75360	-.81558	-.88207	-.95336
17	4	2	-.13123	-.12906	-.12619	-.12255	-.11806	-.11265	-.10621

Coefficients b_i for the BLUE of σ
Shape parameter k

n	s	i	1.4	1.5	1.6	1.7	1.8	1.9	2.0
17	4	3	-.04742	-.04076	-.03339	-.02526	-.01630	-.00645	.00438
17	4	4	-.00438	.00323	.01140	.02018	.02961	.03974	.05063
17	4	5	.02149	.02888	.03667	.04491	.05362	.06285	.07263
17	4	6	.03833	.04502	.05200	.05928	.06690	.07487	.08323
17	4	7	.04978	.05560	.06160	.06780	.07422	.08087	.08778
17	4	8	.05775	.06262	.06760	.07271	.07794	.08333	.08888
17	4	9	.06331	.06724	.07122	.07526	.07939	.08360	.08791
17	4	10	.06714	.07013	.07316	.07621	.07931	.08246	.08567
17	4	11	.06965	.07175	.07387	.07601	.07818	.08038	.08262
17	4	12	.07109	.07235	.07363	.07494	.07627	.07764	.07904
17	4	13	.33648	.33505	.33428	.33411	.33450	.33541	.33680
17	5	1	-.63603	-.68905	-.74600	-.80714	-.87272	-.94303	-1.01839
17	5	2	-.14027	-.13786	-.13474	-.13081	-.12602	-.12026	-.11344
17	5	3	-.05018	-.04308	-.03525	-.02662	-.01714	-.00673	.00468
17	5	4	-.00395	.00411	.01275	.02202	.03196	.04262	.05407
17	5	5	.02382	.03160	.03981	.04847	.05763	.06731	.07757
17	5	6	.04187	.04890	.05621	.06384	.07181	.08015	.08888
17	5	7	.05414	.06020	.06646	.07292	.07960	.08653	.09373
17	5	8	.06265	.06769	.07285	.07814	.08356	.08914	.09489
17	5	9	.06857	.07259	.07668	.08083	.08507	.08941	.09384
17	5	10	.07262	.07564	.07870	.08180	.08495	.08816	.09143
17	5	11	.07522	.07730	.07939	.08152	.08369	.08589	.08814
17	5	12	.43154	.43196	.43314	.43504	.43761	.44081	.44458
17	6	1	-.68878	-.74538	-.80615	-.87136	-.94127	-1.01619	-1.09643
17	6	2	-.15087	-.14818	-.14476	-.14051	-.13534	-.12918	-.12191
17	6	3	-.05326	-.04563	-.03724	-.02804	-.01794	-.00688	.00523
17	6	4	-.00320	.00539	.01459	.02443	.03497	.04627	.05838
17	6	5	.02683	.03509	.04378	.05294	.06261	.07283	.08364
17	6	6	.04633	.05374	.06144	.06947	.07785	.08661	.09579
17	6	7	.05955	.06590	.07245	.07921	.08621	.09345	.10098
17	6	8	.06870	.07394	.07929	.08478	.09042	.09622	.10219
17	6	9	.07503	.07915	.08335	.08762	.09199	.09645	.10103
17	6	10	.07931	.08236	.08545	.08859	.09179	.09505	.09838
17	6	11	.54035	.54363	.54781	.55286	.55872	.56536	.57273
17	7	1	-.75293	-.81393	-.87939	-.94959	-1.02481	-1.10537	-1.19159
17	7	2	-.16344	-.16043	-.15665	-.15200	-.14640	-.13976	-.13196
17	7	3	-.05667	-.04841	-.03936	-.02945	-.01861	-.00677	.00617
17	7	4	-.00198	.00726	.01712	.02765	.03892	.05097	.06386
17	7	5	.03080	.03961	.04888	.05864	.06892	.07978	.09126
17	7	6	.05203	.05989	.06806	.07656	.08543	.09470	.10439
17	7	7	.06639	.07308	.07997	.08709	.09444	.10206	.10997
17	7	8	.07628	.08174	.08733	.09306	.09894	.10499	.11122
17	7	9	.08308	.08732	.09164	.09604	.10054	.10515	.10988
17	7	10	.66645	.67387	.68241	.69201	.70263	.71425	.72680
17	8	1	-.83251	-.89901	-.97034	-1.04679	-1.12866	-1.21628	-1.30998
17	8	2	-.17862	-.17521	-.17098	-.16584	-.15971	-.15247	-.14402
17	8	3	-.06047	-.05143	-.04156	-.03079	-.01903	-.00622	.00773
17	8	4	-.00003	.00999	.02067	.03206	.04421	.05718	.07104
17	8	5	.03611	.04562	.05559	.06608	.07712	.08876	.10104
17	8	6	.05948	.06789	.07662	.08570	.09516	.10504	.11536
17	8	7	.07521	.08231	.08962	.09715	.10494	.11301	.12138
17	8	8	.08599	.09171	.09757	.10358	.10975	.11609	.12264
17	8	9	.81483	.82813	.84281	.85886	.87623	.89489	.91481
17	9	1	-.93376	-1.00733	-1.08619	-1.17067	-1.26109	-1.35777	-1.46109
17	9	2	-.19733	-.19340	-.18860	-.18284	-.17603	-.16804	-.15876
17	9	3	-.06468	-.05465	-.04374	-.03186	-.01895	-.00492	.01034

Coefficients b_i for the BLUE of σ
Shape parameter k

n	s	i	1.4	1.5	1.6	1.7	1.8	1.9	2.0
17	9	4	.00304	.01406	.02578	.03824	.05151	.06566	.08074
17	9	5	.04344	.05381	.06468	.07608	.08807	.10069	.11399
17	9	6	.06947	.07856	.08800	.09780	.10801	.11865	.12976
17	9	7	.08690	.09450	.10232	.11038	.11871	.12733	.13627
17	9	8	.99291	1.01443	1.03776	1.06287	1.08976	1.11840	1.14877
18	0	1	-.46178	-.50360	-.54852	-.59676	-.64852	-.70406	-.76362
18	0	2	-.10770	-.10686	-.10544	-.10338	-.10060	-.09705	-.09266
18	0	3	-.04266	-.03781	-.03234	-.02622	-.01938	-.01177	-.00333
18	0	4	-.00902	-.00312	.00329	.01023	.01775	.02589	.03468
18	0	5	.01136	.01730	.02363	.03035	.03751	.04513	.05323
18	0	6	.02476	.03032	.03615	.04227	.04870	.05545	.06255
18	0	7	.03398	.03898	.04416	.04953	.05510	.06090	.06693
18	0	8	.04050	.04486	.04933	.05391	.05862	.06347	.06847
18	0	9	.04517	.04886	.05260	.05641	.06028	.06424	.06828
18	0	10	.04850	.05152	.05456	.05762	.06071	.06384	.06701
18	0	11	.05083	.05320	.05556	.05792	.06029	.06267	.06506
18	0	12	.05238	.05412	.05584	.05755	.05925	.06095	.06266
18	0	13	.05329	.05443	.05555	.05665	.05776	.05886	.05996
18	0	14	.05368	.05423	.05478	.05534	.05590	.05646	.05704
18	0	15	.05357	.05356	.05359	.05364	.05371	.05382	.05395
18	0	16	.05295	.05241	.05194	.05153	.05119	.05089	.05066
18	0	17	.05168	.05062	.04969	.04888	.04817	.04755	.04702
18	0	18	.04851	.04696	.04564	.04451	.04356	.04276	.04209
18	1	1	-.48250	-.52577	-.57230	-.62230	-.67600	-.73367	-.79555
18	1	2	-.11246	-.11153	-.11001	-.10783	-.10493	-.10124	-.09668
18	1	3	-.04451	-.03944	-.03374	-.02737	-.02026	-.01236	-.00361
18	1	4	-.00936	-.00322	.00343	.01064	.01844	.02688	.03600
18	1	5	.01194	.01810	.02465	.03162	.03904	.04693	.05534
18	1	6	.02593	.03169	.03772	.04406	.05071	.05770	.06506
18	1	7	.03557	.04073	.04608	.05162	.05739	.06338	.06963
18	1	8	.04238	.04687	.05147	.05619	.06105	.06606	.07123
18	1	9	.04725	.05104	.05488	.05880	.06279	.06687	.07104
18	1	10	.05073	.05381	.05692	.06006	.06324	.06646	.06973
18	1	11	.05315	.05556	.05796	.06037	.06280	.06524	.06771
18	1	12	.05476	.05651	.05825	.05998	.06172	.06346	.06521
18	1	13	.05571	.05683	.05794	.05905	.06016	.06128	.06240
18	1	14	.05610	.05661	.05714	.05767	.05822	.05879	.05937
18	1	15	.05596	.05590	.05588	.05589	.05595	.05604	.05616
18	1	16	.05527	.05467	.05414	.05369	.05331	.05300	.05275
18	1	17	.10407	.10165	.09958	.09784	.09638	.09517	.09419
18	2	1	-.50685	-.55176	-.60008	-.65202	-.70783	-.76777	-.83212
18	2	2	-.11799	-.11694	-.11528	-.11296	-.10991	-.10604	-.10129
18	2	3	-.04659	-.04127	-.03531	-.02866	-.02125	-.01304	-.00394
18	2	4	-.00967	-.00327	.00366	.01116	.01927	.02804	.03751
18	2	5	.01270	.01910	.02591	.03315	.04084	.04903	.05774
18	2	6	.02739	.03336	.03961	.04617	.05306	.06030	.06792
18	2	7	.03751	.04284	.04836	.05410	.06005	.06625	.07271
18	2	8	.04465	.04927	.05401	.05888	.06389	.06906	.07440
18	2	9	.04976	.05364	.05759	.06161	.06571	.06990	.07421
18	2	10	.05340	.05655	.05972	.06293	.06618	.06948	.07284
18	2	11	.05594	.05837	.06080	.06325	.06571	.06821	.07073
18	2	12	.05762	.05936	.06109	.06283	.06458	.06634	.06812
18	2	13	.05859	.05967	.06076	.06185	.06295	.06406	.06519
18	2	14	.05896	.05942	.05989	.06039	.06091	.06146	.06203
18	2	15	.05877	.05863	.05854	.05850	.05851	.05857	.05867
18	2	16	.16580	.16304	.16074	.15885	.15734	.15616	.15528

Coefficients b_i for the BLUE of σ
Shape parameter k

n	s	i	1.4	1.5	1.6	1.7	1.8	1.9	2.0
18	3	1	-.53514	-.58196	-.63234	-.68649	-.74468	-.80718	-.87426
18	3	2	-.12431	-.12313	-.12133	-.11885	-.11561	-.11155	-.10657
18	3	3	-.04891	-.04332	-.03706	-.03009	-.02235	-.01377	-.00428
18	3	4	-.00993	-.00323	.00401	.01183	.02029	.02942	.03927
18	3	5	.01367	.02035	.02744	.03497	.04297	.05148	.06054
18	3	6	.02918	.03538	.04187	.04868	.05582	.06333	.07124
18	3	7	.03984	.04537	.05108	.05701	.06318	.06959	.07628
18	3	8	.04738	.05214	.05702	.06204	.06721	.07254	.07805
18	3	9	.05276	.05673	.06078	.06490	.06912	.07343	.07786
18	3	10	.05658	.05978	.06301	.06629	.06961	.07299	.07643
18	3	11	.05924	.06168	.06414	.06661	.06911	.07164	.07421
18	3	12	.06098	.06270	.06442	.06616	.06791	.06968	.07147
18	3	13	.06197	.06300	.06404	.06510	.06618	.06727	.06839
18	3	14	.06231	.06268	.06309	.06353	.06401	.06452	.06507
18	3	15	.23439	.23183	.22982	.22830	.22724	.22659	.22631
18	4	1	-.56806	-.61711	-.66988	-.72659	-.78752	-.85295	-.92316
18	4	2	-.13156	-.13023	-.12826	-.12560	-.12216	-.11787	-.11264
18	4	3	-.05150	-.04558	-.03899	-.03166	-.02353	-.01455	-.00463
18	4	4	-.01012	-.00309	.00451	.01270	.02154	.03108	.04137
18	4	5	.01492	.02191	.02932	.03718	.04553	.05440	.06383
18	4	6	.03137	.03783	.04459	.05167	.05910	.06691	.07512
18	4	7	.04267	.04840	.05433	.06048	.06687	.07352	.08045
18	4	8	.05064	.05556	.06060	.06579	.07113	.07663	.08233
18	4	9	.05633	.06041	.06456	.06880	.07313	.07757	.08212
18	4	10	.06036	.06361	.06690	.07024	.07363	.07709	.08061
18	4	11	.06314	.06560	.06807	.07056	.07309	.07566	.07826
18	4	12	.06495	.06663	.06834	.07006	.07180	.07357	.07537
18	4	13	.06594	.06690	.06789	.06890	.06994	.07100	.07210
18	4	14	.31092	.30916	.30802	.30747	.30746	.30794	.30887
18	5	1	-.60657	-.65826	-.71384	-.77356	-.83769	-.90653	-.98038
18	5	2	-.13990	-.13840	-.13625	-.13338	-.12971	-.12516	-.11964
18	5	3	-.05438	-.04809	-.04111	-.03337	-.02481	-.01536	-.00494
18	5	4	-.01020	-.00277	.00523	.01384	.02313	.03313	.04390
18	5	5	.01653	.02387	.03164	.03988	.04862	.05789	.06775
18	5	6	.03406	.04082	.04788	.05528	.06304	.07118	.07974
18	5	7	.04610	.05206	.05824	.06463	.07128	.07819	.08539
18	5	8	.05458	.05967	.06489	.07025	.07578	.08148	.08738
18	5	9	.06061	.06481	.06908	.07343	.07789	.08246	.08715
18	5	10	.06487	.06818	.07153	.07494	.07840	.08193	.08553
18	5	11	.06779	.07025	.07273	.07524	.07779	.08038	.08303
18	5	12	.06965	.07129	.07296	.07465	.07638	.07814	.07993
18	5	13	.39685	.39657	.39702	.39815	.39991	.40227	.40516
18	6	1	-.65207	-.70688	-.76581	-.82909	-.89703	-.96990	-1.04804
18	6	2	-.14957	-.14787	-.14551	-.14240	-.13847	-.13362	-.12777
18	6	3	-.05759	-.05087	-.04344	-.03522	-.02615	-.01616	-.00517
18	6	4	-.01011	-.00223	.00624	.01535	.02515	.03568	.04702
18	6	5	.01860	.02635	.03454	.04322	.05241	.06215	.07250
18	6	6	.03741	.04451	.05193	.05968	.06781	.07634	.08530
18	6	7	.05031	.05654	.06299	.06967	.07660	.08381	.09131
18	6	8	.05938	.06466	.07008	.07565	.08138	.08730	.09342
18	6	9	.06581	.07012	.07452	.07901	.08360	.08831	.09315
18	6	10	.07032	.07368	.07710	.08057	.08410	.08771	.09140
18	6	11	.07338	.07583	.07831	.08083	.08340	.08602	.08869
18	6	12	.49413	.49616	.49905	.50274	.50719	.51236	.51820
18	7	1	-.70646	-.76505	-.82801	-.89559	-.96810	-1.04583	-1.12912
18	7	2	-.16089	-.15898	-.15637	-.15298	-.14873	-.14354	-.13730

Coefficients b_i for the BLUE of σ
Shape parameter k

n	s	i	1.4	1.5	1.6	1.7	1.8	1.9	2.0
18	7	3	-.06119	-.05396	-.04599	-.03721	-.02754	-.01691	-.00525
18	7	4	-.00976	-.00134	.00768	.01737	.02777	.03894	.05094
18	7	5	.02131	.02954	.03823	.04741	.05713	.06743	.07834
18	7	6	.04163	.04913	.05696	.06514	.07370	.08268	.09209
18	7	7	.05555	.06209	.06885	.07586	.08313	.09068	.09854
18	7	8	.06530	.07080	.07645	.08225	.08823	.09439	.10076
18	7	9	.07218	.07663	.08117	.08581	.09056	.09543	.10043
18	7	10	.07697	.08039	.08387	.08741	.09102	.09471	.09849
18	7	11	.60536	.61075	.61715	.62453	.63283	.64203	.65207
18	8	1	-.77252	-.83575	-.90364	-.97649	-1.05459	-1.13827	-1.22787
18	8	2	-.17434	-.17216	-.16924	-.16552	-.16090	-.15529	-.14859
18	8	3	-.06524	-.05740	-.04878	-.03931	-.02892	-.01753	-.00506
18	8	4	-.00901	.00004	.00974	.02012	.03125	.04318	.05597
18	8	5	.02489	.03370	.04299	.05279	.06314	.07410	.08570
18	8	6	.04704	.05502	.06334	.07202	.08110	.09061	.10057
18	8	7	.06217	.06908	.07622	.08361	.09127	.09923	.10751
18	8	8	.07273	.07849	.08440	.09048	.09673	.10318	.10985
18	8	9	.08013	.08474	.08944	.09424	.09916	.10422	.10941
18	8	10	.73415	.74423	.75554	.76806	.78176	.79658	.81250
18	9	1	-.85439	-.92340	-.99747	-1.07689	-1.16199	-1.25311	-1.35059
18	9	2	-.19059	-.18807	-.18478	-.18064	-.17556	-.16944	-.16217
18	9	3	-.06981	-.06121	-.05180	-.04148	-.03020	-.01787	-.00439
18	9	4	-.00766	.00219	.01271	.02395	.03597	.04883	.06261
18	9	5	.02975	.03927	.04928	.05984	.07097	.08273	.09517
18	9	6	.05413	.06269	.07161	.08090	.09061	.10077	.11140
18	9	7	.07072	.07808	.08568	.09353	.10167	.11012	.11890
18	9	8	.08225	.08832	.09454	.10094	.10753	.11433	.12135
18	9	9	.88560	.90213	.92022	.93986	.96100	.98363	1.00772
19	0	1	-.45093	-.49242	-.53704	-.58500	-.63652	-.69184	-.75122
19	0	2	-.10914	-.10891	-.10814	-.10678	-.10476	-.10202	-.09851
19	0	3	-.04591	-.04157	-.03663	-.03106	-.02480	-.01779	-.00998
19	0	4	-.01302	-.00750	-.00149	.00506	.01217	.01991	.02829
19	0	5	.00703	.01270	.01876	.02522	.03213	.03950	.04738
19	0	6	.02030	.02570	.03137	.03735	.04365	.05029	.05730
19	0	7	.02952	.03444	.03956	.04489	.05044	.05623	.06228
19	0	8	.03611	.04047	.04496	.04958	.05435	.05928	.06439
19	0	9	.04088	.04465	.04849	.05241	.05642	.06052	.06474
19	0	10	.04436	.04753	.05073	.05396	.05724	.06058	.06397
19	0	11	.04687	.04944	.05202	.05461	.05721	.05984	.06249
19	0	12	.04862	.05062	.05260	.05458	.05656	.05854	.06054
19	0	13	.04977	.05121	.05264	.05405	.05545	.05686	.05826
19	0	14	.05042	.05133	.05222	.05311	.05399	.05488	.05577
19	0	15	.05064	.05103	.05143	.05183	.05224	.05267	.05311
19	0	16	.05044	.05034	.05027	.05023	.05022	.05025	.05030
19	0	17	.04981	.04923	.04873	.04829	.04791	.04759	.04732
19	0	18	.04859	.04755	.04664	.04584	.04515	.04455	.04403
19	0	19	.04563	.04416	.04290	.04183	.04093	.04017	.03954
19	1	1	-.46988	-.51273	-.55885	-.60846	-.66179	-.71910	-.78066
19	1	2	-.11367	-.11337	-.11253	-.11108	-.10897	-.10614	-.10250
19	1	3	-.04778	-.04325	-.03812	-.03233	-.02584	-.01858	-.01050
19	1	4	-.01351	-.00778	-.00155	.00523	.01260	.02060	.02928
19	1	5	.00738	.01325	.01952	.02621	.03335	.04098	.04913
19	1	6	.02121	.02678	.03265	.03882	.04533	.05220	.05945
19	1	7	.03081	.03589	.04117	.04666	.05239	.05838	.06463
19	1	8	.03767	.04216	.04678	.05155	.05646	.06155	.06682
19	1	9	.04265	.04652	.05046	.05449	.05861	.06284	.06719

Coefficients b_i for the BLUE of σ
Shape parameter k

n	s	i	1.4	1.5	1.6	1.7	1.8	1.9	2.0
19	1	10	.04627	.04951	.05278	.05610	.05947	.06291	.06641
19	1	11	.04888	.05150	.05413	.05677	.05944	.06214	.06487
19	1	12	.05070	.05272	.05473	.05675	.05876	.06080	.06284
19	1	13	.05189	.05333	.05476	.05619	.05762	.05905	.06049
19	1	14	.05256	.05345	.05433	.05521	.05610	.05700	.05790
19	1	15	.05277	.05313	.05349	.05388	.05428	.05470	.05514
19	1	16	.05255	.05239	.05228	.05221	.05218	.05219	.05224
19	1	17	.05185	.05121	.05066	.05018	.04977	.04944	.04916
19	1	18	.09763	.09529	.09330	.09162	.09022	.08905	.08810
19	2	1	-.49198	-.53635	-.58413	-.63554	-.69083	-.75025	-.81411
19	2	2	-.11889	-.11850	-.11757	-.11602	-.11380	-.11083	-.10704
19	2	3	-.04989	-.04515	-.03979	-.03376	-.02701	-.01948	-.01109
19	2	4	-.01401	-.00805	-.00157	.00547	.01312	.02141	.03040
19	2	5	.00786	.01395	.02045	.02738	.03478	.04267	.05111
19	2	6	.02234	.02810	.03417	.04055	.04728	.05438	.06188
19	2	7	.03238	.03762	.04307	.04874	.05466	.06083	.06729
19	2	8	.03956	.04418	.04894	.05384	.05891	.06415	.06959
19	2	9	.04477	.04873	.05278	.05691	.06115	.06550	.06998
19	2	10	.04855	.05185	.05520	.05860	.06205	.06557	.06916
19	2	11	.05127	.05393	.05660	.05929	.06201	.06477	.06757
19	2	12	.05317	.05520	.05722	.05926	.06131	.06337	.06546
19	2	13	.05440	.05583	.05725	.05867	.06011	.06155	.06300
19	2	14	.05509	.05593	.05678	.05765	.05852	.05941	.06031
19	2	15	.05528	.05557	.05589	.05624	.05661	.05701	.05744
19	2	16	.05500	.05477	.05460	.05448	.05441	.05439	.05442
19	2	17	.15510	.15238	.15011	.14824	.14673	.14554	.14464
19	3	1	-.51747	-.56359	-.61325	-.66669	-.72417	-.78595	-.85233
19	3	2	-.12483	-.12435	-.12331	-.12164	-.11928	-.11616	-.11220
19	3	3	-.05224	-.04726	-.04165	-.03535	-.02831	-.02047	-.01175
19	3	4	-.01450	-.00828	-.00153	.00580	.01375	.02237	.03170
19	3	5	.00850	.01483	.02158	.02878	.03645	.04464	.05339
19	3	6	.02371	.02969	.03598	.04259	.04956	.05691	.06467
19	3	7	.03427	.03969	.04532	.05118	.05729	.06367	.07034
19	3	8	.04101	.04657	.05147	.05652	.06174	.06715	.07275
19	3	9	.04727	.05134	.05549	.05974	.06409	.06856	.07317
19	3	10	.05124	.05461	.05803	.06150	.06503	.06863	.07232
19	3	11	.05408	.05677	.05948	.06222	.06499	.06779	.07065
19	3	12	.05606	.05809	.06013	.06217	.06424	.06633	.06844
19	3	13	.05734	.05873	.06014	.06155	.06297	.06441	.06588
19	3	14	.05802	.05882	.05963	.06045	.06130	.06217	.06306
19	3	15	.05818	.05840	.05866	.05895	.05928	.05965	.06005
19	3	16	.21857	.21594	.21385	.21224	.21107	.21029	.20987
19	4	1	-.54687	-.59501	-.64684	-.70261	-.76258	-.82702	-.89626
19	4	2	-.13159	-.13100	-.12984	-.12804	-.12553	-.12225	-.11809
19	4	3	-.05486	-.04961	-.04371	-.03711	-.02974	-.02155	-.01246
19	4	4	-.01498	-.00846	-.00140	.00625	.01454	.02352	.03324
19	4	5	.00932	.01593	.02297	.03046	.03845	.04696	.05604
19	4	6	.02539	.03161	.03814	.04501	.05224	.05987	.06791
19	4	7	.03653	.04214	.04798	.05404	.06037	.06698	.07388
19	4	8	.04448	.04939	.05445	.05966	.06505	.07063	.07642
19	4	9	.05024	.05441	.05868	.06304	.06752	.07212	.07685
19	4	10	.05441	.05785	.06134	.06488	.06850	.07219	.07596
19	4	11	.05739	.06011	.06285	.06563	.06844	.07130	.07421
19	4	12	.05945	.06147	.06351	.06556	.06764	.06975	.07188
19	4	13	.06076	.06212	.06349	.06488	.06629	.06772	.06918
19	4	14	.06144	.06216	.06291	.06369	.06450	.06534	.06621

Coefficients b_i for the BLUE of σ
Shape parameter k

n	s	i	1.4	1.5	1.6	1.7	1.8	1.9	2.0
19	4	15	.28889	.28688	.28548	.28464	.28432	.28446	.28504
19	5	1	-.58091	-.63141	-.68577	-.74424	-.80708	-.87461	-.94711
19	5	2	-.13931	-.13860	-.13731	-.13536	-.13268	-.12920	-.12483
19	5	3	-.05777	-.05221	-.04599	-.03904	-.03131	-.02272	-.01321
19	5	4	-.01541	-.00856	-.00114	.00688	.01555	.02493	.03508
19	5	5	.01039	.01731	.02467	.03250	.04083	.04971	.05917
19	5	6	.02744	.03393	.04074	.04789	.05542	.06335	.07171
19	5	7	.03925	.04509	.05114	.05744	.06401	.07086	.07802
19	5	8	.04768	.05276	.05798	.06338	.06895	.07472	.08070
19	5	9	.05376	.05806	.06244	.06693	.07154	.07628	.08115
19	5	10	.05816	.06167	.06523	.06886	.07256	.07634	.08021
19	5	11	.06129	.06404	.06681	.06962	.07248	.07538	.07834
19	5	12	.06344	.06545	.06747	.06953	.07161	.07372	.07588
19	5	13	.06478	.06608	.06741	.06876	.07015	.07156	.07301
19	5	14	.36723	.36642	.36630	.36684	.36798	.36968	.37189
19	6	1	-.62065	-.67392	-.73124	-.79287	-.85909	-.93020	-1.00652
19	6	2	-.14818	-.14734	-.14589	-.14378	-.14091	-.13721	-.13260
19	6	3	-.06103	-.05511	-.04851	-.04117	-.03301	-.02398	-.01399
19	6	4	-.01577	-.00853	-.00071	.00773	.01685	.02669	.03732
19	6	5	.01178	.01906	.02679	.03500	.04373	.05302	.06292
19	6	6	.02996	.03676	.04389	.05137	.05924	.06751	.07623
19	6	7	.04256	.04864	.05495	.06152	.06835	.07548	.08292
19	6	8	.05152	.05679	.06221	.06780	.07358	.07956	.08576
19	6	9	.05798	.06241	.06693	.07156	.07631	.08120	.08623
19	6	10	.06264	.06622	.06987	.07358	.07736	.08124	.08520
19	6	11	.06593	.06870	.07151	.07435	.07724	.08020	.08321
19	6	12	.06816	.07015	.07216	.07420	.07628	.07840	.08056
19	6	13	.45508	.45616	.45805	.46070	.46406	.46809	.47274
19	7	1	-.66748	-.72405	-.78489	-.85027	-.92049	-.99584	-1.07667
19	7	2	-.15845	-.15746	-.15585	-.15353	-.15045	-.14650	-.14160
19	7	3	-.06467	-.05834	-.05131	-.04350	-.03485	-.02529	-.01474
19	7	4	-.01600	-.00831	-.00003	.00891	.01853	.02892	.04011
19	7	5	.01359	.02129	.02945	.03811	.04730	.05707	.06747
19	7	6	.03312	.04027	.04776	.05562	.06387	.07255	.08168
19	7	7	.04662	.05299	.05959	.06645	.07359	.08104	.08881
19	7	8	.05621	.06169	.06733	.07315	.07916	.08537	.09181
19	7	9	.06310	.06767	.07234	.07713	.08204	.08709	.09229
19	7	10	.06804	.07171	.07543	.07923	.08312	.08709	.09116
19	7	11	.07151	.07430	.07712	.08000	.08294	.08593	.08899
19	7	12	.55443	.55826	.56303	.56870	.57524	.58258	.59070
19	8	1	-.72341	-.78394	-.84902	-.91891	-.99393	-1.07439	-1.16063
19	8	2	-.17048	-.16932	-.16750	-.16496	-.16161	-.15738	-.15215
19	8	3	-.06878	-.06196	-.05441	-.04605	-.03682	-.02664	-.01544
19	8	4	-.01604	-.00782	.00102	.01053	.02077	.03179	.04365
19	8	5	.01599	.02418	.03285	.04203	.05177	.06210	.07308
19	8	6	.03710	.04467	.05259	.06089	.06959	.07874	.08836
19	8	7	.05167	.05837	.06532	.07253	.08003	.08784	.09599
19	8	8	.06199	.06772	.07362	.07969	.08597	.09246	.09918
19	8	9	.06938	.07411	.07896	.08392	.08901	.09425	.09964
19	8	10	.07464	.07839	.08221	.08611	.09010	.09418	.09837
19	8	11	.66795	.67559	.68436	.69422	.70513	.71706	.72996
19	9	1	-.79128	-.85666	-.92691	-1.00233	-1.08322	-1.16992	-1.26279
19	9	2	-.18477	-.18339	-.18133	-.17852	-.17486	-.17027	-.16465
19	9	3	-.07344	-.06603	-.05785	-.04883	-.03889	-.02797	-.01597
19	9	4	-.01577	-.00691	.00261	.01282	.02378	.03556	.04822
19	9	5	.01921	.02799	.03727	.04708	.05747	.06848	.08016

Coefficients b_i for the BLUE of σ
Shape parameter k

n	s	i	1.4	1.5	1.6	1.7	1.8	1.9	2.0
19	9	6	.04222	.05029	.05872	.06755	.07680	.08650	.09669
19	9	7	.05806	.06516	.07251	.08014	.08807	.09632	.10492
19	9	8	.06925	.07527	.08147	.08785	.09444	.10125	.10830
19	9	9	.07721	.08214	.08718	.09234	.09764	.10310	.10872
19	9	10	.79930	.81212	.82633	.84189	.85878	.87697	.89641
19	10	1	-.87532	-.94675	-1.02346	-1.10577	-1.19400	-1.28850	-1.38964
19	10	2	-.20204	-.20039	-.19802	-.19487	-.19083	-.18581	-.17970
19	10	3	-.07876	-.07062	-.06166	-.05182	-.04102	-.02917	-.01619
19	10	4	-.01500	-.00534	.00500	.01607	.02794	.04066	.05431
19	10	5	.02361	.03312	.04314	.05373	.06492	.07676	.08931
19	10	6	.04895	.05763	.06669	.07616	.08607	.09646	.10736
19	10	7	.06634	.07392	.08177	.08990	.09834	.10712	.11627
19	10	8	.07857	.08494	.09149	.09824	.10520	.11240	.11985
19	10	9	.95365	.97348	.99505	1.01836	1.04338	1.07008	1.09845
20	0	1	-.44078	-.48194	-.52625	-.57392	-.62517	-.68025	-.73942
20	0	2	-.11025	-.11057	-.11039	-.10967	-.10834	-.10635	-.10364
20	0	3	-.04870	-.04483	-.04039	-.03534	-.02962	-.02317	-.01596
20	0	4	-.01653	-.01138	-.00574	.00042	.00715	.01449	.02248
20	0	5	.00319	.00859	.01439	.02059	.02725	.03437	.04201
20	0	6	.01633	.02154	.02706	.03288	.03904	.04555	.05245
20	0	7	.02552	.03034	.03538	.04065	.04615	.05191	.05795
20	0	8	.03214	.03648	.04097	.04560	.05040	.05538	.06055
20	0	9	.03700	.04082	.04472	.04871	.05281	.05703	.06137
20	0	10	.04060	.04387	.04718	.05055	.05398	.05748	.06105
20	0	11	.04324	.04597	.04872	.05148	.05428	.05711	.05998
20	0	12	.04516	.04736	.04955	.05175	.05396	.05618	.05842
20	0	13	.04649	.04818	.04985	.05152	.05318	.05484	.05651
20	0	14	.04735	.04855	.04972	.05089	.05206	.05322	.05438
20	0	15	.04780	.04852	.04924	.04994	.05065	.05136	.05208
20	0	16	.04790	.04816	.04843	.04871	.04901	.04932	.04965
20	0	17	.04764	.04746	.04732	.04721	.04714	.04710	.04709
20	0	18	.04700	.04640	.04587	.04541	.04501	.04467	.04438
20	0	19	.04584	.04482	.04393	.04315	.04247	.04189	.04138
20	0	20	.04307	.04166	.04046	.03945	.03859	.03787	.03727
20	1	1	-.45819	-.50062	-.54633	-.59554	-.64850	-.70545	-.76667
20	1	2	-.11456	-.11483	-.11460	-.11383	-.11244	-.11038	-.10758
20	1	3	-.05058	-.04655	-.04194	-.03670	-.03077	-.02412	-.01666
20	1	4	-.01713	-.01180	-.00596	.00041	.00737	.01495	.02321
20	1	5	.00337	.00895	.01493	.02134	.02821	.03557	.04346
20	1	6	.01702	.02240	.02809	.03409	.04045	.04717	.05429
20	1	7	.02657	.03154	.03673	.04215	.04783	.05377	.06000
20	1	8	.03345	.03792	.04253	.04730	.05224	.05737	.06271
20	1	9	.03851	.04242	.04642	.05052	.05474	.05908	.06356
20	1	10	.04224	.04558	.04898	.05243	.05595	.05955	.06323
20	1	11	.04498	.04777	.05057	.05340	.05627	.05917	.06213
20	1	12	.04697	.04921	.05144	.05368	.05594	.05821	.06051
20	1	13	.04836	.05006	.05175	.05344	.05513	.05683	.05854
20	1	14	.04924	.05043	.05161	.05279	.05396	.05515	.05633
20	1	15	.04971	.05041	.05110	.05180	.05251	.05323	.05395
20	1	16	.04979	.05002	.05026	.05052	.05081	.05111	.05144
20	1	17	.04951	.04928	.04910	.04896	.04887	.04881	.04880
20	1	18	.04881	.04815	.04758	.04708	.04666	.04630	.04601
20	1	19	.09193	.08967	.08775	.08613	.08477	.08365	.08273
20	2	1	-.47837	-.52221	-.56947	-.62036	-.67513	-.73405	-.79742
20	2	2	-.11950	-.11971	-.11942	-.11857	-.11709	-.11494	-.11203
20	2	3	-.05269	-.04848	-.04367	-.03822	-.03208	-.02518	-.01746

Coefficients b_i for the BLUE of σ
Shape parameter k

n	s	i	1.4	1.5	1.6	1.7	1.8	1.9	2.0
20	2	4	-.01777	-.01223	-.00618	.00043	.00764	.01548	.02402
20	2	5	.00363	.00941	.01561	.02224	.02934	.03694	.04509
20	2	6	.01788	.02344	.02931	.03552	.04207	.04901	.05636
20	2	7	.02784	.03297	.03832	.04391	.04976	.05588	.06231
20	2	8	.03503	.03962	.04436	.04927	.05435	.05964	.06513
20	2	9	.04030	.04431	.04841	.05263	.05696	.06142	.06602
20	2	10	.04419	.04761	.05108	.05461	.05822	.06191	.06569
20	2	11	.04705	.04988	.05273	.05562	.05855	.06152	.06455
20	2	12	.04912	.05137	.05363	.05591	.05820	.06052	.06287
20	2	13	.05055	.05225	.05395	.05565	.05736	.05908	.06082
20	2	14	.05147	.05264	.05380	.05497	.05615	.05733	.05853
20	2	15	.05194	.05259	.05326	.05394	.05463	.05533	.05606
20	2	16	.05200	.05217	.05237	.05259	.05285	.05313	.05345
20	2	17	.05166	.05137	.05113	.05095	.05082	.05074	.05071
20	2	18	.14566	.14300	.14076	.13891	.13741	.13623	.13531
20	3	1	-.50148	-.54694	-.59593	-.64869	-.70549	-.76659	-.83229
20	3	2	-.12509	-.12523	-.12487	-.12393	-.12236	-.12010	-.11706
20	3	3	-.05504	-.05062	-.04560	-.03992	-.03352	-.02636	-.01835
20	3	4	-.01843	-.01266	-.00636	.00050	.00797	.01611	.02495
20	3	5	.00400	.01000	.01642	.02330	.03065	.03852	.04695
20	3	6	.01893	.02469	.03076	.03718	.04396	.05113	.05872
20	3	7	.02937	.03466	.04018	.04595	.05199	.05831	.06493
20	3	8	.03689	.04162	.04650	.05155	.05679	.06223	.06789
20	3	9	.04241	.04652	.05073	.05506	.05951	.06409	.06882
20	3	10	.04647	.04997	.05351	.05713	.06082	.06460	.06848
20	3	11	.04946	.05234	.05524	.05818	.06116	.06420	.06729
20	3	12	.05162	.05389	.05617	.05847	.06080	.06315	.06554
20	3	13	.05311	.05480	.05649	.05820	.05992	.06165	.06341
20	3	14	.05405	.05518	.05632	.05747	.05864	.05982	.06101
20	3	15	.05451	.05511	.05574	.05638	.05704	.05773	.05843
20	3	16	.05454	.05464	.05478	.05495	.05517	.05542	.05571
20	3	17	.20470	.20204	.19990	.19822	.19697	.19610	.19557
20	4	1	-.52793	-.57523	-.62621	-.68110	-.74018	-.80373	-.87205
20	4	2	-.13141	-.13148	-.13103	-.13000	-.12833	-.12594	-.12275
20	4	3	-.05765	-.05300	-.04773	-.04179	-.03512	-.02766	-.01933
20	4	4	-.01911	-.01308	-.00651	.00064	.00841	.01686	.02604
20	4	5	.00449	.01075	.01743	.02456	.03220	.04036	.04910
20	4	6	.02020	.02618	.03248	.03913	.04616	.05358	.06144
20	4	7	.03118	.03666	.04237	.04834	.05457	.06110	.06795
20	4	8	.03909	.04396	.04900	.05421	.05961	.06521	.07105
20	4	9	.04488	.04911	.05344	.05788	.06245	.06717	.07203
20	4	10	.04915	.05272	.05635	.06005	.06383	.06770	.07167
20	4	11	.05228	.05520	.05815	.06114	.06417	.06727	.07043
20	4	12	.05453	.05681	.05911	.06143	.06378	.06617	.06859
20	4	13	.05607	.05775	.05943	.06113	.06285	.06459	.06636
20	4	14	.05703	.05812	.05923	.06035	.06149	.06266	.06385
20	4	15	.05747	.05801	.05858	.05917	.05980	.06045	.06114
20	4	16	.26972	.26752	.26592	.26486	.26430	.26419	.26449
20	5	1	-.55829	-.60773	-.66098	-.71832	-.78001	-.84635	-.91764
20	5	2	-.13858	-.13857	-.13803	-.13689	-.13510	-.13257	-.12923
20	5	3	-.06055	-.05564	-.05010	-.04386	-.03688	-.02908	-.02039
20	5	4	-.01980	-.01347	-.00660	.00087	.00899	.01779	.02735
20	5	5	.00516	.01169	.01866	.02610	.03404	.04253	.05161
20	5	6	.02176	.02798	.03454	.04145	.04874	.05645	.06460
20	5	7	.03335	.03904	.04496	.05114	.05760	.06436	.07145
20	5	8	.04170	.04673	.05193	.05732	.06289	.06868	.07470

Coefficients b_i for the BLUE of σ
Shape parameter k

n	s	i	1.4	1.5	1.6	1.7	1.8	1.9	2.0
20	5	9	.04780	.05215	.05660	.06118	.06588	.07073	.07574
20	5	10	.05229	.05594	.05966	.06344	.06732	.07129	.07536
20	5	11	.05558	.05854	.06153	.06458	.06767	.07082	.07404
20	5	12	.05793	.06022	.06253	.06487	.06724	.06965	.07211
20	5	13	.05952	.06117	.06284	.06452	.06624	.06798	.06975
20	5	14	.06049	.06153	.06259	.06367	.06478	.06593	.06710
20	5	15	.34164	.34042	.33987	.33995	.34060	.34179	.34346
20	6	1	-.59337	-.64528	-.70119	-.76136	-.82607	-.89563	-.97034
20	6	2	-.14674	-.14665	-.14601	-.14476	-.14283	-.14014	-.13662
20	6	3	-.06378	-.05858	-.05272	-.04615	-.03881	-.03064	-.02154
20	6	4	-.02047	-.01381	-.00659	.00125	.00974	.01895	.02893
20	6	5	.00604	.01289	.02018	.02796	.03626	.04512	.05459
20	6	6	.02366	.03017	.03701	.04421	.05181	.05983	.06831
20	6	7	.03597	.04188	.04804	.05446	.06117	.06819	.07555
20	6	8	.04481	.05003	.05541	.06099	.06676	.07275	.07898
20	6	9	.05127	.05575	.06034	.06505	.06990	.07491	.08007
20	6	10	.05601	.05975	.06355	.06743	.07140	.07547	.07965
20	6	11	.05946	.06247	.06551	.06860	.07175	.07497	.07825
20	6	12	.06192	.06421	.06653	.06888	.07127	.07371	.07619
20	6	13	.06356	.06517	.06681	.06848	.07018	.07191	.07368
20	6	14	.42166	.42200	.42311	.42494	.42745	.43059	.43431
20	7	1	-.63422	-.68904	-.74806	-.81155	-.87979	-.95311	-1.03182
20	7	2	-.15611	-.15592	-.15517	-.15379	-.15170	-.14884	-.14512
20	7	3	-.06739	-.06185	-.05563	-.04868	-.04094	-.03232	-.02277
20	7	4	-.02110	-.01406	-.00644	.00182	.01075	.02042	.03089
20	7	5	.00721	.01442	.02210	.03026	.03897	.04825	.05815
20	7	6	.02602	.03284	.04001	.04755	.05549	.06387	.07272
20	7	7	.03914	.04532	.05174	.05844	.06544	.07275	.08040
20	7	8	.04855	.05398	.05958	.06536	.07136	.07757	.08404
20	7	9	.05542	.06005	.06479	.06966	.07467	.07984	.08518
20	7	10	.06044	.06427	.06817	.07216	.07624	.08042	.08471
20	7	11	.06408	.06713	.07022	.07337	.07657	.07985	.08320
20	7	12	.06664	.06893	.07125	.07362	.07602	.07848	.08098
20	7	13	.51132	.51393	.51743	.52178	.52692	.53283	.53944
20	8	1	-.68232	-.74059	-.80329	-.87071	-.94314	-1.02090	-1.10434
20	8	2	-.16696	-.16666	-.16578	-.16425	-.16199	-.15893	-.15497
20	8	3	-.07146	-.06552	-.05888	-.05149	-.04326	-.03414	-.02405
20	8	4	-.02166	-.01417	-.00608	.00266	.01211	.02232	.03334
20	8	5	.00878	.01641	.02452	.03314	.04231	.05208	.06249
20	8	6	.02897	.03616	.04371	.05164	.05998	.06877	.07805
20	8	7	.04304	.04952	.05626	.06327	.07059	.07824	.08624
20	8	8	.05312	.05878	.06461	.07064	.07689	.08337	.09009
20	8	9	.06045	.06524	.07016	.07520	.08040	.08575	.09128
20	8	10	.06579	.06972	.07373	.07783	.08202	.08633	.09075
20	8	11	.06963	.07272	.07586	.07906	.08233	.08567	.08909
20	8	12	.61264	.61839	.62519	.63300	.64177	.65146	.66203
20	9	1	-.73969	-.80211	-.86924	-.94138	-1.01883	-1.10194	-1.19105
20	9	2	-.17967	-.17924	-.17820	-.17649	-.17403	-.17074	-.16651
20	9	3	-.07606	-.06965	-.06251	-.05459	-.04580	-.03608	-.02535
20	9	4	-.02208	-.01406	-.00542	.00390	.01396	.02480	.03649
20	9	5	.01087	.01901	.02763	.03678	.04651	.05685	.06786
20	9	6	.03272	.04034	.04833	.05671	.06553	.07481	.08459
20	9	7	.04790	.05474	.06183	.06922	.07692	.08496	.09336
20	9	8	.05875	.06469	.07080	.07712	.08365	.09043	.09747
20	9	9	.06662	.07161	.07672	.08197	.08737	.09294	.09869
20	9	10	.07232	.07637	.08050	.08472	.08905	.09349	.09805

Coefficients b_i for the BLUE of σ
Shape parameter k

n	s	i	1.4	1.5	1.6	1.7	1.8	1.9	2.0
20	9	11	.72832	.73831	.74956	.76203	.77569	.79049	.80640
20	10	1	-.80926	-.87674	-.94928	-1.02718	-1.11078	-1.20042	-1.29645
20	10	2	-.19476	-.19417	-.19295	-.19103	-.18832	-.18474	-.18019
20	10	3	-.08131	-.07433	-.06659	-.05802	-.04856	-.03811	-.02661
20	10	4	-.02227	-.01361	-.00430	.00572	.01650	.02811	.04061
20	10	5	.01373	.02246	.03170	.04150	.05189	.06293	.07466
20	10	6	.03755	.04569	.05421	.06314	.07252	.08239	.09277
20	10	7	.05407	.06133	.06886	.07668	.08484	.09334	.10222
20	10	8	.06584	.07210	.07854	.08519	.09207	.09920	.10660
20	10	9	.07433	.07955	.08489	.09037	.09601	.10183	.10784
20	10	10	.86209	.87773	.89492	.91362	.93382	.95547	.97855
21	0	1	-.43125	-.47207	-.51606	-.56343	-.61440	-.66921	-.72815
21	0	2	-.11110	-.11191	-.11227	-.11213	-.11144	-.11014	-.10817
21	0	3	-.05112	-.04769	-.04371	-.03913	-.03392	-.02801	-.02135
21	0	4	-.01962	-.01483	-.00955	-.00374	.00262	.00958	.01718
21	0	5	-.00023	.00491	.01045	.01640	.02280	.02968	.03707
21	0	6	.01276	.01780	.02314	.02880	.03481	.04118	.04795
21	0	7	.02191	.02663	.03157	.03676	.04219	.04790	.05390
21	0	8	.02855	.03285	.03731	.04193	.04674	.05174	.05695
21	0	9	.03348	.03731	.04124	.04528	.04944	.05374	.05818
21	0	10	.03716	.04050	.04390	.04736	.05091	.05453	.05825
21	0	11	.03991	.04276	.04564	.04855	.05150	.05449	.05755
21	0	12	.04196	.04432	.04669	.04907	.05146	.05388	.05633
21	0	13	.04344	.04533	.04721	.04909	.05097	.05285	.05475
21	0	14	.04447	.04590	.04731	.04872	.05012	.05152	.05293
21	0	15	.04511	.04610	.04707	.04804	.04901	.04997	.05094
21	0	16	.04541	.04598	.04654	.04710	.04767	.04824	.04881
21	0	17	.04542	.04557	.04574	.04593	.04613	.04635	.04658
21	0	18	.04512	.04488	.04468	.04452	.04439	.04430	.04425
21	0	19	.04448	.04386	.04331	.04284	.04242	.04207	.04177
21	0	20	.04337	.04238	.04151	.04075	.04009	.03951	.03902
21	0	21	.04077	.03942	.03828	.03731	.03650	.03581	.03524
21	1	1	-.44731	-.48933	-.53464	-.58346	-.63603	-.69261	-.75348
21	1	2	-.11520	-.11598	-.11632	-.11614	-.11541	-.11407	-.11204
21	1	3	-.05298	-.04941	-.04528	-.04055	-.03516	-.02906	-.02220
21	1	4	-.02031	-.01535	-.00989	-.00390	.00266	.00984	.01769
21	1	5	-.00019	.00511	.01082	.01696	.02356	.03065	.03828
21	1	6	.01328	.01847	.02397	.02980	.03599	.04256	.04954
21	1	7	.02276	.02762	.03271	.03804	.04364	.04952	.05570
21	1	8	.02965	.03407	.03865	.04340	.04835	.05349	.05886
21	1	9	.03476	.03868	.04272	.04687	.05115	.05557	.06014
21	1	10	.03858	.04199	.04547	.04903	.05266	.05639	.06022
21	1	11	.04143	.04434	.04728	.05025	.05328	.05635	.05949
21	1	12	.04355	.04595	.04836	.05079	.05324	.05572	.05823
21	1	13	.04509	.04700	.04890	.05081	.05273	.05466	.05661
21	1	14	.04615	.04758	.04901	.05043	.05186	.05329	.05472
21	1	15	.04681	.04779	.04876	.04973	.05070	.05168	.05266
21	1	16	.04712	.04766	.04820	.04875	.04932	.04989	.05047
21	1	17	.04711	.04723	.04737	.04754	.04772	.04794	.04817
21	1	18	.04678	.04650	.04626	.04607	.04593	.04582	.04576
21	1	19	.04609	.04542	.04483	.04433	.04389	.04352	.04321
21	1	20	.08684	.08466	.08281	.08124	.07993	.07885	.07796
21	2	1	-.46583	-.50917	-.55591	-.60630	-.66057	-.71900	-.78187
21	2	2	-.11988	-.12062	-.12091	-.12069	-.11990	-.11849	-.11638
21	2	3	-.05508	-.05135	-.04705	-.04214	-.03656	-.03025	-.02315
21	2	4	-.02106	-.01591	-.01025	-.00405	.00273	.01015	.01824

Coefficients b_i for the BLUE of σ
Shape parameter k

n	s	i	1.4	1.5	1.6	1.7	1.8	1.9	2.0
21	2	5	-.00010	.00539	.01128	.01762	.02443	.03175	.03961
21	2	6	.01392	.01928	.02496	.03097	.03735	.04412	.05131
21	2	7	.02380	.02880	.03404	.03953	.04529	.05134	.05771
21	2	8	.03097	.03551	.04021	.04510	.05018	.05547	.06099
21	2	9	.03628	.04031	.04444	.04870	.05309	.05763	.06233
21	2	10	.04025	.04374	.04730	.05094	.05466	.05848	.06241
21	2	11	.04322	.04618	.04918	.05221	.05530	.05845	.06166
21	2	12	.04542	.04786	.05030	.05277	.05527	.05779	.06036
21	2	13	.04702	.04894	.05086	.05279	.05473	.05669	.05868
21	2	14	.04811	.04954	.05097	.05239	.05383	.05527	.05673
21	2	15	.04879	.04974	.05070	.05166	.05263	.05360	.05459
21	2	16	.04910	.04960	.05011	.05064	.05118	.05174	.05232
21	2	17	.04907	.04914	.04924	.04936	.04953	.04972	.04994
21	2	18	.04869	.04835	.04806	.04783	.04765	.04753	.04744
21	2	19	.13728	.13467	.13248	.13066	.12918	.12800	.12708
21	3	1	-.48691	-.53174	-.58009	-.63222	-.68836	-.74882	-.81387
21	3	2	-.12514	-.12585	-.12609	-.12581	-.12496	-.12346	-.12126
21	3	3	-.05740	-.05350	-.04901	-.04390	-.03810	-.03156	-.02422
21	3	4	-.02184	-.01649	-.01062	-.00419	.00284	.01051	.01888
21	3	5	.00006	.00574	.01185	.01841	.02545	.03301	.04113
21	3	6	.01471	.02025	.02611	.03232	.03891	.04589	.05331
21	3	7	.02503	.03019	.03559	.04125	.04718	.05342	.05998
21	3	8	.03252	.03719	.04203	.04705	.05228	.05772	.06340
21	3	9	.03806	.04219	.04643	.05080	.05531	.05997	.06480
21	3	10	.04221	.04578	.04942	.05314	.05695	.06086	.06489
21	3	11	.04530	.04831	.05136	.05446	.05761	.06083	.06411
21	3	12	.04759	.05006	.05253	.05504	.05757	.06014	.06276
21	3	13	.04925	.05118	.05311	.05505	.05702	.05900	.06101
21	3	14	.05038	.05179	.05321	.05463	.05607	.05752	.05898
21	3	15	.05107	.05199	.05292	.05386	.05481	.05578	.05676
21	3	16	.05137	.05182	.05229	.05278	.05330	.05384	.05440
21	3	17	.05130	.05131	.05135	.05143	.05156	.05172	.05192
21	3	18	.19245	.18978	.18760	.18588	.18457	.18363	.18302
21	4	1	-.51086	-.55738	-.60756	-.66165	-.71990	-.78261	-.85008
21	4	2	-.13107	-.13172	-.13191	-.13157	-.13064	-.12906	-.12676
21	4	3	-.05998	-.05588	-.05118	-.04585	-.03981	-.03302	-.02539
21	4	4	-.02267	-.01708	-.01097	-.00429	.00300	.01095	.01962
21	4	5	.00030	.00621	.01255	.01935	.02665	.03447	.04287
21	4	6	.01567	.02141	.02748	.03391	.04072	.04794	.05560
21	4	7	.02649	.03182	.03740	.04324	.04937	.05580	.06257
21	4	8	.03434	.03915	.04413	.04931	.05469	.06030	.06614
21	4	9	.04014	.04438	.04874	.05323	.05786	.06265	.06760
21	4	10	.04448	.04813	.05185	.05566	.05957	.06358	.06770
21	4	11	.04772	.05078	.05388	.05704	.06026	.06354	.06690
21	4	12	.05011	.05259	.05510	.05764	.06021	.06282	.06548
21	4	13	.05182	.05375	.05569	.05764	.05962	.06162	.06365
21	4	14	.05299	.05438	.05578	.05719	.05862	.06007	.06154
21	4	15	.05369	.05456	.05545	.05636	.05729	.05824	.05922
21	4	16	.05396	.05435	.05477	.05522	.05570	.05620	.05674
21	4	17	.25288	.25055	.24880	.24757	.24682	.24651	.24659
21	5	1	-.53815	-.58661	-.63887	-.69518	-.75582	-.82108	-.89127
21	5	2	-.13774	-.13835	-.13848	-.13807	-.13706	-.13538	-.13296
21	5	3	-.06284	-.05851	-.05359	-.04800	-.04170	-.03462	-.02669
21	5	4	-.02353	-.01769	-.01131	-.00435	.00324	.01151	.02051
21	5	5	.00066	.00682	.01342	.02049	.02806	.03618	.04489
21	5	6	.01684	.02281	.02911	.03577	.04283	.05030	.05823

Coefficients b_i for the BLUE of σ
Shape parameter k

n	s	i	1.4	1.5	1.6	1.7	1.8	1.9	2.0
21	5	7	.02822	.03374	.03952	.04556	.05190	.05855	.06554
21	5	8	.03648	.04144	.04659	.05193	.05748	.06326	.06929
21	5	9	.04258	.04694	.05142	.05604	.06080	.06572	.07082
21	5	10	.04713	.05087	.05468	.05858	.06259	.06670	.07092
21	5	11	.05052	.05364	.05680	.06002	.06330	.06665	.07008
21	5	12	.05302	.05553	.05806	.06063	.06324	.06589	.06859
21	5	13	.05480	.05672	.05866	.06062	.06261	.06462	.06667
21	5	14	.05600	.05736	.05873	.06012	.06154	.06298	.06445
21	5	15	.05669	.05752	.05836	.05923	.06013	.06105	.06200
21	5	16	.31933	.31778	.31689	.31661	.31688	.31766	.31890
21	6	1	-.56939	-.62009	-.67474	-.73361	-.79698	-.86515	-.93844
21	6	2	-.14529	-.14585	-.14591	-.14543	-.14433	-.14255	-.14000
21	6	3	-.06601	-.06144	-.05625	-.05038	-.04378	-.03638	-.02810
21	6	4	-.02442	-.01829	-.01161	-.00433	.00360	.01222	.02159
21	6	5	.00116	.00761	.01450	.02186	.02975	.03820	.04725
21	6	6	.01827	.02449	.03105	.03798	.04531	.05307	.06130
21	6	7	.03029	.03603	.04202	.04829	.05486	.06175	.06899
21	6	8	.03901	.04415	.04947	.05500	.06074	.06671	.07294
21	6	9	.04545	.04994	.05456	.05931	.06422	.06929	.07454
21	6	10	.05024	.05407	.05798	.06198	.06609	.07030	.07465
21	6	11	.05380	.05697	.06019	.06347	.06682	.07024	.07375
21	6	12	.05641	.05894	.06150	.06410	.06674	.06943	.07218
21	6	13	.05826	.06017	.06211	.06406	.06605	.06808	.07014
21	6	14	.05948	.06080	.06214	.06351	.06490	.06633	.06779
21	6	15	.39274	.39250	.39299	.39418	.39601	.39843	.40141
21	7	1	-.60542	-.65871	-.71614	-.77797	-.84449	-.91602	-.99289
21	7	2	-.15388	-.15438	-.15438	-.15381	-.15262	-.15071	-.14802
21	7	3	-.06955	-.06469	-.05920	-.05302	-.04608	-.03831	-.02964
21	7	4	-.02533	-.01887	-.01184	-.00420	.00411	.01313	.02292
21	7	5	.00186	.00862	.01584	.02356	.03180	.04062	.05006
21	7	6	.02003	.02653	.03338	.04062	.04826	.05635	.06491
21	7	7	.03279	.03877	.04501	.05153	.05836	.06552	.07304
21	7	8	.04203	.04737	.05289	.05862	.06456	.07075	.07720
21	7	9	.04884	.05349	.05826	.06317	.06823	.07347	.07889
21	7	10	.05391	.05784	.06186	.06597	.07019	.07452	.07898
21	7	11	.05766	.06089	.06417	.06752	.07094	.07444	.07802
21	7	12	.06039	.06294	.06553	.06815	.07082	.07355	.07634
21	7	13	.06231	.06420	.06612	.06808	.07006	.07209	.07417
21	7	14	.47435	.47600	.47850	.48179	.48584	.49060	.49602
21	8	1	-.64732	-.70365	-.76433	-.82962	-.89984	-.97530	-1.05633
21	8	2	-.16373	-.16417	-.16409	-.16343	-.16212	-.16009	-.15724
21	8	3	-.07351	-.06833	-.06250	-.05595	-.04861	-.04042	-.03131
21	8	4	-.02623	-.01939	-.01197	-.00391	.00483	.01431	.02459
21	8	5	.00282	.00995	.01755	.02565	.03431	.04355	.05344
21	8	6	.02222	.02905	.03623	.04381	.05181	.06027	.06921
21	8	7	.03583	.04208	.04860	.05542	.06254	.07001	.07784
21	8	8	.04567	.05123	.05697	.06293	.06911	.07554	.08224
21	8	9	.05292	.05772	.06266	.06774	.07299	.07841	.08402
21	8	10	.05828	.06233	.06646	.07069	.07503	.07950	.08409
21	8	11	.06224	.06554	.06889	.07231	.07580	.07937	.08304
21	8	12	.06511	.06767	.07028	.07293	.07563	.07840	.08123
21	8	13	.56571	.56998	.57524	.58143	.58851	.59645	.60518
21	9	1	-.69660	-.75653	-.82105	-.89045	-.96503	-1.04513	-1.13110
21	9	2	-.17514	-.17551	-.17534	-.17457	-.17313	-.17094	-.16791
21	9	3	-.07798	-.07243	-.06619	-.05921	-.05141	-.04274	-.03310
21	9	4	-.02711	-.01982	-.01193	-.00339	.00586	.01587	.02671

Coefficients b_i for the BLUE of σ
Shape parameter k

n	s	i	1.4	1.5	1.6	1.7	1.8	1.9	2.0
21	9	5	.00413	.01169	.01973	.02829	.03742	.04716	.05756
21	9	6	.02497	.03217	.03975	.04773	.05614	.06503	.07441
21	9	7	.03957	.04614	.05299	.06013	.06760	.07542	.08361
21	9	8	.05011	.05592	.06192	.06814	.07460	.08130	.08829
21	9	9	.05785	.06284	.06797	.07325	.07870	.08433	.09015
21	9	10	.06356	.06773	.07199	.07636	.08084	.08545	.09019
21	9	11	.06775	.07111	.07453	.07803	.08160	.08526	.08902
21	9	12	.66889	.67668	.68562	.69569	.70682	.71900	.73217
21	10	1	-.75535	-.81959	-.88872	-.96304	-1.04287	-1.12855	-1.22043
21	10	2	-.18849	-.18878	-.18851	-.18762	-.18603	-.18365	-.18041
21	10	3	-.08306	-.07705	-.07034	-.06285	-.05451	-.04525	-.03500
21	10	4	-.02790	-.02009	-.01165	-.00254	.00732	.01797	.02947
21	10	5	.00593	.01399	.02255	.03166	.04135	.05168	.06269
21	10	6	.02847	.03612	.04416	.05261	.06151	.07090	.08081
21	10	7	.04424	.05118	.05841	.06594	.07381	.08204	.09067
21	10	8	.05560	.06170	.06801	.07453	.08130	.08833	.09565
21	10	9	.06391	.06912	.07447	.07998	.08566	.09152	.09759
21	10	10	.07002	.07433	.07874	.08326	.08789	.09266	.09757
21	10	11	.78664	.79907	.81288	.82806	.84456	.86235	.88139
21	11	1	-.82654	-.89605	-.97080	-1.05113	-1.13735	-1.22984	-1.32895
21	11	2	-.20436	-.20454	-.20415	-.20310	-.20133	-.19874	-.19524
21	11	3	-.08889	-.08233	-.07503	-.06692	-.05793	-.04797	-.03697
21	11	4	-.02854	-.02009	-.01099	-.00118	.00941	.02082	.03313
21	11	5	.00843	.01709	.02629	.03605	.04642	.05745	.06920
21	11	6	.03302	.04120	.04979	.05880	.06829	.07828	.08882
21	11	7	.05018	.05757	.06525	.07325	.08160	.09032	.09945
21	11	8	.06251	.06896	.07562	.08251	.08965	.09707	.10478
21	11	9	.07149	.07696	.08257	.08834	.09430	.10044	.10680
21	11	10	.92271	.94124	.96146	.98337	1.00695	1.03216	1.05898
22	0	1	-.42228	-.46277	-.50644	-.55349	-.60416	-.65870	-.71738
22	0	2	-.11173	-.11299	-.11384	-.11424	-.11413	-.11346	-.11217
22	0	3	-.05322	-.05019	-.04664	-.04252	-.03778	-.03237	-.02623
22	0	4	-.02237	-.01790	-.01296	-.00750	-.00149	.00510	.01233
22	0	5	-.00328	.00160	.00688	.01259	.01874	.02537	.03252
22	0	6	.00956	.01441	.01958	.02507	.03092	.03715	.04378
22	0	7	.01865	.02325	.02809	.03318	.03854	.04418	.05012
22	0	8	.02529	.02954	.03395	.03854	.04333	.04833	.05355
22	0	9	.03026	.03408	.03803	.04209	.04630	.05065	.05516
22	0	10	.03401	.03739	.04085	.04439	.04801	.05174	.05557
22	0	11	.03686	.03979	.04276	.04578	.04886	.05199	.05520
22	0	12	.03901	.04150	.04400	.04652	.04907	.05166	.05428
22	0	13	.04061	.04266	.04471	.04676	.04882	.05090	.05300
22	0	14	.04177	.04339	.04501	.04662	.04822	.04983	.05145
22	0	15	.04256	.04377	.04497	.04617	.04735	.04854	.04973
22	0	16	.04303	.04385	.04466	.04546	.04626	.04706	.04787
22	0	17	.04323	.04367	.04410	.04454	.04499	.04544	.04590
22	0	18	.04316	.04323	.04332	.04342	.04355	.04369	.04385
22	0	19	.04283	.04255	.04230	.04210	.04193	.04180	.04171
22	0	20	.04221	.04158	.04102	.04053	.04011	.03974	.03943
22	0	21	.04115	.04018	.03933	.03859	.03794	.03739	.03691
22	0	22	.03870	.03741	.03632	.03539	.03462	.03396	.03342
22	1	1	-.43716	-.47878	-.52368	-.57211	-.62429	-.68049	-.74100
22	1	2	-.11563	-.11688	-.11772	-.11811	-.11798	-.11728	-.11596
22	1	3	-.05506	-.05191	-.04823	-.04397	-.03908	-.03350	-.02719
22	1	4	-.02312	-.01851	-.01340	-.00778	-.00159	.00521	.01265
22	1	5	-.00336	.00167	.00712	.01299	.01932	.02615	.03351

Coefficients b_i for the BLUE of σ
Shape parameter k

n	s	i	1.4	1.5	1.6	1.7	1.8	1.9	2.0
22	1	6	.00993	.01493	.02024	.02590	.03191	.03832	.04514
22	1	7	.01934	.02407	.02904	.03427	.03978	.04558	.05170
22	1	8	.02622	.03057	.03510	.03982	.04474	.04988	.05525
22	1	9	.03136	.03528	.03932	.04349	.04780	.05227	.05691
22	1	10	.03524	.03870	.04224	.04586	.04958	.05340	.05734
22	1	11	.03818	.04118	.04422	.04731	.05045	.05367	.05696
22	1	12	.04041	.04294	.04549	.04807	.05068	.05333	.05602
22	1	13	.04206	.04414	.04623	.04832	.05042	.05255	.05470
22	1	14	.04326	.04490	.04653	.04817	.04980	.05145	.05310
22	1	15	.04408	.04529	.04650	.04770	.04890	.05011	.05132
22	1	16	.04456	.04537	.04617	.04697	.04778	.04859	.04940
22	1	17	.04476	.04517	.04559	.04602	.04646	.04691	.04738
22	1	18	.04469	.04472	.04478	.04486	.04497	.04511	.04527
22	1	19	.04433	.04400	.04372	.04349	.04330	.04316	.04306
22	1	20	.04365	.04297	.04238	.04186	.04142	.04105	.04073
22	1	21	.08227	.08016	.07837	.07686	.07560	.07455	.07369
22	2	1	-.45423	-.49708	-.54333	-.59322	-.64700	-.70494	-.76734
22	2	2	-.12006	-.12129	-.12211	-.12247	-.12230	-.12157	-.12019
22	2	3	-.05712	-.05384	-.05001	-.04560	-.04054	-.03478	-.02826
22	2	4	-.02394	-.01916	-.01388	-.00807	-.00168	.00533	.01300
22	2	5	-.00341	.00179	.00741	.01346	.01999	.02703	.03461
22	2	6	.01040	.01555	.02103	.02685	.03304	.03964	.04666
22	2	7	.02017	.02504	.03016	.03554	.04120	.04716	.05346
22	2	8	.02732	.03179	.03644	.04128	.04634	.05162	.05714
22	2	9	.03265	.03667	.04081	.04509	.04951	.05410	.05886
22	2	10	.03668	.04022	.04384	.04755	.05136	.05527	.05931
22	2	11	.03974	.04279	.04589	.04904	.05226	.05555	.05892
22	2	12	.04205	.04462	.04721	.04983	.05249	.05520	.05795
22	2	13	.04376	.04586	.04797	.05009	.05223	.05439	.05658
22	2	14	.04500	.04665	.04829	.04993	.05159	.05325	.05494
22	2	15	.04584	.04705	.04825	.04945	.05065	.05187	.05309
22	2	16	.04634	.04712	.04790	.04869	.04949	.05029	.05111
22	2	17	.04653	.04691	.04729	.04770	.04812	.04856	.04902
22	2	18	.04643	.04642	.04644	.04649	.04657	.04669	.04684
22	2	19	.04604	.04565	.04532	.04505	.04484	.04468	.04456
22	2	20	.12980	.12724	.12509	.12330	.12184	.12067	.11976
22	3	1	-.47354	-.51778	-.56553	-.61704	-.67257	-.73240	-.79683
22	3	2	-.12504	-.12624	-.12703	-.12735	-.12715	-.12636	-.12492
22	3	3	-.05941	-.05598	-.05199	-.04740	-.04215	-.03619	-.02946
22	3	4	-.02481	-.01985	-.01438	-.00836	-.00176	.00548	.01340
22	3	5	-.00342	.00197	.00778	.01403	.02077	.02802	.03584
22	3	6	.01098	.01630	.02195	.02795	.03434	.04113	.04836
22	3	7	.02116	.02618	.03145	.03699	.04281	.04895	.05543
22	3	8	.02861	.03321	.03798	.04296	.04816	.05358	.05925
22	3	9	.03417	.03828	.04253	.04692	.05146	.05616	.06105
22	3	10	.03836	.04198	.04568	.04947	.05337	.05738	.06152
22	3	11	.04154	.04465	.04781	.05103	.05431	.05767	.06112
22	3	12	.04394	.04655	.04918	.05185	.05455	.05730	.06011
22	3	13	.04572	.04784	.04996	.05211	.05427	.05647	.05869
22	3	14	.04700	.04864	.05029	.05194	.05360	.05529	.05699
22	3	15	.04787	.04905	.05024	.05143	.05263	.05385	.05508
22	3	16	.04837	.04911	.04987	.05063	.05141	.05221	.05302
22	3	17	.04855	.04887	.04922	.04959	.04999	.05041	.05085
22	3	18	.04842	.04834	.04831	.04832	.04837	.04846	.04858
22	3	19	.18155	.17888	.17669	.17495	.17360	.17261	.17193
22	4	1	-.49537	-.54116	-.59060	-.64392	-.70140	-.76332	-.82999

Coefficients b_i for the BLUE of σ
Shape parameter k

n	s	i	1.4	1.5	1.6	1.7	1.8	1.9	2.0
22	4	2	-.13060	-.13178	-.13254	-.13282	-.13257	-.13173	-.13022
22	4	3	-.06194	-.05833	-.05417	-.04939	-.04393	-.03775	-.03078
22	4	4	-.02575	-.02058	-.01489	-.00865	-.00181	.00568	.01387
22	4	5	-.00337	.00222	.00824	.01471	.02168	.02918	.03725
22	4	6	.01169	.01719	.02303	.02923	.03582	.04283	.05029
22	4	7	.02233	.02751	.03295	.03866	.04466	.05099	.05766
22	4	8	.03012	.03485	.03977	.04489	.05023	.05581	.06164
22	4	9	.03592	.04015	.04451	.04901	.05367	.05850	.06352
22	4	10	.04030	.04400	.04779	.05167	.05566	.05978	.06402
22	4	11	.04362	.04679	.05001	.05329	.05664	.06007	.06360
22	4	12	.04612	.04876	.05143	.05414	.05689	.05969	.06255
22	4	13	.04797	.05010	.05224	.05440	.05659	.05882	.06107
22	4	14	.04929	.05093	.05257	.05422	.05589	.05758	.05930
22	4	15	.05018	.05133	.05250	.05368	.05487	.05608	.05731
22	4	16	.05068	.05138	.05210	.05283	.05359	.05436	.05516
22	4	17	.05084	.05110	.05140	.05173	.05209	.05248	.05290
22	4	18	.23798	.23555	.23368	.23233	.23143	.23096	.23087
22	5	1	-.52006	-.56763	-.61897	-.67434	-.73401	-.79827	-.86745
22	5	2	-.13684	-.13799	-.13872	-.13896	-.13866	-.13775	-.13616
22	5	3	-.06473	-.06094	-.05657	-.05158	-.04590	-.03947	-.03224
22	5	4	-.02673	-.02134	-.01542	-.00893	-.00182	.00594	.01443
22	5	5	-.00324	.00257	.00882	.01553	.02275	.03052	.03886
22	5	6	.01255	.01826	.02431	.03073	.03755	.04480	.05250
22	5	7	.02372	.02908	.03469	.04059	.04679	.05332	.06020
22	5	8	.03188	.03676	.04183	.04711	.05262	.05836	.06437
22	5	9	.03796	.04231	.04679	.05142	.05621	.06117	.06633
22	5	10	.04255	.04634	.05022	.05420	.05829	.06250	.06685
22	5	11	.04602	.04925	.05253	.05588	.05931	.06281	.06641
22	5	12	.04862	.05130	.05401	.05676	.05956	.06241	.06532
22	5	13	.05055	.05269	.05485	.05703	.05924	.06148	.06377
22	5	14	.05192	.05354	.05517	.05682	.05849	.06019	.06191
22	5	15	.05282	.05394	.05508	.05623	.05741	.05861	.05983
22	5	16	.05332	.05396	.05463	.05533	.05605	.05680	.05758
22	5	17	.29969	.29790	.29675	.29617	.29613	.29659	.29749
22	6	1	-.54811	-.59771	-.65123	-.70892	-.77108	-.83799	-.90999
22	6	2	-.14384	-.14497	-.14567	-.14587	-.14551	-.14453	-.14286
22	6	3	-.06782	-.06382	-.05923	-.05400	-.04807	-.04137	-.03384
22	6	4	-.02778	-.02213	-.01594	-.00917	-.00178	.00630	.01511
22	6	5	-.00302	.00304	.00955	.01653	.02403	.03209	.04075
22	6	6	.01361	.01955	.02583	.03250	.03957	.04707	.05505
22	6	7	.02537	.03092	.03674	.04285	.04927	.05602	.06313
22	6	8	.03395	.03900	.04424	.04969	.05537	.06130	.06750
22	6	9	.04035	.04483	.04944	.05421	.05914	.06425	.06955
22	6	10	.04516	.04905	.05303	.05711	.06131	.06563	.07009
22	6	11	.04880	.05209	.05545	.05887	.06237	.06595	.06963
22	6	12	.05153	.05424	.05698	.05977	.06262	.06551	.06848
22	6	13	.05353	.05568	.05784	.06004	.06227	.06454	.06685
22	6	14	.05494	.05654	.05816	.05980	.06147	.06316	.06489
22	6	15	.05586	.05694	.05804	.05916	.06031	.06149	.06269
22	6	16	.36747	.36676	.36677	.36743	.36872	.37058	.37296
22	7	1	-.58017	-.63210	-.68812	-.74848	-.81348	-.88343	-.95865
22	7	2	-.15176	-.15286	-.15352	-.15368	-.15326	-.15220	-.15044
22	7	3	-.07126	-.06702	-.06218	-.05668	-.05046	-.04346	-.03559
22	7	4	-.02887	-.02293	-.01645	-.00937	-.00164	.00679	.01597
22	7	5	-.00268	.00367	.01047	.01775	.02557	.03396	.04297
22	7	6	.01491	.02111	.02766	.03459	.04194	.04974	.05803

Coefficients b_i for the BLUE of σ
Shape parameter k

n	s	i	1.4	1.5	1.6	1.7	1.8	1.9	2.0
22	7	7	.02734	.03312	.03916	.04550	.05216	.05916	.06653
22	7	8	.03641	.04164	.04706	.05271	.05858	.06472	.07112
22	7	9	.04315	.04778	.05254	.05745	.06254	.06781	.07328
22	7	10	.04823	.05222	.05631	.06050	.06481	.06926	.07384
22	7	11	.05205	.05541	.05884	.06233	.06591	.06958	.07334
22	7	12	.05490	.05765	.06044	.06327	.06615	.06910	.07212
22	7	13	.05699	.05914	.06131	.06352	.06577	.06805	.07039
22	7	14	.05845	.06002	.06162	.06324	.06489	.06658	.06831
22	7	15	.44229	.44318	.44487	.44733	.45050	.45434	.45880
22	8	1	-.61708	-.67172	-.73062	-.79407	-.86236	-.93582	-1.01476
22	8	2	-.16075	-.16184	-.16246	-.16256	-.16208	-.16094	-.15907
22	8	3	-.07509	-.07059	-.06546	-.05966	-.05311	-.04576	-.03752
22	8	4	-.03000	-.02374	-.01692	-.00948	-.00138	.00744	.01704
22	8	5	-.00216	.00450	.01163	.01927	.02744	.03621	.04561
22	8	6	.01653	.02301	.02986	.03711	.04478	.05291	.06153
22	8	7	.02972	.03575	.04205	.04865	.05558	.06286	.07051
22	8	8	.03934	.04477	.05041	.05627	.06236	.06872	.07536
22	8	9	.04648	.05126	.05619	.06127	.06653	.07198	.07763
22	8	10	.05184	.05595	.06016	.06448	.06892	.07349	.07821
22	8	11	.05587	.05931	.06281	.06639	.07005	.07381	.07766
22	8	12	.05887	.06165	.06447	.06735	.07028	.07328	.07635
22	8	13	.06104	.06319	.06536	.06758	.06984	.07214	.07450
22	8	14	.52539	.52849	.53251	.53742	.54315	.54968	.55695
22	9	1	-.65997	-.71777	-.78006	-.84712	-.91925	-.99679	-1.08009
22	9	2	-.17107	-.17212	-.17270	-.17275	-.17220	-.17097	-.16898
22	9	3	-.07940	-.07458	-.06913	-.06297	-.05605	-.04829	-.03962
22	9	4	-.03117	-.02454	-.01732	-.00947	-.00093	.00834	.01842
22	9	5	-.00142	.00562	.01313	.02116	.02975	.03895	.04879
22	9	6	.01854	.02536	.03255	.04015	.04819	.05670	.06572
22	9	7	.03262	.03893	.04553	.05243	.05967	.06726	.07525
22	9	8	.04287	.04854	.05441	.06051	.06686	.07348	.08038
22	9	9	.05047	.05543	.06054	.06582	.07127	.07691	.08277
22	9	10	.05616	.06040	.06474	.06919	.07377	.07849	.08336
22	9	11	.06042	.06393	.06752	.07119	.07494	.07879	.08275
22	9	12	.06356	.06638	.06924	.07216	.07514	.07819	.08132
22	9	13	.61838	.62443	.63155	.63970	.64885	.65894	.66994
22	10	1	-.71037	-.77192	-.83820	-.90952	-.98621	-1.06858	-1.15701
22	10	2	-.18300	-.18403	-.18456	-.18455	-.18391	-.18257	-.18046
22	10	3	-.08426	-.07909	-.07325	-.06668	-.05932	-.05109	-.04192
22	10	4	-.03236	-.02528	-.01761	-.00927	-.00023	.00958	.02021
22	10	5	-.00035	.00711	.01507	.02356	.03263	.04233	.05270
22	10	6	.02109	.02830	.03589	.04390	.05236	.06131	.07078
22	10	7	.03620	.04284	.04978	.05703	.06462	.07258	.08095
22	10	8	.04718	.05311	.05926	.06565	.07228	.07920	.08641
22	10	9	.05530	.06047	.06579	.07128	.07695	.08283	.08891
22	10	10	.06136	.06575	.07024	.07485	.07958	.08447	.08950
22	10	11	.06588	.06949	.07316	.07693	.08078	.08474	.08880
22	10	12	.72334	.73325	.74443	.75684	.77046	.78523	.80112
22	11	1	-.77042	-.83644	-.90752	-.98396	-1.06609	-1.15427	-1.24885
22	11	2	-.19698	-.19797	-.19845	-.19836	-.19763	-.19616	-.19389
22	11	3	-.08981	-.08420	-.07790	-.07085	-.06296	-.05417	-.04441
22	11	4	-.03352	-.02592	-.01770	-.00880	.00084	.01128	.02259
22	11	5	.00115	.00913	.01761	.02665	.03630	.04659	.05758
22	11	6	.02436	.03202	.04008	.04858	.05754	.06701	.07702
22	11	7	.04068	.04771	.05504	.06270	.07071	.07911	.08792
22	11	8	.05251	.05876	.06523	.07195	.07892	.08618	.09374

Coefficients b_i for the BLUE of σ
Shape parameter k

n	s	i	1.4	1.5	1.6	1.7	1.8	1.9	2.0
22	11	9	.06125	.06666	.07222	.07796	.08389	.09003	.09639
22	11	10	.06774	.07229	.07695	.08173	.08665	.09172	.09695
22	11	11	.84305	.85798	.87444	.89239	.91182	.93269	.95497
23	0	1	-.41381	-.45397	-.49731	-.54405	-.59442	-.64867	-.70707
23	0	2	-.11217	-.11385	-.11515	-.11604	-.11646	-.11638	-.11572
23	0	3	-.05505	-.05240	-.04924	-.04554	-.04124	-.03630	-.03067
23	0	4	-.02481	-.02066	-.01603	-.01090	-.00524	.00101	.00788
23	0	5	-.00603	-.00139	.00365	.00911	.01502	.02140	.02831
23	0	6	.00666	.01133	.01633	.02165	.02734	.03341	.03989
23	0	7	.01568	.02017	.02490	.02988	.03515	.04071	.04658
23	0	8	.02232	.02650	.03085	.03540	.04016	.04514	.05035
23	0	9	.02732	.03112	.03505	.03913	.04335	.04773	.05229
23	0	10	.03112	.03453	.03802	.04160	.04529	.04909	.05301
23	0	11	.03404	.03703	.04008	.04319	.04636	.04960	.05293
23	0	12	.03627	.03886	.04147	.04411	.04679	.04952	.05229
23	0	13	.03797	.04016	.04234	.04454	.04676	.04900	.05127
23	0	14	.03924	.04103	.04281	.04459	.04637	.04816	.04997
23	0	15	.04015	.04156	.04295	.04433	.04571	.04709	.04848
23	0	16	.04077	.04180	.04282	.04383	.04484	.04584	.04685
23	0	17	.04111	.04179	.04246	.04312	.04378	.04445	.04512
23	0	18	.04122	.04155	.04189	.04223	.04257	.04293	.04330
23	0	19	.04111	.04111	.04112	.04116	.04122	.04130	.04141
23	0	20	.04076	.04044	.04016	.03992	.03972	.03956	.03943
23	0	21	.04015	.03951	.03894	.03845	.03802	.03765	.03734
23	0	22	.03914	.03819	.03736	.03664	.03601	.03547	.03501
23	0	23	.03683	.03559	.03454	.03366	.03291	.03229	.03177
23	1	1	-.42765	-.46887	-.51338	-.56142	-.61321	-.66904	-.72918
23	1	2	-.11589	-.11757	-.11887	-.11976	-.12019	-.12009	-.11942
23	1	3	-.05685	-.05410	-.05084	-.04702	-.04259	-.03751	-.03171
23	1	4	-.02561	-.02132	-.01655	-.01127	-.00544	.00098	.00805
23	1	5	-.00620	-.00142	.00376	.00938	.01545	.02202	.02912
23	1	6	.00691	.01172	.01685	.02233	.02817	.03441	.04107
23	1	7	.01624	.02084	.02570	.03082	.03622	.04193	.04797
23	1	8	.02310	.02738	.03185	.03651	.04140	.04651	.05187
23	1	9	.02826	.03215	.03618	.04036	.04469	.04918	.05386
23	1	10	.03219	.03567	.03924	.04291	.04669	.05059	.05462
23	1	11	.03520	.03826	.04137	.04455	.04779	.05112	.05453
23	1	12	.03751	.04014	.04281	.04550	.04824	.05103	.05388
23	1	13	.03927	.04148	.04371	.04595	.04821	.05050	.05282
23	1	14	.04058	.04238	.04419	.04600	.04781	.04964	.05149
23	1	15	.04152	.04293	.04433	.04573	.04713	.04854	.04996
23	1	16	.04214	.04317	.04419	.04521	.04623	.04725	.04828
23	1	17	.04250	.04316	.04382	.04448	.04514	.04581	.04649
23	1	18	.04261	.04291	.04323	.04356	.04390	.04425	.04462
23	1	19	.04248	.04245	.04244	.04246	.04250	.04258	.04268
23	1	20	.04211	.04175	.04143	.04117	.04095	.04078	.04065
23	1	21	.04145	.04077	.04017	.03965	.03921	.03882	.03850
23	1	22	.07814	.07611	.07438	.07292	.07170	.07069	.06986
23	2	1	-.44344	-.48582	-.53160	-.58101	-.63431	-.69177	-.75369
23	2	2	-.12010	-.12177	-.12307	-.12395	-.12436	-.12424	-.12354
23	2	3	-.05888	-.05601	-.05262	-.04866	-.04409	-.03885	-.03288
23	2	4	-.02648	-.02204	-.01712	-.01167	-.00566	.00095	.00823
23	2	5	-.00636	-.00142	.00392	.00970	.01595	.02271	.03001
23	2	6	.00724	.01219	.01747	.02310	.02911	.03552	.04236
23	2	7	.01691	.02164	.02663	.03189	.03744	.04330	.04951
23	2	8	.02402	.02841	.03300	.03778	.04279	.04804	.05354

Coefficients b_i for the BLUE of σ
Shape parameter k

n	s	i	1.4	1.5	1.6	1.7	1.8	1.9	2.0
23	2	9	.02937	.03336	.03748	.04176	.04620	.05081	.05561
23	2	10	.03344	.03700	.04065	.04440	.04827	.05226	.05639
23	2	11	.03656	.03968	.04285	.04610	.04941	.05281	.05631
23	2	12	.03895	.04163	.04434	.04708	.04988	.05273	.05563
23	2	13	.04077	.04301	.04527	.04754	.04984	.05218	.05455
23	2	14	.04212	.04394	.04576	.04759	.04943	.05129	.05317
23	2	15	.04309	.04450	.04591	.04732	.04873	.05015	.05159
23	2	16	.04374	.04475	.04576	.04678	.04779	.04882	.04986
23	2	17	.04410	.04473	.04537	.04601	.04667	.04734	.04801
23	2	18	.04420	.04447	.04475	.04506	.04538	.04572	.04608
23	2	19	.04405	.04397	.04392	.04391	.04393	.04399	.04408
23	2	20	.04364	.04322	.04287	.04257	.04232	.04213	.04199
23	2	21	.12306	.12056	.11846	.11671	.11527	.11412	.11321
23	3	1	-.46123	-.50490	-.55207	-.60300	-.65794	-.71717	-.78100
23	3	2	-.12480	-.12647	-.12775	-.12862	-.12900	-.12886	-.12812
23	3	3	-.06112	-.05812	-.05459	-.05048	-.04575	-.04033	-.03417
23	3	4	-.02742	-.02282	-.01772	-.01209	-.00589	.00093	.00843
23	3	5	-.00649	-.00139	.00412	.01009	.01653	.02349	.03101
23	3	6	.00765	.01275	.01819	.02399	.03018	.03677	.04382
23	3	7	.01770	.02257	.02770	.03311	.03882	.04485	.05122
23	3	8	.02509	.02961	.03431	.03923	.04437	.04976	.05540
23	3	9	.03065	.03474	.03897	.04335	.04790	.05263	.05755
23	3	10	.03488	.03852	.04226	.04610	.05005	.05414	.05837
23	3	11	.03812	.04130	.04454	.04785	.05124	.05471	.05828
23	3	12	.04061	.04333	.04608	.04887	.05172	.05462	.05759
23	3	13	.04249	.04476	.04704	.04935	.05168	.05405	.05647
23	3	14	.04389	.04572	.04755	.04939	.05125	.05313	.05504
23	3	15	.04489	.04629	.04770	.04910	.05052	.05195	.05340
23	3	16	.04555	.04654	.04754	.04854	.04955	.05057	.05161
23	3	17	.04592	.04651	.04712	.04774	.04838	.04903	.04970
23	3	18	.04600	.04622	.04647	.04674	.04704	.04736	.04770
23	3	19	.04582	.04568	.04559	.04554	.04553	.04556	.04563
23	3	20	.17179	.16913	.16695	.16519	.16382	.16279	.16207
23	4	1	-.48121	-.52633	-.57507	-.62768	-.68443	-.74560	-.81152
23	4	2	-.13004	-.13170	-.13297	-.13382	-.13418	-.13400	-.13322
23	4	3	-.06359	-.06044	-.05676	-.05248	-.04757	-.04196	-.03560
23	4	4	-.02843	-.02364	-.01836	-.01253	-.00611	.00094	.00867
23	4	5	-.00659	-.00131	.00440	.01056	.01721	.02439	.03214
23	4	6	.00815	.01343	.01905	.02503	.03140	.03820	.04545
23	4	7	.01864	.02366	.02895	.03452	.04039	.04659	.05315
23	4	8	.02635	.03099	.03583	.04088	.04616	.05170	.05750
23	4	9	.03214	.03633	.04067	.04517	.04983	.05468	.05973
23	4	10	.03654	.04027	.04409	.04802	.05207	.05626	.06058
23	4	11	.03992	.04316	.04647	.04984	.05330	.05685	.06050
23	4	12	.04250	.04526	.04806	.05090	.05380	.05676	.05978
23	4	13	.04446	.04675	.04906	.05139	.05376	.05616	.05861
23	4	14	.04591	.04774	.04958	.05143	.05331	.05521	.05714
23	4	15	.04694	.04833	.04972	.05112	.05254	.05398	.05543
23	4	16	.04761	.04858	.04955	.05053	.05153	.05254	.05357
23	4	17	.04797	.04852	.04910	.04969	.05030	.05093	.05159
23	4	18	.04804	.04820	.04840	.04863	.04889	.04918	.04951
23	4	19	.22469	.22220	.22025	.21880	.21779	.21720	.21697
23	5	1	-.50368	-.55044	-.60093	-.65543	-.71420	-.77754	-.84577
23	5	2	-.13588	-.13753	-.13879	-.13962	-.13996	-.13974	-.13891
23	5	3	-.06630	-.06300	-.05914	-.05469	-.04958	-.04376	-.03717
23	5	4	-.02950	-.02452	-.01902	-.01297	-.00632	.00097	.00897

Coefficients b_i for the BLUE of σ
Shape parameter k

n	s	i	1.4	1.5	1.6	1.7	1.8	1.9	2.0
23	5	5	-.00665	-.00117	.00475	.01113	.01801	.02543	.03343
23	5	6	.00878	.01424	.02005	.02623	.03281	.03983	.04731
23	5	7	.01974	.02493	.03039	.03613	.04219	.04858	.05534
23	5	8	.02780	.03259	.03757	.04277	.04821	.05390	.05987
23	5	9	.03386	.03817	.04263	.04724	.05204	.05702	.06220
23	5	10	.03846	.04227	.04619	.05022	.05437	.05865	.06309
23	5	11	.04198	.04529	.04866	.05211	.05564	.05927	.06300
23	5	12	.04467	.04748	.05032	.05321	.05616	.05917	.06225
23	5	13	.04671	.04902	.05135	.05371	.05611	.05855	.06104
23	5	14	.04821	.05004	.05188	.05375	.05563	.05755	.05949
23	5	15	.04927	.05064	.05202	.05341	.05483	.05626	.05772
23	5	16	.04995	.05088	.05182	.05278	.05375	.05475	.05577
23	5	17	.05030	.05080	.05133	.05188	.05246	.05307	.05370
23	5	18	.28228	.28030	.27894	.27813	.27785	.27804	.27866
23	6	1	-.52904	-.57765	-.63013	-.68676	-.74781	-.81358	-.88440
23	6	2	-.14240	-.14405	-.14530	-.14611	-.14642	-.14617	-.14528
23	6	3	-.06930	-.06582	-.06178	-.05712	-.05180	-.04574	-.03889
23	6	4	-.03064	-.02544	-.01971	-.01342	-.00651	.00106	.00935
23	6	5	-.00665	-.00094	.00520	.01183	.01896	.02665	.03493
23	6	6	.00955	.01522	.02124	.02764	.03446	.04171	.04944
23	6	7	.02105	.02642	.03207	.03801	.04426	.05086	.05783
23	6	8	.02951	.03445	.03959	.04495	.05056	.05643	.06257
23	6	9	.03585	.04029	.04488	.04963	.05456	.05968	.06501
23	6	10	.04067	.04459	.04860	.05273	.05699	.06139	.06593
23	6	11	.04436	.04774	.05118	.05471	.05832	.06203	.06584
23	6	12	.04717	.05002	.05291	.05585	.05885	.06191	.06505
23	6	13	.04929	.05162	.05397	.05636	.05879	.06126	.06378
23	6	14	.05085	.05267	.05452	.05638	.05828	.06020	.06216
23	6	15	.05193	.05328	.05464	.05601	.05742	.05884	.06030
23	6	16	.05262	.05350	.05440	.05533	.05628	.05725	.05826
23	6	17	.34520	.34412	.34373	.34398	.34483	.34622	.34811
23	7	1	-.55780	-.60852	-.66327	-.72231	-.78595	-.85448	-.92823
23	7	2	-.14972	-.15137	-.15261	-.15341	-.15369	-.15339	-.15245
23	7	3	-.07262	-.06894	-.06469	-.05981	-.05424	-.04793	-.04079
23	7	4	-.03186	-.02641	-.02042	-.01385	-.00665	.00122	.00984
23	7	5	-.00657	-.00061	.00579	.01269	.02010	.02809	.03668
23	7	6	.01049	.01640	.02266	.02931	.03638	.04390	.05190
23	7	7	.02261	.02819	.03404	.04019	.04667	.05350	.06071
23	7	8	.03151	.03662	.04194	.04748	.05328	.05933	.06568
23	7	9	.03818	.04276	.04749	.05239	.05746	.06274	.06823
23	7	10	.04325	.04727	.05139	.05564	.06001	.06452	.06919
23	7	11	.04711	.05057	.05409	.05770	.06139	.06519	.06909
23	7	12	.05005	.05295	.05589	.05888	.06193	.06506	.06826
23	7	13	.05226	.05461	.05699	.05940	.06185	.06436	.06691
23	7	14	.05388	.05569	.05753	.05940	.06130	.06323	.06520
23	7	15	.05499	.05630	.05763	.05899	.06037	.06179	.06324
23	7	16	.41423	.41450	.41555	.41733	.41979	.42288	.42655
23	8	1	-.59062	-.64377	-.70111	-.76293	-.82952	-.90121	-.97831
23	8	2	-.15799	-.15963	-.16087	-.16165	-.16190	-.16156	-.16056
23	8	3	-.07631	-.07241	-.06792	-.06278	-.05694	-.05033	-.04289
23	8	4	-.03315	-.02742	-.02113	-.01425	-.00673	.00149	.01047
23	8	5	-.00638	-.00014	.00656	.01375	.02149	.02980	.03874
23	8	6	.01167	.01783	.02436	.03129	.03864	.04646	.05478
23	8	7	.02448	.03029	.03637	.04277	.04949	.05658	.06405
23	8	8	.03388	.03919	.04470	.05044	.05644	.06271	.06927
23	8	9	.04093	.04566	.05054	.05560	.06084	.06629	.07195

Coefficients b_i for the BLUE of σ
Shape parameter k

n	s	i	1.4	1.5	1.6	1.7	1.8	1.9	2.0
23	8	10	.04626	.05040	.05464	.05901	.06351	.06815	.07296
23	8	11	.05033	.05386	.05747	.06116	.06495	.06884	.07284
23	8	12	.05341	.05635	.05934	.06239	.06550	.06868	.07195
23	8	13	.05572	.05808	.06047	.06291	.06539	.06792	.07051
23	8	14	.05739	.05919	.06102	.06288	.06478	.06672	.06870
23	8	15	.49038	.49253	.49556	.49942	.50407	.50946	.51554
23	9	1	-.62836	-.68432	-.74467	-.80970	-.87971	-.95504	-1.03601
23	9	2	-.16738	-.16903	-.17026	-.17102	-.17124	-.17086	-.16979
23	9	3	-.08044	-.07628	-.07152	-.06609	-.05994	-.05300	-.04519
23	9	4	-.03451	-.02846	-.02184	-.01461	-.00671	.00190	.01129
23	9	5	-.00605	.00051	.00755	.01509	.02319	.03188	.04121
23	9	6	.01314	.01959	.02643	.03367	.04135	.04951	.05817
23	9	7	.02675	.03281	.03916	.04582	.05283	.06020	.06797
23	9	8	.03672	.04223	.04797	.05394	.06017	.06668	.07349
23	9	9	.04418	.04908	.05415	.05938	.06481	.07044	.07630
23	9	10	.04982	.05409	.05847	.06297	.06761	.07240	.07735
23	9	11	.05411	.05774	.06143	.06522	.06910	.07309	.07720
23	9	12	.05735	.06034	.06338	.06649	.06966	.07291	.07624
23	9	13	.05976	.06213	.06454	.06700	.06951	.07207	.07470
23	9	14	.57492	.57956	.58522	.59184	.59939	.60781	.61707
23	10	1	-.67219	-.73143	-.79529	-.86406	-.93807	-1.01764	-1.10313
23	10	2	-.17813	-.17979	-.18102	-.18176	-.18195	-.18151	-.18038
23	10	3	-.08507	-.08062	-.07555	-.06978	-.06327	-.05594	-.04772
23	10	4	-.03594	-.02952	-.02251	-.01487	-.00655	.00251	.01238
23	10	5	-.00551	.00142	.00884	.01678	.02529	.03442	.04421
23	10	6	.01498	.02178	.02896	.03657	.04462	.05317	.06223
23	10	7	.02951	.03586	.04251	.04949	.05682	.06452	.07263
23	10	8	.04014	.04590	.05189	.05811	.06461	.07139	.07848
23	10	9	.04807	.05318	.05844	.06388	.06952	.07536	.08144
23	10	10	.05407	.05848	.06301	.06766	.07246	.07741	.08253
23	10	11	.05861	.06233	.06613	.07002	.07401	.07811	.08234
23	10	12	.06202	.06506	.06816	.07132	.07456	.07787	.08128
23	10	13	.66945	.67736	.68643	.69665	.70796	.72033	.73371
23	11	1	-.72366	-.78678	-.85478	-.92798	-1.00671	-1.09130	-1.18212
23	11	2	-.19058	-.19225	-.19348	-.19420	-.19435	-.19385	-.19263
23	11	3	-.09032	-.08553	-.08008	-.07393	-.06700	-.05922	-.05051
23	11	4	-.03743	-.03057	-.02311	-.01500	-.00617	.00342	.01384
23	11	5	-.00470	.00267	.01054	.01895	.02794	.03758	.04790
23	11	6	.01733	.02452	.03212	.04014	.04863	.05762	.06716
23	11	7	.03292	.03962	.04662	.05396	.06166	.06975	.07825
23	11	8	.04431	.05036	.05664	.06316	.06996	.07705	.08446
23	11	9	.05280	.05813	.06362	.06930	.07517	.08127	.08760
23	11	10	.05920	.06377	.06847	.07330	.07827	.08340	.08871
23	11	11	.06402	.06785	.07176	.07577	.07988	.08411	.08846
23	11	12	.77611	.78821	.80169	.81654	.83271	.85017	.86888
23	12	1	-.78495	-.85271	-.92567	-1.00418	-1.08855	-1.17916	-1.27638
23	12	2	-.20517	-.20684	-.20806	-.20876	-.20886	-.20829	-.20698
23	12	3	-.09632	-.09111	-.08523	-.07861	-.07118	-.06286	-.05358
23	12	4	-.03895	-.03158	-.02358	-.01489	-.00547	.00474	.01583
23	12	5	-.00348	.00440	.01280	.02176	.03134	.04157	.05251
23	12	6	.02036	.02802	.03609	.04461	.05362	.06315	.07323
23	12	7	.03721	.04431	.05173	.05949	.06763	.07616	.08514
23	12	8	.04949	.05587	.06249	.06937	.07652	.08398	.09177
23	12	9	.05863	.06421	.06997	.07592	.08207	.08845	.09508
23	12	10	.06548	.07025	.07514	.08016	.08534	.09069	.09621
23	12	11	.89770	.91518	.93433	.95512	.97755	1.00157	1.02716

Coefficients bi for the BLUE of σ
Shape parameter k

n	s	i	1.4	1.5	1.6	1.7	1.8	1.9	2.0
24	0	1	-.40581	-.44564	-.48866	-.53507	-.58513	-.63908	-.69720
24	0	2	-.11246	-.11452	-.11623	-.11757	-.11849	-.11894	-.11888
24	0	3	-.05664	-.05434	-.05156	-.04825	-.04438	-.03988	-.03471
24	0	4	-.02699	-.02313	-.01881	-.01400	-.00865	-.00274	.00378
24	0	5	-.00851	-.00410	.00070	.00592	.01159	.01774	.02441
24	0	6	.00403	.00853	.01335	.01851	.02403	.02994	.03626
24	0	7	.01299	.01735	.02196	.02683	.03200	.03747	.04326
24	0	8	.01961	.02371	.02800	.03249	.03720	.04214	.04734
24	0	9	.02462	.02839	.03230	.03636	.04058	.04498	.04956
24	0	10	.02846	.03187	.03538	.03900	.04272	.04658	.05057
24	0	11	.03144	.03447	.03757	.04074	.04399	.04732	.05075
24	0	12	.03374	.03640	.03910	.04183	.04462	.04746	.05036
24	0	13	.03552	.03781	.04011	.04243	.04477	.04715	.04957
24	0	14	.03688	.03880	.04072	.04264	.04457	.04652	.04849
24	0	15	.03790	.03946	.04101	.04255	.04410	.04565	.04721
24	0	16	.03862	.03983	.04103	.04222	.04341	.04460	.04579
24	0	17	.03909	.03997	.04083	.04169	.04254	.04340	.04426
24	0	18	.03934	.03989	.04044	.04098	.04153	.04208	.04264
24	0	19	.03938	.03962	.03987	.04012	.04039	.04067	.04096
24	0	20	.03923	.03917	.03913	.03911	.03912	.03915	.03921
24	0	21	.03887	.03852	.03821	.03794	.03771	.03753	.03738
24	0	22	.03827	.03763	.03706	.03657	.03613	.03576	.03544
24	0	23	.03731	.03638	.03557	.03487	.03426	.03374	.03329
24	0	24	.03512	.03394	.03293	.03208	.03137	.03077	.03027
24	1	1	-.41872	-.45955	-.50367	-.55132	-.60273	-.65818	-.71794
24	1	2	-.11601	-.11808	-.11981	-.12116	-.12210	-.12256	-.12249
24	1	3	-.05841	-.05603	-.05315	-.04974	-.04575	-.04113	-.03582
24	1	4	-.02782	-.02384	-.01939	-.01444	-.00895	-.00288	.00382
24	1	5	-.00875	-.00422	.00072	.00608	.01191	.01822	.02506
24	1	6	.00418	.00881	.01375	.01905	.02472	.03078	.03728
24	1	7	.01343	.01790	.02263	.02763	.03293	.03854	.04449
24	1	8	.02026	.02446	.02885	.03346	.03829	.04336	.04869
24	1	9	.02542	.02928	.03329	.03745	.04177	.04628	.05098
24	1	10	.02939	.03288	.03646	.04016	.04398	.04793	.05202
24	1	11	.03246	.03556	.03872	.04196	.04529	.04870	.05221
24	1	12	.03484	.03755	.04029	.04309	.04593	.04884	.05181
24	1	13	.03667	.03900	.04134	.04370	.04610	.04853	.05100
24	1	14	.03808	.04002	.04197	.04392	.04589	.04788	.04990
24	1	15	.03912	.04069	.04226	.04383	.04540	.04699	.04858
24	1	16	.03986	.04108	.04228	.04349	.04469	.04590	.04712
24	1	17	.04034	.04121	.04208	.04294	.04380	.04467	.04554
24	1	18	.04060	.04113	.04167	.04221	.04276	.04332	.04388
24	1	19	.04064	.04086	.04108	.04133	.04158	.04186	.04215
24	1	20	.04048	.04038	.04032	.04028	.04028	.04030	.04035
24	1	21	.04010	.03970	.03936	.03907	.03883	.03863	.03848
24	1	22	.03945	.03877	.03817	.03765	.03721	.03682	.03650
24	1	23	.07440	.07244	.07077	.06935	.06817	.06719	.06639
24	2	1	-.43338	-.47530	-.52062	-.56956	-.62240	-.67939	-.74084
24	2	2	-.12001	-.12208	-.12382	-.12518	-.12611	-.12657	-.12649
24	2	3	-.06039	-.05791	-.05492	-.05139	-.04727	-.04251	-.03705
24	2	4	-.02873	-.02462	-.02003	-.01493	-.00928	-.00303	.00386
24	2	5	-.00900	-.00432	.00076	.00628	.01226	.01875	.02578
24	2	6	.00439	.00915	.01423	.01967	.02549	.03172	.03839
24	2	7	.01396	.01855	.02340	.02854	.03397	.03973	.04583
24	2	8	.02103	.02533	.02984	.03456	.03951	.04471	.05017
24	2	9	.02637	.03032	.03442	.03868	.04311	.04773	.05255

Coefficients b_i for the BLUE of σ
Shape parameter k

n	s	i	1.4	1.5	1.6	1.7	1.8	1.9	2.0
24	2	10	.03047	.03404	.03770	.04149	.04539	.04943	.05362
24	2	11	.03365	.03681	.04004	.04335	.04674	.05023	.05382
24	2	12	.03611	.03886	.04166	.04451	.04741	.05037	.05341
24	2	13	.03801	.04036	.04274	.04514	.04758	.05005	.05258
24	2	14	.03945	.04142	.04338	.04537	.04737	.04939	.05144
24	2	15	.04053	.04211	.04369	.04527	.04686	.04847	.05009
24	2	16	.04129	.04250	.04371	.04492	.04613	.04735	.04858
24	2	17	.04179	.04264	.04349	.04435	.04521	.04608	.04695
24	2	18	.04204	.04255	.04307	.04359	.04413	.04468	.04524
24	2	19	.04208	.04226	.04245	.04267	.04291	.04318	.04346
24	2	20	.04189	.04175	.04165	.04159	.04156	.04157	.04161
24	2	21	.04147	.04103	.04065	.04033	.04006	.03985	.03968
24	2	22	.11698	.11454	.11248	.11076	.10935	.10822	.10732
24	3	1	-.44982	-.49295	-.53958	-.58995	-.64432	-.70297	-.76622
24	3	2	-.12447	-.12654	-.12828	-.12964	-.13057	-.13102	-.13092
24	3	3	-.06258	-.05998	-.05687	-.05321	-.04896	-.04404	-.03842
24	3	4	-.02972	-.02545	-.02071	-.01544	-.00962	-.00319	.00390
24	3	5	-.00923	-.00440	.00084	.00652	.01268	.01935	.02658
24	3	6	.00466	.00956	.01480	.02039	.02638	.03278	.03963
24	3	7	.01459	.01931	.02430	.02957	.03516	.04107	.04733
24	3	8	.02192	.02634	.03096	.03581	.04088	.04622	.05182
24	3	9	.02747	.03152	.03571	.04007	.04461	.04934	.05428
24	3	10	.03172	.03536	.03911	.04298	.04697	.05111	.05540
24	3	11	.03501	.03823	.04153	.04490	.04837	.05193	.05561
24	3	12	.03756	.04036	.04321	.04611	.04906	.05209	.05519
24	3	13	.03953	.04191	.04432	.04676	.04924	.05176	.05432
24	3	14	.04102	.04300	.04499	.04699	.04902	.05107	.05315
24	3	15	.04213	.04372	.04530	.04689	.04849	.05011	.05175
24	3	16	.04292	.04412	.04532	.04652	.04773	.04896	.05020
24	3	17	.04342	.04425	.04509	.04593	.04678	.04764	.04852
24	3	18	.04368	.04415	.04464	.04514	.04566	.04620	.04675
24	3	19	.04370	.04383	.04399	.04418	.04440	.04464	.04490
24	3	20	.04348	.04329	.04315	.04305	.04299	.04297	.04299
24	3	21	.16301	.16037	.15819	.15643	.15504	.15399	.15324
24	4	1	-.46821	-.51268	-.56076	-.61270	-.66876	-.72923	-.79443
24	4	2	-.12941	-.13149	-.13323	-.13459	-.13551	-.13595	-.13583
24	4	3	-.06498	-.06225	-.05901	-.05521	-.05080	-.04572	-.03992
24	4	4	-.03077	-.02635	-.02143	-.01599	-.00998	-.00334	.00396
24	4	5	-.00945	-.00446	.00095	.00682	.01317	.02004	.02748
24	4	6	.00501	.01006	.01546	.02122	.02738	.03397	.04102
24	4	7	.01533	.02020	.02533	.03076	.03649	.04257	.04901
24	4	8	.02296	.02750	.03225	.03723	.04244	.04791	.05367
24	4	9	.02873	.03288	.03718	.04166	.04631	.05116	.05622
24	4	10	.03315	.03688	.04071	.04467	.04876	.05299	.05738
24	4	11	.03657	.03986	.04322	.04666	.05020	.05384	.05760
24	4	12	.03922	.04206	.04496	.04791	.05092	.05400	.05716
24	4	13	.04125	.04367	.04611	.04859	.05110	.05366	.05627
24	4	14	.04280	.04480	.04680	.04882	.05087	.05295	.05506
24	4	15	.04395	.04553	.04712	.04871	.05032	.05196	.05361
24	4	16	.04476	.04594	.04713	.04832	.04953	.05075	.05199
24	4	17	.04527	.04607	.04688	.04770	.04853	.04938	.05025
24	4	18	.04552	.04595	.04640	.04687	.04737	.04788	.04842
24	4	19	.04551	.04559	.04571	.04586	.04604	.04626	.04650
24	4	20	.21279	.21025	.20824	.20671	.20562	.20492	.20458
24	5	1	-.48877	-.53476	-.58447	-.63815	-.69608	-.75857	-.82591
24	5	2	-.13489	-.13698	-.13873	-.14009	-.14100	-.14142	-.14128

Coefficients b_i for the BLUE of σ
Shape parameter k

n	s	i	1.4	1.5	1.6	1.7	1.8	1.9	2.0
24	5	3	-.06762	-.06475	-.06137	-.05741	-.05283	-.04757	-.04157
24	5	4	-.03190	-.02730	-.02220	-.01656	-.01034	-.00348	.00406
24	5	5	-.00965	-.00448	.00113	.00719	.01375	.02084	.02851
24	5	6	.00544	.01066	.01624	.02219	.02854	.03533	.04258
24	5	7	.01621	.02123	.02652	.03211	.03802	.04427	.05089
24	5	8	.02417	.02884	.03373	.03884	.04420	.04983	.05574
24	5	9	.03018	.03444	.03886	.04345	.04823	.05320	.05839
24	5	10	.03479	.03861	.04253	.04659	.05077	.05511	.05960
24	5	11	.03835	.04171	.04514	.04866	.05227	.05599	.05983
24	5	12	.04110	.04400	.04695	.04995	.05302	.05616	.05938
24	5	13	.04322	.04567	.04814	.05065	.05320	.05580	.05845
24	5	14	.04483	.04683	.04885	.05089	.05295	.05505	.05719
24	5	15	.04601	.04759	.04917	.05076	.05238	.05402	.05568
24	5	16	.04684	.04800	.04917	.05035	.05155	.05276	.05400
24	5	17	.04735	.04812	.04889	.04969	.05050	.05133	.05219
24	5	18	.04759	.04797	.04838	.04881	.04927	.04976	.05028
24	5	19	.26675	.26462	.26309	.26210	.26162	.26159	.26198
24	6	1	-.51183	-.55952	-.61106	-.66670	-.72673	-.79146	-.86120
24	6	2	-.14098	-.14309	-.14484	-.14621	-.14712	-.14752	-.14735
24	6	3	-.07052	-.06750	-.06396	-.05983	-.05507	-.04960	-.04338
24	6	4	-.03312	-.02832	-.02301	-.01716	-.01071	-.00361	.00419
24	6	5	-.00982	-.00444	.00137	.00765	.01443	.02177	.02969
24	6	6	.00597	.01139	.01716	.02331	.02987	.03688	.04437
24	6	7	.01725	.02243	.02790	.03367	.03976	.04621	.05303
24	6	8	.02557	.03039	.03543	.04070	.04621	.05200	.05809
24	6	9	.03186	.03624	.04079	.04550	.05041	.05552	.06085
24	6	10	.03667	.04059	.04462	.04877	.05306	.05750	.06211
24	6	11	.04039	.04382	.04733	.05093	.05462	.05843	.06235
24	6	12	.04326	.04621	.04921	.05227	.05539	.05859	.06188
24	6	13	.04547	.04794	.05044	.05298	.05557	.05821	.06091
24	6	14	.04713	.04914	.05117	.05322	.05531	.05743	.05959
24	6	15	.04836	.04992	.05149	.05308	.05470	.05634	.05802
24	6	16	.04920	.05033	.05147	.05264	.05382	.05503	.05626
24	6	17	.04971	.05043	.05117	.05193	.05271	.05352	.05436
24	6	18	.32542	.32405	.32334	.32326	.32374	.32475	.32624
24	7	1	-.53780	-.58743	-.64103	-.69888	-.76128	-.82852	-.90095
24	7	2	-.14778	-.14990	-.15167	-.15304	-.15395	-.15434	-.15414
24	7	3	-.07372	-.07054	-.06681	-.06249	-.05752	-.05184	-.04538
24	7	4	-.03442	-.02940	-.02387	-.01777	-.01106	-.00369	.00439
24	7	5	-.00994	-.00434	.00170	.00822	.01526	.02286	.03106
24	7	6	.00664	.01227	.01825	.02463	.03143	.03867	.04641
24	7	7	.01848	.02385	.02951	.03548	.04178	.04843	.05547
24	7	8	.02721	.03220	.03739	.04283	.04852	.05449	.06076
24	7	9	.03380	.03832	.04300	.04786	.05291	.05817	.06365
24	7	10	.03885	.04287	.04701	.05127	.05568	.06024	.06496
24	7	11	.04274	.04625	.04984	.05352	.05730	.06119	.06521
24	7	12	.04574	.04874	.05179	.05491	.05810	.06136	.06471
24	7	13	.04804	.05054	.05307	.05565	.05827	.06095	.06370
24	7	14	.04977	.05178	.05382	.05589	.05799	.06013	.06231
24	7	15	.05103	.05257	.05414	.05572	.05733	.05898	.06065
24	7	16	.05189	.05298	.05409	.05523	.05639	.05758	.05880
24	7	17	.38946	.38924	.38976	.39098	.39285	.39533	.39837
24	8	1	-.56722	-.61904	-.67499	-.73536	-.80043	-.87054	-.94600
24	8	2	-.15541	-.15755	-.15934	-.16071	-.16162	-.16199	-.16177
24	8	3	-.07726	-.07389	-.06997	-.06544	-.06024	-.05430	-.04757
24	8	4	-.03581	-.03055	-.02476	-.01839	-.01140	-.00373	.00468

Coefficients b_i for the BLUE of σ
Shape parameter k

n	s	i	1.4	1.5	1.6	1.7	1.8	1.9	2.0
24	8	5	-.01000	-.00415	.00215	.00894	.01627	.02416	.03268
24	8	6	.00748	.01334	.01957	.02619	.03325	.04077	.04878
24	8	7	.01995	.02553	.03140	.03759	.04411	.05101	.05829
24	8	8	.02914	.03430	.03969	.04531	.05119	.05736	.06383
24	8	9	.03608	.04074	.04557	.05058	.05579	.06121	.06686
24	8	10	.04138	.04552	.04977	.05416	.05869	.06337	.06823
24	8	11	.04546	.04906	.05273	.05650	.06038	.06437	.06848
24	8	12	.04860	.05166	.05477	.05795	.06120	.06453	.06795
24	8	13	.05100	.05353	.05610	.05871	.06137	.06409	.06687
24	8	14	.05280	.05482	.05686	.05893	.06105	.06320	.06541
24	8	15	.05410	.05562	.05716	.05874	.06034	.06198	.06366
24	8	16	.45969	.46106	.46328	.46629	.47005	.47451	.47962
24	9	1	-.60075	-.65509	-.71374	-.77698	-.84513	-.91850	-.99744
24	9	2	-.16400	-.16618	-.16799	-.16937	-.17028	-.17064	-.17039
24	9	3	-.08120	-.07762	-.07347	-.06870	-.06324	-.05702	-.04999
24	9	4	-.03729	-.03175	-.02568	-.01900	-.01169	-.00368	.00508
24	9	5	-.00997	-.00384	.00276	.00986	.01750	.02573	.03459
24	9	6	.00853	.01465	.02115	.02806	.03540	.04322	.05156
24	9	7	.02171	.02753	.03364	.04008	.04686	.05401	.06157
24	9	8	.03143	.03679	.04238	.04821	.05431	.06070	.06740
24	9	9	.03875	.04358	.04858	.05376	.05914	.06474	.07058
24	9	10	.04434	.04861	.05299	.05751	.06218	.06701	.07201
24	9	11	.04864	.05233	.05610	.05997	.06394	.06804	.07226
24	9	12	.05194	.05505	.05823	.06147	.06479	.06819	.07168
24	9	13	.05444	.05700	.05960	.06224	.06494	.06770	.07053
24	9	14	.05631	.05832	.06037	.06245	.06457	.06674	.06897
24	9	15	.53711	.54062	.54508	.55046	.55671	.56377	.57161
24	10	1	-.63929	-.69653	-.75830	-.82487	-.89656	-.97371	-1.05666
24	10	2	-.17377	-.17598	-.17781	-.17921	-.18012	-.18047	-.18020
24	10	3	-.08560	-.08178	-.07738	-.07233	-.06657	-.06004	-.05267
24	10	4	-.03887	-.03302	-.02661	-.01959	-.01191	-.00352	.00565
24	10	5	-.00981	-.00336	.00357	.01102	.01902	.02763	.03689
24	10	6	.00985	.01627	.02308	.03031	.03798	.04615	.05484
24	10	7	.02386	.02994	.03632	.04303	.05010	.05756	.06542
24	10	8	.03417	.03975	.04557	.05164	.05798	.06462	.07157
24	10	9	.04193	.04694	.05212	.05750	.06307	.06888	.07492
24	10	10	.04784	.05225	.05678	.06144	.06626	.07125	.07642
24	10	11	.05238	.05617	.06004	.06402	.06811	.07231	.07666
24	10	12	.05585	.05902	.06227	.06558	.06897	.07245	.07602
24	10	13	.05847	.06105	.06368	.06636	.06910	.07190	.07478
24	10	14	.62301	.62929	.63667	.64511	.65457	.66500	.67635
24	11	1	-.68400	-.74465	-.81004	-.88049	-.95632	-1.03787	-1.12550
24	11	2	-.18495	-.18720	-.18907	-.19049	-.19141	-.19175	-.19144
24	11	3	-.09056	-.08646	-.08176	-.07639	-.07029	-.06339	-.05562
24	11	4	-.04056	-.03434	-.02755	-.02013	-.01203	-.00318	.00646
24	11	5	-.00948	-.00266	.00466	.01250	.02093	.02997	.03969
24	11	6	.01153	.01829	.02546	.03305	.04111	.04967	.05877
24	11	7	.02648	.03286	.03956	.04659	.05399	.06179	.07001
24	11	8	.03747	.04332	.04940	.05573	.06235	.06928	.07653
24	11	9	.04573	.05095	.05635	.06194	.06774	.07378	.08006
24	11	10	.05202	.05658	.06128	.06611	.07110	.07627	.08161
24	11	11	.05682	.06072	.06472	.06881	.07303	.07736	.08183
24	11	12	.06048	.06373	.06704	.07043	.07390	.07746	.08113
24	11	13	.71903	.72886	.73997	.75233	.76589	.78063	.79649
24	12	1	-.73650	-.80115	-.87083	-.94586	-1.02657	-1.11332	-1.20648
24	12	2	-.19790	-.20020	-.20210	-.20354	-.20446	-.20479	-.20446

Coefficients b_i for the BLUE of σ
Shape parameter k

n	s	i	1.4	1.5	1.6	1.7	1.8	1.9	2.0
24	12	3	-.09618	-.09176	-.08671	-.08097	-.07446	-.06713	-.05890
24	12	4	-.04234	-.03570	-.02846	-.02057	-.01197	-.00260	.00759
24	12	5	-.00890	-.00165	.00612	.01443	.02335	.03290	.04315
24	12	6	.01368	.02085	.02843	.03645	.04495	.05397	.06355
24	12	7	.02973	.03647	.04353	.05094	.05872	.06691	.07554
24	12	8	.04152	.04766	.05404	.06069	.06763	.07488	.08247
24	12	9	.05036	.05582	.06146	.06730	.07336	.07966	.08621
24	12	10	.05706	.06181	.06669	.07172	.07691	.08227	.08783
24	12	11	.06217	.06620	.07033	.07456	.07891	.08339	.08801
24	12	12	.82731	.84165	.85750	.87484	.89364	.91387	.93549
25	0	1	-.39823	-.43773	-.48042	-.52652	-.57626	-.62991	-.68773
25	0	2	-.11263	-.11503	-.11712	-.11888	-.12026	-.12121	-.12169
25	0	3	-.05803	-.05606	-.05363	-.05069	-.04721	-.04313	-.03840
25	0	4	-.02895	-.02536	-.02133	-.01681	-.01178	-.00619	.00000
25	0	5	-.01075	-.00657	-.00199	.00300	.00844	.01435	.02078
25	0	6	.00164	.00596	.01061	.01560	.02096	.02671	.03288
25	0	7	.01052	.01475	.01925	.02401	.02907	.03444	.04014
25	0	8	.01712	.02113	.02535	.02978	.03443	.03933	.04449
25	0	9	.02213	.02586	.02974	.03378	.03799	.04238	.04698
25	0	10	.02601	.02941	.03292	.03655	.04030	.04419	.04823
25	0	11	.02903	.03209	.03523	.03844	.04174	.04514	.04865
25	0	12	.03139	.03411	.03687	.03968	.04255	.04548	.04849
25	0	13	.03324	.03561	.03800	.04042	.04287	.04537	.04791
25	0	14	.03468	.03670	.03873	.04078	.04284	.04492	.04703
25	0	15	.03577	.03746	.03915	.04084	.04253	.04423	.04595
25	0	16	.03659	.03795	.03931	.04065	.04200	.04335	.04471
25	0	17	.03716	.03821	.03925	.04027	.04130	.04233	.04335
25	0	18	.03752	.03826	.03900	.03973	.04045	.04118	.04191
25	0	19	.03769	.03814	.03859	.03903	.03949	.03994	.04040
25	0	20	.03769	.03785	.03802	.03821	.03840	.03861	.03884
25	0	21	.03751	.03739	.03731	.03725	.03721	.03720	.03722
25	0	22	.03715	.03676	.03643	.03614	.03590	.03569	.03552
25	0	23	.03656	.03592	.03535	.03485	.03442	.03404	.03372
25	0	24	.03564	.03473	.03395	.03326	.03267	.03216	.03172
25	0	25	.03357	.03242	.03146	.03064	.02996	.02938	.02891
25	1	1	-.41030	-.45075	-.49449	-.54176	-.59279	-.64786	-.70725
25	1	2	-.11602	-.11844	-.12056	-.12234	-.12375	-.12472	-.12521
25	1	3	-.05977	-.05772	-.05521	-.05218	-.04860	-.04441	-.03957
25	1	4	-.02980	-.02611	-.02196	-.01732	-.01215	-.00642	-.00007
25	1	5	-.01105	-.00675	-.00205	.00307	.00865	.01471	.02130
25	1	6	.00171	.00615	.01092	.01604	.02153	.02742	.03375
25	1	7	.01086	.01520	.01981	.02469	.02987	.03538	.04122
25	1	8	.01766	.02177	.02609	.03062	.03539	.04041	.04570
25	1	9	.02283	.02664	.03061	.03474	.03905	.04355	.04826
25	1	10	.02682	.03029	.03388	.03759	.04143	.04542	.04956
25	1	11	.02993	.03305	.03625	.03954	.04292	.04640	.04999
25	1	12	.03236	.03513	.03794	.04081	.04374	.04674	.04982
25	1	13	.03427	.03667	.03911	.04157	.04408	.04663	.04923
25	1	14	.03575	.03780	.03986	.04194	.04405	.04617	.04833
25	1	15	.03687	.03858	.04029	.04201	.04373	.04546	.04722
25	1	16	.03771	.03908	.04045	.04182	.04319	.04456	.04594
25	1	17	.03830	.03935	.04039	.04143	.04246	.04351	.04455
25	1	18	.03867	.03940	.04013	.04086	.04160	.04233	.04307
25	1	19	.03884	.03927	.03971	.04015	.04060	.04106	.04153
25	1	20	.03883	.03897	.03913	.03930	.03949	.03969	.03992
25	1	21	.03864	.03850	.03839	.03831	.03826	.03824	.03825

Coefficients b_i for the BLUE of σ
Shape parameter k

n	s	i	1.4	1.5	1.6	1.7	1.8	1.9	2.0
25	1	22	.03826	.03784	.03748	.03717	.03691	.03669	.03652
25	1	23	.03763	.03695	.03636	.03584	.03539	.03501	.03468
25	1	24	.07100	.06909	.06748	.06611	.06497	.06402	.06324
25	2	1	-.42396	-.46544	-.51031	-.55880	-.61118	-.66771	-.72870
25	2	2	-.11983	-.12226	-.12440	-.12620	-.12762	-.12860	-.12910
25	2	3	-.06170	-.05957	-.05696	-.05383	-.05014	-.04583	-.04085
25	2	4	-.03073	-.02692	-.02265	-.01787	-.01256	-.00667	-.00015
25	2	5	-.01136	-.00694	-.00210	.00316	.00889	.01511	.02188
25	2	6	.00182	.00638	.01128	.01653	.02217	.02822	.03470
25	2	7	.01128	.01573	.02045	.02546	.03077	.03641	.04241
25	2	8	.01830	.02251	.02693	.03158	.03646	.04160	.04702
25	2	9	.02364	.02754	.03160	.03582	.04024	.04484	.04966
25	2	10	.02776	.03131	.03498	.03877	.04270	.04677	.05100
25	2	11	.03097	.03416	.03742	.04078	.04422	.04778	.05145
25	2	12	.03349	.03630	.03916	.04209	.04508	.04814	.05128
25	2	13	.03545	.03789	.04037	.04287	.04542	.04802	.05068
25	2	14	.03698	.03905	.04115	.04326	.04539	.04755	.04975
25	2	15	.03814	.03986	.04159	.04332	.04506	.04682	.04861
25	2	16	.03900	.04038	.04175	.04312	.04450	.04589	.04730
25	2	17	.03960	.04064	.04168	.04272	.04376	.04481	.04587
25	2	18	.03998	.04070	.04142	.04214	.04286	.04360	.04434
25	2	19	.04016	.04056	.04097	.04140	.04184	.04229	.04275
25	2	20	.04014	.04024	.04037	.04052	.04069	.04088	.04110
25	2	21	.03993	.03974	.03960	.03949	.03942	.03939	.03939
25	2	22	.03951	.03904	.03865	.03831	.03802	.03779	.03760
25	2	23	.11146	.10907	.10706	.10538	.10399	.10288	.10199
25	3	1	-.43922	-.48183	-.52793	-.57776	-.63158	-.68968	-.75236
25	3	2	-.12405	-.12650	-.12866	-.13047	-.13189	-.13288	-.13338
25	3	3	-.06383	-.06160	-.05888	-.05564	-.05183	-.04739	-.04226
25	3	4	-.03174	-.02780	-.02339	-.01847	-.01300	-.00694	-.00024
25	3	5	-.01168	-.00711	-.00214	.00328	.00917	.01556	.02251
25	3	6	.00198	.00667	.01171	.01711	.02289	.02910	.03576
25	3	7	.01177	.01635	.02119	.02633	.03178	.03757	.04372
25	3	8	.01905	.02337	.02790	.03266	.03766	.04293	.04848
25	3	9	.02457	.02857	.03272	.03705	.04156	.04628	.05122
25	3	10	.02884	.03247	.03622	.04009	.04411	.04827	.05260
25	3	11	.03216	.03541	.03874	.04216	.04569	.04932	.05307
25	3	12	.03476	.03763	.04054	.04352	.04657	.04969	.05290
25	3	13	.03680	.03928	.04179	.04433	.04692	.04957	.05227
25	3	14	.03837	.04047	.04259	.04473	.04689	.04909	.05132
25	3	15	.03957	.04130	.04304	.04479	.04655	.04833	.05014
25	3	16	.04046	.04183	.04321	.04459	.04597	.04737	.04879
25	3	17	.04108	.04211	.04313	.04416	.04520	.04625	.04731
25	3	18	.04146	.04215	.04285	.04356	.04427	.04500	.04574
25	3	19	.04163	.04200	.04239	.04279	.04321	.04365	.04410
25	3	20	.04160	.04166	.04175	.04187	.04202	.04219	.04239
25	3	21	.04136	.04112	.04094	.04080	.04070	.04065	.04063
25	3	22	.15506	.15245	.15028	.14853	.14713	.14606	.14528
25	4	1	-.45620	-.50007	-.54753	-.59883	-.65423	-.71404	-.77855
25	4	2	-.12872	-.13119	-.13336	-.13518	-.13662	-.13761	-.13810
25	4	3	-.06616	-.06382	-.06099	-.05763	-.05368	-.04910	-.04382
25	4	4	-.03283	-.02875	-.02418	-.01910	-.01346	-.00722	-.00032
25	4	5	-.01200	-.00728	-.00214	.00344	.00950	.01608	.02322
25	4	6	.00219	.00703	.01221	.01777	.02372	.03010	.03693
25	4	7	.01236	.01707	.02205	.02733	.03292	.03886	.04517
25	4	8	.01991	.02434	.02900	.03388	.03901	.04442	.05011

Coefficients b_i for the BLUE of σ
Shape parameter k

n	s	i	1.4	1.5	1.6	1.7	1.8	1.9	2.0
25	4	9	.02564	.02974	.03399	.03843	.04305	.04789	.05294
25	4	10	.03007	.03378	.03762	.04158	.04569	.04995	.05438
25	4	11	.03352	.03683	.04023	.04373	.04732	.05103	.05486
25	4	12	.03622	.03913	.04210	.04513	.04823	.05141	.05469
25	4	13	.03832	.04083	.04338	.04597	.04860	.05129	.05404
25	4	14	.03995	.04207	.04421	.04637	.04856	.05079	.05306
25	4	15	.04119	.04293	.04467	.04643	.04821	.05001	.05184
25	4	16	.04211	.04347	.04484	.04622	.04761	.04902	.05044
25	4	17	.04274	.04375	.04476	.04578	.04681	.04785	.04892
25	4	18	.04313	.04379	.04446	.04514	.04584	.04656	.04729
25	4	19	.04329	.04361	.04397	.04434	.04473	.04515	.04559
25	4	20	.04323	.04324	.04329	.04337	.04349	.04364	.04382
25	4	21	.20205	.19948	.19743	.19585	.19469	.19391	.19348
25	5	1	-.47510	-.52038	-.56935	-.62227	-.67942	-.74110	-.80762
25	5	2	-.13388	-.13637	-.13856	-.14040	-.14184	-.14284	-.14333
25	5	3	-.06871	-.06626	-.06330	-.05981	-.05571	-.05097	-.04552
25	5	4	-.03400	-.02976	-.02503	-.01977	-.01395	-.00751	-.00040
25	5	5	-.01230	-.00742	-.00212	.00364	.00989	.01667	.02402
25	5	6	.00246	.00746	.01281	.01853	.02466	.03122	.03826
25	5	7	.01305	.01790	.02303	.02846	.03422	.04032	.04681
25	5	8	.02091	.02547	.03025	.03527	.04054	.04609	.05193
25	5	9	.02687	.03108	.03544	.03999	.04473	.04969	.05487
25	5	10	.03148	.03528	.03920	.04326	.04746	.05182	.05636
25	5	11	.03506	.03845	.04192	.04549	.04916	.05295	.05686
25	5	12	.03786	.04083	.04385	.04694	.05010	.05335	.05668
25	5	13	.04005	.04260	.04518	.04781	.05048	.05322	.05601
25	5	14	.04174	.04388	.04604	.04822	.05044	.05270	.05500
25	5	15	.04302	.04476	.04651	.04828	.05007	.05188	.05373
25	5	16	.04396	.04531	.04668	.04805	.04944	.05085	.05228
25	5	17	.04461	.04559	.04658	.04758	.04860	.04964	.05069
25	5	18	.04499	.04561	.04625	.04691	.04759	.04829	.04901
25	5	19	.04514	.04542	.04573	.04606	.04643	.04682	.04724
25	5	20	.25280	.25056	.24889	.24776	.24712	.24692	.24712
25	6	1	-.49619	-.54304	-.59370	-.64843	-.70752	-.77128	-.84001
25	6	2	-.13959	-.14210	-.14432	-.14618	-.14763	-.14863	-.14912
25	6	3	-.07151	-.06893	-.06584	-.06219	-.05794	-.05303	-.04739
25	6	4	-.03526	-.03084	-.02593	-.02048	-.01445	-.00779	-.00045
25	6	5	-.01260	-.00754	-.00204	.00391	.01037	.01737	.02495
25	6	6	.00282	.00799	.01351	.01942	.02574	.03251	.03975
25	6	7	.01387	.01887	.02416	.02976	.03569	.04198	.04864
25	6	8	.02206	.02676	.03169	.03685	.04227	.04797	.05397
25	6	9	.02829	.03261	.03709	.04176	.04663	.05171	.05703
25	6	10	.03309	.03698	.04101	.04516	.04947	.05394	.05858
25	6	11	.03682	.04028	.04383	.04748	.05123	.05510	.05911
25	6	12	.03974	.04276	.04584	.04898	.05221	.05552	.05892
25	6	13	.04201	.04459	.04721	.04988	.05260	.05538	.05822
25	6	14	.04376	.04592	.04810	.05030	.05255	.05483	.05716
25	6	15	.04509	.04683	.04858	.05036	.05215	.05398	.05584
25	6	16	.04605	.04739	.04874	.05011	.05149	.05290	.05433
25	6	17	.04671	.04766	.04862	.04961	.05061	.05163	.05268
25	6	18	.04709	.04767	.04827	.04889	.04954	.05022	.05092
25	6	19	.30775	.30614	.30518	.30481	.30500	.30570	.30685
25	7	1	-.51979	-.56841	-.62097	-.67773	-.73900	-.80507	-.87628
25	7	2	-.14592	-.14847	-.15071	-.15260	-.15407	-.15508	-.15556
25	7	3	-.07459	-.07186	-.06862	-.06482	-.06039	-.05529	-.04944
25	7	4	-.03661	-.03200	-.02689	-.02122	-.01497	-.00807	-.00048

Coefficients b_i for the BLUE of σ
Shape parameter k

n	s	i	1.4	1.5	1.6	1.7	1.8	1.9	2.0
25	7	5	-.01288	-.00761	-.00191	.00426	.01095	.01818	.02602
25	7	6	.00327	.00863	.01435	.02046	.02699	.03398	.04146
25	7	7	.01483	.02001	.02548	.03126	.03738	.04386	.05073
25	7	8	.02341	.02826	.03334	.03866	.04424	.05011	.05629
25	7	9	.02992	.03437	.03898	.04378	.04879	.05402	.05948
25	7	10	.03494	.03894	.04306	.04733	.05175	.05633	.06109
25	7	11	.03884	.04238	.04601	.04974	.05358	.05754	.06164
25	7	12	.04188	.04496	.04809	.05130	.05459	.05797	.06144
25	7	13	.04424	.04686	.04952	.05223	.05499	.05782	.06071
25	7	14	.04606	.04824	.05043	.05266	.05493	.05724	.05960
25	7	15	.04744	.04917	.05093	.05270	.05451	.05635	.05822
25	7	16	.04843	.04974	.05107	.05243	.05380	.05521	.05664
25	7	17	.04909	.05000	.05093	.05189	.05287	.05387	.05491
25	7	18	.36745	.36681	.36690	.36767	.36906	.37103	.37353
25	8	1	-.54634	-.59696	-.65166	-.71072	-.77443	-.84311	-.91709
25	8	2	-.15299	-.15558	-.15785	-.15976	-.16125	-.16227	-.16276
25	8	3	-.07798	-.07510	-.07169	-.06771	-.06309	-.05777	-.05169
25	8	4	-.03806	-.03324	-.02790	-.02200	-.01550	-.00834	-.00046
25	8	5	-.01312	-.00763	-.00170	.00471	.01165	.01915	.02727
25	8	6	.00384	.00941	.01535	.02169	.02846	.03569	.04342
25	8	7	.01598	.02135	.02702	.03300	.03933	.04602	.05312
25	8	8	.02499	.03000	.03525	.04074	.04651	.05256	.05893
25	8	9	.03182	.03640	.04116	.04611	.05126	.05664	.06226
25	8	10	.03707	.04119	.04543	.04981	.05435	.05906	.06395
25	8	11	.04116	.04478	.04850	.05232	.05626	.06032	.06452
25	8	12	.04434	.04747	.05068	.05395	.05731	.06075	.06430
25	8	13	.04680	.04946	.05216	.05491	.05771	.06059	.06353
25	8	14	.04870	.05088	.05309	.05534	.05764	.05997	.06236
25	8	15	.05011	.05184	.05359	.05537	.05718	.05902	.06091
25	8	16	.05113	.05242	.05373	.05506	.05642	.05782	.05924
25	8	17	.43256	.43330	.43485	.43716	.44019	.44388	.44819
25	9	1	-.57637	-.62927	-.68640	-.74806	-.81455	-.88619	-.96332
25	9	2	-.16090	-.16354	-.16585	-.16780	-.16931	-.17035	-.17083
25	9	3	-.08174	-.07868	-.07509	-.07090	-.06606	-.06051	-.05418
25	9	4	-.03962	-.03456	-.02897	-.02280	-.01602	-.00856	-.00037
25	9	5	-.01331	-.00758	-.00139	.00530	.01252	.02032	.02874
25	9	6	.00456	.01037	.01655	.02315	.03018	.03768	.04570
25	9	7	.01736	.02294	.02883	.03503	.04159	.04853	.05588
25	9	8	.02684	.03204	.03748	.04317	.04913	.05539	.06197
25	9	9	.03403	.03877	.04368	.04879	.05411	.05966	.06546
25	9	10	.03955	.04379	.04816	.05268	.05735	.06219	.06722
25	9	11	.04384	.04756	.05138	.05530	.05934	.06350	.06781
25	9	12	.04717	.05038	.05365	.05699	.06042	.06394	.06757
25	9	13	.04975	.05245	.05519	.05798	.06083	.06375	.06675
25	9	14	.05172	.05392	.05615	.05842	.06073	.06309	.06551
25	9	15	.05318	.05490	.05664	.05842	.06023	.06208	.06397
25	9	16	.50392	.50650	.50999	.51434	.51951	.52546	.53213
25	10	1	-.61058	-.66608	-.72600	-.79064	-.86031	-.93533	-1.01606
25	10	2	-.16982	-.17251	-.17487	-.17686	-.17841	-.17946	-.17995
25	10	3	-.08592	-.08266	-.07886	-.07444	-.06936	-.06354	-.05692
25	10	4	-.04129	-.03596	-.03009	-.02362	-.01652	-.00873	-.00019
25	10	5	-.01344	-.00742	-.00094	.00606	.01359	.02173	.03050
25	10	6	.00549	.01156	.01802	.02489	.03222	.04003	.04837
25	10	7	.01902	.02484	.03097	.03743	.04425	.05146	.05909
25	10	8	.02904	.03445	.04010	.04600	.05219	.05868	.06550
25	10	9	.03663	.04154	.04664	.05193	.05743	.06317	.06916

Coefficients b_i for the BLUE of σ
Shape parameter k

n	s	i	1.4	1.5	1.6	1.7	1.8	1.9	2.0
25	10	10	.04246	.04684	.05135	.05600	.06082	.06582	.07101
25	10	11	.04697	.05080	.05472	.05875	.06290	.06718	.07161
25	10	12	.05047	.05375	.05709	.06051	.06402	.06763	.07134
25	10	13	.05317	.05591	.05869	.06153	.06443	.06740	.07046
25	10	14	.05522	.05743	.05967	.06196	.06430	.06668	.06913
25	10	15	.58255	.58751	.59350	.60049	.60843	.61727	.62697
25	11	1	-.64987	-.70838	-.77152	-.83960	-.91293	-.99186	-1.07674
25	11	2	-.17995	-.18270	-.18512	-.18716	-.18874	-.18982	-.19032
25	11	3	-.09059	-.08711	-.08307	-.07839	-.07302	-.06690	-.05996
25	11	4	-.04309	-.03745	-.03125	-.02445	-.01699	-.00881	.00013
25	11	5	-.01346	-.00712	-.00030	.00704	.01494	.02345	.03262
25	11	6	.00666	.01304	.01981	.02701	.03467	.04283	.05153
25	11	7	.02104	.02714	.03354	.04029	.04741	.05492	.06287
25	11	8	.03168	.03732	.04321	.04936	.05580	.06255	.06964
25	11	9	.03972	.04483	.05012	.05562	.06133	.06728	.07349
25	11	10	.04590	.05043	.05509	.05991	.06489	.07006	.07542
25	11	11	.05066	.05460	.05864	.06279	.06707	.07148	.07603
25	11	12	.05435	.05770	.06113	.06463	.06822	.07192	.07572
25	11	13	.05718	.05995	.06278	.06567	.06862	.07165	.07476
25	11	14	.66976	.67775	.68693	.69727	.70872	.72124	.73480
25	12	1	-.69544	-.75745	-.82434	-.89643	-.97404	-1.05752	-1.14724
25	12	2	-.19155	-.19438	-.19686	-.19895	-.20058	-.20169	-.20220
25	12	3	-.09586	-.09212	-.08779	-.08281	-.07712	-.07065	-.06333
25	12	4	-.04502	-.03902	-.03245	-.02525	-.01737	-.00876	.00065
25	12	5	-.01333	-.00662	.00058	.00833	.01665	.02560	.03523
25	12	6	.00817	.01490	.02203	.02960	.03765	.04622	.05533
25	12	7	.02353	.02993	.03666	.04374	.05120	.05906	.06737
25	12	8	.03487	.04078	.04694	.05337	.06010	.06715	.07454
25	12	9	.04344	.04877	.05428	.06001	.06596	.07215	.07861
25	12	10	.05000	.05470	.05955	.06455	.06972	.07507	.08063
25	12	11	.05505	.05912	.06329	.06758	.07199	.07654	.08125
25	12	12	.05895	.06238	.06589	.06949	.07318	.07697	.08088
25	12	13	.76720	.77901	.79222	.80679	.82267	.83985	.85828
25	13	1	-.74891	-.81505	-.88637	-.96319	-1.04584	-1.13469	-1.23013
25	13	2	-.20497	-.20789	-.21045	-.21260	-.21428	-.21543	-.21596
25	13	3	-.10184	-.09780	-.09314	-.08781	-.08173	-.07485	-.06709
25	13	4	-.04710	-.04068	-.03366	-.02600	-.01763	-.00850	.00146
25	13	5	-.01299	-.00585	.00181	.01002	.01884	.02830	.03847
25	13	6	.01013	.01726	.02482	.03282	.04133	.05036	.05996
25	13	7	.02662	.03339	.04049	.04795	.05580	.06408	.07281
25	13	8	.03878	.04500	.05148	.05824	.06530	.07270	.08044
25	13	9	.04795	.05354	.05931	.06530	.07153	.07801	.08475
25	13	10	.05496	.05986	.06491	.07012	.07550	.08108	.08686
25	13	11	.06034	.06455	.06887	.07331	.07788	.08260	.08747
25	13	12	.87704	.89366	.91193	.93182	.95330	.97636	1.00095

Coefficients b_i for the BLUE of σ
Shape parameter k

n	s	i	2.1	2.2	2.3	2.4	2.5
2	0	1	-1.00044	-1.01003	-1.01986	-1.02992	-1.04019
2	0	2	1.00044	1.01003	1.01986	1.02992	1.04019
3	0	1	-1.04889	-1.07728	-1.10620	-1.13567	-1.16569
3	0	2	.58416	.61220	.64045	.66891	.69762
3	0	3	.46473	.46507	.46576	.46676	.46807
3	1	1	-1.92657	-1.97918	-2.03335	-2.08907	-2.14634
3	1	2	1.92657	1.97918	2.03335	2.08907	2.14634
4	0	1	-1.05429	-1.09495	-1.13685	-1.18002	-1.22448
4	0	2	.38704	.41857	.45105	.48451	.51898
4	0	3	.37388	.38400	.39410	.40420	.41432
4	0	4	.29337	.29238	.29170	.29131	.29118
4	1	1	-1.47550	-1.53151	-1.58951	-1.64954	-1.71165
4	1	2	.53510	.57645	.61914	.66323	.70879
4	1	3	.94041	.95506	.97037	.98631	1.00286
5	0	1	-1.04427	-1.09314	-1.14393	-1.19669	-1.25150
5	0	2	.26846	.29912	.33120	.36477	.39989
5	0	3	.29924	.31428	.32953	.34502	.36076
5	0	4	.26530	.26968	.27407	.27847	.28288
5	0	5	.21126	.21006	.20913	.20844	.20797
5	1	1	-1.31309	-1.37359	-1.43665	-1.50235	-1.57078
5	1	2	.33403	.37103	.40979	.45038	.49291
5	1	3	.37322	.39109	.40930	.42787	.44683
5	1	4	.60584	.61147	.61756	.62409	.63104
5	2	1	-1.85227	-1.93538	-2.02205	-2.11237	-2.20646
5	2	2	.46896	.51731	.56788	.62077	.67611
5	2	3	1.38332	1.41807	1.45417	1.49160	1.53035
6	0	1	-1.02802	-1.08258	-1.13960	-1.19921	-1.26152
6	0	2	.18806	.21649	.24659	.27845	.31217
6	0	3	.24279	.25995	.27761	.29579	.31450
6	0	4	.23121	.23930	.24744	.25565	.26392
6	0	5	.20219	.20422	.20629	.20838	.21052
6	0	6	.16378	.16261	.16167	.16094	.16040
6	1	1	-1.22135	-1.28533	-1.35236	-1.42257	-1.49610
6	1	2	.22118	.25401	.28878	.32561	.36461
6	1	3	.28649	.30615	.32642	.34733	.36892
6	1	4	.27307	.28222	.29148	.30087	.31039
6	1	5	.44061	.44296	.44569	.44877	.45218
6	2	1	-1.54574	-1.62467	-1.70739	-1.79402	-1.88473
6	2	2	.27811	.31755	.35925	.40336	.45000
6	2	3	.36043	.38360	.40750	.43216	.45764
6	2	4	.90719	.92353	.94064	.95850	.97709
6	3	1	-2.18957	-2.29874	-2.41306	-2.53268	-2.65779
6	3	2	.39720	.44912	.50387	.56160	.62248
6	3	3	1.79238	1.84962	1.90919	1.97108	2.03531
7	0	1	-1.00934	-1.06790	-1.12941	-1.19401	-1.26185
7	0	2	.12967	.15551	.18313	.21264	.24414
7	0	3	.19881	.21667	.23525	.25457	.27468
7	0	4	.20123	.21155	.22205	.23276	.24368
7	0	5	.18473	.18944	.19417	.19891	.20367
7	0	6	.16179	.16271	.16368	.16470	.16577
7	0	7	.13310	.13202	.13114	.13044	.12990
7	1	1	-1.15803	-1.22454	-1.29451	-1.36812	-1.44553
7	1	2	.14721	.17624	.20727	.24043	.27586
7	1	3	.22672	.24670	.26750	.28917	.31175
7	1	4	.22971	.24116	.25285	.26481	.27704
7	1	5	.21101	.21618	.22140	.22669	.23205

Coefficients b_i for the BLUE of σ
Shape parameter k

n	s	i	2.1	2.2	2.3	2.4	2.5
7	1	6	.34338	.34426	.34549	.34701	.34882
7	2	1	-1.38296	-1.46065	-1.54239	-1.62837	-1.71876
7	2	2	.17435	.20772	.24334	.28136	.32192
7	2	3	.26929	.29199	.31563	.34026	.36591
7	2	4	.27283	.28561	.29869	.31209	.32582
7	2	5	.66649	.67533	.68473	.69467	.70511
7	3	1	-1.75710	-1.85342	-1.95469	-2.06108	-2.17282
7	3	2	.22209	.26230	.30511	.35068	.39918
7	3	3	.34177	.36867	.39664	.42574	.45603
7	3	4	1.19324	1.22246	1.25294	1.28466	1.31761
7	4	1	-2.49516	-2.62917	-2.76989	-2.91757	-3.07244
7	4	2	.32402	.37741	.43405	.49413	.55784
7	4	3	2.17114	2.25176	2.33584	2.42344	2.51460
8	0	1	-.98991	-1.05138	-1.11617	-1.18446	-1.25643
8	0	2	.08532	.10857	.13363	.16061	.18964
8	0	3	.16359	.18140	.20007	.21966	.24022
8	0	4	.17554	.18713	.19905	.21131	.22395
8	0	5	.16744	.17402	.18068	.18742	.19425
8	0	6	.15222	.15510	.15798	.16088	.16380
8	0	7	.13405	.13439	.13480	.13526	.13577
8	0	8	.11176	.11077	.10996	.10932	.10881
8	1	1	-1.10937	-1.17769	-1.24981	-1.32593	-1.40626
8	1	2	.09446	.12009	.14772	.17748	.20949
8	1	3	.18231	.20189	.22244	.24402	.26669
8	1	4	.19584	.20853	.22160	.23508	.24899
8	1	5	.18690	.19406	.20132	.20871	.21622
8	1	6	.17000	.17311	.17625	.17944	.18267
8	1	7	.27985	.28001	.28048	.28121	.28219
8	2	1	-1.27813	-1.35538	-1.43692	-1.52298	-1.61377
8	2	2	.10766	.13634	.16722	.20042	.23611
8	2	3	.20893	.23068	.25350	.27746	.30262
8	2	4	.22457	.23852	.25289	.26773	.28305
8	2	5	.21431	.22204	.22991	.23793	.24612
8	2	6	.52266	.52780	.53340	.53944	.54587
8	3	1	-1.53161	-1.62213	-1.71760	-1.81827	-1.92436
8	3	2	.12878	.16179	.19724	.23529	.27609
8	3	3	.24984	.27460	.30056	.32777	.35631
8	3	4	.26827	.28389	.30000	.31663	.33379
8	3	5	.88473	.90184	.91980	.93859	.95816
8	4	1	-1.95085	-2.06364	-2.18248	-2.30760	-2.43927
8	4	2	.16705	.20702	.24979	.29555	.34448
8	4	3	.31988	.34942	.38032	.41267	.44652
8	4	4	1.46392	1.50720	1.55236	1.59939	1.64827
9	0	1	-.97059	-1.03416	-1.10138	-1.17243	-1.24753
9	0	2	.05053	.07134	.09393	.11842	.14494
9	0	3	.13477	.15210	.17042	.18977	.21021
9	0	4	.15350	.16576	.17846	.19165	.20533
9	0	5	.15148	.15933	.16734	.17552	.18388
9	0	6	.14163	.14601	.15043	.15489	.15939
9	0	7	.12859	.13041	.13223	.13407	.13592
9	0	8	.11398	.11400	.11409	.11424	.11445
9	0	9	.09611	.09522	.09448	.09388	.09341
9	1	1	-1.06955	-1.13915	-1.21283	-1.29080	-1.37331
9	1	2	.05479	.07742	.10199	.12863	.15748
9	1	3	.14772	.16654	.18644	.20748	.22973
9	1	4	.16846	.18174	.19551	.20983	.22471

Coefficients b_i for the BLUE of σ
Shape parameter k

n	s	i	2.1	2.2	2.3	2.4	2.5
9	1	5	.16633	.17480	.18345	.19232	.20140
9	1	6	.15557	.16027	.16503	.16985	.17473
9	1	7	.14133	.14326	.14522	.14723	.14927
9	1	8	.23534	.23513	.23518	.23548	.23599
9	2	1	-1.20270	-1.27972	-1.36126	-1.44755	-1.53883
9	2	2	.06067	.08553	.11249	.14169	.17328
9	2	3	.16522	.18580	.20755	.23053	.25482
9	2	4	.18862	.20303	.21800	.23355	.24972
9	2	5	.18628	.19538	.20469	.21424	.22404
9	2	6	.17422	.17918	.18423	.18937	.19459
9	2	7	.42770	.43080	.43430	.43817	.44238
9	3	1	-1.38974	-1.47699	-1.56928	-1.66687	-1.77002
9	3	2	.06962	.09745	.12756	.16011	.19526
9	3	3	.19035	.21322	.23736	.26284	.28974
9	3	4	.21729	.23315	.24960	.26669	.28446
9	3	5	.21448	.22433	.23442	.24479	.25543
9	3	6	.69800	.70884	.72034	.73244	.74514
9	4	1	-1.66925	-1.77188	-1.88032	-1.99484	-2.11572
9	4	2	.08471	.11682	.15146	.18881	.22904
9	4	3	.22920	.25534	.28288	.31190	.34248
9	4	4	.26104	.27890	.29743	.31666	.33663
9	4	5	1.09431	1.12081	1.14854	1.17747	1.20757
9	5	1	-2.12984	-2.25827	-2.39379	-2.53671	-2.68733
9	5	2	.11347	.15252	.19450	.23960	.28801
9	5	3	.29615	.32755	.36056	.39525	.43172
9	5	4	1.72023	1.77819	1.83874	1.90187	1.96760
10	0	1	-.95176	-1.01688	-1.08589	-1.15903	-1.23652
10	0	2	.02257	.04112	.06141	.08354	.10764
10	0	3	.11075	.12740	.14510	.16392	.18391
10	0	4	.13447	.14700	.16009	.17375	.18803
10	0	5	.13704	.14572	.15465	.16384	.17330
10	0	6	.13122	.13673	.14232	.14800	.15377
10	0	7	.12183	.12485	.12788	.13094	.13401
10	0	8	.11083	.11198	.11315	.11434	.11555
10	0	9	.09886	.09869	.09860	.09858	.09861
10	0	10	.08419	.08336	.08269	.08214	.08171
10	1	1	-1.03564	-1.10613	-1.18092	-1.26025	-1.34438
10	1	2	.02386	.04383	.06565	.08946	.11541
10	1	3	.11988	.13779	.15683	.17708	.19862
10	1	4	.14577	.15922	.17328	.18798	.20335
10	1	5	.14864	.15793	.16751	.17738	.18756
10	1	6	.14237	.14825	.15422	.16031	.16652
10	1	7	.13222	.13542	.13866	.14193	.14525
10	1	8	.12035	.12157	.12282	.12412	.12545
10	1	9	.20256	.20213	.20195	.20199	.20223
10	2	1	-1.14444	-1.22128	-1.30280	-1.38928	-1.48098
10	2	2	.02559	.04724	.07087	.09664	.12469
10	2	3	.13175	.15110	.17167	.19353	.21677
10	2	4	.16043	.17490	.19002	.20583	.22237
10	2	5	.16365	.17358	.18382	.19438	.20528
10	2	6	.15677	.16299	.16932	.17579	.18239
10	2	7	.14559	.14892	.15231	.15575	.15925
10	2	8	.36066	.36255	.36480	.36736	.37023
10	3	1	-1.29018	-1.37529	-1.46554	-1.56121	-1.66257
10	3	2	.02831	.05210	.07802	.10622	.13687
10	3	3	.14798	.16913	.19159	.21543	.24074

Coefficients b_i for the BLUE of σ
Shape parameter k

n	s	i	2.1	2.2	2.3	2.4	2.5
10	3	4	.18030	.19600	.21240	.22955	.24747
10	3	5	.18391	.19458	.20558	.21694	.22866
10	3	6	.17611	.18269	.18940	.19627	.20329
10	3	7	.57357	.58079	.58855	.59681	.60553
10	4	1	-1.49397	-1.59065	-1.69307	-1.80151	-1.91626
10	4	2	.03308	.05975	.08872	.12018	.15429
10	4	3	.17144	.19500	.21996	.24643	.27447
10	4	4	.20865	.22598	.24406	.26293	.28265
10	4	5	.21261	.22422	.23619	.24855	.26131
10	4	6	.86819	.88571	.90413	.92342	.94354
10	5	1	-1.79746	-1.91159	-2.03233	-2.15999	-2.29489
10	5	2	.04219	.07304	.10646	.14264	.18174
10	5	3	.20797	.23501	.26360	.29384	.32582
10	5	4	.25204	.27169	.29216	.31350	.33576
10	5	5	1.29525	1.33185	1.37011	1.41002	1.45156
11	0	1	-.93361	-.99986	-1.07022	-1.14493	-1.22425
11	0	2	-.00032	.01617	.03432	.05424	.07607
11	0	3	.09044	.10630	.12325	.14136	.16070
11	0	4	.11792	.13047	.14365	.15750	.17205
11	0	5	.12404	.13325	.14278	.15265	.16288
11	0	6	.12138	.12771	.13417	.14078	.14755
11	0	7	.11479	.11877	.12279	.12685	.13097
11	0	8	.10638	.10851	.11064	.11279	.11496
11	0	9	.09708	.09781	.09855	.09933	.10012
11	0	10	.08710	.08683	.08663	.08650	.08643
11	0	11	.07481	.07405	.07343	.07293	.07253
11	1	1	-1.00600	-1.07708	-1.15264	-1.23296	-1.31829
11	1	2	-.00091	.01669	.03605	.05732	.08061
11	1	3	.09694	.11386	.13196	.15130	.17196
11	1	4	.12660	.13998	.15405	.16883	.18438
11	1	5	.13326	.14306	.15321	.16374	.17467
11	1	6	.13044	.13716	.14403	.15107	.15830
11	1	7	.12338	.12759	.13186	.13618	.14058
11	1	8	.11438	.11661	.11888	.12117	.12349
11	1	9	.10444	.10520	.10600	.10685	.10773
11	1	10	.17747	.17693	.17661	.17650	.17657
11	2	1	-1.09722	-1.17384	-1.25530	-1.34187	-1.43386
11	2	2	-.00164	.01725	.03802	.06080	.08575
11	2	3	.10512	.12325	.14262	.16331	.18541
11	2	4	.13754	.15182	.16683	.18261	.19920
11	2	5	.14485	.15526	.16605	.17725	.18887
11	2	6	.14182	.14891	.15617	.16363	.17128
11	2	7	.13416	.13855	.14302	.14757	.15220
11	2	8	.12437	.12667	.12901	.13139	.13382
11	2	9	.31100	.31213	.31357	.31531	.31732
11	3	1	-1.21514	-1.29869	-1.38747	-1.48178	-1.58190
11	3	2	-.00234	.01815	.04062	.06524	.09215
11	3	3	.11592	.13549	.15639	.17869	.20248
11	3	4	.15183	.16718	.18330	.20023	.21803
11	3	5	.15994	.17105	.18257	.19452	.20692
11	3	6	.15658	.16407	.17176	.17965	.18776
11	3	7	.14809	.15266	.15732	.16208	.16693
11	3	8	.48512	.49009	.49552	.50138	.50763
11	4	1	-1.37246	-1.46520	-1.56366	-1.66814	-1.77894
11	4	2	-.00268	.01984	.04449	.07143	.10083
11	4	3	.13079	.15222	.17505	.19937	.22529

Coefficients b_i for the BLUE of σ
Shape parameter k

n	s	i	2.1	2.2	2.3	2.4	2.5
11	4	4	.17128	.18796	.20546	.22384	.24312
11	4	5	.18034	.19231	.20471	.21758	.23093
11	4	6	.17644	.18440	.19258	.20098	.20962
11	4	7	.71629	.72847	.74137	.75494	.76916
11	5	1	-1.59175	-1.69739	-1.80942	-1.92814	-2.05389
11	5	2	-.00202	.02327	.05087	.08095	.11369
11	5	3	.15248	.17639	.20183	.22888	.25765
11	5	4	.19911	.21758	.23694	.25722	.27849
11	5	5	.20930	.22240	.23596	.25002	.26460
11	5	6	1.03288	1.05775	1.08382	1.11106	1.13945
12	0	1	-.91623	-.98329	-1.05465	-1.13055	-1.21126
12	0	2	-.01935	-.00474	.01144	.02933	.04903
12	0	3	.07306	.08808	.10422	.12153	.14011
12	0	4	.10340	.11581	.12891	.14274	.15733
12	0	5	.11234	.12185	.13176	.14207	.15281
12	0	6	.11221	.11912	.12623	.13354	.14107
12	0	7	.10786	.11258	.11739	.12227	.12725
12	0	8	.10148	.10441	.10737	.11036	.11338
12	0	9	.09409	.09561	.09714	.09869	.10024
12	0	10	.08617	.08661	.08707	.08757	.08809
12	0	11	.07772	.07739	.07713	.07693	.07679
12	0	12	.06725	.06656	.06599	.06552	.06515
12	1	1	-.97959	-1.05104	-1.12713	-1.20813	-1.29435
12	1	2	-.02115	-.00567	.01149	.03045	.05133
12	1	3	.07768	.09361	.11073	.12910	.14882
12	1	4	.11017	.12332	.13721	.15188	.16738
12	1	5	.11977	.12984	.14033	.15127	.16268
12	1	6	.11967	.12697	.13449	.14225	.15024
12	1	7	.11506	.12004	.12511	.13028	.13557
12	1	8	.10827	.11135	.11447	.11764	.12084
12	1	9	.10041	.10200	.10362	.10526	.10692
12	1	10	.09201	.09247	.09298	.09352	.09411
12	1	11	.15770	.15709	.15670	.15649	.15645
12	2	1	-1.05761	-1.13397	-1.21529	-1.30187	-1.39401
12	2	2	-.02339	-.00689	.01137	.03153	.05372
12	2	3	.08336	.10030	.11849	.13801	.15895
12	2	4	.11849	.13244	.14716	.16272	.17916
12	2	5	.12889	.13954	.15063	.16220	.17428
12	2	6	.12882	.13651	.14443	.15260	.16104
12	2	7	.12388	.12908	.13440	.13983	.14538
12	2	8	.11658	.11977	.12301	.12630	.12965
12	2	9	.10813	.10975	.11140	.11309	.11482
12	2	10	.27285	.27347	.27439	.27558	.27701
12	3	1	-1.15571	-1.23801	-1.32563	-1.41886	-1.51803
12	3	2	-.02603	-.00833	.01124	.03280	.05652
12	3	3	.09064	.10876	.12821	.14906	.17140
12	3	4	.12905	.14392	.15961	.17618	.19367
12	3	5	.14044	.15174	.16351	.17578	.18858
12	3	6	.14038	.14848	.15684	.16545	.17435
12	3	7	.13498	.14042	.14597	.15165	.15747
12	3	8	.12700	.13028	.13362	.13703	.14051
12	3	9	.41925	.42274	.42664	.43092	.43554
12	4	1	-1.28206	-1.37194	-1.46755	-1.56920	-1.67721
12	4	2	-.02907	-.00987	.01130	.03459	.06016
12	4	3	.10032	.11991	.14089	.16335	.18740
12	4	4	.14290	.15890	.17575	.19353	.21229

Coefficients b_i for the BLUE of σ
Shape parameter k

n	s	i	2.1	2.2	2.3	2.4	2.5
12	4	5	.15551	.16758	.18015	.19325	.20690
12	4	6	.15540	.16398	.17283	.18196	.19139
12	4	7	.14936	.15503	.16085	.16680	.17290
12	4	8	.60764	.61641	.62578	.63571	.64617
12	5	1	-1.45013	-1.55012	-1.65637	-1.76920	-1.88894
12	5	2	-.03241	-.01128	.01196	.03747	.06540
12	5	3	.11379	.13527	.15822	.18276	.20897
12	5	4	.16181	.17924	.19758	.21690	.23725
12	5	5	.17592	.18896	.20254	.21667	.23139
12	5	6	.17563	.18480	.19425	.20401	.21408
12	5	7	.85538	.87313	.89182	.91140	.93184
12	6	1	-1.68386	-1.79805	-1.91924	-2.04776	-2.18397
12	6	2	-.03577	-.01200	.01406	.04258	.07373
12	6	3	.13362	.15765	.18329	.21064	.23980
12	6	4	.18897	.20833	.22868	.25008	.27259
12	6	5	.20496	.21930	.23421	.24973	.26587
12	6	6	1.19209	1.22477	1.25899	1.29473	1.33199
13	0	1	-.89963	-.96727	-1.03935	-1.11614	-1.19792
13	0	2	-.03537	-.02248	-.00809	.00791	.02564
13	0	3	.05803	.07221	.08750	.10398	.12173
13	0	4	.09059	.10275	.11564	.12930	.14379
13	0	5	.10179	.11145	.12155	.13213	.14320
13	0	6	.10372	.11104	.11661	.12644	.13455
13	0	7	.10121	.10651	.11192	.11746	.12314
13	0	8	.09647	.10007	.10371	.10741	.11116
13	0	9	.09060	.09281	.09503	.09726	.09952
13	0	10	.08414	.08524	.08635	.08747	.08860
13	0	11	.07733	.07757	.07784	.07815	.07848
13	0	12	.07008	.06972	.06942	.06919	.06901
13	0	13	.06104	.06040	.05987	.05944	.05910
13	1	1	-.95573	-1.02739	-1.10381	-1.18528	-1.27212
13	1	2	-.03797	-.02437	-.00920	.00766	.02635
13	1	3	.06129	.07624	.09238	.10978	.12852
13	1	4	.09592	.10874	.12233	.13676	.15207
13	1	5	.10784	.11802	.12868	.13984	.15154
13	1	6	.10993	.11764	.12561	.13387	.14244
13	1	7	.10729	.11286	.11856	.12440	.13040
13	1	8	.10229	.10606	.10989	.11378	.11774
13	1	9	.09608	.09838	.10071	.10307	.10546
13	1	10	.08925	.09039	.09156	.09274	.09395
13	1	11	.08207	.08233	.08263	.08297	.08335
13	1	12	.14175	.14111	.14067	.14040	.14029
13	2	1	-1.02353	-1.09959	-1.18070	-1.26719	-1.35937
13	2	2	-.04113	-.02673	-.01069	.00713	.02687
13	2	3	.06520	.08101	.09806	.11644	.13623
13	2	4	.10233	.11586	.13020	.14542	.16156
13	2	5	.11514	.12585	.13707	.14882	.16114
13	2	6	.11741	.12549	.13386	.14254	.15154
13	2	7	.11461	.12042	.12638	.13250	.13879
13	2	8	.10928	.11319	.11717	.12123	.12537
13	2	9	.10265	.10502	.10742	.10986	.11234
13	2	10	.09536	.09652	.09770	.09892	.10017
13	2	11	.24268	.24296	.24353	.24433	.24537
13	3	1	-1.10689	-1.18814	-1.27477	-1.36710	-1.46545
13	3	2	-.04491	-.02958	-.01252	.00640	.02734
13	3	3	.07010	.08690	.10499	.12448	.14545

Coefficients b_i for the BLUE of σ
Shape parameter k

n	s	i	2.1	2.2	2.3	2.4	2.5
13	3	4	.11028	.12461	.13980	.15590	.17297
13	3	5	.12416	.13547	.14731	.15971	.17270
13	3	6	.12663	.13513	.14392	.15304	.16250
13	3	7	.12363	.12969	.13592	.14231	.14889
13	3	8	.11787	.12191	.12603	.13024	.13454
13	3	9	.11070	.11311	.11557	.11807	.12062
13	3	10	.36842	.37090	.37375	.37694	.38044
13	4	1	-1.21135	-1.29903	-1.39244	-1.49192	-1.59780
13	4	2	-.04940	-.03295	-.01467	.00556	.02791
13	4	3	.07645	.09443	.11378	.13458	.15694
13	4	4	.12043	.13571	.15189	.16903	.18718
13	4	5	.13561	.14761	.16017	.17332	.18709
13	4	6	.13830	.14726	.15654	.16616	.17613
13	4	7	.13499	.14133	.14785	.15454	.16143
13	4	8	.12866	.13283	.13709	.14145	.14591
13	4	9	.52631	.53280	.53980	.54728	.55520
13	5	1	-1.34555	-1.44147	-1.54356	-1.65217	-1.76764
13	5	2	-.05472	-.03686	-.01708	.00478	.02887
13	5	3	.08501	.10446	.12535	.14778	.17185
13	5	4	.13378	.15024	.16764	.18605	.20553
13	5	5	.15057	.16343	.17686	.19091	.20561
13	5	6	.15349	.16300	.17286	.18306	.19365
13	5	7	.14972	.15638	.16323	.17026	.17751
13	5	8	.72770	.74083	.75471	.76932	.78462
13	6	1	-1.52368	-1.63060	-1.74428	-1.86507	-1.99333
13	6	2	-.06098	-.04131	-.01957	.00438	.03072
13	6	3	.09704	.11840	.14130	.16583	.19211
13	6	4	.15207	.17004	.18902	.20907	.23025
13	6	5	.17088	.18482	.19937	.21457	.23046
13	6	6	.17398	.18419	.19476	.20571	.21706
13	6	7	.99068	1.01445	1.03940	1.06551	1.09274
13	7	1	-1.77096	-1.89333	-2.02328	-2.16117	-2.30739
13	7	2	-.06827	-.04610	-.02168	.00514	.03454
13	7	3	.11496	.13893	.16457	.19198	.22127
13	7	4	.17844	.19848	.21960	.24187	.26536
13	7	5	.19984	.21524	.23129	.24805	.26554
13	7	6	1.34599	1.38678	1.42950	1.47414	1.52068
14	0	1	-.88379	-.95183	-1.02443	-1.10187	-1.18445
14	0	2	-.04900	-.03767	-.02492	-.01065	.00525
14	0	3	.04492	.05826	.07271	.08835	.10525
14	0	4	.07921	.09105	.10364	.11705	.13132
14	0	5	.09224	.10193	.11212	.12282	.13407
14	0	6	.09588	.10347	.11136	.11957	.12810
14	0	7	.09490	.10063	.10652	.11258	.11882
14	0	8	.09154	.09567	.09987	.10415	.10852
14	0	9	.08691	.08970	.09251	.09535	.09823
14	0	10	.08161	.08329	.08498	.08668	.08840
14	0	11	.07595	.07674	.07755	.07837	.07921
14	0	12	.07003	.07014	.07028	.07045	.07066
14	0	13	.06375	.06337	.06305	.06279	.06259
14	0	14	.05585	.05525	.05476	.05436	.05404
14	1	1	-.93396	-1.00570	-1.08229	-1.16406	-1.25132
14	1	2	-.05212	-.04022	-.02685	-.01188	.00481
14	1	3	.04716	.06117	.07635	.09278	.11055
14	1	4	.08343	.09586	.10909	.12318	.13819
14	1	5	.09722	.10740	.11809	.12934	.14118

Coefficients b_i for the BLUE of σ
Shape parameter k

n	s	i	2.1	2.2	2.3	2.4	2.5
14	1	6	.10110	.10906	.11734	.12596	.13494
14	1	7	.10009	.10609	.11227	.11863	.12520
14	1	8	.09656	.10087	.10528	.10977	.11437
14	1	9	.09169	.09459	.09754	.10052	.10355
14	1	10	.08611	.08786	.08963	.09141	.09323
14	1	11	.08015	.08098	.08182	.08269	.08358
14	1	12	.07395	.07407	.07423	.07444	.07468
14	1	13	.12862	.12797	.12750	.12720	.12705
14	2	1	-.99363	-1.06935	-1.15021	-1.23653	-1.32865
14	2	2	-.05585	-.04332	-.02925	-.01351	.00402
14	2	3	.04980	.06454	.08050	.09777	.11645
14	2	4	.08842	.10147	.11537	.13017	.14593
14	2	5	.10313	.11380	.12501	.13681	.14922
14	2	6	.10727	.11561	.12427	.13330	.14270
14	2	7	.10623	.11249	.11893	.12558	.13244
14	2	8	.10249	.10697	.11155	.11623	.12103
14	2	9	.09733	.10033	.10337	.10647	.10962
14	2	10	.09142	.09320	.09502	.09686	.09873
14	2	11	.08511	.08593	.08679	.08768	.08860
14	2	12	.21827	.21832	.21863	.21917	.21992
14	3	1	-1.06567	-1.14599	-1.23175	-1.32326	-1.42089
14	3	2	-.06028	-.04702	-.03215	-.01554	.00294
14	3	3	.05304	.06861	.08546	.10367	.12335
14	3	4	.09449	.10825	.12289	.13847	.15505
14	3	5	.11029	.12150	.13329	.14568	.15872
14	3	6	.11476	.12349	.13257	.14202	.15186
14	3	7	.11366	.12018	.12690	.13384	.14100
14	3	8	.10966	.11430	.11905	.12391	.12889
14	3	9	.10414	.10721	.11034	.11353	.11677
14	3	10	.09780	.09960	.10144	.10331	.10523
14	3	11	.32810	.32987	.33197	.33439	.33709
14	4	1	-1.15399	-1.23986	-1.33149	-1.42921	-1.53337
14	4	2	-.06555	-.05143	-.03562	-.01800	.00158
14	4	3	.05716	.07371	.09159	.11089	.13172
14	4	4	.10206	.11663	.13213	.14861	.16613
14	4	5	.11918	.13101	.14344	.15651	.17025
14	4	6	.12402	.13319	.14273	.15265	.16299
14	4	7	.12282	.12963	.13666	.14390	.15139
14	4	8	.11849	.12329	.12821	.13325	.13842
14	4	9	.11249	.11563	.11883	.12210	.12544
14	4	10	.46332	.46820	.47353	.47929	.48545
14	5	1	-1.26441	-1.35720	-1.45612	-1.56151	-1.67374
14	5	2	-.07183	-.05667	-.03974	-.02089	.00001
14	5	3	.06259	.08031	.09943	.12005	.14226
14	5	4	.11175	.12731	.14383	.16138	.18002
14	5	5	.13047	.14305	.15625	.17011	.18468
14	5	6	.13574	.14543	.15550	.16598	.17689
14	5	7	.13439	.14153	.14889	.15649	.16434
14	5	8	.12959	.13457	.13967	.14491	.15028
14	5	9	.63171	.64167	.65227	.66348	.67527
14	6	1	-1.40596	-1.50765	-1.61594	-1.73118	-1.85376
14	6	2	-.07937	-.06290	-.04456	-.02420	-.00167
14	6	3	.07000	.08920	.10987	.13212	.15605
14	6	4	.12456	.14134	.15914	.17801	.19803
14	6	5	.14527	.15876	.17291	.18775	.20333
14	6	6	.15102	.16134	.17206	.18321	.19481

Coefficients b_i for the BLUE of σ
Shape parameter k

n	s	i	2.1	2.2	2.3	2.4	2.5
14	6	7	.14940	.15693	.16470	.17271	.18099
14	6	8	.84509	.86298	.88182	.90158	.92223
14	7	1	-1.59354	-1.70710	-1.82791	-1.95632	-2.09273
14	7	2	-.08848	-.07031	-.05014	-.02781	-.00316
14	7	3	.08058	.10170	.12439	.14877	.17493
14	7	4	.14216	.16053	.17998	.20058	.22238
14	7	5	.16539	.18007	.19543	.21153	.22841
14	7	6	.17166	.18279	.19434	.20635	.21882
14	7	7	1.12222	1.15232	1.18389	1.21691	1.25134
15	0	1	-.86870	-.93699	-1.00994	-1.08785	-1.17103
15	0	2	-.06070	-.05079	-.03955	-.02688	-.01267
15	0	3	.03339	.04592	.05954	.07434	.09039
15	0	4	.06904	.08051	.09276	.10585	.11982
15	0	5	.08357	.09322	.10339	.11412	.12544
15	0	6	.08864	.09641	.10451	.11297	.12180
15	0	7	.08896	.09501	.10126	.10772	.11440
15	0	8	.08676	.09131	.09596	.10073	.10562
15	0	9	.08318	.08645	.08976	.09312	.09655
15	0	10	.07885	.08103	.08323	.08545	.08770
15	0	11	.07410	.07539	.07669	.07800	.07932
15	0	12	.06911	.06968	.07027	.07087	.07149
15	0	13	.06393	.06394	.06398	.06406	.06417
15	0	14	.05842	.05803	.05771	.05744	.05722
15	0	15	.05145	.05089	.05043	.05006	.04976
15	1	1	-.91394	-.98564	-1.06230	-1.14422	-1.23174
15	1	2	-.06415	-.05379	-.04205	-.02881	-.01395
15	1	3	.03486	.04797	.06223	.07772	.09453
15	1	4	.07239	.08439	.09722	.11092	.12557
15	1	5	.08771	.09779	.10844	.11967	.13154
15	1	6	.09306	.10118	.10965	.11851	.12777
15	1	7	.09341	.09973	.10627	.11303	.12004
15	1	8	.09112	.09586	.10073	.10572	.11084
15	1	9	.08737	.09077	.09423	.09775	.10134
15	1	10	.08283	.08510	.08739	.08972	.09208
15	1	11	.07785	.07919	.08054	.08192	.08331
15	1	12	.07263	.07322	.07383	.07447	.07513
15	1	13	.06722	.06724	.06730	.06740	.06754
15	1	14	.11764	.11699	.11651	.11619	.11600
15	2	1	-.96701	-1.04235	-1.12290	-1.20900	-1.30098
15	2	2	-.06822	-.05736	-.04505	-.03120	-.01565
15	2	3	.03656	.05030	.06523	.08145	.09904
15	2	4	.07630	.08886	.10228	.11662	.13194
15	2	5	.09253	.10307	.11419	.12594	.13834
15	2	6	.09822	.10669	.11553	.12478	.13444
15	2	7	.09862	.10519	.11200	.11905	.12635
15	2	8	.09620	.10113	.10618	.11137	.11670
15	2	9	.09226	.09577	.09935	.10300	.10672
15	2	10	.08747	.08980	.09216	.09456	.09700
15	2	11	.08222	.08358	.08496	.08637	.08780
15	2	12	.07671	.07730	.07792	.07857	.07925
15	2	13	.19815	.19803	.19816	.19851	.19904
15	3	1	-1.03011	-1.10959	-1.19454	-1.28531	-1.38225
15	3	2	-.07302	-.06158	-.04864	-.03409	-.01778
15	3	3	.03862	.05306	.06875	.08577	.10422
15	3	4	.08098	.09415	.10823	.12326	.13931
15	3	5	.09829	.10933	.12097	.13326	.14624

Coefficients b_i for the BLUE of σ
Shape parameter k

n	s	i	2.1	2.2	2.3	2.4	2.5
15	3	6	.10437	.11321	.12245	.13210	.14220
15	3	7	.10481	.11166	.11874	.12608	.13368
15	3	8	.10226	.10736	.11259	.11797	.12351
15	3	9	.09806	.10168	.10536	.10913	.11298
15	3	10	.09297	.09534	.09776	.10021	.10272
15	3	11	.08738	.08874	.09014	.09156	.09302
15	3	12	.29539	.29663	.29820	.30005	.30216
15	4	1	-1.10611	-1.19046	-1.28058	-1.37681	-1.47952
15	4	2	-.07868	-.06657	-.05291	-.03756	-.02038
15	4	3	.04121	.05646	.07300	.09094	.11037
15	4	4	.08670	.10058	.11540	.13122	.14809
15	4	5	.10530	.11690	.12913	.14203	.15564
15	4	6	.11184	.12110	.13077	.14087	.15143
15	4	7	.11232	.11945	.12684	.13449	.14242
15	4	8	.10958	.11486	.12029	.12587	.13161
15	4	9	.10507	.10878	.11257	.11645	.12042
15	4	10	.09959	.10200	.10445	.10695	.10951
15	4	11	.41318	.41690	.42103	.42555	.43042
15	5	1	-1.19907	-1.28936	-1.38574	-1.48857	-1.59822
15	5	2	-.08539	-.07249	-.05797	-.04168	-.02350
15	5	3	.04456	.06077	.07833	.09735	.11791
15	5	4	.09386	.10858	.12427	.14100	.15882
15	5	5	.11402	.12627	.13918	.15278	.16713
15	5	6	.12109	.13083	.14099	.15161	.16270
15	5	7	.12159	.12905	.13678	.14478	.15307
15	5	8	.11859	.12408	.12971	.13551	.14148
15	5	9	.11368	.11749	.12138	.12537	.12946
15	5	10	.55708	.56478	.57305	.58185	.59114
15	6	1	-1.31509	-1.41277	-1.51696	-1.62801	-1.74629
15	6	2	-.09341	-.07955	-.06399	-.04658	-.02717
15	6	3	.04905	.06643	.08523	.10555	.12749
15	6	4	.10307	.11880	.13555	.15337	.17235
15	6	5	.12512	.13816	.15188	.16633	.18155
15	6	6	.13281	.14313	.15388	.16511	.17682
15	6	7	.13330	.14115	.14927	.15767	.16638
15	6	8	.12995	.13566	.14153	.14757	.15379
15	6	9	.73518	.74900	.76360	.77897	.79507
15	7	1	-1.46358	-1.57080	-1.68503	-1.80664	-1.93603
15	7	2	-.10310	-.08804	-.07118	-.05236	-.03144
15	7	3	.05532	.07417	.09452	.11647	.14013
15	7	4	.11529	.13229	.15035	.16956	.18996
15	7	5	.13969	.15371	.16845	.18395	.20025
15	7	6	.14812	.15914	.17062	.18259	.19507
15	7	7	.14852	.15683	.16543	.17433	.18355
15	7	8	.95974	.98270	1.00683	1.03211	1.05850
15	8	1	-1.66008	-1.78003	-1.90767	-2.04341	-2.18764
15	8	2	-.11498	-.09835	-.07979	-.05913	-.03623
15	8	3	.06443	.08521	.10759	.13168	.15759
15	8	4	.13217	.15082	.17060	.19160	.21387
15	8	5	.15955	.17485	.19090	.20776	.22547
15	8	6	.16883	.18076	.19317	.20610	.21958
15	8	7	1.25008	1.28675	1.32520	1.36541	1.40736
16	0	1	-.85431	-.92273	-.99590	-1.07413	-1.15774
16	0	2	-.07082	-.06221	-.05236	-.04116	-.02850
16	0	3	.02319	.03494	.04776	.06174	.07695
16	0	4	.05990	.07098	.08285	.09558	.10921

Coefficients b_i for the BLUE of σ
Shape parameter k

n	s	i	2.1	2.2	2.3	2.4	2.5
16	0	5	.07568	.08522	.09530	.10599	.11730
16	0	6	.08195	.08981	.09804	.10666	.11570
16	0	7	.08337	.08966	.09618	.10295	.10998
16	0	8	.08217	.08705	.09207	.09723	.10255
16	0	9	.07949	.08315	.08688	.09069	.09458
16	0	10	.07598	.07860	.08124	.08392	.08664
16	0	11	.07200	.07373	.07547	.07723	.07900
16	0	12	.06774	.06874	.06975	.07077	.07180
16	0	13	.06333	.06373	.06415	.06459	.06505
16	0	14	.05875	.05869	.05867	.05868	.05872
16	0	15	.05388	.05349	.05316	.05289	.05267
16	0	16	.04767	.04715	.04672	.04638	.04610
16	1	1	-.89539	-.96698	-1.04360	-1.12557	-1.21323
16	1	2	-.07447	-.06551	-.05525	-.04359	-.03041
16	1	3	.02408	.03633	.04970	.06428	.08016
16	1	4	.06257	.07412	.08651	.09979	.11403
16	1	5	.07913	.08907	.09960	.11075	.12256
16	1	6	.08572	.09391	.10249	.11150	.12094
16	1	7	.08722	.09377	.10057	.10764	.11498
16	1	8	.08598	.09106	.09629	.10168	.10724
16	1	9	.08319	.08699	.09088	.09485	.09892
16	1	10	.07952	.08223	.08499	.08779	.09063
16	1	11	.07536	.07715	.07897	.08080	.08266
16	1	12	.07092	.07195	.07300	.07406	.07514
16	1	13	.06631	.06672	.06717	.06763	.06812
16	1	14	.06155	.06150	.06149	.06152	.06159
16	1	15	.10832	.10768	.10720	.10687	.10666
16	2	1	-.94302	-1.01796	-1.09817	-1.18398	-1.27576
16	2	2	-.07874	-.06937	-.05866	-.04650	-.03277
16	2	3	.02507	.03786	.05182	.06704	.08359
16	2	4	.06564	.07769	.09061	.10446	.11931
16	2	5	.08310	.09346	.10443	.11605	.12836
16	2	6	.09006	.09858	.10752	.11689	.12673
16	2	7	.09167	.09847	.10553	.11288	.12052
16	2	8	.09037	.09564	.10106	.10666	.11243
16	2	9	.08745	.09138	.09540	.09951	.10373
16	2	10	.08360	.08639	.08923	.09212	.09506
16	2	11	.07923	.08106	.08292	.08481	.08672
16	2	12	.07456	.07561	.07667	.07776	.07887
16	2	13	.06973	.07014	.07058	.07106	.07156
16	2	14	.18128	.18105	.18105	.18125	.18164
16	3	1	-.99893	-1.07761	-1.16181	-1.25187	-1.34815
16	3	2	-.08372	-.07389	-.06268	-.04996	-.03561
16	3	3	.02626	.03966	.05426	.07017	.08747
16	3	4	.06926	.08186	.09536	.10983	.12532
16	3	5	.08778	.09859	.11003	.12215	.13499
16	3	6	.09517	.10405	.11335	.12311	.13335
16	3	7	.09689	.10395	.11129	.11892	.12686
16	3	8	.09553	.10098	.10660	.11240	.11838
16	3	9	.09244	.09650	.10064	.10489	.10925
16	3	10	.08838	.09123	.09414	.09711	.10013
16	3	11	.08375	.08561	.08750	.08942	.09138
16	3	12	.07882	.07986	.08093	.08202	.08315
16	3	13	.26835	.26922	.27038	.27181	.27348
16	4	1	-1.06525	-1.14827	-1.23708	-1.33201	-1.43345
16	4	2	-.08954	-.07919	-.06741	-.05406	-.03902

Coefficients b_i for the BLUE of σ
Shape parameter k

n	s	i	2.1	2.2	2.3	2.4	2.5
16	4	3	.02776	.04183	.05717	.07385	.09198
16	4	4	.07362	.08684	.10099	.11614	.13236
16	4	5	.09338	.10470	.11667	.12934	.14276
16	4	6	.10127	.11054	.12025	.13043	.14111
16	4	7	.10312	.11047	.11810	.12604	.13430
16	4	8	.10168	.10732	.11314	.11915	.12535
16	4	9	.09839	.10256	.10683	.11121	.11570
16	4	10	.09405	.09697	.09994	.10297	.10608
16	4	11	.08912	.09098	.09289	.09484	.09683
16	4	12	.37239	.37525	.37850	.38209	.38600
16	5	1	-1.14496	-1.23315	-1.32743	-1.42813	-1.53565
16	5	2	-.09638	-.08543	-.07298	-.05891	-.04308
16	5	3	.02968	.04456	.06074	.07832	.09740
16	5	4	.07898	.09292	.10782	.12376	.14081
16	5	5	.10022	.11212	.12469	.13799	.15206
16	5	6	.10870	.11841	.12858	.13923	.15040
16	5	7	.11068	.11835	.12631	.13459	.14320
16	5	8	.10912	.11497	.12101	.12725	.13368
16	5	9	.10557	.10986	.11426	.11878	.12341
16	5	10	.10089	.10385	.10689	.10999	.11316
16	5	11	.49749	.50355	.51011	.51714	.52460
16	6	1	-1.24230	-1.33681	-1.43773	-1.54545	-1.66034
16	6	2	-.10449	-.09282	-.07959	-.06467	-.04791
16	6	3	.03227	.04809	.06527	.08391	.10412
16	6	4	.08573	.10051	.11630	.13317	.15119
16	6	5	.10874	.12132	.13461	.14864	.16348
16	6	6	.11791	.12814	.13884	.15005	.16178
16	6	7	.12002	.12806	.13640	.14506	.15407
16	6	8	.11829	.12439	.13067	.13716	.14386
16	6	9	.11440	.11882	.12336	.12802	.13281
16	6	10	.64944	.66031	.67188	.68410	.69693
16	7	1	-1.36359	-1.46599	-1.57523	-1.69170	-1.81579
16	7	2	-.11419	-.10166	-.08748	-.07152	-.05364
16	7	3	.03583	.05281	.07122	.09116	.11273
16	7	4	.09443	.11025	.12712	.14512	.16432
16	7	5	.11961	.13302	.14717	.16210	.17786
16	7	6	.12960	.14045	.15179	.16366	.17608
16	7	7	.13184	.14031	.14910	.15822	.16771
16	7	8	.12985	.13623	.14280	.14958	.15658
16	7	9	.83662	.85458	.87351	.89338	.91415
16	8	1	-1.51865	-1.63120	-1.75114	-1.87888	-2.01481
16	8	2	-.12598	-.11235	-.09698	-.07972	-.06044
16	8	3	.04095	.05939	.07934	.10091	.12420
16	8	4	.10604	.12316	.14139	.16081	.18148
16	8	5	.13392	.14837	.16359	.17963	.19654
16	8	6	.14488	.15650	.16864	.18133	.19460
16	8	7	.14721	.15621	.16556	.17525	.18532
16	8	8	1.07164	1.09991	1.12960	1.16068	1.19312
17	0	1	-.84058	-.90904	-.98234	-1.06077	-1.14466
17	0	2	-.07963	-.07221	-.06363	-.05379	-.04258
17	0	3	.01412	.02511	.03716	.05034	.06473
17	0	4	.05166	.06233	.07380	.08613	.09938
17	0	5	.06847	.07785	.08781	.09839	.10962
17	0	6	.07576	.08365	.09194	.10066	.10983
17	0	7	.07813	.08459	.09130	.09830	.10560
17	0	8	.07779	.08294	.08824	.09372	.09940

Coefficients b_i for the BLUE of σ
Shape parameter k

n	s	i	2.1	2.2	2.3	2.4	2.5
17	0	9	.07590	.07988	.08396	.08814	.09243
17	0	10	.07310	.07608	.07910	.08218	.08532
17	0	11	.06977	.07188	.07402	.07618	.07837
17	0	12	.06613	.06751	.06891	.07031	.07173
17	0	13	.06232	.06309	.06387	.06467	.06547
17	0	14	.05838	.05866	.05895	.05927	.05961
17	0	15	.05431	.05420	.05413	.05409	.05408
17	0	16	.04997	.04958	.04926	.04898	.04876
17	0	17	.04440	.04391	.04351	.04318	.04292
17	1	1	-.87811	-.94954	-1.02605	-1.10798	-1.19567
17	1	2	-.08340	-.07570	-.06680	-.05658	-.04495
17	1	3	.01454	.02598	.03851	.05222	.06719
17	1	4	.05378	.06487	.07681	.08965	.10345
17	1	5	.07136	.08111	.09148	.10249	.11419
17	1	6	.07899	.08719	.09582	.10490	.11445
17	1	7	.08148	.08819	.09517	.10246	.11006
17	1	8	.08114	.08648	.09200	.09771	.10362
17	1	9	.07917	.08331	.08755	.09190	.09637
17	1	10	.07625	.07934	.08249	.08569	.08897
17	1	11	.07279	.07498	.07720	.07945	.08173
17	1	12	.06900	.07043	.07188	.07334	.07482
17	1	13	.06503	.06583	.06664	.06747	.06832
17	1	14	.06093	.06122	.06154	.06187	.06224
17	1	15	.05672	.05661	.05655	.05653	.05655
17	1	16	.10033	.09970	.09922	.09889	.09867
17	2	1	-.92120	-.99571	-1.07554	-1.16104	-1.25255
17	2	2	-.08776	-.07974	-.07048	-.05986	-.04777
17	2	3	.01500	.02691	.03995	.05421	.06978
17	2	4	.05619	.06773	.08014	.09350	.10784
17	2	5	.07465	.08479	.09556	.10700	.11916
17	2	6	.08267	.09119	.10015	.10957	.11950
17	2	7	.08530	.09225	.09950	.10706	.11495
17	2	8	.08496	.09048	.09620	.10211	.10824
17	2	9	.08291	.08718	.09156	.09606	.10069
17	2	10	.07986	.08304	.08628	.08959	.09297
17	2	11	.07623	.07848	.08076	.08307	.08543
17	2	12	.07227	.07373	.07520	.07671	.07823
17	2	13	.06812	.06892	.06975	.07059	.07146
17	2	14	.06384	.06412	.06443	.06478	.06515
17	2	15	.16697	.16665	.16655	.16665	.16691
17	3	1	-.97119	-1.04913	-1.13261	-1.22199	-1.31763
17	3	2	-.09280	-.08442	-.07476	-.06369	-.05111
17	3	3	.01555	.02797	.04157	.05644	.07265
17	3	4	.05899	.07102	.08395	.09786	.11279
17	3	5	.07848	.08903	.10023	.11213	.12478
17	3	6	.08695	.09580	.10511	.11490	.12520
17	3	7	.08974	.09695	.10446	.11230	.12048
17	3	8	.08939	.09511	.10102	.10714	.11348
17	3	9	.08724	.09164	.09616	.10080	.10559
17	3	10	.08404	.08730	.09063	.09403	.09751
17	3	11	.08022	.08251	.08483	.08720	.08961
17	3	12	.07605	.07752	.07901	.08054	.08209
17	3	13	.07168	.07247	.07330	.07415	.07502
17	3	14	.24566	.24624	.24709	.24820	.24954
17	4	1	-1.02976	-1.11160	-1.19923	-1.29301	-1.39331
17	4	2	-.09865	-.08986	-.07974	-.06818	-.05504

Coefficients b_i for the BLUE of σ
Shape parameter k

n	s	i	2.1	2.2	2.3	2.4	2.5
17	4	3	.01625	.02926	.04348	.05901	.07594
17	4	4	.06233	.07490	.08841	.10292	.11849
17	4	5	.08300	.09401	.10569	.11809	.13127
17	4	6	.09200	.10121	.11090	.12108	.13180
17	4	7	.09497	.10245	.11025	.11838	.12688
17	4	8	.09460	.10052	.10664	.11297	.11953
17	4	9	.09233	.09686	.10152	.10631	.11124
17	4	10	.08894	.09228	.09569	.09918	.10275
17	4	11	.08489	.08721	.08958	.09199	.09445
17	4	12	.08047	.08194	.08344	.08497	.08654
17	4	13	.33861	.34082	.34339	.34628	.34946
17	5	1	-1.09911	-1.18553	-1.27800	-1.37689	-1.48258
17	5	2	-.10545	-.09620	-.08557	-.07344	-.05968
17	5	3	.01718	.03086	.04579	.06207	.07981
17	5	4	.06637	.07956	.09372	.10891	.12521
17	5	5	.08844	.09996	.11218	.12515	.13891
17	5	6	.09805	.10766	.11777	.12839	.13957
17	5	7	.10121	.10900	.11712	.12558	.13441
17	5	8	.10082	.10695	.11329	.11986	.12666
17	5	9	.09839	.10306	.10787	.11281	.11789
17	5	10	.09476	.09818	.10167	.10525	.10891
17	5	11	.09043	.09278	.09517	.09762	.10013
17	5	12	.44890	.45371	.45899	.46468	.47077
17	6	1	-1.18232	-1.27421	-1.37246	-1.47744	-1.58954
17	6	2	-.11344	-.10365	-.09242	-.07964	-.06516
17	6	3	.01847	.03292	.04868	.06584	.08451
17	6	4	.07136	.08527	.10019	.11618	.13331
17	6	5	.09509	.10721	.12006	.13367	.14811
17	6	6	.10541	.11549	.12608	.13721	.14890
17	6	7	.10880	.11693	.12540	.13423	.14344
17	6	8	.10836	.11473	.12132	.12814	.13520
17	6	9	.10572	.11054	.11550	.12060	.12586
17	6	10	.10179	.10528	.10885	.11252	.11627
17	6	11	.58079	.58949	.59880	.60868	.61910
17	7	1	-1.28381	-1.38238	-1.48768	-1.60008	-1.71999
17	7	2	-.12290	-.11247	-.10054	-.08698	-.07166
17	7	3	.02028	.03567	.05241	.07063	.09040
17	7	4	.07767	.09244	.10826	.12520	.14331
17	7	5	.10339	.11622	.12981	.14419	.15942
17	7	6	.11454	.12518	.13634	.14806	.16036
17	7	7	.11818	.12672	.13561	.14487	.15452
17	7	8	.11766	.12430	.13117	.13829	.14565
17	7	9	.11474	.11972	.12486	.13014	.13559
17	7	10	.74026	.75459	.76974	.78569	.80240
17	8	1	-1.41011	-1.51704	-1.63115	-1.75284	-1.88251
17	8	2	-.13425	-.12303	-.11024	-.09575	-.07941
17	8	3	.02293	.03945	.05741	.07690	.09803
17	8	4	.08585	.10169	.11861	.13669	.15602
17	8	5	.11400	.12771	.14219	.15751	.17371
17	8	6	.12616	.13747	.14932	.16175	.17478
17	8	7	.13006	.13909	.14847	.15825	.16843
17	8	8	.12939	.13636	.14357	.15103	.15875
17	8	9	.93597	.95831	.98182	1.00646	1.03220
17	9	1	-1.57140	-1.68909	-1.81455	-1.94819	-2.09043
17	9	2	-.14809	-.13588	-.12200	-.10633	-.08871
17	9	3	.02690	.04487	.06436	.08547	.10831

Coefficients b_i for the BLUE of σ
Shape parameter k

n	s	i	2.1	2.2	2.3	2.4	2.5
17	9	4	.09683	.11399	.13231	.15185	.17269
17	9	5	.12801	.14280	.15841	.17491	.19232
17	9	6	.14137	.15352	.16623	.17955	.19350
17	9	7	.14554	.15516	.16517	.17558	.18642
17	9	8	1.18085	1.21462	1.25006	1.28716	1.32590
18	0	1	-.82747	-.89591	-.96923	-1.04776	-1.13184
18	0	2	-.08734	-.08102	-.07362	-.06503	-.05516
18	0	3	.00599	.01627	.02758	.03999	.05358
18	0	4	.04419	.05444	.06550	.07743	.09028
18	0	5	.06187	.07106	.08085	.09128	.10239
18	0	6	.07003	.07790	.08620	.09496	.10419
18	0	7	.07322	.07978	.08663	.09380	.10129
18	0	8	.07363	.07897	.08451	.09024	.09620
18	0	9	.07242	.07667	.08103	.08553	.09016
18	0	10	.07024	.07353	.07688	.08030	.08381
18	0	11	.06748	.06993	.07241	.07493	.07749
18	0	12	.06438	.06610	.06784	.06960	.07138
18	0	13	.06107	.06218	.06330	.06443	.06558
18	0	14	.05763	.05823	.05885	.05947	.06010
18	0	15	.05411	.05429	.05449	.05471	.05495
18	0	16	.05046	.05032	.05021	.05013	.05009
18	0	17	.04657	.04619	.04587	.04560	.04537
18	0	18	.04154	.04108	.04070	.04039	.04014
18	1	1	-.86195	-.93316	-1.00951	-1.09132	-1.17897
18	1	2	-.09117	-.08463	-.07697	-.06808	-.05787
18	1	3	.00606	.01672	.02846	.04133	.05544
18	1	4	.04586	.05650	.06798	.08037	.09371
18	1	5	.06429	.07383	.08399	.09482	.10637
18	1	6	.07281	.08098	.08959	.09869	.10829
18	1	7	.07614	.08295	.09006	.09750	.10529
18	1	8	.07658	.08212	.08786	.09382	.10001
18	1	9	.07533	.07973	.08426	.08893	.09374
18	1	10	.07307	.07647	.07995	.08351	.08715
18	1	11	.07020	.07274	.07531	.07793	.08059
18	1	12	.06698	.06876	.07057	.07240	.07425
18	1	13	.06354	.06469	.06585	.06703	.06822
18	1	14	.05998	.06060	.06123	.06188	.06255
18	1	15	.05632	.05651	.05672	.05696	.05723
18	1	16	.05256	.05241	.05232	.05226	.05223
18	1	17	.09340	.09278	.09232	.09198	.09175
18	2	1	-.90118	-.97525	-1.05469	-1.13982	-1.23103
18	2	2	-.09556	-.08877	-.08082	-.07160	-.06102
18	2	3	.00611	.01718	.02936	.04272	.05736
18	2	4	.04774	.05878	.07070	.08354	.09739
18	2	5	.06703	.07692	.08746	.09869	.11066
18	2	6	.07595	.08441	.09334	.10276	.11271
18	2	7	.07945	.08649	.09385	.10156	.10963
18	2	8	.07992	.08564	.09158	.09775	.10415
18	2	9	.07862	.08316	.08784	.09266	.09764
18	2	10	.07627	.07977	.08335	.08702	.09079
18	2	11	.07328	.07588	.07853	.08122	.08397
18	2	12	.06992	.07174	.07359	.07547	.07738
18	2	13	.06634	.06750	.06869	.06989	.07112
18	2	14	.06262	.06324	.06388	.06455	.06523
18	2	15	.05882	.05900	.05921	.05946	.05973
18	2	16	.15467	.15429	.15412	.15414	.15431

Coefficients b_i for the BLUE of σ
Shape parameter k

n	s	i	2.1	2.2	2.3	2.4	2.5
18	3	1	-.94626	-1.02347	-1.10626	-1.19498	-1.29000
18	3	2	-.10059	-.09352	-.08524	-.07567	-.06469
18	3	3	.00618	.01770	.03036	.04424	.05944
18	3	4	.04991	.06138	.07376	.08710	.10147
18	3	5	.07018	.08045	.09138	.10303	.11545
18	3	6	.07956	.08833	.09759	.10735	.11766
18	3	7	.08325	.09054	.09816	.10613	.11449
18	3	8	.08376	.08967	.09580	.10217	.10880
18	3	9	.08241	.08709	.09191	.09688	.10201
18	3	10	.07995	.08354	.08722	.09100	.09487
18	3	11	.07682	.07947	.08218	.08494	.08775
18	3	12	.07330	.07514	.07702	.07894	.08088
18	3	13	.06954	.07071	.07190	.07312	.07436
18	3	14	.06564	.06625	.06689	.06755	.06824
18	3	15	.22636	.22672	.22734	.22820	.22927
18	4	1	-.99848	-1.07925	-1.16581	-1.25853	-1.35779
18	4	2	-.10637	-.09898	-.09035	-.08039	-.06896
18	4	3	.00630	.01832	.03151	.04597	.06179
18	4	4	.05246	.06441	.07730	.09118	.10612
18	4	5	.07386	.08454	.09591	.10802	.12091
18	4	6	.08377	.09288	.10249	.11262	.12331
18	4	7	.08768	.09523	.10312	.11138	.12003
18	4	8	.08822	.09433	.10067	.10725	.11410
18	4	9	.08680	.09162	.09659	.10171	.10700
18	4	10	.08421	.08790	.09167	.09555	.09953
18	4	11	.08092	.08362	.08638	.08920	.09208
18	4	12	.07720	.07907	.08097	.08291	.08489
18	4	13	.07324	.07440	.07559	.07681	.07806
18	4	14	.31020	.31191	.31396	.31630	.31892
18	5	1	-1.05956	-1.14442	-1.23531	-1.33261	-1.43671
18	5	2	-.11305	-.10530	-.09627	-.08586	-.07394
18	5	3	.00652	.01910	.03290	.04800	.06451
18	5	4	.05550	.06800	.08146	.09594	.11152
18	5	5	.07823	.08937	.10122	.11384	.12726
18	5	6	.08875	.09823	.10823	.11877	.12988
18	5	7	.09290	.10074	.10893	.11751	.12648
18	5	8	.09348	.09980	.10636	.11317	.12026
18	5	9	.09197	.09694	.10206	.10734	.11280
18	5	10	.08922	.09300	.09687	.10085	.10493
18	5	11	.08572	.08847	.09128	.09415	.09709
18	5	12	.08177	.08364	.08556	.08752	.08952
18	5	13	.40856	.41242	.41671	.42138	.42641
18	6	1	-1.13177	-1.22145	-1.31743	-1.42010	-1.52985
18	6	2	-.12081	-.11265	-.10317	-.09225	-.07978
18	6	3	.00689	.02013	.03462	.05046	.06775
18	6	4	.05922	.07234	.08645	.10162	.11792
18	6	5	.08348	.09515	.10756	.12075	.13478
18	6	6	.09471	.10462	.11506	.12606	.13765
18	6	7	.09914	.10731	.11584	.12476	.13409
18	6	8	.09975	.10631	.11312	.12018	.12752
18	6	9	.09813	.10326	.10854	.11400	.11963
18	6	10	.09518	.09905	.10302	.10711	.11130
18	6	11	.09142	.09421	.09707	.09999	.10299
18	6	12	.52466	.53171	.53931	.54742	.55600
18	7	1	-1.21831	-1.31375	-1.41582	-1.52491	-1.64141
18	7	2	-.12991	-.12127	-.11126	-.09976	-.08665

Coefficients b_i for the BLUE of σ
Shape parameter k

n	s	i	2.1	2.2	2.3	2.4	2.5
18	7	3	.00754	.02153	.03683	.05354	.07174
18	7	4	.06383	.07768	.09256	.10853	.12567
18	7	5	.08992	.10221	.11527	.12913	.14385
18	7	6	.10199	.11239	.12334	.13486	.14700
18	7	7	.10673	.11527	.12419	.13351	.14325
18	7	8	.10736	.11418	.12127	.12862	.13625
18	7	9	.10558	.11089	.11636	.12200	.12783
18	7	10	.10237	.10634	.11042	.11461	.11893
18	7	11	.66292	.67452	.68686	.69987	.71355
18	8	1	-1.32373	-1.42622	-1.53573	-1.65266	-1.77740
18	8	2	-.14069	-.13148	-.12084	-.10865	-.09478
18	8	3	.00859	.02351	.03978	.05751	.07680
18	8	4	.06970	.08442	.10021	.11713	.13527
18	8	5	.09799	.11101	.12483	.13949	.15503
18	8	6	.11104	.12202	.13358	.14573	.15851
18	8	7	.11612	.12511	.13448	.14427	.15449
18	8	8	.11675	.12389	.13130	.13898	.14695
18	8	9	.11476	.12026	.12595	.13181	.13786
18	8	10	.82948	.84748	.86647	.88640	.90726
18	9	1	-1.45479	-1.56611	-1.68493	-1.81165	-1.94671
18	9	2	-.15364	-.14373	-.13233	-.11930	-.10451
18	9	3	.01032	.02636	.04382	.06282	.08345
18	9	4	.07735	.09315	.11006	.12816	.14752
18	9	5	.10833	.12225	.13701	.15264	.16919
18	9	6	.12255	.13426	.14654	.15945	.17302
18	9	7	.12803	.13755	.14748	.15783	.16863
18	9	8	.12861	.13613	.14392	.15200	.16038
18	9	9	1.03324	1.06015	1.08843	1.11806	1.14902
19	0	1	-.81494	-.88329	-.95658	-1.03514	-1.11931
19	0	2	-.09413	-.08882	-.08250	-.07507	-.06645
19	0	3	-.00131	.00829	.01889	.03056	.04338
19	0	4	.03738	.04723	.05788	.06939	.08182
19	0	5	.05579	.06478	.07438	.08462	.09557
19	0	6	.06470	.07253	.08080	.08954	.09880
19	0	7	.06861	.07523	.08218	.08945	.09709
19	0	8	.06968	.07517	.08088	.08682	.09300
19	0	9	.06907	.07353	.07813	.08289	.08780
19	0	10	.06744	.07098	.07461	.07833	.08215
19	0	11	.06518	.06792	.07070	.07353	.07643
19	0	12	.06255	.06457	.06663	.06871	.07081
19	0	13	.05967	.06109	.06252	.06397	.06543
19	0	14	.05666	.05756	.05847	.05939	.06031
19	0	15	.05356	.05402	.05450	.05499	.05549
19	0	16	.05038	.05049	.05062	.05077	.05094
19	0	17	.04710	.04693	.04679	.04669	.04662
19	0	18	.04359	.04321	.04290	.04263	.04241
19	0	19	.03901	.03858	.03822	.03793	.03770
19	1	1	-.84677	-.91772	-.99385	-1.07551	-1.16304
19	1	2	-.09798	-.09250	-.08598	-.07831	-.06941
19	1	3	-.00153	.00841	.01938	.03147	.04475
19	1	4	.03869	.04889	.05992	.07184	.08473
19	1	5	.05783	.06714	.07708	.08770	.09905
19	1	6	.06711	.07520	.08377	.09284	.10243
19	1	7	.07117	.07803	.08522	.09277	.10069
19	1	8	.07229	.07798	.08389	.09005	.09646
19	1	9	.07167	.07629	.08105	.08598	.09108

Coefficients b_i for the BLUE of σ
Shape parameter k

n	s	i	2.1	2.2	2.3	2.4	2.5
19	1	10	.06998	.07365	.07740	.08126	.08523
19	1	11	.06765	.07047	.07335	.07629	.07929
19	1	12	.06491	.06701	.06913	.07129	.07348
19	1	13	.06194	.06340	.06488	.06638	.06790
19	1	14	.05882	.05974	.06068	.06164	.06261
19	1	15	.05560	.05608	.05658	.05709	.05762
19	1	16	.05232	.05243	.05257	.05273	.05292
19	1	17	.04894	.04877	.04864	.04855	.04850
19	1	18	.08734	.08674	.08628	.08594	.08572
19	2	1	-.88269	-.95632	-1.03533	-1.12009	-1.21095
19	2	2	-.10235	-.09667	-.08992	-.08199	-.07280
19	2	3	-.00179	.00850	.01986	.03238	.04612
19	2	4	.04015	.05070	.06212	.07447	.08781
19	2	5	.06012	.06974	.08003	.09102	.10277
19	2	6	.06980	.07817	.08703	.09640	.10633
19	2	7	.07405	.08113	.08856	.09636	.10455
19	2	8	.07523	.08109	.08720	.09356	.10019
19	2	9	.07459	.07934	.08426	.08934	.09461
19	2	10	.07284	.07661	.08048	.08445	.08854
19	2	11	.07041	.07331	.07627	.07930	.08239
19	2	12	.06757	.06971	.07189	.07411	.07636
19	2	13	.06448	.06597	.06748	.06902	.07057
19	2	14	.06123	.06217	.06312	.06410	.06509
19	2	15	.05789	.05837	.05887	.05939	.05993
19	2	16	.05448	.05459	.05472	.05489	.05509
19	2	17	.14399	.14357	.14335	.14330	.14341
19	3	1	-.92362	-1.00015	-1.08227	-1.17034	-1.26475
19	3	2	-.10732	-.10142	-.09442	-.08621	-.07669
19	3	3	-.00209	.00859	.02038	.03334	.04759
19	3	4	.04181	.05275	.06459	.07738	.09119
19	3	5	.06272	.07269	.08334	.09472	.10688
19	3	6	.07286	.08153	.09069	.10039	.11065
19	3	7	.07733	.08465	.09232	.10038	.10884
19	3	8	.07857	.08462	.09092	.09748	.10432
19	3	9	.07791	.08281	.08787	.09311	.09853
19	3	10	.07609	.07996	.08394	.08802	.09223
19	3	11	.07356	.07653	.07956	.08266	.08583
19	3	12	.07059	.07277	.07500	.07726	.07956
19	3	13	.06736	.06887	.07040	.07196	.07355
19	3	14	.06397	.06491	.06587	.06685	.06785
19	3	15	.06048	.06095	.06144	.06196	.06250
19	3	16	.20976	.20995	.21039	.21105	.21192
19	4	1	-.97059	-1.05037	-1.13594	-1.22768	-1.32597
19	4	2	-.11299	-.10684	-.09956	-.09104	-.08116
19	4	3	-.00240	.00871	.02096	.03442	.04920
19	4	4	.04375	.05512	.06740	.08067	.09500
19	4	5	.06573	.07608	.08713	.09892	.11151
19	4	6	.07640	.08538	.09487	.10491	.11554
19	4	7	.08110	.08867	.09661	.10494	.11369
19	4	8	.08242	.08867	.09517	.10194	.10900
19	4	9	.08173	.08678	.09199	.09738	.10297
19	4	10	.07983	.08380	.08788	.09208	.09640
19	4	11	.07717	.08020	.08330	.08647	.08972
19	4	12	.07406	.07627	.07853	.08083	.08318
19	4	13	.07067	.07218	.07373	.07530	.07691
19	4	14	.06710	.06803	.06898	.06996	.07097

Coefficients b_i for the BLUE of σ
Shape parameter k

n	s	i	2.1	2.2	2.3	2.4	2.5
19	4	15	.28601	.28732	.28896	.29088	.29305
19	5	1	-1.02493	-1.10840	-1.19789	-1.29378	-1.39645
19	5	2	-.11949	-.11307	-.10547	-.09660	-.08634
19	5	3	-.00270	.00889	.02165	.03567	.05104
19	5	4	.04604	.05788	.07067	.08448	.09936
19	5	5	.06926	.08003	.09151	.10376	.11684
19	5	6	.08054	.08986	.09971	.11013	.12114
19	5	7	.08551	.09335	.10158	.11020	.11926
19	5	8	.08691	.09336	.10008	.10707	.11436
19	5	9	.08618	.09138	.09675	.10230	.10806
19	5	10	.08417	.08824	.09243	.09674	.10118
19	5	11	.08137	.08446	.08762	.09086	.09418
19	5	12	.07808	.08032	.08260	.08494	.08733
19	5	13	.07449	.07600	.07755	.07913	.08075
19	5	14	.37458	.37770	.38121	.38508	.38928
19	6	1	-1.08838	-1.17614	-1.27016	-1.37084	-1.47856
19	6	2	-.12697	-.12024	-.11230	-.10303	-.09234
19	6	3	-.00296	.00918	.02253	.03717	.05321
19	6	4	.04880	.06118	.07454	.08894	.10446
19	6	5	.07346	.08469	.09667	.10944	.12305
19	6	6	.08543	.09514	.10539	.11623	.12768
19	6	7	.09071	.09886	.10740	.11635	.12574
19	6	8	.09219	.09887	.10583	.11307	.12062
19	6	9	.09142	.09677	.10231	.10805	.11398
19	6	10	.08927	.09345	.09775	.10218	.10674
19	6	11	.08629	.08944	.09266	.09597	.09937
19	6	12	.08278	.08504	.08735	.08972	.09214
19	6	13	.47798	.48375	.49002	.49675	.50391
19	7	1	-1.16331	-1.25612	-1.35549	-1.46179	-1.57544
19	7	2	-.13566	-.12857	-.12023	-.11053	-.09934
19	7	3	-.00313	.00965	.02367	.03903	.05583
19	7	4	.05218	.06519	.07920	.09430	.11054
19	7	5	.07853	.09030	.10284	.11620	.13042
19	7	6	.09131	.10146	.11217	.12348	.13542
19	7	7	.09693	.10543	.11433	.12365	.13342
19	7	8	.09850	.10544	.11266	.12018	.12801
19	7	9	.09765	.10319	.10892	.11485	.12098
19	7	10	.09534	.09964	.10406	.10861	.11330
19	7	11	.09213	.09534	.09863	.10201	.10548
19	7	12	.59954	.60906	.61923	.63001	.64136
19	8	1	-1.25301	-1.35189	-1.45765	-1.57069	-1.69143
19	8	2	-.14584	-.13834	-.12954	-.11931	-.10756
19	8	3	-.00312	.01040	.02521	.04142	.05911
19	8	4	.05641	.07015	.08494	.10084	.11793
19	8	5	.08475	.09716	.11036	.12441	.13935
19	8	6	.09848	.10915	.12040	.13226	.14478
19	8	7	.10450	.11340	.12271	.13246	.14268
19	8	8	.10615	.11339	.12092	.12875	.13691
19	8	9	.10520	.11095	.11689	.12304	.12940
19	8	10	.10267	.10710	.11165	.11634	.12118
19	8	11	.74380	.75852	.77410	.79049	.80767
19	9	1	-1.36218	-1.46847	-1.58205	-1.70334	-1.83275
19	9	2	-.15790	-.14991	-.14055	-.12973	-.11730
19	9	3	-.00281	.01160	.02736	.04458	.06334
19	9	4	.06183	.07645	.09215	.10902	.12712
19	9	5	.09256	.10573	.11972	.13459	.15038

Coefficients b_i for the BLUE of σ
Shape parameter k

n	s	i	2.1	2.2	2.3	2.4	2.5
19	9	6	.10741	.11870	.13058	.14311	.15630
19	9	7	.11389	.12327	.13307	.14332	.15406
19	9	8	.11561	.12320	.13109	.13929	.14783
19	9	9	.11451	.12050	.12668	.13308	.13971
19	9	10	.91707	.93893	.96194	.98607	1.01130
19	10	1	-1.49779	-1.61335	-1.73672	-1.86832	-2.00858
19	10	2	-.17241	-.16380	-.15378	-.14221	-.12896
19	10	3	-.00199	.01352	.03046	.04891	.06899
19	10	4	.06895	.08465	.10149	.11955	.13889
19	10	5	.10260	.11671	.13167	.14754	.16438
19	10	6	.11881	.13085	.14351	.15684	.17087
19	10	7	.12580	.13576	.14616	.15703	.16840
19	10	8	.12758	.13559	.14391	.15256	.16155
19	10	9	1.12846	1.16008	1.19330	1.22810	1.26445
20	0	1	-.80296	-.87117	-.94437	-1.02289	-1.10707
20	0	2	-.10013	-.09576	-.09044	-.08409	-.07662
20	0	3	-.00791	.00105	.01097	.02194	.03402
20	0	4	.03117	.04061	.05084	.06194	.07395
20	0	5	.05019	.05896	.06834	.07839	.08914
20	0	6	.05975	.06749	.07570	.08440	.09363
20	0	7	.06428	.07093	.07792	.08528	.09301
20	0	8	.06593	.07153	.07737	.08347	.08983
20	0	9	.06586	.07049	.07528	.08025	.08540
20	0	10	.06471	.06847	.07233	.07631	.08040
20	0	11	.06291	.06589	.06893	.07204	.07522
20	0	12	.06068	.06297	.06530	.06767	.07008
20	0	13	.05819	.05989	.06160	.06333	.06508
20	0	14	.05555	.05672	.05791	.05910	.06031
20	0	15	.05280	.05353	.05427	.05502	.05577
20	0	16	.04999	.05034	.05071	.05110	.05149
20	0	17	.04711	.04716	.04723	.04733	.04744
20	0	18	.04414	.04395	.04379	.04367	.04358
20	0	19	.04095	.04058	.04027	.04001	.03980
20	0	20	.03677	.03636	.03602	.03575	.03552
20	1	1	-.83246	-.90313	-.97901	-1.06045	-1.14781
20	1	2	-.10396	-.09946	-.09399	-.08745	-.07977
20	1	3	-.00834	.00091	.01116	.02249	.03498
20	1	4	.03218	.04193	.05252	.06399	.07641
20	1	5	.05191	.06097	.07067	.08106	.09219
20	1	6	.06184	.06984	.07832	.08732	.09688
20	1	7	.06654	.07341	.08064	.08825	.09625
20	1	8	.06826	.07405	.08008	.08639	.09298
20	1	9	.06819	.07297	.07793	.08307	.08840
20	1	10	.06701	.07089	.07488	.07899	.08324
20	1	11	.06514	.06822	.07136	.07458	.07788
20	1	12	.06284	.06521	.06761	.07006	.07256
20	1	13	.06027	.06202	.06378	.06557	.06739
20	1	14	.05753	.05874	.05996	.06120	.06246
20	1	15	.05469	.05544	.05621	.05698	.05777
20	1	16	.05178	.05215	.05254	.05294	.05335
20	1	17	.04882	.04887	.04895	.04906	.04919
20	1	18	.04577	.04557	.04543	.04532	.04525
20	1	19	.08199	.08141	.08096	.08063	.08040
20	2	1	-.86552	-.93870	-1.01728	-1.10163	-1.19212
20	2	2	-.10828	-.10363	-.09798	-.09124	-.08331
20	2	3	-.00885	.00071	.01130	.02301	.03591

Coefficients b_i for the BLUE of σ
Shape parameter k

n	s	i	2.1	2.2	2.3	2.4	2.5
20	2	4	.03329	.04337	.05430	.06616	.07899
20	2	5	.05382	.06317	.07320	.08393	.09542
20	2	6	.06415	.07241	.08117	.09046	.10033
20	2	7	.06905	.07614	.08360	.09145	.09971
20	2	8	.07085	.07681	.08304	.08954	.09634
20	2	9	.07078	.07571	.08081	.08611	.09161
20	2	10	.06957	.07356	.07767	.08190	.08627
20	2	11	.06763	.07079	.07402	.07733	.08073
20	2	12	.06525	.06767	.07014	.07265	.07522
20	2	13	.06258	.06436	.06617	.06801	.06987
20	2	14	.05974	.06097	.06222	.06348	.06477
20	2	15	.05680	.05755	.05833	.05912	.05993
20	2	16	.05378	.05415	.05453	.05494	.05536
20	2	17	.05071	.05076	.05084	.05094	.05108
20	2	18	.13465	.13420	.13394	.13384	.13389
20	3	1	-.90291	-.97878	-1.06025	-1.14769	-1.24148
20	3	2	-.11317	-.10834	-.10249	-.09552	-.08734
20	3	3	-.00943	.00047	.01143	.02354	.03688
20	3	4	.03455	.04498	.05628	.06854	.08180
20	3	5	.05598	.06565	.07600	.08709	.09897
20	3	6	.06677	.07529	.08434	.09394	.10412
20	3	7	.07189	.07921	.08690	.09499	.10352
20	3	8	.07378	.07992	.08633	.09304	.10004
20	3	9	.07372	.07878	.08404	.08949	.09515
20	3	10	.07246	.07655	.08077	.08512	.08961
20	3	11	.07045	.07368	.07699	.08038	.08387
20	3	12	.06797	.07044	.07296	.07553	.07815
20	3	13	.06519	.06700	.06884	.07071	.07261
20	3	14	.06223	.06347	.06473	.06601	.06732
20	3	15	.05917	.05992	.06069	.06149	.06230
20	3	16	.05603	.05638	.05676	.05716	.05759
20	3	17	.19534	.19539	.19569	.19620	.19690
20	4	1	-.94546	-1.02432	-1.10897	-1.19979	-1.29717
20	4	2	-.11870	-.11368	-.10761	-.10040	-.09193
20	4	3	-.01007	.00021	.01157	.02411	.03792
20	4	4	.03600	.04681	.05852	.07120	.08493
20	4	5	.05845	.06846	.07918	.09065	.10293
20	4	6	.06976	.07858	.08793	.09784	.10836
20	4	7	.07514	.08269	.09063	.09899	.10778
20	4	8	.07712	.08345	.09006	.09697	.10419
20	4	9	.07707	.08228	.08768	.09329	.09911
20	4	10	.07575	.07995	.08428	.08875	.09336
20	4	11	.07365	.07696	.08034	.08382	.08739
20	4	12	.07106	.07357	.07614	.07876	.08144
20	4	13	.06815	.06998	.07185	.07374	.07568
20	4	14	.06506	.06630	.06756	.06886	.07017
20	4	15	.06185	.06259	.06336	.06415	.06496
20	4	16	.26516	.26617	.26748	.26906	.27087
20	5	1	-.99423	-1.07645	-1.16467	-1.25927	-1.36065
20	5	2	-.12499	-.11976	-.11345	-.10597	-.09720
20	5	3	-.01075	-.00006	.01175	.02477	.03909
20	5	4	.03771	.04893	.06109	.07425	.08847
20	5	5	.06132	.07171	.08282	.09471	.10742
20	5	6	.07322	.08236	.09204	.10230	.11319
20	5	7	.07888	.08669	.09490	.10354	.11263
20	5	8	.08097	.08751	.09433	.10146	.10891

Coefficients b_i for the BLUE of σ
Shape parameter k

n	s	i	2.1	2.2	2.3	2.4	2.5
20	5	9	.08092	.08629	.09185	.09762	.10362
20	5	10	.07955	.08386	.08830	.09288	.09762
20	5	11	.07734	.08071	.08418	.08773	.09138
20	5	12	.07461	.07716	.07978	.08245	.08518
20	5	13	.07155	.07340	.07528	.07720	.07916
20	5	14	.06830	.06953	.07079	.07209	.07341
20	5	15	.34558	.34811	.35101	.35425	.35779
20	6	1	-1.05056	-1.13664	-1.22894	-1.32786	-1.43379
20	6	2	-.13217	-.12671	-.12014	-.11235	-.10324
20	6	3	-.01146	-.00031	.01199	.02555	.04044
20	6	4	.03974	.05144	.06410	.07778	.09256
20	6	5	.06470	.07551	.08706	.09941	.11260
20	6	6	.07727	.08676	.09681	.10746	.11874
20	6	7	.08326	.09135	.09986	.10880	.11821
20	6	8	.08547	.09223	.09928	.10664	.11433
20	6	9	.08541	.09094	.09667	.10262	.10880
20	6	10	.08395	.08838	.09294	.09765	.10251
20	6	11	.08162	.08507	.08860	.09224	.09597
20	6	12	.07873	.08132	.08397	.08669	.08947
20	6	13	.07549	.07734	.07923	.08117	.08315
20	6	14	.43858	.44334	.44856	.45421	.46025
20	7	1	-1.11627	-1.20683	-1.30387	-1.40779	-1.51899
20	7	2	-.14044	-.13472	-.12785	-.11972	-.11024
20	7	3	-.01219	-.00051	.01236	.02652	.04206
20	7	4	.04221	.05444	.06767	.08195	.09737
20	7	5	.06872	.08001	.09206	.10493	.11867
20	7	6	.08207	.09196	.10243	.11351	.12524
20	7	7	.08842	.09684	.10568	.11496	.12472
20	7	8	.09076	.09777	.10508	.11270	.12067
20	7	9	.09069	.09640	.10233	.10847	.11485
20	7	10	.08913	.09368	.09838	.10322	.10822
20	7	11	.08664	.09016	.09378	.09750	.10132
20	7	12	.08355	.08618	.08887	.09162	.09445
20	7	13	.54671	.55461	.56310	.57212	.58166
20	8	1	-1.19380	-1.28965	-1.39227	-1.50208	-1.61948
20	8	2	-.15003	-.14401	-.13680	-.12829	-.11837
20	8	3	-.01290	-.00060	.01293	.02779	.04407
20	8	4	.04526	.05812	.07200	.08697	.10311
20	8	5	.07359	.08543	.09805	.11153	.12589
20	8	6	.08785	.09820	.10914	.12071	.13295
20	8	7	.09462	.10340	.11262	.12229	.13245
20	8	8	.09709	.10438	.11198	.11990	.12817
20	8	9	.09699	.10291	.10904	.11540	.12200
20	8	10	.09530	.09999	.10482	.10981	.11497
20	8	11	.09260	.09621	.09991	.10372	.10763
20	8	12	.67343	.68563	.69858	.71225	.72661
20	9	1	-1.28651	-1.38871	-1.49805	-1.61492	-1.73975
20	9	2	-.16126	-.15489	-.14728	-.13833	-.12792
20	9	3	-.01352	-.00049	.01381	.02949	.04664
20	9	4	.04910	.06270	.07735	.09314	.11013
20	9	5	.07959	.09207	.10538	.11955	.13466
20	9	6	.09490	.10579	.11729	.12944	.14228
20	9	7	.10215	.11137	.12102	.13115	.14179
20	9	8	.10478	.11239	.12032	.12859	.13721
20	9	9	.10463	.11078	.11715	.12376	.13061
20	9	10	.10275	.10760	.11259	.11775	.12307

Coefficients b_i for the BLUE of σ
Shape parameter k

n	s	i	2.1	2.2	2.3	2.4	2.5
20	9	11	.82338	.84140	.86041	.88038	.90129
20	10	1	-1.39926	-1.50923	-1.62676	-1.75227	-1.88620
20	10	2	-.17457	-.16778	-.15970	-.15023	-.13923
20	10	3	-.01395	-.00006	.01517	.03184	.05004
20	10	4	.05406	.06855	.08413	.10089	.11890
20	10	5	.08713	.10040	.11452	.12954	.14552
20	10	6	.10371	.11524	.12741	.14025	.15381
20	10	7	.11151	.12123	.13142	.14209	.15328
20	10	8	.11429	.12229	.13061	.13929	.14833
20	10	9	.11405	.12047	.12712	.13402	.14117
20	10	10	1.00303	1.02889	1.05608	1.08459	1.11438
21	0	1	-.79148	-.85952	-.93258	-1.01101	-1.09514
21	0	2	-.10546	-.10195	-.09756	-.09221	-.08582
21	0	3	-.01388	-.00554	.00373	.01403	.02540
21	0	4	.02548	.03451	.04434	.05502	.06661
21	0	5	.04502	.05355	.06271	.07253	.08308
21	0	6	.05514	.06278	.07090	.07953	.08870
21	0	7	.06022	.06687	.07387	.08126	.08905
21	0	8	.06238	.06806	.07399	.08021	.08671
21	0	9	.06278	.06755	.07249	.07764	.08298
21	0	10	.06207	.06601	.07007	.07426	.07859
21	0	11	.06067	.06386	.06712	.07047	.07392
21	0	12	.05881	.06134	.06391	.06653	.06920
21	0	13	.05667	.05860	.06057	.06256	.06458
21	0	14	.05434	.05577	.05720	.05865	.06012
21	0	15	.05191	.05288	.05387	.05486	.05586
21	0	16	.04940	.04999	.05059	.05120	.05182
21	0	17	.04684	.04710	.04739	.04769	.04800
21	0	18	.04422	.04422	.04425	.04430	.04437
21	0	19	.04152	.04131	.04114	.04100	.04090
21	0	20	.03860	.03825	.03794	.03769	.03748
21	0	21	.03477	.03437	.03405	.03379	.03358
21	1	1	-.81893	-.88929	-.96489	-1.04608	-1.13323
21	1	2	-.10925	-.10565	-.10114	-.09565	-.08909
21	1	3	-.01450	-.00590	.00367	.01428	.02602
21	1	4	.02624	.03556	.04570	.05673	.06869
21	1	5	.04647	.05527	.06472	.07487	.08576
21	1	6	.05695	.06483	.07321	.08212	.09160
21	1	7	.06221	.06907	.07630	.08393	.09199
21	1	8	.06446	.07031	.07644	.08285	.08958
21	1	9	.06488	.06979	.07490	.08021	.08574
21	1	10	.06415	.06821	.07240	.07673	.08120
21	1	11	.06270	.06599	.06936	.07282	.07638
21	1	12	.06079	.06339	.06604	.06875	.07152
21	1	13	.05857	.06057	.06259	.06465	.06674
21	1	14	.05617	.05764	.05912	.06062	.06214
21	1	15	.05366	.05466	.05568	.05671	.05775
21	1	16	.05107	.05168	.05230	.05293	.05358
21	1	17	.04843	.04870	.04900	.04931	.04964
21	1	18	.04573	.04574	.04577	.04583	.04592
21	1	19	.04296	.04276	.04259	.04247	.04238
21	1	20	.07724	.07667	.07624	.07591	.07569
21	2	1	-.84951	-.92222	-1.00036	-1.08429	-1.17439
21	2	2	-.11350	-.10979	-.10514	-.09949	-.09275
21	2	3	-.01520	-.00633	.00354	.01449	.02659
21	2	4	.02707	.03668	.04715	.05852	.07086

Coefficients b_i for the BLUE of σ
Shape parameter k

n	s	i	2.1	2.2	2.3	2.4	2.5
21	2	5	.04806	.05714	.06688	.07735	.08858
21	2	6	.05895	.06707	.07571	.08489	.09467
21	2	7	.06441	.07148	.07893	.08679	.09509
21	2	8	.06675	.07278	.07909	.08570	.09262
21	2	9	.06720	.07225	.07751	.08297	.08867
21	2	10	.06645	.07062	.07493	.07938	.08398
21	2	11	.06495	.06833	.07179	.07535	.07901
21	2	12	.06297	.06564	.06836	.07114	.07398
21	2	13	.06068	.06272	.06480	.06690	.06905
21	2	14	.05820	.05969	.06120	.06274	.06430
21	2	15	.05560	.05661	.05765	.05870	.05976
21	2	16	.05292	.05353	.05416	.05480	.05546
21	2	17	.05019	.05046	.05075	.05107	.05141
21	2	18	.04740	.04740	.04743	.04750	.04759
21	2	19	.12640	.12593	.12564	.12551	.12552
21	3	1	-.88384	-.95907	-1.03991	-1.12673	-1.21991
21	3	2	-.11828	-.11443	-.10964	-.10381	-.09686
21	3	3	-.01599	-.00682	.00337	.01467	.02715
21	3	4	.02800	.03793	.04873	.06046	.07319
21	3	5	.04985	.05922	.06927	.08006	.09164
21	3	6	.06119	.06956	.07847	.08794	.09801
21	3	7	.06689	.07417	.08184	.08994	.09849
21	3	8	.06933	.07553	.08202	.08882	.09595
21	3	9	.06980	.07499	.08039	.08601	.09187
21	3	10	.06903	.07331	.07773	.08230	.08703
21	3	11	.06748	.07093	.07448	.07813	.08188
21	3	12	.06543	.06815	.07093	.07377	.07668
21	3	13	.06305	.06512	.06724	.06939	.07158
21	3	14	.06047	.06198	.06351	.06507	.06666
21	3	15	.05777	.05879	.05983	.06089	.06197
21	3	16	.05498	.05559	.05622	.05686	.05753
21	3	17	.05215	.05241	.05270	.05301	.05335
21	3	18	.18271	.18265	.18283	.18322	.18379
21	4	1	-.92263	-1.00062	-1.08441	-1.17436	-1.27088
21	4	2	-.12365	-.11966	-.11470	-.10868	-.10151
21	4	3	-.01687	-.00738	.00317	.01484	.02774
21	4	4	.02906	.03933	.05050	.06261	.07576
21	4	5	.05189	.06156	.07195	.08309	.09504
21	4	6	.06373	.07238	.08156	.09133	.10172
21	4	7	.06969	.07719	.08511	.09345	.10226
21	4	8	.07225	.07863	.08532	.09232	.09965
21	4	9	.07275	.07808	.08364	.08941	.09543
21	4	10	.07195	.07634	.08087	.08556	.09042
21	4	11	.07034	.07387	.07750	.08123	.08508
21	4	12	.06820	.07097	.07381	.07671	.07969
21	4	13	.06572	.06783	.06997	.07216	.07439
21	4	14	.06303	.06455	.06610	.06768	.06929
21	4	15	.06021	.06123	.06227	.06334	.06443
21	4	16	.05731	.05790	.05852	.05916	.05982
21	4	17	.24703	.24779	.24884	.25014	.25167
21	5	1	-.96672	-1.04780	-1.13485	-1.22828	-1.32848
21	5	2	-.12972	-.12557	-.12043	-.11420	-.10679
21	5	3	-.01783	-.00798	.00295	.01504	.02838
21	5	4	.03030	.04095	.05251	.06505	.07864
21	5	5	.05423	.06425	.07499	.08651	.09886
21	5	6	.06665	.07559	.08508	.09517	.10589

Coefficients b_i for the BLUE of σ
Shape parameter k

n	s	i	2.1	2.2	2.3	2.4	2.5
21	5	7	.07290	.08065	.08881	.09742	.10650
21	5	8	.07559	.08217	.08905	.09626	.10382
21	5	9	.07611	.08160	.08731	.09325	.09944
21	5	10	.07528	.07978	.08444	.08925	.09423
21	5	11	.07360	.07721	.08092	.08474	.08868
21	5	12	.07136	.07418	.07707	.08003	.08307
21	5	13	.06876	.07089	.07307	.07529	.07756
21	5	14	.06594	.06747	.06903	.07062	.07224
21	5	15	.06299	.06399	.06503	.06609	.06718
21	5	16	.32057	.32263	.32504	.32776	.33077
21	6	1	-1.01719	-1.10176	-1.19252	-1.28987	-1.39420
21	6	2	-.13661	-.13229	-.12694	-.12048	-.11281
21	6	3	-.01888	-.00863	.00273	.01528	.02912
21	6	4	.03178	.04284	.05484	.06785	.08194
21	6	5	.05695	.06735	.07849	.09043	.10322
21	6	6	.07003	.07929	.08912	.09956	.11065
21	6	7	.07661	.08462	.09306	.10196	.11134
21	6	8	.07944	.08623	.09334	.10077	.10857
21	6	9	.07999	.08565	.09152	.09764	.10401
21	6	10	.07912	.08374	.08852	.09346	.09857
21	6	11	.07734	.08104	.08484	.08875	.09277
21	6	12	.07498	.07786	.08080	.08381	.08691
21	6	13	.07225	.07440	.07660	.07885	.08115
21	6	14	.06928	.07080	.07237	.07396	.07560
21	6	15	.40490	.40886	.41325	.41803	.42316
21	7	1	-1.07543	-1.16402	-1.25903	-1.36086	-1.46992
21	7	2	-.14447	-.13995	-.13439	-.12768	-.11971
21	7	3	-.02000	-.00930	.00253	.01560	.02999
21	7	4	.03355	.04508	.05758	.07111	.08575
21	7	5	.06017	.07099	.08258	.09498	.10825
21	7	6	.07399	.08361	.09382	.10465	.11615
21	7	7	.08094	.08925	.09800	.10722	.11693
21	7	8	.08393	.09096	.09830	.10599	.11404
21	7	9	.08451	.09034	.09641	.10271	.10927
21	7	10	.08358	.08833	.09324	.09832	.10358
21	7	11	.08170	.08547	.08936	.09337	.09749
21	7	12	.07919	.08211	.08511	.08818	.09133
21	7	13	.07629	.07846	.08068	.08295	.08528
21	7	14	.50205	.50866	.51581	.52344	.53155
21	8	1	-1.14329	-1.23656	-1.33652	-1.44356	-1.55811
21	8	2	-.15350	-.14877	-.14296	-.13596	-.12767
21	8	3	-.02119	-.00998	.00241	.01606	.03107
21	8	4	.03573	.04780	.06086	.07499	.09025
21	8	5	.06401	.07532	.08741	.10034	.11417
21	8	6	.07869	.08872	.09936	.11064	.12260
21	8	7	.08607	.09471	.10381	.11338	.12347
21	8	8	.08923	.09652	.10414	.11211	.12045
21	8	9	.08983	.09586	.10213	.10865	.11543
21	8	10	.08883	.09372	.09878	.10400	.10942
21	8	11	.08681	.09068	.09466	.09876	.10299
21	8	12	.08412	.08709	.09014	.09327	.09649
21	8	13	.61468	.62489	.63578	.64731	.65944
21	9	1	-1.22329	-1.32209	-1.42788	-1.54108	-1.66210
21	9	2	-.16396	-.15899	-.15289	-.14557	-.13692
21	9	3	-.02242	-.01062	.00240	.01674	.03247
21	9	4	.03845	.05113	.06485	.07967	.09566

Coefficients b_i for the BLUE of σ
Shape parameter k

n	s	i	2.1	2.2	2.3	2.4	2.5
21	9	5	.06867	.08054	.09322	.10677	.12123
21	9	6	.08435	.09486	.10599	.11778	.13027
21	9	7	.09221	.10125	.11074	.12073	.13123
21	9	8	.09557	.10316	.11109	.11938	.12805
21	9	9	.09618	.10244	.10894	.11570	.12272
21	9	10	.09508	.10013	.10535	.11075	.11633
21	9	11	.09288	.09685	.10094	.10515	.10950
21	9	12	.74629	.76133	.77724	.79399	.81155
21	10	1	-1.31889	-1.42432	-1.53712	-1.65770	-1.78650
21	10	2	-.17621	-.17095	-.16453	-.15684	-.14776
21	10	3	-.02366	-.01115	.00262	.01775	.03434
21	10	4	.04190	.05533	.06982	.08545	.10230
21	10	5	.07444	.08697	.10034	.11460	.12982
21	10	6	.09128	.10235	.11406	.12645	.13956
21	10	7	.09970	.10919	.11915	.12962	.14063
21	10	8	.10327	.11122	.11951	.12817	.13722
21	10	9	.10388	.11040	.11717	.12420	.13151
21	10	10	.10264	.10787	.11327	.11886	.12465
21	10	11	.90166	.92311	.94571	.96943	.99424
21	11	1	-1.43508	-1.54861	-1.66998	-1.79959	-1.93789
21	11	2	-.19074	-.18514	-.17832	-.17019	-.16062
21	11	3	-.02483	-.01147	.00321	.01930	.03691
21	11	4	.04641	.06072	.07615	.09276	.11064
21	11	5	.08171	.09503	.10924	.12437	.14049
21	11	6	.09994	.11168	.12409	.13720	.15106
21	11	7	.10902	.11904	.12956	.14061	.15221
21	11	8	.11281	.12117	.12990	.13900	.14851
21	11	9	.11339	.12021	.12730	.13466	.14230
21	11	10	1.08739	1.11736	1.14887	1.18189	1.21640
22	0	1	-.78048	-.84831	-.92120	-.99949	-1.08353
22	0	2	-.11020	-.10749	-.10396	-.09955	-.09416
22	0	3	-.01932	-.01156	-.00290	.00675	.01745
22	0	4	.02025	.02889	.03832	.04859	.05976
22	0	5	.04022	.04851	.05744	.06703	.07735
22	0	6	.05083	.05835	.06636	.07490	.08399
22	0	7	.05640	.06302	.07002	.07741	.08523
22	0	8	.05902	.06474	.07074	.07704	.08365
22	0	9	.05984	.06471	.06978	.07506	.08057
22	0	10	.05953	.06361	.06783	.07220	.07673
22	0	11	.05848	.06185	.06530	.06887	.07253
22	0	12	.05695	.05968	.06246	.06531	.06823
22	0	13	.05512	.05727	.05946	.06168	.06395
22	0	14	.05308	.05473	.05639	.05808	.05979
22	0	15	.05092	.05212	.05333	.05456	.05579
22	0	16	.04867	.04949	.05031	.05114	.05198
22	0	17	.04638	.04686	.04735	.04785	.04836
22	0	18	.04403	.04423	.04445	.04468	.04492
22	0	19	.04164	.04161	.04160	.04162	.04165
22	0	20	.03917	.03895	.03877	.03863	.03851
22	0	21	.03650	.03615	.03586	.03561	.03541
22	0	22	.03296	.03259	.03229	.03204	.03184
22	1	1	-.80611	-.87615	-.95144	-1.03235	-1.11924
22	1	2	-.11395	-.11117	-.10756	-.10304	-.09753
22	1	3	-.02007	-.01209	-.00317	.00676	.01778
22	1	4	.02080	.02970	.03942	.05000	.06152
22	1	5	.04144	.04998	.05918	.06907	.07971

Coefficients b_i for the BLUE of σ
Shape parameter k

n	s	i	2.1	2.2	2.3	2.4	2.5
22	1	6	.05241	.06016	.06841	.07721	.08659
22	1	7	.05816	.06498	.07219	.07982	.08789
22	1	8	.06087	.06677	.07295	.07945	.08627
22	1	9	.06173	.06674	.07197	.07742	.08310
22	1	10	.06141	.06561	.06996	.07447	.07915
22	1	11	.06033	.06380	.06737	.07104	.07483
22	1	12	.05877	.06157	.06444	.06738	.07039
22	1	13	.05687	.05909	.06134	.06364	.06598
22	1	14	.05478	.05647	.05818	.05993	.06169
22	1	15	.05255	.05378	.05503	.05629	.05757
22	1	16	.05023	.05107	.05192	.05277	.05364
22	1	17	.04786	.04836	.04886	.04938	.04992
22	1	18	.04545	.04566	.04588	.04612	.04638
22	1	19	.04300	.04296	.04296	.04299	.04303
22	1	20	.04047	.04025	.04008	.03994	.03984
22	1	21	.07300	.07245	.07202	.07170	.07148
22	2	1	-.83450	-.90675	-.98444	-1.06794	-1.15762
22	2	2	-.11811	-.11526	-.11155	-.10691	-.10125
22	2	3	-.02093	-.01270	-.00352	.00671	.01805
22	2	4	.02140	.03057	.04057	.05147	.06332
22	2	5	.04277	.05157	.06103	.07122	.08218
22	2	6	.05414	.06211	.07061	.07967	.08932
22	2	7	.06010	.06712	.07454	.08238	.09069
22	2	8	.06292	.06898	.07534	.08201	.08903
22	2	9	.06381	.06896	.07433	.07993	.08578
22	2	10	.06349	.06780	.07227	.07690	.08171
22	2	11	.06238	.06593	.06959	.07336	.07725
22	2	12	.06076	.06363	.06657	.06958	.07268
22	2	13	.05881	.06107	.06338	.06573	.06813
22	2	14	.05664	.05837	.06012	.06190	.06371
22	2	15	.05433	.05559	.05686	.05815	.05946
22	2	16	.05194	.05279	.05365	.05453	.05541
22	2	17	.04950	.04999	.05050	.05103	.05157
22	2	18	.04701	.04721	.04743	.04768	.04794
22	2	19	.04448	.04444	.04444	.04446	.04452
22	2	20	.11907	.11859	.11828	.11812	.11810
22	3	1	-.86618	-.94079	-1.02101	-1.10722	-1.19980
22	3	2	-.12277	-.11982	-.11599	-.11122	-.10540
22	3	3	-.02189	-.01340	-.00393	.00661	.01829
22	3	4	.02206	.03151	.04182	.05305	.06526
22	3	5	.04426	.05332	.06307	.07356	.08484
22	3	6	.05606	.06427	.07302	.08234	.09228
22	3	7	.06226	.06948	.07711	.08518	.09372
22	3	8	.06519	.07142	.07796	.08482	.09204
22	3	9	.06613	.07142	.07693	.08268	.08869
22	3	10	.06580	.07022	.07480	.07956	.08449
22	3	11	.06465	.06829	.07204	.07590	.07989
22	3	12	.06298	.06591	.06892	.07200	.07517
22	3	13	.06096	.06326	.06562	.06802	.07048
22	3	14	.05871	.06047	.06225	.06406	.06591
22	3	15	.05633	.05759	.05888	.06019	.06152
22	3	16	.05385	.05470	.05556	.05644	.05734
22	3	17	.05131	.05180	.05231	.05284	.05338
22	3	18	.04874	.04893	.04914	.04939	.04965
22	3	19	.17155	.17141	.17150	.17178	.17224
22	4	1	-.90174	-.97891	-1.06187	-1.15100	-1.24670

Coefficients b_i for the BLUE of σ
Shape parameter k

n	s	i	2.1	2.2	2.3	2.4	2.5
22	4	2	-.12797	-.12492	-.12097	-.11605	-.11006
22	4	3	-.02295	-.01418	-.00440	.00647	.01851
22	4	4	.02281	.03257	.04321	.05478	.06737
22	4	5	.04593	.05528	.06533	.07614	.08776
22	4	6	.05823	.06669	.07570	.08530	.09554
22	4	7	.06469	.07213	.07998	.08829	.09707
22	4	8	.06775	.07416	.08088	.08794	.09535
22	4	9	.06873	.07417	.07983	.08574	.09190
22	4	10	.06840	.07293	.07763	.08250	.08756
22	4	11	.06721	.07093	.07477	.07872	.08281
22	4	12	.06547	.06847	.07153	.07468	.07792
22	4	13	.06337	.06572	.06811	.07056	.07306
22	4	14	.06104	.06281	.06462	.06646	.06834
22	4	15	.05856	.05983	.06113	.06245	.06380
22	4	16	.05598	.05682	.05769	.05857	.05947
22	4	17	.05334	.05382	.05432	.05484	.05538
22	4	18	.23113	.23169	.23252	.23360	.23489
22	5	1	-.94186	-1.02188	-1.10786	-1.20020	-1.29930
22	5	2	-.13382	-.13065	-.12657	-.12149	-.11531
22	5	3	-.02412	-.01504	-.00493	.00631	.01874
22	5	4	.02369	.03378	.04477	.05673	.06972
22	5	5	.04784	.05750	.06788	.07904	.09102
22	5	6	.06070	.06943	.07873	.08863	.09917
22	5	7	.06746	.07512	.08321	.09177	.10082
22	5	8	.07066	.07725	.08417	.09143	.09906
22	5	9	.07169	.07727	.08309	.08916	.09549
22	5	10	.07134	.07599	.08081	.08581	.09100
22	5	11	.07011	.07391	.07784	.08189	.08607
22	5	12	.06829	.07135	.07448	.07769	.08099
22	5	13	.06610	.06848	.07092	.07340	.07595
22	5	14	.06367	.06546	.06728	.06914	.07105
22	5	15	.06107	.06235	.06365	.06498	.06633
22	5	16	.05838	.05921	.06007	.06095	.06185
22	5	17	.29879	.30047	.30248	.30478	.30735
22	6	1	-.98742	-1.07063	-1.15999	-1.25592	-1.35880
22	6	2	-.14041	-.13712	-.13289	-.12763	-.12125
22	6	3	-.02540	-.01598	-.00550	.00613	.01899
22	6	4	.02472	.03518	.04657	.05894	.07237
22	6	5	.05005	.06005	.07079	.08232	.09471
22	6	6	.06354	.07256	.08217	.09240	.10328
22	6	7	.07062	.07853	.08689	.09571	.10504
22	6	8	.07399	.08078	.08791	.09539	.10324
22	6	9	.07507	.08081	.08679	.09303	.09954
22	6	10	.07470	.07948	.08442	.08955	.09487
22	6	11	.07341	.07731	.08132	.08546	.08974
22	6	12	.07151	.07462	.07781	.08109	.08446
22	6	13	.06921	.07162	.07409	.07662	.07921
22	6	14	.06665	.06845	.07029	.07218	.07410
22	6	15	.06393	.06520	.06650	.06783	.06919
22	6	16	.37582	.37912	.38283	.38691	.39131
22	7	1	-1.03950	-1.12635	-1.21956	-1.31954	-1.42671
22	7	2	-.14788	-.14445	-.14006	-.13461	-.12801
22	7	3	-.02680	-.01700	-.00611	.00596	.01930
22	7	4	.02596	.03684	.04866	.06149	.07541
22	7	5	.05263	.06301	.07415	.08610	.09892
22	7	6	.06683	.07618	.08613	.09672	.10798

Coefficients b_i for the BLUE of σ
Shape parameter k

n	s	i	2.1	2.2	2.3	2.4	2.5
22	7	7	.07429	.08247	.09111	.10024	.10987
22	7	8	.07783	.08484	.09220	.09992	.10802
22	7	9	.07896	.08488	.09104	.09747	.10417
22	7	10	.07858	.08348	.08856	.09383	.09929
22	7	11	.07721	.08120	.08531	.08955	.09393
22	7	12	.07521	.07837	.08163	.08497	.08841
22	7	13	.07278	.07522	.07773	.08029	.08292
22	7	14	.07008	.07189	.07374	.07563	.07757
22	7	15	.46383	.46940	.47546	.48198	.48893
22	8	1	-1.09956	-1.19058	-1.28822	-1.39286	-1.50494
22	8	2	-.15639	-.15281	-.14824	-.14259	-.13575
22	8	3	-.02832	-.01809	-.00673	.00583	.01970
22	8	4	.02748	.03882	.05113	.06448	.07895
22	8	5	.05568	.06649	.07808	.09050	.10382
22	8	6	.07069	.08042	.09075	.10174	.11342
22	8	7	.07857	.08707	.09603	.10548	.11546
22	8	8	.08231	.08957	.09719	.10517	.11354
22	8	9	.08350	.08961	.09597	.10260	.10952
22	8	10	.08308	.08813	.09336	.09878	.10440
22	8	11	.08163	.08572	.08993	.09428	.09877
22	8	12	.07949	.08272	.08604	.08945	.09296
22	8	13	.07691	.07939	.08192	.08452	.08719
22	8	14	.56492	.57355	.58279	.59261	.60297
22	9	1	-1.16949	-1.26538	-1.36817	-1.47824	-1.59604
22	9	2	-.16616	-.16241	-.15764	-.15176	-.14466
22	9	3	-.02997	-.01924	-.00735	.00578	.02026
22	9	4	.02936	.04124	.05411	.06806	.08315
22	9	5	.05934	.07064	.08274	.09570	.10957
22	9	6	.07528	.08543	.09621	.10765	.11980
22	9	7	.08365	.09249	.10181	.11164	.12201
22	9	8	.08760	.09515	.10305	.11133	.12001
22	9	9	.08885	.09517	.10176	.10862	.11577
22	9	10	.08839	.09359	.09898	.10457	.11037
22	9	11	.08682	.09101	.09534	.09980	.10441
22	9	12	.08452	.08782	.09120	.09469	.09827
22	9	13	.68181	.69449	.70796	.72216	.73708
22	10	1	-1.25186	-1.35351	-1.46237	-1.57886	-1.70341
22	10	2	-.17748	-.17354	-.16855	-.16240	-.15500
22	10	3	-.03172	-.02042	-.00792	.00587	.02105
22	10	4	.03175	.04424	.05777	.07240	.08822
22	10	5	.06379	.07566	.08836	.10194	.11646
22	10	6	.08082	.09146	.10275	.11472	.12742
22	10	7	.08974	.09899	.10873	.11899	.12980
22	10	8	.09393	.10180	.11004	.11866	.12769
22	10	9	.09524	.10181	.10865	.11577	.12319
22	10	10	.09471	.10009	.10567	.11144	.11744
22	10	11	.09299	.09731	.10176	.10635	.11109
22	10	12	.81809	.83610	.85513	.87512	.89605
22	11	1	-1.35023	-1.45879	-1.57496	-1.69914	-1.83179
22	11	2	-.19073	-.18656	-.18131	-.17486	-.16711
22	11	3	-.03357	-.02159	-.00836	.00621	.02221
22	11	4	.03482	.04805	.06235	.07780	.09447
22	11	5	.06931	.08186	.09526	.10957	.12486
22	11	6	.08761	.09883	.11072	.12331	.13666
22	11	7	.09717	.10690	.11713	.12790	.13923
22	11	8	.10164	.10989	.11851	.12753	.13697

Coefficients b_i for the BLUE of σ
Shape parameter k

n	s	i	2.1	2.2	2.3	2.4	2.5
22	11	9	.10299	.10985	.11698	.12440	.13213
22	11	10	.10236	.10795	.11374	.11973	.12595
22	11	11	.97863	1.00363	1.02995	1.05755	1.08642
23	0	1	-.76992	-.83753	-.91021	-.98832	-1.07222
23	0	2	-.11444	-.11247	-.10975	-.10620	-.10176
23	0	3	-.02428	-.01708	-.00899	.00004	.01008
23	0	4	.01542	.02369	.03272	.04259	.05335
23	0	5	.03577	.04382	.05250	.06186	.07194
23	0	6	.04681	.05419	.06208	.07051	.07950
23	0	7	.05280	.05938	.06634	.07372	.08154
23	0	8	.05583	.06157	.06761	.07396	.08065
23	0	9	.05703	.06198	.06714	.07253	.07817
23	0	10	.05707	.06127	.06562	.07015	.07485
23	0	11	.05635	.05986	.06349	.06723	.07109
23	0	12	.05512	.05802	.06099	.06404	.06717
23	0	13	.05357	.05591	.05830	.06073	.06322
23	0	14	.05179	.05364	.05551	.05741	.05935
23	0	15	.04987	.05128	.05270	.05414	.05559
23	0	16	.04786	.04888	.04991	.05095	.05199
23	0	17	.04579	.04647	.04716	.04786	.04856
23	0	18	.04368	.04407	.04447	.04488	.04530
23	0	19	.04153	.04167	.04183	.04201	.04219
23	0	20	.03934	.03928	.03924	.03923	.03923
23	0	21	.03707	.03684	.03666	.03650	.03638
23	0	22	.03461	.03427	.03399	.03375	.03355
23	0	23	.03134	.03098	.03069	.03045	.03026
23	1	1	-.79393	-.86362	-.93859	-1.01920	-1.10581
23	1	2	-.11812	-.11611	-.11334	-.10972	-.10518
23	1	3	-.02514	-.01774	-.00943	-.00015	.01018
23	1	4	.01581	.02430	.03360	.04375	.05482
23	1	5	.03679	.04507	.05400	.06364	.07402
23	1	6	.04818	.05578	.06390	.07257	.08183
23	1	7	.05436	.06113	.06830	.07590	.08395
23	1	8	.05749	.06340	.06962	.07616	.08305
23	1	9	.05874	.06383	.06914	.07470	.08051
23	1	10	.05878	.06310	.06759	.07224	.07709
23	1	11	.05804	.06166	.06539	.06925	.07323
23	1	12	.05679	.05977	.06282	.06596	.06919
23	1	13	.05519	.05759	.06005	.06256	.06513
23	1	14	.05336	.05526	.05718	.05914	.06114
23	1	15	.05138	.05283	.05429	.05577	.05728
23	1	16	.04931	.05036	.05142	.05249	.05357
23	1	17	.04718	.04788	.04859	.04931	.05004
23	1	18	.04501	.04541	.04582	.04625	.04669
23	1	19	.04280	.04295	.04311	.04330	.04350
23	1	20	.04055	.04049	.04046	.04045	.04047
23	1	21	.03823	.03801	.03783	.03769	.03758
23	1	22	.06918	.06865	.06824	.06792	.06770
23	2	1	-.82038	-.89217	-.96940	-1.05246	-1.14172
23	2	2	-.12220	-.12014	-.11729	-.11358	-.10893
23	2	3	-.02612	-.01850	-.00996	-.00042	.01021
23	2	4	.01620	.02495	.03451	.04495	.05633
23	2	5	.03790	.04641	.05560	.06550	.07618
23	2	6	.04968	.05749	.06583	.07475	.08427
23	2	7	.05607	.06303	.07039	.07820	.08648
23	2	8	.05931	.06538	.07177	.07849	.08557

Coefficients b_i for the BLUE of σ
Shape parameter k

n	s	i	2.1	2.2	2.3	2.4	2.5
23	2	9	.06061	.06583	.07129	.07699	.08296
23	2	10	.06066	.06509	.06969	.07447	.07945
23	2	11	.05990	.06361	.06743	.07139	.07548
23	2	12	.05861	.06166	.06479	.06801	.07132
23	2	13	.05696	.05942	.06193	.06451	.06714
23	2	14	.05508	.05701	.05898	.06099	.06303
23	2	15	.05304	.05451	.05600	.05752	.05906
23	2	16	.05090	.05197	.05304	.05413	.05524
23	2	17	.04871	.04941	.05013	.05086	.05161
23	2	18	.04647	.04687	.04728	.04771	.04816
23	2	19	.04419	.04433	.04450	.04469	.04489
23	2	20	.04188	.04181	.04178	.04178	.04180
23	2	21	.11252	.11203	.11170	.11152	.11147
23	3	1	-.84974	-.92374	-1.00336	-1.08897	-1.18096
23	3	2	-.12672	-.12460	-.12167	-.11786	-.11309
23	3	3	-.02720	-.01936	-.01056	-.00075	.01018
23	3	4	.01665	.02564	.03548	.04622	.05793
23	3	5	.03912	.04788	.05733	.06752	.07849
23	3	6	.05134	.05937	.06794	.07711	.08690
23	3	7	.05797	.06511	.07268	.08070	.08921
23	3	8	.06133	.06757	.07412	.08102	.08829
23	3	9	.06268	.06804	.07363	.07949	.08562
23	3	10	.06274	.06728	.07200	.07690	.08200
23	3	11	.06196	.06575	.06967	.07372	.07791
23	3	12	.06063	.06374	.06694	.07024	.07363
23	3	13	.05892	.06143	.06400	.06662	.06932
23	3	14	.05698	.05894	.06095	.06299	.06508
23	3	15	.05487	.05636	.05787	.05941	.06098
23	3	16	.05266	.05373	.05482	.05593	.05705
23	3	17	.05039	.05109	.05182	.05255	.05331
23	3	18	.04807	.04847	.04888	.04931	.04976
23	3	19	.04573	.04586	.04601	.04620	.04640
23	3	20	.16162	.16142	.16144	.16164	.16201
23	4	1	-.88250	-.95890	-1.04107	-1.12941	-1.22432
23	4	2	-.13176	-.12957	-.12655	-.12263	-.11773
23	4	3	-.02840	-.02031	-.01125	-.00113	.01011
23	4	4	.01715	.02642	.03655	.04761	.05965
23	4	5	.04050	.04952	.05925	.06973	.08102
23	4	6	.05319	.06146	.07028	.07971	.08977
23	4	7	.06009	.06743	.07521	.08346	.09220
23	4	8	.06359	.06999	.07673	.08381	.09127
23	4	9	.06500	.07049	.07623	.08224	.08853
23	4	10	.06507	.06972	.07455	.07957	.08480
23	4	11	.06426	.06814	.07215	.07629	.08058
23	4	12	.06288	.06606	.06933	.07269	.07616
23	4	13	.06111	.06367	.06628	.06896	.07170
23	4	14	.05910	.06109	.06313	.06521	.06733
23	4	15	.05691	.05842	.05995	.06151	.06309
23	4	16	.05462	.05569	.05679	.05790	.05903
23	4	17	.05226	.05296	.05368	.05442	.05517
23	4	18	.04986	.05024	.05064	.05107	.05151
23	4	19	.21707	.21747	.21813	.21902	.22012
23	5	1	-.91922	-.99825	-1.08323	-1.17455	-1.27262
23	5	2	-.13739	-.13512	-.13200	-.12796	-.12292
23	5	3	-.02972	-.02136	-.01201	-.00158	.01001
23	5	4	.01773	.02730	.03775	.04915	.06156

Coefficients b_i for the BLUE of σ
Shape parameter k

n	s	i	2.1	2.2	2.3	2.4	2.5
23	5	5	.04206	.05137	.06139	.07219	.08382
23	5	6	.05529	.06381	.07289	.08260	.09295
23	5	7	.06248	.07004	.07805	.08653	.09551
23	5	8	.06614	.07272	.07964	.08692	.09458
23	5	9	.06761	.07325	.07914	.08530	.09175
23	5	10	.06768	.07245	.07740	.08255	.08790
23	5	11	.06685	.07081	.07491	.07915	.08354
23	5	12	.06541	.06865	.07199	.07542	.07896
23	5	13	.06358	.06617	.06883	.07156	.07435
23	5	14	.06147	.06350	.06556	.06767	.06982
23	5	15	.05920	.06071	.06226	.06383	.06544
23	5	16	.05682	.05788	.05898	.06009	.06123
23	5	17	.05436	.05504	.05575	.05648	.05723
23	5	18	.27967	.28103	.28271	.28467	.28688
23	6	1	-.96062	-1.04258	-1.13068	-1.22530	-1.32685
23	6	2	-.14370	-.14134	-.13811	-.13395	-.12876
23	6	3	-.03118	-.02252	-.01285	-.00207	.00988
23	6	4	.01842	.02832	.03912	.05089	.06369
23	6	5	.04385	.05347	.06382	.07496	.08695
23	6	6	.05768	.06647	.07584	.08585	.09651
23	6	7	.06520	.07299	.08124	.08998	.09922
23	6	8	.06902	.07580	.08292	.09041	.09829
23	6	9	.07056	.07636	.08242	.08875	.09537
23	6	10	.07065	.07553	.08061	.08589	.09138
23	6	11	.06978	.07383	.07803	.08236	.08685
23	6	12	.06828	.07158	.07499	.07849	.08210
23	6	13	.06636	.06900	.07170	.07447	.07732
23	6	14	.06416	.06620	.06829	.07043	.07261
23	6	15	.06178	.06330	.06485	.06644	.06805
23	6	16	.05929	.06035	.06143	.06255	.06369
23	6	17	.35046	.35323	.35638	.35988	.36369
23	7	1	-1.00757	-1.09284	-1.18444	-1.28276	-1.38823
23	7	2	-.15079	-.14834	-.14501	-.14071	-.13535
23	7	3	-.03277	-.02379	-.01376	-.00261	.00976
23	7	4	.01924	.02951	.04070	.05287	.06611
23	7	5	.04593	.05588	.06659	.07811	.09050
23	7	6	.06043	.06952	.07921	.08954	.10055
23	7	7	.06832	.07637	.08488	.09389	.10343
23	7	8	.07233	.07932	.08665	.09437	.10249
23	7	9	.07395	.07991	.08614	.09265	.09946
23	7	10	.07403	.07905	.08426	.08968	.09532
23	7	11	.07312	.07727	.08157	.08601	.09060
23	7	12	.07154	.07492	.07839	.08197	.08565
23	7	13	.06953	.07221	.07495	.07777	.08066
23	7	14	.06722	.06928	.07139	.07355	.07576
23	7	15	.06472	.06624	.06779	.06938	.07101
23	7	16	.43077	.43550	.44068	.44629	.45228
23	8	1	-1.06120	-1.15025	-1.24583	-1.34836	-1.45827
23	8	2	-.15882	-.15627	-.15281	-.14837	-.14283
23	8	3	-.03453	-.02518	-.01476	-.00318	.00965
23	8	4	.02026	.03093	.04255	.05518	.06890
23	8	5	.04836	.05869	.06981	.08175	.09457
23	8	6	.06363	.07306	.08310	.09379	.10518
23	8	7	.07194	.08027	.08908	.09839	.10824
23	8	8	.07615	.08337	.09095	.09892	.10729
23	8	9	.07785	.08400	.09042	.09713	.10414

Coefficients b_i for the BLUE of σ
Shape parameter k

n	s	i	2.1	2.2	2.3	2.4	2.5
23	8	10	.07793	.08310	.08845	.09402	.09981
23	8	11	.07697	.08122	.08562	.09018	.09489
23	8	12	.07530	.07874	.08229	.08594	.08971
23	8	13	.07317	.07589	.07868	.08154	.08448
23	8	14	.07073	.07280	.07493	.07711	.07934
23	8	15	.52228	.52962	.53753	.54596	.55489
23	9	1	-1.12300	-1.21639	-1.31656	-1.42394	-1.53894
23	9	2	-.16797	-.16530	-.16171	-.15711	-.15138
23	9	3	-.03644	-.02668	-.01581	-.00376	.00958
23	9	4	.02152	.03265	.04476	.05791	.07217
23	9	5	.05124	.06201	.07357	.08599	.09931
23	9	6	.06739	.07720	.08763	.09873	.11055
23	9	7	.07617	.08482	.09396	.10362	.11382
23	9	8	.08062	.08810	.09595	.10419	.11285
23	9	9	.08240	.08876	.09539	.10232	.10955
23	9	10	.08248	.08780	.09332	.09905	.10501
23	9	11	.08144	.08581	.09033	.09500	.09983
23	9	12	.07966	.08318	.08680	.09053	.09438
23	9	13	.07739	.08015	.08298	.08589	.08888
23	9	14	.62712	.63791	.64941	.66158	.67437
23	10	1	-1.19491	-1.29336	-1.39888	-1.51190	-1.63284
23	10	2	-.17846	-.17568	-.17194	-.16715	-.16122
23	10	3	-.03853	-.02829	-.01692	-.00431	.00961
23	10	4	.02311	.03478	.04744	.06118	.07607
23	10	5	.05471	.06598	.07806	.09102	.10491
23	10	6	.07187	.08211	.09299	.10457	.11687
23	10	7	.08118	.09020	.09971	.10976	.12037
23	10	8	.08590	.09367	.10182	.11038	.11936
23	10	9	.08777	.09436	.10123	.10840	.11589
23	10	10	.08783	.09332	.09902	.10494	.11109
23	10	11	.08669	.09119	.09584	.10064	.10561
23	10	12	.08478	.08838	.09208	.09590	.09984
23	10	13	.74807	.76336	.77954	.79658	.81444
23	11	1	-1.27955	-1.38399	-1.49584	-1.61552	-1.74348
23	11	2	-.19061	-.18769	-.18378	-.17880	-.17263
23	11	3	-.04081	-.03001	-.01804	-.00480	.00980
23	11	4	.02516	.03744	.05076	.06518	.08079
23	11	5	.05895	.07079	.08348	.09707	.11162
23	11	6	.07728	.08802	.09943	.11155	.12442
23	11	7	.08721	.09665	.10660	.11710	.12818
23	11	8	.09222	.10033	.10884	.11776	.12712
23	11	9	.09418	.10104	.10819	.11564	.12342
23	11	10	.09420	.09990	.10580	.11193	.11830
23	11	11	.09295	.09759	.10238	.10733	.11246
23	11	12	.88881	.90992	.93218	.95555	.98001
23	12	1	-1.38059	-1.49220	-1.61164	-1.73933	-1.87572
23	12	2	-.20483	-.20175	-.19765	-.19243	-.18599
23	12	3	-.04326	-.03180	-.01913	-.00514	.01027
23	12	4	.02784	.04086	.05494	.07018	.08665
23	12	5	.06422	.07675	.09015	.10449	.11981
23	12	6	.08393	.09527	.10729	.12005	.13359
23	12	7	.09457	.10451	.11498	.12601	.13763
23	12	8	.09991	.10843	.11735	.12669	.13649
23	12	9	.10197	.10914	.11661	.12439	.13251
23	12	10	.10193	.10785	.11400	.12038	.12700
23	12	11	1.05430	1.08295	1.11310	1.14471	1.17776

Coefficients b_i for the BLUE of σ
Shape parameter k

n	s	i	2.1	2.2	2.3	2.4	2.5
24	0	1	-.75978	-.82714	-.89959	-.97750	-1.06122
24	0	2	-.11823	-.11696	-.11498	-.11225	-.10868
24	0	3	-.02882	-.02214	-.01461	-.00617	.00326
24	0	4	.01097	.01886	.02752	.03699	.04734
24	0	5	.03162	.03943	.04787	.05698	.06682
24	0	6	.04303	.05028	.05803	.06633	.07521
24	0	7	.04940	.05592	.06284	.07018	.07798
24	0	8	.05280	.05855	.06461	.07099	.07773
24	0	9	.05435	.05935	.06458	.07006	.07580
24	0	10	.05470	.05900	.06347	.06811	.07296
24	0	11	.05428	.05792	.06169	.06558	.06962
24	0	12	.05333	.05638	.05951	.06273	.06605
24	0	13	.05203	.05454	.05710	.05973	.06242
24	0	14	.05049	.05251	.05457	.05667	.05881
24	0	15	.04879	.05038	.05199	.05363	.05529
24	0	16	.04699	.04820	.04941	.05065	.05189
24	0	17	.04512	.04599	.04686	.04774	.04864
24	0	18	.04320	.04377	.04435	.04494	.04553
24	0	19	.04126	.04157	.04190	.04223	.04258
24	0	20	.03928	.03938	.03949	.03962	.03976
24	0	21	.03726	.03718	.03712	.03708	.03707
24	0	22	.03517	.03494	.03475	.03459	.03446
24	0	23	.03290	.03257	.03229	.03206	.03186
24	0	24	.02986	.02952	.02924	.02902	.02883
24	1	1	-.78233	-.85167	-.92630	-1.00659	-1.09290
24	1	2	-.12184	-.12055	-.11855	-.11577	-.11214
24	1	3	-.02977	-.02292	-.01519	-.00653	.00316
24	1	4	.01120	.01931	.02820	.03793	.04857
24	1	5	.03247	.04049	.04917	.05854	.06866
24	1	6	.04423	.05168	.05965	.06818	.07731
24	1	7	.05080	.05749	.06460	.07216	.08018
24	1	8	.05430	.06021	.06643	.07300	.07993
24	1	9	.05590	.06103	.06641	.07205	.07796
24	1	10	.05627	.06068	.06527	.07005	.07504
24	1	11	.05583	.05958	.06345	.06745	.07161
24	1	12	.05486	.05799	.06121	.06453	.06795
24	1	13	.05352	.05610	.05874	.06144	.06421
24	1	14	.05194	.05402	.05614	.05830	.06050
24	1	15	.05020	.05183	.05349	.05517	.05688
24	1	16	.04834	.04958	.05084	.05210	.05339
24	1	17	.04642	.04731	.04821	.04912	.05005
24	1	18	.04446	.04504	.04564	.04624	.04686
24	1	19	.04246	.04278	.04311	.04346	.04382
24	1	20	.04043	.04053	.04065	.04078	.04094
24	1	21	.03836	.03827	.03822	.03819	.03819
24	1	22	.03623	.03600	.03581	.03567	.03555
24	1	23	.06574	.06522	.06482	.06451	.06429
24	2	1	-.80706	-.87838	-.95516	-1.03777	-1.12660
24	2	2	-.12582	-.12450	-.12245	-.11961	-.11591
24	2	3	-.03084	-.02379	-.01586	-.00696	.00297
24	2	4	.01144	.01977	.02890	.03890	.04982
24	2	5	.03339	.04163	.05053	.06016	.07055
24	2	6	.04553	.05317	.06136	.07012	.07949
24	2	7	.05231	.05918	.06648	.07423	.08247
24	2	8	.05593	.06199	.06838	.07512	.08223
24	2	9	.05758	.06285	.06837	.07415	.08022

Coefficients b_i for the BLUE of σ
Shape parameter k

n	s	i	2.1	2.2	2.3	2.4	2.5
24	2	10	.05797	.06249	.06720	.07210	.07722
24	2	11	.05753	.06136	.06533	.06944	.07370
24	2	12	.05653	.05973	.06303	.06642	.06993
24	2	13	.05515	.05779	.06048	.06325	.06609
24	2	14	.05352	.05565	.05781	.06002	.06228
24	2	15	.05173	.05339	.05508	.05680	.05856
24	2	16	.04982	.05108	.05236	.05365	.05497
24	2	17	.04784	.04874	.04966	.05059	.05153
24	2	18	.04582	.04641	.04701	.04763	.04825
24	2	19	.04376	.04408	.04442	.04477	.04514
24	2	20	.04167	.04177	.04189	.04203	.04218
24	2	21	.03955	.03946	.03941	.03938	.03938
24	2	22	.10663	.10613	.10580	.10560	.10553
24	3	1	-.83437	-.90778	-.98681	-1.07183	-1.16324
24	3	2	-.13022	-.12886	-.12676	-.12385	-.12005
24	3	3	-.03202	-.02477	-.01662	-.00748	.00273
24	3	4	.01170	.02026	.02964	.03991	.05113
24	3	5	.03440	.04287	.05201	.06190	.07257
24	3	6	.04696	.05481	.06321	.07220	.08183
24	3	7	.05398	.06103	.06852	.07647	.08493
24	3	8	.05772	.06394	.07049	.07741	.08470
24	3	9	.05944	.06484	.07050	.07642	.08264
24	3	10	.05985	.06448	.06930	.07432	.07956
24	3	11	.05940	.06332	.06737	.07158	.07594
24	3	12	.05837	.06164	.06501	.06848	.07207
24	3	13	.05695	.05963	.06239	.06521	.06811
24	3	14	.05527	.05743	.05963	.06189	.06419
24	3	15	.05342	.05511	.05683	.05858	.06036
24	3	16	.05145	.05272	.05402	.05533	.05667
24	3	17	.04941	.05031	.05124	.05217	.05313
24	3	18	.04732	.04790	.04851	.04913	.04976
24	3	19	.04519	.04551	.04584	.04619	.04656
24	3	20	.04304	.04313	.04324	.04337	.04353
24	3	21	.15275	.15249	.15245	.15258	.15288
24	4	1	-.86468	-.94034	-1.02176	-1.10935	-1.20349
24	4	2	-.13510	-.13369	-.13153	-.12854	-.12464
24	4	3	-.03332	-.02586	-.01747	-.00807	.00242
24	4	4	.01199	.02080	.03045	.04101	.05254
24	4	5	.03553	.04423	.05364	.06379	.07475
24	4	6	.04855	.05662	.06525	.07449	.08437
24	4	7	.05583	.06307	.07076	.07893	.08760
24	4	8	.05972	.06610	.07282	.07991	.08740
24	4	9	.06151	.06704	.07284	.07891	.08529
24	4	10	.06194	.06668	.07161	.07676	.08212
24	4	11	.06147	.06548	.06963	.07393	.07839
24	4	12	.06041	.06375	.06719	.07074	.07440
24	4	13	.05894	.06168	.06448	.06736	.07032
24	4	14	.05721	.05940	.06164	.06393	.06628
24	4	15	.05529	.05700	.05874	.06052	.06233
24	4	16	.05325	.05454	.05584	.05717	.05852
24	4	17	.05114	.05205	.05297	.05391	.05487
24	4	18	.04898	.04955	.05015	.05077	.05140
24	4	19	.04678	.04708	.04740	.04775	.04811
24	4	20	.20456	.20482	.20534	.20608	.20702
24	5	1	-.89846	-.97657	-1.06061	-1.15097	-1.24807
24	5	2	-.14051	-.13905	-.13683	-.13375	-.12976

Coefficients b_i for the BLUE of σ
Shape parameter k

n	s	i	2.1	2.2	2.3	2.4	2.5
24	5	3	-.03475	-.02706	-.01841	-.00874	.00205
24	5	4	.01234	.02142	.03135	.04222	.05407
24	5	5	.03680	.04576	.05544	.06589	.07716
24	5	6	.05034	.05864	.06752	.07701	.08716
24	5	7	.05791	.06536	.07326	.08164	.09055
24	5	8	.06196	.06851	.07541	.08269	.09037
24	5	9	.06382	.06949	.07544	.08167	.08820
24	5	10	.06427	.06912	.07418	.07945	.08494
24	5	11	.06379	.06789	.07213	.07653	.08109
24	5	12	.06269	.06610	.06961	.07323	.07697
24	5	13	.06117	.06395	.06681	.06974	.07276
24	5	14	.05937	.06159	.06387	.06619	.06858
24	5	15	.05738	.05911	.06087	.06266	.06450
24	5	16	.05526	.05655	.05786	.05920	.06056
24	5	17	.05307	.05397	.05489	.05583	.05680
24	5	18	.05082	.05138	.05197	.05258	.05321
24	5	19	.26275	.26385	.26525	.26693	.26885
24	6	1	-.93630	-1.01712	-1.10404	-1.19745	-1.29778
24	6	2	-.14655	-.14503	-.14274	-.13957	-.13547
24	6	3	-.03633	-.02838	-.01946	-.00948	.00164
24	6	4	.01275	.02212	.03237	.04357	.05578
24	6	5	.03825	.04749	.05747	.06823	.07983
24	6	6	.05236	.06091	.07006	.07983	.09027
24	6	7	.06026	.06792	.07605	.08467	.09383
24	6	8	.06448	.07121	.07831	.08578	.09367
24	6	9	.06642	.07225	.07835	.08474	.09144
24	6	10	.06690	.07187	.07705	.08245	.08808
24	6	11	.06640	.07059	.07493	.07943	.08410
24	6	12	.06525	.06873	.07231	.07601	.07983
24	6	13	.06367	.06650	.06941	.07240	.07547
24	6	14	.06180	.06405	.06636	.06872	.07114
24	6	15	.05972	.06146	.06324	.06505	.06691
24	6	16	.05752	.05880	.06012	.06146	.06283
24	6	17	.05522	.05611	.05703	.05797	.05893
24	6	18	.32817	.33050	.33319	.33621	.33952
24	7	1	-.97890	-1.06275	-1.15289	-1.24970	-1.35362
24	7	2	-.15330	-.15173	-.14935	-.14610	-.14187
24	7	3	-.03807	-.02983	-.02061	-.01030	.00117
24	7	4	.01325	.02295	.03354	.04510	.05770
24	7	5	.03992	.04947	.05977	.07087	.08283
24	7	6	.05468	.06350	.07293	.08301	.09377
24	7	7	.06293	.07083	.07921	.08809	.09751
24	7	8	.06735	.07428	.08158	.08927	.09738
24	7	9	.06938	.07537	.08164	.08820	.09509
24	7	10	.06987	.07498	.08029	.08583	.09160
24	7	11	.06936	.07364	.07809	.08269	.08747
24	7	12	.06816	.07171	.07536	.07914	.08304
24	7	13	.06650	.06938	.07234	.07538	.07851
24	7	14	.06454	.06682	.06915	.07155	.07400
24	7	15	.06237	.06412	.06590	.06773	.06961
24	7	16	.06005	.06134	.06265	.06399	.06537
24	7	17	.40192	.40594	.41040	.41525	.42047
24	8	1	-1.02719	-1.11446	-1.20822	-1.30887	-1.41682
24	8	2	-.16088	-.15925	-.15680	-.15345	-.14910
24	8	3	-.03997	-.03143	-.02186	-.01119	.00067
24	8	4	.01387	.02393	.03490	.04687	.05989

Coefficients b_i for the BLUE of σ
Shape parameter k

n	s	i	2.1	2.2	2.3	2.4	2.5
24	8	5	.04185	.05175	.06241	.07389	.08625
24	8	6	.05734	.06647	.07622	.08663	.09773
24	8	7	.06600	.07416	.08281	.09198	.10169
24	8	8	.07063	.07778	.08531	.09323	.10158
24	8	9	.07276	.07893	.08538	.09214	.09921
24	8	10	.07328	.07852	.08398	.08966	.09559
24	8	11	.07273	.07713	.08168	.08639	.09129
24	8	12	.07147	.07509	.07883	.08269	.08667
24	8	13	.06973	.07266	.07566	.07876	.08194
24	8	14	.06766	.06997	.07233	.07476	.07724
24	8	15	.06537	.06713	.06893	.07077	.07266
24	8	16	.48535	.49164	.49845	.50575	.51350
24	9	1	-1.08231	-1.17350	-1.27138	-1.37639	-1.48893
24	9	2	-.16946	-.16777	-.16523	-.16177	-.15729
24	9	3	-.04207	-.03317	-.02323	-.01215	.00015
24	9	4	.01466	.02512	.03652	.04893	.06243
24	9	5	.04413	.05441	.06547	.07738	.09018
24	9	6	.06044	.06991	.08002	.09080	.10229
24	9	7	.06956	.07801	.08697	.09644	.10649
24	9	8	.07443	.08183	.08960	.09779	.10640
24	9	9	.07667	.08303	.08969	.09665	.10394
24	9	10	.07720	.08260	.08822	.09406	.10015
24	9	11	.07662	.08113	.08580	.09064	.09565
24	9	12	.07528	.07898	.08280	.08675	.09082
24	9	13	.07343	.07641	.07947	.08262	.08587
24	9	14	.07124	.07357	.07597	.07842	.08094
24	9	15	.58018	.58943	.59933	.60984	.62091
24	10	1	-1.14578	-1.24147	-1.34412	-1.45414	-1.57197
24	10	2	-.17922	-.17746	-.17484	-.17126	-.16664
24	10	3	-.04437	-.03508	-.02471	-.01317	-.00038
24	10	4	.01566	.02658	.03846	.05139	.06543
24	10	5	.04685	.05756	.06908	.08146	.09476
24	10	6	.06409	.07395	.08446	.09566	.10759
24	10	7	.07373	.08252	.09181	.10164	.11205
24	10	8	.07887	.08654	.09460	.10307	.11199
24	10	9	.08122	.08781	.09469	.10188	.10941
24	10	10	.08178	.08734	.09313	.09916	.10543
24	10	11	.08114	.08577	.09057	.09555	.10070
24	10	12	.07970	.08350	.08741	.09144	.09561
24	10	13	.07773	.08076	.08388	.08709	.09039
24	10	14	.68860	.70168	.71558	.73023	.74562
24	11	1	-1.21959	-1.32053	-1.42873	-1.54460	-1.66860
24	11	2	-.19042	-.18858	-.18586	-.18216	-.17738
24	11	3	-.04690	-.03715	-.02629	-.01423	-.00087
24	11	4	.01697	.02840	.04084	.05436	.06902
24	11	5	.05013	.06134	.07338	.08631	.10018
24	11	6	.06845	.07876	.08972	.10140	.11383
24	11	7	.07868	.08784	.09753	.10776	.11859
24	11	8	.08413	.09210	.10048	.10929	.11855
24	11	9	.08660	.09343	.10057	.10803	.11583
24	11	10	.08716	.09292	.09891	.10514	.11162
24	11	11	.08645	.09123	.09618	.10130	.10661
24	11	12	.08490	.08879	.09279	.09693	.10121
24	11	13	.81344	.83145	.85047	.87047	.89142
24	12	1	-1.30643	-1.41357	-1.52833	-1.65112	-1.78240
24	12	2	-.20337	-.20146	-.19863	-.19479	-.18984

Coefficients b_i for the BLUE of σ
Shape parameter k

n	s	i	2.1	2.2	2.3	2.4	2.5
24	12	3	-.04968	-.03940	-.02797	-.01529	-.00128
24	12	4	.01868	.03073	.04382	.05802	.07339
24	12	5	.05415	.06594	.07860	.09216	.10671
24	12	6	.07373	.08455	.09605	.10829	.12130
24	12	7	.08464	.09424	.10437	.11508	.12639
24	12	8	.09043	.09876	.10751	.11670	.12636
24	12	9	.09303	.10015	.10758	.11534	.12346
24	12	10	.09359	.09957	.10578	.11224	.11896
24	12	11	.09279	.09772	.10283	.10812	.11361
24	12	12	.95846	.98277	1.00837	1.03524	1.06334
25	0	1	-.75003	-.81712	-.88932	-.96700	-1.05051
25	0	2	-.12164	-.12101	-.11973	-.11776	-.11502
25	0	3	-.03298	-.02680	-.01980	-.01192	-.00309
25	0	4	.00684	.01438	.02266	.03175	.04169
25	0	5	.02775	.03532	.04351	.05238	.06198
25	0	6	.03949	.04659	.05420	.06236	.07111
25	0	7	.04620	.05265	.05950	.06679	.07455
25	0	8	.04993	.05566	.06172	.06812	.07488
25	0	9	.05178	.05682	.06210	.06765	.07347
25	0	10	.05243	.05680	.06136	.06611	.07107
25	0	11	.05227	.05602	.05991	.06393	.06812
25	0	12	.05157	.05475	.05802	.06140	.06489
25	0	13	.05050	.05316	.05588	.05867	.06154
25	0	14	.04918	.05136	.05359	.05586	.05819
25	0	15	.04768	.04944	.05123	.05305	.05489
25	0	16	.04607	.04745	.04885	.05026	.05169
25	0	17	.04439	.04543	.04648	.04754	.04861
25	0	18	.04265	.04339	.04414	.04489	.04565
25	0	19	.04087	.04135	.04184	.04233	.04283
25	0	20	.03908	.03933	.03959	.03986	.04015
25	0	21	.03725	.03731	.03738	.03747	.03758
25	0	22	.03539	.03528	.03521	.03515	.03512
25	0	23	.03345	.03322	.03302	.03286	.03273
25	0	24	.03134	.03103	.03076	.03053	.03034
25	0	25	.02851	.02818	.02792	.02770	.02753
25	1	1	-.77127	-.84025	-.91453	-.99447	-1.08046
25	1	2	-.12518	-.12455	-.12327	-.12127	-.11850
25	1	3	-.03401	-.02767	-.02050	-.01242	-.00336
25	1	4	.00695	.01468	.02318	.03251	.04272
25	1	5	.02846	.03622	.04463	.05374	.06360
25	1	6	.04054	.04782	.05564	.06402	.07300
25	1	7	.04744	.05406	.06109	.06858	.07655
25	1	8	.05128	.05717	.06339	.06996	.07691
25	1	9	.05319	.05836	.06378	.06948	.07547
25	1	10	.05386	.05834	.06302	.06790	.07301
25	1	11	.05370	.05755	.06154	.06568	.06998
25	1	12	.05299	.05624	.05961	.06308	.06666
25	1	13	.05189	.05461	.05741	.06028	.06323
25	1	14	.05053	.05277	.05506	.05740	.05979
25	1	15	.04900	.05080	.05263	.05450	.05640
25	1	16	.04734	.04876	.05019	.05164	.05312
25	1	17	.04561	.04668	.04775	.04884	.04995
25	1	18	.04382	.04458	.04535	.04613	.04692
25	1	19	.04200	.04249	.04300	.04351	.04403
25	1	20	.04016	.04042	.04069	.04097	.04127
25	1	21	.03829	.03835	.03843	.03853	.03864

Coefficients b_i for the BLUE of σ
Shape parameter k

n	s	i	2.1	2.2	2.3	2.4	2.5
25	1	22	.03638	.03628	.03620	.03616	.03613
25	1	23	.03441	.03418	.03399	.03384	.03372
25	1	24	.06261	.06211	.06171	.06142	.06120
25	2	1	-.79445	-.86532	-.94164	-1.02380	-1.11218
25	2	2	-.12906	-.12842	-.12712	-.12508	-.12225
25	2	3	-.03514	-.02864	-.02129	-.01300	-.00371
25	2	4	.00705	.01498	.02370	.03327	.04374
25	2	5	.02922	.03718	.04581	.05515	.06526
25	2	6	.04167	.04914	.05715	.06574	.07496
25	2	7	.04878	.05556	.06278	.07046	.07863
25	2	8	.05274	.05877	.06515	.07189	.07902
25	2	9	.05471	.06001	.06557	.07141	.07755
25	2	10	.05541	.06000	.06479	.06979	.07502
25	2	11	.05525	.05919	.06327	.06751	.07192
25	2	12	.05452	.05785	.06129	.06484	.06852
25	2	13	.05339	.05617	.05903	.06197	.06499
25	2	14	.05200	.05428	.05662	.05901	.06146
25	2	15	.05042	.05226	.05413	.05604	.05798
25	2	16	.04872	.05016	.05162	.05310	.05461
25	2	17	.04694	.04802	.04911	.05023	.05136
25	2	18	.04510	.04587	.04665	.04744	.04824
25	2	19	.04323	.04372	.04423	.04475	.04528
25	2	20	.04133	.04159	.04186	.04215	.04245
25	2	21	.03941	.03947	.03955	.03965	.03977
25	2	22	.03746	.03735	.03728	.03723	.03721
25	2	23	.10131	.10081	.10046	.10026	.10016
25	3	1	-.81995	-.89279	-.97123	-1.05567	-1.14650
25	3	2	-.13333	-.13267	-.13134	-.12926	-.12636
25	3	3	-.03640	-.02972	-.02217	-.01367	-.00414
25	3	4	.00716	.01530	.02425	.03407	.04481
25	3	5	.03004	.03821	.04707	.05665	.06702
25	3	6	.04290	.05056	.05878	.06760	.07705
25	3	7	.05025	.05721	.06460	.07247	.08085
25	3	8	.05434	.06053	.06706	.07397	.08127
25	3	9	.05639	.06181	.06750	.07348	.07977
25	3	10	.05711	.06181	.06671	.07183	.07719
25	3	11	.05695	.06097	.06515	.06949	.07400
25	3	12	.05620	.05960	.06311	.06675	.07050
25	3	13	.05504	.05788	.06079	.06379	.06688
25	3	14	.05360	.05593	.05831	.06075	.06325
25	3	15	.05198	.05385	.05575	.05769	.05968
25	3	16	.05023	.05168	.05316	.05467	.05621
25	3	17	.04839	.04948	.05059	.05172	.05287
25	3	18	.04650	.04727	.04805	.04885	.04967
25	3	19	.04457	.04506	.04557	.04609	.04662
25	3	20	.04262	.04287	.04313	.04342	.04372
25	3	21	.04064	.04069	.04076	.04086	.04098
25	3	22	.14476	.14447	.14437	.14445	.14469
25	4	1	-.84811	-.92305	-1.00376	-1.09061	-1.18402
25	4	2	-.13804	-.13736	-.13599	-.13386	-.13090
25	4	3	-.03777	-.03091	-.02315	-.01442	-.00464
25	4	4	.00728	.01565	.02484	.03492	.04595
25	4	5	.03096	.03935	.04844	.05828	.06891
25	4	6	.04427	.05213	.06057	.06961	.07930
25	4	7	.05188	.05901	.06660	.07467	.08326
25	4	8	.05611	.06245	.06915	.07623	.08371

Coefficients b_i for the BLUE of σ
Shape parameter k

n	s	i	2.1	2.2	2.3	2.4	2.5
25	4	9	.05824	.06379	.06962	.07574	.08218
25	4	10	.05899	.06379	.06881	.07405	.07953
25	4	11	.05883	.06294	.06721	.07164	.07625
25	4	12	.05805	.06153	.06511	.06882	.07266
25	4	13	.05686	.05975	.06272	.06578	.06893
25	4	14	.05538	.05774	.06017	.06265	.06519
25	4	15	.05370	.05559	.05753	.05950	.06151
25	4	16	.05189	.05336	.05486	.05639	.05794
25	4	17	.04999	.05109	.05221	.05335	.05450
25	4	18	.04804	.04881	.04959	.05040	.05121
25	4	19	.04605	.04653	.04703	.04755	.04808
25	4	20	.04403	.04427	.04453	.04481	.04511
25	4	21	.19336	.19352	.19391	.19452	.19532
25	5	1	-.87932	-.95656	-1.03971	-1.12916	-1.22534
25	5	2	-.14325	-.14254	-.14113	-.13895	-.13591
25	5	3	-.03929	-.03222	-.02424	-.01526	-.00522
25	5	4	.00743	.01604	.02550	.03585	.04718
25	5	5	.03199	.04062	.04996	.06006	.07098
25	5	6	.04579	.05387	.06254	.07182	.08176
25	5	7	.05369	.06101	.06880	.07708	.08589
25	5	8	.05809	.06459	.07146	.07871	.08638
25	5	9	.06029	.06598	.07196	.07823	.08482
25	5	10	.06108	.06600	.07113	.07649	.08210
25	5	11	.06092	.06512	.06948	.07401	.07873
25	5	12	.06012	.06366	.06732	.07111	.07502
25	5	13	.05888	.06183	.06485	.06797	.07118
25	5	14	.05735	.05975	.06221	.06474	.06732
25	5	15	.05561	.05753	.05948	.06148	.06353
25	5	16	.05374	.05522	.05673	.05827	.05985
25	5	17	.05177	.05287	.05399	.05513	.05630
25	5	18	.04975	.05051	.05129	.05209	.05291
25	5	19	.04768	.04815	.04864	.04915	.04968
25	5	20	.24768	.24857	.24974	.25118	.25284
25	6	1	-.91408	-.99383	-1.07966	-1.17196	-1.27114
25	6	2	-.14903	-.14829	-.14684	-.14460	-.14149
25	6	3	-.04095	-.03366	-.02543	-.01620	-.00587
25	6	4	.00762	.01650	.02623	.03688	.04853
25	6	5	.03316	.04204	.05165	.06204	.07327
25	6	6	.04751	.05582	.06473	.07427	.08449
25	6	7	.05573	.06325	.07125	.07976	.08880
25	6	8	.06030	.06698	.07402	.08147	.08934
25	6	9	.06260	.06843	.07456	.08099	.08774
25	6	10	.06342	.06845	.07371	.07920	.08494
25	6	11	.06325	.06755	.07201	.07664	.08146
25	6	12	.06242	.06604	.06977	.07364	.07763
25	6	13	.06114	.06414	.06722	.07039	.07367
25	6	14	.05955	.06199	.06448	.06705	.06968
25	6	15	.05774	.05968	.06166	.06368	.06576
25	6	16	.05579	.05728	.05881	.06036	.06195
25	6	17	.05375	.05485	.05596	.05711	.05828
25	6	18	.05164	.05239	.05316	.05396	.05477
25	6	19	.30843	.31039	.31270	.31531	.31820
25	7	1	-.95297	-1.03552	-1.12432	-1.21975	-1.32225
25	7	2	-.15545	-.15469	-.15319	-.15089	-.14770
25	7	3	-.04278	-.03525	-.02675	-.01723	-.00659
25	7	4	.00787	.01703	.02707	.03805	.05004

Coefficients b_i for the BLUE of σ
Shape parameter k

n	s	i	2.1	2.2	2.3	2.4	2.5
25	7	5	.03449	.04366	.05356	.06426	.07581
25	7	6	.04946	.05803	.06720	.07702	.08752
25	7	7	.05803	.06577	.07400	.08275	.09204
25	7	8	.06280	.06966	.07691	.08455	.09263
25	7	9	.06520	.07119	.07747	.08407	.09100
25	7	10	.06605	.07122	.07660	.08223	.08810
25	7	11	.06588	.07028	.07484	.07958	.08451
25	7	12	.06502	.06871	.07252	.07646	.08055
25	7	13	.06368	.06673	.06987	.07310	.07643
25	7	14	.06202	.06449	.06703	.06963	.07230
25	7	15	.06013	.06209	.06409	.06614	.06823
25	7	16	.05810	.05959	.06112	.06269	.06428
25	7	17	.05597	.05705	.05817	.05931	.06048
25	7	18	.37652	.37997	.38382	.38804	.39261
25	8	1	-.99674	-1.08242	-1.17453	-1.27347	-1.37967
25	8	2	-.16264	-.16184	-.16030	-.15794	-.15467
25	8	3	-.04479	-.03699	-.02820	-.01837	-.00739
25	8	4	.00818	.01766	.02804	.03938	.05175
25	8	5	.03603	.04551	.05574	.06678	.07869
25	8	6	.05169	.06053	.07000	.08012	.09095
25	8	7	.06065	.06864	.07712	.08613	.09569
25	8	8	.06564	.07271	.08017	.08803	.09634
25	8	9	.06815	.07431	.08077	.08755	.09466
25	8	10	.06904	.07434	.07987	.08564	.09166
25	8	11	.06886	.07336	.07803	.08289	.08793
25	8	12	.06795	.07172	.07562	.07965	.08382
25	8	13	.06655	.06966	.07285	.07615	.07954
25	8	14	.06481	.06732	.06989	.07253	.07524
25	8	15	.06283	.06481	.06682	.06889	.07101
25	8	16	.06070	.06220	.06373	.06530	.06690
25	8	17	.45307	.45849	.46440	.47076	.47754
25	9	1	-1.04631	-1.13553	-1.23139	-1.33428	-1.44465
25	9	2	-.17070	-.16988	-.16830	-.16587	-.16252
25	9	3	-.04700	-.03890	-.02979	-.01961	-.00826
25	9	4	.00860	.01844	.02919	.04092	.05372
25	9	5	.03783	.04765	.05824	.06966	.08197
25	9	6	.05426	.06342	.07320	.08366	.09484
25	9	7	.06367	.07192	.08068	.08998	.09985
25	9	8	.06890	.07619	.08388	.09199	.10055
25	9	9	.07152	.07787	.08452	.09150	.09882
25	9	10	.07246	.07791	.08359	.08952	.09570
25	9	11	.07226	.07688	.08167	.08664	.09181
25	9	12	.07131	.07516	.07914	.08326	.08752
25	9	13	.06983	.07299	.07625	.07960	.08306
25	9	14	.06799	.07053	.07314	.07582	.07857
25	9	15	.06591	.06789	.06993	.07201	.07416
25	9	16	.53948	.54746	.55605	.56519	.57485
25	10	1	-1.10287	-1.19613	-1.29626	-1.40366	-1.51878
25	10	2	-.17981	-.17897	-.17734	-.17484	-.17140
25	10	3	-.04943	-.04100	-.03153	-.02096	-.00919
25	10	4	.00916	.01939	.03056	.04275	.05601
25	10	5	.03996	.05016	.06115	.07300	.08575
25	10	6	.05726	.06677	.07691	.08775	.09932
25	10	7	.06717	.07572	.08479	.09441	.10462
25	10	8	.07267	.08022	.08817	.09655	.10538
25	10	9	.07543	.08198	.08885	.09604	.10359

Coefficients b_i for the BLUE of σ
Shape parameter k

n	s	i	2.1	2.2	2.3	2.4	2.5
25	10	10	.07640	.08201	.08786	.09397	.10033
25	10	11	.07618	.08093	.08585	.09095	.09626
25	10	12	.07516	.07911	.08318	.08740	.09176
25	10	13	.07359	.07681	.08014	.08356	.08708
25	10	14	.07164	.07422	.07686	.07958	.08238
25	10	15	.63749	.64878	.66080	.67351	.68688
25	11	1	-1.16794	-1.26587	-1.37092	-1.48353	-1.60411
25	11	2	-.19017	-.18930	-.18763	-.18507	-.18153
25	11	3	-.05212	-.04330	-.03343	-.02241	-.01017
25	11	4	.00991	.02059	.03224	.04493	.05873
25	11	5	.04250	.05314	.06460	.07692	.09018
25	11	6	.06081	.07070	.08126	.09252	.10454
25	11	7	.07127	.08017	.08959	.09958	.11016
25	11	8	.07708	.08492	.09316	.10184	.11099
25	11	9	.07998	.08677	.09387	.10132	.10911
25	11	10	.08099	.08679	.09283	.09912	.10569
25	11	11	.08074	.08563	.09069	.09594	.10140
25	11	12	.07964	.08369	.08787	.09219	.09666
25	11	13	.07796	.08125	.08464	.08813	.09173
25	11	14	.74934	.76483	.78123	.79851	.81662
25	12	1	-1.24358	-1.34694	-1.45774	-1.57641	-1.70338
25	12	2	-.20205	-.20116	-.19943	-.19680	-.19316
25	12	3	-.05509	-.04583	-.03549	-.02397	-.01118
25	12	4	.01093	.02213	.03433	.04760	.06201
25	12	5	.04559	.05673	.06871	.08159	.09543
25	12	6	.06504	.07539	.08642	.09817	.11070
25	12	7	.07615	.08544	.09526	.10567	.11668
25	12	8	.08231	.09046	.09904	.10807	.11758
25	12	9	.08536	.09241	.09979	.10751	.11560
25	12	10	.08641	.09241	.09866	.10518	.11197
25	12	11	.08611	.09115	.09637	.10180	.10742
25	12	12	.08491	.08906	.09336	.09780	.10239
25	12	13	.87792	.89875	.92071	.94379	.96794
25	13	1	-1.33253	-1.44231	-1.55990	-1.68573	-1.82025
25	13	2	-.21580	-.21488	-.21311	-.21039	-.20664
25	13	3	-.05837	-.04860	-.03771	-.02560	-.01218
25	13	4	.01231	.02412	.03696	.05091	.06604
25	13	5	.04939	.06112	.07372	.08724	.10175
25	13	6	.07018	.08105	.09263	.10496	.11808
25	13	7	.08203	.09177	.10207	.11295	.12447
25	13	8	.08857	.09711	.10608	.11551	.12543
25	13	9	.09180	.09916	.10685	.11489	.12331
25	13	10	.09287	.09912	.10561	.11238	.11943
25	13	11	.09251	.09773	.10313	.10874	.11457
25	13	12	1.02705	1.05463	1.08367	1.11413	1.14599

Table 6. Variances and covariance factors, V_1, V_2 and V_3, of the BLUEs of μ and σ for $n = 2(1)25$ and the censoring number $s = 0(1)\left[\frac{n+1}{2}\right]$

Variance-Covariance factors V1,V2 and V3 for
the BLUEs of μ and σ shape parameter k

n	s	0.0	0.1	0.2	0.3	0.4	0.5	0.6
2	0	.50000	.50000	.50000	.50000	.50000	.50000	.50000
2	0	.57080	.57189	.57515	.58058	.58815	.59782	.60956
2	0	.00000	.02499	.04995	.07485	.09964	.12430	.14881
3	0	.33333	.33333	.33333	.33333	.33333	.33332	.33331
3	0	.27548	.27600	.27753	.28009	.28367	.28825	.29383
3	0	.00000	.01666	.03329	.04984	.06629	.08260	.09876
3	1	.44867	.45237	.45594	.45938	.46268	.46583	.46882
3	1	.63783	.63294	.62975	.62830	.62859	.63063	.63443
3	1	.20443	.22279	.24110	.25934	.27751	.29560	.31359
4	0	.25000	.25000	.25000	.25000	.24999	.24998	.24996
4	0	.18005	.18037	.18134	.18296	.18521	.18812	.19166
4	0	.00000	.01249	.02496	.03735	.04966	.06183	.07387
4	1	.28701	.28898	.29090	.29275	.29454	.29625	.29787
4	1	.30208	.29975	.29831	.29777	.29814	.29944	.30167
4	1	.06720	.08071	.09412	.10742	.12058	.13360	.14646
5	0	.20000	.20000	.20000	.20000	.19999	.19998	.19996
5	0	.13332	.13355	.13424	.13539	.13701	.13909	.14165
5	0	.00000	.01000	.01996	.02987	.03969	.04940	.05899
5	1	.21772	.21892	.22009	.22122	.22232	.22337	.22437
5	1	.19476	.19333	.19249	.19224	.19261	.19361	.19523
5	1	.03300	.04362	.05417	.06461	.07493	.08511	.09515
5	2	.28393	.28656	.28903	.29134	.29347	.29542	.29719
5	2	.31809	.31444	.31157	.30952	.30829	.30792	.30841
5	2	.12336	.13413	.14478	.15529	.16565	.17587	.18593
6	0	.16667	.16667	.16667	.16666	.16666	.16665	.16663
6	0	.10570	.10588	.10641	.10729	.10853	.11014	.11211
6	0	.00000	.00833	.01663	.02488	.03306	.04113	.04909
6	1	.17688	.17768	.17846	.17922	.17996	.18066	.18133
6	1	.14277	.14178	.14123	.14112	.14147	.14229	.14358
6	1	.01946	.02821	.03689	.04549	.05399	.06236	.07060
6	2	.20683	.20864	.21038	.21202	.21356	.21499	.21631
6	2	.20436	.20204	.20025	.19902	.19837	.19830	.19883
6	2	.06240	.07141	.08030	.08907	.09771	.10621	.11456
6	3	.29985	.30224	.30437	.30623	.30783	.30917	.31023
6	3	.32920	.32476	.32104	.31807	.31586	.31445	.31385
6	3	.17017	.17858	.18685	.19497	.20296	.21080	.21850
7	0	.14286	.14286	.14286	.14285	.14285	.14284	.14282
7	0	.08750	.08764	.08806	.08877	.08977	.09107	.09266
7	0	.00000	.00714	.01425	.02132	.02832	.03523	.04203
7	1	.14942	.14999	.15055	.15109	.15162	.15212	.15260
7	1	.11233	.11160	.11121	.11117	.11150	.11219	.11327
7	1	.01277	.02021	.02759	.03490	.04212	.04924	.05623
7	2	.16600	.16729	.16854	.16973	.17086	.17191	.17288
7	2	.14928	.14763	.14639	.14557	.14519	.14525	.14577
7	2	.03752	.04518	.05276	.06022	.06758	.07481	.08191
7	3	.20714	.20913	.21098	.21268	.21422	.21559	.21679
7	3	.21136	.20849	.20611	.20424	.20292	.20215	.20195
7	3	.08805	.09564	.10310	.11043	.11762	.12467	.13158
7	4	.32480	.32635	.32757	.32844	.32899	.32921	.32911
7	4	.33753	.33257	.32828	.32467	.32180	.31967	.31831
7	4	.20989	.21624	.22244	.22850	.23442	.24021	.24589
8	0	.12500	.12500	.12500	.12500	.12499	.12498	.12496
8	0	.07461	.07473	.07508	.07567	.07650	.07757	.07890
8	0	.00000	.00625	.01247	.01865	.02477	.03081	.03675
8	1	.12954	.12996	.13038	.13079	.13118	.13156	.13191
8	1	.09242	.09186	.09157	.09157	.09188	.09248	.09340

Variance-Covariance factors V1,V2 and V3 for the BLUEs of μ and σ shape parameter k

n	s	0.0	0.1	0.2	0.3	0.4	0.5	0.6
8	1	.00899	.01547	.02189	.02825	.03453	.04071	.04679
8	2	.13988	.14084	.14177	.14266	.14351	.14431	.14505
8	2	.11707	.11582	.11491	.11432	.11409	.11422	.11472
8	2	.02496	.03161	.03819	.04468	.05108	.05736	.06352
8	3	.16225	.16379	.16525	.16660	.16786	.16900	.17002
8	3	.15419	.15212	.15043	.14914	.14826	.14780	.14779
8	3	.05377	.06048	.06707	.07355	.07992	.08615	.09226
8	4	.21376	.21561	.21727	.21872	.21996	.22098	.22179
8	4	.21679	.21353	.21074	.20843	.20664	.20537	.20464
8	4	.11056	.11689	.12309	.12915	.13507	.14086	.14651
9	0	.11111	.11111	.11111	.11111	.11110	.11109	.11108
9	0	.06502	.06511	.06541	.06591	.06662	.06754	.06867
9	0	.00000	.00555	.01108	.01658	.02201	.02737	.03264
9	1	.11442	.11474	.11507	.11538	.11569	.11598	.11625
9	1	.07842	.07797	.07775	.07778	.07806	.07860	.07941
9	1	.00666	.01239	.01807	.02370	.02926	.03473	.04010
9	2	.12140	.12214	.12285	.12354	.12420	.12482	.12540
9	2	.09605	.09507	.09436	.09393	.09379	.09395	.09442
9	2	.01775	.02363	.02944	.03518	.04083	.04637	.05182
9	3	.13515	.13635	.13749	.13857	.13957	.14049	.14133
9	3	.12075	.11916	.11788	.11693	.11631	.11603	.11612
9	3	.03618	.04213	.04800	.05377	.05943	.06498	.07041
9	4	.16291	.16451	.16598	.16732	.16851	.16955	.17045
9	4	.15810	.15573	.15372	.15208	.15084	.15000	.14960
9	4	.06838	.07422	.07995	.08556	.09104	.09640	.10163
9	5	.22407	.22557	.22681	.22780	.22854	.22903	.22927
9	5	.22116	.21763	.21453	.21190	.20974	.20810	.20697
9	5	.13049	.13570	.14077	.14571	.15051	.15518	.15972
10	0	.10000	.10000	.10000	.10000	.09999	.09998	.09997
10	0	.05760	.05768	.05794	.05837	.05899	.05978	.06077
10	0	.00000	.00500	.00998	.01492	.01981	.02462	.02936
10	1	.10251	.10276	.10302	.10327	.10352	.10375	.10397
10	1	.06806	.06769	.06752	.06757	.06783	.06831	.06902
10	1	.00512	.01026	.01536	.02040	.02539	.03029	.03511
10	2	.10749	.10807	.10864	.10919	.10972	.11021	.11067
10	2	.08129	.08050	.07993	.07961	.07953	.07971	.08015
10	2	.01324	.01850	.02371	.02885	.03390	.03887	.04375
10	3	.11666	.11761	.11852	.11939	.12020	.12095	.12164
10	3	.09892	.09765	.09665	.09591	.09546	.09530	.09544
10	3	.02596	.03130	.03656	.04174	.04683	.05181	.05669
10	4	.13359	.13491	.13614	.13727	.13830	.13923	.14004
10	4	.12373	.12190	.12036	.11912	.11821	.11764	.11741
10	4	.04645	.05177	.05699	.06211	.06712	.07202	.07680
10	5	.16641	.16792	.16926	.17044	.17144	.17226	.17290
10	5	.16131	.15872	.15646	.15456	.15304	.15192	.15120
10	5	.08157	.08664	.09158	.09640	.10110	.10567	.11012
11	0	.09091	.09091	.09091	.09091	.09090	.09089	.09088
11	0	.05169	.05176	.05199	.05237	.05291	.05361	.05449
11	0	.00000	.00454	.00907	.01356	.01800	.02238	.02668
11	1	.09287	.09308	.09329	.09349	.09369	.09388	.09405
11	1	.06009	.05978	.05965	.05970	.05994	.06038	.06102
11	1	.00405	.00871	.01333	.01791	.02243	.02687	.03123
11	2	.09658	.09705	.09751	.09796	.09839	.09879	.09917
11	2	.07040	.06973	.06927	.06902	.06898	.06917	.06958
11	2	.01024	.01500	.01971	.02436	.02894	.03344	.03785
11	3	.10307	.10383	.10457	.10528	.10595	.10656	.10713

Variance-Covariance factors V1,V2 and V3 for
the BLUEs of μ and σ shape parameter k

n	s	0.0	0.1	0.2	0.3	0.4	0.5	0.6
11	3	.08361	.08257	.08176	.08118	.08084	.08075	.08092
11	3	.01950	.02433	.02910	.03380	.03841	.04293	.04735
11	4	.11428	.11536	.11638	.11733	.11820	.11900	.11970
11	4	.10128	.09981	.09859	.09762	.09693	.09652	.09640
11	4	.03358	.03843	.04319	.04787	.05245	.05693	.06131
11	5	.13421	.13553	.13675	.13785	.13882	.13965	.14036
11	5	.12622	.12420	.12246	.12101	.11987	.11906	.11858
11	5	.05587	.06061	.06525	.06978	.07420	.07851	.08271
12	0	.08333	.08333	.08333	.08333	.08333	.08332	.08330
12	0	.04688	.04694	.04715	.04749	.04797	.04859	.04937
12	0	.00000	.00416	.00831	.01243	.01650	.02051	.02444
12	1	.08490	.08507	.08525	.08542	.08558	.08574	.08588
12	1	.05377	.05351	.05340	.05346	.05369	.05409	.05467
12	1	.00329	.00754	.01177	.01596	.02009	.02415	.02814
12	2	.08776	.08815	.08853	.08890	.08926	.08959	.08990
12	2	.06203	.06147	.06109	.06089	.06088	.06106	.06146
12	2	.00815	.01249	.01680	.02104	.02523	.02934	.03336
12	3	.09255	.09318	.09380	.09438	.09493	.09545	.09593
12	3	.07230	.07144	.07076	.07029	.07003	.06999	.07018
12	3	.01516	.01958	.02393	.02822	.03244	.03657	.04061
12	4	.10043	.10132	.10217	.10297	.10371	.10439	.10500
12	4	.08554	.08432	.08332	.08254	.08200	.08170	.08166
12	4	.02537	.02982	.03419	.03848	.04269	.04680	.05082
12	5	.11357	.11471	.11576	.11673	.11760	.11837	.11904
12	5	.10328	.10165	.10026	.09911	.09823	.09762	.09730
12	5	.04064	.04505	.04936	.05358	.05770	.06172	.06564
12	6	.13633	.13760	.13872	.13971	.14054	.14122	.14175
12	6	.12834	.12618	.12428	.12266	.12134	.12033	.11965
12	6	.06452	.06874	.07284	.07684	.08073	.08450	.08817
13	0	.07692	.07692	.07692	.07692	.07692	.07691	.07689
13	0	.04288	.04294	.04312	.04343	.04386	.04443	.04513
13	0	.00000	.00384	.00767	.01147	.01523	.01893	.02256
13	1	.07820	.07835	.07850	.07864	.07878	.07891	.07903
13	1	.04864	.04841	.04833	.04839	.04861	.04898	.04951
13	1	.00271	.00664	.01053	.01439	.01820	.02194	.02561
13	2	.08047	.08079	.08111	.08142	.08172	.08200	.08226
13	2	.05541	.05493	.05461	.05445	.05446	.05464	.05501
13	2	.00663	.01063	.01459	.01850	.02235	.02613	.02983
13	3	.08413	.08465	.08517	.08566	.08613	.08656	.08697
13	3	.06363	.06289	.06232	.06194	.06174	.06173	.06193
13	3	.01211	.01617	.02018	.02413	.02801	.03181	.03554
13	4	.08991	.09066	.09138	.09205	.09269	.09327	.09379
13	4	.07391	.07289	.07205	.07141	.07098	.07076	.07077
13	4	.01983	.02392	.02795	.03191	.03580	.03959	.04330
13	5	.09911	.10008	.10098	.10182	.10259	.10328	.10388
13	5	.08718	.08583	.08468	.08375	.08305	.08258	.08236
13	5	.03088	.03496	.03897	.04289	.04673	.05047	.05412
13	6	.11408	.11522	.11625	.11718	.11800	.11869	.11927
13	6	.10500	.10325	.10172	.10042	.09938	.09861	.09812
13	6	.04721	.05120	.05510	.05889	.06259	.06619	.06969
13	7	.13953	.14067	.14165	.14246	.14311	.14358	.14389
13	7	.13018	.12790	.12587	.12411	.12264	.12148	.12063
13	7	.07252	.07625	.07986	.08337	.08676	.09005	.09323
14	0	.07143	.07143	.07143	.07143	.07142	.07142	.07140
14	0	.03951	.03957	.03973	.04001	.04040	.04091	.04155
14	0	.00000	.00357	.00712	.01065	.01414	.01757	.02094

Variance-Covariance factors V1,V2 and V3 for the BLUEs of μ and σ shape parameter k

n	s	0.0	0.1	0.2	0.3	0.4	0.5	0.6
14	1	.07249	.07262	.07274	.07286	.07298	.07309	.07319
14	1	.04440	.04420	.04413	.04420	.04440	.04474	.04523
14	1	.00228	.00591	.00953	.01310	.01663	.02010	.02351
14	2	.07432	.07460	.07487	.07513	.07539	.07563	.07585
14	2	.05005	.04963	.04935	.04923	.04925	.04944	.04978
14	2	.00549	.00919	.01286	.01648	.02005	.02355	.02698
14	3	.07720	.07764	.07808	.07850	.07890	.07927	.07961
14	3	.05678	.05614	.05565	.05533	.05518	.05520	.05540
14	3	.00989	.01365	.01736	.02101	.02461	.02813	.03158
14	4	.08159	.08223	.08284	.08342	.08396	.08446	.08492
14	4	.06500	.06412	.06341	.06287	.06252	.06236	.06240
14	4	.01590	.01970	.02343	.02711	.03071	.03423	.03768
14	5	.08833	.08915	.08993	.09066	.09133	.09194	.09248
14	5	.07530	.07415	.07319	.07241	.07184	.07147	.07133
14	5	.02423	.02803	.03176	.03542	.03899	.04249	.04589
14	6	.09878	.09977	.10070	.10153	.10228	.10294	.10350
14	6	.08862	.08715	.08589	.08483	.08399	.08338	.08302
14	6	.03603	.03978	.04345	.04704	.05053	.05393	.05724
14	7	.11549	.11658	.11755	.11840	.11911	.11969	.12014
14	7	.10651	.10465	.10301	.10159	.10043	.09952	.09888
14	7	.05333	.05693	.06044	.06385	.06716	.07038	.07349
15	0	.06667	.06667	.06667	.06667	.06666	.06665	.06664
15	0	.03663	.03668	.03683	.03709	.03745	.03791	.03850
15	0	.00000	.00333	.00665	.00994	.01319	.01639	.01954
15	1	.06756	.06767	.06778	.06788	.06798	.06808	.06817
15	1	.04083	.04066	.04060	.04067	.04086	.04118	.04163
15	1	.00194	.00533	.00869	.01203	.01532	.01855	.02172
15	2	.06907	.06931	.06954	.06977	.06999	.07019	.07038
15	2	.04562	.04525	.04501	.04491	.04495	.04513	.04546
15	2	.00462	.00807	.01148	.01486	.01818	.02144	.02464
15	3	.07138	.07176	.07213	.07250	.07284	.07316	.07346
15	3	.05123	.05067	.05025	.04998	.04986	.04990	.05010
15	3	.00822	.01172	.01517	.01857	.02192	.02520	.02841
15	4	.07482	.07536	.07588	.07638	.07685	.07729	.07769
15	4	.05795	.05719	.05658	.05613	.05584	.05572	.05579
15	4	.01303	.01656	.02004	.02346	.02682	.03011	.03332
15	5	.07992	.08063	.08130	.08194	.08252	.08306	.08354
15	5	.06618	.06520	.06437	.06372	.06324	.06295	.06286
15	5	.01951	.02305	.02654	.02996	.03330	.03657	.03975
15	6	.08755	.08842	.08923	.08998	.09065	.09125	.09177
15	6	.07651	.07527	.07420	.07331	.07262	.07213	.07186
15	6	.02839	.03191	.03536	.03874	.04203	.04524	.04836
15	7	.09920	.10019	.10109	.10189	.10259	.10319	.10367
15	7	.08988	.08833	.08696	.08579	.08484	.08412	.08363
15	7	.04087	.04431	.04767	.05093	.05411	.05720	.06019
15	8	.11758	.11858	.11945	.12018	.12076	.12119	.12148
15	8	.10785	.10590	.10416	.10265	.10137	.10035	.09959
15	8	.05904	.06229	.06544	.06849	.07144	.07430	.07706
16	0	.06250	.06250	.06250	.06250	.06249	.06249	.06247
16	0	.03414	.03419	.03433	.03456	.03489	.03532	.03586
16	0	.00000	.00312	.00623	.00932	.01237	.01537	.01831
16	1	.06326	.06336	.06345	.06354	.06363	.06371	.06379
16	1	.03779	.03764	.03759	.03766	.03784	.03814	.03856
16	1	.00167	.00484	.00800	.01112	.01420	.01722	.02019
16	2	.06452	.06473	.06493	.06513	.06532	.06550	.06567
16	2	.04190	.04157	.04137	.04128	.04133	.04150	.04182

Variance-Covariance factors V1,V2 and V3 for
the BLUEs of μ and σ shape parameter k

n	s	0.0	0.1	0.2	0.3	0.4	0.5	0.6
16	2	.00394	.00717	.01036	.01352	.01662	.01968	.02267
16	3	.06641	.06674	.06707	.06738	.06768	.06796	.06822
16	3	.04665	.04616	.04579	.04556	.04547	.04552	.04572
16	3	.00694	.01020	.01343	.01662	.01975	.02282	.02582
16	4	.06916	.06963	.07009	.07052	.07093	.07131	.07166
16	4	.05225	.05158	.05105	.05066	.05042	.05034	.05042
16	4	.01086	.01416	.01742	.02062	.02376	.02684	.02985
16	5	.07314	.07375	.07434	.07490	.07541	.07589	.07631
16	5	.05898	.05812	.05741	.05685	.05645	.05622	.05617
16	5	.01603	.01935	.02262	.02582	.02896	.03203	.03502
16	6	.07890	.07967	.08038	.08105	.08165	.08219	.08267
16	6	.06723	.06615	.06523	.06448	.06390	.06351	.06330
16	6	.02293	.02625	.02949	.03267	.03578	.03881	.04175
16	7	.08740	.08828	.08910	.08984	.09050	.09106	.09154
16	7	.07759	.07627	.07510	.07412	.07333	.07274	.07236
16	7	.03231	.03558	.03877	.04188	.04491	.04786	.05072
16	8	.10020	.10115	.10200	.10274	.10337	.10387	.10426
16	8	.09101	.08938	.08792	.08667	.08562	.08479	.08420
16	8	.04542	.04857	.05163	.05461	.05749	.06028	.06299
17	0	.05882	.05882	.05882	.05882	.05882	.05881	.05880
17	0	.03197	.03201	.03214	.03235	.03266	.03306	.03356
17	0	.00000	.00294	.00587	.00877	.01164	.01446	.01723
17	1	.05948	.05956	.05965	.05973	.05980	.05988	.05994
17	1	.03517	.03503	.03499	.03506	.03523	.03551	.03591
17	1	.00145	.00443	.00740	.01033	.01323	.01608	.01887
17	2	.06055	.06073	.06091	.06108	.06125	.06140	.06155
17	2	.03873	.03844	.03826	.03819	.03824	.03841	.03871
17	2	.00340	.00643	.00943	.01239	.01531	.01818	.02099
17	3	.06212	.06241	.06269	.06297	.06323	.06348	.06370
17	3	.04281	.04237	.04205	.04185	.04178	.04184	.04204
17	3	.00593	.00900	.01203	.01502	.01796	.02085	.02367
17	4	.06436	.06477	.06517	.06555	.06591	.06625	.06655
17	4	.04755	.04696	.04649	.04615	.04595	.04590	.04599
17	4	.00919	.01229	.01534	.01835	.02131	.02420	.02702
17	5	.06753	.06807	.06858	.06907	.06953	.06995	.07033
17	5	.05315	.05239	.05177	.05129	.05095	.05077	.05074
17	5	.01340	.01652	.01959	.02261	.02556	.02844	.03126
17	6	.07201	.07268	.07331	.07390	.07445	.07494	.07537
17	6	.05989	.05895	.05815	.05750	.05701	.05669	.05654
17	6	.01890	.02202	.02509	.02809	.03102	.03388	.03666
17	7	.07842	.07921	.07995	.08062	.08122	.08175	.08221
17	7	.06816	.06701	.06601	.06517	.06451	.06402	.06372
17	7	.02618	.02928	.03231	.03526	.03814	.04095	.04367
17	8	.08774	.08862	.08942	.09013	.09074	.09126	.09167
17	8	.07856	.07716	.07593	.07486	.07399	.07331	.07284
17	8	.03603	.03905	.04200	.04486	.04765	.05035	.05296
17	9	.10164	.10254	.10332	.10397	.10449	.10489	.10516
17	9	.09202	.09032	.08880	.08747	.08634	.08542	.08474
17	9	.04971	.05258	.05537	.05807	.06068	.06320	.06563
18	0	.05556	.05556	.05556	.05555	.05555	.05554	.05553
18	0	.03006	.03009	.03021	.03041	.03070	.03107	.03153
18	0	.00000	.00278	.00554	.00828	.01099	.01366	.01627
18	1	.05613	.05620	.05627	.05635	.05641	.05648	.05653
18	1	.03288	.03276	.03273	.03279	.03296	.03322	.03359
18	1	.00127	.00409	.00688	.00965	.01239	.01507	.01771
18	2	.05704	.05720	.05736	.05751	.05766	.05780	.05793

Variance-Covariance factors V1,V2 and V3 for the BLUEs of μ and σ shape parameter k

n	s	0.0	0.1	0.2	0.3	0.4	0.5	0.6
18	2	.03601	.03575	.03559	.03553	.03559	.03575	.03604
18	2	.00296	.00581	.00864	.01144	.01420	.01690	.01955
18	3	.05836	.05862	.05887	.05911	.05934	.05956	.05976
18	3	.03954	.03915	.03887	.03869	.03864	.03871	.03890
18	3	.00512	.00801	.01087	.01369	.01646	.01918	.02184
18	4	.06022	.06058	.06093	.06127	.06159	.06189	.06216
18	4	.04361	.04308	.04266	.04237	.04220	.04217	.04227
18	4	.00787	.01079	.01367	.01651	.01929	.02202	.02469
18	5	.06279	.06327	.06373	.06416	.06456	.06494	.06528
18	5	.04835	.04767	.04712	.04670	.04641	.04626	.04626
18	5	.01136	.01430	.01720	.02004	.02283	.02556	.02821
18	6	.06636	.06695	.06752	.06804	.06853	.06897	.06936
18	6	.05395	.05312	.05242	.05186	.05143	.05117	.05106
18	6	.01584	.01878	.02168	.02452	.02729	.03000	.03264
18	7	.07134	.07204	.07270	.07331	.07386	.07435	.07477
18	7	.06071	.05970	.05883	.05810	.05753	.05712	.05689
18	7	.02163	.02457	.02744	.03025	.03299	.03566	.03826
18	8	.07837	.07917	.07990	.08056	.08114	.08164	.08206
18	8	.06900	.06779	.06672	.06581	.06507	.06450	.06412
18	8	.02927	.03216	.03498	.03773	.04040	.04300	.04551
18	9	.08848	.08934	.09009	.09075	.09130	.09175	.09209
18	9	.07943	.07798	.07667	.07554	.07459	.07384	.07328
18	9	.03954	.04234	.04506	.04769	.05024	.05271	.05510
19	0	.05263	.05263	.05263	.05263	.05263	.05262	.05261
19	0	.02836	.02839	.02850	.02869	.02896	.02930	.02974
19	0	.00000	.00263	.00525	.00785	.01041	.01293	.01541
19	1	.05313	.05320	.05326	.05333	.05339	.05344	.05349
19	1	.03087	.03076	.03074	.03080	.03096	.03121	.03156
19	1	.00112	.00379	.00644	.00906	.01165	.01419	.01668
19	2	.05392	.05406	.05420	.05434	.05447	.05460	.05471
19	2	.03364	.03340	.03326	.03321	.03327	.03343	.03370
19	2	.00260	.00530	.00798	.01062	.01323	.01579	.01829
19	3	.05505	.05528	.05550	.05572	.05592	.05612	.05630
19	3	.03673	.03638	.03612	.03597	.03593	.03601	.03620
19	3	.00447	.00720	.00990	.01257	.01519	.01777	.02028
19	4	.05661	.05693	.05724	.05754	.05783	.05809	.05834
19	4	.04026	.03978	.03941	.03915	.03901	.03899	.03910
19	4	.00681	.00957	.01230	.01498	.01761	.02019	.02272
19	5	.05874	.05916	.05956	.05995	.06031	.06065	.06095
19	5	.04432	.04371	.04322	.04285	.04260	.04248	.04250
19	5	.00975	.01253	.01527	.01796	.02060	.02318	.02570
19	6	.06163	.06216	.06266	.06313	.06356	.06396	.06432
19	6	.04906	.04832	.04769	.04720	.04683	.04661	.04653
19	6	.01346	.01625	.01899	.02168	.02431	.02688	.02938
19	7	.06559	.06621	.06680	.06735	.06785	.06830	.06869
19	7	.05468	.05378	.05302	.05238	.05189	.05154	.05136
19	7	.01817	.02096	.02369	.02636	.02897	.03151	.03398
19	8	.07104	.07176	.07242	.07303	.07357	.07405	.07445
19	8	.06145	.06038	.05945	.05865	.05801	.05754	.05723
19	8	.02424	.02700	.02970	.03233	.03489	.03737	.03979
19	9	.07866	.07945	.08016	.08079	.08134	.08179	.08216
19	9	.06976	.06850	.06737	.06640	.06558	.06495	.06449
19	9	.03221	.03490	.03753	.04008	.04255	.04495	.04727
19	10	.08954	.09035	.09105	.09164	.09211	.09248	.09272
19	10	.08023	.07872	.07736	.07617	.07515	.07433	.07370
19	10	.04288	.04546	.04796	.05038	.05271	.05496	.05714

Variance-Covariance factors V1,V2 and V3 for the BLUEs of μ and σ shape parameter k

n	s	0.0	0.1	0.2	0.3	0.4	0.5	0.6
20	0	.05000	.05000	.05000	.05000	.05000	.04999	.04998
20	0	.02684	.02687	.02698	.02715	.02740	.02773	.02813
20	0	.00000	.00250	.00499	.00745	.00989	.01229	.01464
20	1	.05044	.05050	.05056	.05062	.05067	.05072	.05077
20	1	.02910	.02900	.02898	.02904	.02919	.02943	.02976
20	1	.00100	.00353	.00604	.00853	.01099	.01340	.01577
20	2	.05113	.05126	.05138	.05151	.05162	.05173	.05184
20	2	.03156	.03134	.03121	.03118	.03124	.03139	.03165
20	2	.00230	.00486	.00740	.00991	.01239	.01481	.01719
20	3	.05210	.05230	.05250	.05270	.05288	.05306	.05322
20	3	.03429	.03397	.03374	.03361	.03358	.03366	.03384
20	3	.00393	.00652	.00908	.01161	.01410	.01654	.01893
20	4	.05343	.05371	.05399	.05426	.05452	.05475	.05497
20	4	.03738	.03694	.03661	.03638	.03626	.03626	.03637
20	4	.00595	.00857	.01115	.01370	.01620	.01864	.02104
20	5	.05521	.05558	.05595	.05629	.05662	.05692	.05719
20	5	.04089	.04035	.03991	.03958	.03936	.03927	.03930
20	5	.00846	.01109	.01369	.01624	.01875	.02120	.02359
20	6	.05760	.05806	.05851	.05894	.05933	.05969	.06001
20	6	.04496	.04429	.04373	.04329	.04298	.04279	.04274
20	6	.01157	.01422	.01682	.01938	.02188	.02432	.02670
20	7	.06080	.06136	.06189	.06239	.06284	.06325	.06361
20	7	.04970	.04890	.04822	.04766	.04723	.04694	.04679
20	7	.01547	.01812	.02072	.02326	.02575	.02817	.03052
20	8	.06513	.06578	.06638	.06694	.06744	.06788	.06826
20	8	.05533	.05439	.05356	.05287	.05231	.05190	.05165
20	8	.02040	.02304	.02561	.02813	.03058	.03296	.03527
20	9	.07103	.07176	.07242	.07301	.07354	.07398	.07435
20	9	.06212	.06100	.06002	.05917	.05847	.05792	.05755
20	9	.02674	.02933	.03186	.03431	.03670	.03902	.04127
20	10	.07923	.08000	.08068	.08127	.08176	.08216	.08246
20	10	.07046	.06915	.06797	.06694	.06607	.06537	.06485
20	10	.03500	.03752	.03996	.04232	.04461	.04682	.04896
21	0	.04762	.04762	.04762	.04762	.04761	.04761	.04760
21	0	.02548	.02551	.02561	.02577	.02600	.02631	.02669
21	0	.00000	.00238	.00475	.00710	.00942	.01170	.01394
21	1	.04801	.04807	.04812	.04817	.04822	.04826	.04830
21	1	.02751	.02742	.02740	.02747	.02761	.02783	.02815
21	1	.00090	.00330	.00570	.00807	.01040	.01270	.01495
21	2	.04862	.04873	.04884	.04896	.04906	.04916	.04925
21	2	.02971	.02952	.02941	.02938	.02944	.02959	.02984
21	2	.00205	.00449	.00690	.00929	.01164	.01395	.01622
21	3	.04946	.04964	.04982	.05000	.05016	.05032	.05046
21	3	.03215	.03185	.03165	.03153	.03151	.03159	.03178
21	3	.00348	.00595	.00838	.01079	.01316	.01548	.01775
21	4	.05060	.05086	.05111	.05135	.05158	.05179	.05199
21	4	.03487	.03448	.03418	.03398	.03387	.03388	.03399
21	4	.00524	.00773	.01019	.01261	.01498	.01731	.01959
21	5	.05211	.05245	.05277	.05308	.05338	.05365	.05390
21	5	.03795	.03746	.03706	.03676	.03658	.03650	.03655
21	5	.00740	.00990	.01238	.01481	.01719	.01952	.02180
21	6	.05411	.05453	.05493	.05531	.05567	.05599	.05629
21	6	.04147	.04087	.04037	.03997	.03970	.03954	.03951
21	6	.01005	.01257	.01505	.01748	.01986	.02219	.02446
21	7	.05674	.05725	.05773	.05818	.05859	.05897	.05930
21	7	.04553	.04481	.04420	.04370	.04333	.04308	.04296

Variance-Covariance factors V1,V2 and V3 for
the BLUEs of μ and σ shape parameter k

n	s	0.0	0.1	0.2	0.3	0.4	0.5	0.6
21	7	.01332	.01585	.01832	.02075	.02312	.02543	.02768
21	8	.06024	.06083	.06138	.06189	.06235	.06276	.06312
21	8	.05029	.04944	.04871	.04809	.04760	.04725	.04704
21	8	.01740	.01992	.02238	.02479	.02713	.02941	.03163
21	9	.06493	.06559	.06620	.06675	.06725	.06767	.06803
21	9	.05593	.05494	.05406	.05332	.05271	.05224	.05192
21	9	.02255	.02503	.02746	.02983	.03213	.03436	.03653
21	10	.07128	.07199	.07264	.07321	.07370	.07411	.07443
21	10	.06274	.06158	.06054	.05964	.05889	.05829	.05785
21	10	.02912	.03155	.03392	.03622	.03844	.04060	.04269
21	11	.08003	.08076	.08140	.08193	.08237	.08270	.08293
21	11	.07110	.06974	.06852	.06744	.06652	.06577	.06519
21	11	.03767	.04001	.04228	.04447	.04658	.04862	.05058
22	0	.04545	.04545	.04545	.04545	.04545	.04544	.04543
22	0	.02425	.02428	.02437	.02452	.02474	.02503	.02539
22	0	.00000	.00227	.00453	.00677	.00899	.01117	.01330
22	1	.04581	.04586	.04590	.04595	.04599	.04603	.04607
22	1	.02609	.02600	.02599	.02605	.02619	.02641	.02670
22	1	.00081	.00310	.00539	.00765	.00988	.01207	.01421
22	2	.04634	.04645	.04655	.04665	.04674	.04683	.04692
22	2	.02807	.02790	.02779	.02777	.02783	.02798	.02822
22	2	.00184	.00416	.00646	.00874	.01099	.01319	.01535
22	3	.04708	.04725	.04741	.04757	.04772	.04786	.04799
22	3	.03025	.02998	.02980	.02970	.02968	.02977	.02994
22	3	.00311	.00545	.00778	.01007	.01233	.01454	.01671
22	4	.04807	.04830	.04852	.04874	.04895	.04915	.04932
22	4	.03268	.03232	.03204	.03186	.03178	.03179	.03191
22	4	.00465	.00702	.00936	.01167	.01394	.01616	.01833
22	5	.04936	.04967	.04996	.05024	.05051	.05075	.05098
22	5	.03539	.03494	.03458	.03432	.03415	.03410	.03415
22	5	.00653	.00891	.01127	.01359	.01586	.01808	.02025
22	6	.05105	.05143	.05179	.05214	.05246	.05276	.05303
22	6	.03847	.03792	.03747	.03712	.03687	.03674	.03673
22	6	.00881	.01121	.01357	.01589	.01817	.02039	.02255
22	7	.05325	.05371	.05414	.05455	.05493	.05527	.05558
22	7	.04199	.04134	.04079	.04034	.04001	.03979	.03970
22	7	.01159	.01400	.01636	.01868	.02095	.02316	.02531
22	8	.05613	.05667	.05717	.05763	.05806	.05844	.05877
22	8	.04606	.04530	.04464	.04409	.04366	.04335	.04318
22	8	.01502	.01742	.01977	.02208	.02432	.02651	.02864
22	9	.05992	.06052	.06109	.06160	.06206	.06246	.06280
22	9	.05083	.04994	.04916	.04849	.04795	.04755	.04729
22	9	.01926	.02165	.02398	.02626	.02847	.03062	.03271
22	10	.06495	.06562	.06622	.06676	.06723	.06764	.06796
22	10	.05649	.05545	.05453	.05374	.05308	.05256	.05219
22	10	.02460	.02695	.02924	.03146	.03362	.03571	.03774
22	11	.07173	.07243	.07304	.07358	.07403	.07439	.07466
22	11	.06331	.06211	.06103	.06008	.05928	.05863	.05814
22	11	.03140	.03368	.03590	.03804	.04011	.04212	.04405
23	0	.04348	.04348	.04348	.04348	.04347	.04347	.04346
23	0	.02313	.02316	.02324	.02339	.02360	.02387	.02421
23	0	.00000	.00217	.00434	.00648	.00860	.01068	.01272
23	1	.04380	.04384	.04388	.04392	.04397	.04400	.04403
23	1	.02481	.02473	.02472	.02478	.02491	.02512	.02540
23	1	.00073	.00293	.00511	.00727	.00940	.01150	.01355
23	2	.04427	.04437	.04446	.04455	.04464	.04472	.04479

Variance-Covariance factors V1,V2 and V3 for
the BLUEs of μ and σ shape parameter k

n	s	0.0	0.1	0.2	0.3	0.4	0.5	0.6
23	2	.02660	.02644	.02635	.02633	.02639	.02654	.02677
23	2	.00166	.00387	.00608	.00825	.01040	.01250	.01457
23	3	.04492	.04507	.04522	.04536	.04550	.04563	.04575
23	3	.02857	.02832	.02815	.02806	.02805	.02814	.02831
23	3	.00279	.00503	.00725	.00944	.01160	.01371	.01578
23	4	.04579	.04600	.04620	.04640	.04659	.04677	.04693
23	4	.03074	.03041	.03016	.03000	.02992	.02994	.03006
23	4	.00415	.00642	.00865	.01086	.01302	.01514	.01722
23	5	.04691	.04718	.04745	.04770	.04794	.04817	.04838
23	5	.03315	.03274	.03241	.03217	.03203	.03199	.03205
23	5	.00580	.00808	.01033	.01254	.01471	.01684	.01891
23	6	.04835	.04869	.04902	.04934	.04963	.04991	.05015
23	6	.03587	.03537	.03495	.03464	.03442	.03431	.03431
23	6	.00778	.01007	.01233	.01455	.01672	.01885	.02092
23	7	.05021	.05063	.05102	.05139	.05174	.05205	.05234
23	7	.03895	.03835	.03785	.03745	.03715	.03697	.03690
23	7	.01017	.01247	.01473	.01695	.01912	.02124	.02330
23	8	.05261	.05310	.05356	.05398	.05437	.05473	.05504
23	8	.04247	.04178	.04118	.04069	.04030	.04004	.03990
23	8	.01308	.01538	.01764	.01985	.02200	.02410	.02614
23	9	.05573	.05628	.05679	.05727	.05770	.05807	.05840
23	9	.04655	.04574	.04504	.04445	.04397	.04362	.04339
23	9	.01665	.01894	.02118	.02336	.02549	.02756	.02957
23	10	.05979	.06040	.06097	.06147	.06192	.06231	.06263
23	10	.05133	.05040	.04957	.04887	.04828	.04783	.04752
23	10	.02105	.02332	.02552	.02767	.02976	.03179	.03375
23	11	.06516	.06582	.06641	.06693	.06737	.06774	.06803
23	11	.05699	.05592	.05497	.05413	.05343	.05286	.05244
23	11	.02657	.02879	.03094	.03303	.03505	.03701	.03890
23	12	.07235	.07302	.07360	.07410	.07450	.07480	.07501
23	12	.06383	.06260	.06148	.06050	.05965	.05896	.05842
23	12	.03358	.03572	.03779	.03979	.04172	.04358	.04537
24	0	.04167	.04167	.04167	.04167	.04166	.04166	.04165
24	0	.02211	.02213	.02222	.02235	.02255	.02281	.02314
24	0	.00000	.00208	.00415	.00621	.00824	.01023	.01219
24	1	.04195	.04199	.04203	.04207	.04211	.04214	.04217
24	1	.02364	.02357	.02356	.02362	.02375	.02395	.02422
24	1	.00066	.00277	.00486	.00693	.00897	.01098	.01294
24	2	.04238	.04247	.04255	.04263	.04271	.04279	.04285
24	2	.02528	.02513	.02504	.02503	.02509	.02523	.02545
24	2	.00150	.00362	.00573	.00782	.00987	.01189	.01386
24	3	.04296	.04309	.04323	.04336	.04348	.04360	.04371
24	3	.02706	.02683	.02667	.02659	.02659	.02668	.02685
24	3	.00251	.00466	.00678	.00888	.01095	.01297	.01495
24	4	.04372	.04391	.04410	.04428	.04445	.04461	.04476
24	4	.02901	.02871	.02848	.02833	.02827	.02830	.02842
24	4	.00373	.00589	.00803	.01014	.01222	.01425	.01624
24	5	.04469	.04494	.04519	.04542	.04564	.04585	.04603
24	5	.03117	.03079	.03049	.03028	.03015	.03012	.03018
24	5	.00518	.00736	.00952	.01164	.01372	.01575	.01774
24	6	.04594	.04625	.04655	.04684	.04711	.04736	.04759
24	6	.03359	.03313	.03275	.03246	.03227	.03218	.03219
24	6	.00692	.00911	.01127	.01340	.01548	.01752	.01950
24	7	.04753	.04791	.04827	.04861	.04893	.04922	.04948
24	7	.03630	.03576	.03530	.03494	.03467	.03451	.03446
24	7	.00900	.01120	.01337	.01549	.01757	.01960	.02157

Variance-Covariance factors V1,V2 and V3 for
the BLUEs of μ and σ shape parameter k

n	s	0.0	0.1	0.2	0.3	0.4	0.5	0.6
24	8	.04956	.05000	.05042	.05081	.05117	.05150	.05179
24	8	.03938	.03875	.03821	.03776	.03742	.03719	.03707
24	8	.01150	.01370	.01587	.01798	.02005	.02207	.02403
24	9	.05215	.05266	.05313	.05357	.05397	.05432	.05463
24	9	.04291	.04218	.04154	.04101	.04058	.04027	.04009
24	9	.01452	.01672	.01887	.02098	.02303	.02502	.02696
24	10	.05549	.05606	.05658	.05705	.05748	.05785	.05816
24	10	.04700	.04616	.04542	.04478	.04426	.04387	.04360
24	10	.01822	.02040	.02253	.02460	.02662	.02858	.03048
24	11	.05983	.06044	.06100	.06149	.06193	.06229	.06258
24	11	.05179	.05082	.04996	.04922	.04859	.04810	.04774
24	11	.02277	.02492	.02701	.02904	.03101	.03292	.03476
24	12	.06553	.06617	.06673	.06722	.06763	.06796	.06821
24	12	.05747	.05637	.05537	.05450	.05375	.05315	.05269
24	12	.02846	.03055	.03258	.03454	.03644	.03827	.04003
25	0	.04000	.04000	.04000	.04000	.04000	.03999	.03998
25	0	.02117	.02120	.02128	.02141	.02160	.02184	.02215
25	0	.00000	.00200	.00399	.00596	.00791	.00982	.01170
25	1	.04026	.04030	.04033	.04037	.04040	.04043	.04046
25	1	.02258	.02252	.02251	.02257	.02269	.02288	.02314
25	1	.00061	.00262	.00463	.00662	.00858	.01050	.01239
25	2	.04064	.04072	.04080	.04088	.04095	.04102	.04108
25	2	.02408	.02394	.02386	.02385	.02392	.02405	.02426
25	2	.00136	.00340	.00542	.00742	.00939	.01133	.01322
25	3	.04116	.04129	.04141	.04153	.04164	.04175	.04185
25	3	.02570	.02549	.02534	.02527	.02528	.02536	.02553
25	3	.00228	.00434	.00637	.00839	.01037	.01231	.01421
25	4	.04183	.04201	.04218	.04234	.04250	.04265	.04279
25	4	.02747	.02718	.02698	.02684	.02679	.02682	.02694
25	4	.00337	.00544	.00749	.00952	.01151	.01346	.01536
25	5	.04269	.04292	.04314	.04336	.04356	.04375	.04392
25	5	.02941	.02906	.02879	.02859	.02848	.02846	.02853
25	5	.00466	.00675	.00881	.01085	.01284	.01479	.01670
25	6	.04378	.04406	.04434	.04460	.04485	.04508	.04529
25	6	.03157	.03115	.03080	.03054	.03037	.03029	.03031
25	6	.00619	.00830	.01037	.01241	.01440	.01636	.01826
25	7	.04515	.04549	.04582	.04613	.04643	.04669	.04693
25	7	.03398	.03348	.03307	.03274	.03250	.03236	.03232
25	7	.00801	.01012	.01220	.01424	.01624	.01818	.02008
25	8	.04688	.04728	.04767	.04803	.04836	.04866	.04893
25	8	.03670	.03612	.03562	.03522	.03491	.03471	.03462
25	8	.01018	.01230	.01437	.01641	.01840	.02034	.02222
25	9	.04907	.04953	.04997	.05037	.05074	.05107	.05136
25	9	.03979	.03912	.03854	.03805	.03767	.03740	.03724
25	9	.01278	.01489	.01696	.01899	.02096	.02288	.02475
25	10	.05185	.05237	.05285	.05330	.05369	.05404	.05434
25	10	.04332	.04256	.04188	.04131	.04085	.04050	.04027
25	10	.01592	.01802	.02007	.02207	.02402	.02592	.02775
25	11	.05541	.05598	.05650	.05697	.05738	.05774	.05803
25	11	.04742	.04654	.04577	.04510	.04454	.04410	.04380
25	11	.01973	.02181	.02383	.02580	.02771	.02957	.03136
25	12	.06001	.06061	.06116	.06163	.06204	.06238	.06264
25	12	.05221	.05122	.05033	.04955	.04889	.04835	.04795
25	12	.02443	.02647	.02844	.03036	.03221	.03401	.03574
25	13	.06602	.06664	.06718	.06763	.06800	.06829	.06848
25	13	.05791	.05678	.05575	.05485	.05407	.05342	.05292

Variance-Covariance factors V1,V2 and V3 for
the BLUEs of μ and σ shape parameter k

n	s	0.0	0.1	0.2	0.3	0.4	0.5	0.6
25	13	.03028	.03225	.03416	.03600	.03777	.03948	.04112

Variance-Covariance factors V1,V2 and V3 for the BLUEs of μ and σ shape parameter k

n	s	0.7	0.8	0.9	1.0	1.1	1.2	1.3
2	0	.50000	.50000	.50000	.50000	.50000	.50000	.50000
2	0	.62333	.63908	.65676	.67632	.69770	.72087	.74575
2	0	.17314	.19728	.22121	.24491	.26838	.29162	.31462
3	0	.33328	.33325	.33321	.33315	.33307	.33298	.33286
3	0	.30040	.30794	.31644	.32588	.33625	.34752	.35968
3	0	.11473	.13049	.14604	.16135	.17643	.19125	.20583
3	1	.47165	.47433	.47685	.47921	.48143	.48350	.48543
3	1	.63998	.64725	.65624	.66692	.67926	.69324	.70881
3	1	.33149	.34928	.36697	.38454	.40201	.41938	.43663
4	0	.24993	.24989	.24983	.24974	.24964	.24951	.24935
4	0	.19585	.20068	.20614	.21223	.21894	.22626	.23418
4	0	.08573	.09741	.10889	.12016	.13120	.14203	.15263
4	1	.29939	.30082	.30215	.30338	.30450	.30552	.30643
4	1	.30483	.30892	.31394	.31989	.32674	.33449	.34312
4	1	.15915	.17166	.18399	.19614	.20811	.21989	.23148
5	0	.19993	.19988	.19981	.19972	.19961	.19946	.19929
5	0	.14467	.14816	.15213	.15657	.16147	.16684	.17267
5	0	.06842	.07768	.08676	.09566	.10436	.11286	.12116
5	1	.22530	.22617	.22697	.22769	.22834	.22891	.22939
5	1	.19750	.20040	.20394	.20812	.21293	.21837	.22442
5	1	.10503	.11474	.12427	.13363	.14281	.15181	.16063
5	2	.29876	.30014	.30133	.30233	.30315	.30379	.30425
5	2	.30977	.31200	.31511	.31907	.32389	.32955	.33602
5	2	.19584	.20560	.21519	.22463	.23392	.24305	.25203
6	0	.16659	.16655	.16648	.16639	.16627	.16613	.16595
6	0	.11445	.11716	.12025	.12371	.12755	.13176	.13634
6	0	.05691	.06458	.07209	.07943	.08660	.09358	.10039
6	1	.18195	.18253	.18305	.18351	.18392	.18426	.18453
6	1	.14535	.14761	.15035	.15358	.15730	.16149	.16615
6	1	.07870	.08664	.09443	.10205	.10951	.11680	.12392
6	2	.21750	.21857	.21951	.22031	.22099	.22153	.22195
6	2	.19997	.20172	.20409	.20706	.21065	.21483	.21960
6	2	.12276	.13080	.13869	.14641	.15398	.16139	.16864
6	3	.31104	.31159	.31190	.31196	.31181	.31143	.31086
6	3	.31407	.31511	.31697	.31964	.32311	.32737	.33240
6	3	.22607	.23350	.24081	.24799	.25504	.26198	.26878
7	0	.14279	.14274	.14267	.14259	.14247	.14233	.14215
7	0	.09455	.09675	.09926	.10209	.10523	.10868	.11244
7	0	.04871	.05525	.06165	.06790	.07398	.07991	.08566
7	1	.15304	.15344	.15380	.15412	.15438	.15459	.15474
7	1	.11473	.11657	.11881	.12144	.12446	.12787	.13166
7	1	.06309	.06982	.07640	.08284	.08912	.09524	.10122
7	2	.17377	.17457	.17527	.17588	.17639	.17680	.17711
7	2	.14676	.14822	.15016	.15257	.15546	.15881	.16262
7	2	.08886	.09568	.10235	.10886	.11523	.12146	.12753
7	3	.21782	.21867	.21935	.21985	.22019	.22036	.22037
7	3	.20234	.20331	.20486	.20700	.20972	.21301	.21686
7	3	.13834	.14496	.15144	.15778	.16399	.17005	.17598
7	4	.32871	.32803	.32707	.32586	.32442	.32275	.32089
7	4	.31772	.31792	.31891	.32066	.32318	.32643	.33041
7	4	.25146	.25692	.26229	.26755	.27273	.27781	.28281
8	0	.12493	.12489	.12482	.12474	.12463	.12449	.12432
8	0	.08048	.08233	.08444	.08682	.08946	.09238	.09556
8	0	.04257	.04828	.05385	.05928	.06456	.06970	.07469
8	1	.13224	.13254	.13279	.13301	.13319	.13332	.13340
8	1	.09464	.09620	.09809	.10031	.10285	.10572	.10891

Variance-Covariance factors V1,V2 and V3 for
the BLUEs of μ and σ shape parameter k

n	s	0.7	0.8	0.9	1.0	1.1	1.2	1.3
8	1	.05274	.05858	.06428	.06984	.07527	.08055	.08570
8	2	.14572	.14633	.14687	.14733	.14771	.14801	.14823
8	2	.11560	.11686	.11850	.12054	.12296	.12576	.12894
8	2	.06955	.07546	.08123	.08686	.09236	.09771	.10293
8	3	.17091	.17169	.17233	.17284	.17323	.17349	.17362
8	3	.14823	.14913	.15048	.15230	.15457	.15729	.16045
8	3	.09823	.10406	.10976	.11533	.12076	.12606	.13122
8	4	.22239	.22278	.22296	.22294	.22274	.22236	.22180
8	4	.20448	.20487	.20583	.20735	.20943	.21205	.21520
8	4	.15203	.15743	.16270	.16785	.17288	.17779	.18258
9	0	.11105	.11100	.11094	.11086	.11076	.11062	.11046
9	0	.07003	.07161	.07342	.07547	.07775	.08026	.08302
9	0	.03781	.04286	.04779	.05260	.05727	.06180	.06620
9	1	.11650	.11673	.11692	.11708	.11719	.11727	.11730
9	1	.08048	.08183	.08346	.08537	.08757	.09004	.09279
9	1	.04536	.05051	.05554	.06044	.06522	.06986	.07437
9	2	.12593	.12640	.12682	.12717	.12746	.12768	.12784
9	2	.09521	.09632	.09775	.09951	.10160	.10401	.10674
9	2	.05714	.06235	.06743	.07239	.07722	.08192	.08650
9	3	.14207	.14272	.14327	.14372	.14407	.14432	.14446
9	3	.11657	.11740	.11861	.12019	.12215	.12448	.12718
9	3	.07572	.08090	.08596	.09089	.09570	.10037	.10493
9	4	.17118	.17177	.17219	.17247	.17260	.17258	.17243
9	4	.14962	.15009	.15100	.15236	.15416	.15639	.15905
9	4	.10673	.11171	.11657	.12130	.12591	.13041	.13478
9	5	.22927	.22905	.22860	.22795	.22709	.22606	.22485
9	5	.20639	.20634	.20684	.20788	.20945	.21155	.21415
9	5	.16416	.16848	.17269	.17680	.18081	.18472	.18853
10	0	.09994	.09990	.09984	.09976	.09966	.09953	.09938
10	0	.06195	.06333	.06491	.06671	.06871	.07092	.07334
10	0	.03400	.03854	.04296	.04726	.05145	.05550	.05943
10	1	.10416	.10433	.10448	.10459	.10467	.10471	.10472
10	1	.06997	.07116	.07259	.07427	.07620	.07837	.08079
10	1	.03982	.04443	.04893	.05331	.05757	.06171	.06573
10	2	.11110	.11147	.11180	.11208	.11230	.11247	.11257
10	2	.08086	.08185	.08313	.08468	.08651	.08863	.09103
10	2	.04851	.05317	.05771	.06214	.06644	.07063	.07470
10	3	.12225	.12279	.12325	.12363	.12392	.12413	.12426
10	3	.09589	.09666	.09774	.09915	.10088	.10293	.10528
10	3	.06146	.06611	.07065	.07506	.07936	.08355	.08761
10	4	.14074	.14131	.14177	.14211	.14233	.14243	.14241
10	4	.11754	.11803	.11889	.12012	.12172	.12367	.12598
10	4	.08146	.08601	.09044	.09475	.09895	.10303	.10699
10	5	.17336	.17365	.17377	.17371	.17350	.17313	.17262
10	5	.15091	.15104	.15160	.15259	.15401	.15585	.15810
10	5	.11446	.11868	.12279	.12679	.13068	.13446	.13815
11	0	.09085	.09081	.09076	.09068	.09059	.09046	.09031
11	0	.05553	.05675	.05816	.05975	.06153	.06350	.06566
11	0	.03089	.03500	.03901	.04291	.04670	.05036	.05391
11	1	.09421	.09435	.09446	.09454	.09459	.09461	.09459
11	1	.06187	.06293	.06421	.06571	.06742	.06936	.07151
11	1	.03551	.03968	.04375	.04771	.05156	.05529	.05892
11	2	.09951	.09982	.10008	.10030	.10047	.10060	.10067
11	2	.07024	.07113	.07227	.07366	.07530	.07719	.07932
11	2	.04217	.04638	.05048	.05448	.05836	.06214	.06580
11	3	.10764	.10809	.10847	.10879	.10904	.10921	.10932

Variance-Covariance factors V1,V2 and V3 for the BLUEs of μ and σ shape parameter k

n	s	0.7	0.8	0.9	1.0	1.1	1.2	1.3
11	3	.08136	.08207	.08306	.08433	.08587	.08770	.08979
11	3	.05167	.05589	.05999	.06399	.06788	.07165	.07532
11	4	.12032	.12084	.12127	.12160	.12183	.12196	.12199
11	4	.09659	.09708	.09789	.09902	.10045	.10220	.10425
11	4	.06557	.06973	.07377	.07771	.08153	.08525	.08886
11	5	.14092	.14135	.14164	.14179	.14182	.14171	.14149
11	5	.11845	.11868	.11926	.12020	.12150	.12315	.12514
11	5	.08679	.09077	.09463	.09839	.10205	.10559	.10904
12	0	.08328	.08324	.08319	.08312	.08302	.08291	.08277
12	0	.05030	.05140	.05266	.05409	.05569	.05747	.05941
12	0	.02830	.03206	.03573	.03929	.04275	.04609	.04933
12	1	.08601	.08612	.08621	.08627	.08630	.08630	.08627
12	1	.05544	.05640	.05755	.05889	.06044	.06218	.06412
12	1	.03204	.03586	.03957	.04318	.04669	.05009	.05339
12	2	.09018	.09044	.09065	.09083	.09096	.09105	.09110
12	2	.06206	.06288	.06391	.06517	.06665	.06835	.07026
12	2	.03730	.04115	.04489	.04853	.05207	.05551	.05884
12	3	.09635	.09673	.09705	.09732	.09752	.09767	.09775
12	3	.07060	.07127	.07218	.07333	.07473	.07638	.07826
12	3	.04456	.04841	.05216	.05581	.05935	.06279	.06613
12	4	.10554	.10600	.10638	.10668	.10689	.10703	.10708
12	4	.08188	.08237	.08313	.08416	.08547	.08705	.08890
12	4	.05474	.05855	.06227	.06588	.06938	.07279	.07609
12	5	.11959	.12004	.12038	.12060	.12072	.12073	.12063
12	5	.09727	.09754	.09813	.09901	.10021	.10171	.10350
12	5	.06945	.07315	.07676	.08026	.08366	.08696	.09016
12	6	.14213	.14235	.14242	.14235	.14214	.14180	.14133
12	6	.11931	.11932	.11967	.12037	.12142	.12281	.12454
12	6	.09173	.09519	.09855	.10181	.10498	.10805	.11103
13	0	.07687	.07683	.07678	.07672	.07663	.07652	.07638
13	0	.04597	.04696	.04810	.04940	.05085	.05247	.05424
13	0	.02611	.02958	.03295	.03623	.03941	.04249	.04546
13	1	.07913	.07922	.07929	.07933	.07935	.07934	.07930
13	1	.05021	.05108	.05213	.05335	.05476	.05634	.05811
13	1	.02921	.03271	.03613	.03945	.04267	.04580	.04882
13	2	.08250	.08271	.08289	.08303	.08314	.08320	.08323
13	2	.05557	.05632	.05727	.05842	.05976	.06131	.06305
13	2	.03346	.03699	.04043	.04378	.04703	.05018	.05323
13	3	.08733	.08765	.08792	.08814	.08831	.08843	.08849
13	3	.06234	.06296	.06380	.06486	.06614	.06763	.06935
13	3	.03917	.04271	.04616	.04951	.05277	.05593	.05899
13	4	.09426	.09466	.09499	.09526	.09545	.09558	.09563
13	4	.07101	.07148	.07220	.07315	.07436	.07580	.07748
13	4	.04692	.05044	.05387	.05720	.06043	.06357	.06661
13	5	.10440	.10482	.10515	.10539	.10554	.10560	.10556
13	5	.08240	.08270	.08327	.08411	.08522	.08659	.08823
13	5	.05767	.06112	.06448	.06774	.07090	.07397	.07694
13	6	.11972	.12005	.12025	.12034	.12030	.12016	.11990
13	6	.09792	.09801	.09840	.09909	.10008	.10137	.10295
13	6	.07309	.07639	.07959	.08270	.08571	.08864	.09147
13	7	.14403	.14401	.14383	.14350	.14303	.14242	.14169
13	7	.12011	.11993	.12009	.12059	.12143	.12261	.12410
13	7	.09632	.09931	.10221	.10502	.10774	.11038	.11294
14	0	.07138	.07134	.07130	.07123	.07115	.07104	.07091
14	0	.04232	.04322	.04427	.04545	.04678	.04826	.04988
14	0	.02423	.02745	.03058	.03362	.03656	.03940	.04215

Variance-Covariance factors V1,V2 and V3 for
the BLUEs of μ and σ shape parameter k

n	s	0.7	0.8	0.9	1.0	1.1	1.2	1.3
14	1	.07328	.07335	.07341	.07344	.07345	.07342	.07337
14	1	.04587	.04667	.04763	.04876	.05004	.05150	.05311
14	1	.02684	.03008	.03325	.03632	.03930	.04218	.04498
14	2	.07605	.07623	.07638	.07649	.07658	.07663	.07663
14	2	.05030	.05100	.05187	.05293	.05416	.05558	.05718
14	2	.03034	.03361	.03679	.03989	.04289	.04580	.04862
14	3	.07992	.08020	.08043	.08062	.08076	.08085	.08090
14	3	.05579	.05637	.05715	.05813	.05931	.06068	.06226
14	3	.03495	.03823	.04142	.04452	.04753	.05045	.05327
14	4	.08532	.08567	.08597	.08620	.08637	.08648	.08653
14	4	.06265	.06311	.06378	.06468	.06579	.06712	.06866
14	4	.04103	.04430	.04748	.05056	.05356	.05646	.05928
14	5	.09294	.09333	.09365	.09388	.09403	.09411	.09411
14	5	.07141	.07173	.07228	.07308	.07411	.07538	.07688
14	5	.04920	.05243	.05556	.05860	.06154	.06440	.06717
14	6	.10396	.10431	.10457	.10472	.10477	.10473	.10458
14	6	.08291	.08305	.08346	.08413	.08507	.08627	.08772
14	6	.06046	.06358	.06661	.06954	.07239	.07515	.07783
14	7	.12045	.12063	.12068	.12059	.12039	.12007	.11964
14	7	.09853	.09847	.09870	.09922	.10004	.10114	.10253
14	7	.07651	.07944	.08227	.08502	.08768	.09026	.09276
15	0	.06662	.06659	.06654	.06648	.06640	.06629	.06617
15	0	.03920	.04003	.04099	.04208	.04330	.04467	.04616
15	0	.02261	.02561	.02852	.03135	.03409	.03674	.03929
15	1	.06824	.06830	.06834	.06836	.06836	.06833	.06828
15	1	.04223	.04296	.04385	.04488	.04607	.04741	.04890
15	1	.02482	.02785	.03079	.03365	.03642	.03910	.04170
15	2	.07056	.07071	.07083	.07093	.07099	.07103	.07102
15	2	.04594	.04659	.04740	.04837	.04952	.05082	.05230
15	2	.02776	.03080	.03376	.03664	.03943	.04214	.04475
15	3	.07373	.07396	.07416	.07432	.07444	.07452	.07455
15	3	.05047	.05102	.05175	.05266	.05375	.05502	.05647
15	3	.03155	.03460	.03757	.04045	.04325	.04596	.04859
15	4	.07804	.07835	.07860	.07881	.07896	.07906	.07910
15	4	.05604	.05648	.05712	.05795	.05898	.06021	.06164
15	4	.03645	.03949	.04245	.04533	.04812	.05082	.05344
15	5	.08395	.08431	.08459	.08481	.08496	.08504	.08505
15	5	.06298	.06330	.06383	.06458	.06555	.06673	.06812
15	5	.04285	.04587	.04880	.05164	.05439	.05706	.05965
15	6	.09221	.09256	.09283	.09301	.09310	.09310	.09302
15	6	.07182	.07200	.07241	.07306	.07395	.07507	.07641
15	6	.05140	.05434	.05720	.05997	.06266	.06526	.06778
15	7	.10404	.10430	.10445	.10450	.10443	.10426	.10400
15	7	.08339	.08340	.08367	.08420	.08499	.08603	.08733
15	7	.06310	.06591	.06864	.07129	.07385	.07633	.07872
15	8	.12163	.12163	.12150	.12123	.12084	.12033	.11971
15	8	.09911	.09891	.09900	.09938	.10005	.10099	.10222
15	8	.07973	.08231	.08481	.08723	.08957	.09183	.09402
16	0	.06245	.06242	.06238	.06232	.06224	.06214	.06202
16	0	.03651	.03728	.03816	.03917	.04031	.04157	.04296
16	0	.02119	.02400	.02672	.02937	.03193	.03441	.03680
16	1	.06385	.06390	.06393	.06395	.06394	.06391	.06385
16	1	.03911	.03980	.04062	.04158	.04268	.04392	.04530
16	1	.02310	.02593	.02868	.03135	.03394	.03645	.03887
16	2	.06582	.06594	.06605	.06613	.06618	.06620	.06619
16	2	.04227	.04288	.04363	.04454	.04560	.04681	.04818

Variance-Covariance factors V1,V2 and V3 for
the BLUEs of μ and σ shape parameter k

n	s	0.7	0.8	0.9	1.0	1.1	1.2	1.3
16	2	.02559	.02843	.03120	.03389	.03650	.03902	.04146
16	3	.06845	.06866	.06883	.06897	.06907	.06913	.06915
16	3	.04608	.04659	.04728	.04813	.04914	.05032	.05167
16	3	.02876	.03161	.03439	.03708	.03970	.04223	.04468
16	4	.07197	.07224	.07247	.07265	.07278	.07286	.07290
16	4	.05067	.05110	.05170	.05249	.05345	.05460	.05592
16	4	.03278	.03563	.03840	.04109	.04370	.04622	.04867
16	5	.07668	.07700	.07726	.07746	.07760	.07768	.07770
16	5	.05630	.05662	.05714	.05785	.05876	.05986	.06115
16	5	.03793	.04076	.04351	.04617	.04875	.05126	.05368
16	6	.08308	.08341	.08367	.08385	.08396	.08399	.08395
16	6	.06330	.06351	.06392	.06455	.06539	.06644	.06769
16	6	.04462	.04740	.05010	.05271	.05525	.05770	.06007
16	7	.09193	.09222	.09241	.09252	.09253	.09245	.09228
16	7	.07221	.07227	.07257	.07309	.07385	.07484	.07605
16	7	.05349	.05618	.05879	.06132	.06376	.06613	.06842
16	8	.10452	.10467	.10470	.10461	.10442	.10412	.10372
16	8	.08385	.08375	.08390	.08430	.08496	.08587	.08702
16	8	.06560	.06814	.07059	.07296	.07526	.07748	.07962
17	0	.05878	.05875	.05871	.05865	.05857	.05848	.05836
17	0	.03416	.03487	.03570	.03663	.03769	.03887	.04016
17	0	.01994	.02257	.02514	.02762	.03003	.03235	.03460
17	1	.06000	.06004	.06006	.06007	.06006	.06002	.05996
17	1	.03642	.03706	.03783	.03872	.03975	.04091	.04220
17	1	.02159	.02425	.02684	.02935	.03178	.03413	.03640
17	2	.06168	.06179	.06188	.06195	.06199	.06200	.06198
17	2	.03914	.03971	.04042	.04126	.04225	.04339	.04466
17	2	.02374	.02641	.02901	.03153	.03398	.03634	.03863
17	3	.06391	.06409	.06424	.06435	.06444	.06449	.06450
17	3	.04238	.04287	.04351	.04431	.04526	.04636	.04762
17	3	.02642	.02910	.03171	.03424	.03669	.03906	.04136
17	4	.06683	.06707	.06727	.06743	.06754	.06761	.06764
17	4	.04624	.04665	.04722	.04796	.04887	.04994	.05118
17	4	.02978	.03246	.03506	.03759	.04004	.04241	.04470
17	5	.07066	.07095	.07118	.07136	.07149	.07156	.07158
17	5	.05089	.05121	.05171	.05238	.05324	.05427	.05548
17	5	.03400	.03667	.03925	.04176	.04419	.04655	.04882
17	6	.07574	.07605	.07630	.07648	.07659	.07663	.07661
17	6	.05657	.05679	.05720	.05780	.05860	.05958	.06076
17	6	.03937	.04200	.04455	.04702	.04942	.05173	.05397
17	7	.08258	.08288	.08309	.08322	.08327	.08324	.08314
17	7	.06362	.06372	.06403	.06455	.06528	.06621	.06735
17	7	.04632	.04888	.05137	.05377	.05610	.05835	.06053
17	8	.09199	.09220	.09231	.09233	.09224	.09207	.09180
17	8	.07258	.07255	.07274	.07315	.07380	.07467	.07576
17	8	.05550	.05795	.06033	.06262	.06484	.06699	.06907
17	9	.10530	.10531	.10521	.10498	.10465	.10421	.10367
17	9	.08429	.08409	.08413	.08443	.08497	.08576	.08679
17	9	.06798	.07025	.07245	.07457	.07662	.07859	.08051
18	0	.05551	.05548	.05544	.05539	.05531	.05522	.05511
18	0	.03209	.03276	.03353	.03440	.03539	.03649	.03770
18	0	.01882	.02131	.02373	.02607	.02834	.03053	.03264
18	1	.05658	.05662	.05664	.05664	.05662	.05658	.05652
18	1	.03408	.03468	.03539	.03623	.03719	.03827	.03948
18	1	.02028	.02279	.02522	.02759	.02988	.03209	.03422
18	2	.05804	.05814	.05821	.05827	.05830	.05830	.05828

Variance-Covariance factors V1,V2 and V3 for
the BLUEs of μ and σ shape parameter k

n	s	0.7	0.8	0.9	1.0	1.1	1.2	1.3
18	2	.03644	.03698	.03764	.03843	.03936	.04043	.04162
18	2	.02214	.02466	.02711	.02948	.03179	.03401	.03617
18	3	.05995	.06010	.06023	.06034	.06041	.06044	.06045
18	3	.03923	.03970	.04030	.04105	.04194	.04298	.04416
18	3	.02444	.02697	.02942	.03181	.03412	.03635	.03851
18	4	.06240	.06262	.06279	.06293	.06303	.06309	.06311
18	4	.04252	.04291	.04345	.04415	.04501	.04601	.04717
18	4	.02728	.02981	.03227	.03465	.03695	.03919	.04134
18	5	.06558	.06583	.06605	.06621	.06633	.06639	.06641
18	5	.04642	.04674	.04721	.04786	.04866	.04964	.05077
18	5	.03080	.03332	.03576	.03813	.04042	.04264	.04479
18	6	.06970	.06999	.07022	.07039	.07050	.07055	.07054
18	6	.05111	.05134	.05174	.05232	.05308	.05401	.05512
18	6	.03520	.03769	.04011	.04245	.04472	.04691	.04903
18	7	.07513	.07541	.07563	.07577	.07584	.07584	.07577
18	7	.05683	.05696	.05728	.05779	.05848	.05937	.06044
18	7	.04077	.04322	.04558	.04787	.05009	.05223	.05431
18	8	.08239	.08263	.08278	.08285	.08283	.08273	.08255
18	8	.06393	.06394	.06416	.06458	.06521	.06604	.06708
18	8	.04795	.05031	.05260	.05481	.05695	.05901	.06101
18	9	.09232	.09244	.09245	.09237	.09218	.09190	.09153
18	9	.07294	.07282	.07291	.07323	.07378	.07454	.07553
18	9	.05741	.05964	.06180	.06388	.06590	.06784	.06972
19	0	.05259	.05256	.05252	.05247	.05240	.05231	.05220
19	0	.03026	.03088	.03160	.03243	.03335	.03439	.03553
19	0	.01783	.02018	.02247	.02469	.02683	.02890	.03090
19	1	.05353	.05356	.05358	.05358	.05356	.05352	.05345
19	1	.03201	.03258	.03325	.03404	.03494	.03596	.03709
19	1	.01911	.02149	.02379	.02603	.02819	.03028	.03230
19	2	.05481	.05490	.05496	.05501	.05503	.05503	.05500
19	2	.03409	.03459	.03522	.03597	.03684	.03784	.03897
19	2	.02074	.02312	.02544	.02769	.02986	.03197	.03400
19	3	.05646	.05660	.05671	.05680	.05686	.05689	.05688
19	3	.03651	.03696	.03753	.03824	.03908	.04005	.04116
19	3	.02274	.02513	.02745	.02970	.03189	.03400	.03603
19	4	.05855	.05874	.05890	.05902	.05911	.05916	.05917
19	4	.03934	.03972	.04024	.04090	.04171	.04266	.04375
19	4	.02517	.02756	.02989	.03214	.03432	.03643	.03847
19	5	.06122	.06145	.06164	.06179	.06190	.06196	.06197
19	5	.04267	.04298	.04344	.04405	.04481	.04573	.04680
19	5	.02815	.03053	.03285	.03509	.03726	.03936	.04139
19	6	.06463	.06489	.06510	.06526	.06537	.06542	.06541
19	6	.04661	.04684	.04723	.04779	.04851	.04940	.05044
19	6	.03182	.03418	.03647	.03869	.04084	.04292	.04493
19	7	.06903	.06930	.06951	.06965	.06973	.06975	.06971
19	7	.05134	.05148	.05180	.05230	.05297	.05381	.05483
19	7	.03638	.03870	.04096	.04314	.04525	.04729	.04927
19	8	.07478	.07503	.07520	.07530	.07532	.07527	.07514
19	8	.05709	.05714	.05738	.05780	.05841	.05920	.06019
19	8	.04213	.04440	.04659	.04871	.05077	.05275	.05466
19	9	.08243	.08261	.08269	.08269	.08259	.08241	.08215
19	9	.06423	.06416	.06430	.06464	.06518	.06592	.06685
19	9	.04952	.05169	.05379	.05582	.05778	.05967	.06150
19	10	.09285	.09287	.09278	.09259	.09230	.09190	.09142
19	10	.07328	.07308	.07309	.07333	.07379	.07446	.07534
19	10	.05924	.06126	.06321	.06510	.06692	.06868	.07037

Variance-Covariance factors V1,V2 and V3 for
the BLUEs of μ and σ shape parameter k

n	s	0.7	0.8	0.9	1.0	1.1	1.2	1.3
20	0	.04996	.04993	.04989	.04984	.04978	.04969	.04959
20	0	.02863	.02921	.02989	.03066	.03154	.03251	.03358
20	0	.01693	.01917	.02134	.02344	.02548	.02744	.02933
20	1	.05080	.05083	.05084	.05083	.05081	.05077	.05070
20	1	.03019	.03072	.03135	.03209	.03295	.03391	.03498
20	1	.01808	.02033	.02251	.02463	.02668	.02866	.03058
20	2	.05193	.05200	.05206	.05210	.05211	.05211	.05207
20	2	.03202	.03250	.03309	.03380	.03462	.03556	.03663
20	2	.01951	.02177	.02397	.02610	.02816	.03016	.03208
20	3	.05336	.05348	.05359	.05366	.05371	.05373	.05373
20	3	.03415	.03457	.03511	.03578	.03658	.03750	.03855
20	3	.02126	.02353	.02573	.02786	.02993	.03193	.03386
20	4	.05517	.05534	.05548	.05559	.05566	.05571	.05571
20	4	.03661	.03697	.03747	.03810	.03886	.03976	.04079
20	4	.02337	.02564	.02784	.02997	.03204	.03404	.03597
20	5	.05744	.05765	.05782	.05795	.05805	.05810	.05811
20	5	.03947	.03977	.04022	.04080	.04153	.04240	.04341
20	5	.02592	.02818	.03038	.03250	.03456	.03656	.03848
20	6	.06030	.06054	.06073	.06088	.06098	.06103	.06103
20	6	.04283	.04306	.04344	.04398	.04467	.04551	.04650
20	6	.02901	.03126	.03344	.03555	.03760	.03957	.04148
20	7	.06392	.06418	.06438	.06452	.06461	.06463	.06460
20	7	.04680	.04696	.04728	.04776	.04840	.04921	.05017
20	7	.03281	.03503	.03718	.03926	.04128	.04322	.04510
20	8	.06858	.06883	.06901	.06912	.06916	.06914	.06905
20	8	.05156	.05163	.05188	.05229	.05289	.05365	.05458
20	8	.03752	.03969	.04180	.04383	.04580	.04770	.04953
20	9	.07464	.07484	.07496	.07501	.07497	.07486	.07467
20	9	.05734	.05732	.05748	.05783	.05836	.05908	.05997
20	9	.04344	.04554	.04757	.04954	.05144	.05327	.05504
20	10	.08266	.08277	.08277	.08269	.08251	.08225	.08190
20	10	.06452	.06438	.06444	.06470	.06516	.06582	.06667
20	10	.05103	.05302	.05495	.05680	.05860	.06033	.06200
21	0	.04758	.04755	.04752	.04747	.04740	.04732	.04722
21	0	.02716	.02771	.02835	.02908	.02991	.03083	.03184
21	0	.01612	.01825	.02032	.02232	.02425	.02612	.02791
21	1	.04833	.04835	.04836	.04835	.04833	.04829	.04822
21	1	.02855	.02906	.02966	.03036	.03116	.03207	.03309
21	1	.01715	.01929	.02137	.02338	.02533	.02721	.02903
21	2	.04933	.04940	.04945	.04948	.04949	.04948	.04944
21	2	.03019	.03064	.03120	.03187	.03265	.03355	.03455
21	2	.01842	.02057	.02266	.02469	.02665	.02854	.03037
21	3	.05059	.05070	.05079	.05086	.05090	.05092	.05090
21	3	.03207	.03247	.03299	.03363	.03438	.03525	.03625
21	3	.01996	.02212	.02421	.02624	.02821	.03011	.03194
21	4	.05217	.05232	.05244	.05254	.05261	.05264	.05265
21	4	.03423	.03458	.03505	.03565	.03638	.03723	.03821
21	4	.02181	.02396	.02606	.02809	.03005	.03195	.03379
21	5	.05412	.05431	.05447	.05459	.05467	.05472	.05472
21	5	.03672	.03701	.03744	.03800	.03869	.03952	.04048
21	5	.02401	.02617	.02825	.03028	.03224	.03413	.03596
21	6	.05655	.05677	.05695	.05708	.05717	.05722	.05722
21	6	.03961	.03984	.04022	.04073	.04139	.04219	.04313
21	6	.02666	.02880	.03088	.03289	.03484	.03672	.03853
21	7	.05959	.05983	.06001	.06015	.06024	.06027	.06025
21	7	.04299	.04316	.04347	.04394	.04456	.04533	.04624

Variance-Covariance factors V1,V2 and V3 for
the BLUEs of μ and σ shape parameter k

n	s	0.7	0.8	0.9	1.0	1.1	1.2	1.3
21	7	.02987	.03199	.03404	.03603	.03795	.03981	.04160
21	8	.06342	.06366	.06384	.06396	.06402	.06401	.06395
21	8	.04698	.04708	.04733	.04774	.04832	.04905	.04994
21	8	.03378	.03587	.03788	.03984	.04172	.04354	.04530
21	9	.06832	.06853	.06868	.06875	.06875	.06868	.06854
21	9	.05177	.05178	.05196	.05231	.05283	.05352	.05438
21	9	.03862	.04065	.04262	.04451	.04634	.04811	.04982
21	10	.07467	.07482	.07488	.07486	.07476	.07459	.07433
21	10	.05759	.05750	.05760	.05787	.05833	.05898	.05980
21	10	.04470	.04665	.04853	.05034	.05210	.05379	.05542
21	11	.08305	.08307	.08300	.08282	.08256	.08221	.08177
21	11	.06480	.06459	.06459	.06478	.06517	.06575	.06652
21	11	.05247	.05430	.05606	.05776	.05940	.06097	.06249
22	0	.04542	.04539	.04536	.04531	.04525	.04517	.04507
22	0	.02583	.02635	.02696	.02765	.02843	.02931	.03027
22	0	.01539	.01742	.01939	.02129	.02314	.02491	.02663
22	1	.04610	.04611	.04612	.04611	.04608	.04604	.04597
22	1	.02709	.02757	.02814	.02880	.02956	.03043	.03139
22	1	.01631	.01835	.02033	.02225	.02411	.02590	.02763
22	2	.04699	.04705	.04709	.04712	.04712	.04711	.04707
22	2	.02855	.02898	.02952	.03015	.03090	.03174	.03270
22	2	.01745	.01950	.02149	.02342	.02529	.02709	.02883
22	3	.04810	.04820	.04828	.04834	.04838	.04839	.04837
22	3	.03023	.03061	.03111	.03171	.03243	.03326	.03420
22	3	.01882	.02087	.02287	.02480	.02667	.02848	.03023
22	4	.04948	.04962	.04974	.04982	.04988	.04991	.04991
22	4	.03213	.03247	.03293	.03350	.03419	.03500	.03593
22	4	.02044	.02250	.02450	.02643	.02830	.03011	.03186
22	5	.05118	.05135	.05150	.05161	.05168	.05172	.05172
22	5	.03432	.03461	.03502	.03556	.03622	.03701	.03792
22	5	.02237	.02442	.02641	.02834	.03021	.03202	.03376
22	6	.05327	.05347	.05363	.05376	.05384	.05389	.05389
22	6	.03683	.03707	.03743	.03793	.03856	.03932	.04022
22	6	.02466	.02670	.02868	.03060	.03246	.03426	.03599
22	7	.05585	.05607	.05624	.05637	.05646	.05649	.05648
22	7	.03974	.03992	.04023	.04069	.04128	.04202	.04289
22	7	.02740	.02942	.03139	.03329	.03513	.03691	.03862
22	8	.05905	.05928	.05946	.05958	.05964	.05965	.05960
22	8	.04315	.04326	.04351	.04392	.04447	.04518	.04603
22	8	.03070	.03270	.03464	.03651	.03832	.04007	.04175
22	9	.06309	.06330	.06346	.06354	.06357	.06353	.06342
22	9	.04717	.04721	.04740	.04775	.04826	.04892	.04975
22	9	.03473	.03668	.03858	.04041	.04217	.04388	.04552
22	10	.06821	.06839	.06849	.06851	.06846	.06834	.06815
22	10	.05198	.05193	.05205	.05234	.05280	.05342	.05421
22	10	.03970	.04159	.04342	.04518	.04689	.04853	.05012
22	11	.07484	.07493	.07493	.07484	.07468	.07443	.07411
22	11	.05783	.05768	.05771	.05793	.05833	.05890	.05965
22	11	.04592	.04772	.04946	.05113	.05275	.05430	.05581
23	0	.04344	.04342	.04338	.04334	.04328	.04320	.04311
23	0	.02463	.02512	.02570	.02635	.02710	.02793	.02884
23	0	.01471	.01665	.01854	.02036	.02212	.02382	.02545
23	1	.04406	.04407	.04407	.04406	.04404	.04399	.04393
23	1	.02577	.02622	.02676	.02739	.02812	.02894	.02985
23	1	.01555	.01750	.01939	.02123	.02300	.02471	.02636
23	2	.04486	.04491	.04495	.04497	.04497	.04495	.04491

Variance-Covariance factors V1,V2 and V3 for
the BLUEs of μ and σ shape parameter k

n	s	0.7	0.8	0.9	1.0	1.1	1.2	1.3
23	2	.02708	.02750	.02800	.02861	.02932	.03012	.03103
23	2	.01658	.01854	.02044	.02228	.02406	.02578	.02744
23	3	.04585	.04594	.04601	.04607	.04609	.04610	.04608
23	3	.02858	.02895	.02943	.03000	.03069	.03148	.03237
23	3	.01780	.01976	.02167	.02351	.02530	.02703	.02869
23	4	.04707	.04720	.04730	.04738	.04743	.04745	.04744
23	4	.03028	.03061	.03105	.03159	.03225	.03303	.03391
23	4	.01924	.02120	.02311	.02496	.02675	.02847	.03014
23	5	.04856	.04872	.04885	.04894	.04901	.04905	.04905
23	5	.03221	.03250	.03289	.03341	.03404	.03479	.03566
23	5	.02093	.02290	.02480	.02664	.02843	.03015	.03182
23	6	.05037	.05055	.05070	.05082	.05090	.05094	.05094
23	6	.03442	.03465	.03501	.03549	.03609	.03682	.03768
23	6	.02293	.02489	.02678	.02862	.03040	.03211	.03377
23	7	.05258	.05279	.05295	.05308	.05315	.05319	.05318
23	7	.03695	.03713	.03744	.03788	.03845	.03916	.03999
23	7	.02530	.02724	.02912	.03094	.03270	.03440	.03605
23	8	.05530	.05552	.05569	.05580	.05587	.05589	.05585
23	8	.03989	.04001	.04026	.04066	.04120	.04187	.04269
23	8	.02812	.03004	.03190	.03370	.03543	.03711	.03873
23	9	.05867	.05888	.05904	.05914	.05917	.05915	.05908
23	9	.04331	.04336	.04357	.04392	.04441	.04506	.04584
23	9	.03152	.03340	.03523	.03699	.03869	.04033	.04192
23	10	.06289	.06308	.06320	.06325	.06323	.06315	.06301
23	10	.04735	.04733	.04747	.04777	.04822	.04882	.04958
23	10	.03565	.03748	.03926	.04097	.04262	.04422	.04575
23	11	.06824	.06837	.06842	.06839	.06829	.06811	.06786
23	11	.05218	.05208	.05215	.05238	.05278	.05334	.05407
23	11	.04073	.04250	.04420	.04584	.04742	.04895	.05042
23	12	.07513	.07515	.07508	.07493	.07469	.07437	.07397
23	12	.05805	.05786	.05784	.05799	.05833	.05884	.05953
23	12	.04709	.04875	.05036	.05190	.05338	.05482	.05619
24	0	.04163	.04161	.04157	.04153	.04147	.04140	.04131
24	0	.02353	.02400	.02455	.02517	.02588	.02667	.02755
24	0	.01410	.01596	.01776	.01950	.02119	.02281	.02438
24	1	.04219	.04220	.04220	.04219	.04216	.04212	.04205
24	1	.02457	.02500	.02552	.02612	.02681	.02759	.02846
24	1	.01486	.01673	.01854	.02029	.02199	.02363	.02521
24	2	.04291	.04296	.04299	.04301	.04301	.04299	.04295
24	2	.02576	.02615	.02664	.02722	.02789	.02866	.02952
24	2	.01579	.01766	.01948	.02124	.02295	.02459	.02618
24	3	.04381	.04389	.04395	.04400	.04402	.04402	.04400
24	3	.02711	.02747	.02792	.02847	.02912	.02988	.03073
24	3	.01688	.01876	.02059	.02235	.02406	.02571	.02731
24	4	.04489	.04500	.04510	.04517	.04521	.04523	.04522
24	4	.02863	.02895	.02937	.02989	.03052	.03126	.03210
24	4	.01817	.02005	.02188	.02364	.02536	.02701	.02860
24	5	.04620	.04635	.04646	.04655	.04662	.04665	.04664
24	5	.03035	.03063	.03101	.03150	.03211	.03283	.03366
24	5	.01967	.02155	.02338	.02514	.02685	.02850	.03009
24	6	.04779	.04796	.04810	.04820	.04827	.04831	.04831
24	6	.03230	.03253	.03288	.03334	.03392	.03462	.03544
24	6	.02143	.02330	.02512	.02688	.02858	.03022	.03181
24	7	.04970	.04989	.05005	.05016	.05024	.05027	.05026
24	7	.03453	.03471	.03501	.03544	.03599	.03667	.03747
24	7	.02349	.02536	.02716	.02891	.03060	.03223	.03380

Variance-Covariance factors V1,V2 and V3 for
the BLUEs of μ and σ shape parameter k

n	s	0.7	0.8	0.9	1.0	1.1	1.2	1.3
24	8	.05203	.05224	.05240	.05251	.05258	.05260	.05257
24	8	.03708	.03721	.03746	.03785	.03837	.03902	.03980
24	8	.02593	.02778	.02956	.03129	.03296	.03457	.03612
24	9	.05489	.05509	.05525	.05535	.05540	.05539	.05533
24	9	.04002	.04010	.04030	.04065	.04113	.04176	.04251
24	9	.02883	.03065	.03241	.03411	.03575	.03733	.03886
24	10	.05841	.05860	.05873	.05880	.05881	.05876	.05864
24	10	.04347	.04347	.04363	.04392	.04437	.04496	.04569
24	10	.03232	.03409	.03581	.03747	.03907	.04061	.04210
24	11	.06281	.06296	.06304	.06305	.06299	.06287	.06268
24	11	.04753	.04746	.04755	.04779	.04819	.04874	.04944
24	11	.03654	.03826	.03992	.04152	.04307	.04456	.04599
24	12	.06837	.06845	.06844	.06836	.06820	.06796	.06766
24	12	.05238	.05223	.05225	.05243	.05277	.05328	.05395
24	12	.04174	.04338	.04496	.04648	.04795	.04937	.05073
25	0	.03997	.03994	.03991	.03987	.03981	.03974	.03965
25	0	.02253	.02298	.02350	.02409	.02477	.02552	.02636
25	0	.01353	.01532	.01704	.01872	.02033	.02189	.02339
25	1	.04048	.04049	.04048	.04047	.04044	.04040	.04033
25	1	.02347	.02389	.02438	.02495	.02562	.02636	.02719
25	1	.01423	.01602	.01776	.01944	.02107	.02263	.02415
25	2	.04113	.04117	.04120	.04122	.04121	.04119	.04115
25	2	.02456	.02494	.02540	.02595	.02660	.02733	.02816
25	2	.01507	.01687	.01861	.02030	.02194	.02351	.02503
25	3	.04194	.04201	.04207	.04211	.04213	.04212	.04210
25	3	.02578	.02612	.02656	.02709	.02771	.02843	.02925
25	3	.01606	.01786	.01961	.02131	.02294	.02452	.02605
25	4	.04291	.04301	.04309	.04316	.04320	.04321	.04320
25	4	.02715	.02746	.02786	.02836	.02897	.02967	.03048
25	4	.01721	.01902	.02077	.02246	.02410	.02569	.02721
25	5	.04407	.04420	.04431	.04439	.04445	.04447	.04447
25	5	.02869	.02896	.02933	.02981	.03039	.03108	.03187
25	5	.01856	.02036	.02211	.02380	.02544	.02702	.02854
25	6	.04547	.04563	.04575	.04585	.04592	.04595	.04595
25	6	.03043	.03066	.03099	.03144	.03200	.03267	.03345
25	6	.02011	.02191	.02366	.02534	.02697	.02855	.03007
25	7	.04714	.04732	.04747	.04757	.04764	.04768	.04767
25	7	.03240	.03258	.03288	.03329	.03382	.03447	.03524
25	7	.02193	.02372	.02545	.02713	.02875	.03032	.03183
25	8	.04916	.04936	.04951	.04962	.04968	.04971	.04968
25	8	.03463	.03477	.03503	.03541	.03591	.03654	.03729
25	8	.02405	.02583	.02755	.02921	.03081	.03236	.03386
25	9	.05161	.05180	.05195	.05206	.05211	.05211	.05206
25	9	.03720	.03728	.03750	.03784	.03831	.03891	.03964
25	9	.02656	.02831	.03000	.03164	.03322	.03475	.03622
25	10	.05459	.05478	.05491	.05499	.05501	.05498	.05489
25	10	.04016	.04019	.04035	.04065	.04109	.04166	.04237
25	10	.02953	.03125	.03291	.03452	.03607	.03756	.03900
25	11	.05826	.05842	.05852	.05856	.05853	.05844	.05829
25	11	.04362	.04358	.04369	.04394	.04434	.04487	.04555
25	11	.03309	.03477	.03638	.03794	.03945	.04090	.04229
25	12	.06283	.06294	.06298	.06294	.06284	.06267	.06243
25	12	.04770	.04759	.04763	.04783	.04818	.04868	.04933
25	12	.03741	.03902	.04057	.04207	.04351	.04490	.04624
25	13	.06859	.06861	.06855	.06841	.06819	.06789	.06753
25	13	.05257	.05238	.05235	.05248	.05278	.05323	.05385

Variance-Covariance factors V1,V2 and V3 for
the BLUEs of μ and σ shape parameter k

n	s	0.7	0.8	0.9	1.0	1.1	1.2	1.3
25	13	.04271	.04423	.04570	.04711	.04847	.04978	.05104

Variance-Covariance factors V1,V2 and V3 for the BLUEs of μ and σ shape parameter k

n	s	1.4	1.5	1.6	1.7	1.8	1.9	2.0
2	0	.50000	.50000	.50000	.50000	.50000	.50000	.50000
2	0	.77232	.80053	.83032	.86166	.89451	.92883	.96460
2	0	.33738	.35990	.38218	.40424	.42606	.44767	.46906
3	0	.33272	.33256	.33238	.33218	.33195	.33169	.33142
3	0	.37271	.38659	.40131	.41684	.43318	.45029	.46817
3	0	.22017	.23425	.24810	.26170	.27507	.28822	.30115
3	1	.48723	.48890	.49046	.49190	.49324	.49449	.49564
3	1	.72596	.74464	.76484	.78651	.80963	.83418	.86012
3	1	.45379	.47085	.48782	.50470	.52149	.53821	.55485
4	0	.24915	.24893	.24867	.24838	.24806	.24770	.24730
4	0	.24270	.25180	.26147	.27170	.28247	.29377	.30560
4	0	.16300	.17315	.18308	.19279	.20229	.21158	.22066
4	1	.30724	.30794	.30855	.30906	.30948	.30982	.31007
4	1	.35261	.36295	.37412	.38610	.39887	.41241	.42671
4	1	.24290	.25414	.26521	.27611	.28684	.29742	.30785
5	0	.19908	.19883	.19855	.19823	.19787	.19747	.19704
5	0	.17894	.18566	.19282	.20040	.20839	.21679	.22557
5	0	.12925	.13714	.14484	.15234	.15965	.16677	.17372
5	1	.22980	.23012	.23036	.23053	.23062	.23063	.23057
5	1	.23107	.23831	.24613	.25451	.26344	.27291	.28289
5	1	.16927	.17773	.18602	.19415	.20211	.20991	.21755
5	2	.30456	.30470	.30469	.30454	.30426	.30386	.30334
5	2	.34330	.35136	.36017	.36972	.37998	.39093	.40254
5	2	.26086	.26955	.27809	.28649	.29475	.30288	.31087
6	0	.16574	.16549	.16520	.16487	.16451	.16410	.16366
6	0	.14129	.14659	.15225	.15825	.16458	.17123	.17820
6	0	.10701	.11345	.11971	.12580	.13172	.13747	.14305
6	1	.18473	.18487	.18494	.18494	.18487	.18474	.18454
6	1	.17128	.17686	.18288	.18934	.19622	.20350	.21118
6	1	.13088	.13767	.14430	.15078	.15710	.16327	.16930
6	2	.22225	.22242	.22247	.22241	.22225	.22198	.22161
6	2	.22495	.23085	.23730	.24428	.25176	.25974	.26820
6	2	.17575	.18269	.18950	.19615	.20266	.20904	.21527
6	3	.31009	.30915	.30804	.30678	.30539	.30386	.30222
6	3	.33817	.34465	.35183	.35967	.36815	.37724	.38691
6	3	.27547	.28204	.28849	.29482	.30103	.30712	.31310
7	0	.14195	.14170	.14142	.14110	.14074	.14034	.13990
7	0	.11651	.12088	.12554	.13049	.13572	.14122	.14699
7	0	.09126	.09669	.10196	.10707	.11203	.11683	.12149
7	1	.15483	.15487	.15484	.15476	.15462	.15441	.15415
7	1	.13583	.14037	.14527	.15051	.15610	.16202	.16825
7	1	.10704	.11271	.11823	.12361	.12884	.13394	.13890
7	2	.17732	.17743	.17745	.17738	.17721	.17696	.17663
7	2	.16688	.17157	.17670	.18224	.18817	.19450	.20119
7	2	.13346	.13924	.14489	.15039	.15577	.16100	.16611
7	3	.22024	.21996	.21954	.21900	.21833	.21755	.21667
7	3	.22125	.22617	.23160	.23752	.24390	.25074	.25801
7	3	.18177	.18743	.19296	.19836	.20364	.20878	.21381
7	4	.31884	.31663	.31426	.31176	.30914	.30642	.30360
7	4	.33508	.34042	.34639	.35298	.36014	.36785	.37608
7	4	.28771	.29252	.29723	.30185	.30638	.31080	.31512
8	0	.12412	.12388	.12361	.12330	.12295	.12256	.12214
8	0	.09901	.10271	.10667	.11088	.11533	.12001	.12491
8	0	.07953	.08422	.08876	.09316	.09741	.10153	.10552
8	1	.13342	.13340	.13332	.13319	.13301	.13277	.13248
8	1	.11242	.11624	.12036	.12478	.12948	.13446	.13970

Variance-Covariance factors V1,V2 and V3 for
the BLUEs of μ and σ shape parameter k

n	s	1.4	1.5	1.6	1.7	1.8	1.9	2.0
8	1	.09070	.09556	.10029	.10488	.10934	.11368	.11789
8	2	.14837	.14843	.14841	.14831	.14813	.14788	.14756
8	2	.13249	.13640	.14066	.14527	.15019	.15544	.16098
8	2	.10802	.11297	.11779	.12248	.12705	.13149	.13581
8	3	.17364	.17354	.17332	.17300	.17258	.17206	.17145
8	3	.16405	.16806	.17248	.17729	.18247	.18801	.19389
8	3	.13626	.14116	.14594	.15060	.15514	.15955	.16385
8	4	.22109	.22022	.21922	.21808	.21683	.21546	.21400
8	4	.21886	.22303	.22767	.23276	.23829	.24424	.25057
8	4	.18726	.19182	.19627	.20061	.20484	.20895	.21295
9	0	.11027	.11004	.10978	.10948	.10914	.10877	.10836
9	0	.08600	.08921	.09265	.09631	.10017	.10424	.10850
9	0	.07045	.07458	.07856	.08242	.08615	.08975	.09323
9	1	.11729	.11723	.11712	.11696	.11676	.11650	.11620
9	1	.09582	.09911	.10267	.10648	.11054	.11483	.11935
9	1	.07876	.08301	.08714	.09115	.09503	.09880	.10245
9	2	.12792	.12794	.12789	.12777	.12758	.12733	.12702
9	2	.10979	.11314	.11679	.12073	.12494	.12943	.13416
9	2	.09095	.09527	.09947	.10356	.10752	.11137	.11511
9	3	.14451	.14446	.14431	.14408	.14376	.14336	.14288
9	3	.13024	.13365	.13739	.14146	.14584	.15051	.15547
9	3	.10936	.11367	.11787	.12194	.12590	.12975	.13349
9	4	.17214	.17172	.17119	.17054	.16979	.16894	.16800
9	4	.16212	.16559	.16944	.17366	.17823	.18314	.18836
9	4	.13904	.14318	.14722	.15114	.15495	.15864	.16224
9	5	.22349	.22198	.22034	.21858	.21671	.21474	.21269
9	5	.21725	.22081	.22483	.22927	.23412	.23935	.24494
9	5	.19225	.19587	.19940	.20282	.20615	.20939	.21252
10	0	.09919	.09897	.09872	.09843	.09811	.09775	.09736
10	0	.07597	.07880	.08184	.08506	.08848	.09207	.09584
10	0	.06323	.06691	.07046	.07389	.07720	.08039	.08347
10	1	.10468	.10459	.10447	.10429	.10408	.10382	.10351
10	1	.08345	.08634	.08947	.09281	.09638	.10015	.10412
10	1	.06963	.07342	.07708	.08063	.08406	.08739	.09061
10	2	.11262	.11260	.11253	.11240	.11220	.11195	.11165
10	2	.09369	.09662	.09982	.10326	.10694	.11085	.11499
10	2	.07865	.08249	.08621	.08983	.09333	.09672	.10001
10	3	.12431	.12427	.12415	.12395	.12368	.12334	.12293
10	3	.10795	.11091	.11417	.11770	.12149	.12555	.12984
10	3	.09156	.09540	.09913	.10274	.10625	.10966	.11296
10	4	.14229	.14205	.14172	.14129	.14077	.14016	.13947
10	4	.12864	.13164	.13495	.13858	.14250	.14670	.15117
10	4	.11085	.11460	.11824	.12177	.12520	.12852	.13174
10	5	.17197	.17120	.17030	.16929	.16818	.16697	.16568
10	5	.16075	.16378	.16717	.17091	.17498	.17937	.18404
10	5	.14172	.14520	.14858	.15186	.15504	.15811	.16109
11	0	.09014	.08993	.08969	.08941	.08910	.08876	.08838
11	0	.06800	.07054	.07325	.07613	.07919	.08241	.08578
11	0	.05734	.06066	.06386	.06694	.06992	.07279	.07555
11	1	.09453	.09444	.09430	.09412	.09390	.09364	.09333
11	1	.07388	.07646	.07924	.08222	.08540	.08876	.09229
11	1	.06243	.06583	.06912	.07230	.07538	.07835	.08123
11	2	.10068	.10065	.10056	.10042	.10022	.09997	.09967
11	2	.08169	.08430	.08713	.09019	.09345	.09693	.10059
11	2	.06936	.07280	.07614	.07938	.08251	.08555	.08849
11	3	.10935	.10931	.10920	.10902	.10878	.10847	.10811

Variance-Covariance factors V1,V2 and V3 for the BLUEs of μ and σ shape parameter k

n	s	1.4	1.5	1.6	1.7	1.8	1.9	2.0
11	3	.09216	.09478	.09766	.10078	.10414	.10771	.11150
11	3	.07888	.08233	.08568	.08893	.09208	.09512	.09808
11	4	.12194	.12179	.12155	.12123	.12083	.12036	.11981
11	4	.10660	.10924	.11216	.11534	.11878	.12247	.12638
11	4	.09237	.09577	.09907	.10226	.10536	.10837	.11127
11	5	.14114	.14069	.14013	.13947	.13872	.13789	.13697
11	5	.12747	.13012	.13308	.13633	.13987	.14367	.14772
11	5	.11239	.11563	.11878	.12183	.12478	.12764	.13041
12	0	.08259	.08239	.08216	.08190	.08160	.08128	.08092
12	0	.06153	.06381	.06626	.06887	.07164	.07455	.07760
12	0	.05246	.05547	.05838	.06119	.06389	.06649	.06899
12	1	.08620	.08609	.08595	.08576	.08554	.08528	.08498
12	1	.06626	.06858	.07109	.07378	.07664	.07967	.08285
12	1	.05658	.05967	.06266	.06554	.06833	.07102	.07362
12	2	.09110	.09104	.09094	.09079	.09060	.09035	.09006
12	2	.07240	.07474	.07729	.08004	.08297	.08609	.08939
12	2	.06207	.06519	.06822	.07115	.07399	.07673	.07938
12	3	.09777	.09772	.09762	.09745	.09722	.09694	.09660
12	3	.08039	.08275	.08533	.08812	.09113	.09433	.09772
12	3	.06937	.07251	.07554	.07849	.08134	.08409	.08676
12	4	.10705	.10694	.10676	.10650	.10617	.10577	.10531
12	4	.09101	.09337	.09598	.09882	.10189	.10517	.10865
12	4	.07930	.08240	.08541	.08833	.09115	.09388	.09652
12	5	.12043	.12014	.11976	.11928	.11873	.11810	.11740
12	5	.10559	.10795	.11058	.11347	.11661	.11997	.12355
12	5	.09327	.09628	.09919	.10201	.10474	.10738	.10993
12	6	.14074	.14004	.13923	.13833	.13734	.13626	.13512
12	6	.12659	.12895	.13161	.13455	.13775	.14121	.14490
12	6	.11392	.11671	.11942	.12205	.12458	.12703	.12939
13	0	.07622	.07602	.07580	.07555	.07527	.07495	.07461
13	0	.05616	.05825	.06048	.06286	.06538	.06804	.07082
13	0	.04833	.05110	.05377	.05634	.05881	.06119	.06347
13	1	.07922	.07911	.07896	.07878	.07856	.07830	.07801
13	1	.06005	.06216	.06444	.06689	.06949	.07225	.07515
13	1	.05175	.05457	.05731	.05994	.06249	.06494	.06731
13	2	.08321	.08315	.08304	.08289	.08269	.08245	.08217
13	2	.06499	.06712	.06944	.07193	.07460	.07743	.08042
13	2	.05619	.05905	.06182	.06449	.06708	.06958	.07199
13	3	.08850	.08845	.08834	.08818	.08797	.08770	.08739
13	3	.07128	.07342	.07576	.07829	.08101	.08391	.08698
13	3	.06196	.06483	.06761	.07030	.07290	.07541	.07784
13	4	.09561	.09553	.09537	.09515	.09486	.09452	.09411
13	4	.07939	.08153	.08389	.08646	.08923	.09219	.09533
13	4	.06955	.07241	.07517	.07784	.08043	.08292	.08534
13	5	.10544	.10524	.10495	.10459	.10415	.10365	.10308
13	5	.09012	.09226	.09463	.09723	.10005	.10307	.10629
13	5	.07982	.08261	.08531	.08793	.09045	.09289	.09525
13	6	.11954	.11909	.11853	.11790	.11718	.11638	.11551
13	6	.10480	.10693	.10932	.11195	.11482	.11791	.12120
13	6	.09421	.09687	.09944	.10192	.10432	.10664	.10887
13	7	.14084	.13988	.13882	.13767	.13643	.13512	.13374
13	7	.12592	.12803	.13043	.13310	.13602	.13918	.14257
13	7	.11542	.11781	.12013	.12237	.12453	.12660	.12860
14	0	.07075	.07057	.07036	.07011	.06984	.06954	.06921
14	0	.05165	.05356	.05561	.05780	.06011	.06256	.06512
14	0	.04481	.04736	.04982	.05219	.05447	.05666	.05877

Variance-Covariance factors V1,V2 and V3 for the BLUEs of μ and σ shape parameter k

n	s	1.4	1.5	1.6	1.7	1.8	1.9	2.0
14	1	.07329	.07318	.07303	.07285	.07263	.07238	.07210
14	1	.05489	.05683	.05892	.06116	.06355	.06608	.06874
14	1	.04768	.05028	.05280	.05523	.05757	.05982	.06200
14	2	.07660	.07653	.07642	.07627	.07607	.07584	.07556
14	2	.05896	.06091	.06303	.06531	.06775	.07034	.07307
14	2	.05135	.05398	.05653	.05899	.06137	.06366	.06587
14	3	.08089	.08084	.08073	.08058	.08037	.08012	.07982
14	3	.06402	.06598	.06812	.07043	.07292	.07557	.07836
14	3	.05601	.05866	.06122	.06369	.06608	.06839	.07062
14	4	.08651	.08644	.08630	.08610	.08585	.08554	.08517
14	4	.07041	.07237	.07452	.07686	.07939	.08209	.08495
14	4	.06200	.06464	.06719	.06965	.07204	.07434	.07656
14	5	.09403	.09387	.09364	.09335	.09299	.09256	.09208
14	5	.07861	.08056	.08273	.08509	.08766	.09040	.09332
14	5	.06985	.07244	.07495	.07738	.07972	.08198	.08416
14	6	.10435	.10403	.10363	.10315	.10259	.10196	.10128
14	6	.08942	.09136	.09353	.09592	.09852	.10132	.10430
14	6	.08041	.08292	.08534	.08768	.08993	.09211	.09421
14	7	.11910	.11847	.11774	.11692	.11603	.11507	.11404
14	7	.10419	.10611	.10829	.11070	.11334	.11619	.11923
14	7	.09517	.09751	.09977	.10195	.10405	.10607	.10801
15	0	.06602	.06584	.06564	.06540	.06514	.06486	.06454
15	0	.04780	.04956	.05146	.05348	.05562	.05788	.06025
15	0	.04176	.04413	.04642	.04862	.05073	.05276	.05470
15	1	.06819	.06808	.06793	.06775	.06754	.06729	.06702
15	1	.05054	.05233	.05426	.05633	.05854	.06087	.06332
15	1	.04420	.04662	.04896	.05120	.05337	.05546	.05746
15	2	.07099	.07091	.07079	.07064	.07045	.07022	.06995
15	2	.05394	.05574	.05769	.05980	.06204	.06443	.06695
15	2	.04728	.04973	.05209	.05437	.05656	.05868	.06073
15	3	.07453	.07447	.07437	.07421	.07402	.07378	.07349
15	3	.05810	.05991	.06188	.06401	.06630	.06873	.07130
15	3	.05113	.05358	.05596	.05825	.06046	.06259	.06465
15	4	.07908	.07901	.07888	.07870	.07847	.07818	.07785
15	4	.06326	.06506	.06704	.06919	.07151	.07399	.07662
15	4	.05597	.05842	.06078	.06307	.06528	.06741	.06946
15	5	.08500	.08487	.08468	.08443	.08412	.08375	.08333
15	5	.06972	.07151	.07350	.07568	.07803	.08054	.08321
15	5	.06215	.06457	.06691	.06916	.07135	.07345	.07548
15	6	.09286	.09263	.09232	.09193	.09148	.09097	.09040
15	6	.07798	.07977	.08176	.08395	.08633	.08889	.09161
15	6	.07022	.07257	.07485	.07704	.07917	.08121	.08318
15	7	.10364	.10319	.10266	.10206	.10137	.10062	.09981
15	7	.08886	.09063	.09262	.09483	.09724	.09983	.10260
15	7	.08104	.08328	.08545	.08754	.08955	.09149	.09335
15	8	.11899	.11817	.11726	.11627	.11520	.11407	.11287
15	8	.10371	.10545	.10744	.10966	.11209	.11473	.11755
15	8	.09614	.09818	.10015	.10205	.10388	.10564	.10732
16	0	.06188	.06171	.06151	.06129	.06104	.06076	.06045
16	0	.04447	.04611	.04787	.04975	.05175	.05385	.05605
16	0	.03910	.04131	.04345	.04549	.04746	.04935	.05116
16	1	.06376	.06365	.06350	.06332	.06312	.06288	.06261
16	1	.04683	.04849	.05028	.05220	.05425	.05641	.05869
16	1	.04121	.04346	.04563	.04773	.04974	.05168	.05355
16	2	.06614	.06606	.06595	.06579	.06561	.06538	.06512
16	2	.04970	.05137	.05318	.05514	.05722	.05943	.06177

Variance-Covariance factors V1,V2 and V3 for the BLUEs of μ and σ shape parameter k

n	s	1.4	1.5	1.6	1.7	1.8	1.9	2.0
16	2	.04382	.04610	.04830	.05042	.05246	.05443	.05633
16	3	.06913	.06906	.06896	.06881	.06862	.06838	.06811
16	3	.05319	.05486	.05668	.05866	.06078	.06303	.06541
16	3	.04705	.04933	.05154	.05368	.05573	.05772	.05963
16	4	.07288	.07281	.07269	.07252	.07230	.07204	.07173
16	4	.05742	.05909	.06093	.06292	.06507	.06736	.06978
16	4	.05103	.05332	.05552	.05765	.05971	.06169	.06361
16	5	.07765	.07755	.07738	.07716	.07689	.07656	.07618
16	5	.06263	.06430	.06614	.06815	.07032	.07264	.07511
16	5	.05602	.05828	.06047	.06258	.06462	.06658	.06848
16	6	.08383	.08365	.08340	.08308	.08271	.08227	.08178
16	6	.06915	.07081	.07265	.07467	.07686	.07922	.08172
16	6	.06237	.06459	.06673	.06880	.07079	.07271	.07456
16	7	.09203	.09170	.09129	.09082	.09027	.08967	.08900
16	7	.07748	.07912	.08096	.08299	.08520	.08759	.09013
16	7	.07063	.07277	.07483	.07682	.07873	.08058	.08235
16	8	.10322	.10264	.10198	.10124	.10043	.09955	.09862
16	8	.08841	.09003	.09187	.09391	.09614	.09856	.10114
16	8	.08169	.08369	.08562	.08748	.08927	.09099	.09264
17	0	.05822	.05806	.05787	.05766	.05742	.05715	.05686
17	0	.04157	.04311	.04475	.04651	.04837	.05033	.05239
17	0	.03675	.03883	.04083	.04275	.04459	.04636	.04805
17	1	.05987	.05976	.05961	.05944	.05924	.05900	.05874
17	1	.04362	.04516	.04684	.04863	.05054	.05256	.05468
17	1	.03859	.04070	.04273	.04469	.04658	.04839	.05013
17	2	.06193	.06185	.06173	.06158	.06139	.06117	.06092
17	2	.04608	.04764	.04933	.05115	.05309	.05515	.05732
17	2	.04084	.04297	.04503	.04701	.04892	.05076	.05253
17	3	.06447	.06440	.06430	.06415	.06396	.06374	.06348
17	3	.04903	.05059	.05229	.05413	.05610	.05820	.06042
17	3	.04358	.04572	.04779	.04978	.05170	.05356	.05534
17	4	.06761	.06755	.06743	.06727	.06707	.06682	.06653
17	4	.05258	.05413	.05584	.05770	.05969	.06182	.06407
17	4	.04692	.04906	.05112	.05312	.05504	.05690	.05868
17	5	.07154	.07145	.07130	.07111	.07086	.07056	.07022
17	5	.05686	.05841	.06013	.06200	.06401	.06617	.06846
17	5	.05102	.05315	.05520	.05718	.05909	.06093	.06270
17	6	.07652	.07637	.07616	.07589	.07557	.07519	.07476
17	6	.06212	.06367	.06538	.06726	.06929	.07148	.07380
17	6	.05614	.05823	.06025	.06220	.06408	.06589	.06763
17	7	.08295	.08270	.08238	.08199	.08154	.08103	.08047
17	7	.06869	.07022	.07193	.07381	.07587	.07807	.08043
17	7	.06264	.06467	.06663	.06852	.07034	.07209	.07378
17	8	.09145	.09102	.09052	.08994	.08930	.08860	.08784
17	8	.07706	.07857	.08027	.08217	.08423	.08647	.08885
17	8	.07107	.07301	.07487	.07667	.07840	.08006	.08166
17	9	.10303	.10231	.10151	.10064	.09970	.09870	.09764
17	9	.08805	.08953	.09122	.09312	.09520	.09745	.09987
17	9	.08235	.08413	.08584	.08749	.08908	.09060	.09205
18	0	.05498	.05482	.05464	.05443	.05420	.05394	.05366
18	0	.03903	.04046	.04201	.04365	.04540	.04724	.04918
18	0	.03468	.03663	.03851	.04031	.04204	.04370	.04529
18	1	.05643	.05632	.05617	.05600	.05581	.05558	.05533
18	1	.04081	.04226	.04383	.04551	.04730	.04919	.05118
18	1	.03629	.03827	.04018	.04202	.04379	.04549	.04712
18	2	.05822	.05814	.05802	.05787	.05769	.05748	.05723

Variance-Covariance factors V1,V2 and V3 for
the BLUEs of μ and σ shape parameter k

n	s	1.4	1.5	1.6	1.7	1.8	1.9	2.0
18	2	.04295	.04441	.04599	.04769	.04951	.05144	.05347
18	2	.03824	.04025	.04218	.04404	.04583	.04755	.04921
18	3	.06041	.06034	.06024	.06009	.05991	.05970	.05945
18	3	.04548	.04694	.04853	.05025	.05209	.05406	.05613
18	3	.04060	.04261	.04455	.04642	.04823	.04996	.05164
18	4	.06309	.06302	.06291	.06275	.06256	.06233	.06205
18	4	.04848	.04994	.05154	.05327	.05514	.05713	.05923
18	4	.04343	.04544	.04739	.04926	.05106	.05280	.05448
18	5	.06637	.06629	.06616	.06597	.06575	.06547	.06516
18	5	.05207	.05352	.05512	.05687	.05875	.06077	.06290
18	5	.04686	.04887	.05080	.05266	.05446	.05619	.05785
18	6	.07047	.07034	.07016	.06992	.06964	.06930	.06892
18	6	.05640	.05784	.05945	.06120	.06310	.06513	.06730
18	6	.05107	.05305	.05496	.05680	.05857	.06028	.06192
18	7	.07564	.07543	.07517	.07484	.07446	.07403	.07354
18	7	.06170	.06313	.06473	.06649	.06840	.07046	.07265
18	7	.05631	.05824	.06010	.06190	.06363	.06529	.06689
18	8	.08230	.08196	.08156	.08110	.08057	.07999	.07936
18	8	.06830	.06972	.07131	.07307	.07500	.07708	.07929
18	8	.06294	.06480	.06659	.06832	.06998	.07158	.07311
18	9	.09108	.09054	.08993	.08925	.08851	.08771	.08686
18	9	.07672	.07811	.07969	.08146	.08339	.08549	.08773
18	9	.07153	.07328	.07496	.07658	.07814	.07963	.08106
19	0	.05208	.05192	.05175	.05155	.05133	.05108	.05080
19	0	.03677	.03812	.03957	.04112	.04277	.04450	.04633
19	0	.03282	.03466	.03644	.03814	.03977	.04133	.04283
19	1	.05337	.05325	.05311	.05295	.05275	.05253	.05229
19	1	.03834	.03971	.04118	.04276	.04444	.04622	.04809
19	1	.03424	.03612	.03792	.03965	.04132	.04292	.04445
19	2	.05494	.05486	.05474	.05459	.05441	.05421	.05397
19	2	.04021	.04158	.04307	.04467	.04638	.04819	.05010
19	2	.03596	.03785	.03967	.04142	.04311	.04473	.04629
19	3	.05685	.05678	.05667	.05653	.05635	.05615	.05591
19	3	.04241	.04378	.04527	.04689	.04862	.05046	.05241
19	3	.03800	.03990	.04173	.04350	.04519	.04683	.04840
19	4	.05915	.05908	.05897	.05882	.05864	.05842	.05816
19	4	.04498	.04635	.04785	.04948	.05123	.05310	.05507
19	4	.04044	.04234	.04417	.04593	.04764	.04927	.05085
19	5	.06194	.06186	.06173	.06156	.06135	.06110	.06081
19	5	.04802	.04939	.05089	.05253	.05430	.05618	.05819
19	5	.04335	.04524	.04707	.04883	.05052	.05215	.05372
19	6	.06536	.06525	.06508	.06487	.06461	.06431	.06396
19	6	.05165	.05300	.05451	.05615	.05793	.05984	.06187
19	6	.04687	.04874	.05055	.05229	.05396	.05558	.05713
19	7	.06960	.06943	.06921	.06893	.06859	.06821	.06778
19	7	.05601	.05736	.05886	.06051	.06230	.06423	.06627
19	7	.05117	.05301	.05478	.05649	.05813	.05972	.06123
19	8	.07495	.07468	.07436	.07397	.07353	.07303	.07249
19	8	.06134	.06267	.06417	.06582	.06763	.06957	.07164
19	8	.05651	.05829	.06001	.06167	.06326	.06479	.06626
19	9	.08182	.08140	.08092	.08038	.07977	.07912	.07841
19	9	.06798	.06929	.07078	.07243	.07424	.07620	.07829
19	9	.06326	.06496	.06660	.06818	.06969	.07114	.07253
19	10	.09086	.09022	.08950	.08872	.08788	.08698	.08604
19	10	.07643	.07772	.07919	.08084	.08266	.08463	.08674
19	10	.07201	.07358	.07509	.07655	.07795	.07928	.08056

Variance-Covariance factors V1,V2 and V3 for
the BLUEs of μ and σ shape parameter k

n	s	1.4	1.5	1.6	1.7	1.8	1.9	2.0
20	0	.04946	.04932	.04915	.04895	.04874	.04850	.04824
20	0	.03476	.03603	.03740	.03887	.04042	.04206	.04378
20	0	.03115	.03290	.03457	.03618	.03773	.03921	.04062
20	1	.05062	.05050	.05037	.05021	.05002	.04980	.04956
20	1	.03616	.03744	.03883	.04032	.04191	.04359	.04536
20	1	.03242	.03419	.03589	.03753	.03911	.04062	.04207
20	2	.05201	.05193	.05181	.05167	.05149	.05129	.05106
20	2	.03780	.03909	.04050	.04201	.04362	.04533	.04713
20	2	.03393	.03572	.03744	.03910	.04069	.04222	.04370
20	3	.05368	.05361	.05351	.05337	.05320	.05300	.05276
20	3	.03972	.04101	.04242	.04395	.04558	.04731	.04915
20	3	.03572	.03752	.03925	.04092	.04252	.04407	.04555
20	4	.05568	.05561	.05551	.05537	.05519	.05498	.05473
20	4	.04195	.04324	.04466	.04619	.04784	.04960	.05146
20	4	.03784	.03963	.04137	.04304	.04465	.04619	.04768
20	5	.05808	.05800	.05789	.05773	.05753	.05729	.05702
20	5	.04456	.04585	.04727	.04881	.05047	.05225	.05413
20	5	.04034	.04213	.04386	.04552	.04713	.04867	.05015
20	6	.06098	.06088	.06073	.06054	.06030	.06003	.05971
20	6	.04764	.04892	.05033	.05188	.05356	.05535	.05726
20	6	.04332	.04510	.04681	.04846	.05005	.05158	.05305
20	7	.06452	.06438	.06418	.06393	.06364	.06330	.06292
20	7	.05129	.05256	.05397	.05553	.05721	.05902	.06095
20	7	.04692	.04867	.05035	.05198	.05354	.05505	.05649
20	8	.06890	.06869	.06841	.06809	.06770	.06727	.06680
20	8	.05568	.05694	.05835	.05990	.06160	.06342	.06536
20	8	.05130	.05301	.05466	.05624	.05777	.05923	.06064
20	9	.07441	.07409	.07370	.07325	.07274	.07218	.07158
20	9	.06104	.06228	.06369	.06524	.06694	.06878	.07074
20	9	.05674	.05838	.05997	.06149	.06295	.06436	.06570
20	10	.08148	.08099	.08042	.07980	.07911	.07838	.07759
20	10	.06771	.06893	.07032	.07187	.07358	.07542	.07740
20	10	.06361	.06516	.06665	.06808	.06946	.07077	.07203
21	0	.04710	.04696	.04680	.04661	.04640	.04617	.04591
21	0	.03295	.03416	.03546	.03685	.03832	.03987	.04150
21	0	.02964	.03130	.03289	.03442	.03589	.03729	.03863
21	1	.04814	.04803	.04789	.04773	.04755	.04734	.04711
21	1	.03420	.03542	.03673	.03815	.03965	.04124	.04291
21	1	.03078	.03246	.03408	.03563	.03712	.03855	.03993
21	2	.04938	.04930	.04918	.04904	.04887	.04867	.04845
21	2	.03566	.03689	.03821	.03964	.04116	.04278	.04449
21	2	.03213	.03382	.03546	.03703	.03854	.03999	.04138
21	3	.05086	.05079	.05068	.05055	.05038	.05019	.04996
21	3	.03735	.03858	.03991	.04135	.04289	.04453	.04627
21	3	.03371	.03541	.03705	.03863	.04015	.04162	.04302
21	4	.05261	.05254	.05244	.05230	.05213	.05193	.05169
21	4	.03931	.04053	.04187	.04332	.04488	.04654	.04829
21	4	.03556	.03726	.03891	.04049	.04202	.04348	.04489
21	5	.05469	.05462	.05451	.05436	.05417	.05394	.05368
21	5	.04157	.04279	.04413	.04559	.04716	.04883	.05061
21	5	.03773	.03943	.04107	.04265	.04417	.04563	.04704
21	6	.05718	.05709	.05695	.05678	.05656	.05630	.05601
21	6	.04421	.04542	.04676	.04822	.04980	.05150	.05329
21	6	.04029	.04198	.04361	.04518	.04669	.04814	.04954
21	7	.06017	.06005	.05988	.05966	.05939	.05909	.05874
21	7	.04731	.04851	.04985	.05131	.05291	.05461	.05643

Variance-Covariance factors V1,V2 and V3 for the BLUEs of μ and σ shape parameter k

n	s	1.4	1.5	1.6	1.7	1.8	1.9	2.0
21	7	.04333	.04500	.04661	.04816	.04965	.05108	.05246
21	8	.06383	.06365	.06342	.06313	.06280	.06242	.06199
21	8	.05098	.05217	.05351	.05498	.05657	.05829	.06012
21	8	.04700	.04863	.05021	.05173	.05319	.05459	.05593
21	9	.06834	.06808	.06775	.06737	.06694	.06646	.06594
21	9	.05540	.05658	.05790	.05937	.06098	.06270	.06455
21	9	.05146	.05305	.05458	.05604	.05746	.05881	.06011
21	10	.07401	.07361	.07316	.07264	.07207	.07146	.07079
21	10	.06079	.06194	.06326	.06473	.06633	.06807	.06993
21	10	.05699	.05850	.05996	.06136	.06270	.06399	.06522
21	11	.08126	.08068	.08003	.07933	.07856	.07775	.07689
21	11	.06748	.06862	.06992	.07138	.07299	.07473	.07661
21	11	.06396	.06537	.06672	.06802	.06927	.07046	.07160
22	0	.04495	.04482	.04466	.04448	.04428	.04405	.04381
22	0	.03133	.03247	.03370	.03502	.03642	.03790	.03945
22	0	.02827	.02985	.03137	.03282	.03421	.03555	.03682
22	1	.04589	.04578	.04565	.04550	.04532	.04511	.04489
22	1	.03245	.03360	.03485	.03619	.03762	.03913	.04071
22	1	.02930	.03090	.03243	.03391	.03533	.03669	.03799
22	2	.04701	.04692	.04681	.04667	.04650	.04631	.04609
22	2	.03375	.03491	.03617	.03752	.03897	.04051	.04212
22	2	.03050	.03212	.03367	.03516	.03660	.03797	.03929
22	3	.04832	.04825	.04815	.04801	.04785	.04766	.04745
22	3	.03525	.03641	.03768	.03904	.04051	.04206	.04370
22	3	.03191	.03353	.03509	.03659	.03804	.03942	.04076
22	4	.04987	.04980	.04970	.04957	.04940	.04921	.04898
22	4	.03698	.03813	.03940	.04078	.04225	.04383	.04549
22	4	.03354	.03517	.03673	.03823	.03968	.04108	.04241
22	5	.05169	.05162	.05151	.05137	.05119	.05098	.05073
22	5	.03895	.04011	.04138	.04276	.04425	.04584	.04752
22	5	.03544	.03706	.03862	.04013	.04157	.04296	.04430
22	6	.05385	.05376	.05364	.05347	.05327	.05303	.05275
22	6	.04124	.04239	.04366	.04505	.04655	.04815	.04985
22	6	.03766	.03927	.04082	.04232	.04376	.04514	.04647
22	7	.05641	.05630	.05615	.05595	.05571	.05543	.05511
22	7	.04390	.04504	.04631	.04770	.04921	.05082	.05254
22	7	.04027	.04187	.04340	.04488	.04631	.04767	.04899
22	8	.05950	.05935	.05915	.05890	.05860	.05826	.05788
22	8	.04702	.04815	.04942	.05081	.05232	.05395	.05568
22	8	.04338	.04495	.04646	.04791	.04930	.05065	.05193
22	9	.06326	.06304	.06277	.06244	.06207	.06164	.06118
22	9	.05072	.05184	.05310	.05449	.05601	.05764	.05939
22	9	.04711	.04863	.05010	.05152	.05288	.05418	.05543
22	10	.06790	.06758	.06720	.06677	.06629	.06575	.06518
22	10	.05516	.05626	.05751	.05890	.06042	.06206	.06382
22	10	.05164	.05311	.05453	.05589	.05719	.05845	.05964
22	11	.07371	.07325	.07273	.07214	.07151	.07083	.07010
22	11	.06057	.06165	.06289	.06427	.06579	.06744	.06921
22	11	.05725	.05864	.05998	.06126	.06249	.06367	.06480
23	0	.04299	.04286	.04271	.04253	.04234	.04212	.04188
23	0	.02985	.03094	.03211	.03337	.03470	.03610	.03758
23	0	.02702	.02853	.02998	.03136	.03269	.03396	.03517
23	1	.04384	.04374	.04361	.04346	.04328	.04309	.04287
23	1	.03086	.03196	.03315	.03442	.03578	.03722	.03873
23	1	.02795	.02948	.03094	.03235	.03370	.03500	.03624
23	2	.04485	.04476	.04465	.04452	.04435	.04417	.04396

Variance-Covariance factors V1,V2 and V3 for the BLUEs of μ and σ shape parameter k

n	s	1.4	1.5	1.6	1.7	1.8	1.9	2.0
23	2	.03203	.03314	.03433	.03562	.03700	.03846	.03999
23	2	.02904	.03058	.03206	.03348	.03484	.03615	.03741
23	3	.04603	.04596	.04586	.04573	.04557	.04538	.04517
23	3	.03338	.03448	.03568	.03698	.03837	.03985	.04141
23	3	.03030	.03184	.03333	.03476	.03613	.03746	.03872
23	4	.04741	.04734	.04724	.04711	.04695	.04676	.04654
23	4	.03490	.03601	.03721	.03852	.03992	.04142	.04300
23	4	.03175	.03329	.03479	.03622	.03760	.03892	.04020
23	5	.04901	.04895	.04884	.04870	.04853	.04833	.04809
23	5	.03665	.03775	.03896	.04027	.04168	.04319	.04479
23	5	.03342	.03497	.03646	.03789	.03927	.04059	.04187
23	6	.05090	.05082	.05070	.05055	.05036	.05013	.04987
23	6	.03865	.03974	.04095	.04227	.04369	.04522	.04683
23	6	.03536	.03690	.03838	.03981	.04118	.04250	.04377
23	7	.05312	.05302	.05288	.05270	.05248	.05222	.05192
23	7	.04096	.04204	.04325	.04457	.04600	.04753	.04916
23	7	.03763	.03915	.04062	.04204	.04340	.04470	.04596
23	8	.05577	.05563	.05545	.05523	.05496	.05465	.05430
23	8	.04364	.04471	.04592	.04724	.04868	.05022	.05186
23	8	.04029	.04179	.04324	.04463	.04597	.04726	.04849
23	9	.05894	.05876	.05852	.05823	.05790	.05753	.05711
23	9	.04677	.04784	.04904	.05036	.05180	.05335	.05501
23	9	.04345	.04492	.04633	.04770	.04900	.05026	.05146
23	10	.06280	.06254	.06222	.06185	.06143	.06097	.06046
23	10	.05049	.05154	.05273	.05405	.05550	.05705	.05872
23	10	.04723	.04866	.05003	.05135	.05262	.05383	.05499
23	11	.06755	.06718	.06675	.06626	.06572	.06514	.06451
23	11	.05495	.05598	.05716	.05848	.05992	.06149	.06316
23	11	.05184	.05320	.05451	.05577	.05697	.05813	.05923
23	12	.07350	.07297	.07238	.07173	.07103	.07029	.06950
23	12	.06038	.06139	.06256	.06387	.06531	.06688	.06856
23	12	.05752	.05880	.06002	.06120	.06232	.06340	.06442
24	0	.04120	.04107	.04092	.04075	.04056	.04035	.04012
24	0	.02850	.02954	.03066	.03186	.03313	.03447	.03588
24	0	.02588	.02732	.02870	.03003	.03129	.03251	.03367
24	1	.04197	.04187	.04174	.04159	.04142	.04123	.04102
24	1	.02942	.03047	.03160	.03282	.03411	.03548	.03692
24	1	.02672	.02818	.02958	.03093	.03222	.03345	.03464
24	2	.04289	.04280	.04269	.04255	.04240	.04221	.04201
24	2	.03048	.03153	.03267	.03390	.03521	.03660	.03807
24	2	.02771	.02918	.03059	.03195	.03325	.03450	.03570
24	3	.04395	.04387	.04377	.04365	.04349	.04331	.04311
24	3	.03169	.03274	.03389	.03512	.03645	.03786	.03934
24	3	.02884	.03032	.03174	.03310	.03441	.03567	.03688
24	4	.04518	.04511	.04502	.04489	.04473	.04455	.04434
24	4	.03305	.03410	.03525	.03650	.03784	.03926	.04076
24	4	.03014	.03162	.03304	.03441	.03573	.03699	.03821
24	5	.04661	.04654	.04644	.04631	.04614	.04595	.04572
24	5	.03460	.03565	.03680	.03805	.03940	.04084	.04236
24	5	.03162	.03310	.03453	.03590	.03721	.03848	.03969
24	6	.04827	.04820	.04808	.04794	.04776	.04754	.04730
24	6	.03637	.03741	.03856	.03982	.04117	.04262	.04416
24	6	.03334	.03481	.03623	.03759	.03890	.04016	.04137
24	7	.05021	.05012	.04999	.04982	.04962	.04937	.04909
24	7	.03839	.03942	.04057	.04183	.04319	.04465	.04620
24	7	.03532	.03678	.03819	.03954	.04084	.04209	.04329

Variance-Covariance factors V1,V2 and V3 for the BLUEs of μ and σ shape parameter k

n	s	1.4	1.5	1.6	1.7	1.8	1.9	2.0
24	8	.05250	.05238	.05222	.05202	.05177	.05149	.05117
24	8	.04071	.04174	.04289	.04415	.04551	.04698	.04854
24	8	.03762	.03907	.04046	.04179	.04308	.04431	.04549
24	9	.05522	.05506	.05485	.05460	.05430	.05396	.05358
24	9	.04340	.04442	.04557	.04683	.04820	.04967	.05124
24	9	.04033	.04175	.04311	.04442	.04568	.04689	.04805
24	10	.05848	.05825	.05798	.05766	.05729	.05688	.05643
24	10	.04656	.04757	.04870	.04996	.05133	.05281	.05439
24	10	.04353	.04492	.04624	.04752	.04874	.04992	.05104
24	11	.06243	.06212	.06175	.06134	.06088	.06037	.05982
24	11	.05029	.05128	.05241	.05366	.05504	.05652	.05811
24	11	.04738	.04871	.04998	.05121	.05239	.05352	.05459
24	12	.06729	.06686	.06637	.06583	.06524	.06460	.06392
24	12	.05477	.05574	.05686	.05810	.05948	.06097	.06256
24	12	.05204	.05330	.05451	.05568	.05679	.05785	.05887
25	0	.03954	.03942	.03927	.03911	.03893	.03872	.03850
25	0	.02727	.02827	.02934	.03048	.03170	.03298	.03433
25	0	.02483	.02621	.02753	.02880	.03001	.03117	.03228
25	1	.04025	.04015	.04003	.03988	.03972	.03953	.03932
25	1	.02811	.02911	.03019	.03136	.03259	.03390	.03528
25	1	.02560	.02700	.02834	.02962	.03086	.03204	.03317
25	2	.04109	.04100	.04089	.04076	.04061	.04043	.04023
25	2	.02907	.03008	.03117	.03234	.03359	.03492	.03632
25	2	.02649	.02790	.02925	.03055	.03180	.03299	.03414
25	3	.04205	.04197	.04188	.04175	.04160	.04142	.04123
25	3	.03016	.03117	.03226	.03344	.03471	.03605	.03747
25	3	.02752	.02893	.03029	.03160	.03285	.03406	.03521
25	4	.04316	.04309	.04299	.04287	.04272	.04254	.04234
25	4	.03139	.03239	.03349	.03468	.03596	.03731	.03875
25	4	.02868	.03010	.03146	.03277	.03403	.03524	.03640
25	5	.04443	.04437	.04427	.04414	.04398	.04380	.04358
25	5	.03277	.03377	.03487	.03607	.03735	.03872	.04018
25	5	.03001	.03143	.03279	.03410	.03536	.03657	.03773
25	6	.04591	.04584	.04573	.04559	.04542	.04522	.04498
25	6	.03434	.03534	.03644	.03764	.03893	.04031	.04177
25	6	.03154	.03295	.03430	.03561	.03687	.03807	.03923
25	7	.04762	.04754	.04742	.04726	.04706	.04683	.04657
25	7	.03612	.03711	.03821	.03941	.04071	.04210	.04358
25	7	.03328	.03468	.03603	.03733	.03858	.03978	.04093
25	8	.04962	.04951	.04937	.04918	.04895	.04869	.04839
25	8	.03815	.03914	.04023	.04144	.04274	.04414	.04563
25	8	.03530	.03668	.03802	.03930	.04054	.04172	.04286
25	9	.05197	.05183	.05164	.05141	.05115	.05084	.05049
25	9	.04049	.04147	.04256	.04376	.04507	.04647	.04797
25	9	.03764	.03901	.04032	.04158	.04280	.04396	.04508
25	10	.05475	.05456	.05432	.05404	.05371	.05334	.05294
25	10	.04320	.04417	.04525	.04645	.04776	.04917	.05068
25	10	.04039	.04172	.04301	.04424	.04543	.04656	.04765
25	11	.05809	.05783	.05752	.05716	.05675	.05630	.05582
25	11	.04637	.04732	.04840	.04960	.05091	.05232	.05384
25	11	.04364	.04493	.04618	.04737	.04852	.04962	.05066
25	12	.06213	.06177	.06136	.06090	.06039	.05984	.05925
25	12	.05012	.05105	.05212	.05331	.05462	.05604	.05757
25	12	.04753	.04877	.04996	.05110	.05220	.05324	.05424
25	13	.06710	.06661	.06606	.06546	.06482	.06413	.06340
25	13	.05461	.05553	.05658	.05777	.05907	.06049	.06202

Variance-Covariance factors V1,V2 and V3 for
the BLUEs of μ and σ shape parameter k

n	s	1.4	1.5	1.6	1.7	1.8	1.9	2.0
25	13	.05225	.05342	.05454	.05561	.05663	.05761	.05854

Variance-Covariance factors V1,V2 and V3 for the BLUEs of μ and σ shape parameter k

n	s	2.1	2.2	2.3	2.4	2.5
2	0	.50000	.50000	.50000	.50000	.50000
2	0	1.00177	1.04032	1.08023	1.12146	1.16400
2	0	.49024	.51122	.53201	.55261	.57303
3	0	.33112	.33080	.33045	.33009	.32971
3	0	.48680	.50617	.52626	.54705	.56854
3	0	.31386	.32637	.33869	.35082	.36276
3	1	.49671	.49770	.49863	.49949	.50029
3	1	.88744	.91612	.94612	.97745	1.01007
3	1	.57143	.58795	.60442	.62083	.63721
4	0	.24688	.24642	.24592	.24540	.24485
4	0	.31793	.33075	.34406	.35784	.37208
4	0	.22956	.23827	.24680	.25515	.26334
4	1	.31024	.31033	.31036	.31031	.31021
4	1	.44174	.45749	.47394	.49108	.50889
4	1	.31813	.32827	.33828	.34815	.35790
5	0	.19656	.19605	.19550	.19492	.19431
5	0	.23474	.24428	.25417	.26442	.27499
5	0	.18049	.18709	.19353	.19981	.20594
5	1	.23043	.23024	.22998	.22965	.22927
5	1	.29337	.30434	.31579	.32769	.34005
5	1	.22505	.23240	.23961	.24669	.25363
5	2	.30271	.30198	.30116	.30026	.29927
5	2	.41480	.42768	.44115	.45520	.46981
5	2	.31873	.32647	.33408	.34157	.34894
6	0	.16317	.16265	.16209	.16150	.16087
6	0	.18547	.19304	.20089	.20901	.21739
6	0	.14849	.15377	.15890	.16389	.16875
6	1	.18429	.18397	.18359	.18316	.18268
6	1	.21924	.22766	.23644	.24556	.25501
6	1	.17518	.18093	.18655	.19203	.19739
6	2	.22115	.22061	.21999	.21929	.21852
6	2	.27711	.28646	.29623	.30640	.31696
6	2	.22138	.22735	.23319	.23891	.24451
6	3	.30047	.29863	.29671	.29471	.29264
6	3	.39715	.40791	.41917	.43092	.44311
6	3	.31895	.32469	.33031	.33582	.34121
7	0	.13942	.13891	.13836	.13777	.13715
7	0	.15300	.15926	.16575	.17246	.17938
7	0	.12601	.13039	.13464	.13876	.14276
7	1	.15384	.15347	.15305	.15258	.15206
7	1	.17479	.18162	.18874	.19612	.20376
7	1	.14374	.14844	.15302	.15748	.16183
7	2	.17622	.17574	.17519	.17457	.17389
7	2	.20823	.21561	.22332	.23134	.23965
7	2	.17110	.17595	.18069	.18531	.18982
7	3	.21569	.21462	.21347	.21224	.21094
7	3	.26568	.27374	.28217	.29095	.30006
7	3	.21871	.22349	.22816	.23270	.23713
7	4	.30070	.29773	.29470	.29162	.28850
7	4	.38480	.39397	.40358	.41358	.42397
7	4	.31934	.32345	.32746	.33136	.33515
8	0	.12167	.12117	.12064	.12007	.11947
8	0	.13003	.13536	.14088	.14659	.15247
8	0	.10938	.11311	.11673	.12022	.12360
8	1	.13214	.13175	.13131	.13083	.13031
8	1	.14520	.15094	.15691	.16310	.16951

Variance-Covariance factors V1,V2 and V3 for the BLUEs of μ and σ shape parameter k

n	s	2.1	2.2	2.3	2.4	2.5
8	1	.12198	.12595	.12981	.13355	.13719
8	2	.14718	.14673	.14621	.14564	.14502
8	2	.16682	.17293	.17931	.18593	.19279
8	2	.14001	.14410	.14807	.15194	.15570
8	3	.17076	.16999	.16914	.16823	.16725
8	3	.20010	.20662	.21342	.22050	.22784
8	3	.16803	.17209	.17604	.17988	.18361
8	4	.21244	.21080	.20909	.20731	.20547
8	4	.25727	.26431	.27168	.27936	.28731
8	4	.21684	.22062	.22429	.22784	.23129
9	0	.10791	.10743	.10692	.10637	.10579
9	0	.11296	.11759	.12239	.12735	.13246
9	0	.09659	.09983	.10297	.10600	.10892
9	1	.11585	.11546	.11502	.11454	.11402
9	1	.12408	.12903	.13417	.13950	.14501
9	1	.10599	.10942	.11274	.11596	.11909
9	2	.12665	.12622	.12574	.12520	.12462
9	2	.13915	.14436	.14980	.15544	.16128
9	2	.11874	.12226	.12568	.12900	.13221
9	3	.14232	.14170	.14101	.14025	.13945
9	3	.16070	.16619	.17191	.17787	.18403
9	3	.13712	.14064	.14406	.14737	.15058
9	4	.16697	.16587	.16469	.16345	.16216
9	4	.19388	.19968	.20574	.21205	.21858
9	4	.16572	.16910	.17237	.17553	.17859
9	5	.21056	.20836	.20610	.20379	.20144
9	5	.25085	.25708	.26360	.27038	.27741
9	5	.21555	.21848	.22131	.22404	.22666
10	0	.09693	.09647	.09598	.09546	.09490
10	0	.09978	.10387	.10811	.11249	.11701
10	0	.08645	.08932	.09208	.09475	.09732
10	1	.10316	.10277	.10234	.10187	.10137
10	1	.10827	.11261	.11713	.12180	.12663
10	1	.09372	.09674	.09965	.10247	.10520
10	2	.11129	.11088	.11042	.10991	.10936
10	2	.11934	.12388	.12862	.13353	.13861
10	2	.10320	.10629	.10929	.11218	.11499
10	3	.12245	.12191	.12132	.12067	.11996
10	3	.13437	.13911	.14406	.14920	.15452
10	3	.11616	.11926	.12226	.12516	.12798
10	4	.13871	.13787	.13697	.13602	.13500
10	4	.15588	.16084	.16601	.17139	.17695
10	4	.13486	.13788	.14080	.14362	.14635
10	5	.16431	.16286	.16136	.15980	.15819
10	5	.18899	.19420	.19964	.20530	.21116
10	5	.16397	.16675	.16944	.17202	.17451
11	0	.08798	.08753	.08706	.08656	.08603
11	0	.08930	.09297	.09677	.10069	.10473
11	0	.07822	.08078	.08325	.08563	.08792
11	1	.09299	.09261	.09219	.09173	.09124
11	1	.09600	.09986	.10388	.10804	.11233
11	1	.08401	.08670	.08929	.09179	.09421
11	2	.09933	.09893	.09849	.09801	.09749
11	2	.10445	.10847	.11267	.11702	.12151
11	2	.09133	.09408	.09673	.09930	.10179
11	3	.10768	.10720	.10666	.10608	.10545

Variance-Covariance factors V1,V2 and V3 for the BLUEs of μ and σ shape parameter k

n	s	2.1	2.2	2.3	2.4	2.5
11	3	.11549	.11967	.12403	.12855	.13323
11	3	.10093	.10370	.10637	.10895	.11145
11	4	.11919	.11851	.11778	.11699	.11614
11	4	.13051	.13484	.13936	.14406	.14892
11	4	.11408	.11680	.11942	.12196	.12440
11	5	.13598	.13492	.13381	.13264	.13141
11	5	.15200	.15650	.16120	.16609	.17115
11	5	.13308	.13565	.13813	.14052	.14282
12	0	.08053	.08010	.07965	.07917	.07867
12	0	.08079	.08411	.08754	.09109	.09475
12	0	.07140	.07372	.07595	.07810	.08016
12	1	.08465	.08428	.08387	.08343	.08296
12	1	.08619	.08967	.09329	.09703	.10090
12	1	.07612	.07854	.08088	.08313	.08530
12	2	.08973	.08935	.08893	.08847	.08798
12	2	.09285	.09646	.10022	.10412	.10815
12	2	.08194	.08441	.08680	.08911	.09133
12	3	.09621	.09577	.09528	.09475	.09418
12	3	.10129	.10502	.10891	.11295	.11713
12	3	.08933	.09183	.09423	.09655	.09879
12	4	.10479	.10421	.10358	.10290	.10217
12	4	.11233	.11618	.12020	.12437	.12869
12	4	.09907	.10153	.10390	.10620	.10840
12	5	.11663	.11579	.11491	.11397	.11298
12	5	.12734	.13131	.13546	.13977	.14423
12	5	.11240	.11477	.11706	.11925	.12137
12	6	.13390	.13262	.13129	.12991	.12848
12	6	.14881	.15293	.15722	.16169	.16631
12	6	.13166	.13384	.13594	.13795	.13987
13	0	.07423	.07383	.07340	.07294	.07245
13	0	.07373	.07676	.07990	.08314	.08647
13	0	.06567	.06779	.06982	.07177	.07364
13	1	.07768	.07732	.07693	.07650	.07605
13	1	.07818	.08135	.08464	.08804	.09155
13	1	.06959	.07179	.07391	.07595	.07792
13	2	.08184	.08148	.08108	.08064	.08017
13	2	.08355	.08683	.09024	.09377	.09742
13	2	.07432	.07657	.07873	.08082	.08284
13	3	.08702	.08661	.08616	.08567	.08514
13	3	.09020	.09358	.09709	.10074	.10451
13	3	.08019	.08245	.08463	.08674	.08877
13	4	.09366	.09315	.09259	.09198	.09134
13	4	.09864	.10211	.10573	.10948	.11337
13	4	.08767	.08992	.09208	.09417	.09618
13	5	.10244	.10176	.10102	.10023	.09939
13	5	.10968	.11325	.11697	.12083	.12482
13	5	.09752	.09971	.10182	.10384	.10579
13	6	.11458	.11360	.11256	.11147	.11033
13	6	.12469	.12835	.13218	.13616	.14027
13	6	.11102	.11308	.11507	.11697	.11879
13	7	.13230	.13081	.12927	.12768	.12607
13	7	.14616	.14994	.15389	.15799	.16224
13	7	.13052	.13236	.13411	.13579	.13738
14	0	.06885	.06846	.06805	.06761	.06715
14	0	.06780	.07058	.07346	.07644	.07950
14	0	.06079	.06273	.06459	.06638	.06810

Variance-Covariance factors V1,V2 and V3 for the BLUEs of μ and σ shape parameter k

n	s	2.1	2.2	2.3	2.4	2.5
14	1	.07178	.07143	.07105	.07064	.07020
14	1	.07152	.07442	.07743	.08055	.08376
14	1	.06409	.06611	.06805	.06991	.07171
14	2	.07525	.07490	.07452	.07410	.07365
14	2	.07594	.07894	.08205	.08528	.08861
14	2	.06801	.07007	.07205	.07396	.07579
14	3	.07948	.07910	.07867	.07821	.07771
14	3	.08131	.08439	.08759	.09092	.09435
14	3	.07277	.07485	.07685	.07877	.08063
14	4	.08476	.08430	.08380	.08325	.08267
14	4	.08796	.09112	.09441	.09782	.10135
14	4	.07870	.08077	.08276	.08467	.08651
14	5	.09154	.09095	.09031	.08963	.08890
14	5	.09640	.09963	.10300	.10650	.11012
14	5	.08627	.08829	.09024	.09211	.09391
14	6	.10053	.09972	.09887	.09797	.09703
14	6	.10745	.11075	.11421	.11779	.12150
14	6	.09623	.09817	.10003	.10182	.10353
14	7	.11295	.11180	.11061	.10938	.10810
14	7	.12245	.12584	.12939	.13307	.13687
14	7	.10988	.11167	.11338	.11502	.11657
15	0	.06419	.06382	.06342	.06300	.06256
15	0	.06273	.06530	.06797	.07072	.07356
15	0	.05657	.05837	.06009	.06174	.06332
15	1	.06671	.06637	.06600	.06561	.06519
15	1	.06589	.06857	.07135	.07422	.07719
15	1	.05939	.06125	.06304	.06476	.06641
15	2	.06965	.06931	.06894	.06854	.06811
15	2	.06959	.07235	.07522	.07819	.08125
15	2	.06269	.06459	.06641	.06817	.06986
15	3	.07317	.07281	.07241	.07197	.07150
15	3	.07401	.07684	.07979	.08284	.08599
15	3	.06664	.06855	.07039	.07216	.07387
15	4	.07747	.07705	.07659	.07609	.07555
15	4	.07938	.08228	.08529	.08842	.09165
15	4	.07145	.07335	.07519	.07696	.07866
15	5	.08286	.08234	.08177	.08117	.08052
15	5	.08604	.08899	.09208	.09528	.09858
15	5	.07743	.07931	.08112	.08286	.08452
15	6	.08977	.08910	.08837	.08761	.08680
15	6	.09448	.09750	.10065	.10392	.10730
15	6	.08507	.08690	.08864	.09032	.09192
15	7	.09895	.09803	.09706	.09605	.09501
15	7	.10553	.10861	.11183	.11517	.11863
15	7	.09514	.09686	.09850	.10007	.10156
15	8	.11163	.11033	.10899	.10762	.10621
15	8	.12054	.12369	.12699	.13041	.13395
15	8	.10894	.11048	.11195	.11334	.11466
16	0	.06012	.05977	.05939	.05898	.05855
16	0	.05836	.06075	.06323	.06579	.06843
16	0	.05290	.05457	.05617	.05770	.05916
16	1	.06231	.06198	.06163	.06125	.06084
16	1	.06107	.06356	.06614	.06880	.07155
16	1	.05534	.05706	.05872	.06031	.06184
16	2	.06483	.06451	.06415	.06377	.06335
16	2	.06422	.06677	.06943	.07218	.07502

Variance-Covariance factors V1,V2 and V3 for the BLUEs of μ and σ shape parameter k

n	s	2.1	2.2	2.3	2.4	2.5
16	2	.05815	.05991	.06160	.06323	.06479
16	3	.06781	.06746	.06708	.06667	.06623
16	3	.06791	.07053	.07326	.07608	.07899
16	3	.06147	.06324	.06495	.06659	.06817
16	4	.07138	.07099	.07056	.07010	.06960
16	4	.07234	.07501	.07779	.08068	.08366
16	4	.06545	.06722	.06892	.07056	.07214
16	5	.07576	.07529	.07478	.07424	.07365
16	5	.07771	.08044	.08328	.08622	.08927
16	5	.07030	.07205	.07374	.07535	.07690
16	6	.08124	.08065	.08002	.07935	.07865
16	6	.08437	.08715	.09004	.09305	.09615
16	6	.07634	.07805	.07969	.08126	.08277
16	7	.08829	.08752	.08671	.08586	.08497
16	7	.09282	.09565	.09860	.10167	.10483
16	7	.08406	.08569	.08726	.08875	.09018
16	8	.09763	.09660	.09552	.09441	.09326
16	8	.10388	.10676	.10977	.11289	.11612
16	8	.09422	.09573	.09717	.09855	.09985
17	0	.05654	.05619	.05583	.05544	.05503
17	0	.05455	.05678	.05910	.06150	.06396
17	0	.04967	.05123	.05272	.05415	.05552
17	1	.05845	.05814	.05779	.05743	.05703
17	1	.05691	.05922	.06163	.06411	.06668
17	1	.05180	.05341	.05495	.05643	.05785
17	2	.06064	.06033	.05998	.05961	.05922
17	2	.05960	.06198	.06446	.06702	.06966
17	2	.05423	.05587	.05744	.05895	.06040
17	3	.06319	.06286	.06250	.06211	.06168
17	3	.06274	.06518	.06771	.07033	.07304
17	3	.05706	.05871	.06030	.06183	.06329
17	4	.06620	.06584	.06544	.06500	.06454
17	4	.06645	.06893	.07151	.07419	.07696
17	4	.06040	.06205	.06364	.06517	.06664
17	5	.06983	.06941	.06894	.06844	.06791
17	5	.07087	.07340	.07603	.07877	.08159
17	5	.06441	.06605	.06762	.06914	.07058
17	6	.07429	.07377	.07321	.07261	.07197
17	6	.07625	.07882	.08150	.08428	.08716
17	6	.06930	.07091	.07245	.07393	.07534
17	7	.07986	.07920	.07851	.07777	.07700
17	7	.08292	.08553	.08826	.09109	.09401
17	7	.07540	.07695	.07843	.07985	.08121
17	8	.08703	.08617	.08528	.08434	.08338
17	8	.09138	.09403	.09680	.09968	.10266
17	8	.08318	.08465	.08604	.08737	.08864
17	9	.09654	.09539	.09421	.09299	.09175
17	9	.10244	.10514	.10796	.11089	.11392
17	9	.09344	.09477	.09602	.09722	.09834
18	0	.05335	.05302	.05267	.05230	.05190
18	0	.05120	.05330	.05547	.05772	.06003
18	0	.04682	.04828	.04967	.05101	.05229
18	1	.05505	.05474	.05441	.05405	.05368
18	1	.05326	.05543	.05769	.06002	.06241
18	1	.04869	.05019	.05163	.05302	.05434
18	2	.05696	.05666	.05633	.05597	.05559

Variance-Covariance factors V1,V2 and V3 for the BLUEs of μ and σ shape parameter k

n	s	2.1	2.2	2.3	2.4	2.5
18	2	.05561	.05783	.06015	.06254	.06501
18	2	.05080	.05234	.05381	.05522	.05657
18	3	.05917	.05885	.05851	.05813	.05773
18	3	.05830	.06058	.06294	.06539	.06792
18	3	.05324	.05479	.05628	.05770	.05907
18	4	.06175	.06140	.06102	.06062	.06018
18	4	.06145	.06376	.06618	.06868	.07126
18	4	.05609	.05764	.05913	.06056	.06193
18	5	.06480	.06441	.06398	.06352	.06302
18	5	.06515	.06751	.06996	.07251	.07514
18	5	.05946	.06100	.06247	.06389	.06525
18	6	.06849	.06802	.06752	.06697	.06640
18	6	.06958	.07198	.07447	.07706	.07973
18	6	.06350	.06501	.06647	.06786	.06919
18	7	.07301	.07243	.07182	.07117	.07049
18	7	.07497	.07740	.07993	.08256	.08528
18	7	.06843	.06990	.07131	.07266	.07395
18	8	.07867	.07795	.07718	.07638	.07554
18	8	.08164	.08410	.08668	.08935	.09211
18	8	.07458	.07598	.07733	.07861	.07982
18	9	.08596	.08502	.08403	.08302	.08198
18	9	.09010	.09260	.09522	.09793	.10073
18	9	.08243	.08374	.08498	.08616	.08727
19	0	.05051	.05019	.04985	.04949	.04911
19	0	.04823	.05021	.05226	.05437	.05655
19	0	.04427	.04564	.04695	.04821	.04941
19	1	.05202	.05172	.05140	.05106	.05069
19	1	.05005	.05210	.05421	.05640	.05866
19	1	.04592	.04734	.04869	.04999	.05124
19	2	.05371	.05341	.05309	.05275	.05238
19	2	.05211	.05420	.05637	.05862	.06094
19	2	.04779	.04923	.05061	.05193	.05320
19	3	.05563	.05533	.05500	.05464	.05426
19	3	.05445	.05658	.05880	.06110	.06347
19	3	.04991	.05136	.05276	.05410	.05538
19	4	.05786	.05754	.05718	.05679	.05638
19	4	.05715	.05932	.06158	.06392	.06634
19	4	.05237	.05382	.05522	.05656	.05785
19	5	.06047	.06011	.05971	.05928	.05882
19	5	.06029	.06250	.06480	.06718	.06964
19	5	.05523	.05668	.05807	.05941	.06068
19	6	.06357	.06315	.06269	.06219	.06166
19	6	.06400	.06624	.06858	.07100	.07350
19	6	.05862	.06005	.06143	.06274	.06400
19	7	.06731	.06680	.06625	.06567	.06505
19	7	.06844	.07071	.07308	.07553	.07807
19	7	.06269	.06409	.06543	.06671	.06793
19	8	.07190	.07127	.07060	.06990	.06916
19	8	.07383	.07613	.07853	.08102	.08359
19	8	.06767	.06901	.07030	.07153	.07270
19	9	.07765	.07685	.07602	.07515	.07425
19	9	.08051	.08284	.08527	.08779	.09040
19	9	.07386	.07513	.07634	.07749	.07859
19	10	.08505	.08402	.08295	.08186	.08074
19	10	.08898	.09134	.09380	.09637	.09901
19	10	.08178	.08294	.08404	.08508	.08606

Variance-Covariance factors V1,V2 and V3 for the BLUEs of μ and σ shape parameter k

n	s	2.1	2.2	2.3	2.4	2.5
20	0	.04795	.04764	.04732	.04697	.04660
20	0	.04558	.04745	.04939	.05139	.05344
20	0	.04198	.04327	.04451	.04570	.04683
20	1	.04930	.04901	.04870	.04837	.04802
20	1	.04721	.04913	.05113	.05320	.05532
20	1	.04346	.04479	.04607	.04729	.04846
20	2	.05080	.05052	.05021	.04988	.04953
20	2	.04902	.05099	.05304	.05516	.05734
20	2	.04511	.04646	.04776	.04901	.05021
20	3	.05250	.05221	.05189	.05155	.05118
20	3	.05107	.05308	.05517	.05733	.05956
20	3	.04698	.04834	.04966	.05092	.05213
20	4	.05445	.05414	.05380	.05343	.05304
20	4	.05341	.05546	.05758	.05979	.06206
20	4	.04911	.05049	.05180	.05307	.05428
20	5	.05671	.05636	.05599	.05558	.05515
20	5	.05612	.05819	.06035	.06259	.06490
20	5	.05158	.05295	.05426	.05552	.05673
20	6	.05935	.05896	.05853	.05807	.05758
20	6	.05926	.06137	.06356	.06584	.06818
20	6	.05447	.05582	.05712	.05836	.05955
20	7	.06249	.06203	.06153	.06100	.06044
20	7	.06298	.06511	.06733	.06964	.07202
20	7	.05788	.05921	.06048	.06170	.06286
20	8	.06628	.06572	.06513	.06450	.06384
20	8	.06742	.06958	.07183	.07416	.07657
20	8	.06198	.06327	.06450	.06568	.06680
20	9	.07093	.07024	.06952	.06876	.06798
20	9	.07281	.07500	.07727	.07964	.08208
20	9	.06699	.06822	.06940	.07051	.07157
20	10	.07676	.07590	.07500	.07406	.07311
20	10	.07950	.08171	.08401	.08640	.08888
20	10	.07324	.07438	.07547	.07650	.07748
21	0	.04564	.04534	.04503	.04469	.04434
21	0	.04321	.04498	.04682	.04871	.05066
21	0	.03991	.04114	.04231	.04343	.04450
21	1	.04686	.04658	.04628	.04595	.04561
21	1	.04466	.04649	.04838	.05033	.05234
21	1	.04124	.04250	.04371	.04487	.04597
21	2	.04820	.04792	.04763	.04731	.04696
21	2	.04627	.04814	.05007	.05208	.05414
21	2	.04271	.04400	.04523	.04640	.04753
21	3	.04971	.04943	.04912	.04879	.04844
21	3	.04808	.04998	.05196	.05400	.05611
21	3	.04437	.04566	.04690	.04809	.04923
21	4	.05142	.05113	.05080	.05045	.05008
21	4	.05014	.05207	.05408	.05615	.05830
21	4	.04624	.04754	.04879	.04998	.05113
21	5	.05339	.05307	.05271	.05233	.05192
21	5	.05248	.05444	.05648	.05859	.06077
21	5	.04839	.04969	.05093	.05212	.05326
21	6	.05567	.05531	.05491	.05448	.05403
21	6	.05519	.05717	.05924	.06138	.06359
21	6	.05088	.05216	.05340	.05458	.05570
21	7	.05835	.05793	.05747	.05698	.05647
21	7	.05834	.06035	.06244	.06461	.06685

Variance-Covariance factors V1,V2 and V3 for the BLUEs of μ and σ shape parameter k

n	s	2.1	2.2	2.3	2.4	2.5
21	7	.05378	.05505	.05626	.05742	.05852
21	8	.06153	.06103	.06049	.05993	.05933
21	8	.06206	.06409	.06621	.06841	.07068
21	8	.05722	.05846	.05963	.06076	.06183
21	9	.06537	.06476	.06412	.06345	.06275
21	9	.06650	.06856	.07070	.07292	.07522
21	9	.06135	.06254	.06367	.06475	.06577
21	10	.07008	.06934	.06856	.06775	.06691
21	10	.07191	.07398	.07614	.07839	.08071
21	10	.06640	.06752	.06858	.06960	.07055
21	11	.07599	.07506	.07409	.07310	.07208
21	11	.07860	.08069	.08288	.08515	.08750
21	11	.07268	.07371	.07469	.07561	.07648
22	0	.04354	.04325	.04295	.04262	.04228
22	0	.04107	.04275	.04449	.04629	.04814
22	0	.03804	.03921	.04032	.04138	.04240
22	1	.04464	.04437	.04408	.04377	.04344
22	1	.04237	.04411	.04590	.04776	.04966
22	1	.03924	.04044	.04158	.04268	.04372
22	2	.04585	.04558	.04530	.04499	.04465
22	2	.04382	.04559	.04742	.04932	.05128
22	2	.04056	.04178	.04294	.04406	.04512
22	3	.04720	.04693	.04663	.04632	.04598
22	3	.04543	.04723	.04910	.05103	.05303
22	3	.04204	.04326	.04444	.04557	.04665
22	4	.04872	.04844	.04813	.04779	.04743
22	4	.04724	.04907	.05097	.05294	.05497
22	4	.04370	.04493	.04611	.04724	.04833
22	5	.05045	.05015	.04981	.04945	.04906
22	5	.04929	.05115	.05308	.05508	.05714
22	5	.04558	.04681	.04799	.04913	.05021
22	6	.05244	.05210	.05173	.05133	.05090
22	6	.05164	.05352	.05547	.05750	.05959
22	6	.04774	.04897	.05014	.05126	.05233
22	7	.05475	.05436	.05394	.05349	.05301
22	7	.05435	.05625	.05823	.06028	.06240
22	7	.05024	.05145	.05261	.05371	.05476
22	8	.05746	.05700	.05652	.05600	.05545
22	8	.05751	.05943	.06143	.06350	.06565
22	8	.05317	.05435	.05548	.05656	.05758
22	9	.06068	.06014	.05956	.05896	.05833
22	9	.06123	.06317	.06519	.06729	.06945
22	9	.05663	.05777	.05887	.05990	.06089
22	10	.06456	.06391	.06323	.06251	.06177
22	10	.06568	.06764	.06968	.07180	.07399
22	10	.06079	.06188	.06292	.06391	.06484
22	11	.06933	.06853	.06770	.06684	.06596
22	11	.07109	.07306	.07512	.07726	.07947
22	11	.06587	.06689	.06786	.06877	.06963
23	0	.04162	.04135	.04105	.04073	.04040
23	0	.03912	.04073	.04239	.04410	.04586
23	0	.03634	.03744	.03850	.03951	.04048
23	1	.04263	.04236	.04208	.04178	.04146
23	1	.04031	.04196	.04366	.04543	.04724
23	1	.03742	.03856	.03965	.04069	.04168
23	2	.04372	.04346	.04318	.04288	.04256

Variance-Covariance factors V1,V2 and V3 for
the BLUEs of μ and σ shape parameter k

n	s	2.1	2.2	2.3	2.4	2.5
23	2	.04161	.04329	.04503	.04684	.04870
23	2	.03862	.03977	.04088	.04194	.04295
23	3	.04493	.04467	.04439	.04408	.04375
23	3	.04305	.04476	.04653	.04837	.05027
23	3	.03994	.04111	.04222	.04329	.04432
23	4	.04629	.04602	.04572	.04540	.04506
23	4	.04466	.04639	.04820	.05007	.05200
23	4	.04142	.04259	.04371	.04479	.04582
23	5	.04783	.04754	.04722	.04687	.04650
23	5	.04647	.04823	.05006	.05196	.05391
23	5	.04309	.04426	.04538	.04646	.04748
23	6	.04958	.04926	.04890	.04853	.04812
23	6	.04853	.05031	.05216	.05408	.05607
23	6	.04498	.04615	.04727	.04833	.04935
23	7	.05159	.05122	.05083	.05041	.04997
23	7	.05088	.05268	.05456	.05650	.05851
23	7	.04716	.04831	.04941	.05047	.05147
23	8	.05392	.05350	.05306	.05258	.05208
23	8	.05359	.05541	.05731	.05927	.06130
23	8	.04967	.05081	.05189	.05292	.05390
23	9	.05666	.05617	.05565	.05510	.05453
23	9	.05675	.05859	.06050	.06249	.06454
23	9	.05262	.05372	.05477	.05577	.05672
23	10	.05992	.05934	.05873	.05809	.05742
23	10	.06048	.06233	.06427	.06627	.06834
23	10	.05610	.05716	.05817	.05913	.06004
23	11	.06385	.06315	.06242	.06166	.06088
23	11	.06494	.06680	.06875	.07078	.07287
23	11	.06029	.06129	.06224	.06314	.06399
23	12	.06868	.06782	.06693	.06602	.06509
23	12	.07035	.07223	.07420	.07624	.07835
23	12	.06540	.06632	.06720	.06802	.06880
24	0	.03987	.03960	.03931	.03901	.03869
24	0	.03735	.03889	.04047	.04211	.04379
24	0	.03478	.03583	.03684	.03781	.03872
24	1	.04078	.04053	.04025	.03996	.03965
24	1	.03843	.04000	.04163	.04331	.04504
24	1	.03577	.03685	.03789	.03888	.03983
24	2	.04178	.04153	.04126	.04097	.04066
24	2	.03961	.04121	.04287	.04459	.04636
24	2	.03685	.03795	.03900	.04001	.04098
24	3	.04288	.04262	.04235	.04205	.04173
24	3	.04090	.04253	.04422	.04597	.04778
24	3	.03804	.03915	.04022	.04124	.04221
24	4	.04410	.04384	.04355	.04324	.04291
24	4	.04234	.04400	.04572	.04749	.04933
24	4	.03937	.04049	.04156	.04258	.04356
24	5	.04547	.04519	.04488	.04455	.04420
24	5	.04396	.04563	.04737	.04918	.05104
24	5	.04086	.04197	.04304	.04407	.04505
24	6	.04702	.04671	.04638	.04602	.04564
24	6	.04577	.04747	.04923	.05106	.05294
24	6	.04253	.04365	.04471	.04573	.04670
24	7	.04879	.04845	.04808	.04768	.04726
24	7	.04783	.04954	.05133	.05317	.05508
24	7	.04444	.04554	.04660	.04760	.04857

Variance-Covariance factors V1,V2 and V3 for the BLUEs of μ and σ shape parameter k

n	s	2.1	2.2	2.3	2.4	2.5
24	8	.05082	.05043	.05002	.04957	.04911
24	8	.05019	.05192	.05372	.05558	.05751
24	8	.04663	.04771	.04875	.04974	.05068
24	9	.05317	.05273	.05225	.05175	.05122
24	9	.05290	.05465	.05647	.05835	.06030
24	9	.04916	.05022	.05123	.05219	.05311
24	10	.05594	.05542	.05487	.05429	.05369
24	10	.05607	.05783	.05966	.06156	.06353
24	10	.05212	.05314	.05412	.05505	.05593
24	11	.05924	.05862	.05797	.05729	.05659
24	11	.05980	.06157	.06342	.06534	.06732
24	11	.05562	.05660	.05753	.05842	.05925
24	12	.06321	.06247	.06169	.06089	.06007
24	12	.06426	.06604	.06791	.06984	.07184
24	12	.05983	.06075	.06162	.06244	.06322
25	0	.03826	.03799	.03772	.03742	.03711
25	0	.03574	.03720	.03872	.04028	.04189
25	0	.03334	.03435	.03532	.03624	.03712
25	1	.03910	.03885	.03858	.03830	.03800
25	1	.03672	.03822	.03978	.04139	.04304
25	1	.03425	.03529	.03628	.03722	.03812
25	2	.04000	.03976	.03950	.03921	.03891
25	2	.03779	.03932	.04091	.04255	.04424
25	2	.03524	.03629	.03729	.03825	.03917
25	3	.04100	.04076	.04049	.04020	.03989
25	3	.03896	.04052	.04213	.04380	.04552
25	3	.03632	.03738	.03840	.03937	.04030
25	4	.04211	.04185	.04158	.04128	.04096
25	4	.04026	.04183	.04347	.04517	.04692
25	4	.03752	.03858	.03960	.04058	.04151
25	5	.04334	.04307	.04278	.04246	.04212
25	5	.04170	.04330	.04496	.04668	.04845
25	5	.03885	.03992	.04094	.04192	.04285
25	6	.04472	.04443	.04411	.04377	.04341
25	6	.04332	.04493	.04661	.04835	.05015
25	6	.04034	.04141	.04242	.04340	.04433
25	7	.04628	.04596	.04562	.04524	.04485
25	7	.04513	.04676	.04846	.05022	.05204
25	7	.04203	.04308	.04409	.04506	.04598
25	8	.04807	.04771	.04732	.04691	.04647
25	8	.04720	.04884	.05056	.05234	.05417
25	8	.04395	.04499	.04598	.04693	.04784
25	9	.05012	.04971	.04927	.04881	.04832
25	9	.04955	.05121	.05295	.05474	.05659
25	9	.04614	.04717	.04814	.04907	.04995
25	10	.05250	.05203	.05152	.05099	.05044
25	10	.05227	.05395	.05569	.05750	.05937
25	10	.04869	.04968	.05063	.05153	.05238
25	11	.05530	.05474	.05416	.05355	.05292
25	11	.05544	.05713	.05889	.06071	.06260
25	11	.05167	.05262	.05353	.05439	.05521
25	12	.05862	.05797	.05728	.05657	.05583
25	12	.05918	.06088	.06265	.06449	.06639
25	12	.05519	.05610	.05696	.05777	.05853
25	13	.06264	.06185	.06103	.06019	.05933
25	13	.06364	.06535	.06714	.06899	.07090

Variance-Covariance factors V1,V2 and V3 for the BLUEs of μ and σ shape parameter k

n	s	2.1	2.2	2.3	2.4	2.5
25	13	.05943	.06027	.06106	.06181	.06251

Table 7. Simulated values of $100p$-percentage points of the correlation goodness-of-fit statistic R for $n = 20$ (based on 5,000 Monte Carlo simulations)

k

p	0.1	0.2	0.3	0.4	0.5	0.6	0.7	0.8	0.9
0.01	0.9260	0.9280	0.9249	0.9229	0.9197	0.9187	0.9151	0.9146	0.9124
0.025	0.9405	0.9404	0.9379	0.9388	0.9376	0.9335	0.9343	0.9331	0.9328
0.05	0.9510	0.9493	0.9488	0.9498	0.9481	0.9462	0.9454	0.9462	0.9444
0.075	0.9556	0.9548	0.9546	0.9553	0.9544	0.9531	0.9530	0.9528	0.9516
0.1	0.9596	0.9592	0.9590	0.9592	0.9585	0.9572	0.9574	0.9570	0.9561
0.15	0.9656	0.9649	0.9643	0.9644	0.9636	0.9634	0.9633	0.9629	0.9623
0.2	0.9696	0.9689	0.9679	0.9683	0.9676	0.9675	0.9675	0.9668	0.9662
0.25	0.9723	0.9716	0.9707	0.9712	0.9707	0.9709	0.9706	0.9701	0.9696
0.3	0.9745	0.9739	0.9734	0.9741	0.9733	0.9733	0.9732	0.9727	0.9721
0.4	0.9778	0.9779	0.9772	0.9780	0.9772	0.9773	0.9771	0.9767	0.9765
0.5	0.9807	0.9810	0.9804	0.9809	0.9805	0.9803	0.9804	0.9800	0.9800
0.6	0.9833	0.9836	0.9832	0.9833	0.9830	0.9829	0.9831	0.9829	0.9828
0.7	0.9857	0.9860	0.9857	0.9858	0.9854	0.9855	0.9856	0.9855	0.9856
0.75	0.9868	0.9872	0.9867	0.9869	0.9867	0.9868	0.9869	0.9867	0.9868
0.8	0.9880	0.9883	0.9879	0.9880	0.9879	0.9881	0.9880	0.9880	0.9880
0.85	0.9892	0.9895	0.9893	0.9892	0.9892	0.9895	0.9892	0.9893	0.9891
0.9	0.9908	0.9909	0.9907	0.9905	0.9906	0.9909	0.9906	0.9906	0.9906
0.925	0.9915	0.9916	0.9915	0.9913	0.9913	0.9918	0.9915	0.9913	0.9915
0.95	0.9924	0.9923	0.9924	0.9923	0.9923	0.9926	0.9925	0.9923	0.9924
0.975	0.9936	0.9934	0.9937	0.9936	0.9934	0.9938	0.9937	0.9937	0.9937
0.99	0.9949	0.9946	0.9946	0.9949	0.9946	0.9950	0.9947	0.9947	0.9949

p \ k	1.0	1.1	1.2	1.3	1.4	1.5	1.6	1.7
0.01	0.9113	0.9088	0.9015	0.9016	0.8981	0.8961	0.8949	0.8947
0.025	0.9281	0.9281	0.9248	0.9224	0.9200	0.9200	0.9170	0.9166
0.05	0.9408	0.9410	0.9376	0.9353	0.9355	0.9344	0.9313	0.9310
0.075	0.9483	0.9473	0.9451	0.9432	0.9442	0.9425	0.9393	0.9393
0.1	0.9530	0.9534	0.9507	0.9480	0.9492	0.9479	0.9464	0.9456
0.15	0.9598	0.9600	0.9584	0.9561	0.9569	0.9563	0.9556	0.9536
0.2	0.9648	0.9644	0.9634	0.9619	0.9619	0.9612	0.9611	0.9598
0.25	0.9685	0.9681	0.9672	0.9661	0.9658	0.9650	0.9652	0.9647
0.3	0.9714	0.9711	0.9704	0.9691	0.9690	0.9681	0.9681	0.9681
0.4	0.9757	0.9756	0.9752	0.9744	0.9740	0.9735	0.9734	0.9735
0.5	0.9793	0.9790	0.9787	0.9783	0.9779	0.9776	0.9771	0.9776
0.6	0.9824	0.9820	0.9821	0.9815	0.9817	0.9811	0.9807	0.9811
0.7	0.9851	0.9848	0.9850	0.9844	0.9847	0.9842	0.9838	0.9842
0.75	0.9865	0.9861	0.9863	0.9858	0.9861	0.9859	0.9854	0.9858
0.8	0.9878	0.9873	0.9877	0.9874	0.9877	0.9874	0.9870	0.9875
0.85	0.9893	0.9888	0.9891	0.9889	0.9892	0.9887	0.9885	0.9890
0.9	0.9907	0.9905	0.9907	0.9906	0.9906	0.9904	0.9902	0.9908
0.925	0.9915	0.9914	0.9915	0.9915	0.9915	0.9915	0.9912	0.9917
0.95	0.9926	0.9923	0.9925	0.9925	0.9925	0.9925	0.9923	0.9929
0.975	0.9939	0.9937	0.9938	0.9939	0.9939	0.9939	0.9938	0.9940
0.99	0.9951	0.9950	0.9951	0.9952	0.9952	0.9953	0.9951	0.9952

p \ k	1.8	1.9	2.0	2.1	2.2	2.3	2.4	2.5
0.01	0.8791	0.8732	0.8776	0.8752	0.8745	0.8804	0.8438	0.8579
0.025	0.9070	0.9020	0.9038	0.8970	0.9001	0.9034	0.8728	0.8845
0.05	0.9270	0.9230	0.9240	0.9181	0.9182	0.9204	0.9002	0.9068
0.075	0.9369	0.9336	0.9337	0.9292	0.9303	0.9302	0.9150	0.9196
0.1	0.9438	0.9402	0.9404	0.9362	0.9374	0.9372	0.9245	0.9285
0.15	0.9524	0.9498	0.9495	0.9474	0.9469	0.9466	0.9399	0.9407
0.2	0.9582	0.9567	0.9558	0.9543	0.9539	0.9534	0.9487	0.9491
0.25	0.9625	0.9612	0.9607	0.9596	0.9590	0.9586	0.9546	0.9544
0.3	0.9662	0.9653	0.9646	0.9636	0.9630	0.9624	0.9599	0.9593
0.4	0.9722	0.9711	0.9709	0.9702	0.9698	0.9691	0.9672	0.9663
0.5	0.9766	0.9759	0.9758	0.9749	0.9749	0.9744	0.9731	0.9724
0.6	0.9804	0.9798	0.9797	0.9790	0.9792	0.9787	0.9777	0.9770
0.7	0.9837	0.9834	0.9835	0.9827	0.9832	0.9826	0.9816	0.9815
0.75	0.9855	0.9851	0.9851	0.9845	0.9850	0.9846	0.9836	0.9836
0.8	0.9871	0.9868	0.9869	0.9863	0.9868	0.9866	0.9856	0.9857
0.85	0.9888	0.9885	0.9887	0.9882	0.9888	0.9885	0.9877	0.9880
0.9	0.9907	0.9904	0.9907	0.9901	0.9907	0.9905	0.9898	0.9902
0.925	0.9916	0.9914	0.9916	0.9913	0.9917	0.9913	0.9909	0.9912
0.95	0.9928	0.9926	0.9927	0.9925	0.9929	0.9925	0.9923	0.9926
0.975	0.9941	0.9940	0.9941	0.9941	0.9945	0.9943	0.9940	0.9939
0.99	0.9954	0.9954	0.9954	0.9954	0.9958	0.9954	0.9955	0.9954

Table 8. Simulated values of $100p$-percentage points of the correlation goodness-of-fit statistic R for $n = 23$ (based on 5,000 Monte Carlo simulations)

p \ k	0.1	0.2	0.3	0.4	0.5	0.6	0.7	0.8	0.9
0.01	0.9331	0.9371	0.9311	0.9323	0.9291	0.9291	0.9220	0.9186	0.9180
0.025	0.9456	0.9481	0.9471	0.9453	0.9433	0.9426	0.9398	0.9365	0.9352
0.05	0.9546	0.9553	0.9548	0.9545	0.9529	0.9528	0.9504	0.9487	0.9484
0.075	0.9601	0.9605	0.9602	0.9602	0.9586	0.9582	0.9569	0.9557	0.9538
0.1	0.9636	0.9643	0.9635	0.9639	0.9625	0.9622	0.9610	0.9601	0.9588
0.15	0.9687	0.9689	0.9685	0.9689	0.9676	0.9673	0.9664	0.9658	0.9647
0.2	0.9719	0.9721	0.9720	0.9725	0.9709	0.9708	0.9698	0.9699	0.9687
0.25	0.9746	0.9748	0.9746	0.9750	0.9737	0.9735	0.9728	0.9726	0.9719
0.3	0.9768	0.9767	0.9768	0.9769	0.9759	0.9757	0.9751	0.9748	0.9743
0.4	0.9802	0.9799	0.9803	0.9801	0.9794	0.9790	0.9789	0.9787	0.9782
0.5	0.9826	0.9825	0.9830	0.9827	0.9822	0.9818	0.9817	0.9817	0.9814
0.6	0.9850	0.9849	0.9850	0.9850	0.9846	0.9844	0.9842	0.9842	0.9840
0.7	0.9870	0.9870	0.9871	0.9872	0.9868	0.9867	0.9863	0.9864	0.9863
0.75	0.9880	0.9880	0.9881	0.9883	0.9878	0.9878	0.9875	0.9875	0.9875
0.8	0.9890	0.9891	0.9892	0.9892	0.9890	0.9888	0.9886	0.9887	0.9886
0.85	0.9901	0.9902	0.9902	0.9904	0.9901	0.9900	0.9898	0.9898	0.9899
0.9	0.9914	0.9913	0.9914	0.9916	0.9914	0.9913	0.9911	0.9912	0.9912
0.925	0.9921	0.9922	0.9920	0.9923	0.9921	0.9920	0.9919	0.9919	0.9920
0.95	0.9929	0.9929	0.9928	0.9931	0.9928	0.9929	0.9930	0.9929	0.9929
0.975	0.9940	0.9940	0.9939	0.9942	0.9942	0.9940	0.9940	0.9939	0.9942
0.99	0.9949	0.9950	0.9950	0.9952	0.9953	0.9952	0.9951	0.9950	0.9953

k

p	1.0	1.1	1.2	1.3	1.4	1.5	1.6	1.7
0.01	0.9201	0.9086	0.9100	0.9053	0.8988	0.9030	0.8998	0.8922
0.025	0.9362	0.9313	0.9331	0.9313	0.9219	0.9243	0.9189	0.9159
0.05	0.9485	0.9449	0.9447	0.9436	0.9374	0.9400	0.9352	0.9331
0.075	0.9542	0.9510	0.9511	0.9509	0.9469	0.9475	0.9446	0.9425
0.1	0.9579	0.9555	0.9561	0.9557	0.9524	0.9525	0.9501	0.9487
0.15	0.9643	0.9627	0.9623	0.9623	0.9595	0.9593	0.9575	0.9563
0.2	0.9682	0.9671	0.9666	0.9667	0.9644	0.9648	0.9628	0.9618
0.25	0.9712	0.9706	0.9701	0.9712	0.9682	0.9682	0.9667	0.9662
0.3	0.9737	0.9737	0.9726	0.9726	0.9710	0.9712	0.9700	0.9694
0.4	0.9778	0.9779	0.9769	0.9770	0.9759	0.9756	0.9750	0.9747
0.5	0.9811	0.9811	0.9802	0.9807	0.9797	0.9794	0.9790	0.9787
0.6	0.9837	0.9838	0.9833	0.9834	0.9827	0.9825	0.9825	0.9822
0.7	0.9862	0.9863	0.9859	0.9862	0.9856	0.9853	0.9853	0.9853
0.75	0.9872	0.9875	0.9871	0.9874	0.9869	0.9868	0.9869	0.9867
0.8	0.9884	0.9887	0.9885	0.9887	0.9884	0.9880	0.9881	0.9882
0.85	0.9897	0.9899	0.9897	0.9899	0.9897	0.9895	0.9894	0.9896
0.9	0.9910	0.9913	0.9912	0.9913	0.9912	0.9909	0.9911	0.9913
0.925	0.9917	0.9922	0.9919	0.9921	0.9922	0.9918	0.9920	0.9923
0.95	0.9927	0.9932	0.9928	0.9932	0.9931	0.9929	0.9930	0.9933
0.975	0.9940	0.9945	0.9941	0.9946	0.9943	0.9942	0.9943	0.9946
0.99	0.9954	0.9957	0.9953	0.9959	0.9955	0.9954	0.9955	0.9956

p \ k	1.8	1.9	2.0	2.1	2.2	2.3	2.4	2.5
0.01	0.8827	0.8850	0.8881	0.8779	0.8748	0.8772	0.8494	0.8530
0.025	0.9110	0.9116	0.9151	0.9042	0.9001	0.9060	0.8795	0.8857
0.05	0.9314	0.9296	0.9317	0.9234	0.9219	0.9263	0.9059	0.9104
0.075	0.9400	0.9383	0.9397	0.9348	0.9344	0.9351	0.9202	0.9242
0.1	0.9461	0.9449	0.9463	0.9428	0.9420	0.9412	0.9298	0.9317
0.15	0.9548	0.9544	0.9549	0.9529	0.9516	0.9501	0.9410	0.9432
0.2	0.9601	0.9600	0.9602	0.9584	0.9577	0.9567	0.9486	0.9514
0.25	0.9645	0.9647	0.9646	0.9630	0.9625	0.9612	0.9546	0.9570
0.3	0.9684	0.9680	0.9677	0.9665	0.9661	0.9650	0.9597	0.9611
0.4	0.9737	0.9736	0.9730	0.9726	0.9722	0.9713	0.9672	0.9681
0.5	0.9778	0.9780	0.9774	0.9769	0.9767	0.9760	0.9729	0.9734
0.6	0.9813	0.9816	0.9810	0.9806	0.9804	0.9800	0.9772	0.9781
0.7	0.9845	0.9846	0.9844	0.9841	0.9842	0.9841	0.9812	0.9822
0.75	0.9861	0.9861	0.9860	0.9858	0.9857	0.9859	0.9833	0.9843
0.8	0.9875	0.9876	0.9876	0.9874	0.9872	0.9875	0.9855	0.9861
0.85	0.9891	0.9892	0.9891	0.9892	0.9892	0.9892	0.9876	0.9881
0.9	0.9910	0.9910	0.9911	0.9910	0.9910	0.9910	0.9897	0.9902
0.925	0.9920	0.9919	0.9921	0.9920	0.9919	0.9920	0.9909	0.9911
0.95	0.9930	0.9930	0.9932	0.9931	0.9929	0.9931	0.9923	0.9922
0.975	0.9943	0.9943	0.9947	0.9944	0.9942	0.9944	0.9939	0.9941
0.99	0.9956	0.9955	0.9957	0.9957	0.9954	0.9957	0.9953	0.9956

Table 9. Simulated values of $100p$-percentage points of the correlation goodness-of-fit statistic R for $n = 25$ (based on 5,000 Monte Carlo simulations)

p \ k	0.1	0.2	0.3	0.4	0.5	0.6	0.7	0.8	0.9
0.01	0.9364	0.9369	0.9359	0.9324	0.9313	0.9300	0.9275	0.9239	0.9211
0.025	0.9497	0.9492	0.9486	0.9477	0.9473	0.9458	0.9437	0.9408	0.9398
0.05	0.9578	0.9579	0.9576	0.9569	0.9559	0.9552	0.9543	0.9531	0.9517
0.075	0.9625	0.9626	0.9621	0.9617	0.9615	0.9612	0.9599	0.9592	0.9581
0.1	0.9662	0.9661	0.9659	0.9654	0.9651	0.9646	0.9633	0.9626	0.9617
0.15	0.9710	0.9709	0.9705	0.9700	0.9697	0.9692	0.9686	0.9680	0.9671
0.2	0.9740	0.9737	0.9735	0.9732	0.9726	0.9724	0.9720	0.9715	0.9708
0.25	0.9762	0.9760	0.9759	0.9755	0.9752	0.9749	0.9744	0.9742	0.9735
0.3	0.9781	0.9779	0.9777	0.9776	0.9773	0.9769	0.9767	0.9763	0.9758
0.4	0.9812	0.9811	0.9809	0.9806	0.9805	0.9803	0.9800	0.9797	0.9793
0.5	0.9837	0.9837	0.9834	0.9833	0.9832	0.9830	0.9828	0.9826	0.9824
0.6	0.9859	0.9858	0.9858	0.9857	0.9856	0.9854	0.9852	0.9851	0.9848
0.7	0.9878	0.9878	0.9878	0.9877	0.9876	0.9876	0.9874	0.9874	0.9873
0.75	0.9888	0.9887	0.9888	0.9887	0.9885	0.9885	0.9885	0.9884	0.9882
0.8	0.9897	0.9897	0.9897	0.9897	0.9896	0.9896	0.9896	0.9895	0.9894
0.85	0.9909	0.9908	0.9907	0.9907	0.9907	0.9908	0.9907	0.9906	0.9906
0.9	0.9919	0.9919	0.9919	0.9920	0.9919	0.9919	0.9919	0.9919	0.9919
0.925	0.9925	0.9926	0.9926	0.9926	0.9927	0.9927	0.9926	0.9925	0.9926
0.95	0.9932	0.9933	0.9933	0.9934	0.9935	0.9935	0.9934	0.9934	0.9934
0.975	0.9943	0.9943	0.9943	0.9944	0.9945	0.9946	0.9946	0.9946	0.9947
0.99	0.9954	0.9954	0.9954	0.9955	0.9956	0.9955	0.9954	0.9956	0.9956

k

p	1.0	1.1	1.2	1.3	1.4	1.5	1.6	1.7
0.01	0.9200	0.9178	0.9151	0.9104	0.9036	0.9060	0.9016	0.8966
0.025	0.9396	0.9372	0.9348	0.9323	0.9269	0.9262	0.9227	0.9212
0.05	0.9504	0.9486	0.9473	0.9452	0.9416	0.9410	0.9387	0.9363
0.075	0.9566	0.9552	0.9542	0.9523	0.9498	0.9487	0.9467	0.9451
0.1	0.9606	0.9596	0.9584	0.9571	0.9552	0.9542	0.9525	0.9511
0.15	0.9660	0.9651	0.9645	0.9636	0.9624	0.9614	0.9601	0.9590
0.2	0.9700	0.9694	0.9685	0.9677	0.9669	0.9662	0.9650	0.9644
0.25	0.9730	0.9725	0.9719	0.9713	0.9705	0.9700	0.9691	0.9685
0.3	0.9754	0.9751	0.9746	0.9740	0.9734	0.9727	0.9719	0.9714
0.4	0.9790	0.9787	0.9786	0.9781	0.9776	0.9772	0.9767	0.9762
0.5	0.9821	0.9819	0.9817	0.9815	0.9810	0.9807	0.9804	0.9799
0.6	0.9846	0.9844	0.9844	0.9842	0.9840	0.9835	0.9834	0.9831
0.7	0.9871	0.9870	0.9868	0.9867	0.9865	0.9863	0.9862	0.9860
0.75	0.9881	0.9880	0.9879	0.9878	0.9877	0.9875	0.9875	0.9874
0.8	0.9892	0.9891	0.9890	0.9890	0.9889	0.9888	0.9888	0.9887
0.85	0.9904	0.9904	0.9904	0.9904	0.9904	0.9903	0.9903	0.9902
0.9	0.9918	0.9917	0.9918	0.9918	0.9918	0.9917	0.9917	0.9916
0.925	0.9925	0.9925	0.9926	0.9926	0.9924	0.9925	0.9924	0.9924
0.95	0.9934	0.9934	0.9934	0.9933	0.9933	0.9933	0.9934	0.9934
0.975	0.9946	0.9946	0.9947	0.9947	0.9947	0.9947	0.9947	0.9947
0.99	0.9957	0.9957	0.9957	0.9957	0.9956	0.9958	0.9956	0.9956

k

p	1.8	1.9	2.0	2.1	2.2	2.3	2.4	2.5
0.01	0.8927	0.8837	0.8887	0.8666	0.8742	0.8803	0.8445	0.8528
0.025	0.9179	0.9122	0.9146	0.9005	0.9068	0.9098	0.8766	0.8835
0.05	0.9341	0.9310	0.9314	0.9233	0.9260	0.9270	0.9035	0.9090
0.075	0.9429	0.9409	0.9407	0.9344	0.9365	0.9366	0.9184	0.9233
0.1	0.9497	0.9468	0.9476	0.9419	0.9433	0.9436	0.9286	0.9318
0.15	0.9581	0.9560	0.9560	0.9521	0.9528	0.9523	0.9425	0.9451
0.2	0.9639	0.9622	0.9619	0.9590	0.9593	0.9585	0.9509	0.9526
0.25	0.9681	0.9663	0.9663	0.9635	0.9640	0.9637	0.9567	0.9583
0.3	0.9708	0.9695	0.9694	0.9670	0.9674	0.9675	0.9611	0.9624
0.4	0.9756	0.9748	0.9746	0.9729	0.9729	0.9729	0.9680	0.9691
0.5	0.9796	0.9792	0.9788	0.9774	0.9777	0.9775	0.9738	0.9746
0.6	0.9828	0.9822	0.9822	0.9811	0.9813	0.9812	0.9785	0.9789
0 7	0.9858	0.9852	0.9854	0.9844	0.9845	0.9846	0.9825	0.9829
0.75	0.9872	0.9868	0.9868	0.9861	0.9863	0.9863	0.9843	0.9847
0.8	0.9886	0.9883	0.9881	0.9876	0.9879	0.9878	0.9861	0.9865
0.85	0.9901	0.9897	0.9899	0.9893	0.9895	0.9896	0.9880	0.9883
0.9	0.9915	0.9914	0.9915	0.9911	0.9913	0.9914	0.9901	0.9904
0.925	0.9923	0.9923	0.9922	0.9920	0.9921	0.9922	0.9912	0.9916
0.95	0.9934	0.9932	0.9934	0.9931	0.9933	0.9935	0.9925	0.9927
0.975	0.9947	0.9945	0.9947	0.9946	0.9946	0.9947	0.9940	0.9942
0.99	0.9955	0.9956	0.9956	0.9959	0.9959	0.9958	0.9958	0.9958

Table 10. Values of the correlation statistic R and the corresponding p-values for Example 1

[illegible]	R	p
0.1	0.9518	0.0544
0.2	0.9566	0.0852
0.3	0.9611	0.1198
0.4	0.9652	0.1786
0.5	0.9689	0.2210
0.6	0.9722	0.2771
0.7	0.9752	0.3513
0.8	0.9778	0.4333
0.9	0.9800	0.5000
1.0	0.9819	0.5839
1.1	0.9835	0.6536
1.2	0.9848	0.6931
1.3	0.9857	0.7464
1.4	0.9864	0.7594
1.5	0.9869	0.7833
1.6	0.9870	0.8000
1.7	0.9870	0.7853
1.8	0.9867	0.7875
1.9	0.9862	0.7824
2.0	0.9856	0.7639
2.1	0.9847	0.7556
2.2	0.9837	0.7139
2.3	0.9826	0.7000
2.4	0.9813	0.6923
2.5	0.9800	0.6667

Table 11. Values of the correlation statistic R and the corresponding p-values for Example 3

[illegible]	R	p
0.1	0.9660	0.124
0.2	0.9703	0.172
0.3	0.9743	0.244
0.4	0.9779	0.331
0.5	0.9810	0.457
0.6	0.9838	0.577
0.7	0.9862	0.695
0.8	0.9882	0.779
0.9	0.9898	0.846
1.0	0.9911	0.904
1.1	0.9919	0.917
1.2	0.9926	0.944
1.3	0.9930	0.946
1.4	0.9930	0.947
1.5	0.9927	0.946
1.6	0.9922	0.930
1.7	0.9915	0.905
1.8	0.9906	0.890
1.9	0.9894	0.863
2.0	0.9881	0.817
2.1	0.9866	0.775
2.2	0.9849	0.723
2.3	0.9831	0.676
2.4	0.9812	0.700
2.5	0.9792	0.627